Lecture Notes in Computer Science 10821

More information about this series at http://www.springer.com/series/7410

Jesper Buus Nielsen · Vincent Rijmen (Eds.)

Advances in Cryptology – EUROCRYPT 2018

37th Annual International Conference on the Theory
and Applications of Cryptographic Techniques
Tel Aviv, Israel, April 29 – May 3, 2018
Proceedings, Part II

Editors
Jesper Buus Nielsen
Aarhus University
Aarhus
Denmark

Vincent Rijmen
University of Leuven
Leuven
Belgium

ISSN 0302-9743 ISSN 1611-3349 (electronic)
Lecture Notes in Computer Science
ISBN 978-3-319-78374-1 ISBN 978-3-319-78375-8 (eBook)
https://doi.org/10.1007/978-3-319-78375-8

Library of Congress Control Number: 2018937382

LNCS Sublibrary: SL4 – Security and Cryptology

Printed on acid-free paper

This Springer imprint is published by the registered company Springer International Publishing AG part of Springer Nature
The registered company address is: Gewerbestrasse 11, 6330 Cham, Switzerland

Preface

Eurocrypt 2018, the 37th Annual International Conference on the Theory and Applications of Cryptographic Techniques, was held in Tel Aviv, Israel, from April 29 to May 3, 2018. The conference was sponsored by the International Association for Cryptologic Research (IACR). Orr Dunkelman (University of Haifa, Israel) was responsible for the local organization. He was supported by a local organizing team consisting of Technion's Hiroshi Fujiwara Cyber Security Research Center headed by Eli Biham, and most notably by Suzie Eid. We are deeply indebted to them for their support and smooth collaboration.

The conference program followed the now established parallel track system where the works of the authors were presented in two concurrently running tracks. Only the invited talks spanned over both tracks.

We received a total of 294 submissions. Each submission was anonymized for the reviewing process and was assigned to at least three of the 54 Program Committee members. Committee members were allowed to submit at most one paper, or two if both were co-authored. Submissions by committee members were held to a higher standard than normal submissions. The reviewing process included a rebuttal round for all submissions. After extensive deliberations, the Program Committee accepted 69 papers. The revised versions of these papers are included in these three-volume proceedings, organized topically within their respective track.

The committee decided to give the Best Paper Award to the papers "Simple Proofs of Sequential Work" by Bram Cohen and Krzysztof Pietrzak, "Two-Round Multiparty Secure Computation from Minimal Assumptions" by Sanjam Garg and Akshayaram Srinivasan, and "Two-Round MPC from Two-Round OT" by Fabrice Benhamouda and Huijia Lin. All three papers received invitations for the *Journal of Cryptology*.

The program also included invited talks by Anne Canteaut, titled "Desperately Seeking Sboxes", and Matthew Green, titled "Thirty Years of Digital Currency: From DigiCash to the Blockchain".

We would like to thank all the authors who submitted papers. We know that the Program Committee's decisions can be very disappointing, especially rejections of very good papers that did not find a slot in the sparse number of accepted papers. We sincerely hope that these works eventually get the attention they deserve.

We are also indebted to the members of the Program Committee and all external reviewers for their voluntary work. The Program Committee work is quite a workload. It has been an honor to work with everyone. The committee's work was tremendously simplified by Shai Halevi's submission software and his support, including running the service on IACR servers.

Finally, we thank everyone else — speakers, session chairs, and rump-session chairs — for their contribution to the program of Eurocrypt 2018. We would also like to thank the many sponsors for their generous support, including the Cryptography Research Fund that supported student speakers.

May 2018

Jesper Buus Nielsen
Vincent Rijmen

Eurocrypt 2018

The 37th Annual International Conference on the Theory and Applications of Cryptographic Techniques

Sponsored by *the International Association for Cryptologic Research*

April 29 – May 3, 2018
Tel Aviv, Israel

General Chair

Orr Dunkelman	University of Haifa, Israel

Program Co-chairs

Jesper Buus Nielsen	Aarhus University, Denmark
Vincent Rijmen	University of Leuven, Belgium

Program Committee

Martin Albrecht	Royal Holloway, UK
Joël Alwen	IST Austria, Austria, and Wickr, USA
Gilles Van Assche	STMicroelectronics, Belgium
Paulo S. L. M. Barreto	University of Washington Tacoma, USA
Nir Bitansky	Tel Aviv University, Israel
Céline Blondeau	Aalto University, Finland
Andrey Bogdanov	DTU, Denmark
Chris Brzuska	TU Hamburg, Germany, and Aalto University, Finland
Jan Camenisch	IBM Research – Zurich, Switzerland
Ignacio Cascudo	Aalborg University, Denmark
Melissa Chase	Microsoft Research, USA
Alessandro Chiesa	UC Berkeley, USA
Joan Daemen	Radboud University, The Netherlands, and STMicroelectronics, Belgium
Yevgeniy Dodis	New York University, USA
Nico Döttling	Friedrich Alexander University Erlangen-Nürnberg, Germany
Sebastian Faust	TU Darmstadt, Germany
Serge Fehr	CWI Amsterdam, The Netherlands
Georg Fuchsbauer	Inria and ENS, France
Jens Groth	University College London, UK
Jian Guo	Nanyang Technological University, Singapore

Martin Hirt	ETH Zurich, Switzerland
Dennis Hofheinz	KIT, Germany
Yuval Ishai	Technion, Israel, and UCLA, USA
Nathan Keller	Bar-Ilan University, Israel
Eike Kiltz	Ruhr-Universität Bochum, Germany
Gregor Leander	Ruhr-Universität Bochum, Germany
Yehuda Lindell	Bar-Ilan University, Israel
Mohammad Mahmoody	University of Virginia, USA
Willi Meier	FHNW, Windisch, Switzerland
Florian Mendel	Infineon Technologies, Germany
Bart Mennink	Radboud University, The Netherlands
María Naya-Plasencia	Inria, France
Svetla Nikova	KU Leuven, Belgium
Eran Omri	Ariel University, Israel
Arpita Patra	Indian Institute of Science, India
David Pointcheval	ENS/CNRS, France
Bart Preneel	KU Leuven, Belgium
Thomas Ristenpart	Cornell Tech, USA
Alon Rosen	IDC Herzliya, Israel
Mike Rosulek	Oregon State University, USA
Louis Salvail	Université de Montréal, Canada
Yu Sasaki	NTT Secure Platform Laboratories, Japan
Thomas Schneider	TU Darmstadt, Germany
Jacob C. N. Schuldt	AIST, Japan
Nigel P. Smart	KU Leuven, Belgium, and University of Bristol, UK
Adam Smith	Boston University, USA
Damien Stehlé	ENS de Lyon, France
Björn Tackmann	IBM Research – Zurich, Switzerland
Dominique Unruh	University of Tartu, Estonia
Vinod Vaikuntanathan	MIT, USA
Muthuramakrishnan Venkitasubramaniam	University of Rochester, USA
Frederik Vercauteren	KU Leuven, Belgium
Damien Vergnaud	Sorbonne Université, France
Ivan Visconti	University of Salerno, Italy
Moti Yung	Columbia University and Snap Inc., USA

Additional Reviewers

Masayuki Abe
Aysajan Abidin
Ittai Abraham
Hamza Abusalah
Divesh Aggarwal
Shashank Agrawal
Shweta Agrawal
Thomas Agrikola
Bar Alon
Abdel Aly
Prabhanjan Ananth
Elena Andreeva

Daniel Apon
Gilad Asharov
Nuttapong Attrapadung
Benedikt Auerbach
Daniel Augot
Christian Badertscher
Saikrishna Badrinarayanan
Shi Bai
Josep Balasch
Marshall Ball
Valentina Banciu
Subhadeep Banik
Zhenzhen Bao
Gilles Barthe
Lejla Batina
Balthazar Bauer
Carsten Baum
Christof Beierle
Amos Beimel
Sonia Belaid
Aner Ben-Efraim
Fabrice Benhamouda
Iddo Bentov
Itay Berman
Kavun Elif Bilge
Olivier Blazy
Jeremiah Blocki
Andrey Bogdanov
Carl Bootland
Jonathan Bootle
Raphael Bost
Leif Both
Florian Bourse
Elette Boyle
Zvika Brakerski
Christian Cachin
Ran Canetti
Anne Canteaut
Brent Carmer
Wouter Castryck
Andrea Cerulli
André Chailloux
Avik Chakraborti
Yilei Chen
Ashish Choudhury
Chitchanok Chuengsatiansup
Michele Ciampi
Thomas De Cnudde
Ran Cohen
Sandro Coretti
Jean-Sebastien Coron
Henry Corrigan-Gibbs
Ana Costache
Geoffroy Couteau
Claude Crépeau
Ben Curtis
Dana Dachman-Soled
Yuanxi Dai
Bernardo David
Alex Davidson
Jean Paul Degabriele
Akshay Degwekar
Daniel Demmler
Amit Deo
Apoorvaa Deshpande
Itai Dinur
Christoph Dobraunig
Manu Drijvers
Maria Dubovitskaya
Léo Ducas
Yfke Dulek
Pierre-Alain Dupont
François Dupressoir
Avijit Dutta
Lisa Eckey
Maria Eichlseder
Maximilian Ernst
Mohammad Etemad
Antonio Faonio
Oriol Farràs
Pooya Farshim
Manuel Fersch
Dario Fiore
Viktor Fischer
Nils Fleischhacker
Christian Forler
Tommaso Gagliardoni
Chaya Ganesh
Juan Garay
Sanjam Garg
Romain Gay
Peter Gaži
Rosario Gennaro
Satrajit Ghosh
Irene Giacomelli
Federico Giacon
Benedikt Gierlichs
Junqing Gong
Dov Gordon
Divya Gupta
Lorenzo Grassi
Hannes Gross
Vincent Grosso
Paul Grubbs
Chun Guo
Siyao Guo
Mohammad Hajiabadi
Carmit Hazay
Gottfried Herold
Felix Heuer
Thang Hoang
Viet Tung Hoang
Akinori Hosoyamada
Kristina Hostáková
Andreas Hülsing
Ilia Iliashenko
Roi Inbar
Vincenzo Iovino
Tetsu Iwata
Abhishek Jain
Martin Jepsen
Daniel Jost
Chiraag Juvekar
Seny Kamara
Chethan Kamath
Bhavana Kanukurthi
Harish Karthikeyan
Suichi Katsumata
Jonathan Katz
John Kelsey
Dakshita Khurana
Eunkyung Kim
Taechan Kim
Elena Kirshanova
Ágnes Kiss
Susumu Kiyoshima

Ilya Kizhvatov
Alexander Koch
Konrad Kohbrok
Lisa Kohl
Stefan Kölbl
Ilan Komargodski
Yashvanth Kondi
Venkata Koppula
Thorsten Kranz
Hugo Krawczyk
Marie-Sarah Lacharite
Kim Laine
Virginie Lallemand
Gaëtan Leurent
Anthony Leverrier
Xin Li
Pierre-Yvan Liardet
Benoît Libert
Huijia Lin
Guozhen Liu
Jian Liu
Chen-Da Liu-Zhang
Alex Lombardi
Julian Loss
Steve Lu
Atul Luykx
Vadim Lyubashevsky
Saeed Mahloujifar
Hemanta Maji
Mary Maller
Umberto Martínez-Peñas
Daniel Masny
Takahiro Matsuda
Christian Matt
Patrick McCorry
Pierrick Méaux
Lauren De Meyer
Peihan Miao
Brice Minaud
Esfandiar Mohammadi
Ameer Mohammed
Maria Chiara Molteni
Tal Moran
Fabrice Mouhartem
Amir Moradi
Pratyay Mukherjee
Marta Mularczyk
Mridul Nandi
Ventzislav Nikov
Tobias Nilges
Ryo Nishimaki
Anca Nitulescu
Ariel Nof
Achiya Bar On
Claudio Orlandi
Michele Orrù
Clara Paglialonga
Giorgos Panagiotakos
Omer Paneth
Louiza Papachristodoulou
Kostas Papagiannopoulos
Sunoo Park
Anat Paskin-Cherniavsky
Alain Passelègue
Kenny Paterson
Michaël Peeters
Chris Peikert
Alice Pellet–Mary
Geovandro C. C. F. Pereira
Leo Perrin
Giuseppe Persiano
Thomas Peters
Krzysztof Pietrzak
Benny Pinkas
Oxana Poburinnaya
Bertram Poettering
Antigoni Polychroniadou
Christopher Portmann
Manoj Prabhakaran
Emmanuel Prouff
Carla Ràfols
Somindu C. Ramanna
Samuel Ranellucci
Shahram Rasoolzadeh
Divya Ravi
Ling Ren
Oscar Reparaz
Silas Richelson
Peter Rindal
Michal Rolinek
Miruna Rosca
Ron Rothblum
David Roubinet
Adeline Roux-Langlois
Vladimir Rozic
Andy Rupp
Yusuke Sakai
Simona Samardjiska
Niels Samwel
Olivier Sanders
Pratik Sarkar
Alessandra Scafuro
Martin Schläffer
Dominique Schröder
Sven Schäge
Adam Sealfon
Yannick Seurin
abhi shelat
Kazumasa Shinagawa
Luisa Siniscalchi
Maciej Skórski
Fang Song
Ling Song
Katerina Sotiraki
Florian Speelman
Gabriele Spini
Kannan Srinathan
Thomas Steinke
Uri Stemmer
Igors Stepanovs
Noah Stephens-Davidowitz
Alan Szepieniec
Seth Terashima
Cihangir Tezcan
Mehdi Tibouchi
Elmar Tischhauser
Radu Titiu
Yosuke Todo
Junichi Tomida
Patrick Towa
Boaz Tsaban
Daniel Tschudi
Thomas Unterluggauer
Margarita Vald
Kerem Varici
Prashant Vasudevan

Philip Vejre
Daniele Venturi
Benoît Viguier
Fernando Virdia
Damian Vizár
Alexandre Wallet
Michael Walter
Haoyang Wang
Qingju Wang
Hoeteck Wee
Felix Wegener
Christian Weinert
Erich Wenger
Daniel Wichs
Friedrich Wiemer
David Wu
Thomas Wunderer
Sophia Yakoubov
Shota Yamada
Takashi Yamakawa
Kan Yasuda
Attila Yavuz
Scott Yilek
Eylon Yogev
Greg Zaverucha
Mark Zhandry
Ren Zhang

Abstract of Invited Talks

Desperately Seeking Sboxes

Anne Canteaut

Inria, Paris, France
anne.canteaut@inria.fr

Abstract. Twenty-five years ago, the definition of security criteria associated to the resistance to linear and differential cryptanalysis has initiated a long line of research in the quest for Sboxes with optimal nonlinearity and differential uniformity. Although these optimal Sboxes have been studied by many cryptographers and mathematicians, many questions remain open. The most prominent open problem is probably the determination of the optimal values of the nonlinearity and of the differential uniformity for a permutation depending on an even number of variables.

Besides those classical properties, various attacks have motivated several other criteria. Higher-order differential attacks, cube distinguishers and the more recent division property exploit some specific properties of the representation of the whole cipher as a collection of multivariate polynomials, typically the fact that some given monomials do not appear in these polynomials. This type of property is often inherited from some algebraic property of the Sbox. Similarly, the invariant subspace attack and its nonlinear counterpart also originate from specific algebraic structure in the Sbox.

Thirty Years of Digital Currency: From DigiCash to the Blockchain

Matthew Green

Johns Hopkins University
mgreen@cs.jhu.edu

Abstract. More than thirty years ago a researcher named David Chaum presented his vision for a cryptographic financial system. In the past ten years this vision has been realized. Yet despite a vast amount of popular excitement, it remains to be seen whether the development of cryptocurrencies (and their associated consensus technologies) will have a lasting positive impact—both on society and on our research community. In this talk I will examine that question. Specifically, I will review several important contributions that research cryptography has made to this field; survey the most promising deployed (or developing) technologies; and discuss the many challenges ahead.

Contents – Part II

Blockchain

Multi-collision Resistance

Signatures

Private Simultaneous Messages

Blockchain

Thunderella: Blockchains with Optimistic Instant Confirmation

Rafael Pass[1(✉)] and Elaine Shi[2]

[1] CornellTech, New York, USA
rafael@cs.cornell.edu
[2] Cornell, Ithaca, USA

Abstract. State machine replication, or "consensus", is a central abstraction for distributed systems where a set of nodes seek to agree on an ever-growing, linearly-ordered log. In this paper, we propose a practical new paradigm called Thunderella for achieving state machine replication by combining a fast, asynchronous path with a (slow) synchronous "fall-back" path (which only gets executed if something goes wrong); as a consequence, we get *simple* state machine replications that essentially are as robust as the best synchronous protocols, yet "optimistically" (if a super majority of the players are honest), the protocol "instantly" confirms transactions.

We provide instantiations of this paradigm in both permissionless (using proof-of-work) and permissioned settings. Most notably, this yields a new blockchain protocol (for the permissionless setting) that remains resilient assuming only that a majority of the computing power is controlled by honest players, yet *optimistically*—if 3/4 of the computing power is controlled by honest players, and a special player called the "accelerator", is honest—transactions are confirmed as fast as the actual message delay in the network. We additionally show the 3/4 optimistic bound is tight for protocols that are resilient assuming only an honest majority.

1 Introduction

State machine replication, also referred to as atomic broadcast, is a core distributed systems abstraction that has been investigated for three decades. In a state machine replication protocol, a set of servers seek to agree on an ever-growing, *linearly-ordered log*, such that two important properties are satisfied: (1) *consistency*, i.e., all servers must have the same view of the log; and (2) *liveness*, i.e., whenever a client submits a transaction, the transaction is incorporated quickly into the log. In this paper, we will also refer to state machine replication as *consensus* for short[1].

The full version of this paper is available at https://eprint.iacr.org/2017/913 [36].

[1] Although the term "consensus" has been used in the distributed systems literature to mean other related abstractions such as single-shot consensus; in this paper, we use "consensus" to specifically refer to "state machine replication".

J. B. Nielsen and V. Rijmen (Eds.): EUROCRYPT 2018, LNCS 10821, pp. 3–33, 2018.
https://doi.org/10.1007/978-3-319-78375-8_1

State machine replication is a fundamental building block for replicated databases. For more than a decade, companies such as Google and Facebook have deployed Paxos-style protocols [5,24,30] to replicate a significant part of their computing infrastructure. These classical deployment scenarios are typically relatively small scale, with fast local-area networking, where crash (rather than byzantine) faults are usually of concern.

Fuelled by decentralized cryptocurrencies, recently the community has been excited about large-scale applications of distributed consensus. Two deployment scenarios are of interest: (1) the *permissionless* setting where anyone can join freely (e.g., decentralized cryptocurrencies); and (2) the *permissioned* setting where only approved participants may join (e.g., a consortium blockchain where multiple banks collaborate to build a distributed ledger). Regardless of which setting, the typical deployment would involve a large number of nodes (e.g., thousands or more) controlled by mutually distrustful individuals and organizations.

Roughly speaking, two broad classes of protocols have been considered for the large-scale setting, each with their own set of deficiencies:

- First, *classical-style protocols* such as PBFT [9] and Byzantine-Paxos [30] confirm transactions quickly in the normal case; but these protocols are notoriously complicated, making implementation, reconfiguration, and maintenance relatively difficult especially in a large-scale setting. Further, these protocols achieve "fast confirmation" by adopting the asynchronous (or partially synchronous) model, and thus inherently they can tolerate at most $\frac{1}{3}$ corruptions [15,38].
- Second, *blockchain-style* protocols, represented by Nakamoto's original blockchain [19,34,35], are a new breakthrough in distributed consensus: these protocols are conceptually simple and tolerate *minority corruptions*. Moreover, it is has been shown how to remove the expensive proof-of-work from blockchain-style consensus [11,26,40] thus solving the energy waste problem. Further, not only has blockchains' robustness been empirically proven, earlier works [11,40] have also shown mathematically that blockchain-style consensus indeed achieves certain robustness properties in the presence of sporadic participation and node churn that none of the classical-style protocols can attain! Unfortunately known blockchain-style protocols suffer from slow transaction confirmation, e.g., Bitcoin's Nakamoto consensus has a 10-minute block interval and it takes several blocks to confirm a transaction with sufficient confidence. Earlier works that mathematically analyze blockchain-style consensus [35,40] have pointed out that such slowness is inherent for blockchain-style protocols since the expected block interval must be set to be sufficiently large for the protocol to retain security.

A natural question that arises is whether there is some way to *simultaneously* reap the benefit of both of these "worlds". Unfortunately, a negative answer was presented by earlier works [38–40] which showed that a natural notion of fast transaction confirmation called "responsiveness" is unattainable against $\frac{1}{3}$

(even static) corruptions in classical or permissionless models. In this paper we consider a new notion called *optimistic responsiveness* that allows us "circumvent" this lower bound such that we can achieve responsiveness most of the time in practice and yet tolerate up to minority corruptions in the worst-case. In our approach, in the *optimistic case* (when e.g., a super majority is honest), we enjoy the fast nature of asynchronous protocols; and yet we retain the resilience of synchronous (e.g., blockchain) protocols as well as their robustness properties (e.g., support for sporadic participation). More precisely, we show how to combine a fast and simple "asynchronous path"—which guarantees consistency but not liveness—with a (slow) synchronous "fall-back" path which only gets executed if something goes wrong.

1.1 The Thunderella Paradigm

To characterize what we mean by "fast" or "instant confirmation", we adopt the same notion of *responsiveness* as proposed in the work by Attiya et al. [1] and later adopted by others [23,38]. A consensus protocol is said to be responsive iff any transaction input to an honest node is confirmed in time that depends only on the *actual network delay*, but not on any a-priori known *upper bound on the network delay*. Henceforth in this paper, we use δ to denote the actual network delay and use Δ to denote an a-priori known upper bound of the network's delay where Δ is possibly provided as input to the protocol.

As shown in [38], achieving responsiveness requires us to assume that 2/3 of the players are honest. (Additionally, all known protocols that are responsive are very complicated, and thus hard to implement.)

Towards overcoming this issue, we here instead consider a notion of **optimistic responsiveness**—where responsiveness is only required to hold whenever some "goodness conditions" are satisfies. More precisely, we consider two sets of conditions:

- worst-case conditions (denoted W) under which the protocol provides worst-case guarantees including consistency and "slow" confirmation (e.g., $W =$ majority honest).
- optimistic-case conditions (denoted $O \subseteq W$) under which the protocol additionally provides responsive confirmation (e.g., $O =$ "more than $\frac{3}{4}$ are honest and online, and some designated player (the "leader") is honest").

Our main result is a paradigm for taking any blockchain protocol (permissioned or permissionless) that satisfies consistency and liveness under conditions W, and transform it into a new protocol that satisfies consistency and liveness under "essentially" the same conditions W (and in many cases, actually the same conditions W), and additionally satisfies optimistic responsiveness under condition O.

The idea in a nutshell. To explain our approach, consider first the following simple protocol:

- We have a designated entity: the leader, or "accelerator".
- Transactions are sent to the leader; the leader signs the transaction (with an increasing sequence number), and sends out the signed transaction to a "committee" of players.
- The committee members "ack" all leader-signed transactions, but at most one per sequence number.
- If a transaction has received more than 3/4 of the committees signatures—we refer to such a transaction as being notarized. Participants, can *directly output* their longest sequence of consecutive (in terms of their sequence numbers) notarized transactions—all those transactions are confirmed.

It is not hard to see that this protocol is consistent under condition W' = "1/2 the committee is honest"); additionally, it satisfies liveness with optimistic responsiveness under condition O = "leader is honest, and 3/4 of the committee is honest". In fact, under these optimistic condition, we only need 2 communication rounds to confirm a transaction! This approach is extremely practical and indeed this protocol is often used in practice—for instance `chain.com` use something very similar as their permissioned blockchain (and manage to handle a very high volume of transactions with fast confirmations).

The problem with this approach, however, is that the protocol does not satisfy liveness (even "slow" liveness) under condition W'. If the leader is cheating (or is simply taken down from the network), the protocol halts. (Indeed, in this case `chain.com` resorts to manually fixing the issue.)

To overcome this problem, we leverage the underlying (slow) blockchain protocol, which satisfies both consistency and liveness under W = "honest majority of players". Roughly speaking, if players notice that transactions are not getting confirmed by the leader/committee, some "evidence" of this is sent to the underlying blockchain. We then enter a "cool-down" period, where committee members stop signing messages from the leader, yet we allow players to broadcast any notarized transactions they have seen so far. The length of the cool-down period is counted in blocks on the underlying blockchain (say κ blocks where κ is a security parameter). Finally, after the cool-down period ends, we can safely enter a "slow period" where transactions only get confirmed in the underlying blockchain. We can next use the blockchain to switch out the leader (if needed) and begin a new epoch of the optimistic protocol.

Let us point out the reason for having a cool-down period: without it, players may disagree on the set of transactions that have been confirmed before entering the "slow mode", and thus may end up with inconsistent views. The cool-down period enables honest players to post all notarized transactions they have seen to the (slow) underlying blockchain, and thus (slowly) reach consistency of this set of transactions; once we have reached this consistent view (at the end of the cool-down), we can finally fully switch over to confirming new transactions on the blockchain.

Collecting evidence of cheating. It only remains to explain how to collect evidence that the leader (and/or committee) is cheating or is simply "unavailable". This turns out to also be simple: if a player notices that his transaction is not getting confirmed by the leader or committee, he can send the transaction to the underlying blockchain. The leader is additionally instructed to confirm all transactions it sees on the blockchain.

Now, if players see some transaction on the blockchain, that has not gotten notarized within a sufficiently long amount of time—counted in blocks in the underlying blockchains (say within n blocks)—they know that the leader/committee must be cheating/unavailable, and thus should enter the cool-down period. (Note that as long as the leader can confirm transactions before n blocks are created on the underlying blockchain, he cannot be "falsely accused"; and, by the security of the underlying blockchains those blocks cannot be created too fast.)

Selecting the committee. So far we have constructed a protocol that satisfies consistency and liveness under conditions $W \cap W'$ (i.e., assuming an honest majority of players, and an honest majority in the committee), and additionally satisfies liveness with optimistic responsiveness under condition O. The question now is how to select the committee. We consider two different approaches:

- **Using all players as the committee:** In a permissioned setting, the simplest approach is to simply use all players as the committee. In this case, $W' = W$ and thus, we trivially have resilience under W. A variant of this approach with improved communication complexity is to subsample a committee among the set of players (for instance, using the approach in [11] which additionally requires a random oracle), and change committees on a regular basis (to ensure adaptive security)—the resulting protocol, however, will only be secure if corruptions are "slow" (to ensure the attacker does not have time to corrupt the whole committee before it gets switched out). If sub-sampling is instead done "secretly" using a VRF and a random oracle (as in [32]), we can also ensure that the resulting protocol is adaptively secure *in a model with erasures*, even with "instantaneous corruption".

  We mention that these approaches may also be used in the permissionless setting if Thunderella is used to construct a crypto currency: then we can use (potentially a sub-sample of) recent "stakeholders" to form a committee.
- **Using "recent miners" as the committee:** A different approach that works in both the permissioned and permissionless setting is to select the committee as the miners of recent blocks (as was done in [38]). We note, however, that to rely on this approach, we need to ensure that the underlying is blockchain is "fair" [37] in the sense that the fraction of honestly mined blocks is close to the fraction of honest players. This is not the case for Nakamoto's original blockchain (see e.g., [17]), but as shown in [37], any blockchain can be turned into a fair one. If we use this approach, the resulting protocol will now be consistent and live under simply the condition W (i.e., honest majority), yet also satisfy optimistic liveness under condition O. (Again, this only gives security under adaptive corruption where corruption is "slow", so the set of recent miners changes sufficiently fast before they can all be corrupted.)

Permissionless Thunderella. For instance, if we apply the second approach (of selecting the committee as the recent miners) to Nakamoto's proof-of-work based blockchain, we get the following theorem:

Theorem 1 (Thunderellafor permissionless environments, informal). *Assume a proof-of-work random oracle. Then, there exists a state machine replication protocol that achieves consistency and (non-responsive) liveness in a permissionless environment as long as the adversary wields no more than $\frac{1}{2} - \epsilon$ the total online computation power in every round where ϵ is an arbitrarily small constant, and moreover it takes a short while for the adversary to adaptively corrupt nodes. Moreover, if more than $\frac{3}{4}$ of the online computation power is honest and online, then the protocol achieves responsiveness (after a short non-responsive warmup period) in any "epoch" in which the leader is honest and online.*

Permissioned Thunderella. Similar theorems can be shown for permissioned environments (in e.g., the "sleepy model" of [40], or even just in the "classic" model of Dolev-Strong [14]).

The classical mode is essentially the standard synchronous model adopted by the existing distributed systems and cryptography literature. In this model, all nodes are spawned upfront, and their identities and public keys are provided to the protocol as input; further, crashed nodes are treated as faulty and count towards the corruption budget. In a classical, synchronous network, we show that the classical Dolev-Strong byzantine agreement protocol [14] can be extended to implement Thunderella's underlying "blockchain". In this case, our Thunderella paradigm (where use the first approach to instantiate the committee) gives rise to the following informal theorem:

Theorem 2 (Thunderellafor permissioned, classical environments (informal)). *Assume the existence of a PKI and one-way functions. There exists a state machine replication protocol that achieves consistency and (non-responsive) liveness in a classical environment under any $f < n$ number of fully adaptive, byzantine corruptions where n denotes the total number of nodes; moreover, the protocol achieves responsiveness as long as the leader is honest and moreover $\lfloor \frac{n+f}{2} + 1 \rfloor$ nodes are honest.*

The "sleepy" model was recently proposed by Pass and Shi [40] to capture the requirements arising from "sporadic participation" in large-scale, permissioned consensus. Specifically, the sleepy model was in fact inspired by permissionless decentralized cryptocurrencies such as Bitcoin, where nodes may come and go frequently during the protocol, and the protocol should nonetheless guarantee consistency and liveness even for players that join late, and for players who might have had a short outage and woke up later to rejoin the protocol.

The sleepy model is "permissioned" in nature in that the set of approved protocol participants and their public keys are a-priori known and provided to the protocol as input. However, unlike the classical setting, (1) nodes are allowed to be non-participating (i.e., *sleeping*); (2) sleeping nodes are not treated as faulty; and (3) the protocol may not know in advance how many players will actually show up. In comparison, in a classical setting, non-participating nodes are regarded as having crashed and count towards the corruption budget; and moreover a classical protocol need not guarantee consistency and liveness for nodes that have crashed but wake up later to rejoin.

In such a sleepy model, Pass and Shi [40] show that roughly speaking, we can achieve consensus when the majority of *online* (i.e., non-sleeping) nodes are honest (interestingly, unlike the classical synchronous model, [40] also prove that no state machine replication protocol can tolerate more than $\frac{1}{2}$ corruption (among online nodes).

Our Thunderella paradigm (again using the first approach for selecting the committee) can be instantiated in the sleepy model using the sleepy consensus protocol as the underlying blockchain. This gives rise to the following informal theorem in a sleepy environment (where we assume that the adversary can adaptively put honest nodes to sleep).

Theorem 3 (Thunderellafor permissioned, sleepy environments (informal)). *Assume the existence of a PKI, enhanced trapdoor permutations, and a common reference string (CRS). There exists a state machine replication protocol that achieves consistency and (non-responsive) liveness in a sleepy environment with static corruptions, as long as $\frac{1}{2} - \epsilon$ of the online nodes are honest in every round for any arbitrarily small constant ϵ; moreover, if more than $\frac{3}{4}$ fraction of nodes are honest and online, the protocol achieves responsiveness (after a short non-responsive warmup period) in any epoch in which leader is honest and online.*

In fact, the above theorem also extends to adaptive corruptions *with erasures* using the adaptively secure version of sleepy consensus [40][2] as Thunderella's underlying blockchain, assuming the existence of a VRF and a random oracle (using the approach from [32]).

Lower bounds on the optimistic honest threshold. We additionally prove that our optimistic bound of 3/4 is tight: no protocol that is (worst-case) resilient for simply an honest majority, can be optimistically responsive when more than 1/4 of the player can be corrupted.

Practical Considerations: Instant Confirmation and Scalability. The *low latency* and *poor scalability* of Nakamoto's blockchain protocol are typically viewed as the main bottlenecks for Bitcoin as well as other cryptocurrencies.

Our paradigm provides a very practical and simple approach for overcoming these issue. The Thunderella paradigm shows how to build on top of currently running blockchains, to enable "optimistic instant confirmation" of transactions. Additionally, note that in our protocol, players only need to send transactions

[2] The paper has multiple adaptively secure versions; here we rely on the one that achieves adaptive security with erasures in the random oracle model (as this protocol has better parameters than the one which satisfies adaptive security without a random oracle).

to the leader, who in turn lead the committee to confirm the transaction. Most notably, the underlying blockchain is essentially only used when something goes wrong, and blocks need not be distributed to the whole network before getting confirmed; thus, Thunderella also solves the scalability issue with Nakamoto's blockchain protocol. Of course, both of these guarantees are only "optimistic"—but arguably, under normal circumstances one would expect 3/4 of the players to act honestly, and the leader could be incentivized (paid) to perform its job (and if it doesn't, will be kicked out). Thus, we believe our approach is a practically viable approach for circumventing the main bottlenecks of today's cryptocurrencies.

Comparison. At the surface, our idea is reminiscent of classical-style protocols such as PBFT and Byzantine-Paxos. In particular, protocols like PBFT also have a very simple normal path that consists of $O(1)$ rounds of voting. However, when the normal path gets stuck, PBFT-style protocols fall back to a "view change" mechanism that is also responsive—and thus these protocols tolerate only $\frac{1}{3}$ corruptions in the worst-case, and are invariably complex due to the need to handle asynchrony. (Furthermore, this approach is not amenable for protocols in the permissionless setting). Our key insight is to instead fall back to a synchronous path in the worst case, thus allowing us to circumvent the $\frac{1}{3}$ lower bound for partial synchrony and yet still be responsive in practice most of the time. Moreover, since our protocol is fundamentally synchronous, we benefit from the simplicity and robustness enjoyed by synchronous protocols (e.g., blockchains).

Interestingly, Thunderella is also a constant factor faster in the fast path than most PBFT- or Paxos-style protocols. PBFT-style protocols typically require multiple rounds of voting even in the normal path (*c.f.* Thunderella has exactly one)—and the latter rounds are necessary to prepare for the possibility of a view change. Although it is possible to compress the normal path to a single round of voting, this is typically achieved either by sacrificing resilience (e.g., tolerating only $\frac{1}{5}$ corruptions) [42] or by adding yet another optimistic layer on top of the normal path—thus further complicating the already complex protocol [29].

Roadmap. In this extended abstract, we simply provide a description and proof of the general Thunderella paradigm (informally described above) assuming the existence of a fixed committee, a majority of which is honest. We defer the formal treatment of how to select the committee to the online full version [36] (although we described it informally above).

2 Definitions

We present informal definitions in this section while deferring the detailed formal definitions to the online full version [36].

We adopt the standard Interactive Turing Machines (ITM) approach to model protocol execution. A protocol refers to an algorithm for a set of interactive Turing Machines (also called nodes) to interact with each other.

The execution of a protocol Π that is directed by an environment $\mathcal{Z}(1^\kappa)$ (where κ is a security parameter), which activates a number of nodes as either *honest* or *corrupt* nodes. Honest nodes would faithfully follow the protocol's prescription, whereas corrupt nodes are controlled by an adversary $\mathcal{A}$ which reads all their inputs/message and sets their outputs/messages to be sent.

A protocol's execution proceeds in *rounds* that model atomic time steps. Henceforth, we use the terms *round* and *time* interchangeably. At the beginning of every round, honest and online nodes receive inputs from an environment $\mathcal{Z}$; at the end of every round, honest and online nodes send outputs to the environment $\mathcal{Z}$.

Corruption model. In the standard distributed systems or cryptography literature, crashed nodes are often treated as faulty and count towards the corruption budget. In this paper, we describe a more general model in which we distinguish *crashed* nodes (also referred to as *sleeping* nodes) and *corrupt* nodes. An honest node may have a short-term or long-term outage during which it is not able to participate in the protocol. However, such a crashed node is not in the control of the adversary—in this case we do not attribute this node as corrupt. Informally, we often refer to the set of honest nodes that have not crashed as being *online*. We also consider all corrupt nodes as being online (since this gives the adversary more advantage).

We stress that the motivation for not treating crashed nodes as corrupt is to allow us to prove a more powerful theorem: our Thunderella paradigm ensures consistency and worst-case liveness when α fraction of the committee are honest but not necessarily online (and assuming that the underlying blockchain is secure). In particular, as we noted, α can be as small as a single member of the committee—but in this case the conditions necessary for instant confirmation are somewhat more stringent (i.e., all committee members must be honest and online for instant confirmation). In a more traditional model where crash is treated as corrupt, all of our theorems still apply—except that "honest" would equate to "honest and online".

More formally, in our model, the environment $\mathcal{Z}$ controls when nodes are spawned, corrupted, put to sleep, or waken up:

- At any time during the protocol execution, the environment $\mathcal{Z}$ can spawn new nodes, and newly spawned nodes can either be *honest* or *corrupt*. The adversary $\mathcal{A}$ has full control of all corrupt nodes.
- At any time during the protocol execution, $\mathcal{Z}$ can issue a `corrupt` instruction to an honest (and possibly sleeping) node. When this happens, its internal states are exposed to $\mathcal{A}$ and $\mathcal{A}$ henceforth controls the node.
- At any time during the protocol execution, $\mathcal{Z}$ can issue a `sleep` instruction to an honest node. When this happens, the node immediately becomes asleep (or sleeping), and it stops sending and receiving protocol messages and performing any computation. Sleeping is similar to the notion of a crash fault in the classical distributed systems terminology. In our paper, though, we treat sleeping nodes as being honest rather than attributing them towards the faulty budget.

- At any time during the protocol execution, $\mathcal{Z}$ can issue a `wake` instruction to an honest, sleeping node. At this point, this node immediately wakes up and continues to participate in the protocol. When an honest, sleeping node wakes up, pending messages that the node should have received while sleeping and additionally some adversarialy inserted messages may be delivered to the waking node.
- At any time during the protocol execution, $\mathcal{Z}$ can issue a `kill` instruction to a corrupt node. At this point, the corrupt node is removed from the protocol execution and is no longer considered as an online node—but note that the adversary $\mathcal{A}$ still knows the internal states of a killed node prior to its being killed.

Formally, we use the terminology *online* nodes to refer to the set of nodes that are *(i)* either honest and not sleeping; or *(ii)* corrupt but not having been killed.

Communication model. We assume that honest and online nodes can send messages to all other honest and online nodes. The adversary $\mathcal{A}$ is in charge of scheduling message delivery. $\mathcal{A}$ cannot modify the contents of messages broadcast by honest nodes, but it can reorder and delay messages sent by honest and online nodes, possibly subject to constraints on the maximum delays to be defined later. The adversary $\mathcal{A}$ is allowed to send messages to a subset of honest and online nodes but not all of them. The identity of the sender is not known to the recipient[3].

Formally, we say that $(\mathcal{A}, \mathcal{Z})$ *respects Δ-bounded delay* iff $\mathcal{Z}$ inputs Δ to all honest nodes when they are spawned, and moreover the following holds:

> *Δ-bounded delay.* Suppose an honest (and online) node sends a message at time t, then in any round $r \geq t + \Delta$, any honest node that is online in round r will have received the message, including nodes that may possibly have been sleeping but just woke up in round r, as well as nodes which may have just been spawned at the beginning of round r.

Throughout this paper, we assume that $\mathcal{Z}$ inputs the maximum delay parameter Δ to all honest nodes upon spawning (as noted in the above definition of Δ-bounded delay)—this means that the protocol has a-priori knowledge of an upper bound on the network's maximum delay. This is akin to the synchronous communication model in the classical distributed systems literature.

2.1 Classical, Sleepy, and Permissionless Models

The above generic model does not impose any constraints on when nodes are spawned, how many nodes are spawned, and which nodes are allowed to join the protocol. Thus the generic model can capture *permissionless* executions.

[3] Later in the paper, for instantiations in the permissioned model under a PKI, authenticated channels are implied by the PKI.

In this generic model, we can also model *classical* and *sleepy* executions by imposing *constraints* on $(\mathcal{A}, \mathcal{Z})$. The classical setting is what the vast majority of distributed systems literature focuses on. In a classical execution, $(\mathcal{A}, \mathcal{Z})$ is required to spawn all nodes, numbered $1..n$, upfront; further, honest nodes are assumed to be always online (i.e., $(\mathcal{A}, \mathcal{Z})$) are not allowed to issue `sleep` or `wake` instructions. The *sleepy* model was first proposed by Pass and Shi [40], which is meant to be "in-between" a fully permissionless and a classical permissioned model. In a sleepy execution, the set of allowed players are determined upfront and number $1..n$; however, nodes can join late, they can also fall asleep and later wake up again. Nodes that fall asleep are not treated as corrupt and when they wake up, the security properties we define such as consistency and liveness must ensue for them.

We use the terms (n, ρ, Δ)-permissionless environment, (n, ρ, Δ)-sleepy environment, or (n, f, Δ)-classical environment to capture the execution environment we care about and the parameters respected by $(\mathcal{A}, \mathcal{Z})$ (where ρ is a corruption fraction but f is the absolute number of corrupt nodes). We defer the formal definition of these terms to the online full version [36].

2.2 State Machine Replication

State machine replication has been a central abstraction in the 30 years of distributed systems literature. In a state machine replication protocol, a set of nodes seek to agree on an ever-growing log over time. We require two critical security properties: (1) *consistency*, i.e., all honest nodes' logs agree with each other although some nodes may progress faster than others; (2) *liveness*, i.e., transactions received by honest nodes as input get confirmed in all honest nodes' logs quickly. We now define what it formally means for a protocol to realize a "state machine replication" abstraction.

Syntax. In a state machine replication protocol, in every round, an honest and online node receives as input a set of transactions txs from $\mathcal{Z}$ at the beginning of the round, and outputs a LOG collected thus far to $\mathcal{Z}$ at the end of the round.

Security definitions. Let $T_{\text{confirm}}(\kappa, n, \rho, \Delta, \delta)$ and $T_{\text{warmup}}(\kappa, n, \rho, \Delta, \delta)$ be polynomial functions in the security parameter κ and possibly other parameters of the view such as the number of nodes n, the corrupt fraction ρ, the actual maximum network delay δ, the network delay upper bound Δ that is provided by $\mathcal{Z}$ to the protocol as input, etc.

Definition 1 (Security of a state machine replication protocol). *We say that a state machine replication protocol Π satisfies consistency (or $(T_{\text{confirm}}, T_{\text{warmup}})$-liveness resp.) w.r.t. some $(\mathcal{A}, \mathcal{Z})$, iff there exists a negligible function* $\mathsf{negl}(\cdot)$, *such that for any $\kappa \in \mathbb{N}$, except with* $\mathsf{negl}(\kappa)$ *probability over the choice of* $\mathsf{view} \leftarrow \mathsf{EXEC}^{\Pi}(\mathcal{A}, \mathcal{Z}, \kappa)$, *consistency (or $(T_{\text{confirm}}, T_{\text{warmup}})$-liveness resp.) is satisfied:*

- *Consistency*: A view satisfies consistency iff the following holds:
  - *Common prefix.* Suppose that in view, an honest node i outputs LOG to $\mathcal{Z}$ at time t, and an honest node j outputs LOG' to $\mathcal{Z}$ at time t' (i and j may be the same or different), it holds that either $\mathsf{LOG} \prec \mathsf{LOG}'$ or $\mathsf{LOG}' \prec \mathsf{LOG}$. Here the relation $\prec$ means "is a prefix of". By convention we assume that $\emptyset \prec x$ and $x \prec x$ for any x.
  - *Self-consistency.* Suppose that in view, a node i is honest and online at time t and $t' \geq t$, and outputs LOG and LOG' at times t and t' respectively, it holds that $\mathsf{LOG} \prec \mathsf{LOG}'$.
- *Liveness*: A view satisfies $(T_{\text{confirm}}, T_{\text{warmup}})$-liveness iff the following holds: if in some round $T_{\text{warmup}} < t \leq |\mathsf{view}| - T_{\text{confirm}}$, some node honest and online in round t either received from $\mathcal{Z}$ an input set txs that contains some transaction m or has m in its output log to $\mathcal{Z}$ in round t, then, for any node i honest and online at any time $t' \geq t + T_{\text{confirm}}$, let LOG be the output of node i at time t', it holds that $\mathsf{m} \in \mathsf{LOG}$.

  Intuitively, liveness says that transactions input to an honest node get included in honest nodes' LOGs within T_{confirm} time; and further, if a transaction appears in some honest node's LOG, it will appear in every honest node's LOG within T_{confirm} time.

2.3 Abstract Blockchain Protocols

A blockchain protocol can be regarded as a way to realize state machine replication. We now formally define what it means for a protocol to realize to a blockchain abstraction. In our paper, our end goal is to realize state machine replication and we leverage an abstract blockchain as an underlying building block. We note that while the blockchain abstraction may superficially resemble that of state machine replication, the blockchain abstraction in fact allows us to additionally express (1) a rough notion of time through chain growth; and (2) fairness properties [37] through chain quality.

Syntax and Security Definitions

Syntax. An abstract blockchain protocol satisfies the following syntax. In each round, every node that is honest and online in this round receives from $\mathcal{Z}$ a set of transactions txs at the beginning of the round; and outputs to $\mathcal{Z}$ an abstract blockchain chain at the end of the round. An abstract blockchain denoted chain is an ordered sequence of blocks of the following format:

$$\mathsf{chain} := \{\mathsf{txs}_i\}_{i \in [|\mathsf{chain}|]}$$

where each txs_i is an application-specific payload such as a set of transactions.

Blockchain notations. We use the notation chain to denote an abstract blockchain. The notation chain[: $-\ell$] denotes the entire chain except the trailing ℓ blocks; chain[: ℓ] denotes the entire chain upto the block at length ℓ; chain[$-\ell$:] denotes the trailing ℓ blocks of chain; and chain[ℓ :] denotes all blocks at length ℓ or greater.

Henceforth we say that chain is "*an honest chain in* view", iff chain is some honest (and online) node's output to the environment $\mathcal{Z}$ in some round in view. We use the notation $\mathsf{chain}_i^t(\mathsf{view})$ to denote node i's chain in round t in view—since the context is clear, we often omit writing the view explicitly in the above notation.

Security definitions. A blockchain protocol should satisfy chain growth, chain quality, and consistency. Intuitively, chain growth requires that honest nodes' blockchains grow steadily, neither too fast nor too slow. Chain quality requires that in any honest node's chain, any sufficiently long window of consecutive blocks contains a certain fraction of blocks that are mined by honest nodes. Consistency requires that all honest nodes' chains agree with each other except for the trailing few blocks. We will formally define these security properties below.

Definition 2 (Security of an abstract blockchain protocol). *We say that a blockchain protocol $\Pi_{\mathsf{blockchain}}$ satisfies (T, g_0, g_1)-chain growth, (T, μ)-chain quality, and T-consistency w.r.t. some $(\mathcal{A}, \mathcal{Z})$, iff there exists a negligible function* $\mathsf{negl}(\cdot)$*, such that for every $\kappa \in \mathbb{N}$, except with* $\mathsf{negl}(\kappa)$ *probability over the choice of view $\leftarrow \mathsf{EXEC}^{\Pi_{\mathsf{blockchain}}}(\mathcal{A}, \mathcal{Z}, \kappa)$, the following hold for view:*

- (T, g_0, g_1)*-chain growth.* A view satisfies (T, g_0, g_1)-chain growth iff the following hold:
  - *Consistent length:* If in round r some honest chain is of length ℓ, then in round $r + \Delta$, all honest chains must be of length at least ℓ.
  - *Growth lower bound:* For any r and t such that $g_0(t - r) \geq T$, let chain^r and chain^t denote two honest chains in round r and t respectively, it holds that
  $$|\mathsf{chain}^t| - |\mathsf{chain}^r| \geq \lfloor g_0(t - r) \rfloor$$
  - *Growth upper bound:* For any r and t such that $g_1(t - r) \geq T$, let chain^r and chain^t denote two honest chains in round r and t respectively, it holds that
  $$|\mathsf{chain}^t| - |\mathsf{chain}^r| \leq \lceil g_1(t - r) \rceil$$
- (T, L, μ)*-chain quality.* A view satisfies (T, L, μ)-chain quality iff the following holds: for any honest chain denoted chain in view, for any T consecutive blocks chain[$j + 1..j + T$], more than μ fraction of the blocks in chain[$j + 1..j + T$] are mined by honest nodes at most L blocks ago—here we say that a block chain[i] is "mined by an honest node at most L blocks ago" iff there is a set txs such that txs $\subseteq$ chain[i] and moreover $\mathcal{Z}$ input txs to some honest node when its last output to $\mathcal{Z}$ contains the prefix chain[: $i - L$] (here if $i - L < 0$, we round it to 0).

– *T-consistency.* A view satisfies T-consistency iff the following hold: for any two honest chains chain^r and chain^t in round r and $t \geq r$ respectively, it holds that

$$\mathsf{chain}^r[: -T] \prec \mathsf{chain}^t$$

We stress that since chain^r and chain^t can possibly belong to the same node, the above definition also implies "future self consistency" except for the trailing T blocks.

Liveness as a derived property. Intuitively, liveness requires that if honest nodes receive a transaction m as input, then m appear in honest chains very soon. More formally, we say that a blockchain protocol $\Pi_{\mathrm{blockchain}}$ satisfies (K, T)-liveness w.r.t. some $(\mathcal{A}, \mathcal{Z})$ iff there exists a negligible function $\mathsf{negl}(\cdot)$ such that for every $\kappa \in \mathbb{N}$, except with $\mathsf{negl}(\kappa)$ probability over the choice of $\mathsf{view} \leftarrow \mathsf{EXEC}^{\Pi_{\mathrm{blockchain}}}(\mathcal{A}, \mathcal{Z}, \kappa)$, the following holds:

– Suppose that in any round $r \geq t$, $\mathcal{Z}$ always inputs a set that contains some m to every honest and online node i unless $\mathsf{m} \in \mathsf{chain}_i^r[: -T]$. Then, for any honest chain denoted chain in view whose length is at least $\ell + K + T$, it holds that $\mathsf{chain}[: \ell + K]$ contains m where ℓ denotes the shortest honest chain length at time t.

The liveness of a blockchain protocol is directly implied by chain growth and chain quality as stated in the following lemma.

Lemma 1 (Liveness). *Suppose that a blockchain protocol $\Pi_{\mathrm{blockchain}}$ satisfies (K, g_0, g_1)-chain growth, (K', L, μ) chain quality and T-consistency w.r.t. some $(\mathcal{A}, \mathcal{Z})$ for positive parameters K, g_0, g_1, K', L, μ and T, then it holds that $\Pi_{\mathrm{blockchain}}$ satisfies $(2K + 2g_1 + K' + L, T)$-liveness w.r.t. $(\mathcal{A}, \mathcal{Z})$.*

Proof. We ignore the negligible fraction of views where relevant bad events take place. Let r' be the earliest round in which some honest chain reaches length at least $\ell + K + g_1 + K' + L + T$, and let chain^* be an honest chain in round r' of length at least $\ell + K + g_1 + K' + L + T$. By chain quality, in the window $\mathsf{chain}^*[\ell + K + g_1 + L + 1 : \ell + K + g_1 + K' + L]$, there must be an honest block denoted B such that $\mathcal{Z}$ input (a subset of) the contents of B to some honest node i in round $r \leq r'$ when its chain contains the prefix $\mathsf{chain}^*[: \ell + K + g_1 + 1]$. By chain growth upper bound, the longest honest chain in round t must be of length at most $\ell + K + g_1$, and thus B must be input to some honest and online node i by $\mathcal{Z}$ in some round r where $t \leq r \leq r'$. By assumption, B must contain m unless $\mathsf{chain}_i^r[: -T]$ already contains m. By consistency, it must be that chain^* and chain_i^r are no longer than $\ell + 2(K + g_1) + K' + L + T$. By consistency, for any honest chain ch in view of length at least $\ell + 2(K + g_1) + K' + L + T$, it must be that $\mathsf{chain}_i^r[: -T] \prec \mathsf{ch}[: \ell + 2(K + g_1) + K' + L]$ and $\mathsf{chain}^*[: -T] \prec \mathsf{ch}[: \ell + 2(K + g_1) + K' + L]$, and thus $\mathsf{ch}[: \ell + 2(K + g_1) + K' + L]$ must contain m.

Blockchain Implies State Machine Replication. Given any blockchain protocol $\Pi_{\text{blockchain}}$, it is easy to construct a state machine replication protocol where (1) nodes run an instance of $\Pi_{\text{blockchain}}$; (2) an honest node broadcasts all newly seen transactions to each other; and (3) in every round, nodes remove the trailing T blocks from the present chain (where T is the consistency parameter) and output the truncated chain to the environment $\mathcal{Z}$ [38,40]. It is not difficult to see that consistency (of the resulting state machine replication protocol) follows directly from consistency of the blockchain; and liveness follows from chain quality and chain growth. The above intuition has been formalized in earlier works [38,40].

2.4 Preliminaries: Responsiveness

Responsiveness. Recall that throughout this paper we always assume that $(\mathcal{A}, \mathcal{Z})$ respects Δ-bounded delay for some Δ, i.e., $\mathcal{Z}$ informs the protocol of a delay upper bound Δ upfront and all honest messages are then delivered within Δ number of rounds. A state machine replication protocol is said to be *responsive* if the transaction confirmation time is independent of the a-priori known upper bound Δ of the network's delay, but depends only on the actual maximum network delay. To put our results in perspective, we formally define the notion of responsiveness below and state a known lower bound result suggesting the impossibility of responsiveness against $\frac{1}{3}$ fraction of corruption. In the remainder of the paper, we will show that if one *optimistically* hopes for responsiveness only in lucky situations, then we can have protocols that retains consistency and liveness even under more than $\frac{1}{3}$ corruption. In practice, this means that we can have protocols that are responsive most of the time, and even when more than $\frac{1}{3}$ nodes are corrupt, the protocol can still guarantee consistency and liveness although performance would degrade when under attaack.

Responsiveness. We define a technical notion called responsiveness for a state machine replication protocol. Intuitively, responsiveness requires that except for a (possibly non-responsive) warmup period in the beginning, all transactions input afterwards will perceive transaction confirmation delay that is independent of the a-priori set upper bound Δ on the network's delay. As shown in earlier works [15,38], responsive state machine replication is impossible if $\frac{1}{3}$ or more fraction of the nodes are corrupt (even in a permissioned, classical environment with static corruptions, and even assuming that a proof-of-work oracle exists).

Definition 3 (Responsive state machine replication [38]**).** *Suppose that $(\mathcal{A}, \mathcal{Z})$ respects Δ-bounded delay for some Δ. We say that a state machine replication protocol Π satisfies $(T_{\text{confirm}}, T_{\text{warmup}}$-responsiveness w.r.t. $(\mathcal{A}, \mathcal{Z})$ iff Π satisfies $(T_{\text{confirm}}, T_{\text{warmup}})$-liveness w.r.t. $(\mathcal{A}, \mathcal{Z})$, and moreover the function T_{confirm} does not depend on the a-prior delay upper bound Δ.*

We say that a protocol Π satisfies consistency (or responsiveness resp.) in (n, f, Δ)-classical, static environments iff for every p.p.t. $(\mathcal{A}, \mathcal{Z})$ pair that respects (n, f, Δ)-classical execution and static corruption, Π satisfies consistency (or responsiveness resp.) w.r.t. $(\mathcal{A}, \mathcal{Z})$. We can similarly define (n, ρ, Δ)-sleepy, static environments and (n, ρ, Δ)-permissionless, static environments.

Theorem 4 (Impossibility of responsiveness against $\frac{1}{3}$ corruption [38]). *For any n and f such that $n \leq 3f$ and for any polynomial T_{confirm} in κ and δ, and T_{warmup} in κ, Δ, and δ, there exists some polynomial function Δ in κ such that no state machine replication protocol no state machine replication protocol can simultaneously achieve consistency and $(T_{\text{confirm}}, T_{\text{warmup}})$-responsiveness in (n, f, Δ)-classical, static environments even assuming the existence of a proof-of-work oracle.*

The proof of the above theorem was presented by Pass and Shi in a recent work [38] where they modified the classical lower bound proof by Dwork et al. [15] and made it work even in the proof-of-work setting.

Recall that permissioned-classical is expressed as constraints on $(\mathcal{A}, \mathcal{Z})$ in our formal framework. This means that a lower bound for $n \leq 3f$ in the classical setting immediately implies a lower bound in more permissive settings where $(\mathcal{A}, \mathcal{Z})$ need not respect the permissioned-classical constraints as long as $n \leq 3f$ (or the equivalent holds). In other words, the above impossibility also applies to sleepy and permissionless settings (we defer formal theorem statements for these settings to the online full version [36]).

Interestingly, how to achieve responsive state machine replication against fewer than $\frac{1}{3}$ fraction of corruption is also known in the in the permissioned setting assuming the existence of a PKI [9], as well as in the permissionless setting assuming proof-of-work [38] (and under additional technical assumptions).

3 Basic Thunderella Protocol with a Static Committee

We first describe the basic Thunderella protocol assuming a static committee that is known a-priori to all nodes. We will discuss how to perform committee reconfiguration in the online full version [36]. For conceptual simplicity, we describe a version of the protocol where the blockchain is also collecting transactions constantly in the background—in practical implementations, it will not be too difficult to optimize our theoretical construction further such that the blockchain need not store an additional copy of all transactions under optimistic conditions.

As mentioned, in general, the Thunderella paradigm can be instantiated with any suitable asynchronous protocol to serve as the optimistic path and any suitable synchronous protocol to serve as the fallback path. However, we believe that a particular attractive instantiation is to use a simple voting-based protocol for the optimistic path and a blockchain as the fallback. Thus for concreteness, we will describe Thunderella for this specific instantiation.

Terminology. Our basic approach assumes three logical entities:

- *miners* of the underlying blockchain $\Pi_{\text{blockchain}}$;
- a *leader*; and
- a *committee* denoted committee.

To retain consistency and worst-case liveness (i.e., confirmation in the speed of the underlying $\Pi_{\text{blockchain}}$), we need to assume that (1) the underlying blockchain $\Pi_{\text{blockchain}}$ retains security (and this would translate to different compliance rules depending on how we instantiate the underlying blockchain); (2) α fraction of the committee are assumed to remain honest (but not necessarily online) where α is a knob that effectively allows us to trade-off security and performance as is explained later. Notably, the leader need not be trusted for consistency and worst-case liveness.

For concreteness, in our description we will often assume that $\alpha = \frac{1}{2}$, but in fact our approach generalizes to any choice where $0 < \alpha < 1$; and whenever appropriate, we will remark how to generalize the scheme's parameters for arbitrary α.

For simplicity, in this section we start out by assuming a static committee. In a permissioned setting, this committee can be the set of all nodes. In a permissionless setting where the set of players are not known in advance, we can elect the committee dynamically from the underlying blockchain using known techniques [37,38] or have stake-holders act as the committee [6,11,32]—however we defer the discussion of committee election and reconfiguration to the online full version [36]. Although we assume a static committee in this section, our basic protocol supports leader reconfiguration. In our presentation below we focus on describing the mechanism that enables leader reconfiguration without specifying concretely what leader re-election policy to adopt—exactly what policy to adopt depends on the application context and we thus defer the discussion of policies to the online full version [36].

3.1 Our Basic Protocol in a Nutshell

We first describe the intuition behind our basic protocol. For simplicity, we focus our description on what happens within a single epoch in which the identity of the leader is common knowledge.

Optimistic Fast Path. The optimistic fast path consists of a single round of voting to confirm each transaction (or batch). The leader serves as a coordinator and sequences transactions in the optimistic path. It tags each freshly seen transaction (or a batch) with a sequence number that increments over time, and the resulting tuple (seq, tx) is referred to as a notarization request. Whenever the committee members hear a notarization request (seq, tx) from the leader, it will sign the tuple (seq, tx) as long as it has not signed any tuple for seq before. For consistency, it is important that an honest committee member signs only one unique tuple (seq, tx) for every sequence number seq.

Whenever an honest node observes that a notarization request $(\mathsf{seq}, \mathsf{tx})$ has collected votes from more than $\frac{3}{4}$ of the committee, $(\mathsf{seq}, \mathsf{tx})$ is considered notarized. Although any notarized transaction is ready to be confirmed, an honest node is only allowed to output a notarized $(\mathsf{seq}, \mathsf{tx})$ tuple iff for every $s < \mathsf{seq}$, a tuple $(s, \_)$ has already been output. In other words, the output sequence is not allowed to skip any sequence numbers (since transactions must be processed in a linearized order). Henceforth, we referred to a sequence of notarized transactions with contiguous, non-skipping sequence numbers as a *lucky sequence*. In other words, honest nodes always output the maximal lucky sequence they have observed.

It is not hard to see that the optimistic, fast path satisfies the following properties as long as *the majority of the online committee members are honest* (below, we focus our discussion for the specific case $\alpha = \frac{1}{2}$, although the argument can easily be generalized to arbitrary choices of α):

- The following *agreement* property is satisfied even when the leader is corrupt and the committee may not be online: if any two honest nodes have output $(\mathsf{seq}, \mathsf{tx})$ and $(\mathsf{seq}, \mathsf{tx}')$ respectively, it must be that $\mathsf{tx} = \mathsf{tx}'$ (except with negligible probability over the choice of view).
- The following *liveness* property is satisfied only when the leader is honest and online and moreover more than $\frac{3}{4}$ of the committee are honest and online (i.e., when the optimistic conditions hold): every transaction input to an honest node will appear in all nodes' output logs in $O(1)$ actual roundtrips—in other words, when optimistic conditions hold, not only do we achieve liveness but we in fact also achieve *responsiveness*.

Note that when the optimistic conditions do not hold, liveness is not guaranteed for the optimistic path. For example, a corrupt leader can propose different transactions to different nodes for the same sequence number, and thus no transaction will collect enough votes to become notarized. Further, progress can also be hampered if not enough committee members are honest and online to vote.

Summarizing the above, if the leader is honest and online and moreover more than $\frac{3}{4}$ fraction of the committee are honest and online, all nodes will confirm transactions responsively in the optimistic path. However, to make our protocol complete, we need to deal with the case when either the leader is corrupt (or not online), or the committee is not more than $\frac{3}{4}$ honest and online—in the latter case, we wish to fall back to the worst-case guarantees offered by the underlying blockchain. Below we describe how such fallback can be achieved.

Falling Back to the Blockchain. In the fallback slow path, nodes will confirm transactions using the slow blockchain. The most non-trivial part of our protocol is how to switch between the optimistic path and the fallback path. To this end, we must answer the following two questions.

1. How do nodes decide when to fall back to the slow path?
2. Once the above decision is made, what is the mechanism for achieving this fallback?

When to fall back. The idea is to use the underlying blockchain to collect evidence of the optimistic path not working (e.g., either due to corrupt or crashed leader or due to not sufficiently many committee members being honest and online). Such evidence must be robust such that the adversary cannot raise false alarms when the optimistic path is actually working.

For conceptual simplicity, we can imagine the following: (1) whenever honest nodes mine a block, they incorporate into the block their entire view so far, including all unnotarized transactions and notarized transactions they have seen—in the actual protocol, the transactions stored in the blockchain can be easily deduplicated and compressed; (2) honest nodes always gossip their views to each other, such that if one honest node sees some (notarized or unnotarized) transaction by round r, then all honest nodes will have seen it by round $r + \Delta$. Thus by the liveness property of the underlying blockchain, if any (notarized or unnotarized) transaction is observed by any honest node in round r, then in roughly $\epsilon\kappa$ blocks of time, the transaction will appear in the blockchain.

Now, we may use the following criterion to detect when the optimistic path is not working:

> *Fallback condition:* Assume that chain is the stabilized prefix of some honest node's blockchain. If some unnotarized transaction tx appears in the block $\mathsf{chain}[\ell]$ but tx still has not become part of a lucky sequence contained in $\mathsf{chain}[: \ell + \kappa]$ where κ is a security parameter[4], then we conclude that the optimistic path has failed, and that a fallback is necessary.

Note that if the optimistic conditions hold, then the leader would have observed the unnotarized tx when its blockchain is roughly ℓ in length, and the committee would have notarized tx quickly; thus tx will soon become part of a lucky sequence contained in every node's blockchain. If this has not happened within κ blocks of time, then the optimistic conditions must no longer hold.

We also note that using the above mechanism, all honest nodes will decide to fall back within Δ rounds from each other. We now reach our next question: what mechanism do we rely on for the falling back?

How to fall back. The challenge is that when honest nodes decide to fall back (within Δ rounds from each other), although their optimistic logs are prefixes of each other, the logs could be of different lengths. One decision to make during the fallback is where (i.e., at which sequence number) to end the optimistic log before switching to blockchain mode—importantly, for consistency, honest nodes must agree on this decision. We point out that agreeing on this decision actually requires a full agreement instance—unlike the optimistic path where we punted on liveness, here this decision must be made with both consistency and liveness.

Thus the most natural idea is to rely on the underlying blockchain to reach agreement regarding this decision. To this end, we introduce the notion of a grace period that serves as a cool-down period before we eventually fall back into slow mode. The grace period consists of κ number of consecutive blocks where κ is

[4] Transactions of a lucky sequence are allowed to appear out of order in the blockchain.

a security parameter. Let chain denote the stabilized part of an honest node's blockchain and suppose that ℓ^* is the first block such that chain$[: \ell^*]$ triggers the "fallback condition" as described above. Then, the grace period will consist of the blocks chain$[\ell^* + 1 : \ell + \kappa]$. Informally speaking, the grace period is utilized in the following manner:

- Let LOG^* be an honest node's output log at the moment that the grace period starts (thus LOG^* must be a lucky sequence);
- Let chain be the stabilized prefix of this honest node's chain:
  - If the grace period has not ended in chain, then the node outputs the longer of (1) LOG^*; and (2) the maximal lucky sequence contained in chain. Note that in this case, the node does not output any additional transactions that are not part of the lucky sequence.
  - Else if the grace period has ended in chain, then the node first outputs the maximal lucky sequence contained in chain; then it outputs every other transaction (notarized or unnotarized) contained in chain (in the order that they are included in chain). In other words, after the grace period is over, the nodes start confirming transactions based on the blockchain.

Let $\mathsf{LOG}_{\max}$ denote the maximal lucky sequence contained in an honest node's blockchain by the end of the grace period. Effectively, in the above mechanism, nodes agree on $\mathsf{LOG}_{\max}$ before falling to blockchain mode. Importantly, the following informal claim must hold:

Claim (Informal). Except with negligible probability, $\mathsf{LOG}_{\max}$ must be at least as long as any honest node's output log when the node detects the start of the grace period.

To see why, recall that as mentioned earlier, all honest nodes gossip always their protocol views to each other; and honest nodes always embed their entire protocol view into any block they mine (in the actual protocol, the messages can be compressed). Thus, by liveness, any honest node's output log when the grace period starts will be in the blockchain κ blocks later.

Initiating a New Optimistic Epoch. So far, we have described our protocol from the perspective of a single epoch in which the leader is common knowledge. Whenever the protocol is in a slow path, however, we would like to allow the nodes to try to reinitiate an optimistic epoch and try to be fast again. This is easy to achieve since our underlying blockchain is always up and live! Thus one can simply rely on the underlying blockchain to implement any policy-based decision to reinitiate a new epoch. For example, the blockchain can be used to agree on (1) at which block length to reinitiate a new epoch; and (2) who will act as the leader in the new epoch. Our Thunderella framework leaves it to the application layer to specify such policy decisions (e.g., such policies can be implemented through generic smart contracts running atop the underlying blockchain).

Our detailed description in the remainder of this section is aware of the existence of multiple epochs, and thus transactions' sequence numbers are tagged with the epoch number to avoid namespace collision.

3.2 Detailed Protocol Description

We now formally describe our basic Thunderella protocol with a static committee. Our description and proofs are modularized. Specifically, we first describe the minimal set of protocol instructions necessary for guaranteeing consistency (Sect. 3.2)—in an actual implementation, security audit should be prioritized for this part of the protocol. We then describe other surrounding mechanisms (e.g., how to concretely instantiate the chain state function and how the leader proposes transactions) that allow us to additionally achieve worst-case liveness (Sect. 3.2) and optimistic responsiveness (Sect. 3.2).

Concrete blockchain parameters. For concreteness, henceforth in this section we assume a blockchain protocol denoted $\Pi_{\mathsf{blockchain}}$ that achieves $(0.05\kappa, g_0, g_1 = \frac{1}{c\Delta})$-chain growth for some positive g_0 and some positive constant c, $(0.05\kappa, 1, \mu)$-chain quality where μ is positive, 0.05κ-consistency, and $(0.05\kappa, 0.05\kappa)$-liveness w.r.t. to any p.p.t. $(\mathcal{A}, \mathcal{Z})$ that is compliant w.r.t. $\Pi_{\mathsf{blockchain}}$. For our later concrete instantiations in permissionless and permissioned settings, existing blockchains constructions [19,35,40] would satisfy the necessary security properties given the above these parameters. Although we assume these concrete parameters, our Thunderella framework can easily be generalized to other parameters.

Useful Definitions. Henceforth, let $\Sigma = (\mathsf{Gen}, \mathsf{Sign}, \mathsf{Vf})$ denote a digital signature scheme.

Notarized transactions. We say that a tuple (e, s, m, V) is a *notarized* transaction for epoch e and sequence number s w.r.t. committee iff

- For each $(\mathsf{pk}, \sigma) \in V$, $\mathsf{pk} \in \mathsf{committee}$ and moreover σ is a valid signature for (e, s, m) under pk—in this case, we also say that (pk, σ) is a *valid vote* for (e, s, m).
- There are more than $\frac{3}{4} \cdot |\mathsf{committee}|$ votes in V with distinct pks.

If (e, s, m, V) is a notarized transaction, we also say that V is a *valid notarization* for (e, s, m).

Remark 1. Note that the above definition works for $\alpha = \frac{1}{2}$. For a general $\alpha \in (0, 1]$, we can simply replace the constant $\frac{3}{4}$ with $1 - \frac{\alpha}{2}$.

Blockchain states. We assume that there is a deterministic and efficiently computable function Γ such that given an abstract blockchain chain, the function Γ divides chain into multiple *epochs* interspersed with *interims*. Each epoch is a sequence of consecutive blocks in chain, and f also outputs a unique epoch number e for each epoch. A sequence of consecutive blocks that do not belong to any epoch are called interim blocks. Each epoch always contains two sub-phases, an *optimistic period* followed by a *grace period*, each of which contains at least κ consecutive blocks and thus each epoch contains at least 2κ consecutive blocks (unless end of chain is reached).

Formally, we say that $\Gamma(\kappa, \cdot, \cdot)$ is a chain-state function iff for any chain and $0 \leq \ell \leq |\mathsf{chain}|$, $\Gamma(\kappa, \mathsf{chain}, \ell)$ outputs one of the following:

- some $(e, \mathtt{optimistic})$: in this case we say that $\mathsf{chain}[\ell]$ is an optimistic block belonging to epoch e (w.r.t. $\Gamma(\kappa, \cdot, \cdot)$);
- some $(e, \mathtt{grace})$: in this case we say that $\mathsf{chain}[\ell]$ is a grace block belonging to epoch e (w.r.t. $\Gamma(\kappa, \cdot, \cdot)$);
- or $\mathtt{interim}$: in this case we say that $\mathsf{chain}[\ell]$ is an interim block (w.r.t. $\Gamma(\kappa, \cdot, \cdot)$).

We say that a chain-state function $\Gamma(\kappa, \cdot, \cdot)$ is admissible iff for any chain:

1. for any $0 \leq \ell \leq \ell' \leq |\mathsf{chain}|$, if $\mathsf{chain}[\ell]$ belongs to epoch e and $\mathsf{chain}[\ell']$ belongs to epoch e', then $e' \geq e$;
2. for every e: all blocks corresponding to epoch e in chain must appear in a consecutive window, and moreover, all optimistic blocks for epoch e must appear before grace blocks for epoch e;
3. for every epoch e appearing in chain: there must be at least κ grace blocks belonging to epoch e in chain unless chain ends at an epoch-e block.
4. for every chain and every $0 \leq \ell \leq |\mathsf{chain}|$, $\Gamma(\kappa, \mathsf{chain}, \ell)$ depends only on $\mathsf{chain}[: \ell]$ but not $\mathsf{chain}[\ell + 1 :]$.

Lucky sequence. A sequence of notarized transactions $\{(e_i, s_i, \mathsf{m}_i, V_i)\}_{i \in [m]}$ is said to be a lucky sequence for epoch e iff for all $i \in [m]$, $e_i = e$ and $s_i = i$.

Blockchain linearization. Given an abstract blockchain chain, we do not simply output all transactions in chain in the most natural way. Instead, we adopt an algorithm denoted $\mathsf{linearize}^{\Gamma(\kappa,\cdot,\cdot)}(\mathsf{chain})$ for chain linearization. Henceforth we often write $\mathsf{linearize}(\mathsf{chain})$ for simplicity without explicitly denoting the chain-state function $\Gamma(\kappa, \cdot, \cdot)$.

Our chain linearization algorithm $\mathsf{linearize}(\mathsf{chain})$ is defined as follows: scan through the chain from left to right, and output the following:

1. For each epoch $\mathsf{chain}[\ell : \ell']$ encountered with the epoch number e, output the following in order:
   - first extract the maximal lucky sequence TXs for epoch e from $\mathsf{chain}[: \ell']$ and output $\mathsf{strip}(\mathsf{TXs})$ where $\mathsf{strip}(\cdot)$ will be defined below;
   - if $\mathsf{chain}[\ell]$ is not the end of chain, let TXs' be all remaining records in $\mathsf{chain}[\ell : \ell']$ not contained in TXs, output $\mathsf{strip}(\mathsf{TXs}')$;
2. For each interim $\mathsf{chain}[\ell : \ell']$ encountered, extract all transactions TXs from $\mathsf{chain}[\ell : \ell']$ and output $\mathsf{strip}(\mathsf{TXs})$.

In the above, the function $\mathsf{strip}(\cdot)$ removes signatures from notarized transactions: for a notarized transaction $\mathsf{strip}(e, s, \mathsf{m}, V) := (e, s, \mathsf{m})$; for an unnotarized transaction we define $\mathsf{strip}(\mathsf{m}) := \mathsf{m}$. If the input to $\mathsf{strip}(\cdot)$ is a sequence of transactions, the same operation is applied to each transaction.

Π_{thunder} : Core Protocol for Consistency

Additional notation. A node's view consists of every message (including blockchains) it has received from $\mathcal{Z}$ or over the network. Henceforth we say that a notarized transaction (e, s, m, V) is in a node's view iff (e, s, m) exists in the node's view, and every $(\mathsf{pk}, \sigma) \in V$ exists in the node's view (not necessarily appearing together in the node's view). Multiple notarized transactions can exist for a unique (e, s, m) by taking different subsets of V—but in our presentation below, we always take V to be all the valid votes for (e, s, m) in a node's view, such that if for some tuple (e, s, m) there is a notarized transaction (e, s, m, V) in a node's view, then the choice is unique.

Assumptions. Although not explicitly noted, henceforth in all of our protocols, we assume that whenever an honest node receives any message on the network, if the message has not been broadcast before, the honest node broadcasts the message.

Protocol Π_{thunder}. Below we describe the $\Pi_{\text{thunder}}^{\Gamma(\kappa,\cdot,\cdot)}$ protocol that is parametrized by an admissible chain-state function $\Gamma(\kappa, \cdot, \cdot)$. Henceforth in our scheme, we often omit explicitly writing the chain-state function $\Gamma(\kappa, \cdot, \cdot)$.

- *Initialize.*
  - Call $(\mathsf{pk}, \mathsf{sk}) \leftarrow \Sigma.\mathsf{Gen}(\kappa)$ to generate a signing key pair. Output pk to $\mathcal{Z}$.
  - Wait to receive $\mathsf{committee}$ from $\mathcal{Z}$, and henceforth, validity of votes and acceptability of chains will be defined w.r.t. $\mathsf{committee}$.
  - Fork an instance of the $\Pi_{\text{blockchain}}$ protocol with appropriate parameters determined by ρ, n and Δ[5].
- *Notarize.* Upon receiving notarization request (e, s, m) from $\mathcal{Z}$: if $\mathsf{pk} \in \mathsf{committee}$ and no signature has been produced for (e, s) earlier, compute $\sigma := \Sigma.\mathsf{Sign}_{\mathsf{sk}}(e, s, \mathsf{m})$ and broadcast $((e, s, \mathsf{m}), \sigma)$.
- *Propose.* Every round, let chain be the output from the $\Pi_{\text{blockchain}}$ instance.
  - Let TXs be a set containing (1) every notarized transaction (e, s, m, V) in the node's view such that no notarized transaction $(e, s, \mathsf{m}, \_)$ has appeared in $\mathsf{chain}[: -0.5\kappa]$; and (2) every unnotarized transaction m in the node's view such that no m or notarized transaction $(e, s, \mathsf{m}, \_)$ has appeared in $\mathsf{chain}[: -0.5\kappa]$.
  - Propose TXs to $\Pi_{\text{blockchain}}$.

[5] Unless otherwise noted, all messages sent from the $\Pi_{\text{blockchain}}$ instance or destined for $\Pi_{\text{blockchain}}$ are automatically passed through, but these messages also count towards the view of the current Π_{thunder} protocol instance.

- *Output.* In every round, let chain be the output from $\Pi_{\mathsf{blockchain}}$.
  - If $\mathsf{chain}[-0.5\kappa]$ is an optimistic block belonging to epoch e:
    (a) let $\mathsf{chain}[-\ell]$ be the starting block for epoch e in chain where $\ell \geq 0.5\kappa$.
    (b) extract the maximal lucky sequence TXs for epoch e from the node's view so far.
    (c) let $\overline{\mathsf{LOG}} := \mathsf{linearize}(\mathsf{chain}[: -(\ell+1)])||\mathsf{strip}(\mathsf{TXs})$.
  - Else, let $\overline{\mathsf{LOG}} := \mathsf{linearize}(\mathsf{chain}[: -0.5\kappa])$.
  - Let LOG be the previous output to $\mathcal{Z}$: if $\overline{\mathsf{LOG}}$ is longer than LOG, output $\overline{\mathsf{LOG}}$; else output LOG to $\mathcal{Z}$.
- *Mempool.* Upon receiving any other message from the network or $\mathcal{Z}$, record the tuple.

Compliant executions. We say that $(\mathcal{A}, \mathcal{Z})$ is compliant w.r.t. $\Pi_{\mathrm{thunder}}^{\Gamma(\kappa,\cdot,\cdot)}$ iff

- $(\mathcal{A}, \mathcal{Z})$ is compliant w.r.t. $\Pi_{\mathsf{blockchain}}$;
- in every view in the support of $\mathsf{EXEC}^{\Pi_{\mathrm{thunder}}^{\Gamma(\kappa,\cdot,\cdot)}}(\mathcal{A}, \mathcal{Z}, \kappa)$, $\mathcal{Z}$ always inputs the same $\mathsf{committee}$ to all honest nodes;
- in every view in the support of $\mathsf{EXEC}^{\Pi_{\mathrm{thunder}}^{\Gamma(\kappa,\cdot,\cdot)}}(\mathcal{A}, \mathcal{Z}, \kappa)$, more than $\frac{1}{2}$ fraction (or in general, more than α fraction) of the distinct public keys in $\mathsf{committee}$ are output by nodes that remain honest (but not necessarily online) forever.

The following theorem says that for any chain-state function f that is admissible, $\Pi_{\mathrm{thunder}}^{f}$ satisfies consistency under compliant executions.

Theorem 5 (Consistency). *Let $\Gamma(\kappa,\cdot,\cdot)$ be any admissible chain-state function. Then, $\Pi_{\mathrm{thunder}}^{\Gamma(\kappa,\cdot,\cdot)}$ satisfies consistency as defined in Sect. 2.2 w.r.t. any p.p.t. $(\mathcal{A}, \mathcal{Z})$ that is compliant w.r.t. $\Pi_{\mathrm{thunder}}^{\Gamma(\kappa,\cdot,\cdot)}$.*

The proof of this theorem is presented in the online full version [36]

Concrete Chain-State Function and Worst-Case Liveness. We will adopt the following chain-state function $\Gamma^{\mathsf{pred}}(\kappa,\cdot,\cdot)$ that is parametrized by a polynomial-time boolean predicate pred henceforth referred to as the "next-epoch" function. Basically, the job of pred is to examine the prefix of some blockchain and decide whether we want to advance to a larger epoch. Specifically, for some chain prefix $\mathsf{chain}[: i]$ if $\mathsf{pred}(\mathsf{chain}[: i], e) = 1$ then the blockchain wants to advance to epoch e if it is not already in epoch e—if there are multiple such e's such that the above holds, then the blockchain wants to go the largest such epoch.

At this moment, we define the chain state function Γ while leaving the pred unspecified. We will show that worst-case liveness is satisfied in compliant executions regardless of the concrete policy pred. Intuitively, our concrete chain state function is very simple: If the blockchain is currently in some epoch e, then the chain will stay in epoch e unless one of the following things happen:

1. either pred (applied to the prefix of the blockchain) wants to go to a larger epoch; or
2. during the current epoch some transaction did not get confirmed for a long time.

If one of the above did happen, then the chain gracefully transitions to an interim ensuring that there are at least κ optimistic blocks for the current epoch e followed by at least κ grace blocks for epoch e. If the blockchain is in an interim and pred wants to go to a next epoch, then we advance to the next epoch immediately. We note that for consistency and worst-case liveness, we in fact only need that there are at least κ grace blocks for each epoch (but not necessarily κ or more optimistic blocks). Here we additionally require that there are at least κ optimistic blocks for each epoch too—this gives the new epoch some time such that the blockchain can pick up possibly stale transactions that ought to have been confirmed such that we do not exit from the current epoch too soon.

More formally, for any chain, $\Gamma^{\mathsf{pred}}(\kappa, \mathsf{chain}, \cdot)$ is inductively defined as the following:

- The $\mathsf{chain}[0] :=$ genesis block is considered an interim block;
- If $\mathsf{chain}[i]$ is an interim block, let e be the largest epoch number such that $\mathsf{pred}(\mathsf{chain}[: i+1], e) = 1$, but no prefix of $\mathsf{chain}[: i]$ was ever in epoch e:
  - If such an epoch e is found: then $\mathsf{chain}[i+1..i+\kappa]$ are all optimistic blocks for epoch e' (and if $|\mathsf{chain}| < i+\kappa$, then all of $\mathsf{chain}[i+1:]$ are optimistic blocks for epoch e').
  - Else $\mathsf{chain}[i+1]$ is also an interim block;
- If $\mathsf{chain}[i]$ is the ℓ-th optimistic block of some epoch e where $\ell \geq \kappa$:
  - If one of the following two conditions C1 or C2 hold, then $\mathsf{chain}[i+1..i+\kappa]$ are all grace blocks for epoch e, and $\mathsf{chain}[i+\kappa+1]$ is an interim block (and if $|\mathsf{chain}| \leq i+\kappa$ then all of $\mathsf{chain}[i+1:]$ are grace blocks for epoch e):
    C1: some m or some notarized transaction $(\_, \_, \mathsf{m}, \_)$ appears in $\mathsf{chain}[: i - 0.5\kappa]$ but $\mathsf{linearize}(\mathsf{chain}[: i])$ does not contain m or $(\_, \_, \mathsf{m})$, i.e., if some transaction has not occurred in any lucky sequence even after a sufficiently long time;
    C2: there exists some $e' > e$ such that $\mathsf{pred}(\mathsf{chain}[: i+1], e') = 1$, i.e., if the next-epoch policy function wants to switch to a larger epoch than the current one.
  - Else $\mathsf{chain}[i+1]$ is an optimistic block of epoch e.

Theorem 6 (Worst-case liveness). *Let $\Gamma(\kappa, \cdot, \cdot) := \Gamma^{\mathsf{pred}}(\cdot, \cdot, \cdot)$ be the chain-state function as specified above for any polynomial-time boolean predicate* pred. *Let g_0 denote the underlying $\Pi_{\mathsf{blockchain}}$'s chain growth lower bound parameter, and let $T_{\mathsf{confirm}}(\kappa) := \frac{3\kappa}{g_0}$. For any p.p.t. $(\mathcal{A}, \mathcal{Z})$ that is compliant w.r.t. $\Pi_{\mathsf{thunder}}^{\Gamma(\kappa,\cdot,\cdot)}$, there exists a negligible function $\mathsf{negl}(\cdot)$ such that for every $\kappa \in \mathbb{N}$, except with $\mathsf{negl}(\kappa)$ probability over the choice of $\mathsf{view} \leftarrow \mathsf{EXEC}^{\Pi_{\mathsf{thunder}}^{\Gamma(\kappa,\cdot,\cdot)}}(\mathcal{A}, \mathcal{Z}, \kappa)$, the following holds: suppose that $\mathcal{Z}$ inputs a transaction m to an honest node in round r, then in any round $r' \geq r + T_{\mathsf{confirm}}(\kappa)$, all honest and online nodes' output LOG to $\mathcal{Z}$ will contain some $(\_, \_, \mathsf{m})$ or m.*

The proof of the above theorem is deferred to the supplemental material.

Coordination Protocol Π_{ella} and Optimistic Responsiveness. We now describe the full protocol $\Pi_{\mathrm{ella}}^{\Gamma(\kappa,\cdot,\cdot)}$ that spells out the leader-based coordination mechanism on top of Π_{thunder} as well as the next-epoch function pred. We will then show under exactly what optimistic conditions our protocol achieves responsiveness.

Description of protocol Π_{ella}. Π_{ella} calls $\Pi_{\mathrm{thunder}}^{\Gamma^{\mathsf{pred}}(\kappa,\cdot,\cdot)}$ where the chain state function $\Gamma(\kappa,\cdot,\cdot) := \Gamma^{\mathsf{pred}}(\kappa,\cdot,\cdot)$ is as defined in Sect. 3.2. We spell out the next-epoch function pred and the rest of Π_{ella} below.

- *Next-epoch function.* The policy function $\mathsf{pred}(\mathsf{chain}, e)$ takes in an abstract blockchain denoted chain and an epoch number e. If there exists a notarized transaction for epoch e in chain, then output 1; else output 0.
- *Initialize*: fork an instance of the $\Pi_{\mathrm{thunder}}^{\Gamma(\kappa,\cdot,\cdot)}$ protocol.
- *Leader switch*: upon input $\mathtt{leader}(e, i)$: if no leader has been recorded for epoch e, record i as the leader for epoch e, and do the following:
  - if current node is i: send a notarization request for a special epoch-start transaction $(e, s = 1, \mathtt{start})$, and let $s = 2$;
  - for every notarization request (e, s, m) received earlier from node i, act as if (e, s, m) has just been received from i.
- *Notarization*: upon receiving notarization request (e, s, m) from i: if i has been recorded as the leader for epoch e, forward the notarization request (e, s, m) to $\Pi_{\mathrm{thunder}}^{\Gamma(\kappa,\cdot,\cdot)}$; else ignore the request.
- *Leader*: every round: let e be the largest epoch recorded thus far and if current node is recorded as the leader for epoch e:
  - for every m in view such that no m or $(\_, \_, \mathsf{m})$ appears in $\mathsf{linearize}(\mathsf{chain}[: -\kappa])$, if a notarization request has not been broadcast for m earlier, then broadcast the notarization request (e, s, m) and let $s := s + 1$.
- *Other messages*: pass through all other messages between $\Pi_{\mathrm{thunder}}^{\Gamma(\kappa,\cdot,\cdot)}$ and $\mathcal{Z}$; similarly pass through all other messages between $\Pi_{\mathrm{thunder}}^{\Gamma(\kappa,\cdot,\cdot)}$ and the network.

Compliant executions. To guarantee consistency and worst-case liveness, basically we just need the same conditions as our earlier $\Pi_{\mathrm{thunder}}^{\Gamma(\kappa,\cdot,\cdot)}$. We say that $(\mathcal{A}, \mathcal{Z})$ is compliant w.r.t. $\Pi_{\mathrm{ella}}^{\Gamma(\kappa,\cdot,\cdot)}$ iff $(\mathcal{A}, \mathcal{Z})$ is compliant w.r.t. $\Pi_{\mathrm{thunder}}^{\Gamma(\kappa,\cdot,\cdot)}$.

Lucky epoch. Below we will describe exactly under what optimistic conditions can we achieve responsiveness. Roughly speaking, whenever a lucky epoch begins, after a short warmup time, we can achieve responsiveness. Specifically, during a lucky epoch, the epoch's leader is online and honest and more than $\frac{3}{4}$ fraction or in general, $1 - \frac{\alpha}{2}$ fraction of the committee remain honest and online.

Formally, given a view, we say that $[T_{\mathrm{start}}, T_{\mathrm{end}}]$ belongs to a lucky epoch corresponding to epoch e and leader i iff the following hold:

- In any round $r \geq T_{\mathrm{start}} + \Delta$, any honest and online node should have received $\mathtt{leader}(e, i)$ where i is the common leader that all honest nodes receive for epoch e. Further, prior to T_{start}, no honest node has received from $\mathcal{Z}$ any $\mathtt{leader}(e', \_)$ instruction where $e' \geq e$.

- the leader (i.e., node i) is honest and online at in any round $t \in [T_{\text{start}}, T_{\text{end}} + 3\Delta]$;
- more than $\frac{3}{4}$ fraction (or in general, more than $1 - \frac{\alpha}{2}$ fraction) of committee are honest and online[6] in any round $t \in [T_{\text{start}}, T_{\text{end}} + 3\Delta]$.

Optimistic responsiveness in lucky epochs. We say that a protocol Π satisfies $(T_{\text{warmup}}, T_{\text{opt}})$-optimistic responsiveness in lucky epochs w.r.t. $(\mathcal{A}, \mathcal{Z})$ iff except with $\mathsf{negl}(\kappa)$ probability over the choice of view $\leftarrow \mathsf{EXEC}^{\Pi_{\text{ella}}^{\Gamma(\kappa,\cdot,\cdot)}}(\mathcal{A}, \mathcal{Z}, \kappa)$: for any duration $[T_{\text{start}}, T_{\text{end}}]$ in view that belongs to a lucky epoch, $[T_{\text{start}} + T_{\text{warmup}}, T_{\text{end}}]$ is a T_{opt}-responsive period in view.

Theorem 7 (Optimistic case responsiveness). *Let g_0 be the underlying $\Pi_{blockchain}$'s chain growth lower bound parameter. For every p.p.t. $(\mathcal{A}, \mathcal{Z})$ that is compliant w.r.t. $\Pi_{\text{ella}}^{\Gamma(\kappa,\cdot,\cdot)}$, $\Pi_{\text{ella}}^{\Gamma(\kappa,\cdot,\cdot)}$ satisfies $(T_{\text{warmup}}, T_{\text{opt}})$-optimistic responsiveness in lucky epochs for $T_{\text{warmup}} = O(\frac{\kappa}{g_0})$, and $T_{\text{opt}} = 3\delta$ where δ is the actual maximum network delay in view.*

The proof of the above theorem is deferred to the online full version [36]. We note that Theorem 7 implies the following: informally speaking, if throughout the execution more than $\frac{3}{4}$ fraction of the committee remain honest and online and moreover, the initial epoch's leader remains honest and online, then once nodes enter the initial epoch, after a short warmup period, our protocol Π_{ella} will achieve responsiveness throughout the remainder of the execution (assuming that the underlying blockchain is secure).

Remark 2 (Leader re-election mechanism). In our scheme earlier, we left it unspecified how the environment $\mathcal{Z}$ will decide when to issue leader-switch instructions of the form `leader(e, i)` that will cause nodes to start a new leader epoch. This is an application-specific policy decision. At this point, our paper focuses on providing a general framework that enables any application-specific policy decisions. In the online full version [36], we will give some suggestions on leader re-election policies that are useful in practice.

Deferred materials. We defer the full proofs, the lower bounds, as well as how to concretely instantiate the Thunderella framework in permissioned and permissionless environments allowing *committee reconfiguration* and leader rotation in the online full version [36]. We now conclude with the related work.

4 Related Work

State machine replication: classical and blockchain-style approaches. State machine replication or atomic broadcast (referred to as consensus for short in this paper) is a central abstraction of distributed systems, and has been extensively

[6] We say that a public key $\mathsf{pk} \in \mathsf{committee}$ is honest and online in round r if some node that is honest and online in round r output pk to $\mathcal{Z}$ earlier.

investigated and widely adopted in real-world systems. Roughly speaking, there are two, technically speaking, fundamentally different approaches towards realizing state machine replication, classical-style consensus [9,14,15,25,30,31], and blockchain-style consensus [11,19,26,34,35,40]. For a while, it has been vaguely understood by the community that blockchain-style protocols and classical ones achieve different properties—but the community has only recently begun to formally understand and articulate these differences.

The recent work by Pass and Shi [40] point out one fundamental difference between classical style and blockchain-style consensus. Most classical protocols [9,14,15,25,30,31], synchronous and asynchronous ones alike, rely on nodes having collected sufficiently many votes to make progress; thus these protocols would fail in a model where *participation is sporadic* and the exact number of players that do show up cannot be predicted upfront. More specifically, classical models of consensus would pessimistically treat nodes that do not show up as faulty (also referred to as crash fault); and if too many nodes do not show up, the protocol fails to make progress. In comparison, blockchain-style protocols can make progress regardless of how many players actually show up. Moreover, blockchain-style consensus has also been shown to be secure in a setting where the number of players can vary over time [18].

Classical deployments of consensus protocols are typically in a relatively small-scale and permissioned setting. Consensus in the permissionless setting was first empirically demonstrated to be possible due to Bitcoin's ingenious Nakamoto blockchain [34]. While the original Nakamoto blockchain relies on proofs-of-work to solve the Sybil attack in the permissionless setting, other proposals have been suggested since then for securely establishing identities in a permissionless setting—for example, proof-of-stake [2,3,7,10,11,26,27,32,41] is a most-oft cited approach where the stake-holders of a cryptocurrency system are responsible for voting on transactions. Recent works [32] have also explored adopting classical style consensus in a permissionless setting where approaches such as proof-of-stake can be used to establish identities.

Other closely related works. Our work is also reminisient of recent works that combine classical consensus and blockchains [12,28,38] although these works are of a different nature as we explain below. Among these works, Hybrid Consensus [38] is the only known formally correct approach, and moreover the only known approach that achieves *responsiveness*. From a theoretical perspective, our results are incomparable to Hybrid Consensus: we tolerate up to $\frac{1}{2}$ corruption in the worst-case and offer responsiveness only in the optimistic case but not in the worst case; in comparison, Hybrid Consensus achieves responsiveness even in the worst case—but in exchange, their protocol can only tolerate up to $\frac{1}{3}$ corruption, and this turns out to be inherent for any worst-case responsive protocol even when assuming proof-of-work [15,38]. From a practical perspective, Thunderella is more likely to be the protocol of choice in a real-world implementation partly due to its simplicity—in comparison, Hybrid Consensus requires a full-fledged classical protocol such as PBFT and Byzantine Paxos as a subroutine, and thus inherits the complexity of these protocols.

A line of research [8,13,16,21,22,33] has investigated Byzantine agreement protocols capable of early-stopping when conditions are more benign than the worst-case faulty pattern: e.g., the actual number of faulty nodes turns out to be smaller than the worst-case resilience bound. However, these works are of a different nature than ours as we explain below. First, these earlier works focus on stopping in a fewer number of synchronous rounds, and it is not part of their goal to achieve *responsiveness*. Second, although some known lower bounds [13] show that the number of actual rounds must be proportional to the actual number of faulty processors—note that these lower bounds work only for deterministic protocols, and thus they are not applicable in our setting.

Finally, the idea of combining asynchrony and synchrony was described in earlier works [4]; other works have also proposed frameworks for composing multiple BFT protocols [20]. However, to the best of our knowledge, none of the earlier works combined a synchronous fallback path and an asynchronous optimistic path in the manner that we do, allowing us to tolerate more than $\frac{1}{3}$ corruptions in the worst-case while still be responsive most of the time in practice.

Acknowledgments. We thank Jian Xie and Youcai Qian for inspiring conversations. We also thank Lorenzo Alvisi and Robbert van Renesse for helpful discussions and moral support. This work is supported in part by NSF grants CNS-1217821, CNS-1314857, CNS-1514261, CNS-1544613, CNS-1561209, CNS-1601879, CNS-1617676, AFOSR Award FA9550-15-1-0262, an Office of Naval Research Young Investigator Program Award, a Microsoft Faculty Fellowship, a Packard Fellowship, a Sloan Fellowship, Google Faculty Research Awards, and a VMWare Research Award.

References

1. Attiya, H., Dwork, C., Lynch, N., Stockmeyer, L.: Bounds on the time to reach agreement in the presence of timing uncertainty. J. ACM **41**(1), 122–152 (1994)
2. Bentov, I., Gabizon, A., Mizrahi, A.: Cryptocurrencies without proof of work. In: Financial Cryptography Bitcoin Workshop (2016)
3. Bentov, I., Lee, C., Mizrahi, A., Rosenfeld, M.: Proof of activity: extending bitcoin's proof of work via proof of stake. In: NetEcon (2014)
4. Birman, K.P., Joseph, T.A.: Exploiting virtual synchrony in distributed systems. In: SOSP (1987)
5. Burrows, M.: The chubby lock service for loosely-coupled distributed systems. In: OSDI (2006)
6. Buterin, V. (2017). https://medium.com/@VitalikButerin/minimal-slashing-conditions-20f0b500fc6c
7. Buterin, V., Zamfir, V.: Casper (2015). https://blog.ethereum.org/2015/08/01/introducing-casper-friendly-ghost/
8. Castañeda, A., Gonczarowski, Y.A., Moses, Y.: Unbeatable consensus. In: DISC (2014)
9. Castro, M., Liskov, B.: Practical byzantine fault tolerance. In: OSDI (1999)
10. User "cunicula", Rosenfeld, M.: Proof of stake brainstorming, August 2011. https://bitcointalk.org/index.php?topic=37194.0
11. Daian, P., Pass, R., Shi, E.: Snow white: provably secure proofs of stake. Cryptology ePrint Archive, Report 2016/919 (2016)

12. Decker, C., Seidel, J., Wattenhofer, R.: Bitcoin meets strong consistency. In: ICDCN (2016)
13. Dolev, D., Reischuk, R., Raymond Strong, H.: Early stopping in byzantine agreement. J. ACM **37**(4), 720–741 (1990)
14. Dolev, D., Raymond Strong, H.: Authenticated algorithms for byzantine agreement. SIAM J. Comput. SIAMCOMP **12**(4), 656–666 (1983)
15. Dwork, C., Lynch, N., Stockmeyer, L.: Consensus in the presence of partial synchrony. J. ACM **35**, 288–323 (1988)
16. Dwork, C., Moses, Y.: Knowledge and common knowledge in a byzantine environment I: crash failures. In: TARK, pp. 149–169 (1986)
17. Eyal, I., Sirer, E.G.: Majority is not enough: bitcoin mining is vulnerable. In: FC (2014)
18. Garay, J.A., Kiayias, A., Leonardos, N.: The bitcoin backbone protocol with chains of variable difficulty. Cryptology ePrint Archive, 2016/1048 (2016)
19. Garay, J., Kiayias, A., Leonardos, N.: The bitcoin backbone protocol: analysis and applications. In: Oswald, E., Fischlin, M. (eds.) EUROCRYPT 2015. LNCS, vol. 9057, pp. 281–310. Springer, Heidelberg (2015). https://doi.org/10.1007/978-3-662-46803-6_10
20. Guerraoui, R., Knežević, N., Quéma, V., Vukolić, M.: The next 700 BFT protocols. In: Proceedings of the 5th European Conference on Computer Systems, EuroSys 2010, pp. 363–376. ACM, New York (2010)
21. Halpern, J.Y., Moses, Y., Waarts, O.: A characterization of eventual Byzantine agreement. SIAM J. Comput. **31**(3), 838–865 (2001)
22. Herlihy, M., Moses, Y., Tuttle, M.R.: Transforming worst-case optimal solutions for simultaneous tasks into all-case optimal solutions. In: PODC (2011)
23. Herzberg, A., Kutten, S.: Early detection of message forwarding faults. SIAM J. Comput. **30**(4), 1169–1196 (2000)
24. Junqueira, F.P., Reed, B.C., Serafini, M.: Zab: high-performance broadcast for primary-backup systems. In: DSN (2011)
25. Katz, J., Koo, C.-Y.: On expected constant-round protocols for Byzantine agreement. J. Comput. Syst. Sci. **75**(2), 91–112 (2009)
26. Kiayias, A., Russell, A., David, B., Oliynykov, R.: Ouroboros: a provably secure proof-of-stake blockchain protocol. In: Katz, J., Shacham, H. (eds.) CRYPTO 2017. LNCS, vol. 10401, pp. 357–388. Springer, Cham (2017). https://doi.org/10.1007/978-3-319-63688-7_12
27. King, S., Nadal, S.: PPCoin: peer-to-peer crypto-currency with proof-of-stake (2012). https://peercoin.net/assets/paper/peercoin-paper.pdf
28. Kokoris-Kogias, E., Jovanovic, P., Gailly, N., Khoffi, I., Gasser, L., Ford, B.: Enhancing bitcoin security and performance with strong consistency via collective signing. CoRR, abs/1602.06997 (2016)
29. Kotla, R., Alvisi, L., Dahlin, M., Clement, A., Wong, E.L.: Zyzzyva: speculative byzantine fault tolerance. In: SOSP (2007)
30. Lamport, L.: Fast paxos. Distrib. Comput. **19**(2), 79–103 (2006)
31. Lamport, L., Malkhi, D., Zhou, L.: Vertical paxos and primary-backup replication. In: PODC, pp. 312–313 (2009)
32. Micali, S.: Algorand: the efficient and democratic ledger (2016). https://arxiv.org/abs/1607.01341
33. Moses, Y., Raynal, M.: No double discount: condition-based simultaneity yields limited gain. Inf. Comput. **214**, 47–58 (2012)
34. Nakamoto, S.: Bitcoin: a peer-to-peer electronic cash system (2008)

35. Pass, R., Seeman, L., Shelat, A.: Analysis of the blockchain protocol in asynchronous networks. In: Coron, J.-S., Nielsen, J.B. (eds.) EUROCRYPT 2017. LNCS, vol. 10211, pp. 643–673. Springer, Cham (2017). https://doi.org/10.1007/978-3-319-56614-6_22
36. Pass, R., Shi, E.: Thunderella: blockchains with optimistic instant confirmation. https://eprint.iacr.org/2017/913
37. Pass, R., Shi, E.: Fruitchains: a fair blockchain. In: PODC (2017)
38. Pass, R., Shi, E.: Hybrid consensus: efficient consensus in the permissionless model. In: DISC (2017)
39. Pass, R., Shi, E.: Rethinking large-scale consensus (invited paper). In: CSF (2017)
40. Pass, R., Shi, E.: The sleepy model of consensus. In: Takagi, T., Peyrin, T. (eds.) ASIACRYPT 2017. LNCS, vol. 10625, pp. 380–409. Springer, Cham (2017). https://doi.org/10.1007/978-3-319-70697-9_14
41. User "QuantumMechanic": Proof of stake instead of proof of work, July 2011. https://bitcointalk.org/index.php?topic=27787.0
42. Song, Y.J., van Renesse, R.: Bosco: one-step byzantine asynchronous consensus. In: Taubenfeld, G. (ed.) DISC 2008. LNCS, vol. 5218, pp. 438–450. Springer, Heidelberg (2008). https://doi.org/10.1007/978-3-540-87779-0_30

But Why Does It Work? A Rational Protocol Design Treatment of Bitcoin

Christian Badertscher[1](✉), Juan Garay[2], Ueli Maurer[1], Daniel Tschudi[3], and Vassilis Zikas[4]

[1] ETH Zurich, Zürich, Switzerland
{christian.badertscher,maurer}@inf.ethz.ch
[2] Texas A&M University, College Station, USA
garay@tamu.edu
[3] Aarhus University, Aarhus, Denmark
tschudi@cs.au.dk
[4] University of Edinburgh and IOHK, Edinburgh, UK
vassilis.zikas@ed.ac.uk

Abstract. An exciting recent line of work has focused on formally investigating the core cryptographic assumptions underlying the security of Bitcoin. In a nutshell, these works conclude that Bitcoin is secure if and only if the majority of the mining power is honest. Despite their great impact, however, these works do not address an incisive question asked by positivists and Bitcoin critics, which is fuelled by the fact that Bitcoin indeed works in reality: Why should the real-world system adhere to these assumptions?

In this work we employ the machinery from the Rational Protocol Design (RPD) framework by Garay *et al.* [FOCS 2013] to analyze Bitcoin and address questions such as the above. We show that under the natural class of incentives for the miners' behavior—i.e., rewarding them for adding blocks to the blockchain but having them pay for mining—we can reserve the honest majority assumption as a fallback, or even, depending on the application, completely replace it by the assumption that the miners aim to maximize their revenue.

Our results underscore the appropriateness of RPD as a "rational cryptography" framework for analyzing Bitcoin. Along the way, we devise significant extensions to the original RPD machinery that broaden its applicability to cryptocurrencies, which may be of independent interest.

1 Introduction

Following a number of informal and/or *ad hoc* attempts to address the security of Bitcoin, an exciting recent line of work has focused on devising a rigorous cryptographic analysis of the system [2,13,14,27]. At a high level, these works

D. Tschudi—Work done while author was at ETH Zurich.
V. Zikas—Work done in part while the author was at RPI.

J. B. Nielsen and V. Rijmen (Eds.): EUROCRYPT 2018, LNCS 10821, pp. 34–65, 2018.
https://doi.org/10.1007/978-3-319-78375-8_2

start by describing an appropriate model of execution, and, within it, an abstraction of the original Bitcoin protocol [23] along with a specification of its security goals in terms of a set of intuitive desirable properties [13,14,27], or in terms of a functionality in a simulation-based composable framework [2]. They then prove that (their abstraction of) the Bitcoin protocol meets the proposed specification under the assumption that the majority of the computing power invested in mining bitcoins is by devices which mine according to the Bitcoin protocol, i.e., *honestly*. This assumption of *honest majority* of computing power—which had been a folklore within the Bitcoin community for years underlying the system's security—is captured by considering the parties who are not mining honestly as controlled by a central adversary who coordinates them trying to disrupt the protocol's outcome.

Meanwhile, motivated by the fact that Bitcoin is an "economic good" (i.e., BTCs are exchangeable for national currencies and goods) a number of works have focused on a rational analysis of the system [7–9,15,22,24,28–32]. In a nutshell, these works treat Bitcoin as a game between the (competing) rational miners, trying to maximize a set of utilities that are postulated as a natural incentive structure for the system. The goal of such an analysis is to investigate whether or not, or under which assumptions on the incentives and/or the level of collaboration of the parties, Bitcoin achieves a stable state, i.e., a game-theoretic equilibrium. However, despite several enlightening conclusions, more often than not the prediction of such analyses is rather pessimistic. Indeed, these results typically conclude that, unless assumptions on the amount of honest computing power—sometimes even stronger than just majority—are made, the induced incentives result in plausibility of an attack to the Bitcoin mining protocol, which yields undesired outcomes such as forks on the blockchain, or a considerable slowdown.

Yet, to our knowledge, no fork or substantial slowdown that is attributed to rational attacks has been observed to date, and the Bitcoin network keeps performing according to its specification, even though mining pools would, in principle, be able to launch collaborative attacks given the power they control.[1] In the game-theoretic setting, this mismatch between the predicted and observed behavior would be typically interpreted as an indication that the underlying assumptions about the utility of miners in existing analysis do not accurately capture the miners' rationale. Thus, two main questions still remain and are often asked by Bitcoin skeptics:

Q1. How come Bitcoin is not broken using such an attack?

Or, stated differently, why does it work and why do majorities not collude to break it?

Q2. Why do honest miners keep mining given the plausibility of such attacks?

[1] We refer to forks of the Bitcoin chain itself, not to forks that spin-off a new currency.

In this work we use a rigorous cryptographic reasoning to address the above questions. In a nutshell, we devise a rational-cryptography framework for capturing the economic forces that underly the tension between honest miners and (possibly colluding) deviating miners, and explain how these forces affect the miners' behavior. Using this model, we show how natural incentives (that depend on the expected revenue of the miners) in combination with a high monetary value of Bitcoin, can explain the fact that Bitcoin is not being attacked in reality *even though* majority coalitions are in fact possible. In simple terms, we show how natural assumptions about the miners' incentives allow to substitute (either entirely or as a fallback assumption) the honest-majority assumption. To our knowledge, this is the first work that formally proves such rational statements that do not rely on assumptions about the adversary's computing power. We stress that the incentives we consider depend solely on costs and rewards for mining—i.e., mining (coinbase) and transaction fees—and, in particular, we make no assumption that implicitly or explicitly deters forming adversarial majority coalitions.

What enables us to address the above questions is utilizing the Rational Protocol Design (RPD) methodology by Garay *et al.* [11] to derive stability notions that closely capture the idiosyncrasies of coordinated incentive-driven attacks on the Bitcoin protocol. To better understand how our model employs RPD to address the above questions, we recall the basic ideas behind the framework.

Instead of considering the protocol participants—in our case, the Bitcoin miners—as rational agents, RPD considers a meta-game, called the *attack game*. The attack game in its basic form is a two-agent zero-sum extensive game of perfect information with a horizon of length two, i.e., two sequential moves.[2] It involves two players, called the *protocol designer* D—who is trying to come up with the best possible protocol for a given (multi-party) task—and the *attacker* A—who is trying to come up with the (polynomial-time) strategy/adversary that optimally attacks the protocol. The game proceeds in two steps: First, (only) D plays by choosing a protocol for the (honest) players to execute; A is informed about D's move and it is now his term to produce his move. The attacker's strategy is, in fact, a cryptographic adversary that attacks the protocol proposed by the designer.

The incentives of both A and D are described by utility functions, and their respective moves are carried out with the goal of maximizing these utilities.[3] In a nutshell, the attacker's utility function rewards the adversary proportionally to how often he succeeds in provoking his intended breach, and depending on its severity. Since the game is zero-sum, the designer's utility is the opposite of the attacker's; this captures the standard goal of cryptographic protocols, namely, "taming" the adversary in the best possible manner.

Based on the above game, the RPD framework introduces the following natural security notion, termed *attack-payoff security*, that captures the quality of

[2] This is often referred to as a *Stackelberg game* in the game theory literature [26].

[3] Notice, however, the asymmetry: The designer needs to come up with a protocol based on speculation of what the adversary's move will be, whereas the attacker plays after being informed about the actual designer's move, i.e., about the protocol.

a protocol Π for a given specification when facing incentive-driven attacks aiming to maximize the attacker's utility. Informally, attack-payoff security ensures that the adversary is not willing to attack the protocol Π in any way that would make it deviate from its ideal specification. In other words, the protocol is secure against the class of strategies that maximize the attacker's utility. In this incentive-driven setting, this is the natural analogue of security against malicious adversaries.[4] For cases where attack payoff security is not feasible, RPD proposes the notion of *attack-payoff optimality*, which ensures that the protocol Π is a best response to the best attack.

A useful feature of RPD (see below) is that all definitions build on Canetti's simulation-based framework (either the standalone framework [5] or the UC framework [6]), where they can be easily instantiated. In fact, there are several reasons, both at the intuitive and technical levels, that make RPD particularly appealing to analyze complex protocols that are already running, such as Bitcoin. First, RPD supports adaptive corruptions which captures the scenario of parties who are currently running their (mining) strategy changing their mind and deciding to attack. This is particularly useful when aiming to address the likelihood of insider attacks against a protocol which is already in operation. For the same reason, RPD is also suitable for capturing attacks induced by compromised hardware/software and/or bribing [4] (although we will not consider bribing here). Second, the use of a central adversary as the attacker's move ensures that, even though we are restricting to incentive-driven strategies, we allow full collaboration of cheaters. This allows, for example, to capture mining pools deciding to deviate from the protocol's specification.

At the technical level, using the attack-game to specify the incentives takes away many of the nasty complications of "rational cryptography" models. For example, it dispenses with the need to define cumbersome computational versions of equilibrium [10,17–19,21,25], since the actual rational agents, i.e., D and A, are not computationally bounded. (Only their actions need to be PPT machines.) Furthermore, as it builds on simulation-based security, RPD comes with a composition theorem allowing for regular cryptographic subroutine replacement. The latter implies that we can analyze protocols in simpler hybrid-worlds, as we usually do in cryptography, without worrying about whether or not their quality or stability will be affected once we replace their hybrids by corresponding cryptographic implementations.

Our contributions. In this work, we apply the RPD methodology to analyze the quality of Bitcoin against incentive-driven attacks, and address the existential questions posted above. As RPD is UC-based, we use the Bitcoin abstraction as a UC protocol and the corresponding Bitcoin ledger functionality from [2] to capture the goal/specification of Bitcoin. As argued in [2], this functionality captures all the properties that have been proposed in [13,27].

We define a natural class of incentives for the attacker by specifying utilities which, on one hand, reward him according to Bitcoin's standard reward

[4] In fact, if we require this for any arbitrary utility function, then the two notions—attack-payoff security and malicious security—coincide.

mechanisms (i.e., block rewards and transaction fees) for blocks permanently inserted in the blockchain by adversarial miners, and, on the other hand, penalize him for resources that he uses (e.g., use of mining equipment and electricity). In order to overcome the inconsistency of rewards being typically in Bitcoins and costs being in real money, we introduce the notion of a *conversion rate* CR converting reward units (such as BTC) into mining-cost units (such as US Dollar) This allows us to make statements about the quality of the protocol depending on its value measured in a national currency.

We then devise a similar incentive structure for the designer, where, again, the honest parties are (collectively) rewarded for blocks they permanently insert into the blockchain, but pay for the resources they use. What differentiates the incentives of the attacker from the designer's is that the latter is utmost interested in preserving the "health" of the blockchain, which we also reflect in its utility definition. Implicit in our formulation is the assumption that the attacker does not gain reward from attacking the system, unless this attack has a financial gain.[5]

Interestingly, in order to apply the RPD methodology to Bitcoin we need to extend it in non-trivial ways, to capture for example non-zero-sum games—as the utility of the designer and the attacker are not necessarily opposites—and to provide stronger notions of security and stability. In more detail, we introduce the notion of *strong attack payoff security*, which mandates that the attacker will stick to playing a passive strategy, i.e., stick to Bitcoin (but might abuse the adversary's power to delay messages in the network). We also introduce the natural notion of *incentive compatibility* (IC) which mandates that both the attacker and the designer will have their parties play the given protocol. Observe that incentive compatibility trivially implies strong attack payoff security, and the latter implies the standard attack payoff security from the original RPD framework assuming the protocol is at least correct when no party deviates. These extensions to RPD widen its applicability and might therefore be of independent interest. We note that although we focus on analysis of Bitcoin here, the developed methodology can be adapted to analyze other main-stream cryptocurrencies.

Having laid out the model, we then use it to analyze Bitcoin. We start our analysis with the simpler case where the utilities do not depend on the messages—i.e., transactions—that are included into the blocks of the blockchain: when permanently inserting a block into the blockchain, a miner is just rewarded with a fixed block-reward value. This can be seen as corresponding to the Bitcoin backbone abstraction proposed in [13], but enriched with incentives to mine blocks. An interpretation of our results for this setting, listed below, is that they address blockchains that are not necessarily intended to be used as cryptocurrency ledgers. Although arguably this is not the case for Bitcoin, our analysis already reveals several surprising aspects, namely, that in this setting one does not need to rely on honest majority of computing power to ensure the quality

[5] In particular, a fork might be provoked by the attacker only if it is expected to increase his revenue.

of the system. Furthermore, these results offer intuition on what is needed to achieve stability in the more complete case, which also incorporates transaction fees. Summarizing, we prove the following statements for this backbone-like setting, where the contents of the blocks do not influence the player's strategies (but the rewards and costs do):

- Bitcoin is strongly attack-payoff secure, i.e., no coordinated coalition has an incentive to deviate from the protocol, provided that the rest of the parties play it. Further, this statement holds no matter how large the coalition (i.e., no matter how large the fraction of corrupt computing power) and no matter how high the conversion rate is. This means that in this backbone-like setting we can fully replace the assumption of honest majority of computing power by the above intuitive rational assumption.[6]
- If the reward for mining a block is high enough so that mining is on average profitable, then the Bitcoin protocol is even incentive-compatible with respect to local deviations. In other words, not only colluding parties (e.g., mining pools) do not have an incentive to deviate, but also the honest miners have a clear incentive to keep mining. Again, this makes no honest-majority assumption. Furthermore, as a sanity check, we also prove that this is not true if the conversion rate drops so that miners expect to be losing revenue by mining. The above confirms the intuition that after the initial bootstrapping phase where value is poured into the system (i.e., CR becomes large enough), such a ledger will keep working according to its specification for as long as the combination of conversion rate and block-reward is high enough.

With the intuition gained from the analysis in the above idealized setting, we next turn to the more realistic setting which closer captures Bitcoin, where block contents are messages that have an associated fee. We refer to these messages as *transactions*, and use the standard restrictions of Bitcoin on the transaction fees: every transaction has a maximum fee and the fee is a multiple of the minimum division.[7] We remark that in all formal analyses [2,13,27] the transactions are considered as provided as inputs by an explicit environment that is supposed to capture the application layer that sits on top of the blockchain and uses it. As such, the environment will also be responsible for the choice of transaction fees and the distribution of transactions to the miners. For most generality, we do not assume as in [13,27] that all transactions are communicated by the environment to all parties via a broadcast-like mechanism, but rather that they are distributed (i.e., input) by the environment to the miners, individually, who might then forward them using the network (if they are honest) or not. This more realistic transaction-submission mechanism is already explicit in [2].

We call this model that incorporates both mining rewards and transaction fees into the reward of the miner for a block as the *full-reward* model. Interestingly, this model allows us to also make predictions about the Bitcoin era when

[6] It should be noted though that our analysis considers, similarly to [2,13,27], a fixed difficulty parameter. The extension to variable difficulty is left as future research.

[7] For Bitcoin the minimum division is 1 satoshi $= 10^{-8}$ BTC, and there is typically a cap on fees [3].

the rewards for mining a block will be much smaller than the transaction fees (or even zero).

We stress that transactions in our work are dealt with as messages that have an explicit fee associated with them, rather than actions which result in transferring BTCs from one miner to another. This means that other than its associated fee, the contents of a transaction does not affect the strategies of the players in the attack game. This corresponds to the assumption that the miners, who are responsible for maintaining the ledger, are different than the users, which, for example, translate the contents of the blocks as exchanges of cryptocurrency value, and which are part of the application/environment. We refer to this assumption as *the miners/users separation principle.* This assumption is explicit in all existing works, and offers a good abstraction to study the incentives for maintaining the ledger—which is the scope of our work—separately from the incentives of users to actually use it. Note that this neither excludes nor trivially deters "forking" by a sufficiently powerful (e.g., 2/3 majority) attacker; indeed, if some transaction fees are much higher than all others, then such an attacker might fork the network by extending both the highest and the second highest chain with the same block containing these high-fee transactions, and keep it forked for sufficiently long until he cashes out his rewards from both forks.

In this full-reward model, we prove the following statements:

- First, we look at the worst-case environment, i.e., the one that helps the adversary maximize its expected revenue. We prove that in this model Bitcoin is still incentive compatible, hence also strongly attack payoff secure. In fact, the same is true if the environment makes sure that there is a sufficient supply of transactions to the honest miners and to the adversary, such that the fees are high enough to build blocks that reach exactly the maximal rewarding value (note that not necessarily the same set of transactions have to be known to the participants). For example, as long as many users submit transactions with the heaviest possible fee (so-called *full-fee transactions*), then the system is guaranteed to work without relying on an honest majority of miners. In a sense, the users can control the stability of the system through transaction fees.
- Next, we investigate the question of whether or not the above is true for arbitrary transaction-fee distributions. Not surprisingly, the answer here is negative, and the protocol is not even attack-payoff secure (i.e, does not even achieve its specification). The proof of this statement makes use of the above sketched forking argument. On the positive side, our proof suggests that in the honest-majority setting where forking is not possible (except with negligible probability), the only way the adversary is incentivized to deviate from the standard protocol is to withhold the transactions he is mining on to avoid risking to lose the fees to honest parties.

Interpreting the above statements, we can relax the assumption for security of Bitcoin from requiring an honest majority to requiring long-enough presence of sufficiently many full-fee transactions, with a fallback to honest majority.

- Finally, observing that the typically large pool of transactions awaiting validation justifies the realistic assumption that there is enough supply to the network (and given the high adoption, this pool will not become small too fast), we can directly use our analysis, to propose a possible modification which would help Bitcoin, or other cryptocurrencies, to ensure incentive compatibility (hence also strong attack-payoff security) in the full-reward model in the long run: The idea is to define an exact cumulative amount on fees (or overall reward) to be allowed for each block. If there are enough high-fee transactions, then the blocks are filled up with transactions until this amount is reached. As suggested by our first analysis with a simple incentive structure, ensuring that this cap is non-decreasing would be sufficient to argue about stability; however, it is well conceivable that such a bound could be formally based on supply-and-demand in a more complex and economy-driven incentive structure and an interesting future research direction is to precisely define such a proposal together with the (economical) assumptions on which the security statements are based. We note that the introduction of such a rule would typically only induce a "soft fork," and would, for a high-enough combination of conversion rate and reward bound, ensure incentive compatibility even when the flat reward per block tends to zero and the main source of rewards would be transaction fees, as it is the plan for the future of Bitcoin.

2 Preliminaries

In this section we introduce some notation and review the basic concepts and definitions from the literature, in particular from [11] and [2] that form the basis of our treatment. For completeness, an expanded version of this review can be found in the full version [1]. Our definitions use and build on the simulation-based security definition by Canetti [6]; we assume some familiarity with its basic principles.

Throughout this work we will assume an (at times implicit) security parameter κ. We use ITM to the denote the set of *probabilistic polynomial time (PPT)* interactive Turing machines (ITMs). We also use the standard notions of *negligible, noticeable*, and *overwhelming* (e.g., see [16]) were we denote negligible (in κ) functions as $\text{negl}(\kappa)$. Finally, using standard UC notation we denote by $\text{EXEC}_{\Pi,\mathcal{A},\mathcal{Z}}$ (resp. $\text{EXEC}_{\mathcal{F},\mathcal{S},\mathcal{Z}}$) the random variable (ensemble if indexed by κ) corresponding to the output of the environment $\mathcal{Z}$ witnessing an execution of protocol Π against adversary $\mathcal{A}$ (resp. an ideal evaluation of functionality $\mathcal{F}$ with simulator $\mathcal{S}$).

2.1 The RPD Framework

The RPD framework [11] captures incentive-driven adversaries by casting attacks as a *meta-game* between two rational players, the protocol designer D and the attacker A, which we now describe. The game is parameterized by a (multi-party) functionality $\mathcal{F}$ known to both agents D and A which corresponds to the ideal goal the designer is trying to achieve (and the attacker to break). Looking ahead, when we analyze Bitcoin, $\mathcal{F}$ will be a ledger functionality (cf. [2]). The designer D chooses a PPT protocol Π for realizing the functionality $\mathcal{F}$ from the set of all probabilistic and polynomial-time (PPT) computable protocols.[8] D sends Π to A who *then* chooses a PPT adversary $\mathcal{A}$ to attack protocol Π. The set of possible terminal histories is then the set of sequences of pairs $(\Pi, \mathcal{A})$ as above.

Consistently with [11], we denote the corresponding attack game by $\mathcal{G}_{\mathcal{M}}$, where $\mathcal{M}$ is referred to as the *attack model*, which specifies all the public parameters of the game, namely: (1) the functionality, (2) the description of the relevant action sets, and (3) the utilities assigned to certain actions (see below).

Stability in RPD corresponds to a refinement of a *subgame-perfect equilibrium* (cf. [26, Definition 97.2]), called *ϵ-subgame perfect equilibrium*, which considers as solutions profiles in which the parties' utilities are ϵ-close to their best-response utilities (see [11] for a formal definition). Throughout this paper, we will only consider $\epsilon = \text{negl}(\kappa)$; in slight abuse of notation, we will refer to negl(κ)-*subgame perfect equilibrium* simply as subgame perfect.

The utilities. The core novelty of RPD is in how utilities are defined. Since the underlying game is zero-sum, it suffices to define the attacker's utility. This utility depends on the goals of the attacker, more precisely, the security breaches which he succeeds to provoke, and is defined, using the simulation paradigm, via the following three-step process:

First, we modify the ideal functionality $\mathcal{F}$ to obtain a (possibly weaker) ideal functionality $\langle\mathcal{F}\rangle$, which explicitly allows the attacks we wish to model. For example, $\langle\mathcal{F}\rangle$ could give its simulator access to the parties' inputs. This allows to score attacks that aim at input-privacy breaches.

Second we describe a scoring mechanism for the different breaches that are of interest to the adversary. Specifically, we define a function $v_{\mathtt{A}}$ mapping the joint view of the relaxed functionality $\langle\mathcal{F}\rangle$ and the environment $\mathcal{Z}$ to a real-valued *payoff*. This mapping defines the random variable (ensemble) $v_{\mathtt{A}}^{\langle\mathcal{F}\rangle,\mathcal{S},\mathcal{Z}}$ as the result of applying $v_{\mathtt{A}}$ to the views of $\langle\mathcal{F}\rangle$ and $\mathcal{Z}$ in a random experiment describing an ideal evaluation with ideal-world adversary $\mathcal{S}$; in turn, $v_{\mathtt{A}}^{\langle\mathcal{F}\rangle,\mathcal{S},\mathcal{Z}}$ defines the *attacker's (ideal) expected payoff* for simulator $\mathcal{S}$ and environment $\mathcal{Z}$, denoted by $U_{I^{\mathtt{A}}}^{\langle\mathcal{F}\rangle}(\mathcal{S},\mathcal{Z})$, so the expected value of $v^{\mathtt{A}}_{\langle\mathcal{F}\rangle,\mathcal{S},\mathcal{Z}}$. The triple $\mathcal{M} = (\mathcal{F}, \langle\mathcal{F}\rangle, v^{\mathtt{A}})$ constitutes the *attack model*.

The third and final step is to use $U_{I^{\mathtt{A}}}^{\langle\mathcal{F}\rangle}(\mathcal{S},\mathcal{Z})$ to define the attackers utility, $u_{\mathtt{A}}(\Pi, \mathcal{A})$, for playing an adversary $\mathcal{A}$ against protocol Π, as the expected payoff of the "best" simulator that successfully simulates $\mathcal{A}$ in its ($\mathcal{A}$'s) favorite environment. This best simulator is the one that translates the adversary's breaches against Π into breaches against the relaxed functionality $\langle\mathcal{F}\rangle$ in a faithful manner, i.e., so that the ideal breaches occur only if the adversary really makes them necessary for the simulator in order to simulate. As argued in [11], this corresponds to the simulator that minimizes the attacker's utility. Formally, for a functionality $\langle\mathcal{F}\rangle$ and a protocol Π, denote by $\mathcal{C}_{\mathcal{A}}$ the class of simulators that

[8] Following standard UC convention, the protocol description includes its hybrids.

are "good" for $\mathcal{A}$, i.e, $\mathcal{C}_{\mathcal{A}} = \{\mathcal{S} \in \mathtt{ITM} \mid \forall \mathcal{Z} : \mathrm{EXEC}_{\Pi,\mathcal{A},\mathcal{Z}} \approx \mathrm{EXEC}_{\langle\mathcal{F}\rangle,\mathcal{S},\mathcal{Z}}\}$.[9] Then the attacker's (expected) utility is defined as:

$$u_{\mathtt{A}}(\Pi, \mathcal{A}) = \sup_{\mathcal{Z}\in\mathtt{ITM}} \left\{ \inf_{\mathcal{S}\in\mathcal{C}_{\mathcal{A}}} \left\{ U_{I^{\mathtt{A}}}^{\langle\mathcal{F}\rangle}(\mathcal{S}, \mathcal{Z}) \right\} \right\}.$$

For $\mathcal{A}$ and Π with $\mathcal{C}_{\mathcal{A}} = \emptyset$, the utility is ∞ by definition, capturing the fact that we only want to consider protocols which at the very least implement the relaxed (i.e., explicitly breachable) functionality $\langle\mathcal{F}\rangle$. Note that as the views in the above experiments are in fact random variable ensembles indexed by the security parameter κ, the probabilities of all the relative events are in fact functions of κ, hence the utility is also a function of κ. Note also that as long as $\mathcal{C}_{\mathcal{A}} = \emptyset$ is non-empty, for each value of κ, both the supremum and the infimum above exist and are finite and reachable by at least one pair $(\mathcal{S}, \mathcal{Z})$, provided the scoring function assigns finite payoffs to all possible transcripts (for $\mathcal{S} \in \mathcal{C}_{\mathcal{A}}$) (cf. [11]).

Remark 1 (Event-based utility [11]). In many applications, including those in our work, meaningful payoff functions have the following, simple representation: Let $(E_1, \ldots, E_\ell)$ denote a vector of (typically disjoint) events defined on the views (of $\mathcal{S}$ and $\mathcal{Z}$) in the ideal experiment corresponding to the security breaches that contribute to the attacker's utility. Each event E_i is assigned a real number γ_i, and the payoff function $v_{\mathtt{A}}^{\vec{\gamma}}$ assigns, to each ideal execution, the sum of γ_i's for which E_i occurred. The ideal expected payoff of a simulator is computed according to our definition as

$$U_{I^{\mathtt{A}}}^{\langle\mathcal{F}\rangle}(\mathcal{S}, \mathcal{Z}) = \sum_{E_i\in\vec{E},\gamma_i\in\vec{\gamma}} \gamma_i \Pr[E_i],$$

where the probabilities are taken over the random coins of $\mathcal{S}$, $\mathcal{Z}$, and $\langle\mathcal{F}\rangle$.

Building on the above definition of utility, [11] introduces a natural notion of security against incentive-driven attackers. Intuitively, a protocol Π is *attack-payoff secure* in a given attack model $\mathcal{M} = (\mathcal{F}, \cdot, v_{\mathtt{A}})$, if the utility of the best adversary against this protocol is the same as the utility of the best adversary in attacking the $\mathcal{F}$-hybrid "dummy" protocol, which only relays messages between $\mathcal{F}$ and the environment.

Definition 1 (Attack-payoff security [11]**).** *Let $\mathcal{M} = (\mathcal{F}, \langle\mathcal{F}\rangle, v_{\mathtt{A}}, v_{\mathtt{D}})$ be an attack model inducing utilities $u_{\mathtt{A}}$ and $u_{\mathtt{D}}$ on the attacker and the designer, respectively,*[10] *and let $\phi^{\mathcal{F}}$ be the dummy $\mathcal{F}$-hybrid protocol. A protocol Π is* attack-payoff secure *for $\mathcal{M}$ if for all adversaries $\mathcal{A} \in \mathtt{ITM}$,*

$$u_{\mathtt{A}}(\Pi, \mathcal{A}) \leq u_{\mathtt{A}}(\phi^{\mathcal{F}}, \mathcal{A}) + \mathrm{negl}(\kappa).$$

[9] This class is finite for every given value of the security parameter, Π, and $\mathcal{A}$.
[10] In [11], by default $u_{\mathtt{D}} = -u_{\mathtt{A}}$ as the game is zero-sum.

Intuitively, this security definition accurately captures security against an incentive-driven attacker, as in simulating an attack against the dummy $\mathcal{F}$-hybrid protocol, the simulator never needs to provoke any of the "breaching" events. Hence, the utility of the best adversary against Π equals the utility of an adversary that does not provoke any "bad event."

2.2 A Composable Model for Blockchain Protocols

In [2], Badertscher *et al.* present a universally composable treatment of the Bitcoin protocol, $\Pi^{\text{₿}}$, in the UC framework. Here we highlight the basic notions and results and refer to the full version [1] for details.

The Bitcoin ledger. The ledger functionality $\mathcal{G}_{\text{LEDGER}}^{\text{₿}}$ maintains a ledger state $\mathtt{state}$, which is a sequence of state blocks. A state block contains (application-specific) content values—the "transactions." For each honest party p_i, the ledger stores a pointer to a state block—the head of the state from p_i's point of view—and ensures that pointers increase monotonically and are not too far away from the head of the state (and that it only moves forward). Parties or the adversary might submit transactions, which are first validated by means of a predicate $\mathsf{ValidTx}_{\text{₿}}$, and, if considered valid, are added to the functionality's buffer. At any time, the $\mathcal{G}_{\text{LEDGER}}^{\text{₿}}$ allows the adversary to propose a candidate next-block for the state. However, the ledger enforces a specific *extend policy* specified by an algorithm $\mathsf{ExtendPolicy}$ that checks whether the proposal is compliant with the policy. If the adversary's proposal does not comply with the ledger policy, $\mathsf{ExtendPolicy}$ rejects the proposal. The policy enforced by the Bitcoin ledger can be succinctly summarized as follows:

- *Ledger's growth.* Within a certain number of rounds the number of added blocks must not be too small or too large.
- *Chain quality.* A certain fraction of the proposed blocks must be mined honestly and those blocks satisfy special properties (such as including all recent transactions).
- *Transaction liveness.* Old enough (and valid) transactions are included in the next block added to the ledger state.

The Bitcoin protocol. In [2] it was proved that (a [13]-inspired abstraction of) Bitcoin as a synchronous-UC protocol [20], called the *ledger protocol* and denoted by $\Pi^{\text{₿}}$, realizes the above ledger. $\Pi^{\text{₿}}$ uses blockchains to store a sequence of transactions. A *blockchain* $\mathcal{C}$ is a (finite) sequence of blocks $\mathbf{B}_1, \ldots, \mathbf{B}_\ell$. Each *block* $\mathbf{B}_i$ consist of a *pointer* $\mathtt{s}_i$, a *state block* $\mathtt{st}_i$, and a *nonce* $\mathtt{n}_i$. string. The chain $\mathcal{C}^{\lceil k}$ is $\mathcal{C}$ with the last k blocks removed. The *state* $\vec{\mathtt{st}}$ of the blockchain $\mathcal{C} = \mathbf{B}_1, \ldots, \mathbf{B}_\ell$ is defined as a sequence of its state blocks, i.e., $\vec{\mathtt{st}} := \mathtt{st}_1 || \ldots || \mathtt{st}_\ell$.

The validity of a blockchain $\mathcal{C} = \mathbf{B}_1, \ldots, \mathbf{B}_\ell$ where $\mathbf{B}_i = \langle \mathtt{s}_i, \mathtt{st}_i, \mathtt{n}_i \rangle$ is decided by a predicate $\mathsf{isvalidchain}_D(\mathcal{C})$. It combines two types of validity: *chain-level*, aka syntactic, validity—which, intuitively requires that valid blocks need to be solving a proof-of-work-type puzzle for a hash function $\mathsf{H} : \{0,1\}^* \to \{0,1\}^\kappa$

and difficulty d—and *state-level*, aka semantic, validity, which specifies whether the block's contents, i.e., transactions, are valid, with respect to a blockchain-specific predicate.

The Bitcoin protocol Π^{B} is executed in a hybrid world where parties have access to a random oracle functionality $\mathcal{F}_{\text{RO}}$ (modeling the hash function H), a multicast asynchronous network using channels with bounded delay $\mathcal{F}_{\text{N-MC}}$, and a global clock $\mathcal{G}_{\text{CLOCK}}$. Each party maintains a (local) current blockchain. It receives the transactions from the environment (and circulates them), and adds newly received valid transactions to a block that is then mined-on using the algorithm $\mathsf{extendchain}_D$. The idea of the algorithm is to find a proof of work—by querying the random oracle $\mathcal{F}_{\text{RO}}$—which allows to extend the local chain with a valid block. After each mining attempt the party uses the network to multicast their current blockchain. Parties always adopt the longest chain that they see starting from a pre-agreed genesis block. The protocol (implicitly) defines the ledger state to be a certain prefix of the contents of the longest chain held by each party. More specifically, if a party holds a valid chain $\mathcal{C}$ that encodes the sequence of state blocks $\vec{\mathbf{st}}$, then the ledger state is defined to be $\vec{\mathbf{st}}^{\lceil T}$, i.e., the party outputs a prefix of the encoded state blocks of its local longest chain. T is chosen such that honest parties output a consistent ledger state.

The flat model of computation. In this paper, we state the results in the synchronous flat model (with fixed difficulty) by Garay *et al.* [13]. This means we assume a number of parties, denoted by n, that execute the Bitcoin protocol Π^{B}, out of which t parties can get corrupted. For simplicity, the network $\mathcal{F}_{\text{N-MC}}$ guarantees delivery of messages sent by honest parties in round r to be available to any other party at the onset of round $r+1$. Moreover, every party will be invoked in every round and can make at most one "calculation" query to the random oracle $\mathcal{F}_{\text{RO}}$ in every round (and an unrestricted number of "verification" queries to check the validity of received chains)[11], and use the above diffusion network $\mathcal{F}_{\text{N-MC}}$ once in a round to send and receive messages. To capture these restrictions in a composable treatment, the real-world assumptions are enforced by means of a "wrapper" functionality, $\mathcal{W}_{\text{flat}}$, which adequately restricts access to $\mathcal{G}_{\text{CLOCK}}, \mathcal{F}_{\text{RO}}$ and $\mathcal{F}_{\text{N-MC}}$ as explained in [2].

Denote by ρ the fraction of dishonest parties (i.e., $t = \rho \cdot n$) and define $p := \frac{\mathsf{d}}{2^\kappa}$ which is the probability of finding a valid proof of work via a fresh query to $\mathcal{F}_{\text{RO}}$ (where d is fixed but sufficiently small, depending on n). Let $\alpha^{\text{flat}} = 1-(1-p)^{(1-\rho)\cdot n}$ be the mining power of the honest parties, and $\beta^{\text{flat}} = p\cdot(\rho\cdot n)$ be the mining power of the adversary.

Theorem 1. *Consider* Π^{B} *in the* $\mathcal{W}_{\text{flat}}(\mathcal{G}_{\text{CLOCK}}, \mathcal{F}_{\text{RO}}, \mathcal{F}_{\text{N-MC}})$*-hybrid world. If, for some* $\lambda > 1$*, the honest-majority assumption*

$$\alpha^{\text{flat}} \cdot (1 - 4\alpha^{\text{flat}}) \geq \lambda \cdot \beta^{\text{flat}}$$

[11] This fine-grained round model with one hash query was already used by Pass et al. [27]. The extension to a larger, constant upper bound of calculation queries per round as in [13] is straightforward for the results in this work.

holds in any real-world execution, then protocol Π^{B} *UC-realizes* $\mathcal{G}^{\text{B}}_{\text{LEDGER}}$ *for some specific range of parameters (given in* [2]*).*

3 Rational Protocol Design of Ledgers

In this section we present our framework for rational analysis of the Bitcoin protocol. It uses as basis the framework for *rational protocol design* (and analysis—RPD framework for short) by Garay *et al.* [11], extending it in various ways to better capture Bitcoin's features. (We refer to Sect. 2 and to the full version for RPD's main components and security definitions.) We note that although our analysis mainly focuses on Bitcoin, several of the extensions have broader applicability, and can be used for the rational analysis of other cryptocurrencies as well.

RPD's machinery offers the foundations for capturing incentive-driven attacks against multi-party protocols for a given specification. In this section we show how to tailor this methodology to the specific task of protocols aimed to securely implement a public ledger. The extensions and generalizations of the original RPD framework we provide add generic features to the RPD framework, including the ability to capture non-zero-sum attack games—which, as we argue, are more suitable for the implementation of a cryptocurrency ledger—and the extension of the class of events which yield payoff to the attacker and the designer.

The core hypothesis of our rational analysis is that the incentives of an attacker against Bitcoin—which affect his actions and attacks—depend only on the possible earnings or losses of the parties that launch the attack. We do not consider, for example, attackers that might create forks just for the fun of it. An attacker might create a "fork" in the blockchain if he expects to gain something by doing so. In more detail, we consider the following events that yield payoff (or inflict a cost) for running the Bitcoin protocol:

- *Inserting a block into the blockchain.* It is typical of cryptocurrencies that when a party manages to insert a block into the ledger's state, then it is rewarded for the effort it invested in doing so. In addition, it is typical in such protocols that the contents of the blocks (usually transactions) have some *transaction fee* associated with them. (For simplicity, in our initial formalization (Sects. 3 and 4) we will ignore transaction fees in our formal statements, describing how they are extended to also incorporate also such fees in Sect. 5.)
- *Spending resources to mine a block.* These resources might be the electricity consumed for performing the mining, the investment on mining hardware and its deterioration with time, etc.

Remark 2 (The miners/users separation principle). We remark that the scope of our work is to analyze the security of cryptocurrencies against incentive-driven attacks by the miners, i.e., the parties that are responsible for maintaining the blockchain. In particular, consistently with [2,13,27] we shall consider the inputs to the protocol as provided by a (not-necessarily rational) environment, which in particular captures the users of the system. As a result, other than the transaction fees, we will assume that the contents of the ledger do not affect the miners' strategies, which we will refer to as the *miners/users separation principle.* This principle captures the case where the users do not collude with the miners—an assumption implicit in the above works. We leave the full rational analysis of the protocol, including application layers for future research.

There are several challenges that one needs to overcome in deriving a formal treatment of incentive-driven attacks against Bitcoin. First, the above reward and cost mechanisms are measured in different "units." Specifically, the block reward is a cryptocurrency convention and would therefore be measured in the specific cryptocurrency's units, e.g., BTCs in the case of the Bitcoin network. On the other hand, the cost for mining (e.g., the cost of electricity, equipment usage, etc.) would be typically measured in an actual currency. To resolve this mismatch—and refrain from adopting a specific currency—we introduce a variable $\mathtt{CR}$ which corresponds to the *conversion rate* of the specific cryptocurrency unit (e.g., BTCs) to the cost unit (e.g., euros or US dollars). As we shall see in the next section, using such an explicit exchange rate allows us to make statements about the quality of the Bitcoin network that depend on its price—as they intuitively should. For example, we can formally confirm high-level statements of the type: "Bitcoin is stable—i.e., miners have incentive to keep mining honestly—as long as its price is high enough" (cf. Sect. 4).

Furthermore, this way we can express all payoffs in terms of cost units: Assume that it takes r rounds for a miner (or a collection of miners) to insert a block into the state. Denote by $\mathtt{mcost}$ the cost for a single mining attempt (in our case a single RO query), and by $\mathtt{breward}$ the fraction of cryptocurrency units (e.g., BTCs) that is given as a reward for each mined block.[12] Then, the payoff for the insertion of a single block is $\mathtt{breward} \cdot \mathtt{CR} - q_r \cdot \mathtt{mcost}$, where q_r is the number of queries to the RO that were required to mine this block during r rounds.

The second challenge is with respect to *when* should a miner receive the reward for mining. There are several reasons why solving a mining puzzle—thereby creating a new block—does not necessary guarantee a miner that he will manage to insert this block into the blockchain, and therefore be rewarded for it, including the possibility of collisions—more than one miner solving the puzzle—or, even worse, adversarial interference—e.g., network delays or "selfish mining." And even if the miner is the only one to solve the puzzle in a given round, he should only be rewarded for it if his block becomes part of the (permanent) state of the blockchain—the so-called blockchain's "common prefix."

To overcome this second challenge we rely on the RPD methodology. In particular, we will use the ideal experiment where parties have access to the global ledger functionality, where we can clearly identify the event of inserting a block into the state, and decide, by looking into the state, which miner added which block.[13]

[12] Currently, for the Bitcoin network, this is 1/4 of the original reward (12.5 BTCs).

[13] In [2], each block of the state includes the identifier of the miner who this block is attributed to.

In order to formalize the above intuitions and apply the RPD methodology to define the utilities in the attack game corresponding to implementing a ledger against an incentive-driven adversary, we need to make some significant adaptations and extensions to the original framework, which is what we do next. We then (Sect. 3.2) use the extended framework to define the attack-model for the Bitcoin protocol, and conclude the section by giving appropriate definitions of security and stability in this model.

3.1 Extending the RPD Framework

We describe how to extend the model from [11] to be able to use it in our context.

Black-box simulators. The first modification is adding more flexibility to how utilities are defined. The original definition of ideal payoff $U_{I^{\mathtt{A}}}^{\langle\mathcal{F}\rangle}(\mathcal{S},\mathcal{Z})$ computes the payoff of the simulator using the joint view of the environment and the functionality. This might become problematic when attempting to assign cost to resources used by the adversary—the RO queries in our scenario, for example. Indeed, these queries are not necessarily in this joint view, as depending on the simulator, one might not be able to extract them.[14] To resolve this we modify the definition to restrict it to black-box simulators, resulting in $\mathcal{C}_{\mathcal{A}}$ being the class of simulators that use the adversary as a black box. This will ensure that the queries to the RO are part of the interaction of the simulator with its adversary, and therefore present in the view of the simulator. Further, we include this part of the simulator's view in the definition of the scoring function $v_{\mathtt{A}}$, which is defined now as a mapping from the joint view of the relaxed functionality $\langle\mathcal{F}\rangle$, the environment $\mathcal{Z}$, and the simulator $\mathcal{S}$ to a real-valued *payoff*.

Non-zero-sum attack games. The second modification is removing the assumption that the attack game is zero-sum. Indeed, the natural incentive of the protocol designer in designing a Ledger protocol is not to optimally "tame" its attacker—as in [11]—but rather to maximize the revenue of the non-adversarially controlled parties while keeping the blockchain healthy, i.e., free of forks. This is an important modification as it captures attacks in which the adversary preserves his rate of blocks inserted into the state, but slows down the growth of the state to make sure that honest miners accrue less revenue in any time interval. For example, the so called "selfish-mining" strategy [9] provokes a slowdown since honest mining power is invested into mining on a chain which is not the longest one (as the longest chain is kept private as long as possible by the party that does the selfish-mining).

To formally specify the utility of the designer in such a non-zero-sum attack game, we employ a similar reasoning as used in the original RPD framework for defining the attacker's utility. The first step, relaxing the functionality, can be omitted provided that we relaxed it sufficiently in the definition of the attacker's utility. In the second step, we define the scoring mechanism for the incentives

[14] Indeed, in the ideal simulation of the Bitcoin protocol presented in [2], there is no RO in the ideal world.

of the designer as a function $v_{\mathtt{D}}$ mapping the joint view of the relaxed functionality $\langle\mathcal{F}\rangle$, the environment $\mathcal{Z}$, and the simulator $\mathcal{S}$ to a real-valued *payoff*, and define the designer's *(ideal) expected payoff* for simulator $\mathcal{S}$ with respect to the environment $\mathcal{Z}$ as

$$U_{I^{\mathtt{D}}}^{\langle\mathcal{F}\rangle}(\mathcal{S},\mathcal{Z}) = E(v_{\mathtt{D}}^{\langle\mathcal{F}\rangle,\mathcal{S},\mathcal{Z}}),$$

where $v_{\mathtt{D}}^{\langle\mathcal{F}\rangle,\mathcal{S},\mathcal{Z}}$ describes (as a random variable) the payoff of D allocated by $\mathcal{S}$ in an execution using directly the functionality $\langle\mathcal{F}\rangle$.

The third and final step is the trickiest. Here we want to use the above ideal expected payoff to define the expected payoff of a designer using protocol Π when the attacker is playing adversary $\mathcal{A}$. In order to ensure that our definition is consistent with the original definition in [11]—which applied to (only) zero-sum games—we need to make sure that the utility of the designer increases as the utility of the attacker decreases and vice versa. Thus, to assign utility for the designer to a strategy profile $(\Pi,\mathcal{A})$, we will use the same simulators and environments that were used to assign the utility for the attacker. Specifically, let $\mathbb{S}_{\mathtt{A}}$ denote the class of simulators that are used to formulate the utility of the adversary, and let $\mathbb{Z}_{\mathtt{A}}$ denote the class of environments that maximize this utility for simulators in $\mathbb{S}_{\mathtt{A}}$[15], then

$$\mathbb{S}_{\mathtt{A}} = \left\{\mathcal{S}\in\mathcal{C}_{\mathcal{A}} \text{ s.t. } \sup_{\mathcal{Z}\in\mathtt{ITM}}\{U_{I^{\mathtt{A}}}^{\langle\mathcal{F}\rangle}(\mathcal{S},\mathcal{Z})\} = u_{\mathtt{A}}(\Pi,\mathcal{A})\right\} \tag{1}$$

and

$$\mathbb{Z}_{\mathtt{A}} = \left\{\mathcal{Z}\in\mathtt{ITM} \text{ s.t. for some } \mathcal{S}\in\mathbb{S}_{\mathtt{A}}: U_{I^{\mathtt{A}}}^{\langle\mathcal{F}\rangle}(\mathcal{S},\mathcal{Z})\} = u_{\mathtt{A}}(\Pi,\mathcal{A})\right\}. \tag{2}$$

It is easy to verify that this choice of simulator respects the utilities being opposite in a zero-sum game as defined in [11], thereby preserving the results following the original RPD paradigm.

Lemma 1. *Let $v_{\mathtt{D}} = -v_{\mathtt{A}}$ and let $U_{I^{\mathtt{D}}}^{\langle\mathcal{F}\rangle}(\mathcal{S},\mathcal{Z})$ defined as above. For some $\mathcal{S}\in\mathbb{S}_{\mathtt{A}}$ and some $\mathcal{Z}\in\mathbb{Z}_{\mathtt{A}}$, define $u_{\mathtt{D}}(\Pi,\mathcal{A}) := U_{I^{\mathtt{D}}}^{\langle\mathcal{F}\rangle}(\mathcal{S},\mathcal{Z})$. Then $u_{\mathtt{D}}(\Pi,\mathcal{A}) = -u_{\mathtt{A}}(\Pi,\mathcal{A})$.*

Proof. Since $v_{\mathtt{D}} = -v_{\mathtt{A}}$, we have that for all $\mathcal{Z},\mathcal{S}\in\mathtt{ITM}$,

$$U_{I^{\mathtt{D}}}^{\langle\mathcal{F}\rangle}(\mathcal{S},\mathcal{Z}) = -U_{I^{\mathtt{A}}}^{\langle\mathcal{F}\rangle}(\mathcal{S},\mathcal{Z}). \tag{3}$$

However, by definition, since $\mathcal{S}\in\mathbb{S}_{\mathtt{A}}$, we have

$$u_{\mathtt{A}}(\Pi,\mathcal{A}) = U_{I^{\mathtt{A}}}^{\langle\mathcal{F}\rangle}(\mathcal{S},\mathcal{Z}) \stackrel{3}{=} -U_{I^{\mathtt{D}}}^{\langle\mathcal{F}\rangle}(\mathcal{S},\mathcal{Z}) = -u_{\mathtt{D}}(\Pi,\mathcal{A}).$$

□

The above lemma confirms that for a zero-sum attack game we can take any pair $(\mathcal{S},\mathcal{Z})\in\mathbb{S}_{\mathtt{A}}\times\mathbb{Z}_{\mathtt{D}}$ in the definition of $u_{\mathtt{D}}(\Pi,\mathcal{A})$ and it will preserve the zero-sum property (and hence all the original RPD results). This is so because all these

[15] Recall that as argued in Sect. 2.1, these sets are non-empty provided $\mathcal{C}_{\mathcal{A}}\neq\emptyset$.

simulators induce the same utility $-u_{\mathtt{A}}(\Pi, \mathcal{A})$ for the designer. However, for our case of non-zero-sum games, each of those simulator/environment combinations might induce a different utility for the designer. To choose the one which most faithfully translates the designer's utility from the real to the ideal world we use the same line of argument as used in RPD for defining the attacker's utility: The best (i.e., the most faithful) simulator is the one which always rewards the designer whenever his protocol provokes some profitable event; in other words, the one that maximizes the designer's expected utility. Similarly, the natural environment is the one that puts the protocol in its worst possible situation, i.e., the one that minimizes its expected gain; indeed, such an environment will ensure that the designer is guaranteed to get his allocated utility. The above leads to the following definition for the designer's utility in non-zero-sum games:

$$u_{\mathtt{D}}(\Pi, \mathcal{A}) := \inf_{\mathcal{Z} \in \mathbb{Z}_{\mathtt{A}}} \left\{ \sup_{\mathcal{S} \in \mathbb{S}_{\mathtt{A}}} \left\{ U_{I^{\mathtt{D}}}^{\langle \mathcal{F} \rangle}(\mathcal{S}, \mathcal{Z}) \right\} \right\}.$$

For completeness, we set $u_{\mathtt{D}}(\Pi, \mathcal{A}) = -\infty$ if $\mathcal{C}_{\mathcal{A}} = \emptyset$, i.e., if the protocol does not even achieve the relaxed functionality. This is not only intutive—as $\mathcal{C}_{\mathcal{A}} = \emptyset$ means that the designer chose a protocol which does not even reach the relaxed goal—but also analogous to how RPD defines the attacker's utility for protocols that do not achieve their relaxed specification.[16]

Finally, the attack model for non-zero-sum games is defined as the quadruple $\mathcal{M} = (\mathcal{F}, \langle \mathcal{F} \rangle, v_{\mathtt{A}}, v_{\mathtt{D}})$.

3.2 Bitcoin in the RPD Framework

Having formulated the above extensions to the RPD framework, we are ready to apply the methodology to analyze Bitcoin.

Basic foundations. We explain in more depth on how to implement the core steps of RPD. First, we define the Ledger functionality from [2] as Bitcoin's ideal goal (see Sect. 2.2). Following the three steps of the methodology, we start by defining the relaxed version of the Ledger, denoted as $\mathcal{G}_{\text{WEAK-LEDGER}}^{\text{B}}$. Informally, the relaxed Ledger functionality operates as the original ledger with the following modifications:

The state is a tree: Instead of storing a single ledger state `state` as a straight-line blockchain-like structure, $\mathcal{G}_{\text{WEAK-LEDGER}}^{\text{B}}$ stores a tree `state-tree` of state blocks where for each node the direct path from the root defines a possible ledger state that might be presented to any of the honest miners. The functionality maintains for each registered party $p_i \in \mathcal{P}$ a pointer $\mathtt{pt}_i$ to a node in the tree which defines p_i's current-state view. Furthermore, instead of restricting the adversary to only be able to set the state "slackness" to be not larger than a specific parameter, $\mathcal{G}_{\text{WEAK-LEDGER}}^{\text{B}}$ offers the command SET-POINTER which allows the adversary to set the pointers of honest parties

[16] Recall that RPD sets $u_{\mathtt{A}}(\Pi, \mathcal{A}) = \infty$ if $\mathcal{A}$ cannot be simulated, i.e., if $\mathcal{C}_{\mathcal{A}} = \emptyset$.

within state-tree with the following restriction: The pointer of an honest party can only be set to a node whose distance to the root is at least the current-pointer node's.

Relaxed validity check of transactions: All submitted transactions are accepted into the buffer buffer without validating against state-tree. Moreover, transactions in buffer which are added to state-tree are not removed as they could be reused at another branch of state-tree.

Ability to create forks: This relaxation gives the simulator the explicit power to create a fork on the ledger's state. This is done as follows: The command NEXT-BLOCK—which, recall, allows the simulator to propose the next block—is modified to allow the simulator to extend an arbitrary leaf of a sufficiently long rooted path of state-tree. Thus, when state-tree is just a single path, this command operates as in the original ledger from [2]. Additionally, in the relaxed ledger, the simulator is also allowed to add the next block to an intermediate, i.e., non-leaf node of state-tree. This is done by using an extra command FORK which, other than extending the chain from the indicated block provides the same functionality as NEXT-BLOCK.

Relaxed state-extension policy: As explained in Sect. 2.2, the extend policy is a compliance check that the ledger functionality performs on blocks that the simulator proposes to be added to the ledger's state. This is to ensure that they satisfy certain conditions. This is the mechanism which the ledger functionality uses to enforce, among others, common generic-ledger properties from the literature, such as the chain quality or the chain growth properties, and for Bitcoin ledgers the transaction-persistence/stability properties [13,27]. of the ledger state, or on transaction persistence/stability [13]. The relaxed ledger uses a much more permissive extend policy, denoted as weakExtendPolicy, derived from ExtendPolicy with the following modifications: Intuitively, in contrast to ExtendPolicy, the weaker version does not check if the adversary inserts too many or too few blocks, and it does not check if all old-enough transactions have been included. There is also no check of whether enough blocks are mined by honest parties, i.e., that there are enough blocks with coin-base transactions from honest parties. In other words, weakExtendPolicy does not enforce any concrete bounds on the chain quality or the chain growth properties of the ledger state, or on transaction persistence/stability. It rather ensures basic validity criteria of the resulting ledger state.

More formally, instead of state, it takes state-tree and a pointer pt as input. It first computes a valid default block $\vec{N}_{\mathtt{df}}$ which can be appended at the longest branch of state-tree. It then checks if the proposed blocks $\vec{N}$ can be safely appended to the node pt (to yield a valid state). If this is the case it returns $(\vec{N}, \mathtt{pt})$; otherwise it returns $\vec{N}_{\mathtt{df}}$ and a pointer to the leaf of the longest branch in state-tree.

The formal description of the relaxed ledger functionality is found in the full version [1]. This completes the first step of the RPD methodology.

The second step is defining the scoring function. This is where our application of RPD considerably deviates from past works [11,12]. In particular, those works consider attacks against generic secure multi-party computation protocols, where the ideal goal is the standard secure function evaluation (SFE) functionality (cf. [6]). The security breaches are breaking correctness and privacy [11] or breaking fairness [12]. These can be captured by relaxing the SFE functionality to allow the simulator to request extra information (breaking privacy), reset the outputs of honest parties to a wrong value (breaking correctness), or cause an abort (breaking fairness.) The payoff function is then defined by looking at events corresponding to whether or not the simulator provokes these events, and the adversary is given payoff whenever the best simulator is forced to provoke them in order to simulate the attack.

However, attacks against the ledger that have as an incentive increasing the revenue of a coalition are not necessarily specific events corresponding to the simulator sending special "break" commands. Rather, they are events that are extracted from the joint views (e.g., which blocks make it to the state and when). Hence, attacks to the ledger correspond to the simulator implicitly "tweeking" its parameters. Therefore, in this work we take the following approach to define the payoffs of the attacker and designer. In contrast to the RPD examples in [11,12], which use explicit events that "downgrade" the ideal functionality for defining utility, we directly use more intuitive events defined on the joint view of the environment, the functionality, and the simulator. The reason is that as we have assumed that the only rationale is to increase one's profit, the incentives in case of cryptocurrencies are as follows: whenever a block is mined, the adversary gets rewarded. A "security breach" is relevant if (and only if) the adversary can get a better reward by doing so.

Defining concrete utility functions. Defining the utility functions lies at the core of a rational analysis of a blockchain protocol like Bitcoin. The number of aspects that one would like to consider steers the complexity of a concrete analysis, the ultimate goal being to reflect exactly the incentive structure of the actual blockchain ecosystem. Our extended RPD framework for blockchain protocols provides a guideline to defining utility functions of various complexity and to conduct the associated formal analysis. Recall that the utility functions are the means to formalize the assumed underlying incentive structure. As such, our approach is extensible: if certain relevant properties or dynamics are identified or believed (such as reflecting a doomsday risk of an attacker or a altruistic motivation of honest miners), one can enrich the incentive structure by reflecting the associated events and rewards in the utility definition, or by making the costs and rewards time-dependent variables. The general goal of this line of research on rational aspects of cryptocurrencies is to eventually arrive at a more detailed model and, if the assumptions are reasonable, to have more predictive models for reality.

Below we define a first, relatively simple incentive model to concretely showcase our methodology. We conduct the associated rational analysis in the next section and observe that, although being a simplified model, we can already draw interesting conclusions from such a treatment.

Utility of the attacker. Informally, this particular utility is meant to capture the average revenue of the attacker. Consider the following sequence of events defined on the views of the environment, the relaxed ledger functionality, and the black-box simulator of the entire experiment (i.e., until the environment halts) for a given adversary $\mathcal{A}$:

1. For each pair $(q, r) \in \mathbb{N}^2$ define event $W^{\mathtt{A}}_{q,r}$ as follows: The simulator makes q mining queries in round r, i.e., it receives q responses on different messages to the RO in round r.[17]
2. For each pair $(b, r) \in \mathbb{N}^2$ define event $I^{\mathtt{A}}_{b,r}$ as follows: The simulator inserts b blocks into the state of the ledger in round r, such that all these blocks were previously queries to the (simulated) random oracle by the adversary. More formally, $I^{\mathtt{A}}_{b,r}$ occurs if the function extend policy (of the weak ledger) is successfully invoked and outputs a sequence of b non-empty blocks (to be added to the state), where for each of these blocks the following properties hold: (1) The block has appeared in the past in the transcript between the adversary and the simulator, and (2) the contents of the block have appeared on this transcript prior to the block's first appearance, as a query from the adversary to its (simulated) RO. We note in passing that this event definition ensures that the simulator (and therefore also the adversary) does not earn reward by adaptively corrupting parties after they have done the work/query to mine a block but before their block is added into the state. In other words, the adversary only gets rewarded for state blocks which corrupted parties mined while they where already under the adversary's control.

Now, using the simplified event-based utility definition (Remark 1) we define the attacker's utility for a strategy profile $(\Pi, \mathcal{A})$ in the attack game as:[18]

$$u^{\text{\bitcoin}}_{\mathtt{A}}(\Pi, \mathcal{A}) = \sup_{\mathcal{Z} \in \mathtt{ITM}} \left\{ \inf_{\mathcal{S}^{\mathcal{A}} \in \mathcal{C}_{\mathcal{A}}} \left\{ \sum_{(b,r) \in \mathbb{N}^2} b \cdot \mathtt{breward} \cdot \mathtt{CR} \cdot \Pr[I^{\mathtt{A}}_{b,r}] - \sum_{(q,r) \in \mathbb{N}^2} q \cdot \mathtt{mcost} \cdot \Pr[W^{\mathtt{A}}_{q,r}] \right\} \right\}.$$

We remark that although the above sums are in principle infinite, in any specific execution these sums will have only as many (non-zero) terms as the number of rounds in the protocol. Indeed, if the experiment finishes in r' rounds then for any $r > r'$, $\Pr[I^{\mathcal{A}}_{b,r}] = \Pr[W^{\mathcal{A}}_{q,r}] = 0$ for all $b \in \mathbb{N}$. Furthermore, we assume that $\mathtt{breward}$, $\mathtt{CR}$ and $\mathtt{mcost}$ are $O(1)$, i.e., independent of the security parameter.

[17] Observe that since our ideal world is the $\mathcal{G}_{\text{CLOCK}}$-hybrid synchronous world, the round structure is trivially extracted from the simulated ideal experiment by the protocol definition and the clock value. Furthermore, the adversary's mining queries can be trivially extracted by its interaction with the black-box simulator.

[18] Recall that we assume synchronous execution as in [2] where the environment gets to decide how many rounds it wishes to witness.

The above expression can be simplified to the following more useful expression. Let $B^{\mathtt{A}}$ denote the random variables corresponding to the number of blocks contributed to the ledger's state by adversarial miners and $Q^{\mathtt{A}}$ denote the number of queries to the RO performed by adversarial miners (throughout the execution of the random experiment). Then the adversary's utility can be described using the expectations of these random variables as follows:

$$u_{\mathtt{A}}^{\text{B}}(\Pi, \mathcal{A}) = \sup_{\mathcal{Z} \in \mathrm{ITM}} \left\{ \inf_{\mathcal{S}^{\mathcal{A}} \in \mathcal{C}_{\mathcal{A}}} \{\mathtt{breward} \cdot \mathtt{CR} \cdot E(B^{\mathtt{A}}) - \mathtt{mcost} \cdot E(Q^{\mathtt{A}})\} \right\}.$$

Utility of the designer. Since the game is not zero-sum we also need to formally specify the utility of the protocol designer. Recall that we have assumed that, analogously to the attacker, the designer accrues utility when honest miners insert a block into the state, and spends utility when mining—i.e., querying the RO. In addition, what differentiates the incentives of the designer from that of an attacker is that his most important goal is to ensure the "health" of the blockchain, i.e., to avoid forks. To capture this, we will assign a cost for the designer to the event the simulator is forced to request the relaxed ledger functionality to fork, which is larger than his largest possible gain. This yields the following events that are relevant for the designer's utility.

1. For each pair $(q, r) \in \mathbb{N}^2$ define $W_{q,r}^{\Pi}$ as follows: The honest parties, as a set, make q mining queries in round r.[19]
2. For each pair $(b, r) \in \mathbb{N}^2$ define $I_{b,r}^{\Pi}$ as follows: The honest parties jointly insert b blocks into the state of the ledger in round r; that is, the simulator inserts b blocks into the state of the ledger in round r, such that for each of these blocks, at least one of the two properties specified in the above definition of $I_{b,r}^{\mathtt{A}}$ does not hold.[20]
3. For each $r \in \mathbb{N}$ define K_r as follows: The simulator uses the FORK command in round r.

The utility of the designer is then defined similarly to the attacker's, where we denote by $\mathbb{S}_{\mathtt{A}}$ the class of simulators that assign to the adversary his actual utility (cf. Eq. 1):

$$u_{\mathtt{D}}^{\text{B}}(\Pi, \mathcal{A}) = \inf_{\mathcal{Z} \in \mathbb{Z}} \left\{ \sup_{\mathcal{S}^{\mathcal{A}} \in \mathbb{S}_{\mathcal{A}}} \left\{ \sum_{(b,r) \in \mathbb{N}^2} b \cdot \mathtt{CR} \cdot (\mathtt{breward} \cdot \Pr[I_{b,r}^{\Pi}] - 2^{\mathsf{polylog}(\kappa)} \cdot \Pr[K_r]) - \sum_{(q,r) \in \mathbb{N}^2} q \cdot \mathtt{mcost} \cdot \Pr[W_{q,r}^{\Pi}] \right\} \right\}.$$

[19] Note that although there is no RO in the ideal model of [2], whenever a miner would make such a query in the Bitcoin protocol, the corresponding dummy party sends a special MAINTAIN-LEDGER command to the Ledger functionality, making it possible for us to count the mining queries also in the ideal world.

[20] By definition, these two properties combined specify when the adversary should be considered the recipient of the reward.

At first glance, the choice of $2^{\mathsf{polylog}(\kappa)}$ might seem somewhat arbitrary. However, it is there to guarantee that if the ledger state forks (recall that this reflects a violation of the common-prefix property) with noticeable probability, then the designer is punished with this super-polynomially high penalty to make his expected payoff negative as κ grows. On the other hand, if the probability of such a fork is sufficiently small (e.g. in the order of $2^{-\Omega(\kappa)}$), then the loss in utility is made negligible. This, combined with the fact that our stability notions will render negligible losses in the utility irrelevant, will allow the designer the freedom to provide slightly imperfect protocols, i.e., protocols where violations of the common-prefix property occur with sufficiently small probability.

We will denote by $\mathcal{M}^{₿}$ the Bitcoin attack model which has $\mathcal{G}^{₿}_{\text{LEDGER}}$ as the goal, $\langle \mathcal{G}^{₿}_{\text{LEDGER}} \rangle$ as the relaxed functionality, and scoring functions for the attacker and designer inducing utilities $u^{₿}_{\mathtt{A}}$ and $u^{₿}_{\mathtt{D}}$, respectively.

3.3 Attack-Payoff Security and Incentive Compatibility

The definition of the respective utilities for designer and attacker completes the specification of an attack game. Next, we define the appropriate notions of security and stability as they relate to Bitcoin and discuss their meaning.

We start with *attack-payoff security* [11], which, as already mentioned, captures that the adversary would have no incentive to make the protocol deviate from a protocol that implements the ideal specification (i.e., from a protocol that implements the ideal [non-relaxed] ledger functionality), and which is useful in arguing about the resistance of the protocol against incentive-driven attacks. However, in the context of Bitcoin analysis, one might be interested in achieving an even stronger notion of incentive-driven security, which instead of restricting the adversary to strategies that yield payoff as much as the ideal ledger $\mathcal{G}^{₿}_{\text{LEDGER}}$ from [2] would, restricts him to play in a coordinated fashion but *passively*, i.e., follow the mining procedure mandated by the Bitcoin protocol, including announcing each block as soon as it is found, but ensure that no two corrupt parties try to solve the same puzzle (i.e., use the same nonce).

One can think of the above strategy as corresponding to cooperating mining-pools which run the standard Bitcoin protocol. Nonetheless, as the adversary has control over message delays, he is able to make sure that whenever he finds a new block in the same round as some other party, his own block will be the one propagated first[21], and therefore the one that will be added to the blockchain. Note that a similar guarantee is not there for honest miners as in the event of collisions—two miners solve a puzzle in the same round—the colliding miners have no guarantee about whose block will make it. We will refer to such an adversary that sticks to the Bitcoin mining procedure but makes sure his blocks are propagated first as *front running*.

Definition 2 (Front-running, passive mining adversary). *The front-running adversarial strategy $\mathcal{A}_{\text{fr}}$ is specified as follows: Upon activation in round*

[21] This can be thought of as a "rushing" strategy with respect to network delays.

$r > 0$, $\mathcal{A}_{\mathtt{fr}}$ activates in a round-robin fashion all its (passively) corrupted parties, say $p_1, \ldots, p_t$. When party p_i generated some new message to be sent through the network, $\mathcal{A}_{\mathtt{fr}}$ immediately delivers m to all its recepients.[22] *In addition, upon any activation, any message submitted to the network $\mathcal{F}_{\text{N-MC}}$ by an honest party is maximally delayed.*

Note that there might be several front-running, passive mining strategies, depending on which parties are corrupted and (in case of adaptive adversaries) when. We shall denote the class of all such adversary strategies by $\mathbb{A}_{\mathtt{fr}}$. We are now ready to provide the definition of (strong) attack-payoff security for Bitcoin. The definition uses the standard notion of *negl-best-response* strategy from game theory: Consider a two-player game with utilities u_1 and u_2, respectively. A strategy for m_1 of p_1 is *best response* to a strategy m_2 of p_2 if for all possible strategies m_1', $u_1(m_1', m_2) \leq u_1(m_1, m_2) + \text{negl}(\kappa)$. For conciseness, in the sequel we will refer to negl-best-response simply as best-response strategies.

Definition 3. *A protocol Π is* strongly attack-payoff secure for attack model $\mathcal{M}^{\text{B}}$ *if for some $\mathcal{A} \in \mathbb{A}_{fr}$ the attacker playing $\mathcal{A}$ is a (negl-)best-response to the designer playing Π.*

Remark 3. It is instructive to see that for such a weak class of adversaries the usual blockchain properties hold with very nice parameters[23]: first, the common-prefix property is satisfied except with negligible probability (as no intentional forks are provoked by anyone). Second, the fraction of honest blocks (in an interval of say k blocks) is roughly $\frac{\alpha}{\alpha+\beta} \overset{p<<1}{\approx} \frac{(1-\rho)np}{(1-\rho)np+\rho np} = (1-\rho)$ and thus, in expectation, the chain quality corresponds to the relative mining power of honest parties. Finally, since the adversary does contribute his mining power to the main chain, the number of rounds it takes for the chain to grow by k blocks is in expectation $\frac{k}{\alpha+\beta} \overset{p<<1}{\approx} \frac{k}{np}$.

Security thus means that if the honest parties stick to their protocol then the adversary has no incentive to deviate. However, unlike in [11], where the game is zero-sum, in a non-zero-sum setting it does not imply that the *designer* has an incentive to stick to the protocol. This means that the definition is useful to answer the question whether, assuming the network keeps mining, some of the miners have an incentive to deviate from the protocol, but it does not address the question of why the *honest miners* would keep mining. To address this question, we adopt the notion of *incentive compatibility* (IC).

Informally, a protocol being incentive-compatible means that both the attacker and the designer are willing to stick to it. In other words, it is strongly attack-payoff secure—i.e., the adversary will run it if the honest parties do—*and*

[22] I.e., $\mathcal{A}_{\mathtt{fr}}$ sets the delay of the corresponding transmissions to 0.

[23] Recall the notation introduced in Sect. 2.2: n denotes the number of parties, ρ the fraction of corrupted parties, α and β denote honest and dishonest mining power, respectively, and p is the probability of a fresh RO-query to return a correct PoW solution.

if the adversary plays it passively (and front-running), then the honest miners will have an incentive to follow the protocol—i.e., the protocol is the designer's best response to a passive front-running adversary. We note that requiring IC for Bitcoin for the class of all possible protocols would imply a proof that Bitcoin is not only a protocol that the miners wish to follow, but also that there is no other protocol that they would rather participate in instead. This is clearly too strong a requirement, even more so in the presence of results [13,28] that argue that there are alternative "fairer" blockchain protocols which improve on the miners' expected revenue. Thus, we can only hope to make such statements for a subclass of possible protocols, and therefore devise a version of IC which is parameterized by the set of all acceptable deviations (i.e., alternative protocols) $\mathbb{\Pi}$. For full generality, we also parameterize it with respect to the class of acceptable adversaries $\mathbb{A}$, but stress that all statements in this work are for the class of all (PPT) adversaries.

Towards providing the formal definition of IC, we first give the straightforward restriction of equilibrium (in our case, subgame-perfect equilibrium) to a subset of strategies.

Definition 4. *Let $\mathbb{\Pi}$ and $\mathbb{A}$ be sets of possible strategies for the designer and the attacker, respectively. We say that a pair $(\Pi, \mathcal{A}) \in (\mathbb{\Pi}, \mathbb{A})$ is a $(\mathbb{\Pi}, \mathbb{A})$*-subgame perfect equilibrium *in the attack game defined by model $\mathcal{M}$, if it is a (negl(κ)-) subgame-perfect equilibrium on the restricted attack game where the set of all possible deviations of the designer (resp., the attacker) is $\mathbb{\Pi}$ (resp., $\mathbb{A}$).*

The formal definition of (parameterized) IC is then as follows:

Definition 5. *Let Π be a protocol and $\mathbb{\Pi}$ be a set of polynomial-time protocols that have access to the same hybrids as Π. We say that Π is $\mathbb{\Pi}$*-incentive compatible *($\mathbb{\Pi}$-IC for short) in the attack model $\mathcal{M}$ iff for some $\mathcal{A} \in \mathbb{A}_{fr}$, $(\Pi, \mathcal{A})$ is a $(\mathbb{\Pi}, \mathtt{ITM})$-subgame-perfect equilibrium in the attack game defined by $\mathcal{M}$.*

4 Analysis of Bitcoin Without Transaction Fees

In this section, we present our RPD analysis of Bitcoin for the concrete incentive structure defined in the previous section. We note that this incentive structure does not, in particular, reflect rewards that stem from transaction fees and hence the reward per block is constant. First, in Sect. 4.1, we prove that Bitcoin is strongly attack-payoff secure—i.e., if the designer plays it, the attacker is better off sticking to it as well (but in a front-running fashion). The result is independent of the distribution of computing power to honest vs adversarial miners and independent of the conversion rate or the values of `breward` and `mcost`.

Subsequently, in Sect. 4.2, we investigate the role of mining costs vs conversion rate vs block rewards for the stability (i.e., IC) of Bitcoin in the presence of such incentive-driven coordinated coalitions (e.g., utility-maximizing mining pools.) We devise conditions on these values that either make the utility of honest parties negative—hence make playing the Bitcoin protocol a sub-optimal choice of the protocol designer, or yield high enough utility for mining that makes Bitcoin *optimal* among all possible deviations from the standard protocol that are still compatible with the Bitcoin network (i.e., produce valid blockchains); combining this with the results from Sect. 4.1, we deduce that for this latter range of parameters Bitcoin is incentive-compatible.

4.1 Attack-Payoff Security of Bitcoin (Without Fees)

The attack-payoff security of Bitcoin without fees is stated in the following theorem.

Theorem 2. *The Bitcoin protocol is strongly attack-payoff secure in the attack model $\mathcal{M}^{B}$.*

Proof. The theorem follows as a direct corollary of the following general lemma.

Lemma 2. *Given any adversarial strategy, there is a front-running, semi-honest mining adversary $\mathcal{A}$ that achieves better utility. In particular, the adversarial strategy $\mathcal{A}$ makes as many RO-queries per round as allowed by the real-world restrictions, and one environment that maximizes its utility is the environment $\mathcal{Z}$ that activates $\mathcal{A}$ as the first ITM in every round until $\mathcal{A}$ halts.*

Proof intuition. The proof of the lemma consists of three steps. First, we analyze Bitcoin in the real world. By invoking the subroutine-replacement theorem from [11, Theorem 6], we are able to work in a hybrid world where we can easily compute the relevant values, such as the number of blocks an adversary can mine in a given interval of rounds (the hybrid world is the so-called state-exchange hybrid world of [2]). Second, we show by a generic argument that this real-world analysis is sufficient to compute the payoffs for the attacker (which is defined on the transcript in the ideal world). Last but not least, we make a case distinction whether the adversary has expected utility smaller than zero (in which he does not corrupt any party and does not participate in the network), or whether mining Bitcoin is profitable for the attacker. In both cases, we prove that for any attacker $\mathcal{A}$, we can devise a front-running and semi-honest mining adversary which gets higher utility. The formal proof of the lemma is found in the full version [1]. □

4.2 Incentive Compatibility of Bitcoin (Without Fees)

We proceed by investigating how the IC of Bitcoin depends on the relation between rewards and the conversion rate. Concretely, we describe a sufficient condition for IC (Theorem 4) and a condition that makes it non-IC (Theorem 3). We start with the negative result, which, informally, says that if the expected costs are too high with respect to the expected rewards, then Bitcoin is not IC (although it is strongly attack-payoff secure as proved above). As above, we denote by p the probability of solving a proof of work (and hence being a candidate to extend the ledger state) using one query to the random oracle (or equivalently, that a query to the state-exchange functionality successfully extends a state).

Theorem 3. *For $n > 0$ and $\mathit{breward} \cdot \mathit{CR} < \frac{\mathit{mcost}}{p}$ the Bitcoin protocol is not incentive compatible.*

The proof is a straightforward calculation of the utility for the designer per round. Under the above condition, this expectation is less than 0, since they spend (on average) more on queries than what the reward compensates. Hence, the best response would be a protocol that does nothing.

While the above condition implies that the Bitcoin protocol is not a stable solution for all choices of the rewards, costs, and CR, we next provide conditions under which the standard Bitcoin protocol is in fact a stable solution in the attack game. For this, we need to compare it to arbitrary alternative strategies that produce valid blocks for the Bitcoin network. Informally, our condition for IC requires that CR and breward are sufficiently higher than the costs.

Theorem 4. *Consider the real world consisting of the random oracle functionality $\mathcal{F}_{\mathrm{RO}}$, the diffusion network $\mathcal{F}_{\mathrm{N\text{-}MC}}$, and the clock $\mathcal{G}_{\mathrm{CLOCK}}$, and let $\mathcal{W}_{\mathrm{flat}}(\cdot)$ be the wrapper that formalizes the restrictions of the flat model.*[24] *Consider the class $\Pi_{\mathsf{isvalidchain}_{H,d}(\cdot)}$ of protocols Π that are defined for the $\mathcal{W}_{\mathrm{flat}}(\mathcal{G}_{\mathrm{CLOCK}}, \mathcal{F}_{\mathrm{RO}}, \mathcal{F}_{\mathrm{N\text{-}MC}})$-hybrid world and which are compatible with the Bitcoin network, i.e., which obey the following two restrictions:*

1. *With probability 1, the real-world transcript (i.e., the real-world UC-execution of Π, any environment and adversary) does not contain a chain $\mathcal{C}$ with $\mathsf{isvalidchain}_{H,d}(\mathcal{C}) = 0$ and this chain was an output to the network from an uncorrupted protocol instance running Π.*
2. *Upon input (READ, sid) to a protocol instance, the return value is (READ, sid, $\vec{st}^{\lceil T}$) (for some integer T), where $\vec{st}^{\lceil T}$ denotes the prefix of the state $\vec{st}$ encoded in the longest valid chain $\mathcal{C}$ received by this protocol instance.*

With respect to the class $\Pi_{\mathsf{isvalidchain}_{H,d}(\cdot)}$, the Bitcoin protocol is an incentive-compatible choice for the protocol designer if Bitcoin is profitable as in Lemma 3, i.e., if we are in the region $\mathit{breward} \cdot \mathit{CR} > \frac{n \cdot \mathit{mcost}}{p}$, and if

$$\mathit{breward} \cdot \mathit{CR} > \frac{\mathit{mcost}}{p \cdot (1-p)^{n-1}}. \tag{4}$$

Remark 4. Formula 4 constitutes a stronger requirement than the mere condition that mining should be profitable (which we treat separately in Lemma 3 for completeness). The theorem says that the probability that a fixed miner is uniquely successful stands in a reasonable relation to the mining cost and block rewards to achieve a stable solution. While Bitcoin would already yield positive utility to the protocol designer in the case of $\mathtt{breward} \cdot \mathtt{CR} > \frac{n \cdot \mathtt{mcost}}{p}$, we have for large n, $\frac{\mathtt{mcost}}{p} \cdot n \leq \frac{\mathtt{mcost}}{p} \cdot (\frac{1}{1-p})^{n-1}$ (for $p \in (0,1)$).

[24] Recall from [2] that we model restrictions by using functionality wrappers. The above implemented restrictions correspond to the so-called flat model of Bitcoin, where each party gets one query to the random oracle per round and can send and receive one vector of messages in each round.

Proof intuition. The proof follows by demonstrating, in a sequence of claims, that the actual choices of the Bitcoin protocol (i.e., our abstraction of it) are optimal under the conditions of the theorem. This includes proving that the assumed resources cannot be employed in a way that would yield better payoff to the protocol designer. Intuitively, if the protocol has to be compatible with the Bitcoin network (i.e., it has to produce valid chains with probability 1), and invest its resources to achieve the optimum reward vs. query ratio in a setting where it knows it is running against front-running adversary running Bitcoin (such as mining pools). Optimality under the theorem's condition follows by deducing a couple of useful properties from the fact that the protocol has to work potentially independently (per round) and by computing (and maximizing) the distribution of the possible query-vs.-reward ratios. The formal proof is found in the full version [1]. □

We note that the above conditions are not necessarily tight. Thus one might wonder whether we can prove or disprove their tightness, and in the latter case investigate tight conditions for the statements to hold. We conclude this section with the following lemma which serves as first partial attempt to investigate this gap. The lemma implies that there might be potential to prove (partial) IC even for values of the parameters that fall in the gap between the above theorems. We leave the thorough investigation of this gap in terms of stability as a future research direction.

Lemma 3. *If* $\mathtt{breward} \cdot \mathit{CR} > \frac{n \cdot \mathtt{mcost}}{p}$ *then the Bitcoin protocol yields, with overwhelming probability, a positive utility for the protocol designer in the presence of front-running adversaries, i.e., the Bitcoin protocol is profitable in such a setting.*

5 Analysis of Bitcoin with Transaction Fees

Recall that in our formal treatment a chain $\mathcal{C}$ encodes a ledger state $\vec{\mathtt{st}}$. A ledger state is a sequence of individual state-blocks, i.e., $\vec{\mathtt{st}} = \mathtt{st}_1 || \ldots || \mathtt{st}_\ell$. In addition, each state-block $\mathtt{st} \in \vec{\mathtt{st}}$ (except the genesis state) of the state encoded in the blockchain has the form $\mathtt{st} = \mathsf{Blockify}(\vec{N})$ where $\vec{N}$ is a vector of transactions, i.e., $\vec{N} = \mathtt{tx}_1, \ldots, \mathtt{tx}_k$. A transaction $\mathtt{tx}_i$ can be seen as the abstract content of a block. Our above analysis assumes that the contents of the blocks do not affect the incentives of the attacker and the designer. In the real-world execution of Bitcoin, however, this is not the case as the contents of the blocks are money-like transactions and have transaction fees associated with them. We model these using positive-valued function $\mathtt{tx} \mapsto f(\mathtt{tx})$ mapping individual transactions to a positive real value that are integer multiples of 1 *Satoshi* (equals 10^{-8} Bitcoin).[25] For sake of brevity, we will also denote by $\hat{f}(\mathtt{st}) := \sum_{\mathtt{tx} \in \mathtt{st}} f(\mathtt{tx})$ the sum of all fees contained in the state block $\mathtt{st}$. The fees have to be considered when defining the utilities in a rational analysis since they are added to the (flat) block reward and the total sum is given as a reward to the miner who inserts the block into

[25] Note that this modeling aspect is not sensitive to the basic unit of measurement.

the ledger state. Hence, this section treats the case where overall block rewards can be a dynamic quantity. In fact, the plan for Bitcoin is to eventually drop the block rewards at which point mining will be incentivized exclusively by the associated transaction fees. In this section we study the security and stability of the Bitcoin network incorporating also such fees.

5.1 Utility Functions with Fees

We first have to change the definition of the utility functions to incorporate that the attacker and the designer receive a different reward when inserting a block into the ledger state. The difference are the transactions fees. To this end, we first introduce a set $\mathcal{T}_\mathcal{Z}$ which contains all transactions that are submitted by the environment (and in particular not by the adversary), and then define the relevant events to capture fees in our model.[26]

- In an execution, let $\mathcal{T}_\mathcal{Z}$ be the set of transactions such that $\mathtt{tx} \in \mathcal{T}_\mathcal{Z}$ if and only if tx first appeared as an input from the environment (i.e., the first occurrence of tx is in a command (SUBMIT, tx) in this execution).
- For each $(\mu, r) \in \mathbb{N}^2$ the event $F^{\mathtt{A}}_{r,\mu}$ is defined as follows: $F^{\mathtt{A}}_{r,\mu}$ denotes the event that the total sum of the transaction fees $f(\mathtt{tx})$ of all $\mathtt{tx} \in \mathcal{T}_\mathcal{Z}$ contained in the blocks that the adversary adds to the state in round r is equal to $\mu \cdot 10^{-8} \cdot \mathtt{CR}$ cost units.[27]
- For each $(\mu, r) \in \mathbb{N}^2$ let the event $F^{\mathtt{D}}_{r,\mu}$ be defined as follows: $F^{\mathtt{D}}_{r,\mu}$ is the event that the total sum of the transaction fees $f(\mathtt{tx})$ of all $\mathtt{tx} \in \mathcal{T}_\mathcal{Z}$ contained in the blocks that the honest miners (jointly) add to the state in round r is equivalent to $\mu \cdot 10^{-8} \cdot \mathtt{CR}$ cost units.

Since it is the environment that decides on the block-content, the sum of the fees in each block is effectively a random variable whose distribution is induced by the environment. The utilities of the attacker and designer that incorporate fees are defined as follows (we use $\hat{u}^{\text{₿}}_{\mathtt{A}}$ and $\hat{u}^{\text{₿}}_{\mathtt{D}}$ to denote the utilities when fees are added to the incentives):

$$\hat{u}^{\text{₿}}_{\mathtt{A}}(\Pi, \mathcal{A}) = \sup_{\mathcal{Z} \in \mathtt{ITM}} \left\{ \inf_{\mathcal{S}^\mathcal{A} \in \mathcal{C}_\mathcal{A}} \left\{ \mathtt{breward} \cdot \mathtt{CR} \cdot E(B^{\mathtt{A}}) - q \cdot \mathtt{mcost} \cdot E(Q^{\mathtt{A}}) \right.\right.$$
$$\left.\left. + \sum_{(\mu,r) \in \mathbb{N}^2} \mu \cdot 10^{-8} \cdot \mathtt{CR} \cdot \Pr[F^{\mathtt{A}}_{r,\mu}] \right\}\right\}$$

[26] Note that we assume that only transactions submitted by the environment can yield fees, since the environment models "the application layer". In particular, if the adversary creates a transaction on his own and includes it in his next mined block, then this should not assign him any payoff.

[27] Recall that CR is the conversion of one cryptocurrency unit (e.g., one bitcoin) to one cost unit (e.g., one US dollar).

and

$$\hat{u}_{\mathtt{D}}^{\text{B}}(\Pi, \mathcal{A}) = \inf_{\mathcal{Z} \in \mathbb{Z}} \left\{ \sup_{\mathcal{S}^{\mathcal{A}} \in \mathbb{S}_{\mathcal{A}}} \left\{ \sum_{(b,r) \in \mathbb{N}^2} b \cdot \mathtt{CR} \cdot (\mathtt{breward} \cdot \Pr[I_{b,r}^{\mathtt{D}}] - 2^{\mathsf{polylog}(\kappa)} \cdot \Pr[K_r]) \right.\right.$$
$$\left.\left. - \sum_{(q,r) \in \mathbb{N}^2} q \cdot \mathtt{mcost} \cdot \Pr[W_{q,r}^{\mathtt{D}}] + \sum_{(\mu,r) \in \mathbb{N}^2} \mu \cdot 10^{-8} \cdot \mathtt{CR} \cdot \Pr[F_{r,\mu}^{\mathtt{D}}] \right\} \right\}.$$

Note that the multiplicative factor 10^{-8} is there to allow us to set μ to the integer multiple of one Satoshi that the fee yields. We will denote by $\hat{\mathcal{M}}^{\text{B}}$ the Bitcoin attack model which has $\mathcal{G}_{\text{LEDGER}}^{\text{B}}$ as the goal, $\langle \mathcal{G}_{\text{LEDGER}}^{\text{B}} \rangle$ as the relaxed functionality, and scoring functions for the attacker and designer inducing utilities $\hat{u}_{\mathtt{A}}^{\text{B}}$ and $\hat{u}_{\mathtt{D}}^{\text{B}}$.

Upper bounds on fees and total reward for blocks. In reality, transaction fees and the overall reward of a block are naturally bounded (either by size limits or by restricting the total value of the system).[28] In the following, we assume that for all $\mathtt{tx}$, $f(\mathtt{tx}) \leq \mathsf{max}_{\mathsf{fee}}$, and that the sum of fees per block is bounded, yielding an upper bound on the total profit per block: For all state blocks $\mathtt{st}$ we require that $\mathtt{breward} + \hat{f}(\mathtt{st}) \leq \mathsf{max}_{\mathsf{block}}$, where $\mathsf{max}_{\mathsf{fee}}$ and $\mathsf{max}_{\mathsf{block}}$ are (strictly) positive multiples of one Satoshi.

Restrictions on the availability of transactions. So far in our treatment, the environment induces a distribution on the available transactions and is in principle unrestricted in doing so. For example, the set $\mathcal{T}_{\mathcal{Z}}$ is not bounded in size except by the running time of $\mathcal{Z}$. As will become apparent below in Theorem 5, putting no restrictions on the set $\mathcal{T}_{\mathcal{Z}}$ can still lead to meaningful statements that apply, for example, to applications that are believed to generate an (a priori) unbounded number of transactions. However, to model different kinds of scenarios that appear in the real world, we have to develop a language that allows us to speak about limited availability of transactions. To this end, we introduce parameterized environments $\mathcal{Z}^{\mathcal{D}}$. More precisely, let $\mathcal{D}$ be an oracle which takes inputs $(\textsc{NextTxs}, r)$ and returns a vector $\vec{T}_r = (\mathtt{tx}_1, p_{i_1}), \ldots, (\mathtt{tx}_k, p_{i_k})$. We say that an environment is $\mathcal{D}$-respecting, if, in every round r, the environment queries the oracle $\mathcal{D}$ and only transactions $\mathtt{tx} \in \vec{T}_r$ are added to $\mathcal{T}_{\mathcal{Z}}$. We further require that $\mathcal{Z}$ submits $(\textsc{submit}, \mathtt{tx}_i)$ to party p_k in round r if and only if $(\mathtt{tx}_i, p_k) \in \vec{T}_r$. For simplicity, we call $\mathcal{D}$ simply a distribution. The utility for the attacker in such environments is taken to be the supremum as above, but only over all $\mathcal{D}$-respecting environments.

5.2 Analysis of Bitcoin (with Fees)

The following theorem says that if we look at unrestricted environments, then Bitcoin is still incentive compatible. This is a consequence of Theorems 2 and 4 and proven formally in the full version [1].

[28] For example, the number of total Bitcoins is limited and the block-size is bounded.

Theorem 5. *Consider arbitrary environments and let the sum of the transaction fees per block be bounded by* $\mathsf{max}_{\mathsf{block}} > 0$. *Then the Bitcoin protocol is strongly attack-payoff secure in the attack model* $\hat{\mathcal{M}}^{\mathcal{B}}$. *It is further incentive-compatible with respect to the class of protocols that are compatible with the Bitcoin network under the same conditions as in Theorem 4), i.e., if*

$$\mathtt{breward} \cdot CR > \frac{\mathtt{mcost}}{p \cdot (1-p)^{n-1}}.$$

The previous statement is void in case the flat block reward is 0. However, for certain types of distributions $\mathcal{D}$, namely, the ones that provide sufficient high-fee transactions to the participants, it will remain in an equilibrium state. The statement is proven in the full version [1].

Theorem 6. *Consider distributions* $\mathcal{D}$ *with the following property: In every round,* $\mathcal{D}$ *outputs a vector of transactions such that any party gets as input a list of transactions to build a valid next state block* $\mathtt{st}$ *to extend the longest chain and such that* $\hat{f}(\mathtt{st}) = \mathsf{max}_{\mathsf{block}}$ *holds (where* $\mathsf{max}_{\mathsf{block}} > 0$*). Then, with respect to* $\mathcal{D}$*-respecting environments, the Bitcoin protocol is strongly attack-payoff secure in the attack model* $\hat{\mathcal{M}}^{\mathcal{B}}$. *It is further incentive compatible with respect to the class of protocols that are compatible with the Bitcoin network (as defined in Theorem 4) if* $\mathsf{max}_{\mathsf{block}} \cdot CR > \frac{\mathtt{mcost}}{p \cdot (1-p)^{n-1}}$.

However, if an application cannot provide enough transactions, it becomes problematic, as the following counterexample shows.

Theorem 7. *There exist distributions* $\mathcal{D}$ *such that the Bitcoin protocol is neither attack-payoff secure nor strongly attack-payoff secure with respect to* $\mathcal{D}$*-respecting environments.*

Proof. The proof is straightforward and follows from a general observation: assume there is just a single transaction in the network which has been received only by a corrupted party p_i. Then, the adversary does not publish this transaction to the network. If he does not, then he will be the one claiming the reward with probability one, which is his best choice. Hence, he does not follow the protocol (as the semi-honest front-running adversary would do) and hence it cannot be strongly attack-payoff secure.

Furthermore, the protocol is also not attack-payoff secure. If the honest-majority assumption does not hold, and thus an adversary can fork the ledger state, he would exercise his power to create a ledger state where it is a corrupted party who mines the block containing the only transaction in the system as this will yield better reward than simply mining on empty blocks. □

Fallback security. Note that because cryptographic security trivially implies attack-payoff security for all possible environments and utilies, we can easily derive a fallback security notion: If the majority of miners mines honestly, then we get attack-payoff security; and even if this fails, we still get attack-payoff security under the assumption that the distribution of the fees and the relation between rewards vs costs vs conversion rate are as in Theorem 5 or 6.

References

1. Badertscher, C., Garay, J., Maurer, U., Tschudi, D., Zikas, V.: But why does it work? A rational protocol design treatment of bitcoin. Cryptology ePrint Archive, Report 2018/138 (2018). https://eprint.iacr.org/2018/138
2. Badertscher, C., Maurer, U., Tschudi, D., Zikas, V.: Bitcoin as a transaction ledger: a composable treatment. In: Katz, J., Shacham, H. (eds.) CRYPTO 2017. LNCS, vol. 10401, pp. 324–356. Springer, Cham (2017). https://doi.org/10.1007/978-3-319-63688-7_11
3. Github: Bitcoin Core Version 0.12.0. Wallet: Transaction Fees. https://github.com/bitcoin/bitcoin/blob/v0.12.0/doc/release-notes.md#wallet-transaction-fees
4. Bonneau, J.: Why buy when you can rent? In: Clark, J., Meiklejohn, S., Ryan, P.Y.A., Wallach, D., Brenner, M., Rohloff, K. (eds.) FC 2016. LNCS, vol. 9604, pp. 19–26. Springer, Heidelberg (2016). https://doi.org/10.1007/978-3-662-53357-4_2
5. Canetti, R.: Security and composition of multiparty cryptographic protocols. J. Cryptol. **13**(1), 143–202 (2000)
6. Canetti, R.: Universally composable security: a new paradigm for cryptographic protocols. In: 42nd FOCS, pp. 136–145. IEEE Computer Society Press, October 2001
7. Carlsten, M., Kalodner, H.A., Weinberg, S.M., Narayanan, A.: On the instability of bitcoin without the block reward. In: Weippl, E.R., Katzenbeisser, S., Kruegel, C., Myers, A.C., Halevi, S. (eds.) ACM CCS 2016, pp. 154–167. ACM Press, October 2016
8. Eyal, I.: The miner's dilemma. In: 2015 IEEE Symposium on Security and Privacy, pp. 89–103. IEEE Computer Society Press, May 2015
9. Eyal, I., Sirer, E.G.: Majority is not enough: bitcoin mining is vulnerable. In: Christin, N., Safavi-Naini, R. (eds.) FC 2014. LNCS, vol. 8437, pp. 436–454. Springer, Heidelberg (2014). https://doi.org/10.1007/978-3-662-45472-5_28
10. Fuchsbauer, G., Katz, J., Naccache, D.: Efficient rational secret sharing in standard communication networks. In: Micciancio, D. (ed.) TCC 2010. LNCS, vol. 5978, pp. 419–436. Springer, Heidelberg (2010). https://doi.org/10.1007/978-3-642-11799-2_25
11. Garay, J.A., Katz, J., Maurer, U., Tackmann, B., Zikas, V.: Rational protocol design: cryptography against incentive-driven adversaries. In: 54th FOCS, pp. 648–657. IEEE Computer Society Press, October 2013
12. Garay, J.A., Katz, J., Tackmann, B., Zikas, V.: How fair is your protocol? A utility-based approach to protocol optimality. In: Georgiou, C., Spirakis, P.G. (eds.) 34th ACM PODC, pp. 281–290. ACM, July 2015
13. Garay, J., Kiayias, A., Leonardos, N.: The bitcoin backbone protocol: analysis and applications. In: Oswald, E., Fischlin, M. (eds.) EUROCRYPT 2015. LNCS, vol. 9057, pp. 281–310. Springer, Heidelberg (2015). https://doi.org/10.1007/978-3-662-46803-6_10
14. Garay, J., Kiayias, A., Leonardos, N.: The bitcoin backbone protocol with chains of variable difficulty. In: Katz, J., Shacham, H. (eds.) CRYPTO 2017. LNCS, vol. 10401, pp. 291–323. Springer, Cham (2017). https://doi.org/10.1007/978-3-319-63688-7_10
15. Gervais, A., Karame, G.O., Wüst, K., Glykantzis, V., Ritzdorf, H., Capkun, S.: On the security and performance of proof of work blockchains. In: Weippl, E.R., Katzenbeisser, S., Kruegel, C., Myers, A.C., Halevi, S. (eds.) ACM CCS 2016, pp. 3–16. ACM Press, October 2016

16. Goldreich, O.: Foundations of Cryptography: Volume 1, Basic Tools. Cambridge University Press, Cambridge (2003)
17. Gradwohl, R., Livne, N., Rosen, A.: Sequential rationality in cryptographic protocols. In: 51st FOCS, pp. 623–632. IEEE Computer Society Press, October 2010
18. Halpern, J.Y., Pass, R., Seeman, L.: Computational extensive-form games. In: EC (2016)
19. Katz, J.: Bridging game theory and cryptography: recent results and future directions. In: Canetti, R. (ed.) TCC 2008. LNCS, vol. 4948, pp. 251–272. Springer, Heidelberg (2008). https://doi.org/10.1007/978-3-540-78524-8_15
20. Katz, J., Maurer, U., Tackmann, B., Zikas, V.: Universally composable synchronous computation. In: Sahai, A. (ed.) TCC 2013. LNCS, vol. 7785, pp. 477–498. Springer, Heidelberg (2013). https://doi.org/10.1007/978-3-642-36594-2_27
21. Kol, G., Naor, M.: Games for exchanging information. In: Ladner, R.E., Dwork, C. (eds.) 40th ACM STOC, pp. 423–432. ACM Press, May 2008
22. Luu, L., Teutsch, J., Kulkarni, R., Saxena, P.: Demystifying incentives in the consensus computer. In: Ray, I., Li, N., Kruegel, C. (eds.) ACM CCS 2015, pp. 706–719. ACM Press, October 2015
23. Nakamoto, S.: Bitcoin: A Peer-to-Peer Electronic Cash System (2008). http://bitcoin.org/bitcoin.pdf
24. Nayak, K., Kumar, S., Miller, A., Shi, E.: Stubborn mining: generalizing selfish mining and combining with an eclipse attack. In: S&P (2016)
25. Ong, S.J., Parkes, D.C., Rosen, A., Vadhan, S.: Fairness with an honest minority and a rational majority. In: Reingold, O. (ed.) TCC 2009. LNCS, vol. 5444, pp. 36–53. Springer, Heidelberg (2009). https://doi.org/10.1007/978-3-642-00457-5_3
26. Osborne, M.J., Rubinstein, A.: A Course in Game Theory. MIT Press, Cambridge (1994)
27. Pass, R., Seeman, L., Shelat, A.: Analysis of the blockchain protocol in asynchronous networks. In: Coron, J.-S., Nielsen, J.B. (eds.) EUROCRYPT 2017. LNCS, vol. 10211, pp. 643–673. Springer, Cham (2017). https://doi.org/10.1007/978-3-319-56614-6_22
28. Pass, R., Shi, E.: FruitChains: a fair blockchain. In: Schiller, E.M., Schwarzmann, A.A. (eds.) 36th ACM PODC, pp. 315–324. ACM, July 2017
29. Rosenfeld, M.: Analysis of bitcoin pooled mining reward systems. CoRR (2011)
30. Sapirshtein, A., Sompolinsky, Y., Zohar, A.: Optimal selfish mining strategies in bitcoin. In: Grossklags, J., Preneel, B. (eds.) FC 2016. LNCS, vol. 9603, pp. 515–532. Springer, Heidelberg (2017). https://doi.org/10.1007/978-3-662-54970-4_30
31. Schrijvers, O., Bonneau, J., Boneh, D., Roughgarden, T.: Incentive compatibility of bitcoin mining pool reward functions. In: Grossklags, J., Preneel, B. (eds.) FC 2016. LNCS, vol. 9603, pp. 477–498. Springer, Heidelberg (2017). https://doi.org/10.1007/978-3-662-54970-4_28
32. Teutsch, J., Jain, S., Saxena, P.: When cryptocurrencies mine their own business. In: Grossklags, J., Preneel, B. (eds.) FC 2016. LNCS, vol. 9603, pp. 499–514. Springer, Heidelberg (2017). https://doi.org/10.1007/978-3-662-54970-4_29

Ouroboros Praos: An Adaptively-Secure, Semi-synchronous Proof-of-Stake Blockchain

Bernardo David[1,2](✉), Peter Gaži[2], Aggelos Kiayias[2,3], and Alexander Russell[4]

[1] Tokyo Institute of Technology, Tokyo, Japan
bernardo.david@iohk.io
[2] IOHK, Hong Kong, China
peter.gazi@iohk.io
[3] University of Edinburgh, Edinburgh, UK
akiayias@inf.ed.ac.uk
[4] University of Connecticut, Mansfield, CT, USA
acr@cse.uconn.edu

Abstract. We present "Ouroboros Praos", a proof-of-stake blockchain protocol that, for the first time, provides security against *fully-adaptive corruption* in the *semi-synchronous setting*: Specifically, the adversary can corrupt any participant of a dynamically evolving population of stakeholders at any moment as long the stakeholder distribution maintains an honest majority of stake; furthermore, the protocol tolerates an adversarially-controlled message delivery delay unknown to protocol participants.

To achieve these guarantees we formalize and realize in the universal composition setting a suitable form of forward secure digital signatures and a new type of verifiable random function that maintains unpredictability under malicious key generation. Our security proof develops a general combinatorial framework for the analysis of semi-synchronous blockchains that may be of independent interest. We prove our protocol secure under standard cryptographic assumptions in the random oracle model.

1 Introduction

The design of *proof-of-stake* blockchain protocols was identified early on as an important objective in blockchain design; a proof-of-stake blockchain substitutes the costly proof-of-work component in Nakamoto's blockchain protocol [20] while still providing similar guarantees in terms of transaction processing in the presence of a dishonest minority of users, where this "minority" is to be understood here in the context of stake rather than computational power.

The basic stability and security properties of blockchain protocols were first rigorously formulated in [12] and further studied in [15,21]; these include common prefix, chain quality and chain growth and refer to resilient qualities of the underlying data structure of the blockchain in the presence of an adversary that attempts to subvert them.

J. B. Nielsen and V. Rijmen (Eds.): EUROCRYPT 2018, LNCS 10821, pp. 66–98, 2018.
https://doi.org/10.1007/978-3-319-78375-8_3

Proof-of-stake protocols typically possess the following basic characteristics. Based on her local view, a party is capable of deciding, in a publicly verifiable way, whether she is permitted to produce the next block. Assuming the block is valid, other parties update their local views by adopting the block, and proceed in this way continuously. At any moment, the probability of being permitted to issue a block is proportional to the relative stake a player has in the system, as reported by the blockchain itself.

A particularly challenging design aspect is that the above probabilistic mechanism should be designed so that the adversary cannot bias it to its advantage. As the stake shifts, together with the evolving population of stakeholders, so does the honest majority assumption, and hence the function that appoints stakeholders should continuously take the ledger status into account. Preventing the biasing of the election mechanism in a context of a blockchain protocol is a delicate task that so far has eluded a practical solution that is secure against all attacks.

Our Results. We present "Ouroboros Praos", a provably secure proof-of-stake protocol that is the first to be secure against adaptive attackers and scalable in a truly practical sense. Our protocol is based on a previous proof-of-stake protocol, Ouroboros [16], as its analysis builds on some of the core combinatorial arguments that were developed to analyze that scheme. Nevertheless, the protocol construction has a number of novel elements that require a significant recasting and generalization of the previous combinatorial analysis. In more detail, our results are as follows.

In Ouroboros Praos, deciding whether a certain participant of the protocol is eligible to issue a block is decided via a private test that is executed locally using a special verifiable random function (VRF) on the current time-stamp and a nonce that is determined for a period of time known as an "epoch". A special feature of this VRF primitive, novel to our approach, is that the VRF must have strong security characteristics even in the setting of malicious key generation: specifically, if provided with an input that has high entropy, the output of the VRF is unpredictable even when an adversary has subverted the key generation procedure. We call such VRF functions "VRF with unpredictability under malicious key generation" and we present a strong embodiment of this notion with a novel Universal Composable (UC) formulation. We also present a very efficient realization of this primitive under the Computational Diffie Hellman (CDH) assumption in the random oracle model. Beyond this VRF notion, we also formalize in a UC fashion key evolving signatures that provide the forward security that is necessary for handling the adaptive corruption setting.

In more detail, we analyze our protocol in the *partial* or *semi-synchronous* model [11,21]. In this setting, we still divide the protocol execution in time units which, as in [16], are called slots, but there is a maximum delay of Δ slots that is applied to message delivery and it is unknown to the protocol participants.[1] In order to cope with the Δ-semisynchronous setting we introduce the

[1] It is worth pointing out that the notion of slots we use in this work can be substantially shorter in terms of real time elapsed compared to the slots of [16], where each slot represented a full round of interaction between all participants.

concept of "empty slots" which occur with sufficient frequency to enable short periods of silence that facilitate synchronization. This feature of the protocol gives also its moniker, "Praos", meaning "mellow", or "gentle". Ensuring that the adversary cannot exploit the stakeholder keys that it possesses to confuse or out-maneuver the honest parties, we develop a combinatorial analysis to show that the simple rule of following the longest chain still enables the honest parties to converge to a unique view with high probability. To accomplish this we revisit and expand the forkable strings and divergence analysis of [16]. We remark that significant alterations are indeed necessary: As we demonstrate in the full version of this paper, the protocol of [16] and its analysis are critically tailored to synchronous operation and is susceptible to a desynchronization attack that can completely violate the common prefix property. Our new combinatorial analysis introduces a new concept of characteristic strings and "forks" that reflects silent periods in protocol execution and network delays. To bound the density of forkable strings in this Δ-semisynchronous setting we establish a syntactic reduction from Δ-semisynchronous characteristic strings to synchronous strings of [16] that preserves the structure of the forks they support. This is followed by a probabilistic analysis that controls the distortion caused by the reduction and concludes that Δ-semisynchronous forkable strings are rare. Finally, we control the effective power of adaptive adversaries in this setting with a stochastic dominance argument that permits us to carry out the analysis of the underlying blockchain guarantees (e.g., common prefix) with a single distribution that provably dominates all distributions on characteristic strings generated by adaptive adversaries. We remark that these arguments yield graceful degradation of the analysis as a function of network delays (Δ), in the sense that the effective stake of the adversary is amplified by a function of Δ.

The above combinatorial analysis is nevertheless only sufficient to provide a proof of the static stake case, i.e., the setting where the stake distribution relevant to the honest majority assumption remains fixed at the onset of the computation and prior to the selection of the random genesis data that are incorporated in the genesis block. For a true proof-of-stake system, we must permit the set of stakeholders to evolve over time and appropriately adapt our honest stakeholder majority assumption. Achieving this requires a bootstrapping argument that allows the protocol to continue unboundedly by revising its stakeholder distribution as it evolves. We bootstrap our protocol in two conceptual steps. First we show how bootstrapping is possible if a randomness beacon is available to all participants. The beacon at regular intervals emits a new random value and the participants can reseed the election process so the stakeholder distribution used for sampling could be brought closer to the one that is current. A key observation here is that our protocol is resilient even if the randomness beacon is weakened in the following two ways: (i) it leaks its value to the adversary ahead of time by a bounded number of time units, (ii) it allows the adversary to reset its value if it wishes within a bounded time window. We call the resulting primitive a "leaky resettable beacon" and show that our bootstrapping argument still holds in this stronger adversarial setting.

In the final refinement of our protocol, we show how it is possible to implement the leaky resettable beacon via a simple algorithm that concatenates the VRF outputs that were contributed by the participants from the blockchain and subjects them to a hash function that is modeled as a random oracle. This implementation explains the reasons behind the beacon relaxation we introduced: leakiness stems from the fact that the adversary can complete the blockchain segment that determines the beacon value before revealing it to the honest participants, while resettability stems from the fact that the adversary can try a bounded number of different blockchain extensions that will stabilize the final beacon value to a different preferred value.

Putting all the above together, we show how our protocol provides a "robust transaction ledger" in the sense that an immutable record of transactions is built that also guarantees that new transactions will be always included. Our security definition is in the Δ-semisynchronous setting with full adaptive corruptions. As mentioned above, security degrades gracefully as Δ increases, and this parameter is unknown to the protocol participants.

Note that implementing the beacon via hashing VRF values will make feasible a type of "grinding attack" where the adversary can trade hashing power for a slight bias of the protocol execution to its advantage. We show how this bias can be controlled by suitably increasing the relevant parameters depending on the hashing power that is available to the adversary.

Comparison to related work. The idea of proof-of-stake protocols has been discussed extensively in the bitcoin forum.[2] The manner that a stakeholder determines eligibility to issue a block is always publicly verifiable and the proof of eligibility is either computed publicly (via a calculation that is verifiable by repeating it) or by using a cryptographic mechanism that involves a secret-key computation and a public-key verification. The first example of the former approach appeared in PPCoin [17], and was followed by others including Ouroboros and Snow White [2,8,16]; while the first example of the latter approach (that we also employ in our work) appeared in NXT (cf. Sect. 2.4.1 of [7]) and was then also used elsewhere, most notably in Algorand [19]. The virtue of the latter approach is exactly in its potential to control adaptive corruptions: due to the fact that the adversary cannot predict the eligibility of a stakeholder to issue a block prior to corrupting it, she cannot gain an advantage by directing its corruption quota to specific stakeholders. Nevertheless, none of these previous works isolated explicitly the properties of the primitives that are required to provide a full proof of security in the setting of adaptive corruptions. Injecting high quality randomness in the PoS blockchain was proposed by Bentov et al. [3,4], though their proposal does not have a full formal analysis. The Ouroboros proof-of-stake protocol [16] is provably secure in a corruption model that excludes fully adaptive attacks by imposing a corruption delay on the corruption requests of the adversary. The Snow White proof-of-stake [8] is the first to prove security in the Δ-semi-synchronous model but—as in the case of Ouroboros—adopts a weak adaptive corruption model.

[2] Refer e.g., to the posts by QuantumMechanic and others from 2011 https://bitcointalk.org/index.php?topic=27787.0 (Last Accessed 19/09/2017).

A recent work close to ours is Algorand [19] that also provides a proof-of-stake ledger that is adaptively secure. It follows an entirely different construction approach that runs a Byzantine agreement protocol for every block and achieves adaptive-corruption security via a novel, appealing concept of player-replaceability. However, Algorand is only secure against a $1/3$ adversary bound; and while the protocol itself is very efficient, it yields an inherently slower block production rate compared to an "eventual consensus" protocol (like Bitcoin, Snow White, and Ouroboros). In principle, proof-of-stake blockchain protocols can advance at the theoretical maximum speed (of one block per communication round), while protocols relying on Byzantine agreement, like Algorand, would require a larger number of rounds to settle each block.

Sleepy consensus [22] puts forth a technique for handling adaptive corruptions in a model that also encompasses fail-stop and recover corruptions; however, the protocol can be applied directly only in a static stake (i.e., permissioned) setting. We note that in fact our protocol can be also proven secure in such mixed corruption setting, where both fail-stop and recover as well as Byzantine corruptions are allowed (with the former occurring at an arbitrarily high rate); nevertheless this is out of scope for the present exposition and we omit further details.

Note that the possibility of adversarial grinding in Ouroboros Praos is also present in previous work that derives randomness by hashing [8,19], as opposed to a dedicated coin-tossing protocol as in [16]. Following the examples of [8,19], we show that security can be guaranteed despite any adversarial bias resulting from grinding. In fact, we show how to use the q-bounded model of [12] to derive a bound that shows how to increase the relevant security parameters given the hashing power that is available to the adversary.

Finally, in the present exposition we also put aside incentives; nevertheless, it is straightforward to adapt the mechanism of input endorsers from the protocol of [16] to our setting and its approximate Nash equilibrium analysis can be ported directly.

2 Preliminaries

We say a function $negl(x)$ is negligible if for every $c > 0$, there exists an $n > 0$ such that $negl(x) < 1/x^c$ for all $x \geq n$. The length of a string w is denoted by $|w|$; ε denotes the empty string. We let $v \,\|\, w$ denote concatenation of strings.

2.1 Transaction Ledger Properties

We adopt the same definitions for transaction ledger properties as [16]. A protocol Π implements a robust transaction ledger provided that the ledger that Π maintains is divided into "blocks" (assigned to time slots) that determine the order with which transactions are incorporated in the ledger. It should also satisfy the following two properties.

Persistence. Once a node of the system proclaims a certain transaction tx in the *stable* part of its ledger, the remaining nodes, if queried, will either report tx in the same position of that ledger or report a stable ledger which is a prefix of that ledger. Here the notion of stability is a predicate that is parameterized by a security parameter k; specifically, a transaction is declared *stable* if and only if it is in a block that is more than k blocks deep in the ledger.

Liveness. If all honest nodes in the system attempt to include a certain transaction then, after the passing of time corresponding to u slots (called the transaction confirmation time), all nodes, if queried and responding honestly, will report the transaction as stable.

In [15,21] it was shown that persistence and liveness can be derived from the following three elementary properties provided that protocol Π derives the ledger from a data structure in the form of a blockchain.

Common Prefix (CP); with parameters $k \in \mathbb{N}$. The chains $\mathcal{C}_1, \mathcal{C}_2$ possessed by two honest parties at the onset of the slots $sl_1 < sl_2$ are such that $\mathcal{C}_1^{\lceil k} \preceq \mathcal{C}_2$, where $\mathcal{C}_1^{\lceil k}$ denotes the chain obtained by removing the last k blocks from $\mathcal{C}_1$, and $\preceq$ denotes the prefix relation.

Chain Quality (CQ); with parameters $\mu \in (0,1]$ and $k \in \mathbb{N}$. Consider any portion of length at least k of the chain possessed by an honest party at the onset of a round; the ratio of blocks originating from the adversary is at most $1 - \mu$. We call μ the chain quality coefficient.

Chain Growth (CG); with parameters $\tau \in (0,1], s \in \mathbb{N}$. Consider the chains $\mathcal{C}_1, \mathcal{C}_2$ possessed by two honest parties at the onset of two slots sl_1, sl_2 with sl_2 at least s slots ahead of sl_1. Then it holds that $\text{len}(\mathcal{C}_2) - \text{len}(\mathcal{C}_1) \geq \tau \cdot s$. We call τ the speed coefficient.

2.2 The Semi-synchronous Model

On a high level, we consider the security model of [16] with simple modifications to account for adversarially-controlled message delays and immediate adaptive corruption. Namely, we allow the adversary $\mathcal{A}$ to selectively delay any messages sent by honest parties for up to $\Delta \in \mathbb{N}$ slots; and corrupt parties without delay.

Time and slots. We consider a setting where time is divided into discrete units called *slots*. A ledger, described in more detail above, associates with each time slot (at most) one ledger *block*. Players are equipped with (roughly) synchronized clocks that indicate the current slot: we assume that any clock drift is subsumed in the slot length. This will permit them to carry out a distributed protocol intending to collectively assign a block to this current slot. In general, each slot sl_r is indexed by an integer $r \in \{1, 2, \ldots\}$, and we assume that the real time window that corresponds to each slot has the following two properties: (1) The current slot is determined by a publicly-known and monotonically increasing function of current time. (2) Each player has access to the current time. Any discrepancies between parties' local time are insignificant in comparison with the length of time represented by a slot.

Security Model. We adopt the model introduced by [12] for analysing security of blockchain protocols enhanced with an ideal functionality $\mathcal{F}$. We note that multiple different "functionalities" can be encompassed by $\mathcal{F}$. In our model we employ the "Delayed Diffuse" functionality, which allows for adversarially-controlled delayed delivery of messages diffused among stakeholders.

The Diffuse Functionality. This functionality is parameterized by $\Delta \in \mathbb{N}$ and denoted as $\mathsf{DDiffuse}_\Delta$. It keeps rounds, executing one round per slot. $\mathsf{DDiffuse}_\Delta$ interacts with the environment $\mathcal{Z}$, stakeholders $U_1, \ldots, U_n$ and an adversary $\mathcal{A}$, working as follows for each round:

1. $\mathsf{DDiffuse}_\Delta$ maintains an incoming string for each party U_i that participates. A party, if activated, is allowed at any moment to fetch the contents of its incoming string, hence one may think of this as a mailbox. Furthermore, parties can give an instruction to the functionality to diffuse a message. Activated parties are allowed to diffuse once in a round.
2. When the adversary $\mathcal{A}$ is activated, it is allowed to: (a) Read all inboxes and all diffuse requests and deliver messages to the inboxes in any order it prefers; (b) For any message m obtained via a diffuse request and any party U_i, $\mathcal{A}$ may move m into a special string $\mathsf{delayed}_i$ instead of the inbox of U_i. $\mathcal{A}$ can decide this individually for each message and each party; (c) For any party U_i, $\mathcal{A}$ can move any message from the string $\mathsf{delayed}_i$ to the inbox of U_i.
3. At the end of each round, the functionality also ensures that every message that was either (a) diffused in this round and not put to the string $\mathsf{delayed}_i$ or (b) removed from the string $\mathsf{delayed}_i$ in this round is delivered to the inbox of party U_i. If any message currently present in $\mathsf{delayed}_i$ was originally diffused at least Δ slots ago, then the functionality removes it from $\mathsf{delayed}_i$ and appends it to the inbox of party U_i.
4. Upon receiving $(\mathsf{Create}, U, \mathcal{C})$ from the environment, the functionality spawns a new stakeholder with chain $\mathcal{C}$ as its initial local chain (as it was the case in [16]).

Modelling Protocol Execution and Adaptive Corruptions. Given the above we will assume that the execution of the protocol is with respect to a functionality $\mathcal{F}$ that incorporates $\mathsf{DDiffuse}$ as well as possibly additional functionalities to be explained in the following sections. The environment issues transactions on behalf of any stakeholder U_i by requesting a signature on the transaction as described in Protocol π_{SPoS} of Fig. 4 and handing the transaction to stakeholders to put them into blocks. Beyond any restrictions imposed by $\mathcal{F}$, the adversary can only corrupt a stakeholder U_i if it is given permission by the environment $\mathcal{Z}$ running the protocol execution. The permission is in the form of a message $(\mathsf{Corrupt}, U_i)$ which is provided to the adversary by the environment. Upon receiving permission from the environment, the adversary immediately corrupts U_i without any delay, differently from [8,16], where corruptions only take place after a given delay. Note that a corrupted stakeholder U_i will relinquish its entire state to $\mathcal{A}$; from this point on, the adversary will be activated in place of the stakeholder U_i. The adversary is able to control transactions and blocks generated by corrupted parties by interacting with $\mathcal{F}_{\mathsf{DSIG}}$, $\mathcal{F}_{\mathsf{KES}}$ and $\mathcal{F}_{\mathsf{VRF}}$, as described in Protocol π_{SPoS} of Sect. 3. In summary, regarding activations we have the following: (a) At each slot sl_j, the environment $\mathcal{Z}$ activates all honest stakeholders.[3] (b) The adversary is activated at least as the last entity in each sl_j (as well as during all adversarial party activations and invocations from the ideal functionalities as prescribed); (c) If a stakeholder does not fetch in a certain slot the messages written to its incoming string from the diffuse functionality they are flushed.

Restrictions imposed on the environment. It is easy to see that the model above confers such sweeping power on the adversary that one cannot establish any significant guarantees on protocols of interest. It is thus important to restrict the environment suitably (taking into account the details of the protocol) so that we may be able to argue security. We require that in every slot, the adversary does not control more than 50% of the stake in the view of any honest stakeholder. This transaction data, including the required signatures by each stakeholder, is obtained by the environment as specified in the protocol. If this is violated, an event $\mathsf{Bad}^{\frac{1}{2}}$ becomes true for the given execution. When the environment spawns a new stakeholder by sending message $(\mathsf{Create}, U, \mathcal{C})$ to the Key and Transaction functionality, the initial local chain $\mathcal{C}$ can be the chain of any honest stakeholder even in the case of "lazy honest" stakeholders without requiring this stakeholder to have been online in the past slot as in [16]. Finally, we note that in all our proofs, whenever we say that a property Q holds with high probability over all executions, we will in fact argue that $Q \vee \mathsf{Bad}^{\frac{1}{2}}$ holds with high probability over all executions. This captures the fact that we exclude environments and adversaries that trigger $\mathsf{Bad}^{\frac{1}{2}}$ with non-negligible probability.

Random Oracle. We also assume the availability of a random oracle. As usually, this is a function $\mathsf{H}\colon \{0,1\}^* \to \{0,1\}^w$ available to all parties that answers every fresh query with an independent, uniformly random string from $\{0,1\}^w$, while any repeated queries are answered consistently.

Erasures. We assume that honest users can do secure erasures, which is argued to be a reasonable assumption in protocols with security against adaptive adversaries, see e.g., [18].

3 The Static Stake Protocol

We first consider the static stake case, where the stake distribution is fixed throughout protocol execution. The general structure of the protocol in the semi-synchronous model is similar to that of (synchronous) Ouroboros [16] but introduces several fundamental modifications to the leader selection process: not all

[3] We assume this to simplify our formal treatment, a variant of our protocol can actually accommodate "lazy honesty" as introduced in [19]. In this variant, honest stakeholders only come online at the beginning of each epoch and at a few infrequent, predictable moments, see the full version.

slots will be attributed a slot leader, some slots might have multiple slot leaders, and slot leaders' identities remain unknown until they act. The first modification is used to deal with delays in the semi-synchronous network as the *empty slots*—where no block is generated—assist the honest parties to synchronize. The last modification is used to deal with adaptive corruptions, as it prevents the adversary from learning the slot leaders' identity ahead of time and using this knowledge to strategically corrupt coalitions of parties with large (future) influence. Moreover, instead of using concrete instantiations of the necessary building blocks, we describe the protocol with respect to *ideal functionalities*, which we later realize with concrete constructions. This difference allows us to reason about security in the ideal model through a combinatorial argument without having to deal with the probability that the cryptographic building blocks fail. Before describing the specifics of the new leader selection process and the new protocol, we first formally define the static stake scenario and introduce basic definitions as stated in [16] following the notation of [12].

In the static stake case, we assume that a fixed collection of n stakeholders $U_1, \ldots, U_n$ interact throughout the protocol. Stakeholder U_i is attributed stake s_i at the beginning of the protocol.

Definition 1 (Genesis Block). *The* genesis block B_0 *contains the list of stakeholders identified by a label* U_i*, their respective public keys and respective stakes*

$$\mathbb{S}_0 = \left((U_1, v_1^{\mathrm{vrf}}, v_1^{\mathrm{kes}}, v_1^{\mathrm{dsig}}, s_1), \ldots, (U_n, v_n^{\mathrm{vrf}}, v_n^{\mathrm{kes}}, v_n^{\mathrm{dsig}}, s_n)\right),$$

and a nonce η.

We note that the nonce η will be used to seed the slot leader election process and that $v_i^{\mathrm{vrf}}, v_i^{\mathrm{kes}}, v_i^{\mathrm{dsig}}$ will be determined by $\mathcal{F}_{\mathsf{VRF}}$, $\mathcal{F}_{\mathsf{KES}}$ and $\mathcal{F}_{\mathsf{DSIG}}$, respectively.

Definition 2 (Epoch, State, Block Proof, Block, Blockchain). *An* epoch *is a set of* R *adjacent slots* $S = \{sl_1, \ldots, sl_R\}$*. (The value* R *is a parameter of the protocol we analyze in this section.) A* state *is a string* $st \in \{0,1\}^\lambda$*. A* block proof *is a value (or set of values)* B_π *containing information that allows stakeholders to verify if a block is valid. A* block $B = (sl_j, st, d, B_{\pi j}, \sigma_j)$ *generated at a slot* $sl_j \in \{sl_1, \ldots, sl_R\}$ *contains the current state* $st \in \{0,1\}^\lambda$*, data* $d \in \{0,1\}^*$*, the slot number* sl_j*, a block proof* $B_{\pi j}$ *and* σ_j*, a signature on* $(st, d, sl_j, B_{\pi j})$ *under the signing key for the time period of slot* sl_j *of the stakeholder* U_i *generating the block.*

A blockchain *(or simply* chain*) relative to the genesis block* B_0 *is a sequence of blocks* $B_1, \ldots, B_n$ *associated with a strictly increasing sequence of slots for which the state* st_i *of* B_i *is equal to* $H(B_{i-1})$*, where* H *is a prescribed collision-resistant hash function. The* length *of a chain* $\mathrm{len}(\mathcal{C}) = n$ *is its number of blocks. The block* B_n *is the* head *of the chain, denoted* $\mathrm{head}(\mathcal{C})$*. We treat the empty string* ε *as a legal chain and by convention set* $\mathrm{head}(\varepsilon) = \varepsilon$*. Let* $\mathcal{C}$ *be a chain of length* *n and k be any non-negative integer. We denote by* $\mathcal{C}^{\lceil k}$ *the chain resulting from removal of the k rightmost blocks of* $\mathcal{C}$*. If* $k \geq \text{len}(\mathcal{C})$ *we define* $\mathcal{C}^{\lceil k} = \varepsilon$*. We let* $\mathcal{C}_1 \preceq \mathcal{C}_2$ *indicate that the chain* $\mathcal{C}_1$ *is a prefix of the chain* $\mathcal{C}_2$*.*

We consider as valid blocks that are generated by a stakeholder in the slot leader set of the slot to which the block is attributed. Later in Sect. 3.3 we discuss slot leader sets and how they are selected.

Definition 3 (Absolute and Relative Stake). *Let* $U_{\mathcal{P}}$*,* $U_{\mathcal{A}}$ *and* $U_{\mathcal{H}}$ *denote the sets of all stakeholders, the set of stakeholders controlled by an adversary* $\mathcal{A}$*, and the remaining (honest) stakeholders, respectively. For any party (resp. set of parties) X we denote by* s_X^+ *(resp.* s_X^-*) the maximum (resp. minimum)* absolute stake *controlled by X in the view of all honest stakeholders at a given slot, and by* $\alpha_X^+ \triangleq s_X^+/s_{\mathcal{P}}$ *and* $\alpha_X^- \triangleq s_X^-/s_{\mathcal{P}}$ *its* relative stake *taken as maximum and minimum respectively across the views of all honest stakeholders. For simplicity, we use* $s_X^{\mathrm{s}}, \alpha_X^{\mathrm{s}}$ *instead of* s_{U_X}, α_{U_X} *for all* $X \in \{\mathcal{P}, \mathcal{A}, \mathcal{H}\}, \mathrm{s} \in \{+, -\}$*. We also call* $\alpha_{\mathcal{A}} \triangleq \alpha_{\mathcal{A}}^+$ *and* $\alpha_{\mathcal{H}} \triangleq \alpha_{\mathcal{H}}^-$ *the* adversarial stake ratio *and* honest stake ratio*, respectively.*

3.1 Forward Secure Signatures and $\mathcal{F}_{\mathsf{KES}}$

In regular digital signature schemes, an adversary who compromises the signing key of a user can generate signatures for any messages it wishes, including messages that were (or should have been) generated in the past. Forward secure signature schemes [1] prevent such an adversary from generating signatures for messages that were issued in the past, or rather allows honest users to verify that a given signature was generated at a certain point in time. Basically, such security guarantees are achieved by "evolving" the signing key after each signature is generated and erasing the previous key in such a way that the actual signing key used for signing a message in the past cannot be recovered but a fresh signing key can still be linked to the previous one. This notion is formalized through *key evolving signature schemes*, which allow signing keys to be evolved into fresh keys for a number of time periods. We remark that efficient constructions of key evolving signature schemes with forward security exist [13] but no previous work has fully specified them in the UC setting.

We present a UC definition of the type of key-evolving signatures that we will take advantage of in our constructions. $\mathcal{F}_{\mathsf{KES}}$ allows us to achieve forward security with erasures (*i.e.*, assuming that parties securely delete old signing keys as the protocol proceeds). This functionality embodies ideal key evolving signature schemes allowing an adversary that corrupts the signer to forge signatures only under the current and future signing keys, but not under a previous signing key that has been updated. Our starting point for $\mathcal{F}_{\mathsf{KES}}$ is the standard digital signature functionality defined in [5] with the difference that packs together with the signing operation a key-evolving operation. During verification, $\mathcal{F}_{\mathsf{KES}}$ lets the adversary set the response to a verification query (taking as input a given time period) only if no key update has been performed since that time period and no entry exists in its internal table for the specific message, signature and time period specified in the query. We present $\mathcal{F}_{\mathsf{KES}}$ in Fig. 1. In the full version, we show that $\mathcal{F}_{\mathsf{KES}}$ can be realized by a construction based on key evolving signature schemes.

Functionality $\mathcal{F}_{\mathsf{KES}}$

$\mathcal{F}_{\mathsf{KES}}$ is parameterized by the total number of signature updates T, interacting with a signer U_S and stakeholders U_i as follows:

- **Key Generation.** Upon receiving a message $(\mathsf{KeyGen}, sid, U_S)$ from a stakeholder U_S , send $(\mathsf{KeyGen}, sid, U_S)$ to the adversary. Upon receiving $(\mathsf{VerificationKey}, sid, U_S, v)$ from the adversary, send $(\mathsf{VerificationKey}, sid, v)$ to U_S, record the triple (sid, U_S, v) and set counter $\mathsf{k_{ctr}} = 1$.
- **Sign and Update.** Upon receiving a message $(\mathsf{USign}, sid, U_S, m, j)$ from U_S, verify that (sid, U_S, v) is recorded for some sid and that $\mathsf{k_{ctr}} \leq j \leq T$. If not, then ignore the request. Else, set $\mathsf{k_{ctr}} = j+1$ and send $(\mathsf{Sign}, sid, U_S, m, j)$ to the adversary. Upon receiving $(\mathsf{Signature}, sid, U_S, m, j, \sigma)$ from the adversary, verify that no entry $(m, j, \sigma, v, 0)$ is recorded. If it is, then output an error message to U_S and halt. Else, send $(\mathsf{Signature}, sid, m, j, \sigma)$ to U_S, and record the entry $(m, j, \sigma, v, 1)$.
- **Signature Verification.** Upon receiving a message $(\mathsf{Verify}, sid, m, j, \sigma, v')$ from some stakeholder U_i do:
  1. If $v' = v$ and the entry $(m, j, \sigma, v, 1)$ is recorded, then set $f = 1$. (This condition guarantees completeness: If the verification key v' is the registered one and σ is a legitimately generated signature for m, then the verification succeeds.)
  2. Else, if $v' = v$, the signer is not corrupted, and no entry $(m, j, \sigma', v, 1)$ for any σ' is recorded, then set $f = 0$ and record the entry $(m, j, \sigma, v, 0)$. (This condition guarantees unforgeability: If v' is the registered one, the signer is not corrupted, and never signed m, then the verification fails.)
  3. Else, if there is an entry (m, j, σ, v', f') recorded, then let $f = f'$. (This condition guarantees consistency: All verification requests with identical parameters will result in the same answer.)
  4. Else, if $j < \mathsf{k_{ctr}}$, let $f = 0$ and record the entry $(m, j, \sigma, v, 0)$. Otherwise, if $j = \mathsf{k_{ctr}}$, hand $(\mathsf{Verify}, sid, m, j, \sigma, v')$ to the adversary. Upon receiving $(\mathsf{Verified}, sid, m, j, \phi)$ from the adversary let $f = \phi$ and record the entry (m, j, σ, v', ϕ). (This condition guarantees that the adversary is only able to forge signatures under keys belonging to corrupted parties for time periods corresponding to the current or future slots.)

  Output $(\mathsf{Verified}, sid, m, j, f)$ to U_i.

Fig. 1. Functionality $\mathcal{F}_{\mathsf{KES}}$.

3.2 UC-VRFs with Unpredictability Under Malicious Key Generation

The usual pseudorandomness definition for VRFs captures the fact that an attacker, seeing a number of VRF outputs and proofs for adversarially chosen inputs under a key pair that is correctly generated by a challenger, cannot distinguish the output of the VRF on a new (also adversarially chosen) input from a truly random string. This definition is too weak for our purposes for two reasons: first, we need a simulation-based definition so that the VRF can be composed directly within our protocol; second, we need the primitive to provide some level of unpredictability even under malicious key generation, *i.e.*, against adversaries who are allowed to generate the secret and pubic key pair.

Our UC formulation of VRFs cannot be implied by the standard VRF security definition or even the simulatable VRF notion of [6]. For instance, the VRF proofs in our setting have to be simulatable without knowledge of the VRF output (which is critical as we would like to ensure that the VRF output is not leaked to the adversary prematurely); it is easy to construct a VRF that is secure in the standard definition, but it is impossible to simulate its proofs without knowledge of the VRF output. Furthermore, if the adversary is allowed to generate its own key pair it is easy to see that the distribution of the VRF outputs cannot be guaranteed. Indeed, even for known constructions (*e.g.* [10]), an adversary that maliciously generates keys can easily and significantly skew the output distribution.

We call the latter property *unpredictability under malicious key generation* and we present, in Fig. 2, a UC definition for VRF's that captures this stronger security requirement.[4] The functionality operates as follows. Given a key generation request from one of the stakeholders, it returns a new verification key v that is used to label a table. Two methods are provided for computing VRF values. The first provides just the VRF output and does not interact with the adversary. In the second, whenever invoked on an input m that is not asked before by a stakeholder that is associated to a certain table labeled by v, the functionality will query the adversary for the value of the proof π, and subsequently sample a random element ρ to associate with m. Verification is always consistent and will validate outputs that have already being inserted in a table. Unpredictability against malicious key generation is captured by imposing the same random selection of outputs even for the function tables that correspond to keys of corrupted stakeholders. Finally, the adversary is allowed to query all function tables maintained by the functionality for which either a proof has been

[4] In fact our UC formulation captures a stronger notion: even for adversarial keys the VRF function will act as a random oracle. We note that while we can achieve this notion in the random oracle model, a weaker condition of mere unpredictability can be sufficient for the security of our protocol. A UC version of the notion of verifiable pseudorandom permutations, cf. [9], could potentially be used towards a standard model instantiation of the primitive.

computed, or they correspond to adversarial keys. In the full version, we show how to realize $\mathcal{F}_{\mathsf{VRF}}$ in the random oracle model under the CDH assumption based on the 2-Hash-DH verifiable oblivious PRF construction of [14].

Functionality $\mathcal{F}_{\mathsf{VRF}}$.

$\mathcal{F}_{\mathsf{VRF}}$ interacts with stakeholders $U_1, \ldots, U_n$ as follows:

- **Key Generation.** Upon receiving a message (KeyGen, sid) from a stakeholder U_i, hand $(\mathsf{KeyGen}, sid, U_i)$ to the adversary. Upon receiving $(\mathsf{VerificationKey}, sid, U_i, v)$ from the adversary, if U_i is honest, verify that v is unique, record the pair (U_i, v) and return $(\mathsf{VerificationKey}, sid, v)$ to U_i. Initialize the table $T(v, \cdot)$ to empty.
- **Malicious Key Generation.** Upon receiving a message $(\mathsf{KeyGen}, sid, v)$ from $\mathcal{S}$, verify that v has not being recorded before; in this case initialize table $T(v, \cdot)$ to empty and record the pair $(\mathcal{S}, v)$.
- **VRF Evaluation.** Upon receiving a message (Eval, sid, m) from U_i, verify that some pair (U_i, v) is recorded. If not, then ignore the request. Then, if the value $T(v, m)$ is undefined, pick a random value y from $\{0,1\}^{\ell_{\mathsf{VRF}}}$ and set $T(v, m) = (y, \emptyset)$. Then output $(\mathsf{Evaluated}, sid, y)$ to P, where y is such that $T(v, m) = (y, S)$ for some S.
- **VRF Evaluation and Proof.** Upon receiving a message $(\mathsf{EvalProve}, sid, m)$ from U_i, verify that some pair (U_i, v) is recorded. If not, then ignore the request. Else, send $(\mathsf{EvalProve}, sid, U_i, m)$ to the adversary. Upon receiving $(\mathsf{Eval}, sid, m, \pi)$ from the adversary, if value $T(v, m)$ is undefined, verify that π is unique, pick a random value y from $\{0,1\}^{\ell_{\mathsf{VRF}}}$ and set $T(v, m) = (y, \{\pi\})$. Else, if $T(v, m) = (y, S)$, set $T(v, m) = (y, S \cup \{\pi\})$. In any case, output $(\mathsf{Evaluated}, sid, y, \pi)$ to P.
- **Malicious VRF Evaluation.** Upon receiving a message $(\mathsf{Eval}, sid, v, m)$ from $\mathcal{S}$ for some v, do the following. First, if $(\mathcal{S}, v)$ is recorded and $T(v, m)$ is undefined, then choose a random value y from $\{0,1\}^{\ell_{\mathsf{VRF}}}$ and set $T(v, m) = (y, \emptyset)$. Then, if $T(v, m) = (y, S)$ for some $S \neq \emptyset$, output $(\mathsf{Evaluated}, sid, y)$ to $\mathcal{S}$, else ignore the request.
- **Verification.** Upon receiving a message $(\mathsf{Verify}, sid, m, y, \pi, v')$ from some party P, send $(\mathsf{Verify}, sid, m, y, \pi, v')$ to the adversary. Upon receiving $(\mathsf{Verified}, sid, m, y, \pi, v')$ from the adversary do:
  1. If $v' = v$ for some (U_i, v) and the entry $T(U_i, m)$ equals (y, S) with $\pi \in S$, then set $f = 1$.
  2. Else, if $v' = v$ for some (U_i, v), but no entry $T(U_i, m)$ of the form $(y, \{\ldots, \pi, \ldots\})$ is recorded, then set $f = 0$.
  3. Else, initialize the table $T(v', \cdot)$ to empty, and set $f = 0$.

  Output $(\mathsf{Verified}, sid, m, y, \pi, f)$ to P.

Fig. 2. Functionality $\mathcal{F}_{\mathsf{VRF}}$.

3.3 Oblivious Leader Selection

As in (synchronous) Ouroboros, for each $0 < j \leq R$, a *slot leader* E_j is a stakeholder who is elected to generate a block at sl_j. However, our leader selection process differs from Ouroboros [16] in three points: (1) potentially, multiple slot leaders may be elected for a particular slot (forming a *slot leader set*); (2) frequently, slots will have *no leaders* assigned to them; and (3) a priori, only a slot leader is aware that it is indeed a leader for a given slot; this assignment is unknown to all the other stakeholders—including other slot leaders of the same slot—until the other stakeholders receive a valid block from this slot leader. The combinatorial analysis presented in Sect. 4 shows (with an honest stake majority) that (i.) blockchains generated according to these dynamics are well-behaved even if multiple slot leaders are selected for a slot and that (ii.) sequences of slots with no leader provide sufficient stability for honest stakeholders to effectively synchronize. As a matter of terminology, we call slots with an associated nonempty slot leader set *active slots* and slots that are not assigned a slot leader *empty slots.*

The idealized slot leader assignment and the active slots coefficient. The fundamental leader assignment process calls for a stakeholder U_i to be independently selected as a leader for a particular slot sl_j with probability p_i depending only on its relative stake. (In this static-stake analysis, relative stake is simply determined by the genesis block B_0.) The exact relationship between p_i and the relative stake α_i is determined by a parameter f of the protocol which we refer to as the *active slots coefficient.* Specifically,

$$p_i = \phi_f(\alpha_i) \triangleq 1 - (1 - f)^{\alpha_i}, \tag{1}$$

where α_i is the relative stake held by stakeholder U_i. We occasionally drop the subscript f and write $\phi(\alpha_i)$ when f can be inferred from context. As the events "U_i is a leader for sl_j" are independent, this process may indeed generate multiple (or zero) leaders for a given slot.

Remarks about $\phi_f(\cdot)$. Observe that $\phi_f(1) = f$; in particular, the parameter f is the probability that a party holding all the stake will be selected to be a leader for given slot. On the other hand, $\phi_f()$ is not linear, but slightly concave. To motivate the choice of the function ϕ_f, we note that it satisfies the "independent aggregation" property:

$$1 - \phi\left(\sum_i \alpha_i\right) = \prod_i (1 - \phi(\alpha_i)). \tag{2}$$

In particular, when leadership is determined according to ϕ_f, the probability of a stakeholder becoming a slot leader in a particular slot is independent of whether this stakeholder acts as a single party in the protocol, or splits its stake among several "virtual" parties. In particular, consider a party U with relative stake α who contrives to split its stake among two virtual subordinate parties with stakes α_1 and α_2 (so that $\alpha_1 + \alpha_2 = \alpha$). Then the probability that one of these virtual parties is elected for a particular slot is $1-(1-\phi(\alpha_1))(1-\phi(\alpha_2))$, as these events are independent. Property (2) guarantees that this is identical to $\phi(\alpha)$. *Thus this selection rule is invariant under arbitrary reapportionment of a party's stake among virtual parties.*

3.4 The Protocol in the $\mathcal{F}_{\mathsf{INIT}}$-Hybrid Model

We will construct our protocol for the static stake case in the $\mathcal{F}_{\mathsf{INIT}}$-hybrid model, where the genesis stake distribution $\mathbb{S}_0$ and the nonce η (to be written in the genesis block B_0) are determined by the ideal functionality $\mathcal{F}_{\mathsf{INIT}}$ defined in Fig. 3. Moreover, $\mathcal{F}_{\mathsf{INIT}}$ also incorporates the diffuse functionality from Sect. 2.2, which is implicitly used by all parties to send messages and keep synchronized with a global clock. $\mathcal{F}_{\mathsf{INIT}}$ also takes stakeholders' public keys from them and packages them into the genesis block at the outset of the protocol. Note that $\mathcal{F}_{\mathsf{INIT}}$ halts if it is not possible to create a genesis block; all security guarantees we provide later in the paper are conditioned on a successful creation of the genesis block.

Functionality $\mathcal{F}_{\mathsf{INIT}}$

$\mathcal{F}_{\mathsf{INIT}}$ incorporates the delayed diffuse functionality from Section 2.2 and is parameterized by the number of initial stakeholders n and their respective stakes $s_1, \ldots, s_n$. $\mathcal{F}_{\mathsf{INIT}}$ interacts with stakeholders $U_1, \ldots, U_n$ as follows:

- In the first round, upon a request from some stakeholder U_i of the form $(\mathsf{ver\_keys}, sid, U_i, v_i^{\mathrm{vrf}}, v_i^{\mathrm{kes}}, v_i^{\mathrm{dsig}})$, it stores the verification keys tuple $(U_i, v_i^{\mathrm{vrf}}, v_i^{\mathrm{kes}}, v_i^{\mathrm{dsig}})$ and acknowledges its receipt. If any of the n stakeholders does not send a request of this form to $\mathcal{F}_{\mathsf{INIT}}$, or if two different stakeholders provide two identical keys, it halts. Otherwise, it samples and stores a random value $\eta \xleftarrow{\$} \{0,1\}^\lambda$ and constructs a genesis block $(\mathbb{S}_0, \eta)$, where $\mathbb{S}_0 = \Big((U_1, v_1^{\mathrm{vrf}}, v_1^{\mathrm{kes}}, v_1^{\mathrm{dsig}}, s_1), \ldots, (U_n, v_n^{\mathrm{vrf}}, v_n^{\mathrm{kes}}, v_n^{\mathrm{dsig}}, s_n)\Big)$.
- In later rounds, upon a request of the form $(\mathsf{genblock\_req}, sid, U_i)$ from some stakeholder U_i, $\mathcal{F}_{\mathsf{INIT}}$ sends $(\mathsf{genblock}, sid, \mathbb{S}_0, \eta)$ to U_i.

Fig. 3. Functionality $\mathcal{F}_{\mathsf{INIT}}$.

Blocks are signed with a forward secure signature scheme modelled by $\mathcal{F}_{\mathsf{KES}}$, while transactions are signed with a regular EUF-CMA secure digital signature modelled by a standard signature functionality $\mathcal{F}_{\mathsf{DSIG}}$, deferred to the full version due to space constraints.

Notice that the implicit leader assignment process described in π_{SPoS} calls for a party U_i to act as a leader for a slot sl_j when $y < T_i$; this is an event that occurs with probability (exponentially close to) $\phi_f(\alpha_i)$ as y is uniform according to the functionality $\mathcal{F}_{\mathsf{VRF}}$.

We are interested in applications where transactions are inserted in the ledger. For simplicity, transactions are assumed to be simple assertions of the form "Stakeholder U_i transfers stake s to Stakeholder $(U_j, v_j^{\mathrm{vrf}}, v_j^{\mathrm{kes}}, v_j^{\mathrm{dsig}})$"

(In an implementation the different public-keys can be hashed into a single value). Protocol π_{SPoS} ensures that the environment learns every stakeholder's public keys and provides an interface for the environment to request signatures on arbitrary transactions. A transaction will consist of a transaction template tx of this format accompanied by a signature of tx by stakeholder U_i. We define a valid transaction as follows:

Definition 4 (Valid Transaction). *A pair (tx, σ) is considered a valid transaction by a verifier* V *if the following holds:*

- *The transaction template tx is of the format "Stakeholder U_i transfers stake s to Stakeholder $(U_j, v_j^{\mathrm{vrf}}, v_j^{\mathrm{kes}}, v_j^{\mathrm{dsig}})$" where U_i and U_j are stakeholders identified by tuples $(U_i, v_i^{\mathrm{vrf}}, v_i^{\mathrm{kes}}, v_i^{\mathrm{dsig}})$ and $(U_j, v_j^{\mathrm{vrf}}, v_j^{\mathrm{kes}}, v_j^{\mathrm{dsig}})$ contained in the current stake distribution $\mathbb{S}$ and $x \in \mathbb{Z}$.*
- *The verifier V obtains $(\mathsf{Verified}, m, 1)$ as answer upon sending $(\mathsf{Verify}, tx, \sigma, v_i^{\mathrm{dsig}})$ to $\mathcal{F}_{\mathsf{DSIG}}$.*
- *Stakeholder U_i possesses x coins at the moment the transaction is issued (or registered in the blockchain) according to the view of the verifier* V.

Given Definitions 2 and 4, we define a *valid chain* as a blockchain (according to Definition 2) where all transactions contained in every block are valid (according to Definition 4). The stakeholders $U_1, \ldots, U_n$ interact among themselves and with $\mathcal{F}_{\mathsf{INIT}}$ through Protocol π_{SPoS} described in Fig. 4. The protocol relies on a $\mathsf{maxvalid}_S(\mathcal{C}, \mathbb{C})$ function that chooses a chain given the current chain $\mathcal{C}$ and a set of valid chains $\mathbb{C}$ that are available in the network. In the static stake case we analyze the simple "longest chain" rule.

Function $\mathsf{maxvalid}(\mathcal{C}, \mathbb{C})$: Returns the longest chain from $\mathbb{C} \cup \{\mathcal{C}\}$. Ties are broken in favor of $\mathcal{C}$, if it has maximum length, or arbitrarily otherwise.

4 Combinatorial Analysis of the Static Stake Protocol

Throughout this section, we focus solely on analysis of the protocol π_{SPoS} using the idealized functionalities $\mathcal{F}_{\mathsf{VRF}}$ and $\mathcal{F}_{\mathsf{KES}}$ for VRFs and digital signatures, respectively—we refer to it as the *hybrid experiment*. Any property of the protocol that we prove true in the hybrid experiment (such as achieving common prefix, chain growth and chain quality) will remain true (with overwhelming probability) in the setting where $\mathcal{F}_{\mathsf{VRF}}$ and $\mathcal{F}_{\mathsf{KES}}$ are replaced by their real-world implementations—in the so-called *real experiment*.

The hybrid experiment yields a stochastic process for assigning slots to parties which we now abstract and study in detail. Our analysis of the resulting blockchain dynamics proceeds roughly as follows: We begin by generalizing the framework of "forks" [16] to our semi-synchronous setting—forks are a natural bookkeeping tool that reflect the chains possessed by honest players during an execution of the protocol. We then establish a simulation rule that associates with each execution of the semi-synchronous protocol an execution of

Protocol π_{SPoS}

The protocol π_{SPoS} is run by stakeholders $U_1, \ldots, U_n$ interacting among themselves and with ideal functionalities $\mathcal{F}_{\text{INIT}}, \mathcal{F}_{\text{VRF}}, \mathcal{F}_{\text{KES}}, \mathcal{F}_{\text{DSIG}}, \mathsf{H}$ over a sequence of slots $S = \{sl_1, \ldots, sl_R\}$. Define $T_i \triangleq 2^{\ell_{\text{VRF}}} \phi_f(\alpha_i)$ as the threshold for a stakeholder U_i, where α_i is the relative stake of U_i, ℓ_{VRF} denotes the output length of $\mathcal{F}_{\text{VRF}}$, f is the active slots coefficient and ϕ_f is the mapping from equation (1). Then π_{SPoS} proceeds as follows:

1. **Initialization.** The stakeholder U_i sends $(\mathsf{KeyGen}, sid, U_i)$ to $\mathcal{F}_{\text{VRF}}$, $\mathcal{F}_{\text{KES}}$ and $\mathcal{F}_{\text{DSIG}}$; receiving $(\mathsf{VerificationKey}, sid, v_i^{\text{vrf}})$, $(\mathsf{VerificationKey}, sid, v_i^{\text{kes}})$ and $(\mathsf{VerificationKey}, sid, v_i^{\text{dsig}})$, respectively. Then, in case it is the first round, it sends $(\mathsf{ver\_keys}, sid, U_i, v_i^{\text{vrf}}, v_i^{\text{kes}}, v_i^{\text{dsig}})$ to $\mathcal{F}_{\text{INIT}}$ (to claim stake from the genesis block). In any case, it terminates the round by returning $(U_i, v_i^{\text{vrf}}, v_i^{\text{kes}}, v_i^{\text{dsig}})$ to $\mathcal{Z}$. In the next round, it sends $(\mathsf{genblock\_req}, sid, U_i)$ to $\mathcal{F}_{\text{INIT}}$, receiving $(\mathsf{genblock}, sid, \mathbb{S}_0, \eta)$ as the answer. If U_i is initialized in the first round, it sets the local blockchain $\mathcal{C} = B_0 = (\mathbb{S}_0, \eta)$ and its initial internal state $st = H(B_0)$. In case U_i is initialized after the first round, it sets its initial state to $st = H(\text{head}(\mathcal{C}))$ where $\mathcal{C}$ is the initial local chain provided by the environment.
2. **Chain Extension.** After initialization, for every slot $sl_j \in S$, every online stakeholder U_i performs the following steps:
   (a) U_i receives from the environment the transaction data $d \in \{0,1\}^*$ to be inserted into the blockchain.
   (b) U_i collects all valid chains received via diffusion into a set $\mathbb{C}$, pruning blocks belonging to future slots and verifying that for every chain $\mathcal{C}' \in \mathbb{C}$ and every block $B' = (st', d', sl', B_\pi{}', \sigma_{j'}) \in \mathcal{C}'$ it holds that the stakeholder who created it is in the slot leader set of slot sl' (by parsing $B_\pi{}'$ as (U_s, y', π') for some s, verifying that $\mathcal{F}_{\text{VRF}}$ responds to $(\mathsf{Verify}, sid, \eta \parallel sl', y', \pi', v_s^{\text{vrf}})$ by $(\mathsf{Verified}, sid, \eta \parallel sl', y', \pi', 1)$, and that $y' < T_s$), and that $\mathcal{F}_{\text{KES}}$ responds to $(\mathsf{Verify}, sid, (st', d', sl', B_\pi{}'), sl', \sigma_{j'}, v_s^{\text{kes}})$ by $(\mathsf{Verified}, sid, (st', d', sl', B_\pi{}'), sl', 1)$. U_i computes $\mathcal{C}' = \mathsf{maxvalid}(\mathcal{C}, \mathbb{C})$, sets $\mathcal{C}'$ as the new local chain and sets state $st = H(\text{head}(\mathcal{C}'))$.
   (c) U_i sends $(\mathsf{EvalProve}, sid, \eta \parallel sl_j)$ to $\mathcal{F}_{\text{VRF}}$, receiving $(\mathsf{Evaluated}, sid, y, \pi)$. U_i checks whether it is in the slot leader set of slot sl_j by checking that $y < T_i$. If yes, it chooses the maximal sequence d' of transactions in d such that adding a block with d' to $\mathcal{C}$ results into a valid chain, and attempts to include d' as follows: It generates a new block $B = (st, d', sl_j, B_\pi, \sigma)$ where st is its current state, $B_\pi = (U_i, y, \pi)$ and σ is a signature obtained by sending $(\mathsf{USign}, sid, U_i, (st, d', sl_j, B_\pi), sl_j)$ to $\mathcal{F}_{\text{KES}}$ and receiving $(\mathsf{Signature}, sid, (st, d', sl_j, B_\pi), sl_j, \sigma)$. U_i computes $\mathcal{C}' = \mathcal{C} \parallel B$, sets $\mathcal{C}'$ as the new local chain and sets state $st = H(\text{head}(\mathcal{C}'))$. Finally, if U_i has generated a block in this step, it diffuses $\mathcal{C}'$.
3. **Signing Transactions.** Upon receiving $(\mathsf{sign\_tx}, sid', tx)$ from the environment, U_i sends $(\mathsf{Sign}, sid, U_i, tx)$ to $\mathcal{F}_{\text{DSIG}}$, receiving $(\mathsf{Signature}, sid, tx, \sigma)$. Then, U_i sends $(\mathsf{signed\_tx}, sid', tx, \sigma)$ back to the environment.

Fig. 4. Protocol π_{SPoS}.

a related "virtual" synchronous protocol. Motivated by the special case of a *static* adversary—which simply corrupts a family of parties at the outset of the protocol—we identify a natural "generic" probability distribution for this simulation theorem which we prove controls the behavior of adaptive adversaries by stochastic domination. Finally, we prove that this simulation amplifies the effective power of the adversary in a controlled fashion and, furthermore, permits forks of the semi-synchronous protocol to be projected to forks of the virtual protocol in a way that preserves their relevant combinatorial properties. This allows us to apply the density theorems and divergence result of [16,23] to provide strong common prefix, chain growth, and chain quality (Sect. 4.4) guarantees for the semi-synchronous protocol with respect to an adaptive adversary.

We begin in Sect. 4.1 with a discussion of characteristic strings, semi-synchronous forks, and their relationship to executions of π_{SPoS} in the hybrid experiment. Section 4.2 then develops the combinatorial reduction from the semi-synchronous to the synchronous setting. The "generic, dominant" distribution on characteristic strings is then motivated and defined in Sect. 4.3, where the effect of the reduction on this distribution is also described. Section 4.4, as described above, establishes various guarantees on the resulting blockchain under the dominant distribution. The full power of adaptive adversaries is considered in Sect. 4.5. Finally, in preparation for applying the protocol in the dynamic stake setting, we formulate a "resettable setting" which further enlarges the power of the adversary by providing some control over the random nonce that seeds the protocol.

4.1 Chains, Forks and Divergence

We begin by suitably generalizing the framework of characteristic strings, forks, and divergence developed in [16] to our semi-synchronous setting.

The leader assignment process given by protocol π_{SPoS} in the hybrid experiment assigns leaders to slots with the following guarantees: *(i.)* a party with relative stake α becomes a slot leader for a given slot with probability $\phi_f(\alpha) \triangleq 1-(1-f)^{\alpha}$; *(ii.)* the event of becoming a slot leader is independent for each party and for each slot (both points follow from the construction of π_{SPoS} and the independent random sampling of every new output in $\mathcal{F}_{\mathsf{VRF}}$). Clearly, these dynamics may lead to slots with multiple slot leaders and, likewise, slots with no slot leader. For a given (adaptive) adversary $\mathcal{A}$ and environment $\mathcal{Z}$, we reflect the outcome of this process with a *characteristic string*, as described below.

Definition 5 (Execution). *For an (adaptive) adversary $\mathcal{A}$ and an environment $\mathcal{Z}$, an* execution $\mathcal{E}$ *of π_{SPoS} is a transcript including the inputs provided by $\mathcal{Z}$, the random coins of the parties, the random coins of the adversary, the responses of the ideal functionalities and the random oracle. This data determines the entire dynamics of the protocol: messages sent and delivered, the internal states of the parties at each step, the set of corrupt parties at each step, etc.*

Definition 6 (Characteristic string). *Let $S = \{sl_1, \ldots, sl_R\}$ be a sequence of slots of length R and $\mathcal{E}$ be an execution (with adversary $\mathcal{A}$ and environment $\mathcal{Z}$).*

For a slot sl_j, let $\mathcal{P}(j)$ denote the set of parties assigned to be slot leaders for slot j by the protocol π_{SPoS} (specifically, those parties U_i for which $y < 2^{\ell_{\mathsf{VRF}}}\phi_f(\alpha_i)$, where $(y, \pi) \leftarrow \mathsf{Prove}_{\mathsf{VRF}.sk_i}(\eta \| sl_j)$). We define the characteristic string $w \in \{0, 1, \bot\}^R$ *of S to be the random variable so that*

$$w_j = \begin{cases} \bot & \text{if } \mathcal{P}(j) = \emptyset, \\ 0 & \text{if } |\mathcal{P}(j)| = 1 \text{ and the assigned party is honest,} \\ 1 & \text{if } |\mathcal{P}(j)| > 1 \text{ or a party in } \mathcal{P}(i) \text{ is adversarial.} \end{cases} \tag{3}$$

For such a characteristic string $w \in \{0, 1, \bot\}^*$ we say that the index j is uniquely honest *if $w_j = 0$,* tainted *if $w_j = 1$, and* empty *if $w_j = \bot$. We say that an index is* active *if $w_j \in \{0, 1\}$. Note that an index is "tainted" according to this terminology in cases where multiple honest parties (and no adversarial party) have been assigned to it.*

We denote by $\mathcal{D}^f_{\mathcal{Z},\mathcal{A}}$ the distribution of the random variable $w = w_1 \dots w_R$ in the hybrid experiment with the active slots coefficient f, adversary $\mathcal{A}$, and environment $\mathcal{Z}$. For a fixed execution $\mathcal{E}$, we denote by $w_{\mathcal{E}}$ the (fixed) characteristic string resulting from that execution.

We emphasize that in an execution of π_{SPoS}, the resulting characteristic string is determined by both the nonce (and the effective leader selection process), the adaptive adversary $\mathcal{A}$, and the environment $\mathcal{Z}$ (which, in particular, determines the stake distribution).

From Executions to Forks. The notion of a "fork", defined in [16], is a bookkeeping tool that indicates the chains broadcast by honest players during an idealized execution of a blockchain protocol. We now adapt the synchronous notion of [16] to reflect the effect of message delays.

An execution of Protocol π_{SPoS} induces a collection of blocks broadcast by the participants. As we now focus merely on the structural properties of the resulting blockchain, for each broadcast block we now retain only two features: the *slot* associated with the block and the *previous block* to which it is "attached" by the idealized digital signature σ_j. (Of course, we only consider blocks with legal structure that meet the verification criteria of π_{SPoS}.) Note that multiple blocks may be associated with a particular slot, either because multiple parties are assigned to the slot or an adversarial party is assigned to a slot (who may choose to deviate from the protocol by issuing multiple blocks). In any case, these blocks induce a natural directed tree by treating the blocks as vertices and introducing a directed edge between each pair of blocks (b, b') for which b' identifies b as the previous block. In the Δ-semisynchronous setting, the maxvalid rule enforces a further critical property on this tree: the depth of any block broadcast by an honest player during the protocol must exceed the depths of any honestly-generated blocks from slots at least Δ in the past. (This follows because such previously broadcast blocks would have been available to the honest player, who always builds on a chain of maximal length.) We call a directed tree with these structural properties a *Δ-fork*, and define them precisely below.

We may thus associate with any execution of π_{SPoS} a fork. While this fork disregards many of the details of the execution, any violations of common prefix are immediately manifested by certain diverging paths in the fork. A fundamental element of our analysis relies on controlling the structure of the forks that can be induced in this way for a given characteristic string (which determines which slots have been assigned to uniquely honest parties). In particular, we prove that common prefix violations are impossible for "typical" characteristic strings generated by π_{SPoS} with an adversary $\mathcal{A}$ by establishing that such diverging paths cannot exist in their associated forks.

Definition 7 (Δ-fork). *Let $w \in \{0,1,\bot\}^k$ and Δ be a non-negative integer. Let $A = \{i \mid w_i \neq \bot\}$ denote the set of active indices, and let $H = \{i \mid w_i = 0\}$ denote the set of uniquely honest indices. A Δ-*fork *for the string w is a directed, rooted tree $F = (V, E)$ with a labeling $\ell : V \rightarrow \{0\} \cup A$ so that (i) the root $r \in V$ is given the label $\ell(r) = 0$; (ii) each edge of F is directed away from the root; (iii) the labels along any directed path are strictly increasing; (iv) each uniquely honest index $i \in H$ is the label of exactly one vertex of F; (v) the function $\mathbf{d} : H \rightarrow \{1, \ldots, k\}$, defined so that $\mathbf{d}(i)$ is the depth in F of the unique vertex v for which $\ell(v) = i$, satisfies the following Δ-*monotonicity *property: if $i, j \in H$ and $i + \Delta < j$, then $\mathbf{d}(i) < \mathbf{d}(j)$.*

As a matter of notation, we write $F \vdash_\Delta w$ to indicate that F is a Δ-fork for the string w. We typically refer to a Δ-fork as simply a "fork".

Also note that our notion of a fork deliberately models honest parties that do not necessarily exploit all the information available to them thanks to the delivery guarantees provided by the $\mathsf{DDiffuse}$ functionality. Nonetheless, it remains true that any execution of the hybrid experiment leads to a fork as we defined it, a relationship that we make fully formal in the full version. Given this relationship, we can later focus on investigating the properties of the distribution $\mathcal{D}^f_{\mathcal{Z},\mathcal{A}}$. Roughly speaking, if we prove that a characteristic string sampled from $\mathcal{D}^f_{\mathcal{Z},\mathcal{A}}$, with overwhelming probability, does not allow for *any* "harmful" forks, then this also implies that a random execution with overwhelming probability results in a "harmless" outcome.

Now we continue with the adaptation of the framework from [16] to the semi-synchronous setting.

Definition 8 (Tines, length, and viability). *A path in a fork F originating at the root is called a* tine. *For a tine t we let* $\mathrm{length}(t)$ *denote its* length, *equal to the number of edges on the path. For a vertex v, we call the length of the tine terminating at v the* depth of v. *For convenience, we overload the notation $\ell(\cdot)$ so that it applies to tines by defining $\ell(t) \triangleq \ell(v)$, where v is the terminal vertex on the tine t. We say that a tine t is Δ-*viable *if* $\mathrm{length}(t) \geq \max_{h+\Delta \leq \ell(t)} \mathbf{d}(h)$, *this maximum extended over all uniquely honest indices h (appearing Δ or more slots before $\ell(t)$). Note that any tine terminating in a uniquely honest vertex is necessarily viable by the Δ-monotonicity property.*

Remarks on viability and divergence. The notion of viability, defined above, demands that the length of a tine t be no less than that of all tines broadcast by uniquely honest slot leaders prior to slot $\ell(t)-\Delta$. Observe that such a tine could, in principle, be selected according to the maxvalid() rule by an honest player online at time $\ell(t)$: in particular, if all blocks broadcast by honest parties in slots $\ell(t)-\Delta,\ldots,\ell(t)$ are maximally delayed, the tine can favorably compete with all other tines that the adversary is obligated to deliver by slot $\ell(t)$. The major analytic challenge, both in the synchronous case and in our semisynchronous setting, is to control the possibility of a *common prefix* violation, which occurs when the adversary can manipulate the protocol to produce a fork with two viable tines with a relatively short common prefix. We define this precisely by introducing the notion of divergence.

Definition 9 (Divergence). *Let F be a Δ-fork for a string $w \in \{0,1,\bot\}^*$. For two Δ-viable tines t_1 and t_2 of F, define their* divergence *to be the quantity*

$$\mathrm{div}(t_1,t_2) \triangleq \min\{\mathrm{length}(t_1),\mathrm{length}(t_2)\} - \mathrm{length}(t_1 \cap t_2),$$

where $t_1 \cap t_2$ denotes the common prefix of t_1 and t_2. We extend this notation to the fork F by maximizing over viable tines: $\mathrm{div}_\Delta(F) \triangleq \max_{t_1,t_2} \mathrm{div}(t_1,t_2)$, taken over all pairs of Δ-viable tines of F. Finally, we define the Δ-divergence of a characteristic string w to be the maximum over all Δ-forks: $\mathrm{div}_\Delta(w) \triangleq \max_{F\vdash_\Delta w} \mathrm{div}_\Delta(F)$.

Our primary goal in this section is to prove that, with high probability, the characteristic strings induced by protocol π_{SPoS} have small divergence and hence provide strong guarantees on common prefix.

The Synchronous Case. The original development of [16] assumed a strictly synchronous environment. Their definitions of characteristic string, fork, and divergence correspond to the case $\Delta = 0$, where characteristic strings are elements of $\{0,1\}^*$. As this setting will play an important role in our analysis—fulfilling the role of the "virtual protocol" described at the beginning of this section—we set down some further terminology for this synchronous case and establish a relevant combinatorial statement based on a result in [16] that we will need for our analysis.

Definition 10 (Synchronous characteristic strings and forks). *A* synchronous characteristic string *is an element of $\{0,1\}^*$. A* synchronous fork F *for a (synchronous) characteristic string w is a 0-fork $F \vdash_0 w$.*

An immediate conclusion of the results obtained in [16,23] is the following bound on the probability that a synchronous characteristic string drawn from the binomial distribution has large divergence.

Theorem 1. *Let $\ell, k \in \mathbb{N}$ and $\epsilon \in (0,1)$. Let $w \in \{0,1\}^\ell$ be drawn according to the binomial distribution, so that $\Pr[w_i = 1] = (1-\epsilon)/2$. Then $\Pr[\mathrm{div}_0(w) \geq k] \leq \exp(\ln \ell - \Omega(k))$.*

4.2 The Semisynchronous to Synchronous Reduction

We will make use of the following mapping, that maps characteristic strings to synchronous characteristic strings.

Definition 11 (Reduction mapping). *For $\Delta \in \mathbb{N}$, we define the function $\rho_\Delta \colon \{0,1,\perp\}^* \to \{0,1\}^*$ inductively as follows: $\rho_\Delta(\varepsilon) = \varepsilon$, $\rho_\Delta(\perp \| w') = \rho_\Delta(w')$,*

$$\rho_\Delta(1 \| w') = 1 \| \rho_\Delta(w'),$$
$$\rho_\Delta(0 \| w') = \begin{cases} 0 \| \rho_\Delta(w') & \text{if } w' \in \perp^{\Delta-1} \| \{0,1,\perp\}^*, \\ 1 \| \rho_\Delta(w') & \text{otherwise.} \end{cases} \quad (4)$$

We call ρ_Δ the reduction mapping for delay Δ.

A critical feature of the map ρ_Δ is that it monotonically transforms Δ-divergence to synchronous divergence. We state this in the following lemma, proven in the full version.

Lemma 1. *Let $w \in \{0,1,\perp\}^*$. Then* $\mathrm{div}_\Delta(w) \leq \mathrm{div}_0(\rho_\Delta(w))$.

4.3 The Dominant Characteristic Distribution

The high-probability results for our desired chain properties depend on detailed information about the distribution on characteristic strings $\mathcal{D}^f_{\mathcal{Z},\mathcal{A}}$ determined by the adversary $\mathcal{A}$, the environment $\mathcal{Z}$, and the parameters f and R. In this section we define a distinguished distribution on characteristic strings which we will see "dominates" the distributions produced by any static adversary. Later in Sect. 4.5 we show that the same is true also for adaptive adversaries. We then study the effect of ρ_Δ on this distribution in preparation for studying common prefix, chain growth, and chain quality.

Motivating the Dominant Distribution: Static Adversaries. To motivate the dominant distribution, consider the distribution induced by a *static* adversary who corrupts—at the outset of the protocol—a set $U_\mathcal{A}$ of parties with total relative stake $\alpha_\mathcal{A}$. (Formally, one can model this by restricting to environments that only allow static corruption.) Recalling Definition 1, a party with relative stake α_i is independently assigned to be a leader for a slot with probability

$$\phi_f(\alpha_i) \triangleq \phi(\alpha_i) \triangleq 1 - (1-f)^{\alpha_i}.$$

The function ϕ_f is concave since

$$\frac{\partial^2 \phi_f}{\partial \alpha^2}(\alpha) = -(\ln(1-f))^2(1-f)^\alpha < 0.$$

Considering that $\phi_f(0) = 0$ and $\phi_f(1) = f$, concavity implies that $\phi_f(\alpha) \geq f\alpha$ for $\alpha \in [0,1]$. As $\phi_f(0) \geq 0$ and ϕ_f is concave, the function ϕ_f is subadditive. This immediately implies the following proposition that will be useful during the analysis.

Proposition 1. *The function $\phi_f(\alpha)$ satisfies the following properties.*

$$\phi_f\left(\sum_i \alpha_i\right) = 1 - \prod_i (1 - \phi_f(\alpha_i)) \leq \sum_i \phi_f(\alpha_i), \qquad \alpha_i \geq 0, \tag{5}$$

$$\frac{\phi_f(\alpha)}{\phi_f(1)} = \frac{\phi_f(\alpha)}{f} \geq \alpha, \qquad \alpha \in [0,1]. \tag{6}$$

Recalling Definition 6, this (static) adversary $\mathcal{A}$ determines a distribution $\mathcal{D}^f_{\mathcal{Z},\mathcal{A}}$ on strings $w \in \{0,1,\perp\}^R$ by independently assigning each w_i so that

$$\begin{aligned} p^{\mathcal{A}}_{\perp} &\triangleq \Pr[w_i = \perp] = \prod_{i \in \mathcal{P}} (1 - \phi(\alpha_i)) = \prod_{i \in \mathcal{P}} (1-f)^{\alpha_i} = (1-f), \\ p^{\mathcal{A}}_0 &\triangleq \Pr[w_i = 0] = \sum_{h \in \mathcal{H}} (1 - (1-f)^{\alpha_h}) \cdot (1-f)^{1-\alpha_i}, \\ p^{\mathcal{A}}_1 &\triangleq \Pr[w_i = 1] = 1 - p^{\mathcal{A}}_{\perp} - p^{\mathcal{A}}_0. \end{aligned} \tag{7}$$

Here $\mathcal{H}$ denotes the set of all honest parties in the stake distribution $\mathcal{S}$ determined by $\mathcal{Z}$. As before, $\mathcal{P}$ denotes the set of all parties.

It is convenient to work with some bounds on the above quantities that depend only on "macroscopic" features of $\mathcal{S}$ and $\mathcal{A}$: namely, the relative stake of the honest and adversarial parties, and the parameter f. For this purpose we note that

$$p^{\mathcal{A}}_0 \geq \sum_{h \in \mathcal{H}} \phi(\alpha_h) \cdot \prod_{i \in \mathcal{P}} (1 - \phi(\alpha_i)) \geq \phi(\alpha_{\mathcal{H}}) \cdot p^{\mathcal{A}}_{\perp} = \phi(\alpha_{\mathcal{H}}) \cdot (1-f), \tag{8}$$

where $\alpha_{\mathcal{H}}$ denotes the total relative stake of the honest parties. Note that this bound applies to all static adversaries $\mathcal{A}$ that corrupt no more than a $1 - \alpha_{\mathcal{H}}$ fraction of all stake. With this in mind, we define the dominant distribution as follows.

Definition 12 (The dominant distribution $\mathcal{D}^f_\alpha$). *For two parameters f and α, define $\mathcal{D}^f_\alpha$ to be the distribution on strings $w \in \{0,1,\perp\}^R$ that independently assigns each w_i so that $p_\perp \triangleq \Pr[w_i = \perp] = 1-f$, $p_0 \triangleq \Pr[w_i = 0] = \phi(\alpha)\cdot(1-f)$, and $p_1 \triangleq \Pr[w_i = 1] = 1 - p_\perp - p_0$.*

The distribution $\mathcal{D}^f_\alpha$ "dominates" $\mathcal{D}^f_{\mathcal{Z},\mathcal{A}}$ for any static adversary $\mathcal{A}$ that corrupts no more than a relative $1-\alpha$ share of the total stake, in the sense that nonempty slots are more likely to be tainted under $\mathcal{D}^f_\alpha$ than they are under $\mathcal{D}^f_{\mathcal{Z},\mathcal{A}}$.

To make this relationship precise, we introduce the partial order $\preceq$ on the set $\{\perp,0,1\}$ so that $x \preceq y$ if and only if $x = y$ or $y = 1$. We extend this partial order to $\{\perp,0,1\}^R$ by declaring $x_1 \ldots x_R \preceq y_1 \ldots y_R$ if and only if $x_i \preceq y_i$ for each i. Intuitively, the relationship $x \prec y$ asserts that y is "more adversarial than" x; concretely, any legal fork for x is also a legal fork for y. We record this in the lemma below, which follows directly from the definition of Δ-fork and div_Δ.

Lemma 2. *Let x and y be characteristic strings in $\{0,1,\perp\}^R$ for which $x \preceq y$. Then (1.) for every fork F, $F \vdash_\Delta x \implies F \vdash_\Delta y$; (2.) for every Δ, $\mathrm{div}_\Delta(x) \leq \mathrm{div}_\Delta(y)$.*

Finally, we define a notion of stochastic dominance for distributions on characteristic strings, and α-dominated adversaries.

Definition 13 (Stochastic dominance). *We say that a subset $E \subseteq \{\perp,0,1\}^R$ is* monotone *if $x \in E$ and $x \preceq y$ implies that $y \in E$. Let $\mathcal{D}$ and $\mathcal{D}'$ be two distributions on the set of characteristic strings $\{\perp,0,1\}^R$. Then we say that $\mathcal{D}'$* dominates *$\mathcal{D}$, written $\mathcal{D} \preceq \mathcal{D}'$, if $\Pr_{\mathcal{D}}[E] \leq \Pr_{\mathcal{D}'}[E]$ for every monotone set E. An adversary $\mathcal{A}$ is called α-dominated if the distribution $\mathcal{D}^f_{\mathcal{Z},\mathcal{A}}$ that it induces on the set of characteristic strings satisfies $\mathcal{D}^f_{\mathcal{Z},\mathcal{A}} \preceq \mathcal{D}^f_\alpha$.*

In our application, the events of interest are $D_\Delta = \{x \mid \mathrm{div}_\Delta(x) \geq k\}$ which are monotone by Lemma 2. We note that any static adversary that corrupts no more than a $1-\alpha$ fraction of stake is α-dominated, and it follows that $\Pr_{\mathcal{D}^f_{\mathcal{Z},\mathcal{A}}}[\mathrm{div}_\Delta(w) \geq k] \leq \Pr_{\mathcal{D}^f_\alpha}[\mathrm{div}_\Delta(w) \geq k]$. This motivates a particular study of the "dominant" distribution $\mathcal{D}^f_\alpha$.

The Induced Distribution $\rho_\Delta(\mathcal{D}^f_\alpha)$. The dominant distribution $\mathcal{D}^f_\alpha$ on $\{0,1,\perp\}^R$ in conjunction with the definition of ρ_Δ of (4) above implicitly defines a family of random variables $\rho_\Delta(w) = x_1 \dots x_\ell \in \{0,1\}^*$, where $w \in \{0,1,\perp\}^R$ is distributed according to $\mathcal{D}^f_\alpha$. Observe that $\ell = R - \#_\perp(w)$ is precisely the number of active indices of w. We now note a few properties of this resulting distribution that will be useful to us later (their proofs are presented in the full version). In particular, we will see that the x_i random variables are roughly binomially distributed, but subject to an exotic stochastic "stopping time" condition in tandem with some distortion of the last Δ variables.

Lemma 3 (Structure of the induced distribution). *Let $x_1 \dots x_\ell = \rho_\Delta(w)$ where $w \in \{0,1,\perp\}^R$ is distributed according to $\mathcal{D}^f_\alpha$. There is a sequence of independent random variables $z_1, z_2, \dots$ with each $z_i \in \{0,1\}$ so that*

$$\Pr[z_i = 0] = \left(\frac{p_0}{p_0 + p_1}\right) p_\perp^{\Delta-1} \geq \alpha \cdot (1-f)^\Delta, \tag{9}$$

$$\text{and} \quad x_1 \dots x_{\ell-\Delta} = \rho_\Delta(w_1 \dots, w_R)^{\lceil \Delta} \quad \text{is a prefix of} \quad z_1 z_2 \dots . \tag{10}$$

(Note that while the z_i are independent with each other, they are not independent with w.)

Divergence for the Dominant Distribution. Our goal is to apply the reduction ρ_Δ, Lemma 1, and Theorem 1 to establish an upper bound on the probability that a string drawn from the dominant distribution $\mathcal{D}^f_\alpha$ has large Δ-divergence. The difficulty is that the distribution resulting from applying ρ_Δ to a string drawn from $\mathcal{D}^f_\alpha$ is no longer a simple binomial distribution, so we cannot apply Theorem 1 directly. We resolve this obstacle in the proof of the following theorem, also given in the full version.

Theorem 2. *Let $f \in (0,1]$, $\Delta \geq 1$, and α be such that $\alpha(1-f)^\Delta = (1+\epsilon)/2$ for some $\epsilon > 0$. Let w be a string drawn from $\{0,1,\bot\}^R$ according to $\mathcal{D}^f_\alpha$. Then we have* $\Pr[\mathrm{div}_\Delta(w) \geq k + \Delta] = 2^{-\Omega(k)+\log R}$.

Remark. Intuitively, the theorem asserts that sampling the characteristic string in the Δ-semisynchronous setting with protocol parameter f according to $\mathcal{D}^f_\alpha$ is, for the purpose of analyzing divergence, comparable to the *synchronous* setting in which the honest stake has been reduced from α to $\alpha(1-f)^\Delta$. Note that this can be made arbitrarily close to α by adjusting f to be small; however, this happens at the expense of longer periods of silence in the protocol.

4.4 Common Prefix, Chain Growth, and Chain Quality

Our results on Δ-divergence from the previous section allow us to easily establish the following three statements, their proofs are again postponed to the full version.

Theorem 3 (Common prefix). *Let $k, R, \Delta \in \mathbb{N}$ and $\varepsilon \in (0,1)$. Let $\mathcal{A}$ be an α-dominated adversary against the protocol π_{SPoS} for some α satisfying $\alpha(1-f)^\Delta \geq (1+\epsilon)/2$. Then the probability that $\mathcal{A}$, when executed in a Δ-semisynchronous environment, makes π_{SPoS} violate the common prefix property with parameter k throughout a period of R slots is no more than* $\exp(\ln R + \Delta - \Omega(k))$. *The constant hidden by the $\Omega(\cdot)$-notation depends on ϵ.*

To obtain a bound on the probability of a violation of the chain growth property, we again consider the Δ-right-isolated uniquely honest slots introduced in Sect. 4.2. Intuitively, we argue that the leader of such a slot has already received all blocks that were created in all previous such slots and therefore the block it creates will be having depth larger than all these blocks. It then follows that the length of the chain grows by at least the number of such slots.

Theorem 4 (Chain growth). *Let $k, R, \Delta \in \mathbb{N}$ and $\varepsilon \in (0,1)$. Let $\mathcal{A}$ be an α-dominated adversary against the protocol π_{SPoS} for some $\alpha > 0$. Then the probability that $\mathcal{A}$, when executed in a Δ-semisynchronous environment, makes π_{SPoS} violate the chain growth property with parameters $s \geq 4\Delta$ and $\tau = c\alpha/4$ throughout a period of R slots, is no more than* $\exp(-c\alpha s/(20\Delta) + \ln R\Delta + O(1))$, *where c denotes the constant $c := c(f,\Delta) = f(1-f)^\Delta$.*

Our chain quality statement of Theorem 5 is a direct consequence of Lemma 4, which observes that a sufficiently long sequence of consecutive blocks in an honest party's chain will most likely contain a block created in a Δ-right-isolated uniquely honest slot.

Lemma 4. *Let $k, \Delta \in \mathbb{N}$ and $\epsilon \in (0,1)$. Let $\mathcal{A}$ be an α-dominated adversary against the protocol π_{SPoS} for some $\alpha > 0$ satisfying $\alpha(1-f)^\Delta = (1+\epsilon)/2$. Let $B_1, \ldots, B_k$ be a sequence of consecutive blocks in a chain C possessed by an honest party. Then at least one block B_i was created in a Δ-right-isolated uniquely honest slot, except with probability* $\exp(-\Omega(k))$.

Theorem 5 (Chain quality). *Let $k, R, \Delta \in \mathbb{N}$ and $\epsilon \in (0,1)$. Let $\mathcal{A}$ be an α-dominated adversary against the protocol π_{SPoS} for some $\alpha > 0$ satisfying $\alpha(1-f)^{\Delta} \geq (1+\epsilon)/2$. Then the probability that $\mathcal{A}$, when executed in a Δ-semisynchronous environment, makes π_{SPoS} violate the chain quality property with parameters k and $\mu = 1/k$ throughout a period of R slots, is no more than* $\exp(\ln R - \Omega(k))$.

4.5 Adaptive Adversaries

The statements in the previous sections give us guarantees on the common prefix, chain growth, and chain quality properties as long as the adversary is α-dominated for some suitable value of α. In Sect. 4.3 we argued that any *static* adversary that corrupts at most $(1-\alpha)$-fraction of stake is α-dominated. In this section we extend this claim also to *adaptive* adversaries, showing that as long as they corrupt no more than $(1-\alpha)$-fraction of stake adaptively throughout the whole execution, they are still α-dominated. The proof is deferred to the full version.

Theorem 6. *Every adaptive adversary $\mathcal{A}$ that corrupts at most $(1-\alpha)$-fraction of stake throughout the whole execution is α-dominated.*

Theorems 3, 4, 5 and 6 together give us the following corollary.

Corollary 1. *Let $\mathcal{A}$ be an adaptive adversary against the protocol Π_{SPoS} that corrupts at most $(1-\alpha)$-fraction of stake. Then the bounds on common prefix, chain growth and chain quality given in Theorems 3, 4, 5 are satisfied for $\mathcal{A}$.*

4.6 The Resettable Protocol

With the analysis of these basic structural events behind us, we remark that the same arguments apply to a modest generalization of the protocol which permits the adversary some control over the nonce. Specifically, we introduce a "resettable" initialization functionality $\mathcal{F}_{\mathsf{INIT}}^{r}$, which permits the adversary to select the random nonce from a family of r independent and uniformly random nonces. Specifically, $\mathcal{F}_{\mathsf{INIT}}^{r}$ is identical to $\mathcal{F}_{\mathsf{INIT}}$, with the following exception:

- Upon receiving the first request of the form $(\mathsf{genblock\_req}, U_i)$ from some stakeholder U_i, $\mathcal{F}_{\mathsf{INIT}}^{r}$ samples a nonce $\eta \xleftarrow{\$} \{0,1\}^{\lambda}$, defines a "nonce candidate" set $H = \{\eta\}$, and permits the adversary to carry out up to $r-1$ *reset events*: each reset event draws an independent element from $\{0,1\}^{\lambda}$, adds the element to the set H, and permits the adversary to replace the current nonce η with any element of H. Finally, $(\mathsf{genblock}, \mathbb{S}_0, \eta)$ is sent to U_i. Later requests from any stakeholder are answered using the same value η.

Looking ahead, our reason to introduce the resettable functionality $\mathcal{F}_{\mathsf{INIT}}^{r}$ is to capture the limited grinding capabilities of the adversary. A simple application of the union bound shows that this selection of η from among a set of size r uniformly random candidate nonces can inflate the probability of events during the run of π_{SPoS} by a factor no more than r. We record this as a corollary below.

Corollary 2 (Corollary to Theorems 3, 4, 5). *The protocol* Π_{SPoS}*, with initialization functionality* $\mathcal{F}^{r}_{\mathsf{INIT}}$*, satisfies the bounds of Theorems 3, 4, 5 with all probabilities scaled by* r*.*

5 The Full Protocol

In this section, we construct a protocol that handles the dynamic case, where the stake distribution changes as the protocol is executed. As in Ouroboros [16], we divide protocol execution in a number of independent *epochs* during which the stake distribution used for sampling slot leaders remains unchanged. The strategy we use to bootstrap the static protocol is, at a high level, similar: we first show how the protocol can accommodate dynamic stake utilizing an ideal "leaky beacon" functionality and then we show this beacon functionality can be simulated via an algorithm that collects randomness from the blockchain.

In order to facilitate the implementation of our beacon, we need to allow the leaky beacon functionality to be adversarially manipulated by allowing a number of "resets" to be performed by the adversary. Specifically, the functionality is parameterized by values τ and r. First, it leaks to the adversary, up to τ slots prior to the end of an epoch, the beacon value for the next epoch. (Looking ahead, we remark that it is essential that the stake distribution used for sampling slot leaders in the next epoch is determined prior to this leakage.) Second, the adversary can *reset* the value returned by the functionality as many as r times. As expected for a beacon, it reports to honest parties the beacon value only once the next epoch starts. After the epoch is started no more resets are allowed for the beacon value. This mimics the functionality $\mathcal{F}_{\mathsf{INIT}}$ and its resettable version $\mathcal{F}^{r}_{\mathsf{INIT}}$. Note that the ability of the adversary to reset the beacon can be quite influential in the protocol execution: for instance, any event that depends deterministically on the nonce of an epoch and happens with probability $1/2$ can be easily forced to happen almost always by the adversary using a small number of resets.

Naturally, we do not want to assume the availability of a randomness beacon in the final protocol, even if it is leaky and resettable. In our final iteration of the protocol we show how it is possible to simulate such beacon using a hash function that is modeled as a random oracle. This hash function is applied to the concatenation of VRF values that are inserted into each block, using values from all blocks up to and including the middle $\approx 8k$ slots of an epoch that lasts approximately $24k$ slots in entirety. (The "quiet" periods before and after this central block of slots that sets the nonce will ensure that the stake distribution, determined at the beginning of the epoch, is stable, and likewise that the nonce is stable before the next epoch begins.) The verifiability of those values is a key property that we exploit in the proof.

Our proof strategy is to reduce any adversary against the basic properties of the blockchain to a resettable-beacon adversary that will simulate the random oracle. The key point of this reduction is that whenever the random oracle adversary makes a query with a sequence of values that is a candidate sequence for determining the nonce for the next epoch, the resettable attacker detects this as a possible reset opportunity and resets the beacon; it obtains the response from the beacon and sets this as the answer to the random oracle query.

The final issue is to bound the number of resets: towards this, note that the adversary potentially controls a constant fraction of the $\approx 8k$ slots associated with nonce selection, and this allows him to explore an a priori large space of independent random potential nonces (and, ultimately, select one as the next epoch nonce). The size of this space is however upper-bounded by the number of random oracle queries that the adversary can afford during the sequence of $\approx 8k$ slots. To formalize this bound we utilize the q-bounded model of [12] that bounds the number of queries the adversary can pose per round: in that model, the adversary is allowed q queries per adversarial party per round ("slot" in our setting).[5] Assuming that the adversary controls t parties, we obtain a bound equal to $\approx 8qtk$.

5.1 The Dynamic Stake Case with a Resettable Leaky Beacon

First we construct a protocol for the dynamic stake case assuming access to a resettable leaky beacon that provides a fresh nonce for each epoch. This beacon is leaky in the sense that it allows the adversary to obtain the nonce for the next epoch before the epoch starts, and resettable in the sense that it allows the adversary to reset the nonce a number of times. We model the resettable leaky randomness beacon in functionality $\mathcal{F}_{RLB}^{\tau,r}$ presented in Fig. 5.

Functionality $\mathcal{F}_{RLB}^{\tau,r}$

$\mathcal{F}_{RLB}^{\tau,r}$ incorporates the diffuse functionality from Section 2.2 and is parameterized by the number of initial stakeholders n and their respective stakes $s_1, \ldots, s_n$, a nonce leakage parameter τ and a number of allowed resets r. $\mathcal{F}_{RLB}^{\tau,r}$ interacts with stakeholders $U_1, \ldots, U_n$ and an adversary $\mathcal{A}$ as follows:

- In the first round, $\mathcal{F}_{RLB}^{\tau,r}$ operates exactly as $\mathcal{F}_{\mathsf{INIT}}$.
- Upon receiving $(\mathsf{genblock\_req}, sid, U_i)$ from stakeholder U_i it operates as functionality $\mathcal{F}_{\mathsf{INIT}}$ on that message.
- Upon receiving $(\mathsf{epochrnd\_req}, sid, U_i, e_j)$ from stakeholder U_i, if $e_j \geq 2$ is the current epoch, $\mathcal{F}_{RLB}^{\tau,r}$ sends $(\mathsf{epochrnd}, sid, \eta_j)$ to U_i.
- For every epoch e_j, at slot $jR-\tau$, $\mathcal{F}_{RLB}^{\tau,r}$ samples the next epoch's nonce $\eta_{j+1} \xleftarrow{\$} \{0,1\}^\lambda$ and leaks it by sending $(\mathsf{epochrnd\_leak}, sid, e_j, \eta_{j+1})$ to the adversary $\mathcal{A}$. Additionally, $\mathcal{F}_{RLB}^{\tau,r}$ sets an internal reset request counter $\mathsf{Resets} = 0$ and sets $\mathbb{P} = \emptyset$.
- Upon receiving $(\mathsf{epochrnd\_reset}, sid, \mathcal{A})$ from $\mathcal{A}$ at epoch e_j, if $\mathsf{Resets} < r$ and if the current slot is past slot $jR - \tau$, $\mathcal{F}_{RLB}^{\tau,r}$ samples a fresh nonce for the next epoch $\eta_{j+1} \xleftarrow{\$} \{0,1\}^\lambda$ and leaks it by sending $(\mathsf{epochrnd\_leak}, sid, \eta_{j+1})$ to $\mathcal{A}$. Finally, $\mathcal{F}_{RLB}^{\tau,r}$ increments Resets and adds η_{j+1} to $\mathbb{P}$.
- Upon receiving $(\mathsf{epochrnd\_set}, sid, \mathcal{A}, \eta)$ from $\mathcal{A}$ at epoch e_j, if the current slot is past slot $jR - \tau$ and if $\eta \in \mathbb{P}$, $\mathcal{F}_{RLB}^{\tau,r}$ sets $\eta_{j+1} = \eta$ and sends $(\mathsf{epochrnd\_leak}, sid, \eta_{j+1})$ to $\mathcal{A}$.

Fig. 5. Functionality $\mathcal{F}_{RLB}^{\tau,r}$.

[5] Note that we utilize the q-bounded model only to provide a more refined analysis; given that the total length of the execution is polynomial in λ one may also use the total execution length as a bound.

We now describe protocol π_{DPoS}, which is a modified version of π_{SPoS} that updates its genesis block B_0 (and thus the assignment of slot leader sets) for every new epoch. The protocol also adopts an adaptation of the static $\mathsf{maxvalid}_S$ function, defined so that it narrows selection to those chains which share common prefix. Specifically, it adopts the following rule, parametrized by a prefix length k:

> Function $\mathsf{maxvalid}(\mathcal{C}, \mathbb{C})$. Returns the longest chain from $\mathbb{C} \cup \{\mathcal{C}\}$ that does not fork from $\mathcal{C}$ more than k blocks (i.e., not more than k blocks of $\mathcal{C}$ are discarded). If multiple exist it returns $\mathcal{C}$, if this is one of them, or it returns the one that is listed first in $\mathbb{C}$.

The protocol π_{DPoS} is described in Fig. 6 and functions in the $\mathcal{F}^{\tau,r}_{RLB}$-hybrid model.

Lazy players. Note that while the protocol π_{DPoS} in Fig. 6 is stated for a stakeholder that is permanently online, this requirement can be easily relaxed. Namely, it is sufficient for an honest stakeholder to join at the beginning of each epoch, determine whether she belongs to the slot leader set for any slots within this epoch (using the Eval interface of $\mathcal{F}_{\mathsf{VRF}}$), and then come online and act on those slots while maintaining online presence at least every k slots. We sketch this variant of the protocol in the full version.

We proceed to the security analysis of the full protocol in the hybrid world where the functionality $\mathcal{F}^{\tau,r}_{RLB}$ is available to the protocol participants. A key challenge is that in the dynamic stake setting, the honest majority assumption that we have in place refers to the stakeholder view of the honest stakeholders in each slot. Already in the first few slots this assumption may diverge rapidly from the stakeholder distribution that is built-in the genesis block.

To accommodate the issues that will arise from the movement of stake throughout protocol execution, we recall the notion of stake shift defined in [16].

Definition 14. *Consider two slots sl_1, sl_2 and an execution $\mathcal{E}$. The* stake shift *between sl_1, sl_2 is the maximum possible statistical distance of the two weighted-by-stake distributions that are defined using the stake reflected in the chain $\mathcal{C}_1$ of some honest stakeholder active at sl_1 and the chain $\mathcal{C}_2$ of some honest stakeholder active at sl_2.*

Finally, the security of π_{DPoS} is stated below and proven in the full version. We slightly abuse the notation from previous sections and denote by $\alpha_{\mathcal{H}}$ a *lower bound* on the honest stake ratio throughout the whole execution.

Theorem 7 (Security of π_{DPoS} with access to $\mathcal{F}^{\tau,r}_{RLB}$). *Fix parameters $k, R, \Delta, L \in \mathbb{N}, \epsilon, \sigma \in (0,1)$ and r. Let $R \geq 16k/f$ be the epoch length, L the total lifetime of the system, and*

$$(\alpha_{\mathcal{H}} - \sigma)(1-f)^{\Delta} \geq (1+\epsilon)/2. \tag{11}$$

The protocol π_{DPoS}, with access to $\mathcal{F}^{\tau,r}_{RLB}$, with $\tau \leq 8k/f$ satisfies persistence with parameters k and liveness with parameters $u = 8k/f$ throughout a period of L slots of Δ-semisynchronous execution with probability $1 - \exp(\ln L + \Delta + \log(r) - \Omega(k))$ assuming that σ is the maximum stake shift over $2R$ slots.

Protocol π_{DPoS}

The protocol π_{DPoS} is run by stakeholders, initially equal to $U_1, \ldots, U_n$ interacting among themselves and with ideal functionalities $\mathcal{F}_{RLB}^{\tau,r}$ (or $\mathcal{F}_{\mathsf{INIT}}$), $\mathcal{F}_{\mathsf{VRF}}, \mathcal{F}_{\mathsf{KES}}, \mathcal{F}_{\mathsf{DSIG}}, \mathsf{H}$ over a sequence of $L = ER$ slots $S = \{sl_1, \ldots, sl_L\}$ consisting of E epochs with R slots each. Define $T_i^j \triangleq 2^{\ell_{\mathsf{VRF}}} \phi_f(\alpha_i^j)$ as the threshold for a stakeholder U_i for epoch e_j, where α_i^j is the relative stake of stakeholder U_i at epoch e_j, ℓ_{VRF} denotes the output length of $\mathcal{F}_{\mathsf{VRF}}$, f is the active slots coefficient and ϕ_f is the mapping from equation (1). Then π_{DPoS} proceeds as follows:

1. **Initialization.** This step is the same as Step 1 in π_{SPoS} except that any messages for $\mathcal{F}_{\mathsf{INIT}}$ are sent to $\mathcal{F}_{RLB}^{\tau,r}$ if it is available instead.
2. **Chain Extension.** After initialization, for every slot $sl \in S$, every online stakeholder U_i performs the following steps:
   (a) This step is the same as Step 2a in π_{SPoS}.
   (b) If a new epoch e_j, with $j \geq 2$, has started, U_i defines $\mathbb{S}_j$ to be the stakeholder distribution drawn from the most recent block with time stamp up to $(j-2)R$ as reflected in $\mathcal{C}$ and sends $(\mathsf{epochrnd\_req}, sid, U_i, e_j)$ to $\mathcal{F}_{RLB}^{\tau,r}$, receiving $(\mathsf{epochrnd}, sid, \eta_j)$ as answer.
   (c) U_i collects all valid chains received via diffusion into a set $\mathbb{C}$, pruning blocks belonging to future slots and verifying that for every chain $\mathcal{C}' \in \mathbb{C}$ and every block $B' = (st', d', sl', B_\pi', \rho', \sigma_{j'}) \in \mathcal{C}'$ it holds that the stakeholder who created it is in the slot leader set of slot sl' (by parsing B_π' as (U_s, y', π') for some s, verifying that $\mathcal{F}_{\mathsf{VRF}}$ responds to $(\mathsf{Verify}, sid, \eta_j \| sl' \| \mathtt{TEST}, y', \pi', v_s^{\mathrm{vrf}})$ by $(\mathsf{Verified}, sid, \eta_j \| sl' \| \mathtt{TEST}, y', \pi', 1)$, and that $y' < T_s^j$ where T_s^j is the threshold of stakeholder U_s for the epoch e_j to which sl' belongs), that $\mathcal{F}_{\mathsf{VRF}}$ responds to $(\mathsf{Verify}, sid, \eta_j \| sl' \| \mathtt{NONCE}, \rho_y', \rho_\pi', v_s^{\mathrm{vrf}})$ (where $\rho' = (\rho_y', \rho_\pi')$) by $(\mathsf{Verified}, sid, \eta_j \| sl' \| \mathtt{NONCE}, \rho_y', \rho_\pi', 1)$, and that $\mathcal{F}_{\mathsf{KES}}$ responds to $(\mathsf{Verify}, sid, (st', d', sl', B_\pi', \rho'), sl', \sigma_{j'}, v_s^{\mathrm{kes}})$ by $(\mathsf{Verified}, sid, (st', d', sl', B_\pi', \rho'), sl', 1)$. U_i computes $\mathcal{C}' = \mathsf{maxvalid}(\mathcal{C}, \mathbb{C})$, sets $\mathcal{C}'$ as the new local chain and sets state $st = H(\mathsf{head}(\mathcal{C}'))$.
   (d) U_i sends $(\mathsf{EvalProve}, sid, \eta_j \| sl \| \mathtt{NONCE})$ to $\mathcal{F}_{\mathsf{VRF}}$, obtaining $(\mathsf{Evaluated}, sid, \rho_y, \rho_\pi)$, Afterwards, U_i sends $(\mathsf{EvalProve}, sid, \eta_j \| sl \| \mathtt{TEST})$ to $\mathcal{F}_{\mathsf{VRF}}$, receiving $(\mathsf{Evaluated}, sid, y, \pi)$. U_i checks whether it is in the slot leader set of slot sl with respect to the current epoch e_j by checking that $y < T_i^j$. If yes, it chooses the maximal sequence d' of transactions in d such that adding a block with d' to $\mathcal{C}$ results into a valid chain, and attempts to include d' as follows: It generates a new block $B = (st, d', sl, B_\pi, \rho, \sigma)$ where st is its current state, $B_\pi = (U_i, y, \pi)$, $\rho = (\rho_y, \rho_\pi)$ and σ is a signature obtained by sending $(\mathsf{USign}, sid, U_i, (st, d', sl, B_\pi, \rho), sl)$ to $\mathcal{F}_{\mathsf{KES}}$ and receiving $(\mathsf{Signature}, sid, (st, d', sl, B_\pi, \rho), sl, \sigma)$. U_i computes $\mathcal{C}' = \mathcal{C} \| B$, sets $\mathcal{C}'$ as the new local chain and sets state $st = H(\mathsf{head}(\mathcal{C}'))$. Finally, if U_i has generated a block in this step, it diffuses $\mathcal{C}'$.
3. **Signing Transactions.** This step is the same as Step 3 in π_{SPoS}.

Fig. 6. Protocol π_{DPoS}

Note that while Theorem 7 (and also Corollary 3 below) formulates the bound (11) in terms of the *overall* upper bound on honest stake ratio $\alpha_{\mathcal{H}}$ and *maximum* stake shift σ over any $2R$-slots interval, one could easily prove more fine-grained statements that would only require inequality (11) to hold for each epoch (with respect to the honest stake ratio in that epoch, and the stake shift occurring for that epoch's stake distribution).

5.2 Instantiating $\mathcal{F}_{RLB}^{\tau,r}$

In this section, we show how to substitute the oracle $\mathcal{F}_{RLB}^{\tau,r}$ of protocol π_{DPoS} with a subprotocol π_{RLB} that simulates $\mathcal{F}_{RLB}^{\tau,r}$. The resulting protocol can then operate directly in the $\mathcal{F}_{\mathsf{INIT}}$-hybrid model as in Sect. 3 (without resets) while utilizing a random oracle $\mathsf{H}(\cdot)$. The sub-protocol π_{RLB} is described in Figure 7.

Protocol π_{RLB}

Let $\mathsf{H}(\cdot)$ be a random oracle. π_{RLB} is a sub-protocol of π_{DPoS} proceeding as follows:

- Upon receiving $(\mathsf{epochrnd\_req}, sid, U_i, e_j)$ from stakeholder U_i, if $e_j \geq 2$ is the current epoch, it performs the following: for every block $B' = (st', d', sl', B_\pi{}', \rho', \sigma_{j'}) \in \mathcal{C}$ (where $\mathcal{C}$ is the callee's U_i's internal chain) belonging to epoch e_{j-1} up to the slot with timestamp up to $(j-2)R + 16k/f$, concatenate the values ρ' into a value v. Compute $\eta_j = \mathsf{H}(\eta_{j-1}||j||v)$ and return $(\mathsf{epochrnd}, sid, \eta_j)$.

Fig. 7. Protocol π_{RLB}.

We will show next that the sub-protocol π_{RLB} can safely substitute $\mathcal{F}_{RLB}^{\tau,r}$ when called from protocol π_{DPoS}. We will perform our analysis in the q-bounded model of [12] assuming that the adversary is capable of issuing q queries per each round of protocol execution per corrupted party and there are t corrupted parties. The proof is deferred to the full version.

Lemma 5. *Consider the event of violating one of common prefix, chain quality, chain growth in an execution of π_{DPoS} using sub-protocol π_{RLB} in the $\mathcal{F}_{\mathsf{INIT}}$-hybrid model with adversary $\mathcal{A}$ and environment $\mathcal{Z}$ with the same parameter choices as Theorem 7. We construct an adversary $\mathcal{A}'$ so that the corresponding event happens with the same probability in an execution of π_{DPoS} in the $\mathcal{F}_{RLB}^{\tau,r}$-hybrid world with adversary $\mathcal{A}'$ and environment $\mathcal{Z}$ assuming that $r = 8tqk/f$.*

Based on the above lemma, it is now easy to revisit Theorem 7, and show that the same result holds for r in the q-bounded model assuming $r = 8tkq/f$ and $\tau \leq 8k/f$ which permits to set our epoch length R to $24k/f$.

Corollary 3 (Security of π_{DPoS} with subprotocol π_{RLB}). *Fix parameters $k, R, \Delta, L \in \mathbb{N}, \epsilon, \sigma \in (0, 1)$. Let $R = 24k/f$ be the epoch length, L the total lifetime of the system, and $(\alpha_{\mathcal{H}} - \sigma)(1-f)^{\Delta} \geq (1+\epsilon)/2$. The protocol π_{DPoS} using* *subprotocol* π_{RLB} *in the* $\mathcal{F}_{\mathsf{INIT}}$*-hybrid model satisfies persistence with parameters* k *and liveness with parameters* $u = 8k/f$ *throughout a period of* L *slots of* Δ*-semisynchronous execution with probability* $1 - \exp(\ln L + \Delta - \Omega(k - \log tkq))$ *assuming that* σ *is the maximum stake shift over* $2R$ *slots.*

Acknowledgements. We thank Christian Badertscher and the anonymous reviewers for several useful suggestions improving the presentation of the paper.

Peter Gaži partly worked on this project while being a postdoc at IST Austria, supported by the ERC consolidator grant 682815-TOCNeT. Aggelos Kiayias was partly supported by H2020 Project #653497, PANORAMIX.

References

1. Bellare, M., Miner, S.K.: A forward-secure digital signature scheme. In: Wiener, M. (ed.) CRYPTO 1999. LNCS, vol. 1666, pp. 431–448. Springer, Heidelberg (1999). https://doi.org/10.1007/3-540-48405-1_28
2. Bentov, I., Gabizon, A., Mizrahi, A.: Cryptocurrencies without proof of work. CoRR, abs/1406.5694 (2014)
3. Bentov, I., Gabizon, A., Mizrahi, A.: Cryptocurrencies without proof of work. In: Clark, J., Meiklejohn, S., Ryan, P.Y.A., Wallach, D., Brenner, M., Rohloff, K. (eds.) FC 2016. LNCS, vol. 9604, pp. 142–157. Springer, Heidelberg (2016). https://doi.org/10.1007/978-3-662-53357-4_10
4. Bentov, I., Lee, C., Mizrahi, A., Rosenfeld, M.: Proof of activity: extending bitcoin's proof of work via proof of stake. SIGMETRICS Perform. Eval. Rev. **42**(3), 34–37 (2014)
5. Canetti, R.: Universally composable signature, certification, and authentication. In: 17th IEEE Computer Security Foundations Workshop, (CSFW-17 2004), p. 219. IEEE Computer Society (2004)
6. Chase, M., Lysyanskaya, A.: Simulatable VRFs with applications to multi-theorem NIZK. In: Menezes, A. (ed.) CRYPTO 2007. LNCS, vol. 4622, pp. 303–322. Springer, Heidelberg (2007). https://doi.org/10.1007/978-3-540-74143-5_17
7. The NXT Community. Nxt whitepaper, July 2014. https://bravenewcoin.com/assets/Whitepapers/NxtWhitepaper-v122-rev4.pdf
8. Daian, P., Pass, R., Shi, E.: Snow white: provably secure proofs of stake. Cryptology ePrint Archive, Report 2016/919 (2016). http://eprint.iacr.org/2016/919
9. Dodis, Y., Puniya, P.: Feistel networks made public, and applications. In: Naor, M. (ed.) EUROCRYPT 2007. LNCS, vol. 4515, pp. 534–554. Springer, Heidelberg (2007). https://doi.org/10.1007/978-3-540-72540-4_31
10. Dodis, Y., Yampolskiy, A.: A verifiable random function with short proofs and keys. In: Vaudenay, S. (ed.) PKC 2005. LNCS, vol. 3386, pp. 416–431. Springer, Heidelberg (2005). https://doi.org/10.1007/978-3-540-30580-4_28
11. Dwork, C., Lynch, N.A., Stockmeyer, L.J.: Consensus in the presence of partial synchrony. J. ACM **35**(2), 288–323 (1988)
12. Garay, J., Kiayias, A., Leonardos, N.: The bitcoin backbone protocol: analysis and applications. In: Oswald, E., Fischlin, M. (eds.) EUROCRYPT 2015. LNCS, vol. 9057, pp. 281–310. Springer, Heidelberg (2015). https://doi.org/10.1007/978-3-662-46803-6_10
13. Itkis, G., Reyzin, L.: Forward-secure signatures with optimal signing and verifying. In: Kilian, J. (ed.) CRYPTO 2001. LNCS, vol. 2139, pp. 332–354. Springer, Heidelberg (2001). https://doi.org/10.1007/3-540-44647-8_20

14. Jarecki, S., Kiayias, A., Krawczyk, H.: Round-optimal password-protected secret sharing and T-PAKE in the password-only model. In: Sarkar, P., Iwata, T. (eds.) ASIACRYPT 2014. LNCS, vol. 8874, pp. 233–253. Springer, Heidelberg (2014). https://doi.org/10.1007/978-3-662-45608-8_13
15. Kiayias, A., Panagiotakos, G.: Speed-security tradeoffs in blockchain protocols. Cryptology ePrint Archive, Report 2015/1019 (2015). http://eprint.iacr.org/2015/1019
16. Kiayias, A., Russell, A., David, B., Oliynykov, R.: Ouroboros: a provably secure proof-of-stake blockchain protocol. In: Katz, J., Shacham, H. (eds.) CRYPTO 2017. LNCS, vol. 10401, pp. 357–388. Springer, Cham (2017). https://doi.org/10.1007/978-3-319-63688-7_12
17. King, S., Nadal, S.: PPCoin: peer-to-peer crypto-currency with proof-of-stake, August 2012. https://peercoin.net/assets/paper/peercoin-paper.pdf
18. Lindell, A.Y.: Adaptively secure two-party computation with erasures. In: Fischlin, M. (ed.) CT-RSA 2009. LNCS, vol. 5473, pp. 117–132. Springer, Heidelberg (2009). https://doi.org/10.1007/978-3-642-00862-7_8
19. Micali, S.: ALGORAND: the efficient and democratic ledger. CoRR, abs/1607.01341 (2016)
20. Nakamoto, S.: The proof-of-work chain is a solution to the byzantine generals' problem. The Cryptography Mailing List, November 2008. https://www.mail-archive.com/cryptography@metzdowd.com/msg09997.html
21. Pass, R., Seeman, L., Shelat, A.: Analysis of the blockchain protocol in asynchronous networks. In: Coron, J.-S., Nielsen, J.B. (eds.) EUROCRYPT 2017. LNCS, vol. 10211, pp. 643–673. Springer, Cham (2017). https://doi.org/10.1007/978-3-319-56614-6_22
22. Pass, R., Shi, E.: The sleepy model of consensus. Cryptology ePrint Archive, Report 2016/918 (2016). http://eprint.iacr.org/2016/918
23. Russell, A., Moore, C., Kiayias, A., Quader, S.: Forkable strings are rare. Cryptology ePrint Archive, Report 2017/241 (2017). http://eprint.iacr.org/2017/241

Sustained Space Complexity

Joël Alwen[1,3], Jeremiah Blocki[2(✉)], and Krzysztof Pietrzak[1]

[1] IST Austria, Klosterneuburg, Austria
jalwen@ist.ac.at
[2] Purdue University, West Lafayette, USA
[3] Wickr Inc., San Francisco, USA

Abstract. Memory-hard functions (MHF) are functions whose evaluation cost is dominated by memory cost. MHFs are egalitarian, in the sense that evaluating them on dedicated hardware (like FPGAs or ASICs) is not much cheaper than on off-the-shelf hardware (like x86 CPUs). MHFs have interesting cryptographic applications, most notably to password hashing and securing blockchains.

Alwen and Serbinenko [STOC'15] define the cumulative memory complexity (cmc) of a function as the sum (over all time-steps) of the amount of memory required to compute the function. They advocate that a good MHF must have high cmc. Unlike previous notions, cmc takes into account that dedicated hardware might exploit amortization and parallelism. Still, cmc has been critizised as insufficient, as it fails to capture possible time-memory trade-offs; as memory cost doesn't scale linearly, functions with the same cmc could still have very different actual hardware cost.

In this work we address this problem, and introduce the notion of sustained-memory complexity, which requires that any algorithm evaluating the function must use a large amount of memory for many steps. We construct functions (in the parallel random oracle model) whose sustained-memory complexity is almost optimal: our function can be evaluated using n steps and $O(n/\log(n))$ memory, in each step making one query to the (fixed-input length) random oracle, while any algorithm that can make arbitrary many parallel queries to the random oracle, still needs $\Omega(n/\log(n))$ memory for $\Omega(n)$ steps.

As has been done for various notions (including cmc) before, we reduce the task of constructing an MHFs with high sustained-memory complexity to proving pebbling lower bounds on DAGs. Our main technical contribution is the construction is a family of DAGs on n nodes with constant indegree with high "sustained-space complexity", meaning that any parallel black-pebbling strategy requires $\Omega(n/\log(n))$ pebbles for at least $\Omega(n)$ steps.

Along the way we construct a family of maximally "depth-robust" DAGs with maximum indegree $O(\log n)$, improving upon the construction of Mahmoody et al. [ITCS'13] which had maximum indegree $O\left(\log^2 n \cdot \mathsf{polylog}(\log n)\right)$.

J. B. Nielsen and V. Rijmen (Eds.): EUROCRYPT 2018, LNCS 10821, pp. 99–130, 2018.
https://doi.org/10.1007/978-3-319-78375-8_4

1 Introduction

In cryptographic settings we typically consider tasks which can be done efficiently by honest parties, but are infeasible for potential adversaries. This requires an asymmetry in the capabilities of honest and dishonest parties. An example are trapdoor functions, where the honest party – who knows the secret trapdoor key – can efficiently invert the function, whereas a potential adversary – who does not have this key – cannot.

1.1 Moderately-Hard Functions

Moderately hard functions consider a setting where there's no asymmetry, or even worse, the adversary has more capabilities than the honest party. What we want is that the honest party can evaluate the function with some reasonable amount of resources, whereas the adversary should not be able to evaluate the function at significantly lower cost. Moderately hard functions have several interesting cryptographic applications, including securing blockchain protocols and for password hashing.

An early proposal for password hashing is the "Password Based Key Derivation Function 2" (PBKDF2) [Kal00]. This function just iterates a cryptographic hash function like SHA1 several times (1024 is a typical value). Unfortunately, PBKDF2 doesn't make for a good moderately hard function, as evaluating a cryptographic hash function on dedicated hardware like ASCIs (Application Specific Integrated Circuits) can be by several orders of magnitude cheaper in terms of hardware and energy cost than evaluating it on a standard x86 CPU. An economic analysis of Blocki et al. [BHZ18] suggests that an attacker will crack *almost all* passwords protected by PBKDF2. There have been several suggestions how to construct better, i.e., more "egalitarian", moderately hard functions. We discuss the most prominent suggestions below.

Memory-Bound Functions. Abadi et al. [ABW03] observe that the time required to evaluate a function is dominated by the number of cache-misses, and these slow down the computation by about the same time over different architectures. They propose *memory-bound* functions, which are functions that will incur many expensive cache-misses (assuming the cache is not too big). They propose a construction which is not very practical as it requires a fairly large (larger than the cache size) incompressible string as input. Their function is then basically pointer jumping on this string. In subsequent work [DGN03] it was shown that this string can also be locally generated from a short seed.

Bandwidth-Hard Functions. Recently Ren and Devadas [RD17] suggest the notion of *bandwidth-hard* functions, which is a refinement of memory-bound functions. A major difference being that in their model computation is not completely free, and this assumption – which of course is satisfied in practice – allows for much more practical solutions. They also don't argue about evaluation time as [ABW03], but rather the more important energy cost; the energy spend for evaluating a function consists of energy required for on chip computation and memory accesses, only the latter is similar on various platforms. In a bandwidth-hard function the memory accesses dominate the energy cost on a standard CPU, and thus the function cannot be evaluated at much lower energy cost on an ASICs as on a standard CPU.

Memory-Hard Functions. Whereas memory-bound and bandwidth-hard functions aim at being egalitarian in terms of time and energy, memory-hard functions (MHF), proposed by Percival [Per09], aim at being egalitarian in terms of hardware cost. A memory-hard function, in his definition, is one where the memory used by the algorithm, multiplied by the amount of time, is high, i.e., it has high space-time (ST) complexity. Moreover, parallelism should not help to evaluate this function at significantly lower cost by this measure. The rationale here is that the hardware cost for evaluating an MHF is dominated by the memory cost, and as memory cost does not vary much over different architectures, the hardware cost for evaluating MHFs is not much lower on ASICs than on standard CPUs.

Cumulative Memory Complexity. Alwen and Serbinenko [AS15] observe that ST complexity misses a crucial point, amortization. A function might have high ST complexity because at some point during the evaluation the space requirement is high, but for most of the time a small memory is sufficient. As a consequence, ST complexity is not multiplicative: a function can have ST complexity C, but evaluating X instances of the function can be done with ST complexity much less than $X \cdot C$, so the amortized ST cost is much less than C. Alwen and Blocki [AB16, AB17] later showed that prominent MHF candidates such as Argon2i [BDK16], winner of the Password Hashing Competition [PHC] do not have high amortized ST complexity.

To address this issue, [AS15] put forward the notion of cumulative-memory complexity (cmc). The cmc of a function is the sum – over all time steps – of the memory required to compute the function by any algorithm. Unlike ST complexity, cmc is multiplicative.

Sustained-Memory Complexity. Although cmc takes into account amortization and parallelism, it has been observed (e.g., [RD16, Cox16]) that it still is not sufficient to guarantee egalitarian hardware cost. The reason is simple: if a function has cmc C, this could mean that the algorithm minimizing cmc uses some T time steps and C/T memory on average, but it could also mean it uses time $100 \cdot T$ and $C/100 \cdot T$ memory on average. In practice this can makes a huge difference because *memory cost doesn't scale linearly.* The length of the wiring required to access memory of size M grows like $\sqrt{M}$ (assuming a two dimensional layout of the circuit). This means for one thing, that – as we increase M – the latency of accessing the memory will grow as $\sqrt{M}$, and moreover the space for the wiring required to access the memory will grow like $M^{1.5}$.

The exact behaviour of the hardware cost as the memory grows is not crucial here, just the point that it's superlinear, and cmc does not take this into account.

In this work we introduce the notion of sustained-memory complexity, which takes this into account. Ideally, we want a function which can be evaluated by a "naïve" sequential algorithm (the one used by the honest parties) in time T using a memory of size S where (1) S should be close to T and (2) any *parallel* algorithm evaluating the function must use memory S' for at least T' steps, where T' and S' should be not much smaller than T and S, respectively.

Property (1) is required so the memory cost dominates the evaluation cost already for small values of T. Property (2) means that even a parallel algorithm will not be able to evaluate the function at much lower cost; any parallel algorithm must make almost as many steps as the naïve algorithm during which the required memory is almost as large as the maximum memory S used by the naïve algorithm. So, the cost of the best parallel algorithm is similar to the cost of the naïve sequential one, even if we don't charge the parallel algorithm anything for all the steps where the memory is below S'.

Ren and Devadas [RD16] previously proposed the notion of "consistent memory hardness" which requires that any *sequential* evaluation algorithm must either use space S' for at least T' steps, or the algorithm must run for a long time e.g., $T \gg n^2$. Our notion of sustained-memory complexity strengthens this notion in that we consider *parallel* evaluation algorithms, and our guarantees are absolute e.g., even if a parallel attacker runs for a very long time he must still use memory S' for at least T' steps. `scrypt` [Per09] is a good example of a MHF that has maximal cmc $\Omega\left(n^2\right)$ [ACP+17] that does not have high sustained space complexity. In particular, for *any* memory parameter M and any running time parameter n we can evaluate `scrypt` [Per09] in time n^2/M and with maximum space M. As was argued in [RD16] an adversary may be able to fit $M = n/100$ space in an ASIC, which would allow the attacker to speed up computation by a factor of more than 100 and may explain the availability of ASICs for `scrypt` despite its maximal cmc.

In this work we show that functions with asymptotically optimal sustained-memory complexity exist in the random oracle model. We note that we must make some idealized assumption on our building block, like being a random oracle, as with the current state of complexity theory, we cannot even prove superlinear circuit lower-bounds for problems in $\mathcal{NP}$. For a given time T, our function uses maximal space $S \in \Omega(T)$ for the naïve algorithm,[1] while any *parallel* algorithm must have at least $T' \in \Omega(T)$ steps during which it uses memory $S' \in \Omega(T/\log(T))$.

Graph Labelling. The functions we construct are defined by directed acyclic graphs (DAG). For a DAG $G_n = (V, E)$, we order the vertices $V = \{v_1, \ldots, v_n\}$ in some topological order (so if there's a path from i to j then $i < j$), with v_1 being the unique source, and v_n the unique sink of the graph. The function is now defined by G_n and the input specifies a random oracle H. The output is the label ℓ_n of the sink, where the label of a node v_i is recursively defined as $\ell_i = H(i, \ell_{p_1}, \ldots, \ell_{p_d})$ where $v_{p_1}, \ldots, v_{p_d}$ are the parents of v_i.

[1] Recall that the naïve algorithm is sequential, so S must be in $O(T)$ as in time T the algorithm cannot even touch more than $O(T)$ memory.

Pebbling. Like many previous works, including [ABW03, RD17, AS15] discussed above, we reduce the task of proving lower bounds – in our case, on sustained memory complexity – for functions as just described, to proving lower bounds on some complexity of a pebbling game played on the underlying graph.

For example, Ren and Devedas [RD17] define a cost function for the so called reb-blue pebbling game, which then implies lower bounds on the bandwidth hardness of the function defined over the corresponding DAG.

Most closely related to this work is [AS15], who show that a lower bound the so called sequential (or parallel) *cumulative (black) pebbling complexity* (cpc) of a DAG implies a lower bound on the sequential (or parallel) cumulative memory complexity (cmc) of the labelling function defined over this graph. Alwen et al. [ABP17] constructed a constant indegree family of DAGs with parallel cpc $\Omega(n^2/\log(n))$, which is optimal [AB16], and thus gives functions with optimal cmc. More recently, Alwen et al. [ABH17] extended these ideas to give the first *practical* construction of an iMHF with parallel cmc $\Omega(n^2/\log(n))$.

The black pebbling game – as considered in cpc – goes back to [HP70, Coo73]. It is defined over a DAG $G = (V, E)$ and goes in round as follows. Initially all nodes are empty. In every round, the player can put a pebble on a node if all its parents contain pebbles (arbitrary many pebbles per round in the parallel game, just one in the sequential). Pebbles can be removed at any time. The game ends when a pebble is put on the sink. The cpc of such a game is the sum, over all time steps, of the pebbles placed on the graph. The sequential (or parallel) cpc of G is the cpc of the sequential (or parallel) black pebbling strategy which minimizes this cost.

It's not hard to see that the sequential/parallel cpc of G directly implies the same upper bound on the sequential/parallel cmc of the graph labelling function, as to compute the function in the sequential/parallel random oracle model, one simply mimics the pebbling game, where putting a pebble on vertex v_i with parents $v_{p_1}, \ldots, v_{p_d}$ corresponds to the query $\ell_i \leftarrow H(i, \ell_{p_1}, \ldots, \ell_{p_d})$. And where one keeps a label ℓ_j in memory, as long as v_j is pebbled. If the labels $\ell_i \in \{0,1\}^w$ are w bits long, a cpc of p translates to cmc of $p \cdot w$.

More interestingly, the same has been shown to hold for interesting notions also for lower bounds. In particular, the *ex-post facto* argument [AS15] shows that any adversary who computes the label ℓ_n with high probability (over the choice of the random oracle H) with cmc of m, translates into a black pebbling strategy of the underlying graph with cpc almost m/w.

In this work we define the *sustained-space complexity* (ssc) of a sequential/parallel black pebbling game, and show that lower bounds on ssc translate to lower bounds on the sustained-memory complexity (smc) of the graph labelling function in the sequential/parallel random oracle model.

Consider a sequential (or parallel) black pebbling strategy (i.e., a valid sequence pebbling configurations where the last configuration contains the sink) for a DAG $G_n = (V, E)$ on $|V| = n$ vertices. For some space parameter $s \leq n$, the s-ssc of this strategy is the number of pebbling configurations of size at least s. The sequential (or parallel) s-ssc of G is the strategy minimizing this value.

For example, if it's possible to pebble G using $s' < s$ pebbles (using arbitrary many steps), then its s-ssc is 0. Similarly as for csc vs cmc, an upper bound on s-ssc implies the same upper bound for $(w \cdot s)$-smc. In Sect. 5 we prove that also lower bounds on ssc translate to lower bounds on smc.

Thus, to construct a function with high parallel smc, it suffices to construct a family of DAGs with constant indegree and high parallel ssc. In Sect. 3 we construct such a family $\{G_n\}_{n\in\mathbb{N}}$ of DAGs where G_n has n vertices and has indegree 2, where $\Omega(n/\log(n))$-ssc is in $\Omega(n)$. This is basically the best we can hope for, as our bound on ssc trivially implies a $\Omega(n^2/\log(n))$ bound on csc, which is optimal for any constant indegree graph [AS15].

Data-Dependent vs Data-Independent MHFs. There are two categories of Memory Hard Functions: data-Independent Memory Hard Functions (iMHFs) and data-dependent Memory Hard Functions (dMHFs). As the name suggests, the algorithm to compute an iMHFs must induce a memory access pattern that is *independent* of the potentially sensitive input (e.g., a password), while dMHFs have no such constraint. While dMHFs (e.g., `scrypt` [PJ12], Argon2d, Argon2id [BDK16]) are potentially easier to construct, iMHFs (e.g., Argon2i [BDK16], DRSample [ABH17]) are resistant to side channel leakage attacks such as cache-timing. For the cumulative memory complexity metric there is a clear gap between iMHFs and dMHFs. In particular, it is known that `scrypt` has cmc at least $\Omega\left(n^2 w\right)$ [ACP+17], while *any* iMHF has cmc *at most* $O\left(\frac{n^2 w \log\log n}{\log n}\right)$. Interestingly, the same gap does *not* hold for smc. In particular, any dMHF can be computed with *maximum space* $O\left(nw/\log n + n\log n\right)$ by recursively applying a result of Hopcroft et al. [HPV77]—see more details in the full version [ABP18].

1.2 High Level Description of Our Construction and Proof

Our construction of a family $\{G_n\}_{n\in\mathbb{N}}$ of DAGs with optimal ssc involves three building blocks:

The first building block is a construction of Paul et al. [PTC76] of a family of DAGs $\{\mathsf{PTC}_n\}_{n\in\mathbb{N}}$ with $\mathsf{indeg}(\mathsf{PTC}_n) = 2$ and space complexity $\Omega(n/\log n)$. More significantly for us they proved that for any sequential pebbling of G_n there is a time interval $[i, j]$ during which at least $\Omega(n/\log n)$ new pebbles are placed on sources of G_n and at least $\Omega(n/\log n)$ are always on the DAG. We extend the proof of Paul et al. [PTC76] to show that the same holds for any *parallel* pebbling of PTC_n; a pebbling game first introduced in [AS15] which natural models parallel computation. We can argue that $j - i = \Omega(n/\log n)$ for any sequential pebbling since it takes at least this many steps to place $\Omega(n/\log n)$ new pebbles on G_n. However, we stress that this argument does not apply to parallel pebblings so this does not directly imply anything about sustained space complexity for parallel pebblings.

To address this issue we introduce our second building block: a family of $\{D_n^\epsilon\}_{n\in\mathbb{N}}$ of extremely depth robust DAGs with $\mathsf{indeg}(D_n) \in O\left(\log n\right)$—for any constant $\epsilon > 0$ the DAG D_n^ϵ is (e,d)-depth robust for any $e + d \leq (1-\epsilon)n$. We remark that our result improves upon the construction of Mahmoody et al. [MMV13] whose construction required $\mathsf{indeg}(D_n) \in O\left(\log^2 n \mathsf{polylog}(\log n)\right)$ and may be of independent interest (e.g., our construction immediately yields a more efficient construction of proofs of sequential work [MMV13]). Our construction of D_n^ϵ is (essentially) the same as Erdos et al. [EGS75] albeit with much tighter analysis. By overlaying an extremely depth-robust DAG D_n^ϵ on top of the sources of PTC_n, the construction of Paul et al. [PTC76], we can ensure that it takes $\Omega(n/\log n)$ steps to pebble $\Omega(n/\log n)$ sources of G_n. However, the resulting graph would have $\mathsf{indeg}(G_n) \in O(\log n)$ and would have sustained space $\Omega(n/\log n)$ for at most $O(n/\log n)$ steps. By contrast, we want a n-node DAG G with $\mathsf{indeg}(G) = 2$ which requires space $\Omega(n/\log n)$ for at least $\Omega(n)$ steps[2].

Our final tool is to apply the indegree reduction lemma of Alwen et al. [ABP17] to $\{D_t^\epsilon\}_{t\in\mathbb{N}}$ to obtain a family of DAGs $\{J_t^\epsilon\}_{t\in\mathbb{N}}$ such that J_t^ϵ has $\mathsf{indeg}\left(J_t^\epsilon\right) = 2$ and $2t \cdot \mathsf{indeg}\left(D_t^\epsilon\right) \in O(t\log t)$ nodes. Each node in D_t^ϵ is associated with a path of length $2\cdot\mathsf{indeg}(D_t^\epsilon)$ in J_t^ϵ and each path p in D_t^ϵ corresponds to a path p' of length $|p'| \geq |p|\cdot\mathsf{indeg}(G_t)$ in J_t^ϵ. We can then overlay the DAG J_t^ϵ on top of the sources in PTC_n where $t = \Omega(n/\log n)$ is the number of sources in PTC_n. The final DAG has size $O(n)$ and we can then show that any legal parallel pebbling requires $\Omega(n)$ steps with at least $\Omega(n/\log n)$ pebbles on the DAG.

2 Preliminaries

In this section we introduce common notation, definitions and results from other work which we will be using. In particular the following borrows heavily from [ABP17, AT17].

2.1 Notation

We start with some common notation. Let $\mathbb{N} = \{0,1,2,\ldots\}$, $\mathbb{N}^+ = \{1,2,\ldots\}$, and $\mathbb{N}_{\geq c} = \{c, c+1, c+2, \ldots\}$ for $c \in \mathbb{N}$. Further, we write $[c] := \{1,2,\ldots,c\}$ and $[b,c] = \{b, b+1, \ldots, c\}$ where $c \geq b \in \mathbb{N}$. We denote the cardinality of a set B by $|B|$.

[2] We typically want a DAG G with $\mathsf{indeg}(G) = 2$ because the compression function H which is used to label the graph typically maps $2w$ bit inputs to w bit outputs. In this case the labeling function would only be valid for graphs with maximum indegree two. If we used tricks such as Merkle-Damgard to build a new compression function G mapping δw bit inputs to w bit outputs then each pebbling step actually corresponds to $(\delta-1)$ calls to the compression function H which means that each black pebbling step actually takes time $(\delta-1)$ on a sequential computer with a single-core. As a consequence, by considering graphs of degree δ, we pay an additional factor $(\delta-1)$ in the gap between the naive and adversarial evaluation of the MHF.

2.2 Graphs

The central object of interest in this work are directed acyclic graphs (DAGs). A DAG $G = (V, E)$ has *size* $n = |V|$. The indegree of node $v \in V$ is $\delta = \mathsf{indeg}(v)$ if there exist δ incoming edges $\delta = |(V \times \{v\}) \cap E|$. More generally, we say that G has indegree $\delta = \mathsf{indeg}(G)$ if the maximum indegree of any node of G is δ. If $\mathsf{indeg}(v) = 0$ then v is called a *source* node and if v has no outgoing edges it is called a *sink*. We use $\mathsf{parents}_G(v) = \{u \in V : (u, v) \in E\}$ to denote the parents of a node $v \in V$. In general, we use $\mathsf{ancestors}_G(v) := \bigcup_{i \geq 1} \mathsf{parents}^i_G(v)$ to denote the set of all ancestors of v—here, $\mathsf{parents}^2_G(v) := \mathsf{parents}_G(\mathsf{parents}_G(v))$ denotes the grandparents of v and $\mathsf{parents}^{i+1}_G(v) := \mathsf{parents}_G\left(\mathsf{parents}^i_G(v)\right)$. When G is clear from context we will simply write $\mathsf{parents}$ ($\mathsf{ancestors}$). We denote the set of all sinks of G with $\mathsf{sinks}(G) = \{v \in V : \nexists(v, u) \in E\}$—note that $\mathsf{ancestors}(\mathsf{sinks}(G)) = V$. The length of a directed path $p = (v_1, v_2, \ldots, v_z)$ in G is the number of nodes it traverses $\mathsf{length}(p) := z$. The depth $d = \mathsf{depth}(G)$ of DAG G is the length of the longest directed path in G. We often consider the set of all DAGs of fixed size n $\mathbb{G}_n := \{G = (V, E) : |V| = n\}$ and the subset of those DAGs at most some fixed indegree $\mathbb{G}_{n,\delta} := \{G \in \mathbb{G}_n : \mathsf{indeg}(G) \leq \delta\}$. Finally, we denote the graph obtained from $G = (V, E)$ by removing nodes $S \subseteq V$ (and incident edges) by $G - S$ and we denote by $G[S] = G - (V \setminus S)$ the graph obtained by removing nodes $V \setminus S$ (and incident edges).

The following is an important combinatorial property of a DAG for this work.

Definition 1 (Depth-Robustness). *For $n \in \mathbb{N}$ and $e, d \in [n]$ a DAG $G = (V, E)$ is (e, d)-depth-robust if*

$$\forall S \subset V \quad |S| \leq e \Rightarrow \mathsf{depth}(G - S) \geq d.$$

The following lemma due to Alwen et al. [ABP17] will be useful in our analysis. Since our statement of the result is slightly different from [ABP17] we include a proof in Appendix A for completeness.

Lemma 1 *[ABP17, Lemma 1]* (**Indegree-Reduction**). *Let $G = (V = [n], E)$ be an (e, d)-depth robust DAG on n nodes and let $\delta = \mathsf{indeg}(G)$. We can efficiently construct a DAG $G' = (V' = [2n\delta], E')$ on $2n\delta$ nodes with $\mathsf{indeg}(G') = 2$ such that for each path $p = (x_1, ..., x_k)$ in G there exists a corresponding path p' of length $\geq k\delta$ in $G'\left[\bigcup_{i=1}^k [2(x_i - 1)\delta + 1, 2x_i\delta]\right]$ such that $2x_i\delta \in p'$ for each $i \in [k]$. In particular, G' is $(e, d\delta)$-depth robust.*

2.3 Pebbling Models

The main computational models of interest in this work are the parallel (and sequential) pebbling games played over a directed acyclic graph. Below we define these models and associated notation and complexity measures. Much of the notation is taken from [AS15, ABP17].

Definition 2 (Parallel/Sequential Graph Pebbling). *Let $G = (V, E)$ be a DAG and let $T \subseteq V$ be a target set of nodes to be pebbled. A* pebbling configuration *(of G) is a subset $P_i \subseteq V$. A legal* parallel *pebbling of T is a sequence $P = (P_0, \ldots, P_t)$ of* pebbling configurations *of G where $P_0 = \emptyset$ and which satisfies conditions 1 & 2 below. A* sequential *pebbling additionally must satisfy condition 3.*

1. *At some step every target node is pebbled (though not necessarily simultaneously).*
$$\forall x \in T\ \exists z \le t \quad : \quad x \in P_z .$$
2. *A pebble can be added only if all its parents were pebbled at the end of the previous step.*
$$\forall i \in [t] \quad : \quad x \in (P_i \setminus P_{i-1}) \ \Rightarrow\ \mathsf{parents}(x) \subseteq P_{i-1} .$$
3. *At most one pebble is placed per step.*
$$\forall i \in [t] \quad : \quad |P_i \setminus P_{i-1}| \le 1 .$$

We denote with $\mathcal{P}_{G,T}$ and $\mathcal{P}^{\parallel}_{G,T}$ the set of all legal sequential and parallel pebblings of G with target set T, respectively. Note that $\mathcal{P}_{G,T} \subseteq \mathcal{P}^{\parallel}_{G,T}$. We will mostly be interested in the case where $T = \mathsf{sinks}(G)$ in which case we write $\mathcal{P}_G$ and $\mathcal{P}^{\parallel}_G$.

Definition 3 (Pebbling Complexity). *The standard notions of* time, space, space-time *and* cumulative (pebbling) complexity *(cc) of a pebbling $P = \{P_0, \ldots, P_t\} \in \mathcal{P}^{\parallel}_G$ are defined to be:*

$$\Pi_t(P) = t \quad \Pi_s(P) = \max_{i \in [t]} |P_i| \quad \Pi_{st}(P) = \Pi_t(P) \cdot \Pi_s(P) \quad \Pi_{cc}(P) = \sum_{i \in [t]} |P_i| .$$

For $\alpha \in \{s, t, st, cc\}$ and a target set $T \subseteq V$, the sequential and parallel pebbling complexities of G are defined as

$$\Pi_\alpha(G, T) = \min_{P \in \mathcal{P}_{G,T}} \Pi_\alpha(P) \qquad \text{and} \qquad \Pi^{\parallel}_\alpha(G, T) = \min_{P \in \mathcal{P}^{\parallel}_{G,T}} \Pi_\alpha(P) .$$

When $T = \mathsf{sinks}(G)$ we simplify notation and write $\Pi_\alpha(G)$ and $\Pi^{\parallel}_\alpha(G)$.

The following defines a sequential pebbling obtained naturally from a parallel one by adding each new pebble on at a time.

Definition 4. *Given a DAG G and $P = (P_0, \ldots, P_t) \in \mathcal{P}^{\parallel}_G$ the* sequential transform $\mathsf{seq}(P) = P' \in \Pi_G$ *is defined as follows: Let difference $D_j = P_i \setminus P_{i-1}$ and let $a_i = |P_i \setminus P_{i-1}|$ be the number of new pebbles placed on G_n at time i. Finally, let $A_j = \sum_{i=1}^{j} a_i$ ($A_0 = 0$) and let $D_j[k]$ denote the kth element of D_j (according to some fixed ordering of the nodes). We can construct $P' = (P'_1, \ldots, P'_{A_t}) \in \mathcal{P}(G_n)$ as follows: (1) $P'_{A_i} = P_i$ for all $i \in [0, t]$, and (2) for $k \in [1, a_{i+1}]$ let $P'_{A_i+k} = P'_{A_i+k-1} \cup D_j[k]$.*

If easily follows from the definition that the parallel and sequential space complexities differ by at most a multiplicative factor of 2.

Lemma 2. *For any DAG G and $P \in \mathcal{P}^{\parallel}_G$ it holds that $\mathsf{seq}(P) \in \mathcal{P}_G$ and $\Pi_s(\mathsf{seq}(P)) \leq 2 * \Pi^{\parallel}_s(P)$. In particular $\Pi_s(G)/2 \leq \Pi^{\parallel}_s(G)$.*

Proof. Let $P \in \mathcal{P}^{\parallel}_G$ and $P' = \mathsf{seq}(P)$. Suppose P' is not a legal pebbling because $v \in V$ was illegally pebbled in P'_{A_i+k}. If $k = 0$ then $\mathsf{parents}_G(v) \not\subseteq P'_{A_{i-1}+a_i-1}$ which implies that $\mathsf{parents}_G(v) \not\subseteq P_{i-1}$ since $P_{i-1} \subseteq P'_{A_{i-1}+a_i-1}$. Moreover $v \in P_i$ so this would mean that also P illegally pebbles v at time i. If instead, $k > 1$ then $v \in P_{i+1}$ but since $\mathsf{parents}_G(v) \not\subseteq P'_{A_i+k-1}$ it must be that $\mathsf{parents}_G(v) \not\subseteq P_i$ so P must have pebbled v illegally at time $i+1$. Either way we reach a contradiction so P' must be a legal pebbling of G. To see that P' is complete note that $P_0 = P'_{A_0}$. Moreover for any sink $u \in V$ of G there exists time $i \in [0, t]$ with $u \in P_i$ and so $u \in P'_{A_i}$. Together this implies $P' \in \mathcal{P}_G$.

Finally, it follows by inspection that for all $i \geq 0$ we have $|P'_{A_i}| = |P_i|$ and for all $0 < k < a_i$ we have $|P'_{A_i+k}| \leq |P_i| + |P_{i+1}|$ which implies that $\Pi_s(P') \leq 2 * \Pi^{\parallel}_s(P)$.

New to this work is the following notion of sustained-space complexity.

Definition 5 (Sustained Space Complexity). *For $s \in \mathbb{N}$ the* s-sustained-space *(s-ss) complexity of a pebbling $P = \{P_0, \ldots, P_t\} \in \mathcal{P}^{\parallel}_G$ is:*

$$\Pi_{ss}(P, s) = |\{i \in [t] : |P_i| \geq s\}|.$$

More generally, the sequential and parallel s-sustained space complexities of G are defined as

$$\Pi_{ss}(G, T, s) = \min_{P \in \mathcal{P}_{G,T}} \Pi_{ss}(P, s) \qquad \text{and} \qquad \Pi^{\parallel}_{ss}(G, T, s) = \min_{P \in \mathcal{P}^{\parallel}_{G,T}} \Pi_{ss}(P, s) \ .$$

As before, when $T = \mathsf{sinks}(G)$ we simplify notation and write $\Pi_{ss}(G, s)$ and $\Pi^{\parallel}_{ss}(G, s)$.

Remark 1 (On Amortization). An astute reader may observe that $\Pi^{\parallel}_{ss}$ is not amortizable. In particular, if we let $G^{\bigotimes m}$ denotes the graph which consists of m independent copies of G then we may have $\Pi^{\parallel}_{ss}\left(G^{\bigotimes m}, s\right) \ll m\Pi^{\parallel}_{ss}(G, s)$. However, we observe that the issue can be easily corrected by defining the *amortized s-sustained-space* complexity of a pebbling $P = \{P_0, \ldots, P_t\} \in \mathcal{P}^{\parallel}_G$:

$$\Pi_{am,ss}(P, s) = \sum_{i=1}^{t} \left\lfloor \frac{|P_i|}{s} \right\rfloor .$$

In this case we have $\Pi^{\parallel}_{am,ss}\left(G^{\bigotimes m}, s\right) = m\Pi^{\parallel}_{am,ss}(G, s)$ where $\Pi^{\parallel}_{am,ss}(G, s) \doteq \min_{P \in \mathcal{P}^{\parallel}_{G,\mathsf{sinks}(G)}} \Pi_{am,ss}(P, s)$. We remark that a lower bound on *s-sustained-space* complexity is a strictly stronger guarantee than an equivalent lower bound for *amortized s-sustained-space* since $\Pi^{\parallel}_{ss}(G, s) \leq \Pi^{\parallel}_{am,ss}(G, s)$. In particular, all of our lower bounds for $\Pi^{\parallel}_{ss}$ also hold with respect to $\Pi^{\parallel}_{am,ss}$.

The following shows that the indegree of any graph can be reduced down to 2 without loosing too much in the parallel sustained space complexity. The technique is similar the indegree reduction for cumulative complexity in [AS15]. The proof is in Appendix A. While we include the lemma for completeness we stress that, for our specific constructions, we will use more direct approach to lower bound $\Pi_{ss}^{\parallel}$ to avoid the δ factor reduction in space.

Lemma 3 (Indegree Reduction for Parallel Sustained Space).

$$\forall G \in \mathbb{G}_{n,\delta},\ \ \exists H \in \mathbb{G}_{n',2} \text{ such that } \forall s \geq 0\ \ \Pi_{ss}^{\parallel}(H, s/(\delta-1)) = \Pi_{ss}^{\parallel}(G, s) \text{ where } n' \in [n, \delta n].$$

3 A Graph with Optimal Sustained Space Complexity

In this section we construct and analyse a graph with very high sustained space complexity by modifying the graph of [PTC76] using the graph of [EGS75]. Theorem 1, our main theorem, states that there is a family of constant indegree DAGs $\{G_n\}_{n=1}^{\infty}$ with maximum possible sustained space $\Pi_{ss}(G_n, \Omega(n/\log n)) = \Omega(n)$.

Theorem 1. *For some constants $c_4, c_5 > 0$ there is a family of DAGs $\{G_n\}_{n=1}^{\infty}$ with $\mathsf{indeg}(G_n) = 2$, $O(n)$ nodes and $\Pi_{ss}^{\parallel}(G_n, c_4 n/\log n) \geq c_5 n$.*

Remark 2. We observe that Theorem 1 is essentially optimal in an asymptotic sense. Hopcroft et al. [HPV77] showed that any DAG G_n with $\mathsf{indeg}(G_n) \in O(1)$ can be pebbled with space at most $\Pi_s^{\parallel}(G_n) \in O(n/\log n)$. Thus, $\Pi_{ss}(G_n, s_n = \omega(n/\log n)) = 0$ for any DAG G_n with $\mathsf{indeg}(G_n) \in O(1)$ since $s_n > \Pi_s(G_n)$.[3]

We now overview the key technical ingredients in the proof of Theorem 1.

Technical Ingredient 1: High Space Complexity DAGs. The first key building blocks is a construction of Paul et al. [PTC76] of a family of n node DAGs $\{\mathsf{PTC}_n\}_{n=1}^{\infty}$ with space complexity $\Pi_s(\mathsf{PTC}_n) \in \Omega(n/\log n)$ and $\mathsf{indeg}(\mathsf{PTC}_n) = 2$. Lemma 2 implies that $\Pi_s^{\parallel}(\mathsf{PTC}_n) \in \Omega(n/\log n)$ since $\Pi_s(\mathsf{PTC}_n)/2 \leq \Pi_s^{\parallel}(\mathsf{PTC}_n)$. However, we stress that this does not imply that the sustained space complexity of PTC_n is large. In fact, by inspection one can easily verify that $\mathsf{depth}(\mathsf{PTC}_n) \in O(n/\log n)$ so we have $\Pi_{ss}(\mathsf{PTC}_n, s) \in O(n/\log n)$ for any space parameter $s > 0$. Nevertheless, one of the core lemmas from [PTC76] will be very useful in our proofs. In particular, PTC_n contains

[3] Furthermore, even if we restrict our attention to pebblings which finish in time $O(n)$ we still have $\Pi_{ss}(G_n, f(n)) \leq g(n)$ whenever $f(n)g(n) \in \omega\left(\frac{n^2 \log\log n}{\log n}\right)$ and $\mathsf{indeg}(G_n) \in O(1)$. In particular, Alwen and Blocki [AB16] showed that for any G_n with $\mathsf{indeg}(G_n) \in O(1)$ then there is a pebbling $P = (P_0, \ldots, P_n) \in \Pi_{G_n}^{\parallel}$ with $\Pi_{cc}^{\parallel}(P) \in O\left(\frac{n^2 \log\log n}{\log n}\right)$. By contrast, the generic pebbling [HPV77] of any DAG with $\mathsf{indeg} \in O(1)$ in space $O(n/\log n)$ can take exponentially long.

$O(n/\log n)$ source nodes (as illustrated in Fig. 1a) and [PTC76] proved that for any sequential pebbling $P = (P_0, \ldots, P_t) \in \Pi_{\mathsf{PTC}_n}$ we can find an interval $[i,j] \subseteq [t]$ during which $\Omega(n/\log n)$ sources are (re)pebbled and at least $\Omega(n/\log n)$ pebbles are always on the graph.

As Theorem 2 states that the same result holds for all parallel pebblings $P \in \Pi^{\parallel}_{\mathsf{PTC}_n}$. Since Paul et al. [PTC76] technically only considered sequential black pebblings we include the straightforward proof of Theorem 2 in the full version of this paper for completeness [ABP18]. Briefly, to prove Theorem 2 we simply consider the sequential transform $\mathsf{seq}(P) = (Q_0, \ldots, Q_{t'}) \in \Pi_{\mathsf{PTC}_n}$ of the parallel pebbling P. Since $\mathsf{seq}(P)$ is sequential we can find an interval $[i', j'] \subseteq [t']$ during which $\Omega(n/\log n)$ sources are (re)pebbled and at least $\Omega(n/\log n)$ pebbles are always on the graph G_n. We can then translate $[i', j']$ to a corresponding interval $[i,j] \subseteq [t]$ during which the same properties hold for P.

Theorem 2. *There is a family of DAGs* $\{\mathsf{PTC}_n = (V_n = [n], E_n)\}_{n=1}^{\infty}$ *with* $\mathsf{indeg}(\mathsf{PTC}_n) = 2$ *with the property that for some positive constants* $c_1, c_2, c_3 > 0$ *such that for each* $n \geq 1$ *the set* $S = \{v \in [n] : \mathsf{parents}(v) = \emptyset\}$ *of sources of* PTC_n *has size* $|S| \leq c_1 n/\log n$ *and for any legal pebbling* $P = (P_1, \ldots, P_t) \in \mathcal{P}^{\parallel}_{\mathsf{PTC}_n}$ *there is an interval* $[i,j] \subseteq [t]$ *such that (1)* $\left|S \cap \bigcup_{k=i}^{j} P_k \setminus P_{i-1}\right| \geq c_2 n/\log n$ *i.e., at least* $c_2 n/\log n$ *nodes in* S *are (re)pebbled during this interval, and (2)* $\forall k \in [i,j], |P_k| \geq c_3 n/\log n$ *i.e., at least* $c_3 n/\log n$ *pebbles are always on the graph.*

One of the key remaining challenges to establishing high sustained space complexity is that the interval $[i,j]$ we obtain from Theorem 2 might be very short for parallel black pebblings. For sequential pebblings it would take $\Omega(n/\log n)$ steps to (re)pebble $\Omega(n/\log n)$ source nodes since we can add at most one new pebble in each round. However, for parallel pebblings we cannot rule out the possibility that all $\Omega(n/\log n)$ sources were pebbled in a single step!

A first attempt at a fix is to modify PTC_n by overlaying a path of length $\Omega(n)$ on top of these $\Omega(n/\log n)$ source nodes to ensure that the length of the interval $j - i + 1$ is sufficiently large. The hope is that it will take now at least $\Omega(n)$ steps to (rep)pebble any subset of $\Omega(n/\log n)$ of the original sources since these nodes will be connected by a path of length $\Omega(n)$. However, we do not know what the pebbling configuration looks like at time $i-1$. In particular, if P_{i-1} contained just $\sqrt{n}$ of the nodes on this path then the it would be possible to (re)pebble all nodes on the path in at most $O(\sqrt{n})$ steps. This motivates our second technical ingredient: extremely depth-robust graphs.

Technical Ingredient 2: Extremely Depth-Robust Graphs. Our second ingredient is a family $\{D_n^{\epsilon}\}_{n=1}^{\infty}$ of highly depth-robust DAGs with n nodes and $\mathsf{indeg}(D_n) \in O(\log n)$. In particular, D_n^{ϵ} is (e,d)-depth robust for *any* $e + d \leq n(1-\epsilon)$. We show how to construct such a family $\{D_n^{\epsilon}\}_{n=1}^{\infty}$ for any constant $\epsilon > 0$ in Sect. 4. Assuming for now that such a family exists we can overlay D_m over the $m = m_n \leq c_1 n/\log n$ sources of PTC_n. Since D_m^{ϵ} is *highly* depth-robust it will take at least $c_2 n/\log n - \epsilon m \geq c_2 n/\log n - \epsilon c_1 n/\log n \in \Omega(n/\log n)$ steps to pebble these $c_2 n/\log n$ sources during the interval $[i, j]$.

Overlaying D_m^ϵ over the $m \in O(n/\log(n))$ sources of PTC_n yields a DAG G with $O(n)$ nodes, $\mathsf{indeg}(G) \in O(\log n)$ and $\Pi_{ss}^{\parallel}(G, c_4 n/\log n) \geq c_5 n/\log n$ for some constants $c_4, c_5 > 0$. While this is progress it is still a weaker result than Theorem 1 which promised a DAG G with $O(n)$ nodes, $\mathsf{indeg}(G) = 2$ and $\Pi_{ss}^{\parallel}(G, c_4 n/\log n) \geq c_5 n$ for some constants $c_4, c_5 > 0$. Thus, we need to introduce a third technical ingredient: indegree reduction.

Technical Ingredient 3: Indegree Reduction. To ensure $\mathsf{indeg}(G_n) = 2$ we instead apply indegree reduction algorithm from Lemma 1 to D_m^ϵ to obtain a graph J_m^ϵ with $2m\delta \in O(n)$ nodes $[2\delta m]$ and $\mathsf{indeg}(J_m^\epsilon) = 2$ before overlaying—here $\delta = \mathsf{indeg}(D_m^\epsilon)$. This process is illustrated in Fig. 1b. We then obtain our final construction G_n, illustrated in Fig. 1, by associating the m sources of PTC_n with the nodes $\{2\delta v \ : v \in [m]\}$ in J_m^ϵ, where $\epsilon > 0$ is fixed to be some suitably small constant.

It is straightforward to show that J_m^ϵ is $(e, \delta d)$-depth robust for any $e + d \leq (1-\epsilon)m$. Thus, it would be tempting that it will take $\Omega(n)$ steps to (re)pebble $c_2 n/\log n$ sources during the interval $[i, j]$ we obtain from Theorem 2. However, we still run into the same problem: In particular, suppose that at some point in time k we can find a set $T \subseteq \{2v\delta : v \in [m]\} \setminus P_k$ with $|T| \geq c_2 n/\log n$ (e.g., a set of sources in PTC_n) such that the longest path running through T in $J_m^\epsilon - P_k$ has length less than $c_5 n$. If the interval $[i, j]$ starts at time $i = k + 1$ then we cannot ensure that it will take time $\geq c_5 n$ to (re)pebble these $c_2 n/\log n$ source nodes.

Claim 1 addresses this challenge directly. If such a problematic time k exists then Claim 1 implies that we must have $\Pi_{ss}^{\parallel}(P, \Omega(n/\log n)) \in \Omega(n)$. At a high level the argument proceeds as follows: suppose that we find such a problem time k along with a set $T \subseteq \{2v\delta : v \in [m]\} \setminus P_k$ with $|T| \geq c_2 n/\log n$ such that $\mathsf{depth}(J_m^\epsilon[T]) \leq c_5 n$. Then for any time $r \in [k - c_5 n, k]$ we know that the length of the longest path running through T in $J_m^\epsilon - P_r$ is at most $\mathsf{depth}(J_m^\epsilon[T] - P_r) \leq c_5 n + (k - r) \leq 2c_5 n$ since the depth can decrease by at most one each round. We can then use the extreme depth-robustness of D_m^ϵ and the construction of J_m^ϵ to argue that $|P_r| = \Omega(n/\log n)$ for each $r \in [k - c_5 n, k]$. Finally, if no such problem time k exists then the interval $[i, j]$ we obtain from Theorem 2 must have length at least $i - j \geq c_5 n$. In either case we have $\Pi_{ss}^{\parallel}(P, \Omega(n/\log n))) \geq \Omega(n)$.

Proof of Theorem 1. We begin with the family of DAGs $\{\mathsf{PTC}_n\}_{n=1}^{\infty}$ from Theorem 2. Fixing $\mathsf{PTC}_n = ([n], E_n)$ we let $S = \{v \in [n] : \mathsf{parents}(v) = \emptyset\} \subseteq V$ denote the sources of this graph and we let $c_1, c_2, c_3 > 0$ be the constants from Theorem 2. Let $\epsilon \leq c_2/(4c_1)$. By Theorem 3 we can find a depth-robust DAG $D_{|S|}^\epsilon$ on $|S|$ nodes which is $(a|S|, b|S|)$-DR for any $a + b \leq 1 - \epsilon$ with indegree $c' \log n \leq \delta = \mathsf{indeg}(D) \leq c'' \log(n)$ for some constants c', c''. We let $J_{|S|}^\epsilon$ denote the indegree reduced version of $D_{|S|}^\epsilon$ from Lemma 1 with $2|S|\delta \in O(n)$ nodes and $\mathsf{indeg} = 2$. To obtain our DAG G_n from J_n^ϵ and PTC_n we associate each of the

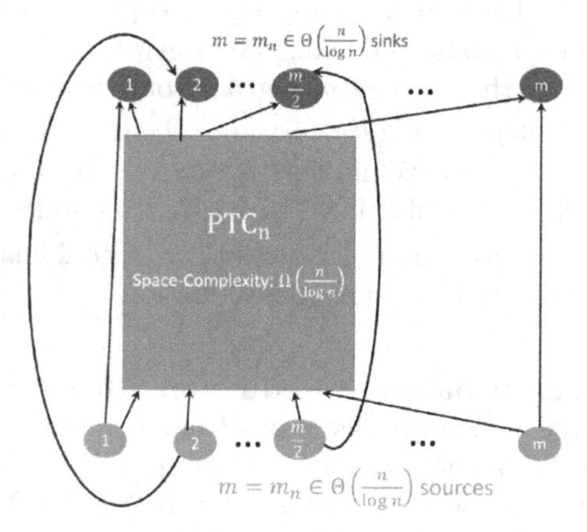

(a) PTC_n: a superconcentrator [PTC76] with $m = \Omega(n/\log n)$ sources and sinks and maximum space complexity $\Pi_s^{\parallel}(\mathsf{PTC}_n) \in \Omega\left(\frac{n}{\log n}\right)$.

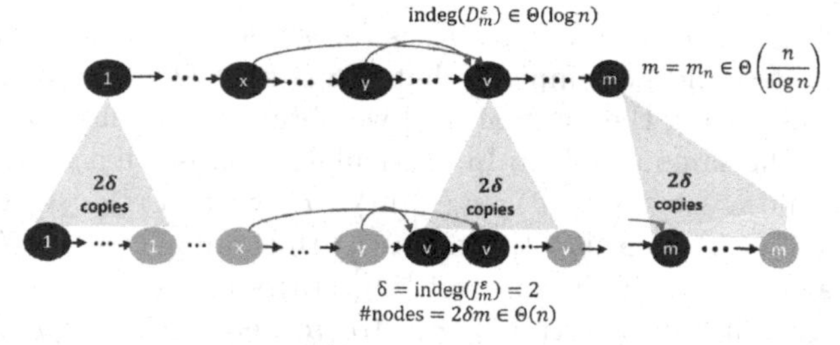

(b) Indegree Recution transforms ϵ-extreme depth robust graph D_m^ϵ with m nodes and $\mathsf{indeg}(D_m^\epsilon) \in O(\log n)$ into indegree reduced graph J_m^ϵ with $2\mathsf{indeg}(D_m^\epsilon) \times m \in O(n)$ nodes and $\mathsf{indeg}(J_m^\epsilon) = 2$.

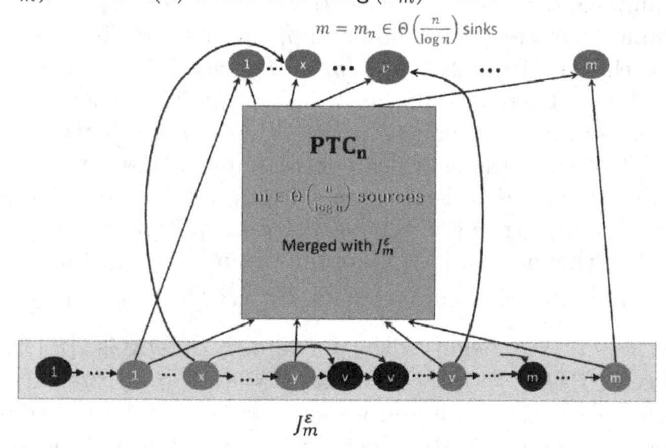

(c) Final Construction G_n. Overlay m nodes J_m^ϵ with m sources in PTC_n.

Fig. 1. Building G_n with $\Pi_{ss}^{\parallel}\left(G_n, \frac{cn}{\log n}\right) \in \Omega(n)$ for some constant $c > 0$.

S nodes $2v\delta$ in J_n^ϵ with one of the nodes in S. We observe that G_n has at most $2|S|\delta + n \in O(n)$ nodes and that $\mathsf{indeg}(G) \le \max\{\mathsf{indeg}(\mathsf{PTC}_n), \mathsf{indeg}(J_n^\epsilon)\} = 2$ since we do not increase the indegree of any node in J_n^ϵ when overlaying and in G_n do not increase the indegree of any nodes other than the sources S from PTC_n (these overlayed nodes have indegree at most 2 in J_n^ϵ).

Let $P = (P_0, \dots, P_t) \in \mathcal{P}_G^{\parallel}$ be given and observe that by restricting $P_i' = P_i \cap V(\mathsf{PTC}_n) \subseteq P_i$ we have a legal pebbling $P' = (P_0', \dots, P_t') \in \mathcal{P}_{\mathsf{PTC}_n}^{\parallel}$ for PTC_n. Thus, by Theorem 2 we can find an interval $[i, j]$ during which at least $c_2 n/\log n$ nodes in S are (re)pebbled and $\forall k \in [i, j]$ we have $|P_k| \ge c_3 n/\log n$. We use $T = S \cap \bigcup_{x=i}^{j} P_x - P_{i-1}$ to denote the source nodes of PTC_n that are (re)pebbled during the interval $[i, j]$. We now set $c_4 = c_2/4$ and $c_5 = c_2 c'/4$ and consider two cases:

Case 1: We have $\mathsf{depth}(\mathsf{ancestors}_{G_n - P_i}(T)) \ge |T|\delta/4$. In other words at time i there is an unpebbled path of length $\ge |T|\delta/4$ to some node in T. In this case, it will take at least $j - i \ge |T|\delta/4$ steps to pebble T so we will have at least $|T|\delta/4 \in \Omega(n)$ steps with at least $c_3 n/\log n$ pebbles. Because $c_5 = c_2 c'/4$ it follows that $|T|\delta/4 \ge c_2 c' n \ge c_5 n$. Finally, since $c_4 \le c_2$ we have $\Pi_{ss}(P, c_4 n/\log n) \ge c_5 n$.

Case 2: We have $\mathsf{depth}(\mathsf{ancestors}_{G_n - P_i}(T)) < |T|\delta/4$. In other words at time i there is no unpebbled path of length $\ge |T|\delta/4$ to any node in T. Now Claim 1 directly implies that $\Pi_{ss}(P, |T| - \epsilon|S| - |T|/2) \ge \delta|T|/4$. This in turn implies that $\Pi_{ss}(P, (c_2/2)n/(\log n) - \epsilon|S|) \ge \delta c_2 n/(2\log n)$. We observe that $\delta c_2 n/(2\log n) \ge c_5 n$ since, we have $c_5 = c_2 c'/4$. We also observe that $(c_2/2)n/\log n - \epsilon|S| \ge (c_2/2 - \epsilon c_1)n/\log n \ge (c_2/2 - c_2/4)n/\log n \ge c_2 n/(4\log n) = c_4 n$ since $|S| \le c_1 n/\log n$, $\epsilon \le c_2/(4c_1)$ and $c_4 = c_2/4$. Thus, in this case we also have $\Pi_{ss}(P, c_4 n/\log n) \ge c_5 n$, which implies that $\Pi_{ss}^{\parallel}(G_n, c_4 n/\log n) \ge c_5 n$. □

Claim 1. *Let D_n^ϵ be an DAG with nodes $V(D_n^\epsilon) = [n]$, indegree $\delta = \mathsf{indeg}(D_n^\epsilon)$ that is (e, d)-depth robust for all $e, d > 0$ such that $e + d \le (1 - \epsilon)n$, let J_n^ϵ be the indegree reduced version of D_n^ϵ from Lemma 1 with 2δ nodes and $\mathsf{indeg}(J_n^\epsilon) = 2$, let $T \subseteq [n]$ and let $P = (P_1, \dots, P_t) \in \mathcal{P}_{J_n^\epsilon, \emptyset}^{\parallel}$ be a (possibly incomplete) pebbling of J_n^ϵ. Suppose that during some round i we have $\mathsf{depth}\left(\mathsf{ancestors}_{J_n^\epsilon - P_i}\left(\bigcup_{v \in T}\{2\delta v\}\right)\right) \le c\delta|T|$ for some constant $0 < c < \frac{1}{2}$. Then $\Pi_{ss}(P, |T| - \epsilon n - 2c|T|) \ge c\delta|T|$.*

Proof of Claim 1. For each time step r we let $H_r = \mathsf{ancestors}_{J_n^\epsilon - P_r}\left(\bigcup_{v \in T}\{2\delta v\}\right)$ and let $k < i$ be the last pebbling step before i during which $\mathsf{depth}(G_k) \ge 2c|T|\delta$. Observe that $k - i \ge \mathsf{depth}(H_k) - \mathsf{depth}(H_i) \ge cn\delta$ since we can decrease the length of any unpebbled path by at most one in each pebbling round. We also observe that $\mathsf{depth}(H_k) = c|T|\delta$ since $\mathsf{depth}(H_k) - 1 \le \mathsf{depth}(H_{k+1}) < 2c|T|\delta$.

Let $r \in [k, i]$ be given then, by definition of k, we have $\mathsf{depth}(H_r) \le 2c|T|\delta$. Let $P_r' = \{v \in V(D_n^\epsilon) : P_r \cap [2\delta(v-1)+1, 2\delta v] \ne \emptyset\}$ be the set of nodes $v \in [n] = V(D_n^\epsilon)$ such that the corresponding path $2\delta(v-1)+1, \dots, 2\delta v$ in J_n^ϵ contains at least one pebble at time r. By depth-robustness of D_n^ϵ we have

$$\mathsf{depth}(D_n^\epsilon[T] - P_r') \ge |T| - |P_r'| - \epsilon n\ . \tag{1}$$

On the other hand, exploiting the properties of the indegree reduction from Lemma 1, we have

$$depth\,(D_n^{\epsilon}[T] - P_r')\,\delta \le \mathsf{depth}\,(H_r) \le 2c|T|\delta\ . \tag{2}$$

Combining Eqs. 1 and 2 we have

$$|T| - |P_r'| - \epsilon n \le \mathsf{depth}\,(D_n^{\epsilon}[T] - P_r') \le 2c|T|\ .$$

It immediately follows that $|P_r| \ge |P_r'| \ge |T| - 2c|T| - \epsilon n$ for each $r \in [k, i]$ and, therefore, $\Pi_{ss}^{\|}(P, |T| - \epsilon n - 2c|T|) \ge c\delta|T|$. □

Remark 3 (On the Explicitness of Our Construction). Our construction of a family of DAGs with high sustained space complexity is explicit in the sense that there is a probabilistic polynomial time algorithm which, except with very small probability, outputs an n node DAG G that has high sustained space complexity. In particular, Theorem 1 relies on an explicit construction of [PTC76], and the extreme depth-robust DAGs from Theorem 3. The construction of [PTC76] in turn uses an object called superconcentrators. Since we have explicit constructions of superconcentrators [GG81] the construction of [PTC76] can be made explicit. While the proof of the existence of a family of extremely depth-robust DAGs is not explicit the proof uses a probabilistic argument and can be adapted to obtain a probabilistic polynomial time which, except with very small probability, outputs an n node DAG G that is extremely depth-robust. In practice, however it is also desirable to ensure that there is a local algorithm which, on input v, computes the set $\mathsf{parents}(v)$ in time $\mathsf{polylog}(n)$. It is an open question whether any DAG G with high sustained space complexity allows for highly efficient computation of the set $\mathsf{parents}(v)$.

4 Better Depth-Robustness

In this section we improve on the original analysis of Erdos et al. [EGS75], who constructed a family of DAGs $\{G_n\}_{n=1}^{\infty}$ with $\mathsf{indeg}(G_n) \in O(\log n)$ such that each DAG G_n is $(e = \Omega(n), d = \Omega(n))$-depth robust. Such a DAG G_n is not sufficient for us since we require that the subgraph $G_n[T]$ is also highly depth robust for any sufficiently large subset $T \subseteq V_n$ of nodes e.g., for any T such that $|T| \ge n/1000$. For any fixed constant $\epsilon > 0$ [MMV13] constructs a family of DAGs $\{G_n^{\epsilon}\}_{n=1}^{\infty}$ which is $(\alpha n, \beta n)$-depth robust for any positive constants α, β such that $\alpha + \beta \le 1 - \epsilon$ but their construction has indegree $O\left(\log^2 n \cdot \mathsf{polylog}\,(\log n)\right)$. By contrast, our results in the previous section assumed the existence of such a family of DAGs with $\mathsf{indeg}\,(G_n^{\epsilon}) \in O(\log n)$.

In fact our family of DAGs is essentially the same as [EGS75] with one minor modification to make the construction for all $n > 0$. Our contribution in this section is an improved analysis which shows that the family of DAGs $\{G_n^{\epsilon}\}_{n=1}^{\infty}$ with indegree $O\,(\log n)$ is $(\alpha n, \beta n)$-depth robust for any positive constants α, β such that $\alpha + \beta \le 1 - \epsilon$.

We remark that if we allow our family of DAGs to have $\mathsf{indeg}(G_n^\epsilon) \in O(\log n \log^* n)$ then we can eliminate the dependence on ϵ entirely. In particular, we can construct a family of DAGs $\{G_n\}_{n=1}^\infty$ with $\mathsf{indeg}(G_n) = O(\log n \log^* n)$ such that for any positive constants such that $\alpha+\beta < 1$ the DAG G_n is $(\alpha n, \beta n)$-depth robust for all suitably large n.

Theorem 3. *Fix $\epsilon > 0$ then there exists a family of DAGs $\{G_n^\epsilon\}_{n=1}^\infty$ with $\mathsf{indeg}(G_n^\epsilon) = O(\log n)$ that is $(\alpha n, \beta n)$-depth robust for any constants α, β such that $\alpha + \beta < 1 - \epsilon$.*

The proof of Theorem 3 relies on Lemmas 4, 5 and 6. We say that G is a δ-local expander if for every node $x \in [n]$ and every $r \leq x, n - x$ and every pair $A \subseteq I_r(x) \doteq \{x - r - 1, \ldots, x\}, B \subseteq I_r^*(x) \doteq \{x + 1, \ldots, x + r\}$ with size $|A|, |B| \geq \delta r$ we have $A \times B \cap E \neq \emptyset$ i.e., there is a directed edge from some node in A to some node in B. Lemma 4 says that for any constant $\delta > 0$ we can construct a family of DAGs $\{\mathsf{LE}_n^\delta\}_{n=1}^\infty$ with $\mathsf{indeg} = O(\log n)$ such that each LE_n^δ is a δ-local expander. Lemma 4 essentially restates [EGS75, Claim 1] except that we require that LE_n is a δ-local expander for *all* $n > 0$ instead of for n sufficiently large. Since we require a (very) minor modification to achieve δ-local expansion for *all* $n > 0$ we include the proof of Lemma 4 in the full version [ABP18] for completeness.

Lemma 4 [EGS75]. *Let $\delta > 0$ be a fixed constant then there is a family of DAGs $\{\mathsf{LE}_n^\delta\}_{n=1}^\infty$ with $\mathsf{indeg} \in O(\log n)$ such that each LE_n^δ is a δ-local expander.*

While Lemma 4 essentially restates [EGS75, Claim 1], Lemmas 5 and 6 improve upon the analysis of [EGS75]. We say that a node $x \in [n]$ is γ-good under a subset $S \subseteq [n]$ if for all $r > 0$ we have $|I_r(x) \backslash S| \geq \gamma |I_r(x)|$ and $|I_r^*(x) \backslash S| \geq \gamma |I_r^*(x)|$. Lemma 5 is similar to [EGS75, Claim 3], which also states that all γ-good nodes are connected by a directed path in $\mathsf{LE}_n - S$. However, we stress that the argument of [EGS75, Claim 3] requires that $\gamma \geq 0.5$ while Lemma 5 has no such restriction. This is crucial to prove Theorem 3 where we will select γ to be very small.

Lemma 5. *Let $G = (V = [n], E)$ be a δ-local expander and let $x < y \in [n]$ both be γ-good under $S \subseteq [n]$ then if $\delta < \min\{\gamma/2, 1/4\}$ then there is a directed path from node x to node y in $G - S$.*

Lemma 6 shows that *almost all* of the remaining nodes in $\mathsf{LE}_n^\delta - S$ will be γ-good. It immediately follows that $\mathsf{LE}_n - S$ contains a directed path running through *almost all* of the nodes $[n] \setminus S$. While Lemma 6 may appear similar to [EGS75, Claim 2] at first glance, we again stress one crucial difference. The proof of [EGS75, Claim 2] is only sufficient to show that at least $n - 2|S|/(1 - \gamma) \geq n - 2|S|$ nodes are γ-good. At best this would allow us to conclude that LE_n^δ is $(e, n - 2e)$-depth robust. Together Lemmas 5 and 6 imply that if LE_n^δ is a δ-local expander ($\delta < \min\{\gamma/2, 1/4\}$) then LE_n^δ is $\left(e, n - e\frac{1+\gamma}{1-\gamma}\right)$-depth robust.

Lemma 6. *For any DAG $G = ([n], E)$ and any subset $S \subseteq [n]$ of nodes at least $n - |S|\frac{1+\gamma}{1-\gamma}$ of the remaining nodes in G are γ-good with respect to S.*

Proof of Theorem 3. By Lemma 4, for any $\delta > 0$, there is a family of DAGs $\{\mathsf{LE}_n^\delta\}_{n=1}^\infty$ with $\mathsf{indeg}\left(\mathsf{LE}_n^\delta\right) \in O(\log n)$ such that for each $n \geq 1$ the DAG LE_n^δ is a δ-local expander. Given $\epsilon \in (0, 1]$ we will set $G_n^\epsilon = \mathsf{LE}_n^\delta$ with $\delta = \epsilon/10 < 1/4$ so that G_n^ϵ is a $(\epsilon/10)$-local expander. We also set $\gamma = \epsilon/4 > 2\delta$. Let $S \subseteq V_n$ of size $|S| \leq e$ be given. Then by Lemma 6 at least $n - e\frac{1+\gamma}{1-\gamma}$ of the nodes are γ-good and by Lemma 5 there is a path connecting all γ-good nodes in $G_n^\epsilon - S$. Thus, the DAG G_n^ϵ is $\left(e, n - e\frac{1+\gamma}{1-\gamma}\right)$-depth robust for any $e \leq n$. In particular, if $\alpha = e/n$ and $\beta = 1 - \alpha\frac{1+\gamma}{1-\gamma}$ then the graph is $(\alpha n, \beta n)$-depth robust. Finally we verify that

$$n - \alpha n - \beta n = -e + e\alpha\frac{1+\gamma}{1-\gamma} = e\frac{2\gamma}{1-\gamma} \leq n\frac{\epsilon}{2-\epsilon/2} \leq \epsilon n\ . \qquad \square$$

The proof of Lemma 5 follows by induction on the distance $|y - x|$ between γ-good nodes x and y. Our proof extends a similar argument from [EGS75] with one important difference. [EGS75] argued inductively that for each good node x and for each $r > 0$ over half of the nodes in $I_r^*(x)$ are reachable from x and that x can be reached from over half of the nodes in $I_r(x)$—this implies that y is reachable from x since there is at least one node $z \in I_{|y-x|}^*(x) = I_{|y-x|}(y)$ such that z can be reached from x and y can be reached from z in $G - S$. Unfortunately, this argument inherently requires that $\gamma \geq 0.5$ since otherwise we may have at least $|I_r^*(x) \cap S| \geq (1-\gamma)r$ nodes in the interval $I_r(x)$ that are not reachable from x. To get around this limitation we instead show, see Claim 2, that more than half of the nodes in the set $I_r^*(x) \setminus S$ are reachable from x and that more than half of the nodes in the set $I_r(x) \setminus S$ are reachable from x—this still suffices to show that x and y are connected since by the pigeonhole principle there is at least one node $z \in I_{|y-x|}^*(x) \setminus S = I_{|y-x|}(y) \setminus S$ such that z can be reached from x and y can be reached from z in $G - S$.

Claim 2. *Let $G = (V = [n], E)$ be a δ-local expander, let $x \in [n]$ be a γ-good node under $S \subseteq [n]$ and let $r > 0$ be given. If $\delta < \gamma/2$ then all but $2\delta r$ of the nodes in $I_r^*(x)\backslash S$ are reachable from x in $G - S$. Similarly, x can be reached from all but $2\delta r$ of the nodes in $I_r(x)\backslash S$. In particular, if $\delta < 1/4$ then more than half of the nodes in $I_r^*(x)\backslash S$ (resp. in $I_r(x)\backslash S$) are reachable from x (resp. x is reachable from) in $G - S$.*

Proof of Claim 2. We prove by induction that (1) if $r = 2^k\delta^{-1}$ for some integer k then all but δr of the nodes in $I_r^*(x)\backslash S$ are reachable from x and, (2) if $2^{k-1} < r < 2^k\delta^{-1}$ then all but $2\delta r$ of the nodes in $I_r^*(x)\backslash S$ are reachable from x. For the base cases we observe that if $r \leq \delta^{-1}$ then, by definition of a δ-local expander, x is directly connected to all nodes in $I_r^*(x)$ so all nodes in $I_r(x)\backslash S$ are reachable.

Now suppose that Claims (1) and (2) holds for each $r' \leq r = 2^k\delta^{-1}$. Then we show that the claim holds for each $r < r' \leq 2r = 2^{k+1}\delta^{-1}$. In particular, let $A \subseteq I_r^*(x)\backslash S$ denote the set of nodes in $I_r^*(x)\backslash S$ that are reachable from x via a directed path in $G - S$ and let $B \subseteq I_{r'-r}^*(x+r)\backslash S$ be the set of all nodes in $I_{r'-r}^*(x+r)\backslash S$ that are *not reachable* from x in $G - S$. Clearly, there are no directed edges from A to B in G and by induction we have $|A| \geq |I_r^*(x)\backslash S| - \delta r \geq r(\gamma - \delta) > \delta r$. Thus, by δ-local expansion $|B| \leq r\delta$. Since, $|I_r^*(x)\backslash(S \cup A)| \leq \delta r$ at most $|I_{r'}^*(x)\backslash(S \cup A)| \leq |B| + \delta r \leq 2\delta r \leq 2\delta r'$ nodes in $I_{2r}^*(x)\backslash S$ are not reachable from x in $G - S$. Since, $r' > r$ the number of unreachable nodes is at most $2\delta r \leq 2\delta r'$, and if $r' = 2r$ then the number of unreachable nodes is at most $2\delta r = \delta r'$.

A similar argument shows that x can be reached from all but $2\delta r$ of the nodes in $I_r(x)\backslash S$ in the graph $G - S$. □

Proof of Lemma 5. By Claim 2 for each r we can reach $|I_r^*(x)\backslash S| - \delta r = |I_r^*(x)\backslash S| \left(1 - \delta \frac{|I_r^*(x)|}{|I_r^*(x)\backslash S|}\right) \geq |I_r^*(x)\backslash S| \left(1 - \frac{\delta}{\gamma}\right) > \frac{1}{2} |I_r^*(x)\backslash S|$ of the nodes in $I_r^*(x)\backslash S$ from the node x in $G - S$. Similarly, we can reach y from more than $\frac{1}{2} |I_r(x)\backslash S|$ of the nodes in $I_r(y)\backslash S$. Thus, by the pigeonhole principle we can find at least one node $z \in I_{|y-x|}^*(x) \setminus S = I_{|y-x|}(y) \setminus S$ such that z can be reached from x and y can be reached from z in $G - S$. □

Lemma 6 shows that almost all of the nodes in $G - S$ are γ-good. The proof is again similar in spirit to an argument of [EGS75]. In particular, [EGS75] constructed a superset T of the set of all γ-bad nodes and then bound the size of this superset T. However, they only prove that $BAD \subset T \subseteq F \cup B$ where $|F|, |B| \leq |S|/(1 - \gamma)$. Thus, we have $|BAD| \leq |T| \leq 2|S|/(1 - \gamma)$. Unfortunately, this bound is not sufficient for our purposes. In particular, if $|S| = n/2$ then this bound does not rule out the possibility that $|BAD| = n$ so that none of the remaining nodes are good. Instead of bounding the size of the superset T directly we instead bound the size of the set $T \setminus S$ observing that $|BAD| \leq |T| \leq |S| + |T \setminus S|$. In particular, we can show that $|T \setminus S| \leq \frac{2\gamma|S|}{1-\gamma}$. We then have $|GOOD| \geq n - |T| = n - |S| - |T\backslash S| \geq n - |S| - \frac{2\gamma|S|}{1-\gamma}$.

Proof of Lemma 6. We say that a γ-bad node x has a forward (resp. backwards) witness r if $|I_r^*(x)\backslash S| > \gamma r$. Let x_1^*, r_1^* be the lexicographically first γ-bad node with a forward witness. Once $x_1^*, r_1^*, \ldots, x_k^*, r_k^*$ have been define let x_{k+1}^* be the lexicographically least γ-bad node such that $x_{k+1}^* > x_k^* + r_k^*$ and x_{k+1}^* has a forward witness r_{k+1}^* (if such a node exists). Let $x_1^*, r_1^*, \ldots, x_k^*, r_{k*}^*$ denote the complete sequence, and similarly define a maximal sequence $x_1, r_1, \ldots, x_k, r_k$ of γ-bad nodes with backwards witnesses such that $x_i - r_i > x_{i+1}$ for each i.

Let

$$F = \bigcup_{i=1}^{k^*} I_{r_i^*}^* (x_i^*), \quad \text{and} \quad B = \bigcup_{i=1}^{k} I_{r_i} (x_i)$$

Note that for each $i \leq k^*$ we have $\left| I_{r_i^*}^* (x_i^*) \setminus S \right| \leq \gamma r$. Similarly, for each $i \leq k$ we have $|I_{r_i} (x_i) \setminus S| \leq \gamma r$. Because the sets $I_{r_i^*}^* (x_i^*)$ are all disjoint (by construction) we have

$$|F\backslash S| \leq \gamma \sum_{i=1}^{k^*} r_i^* = \gamma |F| .$$

Similarly, $|B\backslash S| \le \gamma|B|$. We also note that at least $(1-\gamma)|F|$ of the nodes in $|F|$ are in $|S|$. Thus, $|F|(1-\gamma) \le |S|$ and similarly $|B|(1-\gamma) \le |S|$. We conclude that $|F\backslash S| \le \frac{\gamma|S|}{1-\gamma}$ and that $|B\backslash S| \le \frac{\gamma|S|}{1-\gamma}$.

To finish the proof let $T = F \cup B = S \cup (F\backslash S) \cup (B\backslash S)$. Clearly, T is a superset of all γ-bad nodes. Thus, at least $n - |T| \ge n - |S|\left(1 + \frac{2\gamma}{1-\gamma}\right) = n - |S|\frac{1+\gamma}{1-\gamma}$ nodes are good.

We also remark that Lemma 4 can be modified to yield a family of DAGs $\{\mathsf{LE}_n\}_{n=1}^{\infty}$ with $\mathsf{indeg}(\mathsf{LE}_n) \in O(\log n \log^* n)$ such that each LE_n is a δ_n local expander for some sequence $\{\delta_n\}_{n=1}^{\infty}$ converging to 0. We can define a sequence $\{\gamma_n\}_{n=1}^{\infty}$ such that $\frac{1+\gamma_n}{1-\gamma_n}$ converges to 1 and $2\gamma_n > \delta_n$ for each n. Lemmas 4 and 6 then imply that each G_n is $\left(e, n - e\frac{1+\gamma_n}{1-\gamma_n}\right)$-depth robust for any $e \le n$.

4.1 Additional Applications of Extremely Depth Robust Graphs

We now discuss additional applications of Theorem 3.

Application 1: Improved Proofs of Sequential Work. As we previously noted Mahmoody et al. [MMV13] used extremely depth-robust graphs to construct efficient Proofs-of-Sequential Work. In a proof of sequential work a prover wants to convince a verifier that he computed a hash chain of length n involving the input value x without requiring the verifier to recompute the entire hash chain. Mahmoody et al. [MMV13] accomplish this by requiring the prover computes labels $L_1, \ldots, L_n$ by "pebbling" an extremely depth-robust DAG G_n e.g., $L_{i+1} = H\left(x \| L_{v_1} \| \ldots \| L_{v_\delta}\right)$ where $\{v_1, \ldots, v_\delta\} = \mathsf{parents}(i+1)$ and H is a random oracle. The prover then commits to the labels $L_1, \ldots, L_n$ using a Merkle Tree and sends the root of the tree to the verifier who can audit randomly chosen labels e.g., the verifier audits label L_{i+1} by asking the prover to reveal the values L_{i+1} and L_v for each $v \in \mathsf{parents}(i+1)$. If the DAG is extremely-depth robust then either a (possibly cheating) prover make at least $(1-\epsilon)n$ sequential queries to the random oracle, or the prover will fail to convince the verifier with high probability [MMV13].

We note that the parameter $\delta = \mathsf{indeg}(G_n)$ is crucial to the efficiency of the Proofs-Of-Sequential Work protocol since each audit challenge requires the prover to reveal $\delta + 1$ labels in the Merkle tree. The DAG G_n from [MMV13] has $\mathsf{indeg}(G_n) \in O\left(\log^2 n \cdot \mathsf{polylog}\left(\log n\right)\right)$ while our DAG G_n from Theorem 3 has maximum indegree $\mathsf{indeg}(G_n) \in O\left(\log n\right)$. Thus, we can improve the communication complexity of their Proofs-Of-Sequential Work protocol by a factor of $\Omega(\log n \cdot \mathsf{polylog} \log n)$. However, Cohen and Pietrzak [CP18] found an alternate construction of a Proofs-Of-Sequential Work protocol that does not involve depth-robust graphs and which would almost certainly be more efficient than either of the above constructions in practice.

Application 2: Graphs with Maximum Cumulative Cost. We now show that our family of extreme depth-robust DAGs has the highest possible cumulative pebbling cost even in terms of the *constant* factors. In particular, for any constant $\eta > 0$ and $\epsilon < \eta^2/100$ the family $\{G_n^\epsilon\}_{n=1}^\infty$ of DAGs from Theorem 3 has $\Pi_{cc}^{\parallel}(G_n^\epsilon) \geq \frac{n^2(1-\eta)}{2}$ and $\mathsf{indeg}(G_n) \in O(\log n)$. By comparison, $\Pi_{cc}^{\parallel}(G_n) \leq \frac{n^2+n}{2}$ for *any* DAG $G \in \mathbb{G}_n$—even if G is the complete DAG.

Previously, Alwen et al. [ABP17] showed that any (e, d)-depth robust DAG G has $\Pi_{cc}^{\parallel}(G) > ed$ which implies that there is a family of DAG G_n with $\Pi_{cc}^{\parallel}(G_n) \in \Omega\left(n^2\right)$ [EGS75]. We stress that we need new techniques to prove Theorem 4. Even if a DAG $G \in \mathbb{G}_n$ were $(e, n-e)$-depth robust for every $e \geq 0$ (the only DAG actually satisfying this property is the compete DAG K_n) [ABP17] only implies that $\Pi_{cc}^{\parallel}(G) \geq \max_{e \geq 0} e(n-e) = n^2/4$. Our basic insight is that at time t_i, the first time a pebble is placed on node i in G_n^ϵ, the node $i + \gamma i$ is γ-good and is therefore reachable via an undirected path from all of the other γ-good nodes in $[i]$. If we have $|P_{t_i}| < (1-\eta/2)\, i$ then we can show that at least $\Omega(\eta i)$ of the nodes in $[i]$ are γ-good. We can also show that these γ-good nodes form a depth robust subset and will cost $\Omega\left((\eta-\epsilon)^2 i^2\right)$ to repebble them by [ABP17]. Since, we would need to pay this cost by time $t_{i+\gamma i}$ it is less expensive to simply ensure that $|P_{t_i}| > (1-\eta/2)\, i$. We refer an interested reader to Appendix A for a complete proof.

Theorem 4. *Let* $0 < \eta < 1$ *be a positive constant and let* $\epsilon = \eta^2/100$ *then the family* $\{G_n^\epsilon\}_{n=1}^\infty$ *of DAGs from Theorem 3 has* $\mathsf{indeg}\left(G_n^\epsilon\right) \in O\left(\log n\right)$ *and* $\Pi_{cc}^{\parallel}\left(G_n^\eta\right) \geq \frac{n^2(1-\eta)}{2}$.

Application 3: Cumulative Space in Parallel-Black Sequential-White Pebblings. The black-white pebble game [CS76] was introduced to model nondeterministic computations. White pebbles correspond to nondeterministic guesses and can be placed on any vertex at any time. However, these pebbles can only be removed from a node when all parents of the node contain a pebble (i.e., when we can verify the correctness of this guess). Formally, black white-pebbling configuration $P_i = \left(P_i^W, P_i^B\right)$ of a DAG $G = ([n], E)$ consists of two subsets $P_i^W, P_i^B \subseteq [n]$ where P_i^B (resp. P_i^W) denotes the set of nodes in G with black (resp. white) pebbles on them at time i. For a legal parallel-black sequential-white pebbling $P = (P_0, \ldots, P_t) \in \mathcal{P}_G^{BW}$ we require that we start with no pebbles on the graph i.e., $P_0 = (\emptyset, \emptyset)$ and that all white pebbles are removed by the end i.e., $P_t^W = \emptyset$ so that we verify the correctness of every nondeterministic guess before terminating. If we place a black pebble on a node v during round $i+1$ then we require that all of v's parents have a pebble (either black or white) on them during round i i.e., $\mathsf{parents}\left(P_{i+1}^B \setminus P_i^B\right) \subseteq P_i^B \cup P_i^W$. In the Parallel-Black Sequential-White model we require that at most one new white pebble is placed on the DAG in every round i.e., $\left|P_i^W \setminus P_{i-1}^W\right| \leq 1$ while no such restrict applies for black pebbles.

We can use our construction of a family of extremely depth-robust DAG $\{G_n^\epsilon\}_{n=1}^\infty$ to establish new upper and lower bounds for bounds for parallel-black sequential white pebblings.

Alwen et al. [AdRNV17] previously showed that in the parallel-black sequential white pebbling model an (e,d)-depth-robust DAG G requires cumulative space at least $\Pi_{cc}^{BW}(G) \doteq \min_{P\in\mathcal{P}_G^{BW}} \sum_{i=1}^t \left|P_i^B \cup P_i^W\right| \in \Omega\left(e\sqrt{d}\right)$ or at least $\geq ed$ in the sequential black-white pebbling game. In this section we show that any (e,d)-reducible DAG admits a parallel-black sequential white pebbling with cumulative space at most $O(e^2+dn)$ which implies that any DAG with constant indegree admits a parallel-black sequential white pebbling with cumulative space at most $O(\frac{n^2\log^2\log n}{\log^2 n})$ since any DAG is $(n\log\log n/\log n, n/\log^2 n)$-reducible. We also show that this bound is essentially tight (up to $\log\log n$ factors) using our construction of extremely depth-robust DAGs. In particular, by applying indegree reduction to the family $\{G_n^\epsilon\}_{n=1}^\infty$, we can find a family of DAGs $\{J_n^\epsilon\}_{n=1}^\infty$ with $\mathsf{indeg}\left(J_n^\epsilon\right) = 2$ such that any parallel-black sequential white pebbling has cumulative space at least $\Omega(\frac{n^2}{\log^2 n})$. To show this we start by showing that any parallel-black sequential white pebbling of an extremely depth-robust DAG G_n^ϵ, with $\mathsf{indeg}(G) \in O(\log n)$, has cumulative space at least $\Omega(n^2)$. We use Lemma 1 to reduce the indegree of the DAG and obtain a DAG J_n^ϵ with $n' \in O(n\log n)$ nodes and $\mathsf{indeg}(G) = 2$, such that any parallel-black sequential white pebbling of J_n^ϵ has cumulative space at least $\Omega(\frac{n^2}{\log^2 n})$.

To the best of our knowledge no general upper bound on cumulative space complexity for parallel-black sequential-white pebblings was known prior to our work other than the parallel black-pebbling attacks of Alwen and Blocki [AB16]. This attack, which doesn't even use the white pebbles, yields an upper bound of $O(ne+n\sqrt{nd})$ for (e,d)-reducible DAGs and $O(n^2\log\log n/\log n)$ in general. One could also consider a "parallel-white parallel-black" pebbling model in which we are allowed to place as many white pebbles as he would like in each round. However, this model admits a trivial pebbling. In particular, we could place white pebbles on every node during the first round and remove all of these pebbles in the next round e.g., $P_1 = (\emptyset, V)$ and $P_2 = (\emptyset,\emptyset)$. Thus, any DAG has cumulative space complexity $\Theta(n)$ in the "parallel-white parallel-black" pebbling model.

Theorem 5 shows that (e,d)-reducible DAG admits a parallel-black sequential white pebbling with cumulative space at most $O(e^2+dn)$. The basic pebbling strategy is reminiscent of the parallel black-pebbling attacks of Alwen and Blocki [AB16]. Given an appropriate depth-reducing set S we use the first $e=|S|$ steps to place white pebbles on all nodes in S. Since $G-S$ has depth at most d we can place black pebbles on the remaining nodes during the next d steps. Finally, once we place pebbles on every node we can legally remove the white pebbles. A formal proof of Theorem 5 can be found in the full version of this paper [ABP18].

Theorem 5. *Let $G = (V, E)$ be (e, d)-reducible then $\Pi_{cc}^{BW}(G) \leq \frac{e(e+1)}{2} + dn$. In particular, for any DAG G with* $\mathsf{indeg}(G) \in O(1)$ *we have $\Pi_{cc}^{BW}(G) \in O\left(\left(\frac{n \log\log n}{\log n}\right)^2\right)$.*

Theorem 6 shows that our upper bound is essentially tight. In a nut-shell their lower bound was based on the observation that for any integers i, d the DAG $G - \bigcup_j P_{i+jd}$ has depth at most d since any remaining path must have been pebbled completely in time d—if G is (e, d)-depth robust this implies that $\left|\bigcup_j P_{i+jd}\right| \geq e$. The key difficulty in adapting this argument to the parallel-black sequential white pebbling model is that it is actually possible to pebble a path of length d in $O(\sqrt{d})$ steps by placing white pebbles on every interval of length $\sqrt{d}$. This is precisely why Alwen et al. [AdRNV17] were only able to establish the lower bound $\Omega(e\sqrt{d})$ for the cumulative space complexity of (e, d)-depth robust DAGs—observe that we always have $e\sqrt{d} \leq n^{1.5}$ since $e + d \leq n$ for any DAG G. We overcome this key challenge by using extremely depth-robust DAGs.

In particular, we exploit the fact that extremely depth-robust DAGs are "recursively" depth-robust. For example, if a DAG G is (e, d)-depth robust for any $e+d \leq (1-\epsilon)n$ then the DAG $G-S$ is (e, d)-depth robust for any $e+d \leq (n-|S|) - \epsilon n$. Since $G - S$ is still sufficiently depth-robust we can then show that for some node $x \in V(G-S)$ any (possibly incomplete) pebbling $P = (P_0, P_1, \ldots, P_t)$ of $G - S$ with $P_0 = P_t = (\emptyset, \emptyset)$ either (1) requires $t \in \Omega(n)$ steps, or (2) fails to place a pebble on x i.e. $x \notin \bigcup_{r=0}^{t}\left(P_0^W \cup P_r^B\right)$. By Theorem 3 it then follows that there is a family of DAGs $\{G_n^\epsilon\}_{n=1}^\infty$ with $\mathsf{indeg}\left(G_n^\epsilon\right) \in O\left(\log n\right)$ and $\Pi_{cc}^{BW}(G) \in \Omega(n^2)$. If apply indegree reduction Lemma 1 to the family $\{G_n^\epsilon\}_{n=1}^\infty$ we obtain the family $\{J_n^\epsilon\}_{n=1}^\infty$ with $\mathsf{indeg}(J_n^\epsilon) = 2$ and $O(n)$ nodes. A similar argument shows that $\Pi_{cc}^{BW}\left(J_n^\epsilon\right) \in \Omega(n^2/\log^2 n)$. A formal proof of Theorem 6 can be found in the full version of this paper [ABP18].

Theorem 6. *Let $G = (V = [n], E \supset \{(i, i+1) : i < n\})$ be (e, d)-depth-robust for any $e + d \leq (1-\epsilon)n$ then $\Pi_{cc}^{BW}(G) \geq (1/16 - \epsilon/2)\, n^2$. Furthermore, if $G' = ([2n\delta], E')$ is the indegree reduced version of G from Lemma 1 then $\Pi_{cc}^{BW}(G') \geq (1/16 - \epsilon/2)\, n^2$. In particular, there is a family of DAGs $\{G_n\}_{n=1}^\infty$ with* $\mathsf{indeg}(G_n) \in O\left(\log n\right)$ *and $\Pi_{cc}^{BW}(G) \in \Omega(n^2)$, and a separate family of DAGs $\{H_n\}_{n=1}^\infty$ with* $\mathsf{indeg}(H_n) = 2$ *and $\Pi_{cc}^{BW}(H_n) \in \Omega\left(\frac{n^2}{\log^2 n}\right)$.*

5 A Pebbling Reduction for Sustained Space Complexity

As an application of the pebbling results on sustained space in this section we construct a new type of moderately hard function (MoHF) in the parallel random oracle model pROM. In slightly more detail, we first fix the computational model and define a particular notion of moderately hard function called sustained memory-hard functions (SMHF). We do this using the framework of [AT17] so, beyond the applications to password based cryptography, the results in [AT17] for building provably secure cryptographic applications on top of any MoHF can be immediately applied to SMHFs. In particular this results in a proof-of-work and non-interactive proof-of-work where "work" intuitively means having performed some computation entailing sufficient sustained memory. Finally we prove a "pebbling reduction" for SMHFs; that is we show how to bound the parameters describing the sustained memory complexity of a family of SMHFs in terms of the sustained space of their underlying graphs.[4]

We note that the pebbling reduction below caries over almost unchanged to the framework of [AS15]. That is by defining sustained space in the computational model of [AS15] similarly to the definition below a very similar proof to that of Theorem 7 results the analogous theorem but for the [AT17] framework. Never-the-less we believe the [AT17] framework to result in a more useful definition as exemplified by the applications inherited from that work.

5.1 Defining Sustained Memory Hard Functions

We very briefly sketch the most important parts of the MoHF framework of [AT17] which is, in turn, a generalization of the indifferentiability framework of [MRH04].

We begin with the following definition which describes a family of functions that depend on a (random) oracle.

Definition 6 (Oracle functions). *For (implicit) oracle set* $\mathbb{H}$*, an* oracle function $f^{(\cdot)}$ *(with domain* D *and range* R*), denoted* $f^{(\cdot)} : D \to R$*, is a set of functions indexed by oracles* $h \in \mathbb{H}$ *where each* f^h *maps* $D \to R$*.*

Put simply, an MoHF is a pair consisting of an oracle family $f^{(\cdot)}$ and an honest algorithm $\mathcal{N}$ for evaluating functions in the family using access to a random oracle. Such a pair is secure relative to some computational model M if no adversary $\mathcal{A}$ with a computational device adhering to M (denoted $\mathcal{A} \in M$) can produce output which couldn't be produced simply by called $f^{(h)}$ a limited number of times (where h is a uniform choice of oracle from $\mathbb{H}$). It is assumed that algorithm $\mathcal{N}$ is computable by devices in some (possibly different) computational model $\bar{M}$ when given sufficient computational resources. Usually M is strictly more powerful than $\bar{M}$ reflecting the assumption that an adversary could have a more powerful class of device than the honest party. For example, in this work we will let model $\bar{M}$ contain only sequential devices (say Turing machines which make one call to the random oracle at a time) while M will also include parallel devices.

In this work, both the computational models M and $\bar{M}$ are parametrized by the same space $\mathbb{P}$. For each model, the choice of parameters fixes upperbounds on the power of devices captured by that model; that is on the computational resources available to the permitted devices. For example M_a could be all Turing machines making at most a queries to the random oracle. The security of a given moderatly hard function is parameterized by two functions α and β mapping the

[4] Effectively this does for SMHFs what [AT17] did for MHFs.

parameter space for M to positive integers. Intuitively these functions are used to provide the following two properties.

COMPLETENESS: To ensure the construction is even useable we require that $\mathcal{N}$ is (computable by a device) in model M_a and that $\mathcal{N}$ can evaluate $f^{(h)}$ (when given access to h) on at least $\alpha(a)$ distinct inputs.

SECURITY: To capture how bounds on the resources of an adversary $\mathcal{A}$ limit the ability of $\mathcal{A}$ to evaluate the MoHF we require that the output of $\mathcal{A}$ when running on a device in model M_b (and having access to the random oracle) can be reproduced by some simulator σ using at most $\beta(b)$ oracle calls to $f^{(h)}$ (for uniform randomly sampled $h \leftarrow \mathbb{H}$.

To help build provably secure applications on top of MoHFs the framework makes use of a destinguisher $\mathcal{D}$ (similar to the environment in the Universal Composability [Can01] family of models or, more accurately, to the destinguisher in the indifferentiability framework). The job of $\mathcal{D}$ is to (try to) tell a *real world* interaction with $\mathcal{N}$ and the adversary $\mathcal{A}$ apart from an *ideal world* interaction with $f^{(h)}$ (in place of $\mathcal{N}$) and a simulator (in place of the adversary). Intuitivelly, $\mathcal{D}$'s access to $\mathcal{N}$ captures whatever $\mathcal{D}$ could hope to learn by interacting with an arbitrary application making use of the MoHF. The definition then ensures that even leveraging such information the adversary $\mathcal{A}$ can not produce anything that could not be simulated (by simulator σ) to $\mathcal{D}$ using nothing more than a few calls to $f^{(h)}$.

As in the above description we have ommited several details of the framework we will also use a somewhat simplified notation. We denote the above described real world execution with the pair $(\mathcal{N}, \mathcal{A})$ and an ideal world execution where $\mathcal{D}$ is permitted $c \in \mathbb{N}$ calls to $f^{(\cdot)}$ and simulator σ is permited $d \in \mathbb{N}$ calls to $f^{(h)}$ with the pair $(f^{(\cdot)}, \sigma)_{c,d}$. To denote the statement that no $\mathcal{D}$ can tell an interaction with $(\mathcal{N}, \mathcal{A})$ apart one with $(f^{(\cdot)}, \sigma)_{c,d}$ with more than probability ϵ we write $(\mathcal{N}, \mathcal{A}) \approx_\epsilon (f^{(\cdot)}, \sigma)_{c,d}$.

Finally, to accommodate honest parties with varying amounts of resources we equip the MoHF with a hardness parameter $n \in \mathbb{N}$. The following is the formal security definition of a MoHF. Particular types of MoHF (such as the one we define bellow for sustained memory complexity) differ in the precise notion of computational model they consider. For further intuition, a much more detailed exposition of the framework and how the following definition can be used to prove security for applications we refer to [AT17].

Definition 7 (MoHF security). *Let M and $\bar{M}$ be computational models with bounded resources parametrized by $\mathbb{P}$. For each $n \in \mathbb{N}$, let $f_n^{(\cdot)}$ be an oracle function and $\mathcal{N}(n, \cdot)$ be an algorithm (computable by some device in $\bar{M}$) for evaluating $f_n^{(\cdot)}$. Let $\alpha, \beta : \mathbb{P} \times \mathbb{N} \to \mathbb{N}$, and let $\epsilon : \mathbb{P} \times \mathbb{P} \times \mathbb{N} \to \mathbb{R}_{\geq 0}$. Then, $(f_n^{(\cdot)}, \mathcal{N}_n)_{n \in \mathbb{N}}$ is a $(\alpha, \beta, \epsilon)$-secure* moderately hard function family *(for model M) if*

$$\forall n \in \mathbb{N}, \mathbf{r} \in \mathbb{P}, \mathcal{A} \in M_{\mathbf{r}} \; \exists \sigma \; \forall \mathbf{l} \in \mathbb{P} : \quad (\mathcal{N}(n, \cdot), \mathcal{A}) \approx_{\epsilon(\mathbf{l}, \mathbf{r}, n)} (f_n^{(\cdot)}, \sigma)_{\alpha(\mathbf{l}, n), \beta(\mathbf{r}, n)} , \tag{3}$$

The function family is asymptotically secure *if $\epsilon(\mathbf{l}, \mathbf{r}, \cdot)$ is a negligible function in the third parameter for all values of $\mathbf{r}, \mathbf{l} \in \mathbb{P}$.*

Sustained Space Constrained Computation. Next we define the honest and adversarial computational models for which we prove the pebbling reduction. In particular we first recall (a simplified version of) the pROM of [AT17]. Next we define a notion of sustained memory in that model naturally mirroring the notion of sustained space for pebbling. Thus we can parametrize the pROM by memory threshold s and time t to capture all devices in the pROM with no more sustained memory complexity then given by the choice of those parameters.

In more detail, we consider a resource-bounded computational device $\mathcal{S}$. Let $w \in \mathbb{N}$. Upon startup, $\mathcal{S}^{w\text{-PROM}}$ samples a fresh random oracle $h \leftarrow_\$ \mathbb{H}_w$ with range $\{0,1\}^w$. Now $\mathcal{S}^{w\text{-PROM}}$ accepts as input a pROM algorithm $\mathcal{A}$ which is an oracle algorithm with the following behavior.

A *state* is a pair $(\tau, \mathbf{s})$ where *data* τ is a string and $\mathbf{s}$ is a tuple of strings. The output of step i of algorithm $\mathcal{A}$ is an *output state* $\bar{\sigma}_i = (\tau_i, \mathbf{q}_i)$ where $\mathbf{q}_i = [q_i^1, \ldots, q_i^{z_i}]$ is a tuple of *queries* to h. As input to step $i+1$, algorithm $\mathcal{A}$ is given the corresponding *input state* $\sigma_i = (\tau_i, h(\mathbf{q}_i))$, where $h(\mathbf{q}_i) = [h(q_i^1), \ldots, h(q_i^{z_i})]$ is the tuple of *responses* from h to the queries $\mathbf{q}_i$. In particular, for a given h and random coins of $\mathcal{A}$, the input state σ_{i+1} is a function of the input state σ_i. The initial state σ_0 is empty and the input x_{in} to the computation is given a special input in step 1.

For a given execution of a pROM, we are interested in the following new complexity measure parametrized by an integer $s \geq 0$. We call an element of $\{0,1\}^s$ a *block*. Moreover, we denote the bit-length of a string r by $|r|$. The *length* of a state $\sigma = (\tau, \mathbf{s})$ with $\mathbf{s} = (s^1, s^2, \ldots, s^y)$ is $|\sigma| = |\tau| + \sum_{i \in [y]} |s^i|$. For a given state σ let $b(\sigma) = \lfloor |\sigma|/s \rfloor$ be the number of "blocks in σ". Intuitively, the s-sustained memory complexity (s-SMC) of an execution is the sum of the number of blocks in each state. More precisely, consider an execution of algorithm $\mathcal{A}$ on input x_{in} using coins \$ with oracle h resulting in $z \in \mathbb{Z}_{\geq 0}$ input states $\sigma_1, \ldots, \sigma_z$, where $\sigma_i = (\tau_i, \mathbf{s}_i)$ and $\mathbf{s}_i = (s_i^1, s_i^2, \ldots, s_i^{y_j})$. Then the for integer $s \geq 0$ the *s-sustained memory complexity* (s-SMC) of the execution is

$$s\text{-}\mathsf{smc}(\mathcal{A}^h(x_{\mathsf{in}}; \$)) = \sum_{i \in [z]} b(\sigma_i) ,$$

while the *total number of RO calls* is $\sum_{i \in [z]} y_j$. More generally, the s-SMC (and total number of RO calls) of several executions is the sum of the s-sMC (and total RO calls) of the individual executions.

We can now describe the resource constraints imposed by $\mathcal{S}^{w\text{-PROM}}$ on the pROM algorithms it executes. To quantify the constraints, $\mathcal{S}^{w\text{-PROM}}$ is parametrized by element from $\mathbb{P}^{\text{PROM}} = \mathbb{N}^3$ which describe the limits on an execution of algorithm $\mathcal{A}$. In particular, for parameters $(q, s, t) \in \mathbb{P}^{\text{PROM}}$, algorithm $\mathcal{A}$ is allowed to make a total of q RO calls and have s-SMC at most t (summed across all invocations of $\mathcal{A}$ in any given experiment).

As usual for moderately hard functions, to ensure that the honest algorithm can be run on realistic devices, we restrict the honest algorithm $\mathcal{N}$ for evaluating the SMHF to be a *sequential* algorithms. That is, $\mathcal{N}$ can make only a single call to h per step. Technically, in any execution, for any step j it must be that $y_j \leq 1$.

No such restriction is placed on the adversarial algorithm reflecting the power (potentially) available to such a highly parallel device as an ASIC. In symbols we denote the sequential version of the pROM, which we refer to as the sequential ROM (sROM) by $\mathcal{S}^{w\text{-SROM}}$.

We can now (somewhat) formally define of a sustained memory-hard function for the pROM. The definition is a particular instance of and moderately hard function (c.f. Definition 7).

Definition 8 (Sustained Memory-Hard Function). *For each $n \in \mathbb{N}$, let $f_n^{(\cdot)}$ be an oracle function and $\mathcal{N}_n$ be an sROM algorithm for computing $f^{(\cdot)}$. Consider the function families:*

$$\alpha = \{\alpha_w : \mathbb{P}^{\text{PROM}} \times \mathbb{N} \to \mathbb{N}\}_{w \in \mathbb{N}}, \quad \beta = \{\beta_w : \mathbb{P}^{\text{PROM}} \times \mathbb{N} \to \mathbb{N}\}_{w \in \mathbb{N}},$$

$$\epsilon = \{\epsilon_w : \mathbb{P}^{\text{PROM}} \times \mathbb{P}^{\text{PROM}} \times \mathbb{N} \to \mathbb{N}\}_{w \in \mathbb{N}}.$$

*Then $F = (f_n^{(\cdot)}, \mathcal{N}_n)_{n \in \mathbb{N}}$ is called an $(\alpha, \beta, \epsilon)$-*sustained memory-hard function *(SMHF) if $\forall w \in \mathbb{N}$ F is an $(\alpha_w, \beta_w, \epsilon_w)$-secure moderately hard function family for $\mathcal{S}^{w\text{-PROM}}$.*

5.2 The Construction

In this work $f^{(\cdot)}$ will be a graph function [AS15] (also sometimes called "hash graph"). The following definition is taken from [AT17]. A graph function depends on an oracle $h \in \mathbb{H}_w$ mapping bit strings to bit strings. We also assume the existence of an implicit prefix-free encoding such that h is evaluated on unique strings. Inputs to h are given as distinct tuples of strings (or even tuples of tuples of strings). For example, we assume that $h(0, 00)$, $h(00, 0)$, and $h((0, 0), 0)$ all denote distinct inputs to h.

Definition 9 (Graph function). *Let function $h : \{0,1\}^* \to \{0,1\}^w \in \mathbb{H}_w$ and DAG $G = (V, E)$ have source nodes $\{v_1^{\mathsf{in}}, \ldots, v_a^{\mathsf{in}}\}$ and sink nodes $(v_1^{\mathsf{out}}, \ldots, v_z^{\mathsf{out}})$. Then, for inputs $\mathbf{x} = (x_1, \ldots, x_a) \in (\{0,1\}^*)^{\times a}$, the $(h, \mathbf{x})$-*labeling *of G is a mapping $\mathsf{lab} : V \to \{0,1\}^w$ defined recursively to be:*

$$\forall v \in V \quad \mathsf{lab}(v) := \begin{cases} h(\mathbf{x}, v, x_j) & : v = v_j^{\mathsf{in}} \\ h(\mathbf{x}, v, \mathsf{lab}(v_1), \ldots, \mathsf{lab}(v_d)) & : else \end{cases}$$

where $\{v_1, \ldots, v_d\}$ are the parents of v arranged in lexicographic order.

The graph function *(of G and $\mathbb{H}_w$) is the oracle function*

$$f_G : (\{0,1\}^*)^{\times a} \to (\{0,1\}^w)^{\times z},$$

which maps $\mathbf{x} \mapsto (\mathsf{lab}(v_1^{\mathsf{out}}), \ldots, \mathsf{lab}(v_z^{\mathsf{out}}))$ where lab is the $(h, \mathbf{x})$-labeling of G.

Given a graph function we need an honest (sequential) algorithm for computing it in the pROM. For this we use the same algorithm as already used in [AT17]. The honest oracle algorithm $\mathcal{N}_G$ for graph function f_G computes one label of G at a time in topological order appending the result to its state. If G has $|V| = n$ nodes then $\mathcal{N}_G$ will terminate in n steps making at most 1 call to h per step, for a total of n calls, and will never store more than $n * w$ bits in the data portion of its state. In particular for all inputs $\mathbf{x}$, oracles h (and coins \$) we have that for any $s \in [n]$ if the range of h is in $\{0,1\}^w$ then algorithm $\mathcal{N}$ has sw-SMC of $n - s$.

Recall that we would like to set $\alpha_w : \mathbb{P}^{\mathrm{PROM}} \rightarrow \mathbb{N}$ such that for any parameters (q, s, t) constraining the honest algorithms resources we are still guaranteed at least $\alpha_w(q, s, t)$ evaluations of f_G by $\mathcal{N}_G$. Given the above honest algorithm we can thus set:

$$\forall (q,s,t) \in \mathbb{P}^{\mathrm{PROM}} \quad \alpha_w(q,s,t) := \begin{cases} 0 & : q < n \\ \min(\lfloor q/n \rfloor, \lfloor t/(n - \lfloor s/w \rfloor \rfloor) & : \text{else} \end{cases}$$

It remains to determine how to set β_w and ϵ_w, which is the focus of the remainder of this section.

5.3 The Pebbling Reduction

We state the main theorem of this section which relates the parameters of an SMHF based on a graph function to the sustained (pebbling) space complexity of the underlying graph.

Theorem 7 *(Pebbling reduction). Let $G_n = (V_n, E_n)$ be a DAG of size $|V_n| = n$. Let $F = (f_{G,n}, \mathcal{N}_{G,n})_{n \in \mathbb{N}}$ be the graph functions for G_n and their naïve oracle algorithms. Then, for any $\lambda \geq 0$, F is an $(\alpha, \beta, \epsilon)$-sustained memory-hard function where*

$$\alpha = \{\alpha_w(q,s,t)\}_{w \in \mathbb{N}} ,$$

$$\beta = \left\{ \beta_w(q,s,t) = \frac{\Pi^{\parallel}_{ss}(G,s)(w - \log q)}{1 + \lambda} \right\}_{w \in \mathbb{N}} , \quad \epsilon = \left\{ \epsilon_w(q,m) \leq \frac{q}{2^w} + 2^{-\lambda} \right\}_{w \in \mathbb{N}} .$$

The technical core of the proof follows that of [AT17] closely. The proof can be found in the full version of this paper [ABP18].

6 Open Questions

We conclude with several open questions for future research. The primary challenge is to provide a practical construction of a DAG G with high sustained space complexity. While we provide a DAG G with asymptotically optimal sustained space complexity, we do not optimize for constant factors. We remark that for practical applications to iMHFs it should be trivial to evaluate the function $\mathsf{parents}_G(v)$ without storing the DAG G in memory explicitly. Toward this end it would be useful to either prove or refute the conjecture that any depth-robustness is sufficient for high sustained space complexity e.g., what is the sustained space complexity of the depth-robust DAGs from [EGS75] or [PTC76]? Another interesting direction would be to relax the notion of sustained space complexity and instead require that for any pebbling $P \in \mathcal{P}^{\parallel}(G)$ either (1) P has large cumulative complexity e.g., n^3, or (2) P has high sustained space complexity. Is it possible to design a dMHF with the property for any evaluation algorithm either has (1) sustained space complexity $\Omega(n)$ for $\Omega(n)$ rounds, or (2) has cumulative memory complexity $\omega(n^2)$?

Acknowledgments. This work was supported by the European Research Council under ERC consolidator grant (682815 - TOCNeT) and by the National Science Foundation under NSF Award #1704587. The opinions expressed in this paper are those of the authors and do not necessarily reflect those of the European Research Council or the National Science Foundation.

A Missing Proofs

Reminder of Theorem 4. *Let $0 < \eta < 1$ be a positive constant and let $\epsilon = \eta^2/100$ then the family $\{G_n^\epsilon\}_{n=1}^\infty$ of DAGs from Theorem 3 has* $\mathsf{indeg}(G_n^\epsilon) \in O(\log n)$ *and* $\Pi_{cc}^{\parallel}(G_n^\eta) \geq \frac{n^2(1-\eta)}{2}$.

Proof of Theorem 4. We set $\epsilon = \eta^2/100$ and consider the DAG G_n^ϵ from the proof of Theorem 3. In particular, G_n^ϵ is a $\delta = \epsilon/10$-local expander. We also set $\gamma = \epsilon/4$ when we consider γ-good nodes.

Consider a legal pebbling $P \in \mathcal{P}_{G_n^\epsilon}^{\parallel}$ and let t_i denote the first time that node i is pebbled ($i \in P_{t_i}$, but $i \notin \bigcup_{j<t_i} P_j$). We consider two cases:

Case 1 $|P_{t_i}| \geq (1-\eta/2)\, i$. Observe that if this held for all i then we immediately have $\sum_{j=1}^{t} |P_i| \geq \sum_{j=1}^{n} |P_{t_i}| \geq (1-\eta/2) \sum_{i=1}^{n} i \geq \frac{n^2(1-\epsilon/2)}{2}$.

Case 2 $P_{t_i} < (1-\eta/2)\, i$. Let $GOOD_i$ denote the set of γ-good nodes in $[i]$. We observe that at least $i - (1-\eta/2) i \frac{1-\gamma}{1+\gamma} \geq i\eta/4$ of the nodes in $[i]$ are γ-good by Lemma 6. Furthermore, we note that the subgraph $H_i = G_n^\epsilon[GOOD_i]$ is $(a\,|Good_i|\,, (1-a)\,|Good_i| - \epsilon i)$-depth robust for any constants $a > 0$.[5]

Thus, a result of Alwen et al. [ABP17] gives us $\Pi_{cc}^{\parallel}(H_i) \geq i^2\eta^2/100$ since the DAG H_i is at least $(i\eta/10, i\eta/10)$-depth robust. To see this set $a = 1/2$ and observe that $a|Good_i| \geq i\eta/8$ and that $(1-a)\,|Good_i| - \epsilon i \geq i\eta/8 - \eta i/100 \geq i\eta/10$. Similarly, we note that at time t_i the node $i+\gamma i$ is γ-good. Thus, by Lemma 5 we will have to completely repebble H_i by time $t_{i+\gamma i}$. This means that $\sum_{j=t_i}^{t_{i+\gamma i}} |P_j| \geq \Pi_{cc}^{\parallel}(H_i) \geq i^2\eta^2/100$ and, since $\gamma = \eta^2/400$ we have $i^2\eta^2/100 > 2\gamma i^2 > \sum_{j=i}^{i+\gamma i} j(1-\eta/2)$.

[5] To see this observe that if G_n^ϵ is a δ-local expander then $G_n^\epsilon[\{1,\ldots,i\}]$ is also a δ-local expander. Therefore, Lemmas 5 and 6 imply that $G_n^\epsilon[\{1,\ldots,i\}]$ is (ai, bi)-depth robust for any $a+b \leq 1-\epsilon$. Since, H_i is a subgraph of $G_n^\epsilon[\{1,\ldots,i\}]$ it must be that H_i is $(a\,|Good_i|\,, (1-a)\,|Good_i| - \epsilon i)$-depth robust. Otherwise, we have a set $S \subseteq V(H_i)$ of size $a\,|Good_i|$ such that $\mathsf{depth}(H_i - S) < (1-a)\,|Good_i| - \epsilon i$ which implies that $\mathsf{depth}(G_n^\epsilon[\{1,\ldots,i\}] - S) \leq i - |Good_i| + \mathsf{depth}(Good_i - S) < i - a|Good_i| - \epsilon i$ contradicting the depth-robustness of $G_n^\epsilon[\{1,\ldots,i\}]$.

Let x_1 denote the first node $1 \leq x_1 \leq n - \gamma n$ for which $\left|P_{t_{x_1}}\right| < (1-\eta/2)\, i$ and, once $x_1, \ldots, x_k$ have been defined let x_{k+1} denote the first node such that $n - \gamma n > x_{k+1} > \gamma x_k + x_k$ and $\left|P_{t_{x_{k+1}}}\right| < (1-\eta/2)\, i$. Let $x_1, \ldots, x_{k*}$ denote a maximal such sequence and let $F = \bigcup_{j=1}^{k*} [x_j, x_j + \gamma x_j]$. Let $R = [n - \gamma n] \setminus F$. We have $\sum_{j \in R} |P_j| \geq \sum_{j \in R} j(1-\eta/2)$ and we have $\sum_{j \in F} |P_j| \geq \sum_{j \in R} j(1-\eta/2)$. Thus,

$$\sum_{j=1}^{t} |P_i| \geq \sum_{j \in R} |P_j| + \sum_{j \in F} |P_j| \geq \sum_{j=1}^{n-\gamma n} \frac{n^2\,(1-\eta/2)}{2} \geq \frac{n^2\,(1-\eta/2)}{2} - \gamma n^2 \geq \frac{n^2\,(1-\eta)}{2}\,.$$

□

References

[AB16] Alwen, J., Blocki, J.: Efficiently computing data-independent memory-hard functions. In: Robshaw, M., Katz, J. (eds.) CRYPTO 2016. LNCS, vol. 9815, pp. 241–271. Springer, Heidelberg (2016). https://doi.org/10.1007/978-3-662-53008-5_9

[AB17] Alwen, J., Blocki, J.: Towards practical attacks on Argon2i and balloon hashing. In: Proceedings of the 2nd IEEE European Symposium on Security and Privacy (EuroS&P 2017), pp. 142–157. IEEE (2017). http://eprint.iacr.org/2016/759

[ABH17] Alwen, J., Blocki, J., Harsha, B.: Practical graphs for optimal side-channel resistant memory-hard functions. In: ACM CCS 2017, pp. 1001–1017. ACM Press (2017)

[ABP17] Alwen, J., Blocki, J., Pietrzak, K.: Depth-robust graphs and their cumulative memory complexity. In: Coron, J.-S., Nielsen, J.B. (eds.) EUROCRYPT 2017. LNCS, vol. 10212, pp. 3–32. Springer, Cham (2017). https://doi.org/10.1007/978-3-319-56617-7_1

[ABP18] Alwen, J., Blocki, J., Pietrzak, K.: Sustained space complexity. Cryptology ePrint Archive, Report 2018/147 (2018). https://eprint.iacr.org/2018/147

[ABW03] Abadi, M., Burrows, M., Wobber, T.: Moderately hard, memory-bound functions. In: Proceedings of the Network and Distributed System Security Symposium, NDSS 2003, San Diego, California, USA (2003)

[ACP+17] Alwen, J., Chen, B., Pietrzak, K., Reyzin, L., Tessaro, S.: Scrypt is maximally memory-hard. In: Coron, J.-S., Nielsen, J.B. (eds.) EUROCRYPT 2017. LNCS, vol. 10212, pp. 33–62. Springer, Cham (2017). https://doi.org/10.1007/978-3-319-56617-7_2

[AdRNV17] Alwen, J., de Rezende, S.F., Nordström, J., Vinyals, M.: Cumulative space in black-white pebbling and resolution. In: 8th Innovations in Theoretical Computer Science (ITCS) Conference, Berkeley, 9–11 January 2017

[AS15] Alwen, J., Serbinenko, V.: High parallel complexity graphs and memory-hard functions. In: Proceedings of the Eleventh Annual ACM Symposium on Theory of Computing, STOC 2015 (2015). http://eprint.iacr.org/2014/238

[AT17] Alwen, J., Tackmann, B.: Moderately hard functions: definition, instantiations, and applications. In: Kalai, Y., Reyzin, L. (eds.) TCC 2017. LNCS, vol. 10677, pp. 493–526. Springer, Cham (2017). https://doi.org/10.1007/978-3-319-70500-2_17

[BDK16] Biryukov, A., Dinu, D., Khovratovich, D.: Argon2: new generation of memory-hard functions for password hashing and other applications. In: 2016 IEEE European Symposium on Security and Privacy (EuroS&P), pp. 292–302. IEEE (2016)

[BHZ18] Blocki, J., Harsha, B., Zhou, S.: On the economics of offline password cracking. IEEE Secur. Priv. (2018, to appear)

[Can01] Canetti, R.: Universally composable security: a new paradigm for cryptographic protocols. In: 42nd Annual Symposium on Foundations of Computer Science, Las Vegas, Nevada, pp. 136–145. IEEE, October 2001

[Coo73] Cook, S.A.: An observation on time-storage trade off. In: Proceedings of the Fifth Annual ACM Symposium on Theory of Computing, STOC 1973, pp. 29–33. ACM, New York (1973)

[Cox16] Cox, B.: Re: [Cfrg] Balloon-Hashing or Argon2i. CFRG Mailinglist, August 2016. https://www.ietf.org/mail-archive/web/cfrg/current/msg08426.html

[CP18] Cohen, B., Pietrzak, K.: Simple proofs of sequential work. In: Nielsen, J.B., Rijmen, V. (eds.) EUROCRYPT 2018, Part II. LNCS, vol. 10821, pp. 451–467. Springer, Cham (2018)

[CS76] Cook, S., Sethi, R.: Storage requirements for deterministic polynomialtime recognizable languages. J. Comput. Syst. Sci. **13**(1), 25–37 (1976)

[DGN03] Dwork, C., Goldberg, A., Naor, M.: On memory-bound functions for fighting spam. In: Boneh, D. (ed.) CRYPTO 2003. LNCS, vol. 2729, pp. 426–444. Springer, Heidelberg (2003). https://doi.org/10.1007/978-3-540-45146-4_25

[EGS75] Erdös, P., Graham, R.L., Szemerédi, E.: On sparse graphs with dense long paths. Technical report, Stanford, CA, USA (1975)

[GG81] Gabber, O., Galil, Z.: Explicit constructions of linear-sized superconcentrators. J. Comput. Syst. Sci. **22**(3), 407–420 (1981)

[HP70] Hewitt, C.E., Paterson, M.S.: Record of the project MAC conference on concurrent systems and parallel computation. In: Comparative Schematology, pp. 119–127. ACM, New York (1970)

[HPV77] Hopcroft, J., Paul, W., Valiant, L.: On time versus space. J. ACM **24**(2), 332–337 (1977)

[Kal00] Kaliski, B.: PKCS# 5: password-based cryptography specification version 2.0 (2000)

[MMV13] Mahmoody, M., Moran, T., Vadhan, S.P.: Publicly verifiable proofs of sequential work. In: Kleinberg, R.D. (ed.) Innovations in Theoretical Computer Science, ITCS 2013, Berkeley, CA, USA, 9–12 January 2013, pp. 373–388. ACM (2013)

[MRH04] Maurer, U., Renner, R., Holenstein, C.: Indifferentiability, impossibility results on reductions, and applications to the random Oracle methodology. In: Naor, M. (ed.) TCC 2004. LNCS, vol. 2951, pp. 21–39. Springer, Heidelberg (2004). https://doi.org/10.1007/978-3-540-24638-1_2

[Per09] Percival, C.: Stronger key derivation via sequential memory-hard functions. In: BSDCan 2009 (2009)

[PHC] Password hashing competition. https://password-hashing.net/

[PJ12] Percival, C., Josefsson, S.: The scrypt password-based key derivation function (2012)

[PTC76] Paul, W.J., Tarjan, R.E., Celoni, J.R.: Space bounds for a game on graphs. In: Proceedings of the Eighth Annual ACM Symposium on Theory of Computing, STOC 1976, pp. 149–160. ACM, New York (1976)

[RD16] Ren, L., Devadas, S.: Proof of space from stacked expanders. In: Hirt, M., Smith, A. (eds.) TCC 2016. LNCS, vol. 9985, pp. 262–285. Springer, Heidelberg (2016). https://doi.org/10.1007/978-3-662-53641-4_11

[RD17] Ren, L., Devadas, S.: Bandwidth hard functions for ASIC resistance. In: Kalai, Y., Reyzin, L. (eds.) TCC 2017. LNCS, vol. 10677, pp. 466–492. Springer, Cham (2017). https://doi.org/10.1007/978-3-319-70500-2_16

Multi-collision Resistance

Multi-Collision Resistant Hash Functions and Their Applications

Itay Berman[1(✉)], Akshay Degwekar[1], Ron D. Rothblum[1,2], and Prashant Nalini Vasudevan[1]

[1] MIT, Cambridge, USA
{itayberm,akshayd,ronr,prashvas}@mit.edu
[2] Northeastern University, Boston, USA

Abstract. Collision resistant hash functions are functions that shrink their input, but for which it is computationally infeasible to find a collision, namely two strings that hash to the same value (although collisions are abundant).

In this work we study *multi-collision resistant hash functions* (MCRH) a natural relaxation of collision resistant hash functions in which it is difficult to find a t-way collision (i.e., t strings that hash to the same value) although finding $(t-1)$-way collisions could be easy. We show the following:

- The existence of MCRH follows from the average case hardness of a variant of the *Entropy Approximation* problem. The goal in this problem (Goldreich, Sahai and Vadhan, CRYPTO '99) is to distinguish circuits whose output distribution has high entropy from those having low entropy.
- MCRH imply the existence of *constant-round* statistically hiding (and computationally binding) commitment schemes. As a corollary, using a result of Haitner et al. (SICOMP, 2015), we obtain a blackbox separation of MCRH from any one-way permutation.

1 Introduction

Hash functions are efficiently computable functions that shrink their input and mimic 'random functions' in various aspects. They are prevalent in cryptography, both in theory and in practice. A central goal in the study of the foundations of cryptography has been to distill the precise, and minimal, security requirements necessary from hash functions for different applications.

One widely studied notion of hashing is that of *collision resistant hash functions* (CRH). Namely, hash functions for which it is computationally infeasible to find two strings that hash to the same value, even when such collisions are abundant. CRH have been extremely fruitful and have notable applications

The full version [BDRV17] is available at https://eprint.iacr.org/2017/489.

J. B. Nielsen and V. Rijmen (Eds.): EUROCRYPT 2018, LNCS 10821, pp. 133–161, 2018.
https://doi.org/10.1007/978-3-319-78375-8_5

in cryptography such as digital signatures[1] [GMR88], efficient argument systems for NP [Kil92, Mic00] and (constant-round) statistically hiding commitment schemes [NY89, DPP93, HM96].

In this work we study a natural relaxation of collision resistance. Specifically, we consider hash functions for which it is infeasible to find a t-way collision: i.e., t strings that all have the same hash value. Here t is a parameter, where the standard notion of collision resistance corresponds to the special case of $t = 2$. We refer to such functions as *multi-collision resistant hash functions* (MCRH) and emphasize that, for $t > 2$, it is a *weaker* requirement than that of standard collision resistance.

The property of multi-collision resistance was considered first by Merkle [Mer89] in analyzing a hash function construction based on DES. The notion has also been considered in the context of identification schemes [GS94], micropayments [RS96], and signature schemes [BPVY00]. Joux [Jou04] showed that for *iterated* hash functions, finding a large number of collisions is no harder than finding pairs of highly structured colliding inputs (namely, collisions that share the same prefix). We emphasize that Joux's multi-collision finding attack only applies to certain types of hash functions (e.g., iterated hash functions, or tree hashing) and requires a strong break of collision resistance. In general, it seems that MCRH is a weaker property than CRH.

As in the case of CRH, to obtain a meaningful definition, we must consider keyed functions (since for non keyed functions there are trivial non-uniform attacks). Thus, we define MCRH as follows (here and throughout this work, we use n to denote the security parameter.)

Definition 1.1 (((s,t)-MCRH). *Let* $s = s(n) \in \mathbb{N}$ *and* $t = t(n) \in \mathbb{N}$ *be functions computable in time* $\mathsf{poly}(n)$. *An* (s,t)-Multi-Collision Resistant Hash Function Family ((s,t)-MCRH) *consists of a probabilistic polynomial-time algorithm* Gen *that on input* 1^n *outputs a circuit* h *such that:*

- **s-Shrinkage:** *The circuit* $h : \{0,1\}^n \to \{0,1\}^{n-s}$ *maps inputs of length* n *to outputs of length* $n - s$.
- **t-Collision Resistance:** *For every polynomial size family of circuits* $\mathsf{A} = (\mathsf{A}_n)_{n\in\mathbb{N}}$,

$$\Pr_{\substack{h \leftarrow \mathsf{Gen}(1^n),\\ (x_1,x_2,\ldots,x_t)\leftarrow \mathsf{A}_n(h)}} \left[\begin{array}{c} \text{For all } i \neq j, \\ h(x_i) = h(x_j) \text{ and } x_i \neq x_j \end{array} \right] < \mathsf{negl}(n).$$

Note that the standard notion of CRH simply corresponds to $(1,2)$-MCRH (which is easily shown to be equivalent to $(s,2)$-CRH for any $s = n - \omega(\log n)$). We also remark that Definition 1.1 gives a *non-uniform* security guarantee, which is natural, especially in the context of collision resistance. Note though that all of our results are obtained by *uniform* reductions.

[1] We remark that the weaker notion of universal one-way hash functions (UOWHF) (which is known to be implied by standard one-way functions) suffices for this application [NY89, Rom90].

Remark 1.2 (Shrinkage vs. Collision Resistance). Observe that (s,t)-MCRH are meaningful only when $s \geq \log t$, as otherwise t-way collisions might not even exist (e.g., consider a function mapping inputs of length n to outputs of length $n - \log(t-1)$ in which each range element has exactly $t-1$ preimages).

Moreover, we note that in contrast to standard CRH, it is unclear whether the shrinkage factor s can be trivially improved (e.g., by composition) while preserving the value of t. Specifically, constructions such as Tree Hashing (aka Merkle Tree) inherently rely on the fact that it is computationally infeasible to find *any* collision. It is possible to get some trade-offs between the number of collisions and shrinkage. For example, given an $(s=2, t=4)$-MCRH, we can compose it with itself to get an $(s=4, t=10)$-MCRH. But it is not a priori clear whether there exist transformations that increase the shrinkage s while not increasing t. We remark that a partial affirmative answer to this question was recently given in an independent and concurrent work by Bitansky et al. [BPK17], as long as the hash function is substantially shrinking (see additional details in Sect. 1.2).

Thus, we include both the parameters s and t in the definition of MCRH, whereas in standard CRH the parameter t is fixed to 2, and the parameter s can be given implicitly (since the shrinkage can be trivially improved by composition).

Remark 1.3 (Scaling of Shrinkage vs. Collisions). The shrinkage s is measured in bits, whereas the number of collisions t is just a number. A different definitional choice could have been to put s and t on the same "scale" (e.g., measure the logarithm of the number of collisions) so to make them more easily comparable. However, we refrain from doing so since we find the current (different) scaling of s and t to be more natural.

Remark 1.4 (Public-coin MCRH). One can also consider the stronger *public-coin* variant of MCRH, in which it should be hard to find collisions given not only the description of the hash function, but also the coins that generated the description.

Hsiao and Reyzin [HR04] observed that for some applications of standard collision resistance, it is vital to use the public-key variant (i.e., security can be broken in case the hash function is not public-coin). The distinction is similarly important for MCRH and one should take care of which notion is used depending on the application. Below, when we say MCRH, we refer to the *private-coin* variant (as per Definition 1.1).

1.1 Our Results

The focus of this work is providing a systematic study of MCRH. We consider both the question of constructing MCRH and what applications can we derive from them.

1.1.1 Constructions of MCRH

Since any CRH is in particular also an MCRH, candidate constructions are abundant (based on a variety of concrete computational assumptions). The actual question that we ask, which has a more foundational flavor, is whether we can construct MCRH from assumptions that are not known to imply CRH.

Our first main result is that the existence of MCRH follows from the average-case hardness of a variant of the *Entropy Approximation* problem studied by Goldreich, Sahai and Vadhan [GSV99]. Entropy Approximation, denoted EA, is a promise problem, where YES inputs are circuits whose output distribution (i.e., the distribution obtained by feeding random inputs to the circuit) has entropy at least k, whereas NO inputs are circuits whose output distribution has entropy at most $k-1$ (where k is a parameter that is unimportant for the current discussion). Here by entropy we specifically refer to *Shannon* entropy.[2] Goldreich et al. showed that EA is complete for the class of (promise) problems that have non-interactive statistical zero-knowledge proofs (NISZK).

In this work we consider a variant of EA, first studied by Dvir et al. [DGRV11], that uses different notions of entropy. Specifically, consider the promise problem $\mathsf{EA}_{\min,\max}$, where the goal now is to distinguish between circuits whose output distribution has *min*-entropy[3] at least k from those with *max*-entropy at most $k-1$. It is easy to verify that $\mathsf{EA}_{\min,\max}$ is an easier problem than EA.

Theorem 1.1 (Informal, see Theorem 3.6). *If* $\mathsf{EA}_{\min,\max}$ *is average-case hard, then there exist* (s,t)*-MCRH, where* $s=\sqrt{n}$ *and* $t=6n^2$.

(Note that in the MCRH that we construct there exist $2^{\sqrt{n}}$-way collisions, but it is computationally hard to find even a $6n^2$-way collision.)

In contrast to the original entropy approximation problem, we do not know whether $\mathsf{EA}_{\min,\max}$ is complete for NISZK. Thus, establishing the existence of MCRH based solely on the average-case hardness of NISZK (or SZK) remains open. Indeed such a result could potentially be an interesting extension of Ostrovsky's [Ost91] proof that average-case hardness of SZK implies the existence of one-way functions.

Instantiations. Dvir et al. [DGRV11], showed that the average-case hardness of $\mathsf{EA}_{\min,\max}$ is implied by either the quadratic residuocity (QR) or decisional Diffie Hellman (DDH) assumptions.[4] It is not too hard to see that above extends to any encryption scheme (or even commitment scheme) in which ciphertexts can be perfectly re-randomized.[5]

---

[2] Recall that the Shannon Entropy of a random variable X is defined as $\mathrm{H}_{\mathsf{Shannon}}(X) = \mathrm{E}_{x\leftarrow X}\left[\log\left(\frac{1}{\Pr[X=x]}\right)\right]$.

[3] For a random variable X, the min-entropy is defined as $\mathrm{H}_{\min}(X) = \min_{x\in\mathrm{Supp}(X)}\log\left(\frac{1}{\Pr[X=x]}\right)$ whereas the max-entropy is $\mathrm{H}_{\max}(X)=\log\left(|\mathrm{Supp}(X)|\right)$.

[4] In fact, [DGRV11] show that the same conclusion holds even if we restrict the problem to constant-depth (i.e., NC_0) circuits.

[5] Given such a scheme consider a circuit that has, hard-coded inside, a pair of ciphertexts (c_0,c_1) which are either encryptions of the same bit or of different bits. The circuit gets as input a bit b and random string r and outputs a re-randomization of c_b (using randomness r). If the scheme is perfectly re-randomizing (and perfectly correct) then the min-entropy of the output distribution in case the plaintexts disagree is larger than the max-entropy in case the plaintexts agree.

The hardness of $\mathsf{EA}_{\min,\max}$ can also be shown to follow from the *average-case* hardness of the Shortest Vector Problem or the Closest Vector Problem with approximation factor roughly $\sqrt{n}$.[6] To the best of our knowledge the existence of CRH is not known based on such small approximation factors (even assuming average-case hardness).

We remark that a similar argument establishes the hardness of $\mathsf{EA}_{\min,\max}$ based on the plausible assumption that graph isomorphism is average-case hard.[7]

1.1.2 Applications of MCRH

The main application that we derive from MCRH is a *constant-round* statistically hiding commitment scheme.

Theorem 1.2 (Informally stated, see Theorem 4.4). *Assume that there exists a* $(\log(t), t)$*-MCRH. Then, there exists a* 3*-round statistically-hiding and computationally-binding commitment scheme.*

We note that Theorem 1.2 is optimal in the sense of holding for MCRH that are minimally shrinking. Indeed, as noted in Remark 1.2, (s,t)-MCRH with $s \leq \log(t-1)$ exist trivially and unconditionally.

It is also worthwhile to point out that by a result of Haitner et al. [HNO+09], statistically-hiding commitment schemes can be based on the existence of any one-way function. However, the commitment scheme of [HNO+09] uses a polynomial number of rounds of interaction and the main point in Theorem 1.2 is that we obtain such a commitment scheme with only a *constant* number of rounds.

Moreover, by a result of [HHRS15], any *fully black-box* construction of a statistically hiding commitment scheme from one-way functions (or even one-way permutations) must use a polynomial number of rounds. Loosely speaking, a construction is "fully black-box" [RTV04] if (1) the construction only requires an input-output access to the underlying primitive and (2) the security proof also relies on the adversary in a black-box way. Most constructions in cryptography

[6] The hard distribution for $\mathsf{SVP}_{\sqrt{n}}$ and $\mathsf{CVP}_{\sqrt{n}}$ is the first message from the 2-message honest-verifier SZK proof system of Goldreich and Goldwasser [GG98]. In the case of $\mathsf{CVP}_{\sqrt{n}}$, the input is (B,t,d) where B is the basis of the lattice, t is a target vector and d specifies the bound on the distance of t from the lattice. The distribution is obtained by sampling a random error vector η from the ball of radius $d\sqrt{n}/2$ centered at the origin and outputting $b \cdot t + \eta \mod \mathcal{P}(B)$, where $b \leftarrow \{0,1\}$ and $\mathcal{P}(B)$ is the fundamental parallelopiped of B. When t is far from the lattice, this distribution is injective and hence has high min-entropy while when t is close to the lattice, the distribution is not injective and hence has lower max-entropy. Similarly for $\mathsf{SVP}_{\sqrt{n}}$, on input (B,d), the output is $\eta \mod \mathcal{P}(B)$ where η is again sampled from a ball of radius $d\sqrt{n}/2$.

[7] Note that the graph isomorphism is known to be solvable in polynomial-time for many natural distributions, and the recent breakthrough result of Babai [Bab16] gives a quasi-polynomial worst-case algorithm. Nevertheless, it is still plausible that Graph Isomorphism is average-case quasi-polynomially hard (for some efficiently samplable distribution).

are fully black-box. Since our proof of Theorem 1.2 is via a fully black-box construction, we obtain the following immediate corollary:

Corollary 1.3 (Informally stated). *There does not exist a fully blackbox construction of* MCRH *from one-way permutations.*

Corollary 1.3 can be viewed as an extension of Simon's [Sim98] blackbox separation of CRH from one-way permutations. Due to space limitations, the formal statement and proof of Corollary 1.3 is deferred to the full version of this paper [BDRV17].

1.2 Related Works

Generic Constructions of CRH. Peikert and Waters [PW11] construct CRH from lossy trapdoor functions. Their construction can be viewed as a construction of CRH from $\mathsf{EA}_{\min,\max}$ with a huge gap. (Specifically, the lossy trapdoor function h is either injective (i.e., $\mathrm{H}_{\min}(h) \geq n$) or very shrinking (i.e., $\mathrm{H}_{\max}(h) < 0.5n$).[8] One possible approach to constructing CRH from lossy functions with small 'lossiness' ($\mathrm{H}_{\max}(h)/\,\mathrm{H}_{\min}(h)$) is to first amplify the lossiness and then apply the [PW11] construction. Pietrzak et al. [PRS12] rule out this approach by showing that it is impossible to improve the 'lossiness' in a black-box way.[9] We show that even with distributions where the gap is tiny, we can achieve weaker yet very meaningful notions of collision-resistance.

Applebaum and Raykov [AR16] construct CRH from any average-case hard language with a *Perfect Randomized Encoding* in which the encoding algorithm is one-to-one as a function of the randomness. Perfect Randomized Encodings are a way to encode the computation of a function f on input x such that information-theoretically, the *only* information revealed about x is the value $f(x)$. The class of languages with such randomized encodings PRE is contained in PZK. Their assumption of an average-case hard language with a perfect randomized encoding implies $\mathsf{EA}_{\min,\max}$ as well.

Constant-Round Statistically Hiding Commitments from SZK **Hardness.** The work of Ong and Vadhan [OV08] yields constant-round statistically-hiding commitment schemes from average-case hardness of SZK.[10] Our construction of statistically-hiding commitments via MCRH is arguably simpler, although it relies on a stronger assumption ($\mathsf{EA}_{\min,\max}$) instead of average-case hardness of SZK.

[8] The trapdoor to the lossy function is not used in the construction of CRH.

[9] In contrast, it is easy to see that repetition amplifies the *additive* gap between the min-entropy and the max-entropy. In fact, we use this in our construction.

[10] Actually, Ong and Vadhan [OV08] only construct instance-dependent commitments. Dvir et al. [DGRV11] attribute the construction of constant-round statistically hiding commitments from average-case hardness of SZK to a combination of [OV08] and an unpublished manuscript of Guy Rothblum and Vadhan [RV09].

Distributional CRH. A different weakening of collision resistance was considered by Dubrov and Ishai [DI06]. Their notion, called "distributional collision-resistant" in which it may be feasible to find some specific collision, but it is hard to sample a *random* collision pair. That is, given the hash function h, no efficient algorithm can sample a pair (z_1, z_2) such that z_1 is uniform and z_2 is uniform in the set $\{z : h(z) = h(z_1)\}$. The notions of MCRH and distributional CRH are incomparable and whether one can be constructed from the other is open.

Min-Max Entropy Approximation. The main result of the work of Dvir et al. [DGRV11] (that was mentioned above) was showing that the problem EA for degree-3 polynomial mappings (i.e., where the entropies are measured by Shannon entropy) is complete for $\mathsf{SZK_L}$, a sub-class of SZK in which the verifier and the simulator run in logarithmic space. They also construct algorithms to approximate different notions of entropy in certain restricted settings (but their algorithms do not violate the assumption that $\mathsf{EA}_{\mathrm{min,max}}$ is average-case hard).

1.2.1 Independent Works

MCRH have been recently considered in an independent work by Komargodski et al. [KNY17b] (which was posted online roughly four months prior to the first public posting of our work). Komargodski et al. study the problem, arising from Ramsey theory, of finding either a clique or an independent set (of roughly logarithmic size) in a graph, when such objects are guaranteed to exist. As one of their results, [KNY17b] relate a variant of the foregoing Ramsey problem (for bipartite graphs) to the existence of MCRH. We emphasize that the focus of [KNY17b] is in studying computational problems arising from Ramsey theory, rather than MCRH directly.

Beyond the work of [KNY17b], there are two other concurrent works that specifically study MCRH [BPK17, KNY17a] (and were posted online simultaneously to our work). The main result of [KNY17a] is that the existence of MCRH (with suitable parameters) implies the existence of efficient argument-systems for NP, *á* la Kilian's protocol [Kil92]. Komargodski et al. [KNY17a] also prove that MCRH imply constant-round statistically hiding commitments (similarly to Theorem 1.2), although their result only holds for MCRH who shrink their input by a constant multiplicative factor. Lastly, [KNY17a] also show a blackbox separation between MCRH in which it is hard to find t collisions from those in which it is hard to find $t + 1$ collisions.

Bitansky et al. [BPK17] also study MCRH, with the motivation of constructing efficient argument-systems. They consider both a keyed version of MCRH (as in our work) and an unkeyed version (in which, loosely speaking, the requirement is that adversary cannot produce more collisions than those it can store as non-uniform advice). [BPK17] show a so-called "domain extension" result for MCRH that are sufficiently shrinking. Using this result they construct various succinct and/or zero-knowledge argument-systems, with optimal or close-to-optimal round complexity. In particular, they show the existence of 4 round zero-knowledge arguments for NP based on MCRH, and, assuming unkeyed MCRH, they obtain a similar result but with only 3 rounds of interaction.

1.3 Our Techniques

We provide a detailed overview of our two main results: Constructing MCRH from $\mathsf{EA}_{\mathrm{min,max}}$ and constructing constant-round statistically-hiding commitment scheme from MCRH.

1.3.1 Constructing MCRH from $\mathsf{EA}_{\mathrm{min,max}}$

Assume that we are given a distribution on circuits $\left\{C\colon \{0,1\}^n \to \{0,1\}^{2n}\right\}$ such that it is hard to distinguish between the cases $\mathrm{H}_{\mathsf{min}}(C) \geq k$ or $\mathrm{H}_{\mathsf{max}}(C) \leq k-1$, where we overload notation and let C also denote the output distribution of the circuit when given uniformly random inputs. Note that we have set the output length of the circuit C to $2n$ but this is mainly for concreteness (and to emphasize that the circuit need not be shrinking).

Our goal is to construct an MCRH using C. We will present our construction in steps, where in the first case we start off by assuming a very large entropy gap. Specifically, for the first (over-simplified) case, we assume that it is hard to distinguish between min-entropy $\geq n$ vs. max-entropy $\leq n/2$.[11] Note that having min-entropy n means that C is *injective.*

Warmup: The case of $\mathrm{H}_{\mathsf{min}}(C) \geq n$ vs. $\mathrm{H}_{\mathsf{max}}(C) \ll n/2$. In this case, it is already difficult to find even a 2-way collision in C: if $\mathrm{H}_{\mathsf{min}}(C) \geq n$, then C is injective and no collisions exist. Thus, if one can find a collision, it must be the case that $\mathrm{H}_{\mathsf{max}}(C) \leq n/2$ and so any collision finder distinguishes the two cases.

The problem though is that C by itself is not shrinking, and thus is not an MCRH. To resolve this issue, a natural idea that comes to mind is to hash the output of C, using a pairwise independent hash function.[12] Thus, the first idea is to choose $f : \{0,1\}^{2n} \to \{0,1\}^{n-s}$, for some $s \geq 1$, from a family of pairwise independent hash functions and consider the hash function $h(x) = f(C(x))$.

If $\mathrm{H}_{\mathsf{min}}(C) \geq n$ (i.e., C is injective), then every collision in h is a collision on the hash function f. On the other hand, if $\mathrm{H}_{\mathsf{max}}(C) \leq n/2$, then C itself has many collisions. To be able to distinguish between the two cases, we would like that in the latter case there will be no collisions that originate from f. The image size of C, if $\mathrm{H}_{\mathsf{max}}(C) \ll n/2$, is smaller than $2^{n/2}$. If we set s to be sufficiently small (say constant) than the range of f has size roughly 2^n. Thus, we are hashing a set into a range that is more than quadratic in its size. In such case, we are "below the birthday paradox regime" and a *random function* on this set will be injective. A similar statement can be easily shown also for functions that are merely pairwise independent (rather than being entirely random).

Thus, in case C is injective, all the collisions appear in the second part of the hash function (i.e., the application of f). On the other hand, if C has max-entropy smaller than $n/2$, then all the collisions happen in the first part of the

[11] This setting (and construction) is similar to that of Peikert and Waters's construction of CRH from lossy functions [PW11].

[12] Recall that a collection of functions $\mathcal{F}$ is k-wise independent if for every distinct $x_1, \ldots, x_k$, the distribution of $(f(x_1), \ldots, f(x_k))$ (over the choice of $f \leftarrow \mathcal{F}$) is uniform.

hash function (i.e., in C). Thus, any adversary that finds a collision distinguishes between the two cases and we actually obtain a full-fledged CRH (rather than merely an MCRH) at the cost of making a much stronger assumption.

The next case that we consider is still restricted to circuits that are injective (i.e., have min entropy n) in one case but assumes that it is hard to distinguish injective circuits from circuits having max-entropy $n-\sqrt{n}$ (rather than $n/2$ that we already handled).

The case of $\mathrm{H}_{\min}(C) \geq n$ vs. $\mathrm{H}_{\max}(C) \leq n - \sqrt{n}$. The problem that we encounter now is that in the low max entropy case, the output of C has max-entropy $n - \sqrt{n}$. To apply the above birthday paradox argument we would need the range of f to be of size roughly $(2^{n-\sqrt{n}})^2 \gg 2^n$ and so our hash function would not be shrinking. Note that if the range of f were smaller, than even if f were chosen entirely at random (let alone from a pairwise independent family) we would see collisions in this case (again, by the birthday paradox).

The key observation that we make at this point is that although we will see collisions, there will not be too many of them. Specifically, suppose we set $s \approx \sqrt{n}$. Then, we are now hashing a set of size $2^{n-\sqrt{n}}$ into a range of size $2^{n-\sqrt{n}}$. If we were to choose f entirely at random, this process would correspond to throwing $N = 2^{n-\sqrt{n}}$ balls (i.e., the elements in the range of C) into N bins (i.e., elements in the range of f). It is well-known that in such case, with high probability, the maximal load for any bin will be at most $\frac{\log(N)}{\log\log(N)} < n$. Thus, we are guaranteed that there will be at most n collisions.

Unfortunately, the work of Alon et al. [ADM+99] shows that the same argument does not apply to functions that are merely pairwise independent (rather than entirely random). Thankfully though, suitable derandomizations are known. Specifically, it is not too difficult to show that if we take f from a family of *n-wise independent hash functions*, then the maximal load will also be at most n (see Sect. 2.2 for details).[13]

Similarly to before, in case C is injective, there are no collisions in the first part. On the other hand, in case C has max-entropy at most $n - \sqrt{n}$, we have just argued that there will be less than n collisions in the second part. Thus, an adversary that finds an n-way collision distinguishes between the two cases and we have obtained an (s,t)-MCRH, with $s = \sqrt{n}$ and $t = n$ (i.e., collisions of size $2^{\sqrt{n}}$ exist but finding a collision of size even n is computationally infeasible).

The case of $\mathrm{H}_{\min}(C) \geq k$ vs. $\mathrm{H}_{\max}(C) \leq k - \sqrt{n}$. We want to remove the assumption that when the min-entropy of C is high, then it is in fact injective. Specifically, we consider the case that either C's min-entropy is at least k (for some parameter $k \leq n$) or its max entropy is at most $k - \sqrt{n}$. Note that in the high min-entropy case, C — although not injective — maps at most 2^{n-k} inputs to every output (this is essentially the definition of min-entropy). Our approach is to apply hashing a second time (in a different way), to effectively make C injective, and then apply the construction from the previous case.

[13] We remark that more efficient constructions are known, see Remark 2.4.

Consider the mapping $h'(x) = (C(x), f(x))$, where f will be defined ahead. For h' to be injective, f must be injective over all sets of size 2^{n-k}. Taking f to be pairwise-independent will force to set its output length to be too large, in a way that will ruin the entropy gap between the cases.

As in the previous case, we resolve this difficulty by using many-wise independent hashing. Let $f\colon \{0,1\}^n \to \{0,1\}^{n-k}$ be a $3n$-wise independent hash function. If $\mathrm{H_{min}}(C) \geq k$, then the same load-balancing property of f that we used in the previous case, along with a union bound, implies that with high probability (over the choice of f) there will be no $3n$-way collisions in h'. Our final construction applies the previous construction on h'. Namely,

$$h_{C,f,g}(x) = g(C(x), f(x)),$$

for $f\colon \{0,1\}^n \to \{0,1\}^{n-k}$ and $g\colon \{0,1\}^{3n-k} \to \{0,1\}^{n-\sqrt{n}}$ being $3n$-wise and $2n$-wise independent hash functions, respectively. We can now show that

- If $\mathrm{H_{min}}(C) \geq k$, then there do not exist $3n$ distinct inputs $x_1, \ldots, x_{3n}$ such that they all have the same value of $(C(x_i), f(x_i))$; and
- If $\mathrm{H_{max}}(C) \leq k - \sqrt{n}$, then there do not exist $2n$ distinct inputs $x_1, \ldots, x_{2n}$ such that they all have distinct values of $(C(x_i), f(x_i))$, but all have the same value $g(C(x_i), f(x_i))$.

We claim that $h_{C,f,g}$ is (s,t)-MCRH for $s = \sqrt{n}$ and $t = 6n^2$: First, note that in any set of $6n^2$ collisions for $h_{C,f,g}$, there has to be either a set of $3n$ collisions for (C, f) or a set of $2n$ collisions for g, and so at least one of the conditions in the above two statements is violated. Now, assume that an adversary A finds a $6n^2$-way collision in $h_{C,f,g}$ with high probability. Then, an algorithm D that distinguishes between $\mathrm{H_{min}}(C) \geq k$ to $\mathrm{H_{max}}(C) \leq k - \sqrt{n}$ chooses f and g uniformly at random and runs A on the input $h = h_{C,f,g}$ to get $x_1, \ldots, x_{6n^2}$ with $h(x_1) = \cdots = h(x_{6n^2})$. The distinguisher D now checks which of the two conditions above is violated, and thus can distinguish if it was given C with $\mathrm{H_{min}}(C) \geq k$ or $\mathrm{H_{max}}(C) \leq k - \sqrt{n}$.

We proceed to the case that the entropy gap is 1 (rather than $\sqrt{n}$). This case is rather simple to handle (via a reduction to the previous case).

The case of $\mathrm{H_{min}}(C) \geq k$ vs. $\mathrm{H_{max}}(C) \leq k - 1$. This case is handled by reduction to the previous case. The main observation is that if C has min-entropy at least k, and we take ℓ copies of C, then we get a new circuit with min-entropy at least $\ell \cdot k$. In contrast, if C had max-entropy at most $k-1$, then C' has max-entropy at most $\ell \cdot k - \ell$. Setting $\ell = k$, we obtain that in the second case the max-entropy is $n' - \sqrt{n'}$, where $n' = \ell \cdot k$ is the new input length. Thus, we have obtained a reduction to the $\sqrt{n'}$ gap case that we already handled.

1.3.2 Statistically-Hiding Commitments from MCRH

The fact that MCRH imply constant-round statistically-hiding commitments can be shown in two ways. The first, more direct way, uses only elementary notions such as k-wise independent hashing and is similar to the interactive hashing protocol of Ding et al. [DHRS07]. An alternative method, is to first show that MCRH imply the existence of an ($O(1)$-block) *inaccessible entropy generator* [HRVW09,HV17]. The latter was shown by [HRVW09,HV17] to imply the existence of constant-round statistically-hiding commitments. We discuss these two methods next and remark that in our actual proof we follow the direct route.

1.3.2.1 Direct Analysis

In a nutshell our approach is to follow the construction of Damgård et al. [DPP93] of statistically-hiding commitments from CRH, while replacing the use of pairwise independent hashing, with the interactive hashing of Ding et al. [DHRS07]. We proceed to the technical overview, which does not assume familiarity with any of these results.

Warmup: Commitment from (Standard) CRH. Given a family of collision resistant hash functions $\mathcal{H} = \left\{h\colon \{0,1\}^n \to \{0,1\}^{n-1}\right\}$, a natural first attempt is to have the receiver sample the hash function $h \leftarrow \mathcal{H}$ and send it to the sender. The sender, trying to commit to a bit b, chooses $x \leftarrow \{0,1\}^n$ and $r \leftarrow \{0,1\}^n$, and sends $(y = h(x), r, \sigma = \langle r, x\rangle \oplus b)$ to the receiver. The commitment is defined as $c = (h, y, r, \sigma)$. To reveal, the sender sends (x, b) to the receiver, which verifies that $h(x) = y$ and $\sigma = \langle r, x\rangle \oplus b$. Pictorially, the commit stage is as follows:

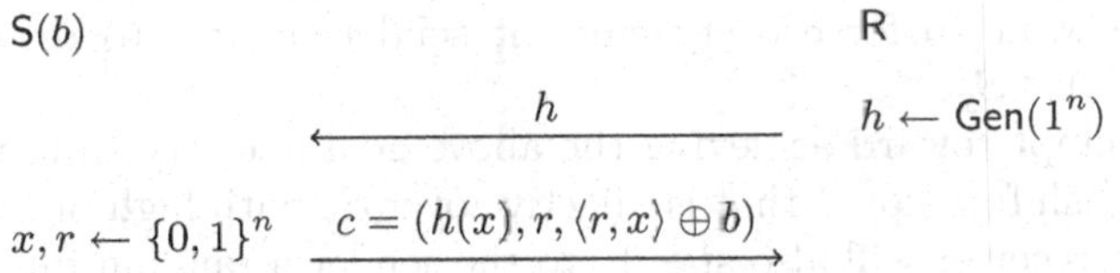

$\mathsf{S}(b)$		R
	$\xleftarrow{\quad h \quad}$	$h \leftarrow \mathsf{Gen}(1^n)$
$x, r \leftarrow \{0,1\}^n$	$\xrightarrow{c = (h(x), r, \langle r, x\rangle \oplus b)}$	

The fact that the scheme is computationaly binding follows immediately from the collision resistance of h: if the sender can find $(x, 0)$ and $(x', 1)$ that pass the receiver's verification, then $x \neq x'$ and $h(x) = h(x')$.

Arguing that the scheme is statistically-hiding is trickier. The reason is that $h(x)$ might reveal a lot of information on x. What helps us is that h is *shrinking*, and thus some information about x is hidden from the receiver. In particular, this means that x has positive min-entropy given $h(x)$. At this point we would like to apply the Leftover Hash Lemma (LHL) to show that for any b, the statistical distance between $(h(x), r, \langle r, x\rangle \oplus b)$ and $(h(x), r, u)$ is small. Unfortunately, the min-entropy level is insufficient to derive anything meaningful from the LHL and indeed the distance between these two distributions is a constant (rather than negligible as required).

To reduce the statistical distance, we increase the min-entropy via repetition. We modify the protocol so that the sender selects k values $\mathbf{x} = (x_1, \ldots, x_k) \leftarrow \{0,1\}^{n \cdot k}$ and $r \leftarrow \{0,1\}^{n \cdot k}$, and sends $\left(h(x_1), \ldots, h(x_k), r, \langle r, \mathbf{x}\rangle \oplus b\right)$ to the receiver. The min-entropy of $\mathbf{x}$, even given $h(x_1), \ldots, h(x_k)$ is now $\Omega(k)$, and the LHL now yields that the statistical distance between the two distributions

$\big(h, h(x_1), \dots, h(x_k), r, \langle r, \mathbf{x}\rangle \oplus 0\big)$ and $\big(h, h(x_1), \dots, h(x_k), r, \langle r, \mathbf{x}\rangle \oplus 1\big)$ is roughly 2^{-k}. Setting k to be sufficiently large (e.g., $k = \mathsf{poly}(n)$ or even $k = \mathsf{poly}\log(n)$) we obtain that the scheme is statistically-hiding. Note that repetition also does not hurt binding: if the sender can find valid decommitments $(\mathbf{x} = (x_1 \dots, x_k), 0)$ and $(\mathbf{x}' = (x'_1, \dots, x'_k), 1)$ that pass the receiver's verification, then there must exist $i \in [k]$ with $x_i \neq x'_i$ and $h(x_i) = h(x'_i)$ (i.e., a collision).

Handling MCRHs. For simplicity, let us focus on the case $t = 4$ (since it basically incorporates all the difficulty encountered when dealing with larger values of t). That is, we assume that $\mathcal{H} = \left\{h \colon \{0,1\}^n \rightarrow \{0,1\}^{n-s}\right\}$ is an (s, t)-MCRH with $s = 2$ and $t = 4$. Namely, it is hard to find 4 inputs that map to the same hash value for a random function from $\mathcal{H}$, even though such 4-way collisions exist. Note however that it might very well be easy to find 3 such colliding inputs. And indeed, the binding argument that we had before breaks: finding $x \neq x'$ with $h(x) = h(x)$ is no longer (necessarily) a difficult task.

The problem comes up because even after the sender 'commits' to $y_1 = h(x_1), \dots, y_k = h(x_k)$, it is no longer forced to reveal $x_1, \dots, x_k$. Intuitively, for every y_i, the sender might know 3 inputs that map to y_i, so, the sender is free to reveal any value in the Cartesian product of these triples. Concretely, let $\mathcal{S}_{y_i}$ be the set of inputs that h maps to y_i that the sender can find efficiently, and let $\mathcal{S}_{\mathbf{y}} = \mathcal{S}_{y_1} \times \dots \times \mathcal{S}_{y_k}$. Since the sender can find at most 3 colliding inputs, it holds that $|\mathcal{S}_{y_i}| \leq 3$ for every i, and thus $|\mathcal{S}_{\mathbf{y}}| \leq 3^k$. To fix the binding argument, we want to force every efficient sender to able to reveal a *unique* $\mathbf{x} = (x_1, \dots, x_k) \in \mathcal{S}_{\mathbf{y}}$.

A first attempt toward achieving the above goal is to try to use a pairwise-independent hash function f that is injective over $\mathcal{S}_{\mathbf{y}}$ with high probability. At a high level, the receiver will also specify to the sender a random function f from the pairwise independent hash function family. The sender in turn sends $f(\mathbf{x})$ as well as $(h(x_1), \dots, h(x_k))$. The receiver adds a check to the verification step to ensure that f maps the decommited input sequence $(x'_1, \dots, x'_k)$ to the value that was pre-specified.

In order for the function f to be injective on the set $\mathcal{S}_{\mathbf{y}}$, the birthday paradox tells us that the range of f must have size at least $|\mathcal{S}_{\mathbf{y}}|^2$ (roughly), which means at least 3^{2k}. Thus, to ensure that f is injective on $\mathcal{S}_{\mathbf{y}}$, we can use a pairwise-independent function $f \colon \{0,1\}^{nk} \rightarrow \{0,1\}^{2k\log(3)}$.

Unfortunately, this scheme is still not binding: f is promised (with high probability) to be injective for *fixed* sets of size 3^k, but the sender can choose $\mathbf{y}$ based on the value of f. Specifically, to choose $\mathbf{y}$ so that f is not injective over $\mathcal{S}_{\mathbf{y}}$. To fix the latter issue, we split the messages that the receiver sends into two rounds. In the first round the receiver sends h and receives $\mathbf{y} = \big(h(x_1), \dots, h(x_k)\big)$ from the sender. Only then the receiver sends f and receives $z_1 = f(\mathbf{x})$. Now, the scheme is binding: since f is chosen *after* $\mathbf{y}$ is set, the pairwise-independence property guarantees that f will be injective over $\mathcal{S}_{\mathbf{y}}$ with high probability. Pictorially, the commit stage of the new scheme is as follows:

S(b)		R
	$\xleftarrow{\quad h \quad}$	$h \leftarrow \mathsf{Gen}(1^n)$
$\mathbf{x} \leftarrow \{0,1\}^{nk}$, $y_i = h(x_i)$	$\xrightarrow{\mathbf{y} = (y_1, y_2 \dots y_k)}$	
	$\xleftarrow{\quad f \quad}$	$f : \{0,1\}^{nk} \to \{0,1\}^{2k\log(3)}$
$r \leftarrow \{0,1\}^{nk}$	$\xrightarrow{f(\mathbf{x}), r, \langle r, \mathbf{x}\rangle \oplus b}$	

But is this scheme statistically-hiding? Recall that previously, to argue hiding, we used the fact that the mapping $(x_1, \dots, x_k) \mapsto (h(x_1), \dots, h(x_k))$ is shrinking. In an analogous manner, here, we need the mapping $(x_1, \dots, x_k) \mapsto (h(x_1), \dots, h(x_k), f(\mathbf{x}))$ to be shrinking. However, the latter mapping maps strings of length $n \cdot k$ bits to strings of length $(n-2) \cdot k + 2\log(3) \cdot k$, which is obviously not shrinking.

One work-around is to simply assume that the given MCRH shrinks much more than we assumed so far. For example, to assume that $\mathcal{H}$ is $(4,4)$-MCRH (or more generally (s,t)-MCRH for $s \gg \log(t)$).[14] However, by adding one more round of interaction we can actually fix the protocol so that it gives statistically-hiding commitments even with tight shrinkage of $\log(t)$.

Overcoming the Birthday Paradox. To guarantee hiding, it seems that we cannot afford the range of f to be as large as $(3^k)^2$. Instead, we set its range size to 3^k (i.e., $f\colon \{0,1\}^{nk} \to \{0,1\}^{k\log(3)}$). Moreover, rather than choosing it from a pairwise independent hash function family, we shall one more use one that is *many-wise-independent*. Again, the important property that we use is that such functions are *load-balanced*[15] with high probability, z_1 — the value that the sender sends in the second round — has at most $\log(3^k) = k \cdot \log(3)$ pre-images from $\mathcal{S}_\mathbf{y}$ under f (i.e., $|\{\mathbf{x} \in \mathcal{S}_\mathbf{y} \colon f(\mathbf{x}) = z_1\}| \le k \cdot \log(3)$). We once more face the problem that the sender can reveal any of these inputs, but now their number is exponentially smaller — it is only $k\log(3)$ (as opposed to 3^k before). We can now choose a pairwise-independent $g\colon \{0,1\}^{nk} \to \{0,1\}^{2(\log(k)+\log\log(3))}$ that is injective over sets of size $k \cdot \log(3)$ (with high probability). For the same reasons that f was sent after h, the receiver sends g only after receiving $f(\mathbf{x})$.

Thus, our final protocol has three rounds (where each round is composed of one message for each of the two parties) and is as follows: In the first round, the receiver selects $h \leftarrow \mathcal{H}$ and sends it to the sender. The sender,

[14] We remark that our construction of MCRH based on $\mathsf{EA}_{\mathbf{min,max}}$ (see Sect. 3) actually supports such large shrinkage.

[15] In a nutshell, the property that we are using is that if $N = 3^k$ balls are thrown into N bins, with high probability the maximal load in every bin will be at most $\log(N)$. It is well-known that hash functions that are $\log(N)$-wise independent also have this property. See Sect. 2.2 for details.

trying to commit to a bit b, chooses $\mathbf{x} = (x_1, \ldots, x_k) \leftarrow \{0,1\}^{nk}$ and sends $\mathbf{y} = (y_1 = h(x_1), \ldots, y_k = h(x_k))$. In the second round, the receiver selects a many-wise-independent hash function $f\colon \{0,1\}^{nk} \rightarrow \{0,1\}^{k\log(3)}$ and sends it to the sender. The sender sends $z_1 = f(\mathbf{x})$ to the receiver. In the third and final round, the receiver selects a pairwise-independent hash function $g\colon \{0,1\}^{n\cdot k} \rightarrow \{0,1\}^{2(\log(k)+\log\log(3))}$ and sends it to the sender. The sender selects $r \leftarrow \{0,1\}^{nk}$, and sends $(z_2 = g(\mathbf{x}), r, \sigma = \langle r, \mathbf{x}\rangle \oplus b)$ to the receiver. The commitment is defined as $c = (h, \mathbf{y}, f, z_1, g, z_2, \sigma)$. To reveal, the sender sends $(\mathbf{x}, b)$ to the receiver, which verifies that $h(x_i) = y_i$ for every i, that $f(\mathbf{x}) = z_1$, $g(\mathbf{x}) = z_2$ and $\sigma = \langle r, \mathbf{x}\rangle \oplus b$. Pictorially, the commit stage is as follows:

$\mathsf{S}(b)$		R
	$\xleftarrow{\quad h \quad}$	$h \leftarrow \mathsf{Gen}(1^n)$
$\mathbf{x} \leftarrow \{0,1\}^{nk}$, $y_i = h(x_i)$	$\xrightarrow{\mathbf{y} = (y_1, y_2 \ldots y_k)}$	
	$\xleftarrow{\quad f \quad}$	$f : \{0,1\}^{nk} \rightarrow \{0,1\}^{k\log(3)}$
	$\xrightarrow{\quad f(\mathbf{x}) \quad}$	
	$\xleftarrow{\quad g \quad}$	$g : \{0,1\}^{nk} \rightarrow \{0,1\}^{2(\log k+\log\log(3))}$
$r \leftarrow \{0,1\}^{nk}$	$\xrightarrow{g(\mathbf{x}), r, \langle r, \mathbf{x}\rangle \oplus b}$	

Intuitively, the scheme is computationally binding since for any computationally bounded sender that committed to c, there is a unique $\mathbf{x}$ that passes the receiver's verification. As for hiding, we need the mapping $(x_1, \ldots, x_k) \mapsto (h(x_1), \ldots, h(x_k), f(\mathbf{x}), g(\mathbf{x}))$ to be shrinking. Observe that we are mapping $n \cdot k$ bits to $(n-2)k + \log(3)k + 2(\log(k) + \log\log(3))$ bits (where all logarithms are to the base 2). Choosing k to be sufficiently large (e.g., $k = \mathsf{poly}(n)$ certainly suffices) yields that the mapping is shrinking.

This completes the high level overview of the direct analysis of our construction of constant-round statistically hiding commitments. The formal proof, done via a reduction from the binding of the scheme to the MCRH property, requires more delicate care (and in particular handling certain probabilistic dependencies that arise in the reduction). See Sect. 4 for details.

1.3.2.2 Analysis via Inaccesible Entropy

Consider the jointly distributed random variables $(h(x), x)$, where h is chosen at random from a family of t-way collision resistant hash functions $\mathcal{H} = \left\{h\colon \{0,1\}^n \rightarrow \{0,1\}^{n-\log(t)}\right\}$ and x is a uniform n-bit string. Since $h(x)$ is only $(n - \log(t))$ bits long, it can reveal only that amount of information about x.

Thus, the entropy of x given $h(x)$ (and h) is at least $\log(t)$. In fact, a stronger property holds: the expected number of pre-images of $h(x)$, over the choice of x, is t. This implies that x given $h(x)$ has $\log(t)$ bits of (a weaker variant of) min-entropy.

While $h(x)$ has t pre-images (in expectation), no *efficient* strategy can find more than $t-1$ of them. Indeed, efficiently finding t such (distinct) pre-images directly violates the t-way collision resistance of h.

In terms of inaccessible entropy, the foregoing discussion establishes that $(h(x), x)$ is a 2-block inaccessible entropy generator where the second block (i.e., x) has real min-entropy $\log(t)$ and accessible max-entropy at most $\log(t-1)$. This block generator is not quite sufficient to get statistically-hiding commitment since the construction of [HRVW09,HV17] requires a larger gap between the entropies. This, however, is easily solved since taking many copies of the same generator increases the entropy gap. That is, the final 2-block generator is $\Big((h(x_1),\ldots,h(x_k)),(x_1,\ldots,x_k)\Big)$, for a suitable choice of k. The existence of constant-round statistically-hiding commitment now follows immediately from [HV17, Lemma 19].[16] The resulting protocol turns out to be essentially the same as that obtained by the direct analysis discussed above (and proved in Sect. 4).

1.4 Organization

In Sect. 2 we provide standard definitions and basic facts. In Sect. 3 we formally state the entropy approximation assumption and present our construction of MCRH based on this assumption. Lastly, In Sect. 4 we describe the construction of constant-round statistically-hiding commitments from MCRH.

As already mentioned, we defer the proof of the blackbox separation of MCRH from one-way permutations to the full version of this paper [BDRV17].

2 Preliminaries

We use lowercase letters for values, uppercase for random variables, uppercase calligraphic letters (e.g., $\mathcal{U}$) to denote sets, boldface for vectors (e.g., $\mathbf{x}$), and uppercase sans-serif (e.g., A) for algorithms (i.e., Turing Machines). All logarithms considered here are in base two. We let poly denote the set of all polynomials. A function $\nu\colon \mathbb{N} \to [0,1]$ is *negligible*, denoted $\nu(n) = \mathsf{negl}(n)$, if $\nu(n) < 1/p(n)$ for every $p \in \mathsf{poly}$ and large enough n.

Given a random variable X, we write $x \leftarrow X$ to indicate that x is selected according to X. Similarly, given a finite set $\mathcal{S}$, we let $s \leftarrow \mathcal{S}$ denote that s is selected according to the uniform distribution on $\mathcal{S}$. We adopt the convention

[16] The general construction of statistically-hiding commitments from inaccessible entropy generators is meant to handle a much more general case than the one needed in our setting. In particular, a major difficulty handled by [HRVW09,HV17] is when the generator has many blocks and it is not known in which one there is a gap between the real and accessible entropies.

that when the same random variable occurs several times in an expression, all occurrences refer to a single sample. For example, $\Pr[f(X) = X]$ is defined to be the probability that when $x \leftarrow X$, we have $f(x) = x$. We write U_n to denote the random variable distributed uniformly over $\{0,1\}^n$. The support of a distribution D over a finite set $\mathcal{U}$, denoted $\mathrm{Supp}(D)$, is defined as $\{u \in \mathcal{U} : D(u) > 0\}$. The *statistical distance* of two distributions P and Q over a finite set $\mathcal{U}$, denoted as $\mathrm{SD}(P, Q)$, is defined as $\max_{\mathcal{S} \subseteq \mathcal{U}} |P(\mathcal{S}) - Q(\mathcal{S})| = \frac{1}{2} \sum_{u \in \mathcal{U}} |P(u) - Q(u)|$.

2.1 Many-Wise Independent Hashing

Many-wise independent hash functions are used extensively in complexity theory and cryptography.

Definition 2.1 (ℓ-wise Independent Hash Functions). *For $\ell \in \mathbb{N}$, a family of functions $\mathcal{F} = \{f : \{0,1\}^n \to \{0,1\}^m\}$ is* ℓ-wise independent *if for every distinct $x_1, x_2, \ldots, x_\ell \in \{0,1\}^n$ and every $y_1, y_2, \ldots, y_\ell \in \{0,1\}^m$, it holds that*

$$\Pr_{f \leftarrow \mathcal{F}}[f(x_1) = y_1 \wedge f(x_2) = y_2 \wedge \cdots \wedge f(x_\ell) = y_\ell] = \frac{1}{M^\ell}.$$

Note that if $\mathcal{H}$ is k-wise independent for $k \geq 2$, it is also universal. The existence of efficient many-wise hash function families is well known.

Fact 2.2 (c.f. [Vad12, Corollary 3.34]). *For every $n, m, \ell \in \mathbb{N}$, there exists a family of ℓ-wise independent hash functions $\mathcal{F}^{(\ell)}_{n,m} = \{f \colon \{0,1\}^n \to \{0,1\}^m\}$ where a random function from $\mathcal{F}^{(\ell)}_{n,m}$ can be selected using $\ell \cdot \max(m, n)$ bits, and given a description of $f \in \mathcal{F}^{(\ell)}_{n.m}$ and $x \in \{0,1\}^n$, the value $f(x)$ can be evaluated in time $\mathsf{poly}(n, m, \ell)$.*

Whenever we only need pairwise independent hash function $\mathcal{F}^{(2)}_{n,m}$, we remove the two from the superscript and simply write $\mathcal{F}_{n,m}$.

2.2 Load Balancing

The theory of load balancing deals with allocating elements into bins, such that no bin has too many elements. If the allocation is done at random, it can be shown that with high probability the max load (i.e., the number of elements in the largest bin) is not large. In fact, allocating via many-wise independent hash function also suffices.

Fact 2.3 (Folklore (see, e.g., [CRSW13])). *Let $n, m, \ell \in \mathbb{N}$ with $\ell \geq 2e$ (where e is the base of the natural logarithm) and let $\mathcal{F}^{(\ell)}_{n,m}$ be an ℓ-wise independent hash function family. Then, for every set $\mathcal{S} \subseteq \{0,1\}^n$ with $|\mathcal{S}| \leq 2^m$ it holds that:*

$$\Pr_{f \leftarrow \mathcal{F}^{(\ell)}_{n,m}}\left[\exists y \in \{0,1\}^m \text{ such that } \left|f^{-1}(y) \cap \mathcal{S}\right| \geq \ell\right] \leq 2^{m-\ell},$$

where $f^{-1}(y) = \{x \in \{0,1\}^n \colon f(x) = y\}$.

Proof. Fix $y \in \{0,1\}^m$. It holds that

$$\begin{aligned}
&\Pr_{f \leftarrow \mathcal{F}^{(\ell)}_{n,m}} \left[\left| f^{-1}(y) \cap \mathcal{S} \right| \geq \ell \right] \\
&\quad \leq \Pr_{f \leftarrow \mathcal{F}^{(\ell)}_{n,m}} \left[\exists \text{ distinct } x_1, \ldots, x_\ell \in \mathcal{S} \colon f(x_1) = y \wedge \cdots \wedge f(x_\ell) = y \right] \\
&\quad \leq \sum_{\text{distinct } x_1,\ldots,x_\ell \in \mathcal{S}} \Pr_{f \leftarrow \mathcal{F}^{\ell}_{n,m}} \left[f(x_1) = y \wedge \cdots \wedge f(x_\ell) = y \right] \\
&\quad \leq \binom{2^m}{\ell} \cdot \left(\frac{1}{2^m} \right)^\ell \\
&\quad \leq \left(\frac{e \cdot 2^m}{\ell} \right)^\ell \cdot \left(\frac{1}{2^m} \right)^\ell \\
&\quad \leq 2^{-\ell},
\end{aligned}$$

where the second inequality is by a union bound, the third inequality follows from the ℓ-wise independence of $\mathcal{F}^{(\ell)}_{n,m}$, the fourth inequality is by a standard bound on binomial coefficients, and the last inequality follows by our assumption that $\ell \geq 2e$.

Fact 2.3 follows from a union bound over all values of $y \in \{0,1\}^m$. □

Remark 2.4 (More Efficient Hash Functions). We remark that more efficient constructions of hash functions guaranteeing the same load balancing performance as in Fact 2.3 are known in the literature.

Specifically, focusing on the setting of $\ell = O(m)$, Fact 2.3 gives a load balancing guarantee for functions whose description size (i.e., key length) is $\Omega(m^2)$ bits. In contrast, a recent result of Celis et al. [CRSW13] constructs such functions that require only $\tilde{O}(m)$ key size. Furthermore, a follow up work of Meka et al. [MRRR14] improves the *evaluation time* of the [CRSW13] hash function to be only poly-logarithmic in m (in the word RAM model).

However, since our focus is not on concrete efficiency, we ignore these optimizations throughout this work.

3 Constructing MCRH Families

In this section, we present a construction of a Multi-Collision Resistant Hash family (MCRH) based on the hardness of estimating certain notions of entropy of a distribution, given an explicit description of the distribution (i.e., a circuit that generates it). We define and discuss this problem in Sect. 3.1, and present the construction of MCRH in Sect. 3.2.

3.1 Entropy Approximation

In order to discuss the problem central to our construction, we first recall some standard notions of entropy.

Definition 3.1. *For a random variable X, we define the following notions of entropy:*

- **Min-entropy:** $\mathrm{H_{min}}(X) = \min_{x \in \mathrm{Supp}(X)} \log\left(\frac{1}{\Pr[X=x]}\right)$.
- **Max-entropy:** $\mathrm{H_{max}}(X) = \log\left(|\mathrm{Supp}(X)|\right)$.
- **Shannon entropy:** $\mathrm{H_{Shannon}}(X) = \mathrm{E}_{x \leftarrow X}\left[\log\left(\frac{1}{\Pr[X=x]}\right)\right]$.

For any random variable, these entropies are related as described below. These relations ensure that the problems we describe later are well-defined.

Fact 3.2. *For a random variable X supported over $\{0,1\}^m$,*

$$0 \leq \mathrm{H_{min}}(X) \leq \mathrm{H_{Shannon}}(X) \leq \mathrm{H_{max}}(X) \leq m.$$

Given a circuit $C : \{0,1\}^n \to \{0,1\}^m$, we overload C to also denote the random variable induced by evaluating C on a uniformly random input from $\{0,1\}^n$. With this notation, the Entropy Approximation problem is defined as below.

Definition 3.3 (Min-Max Entropy Approximation). *Let $g = g(n) \in \mathbb{R}$ be a function such that $0 < g(n) < n$. The* min-max Entropy Approximation *problem with* gap g, *denoted $\mathsf{EA}^{(g)}_{\mathrm{min,max}}$, is a promise problem* (YES, NO) *for* $\mathrm{YES} = \{\mathrm{YES}_n\}_{n \in \mathbb{N}}$ *and* $\mathrm{NO} = \{\mathrm{NO}_n\}_{n \in \mathbb{N}}$, *where we define*

$$\mathrm{YES}_n = \{(1^n, C_n, k) : \ \mathrm{H_{min}}(C_n) \geq k\}, \text{ and}$$
$$\mathrm{NO}_n = \{(1^n, C_n, k) : \ \mathrm{H_{max}}(C_n) \leq k - g(n)\},$$

and where in both cases C_n is a circuit that takes n bits of input, and $k \in \{0, \ldots, n\}$.

We also define $\mathsf{EA}_{\mathrm{min,max}} = \mathsf{EA}^{(1)}_{\mathrm{min,max}}$. That is, when we omit the gap g we simply mean that $g = 1$.

The *Shannon* Entropy Approximation problem (where $\mathrm{H_{min}}$ and $\mathrm{H_{max}}$ above are replaced with $\mathrm{H_{Shannon}}$), with constant gap, was shown by Goldreich et al. [GSV99] to be complete for the class NISZK (promise problems with non-interactive statistical zero knowledge proof systems). For a discussion of generalizations of Entropy Approximation to other notions of entropy, and other related problems, see [DGRV11].

3.1.1 The Assumption: Average-Case Hardness of Entropy Approximation.

Our construction of MCRH is based on the average-case hardness of the Entropy Approximation problem $\mathsf{EA}_{\mathrm{min,max}}$ defined above (i.e., with gap 1). We use the following definition of average-case hardness of promise problems.

Definition 3.4 (Average-case Hardness). *We say that a promise problem* $\Pi = (\mathrm{YES}, \mathrm{NO})$, *where* $\mathrm{YES} = \{\mathrm{YES}_n\}_{n\in\mathbb{N}}$ *and* $\mathrm{NO} = \{\mathrm{NO}_n\}_{n\in\mathbb{N}}$, *is* average-case hard *if there is a probabilistic algorithm* S *such that* $\mathsf{S}(1^n)$ *outputs samples from* $(\mathrm{YES}_n \cup \mathrm{NO}_n)$, *and for every family of polynomial-sized circuits* $\mathsf{A} = (\mathsf{A}_n)_{n\in\mathbb{N}}$,

$$\Pr_{x\leftarrow\mathsf{S}(1^n)}[\mathsf{A}_n(x) = \Pi(x)] \le \frac{1}{2} + \mathsf{negl}(n),$$

where $\Pi(x) = 1$ *if* $x \in \mathrm{YES}$ *and* $\Pi(x) = 0$ *if* $x \in \mathrm{NO}$. *We call* S *a* hard-instance sampler *for* Π. *The quantity* $(\Pr_{x\leftarrow\mathsf{S}(1^n)}[\mathsf{A}_n(x) = \Pi(x)] - 1/2)$ *is referred to as the* advantage *the algorithm* A *has in deciding* Π *with respect to the sampler* S.

In our construction and proofs, it will be convenient for us to work with the problem $\mathsf{EA}_{\min,\max}^{(\lfloor\sqrt{n}\rfloor)}$ rather than $\mathsf{EA}_{\min,\max} = \mathsf{EA}_{\min,\max}^{(1)}$. At first glance $\mathsf{EA}_{\min,\max}^{(\lfloor\sqrt{n}\rfloor)}$ seems to be an easier problem because the gap here is $\lfloor\sqrt{n}\rfloor$, which is much larger. The following simple proposition shows that these two problems are in fact equivalent (even in their average-case complexity). The key idea here is repetition: given a circuit C, we can construct a new circuit C' that outputs C evaluated on independent inputs with a larger gap.

Proposition 3.5. $\mathsf{EA}_{\min,\max}^{(\lfloor\sqrt{n}\rfloor)}$ *is average-case hard if and only if* $\mathsf{EA}_{\min,\max}^{(1)}$ *is average-case hard.*

Proof Sketch. Note that any YES instance of $\mathsf{EA}_{\min,\max}^{(\lfloor\sqrt{n}\rfloor)}$ is itself a YES instance of $\mathsf{EA}_{\min,\max}^{(1)}$, and the same holds for NO instances. So the average-case hardness of $\mathsf{EA}_{\min,\max}^{(\lfloor\sqrt{n}\rfloor)}$ immediately implies that of $\mathsf{EA}_{\min,\max}^{(1)}$, with the same hard-instance sampler. In order to show the implication in the other direction, we show how to use a hard-instance sampler for $\mathsf{EA}_{\min,\max}^{(1)}$ to construct a hard-instance sampler S' for $\mathsf{EA}_{\min,\max}^{(\lfloor\sqrt{n}\rfloor)}$.

S' on input (1^n):

1. Let $\ell = \lfloor\sqrt{n}\rfloor$. S' samples $(1^\ell, C_\ell, k) \leftarrow \mathsf{S}(1^\ell)$.
2. Let $\widehat{C}_n$ be the following circuit that takes an n-bit input x. It breaks x into $\ell+1$ disjoint blocks $x_1, \ldots, x_{\ell+1}$, where $x_1, \ldots, x_\ell$ are of size ℓ, and $x_{\ell+1}$ is whatever remains. It ignores $x_{\ell+1}$, runs a copy of C_ℓ on each of the other x_i's, and outputs a concatenation of all the outputs.
3. S' outputs $(1^n, \widehat{C}_n, k\cdot\ell)$.

As $\widehat{C}_n$ is the ℓ-fold repetition of C_ℓ, its max and min entropies are ℓ times the respective entropies of C_ℓ. So if C_ℓ had min-entropy at least k, then $\widehat{C}_n$ has min-entropy at least $k\cdot\ell$, and if C_ℓ had max-entropy at most $(k-1)$, then $\widehat{C}_n$ has max-entropy at most $(k-1)\cdot\ell = k\cdot\ell - \ell$, where $\ell = \lfloor\sqrt{n}\rfloor$. The proposition follows. □

3.2 The Construction

Our construction of a Multi-Collision Resistant Hash (MCRH) family is presented in Fig. 1. We now prove that the construction is secure under our average-case hardness assumption.

The Construction of MCRH

Let S be a hard-instance sampler for $\mathsf{EA}_{\min,\max}^{(\lfloor\sqrt{n}\rfloor)}$.

Gen(1^n):

1. Sample $(1^n, C_n, k) \leftarrow \mathsf{S}(1^n)$, where C_n maps $\{0,1\}^n \to \{0,1\}^{n'}$.
2. Sample[a] $f \leftarrow \mathcal{F}_{n,(n-k)}^{(3n)}$ and $g \leftarrow \mathcal{F}_{(n'+n-k),(n-\lfloor\sqrt{n}\rfloor)}^{(2n)}$.
3. Output the circuit that computes the function $h_{C_n,f,g} : \{0,1\}^n \to \{0,1\}^{n-\lfloor\sqrt{n}\rfloor}$ that is defined as follows:

$$h_{C_n,f,g}(x) := g\big(C_n(x), f(x)\big).$$

[a]Recall that $\mathcal{F}_{n,m}^{(\ell)} = \{f : \{0,1\}^n \to \{0,1\}^m\}$ is a family of ℓ-wise independent hash functions.

Fig. 1. Construction of MCRH from Entropy Approximation.

Theorem 3.6. *If* $\mathsf{EA}_{\min,\max}^{(\lfloor\sqrt{n}\rfloor)}$ *is average-case hard, then the construction in Fig. 1 is an* (s,t)-MCRH, *where* $s = \lfloor\sqrt{n}\rfloor$ *and* $t = 6n^2$.

The above theorem, along with Proposition 3.5, now implies the following.

Corollary 3.7. *If* $\mathsf{EA}_{\min,\max}$ *is average-case hard, then there exists an* (s,t)-MCRH, *where* $s = \lfloor\sqrt{n}\rfloor$ *and* $t = 6n^2$.

Note that above, the shrinkage being $\lfloor\sqrt{n}\rfloor$ guarantees that there exist $2^{\lfloor\sqrt{n}\rfloor}$-way collisions. But the construction is such that it is not possible to find even a $6n^2$-way collision, (which is sub-exponentially smaller). This is significant because, unlike in the case of standard collision-resistant hash functions (i.e., in which it is hard to find a pair of collisions), shrinkage in MCRHs cannot be easily amplified by composition while maintaining the same amount of collision-resistance (see Remark 1.2).

The rest of this section is dedicated to proving Theorem 3.6.

Proof of Theorem 3.6. Let Gen denote the algorithm described in Fig. 1, and S be the hard-instance sampler used there. Fact 2.2, along with the fact that S runs in polynomial-time ensures that Gen runs in polynomial-time as well. The shrinkage requirement of an MCRH is satisfied because here the shrinkage is $s(n) = \lfloor\sqrt{n}\rfloor$. To demonstrate multi-collision resistance, we show how to use an adversary that finds $6n^2$ collisions in hash functions sampled by Gen to break the average-case hardness of $\mathsf{EA}_{\min,\max}^{(\lfloor\sqrt{n}\rfloor)}$. For the rest of the proof, to avoid cluttering up notations, we will denote the problem $\mathsf{EA}_{\min,\max}^{(\lfloor\sqrt{n}\rfloor)}$ by just EA.

We begin with an informal discussion of the proof. We first prove that large sets of collisions that exist in a hash function output by Gen have different properties depending on whether the instance that was sampled in step 1 of Gen was a YES or NO instance of EA. Specifically, notice that the hash functions that are output by Gen have the form $h_{C_n,f,g}(x) = g(C_n(x), f(x))$; we show that, except with negligible probability:

- In functions $h_{C_n,f,g}$ generated from $(1^n, C_n, k) \in \text{YES}$, with high probability, there do not exist $3n$ distinct inputs $x_1, \ldots, x_{3n}$ such that they all have the same value of $(C_n(x_i), f(x_i))$.
- In functions $h_{C_n,f,g}$ generated from $(1^n, C_n, k) \in \text{NO}$, with high probability, there do not exist $2n$ distinct inputs $x_1, \ldots, x_{2n}$ such that they all have distinct values of $(C_n(x_i), f(x_i))$, but all have the same value $g(C_n(x_i), f(x_i))$.

Note that in any set of $6n^2$ collisions for $h_{C_n,f,g}$, there has to be either a set of $3n$ collisions for (C_n, f) or a set of $2n$ collisions for g, and so at least one of the conclusions in the above two statements is violated.

A candidate average-case solver for EA, when given an instance $(1^n, C_n, k)$, runs steps 2 and 3 of the algorithm Gen from Fig. 1 with this C_n and k. It then runs the collision-finding adversary on the hash function $h_{C_n,f,g}$ that is thus produced. If the adversary does not return $6n^2$ collisions, it outputs a uniformly random answer. But if these many collisions are returned, it checks which of the conclusions above is violated, and thus knows whether it started with a YES or NO instance. So whenever the adversary succeeds in finding collisions, the distinguisher can decide EA correctly with overwhelming probability. As long as the collision-finding adversary succeeds with non-negligible probability, then the distinguisher also has non-negligible advantage, contradicting the average-case hardness of EA.

We now state and prove the above claims about the properties of sets of collisions, then formally write down the adversary outlined above and prove that it breaks the average case hardness of EA.

The first claim is that for hash functions $h_{C_n,f,g}$ generated according to Gen using a YES instance, there is no set of $3n$ distinct x_i's that all have the same value for $C_n(x_i)$ and $f(x_i)$, except with negligible probability.

Claim 3.7.1. *Let $(1^n, C_n, k)$ be a YES instance of EA. Then,*

$$\Pr_{f \leftarrow \mathcal{F}_{n,(n-k)}^{(3n)}} \left[\exists y, y_1 \in \{0,1\}^* : \left|C_n^{-1}(y) \cap f^{-1}(y_1)\right| \geq 3n\right] \leq \frac{1}{2^n}.$$

Intuitively, the reason this should be true is that when C_n comes from a YES instance, it has high min-entropy. This means that for any y, the set $C_n^{-1}(y)$ will be quite small. The function f can now be thought of as partitioning each set $C_n^{-1}(y)$ into several parts, none of which will be too large because of the load-balancing properties of many-wise independent hash functions.

Proof. The above probability can be bounded using the union bound as follows:

$$\Pr_f\left[\exists y, y_1 : \left|C_n^{-1}(y) \cap f^{-1}(y_1)\right| \geq 3n\right] \leq \sum_{y \in \mathrm{Im}(C_n)} \Pr_f\left[\exists y_1 : \left|C_n^{-1}(y) \cap f^{-1}(y_1)\right| \geq 3n\right]. \tag{1}$$

The fact that $(1^n, C_n, k)$ is a YES instance of EA means that $\mathrm{H}_{\mathrm{min}}(C_n) \geq k$. The definition of min-entropy now implies that for any $y \in \mathrm{Im}(C_n)$:

$$\log\left(\frac{1}{\Pr_{x \leftarrow \{0,1\}^n}[C_n(x) = y]}\right) \geq k,$$

which in turn means that $\left|C_n^{-1}(y)\right| \leq 2^{n-k}$. Fact 2.3 (about the load-balancing properties of $\mathcal{F}^{(3n)}_{n,(n-k)}$) now implies that for any $y \in \mathrm{Im}(C_n)$:

$$\Pr_f\left[\exists y_1 : \left|C_n^{-1}(y) \cap f^{-1}(y_1)\right| \geq 3n\right] \leq \frac{2^{n-k}}{2^{3n}} \leq \frac{1}{2^{2n}}. \tag{2}$$

Combining Eqs. (1) and (2), and noting that the image of C_n has at most 2^n elements, we get the desired bound:

$$\Pr_f\left[\exists y, y_1 : \left|C_n^{-1}(y) \cap f^{-1}(y_1)\right| \geq 3n\right] \leq 2^n \cdot \frac{1}{2^{2n}} \leq \frac{1}{2^n}.$$

□

The next claim is that for hash functions $h_{C_n,f,g}$ generated according to Gen using a NO instance, there is no set of $2n$ values of x_i that all have distinct values of $(C_n(x_i), f(x_i))$, but the same value $g(C_n(x_i), f(x_i))$, except with negligible probability.

Claim 3.7.2. *Let* $(1^n, C_n, k)$ *be a* NO *instance of* EA. *Then,*

$$\Pr_{\substack{f \leftarrow \mathcal{F}^{(3n)}_{n,(n-k)} \\ g \leftarrow \mathcal{F}^{(2n)}_{(n'+n-k),(n-\lfloor\sqrt{n}\rfloor)}}}\left[\exists x_1, \ldots, x_{2n} : \begin{array}{c} \text{For all } i \neq j, \\ (C_n(x_i), f(x_i)) \neq (C_n(x_j), f(x_j)) \\ \text{and} \\ g(C_n(x_i), f(x_i)) = g(C_n(x_j), f(x_j)) \end{array}\right] \leq \frac{1}{2^n}.$$

Proof. The fact that $(1^n, C_n, k)$ is a NO instance of EA means that $\mathrm{H}_{\mathrm{max}}(C_n) \leq k - \lfloor\sqrt{n}\rfloor$; that is, C_n has a small range: $|\mathrm{Im}(C_n)| \leq 2^{k-\lfloor\sqrt{n}\rfloor}$.

For any $f \in \mathcal{F}^{(3n)}_{n,(n-k)}$, which is what is sampled by Gen when this instance is used, the range of f is a subset of $\{0,1\}^{n-k}$. This implies that even together, C_n and f have a range whose size is bounded as:

$$|\mathrm{Im}(C_n, f)| \leq 2^{k-\lfloor\sqrt{n}\rfloor} \cdot 2^{n-k} = 2^{n-\lfloor\sqrt{n}\rfloor},$$

where (C_n, f) denotes the function that is the concatenation of C_n and f.

For there to exist a set of $2n$ inputs x_i that all have distinct values for $(C_n(x_i), f(x_i))$ but the same value for $g(C_n(x_i), f(x_i))$, there has to be a y that has more than $2n$ inverses under g that are all in the image of (C_n, f). As g comes from $\mathcal{F}^{(2n)}_{(n'+n-k),(n-\lfloor\sqrt{n}\rfloor)}$, we can use Fact 2.3 along with the above bound on the size of the image of (C_n, f) to bound the probability that such a y exists as follows:

$$\Pr_g\left[\exists y : \left|g^{-1}(y) \cap \mathrm{Im}(C_n, f)\right| \geq 2n\right] \leq \frac{2^{n-\lfloor\sqrt{n}\rfloor}}{2^{2n}} \leq \frac{1}{2^n}.$$

□

Let $\mathsf{A} = (\mathsf{A}_n)_{n\in\mathbb{N}}$ be a polynomial-size family of circuits that given a hash function output by $\mathsf{Gen}(1^n)$ finds a $6n^2$-way collision in it with non-negligible probability. The candidate circuit family $\mathsf{A}' = (\mathsf{A}'_n)_{n\in\mathbb{N}}$ for solving EA on average is described below.

A'_n on input $(1^n, C_n, k)$:

1. Run steps 2 and 3 of the algorithm Gen in Fig. 1 with $(1^n, C_n, k)$ in place of the instance sampled from S there. This results in the description of a hash function $h_{C_n,f,g}$.
2. Run $\mathsf{A}_n(h_{C_n,f,g})$ to get a set of purported collisions $\mathcal{S}$.
3. If $\mathcal{S}$ does not actually contain $6n^2$ collisions under $h_{C_n,f,g}$, output a random bit.
4. If $\mathcal{S}$ contains $3n$ distinct x_i's such that they all have the same value of $(C_n(x_i), f(x_i))$, output 0.
5. If $\mathcal{S}$ contains $2n$ distinct x_i's such that they all have distinct values of $(C_n(x_i), f(x_i))$ but the same value $g(C_n(x_i), f(x_i))$, output 1.

The following claim now states that any collision-finding adversary for the MCRH constructed can be used to break the average-case hardness of EA, thus completing the proof.

Claim 3.7.3. *If* A *finds* $6n^2$ *collisions in hash functions output by* $\mathsf{Gen}(1^n)$ *with non-negligible probability, then* A' *has non-negligible advantage in deciding* EA *with respect to the hard-instance sampler* S *used in* Gen.

Proof. On input $(1^n, C_n, k)$, the adversary A'_n computes $h_{C_n,f,g}$ and runs A_n on it. If A_n does not find $6n^2$ collisions for $h_{C_n,f,g}$, then A'_n guesses at random and is correct in its output with probability 1/2. If A_n does find $6n^2$ collisions, then A'_n is correct whenever one of the following is true:

1. $(1^n, C_n, k)$ is a YES instance and there is no set of $3n$ collisions for (C_n, f).
2. $(1^n, C_n, k)$ is a NO instance and there is no set of $2n$ collisions for g in the image of (C_n, f).

Note that inputs to A'_n are drawn from $\mathsf{S}(1^n)$, and so the distribution over $h_{C_n,f,g}$ produced by A'_n is the same as that produced by $\mathsf{Gen}(1^n)$ itself. With such samples, let E_1 denote the event of (C_n, f) having a set of $3n$ collisions from $\mathcal{S}$ (the set output by A_n), and let E_2 denote the event of g having a set of $2n$ collisions in the image of (C_n, f) from $\mathcal{S}$. Also, let E_Y denote the event of the input to A'_n being a YES instance, E_N that of it being a NO instance, and E_{A} the event that $\mathcal{S}$ contains at least $6n^2$ collisions.

Following the statements above, the probability that A'_n is *wrong* in deciding EA with respect to $(1^n, C_n, k) \leftarrow \mathsf{S}(1^n)$ can be upper-bounded as:

$$\begin{aligned} \Pr\left[\mathsf{A}'_n(1^n, C_n, k) \text{ is wrong}\right] & \\ &= \Pr\left[(\neg E_{\mathsf{A}}) \wedge (\mathsf{A}'_n \text{ is wrong})\right] + \Pr\left[E_{\mathsf{A}} \wedge (\mathsf{A}'_n \text{ is wrong})\right] \\ &\leq \Pr[\neg E_{\mathsf{A}}] \cdot \frac{1}{2} + \Pr[(E_Y \wedge E_1) \vee (E_N \wedge E_2)]. \end{aligned}$$

The first term comes from the fact that if A_n doesn't find enough collisions, A'_n guesses at random. The second term comes from the fact that if both $(E_Y \wedge E_1)$ and $(E_N \wedge E_2)$ are false and E_{A} is true, then since at least one of E_Y and E_N is always true, one of $(E_Y \wedge \neg E_1)$ and $(E_N \wedge \neg E_2)$ will also be true, either of which would ensure that A'_n is correct, as noted earlier.

We now bound the second term above, starting as follows:

$$\begin{aligned} \Pr[(E_Y \wedge E_1) \vee (E_N \wedge E_2)] &\leq \Pr[(E_Y \wedge E_1)] + \Pr[(E_N \wedge E_2)] \\ &= \Pr[E_Y] \Pr[E_1 | E_Y] + \Pr[E_N] \Pr[E_2 | E_N] \\ &\leq \Pr[E_Y] \cdot \mathsf{negl}(n) + \Pr[E_N] \cdot \mathsf{negl}(n) \\ &= \mathsf{negl}(n), \end{aligned}$$

where the first inequality follows from the union bound and the last inequality follows from Claims 3.7.1 and 3.7.2.

Putting this back in the earlier expression,

$$\begin{aligned} \Pr\left[\mathsf{A}'_n(1^n, C_n, k) \text{ is wrong}\right] &\leq \Pr[\neg E_{\mathsf{A}}] \cdot \frac{1}{2} + \mathsf{negl}(n) \\ &= \frac{1}{2} - \frac{\Pr[E_{\mathsf{A}}]}{2} + \mathsf{negl}(n). \end{aligned}$$

In other words,

$$\Pr\left[\mathsf{A}'_n(1^n, C_n, k) \text{ is correct}\right] \geq \frac{1}{2} + \frac{\Pr[E_{\mathsf{A}}]}{2} - \mathsf{negl}(n).$$

So if A succeeds with non-negligible probability in finding $6n^2$ collisions, then A' had non-negligible advantage in deciding EA over S. □

This concludes the proof of Theorem 3.6. □

4 Constant-Round Statistically-Hiding Commitments

In this section we show that multi-collision-resistant hash functions imply the existence of *constant-round* statistically-hiding commitments. Here we follow the "direct route" discussed in the introduction (rather than the "inaccessible entropy route").

For simplicity, we focus on *bit* commitment schemes (in which messages are just single bits). As usual, full-fledged commitment schemes (for long messages) can be obtained by committing bit-by-bit.

Definition 4.1 (Bit Commitment Scheme). *A* bit commitment scheme *is an interactive protocol between two polynomial-time parties — the* sender S *and the* receiver R *— that satisfies the following properties.*

1. *The protocol proceeds in two stages: the* commit *stage and the* reveal *stage.*
2. *At the start of the commit stage both parties get a security parameter* 1^n *as a common input and the sender* S *also gets a private input* $b \in \{0, 1\}$*. At the end of the commit stage the parties have a shared output* c*, which is called the* commitment*, and the sender* S *has an additional private output* d*, which is called the* decommitment.
3. *In the reveal stage, the sender* S *sends* (b, d) *to the receiver* R*. The receiver* R *accepts or rejects based on* c*,* d *and* b*. If both parties follow the protocol, then the receiver* R *always accepts.*

In this section we focus on commitment schemes that are *statistically*-hiding and *computationally*-binding.

Definition 4.2 (Statistically Hiding Bit Commitment). *A bit commitment scheme* (S, R) *is* statistically-hiding *if for every cheating receiver* R^* *it holds that*

$$\mathsf{SD}((\mathsf{S}(0), \mathsf{R}^*)(1^n), (\mathsf{S}(1), \mathsf{R}^*)(1^n)) = \mathsf{negl}(n),$$

where $(\mathsf{S}(b), \mathsf{R}^*)(1^n)$ *denotes the transcript of the interaction between* R^* *and* $\mathsf{S}(b)$ *in the commit stage.*

Definition 4.3 (Computationally Binding Bit Commitment). *A bit commitment scheme* (S, R) *is said to be* computationally-binding *if for every family of polynomial-size circuits sender* $\mathsf{S}^* = (\mathsf{S}^*_n)_{n \in \mathbb{N}}$ *it holds that* S^* *wins in the following game with only with* $\mathsf{negl}(n)$ *probability:*

1. *The cheating sender* S^*_n *interacts with the honest receiver* $\mathsf{R}(1^n)$ *in the commit stage obtaining a commitment* c*.*
2. *Then,* S^*_n *outputs two pairs* $(0, d_0)$ *and* $(1, d_1)$*. The cheating sender* S^* *wins if the honest receiver* R *accepts both* $(c, 0, d_0)$ *and* $(c, 1, d_1)$*.*

We are now ready to state the main result of this section. A *round* of a commitment scheme is a pair of messages, the first sent from the receiver to the sender, and the second the other way.

Theorem 4.4 (MCRH $\Longrightarrow$ **Constant-Round Statistically-Hiding Commitments).** *Let* $t = t(n) \in \mathbb{N}$ *be a polynomial computable in* $\mathsf{poly}(n)$ *time. Assume that there exists a* (s, t)-MCRH *for* $s \geq \log(t)$*, then there exists a* three-round *statistically-hiding computationally-binding commitment scheme.*

As we already mentioned in Sect. 1, constructions of statistically-hiding computationally-binding commitment schemes are known assuming only the minimal assumption that one-way functions exist. Those constructions, however, have a polynomial number of rounds (and this is inherent for black-box constructions [HHRS15]). Theorem 4.4, on the other hand, yields a commitment scheme with only a constant (i.e., three) number of rounds.

Due to space limitation, we defer the complete proof of Theorem 4.4 to the full version of this paper [BDRV17].

Acknowledgments. We thank Vinod Vaikuntanathan for helpful discussions and for his support, and Oded Goldreich, Yuval Ishai and the anonymous reviewers for useful comments. We thank Nir Bitansky, Yael Kalai, Ilan Komargodski, Moni Naor, Omer Paneth and Eylon Yogev for helping us provide a good example of a t-way collision. We also thank Nir Bitansky and an anonymous reviewer for pointing out the connection to inaccessible entropy.

This research was supported in part by NSF Grants CNS-1413920 and CNS-1350619, and by the Defense Advanced Research Projects Agency (DARPA) and the U.S. Army Research Office under contracts W911NF-15-C-0226 and W911NF-15-C-0236. The third author was also partially supported by the SIMONS Investigator award agreement dated 6-5-12 and by the Cybersecurity and Privacy Institute at Northeastern University.

References

[ADM+99] Alon, N., Dietzfelbinger, M., Miltersen, P.B., Petrank, E., Tardos, G.: Linear hash functions. J. ACM **46**(5), 667–683 (1999)

[AR16] Applebaum, B., Raykov, P.: On the relationship between statistical zero-knowledge and statistical randomized encodings. In: Robshaw, M., Katz, J. (eds.) CRYPTO 2016. LNCS, vol. 9816, pp. 449–477. Springer, Heidelberg (2016). https://doi.org/10.1007/978-3-662-53015-3_16

[Bab16] Babai, L.: Graph isomorphism in quasipolynomial time [extended abstract]. In: Wichs, D., Mansour, Y. (eds.) Proceedings of the 48th Annual ACM SIGACT Symposium on Theory of Computing, STOC 2016, Cambridge, MA, USA, 18–21 June 2016, pp. 684–697. ACM (2016)

[BDRV17] Berman, I., Degwekar, A., Rothblum, R.D., Vasudevan, P.N.: Multi collision resistant hash functions and their applications. IACR Cryptology ePrint Archive 2017, 489 (2017)

[BPK17] Bitansky, N., Paneth, O., Kalai, Y.T.: Multi-collision resistance: A paradigm for keyless hash functions. Electron. Colloquium Comput. Complex. (ECCC) **24**, 99 (2017)

[BPVY00] Brickell, E., Pointcheval, D., Vaudenay, S., Yung, M.: Design Validations for Discrete Logarithm Based Signature Schemes. In: Imai, H., Zheng, Y. (eds.) PKC 2000. LNCS, vol. 1751, pp. 276–292. Springer, Heidelberg (2000). https://doi.org/10.1007/978-3-540-46588-1_19

[CRSW13] Elisa Celis, L., Reingold, O., Segev, G., Wieder, U.: Balls and bins: smaller hash families and faster evaluation. SIAM J. Comput. **42**(3), 1030–1050 (2013)

[DGRV11] Dvir, Z., Gutfreund, D., Rothblum, G.N., Vadhan, S.P.: On approximating the entropy of polynomial mappings. In: Proceedings of Innovations in Computer Science - ICS 2010, Tsinghua University, Beijing, China, 7–9 January 2011, pp. 460–475 (2011)

[DHRS07] Ding, Y.Z., Harnik, D., Rosen, A., Shaltiel, R.: Constant-round oblivious transfer in the bounded storage model. J. Cryptol. **20**(2), 165–202 (2007)

[DI06] Dubrov, B., Ishai, Y.: On the randomness complexity of efficient sampling. In: Proceedings of the Thirty-Eighth Annual ACM Symposium on Theory of Computing, pp. 711–720. ACM (2006)

[DPP93] Damgård, I.B., Pedersen, T.P., Pfitzmann, B.: On the existence of statistically hiding bit commitment schemes and fail-stop signatures. In: Stinson, D.R. (ed.) CRYPTO 1993. LNCS, vol. 773, pp. 250–265. Springer, Heidelberg (1994). https://doi.org/10.1007/3-540-48329-2_22

[GG98] Goldreich, O., Goldwasser, S.: On the limits of non-approximability of lattice problems. In: Proceedings of the Thirtieth Annual ACM Symposium on Theory of Computing, pp. 1–9. ACM (1998)

[GMR88] Goldwasser, S., Micali, S., Rivest, R.L.: A digital signature scheme secure against adaptive chosen-message attacks. SIAM J. Comput. **17**(2), 281–308 (1988)

[GS94] Girault, M., Stern, J.: On the length of cryptographic hash-values used in identification schemes. In: Desmedt, Y.G. (ed.) CRYPTO 1994. LNCS, vol. 839, pp. 202–215. Springer, Heidelberg (1994). https://doi.org/10.1007/3-540-48658-5_21

[GSV99] Goldreich, O., Sahai, A., Vadhan, S.: Can statistical zero knowledge be made non-interactive? or on the relationship of SZK and *NISZK*. In: Wiener, M. (ed.) CRYPTO 1999. LNCS, vol. 1666, pp. 467–484. Springer, Heidelberg (1999). https://doi.org/10.1007/3-540-48405-1_30

[HHRS15] Haitner, I., Hoch, J.J., Reingold, O., Segev, G.: Finding collisions in interactive protocols—tight lower bounds on the round and communication complexities of statistically hiding commitments. SIAM J. Comput. **44**(1), 193–242 (2015)

[HM96] Halevi, S., Micali, S.: Practical and provably-secure commitment schemes from collision-free hashing. In: Koblitz, N. (ed.) CRYPTO 1996. LNCS, vol. 1109, pp. 201–215. Springer, Heidelberg (1996). https://doi.org/10.1007/3-540-68697-5_16

[HNO+09] Haitner, I., Nguyen, M.-H., Ong, S.J., Reingold, O., Vadhan, S.P.: Statistically hiding commitments and statistical zero-knowledge arguments from any one-way function. SIAM J. Comput. **39**(3), 1153–1218 (2009)

[HR04] Hsiao, C.-Y., Reyzin, L.: Finding collisions on a public road, or do secure hash functions need secret coins? In: Franklin, M. (ed.) CRYPTO 2004. LNCS, vol. 3152, pp. 92–105. Springer, Heidelberg (2004). https://doi.org/10.1007/978-3-540-28628-8_6

[HRVW09] Haitner, I., Reingold, O., Vadhan, S.P., Wee, H.: Inaccessible entropy. In: Proceedings of the 41st Annual ACM Symposium on Theory of Computing, STOC 2009, Bethesda, MD, USA, 31 May-2 June 2009, pp. 611–620 (2009)

[HV17] Haitner, I., Vadhan, S.: The many entropies in one-way functions. Tutorials on the Foundations of Cryptography. ISC, pp. 159–217. Springer, Cham (2017). https://doi.org/10.1007/978-3-319-57048-8_4

[Jou04] Joux, A.: Multicollisions in iterated hash functions. Application to cascaded constructions. In: Franklin, M. (ed.) CRYPTO 2004. LNCS, vol. 3152, pp. 306–316. Springer, Heidelberg (2004). https://doi.org/10.1007/978-3-540-28628-8_19

[Kil92] Kilian, J.: A note on efficient zero-knowledge proofs and arguments (extended abstract). In: Proceedings of the 24th Annual ACM Symposium on Theory of Computing, Victoria, British Columbia, Canada, 4–6 May 1992, pp. 723–732 (1992)

[KNY17a] Komargodski, I., Naor, M., Yogev, E.: Collision resistant hashing for paranoids: dealing with multiple collisions. IACR Cryptology ePrint Archive 2017, 486 (2017)

[KNY17b] Komargodski, I., Naor, M., Yogev, E.: White-box vs. black-box complexity of search problems: ramsey and graph property testing. Electron. Colloquium Comput. Complex. (ECCC) **24**, 15 (2017)

[Mer89] Merkle, R.C.: One way hash functions and DES. In: Brassard, G. (ed.) CRYPTO 1989. LNCS, vol. 435, pp. 428–446. Springer, New York (1990). https://doi.org/10.1007/0-387-34805-0_40

[Mic00] Micali, S.: Computationally sound proofs. SIAM J. Comput. **30**(4), 1253–1298 (2000)

[MRRR14] Meka, R., Reingold, O., Rothblum, G.N., Rothblum, R.D.: Fast pseudorandomness for independence and load balancing. In: Esparza, J., Fraigniaud, P., Husfeldt, T., Koutsoupias, E. (eds.) ICALP 2014. LNCS, vol. 8572, pp. 859–870. Springer, Heidelberg (2014). https://doi.org/10.1007/978-3-662-43948-7_71

[NY89] Naor, M., Yung, M.: Universal one-way hash functions and their cryptographic applications. In: Proceedings of the 21st Annual ACM Symposium on Theory of Computing, Seattle, Washigton, USA, 14–17 May 1989, pp. 33–43 (1989)

[Ost91] Ostrovsky, R.: One-way functions, hard on average problems, and statistical zero-knowledge proofs. In: Proceedings of the Sixth Annual Structure in Complexity Theory Conference, Chicago, Illinois, USA, 30 June-3 July 1991, pp. 133–138 (1991)

[OV08] Ong, S.J., Vadhan, S.: An equivalence between zero knowledge and commitments. In: Canetti, R. (ed.) TCC 2008. LNCS, vol. 4948, pp. 482–500. Springer, Heidelberg (2008). https://doi.org/10.1007/978-3-540-78524-8_27

[PRS12] Pietrzak, K., Rosen, A., Segev, G.: Lossy functions do not amplify well. In: Cramer, R. (ed.) TCC 2012. LNCS, vol. 7194, pp. 458–475. Springer, Heidelberg (2012). https://doi.org/10.1007/978-3-642-28914-9_26

[PW11] Peikert, C., Waters, B.: Lossy trapdoor functions and their applications. SIAM J. Comput. **40**(6), 1803–1844 (2011)

[Rom90] Rompel, J.: One-way functions are necessary and sufficient for secure signatures. In: Proceedings of the 22nd Annual ACM Symposium on Theory of Computing, Baltimore, Maryland, USA, 13–17 May 1990, pp. 387–394 (1990)

[RS96] Rivest, R.L., Shamir, A.: Payword and micromint: two simple micropayment schemes. In: Proceedings of Security Protocols, International Workshop, Cambridge, United Kingdom, 10–12 April 1996, pp. 69–87 (1996)

[RTV04] Reingold, O., Trevisan, L., Vadhan, S.: Notions of reducibility between cryptographic primitives. In: Naor, M. (ed.) TCC 2004. LNCS, vol. 2951, pp. 1–20. Springer, Heidelberg (2004). https://doi.org/10.1007/978-3-540-24638-1_1

[RV09] Rothblum, G.N., Vadhan, S.P.: Unpublished Manuscript (2009)

[Sim98] Simon, D.R.: Finding collisions on a one-way street: can secure hash functions be based on general assumptions? In: Nyberg, K. (ed.) EUROCRYPT 1998. LNCS, vol. 1403, pp. 334–345. Springer, Heidelberg (1998). https://doi.org/10.1007/BFb0054137

[Vad12] Vadhan, S.P.: Pseudorandomness. Found. Trends Theor. Comput. Sci. **7**(1–3), 1–336 (2012)

Collision Resistant Hashing for Paranoids: Dealing with Multiple Collisions

Ilan Komargodski[1(✉)], Moni Naor[2], and Eylon Yogev[2]

[1] Cornell Tech, NewYork, NY 10044, USA
komargodski@cornell.edu
[2] Weizmann Institute of Science, 76100 Rehovot, Israel
{moni.naor,eylon.yogev}@weizmann.ac.il

Abstract. A collision resistant hash (CRH) function is one that compresses its input, yet it is hard to find a collision, i.e. a $x_1 \neq x_2$ s.t. $h(x_1) = h(x_2)$. Collision resistant hash functions are one of the more useful cryptographic primitives both in theory and in practice and two prominent applications are in signature schemes and succinct zero-knowledge arguments.

In this work we consider a relaxation of the above requirement that we call Multi-CRH: a function where it is hard to find $x_1, x_2, \ldots, x_k$ which are all distinct, yet $h(x_1) = h(x_2) = \cdots = h(x_k)$. We show that for some of the major applications of CRH functions it is possible to replace them by the weaker notion of a Multi-CRH, albeit at the price of adding interaction: we show a constant-round statistically-hiding commitment scheme with succinct interaction (committing to $\mathsf{poly}(n)$ bits requires exchanging $\tilde{O}(n)$ bits) that can be opened locally (without revealing the full string). This in turn can be used to provide succinct arguments for any NP statement.

We formulate four possible worlds of hashing-related assumptions (in the spirit of Impagliazzo's worlds). They are (1) *Nocrypt*, where no one-way functions exist, (2) *Unihash*, where one-way functions exist, and hence also UOWHFs and signature schemes, but no Multi-CRH functions exist, (3) *Minihash*, where Multi-CRH functions exist but no CRH functions exist, and (4) *Hashomania*, where CRH functions exist. We show that these four worlds are distinct in a black-box model: we show a separation of CRH from Multi-CRH and a separation of Multi-CRH from one-way functions.

I. Komargodski—Supported in part by a Packard Foundation Fellowship and AFOSR grant FA9550-15-1-0262. Most work done while the author was a Ph.D. student at the Weizmann Institute of Science, supported in part by a grant from the Israel Science Foundation (no. 950/16) and by a Levzion Fellowship.
M. Naor and E. Yogev—Supported in part by a grant from the Israel Science Foundation (no. 950/16). Moni Naor is the incumbent of the Judith Kleeman Professorial Chair.

J. B. Nielsen and V. Rijmen (Eds.): EUROCRYPT 2018, LNCS 10821, pp. 162–194, 2018.
https://doi.org/10.1007/978-3-319-78375-8_6

1 Introduction

In any function that compresses its input, say from $2n$ bits to n bits, there are many collisions, that is, pairs of distinct inputs whose image is the same. But what is the complexity of finding such a collision? Families of functions where this collision finding task is hard are known as collision resistant hash (CRH) functions.[1] CRH functions have many appealing properties, such as preservation under composition and concatenation. The presumed hardness of finding collisions in such functions is the basis for increased efficiency of many useful cryptographic schemes, in particular signature schemes and succinct (zero-knowledge) arguments, i.e., methods for demonstrating the correctness of a statement that are much shorter than the proof or even the statement itself (see Kilian [37] and Barak and Goldreich [4]). The latter is achieved via a hash tree commitment[2] scheme whose opening is local i.e., opening a bit does not require revealing the full string (known as a "Merkle tree"). Results of this sort enable the construction of efficient delegation of computation where the goal is to offload significant computation to some server but also to verify the computation.

Such a task ("breaking a collision resistant hash function") is indeed hard based on a variety of assumptions such as the hardness of factoring integers, finding discrete logs in finite groups or learning with errors (LWE). There are popular functions (standards) with presumed hardness of collision finding such as SHA-2 and SHA-3 (adopted by NIST[3] in 2015). These functions can be evaluated very quickly; however, their hardness is based on more ad hoc assumptions and some former standards have been shown to be insecure (such as MD4, MD5, SHA-1). On the other hand there is no known construction of CRHs based solely on the existence of one-way functions or even one-way permutations and, furthermore, they were shown to be separated in a black-box model (see Simon [50]).

But a sufficiently compressing function also assures us that there are **multiple collisions**, i.e., k distinct values whose image under the function is equal. What about the problem of finding a k-collision? Assuming such hardness is a *weaker* computational assumption than hardness of finding a single pair of colliding inputs and the question is whether it yields a useful primitive.

In this paper we deal with *multiple collision resistant hash* (MCRH) functions and systematically investigate their properties and applications. We show that for some of the major applications of CRH functions it is possible to replace them by an MCRH, albeit at the price of adding some rounds of interaction:

---

[1] The function $h \in H$ can be sampled efficiently, it is easy to compute $h(x)$ given h and x, however, given h it is hard to find $x_1 \neq x_2$ s.t. $h(x_1) = h(x_2)$.

[2] A commitment scheme is a protocol where a sender commits to a string x in the "commit" phase and that later can be revealed at the opening phase. The two properties are binding and hiding: in what sense is the sender bound to the string x (computationally or information theoretically) and in what sense is x hidden from the receiver before the opening - statistically or computationally.

[3] NIST is the National Institute of Standards and Technology, a US agency.

a constant-round[4] commitment scheme with succinct communication that can be opened locally (without revealing the full string) (see Theorem 2). This implies that it is possible to effectively verify the correctness of computation much more efficiently than repeating it. As an application we get universal arguments [4] (and public-coin zero-knowledge argument systems for NP [3]) with an arbitrary super-constant number of rounds based on MCRH functions. We also provide a constant-round statistically-hiding scheme and thus we can get constant-round statistical zero-knowledge arguments [8].[5]

On the other hand, we show various **black-box separation** results concerning MCRH. First, we separate them from one-way permutations. This follows from the lower bound of Haitner et al. [23] on the number of rounds needed to build a statistically-hiding commitment from one-way permutations. Furthermore, we show a black-box separation from standard CRH: there is no fully black-box construction of a k-MCRH from a $(k+1)$-MCRH for all k with polynomial security loss (see Theorem 4). These results yield an infinite hierarchy of natural cryptographic primitives, each two being separated by a fully black-box construction, between one-way function/permutations and collision-resistant hash function.[6]

One motivation for investigating MCRH functions is the progress in finding collisions in the hash function SHA-1 (that has long been considered insecure [53]) and recently an actual meaningful collision has been found [51]. In general, finding a collision with arbitrary Initialization Vector (IV) allows finding multiple collisions in an iteration of the function (say in a Merkle-Damgård lopsided tree), as shown by Joux [35] (see also Coppersmith and Girault et al. [9,15] for older attacks). However, for the compression function of SHA-1 (or other such functions) there is no non-trivial algorithm for finding multi-collisions.[7] Also, multi-collision resistance is sometimes useful for optimizing concrete parameters, as shown by Girault and Stern [16] for reducing the communication in identification schemes. So the question is what can we do if all we assume about a given hash function is that multi-collision are hard to find, rather than plain collisions.

Our interest in MCRH functions originated in the work of Komargodski et al. [39], where MCRHs were first defined in the context of the bipartite Ramsey problem. They showed that finding cliques or independent sets in succinctly-represented bipartite graphs (whose existence is assured by Ramsey Theory) is equivalent to breaking an MCRH: a hard distribution for finding these

[4] By "constant-round" we mean that for a $c \in \mathbb{N}$ to commit to a string of length n^c using a compression function from $2n$ to n bits the number of rounds is a constant that depends on c.

[5] For constant values of k, we give a 4-round computationally-binding commitment scheme with succinct communication (see Theorem 3).

[6] This hierarchy translates to an infinite hierarchy of natural subclasses in TFNP. See the full version [38] for details.

[7] Beyond the "birthday-like" algorithm that will take time $2^{n \cdot \frac{k-1}{k}}$ [35], where 2^n is the size of the range. See [30] for very recent work in the quantum case.

objects implies the existence of an MCRH and vice-versa (with slightly different parameters).[8]

Families of CRHs compose very nicely, and hence *domain extension* is relatively simple, that is, once we have a CRH that compresses by a single bit we can get any polynomial compression (i.e., from $\mathsf{poly}(n)$ to n bits). In contrast, we do not know how to construct a k'-MCRH on very large domains from a fixed k-MCRH, where k' is not much larger than k. Nevertheless, in Sect. 6, we show how to get such a construction with $k' = k^{O(\log n)}$.

A well-known relaxation of CRHs are Universal One-way Hash Functions (UOWHF) or second pre-image resistance: first the target x is chosen (perhaps adversarially), then a function $h \in_R H$ is sampled and the challenge is to find $x' \neq x$ s.t. $h(X) = h(x')$. Such families are good enough for signatures (at least existentially) and can be used for the hash-and-sign paradigm, if one chooses h per message (see Naor and Yung [46] and Mironov [43]). It is known how to get such family of functions from one-way functions, but the construction is rather involved and inefficient (and there are some inherent reasons for that, see [14]). We show how to go from *any* interactive commitment protocol where the communication is shorter than the string committed to (a succinct commitment) to a UOWHF (see Theorem 5). Together with our commitment schemes, this gives new constructions of UOWHFs on long inputs based on MCRHs with a shorter description than the ones known starting from a UOWHF on fixed input length.

The Four Worlds of Hashing. Impagliazzo's five worlds [32] are a way to characterize the strength of a cryptographic assumption. The worlds he defined are: *Algorithmica* (where $\mathsf{P} = \mathsf{NP}$), *Heuristica* (where NP is hard in the worst case but easy on average, i.e., one simply does not encounter hard problems in NP), *Pessiland* (where hard-on-the-average problems in NP exist, but one-way functions do not exist), *Minicrypt* (where one-way functions exist), and *Cryptomania* (where Oblivious Transfer exists). (Nowadays, it is possible to add a sixth world, *Obfustopia*, where indistinguishability obfuscation for all programs exists.)

In the spirit of Impagliazzo's five worlds of cryptographic assumptions, we define four worlds of hashing-related primitives:

Nocrypt: A world where there are no one-way functions. There are no cryptographic commitments of any kind in this world [34].

Unihash: A world where one-way functions exist (and so do UOWHFs), but there are no MCRH functions. Therefore, signatures exist and hashing applications such as the hash-and-sign paradigm [46]. Also, statistically-hiding commitments exist (albeit with a linear number of rounds) [23,24,26,45]. There are no known short commitment (where the communication is much shorter than the string committed to).

[8] In the bipartite Ramsey problem, the goal is to find a bi-clique or bi-independent set of size $n/4 \times n/4$ in a bipartite graph of size $2^n \times 2^n$. The work of [39] showed that if this problem is hard, then there exists an $(n/4)$-MCRH from n bits to $n/2$ bits. Conversely, If a $\sqrt{n}$-MCRH mapping n bits to $\sqrt{n}/8$ bits exists, then this problem is hard.

Minihash: A world where MCRH exists but there is no CRH: that is, for some polynomial $k(n)$ there exists a k-MCRH that compresses $2n$ to n bits. In this work we give a protocol for short and statistically-hiding commitments with a *constant* number of rounds. Furthermore the string can be opened locally, with little communication and computation, without requiring the full opening of the string.

Hashomania: A world where CRH exists. There is a short commitment protocol that requires only *two* rounds (i.e., two messages) with local opening. This is the famed Merkle-tree.

Note that our separation results imply that these four worlds have black-box separations. Unihash and Minihash are separated by the separation of MCRH from one-way permutations, and Minihash and Hashomania are separated by the separation of CRH from MCRH. Moreover, the separation in Sect. 7 actually implies that the world *Minihash* can be split further into sub-worlds parameterized by k, the number of collisions it is hard to find.

Multi-pair Collision Resistance. A different way to relax the standard notion of collision resistance is what we call *multi-pair-collision-resistance*, where the challenge is to find *arbitrary* k distinct pairs of inputs that collide (possibly to different values). One may wonder what is the difference between these two notions and why we focus on the hardness of finding a k-wise collision rather than hardness of finding k distinct colliding pairs. The answer is that the notion of k-pair-collision-resistance is *existentially equivalent* to the standard notion of collision resistance (see the full version [38] for details).

Concurrent work

In parallel to this work, MCRH functions were studied by two other groups that obtained various related results [6,7] with different motivation and perspective.

Berman et al. [6] showed how to obtain MCRH functions from a specific assumption. Concretely, they construct an n^2-MCRH function compressing inputs of length n to outputs of length $n - \sqrt{n}$ from the average-case hardness of the min-max variant of the entropy approximation problem. This variant is a promise problem where the YES inputs are circuits whose output distribution has *min*-entropy at least κ, whereas NO inputs are circuits whose output distribution has *max*-entropy less than κ.[9] Berman et al. also show how to get a constant-round statistically-hiding (computationally-binding) commitment scheme from any k-MCRH that compresses n bits into $n - \log k$ bits (which implies a black-box separation of MCRH from one-way permutations). However, their commitment scheme is not short and does not support local opening[10].

[9] The original *entropy approximation* problem is the one obtained by replacing the min- and max-entropy with the Shannon entropy. It is known to be complete for the class of languages that have non-interactive statistical zero-knowledge proofs (NISZK) [17].

[10] Starting out with such a weak primitive our methods will not yield a succinct commitment either.

Bitansky et al. [7] replace the CRH assumption with a k-MCRH in several applications related to zero-knowledge and arguments of knowledge for NP with few rounds. They rely on either MCRH that compresses by a polynomial factor (say n^2 bits into n), or alternatively on an MCRH that compresses by a linear factor but lose a quasi-polynomial factor in security. Their main technical component is a two-round (i.e., two-message) short commitments with local opening but with *weak* computational-binding. The latter means that the sender may be able to open the commitment to more than one value, but not to too many values. Their construction of the commitment scheme is related to our construction presented in Theorem 3 but they design a specific code that allows them to get local opening.

Summary of Results and Paper Organization
Our main results are:

1. Any k-MCRH can be used to get a constant round short commitment scheme which is computationally-binding, statistically-hiding and support local opening (à la Merkle commitments). This result in Sect. 5.
2. Any k-MCRH, where k is constant, can be used to get a 4 round short commitment scheme which is computationally-binding and statistically-hiding opening. This appears in Sect. 6.
3. We prove a fully black-box separation between standard collision resistant hash functions and multi-collision resistant ones. This appears in Sect. 7.
4. We present a generic and direct construction of UOWHFs from any short commitment schemes (and thereby from any multi-collision resistant functions). See Sect. 8.

In Sect. 2 we provide an overview of our main ideas and techniques. In Sects. 3 and 4 we provide preliminary standard definitions used throughout the paper and the definition of MCRH functions, respectively.

2 Our Techniques

In this section we present some of our main ideas and techniques used in the construction of the commitment scheme and in the black-box separation result.

2.1 The Main Commitment Scheme

A commitment scheme is a two stage interactive protocol between a sender and a receiver such that after the first stage the sender is bound to at most one value (this is called "binding"). In the second stage the sender can open his committed value to the sender. There are a few security properties one can ask from such a protocol: have statistical/computational binding, and have the committed value of the sender be statistically/computationally hidden *given* the commitment (this is called "hiding"). Our commitment scheme satisfies computational binding and statistical hiding. In this overview we will mostly focus on obtaining computational binding and briefly discuss how we obtain hiding towards the end.

There is a trivial commitment protocol that is (perfectly) binding: let the sender send its value to the receiver. Perhaps the most natural non-trivial property one can ask is that the commitment is *shorter* than the committed string. There are additional useful properties one can require such as *local-opening* which allows the sender to open a small fraction of its input without sending the whole string (local opening is very important in applications of such commitment schemes, e.g., to proof systems and delegation protocols). In our protocol the commitment is short and it supports local-opening; we will focus here on the former and shortly discuss the latter towards the end.

Our goal now is to construct a commitment scheme which is computationally binding and the commitment is shorter than the value of the sender. If we had a standard collision resistant hash function mapping strings of length $2n$ to strings of length n, this task would be easy to achieve: the receiver will sample a hash function h and send it to the sender which will reply with $h(x^*)$, where $x^* \in \{0,1\}^{2n}$ is its value. The commitment thus consists of $(h, h(x^*))$ and its size is n bits.[11] It is easy to verify that for a sender to cheat during the opening phase it actually has to break the collision resistance of h (i.e., come up with a value $x \neq x^*$ such that $h(x) = h(x^*)$).

When h is only a k-MCRH for $k > 2$, the above protocol is clearly insecure: the sender can potentially find two inputs that collide to the same value and cheat when asked to open its commitment. The first observation we make is that even though the sender is not bound to a single value after sending $h(x^*)$, it is bound to a set of values of size at most $k-1$. Otherwise, at least intuitively, he might be able to find k inputs that map to the same output relative to h, which contradicts the security of the k-MCRH. Our first idea is to take advantage of this fact by adding an additional round of communication whose goal is to "eliminate" all but one possible value for the sender. Specifically, after the receiver got $h(x^*)$, it samples a random universal hash[12] function, g, mapping strings of length $2n$ to strings of length m. and sends it to the sender. The sender then responds with $g(x)$. The commitment thus consists of $(h, h(x^*), g, g(x^*))$. To open the commitment the sender just sends x^* (just as before).

One can show that this protocol is binding: Out of the $k-1$ possible values the sender knows that are consistent with $h(x^*)$, with probability roughly $k^2 \cdot 2^{-m}$ there will be no two that agree on $g(x^*)$. Conditioning on this happening, the sender cannot submit $x \neq x^*$ that is consistent with both $h(x^*)$ and $g(x^*)$. To formally show that the protocol is binding we need to show how to find a k-wise collision using a malicious sender. We simulate the protocol between the malicious sender and the receiver *multiple times* (roughly k times) with the same h but with freshly sampled g, by *partially rewinding* the malicious sender. We show that with good probability, every iteration will result with a new collision.

---

[11] For simplicity of presentation here, we ignore the cost of the description of h, so it will not significantly affect the size of the commitment.

[12] A universal hash function is a function of families $\mathcal{G} = \{g\colon \{0,1\}^n \to \{0,1\}^m\}$ such that for any $x, y \in \{0,1\}^n$ such that $x \neq y$ it holds that $\Pr_{g \leftarrow \mathcal{G}}[g(x) = g(y)] \leq 2^{-m}$.

The protocol consists now of 4 rounds, but is the commitment short? Well, it depends on m and on the description size of g. The description size of a universal function is proportional to the input size ($2n$ in our case), which totally ruins the shortness of the protocol. We fix this by sampling g from an *almost* universal family and apply the above protocol. We obtain a 4-round protocol in which the commitment size is of the order roughly $n + m + \log(1/\delta)$, where δ is related to the error probability of the almost universal function. Choosing m and δ appropriately we obtain a protocol with short ($<2n$) commitments.

Handling Longer Inputs. How would we commit on a longer string, say of $10n$ bits or even n^{10} bits? (Recall that all we have is a hash function mapping $2n$ bits into n bits.) One well-known solution is based on a standard collision resistant hash function, and what is known as a Merkle tree (the tree structure will be useful later for a local opening). The input $x \in \{0,1\}^{2^d n}$ (for simplicity think of d as either a large constant or even $O(\log n)$) is partitioned into 2^d blocks each of size n. These blocks are partitioned into pairs and the hash function is applied to each pair resulting in 2^{d-1} blocks. Then, the remaining blocks are partitioned into pairs and the hash function is applied on each pair. This is repeated d times resulting in a binary tree of hash values of depth d. The value associated with the root of the tree is called the root-hash.

If h is a standard CRH sent by the receiver, then it is known that sending the root-hash by the sender is actually a commitment on the input x^* [37,41]. Can we apply the same trick from before to make this a commitment protocol even when h is only a k-MCRH? That is, after sending the root-hash, let the sender sample a good-enough combinatorial hash function $g\colon \{0,1\}^{2^d n} \to \{0,1\}^m$ and send it to the sender that will reply with $g(x^*)$. Is this protocol binding? The answer is "no", even for large values of m. Observe that for every node in the Merkle tree, the sender can potentially provide $k-1$ valid inputs (that hash to the same value). Since the tree is of depth d, one can observe that by a mix-and-match method of different colliding values on different nodes of the tree the sender might be able to come up with as many as $(k-1)^{2^d}$ valid inputs x whose corresponding root-hash is $h(x^*)$. Thus, to satisfy that no two have the same value under g, we have to choose $m \approx 2^d \cdot \log(k-1)$, in which case the almost uniform hash function has a pretty long description. Nevertheless, it is less than $2^d n$ (the input length) so we might hope that some progress has been made. Is this protocol computationally-binding? Not quite. Using the proof technique from above (of partially rewinding and "collecting" collisions) would require running the malicious sender more than $(k-1)^{2^d}$ times until it has to present more than $k-1$ collisions for some value. This is, of course, way too expensive.

The bottom line of the above paragraph is that "mix-and-match" attacks are very powerful for a malicious sender in the context of tree hashing by allowing him to leverage the ability of finding few collisions into an ability to find exponentially many collisions. The reason why this happens is that we compose hash functions but apply the universal hash function on the whole input as a single string. Our next idea is to apply a "small" hash function $g\colon \{0,1\}^{2n} \rightarrow \{0,1\}^n$ per node in the Merkle tree. That is, after the sender sends the root-hash $h(x^*)$, the receiver samples g and sends it to the sender. The sender computes $g(\cdot,\cdot)$ for every pair of siblings along the Merkle tree, concatenates them all and sends this long string back to the receiver. This protocol is more promising since, in some sense, we have a small consistency check per node in the tree which should rule out simple "mix-and-match" attacks. This is our construction and the proof of security works by partially rewinding a malicious sender and "collecting" collisions until we get k collisions with respect to some internal node in the tree (we need to collect roughly $2^d k$ collisions overall so that such a node exists, by the pigeonhole principle). How efficient is the protocol? Details follow.

The protocol still consists of 4 rounds. A commitment consists of the hash function h, the root hash $h(x^*)$, a universal hash function $g\colon \{0,1\}^{2n} \rightarrow \{0,1\}^m$ and the value of g on *every* internal node of the tree. The overall size is thus of order $n + 2^d m$. Notice that $n + 2^d m \ll 2^d n$ whenever d is not too small, so we have made progress! We reduced the size of the commitment by a factor of m/n. The final step is to really get down to a commitment of size roughly n. To achieve this, we apply our protocol *recursively*: Instead of sending the hashes (with respect to g) of all internal nodes, we run our *commit* protocol recursively on this string. Notice that this string is shorter ($2^d m$ compared to $2^d n$) so the recursion does progress. The base of the recursion is when the string length is roughly n bits, then the sender can simply send it to the receiver.

Choosing the parameters carefully, we get various trade-offs between the number of rounds, the commitment size, and the security of the resulting protocol. For example, setting $m = n^{0.99}$, results with a $O(1)$-round protocol in which the commitment size is $O(n)$ (here the big "O" hides constants that depend on $\log_n(|x^*|)$ which is constant for a polynomially long x^*), and whose security is worse than the security of the k-MCRH by an additive factor of $\exp(-n^{0.99})$.

Local Opening. Due to the tree structure of our commitment protocol, it can be slightly modified to support local opening. Recall that the goal here is to allow the receiver to send an index i of a block to the sender, who can reply with the opening of the block, with communication proportional to n but not to the number of blocks 2^d. The idea here is, given an index i of a block, to open the hash values along the path corresponding to the i-block along with the tree sibling of every node in the path. Then, i' is defined to be the index of the block in the shorter string (the string committed to in the next step of the recursion) which containing all the $g(\cdot,\cdot)$ values of the nodes on the path (we make sure that such a block exists). Then, we add the hash values of the path for block i' and continue in a recursive manner.

Statistical Hiding. We show how to transform any short commitment scheme that is computationally binding (but perhaps not hiding) to a new scheme that is short, computationally binding and *statistically hiding*. Moreover, if the original scheme admits a local-opening, then the new scheme admits a local-opening as well. Our transformation is information theoretic, adds no additional assumptions and preserves the security, the number of rounds and communication complexity of the original scheme (up to a small constant factor). The transformation is partially based on ideas originating in the work of Naor and Yung [46, Sect. 5.2] and the follow-up works of Damgård, Pedersen, and Pfitzmann [11,12] giving constructions of statistical-hiding commitments from (standard) collision resistant hash function.

The idea of our transformation is to leverage the fact that the basic commitment protocol is short: when committing to a long string x^*, the communication is very short. Thus, a large portion of x^* is not revealed to the receiver by the protocol so this part of x^* is statistically hidden. The task that remains is to make sure that all of x^* is hidden. Thus, instead of committing to x^* directly, we commit to a random string r that is independent of x^* and slightly longer. Then, we extract from r the remaining randomness r' *given the communication of the protocol* using a strong extractor. Finally, we commit on the string $x^* \oplus r'$. It is not hard to show that if the original scheme was computationally binding, then the new one is as well. The fact that the scheme is statistically-hiding follows from the use of the strong extractor and the fact that the commitment is short.

One problem with the recipe above, is that the protocol (as describe) no longer admits a local-opening. This is because to open an index i, we need the i-th output bit of the extractor, but computing this bit might require reading a large portion of the input of r. Our solution is to break the input to sufficiently small parts such that each part is small enough to fit in a local-opening but is long enough to have enough entropy (given the communication of the protocol) so that we can apply the extractor on it.

2.2 Separating Multi-CRH from Standard CRH

We show barriers of constructing a collision-resistant hash function from a 3-multi-collision-resistant hash function. We rule out fully black-box constructions (see Definition 9). Our proof technique is inspired by the works of Asharov and Segev [2] and Haitner et al. [23], that are based in turn on ideas originating in the works of Simon [50], Gennaro et al. [14] and Wee [54]. However, when trying to adapt their proof to the setting of multi collisions one encounters several obstacles and we explain how to overcome them.

The high-level overview of the proof is to show that there exists an oracle Γ such that relative to Γ there exists a 3-MCRH, however there exist no standard CRH. Our oracle will contain a truly random function f that maps $2n$ bits to n bits. Relative to this oracle, it is clear that 3-MCRH exists, however, also standard CRH exist. We add an oracle $\mathsf{ColFinder}$ that will be used to break any CRH construction. The main difficulty of the proof is to show that this oracle cannot be used to break the 3-MCRH.

The oracle $\mathsf{ColFinder}$ is essentially the same as in Simon [50]. It gets as an input a circuit C, possibly with f gates and it outputs two random elements w, w' such that $C(w) = C(w')$. It is easy to see that no family of hash functions can be collision resistant in the presence of such an oracle. A single call to $\mathsf{ColFinder}$

with the query C (where $C(x) = f(x)$) will find a collision with high probability. The main question is whether this oracle be used to find *multiple* collisions?

Originally, Simon showed that this oracle cannot be used to *invert* a one-way function (or even a permutation). Let $\mathcal{A}$ be an adversary that uses ColFinder to invert f on a random challenge $y = f(x)$. Clearly, if $\mathcal{A}$ make no calls to ColFinder then his chances in inverting y are negligible. Assume, for simplicity, that $\mathcal{A}$ performs only a single query to ColFinder. In order for $\mathcal{A}$ to gain some advantage, it must make an "interesting" query to ColFinder. That is, a query which results in w, w' and the computation of either $C(w)$ or $C(w')$ makes a direct query to some $x \in f^{-1}(y)$. This event is called a hit. An important point is that for any circuit C the marginal distribution of w and of w' is uniform. Therefore, the probability of the event "hit" in the ColFinder query is at most twice that probability when evaluating $C(z)$ for a random z. Thus, we can construct a simulator that replaces $\mathcal{A}$'s query to ColFinder with the evaluation of $C(z)$ and hits an inverse of y with roughly the same probability as $\mathcal{A}$ (while making no queries to ColFinder). The task of inverting y without ColFinder can be shown to be hard, ruling out the existence of such a simulator and in turn of such an adversary $\mathcal{A}$.

Our goal is to extend this proof and show that ColFinder cannot be used to find 3-wise collisions. The above approach above simply does not work: specifically, in our case the event "hit" corresponds to query C to ColFinder that results in w, w' and the computation of $C(w)$ and $C(w')$ *together* make direct queries three elements x_1, x_2, x_3 that collide under f (i.e., $f(x_1) = f(x_2) = f(x_3)$). It might be the case that these three elements are hit by $C(w)$ and $C(w')$ combined, but never by one of them alone. Thus, when simulating the $C(z)$ for a random z we will hit only part of the trio x_1, x_2, x_3 and might never hit all three.

Our main observation is that since $\mathcal{A}$ finds a 3-wise collision, but ColFinder finds only a 2-wise collision (namely w, w'), by the pigeonhole principle, either w or w' will hit two of the three elements of the 3-wise collision. Again, since the marginals of w and w' each are uniform, we can construct a simulator that runs $\mathcal{A}$ and on the query to ColFinder samples a uniform z and compute $C(z)$ and will get a colliding x_1 and x_2 without performing any queries to ColFinder. Then, one can show that such a simulator cannot exist.

Several problems arise with this approach. First, notice that this does not extend to an adversary $\mathcal{A}$ that makes more than one query. In such a case, the resulting simulator finds a 2-wise collision x_1, x_2 without the "hit" event occurring (i.e., finding a 3-wise collision) but while performing several ColFinder queries. Such a simulator (that finds a collision), of course, *trivially* exists, and we do not get the desired contradiction. Nevertheless, we show that the collision found by our simulator is somewhat special, and using ColFinder one can only find "non-special" collisions, ruling out the existence of $\mathcal{A}$ in this case. Second, the event "hit" itself might be spread out within several ColFinder queries, where, for example, one queries finds x_1 and then another query finds x_2, x_3. We tailor a simulator for each case, and show that for each case the resulting simulating cannot exist, completely ruling out the possibility of $\mathcal{A}$ to exist.

3 Preliminaries

Unless stated otherwise, the logarithms in this paper are base 2. For an integer $n \in \mathbb{N}$ we denote by $[n]$ the set $\{1, \ldots, n\}$. We denote by U_n the uniform distribution over n-bit strings. For a distribution $\mathcal{D}$ we denote by $x \leftarrow \mathcal{D}$ an element chosen from $\mathcal{D}$ uniformly at random. We denote by $\circ$ the string concatenation operation. A function $\mathsf{negl}: \mathbb{N} \to \mathbb{R}^+$ is *negligible* if for every constant $c > 0$, there exists an integer N_c such that $\mathsf{negl}(n) < n^{-c}$ for all $n > N_c$.

Definition 1 (Statistical Distance). *The statistical distance between two random variables X, Y is defined by*

$$\Delta(X, Y) \triangleq \frac{1}{2} \cdot \sum_x |\Pr[X = x] - \Pr[Y = x]|$$

We say that X and Y are δ-close (resp. -far) if $\Delta(X, Y) \leq \delta$ (resp. $\Delta(X, Y) \geq \delta$).

3.1 Limited Independence

Definition 2 (k-wise independence). *Fix $n, m, k \in \mathbb{N}$. A function family $\mathcal{G} = \{g: \{0,1\}^n \to \{0,1\}^m\}$ is k-wise independent if for every distinct $x_1, \ldots, x_k \in \{0,1\}^n$ and every $y_1, \ldots, y_k \in \{0,1\}^m$ it holds that*

$$\Pr_{g \leftarrow \mathcal{G}}[\forall i \in [k]: g(x_i) = y_i] = \frac{1}{2^{km}}.$$

It is known that for every $m \leq n$, there exists a k-wise independent family of functions, where each function is described by $k \cdot n$ bits. One well-known construction which is optimal in terms of size is by letting each $g \in \{0,1\}^{k \cdot n}$ describe a degree $k-1$ polynomial over $\mathsf{GF}[2^n]$. The description of the polynomial requires k field elements so $k \cdot n$ bits are enough. Evaluation of such a function is merely an evaluation of the polynomial.

In some applications (including some of ours) the input size n is very large and we prefer that the description size of the hash function to be much shorter. To circumvent this, it is sometimes enough to use *almost* k-wise independent functions.

Definition 3 (Almost k-wise independence). *Fix $n, m, k \in \mathbb{N}$ and $\delta \in \mathbb{R}$. A function family $\mathcal{G} = \{g: \{0,1\}^n \to \{0,1\}^m\}$ is (k, δ)-wise independent if for every distinct $x_1, \ldots, x_k \in \{0,1\}^n$ the distribution of $(g(x_1), \ldots, g(x_k))$ is δ-close to the distribution $(u_1, \ldots, u_k)$, where $g \leftarrow \mathcal{G}$ and each $u_i \leftarrow \{0,1\}^m$ are chosen uniformly at random.*

It is known that for every $m \leq n$, there exists a (k, δ)-wise independent function with each function $g \in \mathcal{G}$ being described by $O(mk + \log(n/\delta))$ bits [1,44] (see also [52]).

3.2 Randomness Extractors

We consider random variables supported on n-bit strings. A random variable X is said to have min-entropy $\mathsf{H}_\infty(X) = k$ if for every $x \in \mathsf{Supp}(X)$ it holds that $\Pr[X = x] \leq 2^{-k}$.

We say that a function $\mathsf{Ext}\colon \{0,1\}^n \times \{0,1\}^d \to \{0,1\}^m$ is a (k, ϵ)-*seeded extractor* if for every distribution X over $\{0,1\}^n$ with min-entropy k, it holds that

$$\Delta(\mathsf{Ext}(X, U_d), U_m) \leq \epsilon.$$

The extractor Ext is said to be *strong* if $\mathsf{Ext}'(x, s) = \mathsf{Ext}(x, s) \circ s$ is a (k, ϵ)-seeded extractor. That is, if

$$\Delta((\mathsf{Ext}(X, U_d) \circ U_d), (U_m \circ U_d)) \leq \epsilon.$$

The famous leftover hash lemma [27,33] says that a pairwise independent function family is a strong extractor.

Proposition 1. *Let $\mathcal{G} = \{g\colon \{0,1\}^n \to \{0,1\}^m\}$ be a pairwise independent family of hash functions where $m = k - 2\log(1/\epsilon)$. Then, $\mathsf{Ext}(x, h) = h(x)$ is a strong (k, ϵ)-seeded extractor.*

Note that the seed length in this extractor equals the number of bits required to sample $g \leftarrow \mathcal{G}$ which is $2n$ bits.

We will also need the following standard proposition that says that conditioning does not reduce entropy by more than the information given by the condition.

Proposition 2. *Let X and Y be random variables. Then, if Y is supported on strings of length k, then $\mathsf{H}_\infty(X \mid Y) \geq \mathsf{H}_\infty(X) - k$.*

3.3 List-Recoverable Codes

The classical notion of error correcting codes ensures that for a code $C \subseteq \mathbb{F}^n$, where $\mathbb{F}$ is a finite field, given a somewhat corrupted version of $c \in C$, it is possible to recover c. The model of allowed corruptions is that some fraction of the symbols in the codeword might be adversarially changed. List recoverable codes were introduced to handle a different model of corruptions: they allow an adversary to submit, for every coordinate $i \in [n]$ a small list $S_i \subseteq \mathbb{F}$ of possible symbols. In this model, it is impossible to completely recover a codeword given the lists, but these codes guarantee that there is only a small list of codewords that are consistent with all the lists.

More precisely, a mapping $C\colon \mathbb{F}^k \to \mathbb{F}^n$ from length k messages to length n codewords, is called (α, ℓ, L)-*list-recoverable* if there is a procedure that is given a sequence of lists $S_1, \ldots, S_n \subseteq \mathbb{F}$ each of size ℓ, and is able to output all messages $x \in \mathbb{F}^k$ such that $C(x)_i \notin S_i$ for at most an α fraction of the coordinates $i \in [n]$. The code guarantees that there are at most L such messages.

Definition 4 (List-recoverable codes). *Let $\alpha \in [0,1]$. We say that a tuple $x \in (\{0,1\}^k)^n$ is* α-consistent *with sets $S_1, \ldots, S_n \subseteq \{0,1\}^k$, if $|\{i : x_i \in S_i|\} \geq \alpha n$.*

A function $C\colon \{0,1\}^v \to (\{0,1\}^k)^n$ is (α, ℓ, L)-list recoverable, if for every set $S_1, \ldots, S_n \subseteq \{0,1\}^k$ each of size at most ℓ, there are at most L strings $x \in \{0,1\}^v$ such that $C(x)$ is α-consistent with $S_1, \ldots, S_n$. For $\alpha = 1$, we omit α in the above notation and call C (ℓ, L)-list recoverable. The strings in the image of C are referred to as codewords.

These code were initially studied in the context of *list-decoding* (and indeed the latter is just a special case of the former with $\ell = 1$) by [18–21]. More recently, they were proven useful in other areas such as compressed sensing [47], non-adaptive domain extension for hashing [25], domain extension for public random functions and MACs [13,40], and more (see Sect. 6, and for example, [29] and references therein).

A natural relaxation of the above codes is to require that $S_1 = \ldots = S_n$. This variant is called *weakly* list-recoverable codes. A list-recoverable code is immediately weakly list-recoverable and the converse also holds albeit with a minor loss in parameters: An (ℓ, L)-weakly list-recoverable code is an (ℓ, nL)-list-recoverable code. Looking ahead, this loss will not make much of a difference for us since our L will be polynomial in n.

For our purposes, we will need a list-recoverable code with $\alpha = 1$. It is well-known (see e.g., [25]) that the notion of weakly list-recoverable codes is equivalent to unbalanced expanders with a certain expansion property. The left set of vertices in the graph is $\{0,1\}^v$, the right set of vertices is $\{0,1\}^k$ and the left degree is n. This graph naturally induces a mapping $C\colon \{0,1\}^v \to (\{0,1\}^k)^n$ which on input $x \in \{0,1\}^v$ (left vertex) outputs n neighbors (right vertices). The mapping C is (ℓ, L)-list-recoverable iff for every set $S \subseteq \{0,1\}^k$ of size larger than L of nodes on the right, the set of left neighbors of S is of size larger than ℓ.

The following instantiation of locally-recoverable codes based on the explicit construction of unbalanced expanders of [22] is taken (with minor modifications) from [25].

Theorem 1 ([22,25]). *For every $\alpha \geq 1/2$, and $k < v$, there exists a* $\mathsf{poly}(n)$-*time computable function $C\colon \{0,1\}^v \to \left(\{0,1\}^k\right)^n$ for $n = O(v \cdot k)^2$ which defines an (α, ℓ, L)-list recoverable code for every $L \leq 2^{k/2}$ and $\ell = \Omega(L)$. The list-recovery algorithm runs in time* $\mathsf{poly}(v, \ell)$.

3.4 Cryptographic Primitives

A function f, with input length $m_1(n)$ and outputs length $m_2(n)$, specifies for every $n \in \mathbb{N}$ a function $f_n\colon \{0,1\}^{m_1(n)} \to \{0,1\}^{m_2(n)}$. We only consider functions with polynomial input lengths (in n) and occasionally abuse notation and write $f(x)$ rather than $f_n(x)$ for simplicity. The function f is computable in polynomial time (efficiently computable) if there exists an algorithm that for any $x \in \{0,1\}^{m_1(n)}$ outputs $f_n(x)$ and runs in time polynomial in n.

A function family ensemble is an infinite set of function families, whose elements (families) are indexed by the set of integers. Let $\mathcal{F} = \{\mathcal{F}_n : \mathcal{D}_n \to \mathcal{R}_n\}_{n \in \mathbb{N}}$ stand for an ensemble of function families, where each $f \in \mathcal{F}_n$ has domain $\mathcal{D}_n$ and range $\mathcal{R}_n$. An efficient function family ensemble is one that has an efficient sampling and evaluation algorithms.

Definition 5 (Efficient function family ensemble). *A function family ensemble $\mathcal{F} = \{\mathcal{F}_n : \mathcal{D}_n \to \mathcal{R}_n\}_{n \in \mathbb{N}}$ is efficient if:*

- *$\mathcal{F}$ is samplable in polynomial time: there exists a probabilistic polynomial-time machine that given 1^n, outputs (the description of) a uniform element in $\mathcal{F}_n$.*
- *There exists a deterministic algorithm that given $x \in \mathcal{D}_n$ and (a description of) $f \in \mathcal{F}_n$, runs in time $\mathsf{poly}(n, |x|)$ and outputs $f(x)$.*

Universal One-Wayness. A one-way function is an efficiently computable function which is hard to invert on a random output for any probabilistic polynomial-time machine. A universal one-way hash function (UOWHF) is a family of compressing functions $\mathcal{H}$ for which any PPT adversary has a negligible chance of winning in the following game: the adversary submits an x and gets back a uniformly chosen $h \leftarrow \mathcal{H}$. The adversary wins if it finds an $x' \neq x$ such that $h(x) = h(x')$. UOWHF were introduced by Naor and Yung [46] and were shown to imply secure digital signature schemes. Rompel [48] (see also [36]) showed how to construct UOWHF based on the minimal assumption that one-way functions exist.

Definition 6 (Universal one-way hash functions (UOWHF)). *An efficient function family ensemble $\mathcal{F} = \{\mathcal{F}_n : \{0,1\}^{m_1(n)} \to \{0,1\}^{m_2(n)}\}_{n \in \mathbb{N}}$ is a universal one-way hash function family if the probability of every probabilistic polynomial-time adversary $\mathcal{A}$ to win in the following game is negligible in n:*

1. *$\mathcal{A}$, given 1^n, submits $x \in \{0,1\}^{m_1(n)}$.*
2. *Challenger responds with a uniformly random $f \leftarrow \mathcal{F}_n$.*
3. *$\mathcal{A}$ (given f) outputs $x' \in \{0,1\}^{m_1(n)}$.*
4. *$\mathcal{A}$ wins iff $x \neq x'$ and $f(x) = f(x')$.*

3.5 Commitment Schemes

A commitment scheme is a two-stage interactive protocol between a sender $\mathcal{S}$ and a receiver $\mathcal{R}$. The goal of such a scheme is that after the first stage of the protocol, called the commit protocol, the sender is bound to at most one value. In the second stage, called the opening protocol, the sender opens its committed value to the receiver. We also require that the opening protocol allows to open only a single bit of the committed string. More precisely, a commitment scheme for a domain of strings $\{0,1\}^\ell$ is defined via a pair of probabilistic polynomial-time algorithms $(\mathcal{S}, \mathcal{R}, \mathcal{V})$ such that:

- The commit protocol: $\mathcal{S}$ receives as input the security parameter 1^n and a string $s \in \{0,1\}^\ell$. $\mathcal{R}$ receives as input the security parameter 1^n. At the end of this stage, $\mathcal{S}$ outputs $\mathsf{decom}_1 \ldots, \mathsf{decom}_\ell$ (the local decommitments) and $\mathcal{R}$ outputs com (the commitment).
- The local-opening procedure: $\mathcal{V}$ receives as input the security parameter 1^n, a commitment com, an index $i \in [\ell]$, a local-decommitment decom_i, and outputs either a bit b or $\bot$.

A commitment scheme is *public coin* if all messages sent by the receiver are independent random coins.

Denote by $(\mathsf{decom}_1, \ldots, \mathsf{decom}_\ell, \mathsf{com}) \leftarrow \langle \mathcal{S}(1^n, s), \mathcal{R} \rangle$ the experiment in which $\mathcal{S}$ and $\mathcal{R}$ interact with the given inputs and uniformly random coins, and eventually $\mathcal{S}$ outputs a list of ℓ decommitment strings and $\mathcal{R}$ outputs a commitment. The completeness of the protocol says that for all $n \in \mathbb{N}$, every string $s \in \{0,1\}^\ell$, every tuple $(\mathsf{decom}_1, \ldots, \mathsf{decom}_\ell, \mathsf{com})$ in the support of $\langle \mathcal{S}(1^n, s), \mathcal{R} \rangle$, and every $i \in [\ell]$, it holds that $\mathcal{V}(i, \mathsf{decom}_i, \mathsf{com}) = s_i$.

Below we define two security properties one can require from a commitment scheme. The properties we list are *statistical-hiding* and *computational-binding*. These roughly say that after the commit stage, the sender is *bound* to a specific value which remains statistically hidden for the receiver.

Definition 7 (ϵ-binding). *A commitment scheme $(\mathcal{S}, \mathcal{R}, \mathcal{V})$ is $(t(n), \epsilon(n))$-binding if for every probabilistic adversary $\mathcal{S}^*$ that runs in time at most $t(n)$, it holds that*

$$\Pr\left[\begin{array}{c}(i, \mathsf{decom}_i, \mathsf{decom}'_i, \mathsf{com}) \leftarrow \langle \mathcal{S}^*(1^n), \mathcal{R} \rangle \text{ and} \\ \bot \neq \mathcal{V}(i, \mathsf{decom}_i, \mathsf{com}) \neq \mathcal{V}(i, \mathsf{decom}'_i, \mathsf{com}) \neq \bot\end{array}\right] \leq \epsilon(n)$$

for all sufficiently large n, where the probability is taken over the random coins of both $\mathcal{S}^*$ and $\mathcal{R}$.

Given a commitment scheme $(\mathcal{S}, \mathcal{R}, \mathcal{V})$ and an adversary $\mathcal{R}^*$, we denote by $\mathsf{view}_{\langle \mathcal{S}(s), \mathcal{R}^* \rangle}(n)$ the distribution on the view of $\mathcal{R}^*$ when interacting with $\mathcal{S}(1^n, s)$. The view consists of $\mathcal{R}^*$'s random coins and the sequence of messages it received from $\mathcal{S}$. The distribution is take over the random coins of both $\mathcal{S}$ and $\mathcal{R}$. Without loss of generality, whenever $\mathcal{R}^*$ has no computational restrictions, we can assume it is deterministic.

Definition 8 (ρ-hiding). *A commitment scheme $(\mathcal{S}, \mathcal{R}, \mathcal{V})$ is $\rho(n)$-hiding if for every (deterministic) adversary $\mathcal{R}^*$ and every distinct $s_0, s_1 \in \{0,1\}^\ell$, it holds that*

$$\Delta\left(\{\mathsf{view}_{\langle \mathcal{S}(s_0), \mathcal{R}^* \rangle}(n)\}, \{\mathsf{view}_{\langle \mathcal{S}(s_1), \mathcal{R}^* \rangle}(n)\}\right) \leq \rho(n)$$

for all sufficiently large $n \in \mathbb{N}$.

Complexity Measures. The parameters of interest are (1) the number of rounds the commit protocol requires, (2) the size of a commitment, and (3) the size of a local opening.

The *size* of a commitment is the size (in bits) of the output of $\mathcal{S}$ denoted above by com. A *short commitment* is such that the size of com is much smaller than ℓ. Preferably, the size of a short commitment depends solely on n, but poly-logarithmic dependence on ℓ is also okay. The size of a local opening is the maximum size of decom_i (in bits). A protocol is said to support local opening if this size depends only on n and at most poly-logarithmically on ℓ.

3.6 Fully Black-Box Constructions

We give a definition of a fully black-box reduction from an MCRH to standard CRH. For this, we generalize the definition of an MCRH to the setting of oracle-aided computation: The generation and evaluation algorithms of an MCRH are given access to an oracle Γ relative to which they can generate a description of a hash function and evaluate an index at a point. The adversary is also given oracle access to Γ in the security game and has to find multiple collisions relative to it.

We focus here on k-MCRH functions with $k = 3$. The following definition of a "black-box construction" is directly inspired by those of [2,23].

Definition 9. *A fully black-box construction of a collision-resistant function family $\mathcal{H}'$ from a 3-MCRH function family $\mathcal{H}$ mapping $2n$ bits to n bits consists of a pair of probabilistic polynomial-time algorithms $(\mathcal{H}.\mathsf{G}, \mathcal{H}.\mathsf{E})$ and an oracle-aided polynomial-time algorithm M such that:*

- **Completeness:** *For any $n \in \mathbb{N}$, for any 3-MCRH function family $\mathcal{H}$ and any function h produced by $h \leftarrow \mathcal{H}'^{\mathcal{H}}(1^n)$, it holds that $h^{\mathcal{H}}\colon \{0,1\}^{2n} \to \{0,1\}^n$.*
- **Black-box proof of security:** *For any collision resistant hash $\mathcal{H}'$, any probabilistic polynomial-time oracle-aided algorithm $\mathcal{A}$, every polynomial $p(\cdot)$, if*

$$\Pr\left[\begin{array}{c} x_1 \neq x_2 \\ h^{\mathcal{H}}(x_1) = h^{\mathcal{H}}(x_2) \end{array} \middle| \begin{array}{c} h \leftarrow h^{\mathcal{H}}(1^n) \\ (x_1, x_2) \leftarrow \mathcal{A}^{\mathcal{H}}(1^n, h) \end{array}\right] \geq \frac{1}{p(n)}$$

for infinitely many values of n, then there exists a polynomial $p'(\cdot)$ such that

$$\Pr\left[\begin{array}{c} x_1, x_2, x_3 \text{are distinct and} \\ h(x_1) = h(x_2) = h(x_3) \end{array} \middle| \begin{array}{c} h \leftarrow \mathcal{H}(1^n) \\ (x_1, x_2, x_3) \leftarrow M^{\mathcal{A},\mathcal{H}}(1^n, h) \end{array}\right] \geq \frac{1}{p'(n)}$$

for infinitely many values of n.

4 Multi-Collision-Resistant Function Families

A multi-collision-resistant hash function is a relaxation of standard collision-resistant hash function in which it is hard to find *multiple* collisions on the same value.

Definition 10 (Multi-Collision-Resistant Hashing). *Let $k = k(n)$ be a polynomial function. An efficient function family ensemble $\mathcal{H} = \{\mathcal{H}_n \colon \{0,1\}^{2n} \to \{0,1\}^n\}_{n \in \mathbb{N}}$ is a (t, ϵ)-secure* k-multi-collision-resistant hash *(MCRH) function family if for any probabilistic algorithm $\mathcal{A}$ that runs in time at most $t(n)$, for large enough $n \in \mathbb{N}$:*

$$\Pr\left[\begin{array}{c} x_1, \ldots, x_k \text{ are distinct and} \\ h(x_1) = \cdots = h(x_k) \end{array} \middle| \begin{array}{c} h \leftarrow \mathcal{H}_n \\ (x_1, \ldots, x_k) \leftarrow \mathcal{A}(h) \end{array}\right] \le \epsilon(n).$$

We call such $x_1, \ldots, x_k$ that map to the same value under h a k-wise collision. *Lastly, we say that $\mathcal{H}$ is a secure k-MCRH if it is $(p, 1/p)$-secure for every polynomial $p(\cdot)$.*

The Compression Ratio. In the definition above we assume that the hash function compresses its input from $2n$ bits into n, where the choice of the constant 2 is somewhat arbitrary. Our choice of linear compression rate (in contrast to, say, a polynomial compression rate) models the basic building blocks in most standards of cryptographic hash functions, such as the ones published by NIST (e.g., most of the SHA-x family).

When considering k-MCRH functions, a compression that eliminates less than $\log k$ bits is not of interest, since such a function exists unconditionally, say by chopping (there simply will be no k-wise collision).

The factor two compression is somewhat arbitrary as any k-MCRH that compresses $(1+\epsilon)n$ bits into n bits can be translated into a $(k^{1/\epsilon})$-MCRH that compresses $2n$ bits into n (e.g. via the Merkle-Damgård iterated construction [10, 42]). It is possible to assume an even stronger hash function that compresses by a polynomial factor, say n^2 bits into n bits (and this is sometimes useful; see the paragraph in the end of Sect. 6 for an example), but this is a strong assumption that we prefer to avoid.

For standard collision resistant hash function (with $k = 2$), it is known that composition allows to translate hash functions that compress by one bit into hash functions that compress by any polynomial factor (from n bits into n^δ bits for any constant $\delta > 0$). Obtaining a similar result for k-MCRH functions (with $k > 2$) without significantly compromising on the value of k in the resulting family is an open problem. In Sect. 6 we give a transformation in which the resulting family is $(k^{O(\log n)})$-MCRH.

Public vs. Private Coins. Our definition above is of a private-coin MCRH, namely, the coins used by the key-generation procedure are not given to the collision finder, but are rather kept secret. One can define the stronger public-coin variant in which the aforementioned coins are given to the attacker. The weaker notion is enough for our applications. There are (other) cases where this distinction matters, see Hsiao and Reyzin [31].

5 Tree Commitments from Multi-CRH

We show how to build a commitment scheme which is computationally-binding, statistically-hiding, round-efficient, has short commitments, and supports local opening. We refer to Sect. 3.5 for the definition of a commitment scheme, computational-binding, statistical-hiding, and the efficiency measures of commitments we consider below.

Theorem 2. *Assume that there exists a (t, ϵ)-secure k-MCRH $\mathcal{H}$ for a polynomial $k = k(n)$ in which every function can be described using $\ell = \ell(n)$ bits. For any parameters $d = d(n)$ and $1 < z \leq n/2d$, there is commitment protocol for strings of length $2^d \cdot n$ with the following properties:*

1. *(t', ϵ')-computationally-binding for $\epsilon' = O\left(2^{\frac{d}{\log(n/(zd))}} \cdot \left(\frac{k^2}{2^{z-d}} + \epsilon\right)\right)$ and $t' = O\left(\frac{\epsilon'^2 \cdot t}{nk2^d \cdot p(n)}\right)$, where $p(\cdot)$ is some fixed polynomial function.*
2. *2^{-n}-statistically-hiding.*
3. *takes $O\left(\frac{d}{\log(n/(zd))}\right)$ rounds.*
4. *the commitment has length $O(d\ell + dn)$.*
5. *supports local opening of size $O(d^2 n)$.*

There are various ways to instantiate z compared to n and d, offering various trade-offs between security and efficiency. We focus here on the case in which we wish to commit on a polynomially-long string, that is, $d = c \log n$ for a constant $c \in \mathbb{N}$. In the following the big "O" notation hides constants that depend on c. Setting $z = n^{1-\delta}$ for a small constant $\delta > 0$, the parameters of our commitment scheme are:

1. $\epsilon' = O\left(\frac{k^2}{2^{n^{1-\delta}}} + \epsilon\right)$ and $t' = \frac{\epsilon'^2 \cdot t}{\mathsf{poly}(n)}$.
2. 2^{-n}-statistically-hiding.
3. takes $O(1)$ rounds.
4. the commitment has length $O(\ell + n \log n)$.
5. supports local opening of size $O(n \log^2 n)$.

This setting is very efficient in terms of rounds (it is a constant that depends solely on c) but suffers in security loss (the resulting scheme is at most $2^{-n^{1-\delta}}$-secure). In the regime where the MCRH is $(p, 1/p)$-secure for every polynomial p, our resulting scheme is as secure (i.e., $(p, 1/p)$-computationally-binding for every polynomial p).

In case ϵ is very small to begin with (e.g., much smaller than $2^{-n^{1-\delta}}$), we can use set z to be $z = n/(2c \log n)$ for a small constant $\delta > 0$. The resulting commitment scheme satisfies:

1. $\epsilon' = O\left(n^{c-1} \cdot \left(\frac{k^2}{2^{n/\log n}} + \epsilon\right)\right)$ and $t' = \frac{t \cdot \epsilon'^2}{\mathsf{poly}(n)}$.
2. 2^{-n}-statistically-hiding.
3. takes $O(\log n)$ rounds.
4. the commitment has length $O(\ell \log n + n \log n)$.
5. supports local opening of size $O(n \log^2 n)$.

This setting has a logarithmic number of rounds, but the security loss is much smaller than before (only of order $2^{-n/\log n}$).

Roadmap. Our protocol is constructed in two main steps. In the first step (given in Sect. 5.1) we construct a protocol with the above properties (i.e., Theorem 2) except that it is *not* statistically hiding (but is computationally-binding, takes few rounds, has short commitment, and supports local opening). In the second step (given in Sect. 5.2), we show how to *generically* bootstrap our commitment scheme, into one that is also statistically-hiding. This reduction is both efficient and security preserving with respect to all parameters.

5.1 A Computationally-Binding Scheme

The main ingredients in our first protocol are an MCRH (Definition 10) and a limited-independent family (Definition 2):

- A (t, ϵ)-secure k-MCRH for a polynomial $k = k(n)$:
$$\mathcal{H} = \{h\colon \{0,1\}^{2n} \to \{0,1\}^n\}.$$
We assume that every $h \in \mathcal{H}$ can be described using $\ell = \ell(n)$ bits.
- A family of pairwise-independent functions mapping strings of length $2n$ to strings of length z:
$$\mathcal{G} = \{g\colon \{0,1\}^{2n} \to \{0,1\}^z\}.$$
Recall that every $g \in \mathcal{G}$ can be described using $4n$ bits.

Description of the Protocol. Our protocol relies on the notion of a Merkle hash tree. This is a method to hash a long string into a short one using a hash function with fixed input length. Let $x \in \{0,1\}^\ell$ be a string. A Merkle hash tree is a binary tree T, associated with a string $x \in \{0,1\}^\ell$ and a hash function $h\colon \{0,1\}^{2n} \to \{0,1\}^n$. Let $x = x_1, \ldots, x_{2^d}$ be the decomposition of x into 2^d blocks, each of length n. Every node v in the tree has a sibling denoted by $N(v)$ (we assume that the sibling of the root is $\perp$). Every node v in the tree is labeled with a string $\pi_v \in \{0,1\}^n$. The tree has 2^d leaves $v_1, \ldots, v_{2^d}$ and the label of v_i are set to $\pi_{v_i} = x_i$. The labels of the rest of the nodes are computed iteratively from the leaves to the root. Given a node v whose both children u_1, u_2 are labeled with π_{u_1}, π_{v_2}, we set the label of v to be $\pi_v = h(\pi_{u_1}, \pi_{v_2})$. The node root has label y and we call it the root-hash.

Given a Merkle hash tree for a string $x = x_1 \ldots x_{2^d} \in \{0,1\}^{2^d \cdot n}$, let path_i be a set of nodes in the tree including the nodes on the path from x_i to the root of the tree and all their siblings along this path. We further let $P_i = \{\pi_v \mid v \in \mathsf{path}_i\}$ be the set of all the labels of the nodes in the set path_i (the labels of the nodes on the path from x_i to the root of the tree and the labels of their siblings). Each set path_i contains $2d$ nodes and thus the description size of each P_i is $2dn$ bits.

The commitment protocol $(\mathcal{S}, \mathcal{R}, \mathcal{V})$ specifies how to commit to a string of length $2^d n$. Our protocol uses a Merkle hash tree with a function h supplied by the receiver and a root hash replied by the sender. In the next round, the receiver chooses a limited-independence function g with z bits of output (z is a tunable parameter) and sends it to the sender. The sender then computes g on the *hash values* of every pair of sibling nodes in the Merkle hash tree (i.e., $g(\pi_v \circ \pi_{N(v)})$). Then it concatenates all of these values into one long string s'. Then, the protocol continues in a recursive manner on this string s'. The length of s' is roughly $z2^d$ bits (there are 2^d internal nodes in the tree and each requires z bits) which is still too long to send as is, however is smaller than the original string s. This allows us to apply the same ideas recursively with the base case, committing on a string of length n, being the trivial protocol of sending the string as is. Choosing parameters carefully, we balance between the efficiency and security of our resulting commitment.

The commitment protocol for strings of length $2^d n$ for $d \geq 1$ is described in Fig. 1.

The commit protocol between $\mathcal{S}$ and $\mathcal{R}$

The sender $\mathcal{S}$ has string $s = s_1 \ldots s_{2^d}$ where $s_i \in \{0,1\}^n$ for all $i \in [2^d]$.

1. $\mathcal{R} \Rightarrow \mathcal{S}$: Sample $h \leftarrow \mathcal{H}$ and send h.
2. $\mathcal{S} \Rightarrow \mathcal{R}$: Compute a Merkle hash-tree T of s using h and send the root-hash y. Let π_v be the hash value in the tree for node $v \in T$, and let $P_i = \{\pi_v \circ \pi_{N(v)}\}_{v \in \mathsf{path}_i}$ for all $i \in [d]$.
3. $\mathcal{R} \Rightarrow \mathcal{S}$: Sample $g \leftarrow \mathcal{G}$ and send g.
4. $\mathcal{S} \Leftrightarrow \mathcal{R}$: Recursively interact to commit on the string $s' = u_1 \circ \ldots \circ u_{2^d}$, where $u_i = \{g(\pi_v \circ \pi_{N(v)})\}_{v \in \mathsf{path}_i}$. Notice that $|s'| = 2^d \cdot dz = 2^{d'} n$, where $d' = d - (\log n - \log z - \log d)$. Denote the outputs of the sender and receiver by
$$((D_1, \ldots, D_{2^{d'}n}), C) \leftarrow \langle \mathcal{S}(1^n, s'), \mathcal{R} \rangle.$$

The output of each party

– $\mathcal{R}$'s output: The receiver $\mathcal{R}$ outputs $\mathsf{com} = (h, y, g, C)$.
– $\mathcal{S}$'s output: The sender $\mathcal{S}$ outputs $(\mathsf{decom}_1, \ldots, \mathsf{decom}_{2^d})$, where decom_i is defined as follows: Let i' be the index in s' of the block containing u_i. Set $\mathsf{decom}_i = (s_i, P_i, D_{i'})$.[a]

The local-opening procedure $\mathcal{V}$

The verifier $\mathcal{V}$ has an index $i \in [2^d]$, a decommitment decom_i, and a commitment com.

1. Verify that s_i appears in P_i in the right location.
2. Verify that the values in P_i are consistent with respect to h and y.
3. Compute $u_i = \{g(\pi_v \circ \pi_{N(v)})\}_{v \in \mathsf{path}_i}$, where $P_i = \{\pi_v \circ \pi_{N(v)}\}_{v \in \mathsf{path}_i}$. Recursively compute $u'_i \leftarrow \mathcal{V}(i', D_{i'}, C)$ and verify that $u'_i = u_i$.
4. If all tests pass, output s_i. Otherwise, output $\perp$.

[a] We assume without loss of generality that each u_i is contained in a single block (otherwise, we pad the string).

Fig. 1. Our commitment protocol for strings of length $2^d \cdot n$.

Rounds and Communication Analysis. Denote by $\mathsf{Size}(2^d n)$, $\mathsf{Rounds}(2^d n)$, and $\mathsf{Decom}(2^d n)$, the *total size* of the commitment, the *number of rounds* of the commit stage, and the *size of a local opening* on a string of length $2^d n$, respectively. The commitment consists of a description of a hash function $h \in \mathcal{H}$ (whose size is denoted by $\ell(n)$), the root-hash value of a Merkle hash-tree (which is of size n), a function $g \in \mathcal{G}$ (which is of size $4n$), and the recursive commitment. The opening for an index $i \in [2^d]$ consists of a block (of size n), the full i-th path (which consists of $2dn$ bits), and the recursive opening.

Recall that the protocol for committing on strings of length $2^d n$ uses (recursively) a protocol for committing on strings of length $2^{d'} n$, where

$$d' = d - (\log n - \log z - \log d) = d - \log(n/(zd)).$$

Moreover, the commitment protocol on strings of length n has communication complexity n, consists of a single round and the opening is n bits. Thus, the total number of recursive call will be

$$\left\lceil \frac{d}{\log(n/(zd))} \right\rceil.$$

We get that the total number of communication rounds is bounded by

$$\mathsf{Rounds}(2^d n) \leq 3 + \mathsf{Rounds}(2^{d'} n) \leq \cdots \leq \left\lceil \frac{3d}{\log(n/(zd))} \right\rceil + O(1)$$

Actually, since each recursive call consists of three messages (except the base of the recursion), we can join the last round of every iteration with the first round in the next one. Therefore, we can get the improved bound

$$\mathsf{Rounds}(2^d n) \leq \frac{2d}{\log(n/(zd))} + O(1)$$

The size of a commitment is bounded by

$$\begin{aligned} \mathsf{Size}(2^d n) &\leq \ell(n) + 5n + \mathsf{Size}(2^{d'} n) \\ &\leq \left\lceil \frac{d \cdot (\ell(n) + 5n)}{\log(n/(zd))} \right\rceil \leq d \cdot (\ell(n) + 5n). \end{aligned}$$

The size of a local opening is bounded by

$$\begin{aligned} \mathsf{Decom}(2^d n) &\leq n + 2dn + \mathsf{Decom}(2^{d'} n) \\ &\leq \left\lceil \frac{3d^2 n}{\log(n/(zd))} \right\rceil \leq 3d^2 n. \end{aligned}$$

Computational-Binding. We show that our protocol is (t_d, ϵ_d)-computationally-binding for strings of length $2^d \cdot n$. We assume that the k-MCRH is (t, ϵ)-secure and that the internal protocol for strings of length $2^{d'} n$ is $(t_{d'}, \epsilon_{d'})$-computationally-binding. We set ϵ_d to satisfy the following recursive relation:

$$\epsilon_d = \frac{4k^2}{2^{z-d}} + 4\epsilon + 4\epsilon_{d'}.$$

Plugging in the number of rounds our recursion takes, we get that

$$\epsilon_d \leq 2^{\frac{6d}{\log n - \log z - \log d}} \cdot \left(\frac{4k^2}{2^{z-d}} + 4\epsilon\right).$$

Let $\mathcal{S}^*$ be a (cheating) adversary that runs in time t_d and breaks the binding of the protocol, namely, with probability ϵ_d, $\mathcal{S}^*$ is able to (locally) open the commitment in two ways without getting caught. That is, after the commitment stage, there is an index $i \in [2^d]$ for which the adversary $\mathcal{S}^*$ is able to open the corresponding block in two *different* ways with probability ϵ_d:

$$\Pr_{\mathcal{S}^*,\mathcal{R}}\left[\begin{array}{c}(i, \mathsf{decom}_i^0, \mathsf{decom}_i^1, \mathsf{com}) \leftarrow \langle \mathcal{S}^*(1^n), \mathcal{R}\rangle \text{ and} \\ \perp \neq \mathcal{V}(i, \mathsf{decom}_i^0, \mathsf{com}) \neq \mathcal{V}(i, \mathsf{decom}_i^1, \mathsf{com}) \neq \perp\end{array}\right] \geq \epsilon_d, \tag{1}$$

where the probability is taken over the random coins of both $\mathcal{S}^*$ and $\mathcal{R}$. The randomness of $\mathcal{R}$ consists of uniformly chosen functions $h \leftarrow \mathcal{H}$ and a function $g \leftarrow \mathcal{G}$, so we can rewrite Eq. (1) as

$$\Pr_{\substack{\mathcal{S}^*, h\leftarrow\mathcal{H}, \\ g\leftarrow\mathcal{G}}}\left[\begin{array}{c}(i, \mathsf{decom}_i^0, \mathsf{decom}_i^1, \mathsf{com}) \leftarrow \langle \mathcal{S}^*(1^n), (h, g)\rangle \\ \text{and} \perp \neq \mathcal{V}(i, \mathsf{decom}_i^0, \mathsf{com}) \neq \mathcal{V}(i, \mathsf{decom}_i^1, \mathsf{com}) \neq \perp\end{array}\right] \geq \epsilon_d. \tag{2}$$

We will show how to construct an adversary $\mathcal{A}$ that runs in time at most t and finds a k-wise collision relative to a randomly chosen hash function $h \leftarrow \mathcal{H}$. The procedure $\mathcal{A}$ will run the protocol $\langle \mathcal{S}^*(1^n), (h, g)\rangle$ with the given h and a function g chosen uniformly at random and get (with good probability) two valid and different openings for some index i. Then, $\mathcal{A}$ will *partially* rewind the adversary $\mathcal{S}^*$ to the stage after he received the hash function h and replied with the root hash y (of s), and execute it again but with a fresh function $g \leftarrow \mathcal{G}$. With noticeable probability, this will again result with two valid openings for some (possibly different) index i. Repeating this process enough times, $\mathcal{A}$ will translate a large number of different (yet valid) decommitments into a large number of collisions relative to h. The fact that these collisions are distinct (with good probability) will follow from the fact that the g's are sampled independently in every repetition.

We introduce some useful notation. Recall that $\mathcal{S}^*$ is the sender that is able to locally open its commitment in two different ways. We slightly abuse notation and also think of $\mathcal{S}^*$ as a distribution over senders (this is without loss of generality since we can assume it chooses all of its randomness ahead of time). For an index $i^* \in [R]$, string $y \in \{0,1\}^n$, we denote by $\mathcal{S}^*|_{h,y}$ a distribution over all senders $\mathcal{S}^*$ in which the first message received is h and the reply is the root-hash y. For a string y that is sampled by choosing $h \leftarrow \mathcal{H}$ and running $\mathcal{S}^*$ for one round to obtain y, the adversary $\mathcal{S}^*|_{h,y}$ is uniformly chosen from all possible continuations of the adversary $\mathcal{S}^*$ given these first two messages. Given this notation, we can write the adversary $\mathcal{S}^*$ as a pair of two distributions $(Y_h, \mathcal{S}^*|_{h,Y_h})$, where Y_h is the distribution of the first message $\mathcal{S}^*$ sends in response to h, and $\mathcal{S}^*|_{h,Y_h}$ is the distribution over the rest of the protocol.

In a high-level, our algorithm $\mathcal{A}$ maintains a binary tree of depth d and 2^d leaves, in which each node v is associated with a set of labels S_v. At the beginning, each such set S_v is initialized to be empty. The algorithm $\mathcal{A}$ uses $\mathcal{S}^*$ to get many valid pairs of openings decom_i^0 and decom_i^1 for an index i:

$$\mathsf{decom}_i^0 = (s_i^0, P_i^0, D_{i'}^0) \text{ and } \mathsf{decom}_i^1 = (s_i^1, P_i^1, D_{i'}^1)$$

that are consistent with a commitment:

$$\mathsf{com} = (h, y, g, C).$$

Since both openings are valid and $s_i \neq s_i'$, it must be that (1) s_i^0 (resp., s_i^1) appears in P_i^0 (resp., P_i^1) and (2) P_i^0 and P_i^1 are consistent with the root-hash y. Thus, it must be that $P_i^0 \neq P_i^1$ and there is a node v on the path path_i that is the first (going from the root to the leaves) non-trivial collision between P_i^0 and P_i^1. Now, by the induction hypothesis, the probability that $\mathcal{S}^*$ cheats in the internal decommitment is small, and since g is sampled uniformly at every iteration, we show that it must be a *new* collision that did not appear before. We will identify the location of the collision at a node v and add this collision to the set S_v. See Fig. 2 for a precise description of $\mathcal{A}$.

The adversary $\mathcal{A}(1^n, h)$:

1. For all nodes v, set $S_v = \emptyset$ to be the empty set.
2. Send to $\mathcal{S}^*$ the function h and receive a root-hash $y \in \{0,1\}^n$.
3. Do the following $T = 50nk2^d/\epsilon_d^2$ times:
   (a) Sample $g \leftarrow \mathcal{G}$.
   (b) Obtain $(i, \mathsf{decom}_i^0, \mathsf{decom}_i^1, \mathsf{com}) \leftarrow \langle \mathcal{S}^*|_{h,y}(1^n), g\rangle$.
   (c) Parse $\mathsf{decom}_i^0 = (s_i^0, P_i^0, D_{i'}^0)$ and $\mathsf{decom}_i^1 = (s_i^1, P_i^1, D_{i'}^1)$. Parse $\mathsf{com} = (h, y, g, C)$.
   (d) If any of the following occurs, continue to the next iteration:
      i. $\perp \neq \mathcal{V}(i, \mathsf{decom}_i^0, \mathsf{com}) \neq \mathcal{V}(i, \mathsf{decom}_i^1, \mathsf{com}) \neq \perp$.
      ii. $\perp \neq \mathcal{V}(i', D_{i'}^0, C) \neq \mathcal{V}(i', D_{i'}^1, C) \neq \perp$.
   (e) Let v the node of the first (from the root to the leaves) non-trivial collision between P_i^0 and P_i^1. Let $X = \pi_v^0, \pi_{N(v)}^0$ and $Y = \pi_v^1, \pi_{N(v)}^1$ be the values of the collision for P_i^0 and P_i^1, respectively. Add X and Y to S_v.
   (f) If there exists a node v for which $|S_v| \geq k$, then output S_v and halt.
4. Output $\perp$.

Fig. 2. The algorithm $\mathcal{A}$ to find a k-wise collision in $\mathcal{H}$.

The analysis of the adversary $\mathcal{A}$ can be found in the full version [38].

Remark 1 (Optimization I: recycling h). In the recursive step of our protocol, we can use the same hash function h as in the first step of the recursion. This saves sending its description (which is of size $\ell(n)$ at every iteration and the resulting size of a commitment is $(O(\ell(n) + d^2n)$.

Remark 2 (Optimization II: almost-universal hashing). In our protocol we used a pairwise independent hash function whose description size is proportional to their input size. This costs us in communication. We could save communication by using almost-universal hash functions whose description size is proportional to their output size.

5.2 Getting Statistical-Hiding Generically

We show how to transform any short commitment scheme Π that is computationally binding (but perhaps not hiding) to a new scheme Π' that is short, computationally binding and *statistically hiding*. Moreover, if Π admits a local-opening, then Π' admits a local-opening as well. Our transformation is information theoretic, adds no additional assumptions and preserves the security, the number of rounds and communication complexity of the original scheme (up to a small constant factor).

High-Level Idea. The underlying idea of our transformation is to leverage the fact that the commitment protocol Π is short. Specifically, when committing to a long string $s \in \{0,1\}^{n^c}$ for some large $c \in \mathbb{N}$, the communication is very short: $\Lambda(n)$ bits for a fixed polynomial function.[13] Thus, when we commit to s, a large portion of s is not revealed to the receiver by the protocol so this part of s is statistically hidden. The question is how to make sure that all of s is hidden.

Our solution takes advantage of the fact that some fraction of s remains hidden. We commit on a random string r that is independent of s and slightly longer. Then, we extract from r the remaining randomness r' *given the communication of the protocol* using a strong extractor (see Sect. 3.2). Finally, we commit on the string $s \oplus r'$. We need to show that this scheme is computationally-binding and statistically-hiding. The former follows from the computational-binding of the original scheme. The latter follows from the fact that r' is completely hidden to the receiver and it masks the value of s.

The details of this transformation appear in the full version [38].

6 Four-Round Short Commitments from Multi-CRH

We show how to construct a *4-round* short commitment protocol based on a family of k-MCRH functions. Compared to the protocol from Sect. 5, this protocol has *no* local opening and is secure only for *constant* values of k. However, it consists only of 4 rounds. Furthermore, using techniques similar to Sect. 5.2 the protocol can be made also statistically-hiding which suffices for some applications such as statistical zero-knowledge arguments [8].

We discuss methods that allow us to prove security even for polynomial values of k towards the end of the section.

[13] Our protocol from Sect. 5.1 has an additional linear dependence on c, but we will ignore this in this section to simplify notation.

Theorem 3. *Assume that there exists a (t, ϵ)-secure k-MCRH $\mathcal{H}$ for a constant k in which every function can be described using $\ell = \ell(n)$ bits. For any $c \in \mathbb{N}$, there is a commitment protocol for strings of length n^c with the following properties:*

1. *(t', ϵ')-computationally-binding for $\epsilon' = 4\epsilon + \frac{O(k^{c \log n})}{2^n}$ and $t' = \frac{t \cdot \epsilon'^2}{O(k^{c \log n}) \cdot p(n)}$, where $p(\cdot)$ is some fixed polynomial function.*
2. *takes 4 rounds (i.e., 4 messages).*
3. *the commitment has length $\ell + O(n)$.*

Proof. We describe a commitment protocol for a string $s = s_1 \ldots s_{2^d}$ of length $2^d \cdot n$ for $d = (c-1) \cdot \log n$. We show that our protocol is computationally-binding and getting statistical-hiding can be done using the generic transformation from Sect. 5.2. Our protocol uses a Merkle hash-tree as described in Sect. 5. The main observation made in this construction is that before using the Merkle hash-tree we can use a special type of encodings, called list-recoverable codes (see Definition 4) to get meaningful security. Let $C\colon \{0,1\}^{2^d n} \to (\{0,1\}^{2n})^{2^{d'}}$ be an (ℓ, L)-list-recoverable code for $\ell = k^d$, $L = O(k^d)$, and $d' = O(\log n)$ with some large enough hidden constants. Such a code exists by Theorem 1 (with efficient encoding and list-recovery). Let $\mathcal{G}$ be a family of (2^{-n})-almost pairwise-independent functions mapping strings of length $2^d n$ to strings of length n (see Definition 3): $\mathcal{G} = \{g\colon \{0,1\}^{2^d n} \to \{0,1\}^n\}$. Recall that every $g \in \mathcal{G}$ can be described using at most $2n + \log(2^d n) + \log(2^n) \leq 4n$ bits.

The commitment protocol for a string $s = s_1 \ldots s_{2^d}$ of length $2^d \cdot n$ for $d = (c-1) \cdot \log n$ works as follows. The receiver first sends a description of an $h \leftarrow \mathcal{H}$ which is a k-MCRH to the sender. The sender then computes the encoding $C(s)$ of s and computes the Merkle hash tree of $s' = C(s)$ (and not of s). The sender sends the root of the hash tree y to the receiver. The receiver replies with a hash function $g \leftarrow \mathcal{G}$ and finally the sender replies with $u = g(s)$. The opening is done is the natural way by letting the sender reveal s to the receiver who then simulates the computation and makes sure that the messages y and u are consistent. See Fig. 3 for the precise description.

By the description of the protocol, one can see that the protocol consists of 4-rounds and has communication complexity of $\ell + n + 4n + n = \ell + 6n$ bits. In addition, for the honest sender the verification succeeds with probability 1.

The analysis of the commitment scheme can be found in the full version [38].

Supporting Arbitrary Larger k. The reason why we could prove security only for constant values of k stems from the fact that our adversary $\mathcal{A}$ for the MCRH runs in time proportional to k^d. The source of this term in the running time is that our Merkle hash tree in the construction is of depth $d = O(\log n)$ and when counting the number of possible openings per coordinate in a leaf, we get k^d possibilities. Our adversary $\mathcal{A}$ basically "collects" this number of different openings for some coordinate and thereby finds a k-wise collision. If k is super-constant the running time of $\mathcal{A}$ becomes super-polynomial.

The commit protocol between $\mathcal{S}$ and $\mathcal{R}$

The sender $\mathcal{S}$ has string $s = s_1 \ldots s_{2^d}$ where $s_i \in \{0,1\}^n$ for all $i \in [2^d]$.

1. $\mathcal{R} \Rightarrow \mathcal{S}$: Samples $h \leftarrow \mathcal{H}$ and sends h.
2. $\mathcal{S} \Rightarrow \mathcal{R}$: Compute $s' = C(s)$ and a Merkle hash-tree T of s' using h and send the root-hash y.
3. $\mathcal{R} \Rightarrow \mathcal{S}$: Sample $g \leftarrow \mathcal{G}$ and send g.
4. $\mathcal{S} \Leftrightarrow \mathcal{R}$: Send $u = g(s)$ to the receiver

The output of the receive is $\mathsf{com} = (h, y, g, u)$**.**

The verifier $\mathcal{V}$ gets the input s simulates the sender to verify y and u.

Fig. 3. Our four-round commitment protocol for strings of length $2^d \cdot n$.

There are two paths we can take to bypass this. One is to assume super-polynomial security of the MCRH and allow our adversary to run in super-polynomial time. The second assumption is to assume that we start with a stronger MCRH that compresses by a polynomial factor (i.e., from $n^{1+\Omega(1)}$ to n) rather than by a factor of 2. This will cause the Merkle hash tree to be of constant depth. Under either of the assumptions, we can support polynomial values of k (in n). However, both assumptions are rather strong on the underlying MCRH and we thus prefer to avoid them; see Sect. 4 for a discussion.

Domain Extension of MCRH Functions. The construction we gave in the proof of Theorem 3 can be viewed as a domain extension method for MCRH functions. Specifically, given a k-MCRH f that maps $2n$ bits into n bits, we constructed a function g that maps $m = m(n)$ bits into n bits for any polynomial $m(\cdot)$ such that g is a $((k-1)^{\log m} + 1)$-MCRH.

Other Uses of List-Recoverable Codes for Domain-Extension. List-recoverable codes have been proven useful in various applications in cryptography. The work of Maurer and Tessaro [40] considers the problem of a extending the domain of a public random function. They show how to use a length-preserving random function f on n input bits, to get a variable-input-length function g that maps arbitrary polynomial-size inputs into arbitrary polynomial-size outputs such that g is indistinguishable from a random function for adversaries making up to $2^{(1-\epsilon)n}$ queries to g. One of their main components in the construction is a list-recoverable code (there referred to as an *input-restricting function family*).

Building on ideas and techniques of [40], Dodis and Steinberger [13] showed that list-recoverable codes are also useful for security preserving domain extension of message-authentication codes (with "beyond-birthday" security).

More recently, Haitner et al. [25] studied the possibility of a *fully parallel* (i.e., non-adaptive) domain-extension scheme that realizes a collision-resistant hash function. Starting with a *random* function f that maps n bits into n bits, they construct a function g that maps $m(n)$ bits into n bits for any polynomial $m(\cdot)$, makes only parallel calls to f, and requiring an attacker to make at least $2^{n/2}$ queries to g to find a collision with high probability. Their construction uses list-recoverable codes and they show that such codes are actually necessary for this task.

7 Separating Multi-CRH from CRH

In this section we rule out fully black-box constructions (see Definition 9) of CRH functions from k-MCRH functions for $k > 2$. Our proof technique is inspired by the works of Asharov and Segev [2] and Haitner et al. [23], that are based in turn on ideas originating in the works of Simon [50] Gennaro et al. [14]) and Wee [54].

We present an oracle Γ relative to which there exists a 3-multi-collision-resistant hash function, but any collision-resistant hash function can be easily broken. The theorem below is stated and proved only for the case of standard CRH and 3-MCRH. The ideas in this proof naturally extend to a separation of k-MCRH from $(k+1)$-MCRH for all fixed values of k.

Theorem 4. *There is no fully black-box construction of a collision-resistant hash function family from a 3-multi-collision-resistant hash function family mapping* $2n$ *bits to* n *bits.*

The proof can be found in the full version [38].

8 UOWHFs, MCRHs and Short Commitments

We explore the relationship between short commitments, MCRHs and universal one-way hash functions (see Definition 6). Our main message is that short commitment protocols (with some requirements listed below) directly imply UOWHFs. The transformation is efficient in the sense that a description of a hash function corresponds to the messages sent by the receiver and evaluation of the function is done by executing the protocol. In some cases this gives a way to construct a UOWHF which is more efficient than the direct construction based on one-way functions or permutations [46, 48] (see comparison below).

Theorem 5. *Any short commitment protocol in which the receiver is public-coin yields a universal one-way hash function with the following properties:*

1. *The key size is the total number of (public) coins sent by the receiver.*
2. *The length of inputs that the UOWHF supports is the length of messages the commitment protocol commits to and the length of the output is the amount of bits sent from the sender to the receiver.*
3. *The evaluation of a hash function amounts to a single execution of the commitment protocol.*

Plugging in our construction of short commitments from MCRH functions from Theorem 2, we obtain a new construction of a UOWHF for messages of length $n2^d$ starting with a k-MCRH for a polynomial $k = k(n)$. The key size in the resulting family is proportional to the number of bits sent from the receiver to the sender: $\ell + O(d \cdot n/\log n)$ bits, where ℓ is the size of an MCRH key.[14]

Using our construction of short commitments from MCRH functions from Theorem 3, we get a new construction of a UOWHF for messages of length $n2^d$ starting from a k-MCRH for any constant k. The key size in the resulting family is $\ell+O(n)$ bits. Notice that this term is independent of d and the hidden constant in the big "O" is pretty small.[15]

Comparison with Previous Constructions. Starting with a standard collision resistant hash function on short inputs, it is known how a collision resistant hash function on long inputs (based on the so called Merkle-Damgård iterated construction) [10,42]. This directly implies a UOWHF. The key size in the resulting construction is optimal: it is just a key for a single collision resistant hash. However, when starting with weaker building blocks the key size grows.

Naor and Yung [46] suggested a solution based on a tree hash (similar to a construction of Wegman and Carter for universal hash functions [55]). In their scheme, hashing a message of length $n2^d$ is done by a balanced tree such that in the i-th level $n2^{d-i}$ bits are hashed into $n2^{d-i-1}$ bits by applying the same basic compression function 2^{d-i-1} times. Each level in the tree requires its own basic compression function which results with a total of d keys for the basic UOWHF. Thus, the total size of a key in the resulting function is $d\ell$ bits, where ℓ is the bit size of a key in the basic UOWHF.

In case that the keys of the basic scheme (ℓ above) are rather long, Shoup [49], following on the XOR tree hash of Bellare and Rogaway [5], offered the following elegant scheme to transform a fixed-input UOWHF into a UOWHF that supports arbitrary long inputs. Given an input of 2^d blocks each of size n, for $i = 1, \dots, 2^d$ we compute $c_i = h((c_{i-1} \oplus s_{\kappa(i)}) \circ m_i)$, where $s_0, \dots, s_d$ are uniformly random "chaining variables", and $\kappa(i)$ chooses one of the elements $s_0, \dots, s_d$.[16] Thus, the total size of a key in the resulting function is $\ell + (d+1)n$ bits.

The proof of Theorem 5 appears in the full version [38].

Acknowledgments. We are grateful to Noga Ron-Zewi for teaching us about list-recoverable codes, for multiple useful discussions, and for sharing with us a preliminary version of [28]. We greatly acknowledge Gilad Asharov and Gil Segev for educating us about black-box separations. We thank Iftach Haitner and Eran Omri for answering questions related to [25]. We also thank Stefano Tessaro for telling us about [13,40] and in particular for explaining the relation of [40] to this work.

[14] The overhead in the key size can be improved if the pairwise hash function is replaced by an almost uniform hash function as described in Remark 2.

[15] The concrete constant in our scheme is roughly 6 but we did not try to optimize it further.

[16] The function $\kappa(i)$ counts the number of times 2 divides i, that is, for $i \geq 1$, $\kappa(i)$ is the largest integer κ such that 2^κ divides i.

References

1. Alon, N., Goldreich, O., Håstad, J., Peralta, R.: Simple construction of almost k-wise independent random variables. Random Struct. Algorithms **3**(3), 289–304 (1992)
2. Asharov, G., Segev, G.: Limits on the power of indistinguishability obfuscation and functional encryption. SIAM J. Comput. **45**(6), 2117–2176 (2016)
3. Barak, B.: How to go beyond the black-box simulation barrier. In: 42nd Annual Symposium on Foundations of Computer Science, FOCS, pp. 106–115. IEEE Computer Society (2001)
4. Barak, B., Goldreich, O.: Universal arguments and their applications. SIAM J. Comput. **38**(5), 1661–1694 (2008)
5. Bellare, M., Rogaway, P.: Collision-resistant hashing: towards making UOWHFs practical. In: Kaliski, B.S. (ed.) CRYPTO 1997. LNCS, vol. 1294, pp. 470–484. Springer, Heidelberg (1997). https://doi.org/10.1007/BFb0052256
6. Berman, I., Degwekar, A., Rothblum, R.D., Vasudevan, P.N.: Multi collision resistant hash functions and their applications. IACR Cryptology ePrint Archive 2017, 489 (2017)
7. Bitansky, N., Kalai, Y.T., Paneth, O.: Multi-collision resistance: A paradigm for keyless hash functions. IACR Cryptology ePrint Archive 2017, 488 (2017)
8. Brassard, G., Chaum, D., Crépeau, C.: Minimum disclosure proofs of knowledge. J. Comput. Syst. Sci. **37**(2), 156–189 (1988)
9. Coppersmith, D.: Another birthday attack. In: Williams, H.C. (ed.) CRYPTO 1985. LNCS, vol. 218, pp. 14–17. Springer, Heidelberg (1986). https://doi.org/10.1007/3-540-39799-X_2
10. Damgård, I.B.: A design principle for hash functions. In: Brassard, G. (ed.) CRYPTO 1989. LNCS, vol. 435, pp. 416–427. Springer, New York (1990). https://doi.org/10.1007/0-387-34805-0_39
11. Damgård, I., Pedersen, T.P., Pfitzmann, B.: On the existence of statistically hiding bit commitment schemes and fail-stop signatures. J. Cryptol. **10**(3), 163–194 (1997)
12. Damgård, I., Pedersen, T.P., Pfitzmann, B.: Statistical secrecy and multibit commitments. IEEE Trans. Inf. Theory **44**(3), 1143–1151 (1998)
13. Dodis, Y., Steinberger, J.: Domain extension for MACs beyond the birthday barrier. In: Paterson, K.G. (ed.) EUROCRYPT 2011. LNCS, vol. 6632, pp. 323–342. Springer, Heidelberg (2011). https://doi.org/10.1007/978-3-642-20465-4_19
14. Gennaro, R., Gertner, Y., Katz, J., Trevisan, L.: Bounds on the efficiency of generic cryptographic constructions. SIAM J. Comput. **35**(1), 217–246 (2005)
15. Girault, M., Cohen, R., Campana, M.: A generalized birthday attack. In: Barstow, D., Brauer, W., Brinch Hansen, P., Gries, D., Luckham, D., Moler, C., Pnueli, A., Seegmüller, G., Stoer, J., Wirth, N., Günther, C.G. (eds.) EUROCRYPT 1988. LNCS, vol. 330, pp. 129–156. Springer, Heidelberg (1988). https://doi.org/10.1007/3-540-45961-8_12
16. Girault, M., Stern, J.: On the length of cryptographic hash-values used in identification schemes. In: Desmedt, Y.G. (ed.) CRYPTO 1994. LNCS, vol. 839, pp. 202–215. Springer, Heidelberg (1994). https://doi.org/10.1007/3-540-48658-5_21
17. Goldreich, O., Sahai, A., Vadhan, S.: Can statistical zero knowledge be made non-interactive? or on the relationship of SZK and *NISZK*. In: Wiener, M. (ed.) CRYPTO 1999. LNCS, vol. 1666, pp. 467–484. Springer, Heidelberg (1999). https://doi.org/10.1007/3-540-48405-1_30

18. Guruswami, V., Indyk, P.: Near-optimal linear-time codes for unique decoding and new list-decodable codes over smaller alphabets. In: Proceedings on 34th Annual ACM Symposium on Theory of Computing, pp. 812–821. ACM (2002)
19. Guruswami, V., Indyk, P.: Linear time encodable and list decodable codes. In: Proceedings of the 35th Annual ACM Symposium on Theory of Computing, pp. 126–135. ACM (2003)
20. Guruswami, V., Indyk, P.: Linear-time list decoding in error-free settings. In: Díaz, J., Karhumäki, J., Lepistö, A., Sannella, D. (eds.) ICALP 2004. LNCS, vol. 3142, pp. 695–707. Springer, Heidelberg (2004). https://doi.org/10.1007/978-3-540-27836-8_59
21. Guruswami, V., Sudan, M.: Improved decoding of Reed-Solomon and algebraic-geometry codes. IEEE Trans. Inf. Theory **45**(6), 1757–1767 (1999)
22. Guruswami, V., Umans, C., Vadhan, S.P.: Unbalanced expanders and randomness extractors from parvaresh-vardy codes. J. ACM **56**(4), 20:1–20:34 (2009)
23. Haitner, I., Hoch, J.J., Reingold, O., Segev, G.: Finding collisions in interactive protocols - tight lower bounds on the round and communication complexities of statistically hiding commitments. SIAM J. Comput. **44**(1), 193–242 (2015)
24. Haitner, I., Horvitz, O., Katz, J., Koo, C., Morselli, R., Shaltiel, R.: Reducing complexity assumptions for statistically-hiding commitment. J. Cryptol. **22**(3), 283–310 (2009)
25. Haitner, I., Ishai, Y., Omri, E., Shaltiel, R.: Parallel hashing via list recoverability. In: Gennaro, R., Robshaw, M. (eds.) CRYPTO 2015. LNCS, vol. 9216, pp. 173–190. Springer, Heidelberg (2015). https://doi.org/10.1007/978-3-662-48000-7_9
26. Haitner, I., Nguyen, M., Ong, S.J., Reingold, O., Vadhan, S.P.: Statistically hiding commitments and statistical zero-knowledge arguments from any one-way function. SIAM J. Comput. **39**(3), 1153–1218 (2009)
27. Håstad, J., Impagliazzo, R., Levin, L.A., Luby, M.: A pseudorandom generator from any one-way function. SIAM J. Comput. **28**, 1364–1396 (1999)
28. Hemenway, B., Ron-Zewi, N., Wootters, M.: Local list recovery of high-rate tensor codes & applications. In: 58th IEEE Annual Symposium on Foundations of Computer Science, FOCS, pp. 204–215. IEEE Computer Society (2017)
29. Hemenway, B., Wootters, M.: Linear-time list recovery of high-rate expander codes. In: Halldórsson, M.M., Iwama, K., Kobayashi, N., Speckmann, B. (eds.) ICALP 2015. LNCS, vol. 9134, pp. 701–712. Springer, Heidelberg (2015). https://doi.org/10.1007/978-3-662-47672-7_57
30. Hosoyamada, A., Sasaki, Y., Xagawa, K.: Quantum multicollision-finding algorithm. IACR Cryptology ePrint Archive 2017, 864 (2017)
31. Hsiao, C.-Y., Reyzin, L.: Finding collisions on a public road, or do secure hash functions need secret coins? In: Franklin, M. (ed.) CRYPTO 2004. LNCS, vol. 3152, pp. 92–105. Springer, Heidelberg (2004). https://doi.org/10.1007/978-3-540-28628-8_6
32. Impagliazzo, R.: A personal view of average-case complexity. In: Proceedings of the Tenth Annual Structure in Complexity Theory Conference, pp. 134–147. IEEE Computer Society (1995)
33. Impagliazzo, R., Levin, L.A., Luby, M.: Pseudo-random generation from one-way functions (extended abstracts). In: Proceedings of the 21st Annual ACM Symposium on Theory of Computing, pp. 12–24. ACM (1989)
34. Impagliazzo, R., Luby, M.: One-way functions are essential for complexity based cryptography (extended abstract). In: 30th Annual Symposium on Foundations of Computer Science, FOCS, pp. 230–235. IEEE Computer Society (1989)

35. Joux, A.: Multicollisions in iterated hash functions. application to cascaded constructions. In: Franklin, M. (ed.) CRYPTO 2004. LNCS, vol. 3152, pp. 306–316. Springer, Heidelberg (2004). https://doi.org/10.1007/978-3-540-28628-8_19
36. Katz, J., Koo, C.: On constructing universal one-way hash functions from arbitrary one-way functions. IACR Cryptology ePrint Archive 2005, 328 (2005)
37. Kilian, J.: A note on efficient zero-knowledge proofs and arguments (extended abstract). In: STOC, pp. 723–732. ACM (1992)
38. Komargodski, I., Naor, M., Yogev, E.: Collision resistant hashing for paranoids: Dealing with multiple collisions. IACR Cryptology ePrint Archive 2017, 486 (2017)
39. Komargodski, I., Naor, M., Yogev, E.: White-box vs. black-box complexity of search problems: ramsey and graph property testing. In: 58th IEEE Annual Symposium on Foundations of Computer Science, FOCS, pp. 622–632 (2017)
40. Maurer, U., Tessaro, S.: Domain extension of public random functions: beyond the birthday barrier. In: Menezes, A. (ed.) CRYPTO 2007. LNCS, vol. 4622, pp. 187–204. Springer, Heidelberg (2007). https://doi.org/10.1007/978-3-540-74143-5_11
41. Merkle, R.C.: A certified digital signature. In: Brassard, G. (ed.) CRYPTO 1989. LNCS, vol. 435, pp. 218–238. Springer, New York (1990). https://doi.org/10.1007/0-387-34805-0_21
42. Merkle, R.C.: One way hash functions and DES. In: Brassard, G. (ed.) CRYPTO 1989. LNCS, vol. 435, pp. 428–446. Springer, New York (1990). https://doi.org/10.1007/0-387-34805-0_40
43. Mironov, I.: Collision-resistant no more: hash-and-sign paradigm revisited. In: Yung, M., Dodis, Y., Kiayias, A., Malkin, T. (eds.) PKC 2006. LNCS, vol. 3958, pp. 140–156. Springer, Heidelberg (2006). https://doi.org/10.1007/11745853_10
44. Naor, J., Naor, M.: Small-bias probability spaces: efficient constructions and applications. SIAM J. Comput. **22**(4), 838–856 (1993)
45. Naor, M., Ostrovsky, R., Venkatesan, R., Yung, M.: Perfect zero-knowledge arguments for NP using any one-way permutation. J. Cryptol. **11**(2), 87–108 (1998)
46. Naor, M., Yung, M.: Universal one-way hash functions and their cryptographic applications. In: Proceedings of the 21st Annual ACM Symposium on Theory of Computing, pp. 33–43. ACM (1989)
47. Ngo, H.Q., Porat, E., Rudra, A.: Efficiently decodable compressed sensing by list-recoverable codes and recursion. In: 29th International Symposium on Theoretical Aspects of Computer Science, STACS. LIPIcs, vol. 14, pp. 230–241. Schloss Dagstuhl - Leibniz-Zentrum fuer Informatik (2012)
48. Rompel, J.: One-way functions are necessary and sufficient for secure signatures. In: Proceedings of the 22nd Annual ACM Symposium on Theory of Computing, pp. 387–394. ACM (1990)
49. Shoup, V.: A composition theorem for universal one-way hash functions. In: Preneel, B. (ed.) EUROCRYPT 2000. LNCS, vol. 1807, pp. 445–452. Springer, Heidelberg (2000). https://doi.org/10.1007/3-540-45539-6_32
50. Simon, D.R.: Finding collisions on a one-way street: can secure hash functions be based on general assumptions? In: Nyberg, K. (ed.) EUROCRYPT 1998. LNCS, vol. 1403, pp. 334–345. Springer, Heidelberg (1998). https://doi.org/10.1007/BFb0054137
51. Stevens, M., Bursztein, E., Karpman, P., Albertini, A., Markov, Y.: The first collision for full SHA-1. In: Katz, J., Shacham, H. (eds.) CRYPTO 2017. LNCS, vol. 10401, pp. 570–596. Springer, Cham (2017). https://doi.org/10.1007/978-3-319-63688-7_19

52. Ta-Shma, A.: Explicit, almost optimal, epsilon-balanced codes. In: Proceedings of the 49th Annual ACM SIGACT Symposium on Theory of Computing, STOC, pp. 238–251 (2017)
53. Wang, X., Yin, Y.L., Yu, H.: Finding collisions in the full SHA-1. In: Shoup, V. (ed.) CRYPTO 2005. LNCS, vol. 3621, pp. 17–36. Springer, Heidelberg (2005). https://doi.org/10.1007/11535218_2
54. Wee, H.: One-way permutations, interactive hashing and statistically hiding commitments. In: Vadhan, S.P. (ed.) TCC 2007. LNCS, vol. 4392, pp. 419–433. Springer, Heidelberg (2007). https://doi.org/10.1007/978-3-540-70936-7_23
55. Wegman, M.N., Carter, L.: New hash functions and their use in authentication and set equality. J. Comput. Syst. Sci. **22**(3), 265–279 (1981)

Signatures

Synchronized Aggregate Signatures from the RSA Assumption

Susan Hohenberger[1](✉) and Brent Waters[2]

[1] Johns Hopkins University, Baltimore, USA
susan@cs.jhu.edu
[2] University of Texas at Austin, Austin, USA
bwaters@cs.utexas.edu

Abstract. In this work we construct efficient aggregate signatures from the RSA assumption in the synchronized setting. In this setting, the signing algorithm takes as input a (time) period t as well the secret key and message. A signer should sign at most once for each t. A set of signatures can be aggregated so long as they were all created for the same period t. Synchronized aggregate signatures are useful in systems where there is a natural reporting period such as log and sensor data, or for signatures embedded in a blockchain protocol.

We design a synchronized aggregate signature scheme that works for a bounded number of periods T that is given as a parameter to a global system setup. The big technical question is whether we can create solutions that will perform well with the large T values that we might use in practice. For instance, if one wanted signing keys to last up to ten years and be able to issue signatures every second, then we would need to support a period bound of upwards of 2^{28}.

We build our solution in stages where we start with an initial solution that establishes feasibility, but has an impractically large signing time where the number of exponentiations and prime searches grows linearly with T. We prove this scheme secure in the standard model under the RSA assumption with respect to honestly-generated keys. We then provide a tradeoff method where one can tradeoff the time to create signatures with the space required to store private keys. One point in the tradeoff is where each scales with $\sqrt{T}$.

Finally, we reach our main innovation which is a scheme where both the signing time and storage scale with $\lg T$ which allows for us to keep both computation and storage costs modest even for large values of T. Conveniently, our final scheme uses the same verification algorithm,

S. Hohenberger—Supported by the National Science Foundation CNS-1228443 and CNS-1414023, the Office of Naval Research N00014-15-1-2778, and a Microsoft Faculty Fellowship.

B. Waters—Supported by NSF CNS-1414082, DARPA SafeWare, Microsoft Faculty Fellowship, and Packard Foundation Fellowship. Any opinions, findings, and conclusions or recommendations expressed in this material are those of the author(s) and do not necessarily reflect the views of the Department of Defense or the U.S. Government.

J. B. Nielsen and V. Rijmen (Eds.): EUROCRYPT 2018, LNCS 10821, pp. 197–229, 2018.
https://doi.org/10.1007/978-3-319-78375-8_7

and has the same distribution of public keys and signatures as the first scheme. Thus we are able to recycle the existing security proof for the new scheme.

We also extend our results to the identity-based setting in the random oracle model, which can further reduce the overall cryptographic overhead. We conclude with a detailed evaluation of the signing time and storage requirements for various settings of the system parameters.

1 Introduction

Aggregate signatures, as introduced by Boneh, Gentry, Lynn and Shacham [13], allow a third party to compress an arbitrary group of signatures $(\sigma_1, \ldots, \sigma_n)$ that verify with respect to a corresponding collection of public key and message pairs $((pk_1, m_1), \ldots, (pk_n, m_n))$ and produce a short aggregated signature that verifies the same collection. There are many applications where reducing the cryptographic overhead is desirable including BGP routing [11,13,29], bundling software updates [1], sensor data [1] and block chain protocols [2].

When exploring a primitive such as aggregate signatures, it is desirable to have multiple realizations under different cryptographic assumptions or constructs. This provides redundancy in the case that one of the assumptions proves to be false. Also different approaches often yield a menu of performance tradeoffs that one can select from in an application-dependent manner.

To date, the design of aggregate signature schemes has mostly been dominated by bilinear (or multilinear) map-based proposals [1,7,10,11,13,14,18,19, 22,23,29,31,36]. Most proposals to aggregate outside of the bilinear setting have required signers to interact either by signing in a sequential chain [4,15,17,27, 28,30,33] or otherwise cooperate interactively on signature creation or verification [3,8]. Here we seek a solution that does not require bilinear maps or signer interaction. We are aware of two prior attempts [20,37] to aggregate RSA-based signatures (without interaction), but to the best of our understanding, both schemes appear to lack basic correctness (that is, each user creates and signs with his own unique modulus, but then the signatures are aggregated and verified with respect to the same modulus).

In this work we construct efficient aggregate signatures from the RSA assumption in the *synchronized* setting of Gentry and Ramzan [19]. In the synchronized setting the signing algorithm will take as input a (time) period t as well the secret key and message. A signer should sign at most once for each t. A set of signatures can be aggregated so long as they were all created for the same period t. Synchronized aggregate signatures are useful in systems where there is a natural reporting period such as log or sensor data. Another example is for use in signatures embedded in a blockchain protocol where the creation of an additional block is a natural synchronization event. For instance, consider a blockchain protocol that records several signed transactions in each new block creation. These signed transactions could use a synchronized aggregate signature scheme with the block iteration as the period number. This would reduce the signature overhead from one per transaction to only one synchronized signature per block iteration.

Ahn, Green and Hohenberger [1] gave a synchronized aggregate signature scheme in bilinear groups from the (standard model) computational Diffie-Hellman assumption by adapting the Hohenberger-Waters [24] short signature scheme. Since Hohenberger and Waters in the same work also provided a similar scheme from the RSA assumption it is natural to wonder why that one could not be adapted as well. Unfortunately, this approach will not work as the HW RSA-based signature scheme requires the signer to have knowledge of $\phi(N)$ and thus the factorization of N. This trapdoor information cannot be securely dispensed among all signers that might work in $\mathbb{Z}_N^*$.

In this work we design a synchronized aggregate signature scheme that works for a bounded number of periods T that is given as a parameter to a global system setup. We believe that such a bound is acceptable in the synchronized setting where a reasonable estimate of it can be derived by first determining a fixed lifetime of keys in the system (e.g., 10 years) and dividing it by the expected frequency that periods will occur (e.g., every minute). The big question is whether we can create solutions that will perform well with the larger T values that we might use in practice. For instance, suppose that we wanted signing keys to last up to ten years and wanted to have the capability of signing on periods as short as a second. In this case we would need to be able to support a period bound of upwards of 2^{28}.

We will build our solution in stages where we start with an initial solution that establishes feasibility of synchronized aggregation in the RSA setting, but has an impractically large signing time where the number of exponentiations and prime searches grows linearly with T. We prove this scheme secure in the standard model under the RSA assumption. We then provide a basic tradeoff that allows one to tradeoff the time to create signatures with the space required to store private keys. One point in the tradeoff is where each scales with $\sqrt{T}$.

We reach our main innovation which is a scheme where both the signing time and storage scale with $\lg(T)$ which allows for us to keep both computation and storage costs modest even for large values of T. Conveniently, our final scheme uses the same verification algorithm, and has the same distribution of public keys and signatures as the first scheme. Thus we are able to recycle the existing security proof for the new scheme.

We continue our exploration of using RSA in the synchronized aggregate setting by demonstrating how to extend our results to be identity-based. Since identity strings are typically much shorter than public keys, this setting can help achieve better overall reduction of cryptographic overhead. Our solution is secure under the standard RSA assumption in the random oracle model.

Finally, we provide a detailed performance evaluation of the various schemes from both a signing time and private key storage perspective, concluding that the $\lg(T)$ construction is relatively practical for realistic settings of the system parameters and far exceeds the performance of the others for most settings.

Overview of the Schemes. In our schemes, messages will be of length L bits which will be broken up into k chunks of ℓ bits each. In our initial scheme a global system setup will first choose an RSA modulus $N = p \cdot q$ where we let g be a generator of the quadratic residues of $\mathbb{Z}_N^*$. Next it picks a key K that is used to define a hash function $H_K(t) = e_t$ that maps a period $t \in [1, T]$ to a prime value e_t. We will defer the details of how this function works to the main body. Finally, the setup computes $E = \prod_{j=1}^{T} e_j \mod \phi(N)$ and $Y = g^E \mod N$ and publishes the public parameters as $\mathsf{pp} = (T, N, g, Y, K)$.

Key generation is performed by choosing random $u_0, u_1, \ldots, u_k$ in $[1, N]$ and setting the secret key as $sk = (u_0, u_1, \ldots, u_k)$ and the public key $pk = (U_0, U_1, \ldots, U_k)$ where $U_j = Y^{u_j} = g^{u_j \prod_{i \in T} e_i}$, for $j = 0$ to k. To sign a message first compute all the primes $e_i \leftarrow H_K(i)$ for $i \neq t$ and then output $\sigma = \left(g^{u_0} \prod_{j=1}^{k} g^{u_j \cdot m_j}\right)^{\prod_{i \in T \setminus \{t\}} e_i} = \left(U_0 \prod_{j=1}^{k} U_j^{m_j}\right)^{1/e_t} \pmod N$. Verification is performed by testing if $\sigma^{e_t} \stackrel{?}{=} U_0 \prod_{i=1}^{k} U_i^{m_i}$. Aggregation is done by simply multiplying individual signatures together (mod N) and testing against the product of the individual verification tests. We remark that our group hash function falls into a more general group hash framework proposed by Hofheinz, Jager and Kiltz [21]. In Sect. 4, we discuss potential future improvements by incorporating their framework.

We give a proof of security under the RSA assumption. Our proof is standard model with respect to honestly-generated keys and uses techniques from [24] for embedding an RSA challenge into the function H_K. The choice of k provides a tradeoff between the secret key storage size which grows linearly with k to the tightness in the reduction which has a loss factor of $2^\ell = 2^{L/k}$.

Taking a step back, our signature scheme involves reconstructing e_t-th roots of a public key and then manipulating these according to the message. Here the secret key simply holds a group element that is root of *all* the e_i values. The underlying structure is reminiscent of earlier RSA-based accumulator schemes (e.g., [6,9]). The problem, however, is that building up this root from the secret key is quite costly and requires $T - 1$ exponentiations and calls to $H_K(\cdot)$ which are roughly equivalent to prime searches. Returning to our example of $T = 2^{28}$, our measured cost of signing one message was more than one day on a common processor. Clearly, we must do better.

We next show how to obtain a basic tradeoff between the time to sign and the size of the private key storage. Very roughly the time to sign will scale linearly with a parameter a and the storage with a parameter b with the constraint that $a \cdot b = T$. Thus we can explore tradeoffs such as setting $a = T, b = 1$ which corresponds to the scheme above, go the opposite direction and set $a = 1, b = T$ to achieve fast signing at the expense of large storage, or try to balance these by choosing $a = b = \sqrt{T}$.

The main technical idea is for the key generation algorithm to organize T into b "windows" each of size a. (We will assume a divides T evenly for ease of exposition.) Each window will be connected with a group element that has g raised to the exponents associated with every period except for a window of a of them. Thus to sign we need to do $a - 1$ exponentiations and prime searches and our private keys roughly grow as b group elements.

While this simple tradeoff technique provides more flexibility, there is still a significant gap from the performance numbers we would like to achieve. Let's return again to our $T = 2^{28}$ example. In setting $a = 1$, we would get very fast signing (a few tens of milliseconds), but with very huge keys of 64 GB. On the other hand, if we aimed for the $\sqrt{T}$ tradeoff we would end up with 4 MB private keys and roughly 9 s per signature. This achieves greater balance, but is still impractical.

This finally moves us to our last solution. Here we wish to find a more intricate way of handling the key storage that allows us to sign efficiently, but without a significant storage penalty. To do this we design a key storage mechanism that has about $2\lg(T)$ group elements and requires $\lg(T)$ exponentiations per signing. Returning to our example of $T = 2^{28}$, we can now achieve the much more practical 16 KB private key storage with 58 ms per signature.

To achieve this, we leverage the fact that the synchronized signatures are performed in sequence over the total number of periods. The goal is to maintain a data structure which (1) is small, (2) is ready to quickly produce a signature for the next period and (3) can perform a small amount of work to update it for future periods. To this end we organize a data structure according to a `levels` parameter where $T = 2^{\texttt{levels}+1} - 2$. In addition, a current `index` value is associated with the structure that indicates how many periods have passed so far. At level i at any time there will be one or two tuples which include a group element which is g raised to all exponents corresponding to periods except those with indices anywhere from 2^i to 2^{i-1}. During each signature the signing algorithm will grab an element from level 1 and use it to sign as well as perform a little bit of work on each level to close the window of exponents further. We defer the details of how this is achieved to Sect. 6. We remark that this approach is conceptually similar to the pebbling optimization used by Itkis and Reyzin [26] to realize efficient forward-secure signatures.

Organization and Summary of the Results. In Sect. 2, we provide the specifications and security definitions. Section 3 covers the algebraic setting, assumptions and related lemmas. Section 4 gives the base construction as well as its proof of security in the standard model under the RSA assumption. Section 5 describes changes to the key generation and signing algorithms that can achieve a tradeoff in private key size versus signing time; one point achieves a balance of $\sqrt{T}$ for both. Section 6 provides a deeper innovation on how change key generation and signing to scale with $\lg(T)$. Recall that the distribution of the public keys and signatures in all of these schemes are the same as are the verification algorithms and thus the security proof in Sect. 4 suffices for all. We then show how to extend these results to the identity-based setting in Sect. 7. Finally, we conclude with a detailed time and space performance analysis of these constructions in Sect. 8 showing that the $\lg(T)$ constructions can be practical even for very large bounds on T.

2 Scheme Specifications and Definitions of Security

In a basic aggregate signature scheme [13], anyone given n signatures on n messages from n users can aggregate all these signatures into a single short signature. This aggregate signature (together with the n public keys and n messages) can be publicly verified to convince anyone that user i authenticated message i for $i = 1$ to n. This is also true for synchronized aggregate signatures except that we assume all signers have a synchronized period identifier (such as a clock) and the following restrictions apply:

1. A signer can issue at most one signature per period and keeps state to ensure this.
2. Only signatures created during the same period can be aggregated.

Gentry and Ramzan [19] were the first to consider this "synchronized" setting in the context of aggregate signatures. In their model, they assumed that signatures were issued using a special tag (which could not be re-used) and only signatures with the same tag could be aggregated. Ahn, Green and Hohenberger [1] formalized this synchronization as a time period, assuming all signers have access to the same clock.[1] Here, we include a bound T on the periods.

Definition 1 (Synchronized Aggregate Signatures [1,19]**).** *A synchronized aggregate signature scheme for a bounded number of periods and message space* $\mathcal{M}(\cdot)$ *is a tuple of algorithms* (Setup, KeyGen, Sign, Verify, Aggregate, AggVerify) *such that*

Setup$(1^\lambda, 1^T)$ *: On input the security parameter* λ *and the period bound* T*, the setup algorithm outputs public parameters* pp.

KeyGen(pp) *: On input the public parameters* pp*, the key generation algorithm outputs a keypair* (pk, sk).

Sign(pp, sk, M, t) *: On input the public parameters* pp*, the signing algorithm takes in a secret key* sk*, a message* $M \in \mathcal{M}(\lambda)$*, the current period* $t \leq T$*, and produces a signature* σ.

Verify$(\mathsf{pp}, pk, M, t, \sigma)$ *: On input the public parameters* pp*, the verification algorithm takes in a public key* pk*, a message* M*, a period* t *and a purported signature* σ*, and returns 1 if and only if the signature is valid and* $t \leq T$*, and 0 otherwise.*

Aggregate$(\mathsf{pp}, t, (pk_1, M_1, \sigma_1), \ldots, (pk_n, M_n, \sigma_n))$ *: On input the public parameters* pp*, a period* t*, a sequence of public keys* $(pk_1, \ldots, pk_n)$*, messages* $(M_1, \ldots, M_n)$*, and purported signatures* $(\sigma_1, \ldots, \sigma_n)$ *for period* $t \leq T$*, it outputs an aggregate signature* σ_{agg} *or error message* $\perp$.

[1] In this work, as in the case of [1], if the signers' clocks become out of sync with each other, this will lead to inefficiencies in the system, as it will not be possible to aggregate some signatures, but this will not open up security issues. As in [1,19], there is a security issue if a tag or period value is reused by the signer, so an adversary's ability to move a user's clock backward could lead to forgeries for that signer.

AggVerify$(\mathsf{pp}, t, (pk_1, \ldots, pk_n), (M_1, \ldots, M_n), \sigma_{agg})$: *On input the public parameters* pp*, a period t, a sequence of public keys $(pk_1, \ldots, pk_n)$ and messages $(M_1, \ldots, M_n)$, and a purported aggregate signature σ_{agg}, the aggregate-verification algorithm outputs 1 if and only if σ_{agg} is a valid aggregate signature and $t \leq T$, and 0 otherwise.*

Efficiency. We require that the setup algorithm run in time polynomial in its inputs and all other algorithms run in time polynomial in λ, T.

Correctness. Let poly(x) *denote the set of polynomials in x. In addition to the standard correctness properties of the basic signature scheme, for a synchronized aggregation scheme, the correctness requirements on* Aggregate *and* AggVerify *stipulate that for all $\lambda \in \mathbb{N}$, $T \in \mathsf{poly}(\lambda), n \in \mathsf{poly}(\lambda)$, $\mathsf{pp} \in \mathsf{Setup}(1^\lambda, 1^T)$, $(pk_1, sk_1), \ldots, (pk_n, sk_n) \in \mathsf{KeyGen}(\mathsf{pp})$, $1 \leq t \leq T$, $M_i \in \mathcal{M}(\lambda)$, $\sigma_i \in \mathsf{Sign}(\mathsf{pp}, sk_i, M_i, t)$ and $\sigma_{agg} \in \mathsf{Aggregate}(\mathsf{pp}, t, (pk_1, M_1, \sigma_1), \ldots, (pk_n, M_n, \sigma_n))$, it holds that*

$$\mathsf{AggVerify}(\mathsf{pp}, t, (pk_1, \ldots, pk_n), (M_1, \ldots, M_n), \sigma_{agg}) = 1.$$

Unforgeability. The definition uses the following game between a challenger and an adversary $\mathcal{A}$ for a given scheme Π = (Setup, KeyGen, Sign, Verify, Aggregate, AggVerify), security parameter λ, and message space $\mathcal{M}(\lambda)$:

Setup: The adversary sends $1^T, 1^n$ to the challenger, who runs $\mathsf{Setup}(1^\lambda, 1^T)$ to obtain the public parameters pp.[2] Then the challenger runs KeyGen(pp) a total of n times to obtain the key pairs $(pk_1, sk_1), \ldots, (pk_n, sk_n)$. The adversary is sent $(\mathsf{pp}, pk_1, (pk_2, sk_2), \ldots, (pk_n, sk_n))$.

Queries: For each period t starting with 1 and incrementing up to T, the adversary can request one signature on a message of its choice in $\mathcal{M}$ under sk_1, or it can choose to skip that period. The challenger responds to a query for M_i during period $t_i \in [1, T]$ as $\mathsf{Sign}(\mathsf{pp}, sk_1, M_i, t_i)$.

Output: Let γ be a function mapping integers to $[1, n]$. Eventually, the adversary outputs a tuple $(t, (pk_{\gamma(1)}, \ldots, pk_{\gamma(k)}), (M_1', \ldots, M_k'), \sigma_{agg})$ and wins the game if:

1. $1 \leq t \leq T$; and
2. there exists an $z^* \in [1, k]$ such that $\gamma(z^*) = 1$; and
3. all $M_i' \in \mathcal{M}$; and
4. M_{z^*}' is not in the set of messages $\mathcal{A}$ queried during the Queries phase[3]; and
5. $\mathsf{AggVerify}(\mathsf{pp}, t, (pk_{\gamma(1)}, \ldots, pk_{\gamma(k)}), (M_1', \ldots, M_k'), \sigma_{agg}) = 1$, where $1 \leq k \leq n$.

---

[2] For any adversary $\mathcal{A}$ that runs in time polynomial in λ will be restricted (by its own running time) to giving T values out that are polynomial in λ.

[3] As observed by [1], one can relax this unforgeability condition to allow the forgery message, M_{z^*}', to have been previously queried to the signing oracle provided that it was not done during the same period used in the forgery. This "stronger" notion can be achieved by any scheme satisfying the above unforgeability definition by having the signer incorporate the period into each message.

We define $\mathsf{SigAdv}_{\mathcal{A},\Pi,\mathcal{M}}(\lambda)$ to be the probability that the adversary $\mathcal{A}$ wins in the above game with scheme Π for message space $\mathcal{M}$ and security parameter λ taken over the coin tosses made by $\mathcal{A}$ and the challenger.

Definition 2 (Unforgeability). *A synchronized aggregate signature scheme Π for message space $\mathcal{M}$ is existentially unforgeable under an adaptive chosen message attack if for all sufficiently large $\lambda \in \mathbb{N}$ and all probabilistic polynomial-time in λ adversaries $\mathcal{A}$, there exists a negligible function* negl, *such that*

$$\mathsf{SigAdv}_{\mathcal{A},\Pi,\mathcal{M}}(\lambda) \leq \mathsf{negl}(\lambda).$$

Discussion. Above, we require that the Setup algorithm is honestly executed, so in practice this could be run by a trusted party or realized via a specialized multiparty protocol (see Sect. 4 for more). We also require that the non-challenge public keys be chosen honestly instead of adversarially. Our later proof requires that the challenger has knowledge of the secret keys corresponding to the non-challenge public keys. This can be realized by working in the Registered Key Model [5] or adding an appropriate NIZK to the user's public key.

3 Number Theoretic Assumptions and Related Lemmas

There are many variants of the RSA assumption [35]. Here we use a variant involving safe primes. A *safe prime* is a prime number of the form $2p+1$, where p is also a prime.

Assumption 1 (RSA). *Let λ be the security parameter. Let integer N be the product of two λ-bit, distinct safe primes primes p, q where $p = 2p' + 1$ and $q = 2q' + 1$. Let e be a randomly chosen prime between 2^{λ} and $2^{\lambda+1} - 1$. Let QR_N be the group of quadratic residues in $\mathbb{Z}_N^*$ of order $p'q'$. Given (N, e) and a random $h \in \mathrm{QR}_N$, it is hard to compute x such that $x^e \equiv h \mod N$.*

We note that a randomly chosen element in $\mathbb{Z}_N^*$ would be a quadratic residue 1/4-th of the time, so the restriction to $h \in \mathrm{QR}_N$ is for convenience and could be relaxed.

In our schemes, we will refer to and require a primality test. For our purposes, it will be sufficient to use the efficient Miller-Rabin test [32,34]. We will also make use of the following lemmas:

Lemma 1 (Cramer-Shoup [16]). *Given $x, y \in \mathbb{Z}_n$ together with $a, b \in \mathbb{Z}$ such that $x^a = y^b$ and $\gcd(a, b) = 1$, there is an efficient algorithm for computing $z \in \mathbb{Z}_n$ such that $z^a = y$.*

Theorem 2 (Prime Number Theorem). *Define $\pi(x)$ as the number of primes $\leq x$. For $x > 1$,*

$$\frac{7}{8} \cdot \frac{x}{\ln x} < \pi(x) < \frac{9}{8} \cdot \frac{x}{\ln x}.$$

4 A Base Scheme for Aggregation from RSA

We begin with a base scheme that assumes a trusted global setup and works in the registered key model, where every signer needs to show their key pair to an authority that certifies their public key. The global setup of our scheme will take as input a security parameter λ and the maximum number of periods T. The message space $\mathcal{M}$ will be $\{0,1\}^L$ where L is some polynomial function of λ. (One can handle messages of arbitrary length by first applying a collision-resistant hash.)

In addition, associated with the scheme will be a "message chunking alphabet" where we break each L-bit message into k chunks each of ℓ bits where $k \cdot \ell = L$ with the restriction that $\ell \leq \lambda$ and thus $2^\ell \leq 2^\lambda$. As we will see the choice of ℓ will effect both the tightness of the security reduction as well as the size of the signatures.[4] We make use of a variant of the hash function in [24] to map integers to primes of an appropriate size.

Setup($1^\lambda, 1^T$) The setup algorithm chooses an integer $N = pq$ as the product of two safe primes where $p-1 = 2p'$ and $q-1 = 2q'$, such that $2^\lambda < \phi(N) < 2^{\lambda+1}$. Let QR_N denote the group of quadratic residues of order $p'q'$ with generator g.

Next, it sets up a hash function $H : [1,T] \rightarrow \{0,1\}^{\lambda+1}$ where H will take as input a period $t \in [1,T]$ and output a prime between 2^λ and $2^{\lambda+1}-1$. It begins by randomly choosing a K' for the PRF function $F : [1,T] \times [1,\lambda^2] \rightarrow \{0,1\}^\lambda$, a random $c \in \{0,1\}^\lambda$ as well as an arbitrary prime e_{default} between 2^λ and $2^{\lambda+1}-1$. We let $K = (K', c, e_{\mathsf{default}})$.

We define how to compute $H_K(t)$. For each $i = 1$ to $\lambda \cdot (\lambda^2 + \lambda)$, let $y_i = c \oplus F_K(t,i)$. If $2^\lambda + y_i$ is prime return it. Else increment i and repeat. If no such $i \leq \lambda \cdot (\lambda^2+\lambda)$ exists, return e_{default}.[5] We note that this computation returns the smallest i such that $2^\lambda + y_i$ is a prime. Notationally, for $t \in [1,T]$ we will let $e_t = H_K(t)$.

The algorithm concludes by computing $E = \prod_{j=1}^{T} e_j \mod \phi(N)$ and $Y = g^E \mod N$.

It publishes the public parameters as $\mathsf{pp} = (T, N, g, Y, K)$.

KeyGen(pp) The algorithm retrieves Y from the pp. It chooses random integers $u_0, u_1, \ldots, u_k$ in $[1,N]$. It sets the secret key as $sk = (u_0, u_1, \ldots, u_k)$ and the public key $pk = (U_0, U_1, \ldots, U_k)$ where

$$U_j = Y^{u_j} = g^{u_j \prod_{i \in T} e_i}, \text{ for } j = 0 \text{ to } k.$$

Sign(pp, sk, M, t) The signing algorithm takes as input a time period $1 \leq t \leq T$ and an $L = (\ell k)$-bit message $M = m_1|m_2|\ldots|m_k$, where each m_i contains

[4] In practice, one might use a collision-resistant hash function to map arbitrarily long messages into $L = 256$ bits and then set $\ell = 32$ and $k = 8$. We discuss the efficiency implications of these choices in Sect. 8.

[5] We expect this default case to be exercised only with negligible probability, but define it so that the function $H_K(t)$ is guaranteed to terminate in a bounded amount of time.

ℓ-bits and these are concatenated together to form M. It computes the primes $(e_1, \ldots, e_{t-1}, e_{t+1}, \ldots, e_T)$ from pp and then outputs

$$\sigma = (g^{u_0} \prod_{j=1}^{k} g^{u_j \cdot m_j})^{\prod_{i \in T \setminus \{t\}} e_i} = (U_0 \prod_{j=1}^{k} U_j^{m_j})^{1/e_t} \pmod N.$$

$Verify(\mathsf{pp}, pk, M, t, \sigma)$ Let $M = m_1|m_2|\ldots|m_k$. The algorithm computes the prime e_t from pp. Output 1 if $1 \le t \le T$ and

$$\sigma^{e_t} \stackrel{?}{=} U_0 \prod_{i=1}^{k} U_i^{m_i} \pmod N$$

or 0 otherwise.

$Aggregate(\mathsf{pp}, t, (pk_1, M_1, \sigma_1), \ldots, (pk_n, M_n, \sigma_n))$ An aggregate signature on signatures from the same time period $1 \le t \le T$ is computed as $\sigma_{agg} = \prod_{j=1}^{n} \sigma_j \pmod N$.

$AggVerify(\mathsf{pp}, t, (pk_1, \ldots, pk_n), (M_1, \ldots, M_n), \sigma_{agg})$ Let $pk_j = (U_{j,0}, U_{j,1}, \ldots, U_{j,k})$ and $M_j = m_{j,1}|m_{j,2}|\ldots|m_{j,k}$. The algorithm computes the prime e_t from pp. Output 1 if $1 \le t \le T$, each public key is unique (i.e., $\forall i \ne j \in [1, n]$, $pk_i \ne pk_j$) and

$$\sigma_{agg}^{e_t} \stackrel{?}{=} \prod_{j=1}^{n} (U_{j,0} \prod_{i=1}^{k} U_{j,i}^{m_{j,i}}) \pmod N$$

or 0 otherwise.

Discussion. Observe that the above group hash function we employ falls into a more general group hash framework proposed by Hofheinz, Jager and Kiltz [21] that uses programmable hash functions. One might use their general framework to explore further concrete efficiency tradeoffs, such as letting the group hash function be more complex and letting the hash function output the product of multiple smaller primes. Our concrete analysis, however, will focus on the core scheme above along with tradeoffs in key storage and signing time that we explore later. We leave open the interesting question of what other tradeoffs can be realized via [21], keeping in mind that some of those instantiations add per signer randomness, which makes aggregation challenging.

Recall from Sect. 2 that **Setup** must be executed honestly. It seems possible that, for this scheme, this might be realized efficiently using a specialized multiparty computation protocol, such as an adaptation of one due to Boneh and Franklin [12] for efficiently allowing a group of parties to generate an RSA modulus, where each party learns N, but no party learns the factorization of N.

4.1 Proof of Security

Theorem 3. *If the RSA assumption (Assumption 1) holds and F is a secure pseudorandom function, then the above synchronized aggregate signature construction is existentially unforgeable under an adaptive chosen message attack.*

Proof. The reduction algorithm receives an RSA challenge (N, e^*, h) and needs to use the attacker to compute $h^{1/e^*} \mod N$. Define a "conforming" attacker as one that will always make a signing query on the period t^* that it forges on. We can assume our attacker is conforming without loss of generality because if there exists an attacker that breaks the scheme, there exits one that breaks it and queries for a signature on period t^* by simply adding a signature query on a random message at that period. Our proof will assume a conforming attacker.

Next, we define a sequence of games.

Game 1: (Security Game) This game is defined to be the same as the security game of the scheme.

Game 2: (Guessing the forgery period and part of its queried message) The same as Game 1, except the game guesses the period the attacker will forge on and a part of the message queried for a signature during the period that will be different from the forgery message, and the adversary only wins if these guesses were correct. Formally, the game chooses random $t' \in [1, T]$, $\alpha \in [1, k]$ and $\beta \in \{0,1\}^\ell$. An adversary wins this game iff: (1) it would have won in Game 1 with a forgery on period t^* for some message $M^* = m_1^*|m_2^*|\ldots|m_k^*$ with some message $M = m_1|m_2|\ldots|m_k$ queried to the signing oracle on period t^*, (2) $t' = t^*$, (3) $\beta = m_\alpha$ and (4) $m_\alpha \neq m_\alpha^*$.

Game 3: (H_K does not default) The attacker wins only if it meets all the conditions to win in Game 2 and $H_K(t^*) \neq e_{\text{default}}$ (that is, the default condition of the hash is not triggered on the forgery message or otherwise equal to the default prime.)

Game 4: (H_K does not collide) The attacker wins only if it meets all the conditions to win in Game 3 and $H_K(t^*) \neq H_K(t)$ for all $t \in [1, T]$ where $t \neq t^*$.

Game 5: (Guess resolving i^* for H_K) The game chooses a random $i^* \in [1, \lambda^3 + \lambda^2]$. Attacker wins only if it meets all the conditions of Game 4 and i^* was the "resolving" index in $H_K(t^*)$; that is, i^* was the smallest i such that $y_i = F_{K'}(t^*, i) \oplus c$ and $(2^\lambda + y_i)$ was a prime.

Game 6: (Programming H_K with random value) The same as Game 5, except that it chooses a random $y' \in \{0,1\}^\lambda$ and set $c = y' \oplus F_{K'}(t^*, i^*)$.

Game 7: (Programming H_K with e^*) The same as Game 6, except choose e^* as a random prime in the range $[2^\lambda, 2^{\lambda+1} - 1]$ and let y' be the λ least significant bits of e^*; that is, drop the leading 1. As before, set $c = y' \oplus F_{K'}(t^*, i^*)$.

We now establish a series of claims that show that if an adversary is successful against the real security game (Game 1) then it will be successful against in Game 7 as well. We will then shortly describe a simulator that can use any adversary successful in Game 7 to solve the RSA challenge.

Define $\mathbf{Adv}_\mathcal{A}[\text{Game } x]$ as the advantage of an adversary $\mathcal{A}$ in Game x.

Claim 4

$$\mathbf{Adv}_\mathcal{A}[\text{Game 2}] \geq \frac{\mathbf{Adv}_\mathcal{A}[\text{Game 1}]}{T \cdot k \cdot 2^\ell}.$$

Proof. Since there is no change to the adversary's view of the game, the probability of the adversary winning in Game 2 is the same as Game 1 times the probability of the game's guesses being correct. There is a $1/T$ probability of guessing the forging period, at least a $1/k$ probability of guessing a message chunk in the signing query that will be different in the forgery (there may be more than one), and a 2^ℓ probability of guessing that chunk's value in the queried message. We note that this gives a polynomial-time reduction for whenever ℓ is polylogarithmic in λ. Recall that any adversary that is polynomial time in λ must give out a 1^T that is polynomially bounded in λ.

Claim 5. *If F is a secure pseudorandom function and $\lambda \geq 4$, then*

$$\mathbf{Adv}_{\mathcal{A}}[\text{Game 3}] = \mathbf{Adv}_{\mathcal{A}}[\text{Game 2}] - \mathsf{negl}(\lambda).$$

Proof. We here need to understand the probability that $H_K(t^*) = e_{\text{default}}$. Using the Prime Number Theorem, we can bound the number of primes in the range $[2^\lambda, 2^{\lambda+1} - 1]$ as follows. Plugging into the formula in Lemma 2, we have that the number of primes less than $2^{\lambda+1} - 1$ is at least $\frac{7}{8} \cdot \frac{2^{\lambda+1}}{(\lambda+1)}$ (the value $2^{\lambda+1}$ is not prime, since it is a power of two, for any $\lambda \geq 1$) and the number of primes less than 2^λ is at most $\frac{9}{8} \cdot \frac{2^\lambda}{\lambda}$. Thus, the total number of primes in our range of interest is at least

$$\frac{7}{8} \cdot \frac{2^{\lambda+1}}{(\lambda+1)} - \frac{9}{8} \cdot \frac{2^\lambda}{\lambda} = \frac{7 \cdot \lambda \cdot 2^{\lambda+1} - 9 \cdot (\lambda+1) \cdot 2^\lambda}{8(\lambda+1)\lambda} \quad (1)$$

$$= \frac{14 \cdot \lambda \cdot 2^\lambda - 9 \cdot (\lambda+1) \cdot 2^\lambda}{8(\lambda+1)\lambda} = \frac{5 \cdot \lambda \cdot 2^\lambda - 9 \cdot 2^\lambda}{8(\lambda+1)\lambda} \quad (2)$$

$$= \frac{(5\lambda - 9) \cdot 2^\lambda}{8(\lambda^2 + \lambda)} > \frac{2^\lambda}{\lambda^2 + \lambda}, \text{ for all } \lambda \geq 4. \quad (3)$$

Let R be a random function that outputs a value in the range $[2^\lambda, 2^{\lambda+1}]$. Then the probability that R outputs a prime is at least:

$$\frac{2^\lambda/(\lambda^2 + \lambda)}{2^{\lambda+1} - 2^\lambda} = \frac{2^\lambda}{2^\lambda(\lambda^2 + 1)} = \frac{1}{\lambda^2 + \lambda} \quad (4)$$

The probability that R fails to output a prime after $\lambda(\lambda^2 + \lambda)$ tries is as follows. We again use the fact that $2^{\lambda+1}$ is not a prime. Recall Chernoff's bound for any $\epsilon \geq 0$, we have $\Pr[X \leq (1 - \epsilon)\mu] \leq e^{-\frac{\epsilon^2 \mu}{2}}$. Here when X is the number of primes output by R in $\lambda(\lambda^2 + \lambda)$ trials, $\epsilon = 1$ and $\mu = \sum^{\lambda(\lambda^2+\lambda)} \Pr[R$ fails to output a prime on one trial$]$, we have that

$$\Pr[R \text{ fails to output a prime in } \lambda^3 + \lambda^2 \text{ trials}] = \Pr[X \leq 0] \leq e^{-\frac{\mu}{2}} \quad (5)$$

$$\leq e^{-\frac{\lambda(\lambda^2+\lambda) \cdot \frac{1}{\lambda^2+\lambda}}{2}} = e^{-\lambda/2} \quad (6)$$

The PRF we employ to sample from this range cannot non-negligibly differ from R in its probability of selecting primes or this provides for a distinguishing attack on the PRF. Thus, the probability that $H_K(t^*) = e_{\text{default}}$ is the probability that the PRF chose the same prime as the setup algorithm, which is negligible at 1 in the number of primes in that range ($> 2^\lambda/(\lambda^2+\lambda)$), plus the probability that H_K triggers the default condition by failing to output a prime, which we also argued was negligibly close to the negligible probability of R doing the same.

Claim 6. *If F is a secure pseudorandom function and $T \in \mathsf{poly}(\lambda)$, then*

$$\mathbf{Adv}_{\mathcal{A}}[\text{Game } 4] = \mathbf{Adv}_{\mathcal{A}}[\text{Game } 3] - \mathsf{negl}(\lambda).$$

Proof. These games differ only in the event that $H_K(t^*) = H_K(t)$ for some $t \in [1,T]$ where $t \neq t^*$. Let R be a random function that outputs a value in the range $[2^\lambda, 2^{\lambda+1}]$. Suppose H_K uses R instead of the PRF. Then the probability of a collision for a single t is one in the number of primes in $[2^\lambda, 2^{\lambda+1}]$ or at most $1/\frac{2^\lambda}{\lambda^2+\lambda} = \frac{\lambda^2+\lambda}{2^\lambda}$, which is negligible. So the probability of a collision for any $t \in [1,T]$ (recall that T is polynomial in λ) is $T \cdot \frac{\lambda^2+\lambda}{2^\lambda} = \frac{\mathsf{poly}(\lambda)(\lambda^2+\lambda)}{2^\lambda} = \frac{\mathsf{poly}(\lambda)}{2^\lambda} = \mathsf{negl}(\lambda)$. When we replace R with the PRF, the probability of a collision cannot non-negligibly differ or this provides a distinguishing attack on the PRF.

Claim 7

$$\mathbf{Adv}_{\mathcal{A}}[\text{Game } 5] = \frac{\mathbf{Adv}_{\mathcal{A}}[\text{Game } 4]}{\lambda^3+\lambda^2}.$$

Proof. The attacker's view in these games is identical. The only difference is whether the game correctly guesses the resolving index i^* for $H_K(t^*)$. Since $i^* \in [1, \lambda^3+\lambda^2]$, the game has a $1/(\lambda^3+\lambda^2)$ chance of guessing this correctly.

Claim 8

$$\mathbf{Adv}_{\mathcal{A}}[\text{Game } 6] = \mathbf{Adv}_{\mathcal{A}}[\text{Game } 5].$$

Proof. In Game 5, c is chosen randomly in $\{0,1\}^\lambda$. In Game 6, c is set by randomly selecting $y' \in \{0,1\}^\lambda$ and setting $c = y' \oplus F_{K'}(t^*, i^*)$, where t^* is the period on which the attacker will attack and i^* is the resolving index for this value. Since y' is chosen randomly and independently of $F_{K'}(t^*, i^*)$, the resulting c will be from the same distribution as Game 5.

Claim 9

$$\mathbf{Adv}_{\mathcal{A}}[\text{Game } 7] = \mathbf{Adv}_{\mathcal{A}}[\text{Game } 6].$$

Proof. An adversary's advantage in these games is the same. In Game 6, the attacker could only win if $2^\lambda + y'$ was a prime, and thus the distributions are the same.

Main Reduction. We now show that if there exists a polynomial-time (in λ) attacker that has advantage $\epsilon = \epsilon(\lambda)$ in Game 7, then there exists a polynomial-time (in λ) attacker for the RSA problem in Assumption 1 with advantage ϵ.

On input an RSA challenge (N, e^*, h), the reduction algorithm proceeds as follows:

Setup.

1. Obtain $1^T, 1^n$ from the aggregate signature adversary $\mathcal{A}$.
2. Make random guesses of $t^* \in [1, T], \alpha \in [1, k], \beta \in \{0,1\}^\ell, i^* \in [1, \lambda^3 + \lambda^2]$.
3. Choose a random PRF key K'. Let y' be the λ least significant bits of the RSA input e^* (note that this is a prime randomly chosen from the appropriate range by the RSA challenger) and set $c = y' \oplus F_{K'}(t^*, i^*)$. Choose a random prime $e_{\text{default}} \in [2^\lambda, 2^{\lambda+1} - 1]$. Set $K = (K', c, e_{\text{default}})$. Thus, note that by construction when i^* is the resolving index for t^*,
$$e_{t^*} = H_K(t^*) = 2^\lambda + (c \oplus F_{K'}(t^*, i^*)) = 2^\lambda + y' = e^*.$$
4. Choose a random $g \in \mathrm{QR}_N$. Compute Y as before.
5. Set the $\mathsf{pp} = (T, N, g, Y, K)$.
6. Set up the "target" user's public key pk_1 as:
   (a) Choose random $u_0, u_1, \dots, u_k \in [1, N]$.
   (b) Set $U_0 = (h^{-\beta})^{\prod_{i \neq t^*}^{T} e_i} \cdot Y^{u_0}$. We note that the reduction algorithm can take the e_t root of U_0 so long as $t \neq t^*$.
   (c) For $j = 1$ to k such that $j \neq \alpha$, compute $U_j = Y^{u_j}$.
   (d) Set $U_\alpha = h^{\prod_{i \neq t^*}^{T} e_i} \cdot Y^{u_\alpha}$.
7. Set $pk_1 = (U_0, U_1, \dots, U_k)$. For $j = 2$ to n, $(pk_j, sk_j) = \mathsf{KeyGen}(\mathsf{pp})$.
8. Send to $\mathcal{A}$ the tuple $(\mathsf{pp}, pk_1, (pk_2, sk_2), \dots, (pk_n, sk_n))$.

Queries. For each period $t = 1$ to T, the adversary can request one signature on a message of its choice in the message space under sk_1 or skip that period. Recall that the adversary must be conforming and thus will request some signature on the forgery period t^*. In our construction, signing during period t requires taking the e_t-th root of each U_j value. By construction, the reduction algorithm can do this so long as: (1) $t \neq t^*$ or (2) for t^*, when the α-th ℓ-bits of the message are the string β. If the reduction is ever asked a query it cannot answer, then it will abort. We note that this only occurs when the guesses of t^*, α, β are incorrect, which is consistent with the attacker not winning in Game 7 anyway. Formally, when asked to sign $M = m_1|m_2|\dots|m_k$ for period $t \neq t^*$, the reduction outputs:

$$\sigma = (h^{-\beta} \cdot h^{m_\alpha})^{\prod_{i \neq t^*, i \neq t}^{T} e_i} \cdot \left(g^{u_0} \prod_{j=1}^{k} g^{u_j m_j}\right)^{\prod_{i \in T \setminus \{t\}} e_i} \tag{7}$$

$$= (h^{-\beta \prod_{i \neq t^*}^{T} e_i} \cdot Y^{u_0})^{1/e_t} \cdot \left(\prod_{j=1, j \neq \alpha}^{k} U_j^{m_j}\right)^{1/e_t} \cdot (h^{\prod_{i \neq t^*}^{T} e_i} \cdot Y^{u_\alpha})^{m_\alpha / e_t} \tag{8}$$

$$= (U_0 \prod_{j=1}^{k} U_j^{m_j})^{1/e_t} \mod N. \tag{9}$$

and when $t = t^*$ and $m_\alpha = \beta$, it outputs the signature:

$$\sigma = (g^{u_0} \prod_{j=1}^{k} g^{u_j m_j})^{\prod_{i \in T \setminus \{t\}} e_i} \quad (10)$$

$$= (1)^{\prod_{i \neq t^*, i \neq t}^{T} e_i} \cdot (g^{u_0} \prod_{j=1}^{k} g^{u_j m_j})^{\prod_{i \in T \setminus \{t\}} e_i} \quad (11)$$

$$= (h^{-\beta} \cdot h^{m_\alpha})^{\prod_{i \neq t^*, i \neq t}^{T} e_i} \cdot (g^{u_0} \prod_{j=1}^{k} g^{u_j m_j})^{\prod_{i \in T \setminus \{t\}} e_i} \quad (12)$$

$$= (h^{-\beta \prod_{i \neq t^*}^{T} e_i} \cdot Y^{u_0})^{1/e_t} \cdot (\prod_{j=1, j \neq \alpha} U_j^{m_j})^{1/e_t} \cdot (h^{\prod_{i \neq t^*}^{T} e_i} \cdot Y^{u_\alpha})^{m_\alpha / e_t} \quad (13)$$

$$= (U_0 \prod_{j=1}^{k} U_j^{m_j})^{1/e_t} \mod N. \quad (14)$$

Output. Eventually $\mathcal{A}$ outputs a tuple $(t_f, (pk_{\gamma(1)}, \ldots, pk_{\gamma(z)}), (M_1, \ldots, M_z), \sigma_{agg})$. Since aggregation order does not matter here[6], we can w.l.o.g. assume that $\gamma(1) = 1$ (corresponding to the target key pk_1); we also drop γ from the subscript below. If the aggregate signature does not verify or if any of the reduction's guesses of t^*, i^*, α, β were incorrect, then abort. These abort conditions are all consistent with the adversary not winning Game 7. Let $E' = \prod_{i \in T \setminus \{t^*\}} e_i$. Otherwise we have that:

$$\sigma_{agg}^{e^*} = \prod_{j=1}^{n} (U_{j,0} \prod_{i=1}^{k} U_{j,i}^{m_{j,i}}) \quad (15)$$

$$= (U_{1,0} \prod_{i=1}^{k} U_{1,i}^{m_{1,i}}) \cdot \prod_{j=2}^{n} (U_{j,0} \prod_{i=1}^{k} U_{j,i}^{m_{j,i}}) \quad (16)$$

$$= (h^{E'(\beta - m_\alpha)} \cdot Y^{u_0} \prod_{j=1}^{k} Y^{u_j m_j}) \cdot \prod_{j=2}^{n} (U_{j,0} \prod_{i=1}^{k} U_{j,i}^{m_{j,i}}) \quad (17)$$

Since the reduction can compute the e^*-th root of all values not in the h term, it can divide them out as:

$$\left(\frac{\sigma_{agg}}{\prod_{j=1}^{n} (g^{u_{j,0}} \prod_{i=1}^{k} g^{u_{j,i} m_{j,i}})^{E'}} \right)^{e^*} \quad (18)$$

$$= \frac{(h^{E'(\beta - m_\alpha)} \cdot Y^{u_0} \prod_{j=1}^{k} Y^{u_j m_j}) \cdot \prod_{j=2}^{n} (U_{j,0} \prod_{i=1}^{k} U_{j,i}^{m_{j,i}})}{\prod_{j=1}^{n} (g^{u_{j,0}} \prod_{i=1}^{k} g^{u_{j,i} m_{j,i}})^{e^* \cdot E'}} \quad (19)$$

$$= h^{E'(\beta - m_\alpha)}. \quad (20)$$

[6] Our scheme has the property that any σ_{agg} that verifies on period t for $pk_1, \ldots, pk_z$ and $M_1, \ldots, M_z$ also verifies on any permutation applied to both sequences.

Now, we have an equation of the form $x^a = y^b$, for $x = \frac{\sigma_{agg}}{\prod_{j=1}^{n}(g^{u_{j,0}} \prod_{i=1}^{k} g^{u_{j,i} m_{j,i}})^{E'}}$, $a = e^*$, $y = h$ and $b = E'(\beta - m_\alpha)$. Recall that the game would have already aborted if e^* was output for any period other than t^* and thus, $\gcd(e^*, E') = 1$. The game would also have aborted if $\beta = m_\alpha$. Finally since the $|\beta| = |m_\alpha| = \ell < \lambda$ and $e^* > 2^\lambda$, we can conclude that $\gcd(a, b) = 1$. This allows the reduction to apply Lemma 1 to efficiently compute $\hat{h} \in \mathbb{Z}_N$ such that $\hat{h}^{e^*} = h \mod N$. The reduction outputs this value as the RSA solution.

Analysis. We argue that the above reduction will succeed in outputting the RSA solution whenever the adversary wins in Game 7. The adversary's view in these scenarios differs only in the way that public key elements U_0 and U_α are chosen. We will first argue that the way they are chosen in Game 7 (and the actual scheme) is statistically close to choosing a random element in QR_N. Next, we argue that the (different) way they are chosen in the reduction above is also statistically close to choosing a random element in QR_N. It follows then that the public key in both Game 7 and the reduction are statistically close and thus cannot be distinguished by our polynomial-time adversary. Moreover, while the signatures are computed via a different method in Game 7 and the reduction, the signature the adversary sees is identical (and unique) given the public information known to the adversary, so there is no information the adversary can use to distinguish. For any given $U \in \mathrm{QR}_N$, prime $e \in [1, N]$, and $m < 2^\lambda$, the values U^{em} and $U^{1/e}$ are unique since each e_i is relatively prime to $\phi(N)$. It remains to support the arguments listed above.

First, recall how U_0, U_α are chosen in Game 7 (and the actual scheme). Here u_0, u_α are randomly chosen from $[1, N]$ and the public key elements are set as:

$$U_0 = Y^{u_0} = g^{u_0 \prod_{i=1}^{T} e_i} \quad , \quad U_\alpha = Y^{u_\alpha} = g^{u_\alpha \prod_{i \in T} e_i}.$$

Observe that the group of QR_N has order $p'q'$. Thus $Y = g^{\prod_{i=1}^{T} e_i}$ is also a generator since all the e_i values are relatively prime to $p'q'$. Since Y is a generator, if we take Y^r for a random $r \in [1, \phi(N)]$ that has the same distribution as choosing a random element in QR_N. Now, the process of raising Y^r for a random $r \in [1, N]$ is statistically close to the process of raising it to a random $r \in [1, \phi(N)]$. The reason is that $N = \phi(N) + p + q - 1$ where the difference $(p + q - 1)$ is negligible. Thus, we achieve our first argument.

Second, recall how U_0, U_α are chosen in the reduction. Here u_0, u_α are randomly chosen from $[1, N]$ and the public key elements are set as:

$$U_0' = (h^{-\beta})^{\prod_{i \neq t^*}^{T} e_i} \cdot Y^{u_0} = h^{-\beta \prod_{i \neq t^*}^{T} e_i} \cdot g^{u_0 \prod_{i=1}^{T} e_i} \quad , \quad U_\alpha' = h^{\prod_{i \neq t^*}^{T} e_i} \cdot Y^{u_\alpha}.$$

We previously argued that the Y^{u_0} and Y^{u_α} components are distributed statistically close to a random element in QR_N. We assume that $h \in \mathrm{QR}_N$; this will be true for a random element in $\mathbb{Z}_N^*$ with 1/4 probability. Each value has an h term that is in QR_N but not necessarily distributed randomly. However, once we multiply this value in the group by a (statistically close to) random element of the group, we have a product that is distributed statistically close to a random element in QR_N. Thus, we achieve our second argument.

Since the adversary cannot distinguish either distribution of public keys from a random distribution, then it cannot distinguish them from each other as well. Thus, whenever the adversary succeeds in Game 7, we can conclude it will also succeed in helping the reduction solve RSA.

5 Trading Off Signing Time with Storage

In this section we show a basic tradeoff between the time to sign and the size of the private key storage. Very roughly the time to sign will scale linearly with a parameter a and the storage with a parameter b with the constraint that $a \cdot b = T$. Thus we can explore tradeoffs such as setting $a = T, b = 1$ as we saw in the last section or go the opposite direction and set $a = 1, b = T$ to achieve fast signing at the expense of large storage, or try to balance these by choosing $a = b = \sqrt{T}$.

Our system will use the same setup, verification and aggregation algorithms as in Sect. 4 and just replace the KeyGen and Sign algorithms. Moreover, the public keys output by the KeyGen algorithm and corresponding signatures output by the Sign algorithm will have an identical distribution to the original Sect. 4 scheme and thus not require a new security proof.

Let the public parameters output from Setup be $\mathsf{pp} = (T, N, g, Y, K)$ as before. Our KeyGeneration algorithm will organize T into b "windows" each of size a. (We assume a divides T evenly for ease of exposition.) Then the private key will be setup to contain a sequence of values R_w which is g raised to all e_i except those in a sliding window of periods. To sign faster during time period t, select these partially computed values where t is in the window and complete its computation for signing by raising to all e_i in that window except e_t.

The new key generation and signing algorithms follow.

KeyGen'(pp, a) It obtains the primes $(e_1, \ldots, e_T)$ and sets $b = T/a$ (we assume it divides evenly for ease of exposition). Next it chooses random integers $u_0, u_1, \ldots, u_k$ in $[1, N]$ and computes $pk = (U_0, U_1, \ldots, U_k)$. For $w = 1$ to b, define Σ_w as the set of integers in $[1, T]$ other than those in the set $\{a(w-1)+1, a(w-1)+2, \ldots, a(w-1)+a\}$.

For $w = 1$ to b, it then computes:

$$R_w = g^{\prod_{i \in \Sigma_w} e_i}$$

where the e_i values are computed using K from pp. It sets the secret key as $sk = (\{R_w\}_{1 \le w \le b}, \{u_i\}_{0 \le i \le k})$. The public key $pk = (U_0, U_1, \ldots, U_k)$ is computed as $U_j = Y^{u_j} = g^{u_j \prod_{i \in T} e_i}$, for $j = 0$ to k as in Sect. 4.

Sign'(pp, sk, M, t) It computes the necessary subset of primes in $(e_1, \ldots, e_T)$ using K in pp and then for period t, selects the window $w = \lceil t/a \rceil$. Let Σ'_w denote the set of periods in the window $\{a(w-1)+1, a(w-1)+2, \ldots, a(w-1)+a\}_{1 \le w \le b}$. It outputs

$$\sigma = \left(R_w^{u_0} \prod_{j=1}^{k} R_w^{u_j \cdot m_j}\right)^{\prod_{i \in \Sigma'_w \setminus \{t\}} e_i} = \left(U_0 \prod_{j=1}^{k} U_j^{m_j}\right)^{1/e_t} \pmod{N}.$$

Analysis. Observe that the public keys and signatures are of the same form and distribution as those of the base system in Sect. 4, as are the verification equations, and thus the security of this tradeoff system follows. We analyze the performance of this system in Sect. 8.

6 Obtaining $O(\lg(T))$ Signing Time and Private Key Size

The previous section showed a basic tradeoff between signing time and private key size. However, it was limited in that the most "balanced" version required both time and storage to grow with the square root of the number of periods.

In this section we show how a more intricate key storage technique can give us much better results with a scheme where the number of exponentiations and prime searches is $\approx \lg(T)$ per signing operation and where we store $\approx \lg(T)$ elements of $\mathbb{Z}_N^*$ in the private key. Unlike the previous schemes where the private key remained static, our method here will require us to update the private key on each signing period. As a consequence a signer will be required to sign using each period in sequence.[7] Again, our new scheme will produce public keys and signatures with exactly the same distribution as the base scheme of Sect. 4. Therefore we will only need to describe and analyze the new method of key generation and storage and are not required to produce a new security proof. As mentioned earlier, this approach has conceptual roots in the pebbling optimization used by Itkis and Reyzin [26] to realize efficient forward-secure signatures.

We present our method by introducing new two algorithms. The first algorithm $\mathtt{StorageInit}(\mathsf{pp}, v)$ takes in the public parameters and an element $v \in \mathbb{Z}_N^*$ and outputs the initial key storage state $\mathtt{store}$. The second algorithm $\mathtt{StorageUpdate}(\mathtt{store})$ takes in the storage $\mathtt{store}$ and outputs an updated storage value $\mathtt{store}$ as well as a group element $s \in \mathbb{Z}_N^*$.

6.1 Storage Algorithms

We assume that there exists an integer '$\mathtt{levels}$' such that $T = 2^{\mathtt{levels}+1} - 2$. (One could always pad T out to match this.) The key storage will be structured as a sequence of sets $S_1, \ldots, S_{\mathtt{levels}}$ where elements of set S_i are of the form

$$w \in \mathbb{Z}_N^*, \mathtt{open} \in [1, T], \mathtt{closing} \in [1, T], \mathtt{count} \in [1, T].$$

Let R be the set of integers $[\mathtt{open}, \mathtt{open} + 2^{i-1} - 1] \cup [\mathtt{closing} + \mathtt{count}, \mathtt{closing} + 2^{i-1} - 1]$. Then $w = v^{\prod_{j \in T \setminus R} e_j}$. Intuitively, w is v raised to all of the e exponents except the sequence of 2^{i-1} values starting at $\mathtt{open}$ and a second sequence of length $2^{i-1} - \mathtt{count}$ starting at $\mathtt{closing} + \mathtt{count}$. When the $\mathtt{StorageUpdate}$ algorithm runs for each i, it will find an element of the set S_i and help "move it forward" by incrementing its counter $\mathtt{count}$ and updating w accordingly. When

[7] We expect this to be the normal mode of operation in a synchronized scheme, however, the previous schemes have the ability to sign for periods in an arbitrary order.

$\texttt{count}$ reaches 2^i the update storage algorithm removes the tuple from the set S_i at level i and then splits it into two parts and puts these in set S_{i-1}. We now describe the algorithms.

$StorageInit(\mathsf{pp}, v)$ Initially, sets $S_1, \ldots, S_{\texttt{levels}}$ are empty. Then for $i = 1$ to $\texttt{levels}$ perform the following:

- Let $R = [2^i - 1, 2^{i+1} - 2]$.
- Compute $w = v^{\prod_{j \in T \setminus R} e_j}$.
- Put in S_i $(w, 2^i - 1, (2^i - 1) + 2^{i-1}, 0)$.
- Put in S_i $(w, (2^i - 1) + 2^{i-1}, 2^i - 1, 0)$.

The storage value $\texttt{store} = \big((S_1, \ldots, S_{\texttt{levels}}), \texttt{index} = 0\big)$ is output.
$StorageUpdate(\mathsf{pp}, \texttt{store})$ For $i = 1$ to $\texttt{levels}$ perform the following:

- Find a tuple (if any exist) in S_i of $(w, \texttt{open}, \texttt{closing}, \texttt{count})$ with the smallest $\texttt{open}$ value.[8]
- Replace it with a new tuple $(w' = w^{e_{\texttt{closing}+\texttt{count}}}, \texttt{open}' = \texttt{open}, \texttt{closing}' = \texttt{closing}, \texttt{count}' = \texttt{count}+1)$ where $(w', \texttt{open}', \texttt{closing}', \texttt{count}')$ is the newly added tuple.

Then for $i = \texttt{levels}$ down to 2

- Find a tuple (if any) of the form $(w, \texttt{open}, \texttt{closing}, \texttt{count} = 2^{i-1})$ in S_i.
- Remove this tuple from the set S_i.
- To the set S_{i-1} add the tuple $(w' = w, \texttt{open}' = \texttt{open}, \texttt{closing}' = \texttt{open} + 2^{i-2}, \texttt{count}' = 0)$ where $(w', \texttt{open}', \texttt{closing}', \texttt{count}')$ is the newly added tuple.
- Also add to the set S_{i-1} the tuple $(w' = w, \texttt{open}' = \texttt{open} + 2^{i-2}, \texttt{closing}' = \texttt{open}, \texttt{count}' = 0)$.

Finally, from S_1 find the tuple $(w, \texttt{open} = \texttt{index} + 1, \texttt{closing}, 1)$. Remove this from S_1 and output $s = w$ which gives $s = v^{\prod_{j \in T \setminus \{(\texttt{index}+1)\}} e_j}$ as needed. Finally, the storage value $\texttt{store} = ((S_1, \ldots, S_{\texttt{levels}}), \texttt{index} = \texttt{index} + 1)$ is output.

6.2 Analysis

We need to show that the storage primitives give the desired correctness and performance properties. To analyze correctness and storage size we consider what the key storage state will look like for each value of $\texttt{index}$ between 0 and T. Recall that in a stored key, $\texttt{index}$ represents the number of signatures generated so far. We describe what each S_i set contains for a particular $\texttt{index}$ value — breaking things into three cases. We will refer to this as our "state description" given below.

[8] In a particular S_i there might be zero, one or two tuples. If there are two, the one with the larger $\texttt{open}$ value is ignored. Ties will not occur, as we will see from the case analysis in Sect. 6.2.

Case 1: $T - \texttt{index} \leq 2^i - 2$. In this case the set S_i will be empty.

Case 2: Not Case 1 and $\texttt{index} = k \cdot 2^i + r$ for $0 \leq r < 2^{i-1}$. S_i will contain two elements. The first is a tuple

$$(w = v^{\prod_{j \in T \setminus R} e_j}, \texttt{open} = (k+1) \cdot 2^i - 1, \texttt{closing} = (k+1) \cdot 2^i - 1 + 2^{i-1}, \texttt{count} = r).$$

Where we let $R = [\texttt{open}, \texttt{open} + 2^{i-1} - 1] \cup [\texttt{closing} + \texttt{count}, \texttt{closing} + 2^{i-1} - 1]$.

The second is a tuple

$$(w = v^{\prod_{j \in T \setminus R} e_j}, \texttt{open} = (k+1) \cdot 2^i - 1 + 2^{i-1}, \texttt{closing} = (k+1) \cdot 2^i - 1, \texttt{count} = 0).$$

Where $R = [\texttt{open}, \texttt{open} + 2^{i-1} - 1] \cup [\texttt{closing} + \texttt{count}, \texttt{closing} + 2^{i-1} - 1]$. (Here $\texttt{count} = 0$.)

Case 3: Not Case 1 and $\texttt{index} = k \cdot 2^i + r$ for $2^{i-1} \leq r < 2^i$. S_i has a single element. A tuple

$$(w = v^{\prod_{j \in T \setminus R} e_j}, \texttt{open} = (k+1) \cdot 2^i - 1 + 2^{i-1}, \texttt{closing} = (k+1) \cdot 2^i - 1, \texttt{count} = r - 2^{i-1}).$$

Where $R = [\texttt{open}, \texttt{open} + 2^{i-1}] \cup [\texttt{closing} + \texttt{count}, \texttt{closing} + 2^{i-1}]$.

Proof of State Description Accuracy.

Theorem 10. *The above state description for variable* index *accurately describes the key storage state after an initial call to* StorageInit(pp, v) *and* index *subsequent calls to* StorageUpdate(pp, store).

Proof. We begin by establishing two claims about when the "pass down" operation can and cannot happen which will be used later on in the proof.

Claim 11. *Suppose that our state description is accurate for period* index. *Consider an update operation where the period moves from* index *to* $\texttt{index}+1$. *This will result in an tuple being "passed down" from S_i to S_{i-1} only if* $\texttt{index} + 1$ *is a multiple of 2^{i-1}, if anything is passed down at all.*

Proof. If $(\texttt{index}, i)$ were in Case 1, then S_i is empty and there is nothing that could be passed down. If in Case 2, then one tuple has a $\texttt{count} = r$ which is the remainder of index mod 2^i. It will trigger a pass down operation only when count increments to $\texttt{count} = 2^{i-1}$. Similarly, in Case 3 there is a tuple with $\texttt{count} = r - 2^{i-1}$. A push down operation is only triggered when it increments to 2^i which means $\texttt{index} + 1$ is a multiple of 2^{i-1}.

Claim 12. *Suppose that our state description is accurate for period* index *and all smaller values. Further suppose that* $\texttt{index} + 1 = 0 \mod 2^i$ *for some i and that set S_{i+1} is in Case 1 at* index. *(I.e.* $T - \texttt{index} \leq 2^{i+1} - 2$.*) Then it will be that at period* $\texttt{index} + 1$ *we have* $T - \texttt{index} \leq 2^i - 2$ *and set S_i is designated as Case 1 by our description.*

Proof. Let z be the value where $T - z = 2^{i+1} - 2$ since $T = 2^{\texttt{levels}+1} - 2$ it follows that $z = y \cdot 2^{i+1}$ for some y. Also note that z must be the smallest value of $\texttt{index}$ where $T - \texttt{index} \leq 2^{i+1} - 2$. It then follows that $z + 2^i - 1$ is the smallest value of $\texttt{index}$ where $T - \texttt{index} \leq 2^{i+1} - 2$ AND $\texttt{index} \mod 2^i$. Now let's consider the next value of $\texttt{index} + 1$ which is equal to $z + 2^i$ and use it to prove that at $\texttt{index} + 1$ the set S_i is assigned to be in Case 1. Then

$$T - (\texttt{index} + 1) = T - (z + 2^i) = (T - z) - 2^i = 2^{i+1} - 2 - 2^i = 2^i - 2.$$

Then we have that at $\texttt{index} + 1$ the set S_i is categorized at Case 1 (and empty) by our description.

We now show that for each $\texttt{index}$ if the state description was valid at $\texttt{index}$ then it is valid at $\texttt{index} + 1$. We break this into three separate claims showing that if a set S_i is in Case 1, 2 and 3 respectively at $\texttt{index}$ that in $\texttt{index} + 1$ it will match the state description.

Claim 13. *Suppose at period* $\texttt{index}$ *the state description is accurate and for a set* S_i *we are in Case 1 where* $T - \texttt{index} \leq 2^i - 2$ *and the set* S_i *is empty. Then at period* $\texttt{index} + 1$ *the state description is accurate for set* S_i.

Proof. For period $\texttt{index} + 1$ we have that $T - (\texttt{index} + 1)$ is also $\leq 2^i - 2$ and therefore it should also be Case 1 and S_i should remain empty. The only way for it not to remain empty would be if the $\texttt{StorageUpdate}$ algorithm "passed down" a new tuple from S_{i+1}. However, if S_i was in Case 1 for period $\texttt{index}$ then S_{i+1} must also be and also be empty. Since S_{i+1} is empty there is nothing to pass down.

Claim 14. *Suppose at period* $\texttt{index}$ *the state description is accurate and for a set* S_i *we are in Case 2 where* $\texttt{index} = k \cdot 2^i + r$ *for* $0 \leq r < 2^{i-1}$. *Then at period* $\texttt{index} + 1$ *the state description is accurate for set* S_i.

Proof. First consider the subcase where $r \neq 2^{i-1} - 1$ which should keep S_i in Case 2 on period $\texttt{index}+1$. We will verify this. Since at period $\texttt{index}$ we are in Case 2 there are two tuples in S_i where the one with the smaller $\texttt{open}$ value is of the form $(w = v^{\prod_{j \in T \setminus R} e_j}, \texttt{open} = (k+1) \cdot 2^i - 1, \texttt{closing} = (k+1) \cdot 2^i - 1 + 2^{i-1}, \texttt{count} = r)$. The update algorithm will increment $\texttt{count}$ to $r + 1$ and update w to $w = w^{e_{\texttt{closing}+\texttt{count}}}$ which gives the needed form to remain in Case 2. The second tuple will is of the form $(w = v^{\prod_{j \in T \setminus R} e_j}, \texttt{open} = (k+1) \cdot 2^i - 1 + 2^{i-1}, \texttt{closing} = (k+1) \cdot 2^i - 1, \texttt{count} = 0)$. The update algorithm will not modify it as the other tuple had the smaller $\texttt{open}$ value. Thus it remains the same which matches the behavior for S_i remaining in Case 2. Finally, we need to check that no new tuples are passed down from S_{i+1}. This follows from the fact (Claim 11) that $\texttt{index} \mod 2^i = r \neq 2^i - 1$ and that a pushdown would only happen as $\texttt{index}$ transfers to being a multiple of 2^i.

We now consider the subcase where $r = 2^{i-1} - 1$ at $\texttt{index}$ and thus at $\texttt{index}+1$ we should be moving into Case 3. In this subcase the set S_i begins with two tuples with one of the form ($w = v^{\prod_{j \in T \setminus R} e_j}$, $\texttt{open} = (k+1) \cdot 2^i - 1$, $\texttt{closing} = (k+1) \cdot 2^i - 1 + 2^{i-1}$, $\texttt{count} = r = 2^{i-1} - 1$). The update operation will first modify the tuple to a new $\texttt{count}$ value of $\texttt{count} = 2^{i-1}$. This will trigger the pushdown operation to move the tuple out of S_i. It then leaves it with one tuple of the needed form which transitions S_i to Case 3 as needed. Again no new elements are pushed onto S_i from S_{i+1} due to Claim 11.

Claim 15. *Suppose at period* $\texttt{index}$ *the state description is accurate and for a set* S_i *we are in Case 3 where* $\texttt{index} = k \cdot 2^i + r$ *for* $2^{i-1} \leq r < 2^i$ *for some* k*. Then at period* $\texttt{index} + 1$ *the state description is accurate for set* S_i*.*

Proof. We first focus on the subcase where $r \neq 2^i - 1$ and thus at $\texttt{index} + 1$ we want to verify that we stay in Case 3. Initially there is one tuple of the form ($w = v^{\prod_{j \in T \setminus R} e_j}$, $\texttt{open} = (k+1) \cdot 2^i - 1 + 2^{i-1}$, $\texttt{closing} = (k+1) \cdot 2^i - 1$, $\texttt{count} = r - 2^{i-1}$). The update algorithm will increment $\texttt{count}$ to $r+1$ and update w to $w = w^{e_{\texttt{closing}+\texttt{count}}}$ which gives the needed form to remain in Case 3. As before no new tuples will be added since $\texttt{index} + 1 \mod 2^i \neq 0$.

We end by considering the subcase where $r = 2^i - 1$. In this subcase there is initially a single tuple with a $\texttt{count}$ value of $\texttt{count} = 2^{i-1} - 1$. The update algorithm will increment this $\texttt{count}$ which triggers its removal from the set. What remains to be seen is whether a new element is added or if it becomes empty.

We now consider two possibilities. If $T - (\texttt{index} + 1) \leq 2^i - 2$, then our description states that set S_i should enter Case 1 on $\texttt{index} + 1$. It is easy to see that if this is true that the set S_{i+1} was already Case 1 and empty on $\texttt{index}$ and nothing new will be added so the set S_i is empty as needed.

The somewhat trickier case is when $T - (\texttt{index}+1) > 2^i - 2$. Here we need to verify that the set S_i ends up in Case 2 with the appropriate tuple at $\texttt{index}+1$. First, since $\texttt{index} + 1 \mod 2^i = 0$ we can apply Claim 12. It states that if set S_{i+1} were in Case 1 (empty) at $\texttt{index}$ then set S_i would be in Case 1 for $\texttt{index} + 1$. Since this is not the case, we have that S_{i+1} must be non empty and in Case 2 or 3.

If S_{i+1} started in Case 2 at $\texttt{index}$, it initially has a tuple of the form:

$$(w = v^{\prod_{j \in T \setminus R} e_j}, \texttt{open} = (\tilde{k} + 1) \cdot 2^{i+1} - 1, \texttt{closing} = (\tilde{k} + 1) \cdot 2^{i+1} - 1 + 2^i, \texttt{count} = 2^i - 1).$$

Where we let $R = [\texttt{open}, \texttt{open}+2^i-1] \cup [\texttt{closing}+\texttt{count}, \texttt{closing}+2^i-1]$. Note by the description $\texttt{index} = 2^{i+1}\tilde{k} + 2^i - 1$. After the update algorithm has its first pass, $\texttt{count}$ is incremented to 2^i and an exponentiation is done that updates w where it is now for $R = [\texttt{open}, \texttt{open} + 2^i - 1]$ as the second half of the range falls off with the new count value. The update algorithm then removes this tuple from S_{i+1} and creates two new tuples from it. One with an $\texttt{open}' = \texttt{open}$ and $\texttt{closing}' = \texttt{open} + 2^i$; the second with $\texttt{open}' = \texttt{open} + 2^i$ and $\texttt{closing}' = \texttt{open}$.

To verify correctness recall that $\texttt{index} = 2^i k + 2^i - 1$ and $\texttt{index} = 2^{i+1}\tilde{k} + 2^i - 1$. It follows that $k = 2 \cdot \tilde{k}$. Second, $\texttt{index} + 1 = 2^i \cdot k'$ where $k' = k + 1$.

To match the description for $\texttt{index}+1$ we must have that the first tuple created has an $\texttt{open}'$ value of $\texttt{open}' = (k'+1)2^i - 1$. Plugging in terms:

$$(k'+1)2^i - 1 = (k+1+1)2^i - 1 = (2\tilde{k}+2)2^i - 1 = (\tilde{k}+1)2^{i+1} - 1.$$

However, this is exactly the value it inherited from open as needed.

The argument that the right tuple is inherited when set S_{i+1} is in Case 3 proceeds in almost the same way as above.

The proof of our theorem now comes via induction. The accuracy of the state description for $\texttt{index} = 0$ can be verified by inspection. We can prove the rest by induction on index. For any index the accuracy of the description $\texttt{index}+1$ follows from its accuracy on period index. In particular, our previous three claims show that for any i if the state S_i is accurate in period index then after the update algorithm executes, S_i will be accurate in period $\texttt{index}+1$ too.

Computational and Storage Efficiency. Analyzing the running time for these storage operations is straightforward. We have that $\texttt{levels} = \lfloor \lg T \rfloor$. In each storage update operation there is at each level at most one prime search operation and at most one exponentiation. This comes from the fact that for each i the algorithm updates a single set element — the one with the smallest open value (if any). Therefore the number of prime searches and exponentiations is bounded by $\lg(T)$ as desired.

The above state description immediately gives us the storage efficiency we desire. There are at most $\lg(T)$ sets i which have at most two tuples. Each tuple has a single group element. As written, a tuple also has three (small) integers (of value at most T), although a program could drop these because they can be inferred from index, so we will not count them in our Sect. 8 analysis.

Sample Snapshot of Storage. To help the reader better understand these storage algorithms, we provide an example of the storage states for $\texttt{levels} = 3$ and $T = 2^{\texttt{levels}+1} - 2 = 2^4 - 2 = 14$ in Appendix A.

6.3 Using the Storage Primitives and Optimizations

We can use the storage primitive above to modify our signing algorithm and key storage of Sect. 4. We describe two slightly different methods to do this.

Method 1. The Setup algorithm will run as before and output the core public parameters as $\mathsf{pp} = (T, N, g, Y, K)$. However, it will also run $\texttt{StorageInit}(\mathsf{pp}, g)$ which outputs a value store which is appended to the public parameters.

The secret key algorithm will choose random integers $u_0, u_1, \ldots, u_k$ in $[1, N]$. It sets the secret key as $sk = (u_0, u_1, \ldots, u_k)$ and the public key $pk = (U_0, U_1, \ldots, U_k)$ where $U_j = Y^{u_j} = g^{u_j \prod_{i \in T} e_i}$, for $j = 0$ to k. Note all of this is identical to the Sect. 4 scheme. However, it additionally appends store from the public parameters to its secret key. The store is the part of the secret key that will be modified at each signing.

During each the t-th signing step, it will call $\texttt{StorageUpdate}(\texttt{pp}, \texttt{store}_{t-1})$ and as output get a new storage value $\texttt{store}_t$ that is uses to replace the previous one as well as $J = Y^{1/e_t}$. It uses this to sign by computing:

$$\sigma = J^{u_0} \prod_{j=1}^{k} J^{u_j \cdot m_j} = \left(U_0 \prod_{j=1}^{k} U_j^{m_j}\right)^{1/e_t} \pmod N.$$

Method 2. This will be similar to Method 1 except that instead of raising to the $u_0, \ldots, u_k$ values at the end of signing the algorithm will keep $k+1$ parallel copies of storage that already have each respective u_i exponent raised. The description below will need to slightly "break into" the abstraction that we gave earlier.

Setup will run as before and output the core public parameters as $\texttt{pp} = (T, N, g, Y, K)$. However, it will also run $\texttt{StorageInit}(\texttt{pp}, g)$ which outputs a value $\texttt{store}$ which is appended to the public parameters.

The secret key algorithm will choose random integers $u_0, u_1, \ldots, u_k$ in $[1, N]$. It sets the public key $pk = (U_0, U_1, \ldots, U_k)$ where $U_j = Y^{u_j} = g^{u_j \prod_{i \in T} e_i}$, for $j = 0$ to k (as in the Sect. 4 scheme). For $j = 0$ to k it computes $\texttt{store}^{(j)}$ by taking each of the group elements in $\texttt{store}$ and raising it to u_j. This process effectively changes $\texttt{store}$ from being a storage of $v = g$ to being a storage of $v_j = g^{u_j}$ for the respective u_j. Note that each conversion takes $2 \cdot \texttt{levels}$ exponentiations since there are $2 \cdot \texttt{levels}$ group elements per storage.

During each t-th signing step, for each $j \in [0, k]$ it will call $\texttt{StorageUpdate}(\texttt{pp}, \texttt{store}^{(j)}_{t-1})$ and as output get a new storage value $\texttt{store}^{(j)}_t$ that is uses to replace the previous one as well as $J_j = U_j^{1/e_t}$. It uses these to sign by computing:

$$\sigma = J_0 \prod_{j=1}^{k} J_j^{m_j} = \left(U_0 \prod_{j=1}^{k} U_j^{m_j}\right)^{1/e_t} \pmod N.$$

Efficiency note: in the scheme above, the update operation will perform $\texttt{levels}$ prime searches for each of the $k+1$ stores. (By prime search we mean computing the relevant e_i values needed in update.) This gives $(k+1) \cdot \texttt{levels}$ total prime searches. However, each of these stores will be computing the same e values. Thus if we slightly break into the abstraction then one can do only $\texttt{levels}$ total prime searches by sharing that part of the computation across all $k+1$ storage updates.

7 Identity-Based Aggregation from RSA

In the full version [25], we provide the definition for synchronized identity-based aggregate signatures. We now give a construction based on the RSA assumption.

Setup$(1^\lambda, 1^T)$ The setup algorithm chooses an integer $N = pq$ as the product of two safe primes where $p - 1 = 2p'$ and $q - 1 = 2q'$, such that $2^\lambda < \phi(N) < 2^{\lambda+1}$.

The scheme assumes a hash function (modeled as a random oracle) $G : \mathcal{I} \rightarrow \mathbb{Z}_N^{*(k+1)}$. It also uses the hash function $H : [1,T] \rightarrow \{0,1\}^{\lambda+1}$ with key K as specified in Sect. 4. It computes:

$$D = \prod_{i=1}^{T} H_K(i)^{-1} \mod \phi(N).$$

It publishes the public parameters as $\mathsf{pp} = (T, N, K)$ and we assume all parties have access to G. The master secret key includes the factorization of N and the value D.

Extract(msk, ID) The algorithm computes $(U_0, \dots, U_k) \leftarrow G(ID)$. For $i = 1$ to k, it computes $d_i = U_i^D \mod N$. It returns the secret key as $sk = (d_0, d_1, \dots, d_k)$.

Sign($\mathsf{pp}, sk_{ID}, M, t$) The signing algorithm takes as input a time period $1 \leq t \leq T$ and an $L = (\ell k)$-bit message $M = m_1|m_2|\dots|m_k$, where each m_i contains ℓ-bits and these are concatenated together to form M. It computes the primes $(e_1, \dots, e_T)$ from pp and then outputs

$$\sigma = (d_0 \prod_{j=1}^{k} d_j^{m_j})^{\prod_{i \in T \setminus \{t\}} e_i} = (U_0 \prod_{j=1}^{k} U_j^{m_j})^{1/e_t} \pmod N.$$

Verify($\mathsf{pp}, ID, M, t, \sigma$) Let $M = m_1|m_2|\dots|m_k$ and $G(ID) = (U_0, \dots, U_k)$ The algorithm computes the prime e_t from pp. Output 1 if $1 \leq t \leq T$ and $\sigma^{e_t} \stackrel{?}{=} U_0 \prod_{i=1}^{k} U_i^{m_i}$ or 0 otherwise.

Aggregate($\mathsf{pp}, t, (ID_1, M_1, \sigma_1), \dots, (ID_n, M_n, \sigma_n)$) As before, $\sigma_{agg} = \prod_{j=1}^{n} \sigma_j \pmod N$.

AggVerify($\mathsf{pp}, t, (ID_1, \dots, ID_n), (M_1, \dots, M_n), \sigma_{agg}$) As before, output 1 if and only if all inputs are in the correct range, each identity is unique and $\sigma_{agg}^{e_t} \stackrel{?}{=} \prod_{j=1}^{n} (U_{j,0} \prod_{i=1}^{k} U_{j,i}^{m_{j,i}})$ where here $G(ID_i) = (U_{i,0}, \dots, U_{i,k})$.

Remarks. We remark that the same performance enhancements explored in Sects. 5 and 6 apply here. For simplicity, we present the identity-based version only for the scheme in Sect. 4.

Theorem 16. *If the RSA assumption (as stated in Assumption 1) holds, F is a secure pseudorandom function and G is modeled as a random oracle, then the above identity-based synchronized aggregate signature construction is existentially unforgeable under an adaptive chosen message attack.*

Proof of this theorem appears in the full version [25] of this work.

Scheme	Signing Operations				
	$\mathbb{P}$	$\mathbb{E}_N$	$\mathbb{E}_{\lvert e \rvert}$	$\mathbb{E}_\ell$	$\mathbb{M}$
Section 4	$T-1$	$k+1$	$T-1$	k	k
Section 5 ($a=\sqrt{T}$)	$\sqrt{T}-1$	$k+1$	$\sqrt{T}-1$	k	k
Section 5 ($a=1$)	0	$k+1$	0	k	k
Section 6 Method 1	$\lg(T)$	$k+1$	$\lg(T)$	k	k
Section 6 Method 2	$\lg(T)$	0	$(k+1)\lg(T)$	k	k

Fig. 1. Signing Operations Evaluation. Let the modulus be N. Let $\mathbb{P}$ be the time for function H_K to output a prime of $|e|$ bits, $\mathbb{E}_j$ be the time to perform a j-bit modular exponentiation, and $\mathbb{M}$ be the time to perform a modular multiplication. For the Sect. 6, we round up and treat $\lg T \approx$ `levels`. For that scheme via Method 2, the results of the prime search from the first store are shared with all other stores.

8 Performance Evaluation

We now analyze the performance of the various RSA-based aggregate signature schemes in this work. In particular we consider: our core signature scheme of Sect. 4, our scheme with $\approx \sqrt{(T)}$ storage and signing time of Sect. 5, our "big storage for fast signing" scheme also of Sect. 5 and our scheme of $\approx \lg(T)$ storage and signing of Sect. 6 via two different methods of implementing signing (which may out perform each other based on the selection of various implementation parameters). The scheme of Sect. 4 has similar performance to that of Sect. 5 when $a = T$ and therefore we do not separately analyze it.

Operation	$\mathbb{P}_{257}$	$\mathbb{P}_{80}$	$\mathbb{E}_{2048}$	$\mathbb{E}_{257}$	$\mathbb{E}_{256}$	$\mathbb{E}_{80}$	$\mathbb{E}_{32}$	$\mathbb{M}$
Time (ms)	0.975	0.311	4.604	0.670	0.629	0.235	0.094	0.00091

Fig. 2. Time recorded in milliseconds for the above operations are averaged over 1,000 iterations for a 2048-bit modulus using NTL v10.5.0 on a modern laptop. Let $\mathbb{P}_x$ denote an x-bit prime search, $\mathbb{E}_x$ be an x-bit modular exponentiation, and $\mathbb{M}$ be a modular multiplication.

For each scheme, we first evaluate its run-time performance with a signing algorithm operations count in Fig. 1. We then proceed to inspect its practical performance using a 2048-bit RSA modulus and a 256-bit message (the latter corresponding to an output of SHA-256). In Fig. 3, we evaluate each scheme with each of the following parameters: 1 message chunk size of 256 bits, 8 message chunks of 32 bits and 256 messages chunks of 1 bit. When message chunks are 256 bits, we use 257-bit prime e values and for chunks of size 32 bits or 1 bit we consider 80-bit e values. Here we make sure that the size of the RSA primes are at least as big as the message chunks, but let them fall no further than 80 bits to avoid collisions.[9] These evaluations will be considered for a maximum number

[9] We remark that the parameters given for this evaluation do not have a total correspondence to the scheme description. For example, using 80-bit e values will techni-

Scheme	Parameters			Time when $T =$					
	k	ℓ	$\|e\|$	2^{12}	2^{16}	2^{20}	2^{24}	2^{28}	2^{32}
Section 4	1	256	257	6.7s	1.8m	28.7m	7.7h	5.1d	81.7d
	8	32	80	2.3s	35.8s	9.5m	2.5h	1.7d	27.1d
	256	1	80	3.4s	37.0s	9.6m	2.5h	1.7d	27.1d
Section 5 ($a = \sqrt{T}$)	1	256	257	113.4ms	0.4s	1.7s	6.7s	27.0s	1.8m
	8	32	80	76.6ms	0.2s	0.6s	2.3s	9.0s	35.8s
	256	1	80	1.2s	1.3s	1.7s	3.4s	10.1s	36.8s
Section 5 ($a = 1$)	1	256	257	9.8ms	9.8ms	9.8ms	9.8ms	9.8ms	9.8ms
	8	32	80	42.2ms	42.2ms	42.2ms	42.2ms	42.2ms	42.2ms
	256	1	80	1.2s	1.2s	1.2s	1.2s	1.2s	1.2s
Section 6 Method 1	1	256	257	29.6ms	36.1ms	42.7ms	49.3ms	55.9ms	62.5ms
	8	32	80	48.8ms	50.9ms	53.1ms	55.3ms	57.5ms	59.7ms
	256	1	80	1.2s	1.2s	1.2s	1.2s	1.2s	1.2s
Section 6 Method 2	1	256	257	28.4ms	37.7ms	47.0ms	56.2ms	65.4ms	74.7ms
	8	32	80	29.9ms	39.6ms	49.3ms	59.1ms	68.8ms	78.5ms
	256	1	80	0.7s	1.0s	1.2s	1.5s	1.7s	1.9s

Fig. 3. Signing Time Evaluations for 90 different performance points; here N is 2048 bits. Times are calculated by taking the average time for an operation (see Fig. 2) and summing up the total times of each operation (see Fig. 1). Let ms denote milliseconds, s denote seconds, m denote minutes, h denote hours, and d denote days.

of periods of $T \in \{2^{12}, 2^{16}, 2^{20}, 2^{24}, 2^{28}, 2^{32}\}$. Technically, for the log scheme the numbers of time periods is $T = 2^{\texttt{levels}+1} - 2$, however for the sake of these comparisons we will ignore the small constants.

To perform the timing evaluations in Fig. 3, we utilized the high-performance NTL number theory library in C++ v10.5.0 by Victor Shoup [38]. Averaged over 1000 iterations, we measured the cost of a prime search of the relevant size as well as the time to compute modular multiplications and modular exponentiations for the relevant exponent sizes using a 2048-bit RSA modulus. We took all time measurements on an early 2015 MacBook Air with a 1.6 GHz Intel Core i5 processor and 8 GB 1600 MHz DDR3 memory. These timing results are recorded in Fig. 2.

We next report on the signer's storage space requirements in Fig. 4 for all of these combinations. And in Fig. 5, we show how to view T in practical terms for how often one can issue signatures according to the synchronized restrictions over the lifespan of a private key.

Some Conclusions. As expected the initial core scheme of Sect. 4 is much too costly for signing. Even for $T = 2^{20}$ (where one signature is permitted every 5 min for 10 years), it takes roughly 10 min to sign a single message, so the processor

cally require a variant of the RSA assumption with smaller exponents. And we do not attempt to set the modulus size to match the security loss of our reduction. (It is unknown whether this loss can actually be utilized by an attacker or not.) Our focus here is to give the reader a sense of the relative performance of the scheme variants for parameters that might be used in practice.

Scheme	SK Elements $\mathbb{Z}_N$	Param. k	Size when $T =$ 2^{12}	2^{16}	2^{20}	2^{24}	2^{28}	2^{32}
S4	$k+1$	1	0.5K	0.5K	0.5K	0.5K	0.5K	0.5K
		8	2.3K	2.3K	2.3K	2.3K	2.3K	2.3K
		256	64.3K	64.3K	64.3K	64.3K	64.3K	64.3K
S5 ($a = \sqrt{T}$)	$(k+1) + \sqrt{T}$	1	16.5K	64.5K	256.5K	1.0M	4.0M	16.0M
		8	18.3K	66.3K	258.3K	1.0M	4.0M	16.0M
		256	80.3K	128.3K	320.3K	1.1M	4.1M	16.1M
S5 ($a = 1$)	$(k+1) + T$	1	1.0M	16.0M	256.0M	4.0G	64.0G	1.0Tb
		8	1.0M	16.0M	256.0M	4.0G	64.0G	1.0Tb
		256	1.1M	16.1M	256.1M	4.0G	64.0G	1.0Tb
S6 Method 1	$(k+1) + 2\lg T$	1	6.5K	8.5K	10.5K	12.5K	14.5K	16.5K
		8	8.3K	10.3K	12.3K	14.3K	16.3K	18.3K
		256	70.3K	72.3K	74.3K	76.3K	78.3K	80.3K
S6 Method 2	$2(k+1)\lg T$	1	12.0K	16.0K	20.0K	24.0K	28.0K	32.0K
		8	54.0K	72.0K	90.0K	108K	106K	144K
		256	1.5M	2.0M	2.5M	3.0M	3.5M	4.0M

Fig. 4. Private Key Size Evaluation. Here the modulus N is 2048 bits. The above numbers are rounded to show one decimal point. Let K denote a kilobyte (2^{10} bytes), M a megabyte (2^{20} bytes), G a gigabyte (2^{30} bytes), and Tb a terabyte (2^{40} bytes). Any of the schemes that compute primes during Signing (all but Sect. 5 when $a = 1$), could instead choose to speed up signing by additionally storing those values at an additional storage cost of T elements of $\mathbb{Z}_{|e|}$. All but the last scheme include $k+1$ elements that are the randomization factors $u_0, \ldots, u_k \in [1, N]$; this space could be shrunk by re-computing these from a PRF.

we took these measurements on could not "break even" by keeping up with the modest pace of one signature every 5 min using the base scheme. At larger time periods, the signing time moves into days. One noticeable aspect is that the $k = 1$ (where k is the number of message chunks) time measurements are about a factor of three greater than when $k \in \{8, 256\}$ for this scheme and the square root one. This is due to the cost difference of searching for and raising to 257-bit primes versus 80-bit primes which dominate these schemes.

The square root tradeoff certainly does better, but cannot break even (on the processor measured) once we hit $T = 2^{28}$. Additionally, the keys are somewhat large on the order of a few megabytes. This could be an issue if we would want to store several keys or a single key on a low memory device.

On the other end of the spectrum when setting $a = 1$, we get relatively fast signatures. Here things flip where it is significantly more expensive to sign for $k = 256$ than $k \in \{1, 8\}$. The reason is that at this point the cost of raising to the u_i values now dominates the computation — whereas in the earlier schemes it was dominated by raising to the e_i values. The main downside of this setting is that the key sizes are huge — breaking into the terabyte range for $T = 2^{32}$.

Setting of T	Frequency of Signatures
2^{12}	76,992 sec (≈ one per day)
2^{16}	4,812 sec (≈ one every 1.5 hours)
2^{20}	300 sec (≈ one every 5 minutes)
2^{24}	19 sec
2^{28}	1.2 sec
2^{32}	0.07 sec (≈ ten per second)

Fig. 5. Approximate view of how to select T based on how often an application needs the ability to issue signatures during a key's 10-year lifespan. (One can approximate a 20-year key lifespan by cutting the above frequencies in half.)

We finally move to our log scheme of Sect. 6 where we start with Method 1. It scales well with the number of time periods where even for $T = 2^{32}$ it is only about 60 ms for $k \in \{1, 8\}$. For $k = 256$ the time is again dominated by the raising to the u_i values at the end. Also, the private keys can be kept in the range of ten to twenty kilobytes for lower k values. (We note that for $k = 256$ one possibility is that the u_i values could be generated from a pseudo random function which could lower the key storage cost.) The second method of using the log storage is more costly in terms of key storage cost. Its performance in signing time is slightly better for smaller values of T, but for values higher than 2^{20} turns worse. For this reason the first method seems to perform better overall than the second.

Altogether, the log storage solution (of Sect. 6 using Method 1) offers practical time/space costs and appears to provide the best overall practical performance of all schemes analyzed.

Acknowledgments. We thank the anonymous reviewers for their helpful comments and Joseph Ayo Akinyele for implementation discussions.

A Sample Snapshot of Storage for Sect. 6 Scheme

To aid the reader, we provide an example of the storage states for $\texttt{levels} = 3$ and $T = 2^{\texttt{levels}+1} - 2 = 2^4 - 2 = 14$ in Fig. 6. This example shows the states after updates; it does not show any intermediate states during an update operation. The example gives just the `open`, `closing` and `count` values. The prior section describes how the corresponding group element w is computed based on these values (see the description of R as the range of indices of e_i values excluded from the product in the exponent.) Initially, we have sets $S_1, \ldots, S_{\texttt{levels}=3}$ that are empty. The values at $\texttt{index} = 0$ show the states after running `StorageInit`. The values at $\texttt{index} > 0$ show the state after a call to `StorageUpdate`.

index	Set S_1			Set S_2			Set S_3		
	`open`	`closing`	`count`	`open`	`closing`	`count`	`open`	`closing`	`count`
0	1 2	2 1	0 0	3 5	5 3	0 0	7 11	11 7	0 0
1	2	1	0	3 5	5 3	1 0	7 11	11 7	1 0
2	3 4	4 3	0 0	5	3	0	7 11	11 7	2 0
3	4	3	0	5	3	1	7 11	11 7	3 0
4	5 6	6 5	0 0	7 9	9 7	0 0	11	7	0
5	6	5	0	7 9	9 7	1 0	11	7	1
6	7 8	8 7	0 0	9	7	0	11	7	2
7	8	7	0	9	7	1	11	7	3
8	9 10	10 9	0 0	11 13	13 11	0 0			
9	10	9	0	11 13	13 11	1 0			
10	11 13	12 11	0 0	12	11	0			
11	12	11	0	13	11	1			
12	13 14	14 13	0 0						
13	14	13	0						
14									

Fig. 6. Storage State Example for `levels` $= 3$, $T = 14$. See above description.

References

1. Ahn, J.H., Green, M., Hohenberger, S.: Synchronized aggregate signatures: new definitions, constructions and applications. In: ACM Conference on Computer and Communications Security, pp. 473–484 (2010)
2. Anonymous. Increasing anonymity in bitcoin (2013). https://bitcointalk.org/index.php?topic=1377298.0
3. Bagherzandi, A., Jarecki, S.: Identity-based aggregate and multi-signature schemes based on RSA. In: Nguyen, P.Q., Pointcheval, D. (eds.) PKC 2010. LNCS, vol. 6056, pp. 480–498. Springer, Heidelberg (2010). https://doi.org/10.1007/978-3-642-13013-7_28
4. El Bansarkhani, R., Mohamed, M.S.E., Petzoldt, A.: MQSAS - a multivariate sequential aggregate signature scheme. In: Bishop, M., Nascimento, A.C.A. (eds.) ISC 2016. LNCS, vol. 9866, pp. 426–439. Springer, Cham (2016). https://doi.org/10.1007/978-3-319-45871-7_25

5. Barak, B., Canetti, R., Nielsen, J.B., Pass, R.: Universally composable protocols with relaxed set-up assumptions. In: Symposium on Foundations of Computer Science, pp. 186–195. IEEE Computer Society (2004)
6. Barić, N., Pfitzmann, B.: Collision-free accumulators and fail-stop signature schemes without trees. In: Fumy, W. (ed.) EUROCRYPT 1997. LNCS, vol. 1233, pp. 480–494. Springer, Heidelberg (1997). https://doi.org/10.1007/3-540-69053-0_33
7. Bellare, M., Namprempre, C., Neven, G.: Unrestricted aggregate signatures. In: Arge, L., Cachin, C., Jurdziński, T., Tarlecki, A. (eds.) ICALP 2007. LNCS, vol. 4596, pp. 411–422. Springer, Heidelberg (2007). https://doi.org/10.1007/978-3-540-73420-8_37
8. Bellare, M., Neven, G.: Identity-based multi-signatures from RSA. In: Abe, M. (ed.) CT-RSA 2007. LNCS, vol. 4377, pp. 145–162. Springer, Heidelberg (2006). https://doi.org/10.1007/11967668_10
9. Benaloh, J., de Mare, M.: One-way accumulators: a decentralized alternative to digital signatures. In: Helleseth, T. (ed.) EUROCRYPT 1993. LNCS, vol. 765, pp. 274–285. Springer, Heidelberg (1994). https://doi.org/10.1007/3-540-48285-7_24
10. Boldyreva, A.: Threshold signatures, multisignatures and blind signatures based on the Gap-Diffie-Hellman-Group signature scheme. In: Desmedt, Y.G. (ed.) PKC 2003. LNCS, vol. 2567, pp. 31–46. Springer, Heidelberg (2003). https://doi.org/10.1007/3-540-36288-6_3
11. Boldyreva, A., Gentry, C., O'Neill, A., Yum, D.H.: Ordered multisignatures and identity-based sequential aggregate signatures, with applications to secure routing. In: ACM Conference on Computer and Communications Security (CCS), pp. 276–285 (2007), http://www.cc.gatech.edu/~amoneill/bgoy.html
12. Boneh, D., Franklin, M.K.: Efficient generation of shared RSA keys. J. ACM **48**(4), 702–722 (2001)
13. Boneh, D., Gentry, C., Lynn, B., Shacham, H.: Aggregate and verifiably encrypted signatures from bilinear maps. In: Biham, E. (ed.) EUROCRYPT 2003. LNCS, vol. 2656, pp. 416–432. Springer, Heidelberg (2003). https://doi.org/10.1007/3-540-39200-9_26
14. Boneh, D., Gentry, C., Lynn, B., Shacham, H.: A survey of two signature aggregation techniques. RSA Cryptobytes **6**(2), 1–9 (2003)
15. Brogle, K., Goldberg, S., Reyzin, L.: Sequential aggregate signatures with lazy verification from trapdoor permutations. Inf. Comput. **239**, 356–376 (2014)
16. Cramer, R., Shoup, V.: Signature schemes based on the strong RSA assumption. ACM Trans. Inf. Syst. Secur. **3**(3), 161–185 (2000)
17. Fischlin, M., Lehmann, A., Schröder, D.: History-free sequential aggregate signatures. In: Visconti, I., De Prisco, R. (eds.) SCN 2012. LNCS, vol. 7485, pp. 113–130. Springer, Heidelberg (2012). https://doi.org/10.1007/978-3-642-32928-9_7
18. Freire, E.S.V., Hofheinz, D., Paterson, K.G., Striecks, C.: Programmable hash functions in the multilinear setting. In: Canetti, R., Garay, J.A. (eds.) CRYPTO 2013. LNCS, vol. 8042, pp. 513–530. Springer, Heidelberg (2013). https://doi.org/10.1007/978-3-642-40041-4_28
19. Gentry, C., Ramzan, Z.: Identity-based aggregate signatures. In: Yung, M., Dodis, Y., Kiayias, A., Malkin, T. (eds.) PKC 2006. LNCS, vol. 3958, pp. 257–273. Springer, Heidelberg (2006). https://doi.org/10.1007/11745853_17
20. Guo, X., Wang, Z.: An efficient synchronized aggregate signature scheme from standard RSA assumption. Int. J. Future Gener. Commun. Netw. **7**(3), 229–240 (2014)

21. Hofheinz, D., Jager, T., Kiltz, E.: Short signatures from weaker assumptions. In: Lee, D.H., Wang, X. (eds.) ASIACRYPT 2011. LNCS, vol. 7073, pp. 647–666. Springer, Heidelberg (2011). https://doi.org/10.1007/978-3-642-25385-0_35
22. Hohenberger, S., Koppula, V., Waters, B.: Universal signature aggregators. In: Oswald, E., Fischlin, M. (eds.) EUROCRYPT 2015. LNCS, vol. 9057, pp. 3–34. Springer, Heidelberg (2015). https://doi.org/10.1007/978-3-662-46803-6_1
23. Hohenberger, S., Sahai, A., Waters, B.: Full domain hash from (leveled) multilinear maps and identity-based aggregate signatures. In: Canetti, R., Garay, J.A. (eds.) CRYPTO 2013. LNCS, vol. 8042, pp. 494–512. Springer, Heidelberg (2013). https://doi.org/10.1007/978-3-642-40041-4_27
24. Hohenberger, S., Waters, B.: Short and stateless signatures from the RSA assumption. In: Halevi, S. (ed.) CRYPTO 2009. LNCS, vol. 5677, pp. 654–670. Springer, Heidelberg (2009). https://doi.org/10.1007/978-3-642-03356-8_38
25. Hohenberger, S., Waters, B.: Synchronized aggregate signatures from the RSA assumption. In: Eurocrypt (This Issue) (2018). The full version appears, https://eprint.iacr.org/2018/082
26. Itkis, G., Reyzin, L.: Forward-secure signatures with optimal signing and verifying. In: Kilian, J. (ed.) CRYPTO 2001. LNCS, vol. 2139, pp. 332–354. Springer, Heidelberg (2001). https://doi.org/10.1007/3-540-44647-8_20
27. Lee, K., Lee, D.H., Yung, M.: Sequential aggregate signatures made shorter. In: Jacobson, M., Locasto, M., Mohassel, P., Safavi-Naini, R. (eds.) ACNS 2013. LNCS, vol. 7954, pp. 202–217. Springer, Heidelberg (2013). https://doi.org/10.1007/978-3-642-38980-1_13
28. Lee, K., Lee, D.H., Yung, M.: Sequential aggregate signatures with short public keys: design, analysis and implementation studies. In: Kurosawa, K., Hanaoka, G. (eds.) PKC 2013. LNCS, vol. 7778, pp. 423–442. Springer, Heidelberg (2013). https://doi.org/10.1007/978-3-642-36362-7_26
29. Lu, S., Ostrovsky, R., Sahai, A., Shacham, H., Waters, B.: Sequential aggregate signatures and multisignatures without random oracles. In: Vaudenay, S. (ed.) EUROCRYPT 2006. LNCS, vol. 4004, pp. 465–485. Springer, Heidelberg (2006). https://doi.org/10.1007/11761679_28. http://cseweb.ucsd.edu/~hovav/dist/agg-sig.pdf
30. Lysyanskaya, A., Micali, S., Reyzin, L., Shacham, H.: Sequential aggregate signatures from trapdoor permutations. In: Cachin, C., Camenisch, J.L. (eds.) EUROCRYPT 2004. LNCS, vol. 3027, pp. 74–90. Springer, Heidelberg (2004). https://doi.org/10.1007/978-3-540-24676-3_5
31. Ma, D., Tsudik, G.: Extended abstract: forward-secure sequential aggregate authentication. In: IEEE Symposium on Security and Privacy, pp. 86–91 (2007)
32. Miller, G.L.: Riemann's hypothesis and tests for primality. J. Comput. Syst. Sci. **13**, 300–317 (1976)
33. Neven, G.: Efficient sequential aggregate signed data. In: Smart, N. (ed.) EUROCRYPT 2008. LNCS, vol. 4965, pp. 52–69. Springer, Heidelberg (2008). https://doi.org/10.1007/978-3-540-78967-3_4
34. Rabin, M.O.: Probabilistic algorithm for testing primality. J. Number Theory **12**, 128–138 (1980)
35. Rivest, R.L., Shamir, A., Adleman, L.: A method for obtaining digital signatures and public-key cryptosystems. Comm. ACM **21**(2), 120–126 (1978)
36. Rückert, M., Schröder, D.: Aggregate and verifiably encrypted signatures from multilinear maps without random oracles. In: Park, J.H., Chen, H.-H., Atiquzzaman, M., Lee, C., Kim, T., Yeo, S.-S. (eds.) ISA 2009. LNCS, vol. 5576, pp. 750–759. Springer, Heidelberg (2009). https://doi.org/10.1007/978-3-642-02617-1_76

37. Sharmila Deva Selvi, S., Sree Vivek, S., Pandu Rangan, C.: Deterministic identity based signature scheme and its application for aggregate signatures. In: Susilo, W., Mu, Y., Seberry, J. (eds.) ACISP 2012. LNCS, vol. 7372, pp. 280–293. Springer, Heidelberg (2012). https://doi.org/10.1007/978-3-642-31448-3_21
38. Shoup, V.: NTL: A Library for doing Number Theory, v10.5.0 (2017). http://www.shoup.net/ntl/

More Efficient (Almost) Tightly Secure Structure-Preserving Signatures

Romain Gay[1,2(✉)], Dennis Hofheinz[3], Lisa Kohl[3], and Jiaxin Pan[3]

[1] Département d'informatique de l'ENS, École normale supérieure, CNRS, PSL Research University, Paris, France
rgay@di.ens.fr
[2] INRIA, Paris, France
[3] Karlsruhe Institute of Technology, Karlsruhe, Germany
{Dennis.Hofheinz,Lisa.Kohl,Jiaxin.Pan}@kit.edu

Abstract. We provide a structure-preserving signature (SPS) scheme with an (almost) tight security reduction to a standard assumption. Compared to the state-of-the-art tightly secure SPS scheme of Abe et al. (CRYPTO 2017), our scheme has smaller signatures and public keys (of about 56%, resp. 40% of the size of signatures and public keys in Abe et al.'s scheme), and a lower security loss (of $\mathbf{O}(\log Q)$ instead of $\mathbf{O}(\lambda)$, where λ is the security parameter, and $Q = \mathsf{poly}(\lambda)$ is the number of adversarial signature queries).

While our scheme is still less compact than structure-preserving signature schemes *without* tight security reduction, it significantly lowers the price to pay for a tight security reduction. In fact, when accounting for a non-tight security reduction with larger key (i.e., group) sizes, the computational efficiency of our scheme becomes at least comparable to that of non-tightly secure SPS schemes.

Technically, we combine and refine recent existing works on tightly secure encryption and SPS schemes. Our technical novelties include a modular treatment (that develops an SPS scheme out of a basic message authentication code), and a refined hybrid argument that enables a lower security loss of $\mathbf{O}(\log Q)$ (instead of $\mathbf{O}(\lambda)$).

Keywords: Structure-preserving signatures · Tight security

R. Gay—Supported by ERC Project aSCEND (639554), and a Google PhD fellowship.

D. Hofheinz—Supported by ERC Project PREP-CRYPTO (724307), and by DFG grants HO 4534/4-1 and HO 4534/2-2.

L. Kohl—Supported by ERC Project PREP-CRYPTO (724307), and by DFG grant HO 4534/2-2.

J. Pan—Supported by DFG grant HO 4534/4-1.

J. B. Nielsen and V. Rijmen (Eds.): EUROCRYPT 2018, LNCS 10821, pp. 230–258, 2018.
https://doi.org/10.1007/978-3-319-78375-8_8

1 Introduction

Structure-Preserving Signatures (SPSs). Informally, a cryptographic scheme (such as an encryption or signature scheme) is called structure-preserving if its operation can be expressed using equations over a (usually pairing-friendly) cyclic group. A structure-preserving scheme has the advantage that we can reason about it with efficient zero-knowledge proof systems such as the Groth-Sahai non-interactive zero-knowledge (NIZK) system [31]. This compatibility is the key to constructing efficient anonymous credential systems (e.g., [10]), and can be extremely useful in voting schemes and mix-nets (e.g., [30]).

In this work, we are concerned with structure-preserving signature (SPS) schemes. Since popular tools such as "structure-breaking" collision-resistant hash functions cannot be used in a structure-preserving scheme, constructing an SPS scheme is a particularly challenging task. Still, there already exist a variety of SPS schemes in the literature [2,4–6,17–19,29,35,37,39,44] (see also Table 1 for details on some of them).

Tight Security for SPS Schemes. A little more specifically, in this work we are interested in *tightly secure* SPS schemes. Informally, a cryptographic scheme is tightly secure if it enjoys a tight security reduction, i.e., a security reduction that transforms any adversary $\mathcal{A}$ on the scheme into a problem-solver with about the same runtime and success probability as $\mathcal{A}$, *independently* of the number of uses of the scheme.[1] A tight security reduction gives security guarantees that do not degrade in the size of the setting in which the scheme is used.

Specifically, tight security reductions allow to give "universal" keylength recommendations that do not depend on the envisioned size of an application. This is useful when deploying an application for which the eventual number of uses cannot be reasonably bounded a priori. Moreover, this point is particularly vital for SPS schemes. Namely, an SPS scheme is usually combined with several other components that all use the same cyclic group. Thus, a keylength increase (which implies changing the group, and which might be necessary for a non-tightly secure scheme for which a secure keylength depends on the number of uses) affects several schemes, and is particularly costly.

In recent years, progress has been made in the construction of a variety of tightly[2] secure cryptographic schemes such as public-key encryption schemes [11, 25,33–35,42,43], identity-based encryption schemes [8,14,20,21,27,36], and signature schemes [3,6,14,16,21,34,35,42]. However, somewhat surprisingly, only few SPS schemes with tight security reductions are known. Moreover, these

[1] We are only interested in reductions to well-established and plausible computational problems here. While the security of any scheme can be trivially (and tightly) reduced to the security of that same scheme, such a trivial reduction is of course not very useful.

[2] Most of the schemes in the literature are only "almost" tightly secure, meaning that their security reduction suffers from a small multiplicative loss (that however is independent of the number of uses of the scheme). In the following, we will not make this distinction, although we will of course be precise in the description and comparison of the reduction loss of our own scheme.

tightly secure SPS schemes [6,35] are significantly less efficient than either "ordinary" SPS or tightly secure signature schemes (see Table 1). One reason for this apparent difficulty to construct tightly secure SPS schemes is that tight security appears to require dedicated design techniques (such as a sophisticated hybrid argument over the bits of an IBE identity [21]), and most known such techniques cannot be expressed in a structure-preserving manner.

Table 1. Comparison of standard-model SPS schemes (in their most efficient variants). We list unilateral schemes (with messages over one group) and bilateral schemes (with messages over both source groups of a pairing) separately. The notation (x_1, x_2) denotes x_1 elements in $\mathbb{G}_1$ and x_2 elements in $\mathbb{G}_2$. $|M|$, $|\sigma|$, and $|pk|$ denote the size of messages, signatures, and public keys (measured in group elements). "Sec. loss" denotes the multiplicative factor that the security reduction to "Assumption" loses, where we omit dominated and additive factors. (Here, "generic" means that only a proof in the generic group model is known.) For the tree-based scheme HJ12, ℓ denotes the depth of the tree (which limits the number of signing queries to 2^ℓ). Q denotes the number of adversarial signing queries, and λ is the security parameter.

Scheme	$\lvert M\rvert$	$\lvert\sigma\rvert$	$\lvert pk\rvert$	Sec. loss	Assumption
HJ12 [35]	1	$10\ell + 6$	13	8	DLIN
ACDKNO16 [2]	$(n_1, 0)$	$(7, 4)$	$(5, n_1 + 12)$	Q	SXDH, XDLIN
LPY15 [44]	$(n_1, 0)$	$(10, 1)$	$(16, 2n_1 + 5)$	$\mathbf{O}(Q)$	SXDH, XDLINX
KPW15 [39]	$(n_1, 0)$	$(6, 1)$	$(0, n_1 + 6)$	$2Q^2$	SXDH
JR17 [37]	$(n_1, 0)$	$(5, 1)$	$(0, n_1 + 6)$	$Q \log Q$	SXDH
AHNOP17 [6]	$(n_1, 0)$	$(13, 12)$	$(18, n_1 + 11)$	80λ	SXDH
Ours (unilateral)	$(n_1, 0)$	$(8, 6)$	$(2, n_1 + 9)$	$6 \log Q$	SXDH
AGHO11 [5]	(n_1, n_2)	$(2, 1)$	$(n_1, n_2 + 2)$	—	Generic
ACDKNO16 [2]	(n_1, n_2)	$(8, 6)$	$(n_2 + 6, n_1 + 13)$	Q	SXDH, XDLIN
KPW15 [39]	(n_1, n_2)	$(7, 3)$	$(n_2 + 1, n_1 + 7)$	$2Q^2$	SXDH
AHNOP17 [6]	(n_1, n_2)	$(14, 14)$	$(n_2 + 19, n_1 + 12)$	80λ	SXDH
Ours (bilateral)	(n_1, n_2)	$(9, 8)$	$(n_2 + 4, n_1 + 9)$	$6 \log Q$	SXDH

1.1 Our Contribution

Overview. We present a tightly secure SPS scheme with significantly improved efficiency and tighter security reduction compared to the state-of-the-art tightly secure SPS scheme of Abe et al. [6]. Specifically, our signatures contain 14 group elements (compared to 25 group elements in [6]), and our security reduction loses a factor of only $\mathbf{O}(\log Q)$ (compared to $\mathbf{O}(\lambda)$), where λ denotes the security parameter, and $Q = \mathsf{poly}(\lambda)$ denotes the number of adversarial signature queries. When accounting for loose reductions through an appropriate keylength increase, the computational efficiency of our scheme even compares favorably to that of state-of-the-art non-tightly secure SPS schemes.

In the following, we will detail how we achieve our results, and in particular the progress we make upon previous techniques. We will also compare our work to existing SPS schemes (both tightly and non-tightly secure).

Central Idea: A Modular Treatment. A central idea in our work (that in particular contrasts our approach to the one of Abe et al.) is a *modular* construction. That is, similar to the approach to tight IBE security of Blazy, Kiltz, and Pan [14], the basis of our construction is a tightly secure message authentication code (MAC). This tightly secure MAC will then be converted into a signature scheme by using NIZK proofs, following (but suitably adapting) the generic MAC-to-signatures conversion of Bellare and Goldwasser [12].

Starting Point: A Tightly Secure MAC. Our tightly secure MAC will have to be structure-preserving, so the MAC used in [14] cannot be employed in our case. Instead, we derive our MAC from the recent tightly secure key encapsulation mechanism (KEM) of Gay, Hofheinz, and Kohl [26] (which in turn builds upon the Kurosawa-Desmedt PKE scheme [41]). To describe their scheme, we assume a group $\mathbb{G} = \langle g \rangle$ of prime order p, and we use the implicit notation $[x] := g^x$ from [24]. We also fix an integer k that determines the computational assumption to which we want to reduce.[3] Now in (a slight simplification of) the scheme of [26], a ciphertext C with corresponding KEM key K is of the form

$$C = ([\mathbf{t}], \pi), \qquad K = [(\mathbf{k}_0 + \mu\mathbf{k}_1)^\top \mathbf{t}] \quad (\text{for } \mu = H([\mathbf{t}])), \tag{1}$$

where H is a collision-resistant hash function, and $\mathbf{k}_0, \mathbf{k}_1, \mathbf{t} \in \mathbb{Z}_p^{2k}$ and π are defined as follows. First, $\mathbf{k}_0, \mathbf{k}_1 \in \mathbb{Z}_p^{2k}$ comprise the secret key. Next, $\mathbf{t} = \mathbf{A}_0\mathbf{r}$ for a fixed matrix $\mathbf{A}_0$ (given as $[\mathbf{A}_0]$ in the public key) and a random vector $\mathbf{r} \in \mathbb{Z}_p^k$ chosen freshly for each encryption. Finally, π is a NIZK proof that proves that $\mathbf{t} \in \text{span}(\mathbf{A}_0) \cup \text{span}(\mathbf{A}_1)$ for another fixed matrix $\mathbf{A}_1$ (also given as $[\mathbf{A}_1]$ in the public key). The original Kurosawa-Desmedt scheme [41] is identical, except that π is omitted, and $k = 1$. Hence, the main benefit of π is that it enables a tight security reduction.[4]

We can view this KEM as a MAC scheme simply by declaring the MAC tag for a message M to be the values (C, K) from (1), only with $\mu := M$ (instead of $\mu = H([\mathbf{t}])$). The verification procedure of the resulting MAC will check π, and then check whether C really decrypts to K. (Hence, MAC verification still requires the secret key $\mathbf{k}_0, \mathbf{k}_1$.) Now a slight adaptation of a generic argument of Dodis et al. [22] reduces the security of this MAC tightly to the security of the underlying KEM scheme. Unfortunately, this resulting MAC is not structure-preserving yet (even if the used NIZK proof π is): the message $M = \mu$ is a scalar (from $\mathbb{Z}_p$).[5]

[3] For $k = 1$, we can reduce to DDH in $\mathbb{G}$, and for $k > 1$, we can reduce to the k-Linear assumption, and in fact even to the weaker Matrix-DDH assumption [24].

[4] Actually, the scheme of [26] uses an efficient designated-verifier NIZK proof π that is however not structure-preserving (and thus not useful for our case), and also induces an additional term in K. For our purposes, we can think of π as a (structure-preserving) Groth-Sahai proof.

[5] A structure-preserving scheme should have group elements (and not scalars) as messages, since Groth-Sahai proofs cannot (easily) be used to prove knowledge of scalars.

Abstracting Our Strategy into a Single "core lemma". We can distill the essence of the security proof of our MAC above into a single "core lemma". This core lemma forms the heart of our work, and shows how to randomize all tags of our MAC. While this randomization follows a previous paradigm called "adaptive partitioning" (used to prove the tight security of PKE [26,33] and SPS schemes [6]), our core lemma induces a much smaller reduction loss. The reason for this smaller reduction loss is that previous works on tightly secure schemes (including [6,26,33]) conduct their reduction along the individual bits of a certain hash value (or message to be signed). Since this hash value (or message) usually has $\mathbf{O}(\lambda)$ bits, this induces a hybrid argument of $\mathbf{O}(\lambda)$ steps, and thus a reduction loss of $\mathbf{O}(\lambda)$. In contrast, we conduct our security argument along the individual bits of the *index* of a signing query (i.e., a value from 1 to Q, where Q is the number of signing queries). This index exists only in the security proof, and can thus be considered as an "implicit" way to structure our reduction.[6]

From MACs to Signatures and Structure-Preserving Signatures. Fortunately, our core lemma can be used to prove not only our MAC scheme, but also a suitable signature and SPS scheme tightly secure. To construct a signature scheme, we can now use an case-tailored (and heavily optimized) version of the generic transformation of Bellare and Goldwasser [12]. In a nutshell, that transformation turns a MAC tag (that requires a secret key to verify) into a publicly verifiable signature by adding a NIZK proof to the tag that proves its validity, relative to a public commitment to the secret key. For our MAC, we only need to prove that the given key K really is of the form $K = [(\mathbf{k}_0 + \mu\mathbf{k}_1)^\top \mathbf{t}]$. This linear statement can be proven with a comparatively simple and efficient NIZK proof π'. For $k = 1$, an optimized Groth-Sahai-based implementation of π, and an implicit π' (that uses ideas from [38,40]), the resulting signature scheme will have signatures that contain 14 group elements.

To turn our scheme into an SPS scheme, we need to reconsider the equation $K = [(\mathbf{k}_0 + \mu\mathbf{k}_1)^\top \mathbf{t}]$ from (1). In our MAC (and also in the signature scheme above), we have set $\mu = M \in \mathbb{Z}_p$, which we cannot afford to do for an SPS scheme. Our solution consists in choosing a different equation that fulfills the following requirements:

(a) it is algebraic (in the sense that it integrates a message $M \in \mathbb{G}$), and
(b) it is compatible with our core lemma (so it can be randomized quickly).

For our scheme, we start from the equation

$$K = [\mathbf{k}_0^\top \mathbf{t} + \mathbf{k}^\top \begin{pmatrix} M \\ 1 \end{pmatrix}] \tag{2}$$

for uniform keys $\mathbf{k}_0, \mathbf{k}$. We note that a similar equation has already been used by Kiltz, Pan, and Wee [39] for constructing SPS schemes, although with a very

[6] A reduction loss of $\mathbf{O}(\log Q)$ has been achieved in the context of IBE schemes [20], but their techniques are different and rely on a composite-order group.

different and non-tight security proof. We can plug this equation into the MAC-to-signature transformation sketched above, to obtain an SPS scheme with only 14 group elements (for $k = 1$) per signature.

Our security proof will directly rely on our core lemma to first randomize the $\mathbf{k}_0^\top \mathbf{t}$ part of (2) in all signatures. After that, similar to [39], an information-theoretic argument (that only uses the pairwise independence of the second part of (2), when viewed as a function of M) shows security.

Our basic SPS scheme is unilateral, i.e., its messages are vectors over only one source group of a given pairing. To obtain a bilateral scheme that accepts "mixed" messages over both source groups of an asymmetric pairing, we can use a generic transformation of [39] that yields a bilateral scheme with signatures of 17 group elements (for $k = 1$).

1.2 Related Work and Efficiency Comparison

In this subsection, we compare our work to the closest existing work (namely, the tightly secure SPS scheme of Abe et al. [6]) and other, non-tightly secure SPS schemes.

Comparison to the Work of Abe et al. The state of the art in tightly secure SPS schemes (and in fact currently the only other efficient tightly secure SPS scheme) is the recent work of Abe et al. [6]. Technically, their scheme also uses a tightly secure PKE scheme (in that case [33]) as an inspiration. However, there are also a number of differences in our approaches which explain our improved efficiency and reduction.

Table 2. Comparison of the computational efficiency of state-of-the-art SPS schemes (in their most efficient, SXDH-based variants) with our SXDH-based schemes in the unilateral (UL) and bilateral (BL) version. With "PPEs" and "Pairings", we denote the number of those operations necessary during verification, where "batched" denotes optimized figures obtained by "batching" verification equations [13]. The "$|M|$" and "Sec. loss" columns have the same meaning as in Table 1. The column "$|\mathbb{G}_1|$" denotes the (bit)size of elements from the first source group in a large but realistic scenario (under some simplifying assumptions), see the discussion in Sect. 1.2. "$|\sigma|$ (bits)" denotes the resulting overall signature size, where we assume that the bitsize of $\mathbb{G}_2$ elements is twice the bitsize of $\mathbb{G}_1$-elements.

Scheme	$\|M\|$	PPEs	Pairings (plain)	Pairings (batched)	Sec. loss	$\|\mathbb{G}_1\|$ (bits)	$\|\sigma\|$ (bits)
KPW [39]	$(n_1, 0)$	3	$n_1 + 11$	$n_1 + 10$	$2Q^2$	322	2576
JR [37]	$(n_1, 0)$	2	$n_1 + 8$	$n_1 + 6$	$Q \log Q$	270	1890
AHNOP [6]	$(n_1, 0)$	15	$n_1 + 57$	$n_1 + 16$	80λ	226	8362
Ours (UL)	$(n_1, 0)$	6	$n_1 + 29$	$n_1 + 11$	$6 \log Q$	216	4320
KPW [39]	(n_1, n_2)	4	$n_1{+}n_2{+}15$	$n_1{+}n_2{+}14$	$2Q^2$	322	4186
AHNOP [6]	(n_1, n_2)	16	$n_1{+}n_2{+}61$	$n_1{+}n_2{+}18$	80λ	226	9492
Ours (BL)	(n_1, n_2)	7	$n_1{+}n_2{+}33$	$n_1{+}n_2{+}15$	$6 \log Q$	216	5400

First, Abe et al.'s scheme involves more (and more complex) NIZK proofs, since they rather closely follow the PKE scheme from [33]. This leads to larger proofs and thus larger signatures. Instead, our starting point is the much simpler scheme of [26] (which only features one comparatively simple NIZK proof in its ciphertext).

Second, while the construction of Abe et al. is rather monolithic, our construction can be explained as a modification of a simple MAC scheme. Our approach thus allows for a more modular exposition, and in particular we can outsource the core of the reduction into a core lemma (as explained above) that can be applied to MAC, signature, and SPS scheme.

Third, like previous tightly secure schemes (and in particular the PKE schemes of [26,33]), Abe et al. conduct their security reduction along the individual bits of a certain hash value (or message to be signed). As explained above, our reduction is more economic, and uses a hybrid argument over an "implicit" counter value.

Efficiency Comparison. We give a comparison to other SPS schemes in Table 1. This table shows that our scheme is still significantly less efficient *in terms of signature size* than existing, non-tightly secure SPS schemes. However, when considering *computational efficiency*, and when accounting for a larger security loss in the reduction with larger groups, things look differently.

The currently most efficient non-tightly secure SPS schemes are due to Jutla and Roy [37] and Kiltz, Pan, and Wee [39]. Table 2 compares the computational complexity of their verification operation with the tightly secure SPSs of Abe et al. and our schemes. Now consider a large scenario with $Q = 2^{30}$ signing queries and a target security parameter of $\lambda = 100$. Assume further that we use groups that only allow generic attacks (that require time about the square root of the group size). This means that we should run a scheme in a group of size at least $2^{2(\lambda+\log L)}$, where L denotes the multiplicative loss of the respective security reduction. Table 2 shows the resulting group sizes in column "$|\mathbb{G}_1|$" (in bits, such that $|\mathbb{G}_1| = 200$ denotes a group of size 2^{200}).

Now very roughly, the computational complexity of pairings can be assumed to be cubic in the (bit)size of the group [7,9,23,28]. Hence, in the unilateral setting, and assuming an optimized verification implementation (that uses "batching" [13]) the computational efficiency of the verification in our scheme is roughly on par with that in the (non-tightly secure) state-of-the-art scheme of Jutla and Roy [37], even for small messages. For larger messages, our scheme becomes preferable. In the bilateral setting, our scheme is clearly the most efficient known scheme.

Roadmap

We fix some notation and recall some preliminaries in Sect. 2. In Sect. 3, we present our basic MAC and prove it secure (using the mentioned core lemma). In Sects. 4 and 5, we present our signature and SPS schemes. Due to lack of space, for some proofs (including the more technical parts of the proof of the core lemma, and a full proof for the signature scheme) we refer to the full version.

2 Preliminaries

In this section we provide the preliminaries which our paper builds upon. First, we want to give an overview of notation used throughout all sections.

2.1 Notation

By $\lambda \in \mathbb{N}$ we denote the security parameter. We always employ $\mathsf{negl}: \mathbb{N} \to \mathbb{R}_{\geq 0}$ to denote a negligible function, that is for all polynomials $p \in \mathbb{N}[X]$ there exists an $n_0 \in \mathbb{N}$ such that $\mathsf{negl}(n) < 1/p(n)$ for all $n \geq n_0$. For any set $\mathcal{S}$, by $s \leftarrow_R \mathcal{S}$ we set s to be a uniformly at random sampled element from $\mathcal{S}$. For any distribution $\mathcal{D}$ by $d \leftarrow \mathcal{D}$ we denote the process of sampling an element d according to the distribution $\mathcal{D}$. For any probabilistic algorithm $\mathcal{B}$ by out $\leftarrow \mathcal{B}(\text{in})$ by out we denote the output of $\mathcal{B}$ on input in. For a deterministic algorithm we sometimes use the notation out $:= \mathcal{B}(\text{in})$ instead. By p we denote a prime throughout the paper. For any element $m \in \mathbb{Z}_p$, we denote by $m_i \in \{0,1\}$ the i-th bit of m's bit representation and by $m_{|i} \in \{0,1\}^i$ the bit string comprising the first i bits of m's bit representation.

It is left to introduce some notation regarding matrices. To this end let $k, \ell \in \mathbb{N}$ such that $\ell > k$. For any matrix $\mathbf{A} \in \mathbb{Z}_p^{\ell \times k}$, we write

$$\mathsf{span}(\mathbf{A}) := \{\mathbf{A}\mathbf{r} \mid \mathbf{r} \in \mathbb{Z}_p^k\} \subset \mathbb{Z}_p^\ell,$$

to denote the *span* of $\mathbf{A}$.

For a full rank matrix $\mathbf{A} \in \mathbb{Z}_p^{\ell \times k}$ we denote by $\mathbf{A}^\perp$ a matrix in $\mathbb{Z}_p^{\ell \times (\ell - k)}$ with $\mathbf{A}^\top \mathbf{A}^\perp = \mathbf{0}$ and rank $\ell - k$. We denote the set of all matrices with these properties as

$$\mathsf{orth}(\mathbf{A}) := \{\mathbf{A}^\perp \in \mathbb{Z}_p^{\ell \times (\ell - k)} \mid \mathbf{A}^\top \mathbf{A}^\perp = \mathbf{0} \text{ and } \mathbf{A}^\perp \text{ has rank } \ell - k\}.$$

For vectors $\mathbf{v} \in \mathbb{Z}_p^{k+n}$ ($n \in \mathbb{N}$), by $\overline{\mathbf{v}} \in \mathbb{Z}_p^k$ we denote the vector consisting of the upper k entries of $\mathbf{v}$ and accordingly by $\underline{\mathbf{v}} \in \mathbb{Z}_p^n$ we denote the vector consisting of the remaining n entries of $\mathbf{v}$.

Similarly, for a matrix $\mathbf{A} \in \mathbb{Z}_p^{2k \times k}$, by $\overline{\mathbf{A}} \in \mathbb{Z}_p^{k \times k}$ we denote the upper square matrix and by $\underline{\mathbf{A}} \in \mathbb{Z}_p^{k \times k}$ the lower one.

2.2 Pairing Groups and Matrix Diffie-Hellman Assumptions

Let GGen be a probabilistic polynomial time (PPT) algorithm that on input 1^λ returns a description $\mathcal{PG} = (\mathbb{G}_1, \mathbb{G}_2, G_T, p, P_1, P_2, e)$ of asymmetric pairing groups where $\mathbb{G}_1$, $\mathbb{G}_2$, $\mathbb{G}_T$ are cyclic group of order p for a 2λ-bit prime p, P_1 and P_2 are generators of $\mathbb{G}_1$ and $\mathbb{G}_2$, respectively, and $e : \mathbb{G}_1 \times \mathbb{G}_2 \to \mathbb{G}_T$ is an efficiently computable (non-degenerate) bilinear map. Define $P_T := e(P_1, P_2)$, which is a generator of $\mathbb{G}_T$. We use implicit representation of group elements. For $i \in \{1, 2, T\}$ and $a \in \mathbb{Z}_p$, we define $[a]_i = aP_i \in \mathbb{G}_i$ as the implicit representation of a in $\mathbb{G}_i$. Given $[a]_1$, $[a]_2$, one can efficiently compute $[ab]_T$ using the pairing e.

For two matrices $\mathbf{A}$, $\mathbf{B}$ with matching dimensions, we define $e([\mathbf{A}]_1, [\mathbf{B}]_2) := [\mathbf{AB}]_T \in \mathbb{G}_T$.

We recall the definitions of the Matrix Decision Diffie-Hellman (MDDH) assumption from [24].

Definition 1 (Matrix distribution). *Let $k, \ell \in \mathbb{N}$, with $\ell > k$ and p be a 2λ-bit prime. We call a PPT algorithm $\mathcal{D}_{\ell,k}$* a matrix distribution *if it outputs matrices in $\mathbb{Z}_p^{\ell \times k}$ of full rank k.*

Note that instantiating $\mathcal{D}_{2,1}$ with a PPT algorithm outputting matrices $\begin{pmatrix} 1 \\ a \end{pmatrix}$ for $a \leftarrow_R \mathbb{Z}_p$, $\mathcal{D}_{2,1}$-MDDH relative to $\mathbb{G}_1$ corresponds to the DDH assumption in $\mathbb{G}_1$. Thus, for $\mathcal{PG} = (\mathbb{G}_1, \mathbb{G}_2, G_T, p, P_1, P_2, e)$, assuming $\mathcal{D}_{2,1}$-MDDH relative to $\mathbb{G}_1$ and relative to $\mathbb{G}_2$, corresponds to the SXDH assumption.

In the following we only consider matrix distributions $\mathcal{D}_{\ell,k}$, where for all $\mathbf{A} \leftarrow_R \mathcal{D}_{\ell,k}$ the first k rows of $\mathbf{A}$ form an invertible matrix. We also require that in case $\ell = 2k$ for any two matrices $\mathbf{A}_0, \mathbf{A}_1 \leftarrow_R \mathcal{D}_{2k,k}$ the matrix $(\mathbf{A}_0 \mid \mathbf{A}_1)$ has full rank with overwhelming probability. In the following we will denote this probability by $1 - \Delta_{\mathcal{D}_{2k,k}}$. Note that if $(\mathbf{A}_0 \mid \mathbf{A}_1)$ has full rank, then for any $\mathbf{A}_0^\perp \in \mathsf{orth}(\mathbf{A}_0)$, $\mathbf{A}_1^\perp \in \mathsf{orth}(\mathbf{A}_1)$ the matrix $(\mathbf{A}_0^\perp \mid \mathbf{A}_1^\perp) \in \mathbb{Z}_p^{2k \times 2k}$ has full rank as well, as otherwise there would exists a non-zero vector $\mathbf{v} \in \mathbb{Z}_p^{2k} \backslash \{\mathbf{0}\}$ with $(\mathbf{A}_0 \mid \mathbf{A}_1)^\top \mathbf{v} = \mathbf{0}$. Further, by similar reasoning $(\mathbf{A}_0^\perp)^\top \mathbf{A}_1 \in \mathbb{Z}_p^{k \times k}$ has full rank.

The $\mathcal{D}_{\ell,k}$-Matrix Diffie-Hellman problem in $\mathbb{G}_i$, for $i \in \{1, 2, T\}$, is to distinguish the between tuples of the form $([\mathbf{A}]_i, [\mathbf{Aw}]_i)$ and $([\mathbf{A}]_i, [\mathbf{u}]_i)$, for a randomly chosen $\mathbf{A} \leftarrow_R \mathcal{D}_{\ell,k}$, $\mathbf{w} \leftarrow_R \mathbb{Z}_p^k$ and $\mathbf{u} \leftarrow_R \mathbb{Z}_p^\ell$.

Definition 2 ($\mathcal{D}_{\ell,k}$-Matrix Diffie-Hellman $\mathcal{D}_{\ell,k}$-MDDH). *Let $\mathcal{D}_{\ell,k}$ be a matrix distribution. We say that the $\mathcal{D}_{\ell,k}$-*Matrix Diffie-Hellman *($\mathcal{D}_{\ell,k}$-*MDDH*)* assumption *holds relative to a prime order group $\mathbb{G}_i$ for $i \in \{1, 2, T\}$, if for all PPT adversaries $\mathcal{A}$,*

$$\begin{aligned} \mathrm{Adv}^{\mathrm{mddh}}_{\mathcal{PG},\mathbb{G}_i,\mathcal{D}_{\ell,k},\mathcal{A}}(\lambda) := |\Pr[&\mathcal{A}(\mathcal{PG}, [\mathbf{A}]_i, [\mathbf{Aw}]_i) = 1] \\ &- \Pr[\mathcal{A}(\mathcal{PG}, [\mathbf{A}]_i, [\mathbf{u}]_i) = 1]| \leq \mathsf{negl}(\lambda), \end{aligned}$$

where the probabilities are taken over $\mathcal{PG} := (\mathbb{G}_1, \mathbb{G}_2, \mathbb{G}_T, p, P_1, P_2) \leftarrow \mathsf{GGen}(1^\lambda)$, $\mathbf{A} \leftarrow_R \mathcal{D}_{\ell,k}, \mathbf{w} \leftarrow_R \mathbb{Z}_p^k, \mathbf{u} \leftarrow_R \mathbb{Z}_p^\ell$.

For $Q \in \mathbb{N}$, $\mathbf{W} \leftarrow_R \mathbb{Z}_p^{k \times Q}$ and $\mathbf{U} \leftarrow_R \mathbb{Z}_p^{\ell \times Q}$, we consider the Q-fold $\mathcal{D}_{\ell,k}$-MDDH assumption, which states that distinguishing tuples of the form $([\mathbf{A}]_i, [\mathbf{AW}]_i)$ from $([\mathbf{A}]_i, [\mathbf{U}]_i)$ is hard. That is, a challenge for the Q-fold $\mathcal{D}_{\ell,k}$-MDDH assumption consists of Q independent challenges of the $\mathcal{D}_{\ell,k}$-MDDH assumption (with the same $\mathbf{A}$ but different randomness $\mathbf{w}$). In [24] it is shown that the two problems are equivalent, where the reduction loses at most a factor $\ell - k$.

Lemma 1 (Random self-reducibility of $\mathcal{D}_{\ell,k}$-MDDH, [24]). *Let $\ell, k, Q \in \mathbb{N}$ with $\ell > k$ and $Q > \ell - k$ and $i \in \{1, 2, T\}$. For any PPT adversary $\mathcal{A}$,* *there exists an adversary* $\mathcal{B}$ *such that* $T(\mathcal{B}) \approx T(\mathcal{A}) + Q \cdot \mathsf{poly}(\lambda)$ *with* $\mathsf{poly}(\lambda)$ *independent of* $T(\mathcal{A})$, *and*

$$\mathrm{Adv}^{Q\text{-mddh}}_{\mathcal{PG},\mathbb{G}_i,\mathcal{D}_{\ell,k},\mathcal{A}}(\lambda) \le (\ell - k) \cdot \mathrm{Adv}^{\mathrm{mddh}}_{\mathcal{PG},\mathbb{G}_i,\mathcal{D}_{\ell,k},\mathcal{B}}(\lambda) + \tfrac{1}{p-1}.$$

Here

$$\begin{aligned}\mathrm{Adv}^{Q\text{-mddh}}_{\mathcal{PG},\mathbb{G}_i,\mathcal{D}_{\ell,k},\mathcal{A}}(\lambda) := |&\Pr[\mathcal{A}(\mathcal{PG},[\mathbf{A}]_i,[\mathbf{AW}]_i) = 1] \\ &- \Pr[\mathcal{A}(\mathcal{PG},[\mathbf{A}]_i,[\mathbf{U}]_i) = 1]|,\end{aligned}$$

where the probability is over $\mathcal{PG} := (\mathbb{G}_1, \mathbb{G}_2, \mathbb{G}_T, p, P_1, P_2) \leftarrow \mathsf{GGen}(1^\lambda)$, $\mathbf{A} \leftarrow_R \mathcal{D}_{\ell,k}$, $\mathbf{W} \leftarrow_R \mathbb{Z}_p^{k \times Q}$ *and* $\mathbf{U} \leftarrow_R \mathbb{Z}_p^{\ell \times Q}$.

For $k \in \mathbb{N}$ we define $\mathcal{D}_k := \mathcal{D}_{k+1,k}$.

The Kernel-Diffie-Hellman assumption $\mathcal{D}_k$-KMDH [45] is a natural *computational analogue* of the $\mathcal{D}_k$-MDDH Assumption.

Definition 3 ($\mathcal{D}_k$-Kernel Diffie-Hellman assumption $\mathcal{D}_k$-KMDH). *Let* $\mathcal{D}_k$ *be a matrix distribution. We say that the* $\mathcal{D}_k$*-Kernel Diffie-Hellman (*$\mathcal{D}_k$*-KMDH) assumption holds relative to a prime order group* $\mathbb{G}_i$ *for* $i \in \{1, 2\}$ *if for all PPT adversaries* $\mathcal{A}$,

$$\begin{aligned}\mathrm{Adv}^{\mathrm{kmdh}}_{\mathcal{PG},\mathbb{G}_i,\mathcal{D}_{\ell,k},\mathcal{A}}(\lambda) &:= \Pr[\mathbf{c}^\top \mathbf{A} = \mathbf{0} \wedge \mathbf{c} \ne \mathbf{0} \mid [\mathbf{c}]_{3-i} \leftarrow_R \mathcal{A}(\mathcal{PG},[\mathbf{A}]_i)] \\ &\le \mathsf{negl}(\lambda),\end{aligned}$$

where the probabilities are taken over $\mathcal{PG} := (\mathbb{G}_1, \mathbb{G}_2, \mathbb{G}_T, p, P_1, P_2) \leftarrow \mathsf{GGen}(1^\lambda)$, *and* $\mathbf{A} \leftarrow_R \mathcal{D}_k$.

Note that we can use a non-zero vector in the kernel of $\mathbf{A}$ to test membership in the column space of $\mathbf{A}$. This means that the $\mathcal{D}_k$-KMDH assumption is a relaxation of the $\mathcal{D}_k$-MDDH assumption, as captured in the following lemma from [45].

Lemma 2 ([45]). *For any matrix distribution* $\mathcal{D}_k$, $\mathcal{D}_k$*-MDDH* $\Rightarrow$ $\mathcal{D}_k$*-KMDH.*

2.3 Signature Schems and Message Authentication Codes

Definition 4 (MAC). *A* message authentication code (MAC) *is a tuple of PPT algorithms* $\mathsf{MAC} := (\mathsf{Gen}, \mathsf{Tag}, \mathsf{Ver})$ *such that:*

$\mathsf{Gen}(1^\lambda)$: *on input of the security parameter, generates public parameters pp and a secret key sk.*

$\mathsf{Tag}(pp, sk, m)$: *on input of public parameters pp, the secret key sk and a message* $m \in \mathcal{M}$, *returns a tag* tag.

$\mathsf{Ver}(pp, sk, m, \mathsf{tag})$: *verifies the tag* tag *for the message* m, *outputting a bit* $b = 1$ *if* tag *is valid respective to* m, *and* 0 *otherwise.*

We say MAC *is* **perfectly correct**, *if for all* $\lambda \in \mathbb{N}$, *all* $m \in \mathcal{M}$ *and all* $(pp, sk) \leftarrow \mathsf{Gen}(1^\lambda)$ *we have*

$$\mathsf{Ver}(pp, sk, m, \mathsf{Tag}(pp, sk, m)) = 1.$$

Definition 5 (UF-CMA **security**). *Let* $\mathsf{MAC} := (\mathsf{Gen}, \mathsf{Tag}, \mathsf{Ver})$ *be a MAC. For any adversary* $\mathcal{A}$, *we define the following experiment:*

$\mathrm{Exp}^{\text{uf-cma}}_{\mathsf{MAC},\mathcal{A}}(\lambda)$:	$\mathrm{TagO}(m)$:
$(pp, sk) \leftarrow \mathsf{Gen}(1^\lambda)$	$\mathcal{Q}_{\mathsf{tag}} := \mathcal{Q}_{\mathsf{tag}} \cup \{m\}$
$\mathcal{Q}_{\mathsf{tag}} := \emptyset$	$\mathsf{tag} \leftarrow \mathsf{Tag}(pp, sk, m)$
$(m^\star, \mathsf{tag}^\star) \leftarrow \mathcal{A}^{\mathrm{TagO}(\cdot)}(pp)$	***return*** tag
if $m^\star \notin \mathcal{Q}_{\mathsf{tag}}$ ***and*** $\mathrm{VerO}(m^\star, \mathsf{tag}^\star) = 1$	
return 1	$\mathrm{VerO}(m, \mathsf{tag})$:
else return 0	$b \leftarrow \mathsf{Ver}(pp, sk, m, \mathsf{tag})$
	return b

The adversary is restricted to one call to VerO. *We say that a MAC scheme* MAC *is* UF-CMA secure, *if for all PPT adversaries* $\mathcal{A}$,

$$\mathrm{Adv}^{\text{uf-cma}}_{\mathsf{MAC},\mathcal{A}}(\lambda) := \Pr[\mathrm{Exp}^{\text{uf-cma}}_{\mathsf{MAC},\mathcal{A}}(\lambda) = 1] \leq \mathsf{negl}(\lambda).$$

Note that in our notion of UF-CMA security, the adversary gets only one forgery attempt. This is due to the fact that we employ the MAC primarily as a building block for our signature. Our notion suffices for this purpose, as an adversary can check the validity of a signature itself.

Definition 6 (Signature). *A* signature scheme *is a tuple of PPT algorithms* $\mathsf{SIG} := (\mathsf{Gen}, \mathsf{Sign}, \mathsf{Ver})$ *such that:*

$\mathsf{Gen}(1^\lambda)$: *on input of the security parameter, generates a pair* (pk, sk) *of keys.*
$\mathsf{Sign}(pk, sk, m)$: *on input of the public key* pk, *the secret key* sk *and a message* $m \in \mathcal{M}$, *returns a signature* σ.
$\mathsf{Ver}(pk, m, \sigma)$: *verifies the signature* σ *for the message* m, *outputting a bit* $b = 1$ *if* σ *is valid respective to* m, *and* 0 *otherwise.*

We say that SIG *is* **perfectly correct**, *if for all* $\lambda \in \mathbb{N}$, *all* $m \in \mathcal{M}$ *and all* $(pk, sk) \leftarrow \mathsf{Gen}(1^\lambda)$,

$$\mathsf{Ver}(pk, m, \mathsf{Sign}(pk, sk, m)) = 1.$$

In bilinear pairing groups, we say a signature scheme SIG *is* structure-preserving *if its public keys, signing messages, signatures contain only group elements and verification proceeds via only a set of pairing product equations.*

Definition 7 (UF-CMA **security**). *For a signature scheme* $\mathsf{SIG} := (\mathsf{Gen}, \mathsf{Sign}, \mathsf{Ver})$ *and any adversary* $\mathcal{A}$, *we define the following experiment:*

$\text{Exp}^{\text{uf-cma}}_{\mathsf{SIG},\mathcal{A}}(\lambda)$:	$\text{SIGNO}(m)$:
$(pk, sk) \leftarrow \mathsf{Gen}(1^\lambda)$ $\mathcal{Q}_{\mathsf{sign}} := \emptyset$ $(m^\star, \sigma^\star) \leftarrow \mathcal{A}^{\text{SIGNO}(\cdot)}(pk)$ ***if*** $m^\star \notin \mathcal{Q}_{\mathsf{sign}}$ ***and*** $\mathsf{Ver}(pk, m^\star, \sigma^\star) = 1$ ***return*** 1 ***else return*** 0	$\mathcal{Q}_{\mathsf{sign}} := \mathcal{Q}_{\mathsf{sign}} \cup \{m\}$ $\sigma \leftarrow \mathsf{Sign}(pk, sk, m)$ ***return*** σ

We say that a signature scheme SIG *is* UF-CMA, *if for all PPT adversaries* $\mathcal{A}$,

$$\text{Adv}^{\text{uf-cma}}_{\mathsf{SIG},\mathcal{A}}(\lambda) := \Pr[\text{Exp}^{\text{uf-cma}}_{\mathsf{SIG},\mathcal{A}}(\lambda) = 1] \le \mathsf{negl}(\lambda).$$

2.4 Non-interactive Zero-Knowledge Proof (NIZK)

The notion of a non-interactive zero-knowledge proof was introduced in [15]. In the following we present the definition from [32]. Non-interactive zero-knowledge proofs will serve as a crucial building block for our constructions.

Definition 8 (Non-interactive zero-knowledge proof [32]). *We consider a family of languages* $\mathcal{L} = \{\mathcal{L}_{pars}\}$ *with efficiently computable witness relation* $\mathcal{R}_{\mathcal{L}}$. *A non-interactive zero-knowledge proof for* $\mathcal{L}$ *is a tuple of PPT algorithms* $\mathsf{PS} := (\mathsf{PGen}, \mathsf{PTGen}, \mathsf{PPrv}, \mathsf{PVer}, \mathsf{PSim})$ *such that:*

$\mathsf{PGen}(1^\lambda, pars)$ *generates a common reference string crs.*
$\mathsf{PTGen}(1^\lambda, pars)$ *generates a common reference string crs and additionally a trapdoor td.*
$\mathsf{PPrv}(crs, x, w)$ *given a word* $x \in \mathcal{L}$ *and a witness* w *with* $\mathcal{R}_{\mathcal{L}}(x, w) = 1$, *outputs a proof* $\Pi \in \mathcal{P}$.
$\mathsf{PVer}(crs, x, \Pi)$ *on input crs,* $x \in \mathcal{X}$ *and* Π *outputs a verdict* $b \in \{0, 1\}$.
$\mathsf{PSim}(crs, td, x)$ *given a crs with corresponding trapdoor td and a word* $x \in \mathcal{X}$, *outputs a proof* Π.

Further we require the following properties to hold.

Completeness: *For all possible public parameters pars, for all words* $x \in \mathcal{L}$, *and all witnesses* w *such that* $\mathcal{R}_{\mathcal{L}}(x, w) = 1$, *we have*

$$\Pr[\mathsf{PVer}(crs, x, \Pi) = 1] = 1,$$

where the probability is taken over $(crs, psk) \leftarrow \mathsf{PGen}\ (1^\lambda, pars)$ *and* $\Pi \leftarrow \mathsf{PPrv}(crs, x, w)$.

Composable zero-knowledge*: *For all PPT adversaries* $\mathcal{A}$ *we have that*

$$\begin{aligned}\text{Adv}^{\text{keygen}}_{\mathsf{PS},\mathcal{A}}(\lambda) := \big|&\Pr[\mathcal{A}(1^\lambda, crs) = 1 \mid crs \leftarrow \mathsf{PGen}(1^\lambda, pars)] \\ &- \Pr[\mathcal{A}(1^\lambda, crs) = 1 \mid (crs, td) \leftarrow \mathsf{PTGen}(1^\lambda, pars)]\big|\end{aligned}$$

is negligible in λ.

Further for all public parameters pars, all pairs (crs, td) in the range of $\mathsf{PTGen}(1^\lambda)$*, all words* $x \in \mathcal{L}$*, and all witnesses* w *with* $\mathcal{R}_\mathcal{L}(x, w) = 1$*, we have that the outputs of*

$$\mathsf{PPrv}(crs, x, w) \textit{ and } \mathsf{PSim}(crs, td, x)$$

are statistically indistinguishable.

Perfect soundness: *For all crs in the range of* $\mathsf{PGen}(1^\lambda, pars)$*, for all words* $x \notin \mathcal{L}$ *and all proofs* Π *it holds* $\mathsf{PVer}(crs, x, \Pi) = 0$.

Remark. We will employ a weaker notion of composable zero-knowledge in the following. Namely:

Composable zero-knowledge: For a PPT adversary $\mathcal{A}$, we define

$$\mathrm{Adv}^{\mathrm{zk}}_{\mathsf{PS},\mathcal{A}}(\lambda) := \left| \Pr\left[b' = b \;\middle|\; \begin{array}{l} crs_0 \leftarrow_R \mathsf{PGen}(1^\lambda, pars); \\ (crs_1, td) \leftarrow_R \mathsf{PTGen}(1^\lambda, pars); \\ b \leftarrow_R \{0,1\}; \\ b' \leftarrow_R \mathcal{A}^{\textsc{Prove}(\cdot,\cdot)}(1^\lambda, crs_b) \end{array} \right] - \frac{1}{2} \right|.$$

<u>$\mathsf{PGen}(1^\lambda, pars)$:</u>

$\mathbf{D} \leftarrow_R \mathcal{D}_k$, $\mathbf{z} \leftarrow_R \mathbb{Z}_p^{k+1} \setminus \mathrm{span}(\mathbf{D})$
//recall $\mathcal{D}_k := \mathcal{D}_{k+1,k}$
$crs := (pars, [\mathbf{D}]_2, [\mathbf{z}]_2)$
return crs

<u>$\mathsf{PPrv}(crs, [\mathbf{x}]_1, \mathbf{r})$:</u>

let $b \in \{0,1\}$ **s.t.** $[\mathbf{x}]_1 = [\mathbf{A}_b]_1 \cdot \mathbf{r}$
$\mathbf{v} \leftarrow_R \mathbb{Z}_p^k$
$[\mathbf{z}_{1-b}]_2 := [\mathbf{D}]_2 \cdot \mathbf{v}$
// $([\mathbf{D}]_2, [\mathbf{z}_{1-b}]_2)$ trapdoor crs
$[\mathbf{z}_b]_2 := [\mathbf{z}]_2 - [\mathbf{z}_{1-b}]_2$
// crs guaranteeing soundness
$\mathbf{S}_0, \mathbf{S}_1 \leftarrow_R \mathbb{Z}_p^{k \times k}$
$[\mathbf{C}_b]_2 := \mathbf{S}_b \cdot [\mathbf{D}]_2^\top + \mathbf{r} \cdot [\mathbf{z}_b]_2^\top$
//commitment to $\mathbf{r}$ with rand. $\mathbf{S}_b$
$[\Pi_b]_1 := [\mathbf{A}_b]_1 \cdot \mathbf{S}_b$
//proof for $\mathbf{x} = \mathbf{A}_b \mathbf{r}$
$[\mathbf{C}_{1-b}]_2 := \mathbf{S}_{1-b} \cdot [\mathbf{D}]_2^\top$
//commitment to $\mathbf{0}$ with rand. $\mathbf{S}_{1-b}$
$[\Pi_{1-b}]_1 := [\mathbf{A}_{1-b}]_1 \cdot \mathbf{S}_{1-b} - [\mathbf{x}]_1 \cdot \mathbf{v}^\top$
//trapdoor proof for $\mathbf{x} = \mathbf{A}_{1-b}\mathbf{r}$
return $([\mathbf{z}_0]_2, ([\mathbf{C}_i]_2, [\Pi_i]_1)_{i \in \{0,1\}})$

<u>$\mathsf{PTGen}(1^\lambda, pars)$:</u>

$\mathbf{D} \leftarrow_R \mathcal{D}_k$, $\mathbf{u} \leftarrow_R \mathbb{Z}_p^k$
$\mathbf{z} := \mathbf{D} \cdot \mathbf{u}$
$crs := (pars, [\mathbf{D}]_2, [\mathbf{z}]_2)$, $td := \mathbf{u}$
return (crs, td)

<u>$\mathsf{PVer}(crs, [\mathbf{x}]_1, ([\mathbf{z}_0]_2, ([\mathbf{C}_i]_2, [\Pi_i]_1)_{i \in \{0,1\}}))$:</u>

$[\mathbf{z}_1]_2 := [\mathbf{z}]_2 - [\mathbf{z}_0]_2$
if for all $i \in \{0,1\}$ **it holds**
 $e([\mathbf{A}_i]_1, [\mathbf{C}_i]_2)$
 $= e([\Pi_i]_1, [\mathbf{D}]_2^\top) + e([\mathbf{x}]_1, [\mathbf{z}_i]_2^\top)$
 //check $\mathbf{C}_i \cdot \mathbf{A}_i \stackrel{?}{=} \Pi_i \cdot \mathbf{D}^\top + \mathbf{x} \cdot \mathbf{z}_i^\top$
  return 1
else return 0

<u>$\mathsf{PSim}(crs, td, [\mathbf{x}]_1)$:</u>

parse $td =: \mathbf{u}$
$\mathbf{v} \leftarrow_R \mathbb{Z}_p^k$
$[\mathbf{z}_0]_2 := [\mathbf{D}]_2 \cdot \mathbf{v}$
$[\mathbf{z}_1]_2 := [\mathbf{z}]_2 - [\mathbf{z}_0]_2$
$\mathbf{S}_0, \mathbf{S}_1 \leftarrow_R \mathbb{Z}_p^{k \times k}$
$[\mathbf{C}_0]_2 := \mathbf{S}_0 \cdot [\mathbf{D}]_2^\top$
$[\Pi_0]_1 := [\mathbf{A}_0]_1 \cdot \mathbf{S}_0 - [\mathbf{x}]_1 \cdot \mathbf{v}^\top$
$[\mathbf{C}_1]_2 := \mathbf{S}_1 \cdot [\mathbf{D}]_2^\top$
$[\Pi_1]_1 := [\mathbf{A}_1]_1 \cdot \mathbf{S}_1 - [\mathbf{x}]_1 \cdot (\mathbf{u} - \mathbf{v})^\top$
return $([\mathbf{z}_0]_2, ([\mathbf{C}_i]_2, [\Pi_i]_1)_{i \in \{0,1\}})$

Fig. 1. NIZK argument for $\mathcal{L}^\vee_{\mathbf{A}_0,\mathbf{A}_1}$ [31,46].

Here $\text{Prove}(x, w)$ returns $\perp$ if $\mathcal{R}_{\mathcal{L}}(x, w) = 0$ or Π_b if $\mathcal{R}_{\mathcal{L}}(x, w) = 1$, where $\Pi_0 \leftarrow_R \mathsf{PPrv}(crs_0, x, w)$ and $\Pi_1 \leftarrow_R \mathsf{PSim}(crs_1, td, x)$. We say that PS satisfies composable zero-knowledge if $\text{Adv}^{\text{zk}}_{\mathsf{PS},\mathcal{A}}(\lambda)$ is negligible in λ for all PPT $\mathcal{A}$.

Note that the original definition of composable zero-knowledge tightly implies our definition of composable zero-knowledge. We choose to work with the latter in order to simplify the presentation of our proofs. Note that for working with this definition in the tightness setting, it is crucial that $\text{Adv}^{\text{zk}}_{\mathsf{PS},\mathcal{A}}(\lambda)$ is independent of the number of queries to the oracle Prove.

2.5 NIZK for Our OR-language

In this section we recall an instantiation of a NIZK for an OR-language implicitly given in [31,46]. This NIZK will be a crucial part of all our constructions, allowing to employ the randomization techniques from [6,26,33] to obtain a tight security reduction.

Public Parameters. Let $\mathcal{PG} \leftarrow \mathsf{GGen}(1^\lambda)$. Let $k \in \mathbb{N}$. Let $\mathbf{A}_0, \mathbf{A}_1 \leftarrow_R \mathcal{D}_{2k,k}$. We define the public parameters to comprise

$$pars := (\mathcal{PG}, [\mathbf{A}_0]_1, [\mathbf{A}_1]_1).$$

We consider $k \in \mathbb{N}$ to be chosen ahead of time, fixed and implicitly known to all algorithms.

OR-Proof ([31,46]). In Fig. 1 we present a non-interactive zero-knowledge proof for the OR-language

$$\mathcal{L}^{\vee}_{\mathbf{A}_0,\mathbf{A}_1} := \{[\mathbf{x}]_1 \in \mathbb{Z}_p^{2k} \mid \exists \mathbf{r} \in \mathbb{Z}_p^k \colon [\mathbf{x}]_1 = [\mathbf{A}_0]_1 \cdot \mathbf{r} \vee [\mathbf{x}]_1 = [\mathbf{A}_1]_1 \cdot \mathbf{r}\}.$$

Note that this OR-proof is implicitly given in [31,46]. We recall the proof in the full version.

Lemma 3. *If the $\mathcal{D}_k$-MDDH assumption holds in the group $\mathbb{G}_2$, then the proof system $\mathsf{PS} := (\mathsf{PGen}, \mathsf{PTGen}, \mathsf{PPrv}, \mathsf{PVer}, \mathsf{PSim})$ as defined in Fig. 1 is a non-interactive zero-knowledge proof for $\mathcal{L}^{\vee}_{\mathbf{A}_0,\mathbf{A}_1}$. More precisely, for every adversary $\mathcal{A}$ attacking the composable zero-knowledge property of PS, we obtain an adversary $\mathcal{B}$ with $T(\mathcal{B}) \approx T(\mathcal{A}) + Q_{\mathsf{prove}} \cdot \mathsf{poly}(\lambda)$ and*

$$\text{Adv}^{\text{zk}}_{\mathsf{PS},\mathcal{A}}(\lambda) \leq \text{Adv}^{\text{mddh}}_{\mathcal{PG},\mathbb{G}_2,\mathcal{D}_k,\mathcal{B}}(\lambda).$$

3 Tightly Secure Message Authentication Code Scheme

Let $k \in \mathbb{N}$ and let $\mathsf{PS} := (\mathsf{PGen}, \mathsf{PTGen}, \mathsf{PPrv}, \mathsf{PSim})$ a non-interactive zero-knowledge proof for $\mathcal{L}^{\vee}_{\mathbf{A}_0,\mathbf{A}_1}$ as defined in Sect. 2.5. In Fig. 2 we provide a MAC $\mathsf{MAC} := (\mathsf{Gen}, \mathsf{Tag}, \mathsf{Ver})$ whose security can be tightly reduced to $\mathcal{D}_{2k,k}$-MDDH and the security of the underlying proof system PS.

$\underline{\mathsf{Gen}(1^\lambda):}$	$\underline{\mathsf{Tag}(pp, sk, \mu \in \mathbb{Z}_p):}$
$\mathcal{PG} \leftarrow \mathsf{GGen}(1^\lambda)$	**parse** $pp =: (\mathcal{PG}, [\mathbf{A}_0]_1, crs)$
$\mathbf{A}_0, \mathbf{A}_1 \leftarrow \mathcal{D}_{2k,k}$	$\mathbf{r} \leftarrow_R \mathbb{Z}_p^k$
$pars := (\mathcal{PG}, [\mathbf{A}_0]_1, [\mathbf{A}_1]_1)$	$[\mathbf{t}]_1 := [\mathbf{A}_0]_1 \mathbf{r}$
$crs \leftarrow \mathsf{PGen}(1^\lambda, pars)$	$\Pi \leftarrow \mathsf{PPrv}(crs, [\mathbf{t}]_1, \mathbf{r})$
$\mathbf{k}_0, \mathbf{k}_1 \leftarrow_R \mathbb{Z}_p^{2k}$	$[u]_1 := (\mathbf{k}_0 + \mu \mathbf{k}_1)^\top [\mathbf{t}]_1$
$pp := (\mathcal{PG}, [\mathbf{A}_0]_1, crs)$	$\mathsf{tag} := ([\mathbf{t}]_1, \Pi, [u]_1)$
$sk := (\mathbf{k}_0, \mathbf{k}_1)$	**return** tag
return (pp, sk)	
	$\underline{\mathsf{Ver}(pp, sk, \mu \in \mathbb{Z}_p, \mathsf{tag}):}$
	parse $\mathsf{tag} =: ([\mathbf{t}]_1, \Pi, [u]_1)$
	$b \leftarrow \mathsf{PVer}(crs, [\mathbf{t}]_1, \Pi)$
	if $b = 1$ **and** $[u]_1 \neq [0]_1$
	and $[u]_1 = (\mathbf{k}_0 + \mu \mathbf{k}_1)^\top [\mathbf{t}]_1$
	return 1
	else return 0

Fig. 2. Tightly secure MAC $\mathsf{MAC} := (\mathsf{Gen}, \mathsf{Tag}, \mathsf{Ver})$ from the $\mathcal{D}_{2k,k}$-MDDH assumption.

Instead of directly proving UF-CMA security of our MAC, we will first provide our so-called core lemma, which captures the essential randomization technique from [6,26,33]. We can employ this lemma to prove the security of our MAC and (structure-preserving) signature schemes. Essentially, the core lemma shows that the term $[\mathbf{k}_0^\top \mathbf{t}]_1$ is pseudorandom. We give the corresponding formal experiment in Fig. 3.

$\underline{\mathrm{Exp}^{\mathrm{core}}_{\beta,\mathcal{A}}(\lambda), \text{ for } \beta \in \{0,1\}:}$	$\underline{\textsc{TagO}():}$
$\mathsf{ctr} := 0$	$\mathsf{ctr} := \mathsf{ctr} + 1$
$\mathcal{PG} \leftarrow \mathsf{GGen}(1^\lambda)$	$\mathbf{r} \leftarrow_R \mathbb{Z}_p^k$
$\mathbf{A}_0, \mathbf{A}_1 \leftarrow_R \mathcal{D}_{2k,k}$	$[\mathbf{t}]_1 := [\mathbf{A}_0]_1 \mathbf{r}$
$pars := (\mathcal{PG}, [\mathbf{A}_0]_1, [\mathbf{A}_1]_1)$	$\Pi \leftarrow \mathsf{PPrv}(crs, [\mathbf{t}]_1, \mathbf{r})$
$crs \leftarrow \mathsf{PGen}(1^\lambda, pars)$	$[u']_1 := (\mathbf{k}_0 + \beta \cdot \mathbf{F}(\mathsf{ctr}))^\top [\mathbf{t}]_1$
$\mathbf{k}_0, \mathbf{k}_1 \leftarrow_R \mathbb{Z}_p^{2k}$	$\mathsf{tag} := ([\mathbf{t}]_1, \Pi, [u']_1)$
$pp := (\mathcal{PG}, [\mathbf{A}_0]_1, crs)$	**return** tag
$\mathsf{tag} \leftarrow \mathcal{A}^{\textsc{TagO}()}(pp)$	
return $\textsc{VerO}(\mathsf{tag})$	$\underline{\textsc{VerO}(\mathsf{tag}):}$
	parse $\mathsf{tag} = ([\mathbf{t}]_1, \Pi, [u']_1)$
	$b \leftarrow \mathsf{PVer}(crs, [\mathbf{t}]_1, \Pi)$
	if $b = 1$ **and** $\exists \mathsf{ctr}' \leq \mathsf{ctr}$:
	$[u']_1 = (\mathbf{k}_0 + \beta \cdot \mathbf{F}(\mathsf{ctr}'))^\top [\mathbf{t}]_1$
	return 1
	else return 0

Fig. 3. Experiment for the core lemma. Here, $\mathbf{F} : \mathbb{Z}_p \to \mathbb{Z}_p^{2k}$ is a random function computed on the fly. We highlight the difference between $\mathrm{Exp}^{\mathrm{core}}_{0,\mathcal{A}}$ and $\mathrm{Exp}^{\mathrm{core}}_{1,\mathcal{A}}$ in gray.

Lemma 4 (Core lemma). *If the $\mathcal{D}_{2k,k}$-MDDH assumption holds in $\mathbb{G}_1$ and the tuple of algorithms* $\mathsf{PS} := (\mathsf{PGen}, \mathsf{PTGen}, \mathsf{PPrv}, \mathsf{PVer})$ *is a non-interactive zero-knowledge proof system for $\mathcal{L}^{\vee}_{\mathbf{A}_0,\mathbf{A}_1}$, then going from experiment $\mathrm{Exp}^{\mathrm{core}}_{0,\mathcal{A}}(\lambda)$ to $\mathrm{Exp}^{\mathrm{core}}_{1,\mathcal{A}}(\lambda)$ can (up to negligible terms) only increase the winning chances of an adversary. More precisely, for every adversary $\mathcal{A}$, there exist adversaries $\mathcal{B}$, $\mathcal{B}'$ with running time $T(\mathcal{B}) \approx T(\mathcal{B}') \approx T(\mathcal{A}) + Q \cdot \mathsf{poly}(\lambda)$ such that*

$$\mathrm{Adv}^{\mathrm{core}}_{0,\mathcal{A}}(\lambda) \leq \mathrm{Adv}^{\mathrm{core}}_{1,\mathcal{A}}(\lambda) + \Delta^{\mathrm{core}}_{\mathcal{A}}(\lambda),$$

where

$$\begin{aligned}\Delta^{\mathrm{core}}_{\mathcal{A}}(\lambda) :=&(4k\lceil \log Q \rceil + 2) \cdot \mathrm{Adv}^{\mathrm{mddh}}_{\mathcal{PG},\mathbb{G}_1,\mathcal{D}_{2k,k},\mathcal{B}}(\lambda)\\ &+ (2\lceil \log Q \rceil + 2) \cdot \mathrm{Adv}^{\mathrm{ZK}}_{\mathsf{PS},\mathcal{B}'}(\lambda)\\ &+ \lceil \log Q \rceil \cdot \Delta_{\mathcal{D}_{2k,k}} + \tfrac{4\lceil \log Q \rceil + 2}{p-1} + \tfrac{\lceil \log Q \rceil \cdot Q}{p}.\end{aligned}$$

Recall that by definition of the distribution $\mathcal{D}_{2k,k}$ (Sect. 2.2), the term $\Delta_{\mathcal{D}_{2k,k}}$ is statistically small.

Proof Outline. Since the proof of Lemma 4 is rather complex, we first outline our strategy. Intuitively, our goal is to randomize the term u' used by oracles TAGO and VERO (i.e., to change this term from $\mathbf{k}_0^\top \mathbf{t}$ to $(\mathbf{k}_0 + \mathbf{F}(\mathsf{ctr}))^\top \mathbf{t}$ for a truly random function $\mathbf{F}$). In this, it will also be helpful to change the distribution of $\mathbf{t} \in \mathbb{Z}_p^{2k}$ in tags handed out by TAGO as needed. (Intuitively, changing $\mathbf{t}$ can be justified with the $\mathcal{D}_{2k,k}$-MDDH assumption, but we can only rely on the soundness of PS if $\mathbf{t} \in \mathrm{span}(\mathbf{A}_0) \cup \mathrm{span}(\mathbf{A}_1)$. In other words, we may assume that $\mathbf{t} \in \mathrm{span}(\mathbf{A}_0) \cup \mathrm{span}(\mathbf{A}_1)$ for any of $\mathcal{A}$'s VERO queries, but only if the same holds for all $\mathbf{t}$ chosen by TAGO.)

We will change u' using a hybrid argument, where in the i-th hybrid we set $u' = (\mathbf{k}_0^\top + \mathbf{F}_i(\mathsf{ctr}_{|i}))^\top \mathbf{t}$ for a random function $\mathbf{F}_i$ on i-bit prefixes, and the i-bit prefix $\mathsf{ctr}_{|i}$ of ctr. (That is, we introduce more and more dependencies on the bits of ctr.) To move from hybrid i to hybrid $i+1$, we proceed again along a series of hybrids (outsourced into the full version), and perform the following modifications:

Partitioning. First, we choose $\mathbf{t} \in \mathrm{span}(\mathbf{A}_{\mathsf{ctr}_{i+1}})$ in VERO, where ctr_{i+1} is the $(i+1)$-th bit of ctr. As noted above, this change can be justified with the $\mathcal{D}_{2k,k}$-MDDH assumption, and we may still assume $\mathbf{t} \in \mathrm{span}(\mathbf{A}_0) \cup \mathrm{span}(\mathbf{A}_1)$ in every TAGO query from $\mathcal{A}$.

Decoupling. At this point, the values u' computed in TAGO and VERO are either of the form $u' = (\mathbf{k}_0^\top + \mathbf{F}_i(\mathsf{ctr}_{|i}))^\top \mathbf{A}_0 \mathbf{r}$ or $u' = (\mathbf{k}_0^\top + \mathbf{F}_i(\mathsf{ctr}_{|i}))^\top \mathbf{A}_1 \mathbf{r}$ (depending on $\mathbf{t}$). Since $\mathbf{F}_i : \{0,1\}^i \to \mathbb{Z}_p^{2k}$ is truly random, and the matrix $\mathbf{A}_0 || \mathbf{A}_1 \in \mathbb{Z}_p^{2k \times 2k}$ has linearly independent columns (with overwhelming probability), the two possible subterms $\mathbf{F}_i(\mathsf{ctr}_{|i})^\top \mathbf{A}_0$ and $\mathbf{F}_i(\mathsf{ctr}_{|i})^\top \mathbf{A}_1$ are independent. Thus, switching to $u' = (\mathbf{k}_0^\top + \mathbf{F}_{i+1}(\mathsf{ctr}_{|i+1}))^\top \mathbf{t}$ does not change $\mathcal{A}$'s view at all.

After these modifications (and resetting $\mathbf{t}$), we have arrived at the $(i+1)$-th hybrid, which completes the proof. However, this outline neglects a number of details, including a proper reasoning of PS proofs, and a careful discussion of the decoupling step. In particular, an additional complication arises in this step from the fact that an adversary may choose $\mathbf{t} \in \mathrm{span}(A_b)$ for an arbitrary bit b not related to any specific ctr. This difficulty is the reason for the somewhat surprising "$\exists \mathsf{ctr}' \leq \mathsf{ctr}$" clause in VERO.

Proof (of Lemma 4). We proceed via a series of hybrid games $\mathsf{G}_0, \ldots, \mathsf{G}_{3.\lceil \log Q \rceil}$, described in Fig. 4, and we denote by ε_i the advantage of $\mathcal{A}$ to win G_i, that is $\Pr[\mathsf{G}_i(\mathcal{A}, 1^\lambda) = 1]$, where the probability is taken over the random coins of G_i and $\mathcal{A}$.

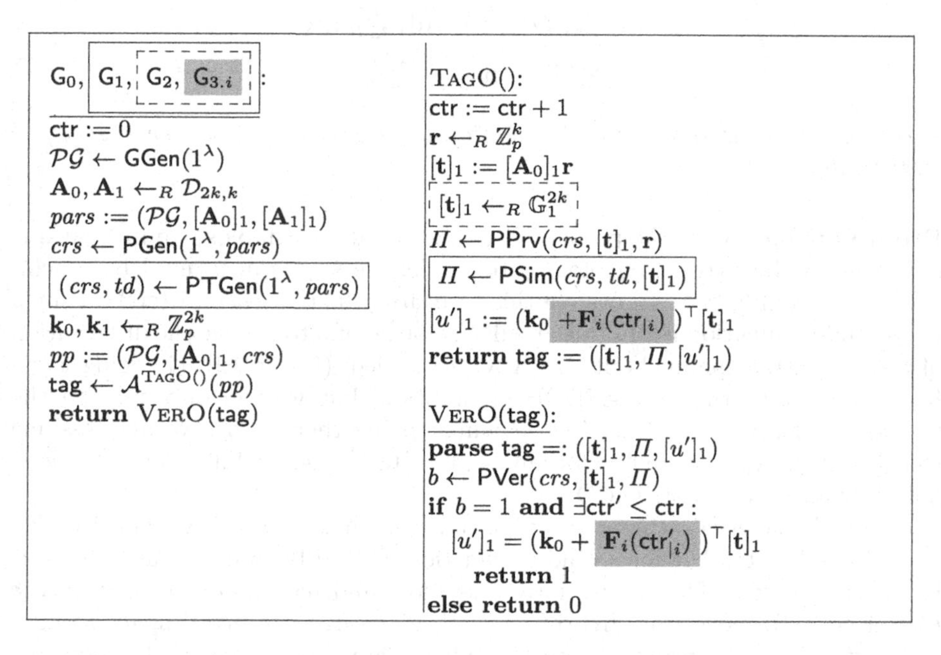

Fig. 4. Games $\mathsf{G}_0, \mathsf{G}_1, \mathsf{G}_2, \mathsf{G}_{3.i}$ for $i \in \{0, \ldots, \lceil \log Q \rceil - 1\}$, for the proof of the core lemma (Lemma 4). $\mathbf{F}_i : \{0,1\}^i \to \mathbb{Z}_p^{2k}$ denotes a random function, and $\mathsf{ctr}_{|i}$ denotes the i-bit prefix of the counter ctr written in binary. In each procedure, the components inside a solid (dotted, gray) frame are only present in the games marked by a solid (dotted, gray) frame.

G_0: We have $\mathsf{G}_0 = \mathrm{Exp}^{\mathrm{core}}_{0,\mathcal{A}}(\lambda)$ and thus by definition:

$$\varepsilon_0 = \mathrm{Adv}^{\mathrm{core}}_{0,\mathcal{A}}(\lambda).$$

$\mathsf{G}_0 \rightsquigarrow \mathsf{G}_1$: Game G_1 is as G_0, except that *crs* is generated by PTGen and the proofs computed by TAGO are generated using PSim instead of PPrv. This change is justified by the zero-knowledge of PS. Namely, let $\mathcal{A}$ be an adversary distinguishing between G_0 and G_1. Then we can construct an adversary

$\mathcal{B}$ on the composable zero-knowledge property of PS as follows. The adversary $\mathcal{B}$ follows G_0, except he uses the *crs* obtained by its own experiment instead of calling PGen. $\mathcal{B}$ answers tag queries following the tag oracle, but instead of computing Π itself it asks its own oracle PROVE. Now $\mathcal{B}$ simulates G_0 in case it was given a real *crs* and it simulates G_1 in case it was given a *crs* generated by PTGen. $\mathcal{B}$ is thus such that $T(\mathcal{B}) \approx T(\mathcal{A}) + Q \cdot \mathsf{poly}(\lambda)$ and

$$|\varepsilon_0 - \varepsilon_1| \leq \mathrm{Adv}^{\mathrm{ZK}}_{\mathsf{PS},\mathcal{B}}(\lambda).$$

$\mathsf{G}_1 \rightsquigarrow \mathsf{G}_2$**:** We can switch $[\mathbf{t}]_1$ to random over $\mathbb{G}_1$ by applying the $\mathcal{D}_{2k,k}$ assumption. More precisely, let $\mathcal{A}$ be an adversary distinguishing between G_1 and G_2 and let $\mathcal{B}$ be an adversary given a Q-fold $\mathcal{D}_{2k,k}$-MDDH challenge $(\mathcal{PG}, [\mathbf{A}_0]_1, [\mathbf{z}_1]_1, \ldots, [\mathbf{z}_Q]_1)$ as input. Now $\mathcal{B}$ sets up the game for $\mathcal{A}$ similar to G_1, but instead choosing $\mathbf{A}_0 \leftarrow_R \mathcal{D}_{2k,k}$, it uses its challenge matrix $[\mathbf{A}_0]_1$ as part of the public parameters *pars*. Further, to answer tag queries $\mathcal{B}$ sets $[\mathbf{t}_i]_1 := [\mathbf{z}_i]_1$ and computes the rest accordingly. This is possible as the proof Π is simulated from game G_1 on. In case $\mathcal{B}$ was given a real $\mathcal{D}_{2k,k}$-challenge, it simulates G_1 and otherwise G_2. Lemma 1 yields the existence of an adversary $\mathcal{B}_1$ with $T(\mathcal{B}_1) \approx T(\mathcal{A}) + Q \cdot \mathsf{poly}(\lambda)$ and

$$|\varepsilon_1 - \varepsilon_2| \leq k \cdot \mathrm{Adv}^{\mathrm{mddh}}_{\mathcal{PG},\mathbb{G}_1,\mathcal{D}_{2k,k},\mathcal{B}_1}(\lambda) + \tfrac{1}{p-1}.$$

$\mathsf{G}_2 \rightsquigarrow \mathsf{G}_{3.0}$**:** As for all $\mathsf{ctr} \in \mathbb{N}$ we have $\mathbf{F}_0(\mathsf{ctr}_{|0}) = \mathbf{F}_0(\epsilon)$ and $\mathbf{k}_0$ is distributed identically to $\mathbf{k}_0 + \mathbf{F}_0(\epsilon)$ for $\mathbf{k}_0 \leftarrow_R \mathbb{Z}_p^{2k}$ we have

$$\varepsilon_2 = \varepsilon_{3.0}.$$

$\mathsf{G}_{3.i} \rightsquigarrow \mathsf{G}_{3.(i+1)}$**:** For the proof of this transition we refer to the full version. We obtain: For every adversary $\mathcal{A}$ there exist adversaries $\mathcal{B}_i$, $\mathcal{B}_i'$ such that $T(\mathcal{B}_i) \approx T(\mathcal{B}_i') \approx T(\mathcal{A}) + Q \cdot \mathsf{poly}(\lambda)$, and

$$\begin{aligned}\varepsilon_{3.i} \leq & \varepsilon_{3.(i+1)} + 4k \cdot \mathrm{Adv}^{\mathrm{mddh}}_{\mathcal{PG},\mathbb{G}_1,\mathcal{D}_{2k,k},\mathcal{B}_i}(\lambda) + 2\mathrm{Adv}^{\mathrm{ZK}}_{\mathsf{PS},\mathcal{B}_i'}(\lambda) \\ & + \Delta_{\mathcal{D}_{2k,k}} + \tfrac{4}{p-1} + \tfrac{Q}{p}.\end{aligned}$$

$\mathsf{G}_{3.\lceil \log Q \rceil} \rightsquigarrow \mathbf{Exp}^{\mathbf{core}}_{1,\mathcal{A}}(\boldsymbol{\lambda})$: It is left to reverse the changes introduced in the transitions from game G_0 to game G_2 to end up at the experiment $\mathrm{Exp}^{\mathrm{core}}_{1,\mathcal{A}}(1^\lambda)$.

In order to do so we introduce an intermediary game G_4, where we set $[\mathbf{t}] := [\mathbf{A}_0]_1\mathbf{r}$ for $\mathbf{r} \leftarrow_R \mathbb{Z}_p^k$. This corresponds to reversing transition $\mathsf{G}_1 \rightsquigarrow \mathsf{G}_2$. By the same reasoning for every adversary $\mathcal{A}$ we thus obtain an adversary $\mathcal{B}_{3.\lceil \log Q \rceil}$ with $T(\mathcal{B}_{3.\lceil \log Q \rceil}) \approx T(\mathcal{A}) + Q \cdot \mathsf{poly}(\lambda)$ such that

$$|\varepsilon_{3.\lceil \log Q \rceil} - \varepsilon_4| \leq k \cdot \mathrm{Adv}^{\mathrm{mddh}}_{\mathcal{PG},\mathbb{G}_1,\mathcal{D}_{2k,k},\mathcal{B}_{3.\lceil \log Q \rceil}}(\lambda) + \tfrac{1}{p-1}.$$

As $[\mathbf{t}]_1$ is now chosen from $\mathrm{span}([\mathbf{A}_0]_1)$ again, we can switch back to honest generation of the common reference string *crs* and proofs Π. As in transition

$\mathsf{G}_0 \rightsquigarrow \mathsf{G}_1$ for an adversary $\mathcal{A}$ we obtain an adversary $\mathcal{B}_4$ with $T(\mathcal{B}_4) \approx T(\mathcal{A}) + Q \cdot \mathsf{poly}(\lambda)$ and

$$|\varepsilon_4 - \mathrm{Adv}^{\mathrm{core}}_{1,\mathcal{A}}(\lambda)| \leq \mathrm{Adv}^{\mathrm{ZK}}_{\mathsf{PS},\mathcal{B}_4}(\lambda).$$

Theorem 1 (UF-CMA security of MAC). *If the $\mathcal{D}_{2k,k}$-MDDH assumptions holds in $\mathbb{G}_1$, and the tuple* $\mathsf{PS} := (\mathsf{PGen}, \mathsf{PTGen}, \mathsf{PPrv}, \mathsf{PVer})$ *is a non-interactive zero-knowledge proof system for $\mathcal{L}^{\vee}_{\mathbf{A}_0,\mathbf{A}_1}$, then the MAC* $\mathsf{MAC} := (\mathsf{Gen}, \mathsf{Tag}, \mathsf{Ver})$ *provided in Fig. 2 is* UF-CMA *secure. Namely, for any adversary $\mathcal{A}$, there exists an adversary $\mathcal{B}$ with running time $T(\mathcal{B}) \approx T(\mathcal{A}) + Q \cdot \mathsf{poly}(\lambda)$, where Q is the number of queries to* TAGO, poly *is independent of Q, and*

$$\mathrm{Adv}^{\mathrm{uf\text{-}cma}}_{\mathsf{MAC},\mathcal{A}}(\lambda) \leq \Delta^{\mathrm{core}}_{\mathcal{B}}(\lambda) + \frac{Q}{p}.$$

Proof. We employ an intermediary game G_0 to prove UF-CMA security of the MAC. By ε_0 we denote the advantage of $\mathcal{A}$ to win game G_0, that is $\Pr[\mathsf{G}_0(\mathcal{A}, 1^\lambda) = 1]$, where the probability is taken over the random coins of G_0 and $\mathcal{A}$.

$\mathrm{Exp}^{\mathrm{uf\text{-}cma}}_{\mathcal{A}}(\lambda) \rightsquigarrow \mathsf{G}_0$: Let $\mathcal{A}$ be an adversary distinguishing between $\mathrm{Exp}^{\mathrm{uf\text{-}cma}}_{\mathcal{A}}(\lambda)$ and G_0. Then we construct an adversary $\mathcal{B}$ with $T(\mathcal{B}) \approx T(\mathcal{A}) + Q \cdot \mathsf{poly}(\lambda)$ allowing to break the core lemma (Lemma 4) as follows. On input pp from $\mathrm{Exp}^{\mathrm{core}}_{\beta}(1^\lambda, \mathcal{B})$ the adversary $\mathcal{B}$ forwards pp to $\mathcal{A}$. Then, $\mathcal{B}$ samples $\mathbf{k}_1 \leftarrow_R \mathbb{Z}_p^{2k}$. Afterwards, on a tag query μ from $\mathcal{A}$, $\mathcal{B}$ queries its own TAGO oracle (which takes no input), receives $([\mathbf{t}]_1, \Pi, [u']_1)$, computes $[u]_1 := [u']_1 + \mu\mathbf{k}_1^\top[\mathbf{t}]_1$, and answers with $([\mathbf{t}]_1, \Pi, [u]_1)$. Finally, given the forgery $(\mu^\star, \mathsf{tag}^\star := ([\mathbf{t}]_1, \Pi, [u^\star]_1))$ from $\mathcal{A}$, if $\mu^\star \notin \mathcal{Q}_{\mathsf{tag}}$ and $[u^\star]_1 \neq [0]_1$, then the adversary $\mathcal{B}$ sends $\mathsf{tag}' := ([\mathbf{t}]_1, \Pi, [u^\star]_1 + \mu\mathbf{k}_1^\top[\mathbf{t}]_1)$ to its experiment (otherwise an invalid tuple). Then we have $\mathrm{Adv}^{\mathrm{uf\text{-}cma}}_{\mathsf{MAC},\mathcal{A}}(\lambda) = \mathrm{Adv}^{\mathrm{core}}_{0,\mathcal{B}}(\lambda)$ and $\varepsilon_0 = \mathrm{Adv}^{\mathrm{core}}_{1,\mathcal{B}}(\lambda)$. The core lemma (Lemma 4) yields

$$\mathrm{Adv}^{\mathrm{core}}_{0,\mathcal{B}}(\lambda) \leq \mathrm{Adv}^{\mathrm{core}}_{1,\mathcal{B}}(\lambda) + \Delta^{\mathrm{core}}_{\mathcal{B}}(\lambda)$$

and thus altogether we obtain

$$\mathrm{Adv}^{\mathrm{uf\text{-}cma}}_{\mathsf{MAC},\mathcal{A}}(\lambda) \leq \varepsilon_0 + \Delta^{\mathrm{core}}_{\mathcal{B}}(\lambda).$$

Game G_0: We now prove that any adversary $\mathcal{A}$ has only negligible chances to win game G_0 using the randomness of $\mathbf{F}$ together with the pairwise independence of $\mu \mapsto \mathbf{k}_0 + \mu\mathbf{k}_1$.

Let $(\mu^\star, \mathsf{tag}^\star)$ be the forgery of $\mathcal{A}$. we can replace $\mathbf{k}_1$ by $\mathbf{k}_1 - \mathbf{v}$ for $\mathbf{v} \leftarrow_R \mathbb{Z}_p^{2k}$, as both are distributed identically. Next, for all $j \leq Q$ we can replace $\mathbf{F}(j)$ by $\mathbf{F}(j) + \mu^{(j)} \cdot \mathbf{v}$ for the same reason. This way, $\mathrm{TAGO}(\mu^{(j)})$ computes

$$\begin{aligned} [u^{(j)}]_1 :&= [(\mathbf{k}_0 + \mu^{(j)}\mathbf{k}_1 \boxed{-\mu^{(j)}\mathbf{v}} + \mathbf{F}(j) \boxed{+\mu^{(j)}\mathbf{v}})^\top \mathbf{t}^{(j)}]_1 \\ &= [(\mathbf{k}_0 + \mu^{(j)}\mathbf{k}_1 + \mathbf{F}(j)^\top \mathbf{t}^{(j)}]_1, \end{aligned}$$

and $\mathrm{VERO}([\mu^\star]_2, \mathsf{tag}^\star := ([\mathbf{t}]_1, \Pi, [u]))$ checks if there exists a counter $i \in \mathcal{Q}_{\mathsf{tag}}$ such that:

$$\begin{aligned} [u]_1 &= [(\mathbf{k}_0 + \mu^\star\mathbf{k}_1 \boxed{-\mu^\star\mathbf{v}} + \mathbf{F}(i) \boxed{+\mu^{(i)}\mathbf{v}})^\top \mathbf{t}]_1 \\ &= [(\mathbf{k}_0 + \mu^\star\mathbf{k}_1 + \mathbf{F}(i)^\top \mathbf{t}^\star]_1 \boxed{+[(\mu^{(i)} - \mu^\star)\mathbf{v}^\top \mathbf{t}]_1}. \end{aligned}$$

$\mathrm{Exp}_{\mathcal{A}}^{\text{uf-cma}}(\lambda)$, $\boxed{\mathsf{G}}$:	$\mathrm{TAGO}(\mu)$:
$\mathcal{Q}_{\mathsf{tag}} := \emptyset$	$\mathcal{Q}_{\mathsf{tag}} := \mathcal{Q}_{\mathsf{tag}} \cup \{\mu\}$
$\boxed{\mathsf{ctr} := 0}$	$\boxed{\mathsf{ctr} := \mathsf{ctr} + 1}$
$\mathcal{PG} \leftarrow \mathsf{GGen}(1^\lambda)$	$\mathbf{r} \leftarrow_R \mathbb{Z}_p^k$
$\mathbf{A}_0, \mathbf{A}_1 \leftarrow_R \mathcal{D}_{2k,k}$	$[\mathbf{t}]_1 := [\mathbf{A}_0]_1 \mathbf{r}$
$pars := (\mathcal{PG}, [\mathbf{A}_0]_1, [\mathbf{A}_1]_1)$	$\Pi \leftarrow \mathsf{PPrv}(crs, [\mathbf{t}]_1, \mathbf{r})$
$crs \leftarrow \mathsf{PGen}(1^\lambda, pars)$	$[u]_1 := (\mathbf{k}_0 + \mu \mathbf{k}_1 \boxed{+\mathbf{F}(\mathsf{ctr})})^\top [\mathbf{t}]_1$
$\mathbf{k}_0, \mathbf{k}_1 \leftarrow_R \mathbb{Z}_p^{2k}$	$\mathsf{tag} := ([\mathbf{t}]_1, \Pi, [u]_1)$
$pp := (\mathcal{PG}, [\mathbf{A}_0]_1, crs)$	**return** tag
$(\mu^\star, \mathsf{tag}^\star) \leftarrow \mathcal{A}^{\mathrm{TAGO}(\cdot)}(pp)$	
if $\mu^\star \notin \mathcal{Q}_{\mathsf{tag}}$	$\mathrm{VERO}(\mu^\star, \mathsf{tag}^\star)$:
and $\mathrm{VERO}(\mu^\star, \mathsf{tag}^\star) = 1$	**parse** $\mathsf{tag}^\star =: ([\mathbf{t}]_1, \Pi, [u]_1)$
return 1	$b \leftarrow \mathsf{PVer}([\mathbf{t}]_1, \Pi)$
else return 0	**if** $b = 1$ **and** $[u]_1 \neq [0]_1$ **and** $\boxed{\exists \mathsf{ctr}' \leq \mathsf{ctr}:}$
	$[u]_1 = (\mathbf{k}_0 + \mu^\star \mathbf{k}_1 \boxed{+\mathbf{F}_i(\mathsf{ctr}')})^\top [\mathbf{t}]_1$
	return 1
	else return 0

Fig. 5. The UF-CMA security experiment and game G for the UF-CMA proof of MAC in Fig. 2. $\mathbf{F} : \{0,1\}^{\lceil \log Q \rceil} \to \mathbb{Z}_p^{2k}$ denotes a random function, applied on ctr written in binary. In each procedure, the components inside a gray frame are only present in the games marked by a gray frame.

For the forgery to be successful, it must hold $\mu^\star \notin \mathcal{Q}_{\mathsf{tag}}$ and $[u] \neq 0$ (and thus $[\mathbf{t}]_1 \neq [\mathbf{0}]_1$). Therefore, each value computed by VERO is (marginally) uniformly random over $\mathbb{G}_1$.

As the verification oracle checks for all counters $i \leq Q$, applying the union bound yields

$$\varepsilon_0 \leq \frac{Q}{p}.$$

4 Tightly Secure Signature Scheme

In this section, we present a signature scheme SIG for signing messages from $\mathbb{Z}_p$, described in Fig. 6, whose UF-CMA security can be tightly reduced to the $\mathcal{D}_{2k,k}$-MDDH and $\mathcal{D}_k$-MDDH assumptions.

SIG builds upon the tightly secure MAC from Sect. 3, and functions as a stepping stone to explain the main ideas of the upcoming structure-preserving signature in Sect. 5. Recall that our MAC outputs $\mathsf{tag} = ([\mathbf{t}]_1, \Pi, [u]_1)$, where Π is a (publicly verifiable) NIZK proof of the statement $\mathbf{t} \in \mathrm{span}(\mathbf{A}_0) \cup \mathrm{span}(\mathbf{A}_1)$, and $u = (\mathbf{k}_0 + \mu \mathbf{k}_1)^\top \mathbf{t}$ has an affine structure. Hence, alternatively, we can also view our MAC as an affine MAC [14] with $\mathbf{t} \in \mathrm{span}(\mathbf{A}_0) \cup \mathrm{span}(\mathbf{A}_1)$ and a NIZK proof for that. Similar to [14], we use (tuned) Groth-Sahai proofs to make $[u]_1$ publicly verifiable. Similar ideas have been used to construct efficient quasi-adaptive NIZK for linear subspace [38,40], structure-preserving signatures [39],

$\underline{\mathsf{Gen}(1^\lambda)}$:	$\underline{\mathsf{Sign}(pk, sk, \mu \in \mathbb{Z}_p)}$:
$\mathcal{PG} \leftarrow \mathsf{GGen}(1^\lambda)$	$\mathbf{r} \leftarrow_R \mathbb{Z}_p^k$
$\mathbf{A}_0, \mathbf{A}_1 \leftarrow \mathcal{D}_{2k,k}$	$[\mathbf{t}]_1 := [\mathbf{A}_0]_1 \mathbf{r}$
$pars := (\mathcal{PG}, [\mathbf{A}_0]_1, [\mathbf{A}_1]_1)$	$\Pi \leftarrow \mathsf{PPrv}(crs, [\mathbf{t}]_1, \mathbf{r})$
$crs \leftarrow \mathsf{PGen}(1^\lambda, pars)$	$[\mathbf{u}]_1 := (\mathbf{K}_0 + \mu \mathbf{K}_1)^\top [\mathbf{t}]_1$
$\mathbf{A} \leftarrow_R \mathcal{D}_k$	$\sigma := ([\mathbf{t}]_1, \Pi, [\mathbf{u}]_1)$
$\mathbf{K}_0, \mathbf{K}_1 \leftarrow_R \mathbb{Z}_p^{2k \times (k+1)}$	**return** σ
$pk := (\mathcal{PG}, [\mathbf{A}_0]_1, crs,$	
$[\mathbf{A}]_2, [\mathbf{K}_0\mathbf{A}]_2, [\mathbf{K}_1\mathbf{A}]_2)$	$\underline{\mathsf{Ver}(pk, \mu \in \mathbb{Z}_p, \sigma)}$:
$sk := (\mathbf{K}_0, \mathbf{K}_1)$	**parse tag** $=: ([\mathbf{t}]_1, \Pi, [\mathbf{u}]_1)$
return (pk, sk)	$b \leftarrow \mathsf{PVer}(crs, [\mathbf{t}]_1, \Pi)$
	if $b = 1$ **and** $[\mathbf{u}]_1 \neq [\mathbf{0}]_1$ **and** $e([\mathbf{u}]_1^\top, [\mathbf{A}]_2)$
	$= e([\mathbf{t}]_1^\top, [\mathbf{K}_0\mathbf{A}]_2 + \mu[\mathbf{K}_1\mathbf{A}]_2)$
	return 1
	else return 0

Fig. 6. Tightly UF-CMA secure signature scheme SIG.

and identity-based encryption schemes [14]. In the following theorem we state the state the security of SIG. For a proof we refer to the full version.

Theorem 2 (Security of SIG). *If* PS := (PGen, PPrv, PVer, PSim) *is a non-interactive zero-knowledge proof system for* $\mathcal{L}^{\vee}_{\mathbf{A}_0,\mathbf{A}_1}$, *then the signature scheme* SIG *described in Fig. 6 is* UF-CMA *secure under the* $\mathcal{D}_{2k,k}$*-MDDH and* $\mathcal{D}_k$*-MDDH assumptions. Namely, for any adversary* $\mathcal{A}$, *there exist adversaries* $\mathcal{B}, \mathcal{B}'$ *with running time* $T(\mathcal{B}) \approx T(\mathcal{B}') \approx T(\mathcal{A}) + Q \cdot \mathsf{poly}(\lambda)$, *where* Q *is the number of queries to* SIGNO, poly *is independent of* Q, *and*

$$\mathrm{Adv}^{\mathrm{uf-cma}}_{\mathsf{SIG},\mathcal{A}}(\lambda) \leq \mathrm{Adv}^{\mathrm{uf-cma}}_{\mathsf{MAC},\mathcal{B}}(\lambda) + \mathrm{Adv}^{\mathrm{mddh}}_{\mathcal{PG},\mathbb{G}_2,\mathcal{D}_k,\mathcal{B}'}(\lambda).$$

5 Tightly Secure Structure-Preserving Signature Scheme

In this section we present a structure-preserving signature scheme SPS, described in Fig. 7, whose security can be tightly reduced to the $\mathcal{D}_{2k,k}$-MDDH and $\mathcal{D}_k$-MDDH assumptions. It builds upon the tightly secure signature presented in Sect. 4 by using a similar idea of [39]. Precisely, we view μ as a label and the main difference between both schemes is that in the proof we do not need to guess which μ the adversary may reuse for its forgery, and thus our security proof is tight.

Theorem 3 (Security of SPS). *If* PS := (PGen, PTGen, PVer, PSim) *is a non-interactive zero-knowledge proof system for* $\mathcal{L}^{\vee}_{\mathbf{A}_0,\mathbf{A}_1}$, *the signature scheme* SPS *described in Fig. 7 is* UF-CMA *secure under the* $\mathcal{D}_{2k,k}$*-MDDH and* $\mathcal{D}_k$*-MDDH assumptions. Namely, for any adversary* $\mathcal{A}$, *there exist adversaries* $\mathcal{B}, \mathcal{B}'$ *with running time* $T(\mathcal{B}) \approx T(\mathcal{B}') \approx T(\mathcal{A}) + Q \cdot \mathsf{poly}(\lambda)$, *where* Q *is the number of queries to* SIGNO, poly *is independent of* Q, *and*

$$\mathrm{Adv}^{\mathrm{uf-cma}}_{\mathsf{SPS},\mathcal{A}}(\lambda) \leq \Delta^{\mathrm{core}}_{\mathcal{B}}(\lambda) + \mathrm{Adv}^{\mathrm{mddh}}_{\mathcal{PG},\mathbb{G}_2,\mathcal{D}_k,\mathcal{B}'}(\lambda) + \frac{Q}{p^k} + \frac{Q}{p}.$$

$\mathsf{Gen}(1^\lambda)$:	$\mathsf{Sign}(pk, sk, [\mathbf{m}]_1 \in \mathbb{G}_1^n)$:
$\mathcal{PG} \leftarrow \mathsf{GGen}(1^\lambda)$	$\mathbf{r} \leftarrow_R \mathbb{Z}_p^k\ [\mathbf{t}]_1 := [\mathbf{A}_0]_1\mathbf{r}$
$\mathbf{A}_0, \mathbf{A}_1 \leftarrow_R \mathcal{D}_{2k,k}$	$\Pi \leftarrow \mathsf{PPrv}(crs, [\mathbf{t}]_1, \mathbf{r})$
$pars := (\mathcal{PG}, [\mathbf{A}_0]_1, [\mathbf{A}_1]_1)$	$[\mathbf{u}]_1 := \mathbf{K}_0^\top [\mathbf{t}]_1 + \mathbf{K}^\top \begin{bmatrix}\mathbf{m}\\1\end{bmatrix}_1$
$crs \leftarrow \mathsf{PGen}(pars, 1^\lambda)$	**return** $\sigma := ([\mathbf{t}]_1, \Pi, [\mathbf{u}]_1)$
$\mathbf{A} \leftarrow_R \mathcal{D}_k$	
$\mathbf{K}_0 \leftarrow_R \mathbb{Z}_p^{2k\times(k+1)}$	
$\mathbf{K} \leftarrow_R \mathbb{Z}_p^{(n+1)\times(k+1)}$	$\mathsf{Ver}(pk, \sigma, [\mathbf{m}]_1)$:
$pk := (\mathcal{PG}, [\mathbf{A}_0]_1, crs, [\mathbf{A}]_2,$	**parse** $\sigma := ([\mathbf{t}]_1, \Pi, [\mathbf{u}]_1)$
$[\mathbf{K}_0\mathbf{A}]_2, [\mathbf{K}\mathbf{A}]_2)$	$b \leftarrow \mathsf{PVer}(pk, [\mathbf{t}]_1, \Pi)$
$sk := (\mathbf{K}_0, \mathbf{K})$	**if** $b = 1$ **and** $e([\mathbf{u}]_1^\top, [\mathbf{A}]_2) = e([\mathbf{t}]_1^\top, [\mathbf{K}_0\mathbf{A}]_2)$
return (pk, sk)	$+e(\begin{bmatrix}\mathbf{m}\\1\end{bmatrix}_1^\top, [\mathbf{K}\mathbf{A}]_2)$
	return 1
	else return 0

Fig. 7. Tightly UF-CMA secure structure-preserving signature scheme SPS with message space $\mathbb{G}_1^n$.

When using PS *from Sect. 2.5, we obtain*

$$\begin{aligned}\mathrm{Adv}^{\mathrm{uf-cma}}_{\mathsf{SPS},\mathcal{A}}(\lambda) \leq &(4k\lceil\log Q\rceil + 2) \cdot \mathrm{Adv}^{\mathrm{mddh}}_{\mathcal{PG},\mathbb{G}_1,\mathcal{D}_{2k,k},\mathcal{B}}(\lambda)\\ &+ (2\lceil\log Q\rceil + 3) \cdot \mathrm{Adv}^{\mathrm{mddh}}_{\mathcal{PG},\mathbb{G}_2,\mathcal{D}_k,\mathcal{B}'}(\lambda) + \lceil\log Q\rceil \cdot \Delta_{\mathcal{D}_{2k,k}}\\ &+ \tfrac{4\lceil\log Q\rceil+2}{p-1} + \tfrac{(Q+1)\lceil\log Q\rceil+Q}{p} + \tfrac{Q}{p^k}.\end{aligned}$$

Strategy. In a nutshell, we will embed a "shadow MAC" in our signature scheme, and then invoke the core lemma to randomize the MAC tags computed during signing queries and the final verification of $\mathcal{A}$'s forgery. A little more specifically, we will embed a term $\mathbf{k}_0^\top\mathbf{t}$ into the $\mathbf{A}$-orthogonal space of each $\mathbf{u}$ computed by SIGNO and VERO. (Intuitively, changes to this $\mathbf{A}$-orthogonal space do not influence the verification key, and simply correspond to changing from one signing key to another signing key that is compatible with the same verification key.) Using our core lemma, we can randomize this term $\mathbf{k}_0^\top\mathbf{t}$ to $(\mathbf{k}_0 + \mathbf{F}(\mathsf{ctr}))^\top\mathbf{t}$ for a random function $\mathbf{F}$ and a signature counter ctr. Intuitively, this means that we use a freshly randomized signing key for each signature query. After these changes, an adversary only has a statistically small chance in producing a valid forgery.

Proof (of Theorem 3). We proceed via a series of hybrid games G_0 to G_2, described in Fig. 8. By ε_i we denote the advantage of $\mathcal{A}$ to win G_i.

$\mathbf{Exp}^{\mathbf{uf\text{-}cma}}_{\mathbf{SPS},\mathcal{A}}(\boldsymbol{\lambda}) \rightsquigarrow \mathbf{G}_0$**:** Here we change the verification oracle as described in Fig. 8.

Note that a pair $(\mu^\star, \sigma^\star)$ that passes VERO in G_0 always passes the VERO check in $\mathrm{Exp}^{\mathrm{uf\text{-}cma}}_{\mathsf{SPS},\mathcal{A}}(\lambda)$. Thus, to bound $|\mathrm{Adv}^{\mathrm{uf\text{-}cma}}_{\mathsf{SPS},\mathcal{A}}(\lambda) - \varepsilon_0|$, it suffices to bound the probability that $\mathcal{A}$ produces a tuple $(\mu^\star, \sigma^\star)$ that passes VERO in $\mathrm{Exp}^{\mathrm{uf\text{-}cma}}_{\mathsf{SPS},\mathcal{A}}(\lambda)$,

G_0, $\boxed{\mathsf{G}_1, \mathsf{G}_2, \mathsf{G}_3, \mathsf{G}_4}$:	$\text{SIGNO}([\mathbf{m}]_1 \in \mathbb{G}_1^n)$:
$\mathcal{Q}_{\mathsf{sign}} := \emptyset$	$\mathcal{Q}_{\mathsf{sign}} := \mathcal{Q}_{\mathsf{sign}} \cup \{[\mathbf{m}]_1\}$
$\boxed{\mathsf{ctr} := 0}$, $\boxed{\widetilde{\mathsf{ctr}} \leftarrow_R [Q]}$	$\boxed{\mathsf{ctr} := \mathsf{ctr} + 1}$
$\mathcal{PG} \leftarrow \mathsf{GGen}(1^\lambda)$	$\mathbf{r} \leftarrow_R \mathbb{Z}_p^k$, $\boxed{\mathbf{r} \leftarrow_R (\mathbb{Z}_p^k)^*}$, $[\mathbf{t}]_1 := [\mathbf{A}_0]_1\mathbf{r}$
$\mathbf{A}_0, \mathbf{A}_1 \leftarrow_R \mathcal{D}_{2k,k}$	$\Pi \leftarrow \mathsf{PPrv}(crs, [\mathbf{t}]_1, \mathbf{r})$
$pars := (\mathcal{PG}, [\mathbf{A}_0]_1, [\mathbf{A}_1]_1)$	$[\mathbf{u}]_1 := \mathbf{K}_0^\top [\mathbf{t}]_1 + \mathbf{K}^\top \begin{bmatrix}\mathbf{m}\\1\end{bmatrix}_1$
$\mathbf{A} \leftarrow_R \mathcal{D}_k$	$\boxed{+\mathbf{a}^\perp(\mathbf{k}_0 \boxed{+\mathbf{F}(\mathsf{ctr})})^\top [\mathbf{t}]_1}$
$\boxed{\mathbf{a}^\perp \in \mathsf{orth}(\mathbf{A})}$	**return** $\sigma := ([\mathbf{t}]_1, \Pi, [\mathbf{u}]_1)$
$crs \leftarrow \mathsf{PGen}(pars, 1^\lambda)$	
$\mathbf{K}_0 \leftarrow_R \mathbb{Z}_p^{2k \times (k+1)}$	$\text{VERO}([\mathbf{m}^\star]_1, \sigma^\star)$:
$\boxed{\mathbf{k}_0 \leftarrow_R \mathbb{Z}_p^{2k}}$	**parse** $\sigma =: ([\mathbf{t}]_1, \Pi, [\mathbf{u}]_1)$
$\mathbf{K} \leftarrow_R \mathbb{Z}_p^{(n+1) \times (k+1)}$	$b \leftarrow \mathsf{PVer}(pk, [\mathbf{t}]_1, \Pi)$
$pk := (crs, pars, [\mathbf{A}]_2,$ $[\mathbf{K}_0\mathbf{A}]_2, [\mathbf{K}\mathbf{A}]_2)$	**if** $b = 1$ **and** $\boxed{\exists \mathsf{ctr}' \leq \mathsf{ctr}:}$
$sk := (\mathbf{K}_0, \mathbf{K})$	$\boxed{\mathsf{ctr}' = \widetilde{\mathsf{ctr}}}$ **and**
$([\mathbf{m}^\star]_1, \sigma^\star) \leftarrow_R \mathcal{A}^{\text{SIGNO}(\cdot)}(pk)$	$[\mathbf{u}]_1 = \mathbf{K}_0^\top [\mathbf{t}]_1 + \mathbf{K}^\top \begin{bmatrix}\mathbf{m}\\1\end{bmatrix}_1$
if $[\mathbf{m}^\star]_1 \notin \mathcal{Q}_{\mathsf{sign}}$ **and** $\text{VERO}([\mathbf{m}^\star]_1, \sigma^\star) = 1$	$\boxed{+\mathbf{a}^\perp(\mathbf{k}_0 \boxed{+\mathbf{F}(\mathsf{ctr}')})^\top [\mathbf{t}]_1}$
return 1	**return** 1
else return 0	**else return** 0

Fig. 8. Games G_0 to G_2 for proving Theorem 3. Here, $\mathbf{F} : \mathbb{Z}_p \to \mathbb{Z}_p^{2k}$ is a random function. In each procedure, the components inside a solid (dotted, double, gray) frame are only present in the games marked by a solid (dotted, double, gray) frame.

but not in G_0. For the signature $\sigma^\star =: ([\mathbf{t}]_1, \Pi, [\mathbf{u}]_1)$ we can write the verification equation in $\mathrm{Exp}^{\text{uf-cma}}_{\mathsf{SPS},\mathcal{A}}(\lambda)$ as

$$e([\mathbf{u}]_1^\top, [\mathbf{A}]_2) = e([\mathbf{t}]_1^\top, [\mathbf{K}_0\mathbf{A}]_2) + e(\begin{bmatrix}\mathbf{m}\\1\end{bmatrix}_1^\top, [\mathbf{K}\mathbf{A}]_2)$$

$$\Leftrightarrow e([\mathbf{u}]_1 - [\mathbf{t}]_1^\top \mathbf{K}_0 - \begin{bmatrix}\mathbf{m}\\1\end{bmatrix}_1^\top \mathbf{K}, [\mathbf{A}]_2) = \mathbf{0}$$

Observe that for any $(\mu^\star, ([\mathbf{t}]_1, \Pi, [\mathbf{u}]_1))$ that passes the verification equation in the experiment $\mathrm{Exp}^{\text{uf-cma}}_{\mathsf{SPS},\mathcal{A}}(\lambda)$, but not the one in G_0, the value

$$[\mathbf{u}]_1 - [\mathbf{t}]_1^\top \mathbf{K}_0 - \begin{bmatrix}\mathbf{m}\\1\end{bmatrix}_1^\top \mathbf{K}$$

is a non-zero vector in the kernel of $\mathbf{A}$. Thus, from $\mathcal{A}$ we can construct an adversary $\mathcal{B}$ against the $\mathcal{D}_k$-KMDH assumption. Finally, Lemma 2 yields an adversary

$\mathcal{B}'$ with $T(\mathcal{B}') \approx T(\mathcal{A}) + Q \cdot \mathsf{poly}(\lambda)$ such that

$$|\mathrm{Adv}^{\text{uf-cma}}_{\mathsf{SPS},\mathcal{A}}(\lambda) - \varepsilon_0| \leq \mathrm{Adv}^{\text{mddh}}_{\mathcal{PG},\mathbb{G}_2,\mathcal{D}_k,\mathcal{B}}(\lambda).$$

$\mathsf{G_0} \rightsquigarrow \mathsf{G_1}$**:** We can replace $\mathbf{K}_0$ by $\mathbf{K}_0 + \mathbf{k}_0(\mathbf{a}^\perp)^\top$ for $\mathbf{a}^\perp \in \mathsf{orth}(\mathbf{A})$ and $\mathbf{k}_i \leftarrow_R \mathbb{Z}_p^{2k}$, as both are distributed identically. Note that this change does not show up in the public key pk. Looking ahead, this change will allow us to use the computational core lemma (Lemma 4). This yields

$$\varepsilon_0 = \varepsilon_1.$$

$\mathsf{G_1} \rightsquigarrow \mathsf{G_2}$**:** Let $\mathcal{A}$ be an adversary playing either G_1 or G_2. We build an adversary $\mathcal{B}$ such that $T(\mathcal{B}) \approx T(\mathcal{A}) + Q \cdot \mathsf{poly}(\lambda)$ and

$$\Pr[\mathrm{Exp}^{\text{core}}_{0,\mathcal{B}}(1^\lambda) = 1] = \varepsilon_1 \text{ and } \Pr[\mathrm{Exp}^{\text{core}}_{1,\mathcal{B}}(1^\lambda) = 1] = \varepsilon_2.$$

This implies, by the core lemma (Lemma 4), that

$$\varepsilon_1 \leq \varepsilon_2 + \Delta^{\text{core}}_{\mathcal{B}}(\lambda).$$

We now describe $\mathcal{B}$ against $\mathrm{Exp}^{\text{core}}_{\beta,\mathcal{B}}(1^\lambda)$ for β equal to either 0 or 1. First, $\mathcal{B}$ receives $pp := (\mathcal{PG}, [\mathbf{A}_0]_1, crs)$ from $\mathrm{Exp}^{\text{core}}_{\beta,\mathcal{B}}(1^\lambda)$, then, $\mathcal{B}$ samples $\mathbf{A} \leftarrow_R \mathcal{D}_k$, $\mathbf{a}^\perp \in \mathsf{orth}(\mathbf{A})$, $\mathbf{K}_0 \leftarrow_R \mathbb{Z}_p^{2k \times (k+1)}$, $\mathbf{K} \leftarrow_R \mathbb{Z}_p^{(n+1)\times(k+1)}$ and forwards $pk := (\mathcal{PG}, [\mathbf{A}_0]_1, crs, [\mathbf{A}]_2, [\mathbf{K}_0\mathbf{A}]_2, [\mathbf{K}\mathbf{A}]_2)$ to $\mathcal{A}$.

To simulate $\textsc{SignO}([\mathbf{m}]_1)$, $\mathcal{B}$ uses its oracle $\textsc{TagO}$, which takes no input, and gives back $([\mathbf{t}]_1, \Pi, [u]_1)$. Then, $\mathcal{B}$ computes $[\mathbf{u}]_1 := \mathbf{K}_0^\top[\mathbf{t}]_1 + \mathbf{a}^\perp[u]_1 + \mathbf{K}^\top \begin{bmatrix} \mathbf{m} \\ 1 \end{bmatrix}_1$, and returns $\sigma := ([\mathbf{t}]_1, \Pi, [\mathbf{u}]_1)$ to $\mathcal{A}$.

Finally, given the forgery $([\mathbf{m}^\star]_1, \sigma^\star)$ with corresponding signature $\sigma^\star := ([\mathbf{t}^\star]_1, \Pi^\star, [\mathbf{u}^\star]_1)$, $\mathcal{B}$ first checks if $[\mathbf{m}^\star]_1 \notin \mathcal{Q}_{\mathsf{sign}}$ and $[\mathbf{u}^\star]_1 \neq [\mathbf{0}]_1$. If it is not the case, then $\mathcal{B}$ returns 0 to $\mathcal{A}$. If it is the case, with the knowledge of $\mathbf{a}^\perp \in \mathbb{Z}_p$, $\mathcal{B}$ efficiently checks whether there exists $[u^\star]_1 \in \mathbb{G}_1$ such that $[\mathbf{u}^\star]_1 - \mathbf{K}_0^\top[\mathbf{t}^\star]_1 - \mathbf{K}^\top \begin{bmatrix} \mathbf{m}^\star \\ 1 \end{bmatrix}_1 = [u^\star]_1\mathbf{a}^\perp$. If it is not the case, $\mathcal{B}$ returns 0 to $\mathcal{A}$. If it is the case, $\mathcal{B}$ computes $[u^\star]_1$ (it can do so efficiently given $\mathbf{a}^\perp$), sets $\mathsf{tag} := ([\mathbf{t}^\star]_1, \Pi^\star, [u^\star]_1)$, calls its verification oracle $\textsc{VerO}(\mathsf{tag})$, and forwards the answer to $\mathcal{A}$.

$\mathsf{G_2} \rightsquigarrow \mathsf{G_3}$**:** In game G_2 the vectors $\mathbf{r}$ sampled by $\textsc{SignO}$ are uniformly random over $\mathbb{Z}_p^k$, while they are uniformly random over $(\mathbb{Z}_p^k)^* = \mathbb{Z}_p^k \backslash \{0\}$ in G_3. Since this is the only difference between the games, the difference of advantage is bounded by the statistical distance between the two distributions of $\mathbf{r}$. A union bound over the number of queries yields

$$\varepsilon_2 - \varepsilon_3 \leq \tfrac{Q}{p^k}.$$

$\mathsf{G_3} \rightsquigarrow \mathsf{G_4}$**:** These games are the same except for the extra condition $\widetilde{\mathsf{ctr}} = \mathsf{ctr}'$ in G_4, which happens with probability $\frac{1}{Q}$ over the choice of $\widetilde{\mathsf{ctr}} \leftarrow_R [Q]$. Since the adversary view is independent of $\widetilde{\mathsf{ctr}}$, we have

$$\varepsilon_4 = \tfrac{\varepsilon_3}{Q}.$$

Game $\mathbf{G}_4$: We prove that $\varepsilon_4 \leq \frac{1}{p}$.

First, we can replace $\mathbf{K}$ by $\mathbf{K} + \mathbf{v}(\mathbf{a}^\perp)^\top$ for $\mathbf{v} \leftarrow_R \mathbb{Z}_p^{n+1}$, and $\{\mathbf{F}(i) : i \in [Q], i \neq \widetilde{\mathsf{ctr}}\}$ by $\{\mathbf{F}(i) + \mathbf{w}_i : i \in [Q], i \neq \widetilde{\mathsf{ctr}}\}$ for $\mathbf{w}_i \leftarrow_R \mathbb{Z}_p^{2k}$. Note that this does not change the distribution of the game.

Thus, for the i-th signing query with $i \neq \widetilde{\mathsf{ctr}}$ the value $\mathbf{u}$ is computed by $\text{SIGNO}([\mathbf{m}_i]_1)$ as

$$[\mathbf{u}]_1 = \mathbf{K}_0^\top [\mathbf{t}]_1 + (\mathbf{K}^\top \boxed{+\mathbf{a}^\perp \mathbf{v}^\top}) \begin{bmatrix} \mathbf{m}_i \\ 1 \end{bmatrix}_1 + \mathbf{a}^\perp (\mathbf{k}_0 + \mathbf{F}(i) \boxed{+\mathbf{w}_i})^\top [\mathbf{t}]_1,$$

with $[\mathbf{t}]_1 := [\mathbf{A}_0]_1 \mathbf{r}$, $\mathbf{r} \leftarrow_R (\mathbb{Z}_p^k)^*$. This is identically distributed to

$$[\mathbf{u}]_1 = \mathbf{K}_0^\top [\mathbf{t}]_1 + \mathbf{K}^\top \begin{bmatrix} \mathbf{m}_i \\ 1 \end{bmatrix}_1 + \gamma_i \cdot \mathbf{a}^\perp, \text{ with } \gamma_i \leftarrow_R \mathbb{Z}_p.$$

For the $\widetilde{\mathsf{ctr}}$'th signing query, we have

$$[\mathbf{u}]_1 = \mathbf{K}_0^\top [\mathbf{t}]_1 + (\mathbf{K}^\top \boxed{+\mathbf{a}^\perp \mathbf{v}^\top}) \begin{bmatrix} \mathbf{m}_{\widetilde{\mathsf{ctr}}} \\ 1 \end{bmatrix}_1 + \mathbf{a}^\perp (\mathbf{k}_0 + \mathbf{F}(\widetilde{\mathsf{ctr}}))^\top [\mathbf{t}]_1.$$

Assuming $\mathcal{A}$ succeeds in producing a valid forgery, VERO computes

$$[\mathbf{u}^\star]_1 = \mathbf{K}_0^\top [\mathbf{t}^\star]_1 + (\mathbf{K}^\top \boxed{+\mathbf{a}^\perp \mathbf{v}^\top}) \begin{bmatrix} \mathbf{m}^\star \\ 1 \end{bmatrix}_1 + \mathbf{a}^\perp (\mathbf{k}_0 + \mathbf{F}(\widetilde{\mathsf{ctr}}))^\top [\mathbf{t}]_1.$$

Since $\mathbf{m}^\star \neq \mathbf{m}_{\widetilde{\mathsf{ctr}}}$ by definition of the security game, we can use the pairwise independence of $\mathbf{m} \mapsto \mathbf{v}^\top \begin{bmatrix} \mathbf{m} \\ 1 \end{bmatrix}_1$ to argue that $\mathbf{v}^\top \begin{bmatrix} \mathbf{m}^\star \\ 1 \end{bmatrix}_1$ and $\mathbf{v}^\top \begin{bmatrix} \mathbf{m}_{\widetilde{\mathsf{ctr}}} \\ 1 \end{bmatrix}_1$ are two independent values, uniformly random over $\mathbb{G}_1$. Thus, the verification equation is satisfied with probability at most $\frac{1}{p}$, that is

$$\varepsilon_4 \leq \frac{1}{p}.$$

Bilateral Structure-Preserving Signature Scheme. Our structure-preserving signature scheme, SPS, defined in Fig. 7 can sign only messages from $\mathbb{G}_1^n$. By applying the generic transformation from [39, Sect. 6], we can transform our SPS to sign messages from $\mathbb{G}_1^{n_1} \times \mathbb{G}_2^{n_2}$ using their two-tier SPS, which is a generalization of [1]. The transformation is tightness-preserving by Theorem 6 of [39] and costs additional k elements from $\mathbb{G}_1$ and $k + 1$ elements from $\mathbb{G}_2$ in the signature. For the SXDH assumption ($k = 1$), our bilateral SPS scheme requires additional 1 element from $\mathbb{G}_1$ and 2 elements from $\mathbb{G}_2$ in the signature.

References

1. Abe, M., Chase, M., David, B., Kohlweiss, M., Nishimaki, R., Ohkubo, M.: Constant-size structure-preserving signatures: generic constructions and simple assumptions. In: Wang, X., Sako, K. (eds.) ASIACRYPT 2012. LNCS, vol. 7658, pp. 4–24. Springer, Heidelberg (2012). https://doi.org/10.1007/978-3-642-34961-4_3
2. Abe, M., Chase, M., David, B., Kohlweiss, M., Nishimaki, R., Ohkubo, M.: Constant-size structure-preserving signatures: generic constructions and simple assumptions. J. Cryptol. **29**(4), 833–878 (2016). https://doi.org/10.1007/s00145-015-9211-7
3. Abe, M., David, B., Kohlweiss, M., Nishimaki, R., Ohkubo, M.: Tagged one-time signatures: tight security and optimal tag size. In: Kurosawa, K., Hanaoka, G. (eds.) PKC 2013. LNCS, vol. 7778, pp. 312–331. Springer, Heidelberg (2013). https://doi.org/10.1007/978-3-642-36362-7_20
4. Abe, M., Fuchsbauer, G., Groth, J., Haralambiev, K., Ohkubo, M.: Structure-preserving signatures and commitments to group elements. J. Cryptol. **29**(2), 363–421 (2016). https://doi.org/10.1007/s00145-014-9196-7
5. Abe, M., Groth, J., Haralambiev, K., Ohkubo, M.: Optimal structure-preserving signatures in asymmetric bilinear groups. In: Rogaway, P. (ed.) CRYPTO 2011. LNCS, vol. 6841, pp. 649–666. Springer, Heidelberg (2011). https://doi.org/10.1007/978-3-642-22792-9_37
6. Abe, M., Hofheinz, D., Nishimaki, R., Ohkubo, M., Pan, J.: Compact structure-preserving signatures with almost tight security. In: Katz, J., Shacham, H. (eds.) CRYPTO 2017. LNCS, vol. 10402, pp. 548–580. Springer, Cham (2017). https://doi.org/10.1007/978-3-319-63715-0_19
7. Acar, T., Lauter, K., Naehrig, M., Shumow, D.: Affine pairings on ARM. In: Abdalla, M., Lange, T. (eds.) Pairing 2012. LNCS, vol. 7708, pp. 203–209. Springer, Heidelberg (2013). https://doi.org/10.1007/978-3-642-36334-4_13
8. Attrapadung, N., Hanaoka, G., Yamada, S.: A framework for identity-based encryption with almost tight security. In: Iwata, T., Cheon, J.H. (eds.) ASIACRYPT 2015. LNCS, vol. 9452, pp. 521–549. Springer, Heidelberg (2015). https://doi.org/10.1007/978-3-662-48797-6_22
9. Barreto, P.S.L.M., Costello, C., Misoczki, R., Naehrig, M., Pereira, G.C.C.F., Zanon, G.: Subgroup security in pairing-based cryptography. In: Lauter, K., Rodríguez-Henríquez, F. (eds.) LATINCRYPT 2015. LNCS, vol. 9230, pp. 245–265. Springer, Cham (2015). https://doi.org/10.1007/978-3-319-22174-8_14
10. Belenkiy, M., Chase, M., Kohlweiss, M., Lysyanskaya, A.: P-signatures and non-interactive anonymous credentials. In: Canetti, R. (ed.) TCC 2008. LNCS, vol. 4948, pp. 356–374. Springer, Heidelberg (2008). https://doi.org/10.1007/978-3-540-78524-8_20
11. Bellare, M., Boldyreva, A., Micali, S.: Public-key encryption in a multi-user setting: security proofs and improvements. In: Preneel, B. (ed.) EUROCRYPT 2000. LNCS, vol. 1807, pp. 259–274. Springer, Heidelberg (2000). https://doi.org/10.1007/3-540-45539-6_18
12. Bellare, M., Goldwasser, S.: New paradigms for digital signatures and message authentication based on non-interactive zero knowledge proofs. In: Brassard, G. (ed.) CRYPTO 1989. LNCS, vol. 435, pp. 194–211. Springer, New York (1990). https://doi.org/10.1007/0-387-34805-0_19

13. Blazy, O., Fuchsbauer, G., Izabachène, M., Jambert, A., Sibert, H., Vergnaud, D.: Batch Groth–Sahai. In: Zhou, J., Yung, M. (eds.) ACNS 2010. LNCS, vol. 6123, pp. 218–235. Springer, Heidelberg (2010). https://doi.org/10.1007/978-3-642-13708-2_14
14. Blazy, O., Kiltz, E., Pan, J.: (Hierarchical) identity-based encryption from affine message authentication. In: Garay, J.A., Gennaro, R. (eds.) CRYPTO 2014. LNCS, vol. 8616, pp. 408–425. Springer, Heidelberg (2014). https://doi.org/10.1007/978-3-662-44371-2_23
15. Blum, M., Feldman, P., Micali, S.: Non-interactive zero-knowledge and its applications (extended abstract). In: 20th ACM STOC, pp. 103–112. ACM Press, May 1988
16. Boneh, D., Mironov, I., Shoup, V.: A secure signature scheme from bilinear maps. In: Joye, M. (ed.) CT-RSA 2003. LNCS, vol. 2612, pp. 98–110. Springer, Heidelberg (2003). https://doi.org/10.1007/3-540-36563-X_7
17. Camenisch, J., Dubovitskaya, M., Haralambiev, K.: Efficient structure-preserving signature scheme from standard assumptions. In: Visconti, I., De Prisco, R. (eds.) SCN 2012. LNCS, vol. 7485, pp. 76–94. Springer, Heidelberg (2012). https://doi.org/10.1007/978-3-642-32928-9_5
18. Cathalo, J., Libert, B., Yung, M.: Group encryption: non-interactive realization in the standard model. In: Matsui, M. (ed.) ASIACRYPT 2009. LNCS, vol. 5912, pp. 179–196. Springer, Heidelberg (2009). https://doi.org/10.1007/978-3-642-10366-7_11
19. Chase, M., Kohlweiss, M.: A new hash-and-sign approach and structure-preserving signatures from DLIN. In: Visconti, I., De Prisco, R. (eds.) SCN 2012. LNCS, vol. 7485, pp. 131–148. Springer, Heidelberg (2012). https://doi.org/10.1007/978-3-642-32928-9_8
20. Chen, J., Gong, J., Weng, J.: Tightly secure IBE under constant-size master public key. In: Fehr, S. (ed.) PKC 2017. LNCS, vol. 10174, pp. 207–231. Springer, Heidelberg (2017). https://doi.org/10.1007/978-3-662-54365-8_9
21. Chen, J., Wee, H.: Fully, (almost) tightly secure IBE and dual system groups. In: Canetti, R., Garay, J.A. (eds.) CRYPTO 2013. LNCS, vol. 8043, pp. 435–460. Springer, Heidelberg (2013). https://doi.org/10.1007/978-3-642-40084-1_25
22. Dodis, Y., Kiltz, E., Pietrzak, K., Wichs, D.: Message authentication, revisited. In: Pointcheval, D., Johansson, T. (eds.) EUROCRYPT 2012. LNCS, vol. 7237, pp. 355–374. Springer, Heidelberg (2012). https://doi.org/10.1007/978-3-642-29011-4_22
23. Enge, A., Milan, J.: Implementing cryptographic pairings at standard security levels. In: Chakraborty, R.S., Matyas, V., Schaumont, P. (eds.) SPACE 2014. LNCS, vol. 8804, pp. 28–46. Springer, Cham (2014). https://doi.org/10.1007/978-3-319-12060-7_3
24. Escala, A., Herold, G., Kiltz, E., Ràfols, C., Villar, J.: An algebraic framework for Diffie-Hellman assumptions. In: Canetti, R., Garay, J.A. (eds.) CRYPTO 2013. LNCS, vol. 8043, pp. 129–147. Springer, Heidelberg (2013). https://doi.org/10.1007/978-3-642-40084-1_8
25. Gay, R., Hofheinz, D., Kiltz, E., Wee, H.: Tightly CCA-secure encryption without pairings. In: Fischlin, M., Coron, J.-S. (eds.) EUROCRYPT 2016. LNCS, vol. 9665, pp. 1–27. Springer, Heidelberg (2016). https://doi.org/10.1007/978-3-662-49890-3_1
26. Gay, R., Hofheinz, D., Kohl, L.: Kurosawa-Desmedt meets tight security. In: Katz, J., Shacham, H. (eds.) CRYPTO 2017. LNCS, vol. 10403, pp. 133–160. Springer, Cham (2017). https://doi.org/10.1007/978-3-319-63697-9_5

27. Gong, J., Chen, J., Dong, X., Cao, Z., Tang, S.: Extended nested dual system groups, revisited. In: Cheng, C.-M., Chung, K.-M., Persiano, G., Yang, B.-Y. (eds.) PKC 2016. LNCS, vol. 9614, pp. 133–163. Springer, Heidelberg (2016). https://doi.org/10.1007/978-3-662-49384-7_6
28. Grewal, G., Azarderakhsh, R., Longa, P., Hu, S., Jao, D.: Efficient implementation of bilinear pairings on ARM processors. In: Knudsen, L.R., Wu, H. (eds.) SAC 2012. LNCS, vol. 7707, pp. 149–165. Springer, Heidelberg (2013). https://doi.org/10.1007/978-3-642-35999-6_11
29. Groth, J.: Simulation-sound NIZK proofs for a practical language and constant size group signatures. In: Lai, X., Chen, K. (eds.) ASIACRYPT 2006. LNCS, vol. 4284, pp. 444–459. Springer, Heidelberg (2006). https://doi.org/10.1007/11935230_29
30. Groth, J., Lu, S.: A non-interactive shuffle with pairing based verifiability. In: Kurosawa, K. (ed.) ASIACRYPT 2007. LNCS, vol. 4833, pp. 51–67. Springer, Heidelberg (2007). https://doi.org/10.1007/978-3-540-76900-2_4
31. Groth, J., Ostrovsky, R., Sahai, A.: New techniques for noninteractive zero-knowledge. J. ACM **59**(3), 1–35 (2012). https://doi.org/10.1145/2220357.2220358. ISSN: 0004-5411. http://doi.acm.org/10.1145/2220357.2220358
32. Groth, J., Sahai, A.: Efficient non-interactive proof systems for bilinear groups. In: Smart, N. (ed.) EUROCRYPT 2008. LNCS, vol. 4965, pp. 415–432. Springer, Heidelberg (2008). https://doi.org/10.1007/978-3-540-78967-3_24
33. Hofheinz, D.: Adaptive partitioning. In: Coron, J.-S., Nielsen, J.B. (eds.) EUROCRYPT 2017. LNCS, vol. 10212, pp. 489–518. Springer, Cham (2017). https://doi.org/10.1007/978-3-319-56617-7_17
34. Hofheinz, D.: Algebraic partitioning: fully compact and (almost) tightly secure cryptography. In: Kushilevitz, E., Malkin, T. (eds.) TCC 2016. LNCS, vol. 9562, pp. 251–281. Springer, Heidelberg (2016). https://doi.org/10.1007/978-3-662-49096-9_11
35. Hofheinz, D., Jager, T.: Tightly secure signatures and public-key encryption. In: Safavi-Naini, R., Canetti, R. (eds.) CRYPTO 2012. LNCS, vol. 7417, pp. 590–607. Springer, Heidelberg (2012). https://doi.org/10.1007/978-3-642-32009-5_35
36. Hofheinz, D., Koch, J., Striecks, C.: Identity-based encryption with (almost) tight security in the multi-instance, multi-ciphertext setting. In: Katz, J. (ed.) PKC 2015. LNCS, vol. 9020, pp. 799–822. Springer, Heidelberg (2015). https://doi.org/10.1007/978-3-662-46447-2_36
37. Jutla, C.S., Roy, A.: Improved structure preserving signatures under standard bilinear assumptions. In: Fehr, S. (ed.) PKC 2017. LNCS, vol. 10175, pp. 183–209. Springer, Heidelberg (2017). https://doi.org/10.1007/978-3-662-54388-7_7
38. Jutla, C.S., Roy, A.: Switching lemma for bilinear tests and constant-size NIZK proofs for linear subspaces. In: Garay, J.A., Gennaro, R. (eds.) CRYPTO 2014. LNCS, vol. 8617, pp. 295–312. Springer, Heidelberg (2014). https://doi.org/10.1007/978-3-662-44381-1_17
39. Kiltz, E., Pan, J., Wee, H.: Structure-preserving signatures from standard assumptions, revisited. In: Gennaro, R., Robshaw, M. (eds.) CRYPTO 2015. LNCS, vol. 9216, pp. 275–295. Springer, Heidelberg (2015). https://doi.org/10.1007/978-3-662-48000-7_14
40. Kiltz, E., Wee, H.: Quasi-adaptive NIZK for linear subspaces revisited. In: Oswald, E., Fischlin, M. (eds.) EUROCRYPT 2015. LNCS, vol. 9057, pp. 101–128. Springer, Heidelberg (2015). https://doi.org/10.1007/978-3-662-46803-6_4
41. Kurosawa, K., Desmedt, Y.: A new paradigm of hybrid encryption scheme. In: Franklin, M. (ed.) CRYPTO 2004. LNCS, vol. 3152, pp. 426–442. Springer, Heidelberg (2004). https://doi.org/10.1007/978-3-540-28628-8_26

42. Libert, B., Joye, M., Yung, M., Peters, T.: Concise multi-challenge CCA-secure encryption and signatures with almost tight security. In: Sarkar, P., Iwata, T. (eds.) ASIACRYPT 2014. LNCS, vol. 8874, pp. 1–21. Springer, Heidelberg (2014). https://doi.org/10.1007/978-3-662-45608-8_1
43. Libert, B., Peters, T., Joye, M., Yung, M.: Compactly hiding linear spans. In: Iwata, T., Cheon, J.H. (eds.) ASIACRYPT 2015. LNCS, vol. 9452, pp. 681–707. Springer, Heidelberg (2015). https://doi.org/10.1007/978-3-662-48797-6_28
44. Libert, B., Peters, T., Yung, M.: Short group signatures via structure-preserving signatures: standard model security from simple assumptions. In: Gennaro, R., Robshaw, M. (eds.) CRYPTO 2015. LNCS, vol. 9216, pp. 296–316. Springer, Heidelberg (2015). https://doi.org/10.1007/978-3-662-48000-7_15
45. Morillo, P., Ràfols, C., Villar, J.L.: The kernel matrix Diffie-Hellman assumption. In: Cheon, J.H., Takagi, T. (eds.) ASIACRYPT 2016. LNCS, vol. 10031, pp. 729–758. Springer, Heidelberg (2016). https://doi.org/10.1007/978-3-662-53887-6_27
46. Ràfols, C.: Stretching Groth-Sahai: NIZK proofs of partial satisfiability. In: Dodis, Y., Nielsen, J.B. (eds.) TCC 2015. LNCS, vol. 9015, pp. 247–276. Springer, Heidelberg (2015). https://doi.org/10.1007/978-3-662-46497-7_10

Private Simultaneous Messages

The Communication Complexity of Private Simultaneous Messages, Revisited

Benny Applebaum[1(✉)], Thomas Holenstein[2], Manoj Mishra[1], and Ofer Shayevitz[1]

[1] Tel Aviv University, Tel Aviv, Israel
benny.applebaum@gmail.com, mishra.m@gmail.com, ofersha@gmail.com
[2] Google, Zurich, Switzerland
thomas.holenstein@gmail.com

Abstract. Private Simultaneous Message (PSM) protocols were introduced by Feige, Kilian and Naor (STOC '94) as a minimal non-interactive model for information-theoretic three-party secure computation. While it is known that every function $f : \{0,1\}^k \times \{0,1\}^k \rightarrow \{0,1\}$ admits a PSM protocol with exponential communication of $2^{k/2}$ (Beimel et al., TCC '14), the best known (non-explicit) lower-bound is $3k - O(1)$ bits. To prove this lower-bound, FKN identified a set of simple requirements, showed that any function that satisfies these requirements is subject to the $3k - O(1)$ lower-bound, and proved that a random function is likely to satisfy the requirements.

We revisit the FKN lower-bound and prove the following results:

(Counterexample) We construct a function that satisfies the FKN requirements but has a PSM protocol with communication of $2k + O(1)$ bits, revealing a gap in the FKN proof.

(PSM lower-bounds) We show that, by imposing additional requirements, the FKN argument can be fixed leading to a $3k - O(\log k)$ lower-bound for a random function. We also get a similar lower-bound for a function that can be computed by a polynomial-size circuit (or even polynomial-time Turing machine under standard complexity-theoretic assumptions). This yields the first non-trivial lower-bound for an explicit Boolean function partially resolving an open problem of Data, Prabhakaran and Prabhakaran (Crypto '14, IEEE Information Theory '16). We further extend these results to the setting of imperfect PSM protocols which may have small correctness or privacy error.

(CDS lower-bounds) We show that the original FKN argument applies (as is) to some weak form of PSM protocols which are strongly related to the setting of Conditional Disclosure of Secrets (CDS). This connection yields a simple combinatorial criterion for establishing linear $\Omega(k)$-bit CDS lower-bounds. As a corollary, we settle the complexity of the Inner Product predicate resolving an open problem of Gay, Kerenidis, and Wee (Crypto '15).

T. Holenstein—This work was done while the author was at ETH Zurich.

J. B. Nielsen and V. Rijmen (Eds.): EUROCRYPT 2018, LNCS 10821, pp. 261–286, 2018.
https://doi.org/10.1007/978-3-319-78375-8_9

1 Introduction

Information theoretic cryptography studies the problem of secure communication and computation in the presence of computationally unbounded adversaries. Unlike the case of computational cryptography whose full understanding is closely tied to basic open problems in computational complexity, information theoretic solutions depend "only" on non-computational (typically combinatorial or algebraic) objects. One may therefore hope to gain a full understanding of the power and limitations of information theoretic primitives. Indeed, Shannon's famous treatment of perfectly secure symmetric encryption [30] provides an archetypical example for such a study.

Unfortunately, for most primitives, the picture is far from being complete. This is especially true for the problem of *secure function evaluation* (SFE) [33], in which a set of parties $P_1, \ldots, P_m$ wish to jointly evaluate a function f over their inputs while keeping those inputs private. Seminal completeness results show that *any function* can be securely evaluated with information theoretic security [10,13] (or computational security [19,33]) under various adversarial settings. However, the *communication complexity* of these solutions is tied to the *computational complexity* of the function (i.e., its circuit size), and it is unknown whether this relation is inherent. For instance, as noted by Beaver, Micali, and Rogaway [8] three decades ago, we cannot even rule out the possibility that any function can be securely computed by a constant number of parties with communication that is polynomial in the input length, even in the simple setting where the adversary passively corrupts a single party. More generally, the communication complexity of securely computing a function (possibly via an inefficient protocol) is wide open, even in the most basic models.

1.1 A Minimal Model for Secure Computation

In light of the above, it makes sense to study the limitation of information theoretic secure computation in its simplest form. In [16] Feige, Kilian and Naor (hereinafter referred to as FKN) presented such a "Minimal Model for Secure Computation". In this model, Alice and Bob hold private inputs, x and y, and they wish to let Charlie learn the value of $f(x, y)$ without leaking any additional information. The communication pattern is minimal. Alice and Bob each send to Charlie a single message, a and b respectively, which depends on the party's input and on a random string r which is shared between Alice and Bob but is hidden from Charlie. Given (a, b) Charlie should be able to recover $f(x, y)$ without learning additional information. The parties are assumed to be computationally unbounded, and the goal is to minimize the communication complexity of the protocol (i.e., the total number of bits sent by Alice and Bob). Following [23], we refer to such a protocol as a *private simultaneous message* protocol (PSM) (Fig. 1).

Definition 1 (Private Simultaneous Messages). *A private simultaneous message (PSM) protocol* $\Pi = (\Pi_A, \Pi_B, g)$ *for a function* $f : \mathcal{X} \times \mathcal{Y} \to \mathcal{Z}$ *is a triple of functions* $\Pi_A : \mathcal{X} \times \mathcal{R} \to \mathcal{A}$, $\Pi_B : \mathcal{Y} \times \mathcal{R} \to \mathcal{B}$, *and* $g : \mathcal{A} \times \mathcal{B} \to \mathcal{Z}$ *that satisfy the following two properties.*

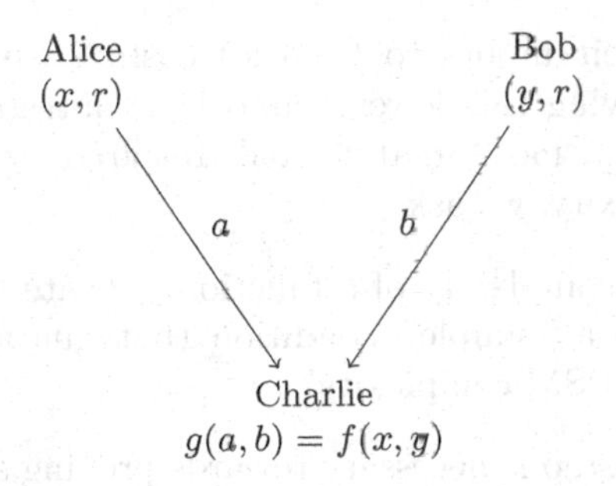

Fig. 1. Schematic of a PSM protocol.

- *(δ-Correctness) The protocol has correctness error of δ if for every $(x, y) \in \mathcal{X} \times \mathcal{Y}$ it holds that*

$$\Pr_{r \xleftarrow{\$} \mathcal{R}} [f(x, y) \neq g(\Pi_A(x, r), \Pi_B(y, r))] \leq \delta$$

- *(ϵ-Privacy) The protocol has privacy error of ϵ if for every pair of inputs $(x, y) \in \mathcal{X} \times \mathcal{Y}$ and $(x', y') \in \mathcal{X} \times \mathcal{Y}$ for which $f(x, y) = f(x', y')$ the random variables*

$$(\Pi_A(x, r), \Pi_B(y, r)) \quad \text{and} \quad (\Pi_A(x', r), \Pi_A(y', r)), \tag{1}$$

induced by a uniform choice of $r \xleftarrow{\$} \mathcal{R}$, are ϵ-close in statistical distance.

We mainly consider perfect *protocols which enjoy both* perfect correctness *($\delta = 0$) and* perfect privacy *($\epsilon = 0$). We define the* communication complexity *of the protocol to be* $\log |\mathcal{A}| + \log |\mathcal{B}|$.

The correctness and privacy conditions assert that, for every pair of inputs (x, y) and (x', y'), the transcript distributions are either close to each other when $f(x, y) = f(x', y')$, or far apart when $f(x, y) \neq f(x', y')$. Hence, the joint computation of Alice and Bob, $C_r(x, y) = (\Pi_A(x, r), \Pi_B(y, r))$, can be also viewed as a "randomized encoding" [5,24] (or "garbled version") of the function $f(x, y)$ that has the property of being 2-decomposable into an x-part and a y-part. Being essentially non-interactive, such protocols (and their multiparty variants [23]) have found various applications in cryptography (cf. [2,22]). Moreover, it was shown in [6,9] that PSM is the strongest model among several other non-interactive models for secret-sharing and zero-knowledge proofs.

FKN showed that any function $f : \{0, 1\}^k \times \{0, 1\}^k \rightarrow \{0, 1\}$ admits a PSM protocol [16]. The best known communication complexity is polynomial for log-space computable functions [16] and $O(2^{k/2})$ for general functions [9]. While it seems likely that some functions require super-polynomial communication, the best known lower-bound, due to the original FKN paper, only shows that a random function requires $3k - O(1)$ bits of communication. This lower-bound is somewhat weak but still non-trivial since an insecure solution (in which Alice and Bob just send their inputs to Charlie) costs $2k$ bits of communication. The question of improving this lower-bound is an intriguing open problem. In this paper, we aim for a more modest goal. Inspired by the general theory of communication complexity, we ask:

> How does the PSM complexity of a function f relate to its combinatorial properties? Is there a "simple" condition that guarantees a non-trivial lower-bound on the PSM complexity?

We believe that such a step is necessary towards proving stronger lower-bounds. Additionally, as we will see, this question leads to several interesting insights for related information-theoretic tasks.

1.2 Revisiting the FKN Lower-Bound

Our starting point is the original proof of the $3k$ lower-bound from [16]. In order to prove a lower-bound FKN relax the privacy condition by requiring that Charlie will not be able to recover the last bit of Alice's input. Formally, let us denote by $\bar{x}$ the string obtained by flipping the last bit of x. Then, the privacy condition (Eq. 1) is relaxed to hold only over *sibling* inputs (x, y) and $(\bar{x}, y)$ for which $f(x, y) = f(\bar{x}, y)$. We refer to this relaxation as *weak privacy.* Since (standard) privacy implies weak privacy, it suffices to lower-bound the communication complexity of weakly private PSM protocols.

To prove a lower-bound for random functions, FKN (implicitly) identify three conditions which hold for most functions and show that if a function $f : \{0,1\}^k \times \{0,1\}^k \rightarrow \{0,1\}$ satisfies these conditions then any weak PSM for f has communication complexity of at least $3k - O(1)$. The FKN conditions are:

1. The function f is *non-degenerate*, namely, for every $x \neq x'$ there exists y for which $f(x, y) \neq f(x', y)$ and similarly, for every $y \neq y'$ there exists x for which $f(x, y) \neq f(x, y')$.
2. The function is *useful* in the sense that for at least $\frac{1}{2} - o(1)$ of the inputs (x, y) it holds that $f(x, y) = f(\bar{x}, y)$ where $\bar{x}$ denotes the string x with its last bit flipped. (An input (x, y) for which the equation holds is referred to as being *useful.*[1])
3. We say that $(x_1, \ldots, x_m) \times (y_1, \ldots, y_n)$ is a *complement similar* rectangle of f if $f(x_i, y_j) = f(\bar{x}_i, y_j)$ for every $1 \leq i \leq m$ and $1 \leq j \leq n$. Then, f has no complement similar rectangle of size mn larger than $M = 2^{k+1}$. Equivalently, the function $f'(x, y) = f(x, y) - f(\bar{x}, y)$, which can be viewed as a partial derivative of f with respect to its last coordinate, has no 0-monochromatic rectangle of size M.

We observe that the above conditions are, in fact, insufficient to prove a non-trivial lower-bound. As a starting point, we note that the inner-product function

[1] In the FKN terminology such an input (x, y) is referred to as being dangerous.

has low PSM complexity and has no large monochromatic rectangles. While the inner-product function cannot be used directly as a counterexample (since it has huge complement similar rectangles), we can construct a related function f such that: (1) the derivative f' is (a variant of) the inner product function and so f' has no large monochromatic rectangles; and (2) by applying some local preprocessing on Alice's input, the computation of $f(x, y)$ reduces to the computation of the inner product function. Altogether, we prove the following theorem (see Sect. 3).

Theorem 1 (FKN counterexample). *There exists a function* $f : \{0,1\}^k \times \{0,1\}^k \to \{0,1\}$ *that satisfies the FKN conditions but has a (standard) PSM of communication complexity of* $2k + O(1)$.

Let us take a closer look at the proof of the FKN lower-bound to see where the gap is. The FKN proof boils down to showing that the set S_r of all possible transcripts (a, b) sent by Alice and Bob under a random string r, has relatively small intersection with the set $S_{r'}$ of all possible transcripts (a, b) sent by Alice and Bob under a different random string r'. Such a collision, $c = (a, b) \in S_r \cap S_{r'}$, is counted as a *trivial collision* if the inputs (x, y) that generate c under r are the same as the inputs (x', y') that generate c under r'. Otherwise, the collision is counted as *non-trivial*. The argument mistakenly assumes that all non-trivial collisions are due to sibling inputs, i.e., $(x', y') = (\bar{x}, y)$. In other words, it is implicitly assumed that the transcript (a, b) *fully reveals* all the information about (x, y) except for the last input of x. (In addition to the value of $f(x, y)$ which is revealed due to the correctness property.) Indeed, we show that the FKN argument holds if one considers fully-revealing PSM protocols. (See Theorem 8 for a slightly stronger version.)

Theorem 2 (LB's against weakly private fully revealing PSM). *Let* $f : \{0,1\}^k \times \{0,1\}^k \to \{0,1\}$ *be a non-degenerate function. Let* M *be an upper-bound on size of the largest complement similar rectangle of* f *and let* U *be a lower-bound on the number of useful inputs of* f. *Then, any weakly-private fully-revealing PSM for* f *has communication complexity of at least* $2 \log U - \log M - O(1)$. *In particular, for all but* $o(1)$ *fraction of the functions* $f : \{0,1\}^k \times \{0,1\}^k \to \{0,1\}$, *we get a lower-bound of* $3k - O(1)$.

A lower-bound of c bits against fully-revealing weakly-private PSM easily yields a lower-bound of $c - 2k + 1$ bits for PSM. (Since a standard PSM can be turned into a fully-revealing weakly-private PSM by letting Alice/Bob append $x[1 : k - 1]$ and y to their messages.) Unfortunately, this loss (of $2k$ bits) makes the $3k$ bit lower-bound useless. Moreover, Theorem 1 shows that this loss is unavoidable. Put differently, fully-revealing weakly-private PSM may be more expensive than standard PSM. Nevertheless, as we will see in Sect. 1.4, lower-bounds for fully-revealing weakly-private PSM have useful implications for other models.

1.3 Fixing the PSM Lower-Bound

We show that the FKN argument can be fixed by posing stronger requirements on f. Roughly speaking, instead of limiting the size of complement similar rectangles, we limit the size of any pair of *similar* rectangles by a parameter M. That is, if the restriction of f to the ordered rectangle $R = (x_1, \ldots, x_m) \times (y_1, \ldots, y_\ell)$ is equal to the restriction of f to the ordered rectangle $R' = (x'_1, \ldots, x'_m) \times (y'_1, \ldots, y'_\ell)$ and the rectangles are disjoint in the sense that either $x_i \neq x'_i$ for every i, or $y_j \neq y'_j$ for every j, then the size $m\ell$ of R should be at most M. (See Sect. 2 for a formal definition.)

Theorem 3 (perfect-PSM LB's). *Let $\mathcal{X}, \mathcal{Y}$ be sets of size at least 2, and let $f : \mathcal{X} \times \mathcal{Y} \to \{0,1\}$ be a non-degenerate function for which any pair of disjoint similar rectangles (R, R') satisfies $|R| \leq M$. Then, any perfect PSM for f has communication of at least $2(\log|\mathcal{X}| + \log|\mathcal{Y}|) - \log M - 3$.*

The theorem is proved by a distributional version of the FKN argument which also implies Theorem 2. (See Sect. 4.) As a corollary, we recover the original lower-bound claimed by FKN.

Corollary 1. *For a $1 - o(1)$ fraction of the functions $f : \{0,1\}^k \times \{0,1\}^k \to \{0,1\}$ any perfect PSM protocol for f requires $3k - 2\log k - O(1)$ bits of total communication.*[2]

Proof. It is not hard to verify that $1 - o(1)$ fraction of all functions are non-degenerate. In Sect. 6 we further show that, for $1 - o(1)$ of the functions, any pair of disjoint similar rectangles (R, R') satisfies $|R| \leq k^2 \cdot 2^k$. The proof follows from Theorem 3. □

By partially de-randomizing the proof, we show that the above lower-bound applies to a function that is computable by a family of polynomial-size circuits, or, under standard complexity-theoretic assumptions, by a polynomial-time Turing machine. This resolves an open question of Data, Prabhakaran and Prabhakaran [15] who proved a similar lower-bound for an explicit *non-boolean* function $f : \{0,1\}^k \times \{0,1\}^k \to \{0,1\}^{k-1}$. Prior to our work, we could not even rule out the (absurd!) possibility that all efficiently computable functions admit a perfect PSM with communication of $2k + o(k)$.

Theorem 4. *There exists a sequence of polynomial-size circuits*

$$f = \{f_k : \{0,1\}^k \times \{0,1\}^k \to \{0,1\}\}$$

such that any perfect PSM for f_k has communication complexity of at least $3k - O(\log k)$ bits. Moreover, assuming the existence of a hitting-set generator against co-nondeterministic uniform algorithms, f is computable by a polynomial-time Turing machine.[3]

[2] The constant 2 can be replaced by any constant larger than 1.

[3] It is worth mentioning that the proof of Theorem 4 strongly relies on the explicit combinatorial condition given in Theorem 3 (and we do not know how to obtain it directly from Corollary 1). This illustrates again the importance of relating PSM complexity to other more explicit properties of functions.

Remark 1 (On the hitting-set generator assumption). The exact definition of a hitting-set generator against co-nondeterministic uniform algorithms is postponed to Sect. 6. For now, let us just say that the existence of such a generator follows from standard Nissan-Wigderson type complexity-theoretic assumptions. In particular, it suffices to assume that the class E of functions computable in $2^{O(n)}$-deterministic time contains a function that has no sub-exponential nondeterministic circuits [28], or, more liberally, that some function in E has no sub-exponential time Arthur-Merlin protocol [21]. (See also the discussion in [7].)

Lower-bounds for imperfect PSM's. We extend Theorem 3 to handle imperfect PSM protocols by strengthening the non-degeneracy condition and the non self-similarity condition. This can be used to prove an imperfect version of Corollary 1 showing that, for almost all functions, an imperfect PSM with correctness error δ and privacy error ϵ must communicate at least

$$\min\{3k - 2\log(k), 2k + \log(1/\epsilon), 2k + \log(1/\delta)\} - O(1)$$

bits. An analogous extension of Theorem 4, yields a similar bound for an explicit function. (See Sect. 5.)

1.4 Applications to Conditional Disclosure of Secrets

We move on to the closely related model of *Conditional Disclosure of Secrets* (CDS) [18]. In the CDS model, Alice holds an input x and Bob holds an input y, and, in addition, Alice holds a secret bit s. The referee, Charlie, holds both x and y, but does not know the secret s. Similarly to the PSM case, Alice and Bob use shared randomness to compute the messages a and b that are sent to Charlie. The CDS requires that Charlie can recover s from (a, b) if and only if the predicate $f(x, y)$ evaluates to one.[4]

Definition 2 (Conditional Disclosure of Secrets). *A conditional disclosure of secrets (CDS) protocol $\Pi = (\Pi_A, \Pi_B, g)$ for a predicate $f : \mathcal{X}\times\mathcal{Y} \to \{0,1\}$ and domain $\mathcal{S}$ of secrets is a triple of functions $\Pi_A : \mathcal{X}\times\mathcal{S}\times\mathcal{R} \to \mathcal{A}$, $\Pi_B : \mathcal{Y}\times\mathcal{R} \to \mathcal{B}$ and $g : \mathcal{X} \times \mathcal{Y} \times \mathcal{A} \times \mathcal{B} \to \mathcal{S}$ that satisfy the following two properties:*

1. *(Perfect Correctness) For every (x, y) that satisfies f and any secret $s \in \mathcal{S}$ we have that:*

$$\Pr_{r \xleftarrow{\$} \mathcal{R}} [g(x, y, \Pi_A(x, s, r), \Pi_B(y, r)) \neq s] = 0.$$

2. *(Perfect Privacy) For every input (x, y) that does not satisfy f and any pair of secrets $s, s' \in \mathcal{S}$ the distributions*

$$(x, y, \Pi_A(x, s, r), \Pi_B(y, r)) \quad \textit{and} \quad (x, y, \Pi_A(x, s', r), \Pi_B(y, r)),$$

induced by $r \xleftarrow{\$} \mathcal{R}$ are identically distributed.

[4] Usually, it is assumed that both Alice and Bob hold the secret s. It is not hard to see that this variant and our variant (in which only Alice knows the secret) are equivalent up to at most 1-bit of additional communication.

The communication complexity *of the CDS protocol is* $(\log |\mathcal{A}| + \log |\mathcal{B}|)$ *and its* randomness complexity *is* $\log |\mathcal{R}|$. *By default, we assume that the protocol supports single-bit secrets* ($\mathcal{S} = \{0,1\}$).[5]

Intuitively, CDS is weaker than PSM since it either releases s or keeps it private but it cannot *manipulate the secret data.*[6] Still, this notion has found useful applications in various contexts such as information-theoretically private information retrieval (PIR) protocols [14], priced oblivious transfer protocols [1], secret sharing schemes for graph-based access structures (cf. [11,12,31]), and attribute-based encryption [20,29].

The communication complexity of CDS. In light of the above, it is interesting to understand the communication complexity of CDS. Protocols with communication of $O(t)$ were constructed for t-size Boolean formula by [18] and were extended to t-size (arithmetic) branching programs by [25] and to t-size (arithmetic) span programs by [6]. Until recently, the CDS complexity of a general predicate $f : \{0,1\}^k \times \{0,1\}^k \to \{0,1\}$ was no better than its PSM complexity, i.e., $O(2^{k/2})$ [9]. This was improved to $2^{O(\sqrt{k \log k})}$ by Liu, Vaikuntanathan and Wee [27]. Moreover, Applebaum et al. [4] showed that, for very long secrets, the amortized complexity of CDS can be reduced to $O(\log k)$ bits per bit of secret. Very recently, the amortized cost was further reduced to $O(1)$ establishing the existence of general CDS with constant rate [3].

Lower-bounds for the communication complexity of CDS were first established by Gay et al. [17]. Their main result shows that the CDS communication of a predicate f is at least logarithmic in its randomized one-way communication complexity, and leads to an $\Omega(\log k)$ lower-bound for several explicit functions. Applebaum et al. [4] observed that weakly private PSM reduces to CDS. This observation together with the $3k$-bit FKN lower-bound for weakly private PSM has lead to a CDS lower-bound of $k - o(k)$ bits for some non-explicit predicate. (The reduction loses about $2k$ bits.)

In this paper, we further exploit the connection between CDS and PSM by observing that CDS protocols for a predicate $h(x,y)$ give rise to weakly private fully revealing PSM for the function $f((x \circ s), y) = h(x,y) \wedge s$, where $\circ$ denotes concatenation. By using our lower-bounds for weakly private fully revealing PSM's we get the following theorem. (See Sect. 7 for a proof.)

Theorem 5. *Let* $h : \mathcal{X} \times \mathcal{Y} \to \{0,1\}$ *be a predicate. Suppose that* M *upper-bounds the size of the largest 0-monochromatic rectangle of* h *and that for every* $x \in \mathcal{X}$, *the residual function* $h(x, \cdot)$ *is not the constant zero function. Then, the communication complexity of any perfect CDS for* h *is at least*

$$2 \log |f^{-1}(0)| - \log M - \log |\mathcal{X}| - \log |\mathcal{Y}| - 1,$$

where $|f^{-1}(0)|$ *denotes the number of inputs* (x,y) *that are mapped to zero.*

[5] One may consider imperfect variants of CDS. In this paper we restrict our attention to the (more common) setting of perfect CDS.

[6] This is analogous to the relation between Functional Encryption and Attribute Based Encryption. Indeed, CDS can be viewed as an information-theoretic one-time variant of Attribute Based Encryption.

Unlike the non-explicit lower-bound of [4], the above theorem provides a simple and clean sufficient condition for proving non-trivial CDS lower-bounds. For example, we can easily show that a random function has at least linear CDS complexity.

Corollary 2. *For all but a* $o(1)$ *fraction of the predicates* $h : \{0,1\}^k \times \{0,1\}^k \rightarrow \{0,1\}$, *any perfect CDS for* h *has communication of at least* $k - 4 - o(1)$.

Proof. Let $h : \{0,1\}^k \times \{0,1\}^k \rightarrow \{0,1\}$ be a randomly chosen predicate. Let $K = 2^k$ and let $\epsilon = 1/\sqrt{K}$. There are exactly $2^K \cdot 2^K = 2^{2K}$ rectangles. Therefore, by a union-bound, the probability of having a 0-monochromatic rectangle of size $M = 2K(1+\epsilon)$ is at most

$$2^{2K} \cdot 2^{-M} = 2^{-2\epsilon K} = 2^{-\Omega(\sqrt{K})}.$$

Also, since h has K^2 inputs, the probability of having less than $(\frac{1}{2} - \epsilon) \cdot K^2$ unsatisfying inputs is, by a Chernoff bound, $2^{-\Omega(\epsilon^2 K^2)} = 2^{-\Omega(K)}$. Finally, by the union bound, the probability that there exists $x \in \mathcal{X}$ for which $h(x, \cdot)$ is the all-zero function is at most $K \cdot 2^{-K}$. It follows, by Theorem 5, that with probability of $1 - 2^{-\Omega(\sqrt{K})}$, the function h has a CDS complexity of at least $k - 4 - o(1)$. □

We can also get lower-bounds for explicit functions. For example, Gay et al. [17] studied the CDS complexity of the binary inner product function $h(x, y) = \langle x, y \rangle$. They proved an upper-bound of $k+1$ bits and a lower-bound of $\Omega(\log k)$ bits, and asked as an open question whether a lower-bound of $\Omega(k)$ can be established. (The question was open even for the special case of linear CDS for which [17] proved an $\Omega(\sqrt{k})$ lower-bound). By plugging the inner-product predicate into Theorem 5, we conclude:

Corollary 3. *Any perfect CDS for the inner product predicate* $h_{ip} : \{0,1\}^k \times \{0,1\}^k \rightarrow \{0,1\}$ *requires at least* $k - 3 - o(1)$ *bits of communication.*

Proof. It suffices to prove the lower bound for the restriction of inner-product in which $x \neq 0^n$. It is well known (cf. [26]) that the largest monochromatic rectangle is of size $M = 2^k$, and the number of "zero" inputs is exactly $S = 2^{2k-1} - 2^k$. Hence, Theorem 5 yields a lower-bound of $k - 3 - o(1)$. □

This lower-bound matches the $k+1$ upper-bound up to a constant additive difference (of 4 bits). It also implies that in any ABE scheme for the inner-product function which is based on the dual system methodology [32] either the ciphertext or the secret-key must be of length $\Omega(k)$. (See [17] for discussion.)

Organization. Following some preliminaries (Sect. 2), we present the counter example for the FKN lower-bound (Sect. 3). We then analyze the communication complexity of perfect PSM (Sect. 4) and imperfect PSM (Sect. 5). Based on these results, we obtain PSM lower-bounds for random and explicit functions (Sect. 6), as well as CDS lower-bounds (Sect. 7).

2 Preliminaries

For a string (or a vector) x of length n, and indices $1 \leq i \leq j \leq n$, we let $x[i]$ denote the i-th entry of x, and let $x[i:j]$ denote the string $(x[i], x[i+1] \dots, x[j])$. By convention, all logarithms are taken base 2.

Rectangles. An *(ordered) rectangle* of size $m \times n$ over some finite domain $\mathcal{X} \times \mathcal{Y}$ is a pair $\rho = (\mathbf{x}, \mathbf{y})$, where $\mathbf{x} = (x_1, \dots, x_m) \subseteq \mathcal{X}^m$ and $\mathbf{y} = (y_1, \dots, y_n) \subseteq \mathcal{Y}^n$ satisfy $x_i \neq x_j$ and $y_i \neq y_j$ for all $i \neq j$. We say that (x, y) belongs to ρ if $x = x_i$ and $y = y_j$ for some i, j (or by abuse of notation we simply write $x \in \mathbf{x}$ and $y \in \mathbf{y}$). The size of an $m \times n$ rectangle ρ is mn, and its density with respect to some probability distribution μ over $\mathcal{X} \times \mathcal{Y}$, is $\sum_{x \in \mathbf{x}, y \in \mathbf{y}} \mu(x, y)$. Let $\rho = (\mathbf{x}, \mathbf{y})$ and $\rho' = (\mathbf{x}', \mathbf{y}')$ be a a pair of $m \times n$-rectangles. We say that ρ and ρ' are *x-disjoint* (resp., *y-disjoint*) if $x_i \neq x'_i$ for all $i \in \{1, \dots, m\}$ (resp., if $y_j \neq y'_j$ for all $j \in \{1, \dots, n\}$). We say that ρ and ρ' are disjoint if they are either x-disjoint or y-disjoint.

As an example, consider the three 2×3 rectangles $\rho_1 = ((1,2),(5,6,7))$, $\rho_2 = ((2,1),(6,5,4))$, and $\rho_3 = ((1,3),(7,5,6))$. Among those, ρ_1 and ρ_3 are y-disjoint but not x-disjoint, ρ_2 and ρ_3 are x-disjoint but not y-disjoint, and ρ_1 and ρ_2 are both x-disjoint and y-disjoint. Therefore, each of these pairs is considered to be disjoint.

If $f : \mathcal{X} \times \mathcal{Y} \to \mathcal{Z}$ is a function and ρ a rectangle of size $m \times n$, we let $f_{[\rho]}$ be the matrix M of size $m \times n$ whose entry M_{ij} is $f(x_i, y_j)$. A rectangle ρ is *0-monochromatic* (resp., 1-monochromatic) if $f_{[\rho]}$ is the all-zero matrix (resp., all-one matrix). A rectangle ρ is *similar* to a rectangle ρ' (with respect to f) if $f_{[\rho]} = f_{[\rho']}$. A rectangle $(\mathbf{x} = (x_1, \dots, x_m), \mathbf{y})$ is *complement similar* if it is similar to the rectangle $((\bar{x}_1, \dots, \bar{x}_m), \mathbf{y})$, where $\bar{x}$ denotes the string x with its last bit flipped.

Probabilistic notation. We will use calligraphic letters $\mathcal{A}$, $\mathcal{B}$, ..., to denote finite sets. Lower case letters denote values from these sets, i.e., $x \in \mathcal{X}$. Upper case letters usually denote random variables (unless the meaning is clear from the context).

Given two random variables A and B over the same set $\mathcal{A}$, we use $\|A - B\|$ to denote their *statistical distance* $\|A - B\| = \frac{1}{2} \sum_{a \in \mathcal{A}} |\Pr[A = a] - \Pr[B = a]|$. The min-entropy of A, denoted by $H_\infty(A)$, is minus the logarithm of the probability of the most likely value of A, i.e., $-\log \max_{a \in \mathcal{A}} \Pr[A = a]$.

3 A Counterexample to the FKN Lower-Bound

Let $\mathbf{T}_0, \mathbf{T}_1$ be a pair of $(k-1) \times (k-1)$ non-singular matrices (over the binary field $\mathbb{F} = \mathrm{GF}[2]$) with the property that $\mathbf{T} = \mathbf{T}_0 + \mathbf{T}_1$ is also non-singular. (The existence of such matrices is guaranteed via a simple probabilistic argument.[7]) Define the mapping $L : \mathbb{F}^k \to \mathbb{F}^k$ by

$$x \mapsto (\mathbf{T}_{x[k]} \cdot x[1:k-1]) \circ x[k],$$

where $\circ$ denotes concatenation. That is, if the last entry of x is zero then L applies $\mathbf{T}_0$ to the $k-1$ prefix $x' = x[1:k-1]$ and extends the resulting $k-1$ vector by an additional 0 entry, and if $x[k] = 1$ then the prefix x' is sent to $\mathbf{T}_1 x'$ and the vector is extended by an additional 1 entry. Note that L is a bijection (since $\mathbf{T}_0, \mathbf{T}_1$ are non-singular). The function $f : \mathbb{F}^k \times \mathbb{F}^k \to \mathbb{F}^k$ is defined by

$$(x, y) \mapsto \langle L(x), y \rangle,$$

where $\langle \cdot, \cdot \rangle$ denotes the inner-product function over $\mathbb{F}$.

In Sect. 3.1, we will prove that f satisfies the FKN conditions (described in Sect. 1.2).

Lemma 1. *The function f is (1) non-degenerate, (2) useful, and (3) its largest complement similar rectangle is of size at most $M = 2^{k+1}$.*

Recall that f is non-degenerate if for every distinct $x \neq x'$ (resp., $y \neq y'$) the residual functions $f(x, \cdot)$ and $f(x', \cdot)$ (resp., $f(\cdot, y')$ and $f(\cdot, y')$) are distinct. It is useful if $\Pr_{x,y}[f(x, y) \neq f(\bar{x}, y)] \geq \frac{1}{2}$, where $\bar{x}$ denotes the string x with its last entry flipped. Also, a rectangle $R = (\mathbf{x}, \mathbf{y})$ is complement similar if $f(x, y) = f(\bar{x}, y)$ for every $x \in \mathbf{x}, y \in \mathbf{y}$.

In Sect. 3.2 we will show that f admits a PSM with communication complexity of $2k + O(1)$.

Lemma 2. *The function f has a PSM protocol with communication complexity of $2k + 2$.*

Theorem 1 follows from Lemmas 1 and 2.

---

[7] When $k-1$ is even, there is a simple deterministic construction: Take $\mathbf{T}_0$ (resp., $\mathbf{T}_1$) to be the upper triangular matrix (resp., lower triangular matrix) whose entries on and above main diagonal (resp., on and below the diagonal) are ones and all other entries are zero. It is not hard to verify that both matrices are non-singular. Also $\mathbf{T} = \mathbf{T}_0 + \mathbf{T}_1$ has a zero diagonal and ones in all other entries and so $\mathbf{T}$ has full rank if $k-1$ is even. The same construction can be used when $k-1$ is odd, at the expense of obtaining a matrix $\mathbf{T}$ with an almost full rank that has only minor affect on the parameter M obtained in Lemma 1.

3.1 f Satisfies the FKN Properties (Proof of Lemma 1)

(1) f is non-degenerate. Fix $x_1 \neq x_2 \in \mathbb{F}^k$ and observe that $L(x_1) \neq L(x_2)$ (since L is a bijection). Therefore there exists y for which $f(x_1, y) = \langle L(x_1), y\rangle \neq \langle L(x_2), y\rangle = f(x_2, y)$. (In fact this holds for half of y's). Similarly, for every $y_1 \neq y_2$ there exists $v \in \mathbb{F}^k$ for which $\langle v, y_1\rangle \neq \langle v, y_2\rangle$, and since L is a bijection we can take $x = L^{-1}(v)$ and get that $f(x, y_1) = \langle v, y_1\rangle \neq \langle v, y_2\rangle = f(x, y_2)$.

(2) f is useful. Choose $x' \xleftarrow{\$} \mathbb{F}^{k-1}$ and $y \xleftarrow{\$} \mathbb{F}^k$ and observe that $f(x' \circ 0, y) = f(x' \circ 1, y)$ if and only if

$$\langle \mathbf{T}x', y[1 : k-1]\rangle + y_k = 0,$$

which happens with probability $\frac{1}{2}$.

(3) The largest complement similar rectangle is of size at most 2^{k+1}. Fix some rectangle $R = (\mathbf{x}, \mathbf{y})$, where $\mathbf{x} = (x_1, \ldots, x_m) \in (\mathbb{F}^k)^m$ and $\mathbf{y} = (y_1, \ldots, y_n) \in (\mathbb{F}^k)^n$. We show that if R is complement similar then $mn \leq 2 \cdot 2^k$. Since R is complement similar for every $x \in \mathbf{x}, y \in \mathbf{y}$ it holds

$$f(x, y) = f(\bar{x}, y),$$

which by definition of f implies that

$$\langle \mathbf{T}x' \circ 1, y\rangle = 0,$$

where x' is the $(k-1)$ prefix of x. Let d be the dimension of the linear subspace spanned by the vectors in $\mathbf{x}$, and so $m \leq 2^d$. Since $\mathbf{T}$ has full rank, the dimension of the subspace V spanned by $\{(\mathbf{T}x[1 : k-1] \circ 1) : x \in \mathbf{x}\}$ is at least $d-1$. (We may lose 1 in the dimension due to the removal of the last entry of the vectors $x \in \mathbf{x}$.) Noting that every $y \in \mathbf{y}$ is orthogonal to V, we conclude that the dimension of the subspace spanned by $\mathbf{y}$ is at most $k - (d-1)$. It follows that $n \leq 2^{k-(d-1)}$ and so $mn < 2 \cdot 2^k$. □

3.2 PSM for f (Proof of Lemma 2)

Note that f can be expressed as applying the inner product to v and y where v can be locally computed based on x. Hence it suffices to construct a PSM for the inner-product function and let Alice compute v and apply the inner-product protocol to v. (This reduction is a special instance of the so-called substitution lemma of randomize encoding, cf. [2,22].) Lemma 2 now follows from the following lemma.

Lemma 3. *The inner product function $h_{ip} : \mathbb{F}^k \times \mathbb{F}^k \rightarrow \mathbb{F}$ has a PSM protocol with communication complexity of* $2k + 2$.

A proof of the lemma appears[8] in [27, Corollary 3]. For the sake of self-containment we describe here an alternative proof.

[8] We thank the anonymous reviewer for pointing this out.

Proof. We show a PSM $\Pi = (\Pi_A, \Pi_B, g)$ with communication $2k$ under the promise that the inputs of Alice and Bob, x, y, are both not equal to the all zero vector. To get a PSM for the general case, let Alice and Bob locally extend their inputs x, y to $k+1$-long inputs $x' = x \circ 1$ and $y' = y \circ 1$. Then run the protocol Π and at the end let Charlie flip the outcome. It is easy to verify that the reduction preserves correctness and privacy. Since the inputs are longer by a single bit the communication becomes $2(k+1)$ as promised.

We move on to describe the protocol Π. The common randomness consists of a random invertible matrix $\mathbf{R} \in \mathbb{F}^{k\times k}$. Given non-zero $x \in \mathbb{F}^k$, Alice outputs $a = \mathbf{R}x$ where x is viewed as a column vector. Bob, who holds $y \in \mathbb{F}^k$, outputs $b = y^T\mathbf{R}^{-1}$. Charlie outputs ba.

Prefect correctness is immediate: $(y^T\mathbf{R}^{-1}) \cdot (\mathbf{R}x) = y^T x$, as required. To prove perfect privacy, we use the following claim.

Claim 6. *Let $x, y \in \mathbb{F}^k$ be non-zero vectors and denote their inner-product by z. Then, there exists an invertible matrix $\mathbf{M} \in \mathbb{F}^{k\times k}$ for which $\mathbf{M}e_1 = x$ and $v_z^T\mathbf{M}^{-1} = y^T$ where e_i is the i-th unit vector, and v_z is taken to be e_1 if $z = 1$ and e_k if $z = 0$.*

Proof. Let us first rewrite the condition $v_z^T\mathbf{M}^{-1} = y^T$ as $v_z^T = y^T\mathbf{M}$. Let $V \subset \mathbb{F}^k$ be the linear subspace of all vectors that are orthogonal to y. Note that the dimension of V is $k-1$. We distinguish between two cases based on the value of z.

Suppose that $z = 0$, that is, $x \in V$ and $v_z = e_k$. Then set the first column of $\mathbf{M}$ to be x and choose the next $k-2$ columns $\mathbf{M}_2, \ldots, \mathbf{M}_{k-1}$ so that together with x they form a basis for V. Let the last column $\mathbf{M}_k$ be some vector outside V. Observe that the columns are linearly independent and so $\mathbf{M}$ is invertible. Also, it is not hard to verify that $\mathbf{M}e_1 = x$ and that $y^T\mathbf{M} = e_k^T$.

Next, consider the case where $z = 1$, that is, $x \notin V$ and $v_z = e_1$. Then, take $\mathbf{M}_1 = x$ and let the other columns $\mathbf{M}_2, \ldots, \mathbf{M}_k$ to be some basis for V. Since x is non-zero the columns of $\mathbf{M}$ are linearly independent. Also, $\mathbf{M}e_1 = x$ and $y^T\mathbf{M} = e_1^T$. The claim follows. □

We can now prove perfect privacy. Fix some non-zero $x, y \in \mathbb{F}^k$ and let $z = \langle x, y\rangle$. We show that the joint distribution of the messages (A, B) depends only on z. In particular, (A, B) is distributed identically to $(\mathbf{R}e_1, v_b^T\mathbf{R}^{-1})$ where $\mathbf{R}$ a random invertible matrix. Indeed, letting $\mathbf{M}$ be the matrix guaranteed in Claim 6 we can write

$$(\mathbf{R}x, y^T\mathbf{R}^{-1}) = (\mathbf{R}(\mathbf{M}e_1), (v_z^T\mathbf{M}^{-1})\mathbf{R}^{-1}).$$

Noting that $\mathbf{T} = \mathbf{RM}$ is also a random invertible matrix (since the the set of invertible matrices forms a group) we conclude that the RHS is identically distributed to $\mathbf{T}e_1, v_z^T\mathbf{T}^{-1}$, as claimed. □

Remark 2. Overall the PSM for f has the following form: Alice sends $a = \mathbf{R} \cdot (L(x) \circ 1)$ and Bob sends $b = (y \circ 1)^T\mathbf{R}$ where $\mathbf{R} \in \mathbb{F}^{(k+1)\times(k+1)}$ is a random invertible matrix. The privacy proof shows that if the input (x, y) is mapped to (a, b) for some $\mathbf{R}$ then for every (x', y') for which $f(x, y) = f(x', y')$, there exists

$\mathbf{R}'$ under which the input (x', y') is mapped to (a, b) as well. Hence, there are collisions between non-sibling inputs. As explained in the introduction, this makes the FKN lower-bound inapplicable.

4 Lower Bound for Perfect PSM Protocols

In this Section we will prove a lower bound for perfect PSM protocols.

Definition 3. *For a function $f : \mathcal{X} \times \mathcal{Y} \rightarrow \mathcal{Z}$ and distribution μ over the domain $\mathcal{X} \times \mathcal{Y}$ with marginals μ_A and μ_B, define*

$$\alpha(\mu) = \max_{(R_1, R_2)} \min(\mu(R_1), \mu(R_2)),$$

where the maximum ranges over all pairs of similar disjoint rectangles (R_1, R_2). We also define

$$\beta(\mu) = \Pr[\,(X, Y) \neq (X', Y') \mid f(X, Y) = f(X', Y')\,],$$

where (X, Y) and (X', Y') represent two independent samples from μ. Finally, we say that f is non-degenerate *with respect to μ if for every $x \neq x'$ in the support of μ_A there exists some $y \in \mathcal{Y}$ for which $f(x, y) \neq f(x', y)$, and similarly for every $y \neq y'$ in the support of μ_B there exists some $x \in \mathcal{X}$ for which $f(x, y) \neq f(x, y')$.*

We prove the following key lemma.

Lemma 4. *Let $f : \mathcal{X} \times \mathcal{Y} \rightarrow \mathcal{Z}$. Then the communication complexity of any perfect PSM protocol is at least*

$$\max_{\mu} \log(1/\alpha(\mu)) + H_\infty(\mu) - \log(1/\beta(\mu)) - 1,$$

where the maximum is taken over all (not necessarily product) distribution μ under which f is non-degenerate.

The lower-bound is meaningful as long as β is not too small. Intuitively, this makes sure that the privacy requirement (which holds only over inputs on which the function agrees) is not trivial to achieve under μ.

For the special case of a Boolean function f, we can use the uniform distribution over $\mathcal{X} \times \mathcal{Y}$ and prove Theorem 3 from the introduction (restated here for the convenience of the reader).

Theorem 7 (Theorem 3 restated). *Let $\mathcal{X}, \mathcal{Y}$ be sets of size at least 2. Let $f : \mathcal{X} \times \mathcal{Y} \rightarrow \{0, 1\}$ be a non-degenerate function for which any pair of disjoint similar rectangles (R, R') satisfies $|R| \leq M$. Then, any perfect PSM for f has communication of at least $2(\log|\mathcal{X}| + \log|\mathcal{Y}|) - \log M - 3$.*

Proof. For the uniform distribution μ we have $\alpha(\mu) \leq M/(|\mathcal{X}||\mathcal{Y}|)$, $H_\infty(\mu) = \log|\mathcal{X}| + \log|\mathcal{Y}|$ and

$$\beta(\mu) \geq \Pr[(X,Y) \neq (X',Y')] - \Pr[f(X,Y) \neq f(X',Y')],$$

where X,Y and X',Y' are two independent copies of uniformly distributed inputs. The minuend is $1 - 1/(|\mathcal{X}||\mathcal{Y}|)$ and the subtrahend is at most $\frac{1}{2}$ (since f is Boolean). For $|\mathcal{X}||\mathcal{Y}| \geq 4$, we get $\beta(\mu) \geq 1/4$, and the proof follows from the key lemma (Lemma 4). □

We note that the constant 3 can be replaced by $2 + o_k(1)$ when the size of the domain $\mathcal{X} \times \mathcal{Y}$ grows with k.

Weakly Private Fully Revealing PSM. We can also derive a lower-bound on the communication complexity of weakly private fully revealing PSM. We begin with a formal definition.

Definition 4 (Weakly Private Fully Revealing PSM). *A weakly private fully revealing PSM $\Pi = (\Pi_A, \Pi_B, g)$ for a function $f : \mathcal{X} \times \mathcal{Y} \to \mathcal{Z}$ is a perfect PSM for the function $f' : \{0,1\}^{k_1} \times \{0,1\}^{k_2} \to \{0,1\}^{k_1-1} \times \{0,1\}^{k_2} \times \{0,1\}$ that takes (x,y) and outputs $(x[1:k_1-1], y, f(x,y))$, where $x[1:k_1-1]$ is the k_1-1 prefix of x.*

In the following, we say that f is *weakly non-degenerate* if for every x there exists y such that $f(x,y) \neq f(\bar{x},y)$. Recall that an input (x,y) is useful if $f(x,y) = f(\bar{x},y)$. We prove the following (stronger) version of Theorem 2 from the introduction.

Theorem 8. *Let $f : \{0,1\}^{k_1} \times \{0,1\}^{k_2} \to \{0,1\}$ be a weakly non-degenerate function. Let M be an upper-bound on size of the largest complement similar rectangle of f and let U be a lower-bound on the number of useful inputs of f. Then, any weakly-private fully-revealing PSM for f has communication complexity of at least $2\log U - \log M - 2$. In particular, for all but an $o(1)$ fraction of the predicates $f : \{0,1\}^k \times \{0,1\}^k \to \{0,1\}$ we get a lower-bound of $3k - 4 - o(1)$.*

Proof. Let f' be the function defined in Definition 4 based on f. We will prove a lower-bound on the communication complexity of any perfect PSM for f'. Let μ be the uniform distribution over the set of useful inputs. Since f is weakly non-degenerate the function f' is non-degenerate under μ. Also, observe that

$$\alpha(\mu) \leq M/U, \quad \beta(\mu) = 1/2, \quad \text{and} \quad H_\infty(\mu) \geq \log U.$$

The first part of the theorem follows from Lemma 4.

To prove the second ("in particular") part observe that, for a random function f, each pair of inputs (x,y) and $(\bar{x},y)$ gets the same f-value with probability $\frac{1}{2}$ independently of other inputs. Hence, with all but $o(1)$ probability, a fraction of $\frac{1}{2} - o(1)$ of all 2^{2k-1} of the pairs is mapped to the same value, and so there will be $2^{2k-1}(1-o(1))$ useful inputs. (Since each successful pair contributes two useful inputs.) Also, each M-size rectangle R is complement similar with probability 2^{-M}. By taking a union bound over all $2^{2^{k+1}}$ rectangles, we conclude that f has an $M = 2^{k+1}(1+o(1))$-size complement similar rectangle with probability at most $2^{2^{k+1}-M} = o(1)$. We conclude that, all but an $o(1)$ fraction of the functions, do not have weakly-private fully-revealing PSM with complexity smaller than $3k - 4 - o(1)$. □

4.1 Proof of the Key Lemma (Lemma 4)

Fix some function $f : \mathcal{X} \times \mathcal{Y} \to \mathcal{Z}$ and let $\Pi = (\Pi_A, \Pi_B, g)$ be a perfect PSM protocol for f. Let μ denote some distribution over the domain $\mathcal{X} \times \mathcal{Y}$ and assume that f is non-degenerate with respect to μ.

We will use a probabilistic version of the FKN proof. In particular, consider two independent executions of Π on inputs that are sampled independently from μ. We let X, Y and R (resp., X', Y' and R') denote the random variables that represent the inputs of Alice and Bob and their shared randomness in the first execution (resp., second execution). Thus, we can for example write $\Pr[(A, B) = (A', B') \wedge X \neq X']$ to denote the probability that the messages in the two executions match while the two inputs for Alice are different.

To simplify notation somewhat, we define the following events:

$$\begin{aligned}
\mathcal{P}^{(=)} &:\equiv (A = A') \wedge (B = B') \\
\mathcal{I}^{(=)} &:\equiv (X = X') \wedge (Y = Y') \\
\mathcal{I}^{(\neq)} &:\equiv (X \neq X') \vee (Y \neq Y') \equiv \neg \mathcal{I}^{(=)} \\
\mathcal{F}^{(=)} &:\equiv f(X, Y) = f(X', Y')
\end{aligned}$$

(The notation $\mathcal{P}$ is chosen to indicate equivalence/inequivalence of *Protocol* message and $\mathcal{I}$ to indicate equivalence/inequivalence of the *Inputs*.) Our lower-bound follows from the following claims.

Claim 9. *The communication complexity of Π is at least* $\log(1/\Pr[\mathcal{I}^{(\neq)} \wedge \mathcal{P}^{(=)}]) - \log(1/\beta)$.

Proof. We will compute the collision probability $\Pr[(A, B) = (A', B')]$ of two random executions by showing that

$$\Pr[\mathcal{P}^{(=)}] = \frac{\Pr[\mathcal{I}^{(\neq)} \wedge \mathcal{P}^{(=)}]}{\Pr[\mathcal{I}^{(\neq)} | \mathcal{F}^{(=)}]} = \frac{\Pr[\mathcal{I}^{(\neq)} \wedge \mathcal{P}^{(=)}]}{\beta}. \tag{2}$$

Because the collision probability of two independent instances of a random variable is at least the inverse of the alphabet size, the alphabet of A and B must have size at least $\beta / \Pr[\mathcal{I}^{(\neq)} \wedge \mathcal{P}^{(=)}]$. Thus, in total the protocol requires

$$\log(1/\Pr[\mathcal{I}^{(\neq)} \wedge \mathcal{P}^{(=)}]) - \log(1/\beta)$$

bits of communication.

We move on to prove (2). By perfect correctness, $\mathcal{P}^{(=)}$ can only happen if $\mathcal{F}^{(=)}$ happens, therefore

$$\frac{\Pr[\mathcal{P}^{(=)}]}{\Pr[\mathcal{I}^{(\neq)} \wedge \mathcal{P}^{(=)}]} = \frac{\Pr[\mathcal{F}^{(=)}]\Pr[\mathcal{P}^{(=)}|\mathcal{F}^{(=)}]}{\Pr[\mathcal{I}^{(\neq)} \wedge \mathcal{P}^{(=)}]}. \tag{3}$$

By the same reasoning, we can express the denominator of the RHS by

$$\Pr[\mathcal{I}^{(\neq)} \wedge \mathcal{P}^{(=)} \wedge \mathcal{F}^{(=)}] = \Pr[\mathcal{F}^{(=)}]\Pr[\mathcal{I}^{(\neq)}|\mathcal{F}^{(=)}]\Pr[\mathcal{P}^{(=)}|\mathcal{F}^{(=)} \wedge \mathcal{I}^{(\neq)}].$$

It follows that (3) equals to

$$\frac{\Pr[\mathcal{F}^{(=)}]\Pr[\mathcal{P}^{(=)}|\mathcal{F}^{(=)}]}{\Pr[\mathcal{F}^{(=)}]\Pr[\mathcal{I}^{(\neq)}|\mathcal{F}^{(=)}]\Pr[\mathcal{P}^{(=)}|\mathcal{F}^{(=)} \wedge \mathcal{I}^{(\neq)}]} = \frac{1}{\Pr[\mathcal{I}^{(\neq)}|\mathcal{F}^{(=)}]}, \tag{4}$$

where equality follows by noting that $\Pr[\mathcal{P}^{(=)}|\mathcal{F}^{(=)}] = \Pr[\mathcal{P}^{(=)}|\mathcal{F}^{(=)} \wedge \mathcal{I}^{(\neq)}]$ (due to perfect privacy). Multiplying the LHS of (3) and the RHS of (4) by $\Pr[\mathcal{I}^{(\neq)} \wedge \mathcal{P}^{(=)}]$, we conclude (2). □

Claim 10. *For any pair of strings* $r \neq r'$,

$$\Pr[\mathcal{P}^{(=)} \wedge \mathcal{I}^{(\neq)}|R = r, R' = r'] \leq 2\alpha(\mu)2^{-H_\infty(\mu)}.$$

Proof. We see that

$$\Pr\left[\mathcal{P}^{(=)} \wedge \mathcal{I}^{(\neq)}|R = r \wedge R' = r'\right] \leq \Pr\left[\mathcal{P}^{(=)} \wedge (X \neq X')|R = r \wedge R' = r'\right] + \Pr\left[\mathcal{P}^{(=)} \wedge (Y \neq Y')|R = r \wedge R' = r'\right].$$

Due to symmetry it suffices to bound the first summand by $\alpha(\mu)2^{-H_\infty(\mu)}$.

Say that x *collides* with x' if $\Pi_A(x,r) = \Pi_A(x',r')$. Restricting our attention to x's in the support of μ_A, we claim that every x can collide with at most a single x'. Indeed, if this is not the case, then $\Pi_A(x,r) = \Pi_A(x',r') = \Pi_A(x'',r')$. The second equality implies that when the randomness is r', for every y, the messages (a,b) communicated under (x',y) are equal to the ones communicated under (x'',y). By perfect correctness, this implies that $f(x',y) = f(x'',y)$ for every y, contradicting the non-degeneracy of f under μ. Analogously, let us say that y collides with y' if $\Pi_B(y,r) = \Pi_B(y',r')$. The same reasoning shows that every y in the support of μ_B can collide with at most a single y' in the support of μ_B.

Let $\mathbf{x} = (x_1,\ldots,x_m)$ and $\mathbf{x}' = (x'_1,\ldots,x'_m)$ be a complete list of entries for which x_i collides with x'_i and $x_i \neq x'_i$ and $\mu_A(x_i), \mu_A(x'_i) > 0$. Analogously let $\mathbf{y} = (y_1,\ldots,y_n)$ and $\mathbf{y}' = (y'_1,\ldots,y'_n)$ be a complete list for which y_i collides with y'_i and $\mu_B(y_i), \mu_B(y'_i) > 0$. (Note that we do not require $y_i \neq y'_i$.) Since collisions are unique (as explained above), the tuples $\mathbf{x}, \mathbf{x}', \mathbf{y}, \mathbf{y}'$ are uniquely determined up to permutation.

By definition, the tuples (x,y,x',y') with $x \neq x'$, and $(a,b) = (a',b')$ are exactly those of the form (x_i, y_j, x'_i, y'_j) for some i and j.

Now, consider the two x-disjoint rectangles $\rho = (\mathbf{x}, \mathbf{y})$ and $\rho' = (\mathbf{x}', \mathbf{y}')$ and assume, without loss of generality, that $\mu(\rho) \leq \mu(\rho')$. Since Alice and Bob both send the same messages with randomness r on inputs (x_i, y_j) as they send with randomness r' on inputs x'_i, y'_j, we see that it must be that $f(x_i, y_j) = f(x'_i, y'_j)$ if the protocol is correct. Therefore, $f_{[\rho]} = f_{[\rho']}$, and so $\mu(\rho) \leq \alpha(\mu)$.

To complete the argument, note that $\mathcal{P}^{(=)} \wedge (X \neq X')$ can only happen if we pick $(X, Y) = (x_i, y_j)$ and $(X', Y') = (x'_i, y'_j)$ for some i, j. The event that there exists i, j for which $(X, Y) = (x_i, y_j)$ has probability at most $\alpha(\mu)$. The event that $(X', Y') = (x'_i, y'_j)$ for the same (i, j) has probability at most $\max_{x,y} \mu(x, y) = 2^{-H_\infty(\mu)}$. □

Combining Claims 9 and 10, we derive Lemma 4. □

5 Lower Bounds for Imperfect PSM Protocols

In this section we state a lower-bound on the communication complexity of imperfect PSM protocols. For this, we will have to strengthen the requirements from the function f.

We call f *strongly non-degenerate* if for any $x \neq x'$ we have $|\{y | f(x, y) = f(x', y)\}| \leq 0.9|\mathcal{Y}|$ and for any $y \neq y'$ we have $|\{x | f(x, y) = f(x, y')\}| \leq 0.9|\mathcal{X}|$. A pair of ordered $m \times n$ rectangles $R = (\mathbf{x}, \mathbf{y})$ and $R' = (\mathbf{x}', \mathbf{y}')$ in which either $x_i \neq x'_i$ for all $i \in [m]$, or $y_i \neq y'_i$ for all $i \in [n]$ are called *approximately similar* if for 0.99 of the pairs (i, j) we have $f(x_i, y_j) = f(x'_i, y'_j)$. (The constants 0.9 and 0.99 are somewhat arbitrary and other constants may be chosen.)

In the full version we prove the following theorem:

Theorem 11. *Let $f : \mathcal{X} \times \mathcal{Y} \to \mathcal{Z}$ be a strongly non-degenerate function whose largest approximately similar pair of rectangles is of size at most M. Then, any PSM for f with privacy error of ϵ and correctness error of $\delta < \frac{1}{100}$, requires at least*

$$\log|\mathcal{X}| + \log|\mathcal{Y}| + \min\left\{\begin{array}{l} \log|\mathcal{X}| + \log|\mathcal{Y}| - \log\left(\frac{1}{\Pr[\mathcal{F}^{(=)}]}\right), \\ \log|\mathcal{X}| + \log|\mathcal{Y}| - \log M, \\ \log(1/\epsilon), \\ \log(1/\delta) - \log\left(\frac{1}{\Pr[\mathcal{F}^{(=)}]}\right) \end{array}\right\} - c \quad (5)$$

bits of communication, where c is some universal constant (that does not depend on f) and $\Pr[\mathcal{F}^{(=)}] = \Pr[f(X, Y) = f(X', Y')]$ when (X, Y) and (X', Y') are picked independently and uniformly at random from $\mathcal{X} \times \mathcal{Y}$.

In the special case of a Boolean function f, it holds that $\Pr[\mathcal{F}^{(=)}] = \Pr[f(X, Y) = f(X', Y')] \geq 1/2$, and the communication lower-bound simplifies to

$$\log|\mathcal{X}| + \log|\mathcal{Y}| + \min\{\log|\mathcal{X}| + \log|\mathcal{Y}| - \log M, \log(1/\epsilon), \log(1/\delta)\} - c$$

where c is some universal constant. In Sect. 6, we will use Theorem 11 to prove imperfect PSM lower-bounds for random functions and for efficiently computable functions.

6 Imperfect PSM Lower-Bounds for Random and Explicit Functions

In this section we will show that most functions have non-trivial imperfect PSM complexity, and establish the existence of an explicit function that admits a non-trivial imperfect PSM lower-bound. Formally, in Sect. 6.1 we will prove the following theorem (which strengthens Corollary 1 from the introduction).

Theorem 12. *For a $1 - o(1)$ fraction of the functions $f : \{0,1\}^k \times \{0,1\}^k \to \{0,1\}$ any PSM protocol for f with privacy error of ϵ and correctness error of δ, $\delta < \frac{1}{100}$, requires at least*

$$\ell(k, \epsilon, \delta) = \min\{3k - 2\log(k), 2k + \log(1/\epsilon), 2k + \log(1/\delta)\} - c \tag{6}$$

bits of communication, where c is some universal constant.

By de-randomizing the proof, we derive (in Sect. 6.2) the following theorem (which strengthens Theorem 4 from the introduction).

Theorem 13. *There exists a sequence of polynomial-size circuits*

$$f = \{f_k : \{0,1\}^k \times \{0,1\}^k \to \{0,1\}\}$$

such that any δ-correct ϵ-private PSM for f_k has communication complexity of at least $\ell(k, \epsilon, \delta)$ bits (as defined in (6)). Moreover, assuming the existence of a hitting-set generator against co-nondeterministic uniform algorithms, there exists an explicit family f which is computable by a polynomial-time Turing machine whose imperfect PSM communication complexity is at least $\ell(k, \epsilon, \delta) - O(\log k)$.

The reader is advised to read the following subsections sequentially since the proof of Theorem 13 builds over the proof of Theorem 12.

6.1 Lower Bounds for Random Functions (Proof of Theorem 12)

We will need the following definition.

Definition 5 (good function). *We say that a function $f : \{0,1\}^k \times \{0,1\}^k \to \{0,1\}$ is* good *if it satisfies the following conditions:*

1. *For every $x \neq x'$ and every set $\mathbf{y}$ of k^2 consecutive strings (according to some predefined order over $\{0,1\}^k$), it holds that $f(x, y) = f(x', y)$ for at most 0.9-fraction of the elements $y \in \mathbf{y}$.*
2. *Similarly, for every $y \neq y'$ and set $\mathbf{x}$ of k^2 consecutive strings (according to some predefined order over $\{0,1\}^k$), it holds that $f(x, y) = f(x, y')$ for at most 0.9-fraction of $x \in \mathbf{x}$.*
3. *For every pair of $k^2 \times k^2$ x-disjoint or y-disjoint rectangles R, R', it holds that $f_{[R]}$ disagrees with $f_{[R']}$ on at least 0.01 fraction of the entries.*

Claim 14. *Any good* $f : \{0,1\}^k \times \{0,1\}^k \to \{0,1\}$ *satisfies the conditions of Theorem 11 with* $M = 2^k \cdot k^2$, *and therefore any* δ*-correct* ϵ*-private PSM for* f, $\delta < \frac{1}{100}$, *requires communication of*

$$\ell(k, \epsilon, \delta) = \min\{3k - 2\log(k), 2k + \log(1/\epsilon), 2k + \log(1/\delta)\} - c,$$

for some universal constant c.

Proof. Fix some good f. Condition (1) guarantees that $f(x, \cdot)$ and $f(x', \cdot)$ differ on 0.1 fraction of each k^2 block of consecutive y's, and therefore, overall, they must differ on a 0.1 fraction of all possible y's. Applying the same argument on the y-axis (using condition (2)), we conclude that a good f must be strongly non-degenerate.

Similarly, a good f cannot have a pair of x-disjoint approximately similar $m \times n$ rectangles R, R' of size $mn \geq 2^k \cdot k^2$. To see this, observe that the latter condition implies that m, n are both larger than k^2, and therefore, again by an averaging argument, there must exists a pair of $k^2 \times k^2$ x-disjoint sub-rectangles $R_0' \subseteq R_0, R_1' \subseteq R_1$ which are also approximately similar. Applying the same argument to y-disjoint rectangles we conclude that any good f satisfies the conditions of Theorem 11. □

We say that a family of functions $\{f_z : \mathcal{A} \to \mathcal{B}\}_{z \in \mathcal{Z}}$ is t-wise independent functions if for any t-tuple of distinct inputs $(a_1, \ldots, a_t)$ and for a uniformly chosen $z \xleftarrow{\$} \mathcal{Z}$, the joint distribution of $(f_z(a_1), \ldots, f_z(a_t))$ is uniform over $\mathcal{B}^t$.

Claim 15. *Pick* $f : \{0,1\}^k \times \{0,1\}^k \to \{0,1\}$ *uniformly at random among all such functions. Then, with probability* $1 - o(1)$, *the resulting function is good. Moreover, this holds even if* f *is chosen from a family of* k^4*-wise independent functions.*

Proof. Choose f randomly from a family of k^4-wise independent hash functions. Fix a pair of $x \neq x'$ and a k^2-subset $\mathbf{y} \subset \{0,1\}^k$ of consecutive y's. By a Chernoff bound, the probability that $f(x, y) = f(x', y)$ for more than 0.9 of $y \in \mathbf{y}$ is at most $2^{-\Omega(k^2)}$. There are at most 2^{2k} pairs of x, x', and at most 2^k different sets $\mathbf{y}$ of consecutive y's, therefore by a union bound the probability that condition (1) does not hold is $2^{3k} 2^{-\Omega(k^2)} = 2^{-\Omega(k^2)}$. A similar argument, shows that (2) fails with a similar probability.

We move on to prove there is no pair of approximately similar x-disjoint rectangles of size exactly $k^2 \times k^2$. (Again, the case of y-disjoint rectangles is treated similarly.)

Let $m = k^2$. Fix two x-disjoint $m \times m$-rectangles $R = (\mathbf{x}, \mathbf{y})$ and $R' = (\mathbf{x}', \mathbf{y}')$. We want to give an upper bound on the probability that $f_{[R]}$ agrees with $f_{[R']}$ on 99% of their entries. This event happens only if the entries of f satisfy all but 1% of the the m^2 equations $f(x_i, y_j) = f(x_i', y_j')$ for $(i,j) \in \{1, \ldots, m\} \times \{1, \ldots, m\}$. The probability that any such equation is satisfied is $\frac{1}{2}$: since the rectangles are x-disjoint the equation is non-trivial. We can further find a subset T of at least $m^2/2$ such equations such that each equation in the subset uses an entry

$f(x, y)$ that is not used in any other equation. Let us fix some $0.01m^2$ subset S of equations that are allowed to be unsatisfied. After removing S from T, we still have at least $0.49m^2$ equations that are simultaneously satisfied with probability of at most $2^{-0.49m^2}$. There are at most $2^{H_2(0.01)m^2}$ sets S (where H_2 is the binary entropy function), and at most 2^{2mk} choices for R and 2^{2mk} choices for R'. Hence, by a union bound, the probability that (3) fails is at most

$$2^{-0.49m^2+0.081m^2+4m^{3/2}} < 2^{-\Omega(m^2)},$$

the claim follows. □

Theorem 12 follows from Claims 14 and 15. □

6.2 Explicit Lower-Bound (Proof of Theorem 13)

Our next goal is to obtain an explicit lower-bound. We begin by noting that good functions (as per definition 5) can be identified by efficient co-nondeterministic algorithms.

Definition 6. *A co-nondeterministic algorithm $M(x, y)$ is a Turing machine that takes z as its primary input and v as a witness. For each $z \in \{0, 1\}^*$ we define $M(z) = 1$ if there exist a witness v such that $M(z, v) = 0$.*

Claim 16. *There exists a co-nondeterministic algorithm that given some s-bit representation of a function $f : \{0, 1\}^k \times \{0, 1\}^k \to \{0, 1\}$ accepts f if and only if f is good with complexity of $O(k^4 t)$ where t is the time complexity of evaluating f on a given point.*

Proof. It suffices to describe a polynomial-time verifiable witness for the failure of each of the goodness conditions. If f is not good due to (1), then the witness is a pair $x \neq x'$ and a k^2-set $\mathbf{y}$ of consecutive y's. Since f_z can be efficiently evaluated we can verify that $f(x, y) = f(x', y)$ for more than 0.9-fraction of the y's in $\mathbf{y}$ in times $O(k^2 t)$. A violation of (2) is treated similarly. If f is not good due to (3), then the witness is a pair of x-disjoint or y-disjoint $k^2 \times k^2$ rectangles R, R' that are approximately similar. Again, we can verify the validity of this witness in time $O(k^4 t)$. □

Let $s(k) = \text{poly}(k)$ and let $\{f_z : \{0, 1\}^k \times \{0, 1\}^k \to \{0, 1\}\}_{z \in \{0,1\}^s}$ be a family of k^4-wise independent functions with an *evaluator* algorithm F which takes an index $z \in \{0, 1\}^s$ and input $(x, y) \in \{0, 1\}^k \times \{0, 1\}^k$ and outputs in time $t(k)$ the value of $f_z(x, y)$. (Such an F can be based on k^4-degree polynomials over a field of size $\Theta(k^4)$). Claims 14 and 15 imply that for most choices of z, the function f_z has an imperfect PSM complexity of at least $\ell(k, \epsilon, \delta)$. Since F is efficiently computable, for every z there is a polynomial-size circuit that computes f_z. Hence, there exists a polynomial-size computable function for which the $\ell(k, \epsilon, \delta)$ lower-bound holds, and the first part of Theorem 13 follows.

To prove the second part, we use a properly chosen pseudorandom generator (PRG) $G : \{0, 1\}^{O(\log k)} \to \{0, 1\}^s$ to "derandomize" the family $\{f_z\}$.

That is, we define the function $g : \{0,1\}^{O(\log k)} \times \{0,1\}^k \times \{0,1\}^k \to \{0,1\}$ which takes (w, x, y) and outputs $f_z(x, y)$ where $z = G(w) \in \{0,1\}^s$. Concretely, we require G to "hit" the image of any co-nondeterministic algorithms of complexity $T = O(k^4 t)$. Formally, this means that for every T-time co-nondeterministic algorithm M it holds that if $\Pr_z[M(z) = 1] \geq \frac{1}{2}$ then there exists a "seed" r for which $M(G(r)) = 1$.

Taking M to be the algorithm from Claim 16, we conclude, by Claims 15 and 14, that for some seed w, the function $f_{G(w)}$ has an imperfect PSM complexity of at least $\ell(k, \epsilon, \delta)$. Let us parse g as a two-party function, say by partitioning w to two halves w_A, w_B and giving (x, w_A) to Alice, and y, w_B to Bob. We conclude that g must have an imperfect PSM complexity of at least $\ell(k, \epsilon, \delta)$. Since the input length k' of Alice and Bob becomes longer by an additional $O(\log k)$ bits, the lower-bound becomes at least $\ell(k', \epsilon, \delta) - O(\log k')$, as claimed. The part of Theorem 13 follows. □

7 Lower-Bounds for Conditional Disclosure of Secrets

In this section we derive CDS lower bounds. We begin with a reduction from fully revealing weakly hiding PSM (Definition 4) to CDS.

Claim 17. *Let $h : \mathcal{X} \times \mathcal{Y} \to \{0,1\}$ be a predicate. Define the function $f : \mathcal{X}' \times \mathcal{Y} \to \{0,1\}$ where $\mathcal{X}' = \mathcal{X} \times \{0,1\}$ by $f((x, s), y) = s \wedge h(x, y)$. If h has a perfect CDS with communication complexity of c then f has a weakly-private fully-revealing PSM with complexity of $c + \log |\mathcal{X}| + \log |\mathcal{Y}|$.*

Proof. Given a CDS protocol $\Pi = (\Pi_A, \Pi_B, g)$ for h we construct a weakly-private fully-revealing PSM for f as follows. Given an input (x, s), Alice sends $(x, a = \Pi_A(x, s, r))$ where x plays the role of the Alice's input in the CDS, s plays the role of the secret, and r is a shared string uniformly sampled from $\mathcal{R}$. Bob takes his input y, and sends $(y, b = \Pi_B(y, r))$. Charlie outputs $h(x, y) \wedge g(x, y, a, b)$.

It is not hard to verify that the protocol is perfectly correct and fully revealing. Indeed, a PSM decoding error happens only if $g(x, y, a, b)$ fails to decode the secret s (which happens with probability zero). To prove weak privacy observe that if f agrees on a pair of inputs, $((x, 0), y)$ and $((x, 1), y)$, then $h(x, y)$ must be zero. By CDS privacy, for $R \xleftarrow{\$} \mathcal{R}$ the distribution $(x, y, \Pi_A(x, 0, R), \Pi_B(y, R))$ is identical to the distribution $(x, y, \Pi_A(x, 1, R), \Pi_B(y, R))$, as required. □

Next, we show that the properties of f needed for applying Theorem 8, follow from simple requirements on h. In the following, we say that $x \in \mathcal{X}$ is a *null input* if the residual function $h(x, \cdot)$ is the constant zero function.

Claim 18. *Let h and f be as in Claim 17. Then*

1. *The size of the largest complement similar rectangle of f equals to the size of the largest 0-monochromatic rectangle of h.*

2. *The number U of useful inputs of f is exactly two times larger than the number of inputs that are mapped by h to zero.*
3. *If h has no input x for which the residual function $h(x,\cdot)$ is the constant zero function, then f is weakly non-degenerate.*

Proof. The claim follows immediately by noting that for every (x,y) it holds that $f((x,1),y) = f((x,0),y)$ if and only if $h(x,y) = 0$. We proceed with a formal argument.

1. Consider some complement similar rectangle $R = (\mathbf{x}' \times \mathbf{y})$ of f. For every $(x,b) \in \mathbf{x}'$ and $y \in \mathbf{y}$, it holds that
$$f((x,b),y) = f((x,1-b),y),$$
and therefore $h(x,y) = 0$ and R is a 0-monochromatic rectangle of h.
2. Every input (x,y) that does not satisfy h induces an unordered pair, $((x,1),y)$ and $((x,0),y)$, of useful inputs for f. Therefore, the number of (ordered) useful inputs of f is exactly $2|h^{-1}(0)|$.
3. Fix some $(x,s) \in \mathcal{X}'$ and assume, towards a contradiction, that for every y it holds that $f((x,s),y) = f((x,1-s),y)$. By the definition of f this means that $h(x,y) = 0$ for every y, contradicting our assumption on h. □

Theorem 5 (restated here for convenience) now follows immediately from the lower-bound on weakly-private fully revealing PSM (Theorem 8).

Theorem 19 (Theorem 5 restated). *Let $h : \mathcal{X} \times \mathcal{Y} \to \{0,1\}$ be a predicate. Suppose that M upper-bounds the size of the largest 0-monochromatic rectangle of h and that for every $x \in \mathcal{X}$, the residual function $h(x,\cdot)$ is not the constant zero function. Then, the communication complexity of any perfect CDS for h is at least*
$$2\log|f^{-1}(0)| - \log M - \log|\mathcal{X}| - \log|\mathcal{Y}| - 1,$$
where $|f^{-1}(0)|$ denotes the number of inputs (x,y) that are mapped to zero.

Proof. Let $h : \mathcal{X} \times \mathcal{Y} \to \{0,1\}$ be a predicate that satisfies the theorem requirement. That is, M upper-bounds the size of the largest 0-monochromatic rectangle of h, there at least S inputs that are mapped to zero, and for every $x \in \mathcal{X}$, the residual function $h(x,\cdot)$ is not the constant zero function.

Suppose that h has a perfect CDS with communication complexity of c. By Claim 17, the function f (defined in the claim) has a weakly-private fully-revealing PSM with complexity of at most
$$c + \log|\mathcal{X}| + \log|\mathcal{Y}|,$$
which, by Claim 18 and Theorem 8, is at least
$$2\log U - \log M - 2 = 2\log S - \log M - 1.$$
It follows that
$$c \geq 2\log S - \log M - 1 - (\log|\mathcal{X}| + \log|\mathcal{Y}|),$$
as required. □

Example 1 (The index predicate). As a sanity check, consider the index predicate $f_{ind} : [k] \times \{0,1\}^k \to \{0,1\}$ which given an index $i \in [k]$ and a string $y \in \{0,1\}^k$ outputs $y[i]$, the i-th bit of y. Clearly exactly half of all inputs are mapped to 0. Also, for every i the residual function $f(i, \cdot)$ is not the constant zero. Finally, every zero rectangle is of the form $I \times \{y : y[i] = 0, \forall i \in I\}$ where $I \subseteq [k]$. This implies that the size of any such rectangle is exactly $|I| \cdot 2^{k-|I|} \leq 2^{k-1}$. Plugging this into Theorem 19, we get a lower-bound of

$$2(k + \log k - 1) - (k - 1) - k - \log k - 1 \geq \log k - 2.$$

References

1. Aiello, B., Ishai, Y., Reingold, O.: Priced oblivious transfer: how to sell digital goods. In: Pfitzmann, B. (ed.) EUROCRYPT 2001. LNCS, vol. 2045, pp. 119–135. Springer, Heidelberg (2001). https://doi.org/10.1007/3-540-44987-6_8
2. Applebaum, B.: Garbled circuits as randomized encodings of functions: a primer. Tutorials on the Foundations of Cryptography. ISC, pp. 1–44. Springer, Cham (2017). https://doi.org/10.1007/978-3-319-57048-8_1
3. Applebaum, B., Arkis, B.: Conditional disclosure of secrets and d-uniform secret sharing with constant information rate. In: Electronic Colloquium on Computational Complexity (ECCC), vol. 24, p. 189 (2017)
4. Applebaum, B., Arkis, B., Raykov, P., Vasudevan, P.N.: Conditional disclosure of secrets: amplification, closure, amortization, lower-bounds, and separations. In: Katz, J., Shacham, H. (eds.) CRYPTO 2017. LNCS, vol. 10401, pp. 727–757. Springer, Cham (2017). https://doi.org/10.1007/978-3-319-63688-7_24
5. Applebaum, B., Ishai, Y., Kushilevitz, E.: Cryptography in NC^0. In: FOCS, pp. 166–175 (2004)
6. Applebaum, B., Raykov, P.: From private simultaneous messages to zero-information arthur-merlin protocols and back. In: Kushilevitz, E., Malkin, T. (eds.) TCC 2016. LNCS, vol. 9563, pp. 65–82. Springer, Heidelberg (2016). https://doi.org/10.1007/978-3-662-49099-0_3
7. Barak, B., Jinong, S., Vadhan, S.P.: Derandomization in cryptography. SIAM J. Comput. **37**(2), 380–400 (2007)
8. Beaver, D., Micali, S., Rogaway, P.: The round complexity of secure protocols (extended abstract). In: STOC, pp. 503–513 (1990)
9. Beimel, A., Ishai, Y., Kumaresan, R., Kushilevitz, E.: On the cryptographic complexity of the worst functions. In: Lindell, Y. (ed.) TCC 2014. LNCS, vol. 8349, pp. 317–342. Springer, Heidelberg (2014). https://doi.org/10.1007/978-3-642-54242-8_14
10. Ben-or, M., Goldwasser, S., Wigderson, A.: Completeness theorems for non-cryptographic fault-tolerant distributed computation (extended abstract). In: STOC, pp. 1–10 (1988)
11. Brickell, E.F., Davenport, D.M.: On the classification of ideal secret sharing schemes. J. Cryptol. **4**(2), 123–134 (1991)
12. Capocelli, R.M., De Santis, A., Gargano, L., Vaccaro, U.: On the size of shares for secret sharing schemes. J. Cryptol. **6**(3), 157–167 (1993)
13. Chaum, D., Crépeau, C., Damgård, I.: Multiparty unconditionally secure protocols (extended abstract). In: STOC, pp. 11–19 (1988)

14. Chor, B., Kushilevitz, E., Goldreich, O., Sudan, M.: Private information retrieval. J. ACM **45**(6), 965–981 (1998)
15. Data, D., Prabhakaran, V.M., Prabhakaran, M.M.: Communication and randomness lower bounds for secure computation. IEEE Trans. Inf. Theor. **62**(7), 3901–3929 (2016)
16. Feige, U., Kilian, J., Naor, M.: A minimal model for secure computation (extended abstract). In: STOC, pp. 554–563 (1994)
17. Gay, R., Kerenidis, I., Wee, H.: Communication complexity of conditional disclosure of secrets and attribute-based encryption. In: Gennaro, R., Robshaw, M. (eds.) CRYPTO 2015. LNCS, vol. 9216, pp. 485–502. Springer, Heidelberg (2015). https://doi.org/10.1007/978-3-662-48000-7_24
18. Gertner, Y., Ishai, Y., Kushilevitz, E., Malkin, T.: Protecting data privacy in private information retrieval schemes. J. Comput. Syst. Sci. **60**(3), 592–629 (2000)
19. Goldreich, O., Micali, S., Wigderson, A.: How to play any mental game or a completeness theorem for protocols with honest majority. In: STOC (1987)
20. Goyal, V., Pandey, O., Sahai, A., Waters, B.: Attribute-based encryption for fine-grained access control of encrypted data. In: Juels, A., Wright, R.N., De Capitani di Vimercati, S., (eds.), Proceedings of the 13th ACM Conference on Computer and Communications Security, CCS 2006, Alexandria, VA, USA, 30 October–3 November 2006, vol. 1, pp. 89–98. ACM (2006)
21. Gutfreund, D., Shaltiel, R., Ta-Shma, A.: Uniform hardness versus randomness tradeoffs for arthur-merlin games. Comput. Complex. **12**(3–4), 85–130 (2003)
22. Ishai, Y.: Randomization techniques for secure computation. In: Prabhakaran, M., Sahai, A., (eds), Secure Multi-Party Computation of Cryptology and Information Security Series, vol. 10, pp. 222–248. IOS Press (2013)
23. Ishai, Y., Kushilevitz, E.: Private simultaneous messages protocols with applications. In: ISTCS (Israel Symposium on Theory of Computing and Systems), pp. 174–184 (1997)
24. Ishai, Y., Kushilevitz, E.: Randomizing polynomials: a new representation with applications to round-efficient secure computation. In: FOCS, pp. 294–304 (2000)
25. Ishai, Y., Wee, H.: Partial garbling schemes and their applications. In: Esparza, J., Fraigniaud, P., Husfeldt, T., Koutsoupias, E. (eds.) ICALP 2014. LNCS, vol. 8572, pp. 650–662. Springer, Heidelberg (2014). https://doi.org/10.1007/978-3-662-43948-7_54
26. Kushilevitz, E., Nisan, N.: Communication Complexity. Cambridge University Press, Cambridge (1997)
27. Liu, T., Vaikuntanathan, V., Wee, H.: Conditional disclosure of secrets via non-linear reconstruction. In: Katz, J., Shacham, H. (eds.) CRYPTO 2017. LNCS, vol. 10401, pp. 758–790. Springer, Cham (2017). https://doi.org/10.1007/978-3-319-63688-7_25
28. Miltersen, P.B., Vinodchandran, N.V.: Derandomizing Arthur-Merlin games using hitting sets. In: FOCS, pp. 71–80 (1999)
29. Sahai, A., Waters, B.: Fuzzy identity-based encryption. In: Cramer, R. (ed.) EUROCRYPT 2005. LNCS, vol. 3494, pp. 457–473. Springer, Heidelberg (2005). https://doi.org/10.1007/11426639_27
30. Shannon, C.E.: Communication theory of secrecy systems. Bell Syst. Tech. J. **28**, 656–715 (1949)

31. Sun, H.-M., Shieh, S.-P.: Secret sharing in graph-based prohibited structures. In: Proceedings IEEE INFOCOM 1997, The Conference on Computer Communications, Sixteenth Annual Joint Conference of the IEEE Computer and Communications Societies, Driving the Information Revolution, Kobe, Japan, pp. 718–724. IEEE, 7–12 April 1997
32. Waters, B.: Ciphertext-policy attribute-based encryption: an expressive, efficient, and provably secure realization. In: Catalano, D., Fazio, N., Gennaro, R., Nicolosi, A. (eds.) PKC 2011. LNCS, vol. 6571, pp. 53–70. Springer, Heidelberg (2011). https://doi.org/10.1007/978-3-642-19379-8_4
33. Yao, A.C.-C.: Protocols for secure computations (extended abstract). In: FOCS, pp. 160–164 (1982)

The Complexity of Multiparty PSM Protocols and Related Models

Amos Beimel[1(✉)], Eyal Kushilevitz[2], and Pnina Nissim[1]

[1] Department of Computer Science, Ben Gurion University, Beer Sheva, Israel
amos.beimel@gmail.com, pninani@post.bgu.ac.il
[2] Department of Computer Science, Technion, Haifa, Israel
eyalk@cs.technion.ac.il

Abstract. We study the efficiency of computing *arbitrary* k-argument functions in the *Private Simultaneous Messages* (PSM) model of [10,14]. This question was recently studied by Beimel et al. [6], in the two-party case ($k = 2$). We tackle this question in the general case of PSM protocols for $k \geq 2$ parties. Our motivation is two-fold: On one hand, there are various applications (old and new) of PSM protocols for constructing other cryptographic primitives, where obtaining more efficient PSM protocols imply more efficient primitives. On the other hand, improved PSM protocols are an interesting goal on its own. In particular, we pay a careful attention to the case of small number of parties (e.g., $k = 3, 4, 5$), which may be especially interesting in practice, and optimize our protocols for those cases.

Our new upper bounds include a k-party PSM protocol, for any $k > 2$ and any function $f : [N]^k \rightarrow \{0, 1\}$, of complexity $O(\text{poly}(k) \cdot N^{k/2})$ (compared to the previous upper bound of $O(\text{poly}(k) \cdot N^{k-1})$), and even better bounds for small values of k; e.g., an $O(N)$ PSM protocol for the case $k = 3$. We also handle the more involved case where different parties have inputs of different sizes, which is useful both in practice and for applications.

As applications, we obtain more efficient Non-Interactive secure Multi-Party (NIMPC) protocols (a variant of PSM, where some of the parties may collude with the referee [5]), improved ad-hoc PSM protocols (another variant of PSM, where the subset of participating parties is not known in advance [4,7]), secret-sharing schemes for uniform access structures with smaller share size than previously known, and better homogeneous distribution designs [4] (a primitive with many cryptographic applications on its own).

1 Introduction

Private simultaneous messages (PSM) protocols, introduced by Feige, Kilian, and Naor [10] and further studied by Ishai and Kushilevitz [14], are secure multi-party computation (MPC) protocols with a minimal communication pattern. In

A. Beimel and P. Nissim—Supported by ISF grants 544/13 and 152/17.
E. Kushilevitz—Supported by ISF grant 1709/14, BSF grant 2012378, and NSF-BSF grant 2015782.

J. B. Nielsen and V. Rijmen (Eds.): EUROCRYPT 2018, LNCS 10821, pp. 287–318, 2018.
https://doi.org/10.1007/978-3-319-78375-8_10

a PSM protocol for a function f, there are k parties, each of them holds a common random string r and a private input x_i. Each party computes a message based on its input and the common random string and sends it to a referee. The referee, which gets the k messages but does not know the common random string, should be able to compute $f(x_1, \ldots, x_k)$ without learning any additional information about the inputs. This model, beside being interesting for its simplicity, implies many cryptographic primitives e.g., constant round secure multi-party protocols (without a common random string) [14,15], protocols for conditional disclosure of secrets [6], generalized oblivious transfer protocols [14], and zero-information Arthur-Merlin protocols [1]. Several generalizations of PSM protocols have been studied, e.g., non-interactive (or robust) MPC [5], in which security is guaranteed even when the referee colludes with some parties, and ad-hoc PSM protocols [4,7], in which only a subset of the parties take part in computing the function. It was shown that PSM protocols imply these generalizations [7,8].

The common random string is crucial for this model – without it only very simple functions can be securely computed against an unbounded adversary; however, given the common random string, there is a PSM protocol (i.e., without any interaction) for any function [10]. Furthermore, many functions can be computed by PSM protocols with short messages, e.g., functions that have small non-deterministic branching programs (i.e., NL) [10] and even functions that have small modular branching programs [14]. Beimel et al. [6] presented improved upper-bounds for computing *arbitrary* functions in the 2-party PSM model. In this paper, we present improved upper-bounds for computing arbitrary functions in the *k-party* case. Then, we show that these improvements imply better complexity for various other primitives, including homogenous distribution designs, t-robust non-interactive secure multi-party computation protocols, and secret-sharing schemes for, so-called, uniform access structures. We elaborate below.

1.1 Our Results

Our main contributions are constructions of multi-party PSM protocols for every function. Prior to our work, the length of the messages in the best known PSM protocol for an arbitrary function $f : [N]^k \to \{0,1\}$ was $O(N^{k-1})$ for $k \geq 3$ [10] and $O(N^{1/2})$ for $k = 2$ [6]. We present, for every $k > 2$, a PSM protocol with messages of length $O(\text{poly}(k) N^{k/2})$; for $k = 3, 4, 5$ we present better protocols (see Fig. 1). Understanding the complexity of secure computation with small number of parties is motivated by practical systems and was done in other secure computation contexts (see, e.g., [13]). We also design PSM protocols for functions in which the inputs have different lengths. For example, for any 3-argument function $f : [N^\alpha] \times [N^\alpha] \times [N] \to [N]$ for some $\alpha \geq 1$ (i.e., a function in which the largest two input domains are the same), we design a PSM protocol whose communication complexity is $O(N^{(2\alpha+1)/3})$ (that is, proportional to the geometric average of the domains).

Num. of parties	Complexity	citation
2	$O(N^{1/2})$	[6]
3	$O(N)$	this paper
4	$O(N^{5/3})$	this paper
5	$O(N^{7/3})$	this paper
$k \geq 6$	$O(k^3 \cdot N^{k/2})$	this paper

Fig. 1. Complexity of PSM protocols for an arbitrary functions.

There are two additional advantages for our protocols for $k \geq 6$:

- They can handle long outputs with the same message length, that is for every function $f : [N]^k \to [N^k]$, we construct a PSM protocol with complexity $O(k^3 \cdot N^{k/2})$.[1]
- By increasing the complexity by a poly$(\log N)$ factor, the protocol can be made *locally computable*; that is, each party, holding an input from $[N]$, can compute $\log N$ messages, where each message depends on a *single* bit of the input. This property is useful when we design PSM protocols for some of the applications (e.g., when constructing NIMPC protocols from our PSM protocols; see below).

Following [6], our protocols use techniques from private information retrieval (PIR) [9]; specifically, we use the cube approach of [9], where a function $f : [N]^k \to \{0,1\}$ is represented by a d-dimensional cube, for some integer d. The 2-party PSM protocol of [6] uses a 4-dimensional cube. For $k = 3, 4, 5$ we use a 3-dimensional cube, and for $k \geq 6$ we use a 2-dimensional cube. These turn out to be the optimal values of d for *our* approach. The fact that we can only use integral values for d results in the "somewhat unnatural" exponents in the Fig. 1. As the number of dimensions in our protocols for $k \geq 4$ is smaller than the number of parties, our protocols have to address a few problems that were not relevant in the 2-party protocol of [6].

We note that, by simulation arguments, if for every N, every function $f : [N]^k \to \{0,1\}$ has a k-party PSM protocol with message length $O(N^\alpha)$ for some constant α, then every function $g : [N]^2 \to \{0,1\}$ has a 2-party PSM protocol with message length $O(N^{\alpha/\lfloor k/2 \rfloor})$. Thus, if one can improve the message length for k-party PSM protocols for an arbitrary function beyond $O(N^{k/4})$ for an even k, then this would yield 2-party PSM protocols with message length $O(N^\alpha)$ for $\alpha < 1/2$. Similarly, any improvement for k-party PSM protocols for $k > 6$, would imply an improvement for 6-party PSM protocols compared to our protocols. Thus, to improve the complexity of k-party PSM protocols for arbitrary functions, one might want to start with designing k-party PSM protocols for small values of k.

[1] As the inputs are from $[N]^k$, we can assume without loss of generality that the size of the domain of the outputs of f is at most N^k.

1.2 Applications

We show that our PSM protocols imply the following constructions.

Non-Interactive secure Multi-Party (NIMPC) protocols. A non-interactive MPC protocol [5] is a PSM protocol in which the security is guaranteed even when the referee colludes with some parties. Specifically, a k-party NIMPC protocol is t*-robust* if it is secure against any coalition of the referee and at most t parties, and it is *fully robust* if it is k-robust. Prior to our work, the length of messages in the best known fully robust NIMPC protocol for an arbitrary function $f : [N]^k \rightarrow \{0, 1\}$ was $O(\text{poly}(\log N, k) \cdot N^k)$ [19] (improving on [5]). No better t-robust protocols were known for any $t > 0$. We construct t-robust NIMPC protocols for any function $f : [N]^k \rightarrow \{0, 1\}$ with complexity $\tilde{O}(N^{k/2+t})$; that is, we improve the complexity when $t < k/2$. Our construction is based on an information-theoretic transformation of [8] that takes any PSM (i.e., 0-robust NIMPC) protocol and transforms it into a t-robust NIMPC protocol. An immediate application of this transformation yields a t-robust NIMPC protocol with complexity $\tilde{O}(N^{k/2+t+1})$. We use properties of our protocols and of the transformation of [8] to improve the complexity by a factor of N. For example, we construct a fully-robust 3-party NIMPC protocol with complexity $\tilde{O}(N^{2.5})$ (compared to $O(N^3)$ using the transformation as is). Thus, for 3 parties, we improve the complexity of fully-robust NIMPC protocols compared to [19].

Ad-hoc PSM protocols. A k-out-of-n ad-hoc PSM protocol [4,7] is a PSM protocol with n parties, where only k parties, whose identity is not known in advance, actually participate in the execution of the protocol. For example, think of an election, where only some of the potential voters will end up voting. Using a transformation, presented in [7], from a t-party PSM protocol for a symmetric function to an ad-hoc PSM protocol for the same function, we construct a k-out-of-n ad-hoc PSM protocol for any symmetric function $f : [N]^k \rightarrow \{0, 1\}$ with communication complexity $O(e^k \cdot k^6 \cdot \log n \cdot N^{k/2})$.

Distribution designs. The goal of a distribution design, introduced in [4], is to find a joint distribution on N random variables that satisfies a given set of constraints on the marginal distributions. Each constraint can either require that two sets of variables are identically distributed or, alternatively, that two sets of variables have disjoint supports. Distribution design generalizes many cryptographic primitives, such as secret-sharing, PSM protocols, and NIMPC protocols. We consider k-homogeneous sets of constraints, where all sets in the constraints are of size exactly k. In [4], it was shown that every k-homogeneous set of constraints without contradictions can be realized by a distribution design such that the size of the support of each variable is $O(\binom{N}{k} k \log N)$. We show that, for every k-homogeneous set of constraints, we can define a symmetric function $f : [N]^k \rightarrow \binom{N}{[k]}$ such that any ad-hoc PSM protocol Π for f can be used to construct a distribution design realizing the constraints, where each variable is a message of a party in Π. Using the 2-party PSM of [6] and the two transformations described above, we get a distribution design for every 2-homogeneous set of constraints in which the size of the support of each variable is $O(\log^2 N \cdot \sqrt{N})$. Using our constructions of PSM protocols, we get a distribution design for every k-homogeneous set of constraints in which the support of each variable is of size $O(k^6 e^k \log N \cdot N^{k/2})$. For $3 \leq k \leq 5$, we get better distribution designs.

Conditional Disclosure of Secrets. In Conditional Disclosure of Secrets (CDS) protocols, introduced in [12], there are k parties, a referee, and a function $f : [N]^k \rightarrow \{0,1\}$. As in the PSM model, each party gets a common random string r and an input x_i. In addition, all parties (excluding the referee) have a secret s. Each party P_i sends one message to the referee, based on r, x_i and s. The referee, which in the CDS setting knows the inputs $x_1, \ldots, x_k$, should learn s if and only if $f(x_1, \ldots, x_k) = 1$. It was shown in [2] that every PSM protocol for f implies a CDS protocol for f with the same complexity. Thus, our PSM protocols imply CDS protocols. However, there are direct constructions of CDS protocols that are much more efficient than the known PSM protocols. Liu, Vaikuntanathan, and Wee [16] showed a CDS protocol for an arbitrary 2-party function $f : [N] \times [N] \rightarrow \{0,1\}$ with communication complexity $2^{O(\sqrt{\log N \log \log N})} = N^{o(1)}$. Very recently, Liu, Vaikuntanathan, and Wee [17] have shown a construction of k-party CDS protocols for any function $f : [N]^k \rightarrow \{0,1\}$ with complexity $2^{\tilde{O}(\sqrt{k \log N})}$.

We show that CDS protocols imply secret-sharing schemes for uniform access structures. An access structure is t-uniform if all sets of size less than t are unauthorized, all sets of size greater that t are authorized, and sets of size t can be either authorized or unauthorized. 2-uniform access structures are called forbidden graph access structures [18] and were studied in, e.g., [2,3]. Secret-sharing schemes for forbidden bipartite graph access structures with N parties are equivalent to 2-party CDS protocols for functions $f : [N]^2 \rightarrow \{0,1\}$. We show that if every k-party function $f : \{0,1\}^k \rightarrow \{0,1\}$ has a CDS protocol with communication complexity $\mathrm{Com}(k)$, then every t-uniform access structure with k parties can be realized by a secret-sharing scheme with share size $k \cdot \mathrm{Com}(k)$. Combined with the result of [17], we get that every t-uniform access structures with k parties can be realized by a secret-sharing scheme with share size $2^{\tilde{O}(\sqrt{k})}$.

1.3 Discussion

CDS protocols vs. PSM protocols. The models of CDS and PSM look similar except that, in CDS protocols, the referee knows the inputs and should learn the secret if and only if some function of the inputs returns 1, while in PSM protocols the referee should learn the value of the function without learning additional information about the inputs. It was shown in [6,12] that a PSM protocol for f implies a CDS protocol for f with the same complexity. This similarity and the recent dramatic efficiency improvements for CDS protocols [16,17] may indicate that better PSM protocols also exist.

There are, however, some differences between the models. In [11], it was shown that a CDS protocol for a function $f : [N] \times [N] \rightarrow \{0,1\}$ can be constructed from a CDS for the index function – a function where P_1 holds a list of length N, party P_2 holds an index i and the referee should reconstruct the secret if and only if the ith element in the list is 1. The CDS protocols of [16] for an arbitrary function build on a construction of a CDS protocol for the index function. The PSM of [10] for an arbitrary function can be seen as, implicitly, constructing a PSM for the index function. However, the correctness of PSM protocols (even without the security requirement) implies that the communication complexity of any PSM protocol for the index function is $\Omega(N)$. Thus, for non-binary functions there is a separation between the CDS and PSM models. It is open if such huge separation exists for Boolean functions. We note that our constructions of PSM protocols use PSM protocols for the (k-dimensional) index function, however with shorter lists.

An alternative approach. We next describe an alternative approach for constructing PSM protocols with communication complexity that is better than that of previously known protocols, but is worse than the complexity of the protocols constructed in this paper. We explain some difficulties in applying this approach. Suppose we want to construct a 4-party PSM protocol for a function $f : [N]^4 \rightarrow \{0,1\}$ using a 2-party PSM protocol. Viewing the function f as a two-argument function with domain $[N^2] \times [N^2]$, by [6], it has a 2-party PSM protocol with complexity $O(N)$, where the first message m_1 depends on the inputs of the first two parties and the second message m_2 depends on the inputs of the other two parties. Thus, the first two parties can execute a PSM protocol for computing m_1. This can be done by $O(N)$ invocations of a (2-party) PSM protocol with a binary output. If the parties could have used the PSM of [6], which has complexity $O(N^{0.5})$, then the resulting PSM protocol would have complexity $O(N^{1.5})$. However, we do not know how to use the PSM protocol of [6] here, since it only applies to deterministic functions and the messages of the 2-party PSM protocol depend on the common randomness.[2] Instead, one can apply the protocol of [10], which can be used for randomized functions as well, but has complexity $O(N)$. This gives a protocol with complexity $O(N^2)$. More generally, for k-party functions this approach results in a PSM with complexity $O(N^{3k/4-1})$. This approach is described in more details in Sect. 6.

Our protocols can be viewed as a generalization of the above approach. Instead of using the PSM protocol of [10] to compute the message of a 2-party protocol, we use a special purpose PSM protocol to compute the message. For example, our k-party protocol from Sect. 4 can be viewed as simulating the 2-party PSM protocol from Sect. 3.2.

[2] We can treat the random string generating the message as a common input (or as an input of, e.g, P_1). However, this increases the length of the inputs, resulting in a non-efficient protocol.

2 Preliminaries

In this section we define PSM protocols and describe two PSM protocols that will be used in this paper.

2.1 Private Simultaneous Messages Protocols

In a PSM protocol, k parties $P_1, \ldots, P_k$ hold a common random string r and inputs $x_1, \ldots, x_k$, respectively; each party P_i sends a single message to a referee, based on r and its x_i, so that the referee learns the value of a function $f(x_1, \ldots, x_k)$ but nothing else. It is formally defined as follows:

Definition 2.1 (PSM protocols – Syntax and correctness). *Let $X_1, \ldots, X_k$, and Z be finite domains. A private simultaneous messages (PSM) protocol $\mathcal{P}$, computing a k-argument function $f : X_1 \times \cdots \times X_k \to Z$, consists of:*

- *A finite domain R of common random inputs, and k finite message domains $M_1, \ldots, M_k$.*
- *Message computation algorithms* $\textsc{Enc}_1, \ldots, \textsc{Enc}_k$, *where* $\textsc{Enc}_i : X_i \times R \to M_i$.
- *A reconstruction algorithm* $\textsc{Dec} : M_1 \times \cdots \times M_k \to Z$.

We say that the protocol $\mathcal{P}$ is correct (with respect to f) if $\textsc{Dec}(\textsc{Enc}_1(x_1, r), \ldots, \textsc{Enc}_k(x_k, r)) = f(x_1, \ldots, x_k)$, *for every input* $(x_1, \ldots, x_k) \in X_1 \times \cdots \times X_k$ *and every random input $r \in R$. The communication complexity of PSM protocol $\mathcal{P}$ is defined as* $\sum_{i=1}^{k} \log |M_i|$. *The randomness complexity of PSM protocol $\mathcal{P}$ is defined as* $\log |R|$.

The security of a PSM protocol requires that the message distribution seen by the referee on input $x_1, \ldots, x_k$ can be generated by a simulator that has access only to $f(x_1, \ldots, x_k)$; that is, everything that can be learned from the PSM protocol can be learned from the output of f.

Definition 2.2 (PSM protocols – Security). *A PSM protocol $\mathcal{P}$ is secure with respect to f if there exists a randomized algorithm* $\textsc{Sim}$ *such that, for every input* $(x_1, \ldots, x_k) \in X_1 \times \cdots \times X_k$, *the distribution of messages* $(\textsc{Enc}_1(x_1, r), \ldots, \textsc{Enc}_k(x_k, r))$ *induced by uniformly choosing a common random string $r \in R$ and the distribution of the output of* $\textsc{Sim}(f(x_1, \ldots, x_k))$ *are identical.*

Ishai and Kushilevitz [14] have shown that every function that has a small modular or non-deterministic branching program can be computed by an efficient PSM protocol. We will use their result for a deterministic branching program.

Theorem 2.3 ([10,14]). *Let $BP = (V, E, \phi, s, t)$ be a deterministic branching program of size $\alpha(k)$ computing a function $f : \{0,1\}^k \to \{0,1\}$. Then, there exist a PSM protocol for f with communication and randomness complexity $O(k \cdot \alpha(k)^2)$.*

Notation. Denote by $[N]$ the set $\{1, 2, \ldots, N\}$. For a finite set S, denote choosing a random element i from S with a uniform distribution by $i \in_R S$; similarly, denote choosing a random subset T of S with a uniform distribution by $T \subseteq_R S$.

2.2 A PSM Protocol for the Index Function

In this section, we show a construction of [10] of a PSM protocol for the index function defined below.

Definition 2.4. *We represent a string $D \in \{0,1\}^{N^{k-1}}$ as a $(k-1)$-dimensional cube (array), that is, $D = (D_{x_2,\ldots,x_k})_{x_2,\ldots,x_k \in [N]}$, where each $D_{x_2,\ldots,x_k}$ is a bit. For a $(k-1)$-dimensional cube D, and a position $(i_2, \ldots, i_k)$ let $D_{i_2,\ldots,i_k}$ denote the value in position $(i_2, \ldots, i_k)$ in the cube D. We define a k-argument function $\mathrm{ind}_{N,k} : \{0,1\}^{N^{k-1}} \times [N] \times \cdots \times [N] \rightarrow \{0,1\}$, whose inputs are a $(k-1)$-dimensional cube and $k-1$ indices, by $\mathrm{ind}_{N,k}(D, x_2, \ldots, x_k) = D_{x_2,\ldots,x_k}$; that is, $\mathrm{ind}_{N,k}(D, x_2, \ldots, x_k)$ returns the value of D in the position indexed by $x_2, \ldots, x_k$. When N, k will be clear from the context, we will write* ind *instead of* $\mathrm{ind}_{N,k}$.

For sake of completeness, we next present a PSM protocol for the index function, based on a PSM from [10]; for simplicity, we assume here that N is a power of 2 (and so values in $[N]$ can be represented by $\log N$-bit strings).

Claim 2.5 ([10]). *There is a k-party PSM protocol $\mathcal{P}_{\mathrm{ind}_{N,k}}$ computing the function $\mathrm{ind}_{N,k}$ with communication and randomness complexity is $O(kN^{k-1})$.*

Proof. In a non-secure protocol for $\mathrm{ind}_{N,k}$, party P_1 would send its input D and parties $P_2, \ldots, P_k$ would send their inputs, $x_2, \ldots, x_k$ respectively, to the referee who could easily compute $D_{x_2,\ldots,x_k}$. However, in a secure protocol, the referee should not learn information on D and $x_2, \ldots x_k$ except for $D_{x_2,\ldots,x_k}$.

To hide the indices $x_2, \ldots, x_k$ we permute D: we give $(k-1)$ random strings $r_2, \ldots, r_k \in [N]$ to the parties as their common randomness. Now, party P_1 creates a new cube D' such that $D'_{x_2,\ldots,x_k} = D_{x_2 \oplus r_2,\ldots,x_k \oplus r_k}$ for every $x_2, \ldots, x_k \in [N]$. Party P_1 sends to the referee D' and parties $P_2, \ldots, P_k$ send to the referee $x_2 \oplus r_2, \ldots, x_k \oplus r_k$ respectively. The referee computes $D'_{x_2 \oplus r_2,\ldots,x_k \oplus r_k} = D_{(x_2 \oplus r_2) \oplus r_2,\ldots,(x_k \oplus r_k) \oplus r_k} = D_{x_2,\ldots,x_k}$ as required.

In the above protocol the referee does not learn information on $x_2, \ldots x_k$; however, it learns information on the cube D because party P_1 sends D', which is a shift of the cube D. We fix this protocol by masking D' and revealing to the referee only the mask of position $x_2, \ldots, x_k$. Specifically, we choose $(k-1)$ random strings $r^2, \ldots, r^k \in \{0,1\}^{N^{k-1}}$; each string is viewed as a $(k-1)$-dimensional cube. Party P_1 computes $D'' = D' \oplus r^2 \oplus \cdots \oplus r^k$ and sends D'', which is now a random string. Each party P_j, for $j = 2, \ldots, k$, sends $x_j \oplus r_j$ and also $r^j_{i_2,\ldots,i_{j-1},x_j \oplus r_j,i_{j+1},\ldots,i_k}$ for every $i_2, \ldots, i_{j-1}, i_{j+1}, \ldots, i_k \in [N]$ (the length of the message of P_j is $\log N + N^{k-2}$). The referee computes $D_{x_2,\ldots,x_k}$ as $D''_{x_2 \oplus r_2,\ldots,x_k \oplus r_k} \oplus (\bigoplus_{j=2}^{k} r^j_{x_2 \oplus r_2,\ldots,x_k \oplus r_k})$. To see that the protocol is secure note that for each entry in the cube the referee gets at most $k-2$ masks (except for the entry $(x_2 \oplus r_2, \ldots, x_k \oplus r_k)$ for which the referee gets all $k-1$ masks). The communication complexity of this protocol is $O(N^{k-1} + k \cdot N^{k-2}) = O(k \cdot N^{k-1})$ and the randomness complexity is $O(k \cdot N^{k-1})$. □

Remark: The dominant contribution to the complexity of the above protocol comes from the size of the cube (i.e., N^{k-1}). We will sometimes need a natural extension of $\mathcal{P}_{\text{ind}}$, where the dimensions are not necessarily of the same size. It is not hard to see that the complexity of this variant remains proportional to the size of the cube.

2.3 A PSM Protocol for $S \oplus \{x\}$

For a set S and an element i, let $S \oplus \{i\}$ denote the set $S \setminus \{i\}$ if $i \in S$, and the set $S \cup \{i\}$ otherwise.

Definition 2.6. *Define the function* Sxor : $\{0,1\}^{N^k} \times [N] \times \cdots \times [N] \to \{0,1\}^{N^k}$ *as the function, whose inputs are a string of length N^k (interpreted as a set contained in $[N^k]$) and k elements from $[N]$, where* Sxor *outputs a string of length N^k (again, interpreted as a set contained in $[N^k]$) such that* $\text{Sxor}(S, x_1, \ldots, x_k) = S \oplus \{x_1 \circ x_2 \circ \cdots \circ x_k\}$ *(where $x_1 \circ x_2 \circ \cdots \circ x_k$ is the concatenation of the k strings, interpreted as an element of $[N^k]$).*

We construct a k-party PSM protocol for Sxor, where P_1 holds S and x_1 and $P_2, \ldots, P_k$ hold $x_2, \ldots, x_k$, respectively.[3]

Claim 2.7. *There exists a PSM protocol $\mathcal{P}_{\text{Sxor}}$ computing the function* Sxor *with communication and randomness complexity $O(k^3 \cdot N^k)$.*

Proof. Let $S' = \text{Sxor}(S, x_1, \ldots, x_k)$ and for every $i_1, \ldots, i_k \in [N]$ denote $S_{i_1,\ldots,i_k}$ and $S'_{i_1,\ldots,i_k}$ as the $i_1 \circ \cdots \circ i_k$th bits of the strings S and S' respectively.

To compute Sxor, for every bit of S' we execute a PSM protocol computing the bit as explained below. For each $(i_1, \ldots, i_k)$ such that $(i_1, \ldots, i_k) \neq (x_1, \ldots, x_k)$, it holds that $S'_{i_1,\ldots,i_k} = S_{i_1,\ldots,i_k}$, and $S'_{x_1,\ldots,x_k} = S_{x_1,\ldots,x_k} \oplus 1$. Let ℓ_j be 1 if $i_j = x_j$ and 0 otherwise. Notice that $S'_{i_1,\ldots,i_k} = S_{i_1,\ldots,i_k} \oplus (\ell_1 \wedge \ell_2 \wedge \ldots \wedge \ell_k)$. Thus, every bit of $S'_{i_1,\ldots,i_k}$ depends only on one bit of S and on $\ell_1, \ldots, \ell_k$, where party P_j can locally compute ℓ_j from x_j and i_j. Define the function $g : \{0,1\}^{k+1} \to \{0,1\}$ such that $g(s, \ell_1, \ldots, \ell_k) = s \oplus (\ell_1 \wedge \ell_2 \wedge \ldots \wedge \ell_k)$ for every $s, \ell_1, \ldots, \ell_k \in \{0,1\}$. We have shown that $S'_{i_1,\ldots,i_k}$ could be computed using N^k copies of a PSM for g.

Next, we show the existence of an efficient k-party PSM protocol $\mathcal{P}_g$ for g, where P_1 holds the inputs s and ℓ_1 and each party P_j, for $2 \leq j \leq k$, holds the input ℓ_j. There is a simple deterministic branching program of size $O(k)$ computing g, thus, by Theorem 2.3, we get that g has an efficient PSM protocol $\mathcal{P}_g$ with complexity $O(k^3)$.

[3] When we use this PSM protocol, all parties know S. We do not use this advantage as it cannot significantly improve the complexity of the protocol.

Protocol $\mathcal{P}_{\text{Sxor}}$ computing Sxor executes the protocol $\mathcal{P}_g$ for every bit of S', namely N^k times. The correctness and privacy of protocol $\mathcal{P}_{\text{Sxor}}$ follows immediately from the correctness and privacy of the PSM protocol $\mathcal{P}_g$. The complexity of $\mathcal{P}_{\text{Sxor}}$ is $O(k^3 N^k)$. □

3 A 3-Party PSM Protocol for an Arbitrary Function

In this section we show that every function $f : [N]^3 \to \{0,1\}$ has a PSM protocol with communication and randomness complexity of $O(N)$. Our construction is inspired by the cubes approach of [9]. We describe this approach in Sect. 3.1. Next, as a warm-up, we construct a 2-party PSM protocol using this approach in Sect. 3.2. We describe the 3-party PSM protocol in Sect. 3.3.

3.1 The Cube Approach

We start with a high level description of the cube approach; specifically, for the case of 2-dimensional cubes, we present a PIR protocol with 4 servers from [9]. Recall that in a PIR protocol, a client holds an index x, each server holds a copy of database D, and the goal of the client is to retrieve D_x without disclosing information about x.

The starting point of the cube approach [9] (restricted here to 2 dimensions) is viewing the database D as a 2-dimensional cube containing N^2 bits, that is $D = (D_{i_1,i_2})_{i_1,i_2 \in [N]}$. Correspondingly, the index that the client wishes to retrieve is viewed as a pair $x = (x_1, x_2)$. The protocol starts by the client choosing a random subset for each dimension, i.e. $S_1, S_2 \subseteq_R [N]$. The client then creates 4 queries of the form (T_1, T_2) where each T_j is either S_j itself or $S_j \oplus \{x_j\}$; i.e. (S_1, S_2), $(S_1 \oplus \{x_1\}, S_2)$, $(S_1, S_2 \oplus \{x_2\})$, and $(S_1 \oplus \{x_1\}, S_2 \oplus \{x_2\})$; we denote these 4 queries by $q_{00}, q_{10}, q_{01}, q_{11}$, respectively. The client sends each query to a different server ($2 \cdot N$ bits to each server). A server, which gets a query (T_1, T_2), replies with a single bit which is the XOR of all bits of D in the sub-cube $T_1 \otimes T_2$, i.e. $\bigoplus_{i_1 \in T_1, i_2 \in T_2} D_{i_1,i_2}$. The observation made in [9] is that each element of the cube appears in an even number of those 4 sub-cubes except the entry $x = (x_1, x_2)$ that appears exactly once. Therefore, taking the XOR of the 4 answer bits, all elements of the cube are canceled except for the desired element in position x. Each server gets no information about x from its query. For example, in the query q_{01}, the sets S_1 and $S_2 \oplus \{x_2\}$ are uniformly distributed, independently of x_1, x_2.

The above approach can be generalized to any number of dimensions. Specifically, for 3 dimensions, the client chooses 3 sets $S_1, S_2, S_3 \subseteq [N]$ and generates 8 queries $q_{000}, \ldots, q_{111}$.

3.2 A 2-Party PSM Protocol

Given a function $f : [N] \times [N] \to \{0,1\}$, we construct a 2-party PSM protocol $\mathcal{P}_2$ for f using the above approach. This PSM protocol is not as efficient as the

PSM protocol of [6]; we present it to introduce the ideas we use in our PSM protocol for $k > 2$ parties. The protocol is formally described in Fig. 2.

Next, we give an informal description of an insecure protocol, and then we fix it so it will be secure. We associate the function f with an N^2-bit database, viewed as a 2-dimensional cube (that is, the (x, y) entry in the database is $f(x, y)$). The common randomness of the two parties is viewed as two random subsets, one for each dimension, i.e. $S_1, S_2 \subseteq_R [N]$. In the PSM protocol, party P_1 holds x_1, party P_2 holds x_2, and the referee wishes to compute $f(x_1, x_2)$, i.e. the XOR of the answers to the same 4 queries mentioned above. This should be done without learning information about x_1, x_2 (besides what follows from $f(x_1, x_2)$). In the protocol, P_1 computes the answers, denoted a_{00}, a_{10}, to the queries q_{00}, q_{10} (using S_1, S_2 and its input x_1). For example, the answer to the query q_{10} is

$$a_{10} = \oplus_{i_1 \in S_1 \oplus \{x_1\}, i_2 \in S_2} f(i_1, i_2).$$

It then sends these answers to the referee. Similarly, P_2 computes the answer, denoted a_{01}, to the query q_{01} and sends it to the referee. Now, the referee has the answers to 3 of the 4 queries and the only query that remains unanswered is q_{11}. To compute the answer to q_{11}, party P_1 sends to the referee $S_1 \oplus \{x_1\}$ and P_2 sends $S_2 \oplus \{x_2\}$, i.e. the referee gets the query q_{11} and can compute the corresponding answer a_{11} (as it knows the function f). Now, the referee has the answers to all 4 queries and it can compute $f(x_1, x_2) = a_{00} \oplus a_{10} \oplus a_{01} \oplus a_{11}$. The correctness of this protocol follows immediately from the correctness of the cube approach.

This protocol is not secure because the referee learns the answers to all 4 queries and this could leak information about x_1 and x_2. In order to deal with this problem, we add one more random bit to the common randomness of the two parties, i.e., $b \in_R \{0, 1\}$. Party P_1 sends to the referee $a_{00} \oplus a_{10} \oplus b$ instead of a_{00}, a_{10}, and P_2 sends $a_{01} \oplus b$ instead of a_{01}. Now, the referee only gets random bits from the parties (from the randomness of S_1, S_2, b) and it learns only $f(x_1, x_2)$. The communication complexity and randomness complexity of this protocol are $O(N)$.

More formally, to argue that the PSM protocol $\mathcal{P}_2$ described in Fig. 2 is secure, we construct a simulator whose input is $f(x_1, x_2)$ and whose output is two messages distributed identically to the messages in $\mathcal{P}_2$. The simulator first chooses with uniform distribution two sets $T_1, T_2 \subseteq_R [N]$ and a random bit $c \in \{0, 1\}$. It then computes the answer to the query q_{11}, namely, $a_{11} = \oplus_{i_1 \in T_1, i_2 \in T_2} f(i_1, i_2)$. Finally, it outputs $(c, T_1), (c \oplus f(x_1, x_2) \oplus a_{11}, T_2)$. Note that in the messages of $\mathcal{P}_2$ the sets $S_1 \oplus \{x_1\}$ and $S_2 \oplus \{x_2\}$ are distributed independently with uniform distribution. Furthermore, m_1 is uniformly distributed given the two sets, since we mask m_1 with a random bit b. Finally, by the correctness of $\mathcal{P}_2$, given $m_1, S_1 \oplus \{x_1\}, S_2 \oplus \{x_2\}$, and $f(x_1, x_2)$, the value of m_2 is fully determined (as the sets determine a_{11}). Thus, the simulator's output is distributed as the messages generated in protocol $\mathcal{P}_2$ on inputs x_1, x_2.

Protocol $\mathcal{P}_2$

Common randomness: Both parties share uniform random strings:

- $S_1, S_2 \subseteq_R [N]$ and $b \in_R \{0,1\}$.

The protocol:

1. P_1, holding x_1, computes the answers a_{00}, a_{10} to the queries $q_{00} = (S_1, S_2)$ and $q_{10} = (S_1 \oplus \{x_1\}, S_2)$ respectively. It then computes $q^1_{11} = S_1 \oplus \{x_1\}$. Finally, it sends $m_1 = a_{00} \oplus a_{10} \oplus b$ and q^1_{11} to the referee.
2. P_2, holding x_2, computes the answer a_{01} to the query $q_{01} = (S_1, S_2 \oplus \{x_2\})$. It then computes $q^2_{11} = S_2 \oplus \{x_2\}$. Finally, it sends $m_2 = a_{01} \oplus b$ and q^2_{11} to the referee.
3. The referee computes the answer a_{11} to the query $q_{11} = (q^1_{11}, q^2_{11}) = (S_1 \oplus \{x_1\}, S_2 \oplus \{x_2\})$. The referee computes $f(x_1, x_2) = m_1 \oplus m_2 \oplus a_{11}$.

Fig. 2. A 2-party PSM protocol $\mathcal{P}_2$ for a function $f : [N] \times [N] \to \{0,1\}$.

3.3 A 3-Party PSM Protocol

In this section, we show how to construct a PSM protocol $\mathcal{P}_3$ for any function $f : [N]^3 \to \{0,1\}$ with communication and randomness complexity $O(N)$. As in the 2-party case above, our construction is inspired by the cube approach [9], as described in Sect. 3.1, using 3-dimensional cubes.

Again, we first give an informal description of an insecure protocol, and then we fix it so it will be secure. The protocol $\mathcal{P}_3$ is formally described in Fig. 3. We associate f with an N^3-bit database that is viewed as a 3-dimensional cube. The common randomness of the 3 parties consists of 3 random subsets, one for each dimension, i.e., $S_1, S_2, S_3 \subseteq_R [N]$. The referee wishes to compute $f(x_1, x_2, x_3)$, i.e. the XOR of the answers to 8 queries. Each query is of the form (T_1, T_2, T_3) where each T_j is either S_j or $S_j \oplus \{x_j\}$. Party P_1 computes the answers a_{000}, a_{100} to the queries q_{000}, q_{100} respectively, and sends these answers to the referee. Similarly, party P_2 (resp. P_3) computes the answer a_{010} (resp. a_{001}) to the query q_{010} (resp. q_{001}) and sends the answer to the referee. There is no party that knows the values of two inputs from $\{x_1, x_2, x_3\}$ and therefore, no party can answer queries of weight 2; e.g. q_{110}. However, using an idea of [9], party P_1 can provide the answers to the queries $(S_1 \oplus \{x_1\}, S_2 \oplus \{\ell\}, S_3)$ for all possible values of $\ell \in [N]$. This is a list of length N in which the entry corresponding to $\ell = x_2$ is the desired answer for the query q_{110}. Party P_1 computes the N-bit list and represents it as a 1-dimension cube; party P_2 holds x_2, which is the position of the answer a_{110} in the list. Now, P_1 and P_2 execute the (2-party) PSM protocol $\mathcal{P}_{\text{ind}}$ for the index function, described in Sect. 2.2, which enables the referee to compute the answer to the query q_{110} without leaking any additional information. Similarly, P_1 and P_3 (resp. P_2 and P_3) execute the PSM protocol $\mathcal{P}_{\text{ind}}$ that enables the referee to compute the answer to the query q_{101} (resp. q_{011}). Finally, party P_1

Protocol $\mathcal{P}_3$

Common randomness: The three parties share uniform random strings:

- $S_1, S_2, S_3 \subseteq_R [N]$.
- $r_{110}, r_{101}, r_{011} \in_R \{0,1\}^{O(N)}$ required for the PSM protocol $\mathcal{P}_{\text{ind}}$ with a list of length N.
- $b_{100}, b_{010}, b_{001}, b_{110}, b_{101}, b_{011} \in_R \{0,1\}$, $b_{000} = b_{100} \oplus b_{010} \oplus b_{001} \oplus b_{110} \oplus b_{101} \oplus b_{011}$.

The protocol:

1. Party P_1:
   - Computes the answers a_{000}, a_{100} to the queries $q_{000} = (S_1, S_2, S_3)$ and $q_{100} = (S_1 \oplus \{x_1\}, S_2, S_3)$, respectively.
   - Computes the answers to all queries $(S_1 \oplus \{x_1\}, S_2 \oplus \{\ell\}, S_3)$ for all possible values of $\ell \in [N]$ and represents the answers as a 1-dimension database $(a^1_{110}, \ldots, a^N_{110})$. Using $D^{110} = (a^1_{110} \oplus b_{110}, \ldots, a^N_{110} \oplus b_{110})$ and common randomness r_{110}, party P_1 computes m^1_{110} – the message of the first party in the 2-party PSM protocol $\mathcal{P}_{\text{ind}}$.
   - Computes the answers to all queries $(S_1 \oplus \{x_1\}, S_2, S_3 \oplus \{\ell\})$ for all possible values of $\ell \in [N]$, and represents the answers as a 1-dimension database $(a^1_{101}, \ldots, a^N_{101})$. Using $D^{101} = (a^1_{101} \oplus b_{101}, \ldots, a^N_{101} \oplus b_{101})$ and common randomness r_{101}, computes m^1_{101} – the message of the first party in the 2-party PSM protocol $\mathcal{P}_{\text{ind}}$.
   - Sends $m_{000} = a_{000} \oplus b_{000}$, $m_{100} = a_{100} \oplus b_{100}$, m^1_{110}, m^1_{101}, and $S_1 \oplus \{x_1\}$ to the referee.
2. Party P_2:
   - Computes the answer a_{010} to the query $q_{010} = (S_1, S_2 \oplus \{x_2\}, S_3)$.
   - Using x_2 and common randomness r_{110}, computes m^2_{110} – the message of the second party in the 2-party PSM protocol $\mathcal{P}_{\text{ind}}$.
   - Computes the answers to all queries $(S_1, S_2 \oplus \{x_2\}, S_3 \oplus \{\ell\})$ for all possible values of $\ell \in [N]$ and represents the answers as a 1-dimension database $(a^1_{011}, \ldots, a^N_{011})$. Using $D^{011} = (a^1_{011} \oplus b_{011}, \ldots, a^N_{011} \oplus b_{011})$ and common randomness r_{011}, computes m^2_{011} – the message of the first party in the PSM protocol $\mathcal{P}_{\text{ind}}$.
   - Sends $m_{010} = a_{010} \oplus b_{010}$, m^2_{110}, m^2_{011}, and $S_2 \oplus \{x_2\}$ to the referee.
3. Party P_3:
   - Computes the answer a_{001} to the query $q_{001} = (S_1, S_2, S_3 \oplus \{x_3\})$.
   - Using x_3 and common randomness r_{101}, computes m^3_{101} – the message of the second party in the 2-party PSM protocol $\mathcal{P}_{\text{ind}}$.
   - Using x_3 and common randomness r_{011} it computes m^2_{011} – the message of the second party in the 2-party PSM protocol $\mathcal{P}_{\text{ind}}$.
   - Sends $m_{001} = a_{001} \oplus b_{001}$, m^3_{101}, m^3_{011}, and $S_3 \oplus \{x_3\}$ to the referee.
4. The referee:
   - Computes the answer a_{111} to the query $q_{111} = (S_1 \oplus \{x_1\}, S_2 \oplus \{x_2\}, S_3 \oplus \{x_3\})$ (using the sets received from P_1, P_2, P_3 and the truth table of f).
   - Using the PSM messages m^1_{110}, m^2_{110} the referee computes $D^{110}_{x_2}$. Using the PSM messages m^1_{101}, m^3_{101} the referee computes $D^{101}_{x_3}$. Using the PSM messages m^2_{011}, m^3_{011} the referee computes $D^{011}_{x_3}$.
   - Output $f(x) = m_{000} \oplus m_{100} \oplus m_{010} \oplus m_{001} \oplus D^{110}_{x_2} \oplus D^{101}_{x_3} \oplus D^{011}_{x_3} \oplus a_{111}$.

Fig. 3. A 3-party PSM protocol $\mathcal{P}_3$ for a function $f : [N]^3 \to \{0,1\}$.

sends to the referee $S_1 \oplus \{x_1\}$, party P_2 sends $S_2 \oplus \{x_2\}$, and party P_3 sends $S_3 \oplus \{x_3\}$. The referee gets the query q_{111} and computes the answer a_{111}. It now has the answers to all 8 queries and it can compute $f(x_1, x_2, x_3)$ as the XOR of these 8 answers.

As in the 2-party case, this protocol is not secure because the referee learns the answers to all 8 queries, which could leak information about x_1, x_2, and x_3. To deal with that, we mask these answers so that the referee gets random bits from the parties whose sum is $f(x_1, x_2, x_3)$ (the details of applying these masks appear in Fig. 3).

Theorem 3.1. *Let $f : [N]^3 \to \{0,1\}$ be a function. The protocol $\mathcal{P}_3$ is a secure PSM protocol for f with communication and randomness complexity $O(N)$.*

Proof. First, we argue that the protocol is correct. The output of the referee is

$$
\begin{aligned}
& m_{000} \oplus m_{100} \oplus m_{010} \oplus m_{001} \oplus D^{110}_{x_2} \oplus D^{101}_{x_3} \oplus D^{011}_{x_3} \oplus a_{111} \\
&= (a_{000} \oplus b_{000}) \oplus (a_{100} \oplus b_{100}) \oplus (a_{010} \oplus b_{010}) \oplus (a_{001} \oplus b_{001}) \oplus (a^{x_2}_{110} \oplus b_{110}) \\
&\quad \oplus (a^{x_3}_{101} \oplus b_{101}) \oplus (a^{x_3}_{011} \oplus b_{011}) \oplus a_{111} \\
&= a_{000} \oplus a_{100} \oplus a_{010} \oplus a_{001} \oplus a^{x_2}_{110} \oplus a^{x_3}_{101} \oplus a^{x_3}_{011} \oplus a_{111} && (1) \\
&= f(x_1, x_2, x_3), && (2)
\end{aligned}
$$

where the equality in (1) follows from the fact that the exclusive or of the b's is zero, and the equality in (2) follows from the correctness of the cube approach.

To argue that protocol $\mathcal{P}_3$ is secure, we construct a simulator whose input is $f(x_1, x_2, x_3)$ and whose output is three messages distributed as the messages in $\mathcal{P}_3$. The simulator on input $f(x_1, x_2, x_3)$ does the following:

1. Chooses three random sets $T_1, T_2, T_3 \subseteq_R [N]$ and 6 random bits $c_{100}, c_{010}, c_{001}, c_{110}, c_{101}, c_{011} \in_R \{0,1\}$.
2. Computes the answer to the query q_{111}, namely,
$$a_{111} = \oplus_{i_1 \in T_1, i_2 \in T_2, i_3 \in T_3} f(i_1, i_2, i_3).$$
3. Computes $c_{000} = f(x_1, x_2, x_3) \oplus a_{111} \oplus c_{100} \oplus c_{010} \oplus c_{001} \oplus c_{110} \oplus c_{101} \oplus c_{011}$.
4. Invokes the simulator SIM_{ind} of protocol $\mathcal{P}_{\text{ind}}$ 3 times:
   - $(m^1_{110}, m^2_{110}) \leftarrow \text{SIM}_{\text{ind}}(c_{110})$,
   - $(m^1_{101}, m^3_{101}) \leftarrow \text{SIM}_{\text{ind}}(c_{101})$,
   - $(m^2_{011}, m^3_{011}) \leftarrow \text{SIM}_{\text{ind}}(c_{011})$.
5. Outputs
$$(c_{000}, c_{100}, m^1_{110}, m^1_{101}, T_1), (c_{010}, m^2_{110}, m^2_{011}, T_2), (c_{001}, m^3_{101}, m^3_{011}, T_3).$$

Note that in the messages of protocol $\mathcal{P}_3$ the sets $S_1 \oplus \{x_1\}, S_2 \oplus \{x_2\}$, and $S_3 \oplus \{x_3\}$ are distributed independently with uniform distribution. Furthermore, m_{100}, m_{010}, m_{001}, $D^{110}_{x_2}, D^{101}_{x_3}, D^{011}_{x_3}$ are uniformly distributed given these 3 sets, since we mask them using independent random bits. Since $D^{110}_{x_2}$ and c_{110} are both random bits, the output of $\text{SIM}_{\text{ind}}(c_{110})$ is distributed as the messages of $\mathcal{P}_{\text{ind}}(D^{110}, x_2)$. The same holds for the other two invocations of SIM_{ind}. Finally, given the sets $S_1 \oplus \{x_1\}, S_2 \oplus \{x_2\}, S_3 \oplus \{x_3\}$, the bits $m_{100}, m_{010}, m_{001}, D^{110}_{x_2}, D^{101}_{x_3}, D^{011}_{x_3}, a_{111}$, and $f(x_1, x_2, x_3)$, the value of m_{000} is fully determined (by the correctness of $\mathcal{P}_3$ and by the fact that the above sets determine a_{111}). Thus, the simulator's output is distributed as the messages generated in protocol $\mathcal{P}_3$ on inputs x_1, x_2, x_3.

We next analyze the complexity of $\mathcal{P}_3$. The communication and randomness complexity of each invocation of $\mathcal{P}_{\text{ind}}$ is linear in the length of the list, i.e., it is $O(N)$. In addition, parties P_1, P_2, P_3 send the subsets $S_1 \oplus \{x_1\}, S_2 \oplus \{x_2\}, S_3 \oplus \{x_3\}$ respectively, which are of size N, and $O(1)$ bits each for the answers of the queries $q_{000}, q_{100}, q_{010}, q_{001}$. Therefore, the communication complexity and randomness complexity of our PSM protocol $\mathcal{P}_3$ is $O(N)$. □

4 A k-Party PSM Protocol for an Arbitrary Function and Some Extensions

In this section, we show how to construct a k-party PSM protocol $\mathcal{P}_k$ for a function $f : [N]^k \rightarrow [N^k]$ with communication and randomness complexity $O(k^3 \cdot N^{k/2})$. The above complexity is achieved even when the output is of length $k \log N$ (as the input length is $k \log N$, we can assume, without loss of generality, that the output length is at most $k \log N$). In general, when the output of f is an L-bit string, one can execute a PSM protocol for every bit of the output, and the complexity of the PSM protocol for f is L times the complexity of a PSM protocol for a Boolean function. In protocol $\mathcal{P}_k$, we do not pay any penalty for long outputs. We also present better protocols for 4 and 5 parties.

We first describe our construction for an even k. Our construction is inspired by our 2-party PSM protocol presented in Sect. 3.2. The protocol is formally described in Fig. 4. Next, we give an informal description of the protocol for the case that the range of f is Boolean. We associate f with an N^k-bit database that is viewed as a 2-dimensional cube, where each dimension of the cube is of size $N^{k/2}$. Correspondingly, each input $x = (x_1, \ldots, x_k) \in [N]^k$ is viewed as a pair $(y_1, y_2) \in [N^{k/2}]^2$, where $y_1 = (x_1, \ldots, x_{k/2})$ and $y_2 = (x_{k/2+1}, \ldots, x_k)$. The common randomness of the k parties contains two random subsets, one for each dimension, i.e. $S_1, S_2 \subseteq_R [N^{k/2}]$. The referee wishes to compute $f(x_1, \ldots, x_k)$, i.e., the XOR of answers to the same 4 queries of the cube approach described in Sect. 3.1. Party P_1 can easily compute the answer to the query (S_1, S_2) (i.e., q_{00}) and send the answer to the referee. However, as the inputs y_1, y_2 are distributed among the parties, there are two problems we have to address: (1) how to answer the queries $(S_1 \oplus \{y_1\}, S_2), (S_1, S_2 \oplus \{y_2\})$, and (2) how to send $S_1 \oplus \{y_1\}$ and $S_2 \oplus \{y_2\}$ to the referee.

We first address the first problem. The answer to query $(S_1 \oplus \{y_1\}, S_2)$ depends on y_1, i.e., on inputs $(x_1, \ldots, x_{k/2})$, and there is no party that knows all these inputs. We solve this problem in a similar way to what is done in [9] and in our protocol $\mathcal{P}_3$, that is, by using the PSM protocol $\mathcal{P}_{\text{ind}}$. Although party P_1 does not know the exact value of $y_1 = (x_1, \ldots, x_{k/2})$, it can compute the answers to all queries $(S_1 \oplus \{\ell\}, S_2)$ for $\ell = (x_1, i_2, \ldots, i_{k/2})$ for all possible values $i_2, \ldots, i_{k/2} \in [N]^{k/2-1}$. This is a list of length $N^{k/2-1}$ in which the entry corresponding to $\ell = y_1$ is the desired answer for the query q_{10}. We view this answer as a $(k/2-1)$-dimensional cube such that the answer corresponding to the values $(x_1, i_2, i_3, \ldots, i_{k/2})$ is in position $(i_2, i_3, \ldots, i_{k/2})$. Specifically, the answer to the query $(S_1 \oplus \{y_1\}, S_2)$ is in position $(x_2, \ldots, x_{k/2})$ in the cube. Parties $P_1, \ldots, P_{k/2}$ use the PSM protocol $\mathcal{P}_{\text{ind}}$ for the index function described in Sect. 2.2, from which the referee learns the answer to the query $(S_1 \oplus \{y_1\}, S_2)$ and nothing else. Similarly, P_k can compute an $N^{k/2-1}$-bit cube, corresponding to all choices of $i_{k/2+1}, \ldots, i_{k-1}$, such that the answer to the query $(S_1, S_2 \oplus \{y_2\})$ is in position $x_{k/2+1}, \ldots, x_{k-1}$ in this cube. Parties $P_k, P_{k/2+1} \ldots, P_{k-1}$ use the PSM protocol $\mathcal{P}_{\text{ind}}$, from which the referee learns the answer to the query $(S_1, S_2 \oplus \{y_2\})$ and nothing else.

The only query that remained unanswered is the query q_{11}. Parties $P_1, \ldots, P_{k/2}$ execute the PSM protocol $\mathcal{P}_{\text{Sxor}}$ described in Sect. 2.3 that enables the referee to compute $S_1 \oplus \{y_1\}$ without learning any information about y_1. Similarly, parties $P_{k/2+1}, \ldots, P_k$ execute the PSM protocol $\mathcal{P}_{\text{Sxor}}$ that enables the referee to compute $S_2 \oplus \{y_2\}$. The referee learns the query $q_{11} = (S_1 \oplus \{y_1\}, S_2 \oplus \{y_2\})$ and computes the corresponding answer a_{11}. Now, the referee has the answers to all 4 queries and it can XOR them to compute $f(x_1, \ldots, x_k)$.

The main contributions to the communication complexity of the above protocol is the invocations of the PSM protocols $\mathcal{P}_{\text{ind}}$ and $\mathcal{P}_{\text{Sxor}}$. We invoke $\mathcal{P}_{\text{ind}}$ with $k/2$ parties and a database containing $N^{k/2-1}$ bits, thus, the complexity of this protocol is $O(kN^{k/2-1})$. We invoke $\mathcal{P}_{\text{Sxor}}$ with $k/2$ parties and a set contained in $[N^{k/2}]$, thus, its complexity is $O(k^3 \cdot N^{k/2})$.

Note that the complexity of invoking $\mathcal{P}_{\text{Sxor}}$ dominates the complexity of invoking $\mathcal{P}_{\text{ind}}$. We capitalize on this gap and construct a PSM protocol for any function with output range $[N^k]$. Again, we represent $f : [N]^k \to [N^k]$ as a two dimensional cube, where the size of each dimension is $N^{k/2}$, however now every entry in the cube is from $[N^k]$. The protocol proceeds as above, where the only difference is that we invoke $\mathcal{P}_{\text{ind}}$ with a database containing $N^{k/2-1}$ elements from $[N^k]$.[4] Thus, the complexity of this protocol is $O(kN^{k/2-1} \cdot k \log N)$. The resulting PSM protocol has complexity $O(k^3 \cdot N^{k/2})$.

Next, we give an informal description of protocol $\mathcal{P}_k$ for an odd k, i.e., $k = 2t+1$ for some t. Here, we partition the input x_{t+1} of party P_{t+1} to two parts, i.e. $x_{t+1} = (x^1_{t+1}, x^2_{t+1})$ such that $x^1_{t+1}, x^2_{t+1} \in [N^{1/2}]$. Again, we associate f with an N^k-bit database that is viewed as a 2-dimensional cube (i.e., the size of each dimension is $N^{k/2}$). Correspondingly, each input $x = (x_1, \ldots, x_k) \in [N]^k$ is viewed as a pair $(y_1, y_2) \in [N^{k/2}]^2$, where $y_1 = (x_1, \ldots, x_t, x^1_{t+1})$ and $y_2 = (x^2_{t+1}, x_{t+2}, \ldots, x_k)$. The referee needs the answers to the same 4 queries in order to compute $f(x_1, \ldots, x_k)$. The rest of the protocol is similar to the protocol for an even k, just that in this case, party P_{t+1} participates in the PSM protocols

[4] Protocol $\mathcal{P}_{\text{ind}}$ (described in Sect. 2.3) can deal with L-bit entries and its complexity, for a list of N entries, is $O(L \cdot N)$.

Protocol $\mathcal{P}_k$

Common randomness: The k parties share uniform random strings:

- $S_1, S_2 \subseteq_R [N^{k/2}]$.
- $r_{10}, r_{01} \in_R \{0,1\}^{O(kN^{k/2-1} \log N)}$ required for the PSM protocol $\mathcal{P}_{\text{ind}}$ with a list of length $[N]^{k/2-1}$, where each element is from $[N^k]$.
- $\rho_{10}, \rho_{01} \in_R \{0,1\}^{O(N^{k/2})}$ required for the $k/2$-party PSM protocol $\mathcal{P}_{\text{Sxor}}$ with a set $S \subseteq [N^{k/2}]$.
- $b_{10}, b_{01} \in_R \{0,1\}^{k \log N}$, $b_{00} = b_{10} \oplus b_{01}$.

The protocol:

1. Party P_1:
   - Compute the answer a_{00} to the query $q_{00} = (S_1, S_2)$.
   - Compute the answers to all queries $(S_1 \oplus \{\ell\}, S_2)$ where $\ell = x_1 \circ i_2 \circ \ldots \circ i_{k/2}$, for all possible values $i_2, \ldots, i_{k/2} \in [N]$, and represent the answers as a $(k/2-1)$-dimensional cube $(a_{10}^{i_2,\ldots,i_{k/2}})_{i_2,\ldots,i_{k/2} \in [N]}$. Using $D^{10} = (a_{10}^{i_2,\ldots,i_{k/2}} \oplus b_{10})_{i_2,\ldots,i_{k/2} \in [N]}$ and common randomness r_{10}, compute m_{10}^1 – the message of the first party in $\mathcal{P}_{\text{ind}}$.
   - Using x_1, S_1 and common randomness ρ_{10}, compute s_{10}^1 – the message of the first party in $\mathcal{P}_{\text{Sxor}}$.
   - Send $m_{00} = a_{00} \oplus b_{00}$, m_{10}^1, and s_{10}^1 to the referee.
2. Party P_j, where $j \in \{2, \ldots, k/2\}$:
   - Using x_j and common randomness r_{10}, compute m_{10}^j – the message of the jth party in $\mathcal{P}_{\text{ind}}$.
   - Using x_j and common randomness ρ_{10}, compute s_{10}^j – the message of the jth party in $\mathcal{P}_{\text{Sxor}}$.
   - Send m_{10}^j and s_{10}^j to the referee.
3. Party P_k:
   - Compute the answers to all queries $(S_1, S_2 \oplus \{\ell\})$ where $\ell = i_{k/2+1} \circ \ldots \circ i_{k-1} \circ x_k$, for all possible values $i_{k/2-1}, \ldots, i_{k-1} \in [N]$, and represent the answers as a $(k/2-1)$-dimensional cube $(a_{01}^{i_{k/2+1},\ldots,i_{k-1}})_{i_{k/2+1},\ldots,i_{k-1} \in [N]}$. Using $D^{01} = (a_{01}^{i_{k/2+1},\ldots,i_{k-1}} \oplus b_{01})_{i_{k/2+1},\ldots,i_{k-1} \in [N]}$ and common randomness r_{01}, compute m_{01}^k – the message of the first party in $\mathcal{P}_{\text{ind}}$.
   - Using x_k, S_2 and common randomness ρ_{01}, compute s_{01}^k – the message of the first party in $\mathcal{P}_{\text{Sxor}}$.
   - Send m_{01}^k and s_{01}^k to the referee.
4. Party P_j, where $j \in \{k/2+1, \ldots, k-1\}$:
   - Using x_j and common randomness r_{01}, compute m_{01}^j – the message of the $(j-k/2+1)$th party in $\mathcal{P}_{\text{ind}}$.
   - Using x_j and common randomness ρ_{01}, compute s_{01}^j – the message of the $(j-k/2+1)$th party in $\mathcal{P}_{\text{Sxor}}$.
   - Send m_{01}^j and s_{01}^j to the referee.
5. The referee:
   - Using the messages $s_{10}^1, \ldots, s_{10}^{k/2}$ of $\mathcal{P}_{\text{Sxor}}$, compute $q_{11}^1 = S_1 \oplus \{y_1\}$. Using the messages $s_{01}^k, s_{01}^{k/2+1}, \ldots, s_{01}^{k-1}$ of $\mathcal{P}_{\text{Sxor}}$, compute $q_{11}^2 = S_2 \oplus \{y_2\}$. Compute the answer a_{11} to the query $q_{11} = (q_{11}^1, q_{11}^2) = (S_1 \oplus \{y_1\}, S_2 \oplus \{y_2\})$.
   - Using the messages $m_{10}^1, \ldots, m_{10}^{k/2}$ of $\mathcal{P}_{\text{ind}}$, compute $D_{x_2,\ldots,x_{k/2}}^{10}$. Using the messages $m_{01}^k, m_{01}^{k/2+1}, \ldots, m_{01}^{k-1}$ of $\mathcal{P}_{\text{ind}}$, compute $D_{x_{k/2+1},\ldots,x_{k-1}}^{01}$.
   - Output $f(x) = m_{00} \oplus D_{x_2,\ldots,x_{k/2}}^{10} \oplus D_{x_{k/2+1},\ldots,x_{k-1}}^{01} \oplus a_{11}$.

Fig. 4. A k-party PSM protocol $\mathcal{P}_k$ for a function $f : [N]^k \to [N^k]$ for an *even* k.

for both queries q_{10}, q_{01}, each time only with half of its input, as well as in both PSM protocols for $S_1 \oplus \{y_1\}$ and for $S_2 \oplus \{y_2\}$. The communication complexity and randomness complexity of this protocol are $O(k^3 N^{k/2})$.

Theorem 4.1. *Let $f : [N]^k \to [N^k]$ be a function. The protocol $\mathcal{P}_k$ is a secure PSM protocol for f with communication and randomness complexity $O(k^3 N^{k/2})$.*

Proof. Correctness follows from the cube approach, where $f(x_1, \ldots, x_k)$ is the XOR of the answers for the 4 queries $q_{00}, q_{10}, q_{01}, q_{11}$. The referee computes these 4 answers (each answer is a string in $[N^k]$); however, the first 3 answers are masked. Nevertheless, the XOR of the 3 masks b_{00}, b_{10}, b_{01} is zero, so when the referee computes the XOR of the 4 masked answers, the masks are canceled and the referee gets the correct answer.

To argue that the PSM protocol $\mathcal{P}_k$ is secure, we construct a simulator whose input is $f(x_1, \ldots, x_k)$ and whose output is k messages distributed as the messages in $\mathcal{P}_k$. To simplify the indices, we only construct a simulator for an even k; however, the simulator (with minor changes) remains valid also for an odd k. The simulator on input $f(x_1, \ldots, x_k)$ does the following:

1. Chooses two random sets $T_1, T_2 \subseteq_R [N^{k/2}]$ and two random strings $c_{10}, c_{01} \in_R \{0,1\}^{k \log N}$.
2. Computes the answer to the query q_{11}, namely, $a_{11} = \oplus_{i_1 \in T_1, i_2 \in T_2} f(i_1, i_2)$ (where i_1, i_2 are considered as $k/2$ elements from $[N]$).
3. Computes $c_{00} = f(x_1, \ldots, x_k) \oplus a_{11} \oplus c_{10} \oplus c_{01}$.
4. Invokes the simulator Sim_{ind} of protocol $\mathcal{P}_{\text{ind}}$ twice:
   - $(m_{10}^1, \ldots, m_{10}^{k/2}) \leftarrow \text{Sim}_{\text{ind}}(c_{10})$,
   - $(m_{01}^k, m_{01}^{k/2+1}, \ldots, m_{01}^{k-1}) \leftarrow \text{Sim}_{\text{ind}}(c_{01})$.
5. Invokes the simulator Sim_{Sxor} of protocol $\mathcal{P}_{\text{Sxor}}$ twice:
   - $(s_{10}^1, \ldots, s_{10}^{k/2}) \leftarrow \text{Sim}_{\text{Sxor}}(T_1)$,
   - $(s_{01}^k, s_{01}^{k/2+1}, \ldots, s_{01}^{k-1}) \leftarrow \text{Sim}_{\text{Sxor}}(T_2)$.
6. Outputs

$$(c_{00}, m_{10}^1, s_{10}^1), (m_{10}^2, s_{10}^2), \ldots, (m_{10}^{k/2}, s_{10}^{k/2}), (m_{01}^{k/2+1}, s_{01}^{k/2+1}), \ldots, (m_{01}^k, s_{01}^k).$$

Note that, in the messages of $\mathcal{P}_k$, the sets $S_1 \oplus \{y_1\}, S_2 \oplus \{y_2\}$ are distributed independently with uniform distribution. Furthermore, $D^{10}_{x_2,\ldots,x_{k/2}}, D^{10}_{x_{k/2+1},\ldots,x_{k-1}}$ are uniformly distributed given these 2 sets, since we mask them using independent random strings. Since $D^{10}_{x_2,\ldots,x_{k/2}}$ and c_{10} are both random strings, the output of $\text{Sim}_{\text{ind}}(c_{10})$ is distributed as the messages of $\mathcal{P}_{\text{ind}}(D^{10}, x_2, \ldots, x_{k/2})$. Since $S_1 \oplus \{y_1\}$ and T_1 are both random sets, the output of $\text{Sim}_{\text{Sxor}}(T_1)$ is distributed as the messages in $\mathcal{P}_{\text{Sxor}}((S_1, x_1), x_2, \ldots, x_{k/2})$. The same holds for the other invocation of Sim_{ind} and Sim_{Sxor}. Finally, given the sets $S_1 \oplus \{y_1\}, S_2 \oplus \{y_2\}$, the strings $m_{10}, m_{01}, D^{10}_{x_2,\ldots,x_{k/2}}, D^{01}_{x_{k/2+1},\ldots,x_{k-1}}$, and $f(x_1, \ldots, x_k)$, the value of m_{00} is fully determined (by the correctness of $\mathcal{P}_k$ and by the fact that the above sets determine a_{11}). Thus, the simulator's output is distributed in the same way as the messages generated in protocol $\mathcal{P}_k$ on inputs $x_1, \ldots, x_k$.

The communication and randomness complexity of each invocation of the PSM protocol $\mathcal{P}_{\text{ind}}$ are $O(k \cdot N^{k/2-1} \cdot k \log N)$. The communication and randomness complexity of each invocation of the PSM protocol $\mathcal{P}_{\text{Sxor}}$ is $O(k^3 \cdot N^{k/2})$. Party P_1 also sends a string of length $k \log N$ to the referee (that is, m_{00}). Therefore, the communication complexity and randomness complexity of the PSM protocol $\mathcal{P}_k$ are $O(k^3 \cdot N^{k/2})$. □

4.1 PSM Protocols for 4 and 5 Parties

We next show how to use the ideas of our previous protocols to construct more efficient k-party PSM protocols, for $k = 4, 5$.

Theorem 4.2.

- *Let $f : [N]^4 \to \{0,1\}$ be a function. There is a secure 4-party PSM protocol $\mathcal{P}_4$ for f with communication and randomness complexity $O(N^{5/3})$.*
- *Let $f : [N]^5 \to \{0,1\}$ be a function. There is a secure 5-party PSM protocol $\mathcal{P}_5$ for f with communication and randomness complexity $O(N^{7/3})$.*

Proof Sketch. The protocols $\mathcal{P}_4$ and $\mathcal{P}_5$ are similar to protocol $\mathcal{P}_k$, except that we view f as a 3-dimensional cube. Specifically, in $\mathcal{P}_4$ the size of each dimension of the cube is $N^{4/3}$. We partition the inputs x_2, x_3 as follows: $x_2 = (x_2^1, x_2^2)$ and $x_3 = (x_3^2, x_3^3)$ such that $x_2^1, x_3^3 \in [N^{1/3}]$ and $x_2^2, x_3^2 \in [N^{2/3}]$. We view each input (x_1, x_2, x_3, x_4) as a 3-tuple $(y_1, y_2, y_3) \in [N^{4/3}]^3$, where $y_1 = (x_1, x_2^1), y_2 = (x_2^2, x_3^2)$, and $y_3 = (x_3^3, x_4)$. The common randomness of the parties contains 3 random sets $S_1, S_2, S_3 \subseteq_R [N^{4/3}]$, random strings for protocols $\mathcal{P}_{\text{ind}}$ and $\mathcal{P}_{\text{Sxor}}$, and random masks. The referee should get the answers of 8 queries $q_{000}, \ldots, q_{111}$. The (masked) answer to q_{000} is computed by P_1 and sent to the referee. The answers to the queries of weight 2 and 3 are computed using protocol $\mathcal{P}_{\text{ind}}$, where the more expensive queries are queries of weight 2. As an example, we explain how to answer q_{110}. The answer to this query requires knowing $(y_1, y_2) \in [N^{4/3}]^2$. As $y_1 = (x_1, x_2^1)$ and party P_1 has x_1, party P_1 can prepare a list of length $N^{5/3}$ (one entry for each possible value of x_2^1, y_2) and P_1, P_2, and P_3 use the 3-party PSM protocol $\mathcal{P}_{\text{ind}}$ to send the answer of q_{110} to the referee. As the list contains $N^{5/3}$ entries (where each entry is a bit), the complexity of invoking $\mathcal{P}_{\text{ind}}$ is $O(N^{5/3})$ (the input of P_2 is from $[N]$ and of P_3 is from $[N^{2/3}]$). Queries q_{101}, q_{011} are dealt in a similar way, where P_1 and P_4, respectively, construct the list. To answer query q_{111}, the sets $S_1 \oplus \{y_1\}, S_2 \oplus \{y_2\}, S_3 \oplus \{y_3\}$ are sent to the referee using $\mathcal{P}_{\text{Sxor}}$. As each set is contained in $[N^{4/3}]$, the complexity of invoking $\mathcal{P}_{\text{Sxor}}$ is $O(N^{4/3})$. The total communication and randomness complexity of $\mathcal{P}_4$ is $O(N^{5/3})$.

In $\mathcal{P}_5$, the size of each dimension of the cube is $N^{5/3}$ and the 5 inputs are partitioned into 3 inputs $y_1, y_2, y_3 \in [N^{5/3}]$. The details are similar to the PSM protocol $\mathcal{P}_4$. As for the complexity analysis, to answer q_{110}, party P_1 needs $(y_1, y_2) \in [N^{5/3}]^2$ and it knows $x_1 \in [N]$ which is part of y_1, thus, it creates a list of length $N^{7/3}$, and the complexity of invoking $\mathcal{P}_{\text{ind}}$ is $O(N^{7/3})$. As each set in $\mathcal{P}_5$ is contained in $[N^{5/3}]$, the complexity of invoking $\mathcal{P}_{\text{Sxor}}$ is $O(N^{5/3})$. The total communication and randomness complexity of $\mathcal{P}_5$ is $O(N^{7/3})$. □

Discussion. Our protocols, and the 2-party PSM protocol of [6], use cubes of different dimensions for different values of k; i.e., in [6], 4 dimensions are used for 2 parties, and in this work 3 dimensions are used for 3, 4, and 5 parties, and 2 dimensions are used for more than 5 parties. These are the optimal dimensions, when using our approach, as we next explain. If there are k inputs (each from the domain $[N]$) and d dimensions, then the size of at least one dimension is $N^{k/d}$. Thus, communicating each set $S_j \oplus \{y_j\}$ (either directly, as in the 2-party and 3-party protocols, or using the PSM protocol $\mathcal{P}_{\text{Sxor}}$, as done for $k > 3$ parties) requires $\Omega(N^{k/d})$ bits. The parties also need to send the answers to the $2^d - 1$ queries of weight at most $d-1$ using protocol $\mathcal{P}_{\text{ind}}$. The most expensive queries are queries of weight $d-1$, which involve $d-1$ "virtual" inputs y_j, each one is taken from $[N^{k/d}]$. As each party P_j knows only x_j, the parties will invoke the PSM protocol $\mathcal{P}_{\text{ind}}$ with a list of length at least $N^{(k/d)(d-1)-1} = N^{k-k/d-1}$;[5] the cost of this invocation will be at least $N^{k-k/d-1}$. Thus, for a given number of parties k, we need to choose d that will minimize $\max\{k/d, k-k/d-1\}$. If, hypothetically, we could choose a non-integral dimension, we would take $d = 2k/(k-1)$ and the complexity of our protocol would have been $O(N^{(k-1)/2})$. This matches the complexity of the PSM protocols for 2 and 3 parties. For $k > 3$, we achieve a slightly worst complexity, since we need to round $2k/(k-1)$ to the nearest integer.

5 PSM Protocols with Inputs of Different Sizes

In this section, we construct PSM protocols for functions in which the domains of inputs are not necessarily the same. That is, we consider functions $f : [N^{\alpha_1}] \times [N^{\alpha_2}] \times \cdots \times [N^{\alpha_k}] \to \{0,1\}$, for some integer N and positive numbers $\alpha_1, \ldots, \alpha_k$. By reordering the parties and normalization, we can assume that $\alpha_1 \geq \alpha_2 \geq \cdots \geq \alpha_k = 1$.

We first observe that the complexity of the PSM of [10] for an arbitrary function does not depend on the size of the largest domain (i.e., of party P_1).

Claim 5.1. *Let $f : [N^{\alpha_1}] \times [N^{\alpha_2}] \times \cdots \times [N^{\alpha_k}] \to \{0,1\}$ be a function. Then, there is a PSM protocol for f with communication complexity $O(N^{\sum_{i=2}^{k} \alpha_i})$ and randomness complexity $O(k \cdot N^{\sum_{i=2}^{k} \alpha_i})$.*

Proof. We describe a protocol with the desired complexity. Party P_1 prepares a list of length $N^{\sum_{i=2}^{k} \alpha_i}$, which contains $f(x_1, i_2, \ldots, i_k)$ for every $(i_2, i_3, \ldots, i_k) \in [N^{\alpha_2}] \times \cdots \times [N^{\alpha_k}]$. The parties invoke the PSM $\mathcal{P}_{\text{ind}}$, where the input of P_1 is this list and the input of P_j, for $2 \leq j \leq k$, is x_j. □

We next construct more efficient k-party protocols. Again, we use the cube approach, with 2 differences compared to the previous protocols. First, the number of dimensions d will be bigger than in our previous protocols (the number

[5] For the interesting parameters (specifically, when $d \leq k$), there will always be a party whose entire input x_i is part of the query and the length of the list would be exactly $N^{k-k/d-1}$.

of dimensions grows as $\sum_{i=2}^{k} \alpha_i$ grows). Thus, the k inputs are partitioned to d virtual inputs. Unlike previous protocols, each P_i, where $1 \leq i \leq k$, holds part of each virtual input.

Lemma 5.2. *Let $f : [N^{\alpha_1}] \times [N^{\alpha_2}] \times \cdots \times [N^{\alpha_k}] \to \{0,1\}$ be a function where $\alpha_1 \geq \alpha_2 \geq \cdots \geq \alpha_k = 1$. Then, there is a PSM protocol for f with communication and randomness complexity*

$$O(\min\{k \cdot N^{(\sum_{i=2}^{k} \alpha_i)(1-1/\lceil a \rceil)}, k^3 \cdot N^{(\sum_{i=1}^{k} \alpha_i)/\lfloor a \rfloor}\})$$

where $a = \alpha_1/(\sum_{i=2}^{k} \alpha_i) + 2$.

Proof. We view f as a d-dimensional cube, where the size of each dimension is $N^{(\sum_{i=1}^{k} \alpha_i)/d}$ and d will be fixed later. We partition the inputs as follows: $x_j = (x_j^1, x_j^2, ..., x_j^d)$ for $1 \leq j \leq k$, where $x_j^1, x_j^2, ..., x_j^d \in [N^{\alpha_j/d}]$. We define $y_\ell = (x_1^\ell, x_2^\ell, \dots, x_k^\ell)$ for each $1 \leq \ell \leq d$. The common randomness of the parties contains d random sets $S_1, ..., S_d \subseteq_R [N^{(\sum_{i=1}^{k} \alpha_i)/d}]$, random strings for protocols $\mathcal{P}_{\text{ind}}$ and $\mathcal{P}_{\text{Sxor}}$, and random masks. Using the cube approach, the referee should get the answers to 2^d queries. To answer query $q_{11\dots1}$ (the query of weight d), the sets $S_1 \oplus \{y_1\}, ..., S_d \oplus \{y_d\}$ are sent to the referee using $\mathcal{P}_{\text{Sxor}}$. As each set is contained in $[N^{(\sum_{i=1}^{k} \alpha_i)/d}]$, the complexity of invoking $\mathcal{P}_{\text{Sxor}}$ is $O(k^3 \cdot N^{(\sum_{i=1}^{k} \alpha_i)/d})$. The (masked) answer to $q_{00\dots0}$ (the query of weight 0) is computed by P_1 and sent to the referee. The answers to the queries of weight m for any $1 \leq m \leq d-1$ are computed using protocol $\mathcal{P}_{\text{ind}}$, where the more expensive queries are queries of weight $d-1$. To answer the query $q_{11\dots10}$, party P_1, which has $x_1^1, ..., x_1^{d-1}$, prepares a list of length $N^{(\sum_{i=2}^{k} \alpha_i)(d-1)/d}$ (one entry for each possible value of $(x_2^1, \dots, x_k^1), ..., (x_2^{d-1}, \dots, x_k^{d-1})$) and the parties $P_1, \dots, P_k$ use the PSM protocol $\mathcal{P}_{\text{ind}}$ to send the masked answer of $q_{11\dots10}$ to the referee. The complexity of invoking $\mathcal{P}_{\text{ind}}$ is $O(k \cdot N^{(\sum_{i=2}^{k} \alpha_i)(d-1)/d})$. All other queries are dealt in a similar way, where P_1 constructs the list. The total communication and randomness complexity of the protocol are

$$O(\max\{k \cdot N^{(\sum_{i=2}^{k} \alpha_i)(d-1)/d}, k^3 \cdot N^{(\sum_{i=1}^{k} \alpha_i)/d}\}).$$

If we could choose a non-integral value for d then the optimal value of d in this approach would be $d = a = \alpha_1/(\sum_{i=2}^{k} \alpha_i) + 2$. Since d must be an integral value the communication and randomness complexity are minimized either on $\lfloor a \rfloor$ or on $\lceil a \rceil$. Thus, we get a protocol with the following complexity.

$$O(\min\{k \cdot N^{(\sum_{i=2}^{k} \alpha_i)(1-1/\lceil a \rceil)}, k^3 \cdot N^{(\sum_{i=1}^{k} \alpha_i)/\lfloor a \rfloor}\}).$$

□

Corollary 5.3. *Let $f : [N^{\alpha_1}] \times [N^{\alpha_2}] \times \cdots \times [N^{\alpha_k}] \to \{0,1\}$ be a function where $\alpha_1 \geq \alpha_2 \geq \cdots \geq \alpha_k = 1$. Then, there is a PSM protocol for f with communication and randomness complexity $O(k \cdot N^{(\sum_{i=2}^{k} \alpha_i) \cdot c})$ where $c < 1$.*

5.1 2-Party PSM Protocols

In this section, we describe results for 2-party PSM protocols summarized in the following lemma.

Lemma 5.4. *Let $f : [N^{\alpha}] \times [N] \to \{0,1\}$, where $\alpha \geq 1$. Then, there is a PSM for f with communication and randomness complexity:*

- $O(N^{\alpha/2})$ *if* $1 \leq \alpha \leq 3/2$,
- $O(N^{(\alpha+1)/(\lfloor\alpha+2\rfloor)})$ *if* $\alpha - \lfloor\alpha\rfloor \leq 1/\lceil\alpha+2\rceil$ *and* $\alpha > 3/2$,
- $O(N^{1-\frac{1}{\lceil\alpha+2\rceil}})$ *otherwise.*

Proof. For the first item, we use the protocol of [6], which has complexity $O(N^{\alpha/2})$ (where we consider both domains to be of size N^{α}). The second and third items follow from Lemma 5.2. □

5.2 3-Party PSM Protocols

In this section, we consider 3-party PSM protocols and show that, for many values of the parameters α_1, α_2, there is a PSM protocol whose complexity is the geometric mean of the sizes of domains.

Claim 5.5. *Let $f : [N^{\alpha_1}] \times [N^{\alpha_2}] \times [N] \to \{0,1\}$, where $\alpha_1 \geq \alpha_2 \geq 1$ and $\alpha_1 \leq 2\alpha_2 - 1$. Then, there is a PSM for f with communication and randomness complexity $O(N^{(\alpha_1+\alpha_2+1)/3})$.*

Proof. The idea of the protocol is similar to the protocols of Theorem 4.2. We view f as a 3-dimensional cube, where the size of each dimension is $N^{(\alpha_1+\alpha_2+1)/3}$. We partition the inputs $x_1 \in [N^{\alpha_1}], x_2 \in [N^{\alpha_2}], x_3 \in [N]$ to equal size inputs $y_1, y_2, y_3 \in [N^{(\alpha_1+\alpha_2+1)/3}]$. The way we partition the inputs will be described below. The common randomness of the parties contains 3 random sets $S_1, S_2, S_3 \subseteq_R [N^{(\alpha_1+\alpha_2+1)/3}]$, random strings for protocols $\mathcal{P}_{\text{ind}}$ and $\mathcal{P}_{\text{Sxor}}$, and random masks. The referee should get the answers to 8 queries $q_{000}, \ldots, q_{111}$. To answer query q_{111}, the sets $S_1 \oplus \{y_1\}, S_2 \oplus \{y_2\}, S_3 \oplus \{y_3\}$ are sent to the referee using $\mathcal{P}_{\text{Sxor}}$. As each set is contained in $[N^{(\alpha_1+\alpha_2+1)/3}]$, the complexity of invoking $\mathcal{P}_{\text{Sxor}}$ is $O(N^{(\alpha_1+\alpha_2+1)/3})$. The (masked) answer to q_{000} is computed by P_1 and sent to the referee. The answers to the queries of weight 1 and 2 are computed using protocol $\mathcal{P}_{\text{ind}}$, where the more expensive queries are queries of weight 2. The details of how to answer these queries depends on the partition of the inputs into y_1, y_2, y_3.

We partition x_1 and x_2 as follows: $x_1 = (x_1^1, x_1^3)$ and $x_2 = (x_2^2, x_2^3)$, where $x_1^1, x_2^2 \in [N^{(\alpha_1+\alpha_2+1)/3}]$, $x_1^3 \in [N^{(2\alpha_1-\alpha_2-1)/3}]$ (note that $2\alpha_1 - \alpha_2 - 1 \geq 0$ since $\alpha_1 \geq \alpha_2 \geq 1$), and $x_2^3 \in [N^{(2\alpha_2-\alpha_1-1)/3}]$ (note that $2\alpha_2 - \alpha_1 - 1 \geq 0$ by our assumption). We define $y_1 = x_1^1, y_2 = x_2^2$, and $y_3 = (x_3, x_1^3, x_2^3)$. To answer the query q_{110}, party P_1, which has $y_1 = x_1^1$, prepares a list of length $N^{(\alpha_1+\alpha_2+1)/3}$ (one entry for each possible value of y_2) and P_1, P_2 use the 2-party PSM protocol $\mathcal{P}_{\text{ind}}$ to send the masked answer of q_{110} to the referee. The complexity of invoking

$\mathcal{P}_{\text{ind}}$ is $O(N^{(\alpha_1+\alpha_2+1)/3})$. Queries q_{101}, q_{011} are dealt in a similar way, where P_1 and P_2, respectively, construct the list and all 3 parties participate (as each has a part of y_3). The total communication and randomness complexity of the protocol are $O(N^{(\alpha_1+\alpha_2+1)/3})$. □

Claim 5.6. *Let $f : [N^{\alpha_1}] \times [N^{\alpha_2}] \times [N] \to \{0,1\}$, where $\alpha_1 \geq \alpha_2 \geq 1$ and $\alpha_1 \geq 2\alpha_2 - 1$. Then, there is a PSM for f with communication and randomness complexity $O(N^{(\alpha_1+1)/2})$.*

Proof. The idea of the protocol is similar to the protocols of Theorem 4.2. We view f as a 3-dimensional cube, where the size of each dimension is $N^{(\alpha_1+\alpha_2+1)/3}$. We define $x_1 = (x_1^1, x_1^3)$, where $x_1^1 \in [N^{(\alpha_1+1)/2}]$ and $x_1^3 \in [N^{(\alpha_1-1)/2}]$, and $y_1 = x_1^1, y_2 = x_2$, and $y_3 = (x_3, x_1^3)$. The common randomness of the parties contains 3 random sets $S_1, S_3 \subseteq_R [N^{(\alpha_1+1)/2}]$ and $S_2 \subseteq_R [N^{\alpha_2}]$, random strings for protocols $\mathcal{P}_{\text{ind}}$ and $\mathcal{P}_{\text{Sxor}}$, and random masks. The referee should get the answers to 8 queries $q_{000}, \ldots, q_{111}$. To answer query q_{111}, the sets $S_1 \oplus \{y_1\}, S_2 \oplus \{y_2\}, S_3 \oplus \{y_3\}$ are sent to the referee using $\mathcal{P}_{\text{Sxor}}$. The complexity of invoking $\mathcal{P}_{\text{Sxor}}$ is $O(N^{(\alpha_1+1)/2})$ (where, for S_2, we use the assumption that $\alpha_1 \geq 2\alpha_2 - 1$). The (masked) answer to q_{000} is computed by P_1 and sent to the referee. The answers to the queries of weight 2 and 3 are computed using protocol $\mathcal{P}_{\text{ind}}$, where the most expensive query is q_{011}, for which P_2 has to prepare a list of length $N^{(\alpha_1+1)/2}$ (one entry for every possible value of y_3) and parties P_1, P_2, and P_3 use the 3-party PSM protocol $\mathcal{P}_{\text{ind}}$ to send the answer of q_{011} to the referee. The complexity of invoking $\mathcal{P}_{\text{ind}}$ is $O(N^{(\alpha_1+1)/2})$. Queries q_{101}, q_{011} are dealt in a similar way. The total communication and randomness complexity of the protocol are $O(N^{(\alpha_1+1)/2})$. □

The next two lemmas summarize the various cases of 3-argument functions; the first claim deals with the case $\alpha_2 \geq 2$ and the second claim with the case $\alpha_2 < 2$.

Lemma 5.7. *Let $f : [N^{\alpha_1}] \times [N^{\alpha_2}] \times [N] \to \{0,1\}$, where $\alpha_1 \geq \alpha_2 \geq 1$. If $\alpha_2 \geq 2$ then there is a PSM for f with communication and randomness complexity:*

- $O(N^{(\alpha_1+\alpha_2+1)/3})$ *if* $\alpha_1 < \alpha_2 + 1$.
- $O(\min\{N^{(\alpha_2+1)(1-1/\lceil a \rceil)}, N^{(\alpha_1+\alpha_2+1)/\lfloor a \rfloor}\})$ *where* $a = (\alpha_1 + 2\alpha_2 + 2)/(\alpha_2 + 1)$ *if* $\alpha_1 \geq \alpha_2 + 1$.

Proof. If $\alpha_1 < \alpha_2 + 1$, then $\alpha_1 < 2\alpha_2 - 1$ (since $\alpha_2 \geq 2$). Thus, the first item is implied by Claim 5.5. The second item follows from Lemma 5.2. □

Lemma 5.8. *Let $f : [N^{\alpha_1}] \times [N^{\alpha_2}] \times [N] \to \{0,1\}$, where $\alpha_1 \geq \alpha_2 \geq 1$. Assume $1 \leq \alpha_2 < 2$. Then there is a PSM for f with communication and randomness complexity:*

- $O(N^{(\alpha_1+\alpha_2+1)/3})$ *if* $\alpha_2 \leq \alpha_1 \leq 2\alpha_2 - 1$.
- $O(N^{(\alpha_1+1)/2})$ *if* $2\alpha_2 - 1 \leq \alpha_1 \leq (4\alpha_2 + 1)/3$.
- $O(\min\{N^{(\alpha_2+1)(1-1/\lceil a \rceil)}, N^{(\alpha_1+\alpha_2+1)/\lfloor a \rfloor}\})$ *where* $a = (\alpha_1 + 2\alpha_2 + 2)/(\alpha_2 + 1)$ *if* $(4\alpha_2 + 1)/3 \leq \alpha_1$.

Proof. The first item follows from Claim 5.5. The second item follows from Claim 5.6. The third item follows from Lemma 5.2. □

6 A PSM for k Parties from a PSM for t Parties

In this section, we show how to construct a k-party PSM protocol for a function from a t-party PSM protocol for a related function where $k > t$. This is a generic transformation, which does not result in better protocols than the protocols presented in this paper. However, it shows that improvements in the complexity of t-party PSM protocols for small values of t will results in better k-party PSM protocols for all values k.

Claim 6.1. *Let k, t, N be integers such that $k > t$ and $N \geq 2^t$ and let $g : [N]^k \rightarrow \{0,1\}$ be a function. If every function $f : [n]^t \rightarrow \{0,1\}$ has a t-party PSM protocol with communication and randomness complexity $O(n^\alpha)$ for $n = N^{k/t}$, then there is a PSM protocol for g with communication and randomness complexity $O(k \cdot t \cdot N^{(\alpha+1)k/t-1})$.*

Proof. We construct the following protocol $\mathcal{P}_k$ for g. We partition the inputs $x_1, ..., x_k$ to equal size inputs $y_1, ..., y_t$ where $y_j \in [N^{k/t}]$ for $j = 1, ..., t$. Let $x_i = (x_i^1, ..., x_i^t)$ for all $i = t+1, ..., k$ where $x_i^1, ..., x_i^t \in [N^{1/t}]$. We define $y_j = (x_j, x_{t+1}^j, ..., x_k^j)$ for $j = 1, ..., t$. Furthermore, we define $g' : [N^{k/t}]^t \rightarrow \{0,1\}$ such that $g'(y_1, ..., y_t) = g(x_1, ..., x_k)$. By the assumption of the claim, there is a PSM protocol $\mathcal{P}_t$ for g' with communication and randomness complexity $O(n^\alpha) = O(N^{\alpha k/t})$.

In protocol $\mathcal{P}_k$, for every $1 \leq j \leq t$, party P_j together with parties $P_{t+1}, ..., P_k$ simulate the j-th party of protocol $\mathcal{P}_t$ as follows. For this simulation, party P_j prepares a list of length $N^{k/t-1}$ of the possible messages of the j-th party in $\mathcal{P}_t$ with input $y_j = (x_j, x_{t+1}^j, ..., x_k^j)$. As P_j knows x_j and does not know $x_{t+1}^j, ..., x_k^j$ the list contains one entry for every possible value of $(x_{t+1}^j, ..., x_k^j) \in [N^{k/t-1}]$. Each entry in the list is a message taken from $[N^{\alpha k/t}]$. Parties $P_j, P_{t+1}, ..., P_k$ use the PSM protocol $\mathcal{P}_{\text{ind}}$ to send the message corresponding to their inputs to the referee. The referee computes the t messages and then computes $g(x_1, ..., x_k) = g'(y_1, ..., y_t)$ according to the protocol $\mathcal{P}_t$. The communication and randomness complexities of the PSM protocol $\mathcal{P}_{\text{ind}}$ are $O(k \cdot N^{(\alpha+1)k/t-1})$ and the parties $P_{t+1}, ..., P_k$ invoke this protocol t times. Therefore, the total communication and randomness complexity of this protocol are $O(k \cdot t \cdot N^{(\alpha+1)k/t-1})$. □

For example, for $t = 2$ there is a 2-party PSM protocol with complexity $O(N^{0.5})$ [6] (here $\alpha = 0.5$). This implies that any function $f : [N]^k \rightarrow \{0,1\}$ has a k-party PSM protocol with complexity $O(N^{3k/4-1})$. This is inferior to our protocols from Sects. 3 and 4. For $t = 3$, we get that any function $f : [N]^k \rightarrow \{0,1\}$ has a k-party PSM protocol with complexity $O(N^{2k/3-1})$ (using our 3-party PSM protocol which has complexity $O(N)$, i.e. $\alpha = 1$). In particular, for $k = 4$, we can construct a 4-party PSM protocol with complexity $O(N^{5/3})$ from our 3-party PSM protocol, matching the complexity of our protocol from Sect. 4.1. Thus, if one can improve the message length of 3-party PSM protocols for an arbitrary function to $o(N)$, then this would yield 4-party PSM protocol with message length $o(N^{5/3})$ improving on our construction.

To conclude, to improve the complexity of k-party PSM protocols for an arbitrary functions, one might want to start with designing k-party PSM protocols for small values of k. For example, if the complexity of the 2-party PSM protocols will be improved to $O(N^{\beta})$ for $\beta < 2/k$, then we will get k-party PSM protocols with complexity better than $O(N^{k/2})$.

7 Applications

In this section, we present some applications of our PSM protocols for several cryptographic primitives.

7.1 t-robust NIMPC Protocols

A Non-Interactive secure Multi-Party Computation (NIMPC) protocol, defined in [5], is a PSM protocol that is secure even if some parties collude with the referee. Such parties may send the referee their messages for *every* possible input and therefore the referee can always compute the function on many inputs. The model is defined with *correlated* randomness $r_1, \dots, r_k$ between the k parties, rather than common randomness, and the dishonest parties can, alternatively, send the referee their r_i's.

In such setting, one may only hope for a so-called "best possible security"; that is, in a t-robust NIMPC protocol, an adversary controlling at most t parties and seeing all the messages sent by honest parties learns no information that is not implied by the unavoidable information – the restriction of f fixing the inputs of the honest parties.

The communication complexity of the best known fully-robust (i.e., k-robust) NIMPC protocol, for an arbitrary function $f : [N]^k \to \{0,1\}$, was $O(\text{poly}(\log N, k) \cdot N^k)$ [19] (improving on [5]). We show that every function $f : [N]^k \to \{0,1\}$ has a t-robust NIMPC with complexity $\tilde{O}(N^{k/2+t})$, which improves on previous constructions when $t < k/2$. Our construction is based on an information-theoretic transformation of [8], which constructs a t-robust NIMPC protocol for f from any PSM (that is, 0-robust NIMPC) protocol for f.

Theorem 7.1 ([8]). *Let t be a positive integer and $\mathcal{P}$ be a PSM protocol for a Boolean function $f : [N]^k \to \{0,1\}$ with randomness and communication complexity $\alpha(N)$. Then, there exists a t-robust NIMPC protocol for f with randomness and communication complexity $O((2\max\{N,k\})^{t+1}k(t\log(N+k)+\alpha(N)))$.*

Using Theorem 7.1 and our PSM protocol $\mathcal{P}_k$, we get a t-robust NIMPC protocol for $f : [N]^k \to \{0,1\}$ with randomness and communication complexity $O(k^4 2^t N^{k/2+t+1})$ (assuming that $N \geq k$). We next construct a t-robust NIMPC protocol where we improve the complexity by a factor of N. This optimization is more significant when k and t are small. For example, consider a 3-argument function $f : [N]^3 \to \{0,1\}$. By Theorem 3.1, the function f has a PSM protocol with complexity $O(N)$. Thus, by Theorem 7.1, we get a 1-robust PSM for f with complexity $O(N^3)$. Notice that a 1-robust 3-party NIMPC protocol is a fully-robust protocol (as even in an ideal world 2 parties can basically learn the input of the third party). Our optimized fully robust NIMPC protocol for f will have complexity $O(N^{2.5} \log^4 N)$.

Theorem 7.2. *Let t, k, N be positive integers such that $t < k/2$ and $k \leq N$ and $f : [N]^k \to \{0,1\}$ be a Boolean function. Then, there exists a t-robust NIMPC protocol for f with randomness and communication complexity $O(k^4 \log^4 N \cdot (2N)^{k/2+t})$.*

Proof. The basic idea is that, instead of viewing f as a k-argument function, where each party has an input from $[N]$, we consider a $(k \log N)$-argument function f', where each party has an input from $\{0,1\}$ and construct a $(t \log N)$-robust protocol for f'. For simplicity of notation we assume that N is a power of 2. (This results in multiplying N by at most 2, yielding the term $2N$ in Theorem 7.2.) We define $f' : \{0,1\}^{k \log N} \to \{0,1\}$, where

$$\begin{aligned} &f'(x_{1,1}, \ldots, x_{1,\log N}, \ldots, x_{k,1}, \ldots, x_{k,\log N}) \\ &= f((x_{1,1}, \ldots, x_{1,\log N}), \ldots, (x_{k,1}, \ldots, x_{k,\log N})). \end{aligned}$$

By Theorem 4.1, the function f' has a PSM protocol with communication and randomness complexity $O((k \log N)^3 2^{(k \log N)/2}) = O(k^3 \log^3 N \cdot N^{k/2})$. We cannot apply Theorem 7.1 directly, as it will not result in the desired complexity. We observe that the NIMPC that we construct needs to be robust only against sets T that contain t blocks of size $\log N$ (where a block is a set of $\log N$ parties of f' holding the bits of some party P_j in f). By [8, Theorem 6.4], we need a matrix H' that is T-admissible only for such sets T.[6] If we have such matrix with ℓ rows over a finite field $\mathbb{F}_q$, then we can replace the term $(2 \max\{N, k\})^{t+1}$ with $q^{\ell+1}$ in Theorem 7.1.

We next explain how to construct such H'. In [8], it was shown that a $t \times k$ parity-check matrix H of the Reed-Solomon code over $\mathbb{F}_q$, where $q \geq \max\{k, N\}$ is a prime-power, is T-admissible for every set T of size t. We take such matrix H over the field $\mathbb{F}_{2^{\log N}}$ and construct a matrix H' over $\mathbb{F}_2$ with $t \log N$ rows and $k' = k \log N$ columns by replacing each entry $a \in \mathbb{F}_{2^{\log N}}$ with a $\log N \times \log N$ matrix A over $\mathbb{F}_2$ such that for every $b \in \mathbb{F}_{2^{\log N}}$ it holds that ab, viewed as a vector over $\mathbb{F}_2$, is the same as $A(b_0, \ldots, b_{\log N-1})^T$ (when viewing elements of $\mathbb{F}_{2^{\log N}}$ as polynomials, the matrix A simulates the multiplication of the polynomial by the polynomial representing a modulo the irreducible polynomial). The matrix H' is T-admissible for every T that contains at most t blocks.

By [8, Theorem 6.4], the function f' has an NIMPC protocol that is T-robust, for every T that contains at most t blocks, and has complexity

$$O(q^{\ell+1}(k')^4 2^{k'/2})) = O(2^{t \log N+1}(k \log N)^4 \cdot 2^{(k \log N)/2}) = O(k^4 \log^4 N \cdot N^{k/2+t}).$$

If N is not a power of 2, we need to replace N by $2N$ in the complexity. To construct a PSM protocol for f, the parties $P_1, \ldots, P_k$ execute the PSM for f',

[6] A matrix H' is T-admissible if $H'\boldsymbol{u} \neq H'\boldsymbol{v}$ for any two vectors $\boldsymbol{u}, \boldsymbol{v}$ that agree on all entries not indexed by T.

where each party P_j sends the messages of the block of parties holding the bits of its input. □

If we consider a 3-argument function $f : [N]^3 \to \{0,1\}$, we get a 1-private NIMPC protocol with communication and randomness complexity $O(\log^4 N \cdot N^{5/2})$. As discussed above, this protocol is fully robust. This improves on the previously best known 3-party NIMPC protocol of [19] whose complexity is $O(N^3 \log^2 N)$. While the protocol of [19] hides the function f, in our protocol the referee needs to know f in order to reconstruct its output. In [19], it is proved that in any NIMPC protocol for every k-argument function $f : [N]^k \to \{0,1\}$ that hides the function, the size of the common randomness is $\Omega(N^k)$. The proof of this lower bound actually holds for any PSM protocol in which the reconstruction of f's value is independent of f. Thus, in a fully robust 3-party NIMPC protocol with randomness complexity $o(N^3)$ for an arbitrary function $f : [N]^3 \to \{0,1\}$, the referee has to know f in advance.

Example 7.3. For the 3-party case, the construction of the admissible matrix H' is much simpler starting from the 1-admissible matrix $H = (1,1,1)$. The resulting matrix H' is $H' = (I_{\log N} I_{\log N} I_{\log N})$ (that is, H' contains 3 copies of the $(\log N) \times (\log N)$ identity matrix). Consider two vectors $\boldsymbol{u}, \boldsymbol{v}$ that differ only in the first block i.e., $\boldsymbol{u} = (\boldsymbol{u_1}, \boldsymbol{u_2}, \boldsymbol{u_3})$ and $\boldsymbol{v} = (\boldsymbol{v_1}, \boldsymbol{v_2}, \boldsymbol{v_3})$ where $\boldsymbol{u_1} \neq \boldsymbol{v_1}$ while $\boldsymbol{u_2} = \boldsymbol{v_2}$ and $\boldsymbol{u_3} = \boldsymbol{v_3}$. Then, $H'\boldsymbol{u} - H'\boldsymbol{v} = H'(\boldsymbol{u} - \boldsymbol{v}) = I_{\log N}(\boldsymbol{u_1} - \boldsymbol{v_1}) \neq \boldsymbol{0}$. Thus, H' is T-admissible for the 3 blocks T.

7.2 Ad-hoc PSM Protocols and Homogeneous Distribution Designs

A k-out-of-n ad-hoc PSM protocol is a PSM protocol with n parties, where only k parties, whose identity is not known in advance, actually participate in the protocol. For a formal definition of ad-hoc PSM protocols see [7].

We obtain improved k-out-of-n ad-hoc PSM protocols for symmetric functions[7] $f : [N]^k \to \{0,1\}$ with communication complexity $O(e^k \cdot k^6 \cdot \log n \cdot N^{k/2})$ and randomness complexity $O(e^k \cdot k^8 \cdot \log n \cdot N^{k/2})$, based on a transformation from k-party PSM protocols to ad-hoc PSM protocols from [7] and on our new PSM protocols.

Theorem 7.4 ([7]). *Assume that there is a k-party PSM protocol Π for a symmetric function f with randomness complexity* $\mathrm{Rnd}(\Pi)$ *and communication complexity* $\mathrm{Com}(\Pi)$. *Then, there is a k-out-of-n ad-hoc PSM protocol for f with randomness complexity $O(e^k \cdot k^3 \cdot \log n \cdot (\mathrm{Rnd}(\Pi) + k^2 \cdot \max\{\mathrm{Com}(\Pi), \log n\}))$ and communication complexity $O(e^k \cdot k^3 \cdot \log n \max\{\mathrm{Com}(\Pi), \log n\})$.*

Using Theorem 7.4 and our PSM protocol $\mathcal{P}_k$, whose communication and randomness complexity are $O(k^3 N^{k/2})$, we get:

[7] A function f is symmetric if for every input $(x_1, \ldots, x_k) \in [N]^k$ and every permutation $\pi : [k] \to [k]$, it holds that $f(x_1, \ldots, x_k) = f(x_{\pi(1)}, \ldots, x_{\pi(k)})$.

Theorem 7.5. *Let $f : [N]^k \to [N^k]$ be a symmetric function. Then, there is a k-out-of-n ad-hoc PSM protocol for f with communication complexity $O(e^k \cdot k^6 \cdot \log n \cdot N^{k/2})$ and randomness complexity $O(e^k \cdot k^8 \cdot \log n \cdot N^{k/2})$.*

We next show that ad-hoc PSM protocols imply distribution design. The goal of *distribution design*, introduced in [4], is to find a joint distribution on N random variables $(X_1, X_2, \ldots, X_N)$ that satisfies a given set of constraints on the marginal distributions. If such a distribution exists we say that the set of constraints is realizable. Each constraint involves two equal-size sets $\{i_1, \ldots, i_d\}$ and $\{j_1, \ldots, j_d\}$ and can either be an equality constraint of the form $\{i_1, \ldots, i_d\} \equiv \{j_1, \ldots, j_d\}$, in which case the two marginal distributions should be identical, or a disjointness constraint of the form $\{i_1, \ldots, i_d\} \parallel \{j_1, \ldots, j_d\}$, in which case the two marginal distributions should have disjoint supports.[8] A k-homogenous set of constraints is one where all sets in the constraints have the same size k. Borrowing terminology from secret sharing (which is one of the applications of distribution design), we refer to the value of a random variable X_i as the ith share.

We obtain distribution designs for 2-homogeneous sets of constraints with share size $O(\sqrt{N} \cdot \log^2 N)$. Previously, the best known distribution design had share size of $O(N \log N)$. We also obtain distribution designs for k-homogeneous sets of constraints, for $k > 2$, with share size $O(e^k \cdot k^6 \cdot N^{k/2} \cdot \log N)$, improving on the best known distribution design whose share size was $O(\binom{N}{k} \cdot \min\{d \log N, N\})$. Our construction is based on a PSM protocol for a function that represents the constraints. We use the messages of this protocol to define the random variables.

Definition 7.6. *Given a k-homogenous set of constraints $\mathcal{R}$ on N variables, define the following symmetric function $f_{\mathcal{R}}$: Construct an undirected graph, whose vertices are all subsets of size k. Connect two vertices (sets of size k) A and B if and only if "$A \equiv B$" $\in \mathcal{R}$ and let $\mathcal{A}_1, \ldots, \mathcal{A}_\ell$ be the connected components of this graph.*[9] *Define a symmetric function $f_{\mathcal{R}} : [N]^k \to \binom{N}{k}$ such that, for every set $\{i_1, \ldots, i_k\}$ (i.e. $i_1 < i_2 < \ldots < i_k$), define $f_{\mathcal{R}}(i_1, \ldots, i_k) = j$ where $(i_1, \ldots, i_k) \in \mathcal{A}_j$. To make this function symmetric, define $f_{\mathcal{R}}(\sigma(i_1), \ldots, \sigma(i_k)) = f_{\mathcal{R}}(i_1, \ldots, i_k)$ for every permutation $\sigma : [k] \to [k]$.*

Theorem 7.7. *Let $\mathcal{R}$ be a k-homogenous set of constraints on N variables. Assume that there is a k-out-of-N ad-hoc PSM protocol $\mathcal{P}_{k,N}$ for $f_{\mathcal{R}} : [N]^k \to \binom{N}{k}$ with communication complexity $\mathrm{Com}(\mathcal{P}_{k,N})$. If $\mathcal{R}$ is realizable, then there is a distribution design X realizing $\mathcal{R}$ with share size $\mathrm{Com}(\mathcal{P}_{k,N}) + \log N$.*

---

[8] We only consider the projective case, in which the constraints are restricted to be on sets of variables; that is, the elements in the sets are sorted. In [4], also constraints for non-sorted elements are considered.

[9] By [4], a distribution design exists if and only if for every constraint "$A \| B$" $\in \mathcal{R}$, the sets A and B are in different connected components. Furthermore, it is enough to construct a distribution design X where $X_A \equiv X_B$ whenever A and B are in the same component $\mathcal{A}_i$, and $X_A \parallel X_B$ whenever A and B are in different components.

Proof. To construct the distribution design, we first choose a random permutation $\pi : [N] \to [N]$. Let $M_{\pi(i)}$ be the messages in $\mathcal{P}_{k,N}$ of party $P_{\pi(i)}$ with input i. We set X_i to be $(M_{\pi(i)}, \pi(i))$ for all $i \in [N]$. The reason that we choose the permutation π is that an ad-hoc PSM protocol does not hide the identities of the parties sending messages. We claim that $X = (X_1, \ldots, X_N)$ realizes $\mathcal{R}$:

Equivalence constraints. Let $A = \{i_1, \ldots, i_k\}$ and let $B = \{j_1, \ldots, j_k\}$ be two sets such that "$(i_1, \ldots, i_k) \equiv (j_1, \ldots, j_k)$" $\in \mathcal{R}$. Vertices A and B are in the same connected component and therefore $f_{\mathcal{R}}(i_1, \ldots, i_k) = f_{\mathcal{R}}(j_1, \ldots, j_k)$. From the construction, $X_A = (X_{i_1}, \ldots, X_{i_k}) = ((M_{\pi(i_1)}, \pi(i_1)), \ldots, (M_{\pi(i_k)}, \pi(i_k)))$, and $X_B = (X_{j_1}, \ldots, X_{j_k}) = ((M_{\pi(j_1)}, \pi(j_1)), \ldots, (M_{\pi(j_k)}, \pi(j_k)))$. We argue that X_A and X_B are equally distributed. First, π is a random permutation, thus $\{\pi(i_1), \ldots, \pi(i_k)\} \equiv \{\pi(j_1), \ldots, \pi(j_k)\}$. We next fix a set $C = \{c_1, ..., c_k\} \subseteq [N]$ of size k and show that the distribution of X_A conditioned on $\pi(i_\ell) = c_\ell$ for every $\ell \in [k]$ is equal to the distribution of X_B conditioned on $\pi(j_\ell) = c_\ell$ for every $\ell \in [k]$. That is, X_A and X_B contain the messages of the same set C with inputs $i_1, \ldots, i_k$ and $j_1, \ldots, j_k$ respectively. Since $f_{\mathcal{R}}(i_1, \ldots, i_k) = f_{\mathcal{R}}(j_1, \ldots, j_k)$, by the security of the ad-hoc PSM protocol, these messages are equally distributed.

Disjointness constraints. Let $A = \{i_1, \ldots, i_k\}$ and let $B = \{j_1, \ldots, j_k\}$ be two subsets such that "$(i_1, \ldots, i_k) \parallel (j_1, \ldots, j_k)$" $\in \mathcal{R}$. Vertices A and B are in different connected components and therefore $f_{\mathcal{R}}(i_1, \ldots, i_k) \neq f_{\mathcal{R}}(j_1, \ldots, j_k)$. Let X'_A and X'_B be the messages in X_A and X_B respectively sorted according to π; i.e. $X'_A = (M_{\pi(i_{t_1})}, \ldots, M_{\pi(i_{t_k})})$ such that $\pi(i_{t_1}) < \pi(i_{t_2}) < \ldots < \pi(i_{t_k})$ and, similarly, $X'_B = (M_{\pi(j_{m_1})}, \ldots, M_{\pi(j_{m_k})})$ such that $\pi(j_{m_1}) < \pi(j_{m_2}) < \ldots < \pi(j_{m_k})$. By the correctness of the ad-hoc PSM protocol $\mathcal{P}_{k,N}$, we can reconstruct $f_{\mathcal{R}}(i_1, \ldots, i_k)$ from the messages $(M_{\pi(i_{t_1})}, \ldots, M_{\pi(i_{t_k})})$ and, similarly, we can reconstruct $f_{\mathcal{R}}(j_1, \ldots, j_k)$ from the messages $(M_{\pi(j_{m_1})}, \ldots, M_{\pi(j_{m_k})})$ and therefore, $X_A \parallel X_B$. □

Corollary 7.8. *Let $\mathcal{R}$ be a k-homogenous set of constraints on N variables. If $\mathcal{R}$ is realizable, then there is a distribution design X realizing $\mathcal{R}$ with share size $O(k^3 N^{k/2})$. If $k = 2$, the size of the shares is $O(N^{0.5})$, if $k = 3$, the size of the shares is $O(N)$, if $k = 4$, the size of the shares is $O(N^{5/3})$, and if $k = 5$, the size of the shares is $O(N^{7/3})$.*

The size of the shares in our distribution design is smaller than the size of the shares in the distribution designs of [6] for homogeneous sets of constraints when the number of constraints is large. However, the distribution designs in [6] have extra properties that can be used to construct distribution designs for non-homogeneous sets of constraints. We do not know how to use our distribution designs to realize non-homogeneous sets of constraints.

7.3 Conditional Disclosure of Secrets and Secret-Sharing Schemes for Uniform Access Structures

A CDS protocol allows a set of parties $P_1, \ldots, P_k$ to disclose a secret to a referee, subject to a given condition on their inputs. In such a protocol, each party P_i

holds an input $x_i \in [N]$, a joint secret s, and a common random string, and there is a public function $f : [N]^k \to \{0,1\}$. The referee knows $x_1, \ldots, x_k$. The protocol involves only a unidirectional communication from the parties to the referee, which should learn s if and only if $f(x_1, \ldots, x_k) = 1$.

Using the transformations of [6,12] from PSM protocols to CDS protocols, our PSM protocols from Sect. 4 imply k-party CDS protocols with complexity $O(k^3 \cdot N^{k/2})$ for an arbitrary function. However, in a very recent result, Liu, Vaikuntanathan, and Wee [17] have constructed much more efficient CDS protocols; they construct a k-party CDS protocols for any function $f : [N]^k \to \{0,1\}$ with complexity $2^{\tilde{O}(\sqrt{k \log N})}$.

We show that a CDS protocol implies a secret-sharing scheme for t-uniform access structures in which the share size is the communication complexity of the CDS protocols. An access structure is said to be t-uniform if the size of every minimal authorized sets is either t or $t+1$ and all sets of size at least $t+1$ are authorized. That is, every set of size less than t is unauthorized, every set of size greater than t is authorized, and some sets of size t are authorized. Our construction realizing a uniform access structure takes a CDS protocol for a function $f : \{0,1\}^k \to \{0,1\}$ and transforms it to a secret-sharing scheme realizing a t-uniform access structure (with k parties).

Definition 7.9. *Let $\mathcal{P} = \{P_1, \ldots, P_k\}$ be a set of parties. We represent a subset of parties $A \subseteq \mathcal{P}$ by its characteristic string $x_A = (x_1, \ldots, x_k) \in \{0,1\}^k$ where for each $i \in [k]$, $x_i = 1$ if and only if $P_i \in A$. For a t-uniform access structure $\mathcal{A}$, we define a function $f_{\mathcal{A}} : \{0,1\}^k \to \{0,1\}$ such that for every subset of parties $A \subseteq \mathcal{P}$, $f(x_A) = 1$ if and only if $A \in \mathcal{A}$. In particular, if $|A| > t$ then $f(x_A) = 1$.*

Theorem 7.10. *Let $\mathcal{A}$ be a t-uniform access structure on a set of parties $\mathcal{P} = \{P_1, \ldots, P_k\}$, and let Π be a CDS protocol for the function $f_{\mathcal{A}} : \{0,1\}^k \to \{0,1\}$ with a secret s. If Π has communication complexity* $\mathrm{Com}(\Pi)$*, then there is a secret-sharing scheme for $\mathcal{A}$ with share size*

$$O(k \cdot \max\{\mathrm{Com}(\Pi), \log k\} + \log k).$$

Proof. Assume the dealer wants to share a secret $s \in \{0,1\}$. The dealer chooses at random a bit $s_1 \in \{0,1\}$ and shares s_1 using Shamir's t-out-of-k secret-sharing scheme. Let $s_2 = s \oplus s_1$. The dealer chooses the randomness r, required for the CDS protocol Π. Let $M_{i,0}, M_{i,1}$ be the message of party P_i on inputs 0 and 1 respectively in the CDS protocol Π with secret s_2 and randomness r. For each $i \in [k]$ the dealer shares $M_{i,0}$ in Shamir's t-out-of-$(k-1)$ threshold secret-sharing scheme among all the parties except for party P_i. Next, the dealer gives to each party P_i the share $M_{i,1}$. The size of $M_{i,b}$ is $\mathrm{Com}(\Pi)$ for each $i \in [k]$, and $b \in \{0,1\}$. The share of each party P_i is $M_{i,1}$ and $k-1$ additional shares created in the t-out-of-$(k-1)$ Shamir's threshold secret-sharing scheme for messages of the CDS, and a share of the bit s_1. Thus, the total share size is $k \cdot \mathrm{Com}(\Pi) + \log k$ (assuming that $\mathrm{Com}(\Pi) \geq \log k$). We claim that this secret-sharing scheme realizes $\mathcal{A}$.

First, let A be an authorized set of parties such that $|A| \geq t$, thus, $f(x_A) = 1$ and the parties in A can reconstruct s_1. By the correctness of Π, if the parties have the messages $M_{i,1}$ for each $P_i \in A$ and $M_{i,0}$ for each $P_i \notin A$, then they can reconstruct the secret s_2. As $|A| \geq t$, the parties in A hold at least t shares of $M_{i,0}$ for every $P_i \notin A$ and therefore, they can reconstruct $M_{i,0}$. The parties in A also have their message on input 1; i.e. $M_{i,1}$ for each $P_i \in A$. Therefore, they can reconstruct the secret s_2, and therefore, reconstruct s.

Second, let A be a set of parties such that $|A| < t$. The parties have no information on s_1, hence no information on s. Finally, let A be an unauthorized subset of parties such that $|A| = t$. From the correctness of Shamir's secret-sharing scheme, the parties in A can reconstruct $M_{i,0}$ for every $P_i \notin A$ and have no information on $M_{i,0}$ for $P_i \in \mathcal{A}$ (since P_i does not get a share of $M_{i,0}$). The parties in A also have their message on input 1; i.e. $M_{i,1}$ for each $P_i \in A$. The set A is unauthorized, thus, $f(x_A) = 0$. From the security of Π, the parties in A do not learn any information on the secret s_2 from the messages on input $f(x_A) = 0$, therefore, they learn no information about s. □

Corollary 7.11. *Let $\mathcal{A}$ be a t-uniform access structure with k parties. Then, there exists a secret-sharing scheme realizing $\mathcal{A}$ with shares of size $2^{\tilde{O}(\sqrt{k})}$.*

References

1. Applebaum, B., Raykov, P.: From private simultaneous messages to zero-information Arthur-Merlin protocols and back. In: Kushilevitz, E., Malkin, T. (eds.) TCC 2016. LNCS, vol. 9563, pp. 65–82. Springer, Heidelberg (2016). https://doi.org/10.1007/978-3-662-49099-03
2. Beimel, A., Farràs, O., Mintz, Y., Peter, N.: Linear secret-sharing schemes for forbidden graph access structures. In: Kalai, Y., Reyzin, L. (eds.) TCC 2017. LNCS, vol. 10678, pp. 394–423. Springer, Cham (2017). https://doi.org/10.1007/978-3-319-70503-313
3. Beimel, A., Farràs, O., Peter, N.: Secret sharing schemes for dense forbidden graphs. In: Zikas, V., De Prisco, R. (eds.) SCN 2016. LNCS, vol. 9841, pp. 509–528. Springer, Cham (2016). https://doi.org/10.1007/978-3-319-44618-927
4. Beimel, A., Gabizon, A., Ishai, Y., Kushilevitz, E.: Distribution design. In: ITCS 2016, pp. 81–92 (2016)
5. Beimel, A., Gabizon, A., Ishai, Y., Kushilevitz, E., Meldgaard, S., Paskin-Cherniavsky, A.: Non-interactive secure multiparty computation. In: Garay, J.A., Gennaro, R. (eds.) CRYPTO 2014. LNCS, vol. 8617, pp. 387–404. Springer, Heidelberg (2014). https://doi.org/10.1007/978-3-662-44381-122
6. Beimel, A., Ishai, Y., Kumaresan, R., Kushilevitz, E.: On the cryptographic complexity of the worst functions. In: Lindell, Y. (ed.) TCC 2014. LNCS, vol. 8349, pp. 317–342. Springer, Heidelberg (2014). https://doi.org/10.1007/978-3-642-54242-814
7. Beimel, A., Ishai, Y., Kushilevitz, E.: Ad hoc PSM protocols: secure computation without coordination. In: Coron, J.-S., Nielsen, J.B. (eds.) EUROCRYPT 2017. LNCS, vol. 10212, pp. 580–608. Springer, Cham (2017). https://doi.org/10.1007/978-3-319-56617-720

8. Benhamouda, F., Krawczyk, H., Rabin, T.: Robust non-interactive multiparty computation against constant-size collusion. In: Katz, J., Shacham, H. (eds.) CRYPTO 2017. LNCS, vol. 10401, pp. 391–419. Springer, Cham (2017). https://doi.org/10.1007/978-3-319-63688-713
9. Chor, B., Goldreich, O., Kushilevitz, E., Sudan, M.: Private information retrieval. J. ACM **45**, 965–981 (1998)
10. Feige, U., Kilian, J., Naor, M.: A minimal model for secure computation. In: 26th STOC, pp. 554–563 (1994)
11. Gay, R., Kerenidis, I., Wee, H.: Communication complexity of conditional disclosure of secrets and attribute-based encryption. In: Gennaro, R., Robshaw, M. (eds.) CRYPTO 2015. LNCS, vol. 9216, pp. 485–502. Springer, Heidelberg (2015). https://doi.org/10.1007/978-3-662-48000-724
12. Gertner, Y., Ishai, Y., Kushilevitz, E., Malkin, T.: Protecting data privacy in private information retrieval schemes. JCSS **60**(3), 592–629 (2000)
13. Ishai, Y., Kumaresan, R., Kushilevitz, E., Paskin-Cherniavsky, A.: Secure computation with minimal interaction, revisited. In: Gennaro, R., Robshaw, M. (eds.) CRYPTO 2015. LNCS, vol. 9216, pp. 359–378. Springer, Heidelberg (2015). https://doi.org/10.1007/978-3-662-48000-718
14. Ishai, Y., Kushilevitz, E.: Private simultaneous messages protocols with applications. In: 5th Israel Symposium on Theory of Computing and Systems, pp. 174–183 (1997)
15. Ishai, Y., Kushilevitz, E., Paskin, A.: Secure multiparty computation with minimal interaction. In: Rabin, T. (ed.) CRYPTO 2010. LNCS, vol. 6223, pp. 577–594. Springer, Heidelberg (2010). https://doi.org/10.1007/978-3-642-14623-731
16. Liu, T., Vaikuntanathan, V., Wee, H.: Conditional disclosure of secrets via non-linear reconstruction. In: Katz, J., Shacham, H. (eds.) CRYPTO 2017. LNCS, vol. 10401, pp. 758–790. Springer, Cham (2017). https://doi.org/10.1007/978-3-319-63688-725
17. Liu, T., Vaikuntanathan, V., Wee, H.: Towards breaking the exponential barrier for general secret sharing. In: Nielsen, J.B., Rijmen, V. (eds.) EUROCRYPT 2018, Part II. LNCS, vol. 10821, pp. 567–596. Springer, Cham (2018)
18. Sun, H.-M., Shieh, S.-P.: Secret sharing in graph-based prohibited structures. In: INFOCOM 1997, pp. 718–724 (1997)
19. Yoshida, M., Obana, S.: On the (in)efficiency of non-interactive secure multiparty computation. In: Kwon, S., Yun, A. (eds.) ICISC 2015. LNCS, vol. 9558, pp. 185–193. Springer, Cham (2016). https://doi.org/10.1007/978-3-319-30840-112

Masking

Formal Verification of Masked Hardware Implementations in the Presence of Glitches

Roderick Bloem([✉]), Hannes Gross, Rinat Iusupov, Bettina Könighofer, Stefan Mangard, and Johannes Winter

Institute for Applied Information Processing and Communications (IAIK), Graz University of Technology, Inffeldgasse 16a, 8010 Graz, Austria
{roderick.bloem,hannes.gross,rinat.iusupov,bettina.konighofer, stefan.mangard,johannes.winter}@iaik.tugraz.at

Abstract. Masking provides a high level of resistance against side-channel analysis. However, in practice there are many possible pitfalls when masking schemes are applied, and implementation flaws are easily overlooked. Over the recent years, the formal verification of masked software implementations has made substantial progress. In contrast to software implementations, hardware implementations are inherently susceptible to glitches. Therefore, the same methods tailored for software implementations are not readily applicable.

In this work, we introduce a method to formally verify the security of masked hardware implementations that takes glitches into account. Our approach does not require any intermediate modeling steps of the targeted implementation. The verification is performed directly on the circuit's netlist in the probing model with glitches and covers also higher-order flaws. For this purpose, a sound but conservative estimation of the Fourier coefficients of each gate in the netlist is calculated, which characterize statistical dependence of the gates on the inputs and thus allow to predict possible leakages. In contrast to existing practical evaluations, like t-tests, this formal verification approach makes security statements beyond specific measurement methods, the number of evaluated leakage traces, and the evaluated devices. Furthermore, flaws detected by the verifier are automatically localized. We have implemented our method on the basis of a SAT solver and demonstrate the suitability on a range of correctly and incorrectly protected circuits of different masking schemes and for different protection orders. Our verifier is efficient enough to prove the security of a full masked first-order AES S-box, and of the Keccak S-box up to the third protection order.

Keywords: Masking · Formal verification
Threshold implementations · Hardware security
Side-channel analysis · Private circuits

J. B. Nielsen and V. Rijmen (Eds.): EUROCRYPT 2018, LNCS 10821, pp. 321–353, 2018.
https://doi.org/10.1007/978-3-319-78375-8_11

1 Introduction

Security critical embedded systems rely on the protection of sensitive information against exposure. While the transmission of sensitive information over conventional communication channels can be protected by means of strong cryptography, the protection against unintentionally created side channels, like power consumption [28] or electromagnetic emanation [33], requires additional countermeasures.

Since the risk of side-channel analysis (SCA) is inevitable in many applications, different countermeasures have been proposed in the past. One of the best researched and most effective countermeasures against SCA is masking. Many different masking schemes have been introduced [24,26,31,35,37] over the years. The history of masking, however, is also a history of failure and learning. Some of the first masking schemes were shown to be insecure in practical implementations because glitches in the combinatorial logic (temporary logic states caused by propagation time differences of the driving signals) were not considered in the formal models [26,37]. The first provably secure masking scheme with inherent resistance to glitches was the threshold implementation (TI) scheme of Nikova et al. [31]. Over the last years the TI scheme has been extended further by Bilgin et al. [12], and other schemes have been proposed like the consolidated masking scheme (CMS) of Reparaz et al. [35], and the domain-oriented masking scheme (DOM and the low-randomness variant UMA) of Gross et al. [23,24].

Even if the used masking scheme is secure, this is not automatically true for the implementations that rely on this scheme. One problem that thus still remains is the verification of masked implementations. There are basically two approaches that are used in practice to verify the resistance against SCA, namely formal verification and empirical leakage assessment. The predominant approach for the verification of masked hardware implementations is still the empirical leakage assessment in form of statistical significance tests [21] or by attacking the devices by using state-of-the-art side-channel analysis techniques. However, such practical evaluations are never complete, in a sense that if no flaw is found it remains uncertain whether or not the implementation could be broken with a better measurement setup or more leakage traces.

Recently there has been some substantial development towards the formal verification for masked software implementations [3,7,17]. However, these verification methods are tailored to software and do not take glitches into account. Therefore, they cannot readily be applied to hardware implementations. In terms of non-empirical verification of hardware implementations, there exist tools to test for leakage by either modeling of the circuit in software [34] or approaches that simulate possible leakages by assuming a specific power model [9]. To the best of our knowledge there exist no formal tools that take glitches into account and directly prove the security of masked hardware implementations on its netlist.

Our contribution. In this work, we introduce a method to formally prove the security of masked hardware implementations in the presence of glitches. In contrast to existing formal or non-empirical verification approaches for hardware designs, the introduced approach does not require any additional modeling of the circuit or the leakage source and proves the security of a circuit directly on its netlist. Compared to empirical verification methods based on the statistical analysis of leakage traces, our formal approach allows direct localization of the detected flaws, and gives conclusive security statements that are independent of device- or measurement-specific conditions, or the amount of gathered leakage information.

We base our approach on the probing model of Ishai et al. [26] and take the effects of glitches into account. We introduce a circuit verification method that performs a conservative estimate of the data that an attacker can learn by probing different gates and wires. The verification works directly on the gate-level representation of the circuit. It uses the Fourier expansion (or Walsh expansion) of the functions that are computed and uses the fact that a non-zero Fourier coefficient for a linear combination of variables indicates a correlation between the function and these variables (cf. [10]). A correlation with a linear combination of variables that contains secrets but no uniformly distributed masking variables corresponds to an information leak. By only keeping track of whether coefficients are zero or not, we circumvent the complexity of a full Fourier representation of all functions computed by all gates of the circuit, at the cost of a loss of precision that may lead to false alarms. The information of whether a coefficient is zero or not can be encoded as a propositional logic formula whose size is linear in the size of the circuit and vulnerability can be computed efficiently by a SAT solver.

To show the practicality of this approach, we check a variety of masked circuits that originate from different masking schemes. We focus on acyclic (feedback free) pipelined masked circuits, like the S-boxes of symmetric primitives which are usually the parts of a circuit that are the hardest to protect in practice and therefore most susceptible to flaws. The security of the linear circuits parts, on the other hand, can be established and verified more easily in practice, for instance by ensuring that only one masked value or mask of one variable is used inside each of the linear circuit parts [24]. For the same reason multiple cipher rounds and S-box lookups can be analyzed separately, as long as it is ensured that the masked outputs of the nonlinear parts are always independently and freshly masked (which is the case for most masking schemes).

We ran our tool on a set of example circuits including the S-boxes of Keccak, Fides and AES. Our verifier is efficient enough to formally prove the resistance of a full first-order masked AES S-box. Because of the circuit size of the AES S-box, which consumes about 40% of the entire AES area [24], the parallelized evaluation takes about 10 h. We also prove a Keccak S-box up to order three, and the $GF(2)$ multipliers of DOM up to order four. Furthermore, we show that our approach correctly detects flaws in masked circuits that are known to be flawed in the presence of glitches e.g. [26,37]. The implementation of our tool and some example circuits are available on github [27].

This paper is organized as follows. In Sect. 2 we give a short overview of existing verification approaches and discuss the differences to our approach. In Sect. 3, we introduce the used notation and the Fourier expansion. We give an introduction to masked circuits and the probing model in Sect. 4, and show how to leverage the Fourier expansion to test for probing security. We start the introduction of our verification approach in Sect. 5, at first without taking signal timings into account. Before we complete the description of our general verification approach in Sect. 7, we first discuss in Sect. 6 how we model timing effects i.e. glitches. A concrete instantiation of our verification approach based on a SAT solver is then introduced in Sect. 8. Evaluation results for various masked circuits are discussed in Sect. 9. We conclude this work in Sect. 10.

2 Related Work

Automated verification of masked implementations has been intensively researched over the last years and recently many works targeting this topic have been published [1,6,7,9,16–18,30]. Most of the existing work, however, targets software based masking which does not include the effects of glitches.

Verification of masked software. One of the most groundbreaking works towards the efficient verification of masked software implementations is the work of Barthe et al. [3]. Instead of proving the security of a whole implementation at once, this work introduces the notion of strong non-interference (SNI). SNI is an extension to the more general non-interference (NI) notion introduced in [2]. The SNI notion allows to prove the security of smaller code sequences (called gadgets) in terms of composability with other code parts. Gadgets fulfilling this SNI notion can be freely composed with other gadgets without interfering with their SCA resistance.

Verification of algorithms that fulfill this notion scale much better than other approaches but, on the other hand, not all masking algorithms that are secure are also directly composable. As a matter of fact the most efficient software masking algorithms in terms of randomness of Belaid et al. [7,8], Barthe et al. [4], and Gross et al. [23], for example, do not achieve SNI directly.

In contrast to Barthe et al.'s work on SNI [3], our approach does not check for composability and is therefore less restrictive to the circuits and masking schemes that can be proven (similar to the NI approach of Barthe et al.' [2]). Since Barthe et al.'s work is designed to prove masked software implementations it does not cover glitches. In our work we introduce the necessary formal groundwork for the verification of masked circuits and in particular the propagation of glitches. Our approach is thereby not bound to our SAT realization but is also compatible with existing tools like easycrypt which is developed by Barthe et al. [5].

Most recently another formal verification approach by Coron [14] was introduced that builds on the work of Barthe et al.. Essentially two approaches are discussed in this work. The first approach is basically the same as the approach in [2] but written in Common Lisp language. The second approach is quite different and works by using elementary transformations in order to make the targeted program verifiable using the NI and SNI properties. The work of Coron targets again only software based masking and does not take glitches into account.

Eldib et al. [17] present an approach to verify masked software implementations. Similar to our approach, the verification problem is encoded into SMT and verified by checking the constraints for individual nodes (operations) inside the program. This approach allows direct localization of the vulnerable code parts.

However, their approach targets software and therefore does not cover glitches. It also produces SMT formulas that are exponential in the number of secret variables, whereas the formulas that are produced by our approach are only linear.

Bhasin et al. [10] also use Fourier transforms to estimate the side channel attack resistance of circuits. Their approach uses a SAT solver to construct low-weight functions of a certain resistance order. They have not used their approach to evaluate existing implementations of cryptographic functions, and they do not take glitching behavior into account.

Verification of masked hardware. Similar to our approach, Bertoni et al. [9] address verification of masked hardware implementations in the presence of glitches. In this work all possible transients at the input of a circuit are considered and all resulting glitches that could occur at the gates are modeled. However, this approach focuses on first-order masking of purely combinatorial logic and uses a rather simple power model to measure the impact (transitions from 0 to 1 result in the same power consumption as transitions from 1 to 0). We note that focusing on combinatorial logic only, leaves out most of the existing hardware-based masking schemes such as [23,24,31,35]. Bertoni et al. demonstrated their approach on a masked implementation of Keccak based on a masking scheme that is known to be insecure in the presence of glitches.

In contrast to Bertoni et al.'s work, our approach considers combinatorial logic as well as sequential gates (registers), covers also higher-order leakages, and is not restricted to circuits with only one output bit.

In the work of Reparaz [34], a leakage assessment approach is introduced that works by simulating leakages of a targeted hardware implementation in software. At first, a high-level model of the hardware implementation is created, and the verification then works by simulating the model with different inputs and extracting leakage traces. The verification result is gathered by applying statistical significance tests (t-tests) to the simulated leakage traces. Compared to our approach, the leakage detection approach of Reparaz does not perform a formal verification but an empirical leakage assessment. Furthermore, the verification is not directly performed on the targeted hardware implementation but requires to model its (leakage) behavior in software.

Most recently a paper by Faust et al. [19] was published that introduces the so-called robust-probing model as extension to the original probing model with regard to glitches. They build upon the work of Barthe et al. [3] and target the verification of the SNI notion in their extended probing model. In contrast to Faust et al.'s approach, our formal verification approach does not strife for the verification of the SNI notion which trades higher randomness and implementations costs against faster verification. Furthermore, to the best of our knowledge, there exists no implementation of their verification approach in form of a tool to check real hardware circuits.

3 Preliminaries

In the following we make extensive use of the usual set notation, where $S \triangle T = S \setminus T \cup T \setminus S$ denotes the symmetric difference of S and T and for two sets of sets $\mathcal{S}$ and $\mathcal{T}$, we define $\mathcal{S} \triangle \mathcal{T} = \{S \triangle T \mid S \in \mathcal{S}, T \in \mathcal{T}\}$ to be the pointwise set difference of all elements. We write $\mathbb{B} = \{\mathsf{true}, \mathsf{false}\}$ for the set of Booleans. For a set X of Boolean variables, we identify an assignment $f : X \to \mathbb{B}$ with the set of variables x for which $f(x) = \mathsf{true}$. For a Boolean function $f(X, Y)$ and an assignment $x \subseteq X$, we write $f|_x$ to denote the function $f|_x(y) = f(x, y)$.

Fourier expansion of Boolean functions. There is a close connection between statistical dependence and the Fourier expansion of Boolean functions. First we formally define statistical independence.

Definition 1 (Statistical independence). *Let X, Y, and Z be sets of Boolean variables and let $f : 2^X \times 2^Y \to 2^Z$. We say that* f *is statistically independent of X if for all z there is a c such that for all x we have $|\{y \mid f(x, y) = z\}| = c$.*

Lemma 2. *Let $F : \mathbb{B}^X \times \mathbb{B}^Y \to \mathbb{B}^Z$. Function* F *is statistically independent of X iff for all functions $f : \mathbb{B}^Z \to \mathbb{B}$ we have that $f \circ F$ is statistically independent of X.*

Please find the proof in Appendix A. To define Fourier expansions, we follow the exposition of [32] and associate true with -1 and false with 1. We can then represent a Boolean function as a multilinear polynomial over the rationals.

Definition 3 (Fourier expansion). *A Boolean function $f : \{-1, 1\}^n \to \{-1, 1\}$ can be* uniquely *expressed as a multilinear polynomial in the n-tuple of variables $X = (x_1, x_2, \ldots, x_n)$ with $x_i \in \{\pm 1\}$, i.e., the multilinear polynomial of f is a linear combination of monomials, called* Fourier characters, *of the form $\chi_T(X) = \prod_{x_i \in T} x_i$ for every subset $T \subseteq X$. The coefficient of $\chi_T \in \mathbb{Q}$ is called the* Fourier coefficient $\widehat{f}(T)$ *of the subset* T. *Thus we have the* Fourier representation *of f:*

$$f(X) = \sum_{T \subseteq X} \widehat{f}(T) \chi_T(X) = \sum_{T \subseteq X} \widehat{f}(T) \prod_{x_i \in T} x_i.$$

The Fourier characters $\chi_T : \{-1, 1\}^n \to \{-1, 1\}$ form an orthonormal basis for the vector space of functions in $f : \{-1, 1\}^n \to \{-1, 1\}$. The Fourier coefficients are given by the projection of the function to its basis, i.e., for $f : \{-1, 1\}^n \to \{-1, 1\}$ and $T \subseteq X = (x_1, x_2, \ldots, x_n)$, the coefficient $\widehat{f}(T)$ is given by $\widehat{f}(T) = 1/2^n \cdot \sum_{X \in \{\pm 1\}^n} (f(X) \cdot \chi_T(X))$. In order to prevent confusion between multiplication and addition on rationals and conjuction and XOR on Booleans, we write $\cdot$ and $+$ for the former and $\wedge$ and $\oplus$ for the latter.

As an example, the Fourier expansion of $x \wedge y$ is

$$1/2 + 1/2 \cdot x + 1/2 \cdot y - 1/2 \cdot x \cdot y. \tag{1}$$

If $x = \mathsf{false} = 1$ and $y = \mathsf{true} = -1$, for example, the polynomial evaluates to $1/2 + 1/2 - 1/2 + 1/2 = 1 = \mathsf{false}$ as expected for an AND function.

Let us note some simple facts. (1) the Fourier expansion uses the exclusive or of variables as the basis: $x \oplus y = x \cdot y$. (2) $f^2 = 1$ for the Fourier expansion of any Boolean function f [32]. (3) there are two linear functions of two arguments: $f = x \cdot y$ (XOR) and $f = -(x \cdot y)$ (XNOR). All other functions f are nonlinear and for them, each of $\widehat{f}(\emptyset)$, $\widehat{f}(\{x\})$, $\widehat{f}(\{y\})$, and $\widehat{f}(\{x, y\})$ is nonzero. (We are ignoring the constant and unary functions.) (4) the statistical dependence of the functions can be read off directly from the Fourier expansion: the conjunction has a constant bias, positively correlates with x and y, and negatively with its $x \oplus y$. This last fact can be generalized to the following lemma.

Lemma 4 (Xiao-Massey [39]). *A Boolean function $f : \{-1, 1\}^n \rightarrow \{-1, 1\}$ is statistically independent of a set of variables $X' \subseteq X$ iff $\forall T \subseteq X'$ it holds that if $T \neq \emptyset$ then $\widehat{f}(T) = 0$.*

4 Masking and the Probing Model

The intention of masking is to harden side-channel analysis attacks (like differential power analysis or electromagnetic emanation analysis) by making side-channel information independent of the underlying security sensitive information. This independence is achieved through the randomization of the representation of security sensitive variables inside the circuit. For this purpose, randomly produced and uniformly distributed masks are added (XOR) to the security sensitive variables on beforehand of a security critical computation. The number of used masks depends on the used masking scheme and is a function of the security order.

As a simple example, we consider the security sensitive 1-bit variable s in Eq. 2 that is protected by adding a uniformly random mask m_s, resulting in the masked representation s_m.

$$s_m = s \oplus m_s. \tag{2}$$

The masked value s_m is again uniformly distributed and statistically independent of s, i.e., it has the same probability to be 0 or 1 regardless of the value of s. Any operation that is performed only on s_m is statistically independent of s and thus also the produced side-channel information. Since the mask m_s is randomly produced, operations on the mask are uncritical. However, the combination of side-channel information on s_m and m_s can reveal information on s. The independence achieved through masking is thus only given up to a certain degree (the number of fresh masks used for masking s), and it is important to ensure this degree of independence throughout the entire circuit. The degree of independence is usually refereed to as the protection order d.

Masked circuits. For the remainder of the paper, let us fix an ordered set $X = \{x_0, \ldots, x_n\}$ of input variables. We partition the input variables X into three categories:

- $S = \{s_1, \ldots s_j\}$ are security sensitive variables such as key material and intermediate values of cryptographic algorithms that must be protected against an attacker by means of masking.

- $M = \{m_1, \ldots m_k\}$ are masks that are used to break the statistical dependency between the secrets S and the information carried on the wires and gates. Masks are assumed to be fresh random variables with uniform distribution and with no statistical dependency to any other variable of the circuit.
- $P = \{p_1, \ldots p_l\}$ are all other variables including publicly known constants, control signals, et cetera. Unlike secret variables, these signals do not need to be protected by masks and are unsuitable to protect secret variables.

We define a circuit $C = (\mathcal{G}, \mathcal{W}, R, f, \mathcal{I})$, where $(\mathcal{G}, \mathcal{W})$ is an acyclic directed graph with vertices $\mathcal{G}$ (gates) and edges $\mathcal{W} \subseteq \mathcal{G} \times \mathcal{G}$ (wires). Gates with indegree zero are called inputs I, gates with outdegree zero are called outputs O. Furthermore, $R \subseteq \mathcal{G}$ is a set of registers, f is a function that associates with any gate $g \in \mathcal{G} \setminus I$ with indegree k a function $f(g) : \mathbb{B}^k \rightarrow \mathbb{B}$, and $\mathcal{I} : I \rightarrow (2^X \rightarrow \mathbb{B})$ associates an externally computed Boolean function over X to each input. We require that registers have indegree one and that the associated function is the identity. In the following, we assume, wlog, that all gates, except inputs and registers, have indegree 2 and we partition these gates into a set L of linear gates (XOR, XNOR) and a set N of nonlinear gates (AND, NAND, OR, NOR, the two implications and their negations). We also require that for any gate g, any path from some input to g has the same number of registers.

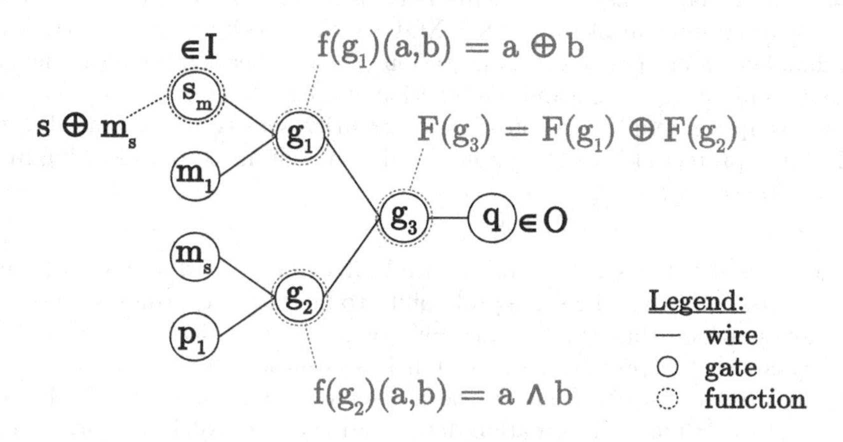

Fig. 1. Circuit graph of circuit in Fig. 2

The intuitive meaning of f is the local function computed by a gate. For instance, if g is an AND gate, $f(g)(x, y) = x \wedge y$. We associate with every gate another function $F(g) : 2^X \rightarrow \mathbb{B}$, which defines the function computed by the output of the gates in terms of the circuit inputs. The function $F(g)$ is defined by the functions of the predecessor gates and $f(g)$ in the obvious way. Given a sequence of gates $(g_1, \ldots, g_d)$, we extend F pointwise to $F(g_1, \ldots, g_d) : 2^X \rightarrow \mathbb{B}^d$: $F(g_1, \ldots, g_d)(x) = (g_1(x), \ldots, g_d(x))$. We often identify a gate with its function.

As an example, consider the circuit graph in Fig. 1 (which corresponds to the circuit depicted in Fig. 2). We have $f(g_3)(a, b) = a \oplus b$ and $F(g_3) = (s_m \oplus m_1) \oplus (m_s \wedge p_1)$.

For a circuit C, a sequence of gates $G = (g_1, \ldots, g_n)$, and a sequence of functions $F = (f_1, \ldots, f_n)$ with $f_i \in \mathbb{B}^2 \to \mathbb{B}$, we write $C[G \mapsto F]$ for the circuit C in which gate g_i is replaced by a gate with the Boolean function f_i.

Security of masked circuits. The security of various masking schemes is often analyzed in the so-called probing model that was introduced by Ishai et al. [26]. It was shown by Faust et al. [20] and Rivain et al. [36] that the probing model is indeed suitable to model side-channel attacks and to describe the resistance of an implementation in relation to the protection order d. As it was shown by Chari et al. [13], there is an exponential relation between d and the number of leakage traces required to exploit the side-channel information.

In the probing model, an attacker is bound to d probing needles which can be freely placed on arbitrary circuit gates (or wires). Probes are placed permanently on these gates and monitor all signals states and signal transitions that occur at the probed circuit gate from the circuit reset onwards. Thus one probe records the probed signals at all time instances. The probing model quantifies the level of side-channel resistance of a circuit over the minimum number of probing needles an attacker requires to extract any secret information. More specifically, a circuit is secure in the probing model if an attacker cannot combine the information gathered from d probes over all points in time in an arbitrary function F such that F statistically depends on any of the secret variables in S. We model a probe as the ability to read the Boolean function produced by the probed gate or its associated wire. Since we assume that the masking variables are uniformly distributed, and the public variables are known, the circuit leaks information iff F is statistically dependent on S regardless of the values that the public variables take.

Definition 5 (secure functions). *A function $f : 2^X \to \mathbb{B}^d$ is secure if* f *is for any assignment $p \subseteq P$ to the public variables, $f|_p$ is statistically independent of* S.

Definition 6 (d-probing security [26]). *A circuit $C = (\mathcal{G}, \mathcal{W}, f, \mathcal{I})$ is order d probing secure (d-probing secure) iff for any gates $g_1, \ldots, g_d \in \mathcal{G}$, $F(g_1, \ldots, g_d)$ is secure.*

Verification example using the Fourier expansion. According to Lemma 4, we can decide whether the values computed by a circuit are secure by computing the Fourier expansion of all its gates and checking whether there is a coefficient that contains only secret variables without a mask (and with or without public variables). Formally we check that $\emptyset \neq S' \subseteq S \cup P$ such that $\widehat{F}(g)(S') \neq 0$. The first-order security of a circuit can thus be verified using the probing model by calculating the Fourier expansion of the whole circuit. As an example consider the Fourier expansion of the circuit in Fig. 2 for which we have:

$$\begin{aligned}
F(g_1) &= s_m \cdot m_1, \\
F(g_2) &= 1/2 + 1/2 \cdot m_s + 1/2 \cdot p_1 - 1/2 \cdot m_s p_1, \text{ and} \\
F(g_3) &= F(g_1) \cdot F(g_2) \\
&= 1/2 \cdot s_m m_1 + 1/2 \cdot m_s s_m m_1 + 1/2 \cdot p_1 s_m m_1 - 1/2 \cdot m_s p_1 s_m m_1.
\end{aligned}$$

Assuming that $s_m = s \oplus m_s$ and using the properties of the Fourier expansion this implies that

$$F(g_3) = 1/2 \cdot s m_s m_1 + 1/2 \cdot s m_1 + 1/2 \cdot s p_1 m_s m_1 - 1/2 \cdot s p_1 m_1. \quad (3)$$

For the example circuit in Fig. 2, if s is a secret and m_1 is a uniformly distributed random mask, then g_3 in Eq. 3 computes a function that does not reveal any secret information. This follows from the fact that in $F(g_3)$ there are only (non-zero) Fourier coefficients for terms that contain s and at least one masked value.

Since the exact computation of Fourier coefficients is very expensive and the extension to higher-order probing security nontrivial, in the following we develop a method to estimate the Fourier coefficients of each gate and to check for higher-order security.

5 Our Verification Approach for Stable Signals

In this section, we present a sound verification method for (d-)probing security for the steady-state of a digital circuit. It is assumed that the signals at the circuit input are fixed to a certain value and that all intermediate signals at the gates and the circuit output have reached their final (stable) state. This approach is later on extended in Sects. 6 and 7 to cover transient signals and glitches.

Since the formal verification of the security order of masked circuits has proven to be a non-trivial problem in practice, the intuition behind a circuit verifier is to have a method that correctly classifies a wide range of practically relevant and securely masked circuits but rejects all insecure circuits. Any circuit that is not secure according to Definition 6 is rejected. Our verification approach can be subdivided into three parts: (1) the labeling system, (2) the propagation rules, and (3) the actual verification.

5.1 Labeling

In order to check the security of a circuit we introduce a labeling over the set of input variables X for the stable signals $\mathcal{S} : \mathcal{G} \rightarrow 2^{2^X}$ that associates a set of sets of variables to every gate. This labeling system is based on the Fourier representation of Boolean functions (see Sect. 3) and intuitively, a label contains at least those sets $X' \subseteq X$ for which $\widehat{f}(X') \neq 0$ (the sets that correlate with the Boolean functions).

The initial labeling is derived from $\mathcal{I}$. For an input g which is fed by function $f_g = \mathcal{I}(g)$, we have $\mathcal{S}(g) = \{X' \subset X \mid \widehat{f_g}(X') \neq 0\}$. In practice, the initial labeling of the circuits is easy to determine as inputs are typically provided with either a single variable m or a masked secret $x \oplus m$. An example for the labeling of an example circuit is shown in Fig. 2 (blue). Inputs containing security sensitive variables contain a single set listing all security sensitive variables and masks that protect this sensitive variables. For the masked signal $s_m = s \oplus m_s$, for example, the initial label is $\mathcal{S}(s_m) = \{\{s, m_s\}\}$. The meaning of the label is that by probing this input the attacker does not learn anything about s. In order to reveal any information on s, also some information on m_s needs to be combined with this wire in, either by the circuit itself (which would be a first-order flaw) or by the attacker by probing an according wire. If the attacker is assumed to be restricted to a single probing needle ($d = 1$) the signal s_m is secure against first-order attacks. Finally, the masked inputs m_s and m_1 in Fig. 2 contain only the mask variables. Formally, for inputs $g \in I$ with function $\mathcal{I}(g) = f(X)$, we set $\mathcal{S}(g) = \{X' | X' = X\}$.

5.2 Propagation Rules

To estimate the information that an attacker can learn by probing the output of a gate, we propagate the input labels through the circuit. For the verification we conservatively estimate which coefficients of the Fourier representation are different from zero and correlate with the variables. We prove at the end of this section that our estimation is sufficient to point out all security relevant information.

Nonlinear gates. To generate the labels for the outputs of each gate of the circuit, we introduce the *nonlinear gate* rule. The nonlinear gate rule corresponds to a worst-case estimation of the concrete Fourier spectrum of the signals and trivially catches all flaws. The labeling for the output of the nonlinear gate $g \in N$, with inputs g_a and g_b is:

$$\mathcal{S}(g) = \{\emptyset\} \cup \mathcal{S}(g_a) \cup \mathcal{S}(g_b) \cup \mathcal{S}(g_a) \,\Delta\, \mathcal{S}(g_b).$$

See gate g_2 in Fig. 2 for a simple example of an AND gate calculating $m_s \wedge p_1$. The resulting labels denote the information that can be learned by probing this gate which could be either m_s or p_1 alone, or together. The labeling reflects the Fourier spectrum of the AND gate (see Eq. 1). In particular the labeling shows all terms of the Fourier polynomial which coefficients are different from zero and are therefore statistical dependent.

Linear gate rule. By following the Definition 3 of the Fourier expansions further we can also model linear gates which have a reduced spectrum compared to nonlinear gates. We model this circumstance by introducing a new rule for labeling a linear gate $g \in L$ with inputs g_a and g_b:

$$\mathcal{S}(g) = \mathcal{S}(g_a) \,\Delta\, \mathcal{S}(g_b).$$

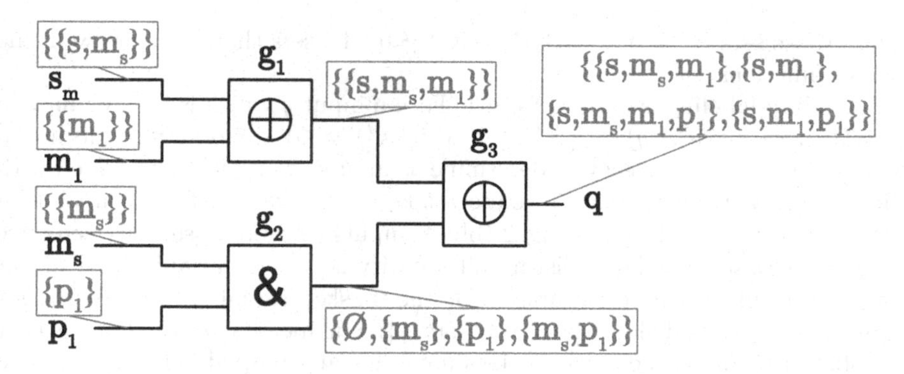

Fig. 2. Masked circuit example with according labels after the propagation step (Color figure online)

Combined example. To demonstrate how the propagation step works in practice, we applied the propagation rules (summarized in Table 1) to an example circuit. The result is shown in Fig. 2. The AND gate g_2 is a nonlinear gate, and the propagation rules are then iteratively applied to the gates g_1 to g_3. The output labeling of g_1 indicates that the security critical variable s is here not only protected by m_s but also by the mask m_1. Combining the public signal p_1 with the mask m_s in the nonlinear gate results in a nonuniform output signal which is indicated by the $\{\emptyset\}$ label at the output of g_2. For the calculation of the labels of g_3, the linear rule is used on the output labels of g_1 and g_2 which results in a labeling that indicates that s is even in the worst-case still protected by m_s, or m_1, or both.

5.3 Verification

For the verification step, in the first-order case, the circuit verifier checks if any of the sublabels created in the propagation step contains one or more secret variables without any masking variables (public variables are ignored since they are unable to mask secret data). If this is the case, the verifier rejects the circuit. In the example circuit in Fig. 2, all of the labels that contain s also contain m_1 or m_s and therefore the circuit is accepted by the verifier.

Table 1. Propagation rules for the stable set $\mathcal{S}(g)$ connected to the gates g_a and g_b

Gate type of g	Stable set rule
Input $\mathcal{I}(g) = f(X)$	$\mathcal{S}(g) = \{X' \mid X' = X\}$
Nonlinear gate	$\mathcal{S}(g) = \{\emptyset\} \cup \mathcal{S}(g_a) \cup \mathcal{S}(g_b) \cup \mathcal{S}(g_a) \bigtriangleup \mathcal{S}(g_b)$
Linear gate	$\mathcal{S}(g) = \mathcal{S}(g_a) \bigtriangleup \mathcal{S}(g_b)$
Register	$\mathcal{S}(g) = \mathcal{S}(g_a)$

Higher-order verification. For the generalization to d-order verification it is quite tempting to model the attackers abilities by letting the attacker pick multiple labels from any gate and combining them in an arbitrary manner. However, we note that the labeling does not reflect the relation of the probed information among each other and thus does not give a suitable approximation of what can be learned when multiple gates are probed. As a trivial example consider a circuit that calculates $q = (a \wedge b) \oplus c$ where all inputs are uniformly distributed. The labeling of the output q after the propagation step consists of the labels $\{c\}$, $\{a, c\}$, $\{b, c\}$, and $\{a, b, c\}$ for all of which an attacker probing q would indeed see a correlation. If an attacker restricted to two probes would probe q with the first probe, she obviously would not learn anything more by probing q a second time. In other words, if we would model a higher-order attacker by the ability to combine multiple labels, she could combine the label $\{c\}$ with any other label of q, e.g. $\{a, b, c\}$, in order to get information on a or b which is of course not the case.

Instead of modeling higher-order verification by the straight-forward combination of labels, we instead check the nonlinear combination of any tuple of d gates. An attacker can thus pick any number of up to d gates and combines them in an arbitrary function. We then need to check that even the worst case function over the gates could never contain a secret variable without a mask. This causes an obvious combinatorial blowup. In Sect. 8, we show how to harness a SAT solver to combat this problem. A proof for the correctness of the verification without glitches is provided in Appendix B.

In the next two sections we extend the verifier to cover glitches which shows that the example circuit is actually insecure.

6 Modeling Transient Timing Effects

So far, we have only considered the circuit's stable signals. We now discuss signal timing effects inside one clock cycle i.e. glitches and formalize how we model glitches in the probing model. Subsequently, we discuss how we model information that is collected from multiple clock cycles.

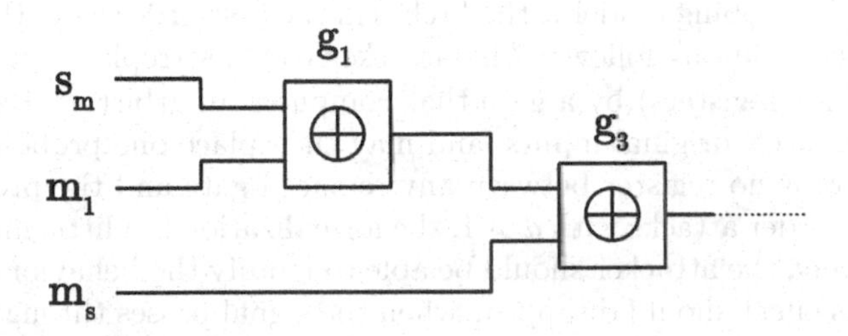

Fig. 3. Masked circuit example, insecure due to glitches

6.1 Glitches

As an example of what can go wrong when differences in the signal propagation times are not taken into account [29], consider the circuit in Fig. 3. The depicted circuit is secure in the original probing model as introduced in [26].

The information on the outputs of the XOR gates is ($s_m = s \oplus m_s$):

$$g_1 = s_m \oplus m_1 = s \oplus m_s \oplus m_1 \text{ and}$$
$$g_3 = s_m \oplus m_1 \oplus m_s = s \oplus m_1.$$

Since the other circuit gates (input terminals are modeled as gates) only carry information on the masked value s_m or the masks m_s and m_1, a single probe on any parts of the circuit does not reveal s and the circuit is thus first-order secure in the original probing model.

However, if we assume that in a subsequent clock cycle (Cycle 2 in Fig. 4) a different secret s' is processed, the circuit inputs change accordingly from s_m, m_s, and m_1 to s'_m, m'_s, and m'_1, respectively. Figure 4 shows an example on how these changes propagate through the circuit. Due to signal timing variance caused by physical circumstances, like different wire lengths or different driving strengths of transistors, so-called glitches arise. As a result of this timing variance m_1 changes its value later (t_2) than the other inputs (t_1) thus creating a temporary information leak (glitch). An attacker who places one probe on the output of g_3 firsts observes the original value $s \oplus m_1$ (at time t_0) and then $s' \oplus m_1$ (between t_1 and t_2). By combining the information the attacker obtains the information $(s \oplus m_1) \oplus (s' \oplus m_1)$ which is equivalent to $s \oplus s'$. Thus, she learns the relation of two secret bits. This information could not be obtained by combining the stable signals in the two clock cycles. Indeed, the leakage critically depends on the temporary information provided by the glitch in the circuit. To verify the security of a circuit in the probing model with glitches, all possible signal combinations that could arise because of propagation delays of signals need to be considered.

6.2 Formalization of Probing Security with Glitches

To formalize the probing model with glitches in the first-order case, the attacker's abilities are extended as follows: The attacker can first replace any number of gates (except for registers) by a gate that computes an arbitrary Boolean function from the gate's original inputs, and may then place one probe on any wire such that there is no register between any replaced gate and the probe.

For higher-order attacks with $d > 1$, the formalization is a little more cumbersome. Intuitively, the attacker should be able to modify the behavior of arbitrary gates, but this effect should disappear when the signal passes through a register. We model this by copying the combinational parts of the circuit and allowing

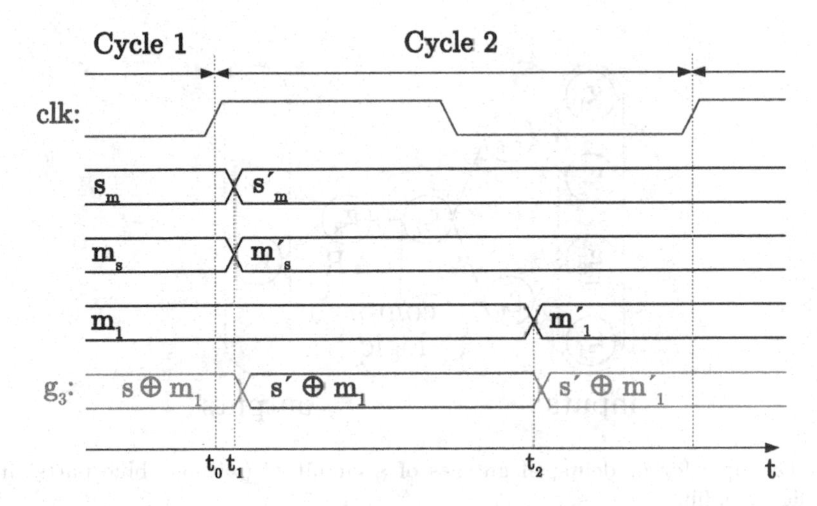

Fig. 4. Waveform example for the circuit in Fig. 3, showing security critical glitch (red) (Color figure online)

the attacker to change gates in the copy, whereas the original, unmodified signals are propagated by the unmodified gates. Figure 5 illustrates an example for the modeling of the glitches. The copied gates, which the attacker may modify, are drawn in blue. Note in particular that gate g_7 feeds into register g_8, but the copy g_7' becomes a new primary output.

Formally, given a circuit $C = (\mathcal{G}, \mathcal{W}, R, f, \mathcal{I})$, we do the following.

(1) We define a circuit $C' = (\mathcal{G}', \mathcal{W}', R, f', \mathcal{I})$. We copy all the gates except inputs and registers: $\mathcal{G}' = \mathcal{G} \cup \{g' \mid g \in \mathcal{G} \setminus R \setminus I\}$. We introduce wires from the inputs and registers to the copied gates and introduce wires between the copied gates: $\mathcal{W}' = \mathcal{W} \cup \{(g, h') \mid (g, h) \in \mathcal{W}, g \in I \cup R\} \cup \{(g', h') \mid (g, h) \in \mathcal{W}, g \notin I \cup R, h \notin R\}$. Finally, the functions of the copied gates are the same as those of the originals: $f'(g') = f(g)$ for $g \in \mathcal{G}' \setminus \mathcal{G}$.

(2) The attacker may replace any gate copy g' by a gate that computes an arbitrary Boolean function. We model this by defining a set of circuits, one for any set of gates that the attacker may modify:

$$\mathcal{C}_{\text{glitch}}(C) = \{C'[(g_1', \ldots, g_n') \mapsto (f_1, \ldots, f_n)] \mid (g_1, \ldots, g_n) \in \mathcal{G}^n, \forall i. f_i : \mathbb{B}^2 \to \mathbb{B}\}.$$

Definition 7 (d-probing security with glitches). *A circuit C is order d probing secure with glitches iff for any $C_{glitch} = (\mathcal{G}', \mathcal{W}', R, f', \mathcal{I}) \in \mathcal{C}_{glitch}$ and any gates $g_1, \ldots, g_d \in \mathcal{G}'$, $F(g_1, \ldots, g_d)$ is secure.*

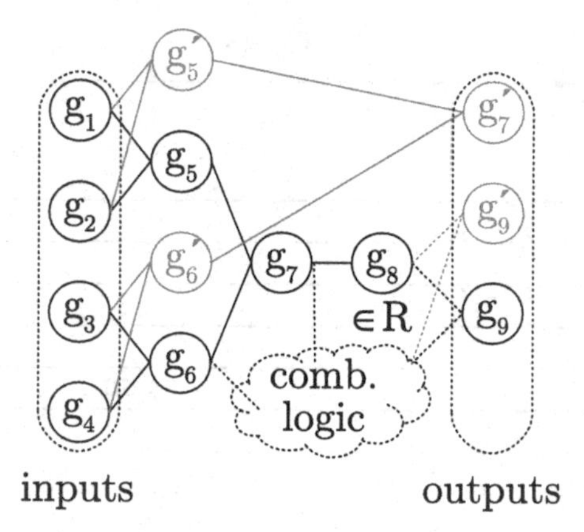

Fig. 5. Example for modeling of glitches of a circuit C (without blue parts) in C' (Color figure online)

6.3 Modeling Information from Multiple Clock Cycles

The verification of higher-order probing security requires to model information that is obtained and combined over different clock cycles. In our verification approach we consider dependencies between variables rather than concrete instantiation of these variables. The way we model glitches allows an attacker to exploit the worst case dependencies between the variables in between two register stages. We now state assumptions on masked circuit that ensure that the worst case dependencies are the same in each clock cycle.

Assumptions on masked circuits. Without loss of relevance for masked circuits we make the following assumptions which are inspired by practical masked circuits: (1) We assume that the values on the inputs remain the same throughout a clock cycle, they toggle only once at the beginning of a new clock cycle (registered inputs). (2) The class of the variables that are used in the input functions and the functions themselves do not change over time. For the circuit in Fig. 3, for example, the input s_m always contains a variable $s \in S$ and the associated mask $m_s \in M$ even though in each clock cycle the variables may change (e.g. from s to s'). (3) Mask variables are fresh random and uniform distributed at each clock cycle. (4) The circuits are feedback free and loop free, except the inherent feedback loops of registers. (5) The register depth (number of registers passed, counting from the input of the circuit) for each variable combined in a gate function is the same. No information resulting from different clock cycles is thus combined apart from the effects of delays and glitches which may temporarily combine information from two successive clock cycles. This assumption is motivated by the fact that most of the masked hardware designs, e.g. common S-box designs, are designed in a pipelined way.

From these assumptions it follows that all variables change in each cycle (e.g. from s to s', and so on), however, at varying times and in an arbitrary order. The variable classes and functions remain the same, and as a result from the assumptions 4 and 5 it is ensured that only variables that are fed into the circuit at the same cycle or from the cycle before are combined. It is therefore enough to consider the propagation of dependencies instead of concrete instantiation of variables.

7 Extending the Verification Approach to Transient Signals

In this section we use the modeling of the transient timing effects from the previous section to complete our verification approach. We take glitches into account by extending the propagation rules accordingly. The modeling of information from different clock cycles, on the other hand, does not require any changes in the verification approach from Sect. 5.

The nonlinear gate rule in Table 1 already inherently covers glitches by propagating the labels of the inputs and all possible combinations of these labels directly to the output. To hinder the propagation of glitches, circuit designers use registers that propagate their input only on a specific clock event, and thus isolate the register input from the output during the evaluation phase. We model the glitching behavior of a circuit by introducing an additional transient set of labels $\mathcal{T}$ per gate. Each gate thus has two associated sets: $\mathcal{S}$ carries the information of the stable state of the circuit as before, and the transient set $\mathcal{T}$ describes the transient information that is only accessible to an attacker in between two registers (or an input and a register, or a register and an output). In between two registers we also apply the nonlinear gate rule to linear gates to ensure we cover all possible effects of glitches.

Figure 6 illustrates the new linear gate rule for the stable (blue) and the transient (red) set of labels. The stable and transient sets of the inputs are equal at the beginning because the inputs are either circuit inputs or outputs of a register. When the signals propagate through the linear XOR gate, the transient set is calculated by applying the linear rule from Table 2 and the stable set with the linear rule from Table 1. After the signal passes the register, only the stable information remains and the transient set carries thus the same information as the stable set. Table 2 summarizes the rules for creating the transient-set labels $\mathcal{T}(g)$. Please note that introducing the transient set and the transient gate rules corresponds to the modeling of glitches from Sect. 6 as depicted in Fig. 5 (blue), where the gates in between two registers are copied and their function can be changed in an arbitrary manner by the attacker. Replacing the transient labels with the stable labels at a register corresponds to connecting the copied gates to the circuit output to hinder the propagation of glitches.

Aside from the introduction of the transient set and the according propagation rules, the verification work as described in Sect. 5. The circuit inputs are initially labeled according to their input variables where both the stable and transient sets hold the same labels. Then for all possible combinations of up to d gates the propagation of the labels is performed according to the stable and transient propagation rules. The circuit is order-d probing secure if for no combination of gates produces a label that only consists of secrets and public variable

Table 2. Propagation rules for the transient set $\mathcal{T}(g)$ fed by the gates g_a and g_b

Gate type of g	Transient set rule
Input	$\mathcal{T}(g) = \mathcal{S}(g)$
Nonlinear gate	$\mathcal{T}(g) = \{\emptyset\} \cup \mathcal{T}(g_a) \cup \mathcal{T}(g_b) \cup \mathcal{T}(g_a) \Delta \mathcal{T}(g_b)$
Linear gate	$\mathcal{T}(g) = \{\emptyset\} \cup \mathcal{T}(g_a) \cup \mathcal{T}(g_b) \cup \mathcal{T}(g_a) \Delta \mathcal{T}(g_b)$
Register	$\mathcal{T}(g) = \mathcal{S}(g_a)$

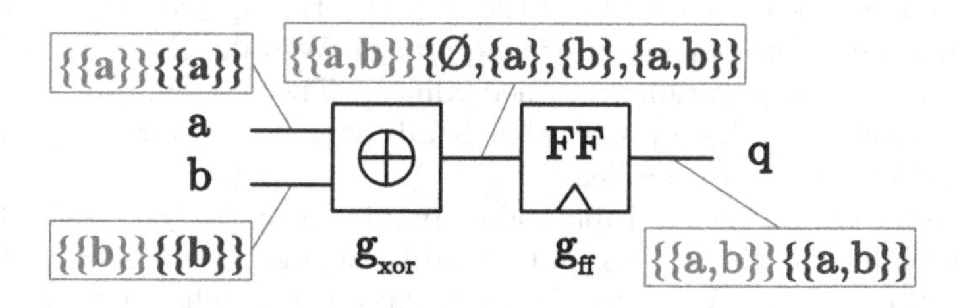

Fig. 6. XOR gate rules for stable (blue) and transient (red) signal sets (Color figure online)

without masks. A proof for the verification approach for transient signals is provided in Appendix C.

Example. The transient labels $\mathcal{T}$ of the circuit in Fig. 2 are shown in Fig. 7 (the stable sets are omitted since they do not carry any additional information). Due to the transient set propagation rules, the functionality of the gates g_1 and g_3, which are linear gates in the underlying circuit in Fig. 2, are replaced with nonlinear gates. As can be observed at the output of the circuit, the verification under the consideration of glitches leads to a rejection of the circuit because the s variable (black labels) is in the output labeling without being masked by either m_s or m_1.

To make it clear that the circuit is indeed insecure, we assume that $p_1 = \mathsf{true}$ and that s_m and m_s change their values to s'_m and m'_s, resp., but the value of m_1 and p_1 temporarily remains unchanged. Then, g_1 transitions from $s \oplus m_s \oplus m_1$ to $s' \oplus m'_s \oplus m_1$ and as a result g_3 transitions from $s \oplus m_1$ to $s' \oplus m_1$, thus leaking information about the relation of s and s'. (Cf. Fig. 4). The flaw can be easily repaired by adding a register after g_1 which ensures that s_m is always remasked before m_s is combined with s_m in g_3, and the same labels as in Fig. 2 for g_1 would thus be propagated.

8 SAT Based Circuit Verification

In this section, we introduce one concrete instantiation of our verification approach based on a SAT solver. The verification approach introduced in the previous sections is thus encoded as formulas in propositional logic. We start with the stable set rules and verification before we extend the encoding to the transient set rules.

Verification of stable signals. The SAT based verification works as follows. Intuitively, for every gate g, we pick one set $X' \subseteq \mathcal{S}(g)$, i.e., we pick one Fourier character with a possibly nonempty coefficient. We then encode the rules for the linear and nonlinear gates of Tables 2 and 1, respectively. To check for higher-order security we connect an XOR gate (checking gate) to any possible subset of up to d gates and check that the label of this gate does not contain a label with just secrets and no masks.

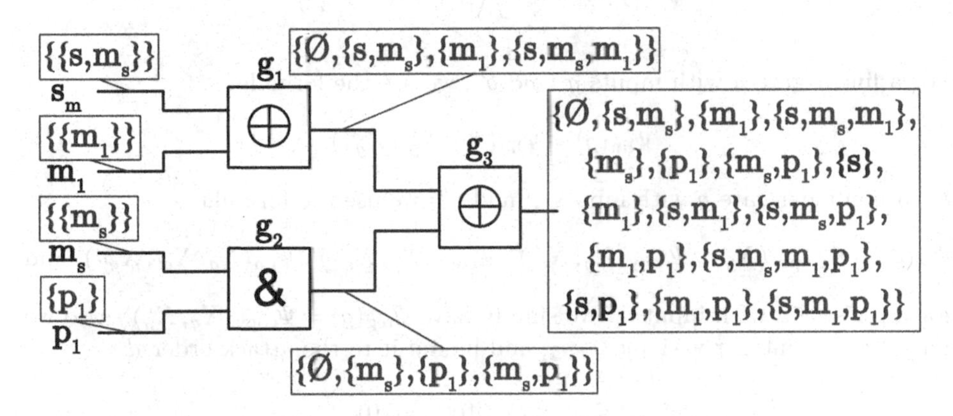

Fig. 7. Masked circuit example from Fig. 2 reevaluated with the transient rules (red) which leads to a flaw due to glitches (black labels) (Color figure online)

Let $C = (\mathcal{G}, \mathcal{W}, R, f, \mathcal{I})$ be a circuit. For each gate g we introduce a set of Boolean variables $X_g = \{x_g \mid x \in X\}$ and a Boolean *activation variable* a_g. For a checking gate g_c we introduce a set of Boolean variables X_{g_c}. We define a formula Ψ to check whether the masking scheme is secure. Recall that L and N are the sets of linear gates and nonlinear gates, resp. Formula Ψ consist of multiple parts:

$$\Psi = \Psi_{\text{gates}} \wedge \Psi_{\text{unsafe}}, \text{ where}$$

$$\Psi_{\text{gates}} = \bigwedge_{g \in I} \Psi_{\text{inp}}(g) \wedge \bigwedge_{g \in N} \Psi_{\text{nl}}(g) \wedge \bigwedge_{g \in L} \Psi_{\text{lin}}(g) \wedge \bigwedge_{g \in R} \Psi_{\text{reg}}(g).$$

The labeling of the inputs is determined by $\mathcal{I}$. For $X' \subseteq X$, we define

$$\psi_g(X') = \bigwedge_{x \in X} \begin{cases} x_g(X') & \text{if } x \in X', \\ \neg x_g(X') & \text{if } x \notin X', \text{ and} \end{cases}$$

$$\Psi_{\text{inp}}(g) = \bigvee_{X' \subseteq X : \widehat{\mathcal{I}(g)}(X') \neq 0} \psi_g(X').$$

To define the behavior of linear and nonlinear gates we define the following auxiliary formulas using the rules from Table 1, where $T = (t_1, \ldots, t_n)$,

$U = (u_1, \ldots, u_n)$, and $V = (v_1, \ldots, v_n)$ are ordered sets of variables, and define $\leftrightarrow$ to denote equality.

$$\Psi_{\text{empty}}(T) = \bigwedge_i \neg t_i,$$

$$\Psi_{\text{copy}}(T, U) = \bigwedge_i (t_i \leftrightarrow u_i), \text{ and}$$

$$\Psi_{\text{lin}}(T, U, V) = \bigwedge_i (t_i \leftrightarrow (u_i \oplus v_i)).$$

For a linear gate g with inputs g' and g'', we use the formula

$$\Psi_{\text{lin}}(g) = \Psi_{\text{lin}}(X_g, X_{g'}, X_{g''}),$$

for a nonlinear gate g with inputs g' and g'', we use the formula

$$\Psi_{\text{nl}}(g) = \Psi_{\text{empty}}(X_g) \vee \Psi_{\text{copy}}(X_g, X_{g'}) \vee \Psi_{\text{copy}}(X_g, X_{g''}) \vee \Psi_{\text{lin}}(X_g, X_{g'}, X_{g''}), \text{ and}$$

for a register g with input g', we simply have $\Psi_{\text{reg}}(g) = \Psi_{\text{copy}}(X_g, X'_g)$. Also we introduce an integer variable a_{sum}, and bound it to the attack order d:

$$a_{sum} = \sum_g \text{ite}(a_g, 1, 0)$$

$$a_{sum} \leq d.$$

The function $Ite(a_g, 1, 0)$ (if-then-else) converts a Boolean variable to Integer.

For the checking gate we xor the corresponding inputs:

$$\Psi(g_c) = \bigwedge_{x \in X} x_{g_c} \leftrightarrow \bigoplus_{g \in \mathcal{G}} a_g \wedge x_g.$$

Finally, for the checking gate g_c we define a constraint to check whether security is violated, that is, whether there is a non-zero Fourier coefficient which contains secrets and no masks:

$$\Psi_{\text{unsafe}}(g_c) = \bigvee_{s \in S} s_g \wedge \bigwedge_{m \in M} \neg m_g.$$

Formula Ψ contains $|X| \cdot |\mathcal{G}|$ propositional variables and $O(|X| \cdot |\mathcal{G}|)$ constraints.

An example for the SAT encoding is provided in Appendix F along with a proof for its correctness in Appendix D.

Extension to transient signals. The encoding for the transient rules follows the exposition in Sect. 7 and in particular the rules from Table 2. We introduce a second set of variables $X'_g = \{x'_g \mid x \in X\}$, which represent the Fourier characters on the "copied" gates in Definition 7. We introduce a slightly modified set of constraints, where we write Φ' to denote a formula Φ in which each variable x_g has been replaced by x'_g.

$$\Phi = \Phi_{\text{gates}} \wedge \Phi'_{\text{unsafe}}, \text{ where}$$

$$\Phi_{\text{gates}} = \bigwedge_{g \in I} (\Psi_{\text{inp}}(g) \wedge \Psi'_{\text{inp}}(g)) \wedge \bigwedge_{g \in N} (\Psi_{\text{nl}}(g) \wedge \Psi'_{\text{nl}}(g)) \wedge$$
$$\bigwedge_{g \in L} (\Psi_{\text{lin}}(g) \wedge \Psi'_{\text{nl}}(g)) \wedge \bigwedge_{g \in R} \Phi_{\text{reg}}(g),$$

where for a register g with input g', we copy only the original (glitch-free) signals:

$$\Phi_{\text{reg}}(g) = \Psi_{\text{copy}}(X_g, X_{g'}) \wedge \Psi_{\text{copy}}(X_g, X'_{g'}).$$

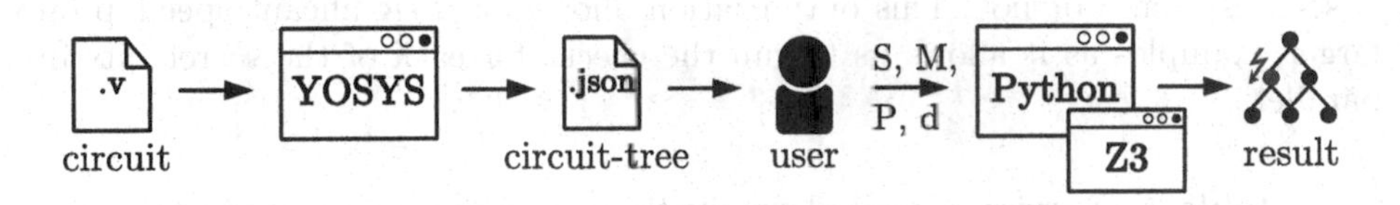

Fig. 8. Illustration of the verification flow

Note the use of the constraint for nonlinear gates for the copy of linear gates, which corresponds to the attacker's ability to replace such a gate by any other gate in $\mathcal{C}_{\text{glitch}}(C)$. Finally, we check for leakage only on the gate copies:

$$\Phi'_{\text{unsafe}}(g_c) = \bigvee_{s \in S} s'_{g_c} \wedge \bigwedge_{m \in M} \neg m'_{g_c}.$$

Formula Φ contains $2 \cdot |X| \cdot |\mathcal{G}|$ propositional variables and $O(|X| \cdot |\mathcal{G}|)$ constraints. A proof for the correctness of the encoding for transient signals is given in Appendix E.

9 Practical Results

Figure 8 illustrates the implemented verification flow that is used to gather the results presented in this section. At first the circuit description is parsed using Yosys 0.7 [38] open synthesis suite. The resulting circuit tree is stored in JavaScript Object Notation (JSON). The user then needs to provide the circuit's input labels by telling the JSON parser (written in Python) which signals are secrets (S), masks (M), or other signals (P), and for which security order (d) the circuit needs to be tested. The construction of the SAT formulas is then performed in about 1,000 lines of Python code, and checked by the Z3 Theorem

Prover 4.5.1 [15] (initial experiments with other SAT solvers, including Crypto-Minisat 5.0.1 were not encouraging). All results are gathered on a Intel Xeon E5-2699v4 CPU with a clock frequency of 3.6 GHz and 512 GB of RAM running in a 64-bit Linux OS environment (Debian 9).

Optimizations. There are a two simple optimizations that we use to speed up the verification. First, we can treat public variables at the inputs as constants. We can easily prove by induction that if $P' \cup S' \cup M' \in \mathcal{S}(q)$ for some gate g and $P' \subseteq P$, $S' \subseteq S$, and $M' \subseteq M$ and we compute a new labeling $\mathcal{S}'$ by treating the public variables as constants, then $S' \cup M' \in \mathcal{S}'(q)$ and thus, if $\mathcal{S}(q)$ is insecure, so is $\mathcal{S}'(q)$. A similar argument holds for $\mathcal{T}$ and for combinations of signals.

Second, we can treat secret bits one at a time, treating the other secret bits as constants. The argument is much the same as for the first optimization, if a function is insecure then it has a label with at least one secret bit and no masks. Removing any other secret bits from the label does not affect whether the label is seen as secure or not. This optimization allows for a significant speedup on larger examples as it allows us to run the checks for each of the secret bits in parallel.

Table 3. Overview of masked circuits the first order verification results

Name	Gates			Variables			Part	w/o glitches		w/glitches	
	Linear	Nonlin	Reg	Secret	Mask	Pub		Time	Result	Time	Result
Trichina gate [37]	4	4	0	2	3	1		≤1 s	✗	≤2 s	✗
ISW AND [26]	4	4	0	2	3	0		≤1 s	✓	≤2 s	✗
TI AND [31]	6	9	0	2	4	0		≤1 s	✓	≤3 s	✓
DOM AND [24]	4	4	2	2	3	1		≤1 s	✓	≤2 s	✓
DOM Keccak S-box [25]	30	20	10	5	10	1		≤1 s	✓	≤20 s	✓
DOM AES S-box [24]	392	144	208	8	26	1	1–8	≤30 s	✓	≤5–10 h	✓
TI Fides-160 S-box [11]	128	60	0	5	15	0	1–4	≤1–2 s	✓	≤1–3 s	✓
TI Fides-192 APN [11]	4,035	3,046	0	6	24	0					
	134	44	0	0	24	0	1	≤2 s	✓	≤5 s	✓
	649	314	0	6	24	0	2	≤1 m	✓	≤15 m	✓
	1,697	1,098	0	6	24	0	3	≤20 m	✓	≤2 h	✓
	1,186	1,086	0	6	24	0	4	≤10 m	✓	≤40 m	✓
	369	504	0	6	24	0	5	≤2 m	✓	≤3 m	✓

Evaluation. An overview of the experiments is given in Table 3. The table states the number of linear and nonlinear gates of the circuits as well as the number of variables classified as secret, mask, and public, resp. Furthermore, the verification results are given for the stable set (without glitches) and transient set (with glitches) verification separately. The table states whether the circuit is secure in the given model (✓ for secure and ✗ for insecure) and the time needed for checking and generation of the constraints.

The selection of masked circuits cover different masked $GF(2)$ multiplier constructions (masked AND gates) of the Trichina gate [37], the ISW scheme [26], the threshold implementation (TI) scheme [31], and the domain-oriented masking scheme (DOM) [24]. We also check larger circuits including the AES S-box constructions of Gross et al. [24] using the domain-oriented masking (DOM) scheme. Furthermore, we verify a FIDES S-box implementation by Bilgin et al. [11], and a Keccak S-box by Gross et al. [25].

9.1 Verification of First-Order Masked Implementations

Table 3 shows the verification results of the first-order masked hardware implementations. For larger circuits, like the AES S-box, we checked each of the secret bits separately. If multiple CPU's are available, these verifications can run simultaneously and we thus split the verification up into multiple parts.

Masked AND gates. The first masked AND we verify is the so-called Trichina gate which was originally designed to resist first-order attacks. Equation 4 shows the underlying logic function. The Trichina gate was designed without considering the effect of glitches. As a result, if the inputs are correctly and independently masked ($a_m = a \oplus m_a$ and $b_m = b \oplus m_b$), the stable state of the output of the circuit is also correctly masked.

$$q = a_m \wedge b_m \oplus a_m \wedge m_b \oplus m_a \wedge b_m \oplus m_a \wedge m_b \oplus m_q \tag{4}$$

However, due to timing differences in the propagation of the signals, glitches may occur in one of the XOR gates. This makes the design vulnerable unless additional measures are taken which is also indicated by our verification results. Interestingly also the result of the stable verification already shows the vulnerability of the Trichina gate. This is due to timing optimizations of the synthesizer that change the sequence in which the AND gate outputs are XORed together which is a common problem in masked circuit designs and is easily overseen.

The masked AND gate from the ISW scheme is similar to the Trichina gate but scalable to any protection order. This gate suffers from the same vulnerability to glitches, which makes any straightforward implementation of the original proposed circuit construction insecure against first-order attacks. This time the flaw is not detected in the stable set because the gates are arranged in a way that the secrets are always masked in the stable analysis of the circuit. However, the circuit is nevertheless vulnerable to glitches which is shown in the transient analysis of the circuit.

To overcome the issue of glitches, the threshold implementation (TI) scheme proposed a masking with two fresh masks per sensitive variable (e.g., $a_m = a \oplus m_{a0} \oplus m_{a1}$). The resistance to glitches is then achieved by ensuring that in no part of the circuit the masked value and all its masks come together.

A different approach, which requires fewer masks is provided e.g. by the domain-oriented masking AND. For the security of this masked AND a separation of the terms is required by using a register for the combination of the terms with a fresh random mask. The verifier also correctly labels the DOM AND to be secure for the stable and in the transient verification. The verification without glitches takes less than a second for all masked AND gate constructions, and less than three seconds for the verification with glitches.

Verification of masked S-box circuits and permutations. For the remaining circuits, we either used the source code which is available online [22] for the DOM Keccak S-box and the DOM AES S-box, or in case of Fides kindly received the circuit design from the designers. In order to check the circuits in a more efficient manner, we used different optimizations. For the TI S-box of the Fides-160 design, we checked the individual S-box functions in parallel but for a fixed assignment of the secrets and masks. The result for all Fides-160 TI functions is computed in less than three seconds with and without glitches. For the TI Fides-192 design not only the S-Box but the whole APN permutation is split into five functions. Again we assumed a fixed masking and checked the functions individually, which makes the verification of the first TI function very fast (because no secrets are fed into this part). For the other circuit parts, the verification takes between 20 min for the verification without glitches verification and 2 h for the verification with glitches. Please note that the differences in the verification timings for the different parts of Fides-192 result from the varying gate counts. All circuit parts are labeled to be secure. Finally, we also checked a DOM AES S-box design for which we checked the whole circuit for the eight individual secret bits separately. The stable set verification takes less than 30 s, and the verification of the transient sets between 5 and 10 h for each part. Again the verification result indicates a securely masked first-order protected circuit.

9.2 Verification of Higher-Order Masked Implementations

To evaluate the performance of our verification approach for higher-order masked circuits, we run our tool on the generically masked DOM AND gate [24] and the Keccak S-box [25]. The results are shown in Table 4 where the protection order of the circuit and the verification order are always set equal and are summarized in a single column (order).

The verification of the second-order masked DOM AND takes less than a second. For the fourth-order protected AND the verification time increases to about 7 min. The influence of the protection order at varying verification orders is depicted in Fig. 9. We evaluated each masked DOM AND from first-order up to its claimed protection order plus one additional order. This figure underlines the intuition that finding a flaw takes less time than ensuring that the circuit is free from flaws.

For the Keccak S-box circuit we again split the verification for the five secrets into five separate verificaiton runs. The verification for the second order than takes about 10 s per verification run without glitches and about 40 s when glitches are considered. For the third-order verification the times increase to 4 min and 25 min, respectively.

Table 4. Overview of masked circuits and the higher order verification results

Name	Order	Gates			Variables			w/o glitches		w/glitches	
		Linear	Nonlin	Reg	Secret	Mask	Pub	Time	Result	Time	Result
DOM AND [24]	2	12	9	9	2	11	1	≤1 s	✓	≤1 s	✓
	3	24	16	16	2	17	1	≤4 s	✓	≤20 s	✓
	4	40	25	25	2	24	1	≤2 m	✓	≤7 m	✓
Keccak S-box[a] [25]	2	75	45	45	5	35	7	≤10 s	✓	≤40 s	✓
	3	140	80	80	5	60	7	≤4 m	✓	≤25 m	✓

[a] For the Keccak S-box we performed the verification for the five secrets separately.

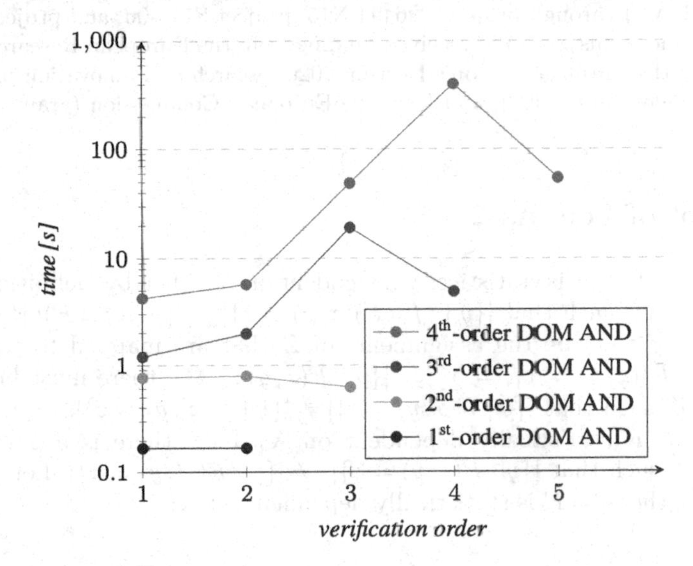

Fig. 9. Verification time for the DOM ANDs with varying protection and verification order

10 Conclusions

In this paper we introduced the formal groundwork for the verification of masked hardware implementations in the presence of glitches. We built upon the probing model of Ishai et al. and presented a method to conservatively estimate the security of circuits under this model for the worst case signal timings. Our approach is based on an estimation of the non-zero Fourier coefficients of the functions computed by the circuit, and we have provided a proof of its correctness. To demonstrate the practicality, we have implemented our formal approach on top of the Z3 theorem prover to verify the masking properties of a side-channel protected hardware implementation directly on the gate-level netlist. We have shown the suitability of our approach to verify masked circuits on practical examples from different masking schemes and different sources.

The advantages of this approach are evident. Circuits deemed secure do not leak secret information under any possible signal timings, which includes the effects of glitches in the combinatorial logic of the circuit, and even for higher-order attacks. If a circuit is rejected, we can pinpoint the gate that causes the potential leakage, which makes checking and fixing of the flaw much easier than by conventional approaches. Furthermore, the verifier can be used at different development stages of the masked circuit or for testing new masking schemes. This makes it a useful method for both practical applications as well as for research purposes.

Acknowledgements. The work has been supported in part by the Austrian Science Fund (FWF) through project P26494-N15, project S114-06, and project W1255-N23. Furthermore this work has received funding from the European Research Council (ERC) under the European Union's Horizon 2020 research and innovation programme (grant agreement No 681402), and from the European Commission (grant agreement No 644905).

A Proof of Lemma 2

1. Suppose that $f \circ F$ is statistically dependent on X. Then by Definition 1 there are $x, x' \in 2^X$ such that $|\{y \mid (f \circ F)(x,y) = 1\}| \neq |\{y \mid (f \circ F)(x',y) = 1\}|$. Let $Z' = f^{-1}(1)$ be the assignments of Z that are mapped to true. Since $|\{y \mid (f \circ F)(x,y) = 1\}| = \sum_{z \in Z} |\{y \mid F(x,y) = z\}|$, there must be at least one $z \in Z'$ such that $|\{y \mid F(x,y) = z\}| \neq |\{y \mid F(x',y) = z\}|$.
2. Suppose F is statistically dependent on X. Then there is a $z \in 2^Z$ and $x, x' \in 2^X$ such that $|\{y \mid F(x,y) = z\}| \neq |\{y \mid F(x',y) = z\}|$. Let $f(z') = 1$ iff $z' = z$, then $f \circ F$ is statistically dependent on X.

□

B Proof of the Stable Verification Approach

Lemma 8. *For any circuit C, any gate* g *and any $T \subseteq X$, if the Fourier coefficient $\widehat{g}(T) \neq 0$, then $T \in \mathcal{S}(g)$.*

Proof. We prove the lemma by induction on the depth k of the circuit.

Base. $k = 0$. For an input g the lemma holds by the definition of the initial label.

Inductive step. Let $k \geq 1$ and suppose that a gate g at depth k is the output of a gate with two input gates u and v, with Fourier representations $u(X) = \sum_{T \subseteq X} \widehat{u}(T)\chi_T(X)$, and $v(X) = \sum_{T \subseteq X} \widehat{v}(T)\chi_T(X)$.

We distinguish two cases: (1) g is linear or (2) g is nonlinear. (The case of registers is trivial when we do not consider glitching.)

Case 1: $g = u \oplus v$. The Fourier representation of g is

$$g(X) = u(X) \cdot v(X) = \sum_{T \subseteq X} \widehat{g}(T) \cdot \chi_T(X),$$

where

$$\widehat{g}(T) = \sum_{T_1 \subseteq X} \widehat{u}(T_1) \cdot \widehat{v}(T \triangle T_1).$$

Assume that $\widehat{g}(T) \neq 0$. If $\widehat{g}(T) \neq 0$, then there exists a set $T_1 \subseteq X$ such that $\widehat{u}(T_1) \cdot \widehat{v}(T \triangle T_1) \neq 0$. Therefore, $\widehat{u}(T_1) \neq 0$ and $\widehat{v}(T \triangle T_1) \neq 0$. By the inductive hypothesis, it holds that $T_1 \in \mathcal{S}(u)$ and $T \triangle T_1 \in \mathcal{S}(v)$, which, by the linear rule means that $T_1 \triangle (T \triangle T_1) = T \in \mathcal{S}(g)$.

Case 2: g is a nonlinear gate. In this case, the Fourier representation of g is

$$g(X) = \alpha_{00} + \alpha_{01} \cdot u(X) + \alpha_{10} \cdot v(X) + \alpha_{11} \cdot u(X) \cdot v(X)$$

for some α_{ij}. Consequently, $\widehat{g}(T) \neq 0$ implies that either (1) $\widehat{u}(T) \neq 0$, (2) $\widehat{v}(T) \neq 0$, or (3) $\exists T' \subseteq X.\widehat{u}(T) \neq 0$ and $\widehat{v}(T' \triangle T) \neq 0$. (The converse does not hold.) In each of these three conditions, $T \in \mathcal{S}(w)$. □

The next lemma shows that an arbitrary function for d gates corresponds to a generalization of the non-linear rule from Table 1. (Note the use of $\triangle_{d \in D} T_d$ to denote the symmetric set difference of all T_ds).

Lemma 9. *Let $F_1, \ldots, F_d : \mathbb{B}^X \to \mathbb{B}$, let $f : \mathbb{B}^d \to \mathbb{B}$, and let $F(x) = f(F_1(x), \ldots, F_d(x))$. For any $T \subseteq X$, we have that $\widehat{F}(T) \neq 0$ implies that there is a $D \subseteq \{1, \ldots, d\}$ and $T_1 \ldots T_d \subseteq X$ such that $T = \triangle_{i \in D} T_i$ and for all i, $\widehat{F}_i(T_i) \neq 0$.*

Proof. Let $f(a_1, \ldots, a_d) = \sum_{D \subseteq \{1,\ldots,d\}} \alpha_D \prod_{i \in D} a_i$ be the Fourier expansion of f. We have that

$$\begin{aligned}
F(x) &= \sum_{D \subseteq \{1,\ldots,d\}} \alpha_D \prod_{i \in D} F_i(x) \\
&= \sum_{D \subseteq \{1,\ldots,d\}} \alpha_D \prod_{i \in D} \sum_{T \subseteq X} \widehat{F}_i(T) \chi_T(X) \\
&= \sum_{D \subseteq \{1,\ldots,d\}} \alpha_D \sum_{\substack{T_1 \subseteq X \\ \vdots \\ T_d \subseteq X}} \prod_{i \in D} \widehat{F}_i(T_i) \chi_{T_i}(X) \\
&= \sum_{D \subseteq \{1,\ldots,d\}} \alpha_D \sum_{\substack{T_1 \subseteq X \\ \vdots \\ T_d \subseteq X}} \chi_{\triangle_{i \in D} T_i}(X) \prod_{i \in D} \widehat{F}_i(T_i).
\end{aligned}$$

Consequently, if $\widehat{F}(T) \neq 0$, then there is a $D \subseteq \{1, \ldots, d\}$ and $T_1 \ldots T_d \subseteq X$ such that for all $i \in D$ $\widehat{F}_i(T_i) \neq 0$ and $T = \triangle_{i \in K} T_i$. □

Note that for $d = 2$, the lemma specializes to $\widehat{F}(T) \neq 0$ implies $T = \emptyset$, $\widehat{F_1(T)} \neq 0$, $\widehat{F_2(T)} \neq 0$, or $T = T_1 \triangle T_2$, $\widehat{F_1(T_1)} \neq 0$, and $\widehat{F_2(T_2)} \neq 0$, which reflects the nonlinear rule.

Theorem 10. *Let C be a circuit. If for all $S' \subseteq S \cup P$ such that $S' \cap S \neq \emptyset$, for all sets $g_1, \ldots, g_d$ of gates, for all $D \subseteq \{1, \ldots, d\}$, for all $T_1 \ldots T_d \subseteq X$ such that $\triangle_{d \in D} T_d = S'$, and for all $i \in D$, we have that $T_i \notin \mathcal{S}(g_i)$, then C is order d secure.*

Proof. (By contradiction.) Suppose that C is not order d secure. By Definitions 5 and 6, and Lemma 2, this implies that there are gates $g_1, \ldots, g_d$, a function$f$ and an assignment $p \subseteq P$ such that $f \circ F|_p(g_1, \ldots, g_d)$ is statistically dependent on S. By Lemma 4, this implies that for some $S' \subseteq S$, $S' \neq \emptyset$, $\widehat{f \circ F|_p(g)}(S') \neq \emptyset$. The Fourier expansion of $F|_p$ is obtained from the expansion of p by substituting -1 or 1 for each public variable, so if $\widehat{F|_p}(S') \neq 0$, then $\widehat{F}(S' \cup p') \neq 0$ for some $p' \subseteq p$. By Lemmas 8 and 9, this means that there is a $D \subseteq \{1, \ldots, d\}$, $T_1 \ldots T_d \subseteq X$ such that $\triangle_{d \in D} T_k = S'$, and $i \in K$, such that that $T_i \in \mathcal{S}(g_i)$. □

Note that for the first order attacks, we just need to consider the labels of all gates (see Lemma 8. For order d attacks, conceptually we connect a d-ary AND gate to any set of d gates and check that the label of this gate does not contain a label with secrets and no masks.

It is worth pointing out that the converse of the theorem does not hold. As an example of imprecision, one can construct $a \oplus b$ as a combination of three nonlinear gates. In this case, the output would be labeled $\{\emptyset, \{a\}, \{b\}, \{a, b\}\}$ although the Fourier expansion of the circuit is $a \cdot b$.

C Proof of the Transient Verification Approach

Lemma 11. *For any gate g of C and any $T \subseteq X$, if for any $C_{glitch} \in \mathcal{C}_{glitch}$, either $\widehat{g}(T) \neq 0$ or $\widehat{g}'(T) \neq 0$, then $T \in \mathcal{T}(g)$.*

Proof. The proof follows that of Lemma 8 with the modification that regardless of the function of a gate g', if g' has inputs g_1 and g_2, then $\widehat{g'}(T) \neq 0$ implies that either (1) $\widehat{g_1}(T) \neq 0$, (2) $\widehat{g_2}(T) \neq 0$, or (3) $\exists T' \subseteq X . \widehat{g_1}(T') \neq 0$ and $\widehat{g_2}(T' \triangle T) \neq 0$. Thus, we can overapproximate the set of non-zero Fourier coefficients for a copied gate g' with inputs g_1 and g_2 by the set that consists of the union of the nonzero coefficients $\mathcal{T}(g_1)$ of g_1, the nonzero coefficients $\mathcal{T}(g_2)$ of g_2, and the set $\mathcal{T}(g_1) \triangle \mathcal{T}(g_2)$. □

Theorem 12. *Let C be a circuit. If for all $S' \subseteq S \cup P$ such that $S' \cap S \neq \emptyset$, for all sets $g_1, \ldots, g_d$ of gates, for all $D \subseteq \{1, \ldots, d\}$, for all $T_1 \ldots T_d \subseteq X$ such that $\triangle_{d \in D} T_d = S'$, and for all $i \in D$, we have that $T_i \notin \mathcal{T}(g_i)$, then C is order d secure.*

Proof. The proof proceeds along the lines Theorem 12 by using Lemma 11 instead of Lemma 8. □

D Proof of the SAT Based Verification for Stable Signals

Lemma 13. *For any gate $g \in \mathcal{G}$ and any $X' \subseteq X$, if $\widehat{F}(g)(X') \neq 0$ then there is a satisfying assignment χ of Ψ_{gates} with $\chi(x_g) = \mathsf{true}$ iff $x \in X'$.*

Proof. The Fourier expansion of any gate g can be obtained recursively from the Fourier expansion of its inputs, where the coefficient of every Fourier character is obtained by multiplying out the coefficients of the inputs. For a linear gate, this is a simple multiplication. A nonlinear gate results in more Fourier coefficients: those for each input separately, and those that result from the multiplication of the inputs. We can represent the way that each Fourier character is derived as a subgraph of the circuit.

Suppose that a coefficient $\widehat{f}(g)(X')$ is nonzero. Then there is a set of wires W' and a set of gates G' that fulfills the following constraints. (1) $g \in G'$. (2) If $g \in G'$ and g is a linear gate, then both incoming wires are in W'; if g' in nonlinear, than 0, 1, or 2 of the incoming wires are in W'. (3) if $w \in W'$ then the gate g' that feeds w is in G'. (4) X' is the symmetric set difference of the nonzero Fourier coefficients of the inputs that feed into an *odd* number of paths to g. The latter observation follows from the fact that in Fourier representation $f^2 = 1$ for any Boolean function f.

The choice of W' and G' is dictated by the choices made at the nonlinear gates, which corresponds to the disjuncts in the definition of $\Psi_{\text{nl}}(g)$. The satisfying assignments of the formula then follow the paths described above, where the cancellation of coefficients that occur an even number of times is ensured by the XORs in the formulas for linear and nonlinear gates. □

Theorem 14. *If C is not order d secure without glitches then Ψ is satisfiable.*

Proof. The theorem follows easily from the Lemma 13 by the fact that information leakage occurs iff there is a gate which is statistically dependent on a set of secret variables (Lemma 4, Definition 6). □

Note if the formula Ψ is satisfiable, i.e. the circuit is not secure, we can easily see what gates the solver picked for the probes. The activation variables for those gates are equal to *true*.

E Proof of the SAT Based Approach for Transient Signals

Lemma 15. *For any circuit $C' \in \mathcal{C}_{glitch}(C)$, gate* g *in C', and $X' \subseteq X$, if $\widehat{F}(g)(X') \neq 0$, then there is a satisfying assignment χ of Φ_{gates} with $\chi(x'_g) = \mathsf{true}$ iff $x \in X'$.*

Proof. The proof follows that of Lemma 13 with the modification that for a gate g, an assignment to the variables $\{x'_g \mid x \in X\}$ is part of a satisfying assignment if the corresponding Fourier coefficient is non-zero for *any* $C' \in \mathcal{C}_{\text{glitch}}(G)$. Intuitively, a nonlinear gate presents the "worst-case" scenario that subsumes the behavior of an arbitrary gate. □

Theorem 16. *If C is not order d secure with glitches then Φ is satisfiable.*

Proof. The proof follows from Lemmas 4, 15, and Definition 7. □

F Example for the SAT Encoding

To illustrate the encoding let us consider the example on the Fig. 2. For this circuit we have one secret variable s and two mask variables m_s and m_1. Since input s_m is driven by the function $\mathcal{I}(s_m) = s \oplus m_s = s \cdot m_s$, we have the constraint

$$\psi_{\text{inp}}(s_m) = s_{s_m} \wedge m_{s,s_m} \wedge \neg m_{1,s_m} \wedge \neg p_{1,s_m}.$$

For the other inputs we have:

$$\psi_{\text{inp}}(m_1) = \neg s_{m_1} \wedge \neg m_{s,m_1} \wedge m_{1,m_1} \wedge \neg p_{1,m_1}$$

$$\psi_{\text{inp}}(m_s) = \neg s_{m_s} \wedge m_{s,m_s} \wedge \neg m_{1,m_s} \wedge \neg p_{1,m_s}$$

$$\psi_{\text{inp}}(p_1) = \neg s_{p_1} \wedge \neg m_{s,p_1} \wedge \neg m_{1,p_1} \wedge p_{1,p_1}.$$

For the linear gates g_1 and g_3 we use the linear rule

$$\begin{aligned}\Psi_{\text{lin}}(g_1) = (s_{g_1} \leftrightarrow (s_{s_m} \oplus s_{m_1})) \wedge (m_{s,g_1} \leftrightarrow (m_{s,s_m} \oplus m_{s,m_1}))\\ \wedge (m_{1,g_1} \leftrightarrow (m_{1,s_m} \oplus m_{1,m_1}) \wedge (p_{1,g_1} \leftrightarrow (p_{1,s_m} \oplus p_{1,m_1}))\end{aligned}$$

$$\begin{aligned}\Psi_{\text{lin}}(g_3) = (s_{g_3} \leftrightarrow (s_{g_2} \oplus s_{g_1})) \wedge (m_{s,g_3} \leftrightarrow (m_{s,g_1} \oplus m_{s,g_2}))\\ \wedge (m_{1,g_3} \leftrightarrow (m_{1,g_1} \oplus m_{1,g_2}) \wedge (p_{1,g_3} \leftrightarrow (p_{1,g_1} \oplus p_{1,g_2})).\end{aligned}$$

For the non-linear gate g_2 we use the non-linear rule:

$$\begin{aligned}\Psi_{nl}(g_2) = (\neg s_{g_2} \wedge \neg m_{s,g_2} \wedge \neg m_{1,g_2} \wedge \neg p_{1,g_2})\\ \vee (s_{g_2} \leftrightarrow s_{m_s} \wedge m_{s,g_2} \leftrightarrow m_{s,m_s} \wedge m_{1,g_2} \leftrightarrow m_{1,m_s} \wedge p_{1,g_2} \wedge p_{1,m_s})\\ \vee (s_{g_2} \leftrightarrow s_{p_1} \wedge m_{s,g_2} \leftrightarrow m_{s,p_1} \wedge m_{1,g_2} \leftrightarrow m_{1,p_1} \wedge p_{1,g_2} \wedge p_{1,p_1})\\ \vee (s_{g_2} \leftrightarrow (s_{m_s} \oplus s_{p_1})) \wedge (m_{s,g_2} \leftrightarrow (m_{s,m_s} \oplus m_{s,p_1}))\\ \wedge (m_{1,g_2} \leftrightarrow (m_{1,m_s} \oplus m_{1,p_1}) \wedge (p_{1,g_2} \leftrightarrow (p_{1,m_s} \oplus p_{1,p_1})).\end{aligned}$$

For the checking gate we have:

$$\begin{aligned}\Psi(g_c) = (s_{g_c} \leftrightarrow (a_{s_m} \wedge s_{s_m} \oplus a_{m_1} \wedge s_{m_1} \oplus a_{m_s} \wedge s_{m_s} \oplus a_{p_1} \wedge s_{p_1}))\\ \wedge (m_{1,g_c} \leftrightarrow (a_{s_m} \wedge m_{1,s_m} \oplus a_{m_1} \wedge m_{1,m_1} \oplus a_{m_s} \wedge m_{1,m_s} \oplus a_{p_1} \wedge m_{1,p_1}))\\ \wedge (m_{s,g_c} \leftrightarrow (a_{s_m} \wedge m_{s,s_m} \oplus a_{m_1} \wedge m_{s,m_1} \oplus a_{m_s} \wedge m_{s,m_s} \oplus a_{p_1} \wedge m_{s,p_1}))\\ \wedge (p_{1,g_c} \leftrightarrow (a_{s_m} \wedge p_{1,s_m} \oplus a_{m_1} \wedge p_{1,m_1} \oplus a_{m_s} \wedge p_{1,m_s} \oplus a_{p_1} \wedge p_{m,p_1})).\end{aligned}$$

To check first order security we bound a_{sum} to 1:

$$a_{sum} = Ite(a_{s_m}, 1, 0) + Ite(a_{m_1}, 1, 0) + Ite(a_{m_s}, 1, 0) + Ite(a_{p_1}, 1, 0) + Ite(a_{g_1}, 1, 0) + Ite(a_{g_2}, 1, 0) + Ite(a_{g_3}, 1, 0)$$

The unsafety constraint is

$$\Psi_{\text{unsafe}}(g_c) = s_{g_c} \wedge \neg m_{s,g_c} \wedge \neg m_{1,g_c}.$$

In this example the formula Ψ for the entire circuit is unsatisfiable meaning that it is first secure in the probing model without glitches.

References

1. Barthe, G., Belaïd, S., Dupressoir, F., Fouque, P., Grégoire, B.: Compositional verification of higher-order masking: application to a verifying masking compiler. IACR Cryptology ePrint Archive, 2015:506 (2015)
2. Barthe, G., Belaïd, S., Dupressoir, F., Fouque, P.-A., Grégoire, B., Strub, P.-Y.: Verified proofs of higher-order masking. In: Oswald, E., Fischlin, M. (eds.) EUROCRYPT 2015. LNCS, vol. 9056, pp. 457–485. Springer, Heidelberg (2015). https://doi.org/10.1007/978-3-662-46800-5_18
3. Barthe, G., Belaïd, S., Dupressoir, F., Fouque, P., Grégoire, B., Strub, P., Zucchini, R.: Strong non-interference and type-directed higher-order masking. In: Proceedings of the 2016 ACM SIGSAC CCS, Vienna, Austria, 24–28 October 2016, pp. 116–129 (2016)
4. Barthe, G., Dupressoir, F., Faust, S., Grégoire, B., Standaert, F., Strub, P.: Parallel implementations of masking schemes and the bounded moment leakage model. IACR Cryptology ePrint Archive, 2016:912 (2016)
5. Barthe, G., Dupressoir, F., Grégoire, B., Stoughton, A., Strub, P.: EasyCrypt: Computer-Aided Cryptographic Proofs (2017). https://github.com/EasyCrypt/easycrypt
6. Bayrak, A.G., Regazzoni, F., Novo, D., Ienne, P.: Sleuth: automated verification of software power analysis countermeasures. In: Bertoni, G., Coron, J.-S. (eds.) CHES 2013. LNCS, vol. 8086, pp. 293–310. Springer, Heidelberg (2013). https://doi.org/10.1007/978-3-642-40349-1_17
7. Belaïd, S., Benhamouda, F., Passelègue, A., Prouff, E., Thillard, A., Vergnaud, D.: Randomness complexity of private circuits for multiplication. In: Fischlin, M., Coron, J.-S. (eds.) EUROCRYPT 2016. LNCS, vol. 9666, pp. 616–648. Springer, Heidelberg (2016). https://doi.org/10.1007/978-3-662-49896-5_22
8. Belaïd, S., Benhamouda, F., Passelègue, A., Prouff, E., Thillard, A., Vergnaud, D.: Private multiplication over finite fields. In: Katz, J., Shacham, H. (eds.) CRYPTO 2017. LNCS, vol. 10403, pp. 397–426. Springer, Cham (2017). https://doi.org/10.1007/978-3-319-63697-9_14
9. Bertoni, G., Martinoli, M.: A methodology for the characterisation of leakages in combinatorial logic. In: Carlet, C., Hasan, M.A., Saraswat, V. (eds.) SPACE 2016. LNCS, vol. 10076, pp. 363–382. Springer, Cham (2016). https://doi.org/10.1007/978-3-319-49445-6_21

10. Bhasin, S., Carlet, C., Guilley, S.: Theory of masking with codewords in hardware: low-weight dth-order correlation-immune boolean functions. IACR Cryptology ePrint Archive, 2013:303 (2013)
11. Bilgin, B., Bogdanov, A., Knežević, M., Mendel, F., Wang, Q.: FIDES: lightweight authenticated cipher with side-channel resistance for constrained hardware. In: Bertoni, G., Coron, J.-S. (eds.) CHES 2013. LNCS, vol. 8086, pp. 142–158. Springer, Heidelberg (2013). https://doi.org/10.1007/978-3-642-40349-1_9
12. Bilgin, B., Gierlichs, B., Nikova, S., Nikov, V., Rijmen, V.: Higher-order threshold implementations. In: Sarkar, P., Iwata, T. (eds.) ASIACRYPT 2014. LNCS, vol. 8874, pp. 326–343. Springer, Heidelberg (2014). https://doi.org/10.1007/978-3-662-45608-8_18
13. Chari, S., Jutla, C.S., Rao, J.R., Rohatgi, P.: Towards sound approaches to counteract power-analysis attacks. In: Wiener, M. (ed.) CRYPTO 1999. LNCS, vol. 1666, pp. 398–412. Springer, Heidelberg (1999). https://doi.org/10.1007/3-540-48405-1_26
14. Coron, J.-S.: Formal verification of side-channel countermeasures via elementary circuit transformations. Cryptology ePrint Archive, Report 2017/879
15. de Moura, L., Bjørner, N.: Z3: an efficient SMT solver. In: Ramakrishnan, C.R., Rehof, J. (eds.) TACAS 2008. LNCS, vol. 4963, pp. 337–340. Springer, Heidelberg (2008). https://doi.org/10.1007/978-3-540-78800-3_24
16. Eldib, H., Wang, C.: Synthesis of masking countermeasures against side channel attacks. In: Biere, A., Bloem, R. (eds.) CAV 2014. LNCS, vol. 8559, pp. 114–130. Springer, Cham (2014). https://doi.org/10.1007/978-3-319-08867-9_8
17. Eldib, H., Wang, C., Schaumont, P.: SMT-based verification of software countermeasures against side-channel attacks. In: Ábrahám, E., Havelund, K. (eds.) TACAS 2014. LNCS, vol. 8413, pp. 62–77. Springer, Heidelberg (2014). https://doi.org/10.1007/978-3-642-54862-8_5
18. Eldib, H., Wang, C., Taha, M.M.I., Schaumont, P.: QMS: evaluating the side-channel resistance of masked software from source code. In: DAC 2014, San Francisco, CA, USA, 1–5 June 2014, pp. 209:1–209:6 (2014)
19. Faust, S., Grosso, V., Pozo, S.M.D., Paglialonga, C., Standaert, F.: Composable masking schemes in the presence of physical defaults and the robust probing model. IACR Cryptology ePrint Archive, 2017:711 (2017)
20. Faust, S., Rabin, T., Reyzin, L., Tromer, E., Vaikuntanathan, V.: Protecting circuits from leakage: the computationally-bounded and noisy cases. In: Gilbert, H. (ed.) EUROCRYPT 2010. LNCS, vol. 6110, pp. 135–156. Springer, Heidelberg (2010). https://doi.org/10.1007/978-3-642-13190-5_7
21. Goodwill, G., Jun, B., Jaffe, J., Rohatgi, P.: A Testing Methodology for Side-Channel Resistance Validation. In: NIST Non-Invasive Attack Testing Workshop (2011)
22. Gross, H.: Collection of protected hardware implementations. https://github.com/hgrosz
23. Gross, H., Mangard, S.: Reconciling d+1 masking in hardware and software. In: CHES 2017 (2017)
24. Gross, H., Mangard, S., Korak, T.: An efficient side-channel protected AES implementation with arbitrary protection order. In: Handschuh, H. (ed.) CT-RSA 2017. LNCS, vol. 10159, pp. 95–112. Springer, Cham (2017). https://doi.org/10.1007/978-3-319-52153-4_6
25. Gross, H., Schaffenrath, D., Mangard, S.: Higher-order side-channel protected implementations of keccak. Cryptology ePrint Archive, Report 2017/395

26. Ishai, Y., Sahai, A., Wagner, D.: Private circuits: securing hardware against probing attacks. In: Boneh, D. (ed.) CRYPTO 2003. LNCS, vol. 2729, pp. 463–481. Springer, Heidelberg (2003). https://doi.org/10.1007/978-3-540-45146-4_27
27. Iusupov, R.: REBECCA - Masking verification tool. https://github.com/riusupov/rebecca
28. Kocher, P., Jaffe, J., Jun, B.: Differential power analysis. In: Wiener, M. (ed.) CRYPTO 1999. LNCS, vol. 1666, pp. 388–397. Springer, Heidelberg (1999). https://doi.org/10.1007/3-540-48405-1_25
29. Mangard, S., Schramm, K.: Pinpointing the side-channel leakage of masked AES hardware implementations. In: Goubin, L., Matsui, M. (eds.) CHES 2006. LNCS, vol. 4249, pp. 76–90. Springer, Heidelberg (2006). https://doi.org/10.1007/11894063_7
30. Moss, A., Oswald, E., Page, D., Tunstall, M.: Compiler assisted masking. In: Prouff, E., Schaumont, P. (eds.) CHES 2012. LNCS, vol. 7428, pp. 58–75. Springer, Heidelberg (2012). https://doi.org/10.1007/978-3-642-33027-8_4
31. Nikova, S., Rechberger, C., Rijmen, V.: Threshold implementations against side-channel attacks and glitches. In: Ning, P., Qing, S., Li, N. (eds.) ICICS 2006. LNCS, vol. 4307, pp. 529–545. Springer, Heidelberg (2006). https://doi.org/10.1007/11935308_38
32. O'Donnell, R.: Analysis of Boolean Functions. Cambridge University Press, Cambridge (2014)
33. Quisquater, J.-J., Samyde, D.: ElectroMagnetic analysis (EMA): measures and counter-measures for smart cards. In: Attali, I., Jensen, T. (eds.) E-smart 2001. LNCS, vol. 2140, pp. 200–210. Springer, Heidelberg (2001). https://doi.org/10.1007/3-540-45418-7_17
34. Reparaz, O.: Detecting flawed masking schemes with leakage detection tests. In: Peyrin, T. (ed.) FSE 2016. LNCS, vol. 9783, pp. 204–222. Springer, Heidelberg (2016). https://doi.org/10.1007/978-3-662-52993-5_11
35. Reparaz, O., Bilgin, B., Nikova, S., Gierlichs, B., Verbauwhede, I.: Consolidating masking schemes. In: Gennaro, R., Robshaw, M. (eds.) CRYPTO 2015. LNCS, vol. 9215, pp. 764–783. Springer, Heidelberg (2015). https://doi.org/10.1007/978-3-662-47989-6_37
36. Rivain, M., Prouff, E.: Provably secure higher-order masking of AES. In: Mangard, S., Standaert, F.-X. (eds.) CHES 2010. LNCS, vol. 6225, pp. 413–427. Springer, Heidelberg (2010). https://doi.org/10.1007/978-3-642-15031-9_28
37. Trichina, E.: Combinational logic design for AES subbyte transformation on masked data. IACR Cryptology ePrint Archive (2003)
38. Wolf, C., Glaser, J.: Yosys - a free verilog synthesis suite. In: Proceedings of Austrochip 2013 (2013)
39. Xiao, G., Massey, J.L.: A spectral characterization of correlation-immune combining functions. IEEE Trans. Inf. Theory **34**(3), 569–571 (1988)

Masking the GLP Lattice-Based Signature Scheme at Any Order

Gilles Barthe[1], Sonia Belaïd[2], Thomas Espitau[3], Pierre-Alain Fouque[4], Benjamin Grégoire[5], Mélissa Rossi[6,7], and Mehdi Tibouchi[8](✉)

[1] IMDEA Software Institute, Madrid, Spain
gilles.barthe@imdea.org
[2] CryptoExperts, Paris, France
sonia.belaid@cryptoexperts.com
[3] UPMC, Paris, France
thomas.espitau@lip6.fr
[4] Univ Rennes, Rennes, France
pierre-alain.fouque@univ-rennes1.fr
[5] Inria Sophia Antipolis, Sophia Antipolis, France
benjamin.gregoire@sophia.inria.fr
[6] Thales, Paris, France
[7] Département d'informatique de l'École normale supérieure de Paris, CNRS, PSL Research University, INRIA, Paris, France
melissa.rossi@ens.fr
[8] NTT Secure Platform Laboratories, Tokyo, Japan
tibouchi.mehdi@lab.ntt.co.jp

Abstract. Recently, numerous physical attacks have been demonstrated against lattice-based schemes, often exploiting their unique properties such as the reliance on Gaussian distributions, rejection sampling and FFT-based polynomial multiplication. As the call for concrete implementations and deployment of postquantum cryptography becomes more pressing, protecting against those attacks is an important problem. However, few countermeasures have been proposed so far. In particular, masking has been applied to the decryption procedure of some lattice-based encryption schemes, but the much more difficult case of signatures (which are highly non-linear and typically involve randomness) has not been considered until now.

In this paper, we describe the first masked implementation of a lattice-based signature scheme. Since masking Gaussian sampling and other procedures involving contrived probability distribution would be prohibitively inefficient, we focus on the GLP scheme of Güneysu, Lyubashevsky and Pöppelmann (CHES 2012). We show how to provably mask it in the Ishai–Sahai–Wagner model (CRYPTO 2003) at any order in a relatively efficient manner, using extensions of the techniques of Coron et al. for converting between arithmetic and Boolean masking. Our proof relies on a mild generalization of probing security that supports the notion of public outputs. We also provide a proof-of-concept implementation to assess the efficiency of the proposed countermeasure.

Keywords: Side-channel · Masking · GLP lattice-based signature

J. B. Nielsen and V. Rijmen (Eds.): EUROCRYPT 2018, LNCS 10821, pp. 354–384, 2018.
https://doi.org/10.1007/978-3-319-78375-8_12

1 Introduction

As the demands for practical implementations of postquantum cryptographic schemes get more pressing ahead of the NIST postquantum competition and in view of the recommendations of various agencies, understanding the security of those schemes against physical attacks is of paramount importance. Lattice-based cryptography, in particular, is an attractive option in the postquantum setting, as it allows to design postquantum implementations of a wide range of primitives with strong security guarantees and a level of efficiency comparable to currently deployed RSA and elliptic curve-based schemes. However, it poses new sets of challenges as far as side-channels and other physical attacks are concerned. In particular, the reliance on Gaussian distributions, rejection sampling or the number-theoretic transform for polynomial multiplication have been shown to open the door to new types of physical attacks for which it is not always easy to propose efficient countermeasures.

The issue has in particular been laid bare in a number of recent works for the case of lattice-based signature schemes. Lattice-based signature in the random oracle model can be roughly divided into two families: on the one hand, constructions following Lyubashevsky's "Fiat–Shamir with aborts" paradigm [23], and on the other hand, hash-and-sign signatures relying on lattice trapdoors, as introduced by Gentry, Peikert and Vaikuntanathan [19]. Attempts have been made to implement schemes from both families, but Fiat–Shamir signatures are more common (although their postquantum security is admittedly not as well grounded). The underlying framework is called Fiat–Shamir *with aborts* because, unlike RSA and discrete logarithm-based constructions, lattice-based constructions involve sampling from sets that do not admit a nice algebraic structure. A naïve sampling algorithm would leak partial key information, in much the same way as it did in early heuristic schemes like GGH and NTRUSign; this is avoided by forcing the output signature to be independent of the secret key using rejection sampling. Many instantiations of the framework have been proposed [15,21,23,24,27], some of them quite efficient: for example, the BLISS signature scheme [15] boasts performance and key and signature sizes roughly comparable to RSA and ECDSA signatures.

However, the picture becomes less rosy once physical attacks are taken into account. For instance, Groot Bruinderink et al. [20] demonstrated a cache attack targetting the Gaussian sampling of the randomness used in BLISS signatures, which recovers the entire secret key from the side-channel leakage of a few thousand signature generations. Fault attacks have also been demonstrated on all kinds of lattice-based signatures [6,17]. In particular, Espitau et al. recover the full BLISS secret key using a single fault on the generation of the randomness (and present a similarly efficient attack on GPV-style signatures). More recently, ACM CCS 2017 has featured several papers [18,26] exposing further side-channel attacks on BLISS, its variant BLISS–B, and their implementation in the strongSwan VPN software. They are based on a range of different side channels (cache attacks, simple and correlation electromagnetic analysis, branch tracing, etc.), and some of them target new parts of the signature generation algorithm, such as the rejection sampling.

In order to protect against attack such as these, one would like to apply powerful countermeasures like masking. However, doing so efficiently on a scheme like BLISS seems hard, as discussed in [18]. Indeed, the sampling of the Gaussian randomness in BLISS signature generation involves either very large lookup tables, which are expensive to mask efficiently, or iterative approaches that are hard to even implement in constant time–let alone mask. Similarly, the rejection sampling step involves transcendental functions of the secret data that have to be computed to high precision; doing so in masked form seems daunting.

However, there exist other lattice-based signatures that appear to support side-channel countermeasures like masking in a more natural way, because they entirely avoid Gaussians and other contrived distributions. Both the sampling of the randomness and the rejection sampling of signatures target uniform distributions in contiguous intervals. Examples of such schemes include the GLP scheme of Güneysu, Lyubashevsky and Pöppelmann [21], which can be seen as the ancestor of BLISS, and later variants like the Dilithium scheme of Ducas et al. [16] (but not Dilithium-G).

In this paper, we show how to efficiently mask the GLP scheme at any masking order, so as to achieve security against power analysis and related attacks (both simple power analysis and higher-order attacks like differential/correlation power analysis). This is to the best of our knowledge the first time a masking countermeasure has been applied to protect lattice-based signatures.

Related Work. Masking is a well-known technique introduced by Chari, Rao and Rohatgi at CHES 2002 [7] and essentially consists in splitting a secret value into $d+1$ ones (d is thus the masking order), using a secret sharing scheme. This will force the adversary to read many internal variables if he wants to recover the secret value, and he will gain no information if he observes fewer than d values. The advantage of this splitting is that linear operations cost nothing, but the downside is that non-linear operations (such as the AES S-box) can become quite expensive. Later, Ishai, Sahai and Wagner [22] developed a technique to prove the security of masking schemes in the threshold probing model (ISW), in which the adversary can read off at most d wires in a circuit. Recently, Duc, Dziembowski and Faust [14] proved the equivalence between this threshold model and the more realistic noisy model, in which the adversary acquires leakage on *all* variables, but that leakage is perturbed with some noise distribution, as is the case in practical side-channel attacks. Since the ISW model is much more convenient for designing and proving masking countermeasures, it is thus preferred, as the equivalence results of Duc et al. ultimately ensure that a secure implementation in the ISW model at a sufficiently high masking order is going to be secure against practical side-channel attacks up to a given signal-to-noise ratio.

Masking has been applied to lattice-based encryption schemes before [28,29]. However, in these schemes, only the decryption procedure needs to be protected, and it usually boils down to computing a scalar product between the secret key and the ciphertext (which is a linear operation in the secret data) followed by a comparison (which is non-linear, but not very difficult to mask). Oder et al. [25] point out a number of issues with those masked decryption algorithms, and describe another one, for a CCA2-secure version of Ring-LWE public-key encryption.

Our Results. Masking lattice-based signatures, even in the comparatively simple case of GLP, turns out to be surprisingly difficult—possibly more so than any of the previous masking countermeasures considered so far in the literature. The probabilistic nature of signature generation, as well as its reliance on rejection sampling, present challenges (both in terms of design and of proof techniques) that had not occurred in earlier schemes, most of them deterministic. In addition, for performance reasons, we are led to require a stronger security property of the original, unprotected signature scheme itself, which we have to establish separately. More precisely, the following issues arise.

Conversion between Boolean and mod-p arithmetic masking. Most steps of the signing algorithm involve linear operations on polynomials in the ring $\mathcal{R} = \mathbb{Z}_p[x]/(x^n+1)$. They can thus be masked very cheaply using mod-p arithmetic masking: each coefficient is represented as a sum of $d+1$ additive shares modulo p. For some operations, however, this representation is less convenient.

This is in particular the case for the generation of the randomness at the beginning of the algorithm, which consists of two polynomials $\mathbf{y}_1, \mathbf{y}_2$ with uniformly random coefficients in a subinterval $[-k, k]$ of $\mathbb{Z}_p$. Generating such a random value in masked form is relatively easy with Boolean masking, but seems hard to do efficiently with arithmetic masking. Therefore, we have to carry out a conversion from Boolean masking to mod-p arithmetic masking. Such conversions have been described before [11,13], but only when the modulus p was a power of 2. Adapting them to our settings requires some tweaks.

Similarly, the rejection sampling step amounts to checking whether the polynomials in the signature have their coefficients in another interval $[-k', k']$. Carrying out the corresponding comparison is again more convenient with Boolean masking, and hence we need a conversion algorithm in the other direction, from mod-p arithmetic masking to Boolean masking. We are again led to adapt earlier works on arithmetic-to-Boolean masking conversion [12,13] to the case of a non-prime modulus.

Security of the signature scheme when revealing the "commitment" value. One of the operations in signature generation is the computation of a hash function mapping to polynomials in $\mathcal{R}$ of a very special shape. Masking the computation of this hash function would be highly inefficient and difficult to combine with the rest of the algorithm. Indeed, the issue with hashing is not obtaining a masked bit string (which could be done with something like SHA-3), but expanding that bit string into a random-looking polynomial $\mathbf{c}$ of fixed, low Hamming weight in masked form. The corresponding operation is really hard to write down as a circuit. Moreover, even if that could be done, it would be terrible for performances because subsequent multiplications by c are no longer products by a known sparse constant, but full-blown ring operations that have to be fully masked.

But more importantly, this masking *should* intuitively be unnecessary. Indeed, when we see the signature scheme as the conversion of an identification protocol under the Fiat–Shamir transform, the hash function computation corresponds to the verifier's sampling of a random challenge $\mathbf{c}$ after it receives the commitment value $\mathbf{r}$ from the prover. In particular, the verifier always learns the commitment value $\mathbf{r}$ (corresponding to the input of the hash function), so if the identification protocol is "secure", one should always be able to reveal this value without compromising security. But the security of the signature scheme only offers weak guarantees on the security of the underlying identification protocol, as discussed by Abdalla et al. [1].

In usual Fiat–Shamir signatures, this is never an issue because the commitment value can always be publicly derived from the signature (as it is necessary for signature verification). However, things are more subtle in the Fiat–Shamir with aborts paradigm, since the value $\mathbf{r}$ is not normally revealed in executions of the signing algorithm that do not pass the rejection sampling step. In our setting, though, we would like to unmask the value to compute the hash function in all cases, before knowing whether the rejection sampling step will be successful. If we do so, the side-channel attacker can thus learn the pair $(\mathbf{r}, \mathbf{c})$ corresponding to rejected executions as well, and this is not covered by the original security proof, nor does security with this additional leakage look reducible to the original security assumption.

However, it is heuristically a hard problem to distinguish those pairs from uniform (an LWE-like problem with a rather unusual distribution), so one possible approach, which requires no change at all to the algorithm itself, is to redo the security proof with an additional, ad hoc hardness assumption. This is the main approach that we suggest in this paper. Although heuristically safe, it is rather unsatisfactory from a theoretical standpoint, so we additionally propose another approach:[1] compute the hash function not in terms of $\mathbf{r}$ itself, but of $f(\mathbf{r})$ where f is a statistically-hiding commitment scheme whose opening information is added to actual signatures, but not revealed in executions of the algorithm that do not pass the rejection sampling. Using a suitable f, $f(\mathbf{r})$ can be efficiently computed in masked form, and only the result needs to be unmasked. It is then clear that the leakage of $\big(f(\mathbf{r}), \mathbf{c}\big)$ is innocuous, and the modified scheme can be proved entirely with no additional hardness assumption. The downside of this approach is of course that the commitment key increases the size of the public key, the opening information increases the size of signatures, and the masked computation of the commitment itself takes a not insignificant amount of time. For practical purposes, we therefore recommend the heuristic approach.

Security of masking schemes with output-dependent probes. In order to prove the security of our masked implementation we see that we reveal some public value $\mathbf{r}$ or a commitment of it. Consequently, we must adapt the notion of security

[1] We are indebted to Vadim Lyubashevsky for suggesting this approach.

from the threshold probing model to account for public outputs; the idea here is not to state that public outputs do not leak relevant information, but rather that the masked implementation does not leak more information than the one that is released through public outputs. We capture this intuition by letting the simulator depend on the distribution of the public outputs. This extends the usual "non-interference" (NI) security notion to a new, more general notion of "non-interference with public outputs" (NIo).

Security proofs. The overall security guarantee for the masked implementation is established by proving the security of individual gadgets and asserting the security of their combination. For some gadgets, one establishes security in the usual threshold probing model, opening the possibility to resort to automated tools such as maskComp [4] to generate provably secure masked implementations. For other gadgets, the proofs of security are given by exhibiting a simulator, and checking its correctness manually. Finally, the main theorem is deduced from the proof of correctness and security in the threshold probing model with public outputs for the masked implementation, and from a modified proof of security for the GLP scheme.

Organization of the Paper. In Sect. 2, we describe the GLP signature scheme and the security assumption on which its security is based. In Sect. 3, we present the new security notions used in our proofs. Then, in Sect. 4, we describe how to mask the GLP algorithm at any masking order. Finally, in Sect. 5, we describe an implementation of this masking countermeasure, and suggest some possible efficiency improvements.

2 The GLP Signature Scheme

2.1 Parameters and Security

Notations. Throughout this paper, we will use the following notations: n is a power of 2, p is a prime number congruent to 1 modulo $2n$, $\mathcal{R}$ is the polynomial ring modulo x^n+1, $\mathcal{R} = \mathbb{Z}_p[x]/(x^n+1)$. The elements of $\mathcal{R}$ can be represented by polynomials of degree $n-1$ with coefficients in the range $[-\frac{p-1}{2}, \frac{p-1}{2}]$. For an integer k such that $0 < k \leq (p-1)/2$, we denote by $\mathcal{R}_k$ the elements of $\mathcal{R}$ with coefficients in the range $[-k, k]$. We write $\xleftarrow{\$} S$ for picking uniformly at random in a set S or $\xleftarrow{\$} \mathcal{D}$ for picking according to some distribution $\mathcal{D}$.

The key generation algorithm for the GLP signature scheme is as follows:

Algorithm 1. GLP key derivation

Result: Signing key sk, verification key pk

1 $\mathbf{s_1}, \mathbf{s_2} \xleftarrow{\$} \mathcal{R}_1$ //$\mathbf{s_1}$ and $\mathbf{s_2}$ have coefficients in $\{-1, 0, 1\}$
2 $\mathbf{a} \xleftarrow{\$} \mathcal{R}$
3 $\mathbf{t} \leftarrow \mathbf{a}\mathbf{s_1} + \mathbf{s_2}$
4 $sk \leftarrow (\mathbf{s_1}, \mathbf{s_2})$
5 $pk \leftarrow (\mathbf{a}, \mathbf{t})$

Given the verification key $pk = (\mathbf{a}, \mathbf{t})$, if an attacker can derive the signing key, he can be used to also solve a $\mathbf{DCK}_{p,n}$ problem defined in [21].

Definition 1. *The* $\mathbf{DCK}_{p,n}$ problem (Decisional Compact Knapsack problem) *is the problem of distinguishing between the uniform distribution over* $\mathcal{R} \times \mathcal{R}$ *and the distribution* $(\mathbf{a}, \mathbf{a}\mathbf{s}_1 + \mathbf{s}_2)$ *with* $\mathbf{s}_1$, $\mathbf{s}_2$ *uniformly random in* $\mathcal{R}_1$.

In the security proof of our variant of the signature scheme, we introduce a new computational problem.

Definition 2. *The* $\mathbf{R\text{-}DCK}_{p,n}$ problem (Rejected-Decisional Compact Knapsack problem) *is the problem of distinguishing between the uniform distribution over* $\mathcal{R} \times \mathcal{R} \times \mathcal{D}_\alpha^n$ *and the distribution* $(\mathbf{a}, \mathbf{a}\mathbf{y}_1 + \mathbf{y}_2, \mathbf{c})$ *where* $(\mathbf{a}, \mathbf{c}, \mathbf{y}_1, \mathbf{y}_2)$ *is uniformly sampled in* $\mathcal{R} \times \mathcal{D}_\alpha^n \times \mathcal{R}_k^2$*, conditioned by the event* $\mathbf{s}_1\mathbf{c} + \mathbf{y}_1 \notin \mathcal{R}_{k-\alpha}$ *or* $\mathbf{s}_2\mathbf{c} + \mathbf{y}_2 \notin \mathcal{R}_{k-\alpha}$.

As shown in the full version of this paper [5], assuming the hardness of $\mathbf{R\text{-}DCK}_{p,n}$ can be avoided entirely by computing the hash value $\mathbf{c}$ not in terms of $\mathbf{r} = \mathbf{a}\mathbf{y}_1 + \mathbf{y}_2$, but of a statistically hiding commitment thereof. This approach shows that masking can be done based on the exact same assumptions as the original scheme, but at some non-negligible cost in efficiency.

To obtain a scheme that more directly follows the original one and to keep the overhead reasonable, we propose to use $\mathbf{R\text{-}DCK}_{p,n}$ as an extra assumption, which we view as a pragmatic compromise. The assumption is admittedly somewhat artificial, but the same can be said of $\mathbf{DCK}_{p,n}$ itself to begin with, and heuristically, $\mathbf{R\text{-}DCK}_{p,n}$ is similar, except that it removes smaller (hence "easier") instances from the distribution: one expects that this makes distinguishing harder, even though one cannot really write down a reduction to formalize that intuition.

2.2 The Signature Scheme

This part describes the signature scheme introduced in [21]. Additional functions like transform and compress introduced in [21] can be used to shorten the size of the signatures. Note however that for masking purposes, we only need to consider the original, non-compressed algorithm of Güneysu et al., which we describe below. Indeed, signature compression does not affect our masking technique at all, because it only involves unmasked parts of the signature generation algorithm (the input of the hash function and the returned signature itself). As a result, although this paper only discusses the non-compressed scheme, we can directly apply our technique to the compressed GLP scheme with no change, and in fact this is what our proof-of-concept implementation in Sect. 5 actually does.

The signature scheme needs a particular cryptographic hash function, $H : \{0,1\}^* \to \mathcal{D}_\alpha^n$, where $\mathcal{D}_\alpha^n$ is the set of polynomials in $\mathcal{R}$ that have all zero coefficients except for at most $\alpha = 32$ coefficients that are in $\{-1, +1\}$ (or $\alpha = 16$ when using the updated parameters presented in [8]).

Let k be a security parameter. Algorithms 2 and 3 respectively describe the GLP signature and verification. Here is the soundness equation for the verification: $\mathbf{a}\mathbf{z}_1 + \mathbf{z}_2 - \mathbf{t}\mathbf{c} = \mathbf{a}\mathbf{y}_1 + \mathbf{y}_2$.

The parameter k controls the trade-off between the security and the runtime of the scheme. The smaller k gets, the more secure the scheme becomes and the shorter the signatures get but the time to sign will increase. The authors of the implementation of [21] suggest $k = 2^{14}$, $n = 512$ and $p = 8383489$ for ≈ 100 bits of security and $k = 2^{15}$, $n = 1024$ and $p = 16760833$ for > 256 bits of security.

2.3 Security Proof of the r-GLP Variant

As mentioned above, masking the hash function of the GLP signature directly has a prohibitive cost, and it is thus preferable to unmask the input $\mathbf{r} = \mathbf{a}\mathbf{y}_1 + \mathbf{y}_2$ to compute the hash value $\mathbf{c} = H(\mathbf{r}, \mathbf{m})$. Doing so allows a side-channel attacker to learn the pair $(\mathbf{r}, \mathbf{c})$ corresponding to rejected executions as well, and since that additional information is not available to the adversary in the original setting, we need to show that it does not affect the security of the scheme.

This stronger security requirement can be modeled as the unforgeability under chosen message attacks of a modified version of the GLP signature scheme in which the pair $(\mathbf{r}, \mathbf{c})$ is made public when a rejection occurs. We call this modified scheme **r**-GLP, and describe it as Algorithm 4. The modification means that, in the EUF-CMA security game, the adversary gets access not only to correctly generated GLP signatures, but also to pairs $(\mathbf{r}, \mathbf{c})$ when rejection occurs, which is exactly the setting that arises as a result of unmasking the value $\mathbf{r}$. The following theorem, proved in the full version of this paper [5], states that the modified scheme is indeed secure, at least if we are willing to assume the hardness of the additional $\mathbf{DCK}_{p,n}$ assumption.

Theorem 1. *Let n, p, $\mathcal{R}$ and $\mathcal{D}^n_\alpha$ as defined in Sect. 2.1. Assuming the hardness of the $\mathbf{DCK}_{p,n}$ and $\mathbf{R\text{-}DCK}_{p,n}$ problems, the signature* **r**-*GLP is EUF-CMA secure in the random oracle model.*

Remark 1. As mentioned previously, we can avoid the non-standard assumption $\mathbf{R\text{-}DCK}_{p,n}$ by hashing not $\mathbf{r}$ but $f(\mathbf{r})$ for some statistically hiding commitment f (which can itself be constructed under $\mathbf{DCK}_{p,n}$, or standard lattice assumptions). See the full version of this paper for details [5]. The downside of that

Algorithm 2. GLP signature

Data: $\mathbf{m}$, pk, sk
Result: Signature σ
1 $\mathbf{y}_1, \mathbf{y}_2 \xleftarrow{\$} \mathcal{R}_k$
2 $\mathbf{c} \leftarrow H(\mathbf{r} = \mathbf{a}\mathbf{y}_1 + \mathbf{y}_2, \mathbf{m})$
3 $\mathbf{z}_1 \leftarrow \mathbf{s}_1\mathbf{c} + \mathbf{y}_1$
4 $\mathbf{z}_2 \leftarrow \mathbf{s}_2\mathbf{c} + \mathbf{y}_2$
5 **if** $\mathbf{z}_1$ *or* $\mathbf{z}_2 \notin \mathcal{R}_{k-\alpha}$ **then**
6 | restart
7 **end**
8 **return** $\sigma = (\mathbf{z}_1, \mathbf{z}_2, \mathbf{c})$

Algorithm 3. GLP verification

Data: $\mathbf{m}$, σ, pk
1 **if** $\mathbf{z}_1$, $\mathbf{z}_2 \in \mathcal{R}_{k-\alpha}$ *and* $\mathbf{c} = H(\mathbf{a}\mathbf{z}_1 + \mathbf{z}_2 - \mathbf{t}\mathbf{c}, \mathbf{m})$ **then**
2 | accept
3 **else**
4 | reject
5 **end**

Algorithm 4. Tweaked signature with public $\mathbf{r}$

Data: $\mathbf{m}$, $pk = (\mathbf{a}, \mathbf{t})$, $sk = (\mathbf{s}_1, \mathbf{s}_2)$
Result: Signature σ

```
1  y1 <-$ R_k
2  y2 <-$ R_k
3  r <- a y1 + y2
4  c <- H(r, m)
5  z1 <- s1 c + y1
6  z2 <- s2 c + y2
7  if z1 or z2 ∉ R_{k-α} then
8  |  (z1, z2) <- (⊥, ⊥)
9  end
10 return σ = (z1, z2, c, r)
```

approach is that it has a non negligible overhead in terms of key size, signature size, and to a lesser extent signature generation time.

3 Threshold Probing Model with Public Outputs

In this section, we briefly review the definition of the threshold probing model, and introduce an extension to accommodate public outputs.

3.1 Threshold Probing Model

The threshold probing model introduced by Ishai, Sahai and Wagner considers implementations that operate over shared values [22].

Definition 3. *Let d be a masking order. A shared value is a $(d+1)$-tuple of values, typically integers or Booleans.*

A (u, v)-gadget is a probabilistic algorithm that takes as inputs u shared values, and returns distributions over v-tuples of shared values. (u, v)-gadgets are typically used to implement functions that take u inputs and produce v outputs.

Gadgets are typically written in pseudo-code, and induce a mapping from u-tuples of shared values (or equivalently $u(d+1)$-tuples of values) to a distribution over v-tuples of values, where the output tuple represents the joint distribution of the output shared values as well as all intermediate values computed during the execution of the gadget.

We now turn to the definition of probing security. Informally, an implementation is d-probing secure if and only if an adversary that can observe at most d intermediate values cannot recover information on secret inputs.

Definition 4. d-non-interference (d-NI): *A gadget is d-non-interfering if and only if every set of at most d intermediate variables can be perfectly simulated with at most d shares of each input.*

Definition 5. d-strong-non-interference (d-SNI): *A gadget is d-strongly non interfering if and only if every set of size $d_0 \leq d$ containing d_1 intermediate variables and $d_2 = d_0 - d_1$ returned values can be perfectly simulated with at most d_1 shares of each input.*

This notion of security is formulated in a simulation-based style. It is however possible to provide an equivalent notion as an information flow property in the style of programming language security and recent work on formal methods for proving security of masked implementations.

The maskComp tool. For certain composition proofs, we will use the maskComp tool from Barthe et al. [4]. It uses a type-based information flow analysis with cardinality constraints and ensures that the composition of gadgets is d-NI secure at arbitrary orders, by inserting refresh gadgets when required.

3.2 Threshold Probing Model with Public Outputs

The security analysis of our masked implementation of GLP requires an adaptation of the standard notion of security in the threshold probing model. Specifically, our implementation does not attempt to mask the computation of $H(\mathbf{r}, \mathbf{m})$ at line 2 of Algorithm 2; instead, it recovers $\mathbf{r}$ from its shares and then computes $H(\mathbf{r}, \mathbf{m})$. This optimization is important for the efficiency of the masked algorithm, in particular because it is not immediately clear whether one can mask the hash function H efficiently—note that this kind of optimization is also reminiscent of the method used to achieve efficient sorting algorithms in multi-party computations.

From a security perspective, recombining $\mathbf{r}$ in the algorithm is equivalent to making $\mathbf{r}$ a public output. In contrast with "return values", we will refer to "outputs" as values broadcast on a public channel during the execution of the masked algorithm. The side-channel attacker can therefore use outputs in attacks. Since the usual notions of NI and SNI security do not account for outputs in that sense, we need to extend those notions of security to support algorithms that provide such outputs. The idea here is not to state that public outputs do not leak relevant information, but rather that the masked implementation does not leak more information than the one that is released through public outputs. We capture this intuition by letting the simulator depend on the distribution of the public outputs.

Definition 6. *A gadget with public outputs is a gadget together with a distinguished subset of intermediate variables whose values are broadcast during execution.*

We now turn to the definition of probing security for gadgets with public outputs.

Definition 7. d-non-interference for gadgets with public outputs (d-NIo): *A gadget with public outputs X is d-NIo if and only if every set of at most d*

intermediate variables can be perfectly simulated with the public outputs and at most d shares of each input.

Again, it is possible to provide an equivalent notion as an information flow property in the style of programming language security.

Note that the use of public outputs induces a weaker notion of security.

Lemma 1. *Let G be a d-NI-gadget. Then G is d-NIo secure for every subset X of intermediate variables.*

Informally, the lemma states that a gadget that does not leak any information also does not leak more information than the one revealed by a subset of its intermediate variables. The lemma is useful to resort to automated tools for proving NI security of some gadgets used in the masked implementations of GLP. In particular, we will use the maskComp tool.

Since d-NIo security is weaker than d-NI security, we must justify that it delivers the required security guarantee. This is achieved by combining the proofs of security for the modified version of GLP with public outputs, and the proofs of correctness and security for the masked implementations of GLP.

4 Masked Algorithm

In this section, the whole GLP scheme is turned into a functionally equivalent scheme secure in the d-probing model with public outputs. Note that it suffices to mask the key derivation in the d-probing model and the signature in the d-probing model with public output $\boldsymbol{r}$, since the verification step does not manipulate sensitive data.

Remark 2. The masked version of GLP scheme with commitment has also been turned into a functionally equivalent scheme proved secure in the d-probing model with public output $\boldsymbol{r}$. Its masked version is a little more complex, it is detailed in the full version of this paper [5].

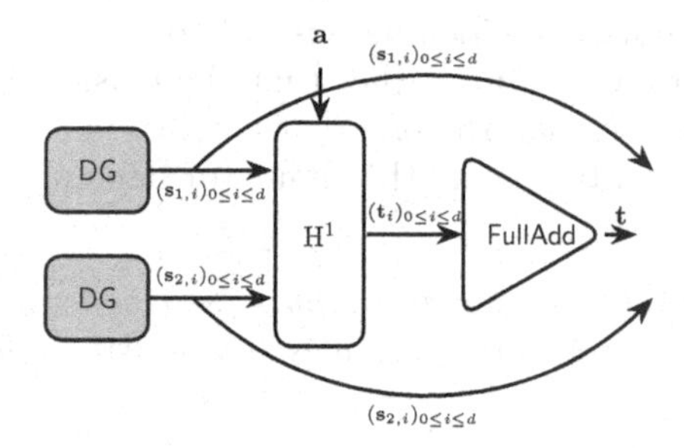

Fig. 1. Composition of mKD (The blue gadgets will be proved d-NIo, the white ones will be proved d-NI) (Color figure online)

4.1 Overall Structure

For simplicity, we will show the masking on a single iteration version of the signature. The masking can be generalized by calling the masked signature again if it fails.

To ensure protection against d-th order attacks, we suggest a masking countermeasure with $d+1$ shares for the following sensitive data: $\mathbf{y}_1$, $\mathbf{y}_2$, $\mathbf{s}_1$ and $\mathbf{s}_2$. All the public variables are $(\mathbf{a}, \mathbf{t})$ (i.e., the public key), $\mathbf{m}$ (i.e., the message), $RejSp$ (i.e., the bit corresponding to the success of the rejection sampling), $(\mathbf{z}_1, \mathbf{z}_2, \mathbf{c})$ (i.e., the signature). As mentioned before, because of the need of $\mathbf{r}$ recombination, even if $\mathbf{r}$ is an intermediate value, it is considered as a public output.

Most operations carried out in the GLP signing algorithm are arithmetic operations modulo p, so we would like to use arithmetic masking. It means for example that $\mathbf{y}_1$ will be replaced by $\mathbf{y}_{1,0}, \ldots \mathbf{y}_{1,d} \in \mathcal{R}$ such that

$$\mathbf{y}_1 = \mathbf{y}_{1,0} + \ldots + \mathbf{y}_{1,d} \mod p.$$

The issue is that at some points of the algorithm, we need to perform operations that are better expressed using Boolean masking. Those parts will be extracted from both the key derivation and the signature to be protected individually and then securely composed. The different new blocks to achieve protection against d-th order attacks are depicted hereafter and represented in Figs. 1 and 2:

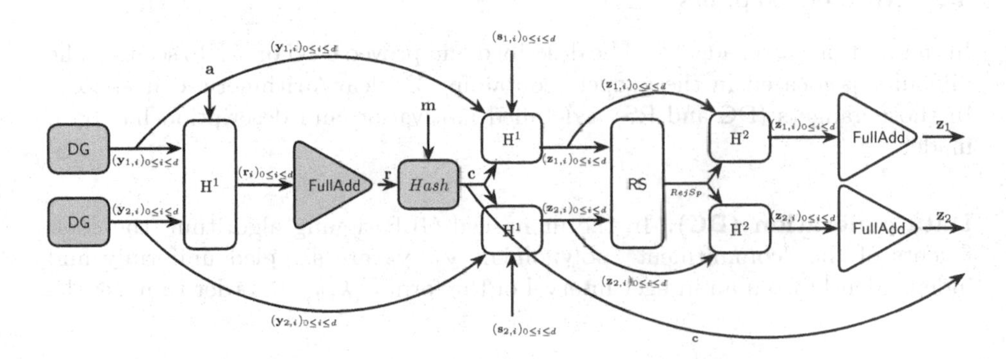

Fig. 2. Composition of mSign (The blue gadgets will be proved d-NIo, the white ones will be proved d-NI and the red one won't be protected) (Color figure online)

- Generation of the shared data (DG), masked version of line 1 in Algorithm 2 and line 1 in Algorithm 1, is a function to generate shares of $\mathbf{y}_1$, $\mathbf{y}_2$, $\mathbf{s}_1$ and $\mathbf{s}_2$. It will be described in Algorithm 7, decomposed and proved d-NIo secure by decomposition.
- Rejection Sampling (RS), masked version of line 5 in Algorithm 2, is a test to determine if $\mathbf{z}_1$ and $\mathbf{z}_2$ belong to the set $\mathcal{R}_{k-\alpha}$. It will be detailed in Algorithm 15 and proved d-NI secure by decomposition.

- Refresh and unmask (FullAdd) is a function that unmasks securely a variable by adding together its shares modulo p without leaking the partial sums. It will be described in Algorithm 16 and proved d-NIo secure and d-NI secure when used at the end.
- H^1 and H^2 are the elementary parts, masked versions of line 2, 3–4 and then 5–6 in Algorithm 2. H^1 is also the masked version of the instruction called in line 3 of the key derivation algorithm (Algorithm 1). They are made of arithmetic computations. They are depicted in Algorithms 17 and 18. They will be proved d-NI secure.
- Hash function, line 2 in Algorithm 2. As mentioned before, is left unmasked because it manipulates only public data.

Algorithm 6 shows a high level picture of mSign with all these blocks and Algorithm 5 shows mKD.

The proofs of dNI or d-NIo security will be given in the following subsection. Then, the composition will be proved in Sect. 4.3 to achieve global security in the d-probing model with public outputs. This yields the d-NIo security of the masked signature and masked key generation algorithms in Theorems 2 and 3, respectively. By combining these results with the refined analysis of the GLP signature in Theorem 1, one obtains the desired security guarantee, as discussed in Sect. 3.

4.2 Masked Gadgets

In this section each gadget will be described and proved d-NI or d-NIo secure. The difficulty is located in the gadgets containing Boolean/arithmetic conversions. In those gadgets (DG and RS) a detailed motivation and description has been made.

Data generation (DG). In the unmasked GLP signing algorithm, the coefficients of the "commitment" polynomials $\mathbf{y}_1$, $\mathbf{y}_2$ are sampled uniformly and independently from an integer interval of the form $[-k, k]$. In order to mask the

Algorithm 5. mKD

Result: Signing key sk, verification key pk

1 $(\mathbf{s}_{1,i})_{0 \leq i \leq d} \leftarrow \mathsf{DG}(1, d)$
2 $(\mathbf{s}_{2,i})_{0 \leq i \leq d} \leftarrow \mathsf{DG}(1, d)$
3 $\mathbf{a} \xleftarrow{\$} \mathcal{R}$
4 $(\mathbf{t}_i)_{0 \leq i \leq d} \leftarrow \mathrm{H}^1(\mathbf{a}, (\mathbf{s}_{1,i})_{0 \leq i \leq d}, (\mathbf{s}_{2,i})_{0 \leq i \leq d})$
5 $\mathbf{t} \leftarrow \mathsf{FullAdd}((\mathbf{t}_i)_{0 \leq i \leq d})$
6 $sk \leftarrow ((\mathbf{s}_{1,i})_{0 \leq i \leq d}, (\mathbf{s}_{2,i})_{0 \leq i \leq d})$
7 $pk \leftarrow (\mathbf{a}, \mathbf{t})$
8 **return as public key** $(\mathbf{a}, \mathbf{t})$
9 **return as secret key** $((\mathbf{s}_{1,i})_{0 \leq i \leq d}, (\mathbf{s}_{2,i})_{0 \leq i \leq d})$

Algorithm 6. mSign

Data: m, $pk = (\mathbf{a}, \mathbf{t})$, $sk = ((\mathbf{s}_{1,i})_{0 \le i \le d}, (\mathbf{s}_{2,i})_{0 \le i \le d})$
Result: Signature σ
1 $(\mathbf{y}_{1,i})_{0 \le i \le d} \leftarrow \mathsf{DG}(k, d)$
2 $(\mathbf{y}_{2,i})_{0 \le i \le d} \leftarrow \mathsf{DG}(k, d)$
3 $(\mathbf{r}_i)_{0 \le i \le d} \leftarrow \mathrm{H}^1(\mathbf{a}, (\mathbf{y}_{1,i})_{0 \le i \le d}, (\mathbf{y}_{2,i})_{0 \le i \le d})$
4 $\mathbf{r} \leftarrow \mathsf{FullAdd}((\mathbf{r}_i)_{0 \le i \le d})$
5 $\mathbf{c} \leftarrow hash(\mathbf{r}, \mathbf{m})$
6 $(\mathbf{z}_{1,i})_{0 \le i \le d} \leftarrow \mathrm{H}^1(\mathbf{c}, (\mathbf{s}_{1,i})_{0 \le i \le d}, (\mathbf{y}_{1,i})_{0 \le i \le d})$
7 $(\mathbf{z}_{2,i})_{0 \le i \le d} \leftarrow \mathrm{H}^1(\mathbf{c}, (\boldsymbol{ts}_{2,i})_{0 \le i \le d}, (\mathbf{y}_{2,i})_{0 \le i \le d})$
8 $RejSp \leftarrow \mathsf{RS}((\mathbf{z}_{1,i})_{0 \le i \le d}, (\mathbf{z}_{2,i})_{0 \le i \le d}, k - \alpha)$
9 $(\mathbf{z}_{1,i})_{0 \le i \le d} \leftarrow \mathrm{H}^2(RejSp, (\mathbf{z}_{1,i})_{0 \le i \le d})$
10 $(\mathbf{z}_{2,i})_{0 \le i \le d} \leftarrow \mathrm{H}^2(RejSp, (\mathbf{z}_{2,i})_{0 \le i \le d})$
11 $\mathbf{z}_1 \leftarrow \mathsf{FullAdd}((\mathbf{z}_{1,i})_{0 \le i \le d})$
12 $\mathbf{z}_2 \leftarrow \mathsf{FullAdd}((\mathbf{z}_{2,i})_{0 \le i \le d})$
13 **return** $\sigma = (\mathbf{z}_1, \mathbf{z}_2, \mathbf{c})$

Algorithm 7. Data Generation (DG)

Data: k and d
Result: A uniformly random $\mathbf{y}$ integer in $\mathcal{R}_k$ in arithmetic masked form $(\mathbf{y}_i)_{0 \le i \le d}$.
1 $(\mathbf{y}_i)_{0 \le i \le d} \leftarrow \{0\}^d$
2 **for** $j = 1$ *to* n **do**
3 $\quad (a_i)_{0 \le i \le d} \leftarrow \mathsf{RG}(k, d)$
4 $\quad (\mathbf{y}_i)_{0 \le i \le d} \leftarrow (\mathbf{y}_i + a_i x^j)_{0 \le i \le d}$
5 **end**
6 **return** $(\mathbf{y}_i)_{0 \le i \le d}$

signing algorithm, one would like to obtain those values in masked form, using order-d arithmetic masking modulo p. Note that since all of these coefficients are completely independent, the problem reduces to obtaining an order-d mod-p arithmetic masking of a single random integer in $[-k, k]$.

Accordingly, we will first create an algorithm called Random Generation (RG) which generates an order-d mod-p arithmetic masking of a single random integer in $[-k, k]$. Next, we will use RG in an algorithm called Data Generation (DG) which generates a sharing of a value in $\mathcal{R}_k$. DG is calling RG n times and is described in Algorithm 7. RG is described hereafter and will be given in Algorithm 14.

Let us now build RG. Carrying out this masked random sampling in arithmetic form directly and securely seems difficult. On the other hand, it is relatively easy to generate a Boolean masking of such a uniformly random value. We can then convert that Boolean masking to an arithmetic masking using Coron et al.'s higher-order Boolean-to-arithmetic masking conversion technique [13]. The technique has to be modified slightly to account for the fact that the modulus p of the arithmetic masking is not a power of two, but the overall structure of the algorithm remains the same. To obtain a better complexity, we also use the Kogge–Stone adder based addition circuit already considered in [12].

A more precise description of our approach is as follows. Let $K = 2k+1$, and w_0 be the smallest integer such that $2^{w_0} > K$. Denote also by w the bit size of the Boolean masking we are going to use; we should have $w > w_0 + 1$ and $2^w > 2p$. For GLP masking, a natural choice, particularly on a 32-bit architecture, would be $w = 32$.

Now the first step of the algorithm is to generate w_0-bit values $(x_i^0)_{0 \le i \le d}$ uniformly and independently at random, and apply a multiplication-based share refreshing algorithm Refresh, as given in Algorithm 8, to obtain a fresh w-bit Boolean masking $(x_i)_{0 \le i \le d}$ of the same value x:

$$x = \bigoplus_{i=0}^{d} x_i^0 = \bigoplus_{i=0}^{d} x_i.$$

Note that x is then a uniform integer in $[0, 2^{w_0} - 1]$.

Algorithm 8. Multiplication-based refresh algorithm for Boolean masking (Refresh)

Data: A Boolean masking $(x_i)_{0 \le i \le d}$ of some value x; the bit size w of the returned masks
Result: An independent Boolean masking $(x'_i)_{0 \le i \le d}$ of x

1 $(x'_i)_{0 \le i \le d} \leftarrow (x_i)_{0 \le i \le d}$
2 **for** $i = 0$ *to* d **do**
3 | **for** $j = i + 1$ *to* d **do**
4 | | pick a uniformly random w-bit value r
5 | | $x'_i \leftarrow x'_i \oplus r$
6 | | $x'_j \leftarrow x'_j \oplus r$
7 | **end**
8 **end**
9 **return** $(x'_i)_{0 \le i \le d}$

We then carry out a rejection sampling on x: if $x \ge K$, we restart the algorithm. If this step is passed successfully, x will thus be uniformly distributed in $[0, K - 1] = [0, 2k]$. Of course, the test has to be carried out securely at order d. This can be done as follows: compute a random w-bit Boolean masking $(k_i)_{0 \le i \le d}$ of the constant $(-K)$ (the two's complement of K over w bits; equivalently, one can use $2^w - K$), and carry out the d-order secure addition $\mathsf{SecAdd}((x_i)_{0 \le i \le d}, (k_i)_{0 \le i \le d})$, given in Algorithm 9 (where Refresh denotes the d-SNI multiplication-based refresh as proven in [4] and recalled in Algorithm 8). The result is a Boolean masking $(\delta_i)_{0 \le i \le d}$ of the difference $\delta = x - K$ in two's complement form. In particular, the most significant bit b of δ is 0 if and only if $x \ge K$. Since computing the most significant bit is an $\mathbb{F}_2$-linear operation,

Algorithm 9. Integer addition of Boolean maskings (SecAdd), as generated by the maskComp tool from the Kogge–Stone adder of [12]

Data: Boolean maskings $(x_i)_{0\le i\le d}$, $(y_i)_{0\le i\le d}$ of integers x, y; the bit size w of the masks
Result: A Boolean masking $(z_i)_{0\le i\le d}$ of $x+y$

1 $(p_i)_{0\le i\le d} \leftarrow (x_i \oplus y_i)_{0\le i\le d}$
2 $(g_i)_{0\le i\le d} \leftarrow \mathsf{SecAnd}\big((x_i)_{0\le i\le d}, (y_i)_{0\le i\le d}, w\big)$
3 **for** $j = 1$ *to* $W := \lceil \log_2(w-1)\rceil - 1$ **do**
4 $\quad$ pow $\leftarrow 2^{j-1}$
5 $\quad$ $(a_i)_{0\le i\le d} \leftarrow (g_i \ll \text{pow})_{0\le i\le d}$
6 $\quad$ $(a_i)_{0\le i\le d} \leftarrow \mathsf{SecAnd}\big((a_i)_{0\le i\le d}, (p_i)_{0\le i\le d}, w\big)$
7 $\quad$ $(g_i)_{0\le i\le d} \leftarrow (g_i \oplus a_i)_{0\le i\le d}$
8 $\quad$ $(a'_i)_{0\le i\le d} \leftarrow (p_i \ll \text{pow})_{0\le i\le d}$
9 $\quad$ $(a'_i)_{0\le i\le d} \leftarrow \mathsf{Refresh}\big((a_i)_{0\le i\le d}, w\big)$
10 $\quad$ $(p_i)_{0\le i\le d} \leftarrow \mathsf{SecAnd}\big((p_i)_{0\le i\le d}, (a'_i)_{0\le i\le d}, w\big)$
11 **end**
12 $(a_i)_{0\le i\le d} \leftarrow (g_i \ll 2^W)_{0\le i\le d}$
13 $(a_i)_{0\le i\le d} \leftarrow \mathsf{SecAnd}\big((a_i)_{0\le i\le d}, (p_i)_{0\le i\le d}, w\big)$
14 $(g_i)_{0\le i\le d} \leftarrow (g_i \oplus a_i)_{0\le i\le d}$
15 $(z_i)_{0\le i\le d} \leftarrow \big(x_i \oplus y_i \oplus (g_i \ll 1)\big)_{0\le i\le d}$
16 **return** $(z_i)_{0\le i\le d}$

we can carry it out componentwise to obtain a masking $(b_i)_{0\le i\le d}$ of b with $b_i = \delta_i \gg (w-1)$. The resulting bit b is non-sensitive, so we can unmask it to check whether to carry out the rejection sampling.

After carrying out these steps, we have obtained a Boolean masking of a uniformly random integer in $[0, 2k]$. What we want is a *mod-p arithmetic* masking of a uniformly random integer in the interval $[-k, k]$, which is of the same length as $[0, 2k]$. If we can convert the Boolean masking to an arithmetic masking, it then suffices to subtract k from one of the shares and we obtain the desired result. To carry out the Boolean-to-arithmetic conversion itself, we essentially follow the approach of [13, Sect. 5], with a few changes to account for the fact that p is not a power of two.

The main change is that we need an algorithm for the secure addition modulo p of two values y, z in Boolean masked form $(y_i)_{0\le i\le d}$, $(z_i)_{0\le i\le d}$ (assuming that $y, z \in [0, p)$). Such an algorithm SecAddModp is easy to construct from SecAdd (see Algorithm 10 with SecAnd the d-order secure bitwise AND operation from [22,30] and recalled in Algorithm 11) and the comparison trick described earlier. More precisely, the approach is to first compute $(s_i)_{0\le i\le d} = \mathsf{SecAdd}\big((y_i)_{0\le i\le d}, (z_i)_{0\le i\le d}\big)$, which is a Boolean sharing of the sum $s = y + z$ without modular reduction, and then $(s'_i)_{0\le i\le d} = \mathsf{SecAdd}\big((s_i)_{0\le i\le d}, (p_i)_{0\le i\le d}\big)$ for a Boolean masking $(p_i)_{0\le i\le d}$ of the value $-p$ in two's complement form (or equivalently $2^w - p$). The result is a masking of $s' = s - p$ in two's complement form. In particular, we have $s \ge p$ if and only if the most significant bit b of s' is 0. Denote by r the desired modular addition $y + z \bmod p$. We thus have:

Algorithm 10. Mod-p addition of Boolean maskings (SecAddModp)

Data: Boolean maskings $(x_i)_{0\leq i\leq d}$, $(y_i)_{0\leq i\leq d}$ of integers x, y; the bit size w of the masks (with $2^w > 2p$)
Result: A Boolean masking $(z_i)_{0\leq i\leq d}$ of $x + y \bmod p$

1 $(p_i)_{0\leq i\leq d} \leftarrow (2^w - p, 0, \ldots, 0)$
2 $(s_i)_{0\leq i\leq d} \leftarrow \mathsf{SecAdd}\big((x_i)_{0\leq i\leq d}, (y_i)_{0\leq i\leq d}, w\big)$
3 $(s'_i)_{0\leq i\leq d} \leftarrow \mathsf{SecAdd}\big((s_i)_{0\leq i\leq d}, (p_i)_{0\leq i\leq d}, w\big)$
4 $(b_i)_{0\leq i\leq d} \leftarrow \big(s'_i \gg (w-1)\big)_{0\leq i\leq d}$
5 $(c_i)_{0\leq i\leq d} \leftarrow \mathsf{Refresh}\big((b_i)_{0\leq i\leq d}, w\big)$
6 $(z_i)_{0\leq i\leq d} \leftarrow \mathsf{SecAnd}\big((s_i)_{0\leq i\leq d}, (\widetilde{c_i})_{0\leq i\leq d}, w\big)$
7 $(c_i)_{0\leq i\leq d} \leftarrow \mathsf{Refresh}\big((b_i)_{0\leq i\leq d}, w\big)$
8 $(z_i)_{0\leq i\leq d} \leftarrow (z_i)_{0\leq i\leq d} \oplus \mathsf{SecAnd}\big((s'_i)_{0\leq i\leq d}, (\widetilde{\neg c_i})_{0\leq i\leq d}, w\big)$
9 **return** $(z_i)_{0\leq i\leq d}$

Algorithm 11. Bitwise AND of Boolean maskings (SecAnd) from [22,30]

Data: Boolean maskings $(x_i)_{0\leq i\leq d}$, $(y_i)_{0\leq i\leq d}$ of integers x, y; the bit size w of the masks
Result: A Boolean masking $(r_i)_{0\leq i\leq d}$ of $x \wedge y$

```
1  (r_i)_{0≤i≤d} ← (x_i ∧ y_i)_{0≤i≤d}
2  for i = 0 to d do
3  |   for j = i + 1 to d do
4  |   |   pick a uniformly random w-bit value z_ij
5  |   |   z_ji ← (x_i ∧ y_j) ⊕ z_ij
6  |   |   z_ji ← z_ji ⊕ (x_j ∧ y_i)
7  |   |   r_i ← r_i ⊕ z_ij
8  |   |   r_j ← r_j ⊕ z_ji
9  |   end
10 end
11 return (r_i)_{0≤i≤d}
```

$$r = \begin{cases} s & \text{if } b = 1; \\ s' & \text{if } b = 0. \end{cases}$$

As a result, we can obtain the masking of r in a secure way as:

$$(r_i)_{0\leq i\leq d} = \mathsf{SecAnd}\big((s_i)_{0\leq i\leq d}, (\widetilde{b_i})_{0\leq i\leq d}\big) \oplus \mathsf{SecAnd}\big((s'_i)_{0\leq i\leq d}, (\widetilde{\neg b_i})_{0\leq i\leq d}\big),$$

where we denote by $\widetilde{b}$ the extension of the bit b to the entire w-bit register (this is again an $\mathbb{F}_2$-linear operation that can be computed componentwise). This concludes the description of SecAddModp.

Using SecAddModp instead of SecAdd in the algorithms of [13, Sect. 4], we also immediately obtain an algorithm SecArithBoolModp for converting a mod-p arithmetic masking $a = \sum_{i=0}^{d} a_i \bmod p$ of a value $a \in [0, p)$ into a Boolean masking $a = \bigoplus_{i=0}^{d} a'_i$ of the same value. The naive way of doing so (see Algorithm 12),

Algorithm 12. Secure conversion from mod-p arithmetic masking to Boolean masking (SecArithBoolModp); this is the simple version (cubic in the masking order)

Data: Arithmetic masking $(a_i)_{0\leq i\leq d}$ modulo p of an integer a; the bit size w of the returned masks (with $2^w > 2p$)
Result: A Boolean masking $(a'_i)_{0\leq i\leq d}$ of a
1 $(a'_i)_{0\leq i\leq d} \leftarrow (0,\ldots,0)$
2 **for** $j = 0$ *to* d **do**
3 | $(b_i)_{0\leq i\leq d} \leftarrow (a_j, 0, \ldots, 0)$
4 | $(b_i)_{0\leq i\leq d} \leftarrow \mathsf{Refresh}((b_i)_{0\leq i\leq d}, w)$
5 | $(a'_i)_{0\leq i\leq d} \leftarrow \mathsf{SecAddModp}((a'_i)_{0\leq i\leq d}, (b_i)_{0\leq i\leq d}, w)$
6 **end**
7 **return** $(a'_i)_{0\leq i\leq d}$

Algorithm 13. Refresh-and-unmask algorithm for Boolean masking (FullXor) from [13]

Data: A Boolean masking $(x_i)_{0\leq i\leq d}$ of some value x; the bit size w of the masks
Result: The value x
1 $(x'_i)_{0\leq i\leq d} \leftarrow \mathsf{FullRefresh}((x_i)_{0\leq i\leq d}, w)$
2 $x \leftarrow x'_0$ **for** $i = 1$ *to* d **do**
3 | $x \leftarrow x \oplus x'_i$
4 **end**
5 **return** x

which is the counterpart of [13, Sect. 4.1], is to simply construct a Boolean masking of each of the shares a_i, and to iteratively apply SecAddModp to those masked values. This is simple and secure, but as noted by Coron et al., this approach has cubic complexity in the masking order d (because SecAdd and hence SecAddModp are quadratic). A more advanced, recursive approach allows to obtain quadratic complexity for the whole conversion: this is described in [13, Sect. 4.2], and directly applies to our setting.

With both algorithms SecAddModp and SecArithBoolModp in hand, we can easily complete the description of our commitment generation algorithm by mimicking [13, Algorithm 6]. To convert the Boolean masking $(x_i)_{0\leq i\leq d}$ of x to a mod-p arithmetic masking, we first generate random integer shares $a_i \in [0, p)$, $1 \leq i \leq d$, uniformly at random. We then define $a'_i = -a_i \bmod p = p - a_i$ for $1 \leq i \leq d$ and $a'_0 = 0$. The tuple $(a'_i)_{0\leq i\leq d}$ is thus a mod-p arithmetic masking of the sum $a' = -\sum_{1\leq i\leq d} a_i \bmod p$. Using SecArithBoolModp, we convert this arithmetic masking to a Boolean masking $(y_i)_{0\leq i\leq d}$, so that $\bigoplus_{i=0}^{d} y_i = a'$. Now, let $(z_i)_{0\leq i\leq d} = \mathsf{SecAddModp}((x_i)_{0\leq i\leq d}, (y_i)_{0\leq i\leq d})$; this is a Boolean masking of:

$$z = (x + a') \bmod p = \left(x - \sum_{i=1}^{d} a_i\right) \bmod p.$$

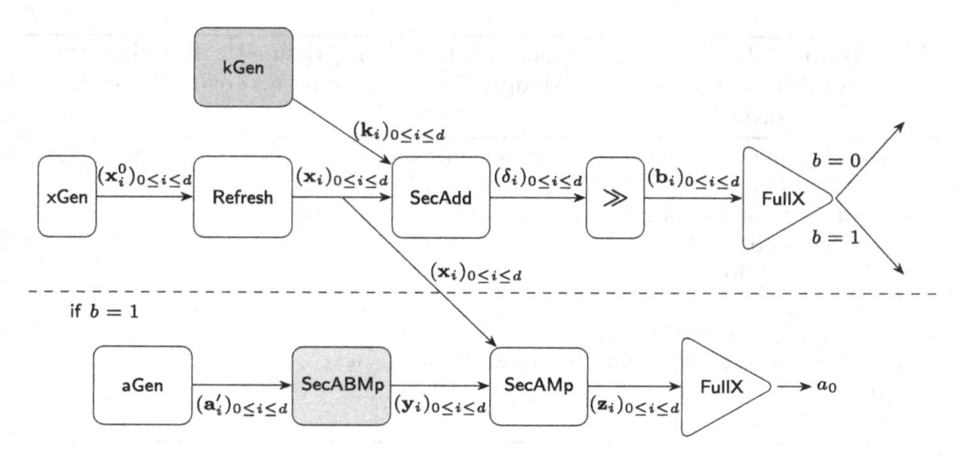

Fig. 3. Randomness Generation RG (The green (resp. white, red) gadgets will be proved d-SNI (resp. d-NI, unmasked)) (Color figure online)

Algorithm 14. Randomness generation (RG)

Data: k and d

Result: A uniformly random a integer in $[-k, k]$ in mod-p arithmetic masked form $(a_i)_{0\le i\le d}$.

1 generate uniformly random w_0-bit values $(x_i^0)_{0\le i\le d}$
2 $(x_i)_{0\le i\le d} \leftarrow \mathsf{Refresh}((x_i^0)_{0\le i\le d})$
3 initialize $(k_i)_{0\le i\le d}$ to a w-bit Boolean sharing of the two's complement value $-K = -2k - 1$
4 $(\delta_i)_{0\le i\le d} \leftarrow \mathsf{SecAdd}((x_i)_{0\le i\le d}, (k_i)_{0\le i\le d})$
5 $(b_i)_{0\le i\le d} \leftarrow (\delta_i)_{0\le i\le d} \gg (w-1)$
6 $b \leftarrow \mathsf{FullXor}((b_i)_{0\le i\le d})$
7 output b
8 **if** $b = 0$ **then**
9 | **restart**
10 **end**
11 generate uniform integers $(a_i)_{1\le i\le d}$ in $[0, p)$
12 $a_i' \leftarrow -a_i \bmod p$ for $i = 1, \ldots, d$
13 $a_0' \leftarrow 0$
14 $(y_i)_{0\le i\le d} \leftarrow \mathsf{SecArithBoolModp}((a_i')_{0\le i\le d})$
15 $(z_i)_{0\le i\le d} \leftarrow \mathsf{SecAddModp}((x_i)_{0\le i\le d}, (y_i)_{0\le i\le d})$
16 $a_0 \leftarrow \mathsf{FullXor}((z_i)_{0\le i\le d})$
17 **return** $(a_i)_{0\le i\le d}$

We then securely unmask this value using Coron et al. FullXor procedure, recalled in Algorithm 13, and set $a_0 = z - k \bmod p$. Then, we have:

$$\sum_{i=0}^{d} a_i \bmod p = z - k + \sum_{i=1}^{d} a_i \bmod p = x - k - \sum_{i=1}^{d} a_i + \sum_{i=1}^{d} a_i \bmod p = x - k \bmod p.$$

Thus, $(a_i)_{0 \le i \le d}$ is a correct mod-p arithmetic masking of a uniformly random value in $[-k, k]$ as required. The whole procedure is summarized in Algorithm 14 and described in Fig. 3 where xGen stands for the generation of x^0's shares, Refresh for the multiplication-based refreshing from [4,22], kGen for the generation of k's shares, $\gg$ for the right shift of δ's shares, FullX for FullXor, aGen for the generation of a's shares, SecABM for SecArithBoolModp, and SecAMp for SecAddModp.

The success probability of the rejection sampling step (the masked comparison to K) is $K/2^{w_0}$, and hence is at least 1/2 by definition of w_0. Therefore, the expected number of runs required to complete is at most 2 (and in fact, a judicious choice of k, such as one less than a power of two, can make the success probability very close to 1). Since all the algorithms we rely on are at most in the masking order and (when using the masked Kogge–Stone adder of [12]) logarithmic in the size w of the Boolean shares, the overall complexity is thus $O(d^2 \log w)$.

Now that the randomness generation is decribed, each intermediate gadget will be proven either d-NI or d-NIo secure. Then, the global composition is proven d-NIo secure as well.

Lemma 2. *Gadget SecAdd is d-NI secure.*

Proof. Gadget SecAdd is built from the Kogge-Stone adder of [12] with secure AND and secure linear functions such as exponentiations and Boolean additions. As to ensure its security with the combination of these atomic masked functions, the tool maskComp was used to properly insert the mandatory d-SNI refreshings, denoted as Refresh in Algorithm 9. As deeply explained in its original paper, maskComp provides a formally proven d-NI secure implementation. □

Lemma 3. *Gadget SecAddModp is d-NI secure.*

Proof. Gadget SecAddModp is built from the gadget SecAdd and SecAnd and linear operations (like $\oplus$). We use the tool maskComp to generate automatically a verified implementation. Note that the tool automatically adds the two refreshs (line 5 and 7) and provides a formally proven d-NI secure implementation. □

Lemma 4. *Gadget SecArithBoolModp is d-SNI secure.*

Proof. A graphical representation of SecArithBoolModp is in Fig. 4. Let O be a set of observations performed by the attacker on the final returned value, let I_{A_j} be the set of internal observations made in step j in the gadget SecAddModp (line 5), and I_{R_j} be the set of internal observations made in the step j in the initialisation of b (line 3) or in the Refresh (line 4). Assuming that $|O| + \sum(|I_{A_j}| + |I_{R_j}|) \le d$, the gadget is d-SNI secure, if we can build a simulator allowing to simulate all the internal and output observations made by the attacker using a set S of shares of a such that $|S| \le \sum(|I_{A_j}| + |I_{R_j}|)$.

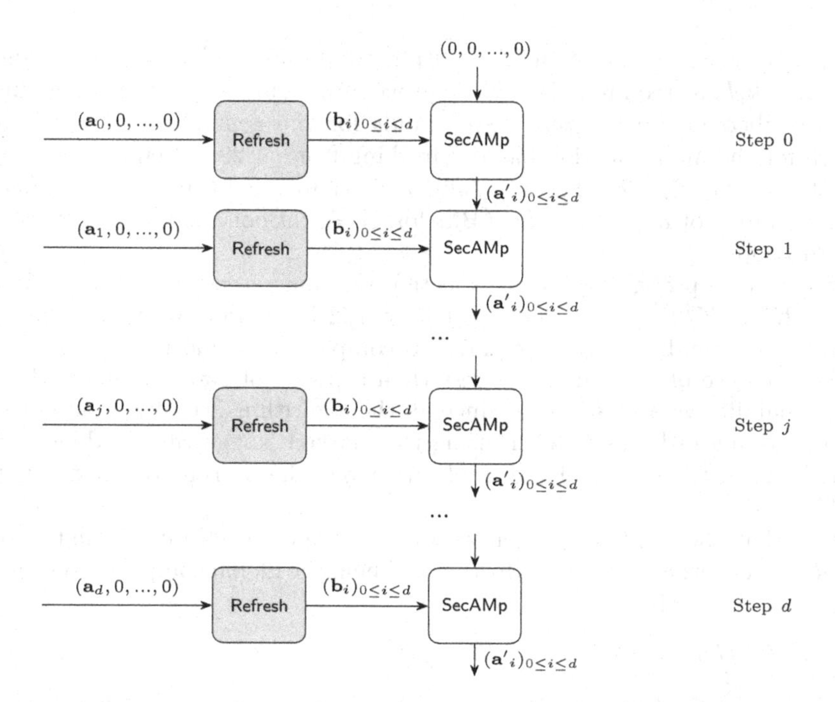

Fig. 4. Graphical Representation of SecArithBoolModp (The green (resp. white) gadgets will be proved d-SNI (resp. d-NI)) (Color figure online)

At the last iteration (see Fig. 5), the set of observations $O \cup I_{A_d}$ can be simulated using a set $S_{a'_{d-1}}$ of shares of a' and $S_{b_{d-1}}$ of shares of b with $|S_{a'_{d-1}}| \leq |O| + |I_{A_d}|$ and $|S_{b_{d-1}}| \leq |O| + |I_{A_d}|$ (because the gadget SecAddModp is d-NI secure). Since the Refresh is d-SNI secure, the sets $S_{b_{d-1}}$ and I_{R_d} can be simulated using a set $S_{b'_{d-1}}$ of input share with $|S_{b'_{d-1}}| \leq |I_{R_d}|$. If I_{R_d} is not empty, then $S_{b'_{d-1}}$ may contain a_d, so we add a_d to S. For each iteration of the loop this process can be repeated. At the very first iteration, several shares of a' may be necessary to simulate the set of observations. However, there are all initialized to 0, nothing is added to S.

At the end we can conclude that the full algorithm can be simulated using the set S of input shares. Furthermore we have $|S| \leq \sum |I_{R_j}|$ (since a_j is added

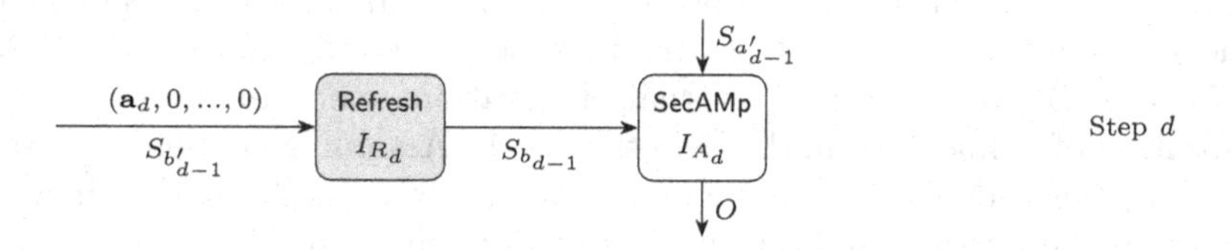

Fig. 5. Last step of SecArithBoolModp with probes (The green (resp. white) gadgets will be proved d-SNI (resp. d-NI)) (Color figure online)

in S only if I_{R_j} is not empty), so we can conclude that $|S| \leq \sum |I_{A_j}| + |I_{R_j}|$ which concludes the proof. □

Lemma 5. *Gadget RG is d-NIo secure with public output b.*

Proof. Here we need to ensure that the returned shares of a cannot be revealed to the attacker through a d-order side-channel attack. Since xGen and aGen are just random generation of shares, the idea is to prove that any set of $t \leq d$ observations on RG including these inputs can be perfectly simulated with at most t shares of x and t shares of a.

Gadget RG is built with no cycle. In this case, from the composition results of [4], it is enough to prove that each sub-gadget is d-NI to achieve global security. In our case, it is enough to prove that each sub-gadget is d-NIo with the knowledge of b to achieve global security.

From Lemmas 2, 4, and 3, SecAdd, SecArithBoolModp, and SecAddModp are d-NI secure. $\gg$ is trivially d-NI secure as well since it applies a linear function, gadget FullRefresh is d-SNI secure thus d-NI secure by definition, and gadget kGen is generating shares of a non sensitive value.

At this point, both gadgets FullXor have to be analyzed to achieve the expected overall security. We start with the gadget computing b. After its execution, b is broadcasted. Since b have to be public, its knowledge does not impact the security but because of this output, the security of RG will be d-NIo with public output b and not d-NI. FullXor is composed of a d-SNI secure refreshing (made of $d+1$ linear refreshing) of the shares and of a Boolean addition of these resulting shares. The attacker is not able to observe intermediate variable of all the linear refreshings (since he only has $\delta \leq d$ available observations), thus we consider that the ith refreshing is left unobserved. As a consequence, all the previous observations involve only one b's share and all the following observations are independent from b's share except for their sum. That is, FullXor is d-NI secure. As for its second instance to compute a_0, FullXor is still d-NI secure but a_0 is not revealed after its execution. While the attacker is able to observe its value, it is not returned for free. All the $\delta_0 \leq d$ observations made by the attacker of this last instance of FullXor can be perfectly simulated with a_0 (for the observations performed after the unobserved linear refreshing) and at most $\delta_0 - 1$ shares of z (for the observations made before the unobserved linear refreshing). □

Remark 3. The knowledge of b (ie. the success of the randomness generation) is not sensitive and we decided to consider it as a public output. To simplify the notation when we report the security on the whole scheme, we will omit b in the public outputs.

Lemma 6. *Gadget DG is d-NIo secure with public output b.*

Proof. From Lemma 5, Gadget DG is d-NIo secure since it only consists in the linear application of Gadget RG to build the polynomial coefficients. □

Algorithm 15. Rejection sampling (RS)

Data: The $4n$ values $a^{(j)}$ to check, in mod-p arithmetic masked representation $(a_i^{(j)})_{0\leq i\leq d}$.

Result: The bit r equal to 1 if all values satisfy that $a^{(j)} + k - \alpha \geq 0$, and 0 otherwise.

```
1  initialize (r_i)_{0≤i≤d} as a single-bit Boolean masking of 1
2  initialize (p_i)_{0≤i≤d} as a w-bit Boolean masking of −p
3  initialize (p'_i)_{0≤i≤d} as a w-bit Boolean masking of −(p+1)/2
4  initialize (k'_i)_{0≤i≤d} as a w-bit Boolean masking of k − α
5  for j = 1 to 4n do
6  |  (a'_i)_{0≤i≤d} ← SecArithBoolModp((a_i^{(j)})_{0≤i≤d})
7  |  (δ_i)_{0≤i≤d} ← SecAdd((a'_i)_{0≤i≤d}, (p'_i)_{0≤i≤d})
8  |  (b_i)_{0≤i≤d} ← (δ_i)_{0≤i≤d} ≫ (w − 1)
9  |  (s_i)_{0≤i≤d} ← SecAdd((a'_i)_{0≤i≤d}, (p_i)_{0≤i≤d})
10 |  (c_i)_{0≤i≤d} ← Refresh((b_i)_{0≤i≤d})
11 |  (a'_i)_{0≤i≤d} ← SecAnd((a'_i)_{0≤i≤d}, (c̃_i)_{0≤i≤d})
12 |  (c_i)_{0≤i≤d} ← Refresh((b_i)_{0≤i≤d})
13 |  (a'_i)_{0≤i≤d} ← (a'_i)_{0≤i≤d} ⊕ SecAnd((s_i)_{0≤i≤d}, ¬(c̃_i)_{0≤i≤d})
14 |  (δ_i)_{0≤i≤d} ← SecAdd((a'_i)_{0≤i≤d}, (k'_i)_{0≤i≤d})
15 |  (b_i)_{0≤i≤d} ← (δ_i)_{0≤i≤d} ≫ (w − 1)
16 |  (r_i)_{0≤i≤d} ← SecAnd((r_i)_{0≤i≤d}, ¬(b_i)_{0≤i≤d})
17 end
18 r ← FullXor((r_i)_{0≤i≤d})
19 return r
```

Rejection Sampling (RS). Right before the rejection sampling step of the masked signing algorithm, the candidate signature polynomials $\mathbf{z}_1$ and $\mathbf{z}_2$ have been obtained as sums of $d+1$ shares modulo p, and we want to check whether the coefficients in $\mathbb{Z}/p\mathbb{Z}$ represented by those shares are all in the interval $[-k+\alpha, k-\alpha]$. Again, carrying out this check using mod-p arithmetic masking seems difficult, so we again resort to Boolean masking.

For each coefficient $z_{i,j}$ of $\mathbf{z}_1$ and $\mathbf{z}_2$, one can trivially obtain a mod-p arithmetic masked representation of both $z_{i,j}$ and $-z_{i,j}$, and the goal is to check whether those values, when unmasked modulo p in the interval $[(-p+1)/2, (p-1)/2)]$, are all greater than $-k+\alpha$.

Let a be one of those values, and $a = a_0 + \cdots + a_d \bmod p$ its masked representation. Using SecArithBoolModp as above, we can convert this mod-p arithmetic masking to a w-bit Boolean masking $(a'_i)_{0\leq i\leq d}$. From this masking, we first want to obtain a masking of the centered representative of $a \bmod p$, i.e. the value a'' such that:

$$a'' = \begin{cases} a & \text{if } a \leq (p-1)/2, \\ a - q & \text{otherwise.} \end{cases}$$

This can be done using a similar approach as the one taken for randomness generation: compute a Boolean masking $(b_i)_{0\leq i\leq d}$ of the most significant bit $a-(p+1)/2$ (which is 1 in the first case and 0 in the second case), and a Boolean masking $(s_i)_{0 \le i \le d}$ of the sum $a - q$. Then, a Boolean masking of $(a''_i)_{0 \le i \le d}$ is obtained as:

$$(a''_i)_{0 \le i \le d} = \mathsf{SecAnd}\big((a'_i)_{0 \le i \le d}, (\widetilde{b_i})_{0 \le i \le d}\big) \oplus \mathsf{SecAnd}\big((s_i)_{0 \le i \le d}, \neg(\widetilde{b_i})_{0 \le i \le d}\big).$$

Finally, once this Boolean masking is obtained, it suffices to add $k - \omega$ to it and check the most significant bit to obtain the desired test.

We cannot directly unmask that final bit, but we can compute it in masked form for all the $4n$ values to be tested, and apply SecAnd iteratively on all of these values to compute a Boolean masked representation of the bit equal to 1 if all the coefficients are in the required intervals, and 0 otherwise. This bit can be safely unmasked, and is the output of our procedure. The whole algorithm is summarized as Algorithm 15.

Since both SecArithBoolModp and SecAnd have quadratic complexity in the masking order (and SecArithBoolModp has logarithmic complexity in the size w of the Boolean shares), the overall complexity of this algorithm is $O(nd^2 \log w)$.

Lemma 7. *Gadget RS is d-NI secure.*

Proof. From Lemmas 2 and 4, Gadgets SecArithBoolModp and SecAdd are d-NI secure. Gadget SecAnd is d-SNI secure from [4,30] and $\gg$ is linear, thus trivially d-NI secure as well.

As done for Gadget SecAdd, the tool maskComp was called to generate a d-NI circuit from the initial sequence of gadgets. It inserted gadgets Refresh (as shown in Algorithm 15) at specific locations so that the overall circuit is formally proven to be d-NI secure. □

Refresh and Unmask (FullAdd). This part provides a computation of the sensitive value as the sum of all its shares. It is a gadget with public output because the final value is returned and also output. This output is useful when FullAdd is used to recombine the intermediate value r.

Before summing, the sharing is given as input for FullRefresh [10, Algorithm 4], which is made of a succession of $d+1$ linear refresh operations. Those linear refreshing modify the sharing by adding randoms elements to each share while keeping constant the value of the sum. Their number is strictly superior to d which is useful to consider that any share or strictly partial sum of shares at the output of the final linear refreshing is independent from the original sharing. Then, the following partial sums do not give any information about the original sharing which is dependent of the sensitive values. The whole algorithm, given in Algorithm 16 has a quadratic complexity in d.

Lemma 8. *Gadget FullAdd is d-NIo secure with public output r.*

Proof. Let $\delta \le d$ be the number of observations made by the attacker. We use a combination of $d+1$ linear refresh operations. That is, there is at least one of the linear refreshing (we call it the ith refreshing) which is not observed by the attacker. For all the $\delta_1 \le \delta$ observations preceding the ith refreshing in FullAdd,

Algorithm 16. FullAdd

```
  Data: (r_i)_{0≤i≤d}
  Result: r
1 if (r_i)_{0≤i≤d} = ⊥ then
2 |  return ⊥
3 end
4 (r_i)_{0≤i≤d} ← FullRefresh ((r_i)_{0≤i≤d})
5 r ← (r_0 + ... + r_d)
6 output (r)
7 return (r)
```

Algorithm 17. H^1

```
  Data: a, (y_{1,i})_{0≤i≤d}
1 for 0 ≤ i ≤ d do
2 |  r_i ← a y_{1,i} + y_{2,i}
3 end
4 return (r_i)_{0≤i≤d}
```

Algorithm 18. H^2

```
  Data: RejSp, (z_{1,i})_{0≤i≤d}
1 if RejSp = 0 then
2 |  (z_{1,i})_{0≤i≤d} ← ⊥
3 end
4 return (z_{1,i})_{0≤i≤d}
```

they can be perfectly simulated with at most δ_1 shares of $\mathbf{r}$ since each one of them involves at most one $\mathbf{r}_i$. As for the observations performed after the ith refreshing, each one of them is independent from the $\mathbf{r}_i$ inside the refresh mask and each intermediate sum of the unmask part is independent of the $\mathbf{r}_i$ as well with the knowledge of $\mathbf{r}$. Then, during the sum computation, all the $\mathbf{r}_i$ can be simulated with fresh random that sum to $\mathbf{r}$ (the public output). Thus, at most δ shares of $\mathbf{r}$ and $\mathbf{r}$ itself are enough to simulate further probes. □

Remark 4. When FullAdd is used at the very end of the whole algorithm (mKD or mSign), the public outputs are also among the returned values. Then, in those cases, it can be considered as d-NI.

Transition Parts. The elementary parts H^1 and H^2 are quite easy to build since they perform only linear operations on the input data. A masked implementation only performs these linear operations on each share to securely compute the returned shares. H^1 and H^2 are described in Algorithms 17 and 18.

Lemma 9. *Gadget* H^2 *and* H^1 *are d-NI secure.*

The straightforward proof is given in the full version of this paper.

Hash Function. The hash function does not manipulate any sensitive data. Thus, it is left unmasked.

4.3 Proofs of Composition

Theorem 2. *The masked GLP sign in Algorithm 6 is d-NIo secure with public output* $\{r, b\}$.

Proof. From Lemmas 6, 7 and 9, Algorithms DG, RS, H^1, and H^2 are all d-NI. From Lemma 8, FullAdd is d-NIo secure.

Let us assume that an attacker has access to $\delta \leq d$ observations on the whole signature scheme. Then, we want to prove that all these δ observations can be perfectly simulated with at most δ shares of each secret among $\mathbf{y}_1$, $\mathbf{y}_2$, $\mathbf{s}_1$ and $\mathbf{s}_2$ and the public variables. With such a result, the signature scheme is then secure in the d-probing model since no set of at most d observations would give information on the secret values.

In the following, we consider the following distribution of the attacker's δ observations: δ_1 (resp. δ_2) on the instance of DG that produces shares of $\mathbf{y}_1$ (resp. $\boldsymbol{y}_2$), δ_3 on H^1, δ_4 on FullAdd of $\mathbf{r}$, δ_5 (resp. δ_6) on H^1 which produces $\mathbf{z}_1$ (resp. $\mathbf{z}_2$), δ_7 on the instance of RS, δ_8 (resp. δ_9) on H^2 applied on $\mathbf{z}_1$ (resp. $\mathbf{z}_2$), and δ_{10} (resp. δ_{11}) on FullAdd of $\mathbf{z}_1$ (resp. $\mathbf{z}_2$). Some other observations can be made on the *Hash* function, their number won't matter during the proof. Finally, we have $\sum_{i=1}^{11} \delta_i \leq \sum_{i=1}^{11} + \delta_{Hash} \leq \delta$.

Now, we build the proof from right to left as follows.

Both last FullAdd blocks in the very end of mSign are d-NI secure, then all the observations performed during the execution of FullAdd on $\mathbf{z}_1$ (resp. $\mathbf{z}_2$) can be perfectly simulated with at most δ_{10} (resp. δ_{11}) shares of $\mathbf{z}_1$ (resp. $\mathbf{z}_2$).

H^2 is d-NI secure, then all the observations from the call of H^2 on $\mathbf{z}_1$ (resp. $\mathbf{z}_2$) can be perfectly simulated with $\delta_8 + \delta_{10}$ (resp. $\delta_9 + \delta_{11}$) shares of the sensitive input $\mathbf{z}_1$ (resp. $\mathbf{z}_2$). The inputs $\mathbf{z}_1$ and $\mathbf{z}_2$ do not come from RS which do not act on them. They are directly taken from the returned values of H^1.

RS is d-NI secure and do not return any sensitive element, then all the observations performed in gadget RS can be perfectly simulated with at most δ_7 shares of $\mathbf{z}_1$ and $\mathbf{z}_2$. So, after H^1, the observations can be simulated with $\delta_7 + (\delta_8 + \delta_{10})$ shares of $\mathbf{z}_1$ and $\delta_7 + (\delta_9 + \delta_{11})$ shares of $\mathbf{z}_2$.

H^1 is d-NI secure as well, thus all the observations from the call of H^1 on $\mathbf{y}_1$ can be perfectly simulated with $\delta_5 + \delta_7 + \delta_8 + \delta_{10} \leq \delta$ shares of $\mathbf{y}_1$ and $\mathbf{s}_1$. Respectively, on $\mathbf{y}_2$, the observations can be perfectly simulated from $\delta_6 + \delta_7 + \delta_9 + \delta_{11} \leq \delta$ shares of $\mathbf{y}_2$ and $\mathbf{s}_2$.

The left FullAdd gadget is d-NIo secure and do not return any sensitive element, then all the observations performed in this gadget can be perfectly simulated with at most δ_4 shares of $\mathbf{r}$.

The left H^1 gadget is d-NI secure, thus all the observations from its call can be perfectly simulated with at most $\delta_3 + \delta_4$ shares of each one of the inputs $\mathbf{y}_1$ and $\mathbf{y}_2$.

DG is also d-NI secure, thus we need to ensure that the number of reported observations does not exceed δ. At the end of DG, the simulation relies on $(\delta_3+\delta_4)+(\delta_5+\delta_7+\delta_8+\delta_{10}) \leq \delta$ shares of $\mathbf{y}_1$ and $(\delta_3+\delta_4)+(\delta_6+\delta_7+\delta_9+\delta_{11}) \leq \delta$ shares of $\mathbf{y}_2$. With the additional δ_1 (resp. δ_2) observations performed on the first (resp. the second) instance of DG, the number of observations remains below δ which is sufficient to ensure security of the whole scheme in the d-probing model. □

Theorem 3. *The masked GLP key derivation in Algorithm 5 is d-NIo secure with public output b.*

Proof. From Lemmas 6 and 9, Algorithms DG, H^1 are all d-NI. From Lemma 8, FullAdd is d-NIo secure.

Here too, let us assume that an attacker has access to $\delta \leq d$ observations on the whole signature scheme. Then, we want to prove that all these δ observations can be perfectly simulated with at most δ shares of each secret among $\mathbf{s}_1$ and $\mathbf{s}_2$.

We now consider the following distribution of the attacker's δ observations: δ_1 (resp. δ_2) on the instance of DG that produces shares of $\mathbf{s}_1$ (resp. $\mathbf{s}_2$), δ_3 on H^1, and δ_4 on FullAdd, such that $\sum_{i=1}^{4} \delta_i = \delta$.

Now, we build the proof from right to left: FullAdd is used at the very end of mKD, so it is d-NI secure. Thus, all the observations from the call of FullAdd can be perfectly simulated with $\delta_4 \leq \delta$ sensitive shares of the input $\mathbf{t}$.

H^1 is d-NI, thus all the observations from its call can be perfectly simulated with at most $\delta_3 + \delta_4 \leq \delta$ shares of each one of the inputs $\mathbf{s}_1$ and $\mathbf{s}_2$.

DG is d-NIo, thus we need to ensure that the number of reported observations does not exceed δ. At the end of DG, the simulation relies on $(\delta_3 + \delta_4) \leq \delta$ shares of $\mathbf{s}_1$ and $\mathbf{s}_2$. With the additional δ_1 (resp. δ_2) observations performed on the first (resp. the second) instance of DG, the number of observations on each block remains below δ. All the observations can thus be perfectly simulated with the only knowledge of the outputs, that is, the key derivation algorithm is this d-NIo secure. □

5 Implementation of the Countermeasure

We have carried out a completely unoptimized implementation of our masking countermeasure based on a recent, public domain implementation of the GLP signature scheme called GLYPH [8,9]. The GLYPH scheme actually features a revised set of parameters supposedly achieving a greater level of security (namely, $n = 1024$, $p = 59393$, $k = 16383$ and $\alpha = 16$), as well as a modified technique for signature compression. We do not claim to vouch for those changes, but stress that, for our purposes, they are essentially irrelevant. Indeed, the overhead of our countermeasure only depends on the masking order d, the bit size w of Boolean masks (which should be chosen as $w = 32$ both for GLYPH and the original GLP parameters) and the degree n of the ring $\mathcal{R}$ (which is the same in GLYPH as in the high-security GLP parameters). Therefore, our results on GLYPH should carry over to a more straightforward implementation of GLP as well.

Implementation results on a single core of an Intel Core i7-3770 CPU are provided in Table 1. In particular, we see that the overhead of our countermeasure with 2, 3 and 4 shares (secure in the d-probing model for $d = 1, 2, 3$ respectively) is around 15×, 30× and 73×. In view of the complete lack of optimizations

Table 1. Implementation results. Timings are provided for 100 executions of the signing and verification algorithms, on one core of an Intel Core i7-3770 CPU-based desktop machine.

Number of shares $(d+1)$	Unprotected	2	3	4	5	6
Total CPU time (s)	0.540	8.15	16.4	39.5	62.1	111
Masking overhead	—	×15	×30	×73	×115	×206

of this implementation, we believe that those results are quite promising. The memory overhead is linear in the masking order, so quite reasonable in practice (all masked values are simply represented as a vector of shares).

For future work, we mention several ways in which our implementation could be sped up:

- For simplicity, we use a version of SecArithBoolModp with cubic complexity in the masking order, as in [13, Sect. 4.1]. Adapting the quadratic algorithm of [13, Sect. 4.2] should provide a significant speed-up. Moreover, for small values of d, Coron's most recent algorithm [11] should be considerably faster. However, the technique from [11] unfortunately has an overhead exponential in the masking order, so it is not suitable for our purpose of masking GLP at any order.
- Several of our algorithms call SecAdd on two masked values one of which is actually a public constant. One could use a faster SecAddConst procedure that only protect the secret operand instead.
- Our algorithms are generic, and do not take advantage of the special shape of k for example. In the case of GLYPH, a comparison to $k = 2^{14} - 1$ could be greatly simplified.
- One key way in which masking affects the efficiency of GLP signing is in the computation of the product $\boldsymbol{a} \cdot \boldsymbol{y}_1$. This product is normally carried out using a number-theoretic transform (NTT), with $O(n \log n)$ complexity. However, the NTT is not linear, and is thus inconvenient to use when $\boldsymbol{y}_1$ is masked. In our implementation, we use the schoolbook $O(n^2)$ polynomial multiplication instead. However, one could consider other approaches: either use a faster linear algorithm, like Karatsuba or Toom–Cook, or try and mask the NTT itself.
- Many other more technical improvements are also possible: for example, we have made no attempt to reduce the number of unnecessary array copies.

6 Conclusion

In this paper, we have described a provably secure masking of the GLP lattice-based signature scheme, as well as a proof-of-concept implementation thereof. The security proof itself involved a number of new techniques in the realm of masking countermeasures. Our method should apply almost identically to other lattice-based Fiat–Shamir type signature schemes using uniform distributions in intervals (as opposed to Gaussian distributions). This includes the Bai–Galbraith signature scheme [2], as well as the recently proposed Dilithium signature [16].

We have mostly ignored the issue of signature compression, which is an important one in all of these constructions, GLP included. However, it is easy to see that compression can be securely applied completely separately from our countermeasure: this is because it only affects already generated signatures (which are non-sensitive) as well as the input to the hash function (which is already unmasked in our technique).

On the other hand, extending our approach to schemes using Gaussian distributions appears to be really difficult: neither Boolean masking nor arithmetic masking with uniform masks seems particularly well-suited to address the problem. One way to tackle the problem might be to consider masking with non-uniform noise, and only achieving statistically close instead of perfect simulatability. Developing such a framework, however, is certainly a formidable challenge.

Masking hash-and-sign type signatures in using GPV lattice trapdoors is probably even harder, as they involve Gaussian sampling not only in $\mathbb{Z}$ but on arbitrary sublattices of $\mathbb{Z}^n$, with variable centers. It seems unlikely that a masked GPV signature scheme can achieve a reasonable level of efficiency.

Finally, while we have used the maskComp tool to securely instantiate the masked versions of some of the gadgets we use in our construction, it would be interesting to leverage recent advances in verification [3] and synthesis [4] of masked implementations in a more systematic way in the lattice-based setting. Even for verification, the sheer size of the algorithms involved poses significant challenges in terms of scalability; however, automated tool support would be invaluable for the further development of masking in the postquantum setting.

Acknowledgements. We are indebted to Vadim Lyubashevsky for fruitful discussions, and to the reviewers of EUROCRYPT for their useful comments. We acknowledge the support of the French Programme d'Investissement d'Avenir under national project RISQ. This work is also partially supported by the European Union PROMETHEUS project (Horizon 2020 Research and Innovation Program, grant 780701) and ONR Grant N000141512750.

References

1. Abdalla, M., An, J.H., Bellare, M., Namprempre, C.: From identification to signatures via the fiat-shamir transform: minimizing assumptions for security and forward-security. In: Knudsen, L.R. (ed.) EUROCRYPT 2002. LNCS, vol. 2332, pp. 418–433. Springer, Heidelberg (2002). https://doi.org/10.1007/3-540-46035-7_28
2. Bai, S., Galbraith, S.D.: An improved compression technique for signatures based on learning with errors. In: Benaloh, J. (ed.) CT-RSA 2014. LNCS, vol. 8366, pp. 28–47. Springer, Cham (2014). https://doi.org/10.1007/978-3-319-04852-9_2
3. Barthe, G., Belaïd, S., Dupressoir, F., Fouque, P.-A., Grégoire, B., Strub, P.-Y.: Verified proofs of higher-order masking. In: Oswald, E., Fischlin, M. (eds.) EUROCRYPT 2015. LNCS, vol. 9056, pp. 457–485. Springer, Heidelberg (2015). https://doi.org/10.1007/978-3-662-46800-5_18

4. Barthe, G., Belaïd, S., Dupressoir, F., Fouque, P.-A., Grégoire, B., Strub, P.-Y., Zucchini, R.: Strong non-interference and type-directed higher-order masking. In: Weippl, E.R., Katzenbeisser, S., Kruegel, C., Myers, A.C., Halevi, S. (eds.) ACM CCS 16, pp. 116–129. ACM Press, October 2016
5. Barthe, G., Belaïd, S., Espitau, T., Fouque, P.-A., Grégoire, B., Rossi, M., Tibouchi, M.: Masking the GLP lattice-based signature scheme at any order. Cryptology ePrint Archive (2018). http://eprint.iacr.org/. Full version of this paper
6. Bindel, N., Buchmann, J.A., Krämer, J.: Lattice-based signature schemes and their sensitivity to fault attacks. In: Maurine, P., Tunstall, M. (eds.) FDTC 2016, pp. 63–77. IEEE Computer Society (2016)
7. Chari, S., Rao, J.R., Rohatgi, P.: Template attacks. In: Kaliski Jr., B.S., Koç, K., Paar, C. (eds.) CHES 2002. LNCS, vol. 2523, pp. 13–28. Springer, Heidelberg (2003). https://doi.org/10.1007/3-540-36400-5_3
8. Chopra, A.: GLYPH: a new insantiation of the GLP digital signature scheme. Cryptology ePrint Archive, Report 2017/766 (2017). http://eprint.iacr.org/2017/766
9. Chopra, A.: Software implementation of GLYPH. GitHub repository (2017). https://github.com/quantumsafelattices/glyph
10. Coron, J.-S.: Higher order masking of look-up tables. In: Nguyen, P.Q., Oswald, E. (eds.) EUROCRYPT 2014. LNCS, vol. 8441, pp. 441–458. Springer, Heidelberg (2014). https://doi.org/10.1007/978-3-642-55220-5_25
11. Coron, J.-S.: High-order conversion from Boolean to arithmetic masking. Cryptology ePrint Archive, Report 2017/252 (2017). http://eprint.iacr.org/2017/252
12. Coron, J.-S., Großschädl, J., Tibouchi, M., Vadnala, P.K.: Conversion from arithmetic to Boolean masking with logarithmic complexity. In: Leander, G. (ed.) FSE 2015. LNCS, vol. 9054, pp. 130–149. Springer, Heidelberg (2015). https://doi.org/10.1007/978-3-662-48116-5_7
13. Coron, J.-S., Großschädl, J., Vadnala, P.K.: Secure conversion between Boolean and arithmetic masking of any order. In: Batina, L., Robshaw, M. (eds.) CHES 2014. LNCS, vol. 8731, pp. 188–205. Springer, Heidelberg (2014). https://doi.org/10.1007/978-3-662-44709-3_11
14. Duc, A., Dziembowski, S., Faust, S.: Unifying leakage models: from probing attacks to noisy leakage. In: Nguyen, P.Q., Oswald, E. (eds.) EUROCRYPT 2014. LNCS, vol. 8441, pp. 423–440. Springer, Heidelberg (2014). https://doi.org/10.1007/978-3-642-55220-5_24
15. Ducas, L., Durmus, A., Lepoint, T., Lyubashevsky, V.: Lattice signatures and bimodal Gaussians. In: Canetti, R., Garay, J.A. (eds.) CRYPTO 2013. LNCS, vol. 8042, pp. 40–56. Springer, Heidelberg (2013). https://doi.org/10.1007/978-3-642-40041-4_3
16. Ducas, L., Lepoint, T., Lyubashevsky, V., Schwabe, P., Seiler, G., Stehle, D.: CRYSTALS - dilithium: digital signatures from module lattices. Cryptology ePrint Archive, Report 2017/633 (2017). http://eprint.iacr.org/2017/633
17. Espitau, T., Fouque, P.-A., Gérard, B., Tibouchi, M.: Loop-abort faults on lattice-based Fiat-Shamir and hash-and-sign signatures. In: Avanzi, R., Heys, H. (eds.) SAC 2016. LNCS, vol. 10532, pp. 140–158. Springer, Cham (2017). https://doi.org/10.1007/978-3-319-69453-5_8
18. Espitau, T., Fouque, P.-A., Gérard, B., Tibouchi, M.: Side-channel attacks on BLISS lattice-based signatures: exploiting branch tracing against strongSwan and electromagnetic emanations in microcontrollers. In: Thuraisingham, B.M., Evans, D., Malkin, T., Xu, D. (eds.) ACM CCS 17, pp. 1857–1874. ACM Press, October/November 2017

19. Gentry, C., Peikert, C., Vaikuntanathan, V.: Trapdoors for hard lattices and new cryptographic constructions. In: Ladner, R.E., Dwork, C. (eds.) 40th ACM STOC, pp. 197–206. ACM Press, May 2008
20. Groot Bruinderink, L., Hülsing, A., Lange, T., Yarom, Y.: Flush, gauss, and reload – a cache attack on the BLISS lattice-based signature scheme. In: Gierlichs, B., Poschmann, A.Y. (eds.) CHES 2016. LNCS, vol. 9813, pp. 323–345. Springer, Heidelberg (2016). https://doi.org/10.1007/978-3-662-53140-2_16
21. Güneysu, T., Lyubashevsky, V., Pöppelmann, T.: Practical lattice-based cryptography: a signature scheme for embedded systems. In: Prouff, E., Schaumont, P. (eds.) CHES 2012. LNCS, vol. 7428, pp. 530–547. Springer, Heidelberg (2012). https://doi.org/10.1007/978-3-642-33027-8_31
22. Ishai, Y., Sahai, A., Wagner, D.: Private circuits: securing hardware against probing attacks. In: Boneh, D. (ed.) CRYPTO 2003. LNCS, vol. 2729, pp. 463–481. Springer, Heidelberg (2003). https://doi.org/10.1007/978-3-540-45146-4_27
23. Lyubashevsky, V.: Fiat-Shamir with aborts: applications to lattice and factoring-based signatures. In: Matsui, M. (ed.) ASIACRYPT 2009. LNCS, vol. 5912, pp. 598–616. Springer, Heidelberg (2009). https://doi.org/10.1007/978-3-642-10366-7_35
24. Lyubashevsky, V.: Lattice signatures without trapdoors. In: Pointcheval, D., Johansson, T. (eds.) EUROCRYPT 2012. LNCS, vol. 7237, pp. 738–755. Springer, Heidelberg (2012). https://doi.org/10.1007/978-3-642-29011-4_43
25. Oder, T., Schneider, T., Pöppelmann, T., Güneysu, T.: Practical CCA2-secure and masked ring-LWE implementation. Cryptology ePrint Archive, Report 2016/1109 (2016). http://eprint.iacr.org/2016/1109
26. Pessl, P., Bruinderink, L.G., Yarom, Y.: To BLISS-B or not to be: attacking strongSwan's implementation of post-quantum signatures. In: Thuraisingham, B.M., Evans, D., Malkin, T., Xu, D. (eds.) ACM CCS 17, pp. 1843–1855. ACM Press, October/November 2017
27. Pöppelmann, T., Ducas, L., Güneysu, T.: Enhanced lattice-based signatures on reconfigurable hardware. In: Batina, L., Robshaw, M. (eds.) CHES 2014. LNCS, vol. 8731, pp. 353–370. Springer, Heidelberg (2014). https://doi.org/10.1007/978-3-662-44709-3_20
28. Reparaz, O., de Clercq, R., Roy, S.S., Vercauteren, F., Verbauwhede, I.: Additively homomorphic ring-LWE masking. In: Takagi, T. (ed.) PQCrypto 2016. LNCS, vol. 9606, pp. 233–244. Springer, Cham (2016). https://doi.org/10.1007/978-3-319-29360-8_15
29. Reparaz, O., Sinha Roy, S., Vercauteren, F., Verbauwhede, I.: A masked ring-LWE implementation. In: Güneysu, T., Handschuh, H. (eds.) CHES 2015. LNCS, vol. 9293, pp. 683–702. Springer, Heidelberg (2015). https://doi.org/10.1007/978-3-662-48324-4_34
30. Rivain, M., Prouff, E.: Provably secure higher-order masking of AES. In: Mangard, S., Standaert, F.-X. (eds.) CHES 2010. LNCS, vol. 6225, pp. 413–427. Springer, Heidelberg (2010). https://doi.org/10.1007/978-3-642-15031-9_28

Masking Proofs Are Tight and How to Exploit it in Security Evaluations

Vincent Grosso[1(✉)] and François-Xavier Standaert[2]

[1] Digital Security Group, Radboud University Nijmegen,
Nijmegen, The Netherlands
v.grosso@cs.ru.nl

[2] ICTEAM - Crypto Group, Université catholique de Louvain,
Louvain-la-Neuve, Belgium

Abstract. Evaluating the security level of a leaking implementation against side-channel attacks is a challenging task. This is especially true when countermeasures such as masking are implemented since in this case: (*i*) the amount of measurements to perform a key recovery may become prohibitive for certification laboratories, and (*ii*) applying optimal (multivariate) attacks may be computationally intensive and technically challenging. In this paper, we show that by taking advantage of the tightness of masking security proofs, we can significantly simplify this evaluation task in a very general manner. More precisely, we show that the evaluation of a masked implementation can essentially be reduced to the one of an unprotected implementation. In addition, we show that despite optimal attacks against masking schemes are computationally intensive for large number of shares, heuristic (soft analytical side-channel) attacks can approach optimality efficiently. As part of this second contribution, we also improve over the recent multivariate (aka horizontal) side-channel attacks proposed at CHES 2016 by Battistello et al.

1 Introduction

Say you design a new block cipher and want to argue about its resistance against linear cryptanalysis [44]. One naive approach for this purpose would be to launch many experimental attacks. Yet, such a naive approach rapidly turns out to be unsuccessful if the goal is to argue about security levels beyond the computational power of the designer (e.g., 80-bit or 128-bit security for current standards). Hence, symmetric cryptographers have developed a variety of tools allowing them to bound the security of a block cipher against linear cryptanalysis, under sound and well-defined assumptions. As a typical example of these tools, one can cite the wide-trail strategy that has been used in the design of the AES Rijndael [17]. Its main idea is to minimize the bias (i.e., the informativeness) of the best linear characteristics within the cipher, which can be estimated under some independence assumptions thanks to the piling-up lemma.

Interestingly, the last years have shown a similar trend in the field of side-channel security evaluations. That is, while certification practices are still heavily dominated by "attack-based evaluations", solutions have emerged in order

J. B. Nielsen and V. Rijmen (Eds.): EUROCRYPT 2018, LNCS 10821, pp. 385–412, 2018.
https://doi.org/10.1007/978-3-319-78375-8_13

to both extend the guarantees and reduce the cost of these evaluations. More precisely, current certification practices focus either on the automatic verification of some minimum (non-quantitative) properties based on so-called leakage detection tools (e.g., [13,22,30,42,55]), or on the exhibition of concrete attack paths exploiting the detected leakages (typically taking advantages of standard distinguishers such as [9,11,28,54]). But they are anyway unable to claim security levels beyond the measurement efforts of the evaluation laboratory. In order to mitigate this limitation, one first intuitive line of papers proposed tools allowing to easily predict the success rate of some specialized distinguishers, based on parameters such as the noise level of the implementation [18,27,38,51]. In parallel, and following a more standard cryptographic approach trying to be independent of the adversarial strategy, significant progresses have been made in the mathematical treatment of physical security. In particular the masking countermeasure, which is one of the most common methods to improve the security of leaking cryptographic implementations, has been analyzed in several more or less formal models [10,19,34,50,58]. These works suggest that physical security via masking has strong analogies with the case of linear cryptanalysis. Namely, security against linear cryptanalysis is obtained by ensuring that the XOR of many (local) linear approximations has low bias. Similarly, masking ensures that every sensitive variable within an implementation is split (e.g., XORed) into several shares that the adversary has to recombine. So intuitively, masking security proofs can be viewed as a noisy version of the piling-up lemma.

Following these advances, the integration of masking proofs as a part of concrete security evaluation practices, undertaken in [20], appears as a necessary next step. And this is especially true when envisioning future cryptographic implementations with high (e.g., >80-bit) security levels, for which an attack-based certification process is unlikely to bring any meaningful conclusion. So the main objective of this paper is to follow such an approach and to show how masking security proofs can be used to gradually simplify side-channel security evaluations, at the cost of some conservative assumptions, but also some more critical ones (e.g., related to the independence of the shares' leakages).

More precisely, we start from the observation that a so far under-discussed issue in physical security evaluations is the case of attacks taking advantage of multiple leaking intermediate variables (e.g., see [33,43,59] for recent references).[1] As put forward in a recent CHES 2016 paper, this issue gains relevance in the context of masked implementations, in view of the (quadratic) cost overheads those implementations generally imply [7]. In this respect, our first contribution is to extend the analysis of masking security proofs from [20] and to show that these proofs remain essentially tight also for multi-target attacks.

Next, and since we aim to discuss the cost of side-channel security evaluations, we propose a simple metric for the evaluation complexity, and use it to extensively discuss the tradeoff between the time needed for a (worst-case) security evaluation and the risks related to the (e.g., independence) assumptions

[1] Which is an orthogonal concern to the more studied one of exploiting multiple leakage samples per intermediate variable (e.g., see [1] and follow up works).

it exploits. As our investigations suggest that the time complexity of optimal side-channel attacks can become a bottleneck when the security levels of masked implementations increase, we also study efficient (heuristic) multi-target attacks against masked implementations. Our best attack significantly improves the multivariate (aka horizontal) iterative attack proposed at CHES 2016 by Battistello et al., that we re-frame as a Soft Analytical Side-Channel Attack [33,59]. Note that our results provide a complementary view to those of Battistello et al., since they typically fix the masking noise parameter and look for the number of masking shares such that their attack is feasible, while we rather fix the number of shares and estimate the resulting security level in function of the noise.

Eventually, we show that the security evaluation of a leaking implementation against worst-case attacks taking advantage of all the target intermediate variables that can be enumerated by an adversary (so still limited to the first/last cipher rounds) boils down to the information theoretic analysis of a couple of its samples, for which good tools exist to guarantee a sound treatment [23,24]. By combining information theoretic evaluations with metric-based bounds for the complexity of key enumeration [48], we can obtain security graphs for optimal attacks, plotting the success rate in function of the measurement and time complexity, within seconds of computation on a computer. We argue that such tools become increasingly necessary for emerging high-security implementations.

2 Cautionary Remarks

Admittedly, the more efficient evaluations we discuss next are based on a number of simplifying assumptions. In this respect, we first recall that secure masking depends on two conditions: sufficient noise and independent leakages. This paper is about the first condition only. That is, we assume that the independence condition is fulfilled (to a sufficient extent), and study how exploiting all the leakage samples in an implementation allows reducing its noise.

We note that tools to ensure (or at least test empirically) the independence condition are already widely discussed in the literature. Concretely, there are two main issues that can break this assumption. First, imperfect refreshing schemes can cause d'-tuples of leakage samples to be key-dependent with d' lower than the number of shares used in the masking scheme d. For example, such an issue was put forward in [16]. It can be provably avoided by using "composable" (e.g., SNI [4]) gadgets or testing the security of the masking description code (i.e., the instructions defining an algorithm) thanks to formal methods [3].

Second, and more critically, different case studies have shown that actual leakage functions can break the independence assumption and recombine (a part of) the shares, e.g., because of transitions in software implementations [14] or glitches in hardware implementations [40]. Nevertheless, in practice such (partial) recombinations typically reduce the (statistical) "security order" of the implementations, captured by the lowest statistical moment of the leakage distribution that is key-dependent (minus one) [5], to some value d'' below the optimal $(d-1)$, while leaving security margins (i.e., $d'' > 1$). As a result, by increasing the number of shares d, one can generally mitigate these physical defaults to a good extent [2,46]. Furthermore, simple leakage detection tools such as [13,22,30,42,55] can be used to (empirically) assess the security order of an implementation, and these non-independence issues can be reflected in information theoretic evaluations (see [20], Sect. 4.2). So overall, ensuring the independence of the shares' leakages in a masked implementation is an orthogonal concern to ours. While non-independence issues may indeed increase the information leakage of the tuples of samples exploited in an high-order side-channel attack, it does not affect the importance/relevance of taking all the exploitable tuples into account in a (worst-case) security evaluation, which is our main concern.

Eventually, we insist that this work is prospective in the sense that our typical targets are masked implementations with (very) large number of shares, aimed at (very) high security levels (e.g., no key recovery with less than 2^{40} measurements). In this respect, we refer to two recently accepted papers (to Eurocrypt 2017) as an excellent motivation for our purposes [5,31]. In particular, [31] describes AES implementations masked with 5 to 10 shares, for which the security evaluation was left as an open problem by the authors and that are typical targets for which attack-based evaluations are unlikely to bring meaningful conclusions. Our following discussions describe theoretical tools allowing one to state sound security claims for such implementations. The important message they carry is that even when the independent shares' leakage assumption is guaranteed, one also needs to pay attention to noise. Simple univariate tests are not enough for this purpose. Performing highly multivariate attacks is (very) expensive. We introduce an intermediate path that allows principled reasoning and to assess the risks of overstated security based on well identified parameters. Quite naturally, this intermediate path also comes with limitations. Namely, since we focus on (very) high security levels, the bounds we provide are also less accurate, and reported as log-scaled plots for convenience (i.e., we typically ignore the impact of small constants as a first step). We will conclude the paper by referring to a recent CHES 2017 work that demonstrated to applicability of our tools based on a 32-share masked AES implementation in an ARM Cortex M4 [35].

3 Background

3.1 S-box Implementations

Our investigations consider both the unprotected and the masked implementation of an m-bit S-box S taking place in the first round of a block cipher.

For the unprotected case, we denote the input plaintext with x and the secret key with k. We define $y_a = x \oplus k$ as the result of a key addition between x and k, and $y_b = \mathsf{S}(y_a)$ as the S-box output. The vector of the target intermediate variables is further denoted with $\boldsymbol{y} = [y_a, y_b]$ and the leakage vector corresponding to these variables with $\boldsymbol{L} = [L_a, L_b] + \boldsymbol{N}$, where $\boldsymbol{N}$ is a bivariate random variable representing an additive Gaussian noise. We make the usual assumption that the noise covariance matrix is diagonal and each sample L_i has a similar noise

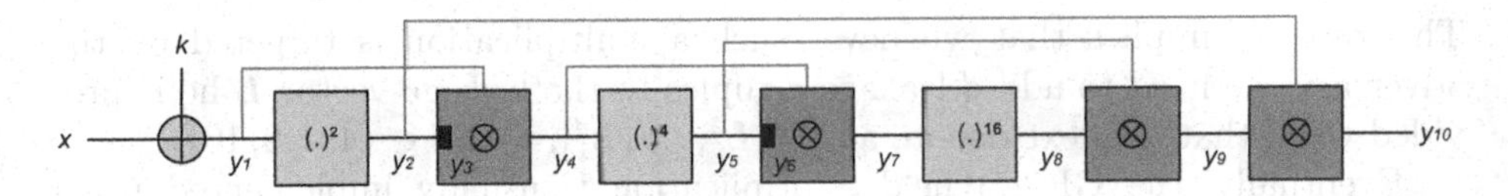

Fig. 1. Multiplication chain for the inversion in GF(2^8) from [52].

variance σ_n^2.[2] Eventually, the deterministic part of the leakage samples are the output of a leakage function L such that $L_i = \mathsf{L}_i(y_i)$, $\forall i \in \{a, b\}$. For simplicity, our experiments will consider L_i to be the Hamming weight function $\forall i$'s. We detail in Sect. 3.2 why this choice does not affect our conclusions.

For the masked case, we will focus on the secure inversion in GF(2^8) proposed in [52], which is the core of the AES S-box and illustrated in Fig. 1. More precisely, we use a slightly modified version of the algorithms of [52], with a secure refreshing R (e.g., from [4,34]) represented by black rectangles on the figure, in order to avoid the attack put forward in [16].[3] Next, we define the notations $y_1 = y_a = x \oplus k$, $y_2 = (y_1)^2$, $y_3 = \mathsf{R}(y_2)$, $y_4 = y_1 \otimes y_3 = (y_1)^3$, $y_5 = (y_4)^4 = (y_1)^{12}$, $y_6 = \mathsf{R}(y_5)$, $y_7 = y_4 \otimes y_6 = (y_1)^{15}$, $y_8 = (y_7)^{16} = (y_1)^{240}$, $y_9 = y_5 \otimes y_8 = (y_1)^{252}$, $y_{10} = y_2 \otimes y_9 = (y_1)^{254}$, with $\otimes$ the field multiplication. This leads to a vector of target intermediate variables $\boldsymbol{y} = [y_1, y_2, \ldots, y_{10}]$. For an implementation masked with d shares, we additionally have a vector of shares $\bar{\boldsymbol{y}} = [\bar{y}_1, \bar{y}_2, \ldots, \bar{y}_{10}]$ such that $\bar{y}_i = [y_i(1), y_i(2), \ldots, y_i(d)]$ $\forall i \in \{1, 2, \ldots, 10\}$. It leads to a leakage vector $\bar{\boldsymbol{L}} = [\bar{L}_1, \bar{L}_2, \ldots, \bar{L}_{10}] + \boldsymbol{N}$, where each leakage d-tuple is denoted as $\bar{L}_i = [L_i(1), L_i(2), \ldots, L_i(d)]$ and made of d samples, the multivariate noise variable is defined as in the unprotected case (with more dimensions) and $L_i(j) = \mathsf{L}_{i,j}(y_i(j))$ $\forall i \in \{1, 2, \ldots, 10\}, j \in \{1, 2, \ldots, d\}$. Such a masking scheme has security order $(d-1)$, meaning that any $(d-1)$-tuple of leakage samples is independent of k, *given that each leakage sample depends on a single share.* We call this assumption the Independent Shares' Leakage (ISL) assumption.

Concretely, the multiplication chain of Fig. 1 is made of squarings, that are GF(2)-linear, and multiplications. In order to evaluate them securely, we use Algorithms 1 and 2 given in Appendix A. For the squarings, the operations are applied to each share independently and therefore can be tabulized. For the multiplications, the different shares need to interact and the algorithm has quadratic overheads that correspond to the computation of all the partial products and their refreshing. For example, for $x = x(1) \oplus \cdots \oplus x(d)$ and $y = y(1) \oplus \cdots \oplus y(d)$, producing the shares of $x \otimes y$ requires to compute (for $d = 3$):

$$\begin{pmatrix} x(1) \otimes y(1) & x(1) \otimes y(2) & x(1) \otimes y(3) \\ x(2) \otimes y(1) & x(2) \otimes y(2) & x(2) \otimes y(3) \\ x(3) \otimes y(1) & x(3) \otimes y(2) & x(3) \otimes y(3) \end{pmatrix} \oplus \begin{pmatrix} 0 & r_{1,2} & r_{1,3} \\ -r_{1,2} & 0 & r_{2,3} \\ -r_{1,3} & -r_{2,3} & 0 \end{pmatrix}. \quad (1)$$

[2] The impact of this noise assumption is specifically discussed in Sect. 5.3.

[3] Note that more efficient solutions for this secure inversion exist, such as [32]. We kept the chain of Rivain and Prouff because for its simpler description.

This directly implies that whenever such a multiplication is targeted by the adversary, we need to add d leakage d-tuples to the leakage vector $\bar{\boldsymbol{L}}$ he is provided with, that we next denote as $[\bar{L}_i^1, \bar{L}_i^2, \ldots, \bar{L}_i^d]$, with $i \in \{4, 7, 9, 10\}$.

Eventually, the GF(2^8) field multiplication is usually implemented using log/alog tables, as described in Appendix A, Algorithm 3. In case the adversary additionally targets these operations, another set of d leakage d-tuples must be added to $\bar{\boldsymbol{L}}$, next denoted as $[\bar{L}_i^{d+1}, \bar{L}_i^{d+2}, \ldots, \bar{L}_i^{2d}]$, with $i \in \{4, 7, 9, 10\}$.

In the following, we will consider different (more or less powerful) attack cases:

C1. The adversary targets only a single d-tuple (e.g., the S-box output one).
C2. The adversary exploits the ten d-tuples of the multiplication chain.
C3. The adversary additionally exploits the leakage of the four secure multiplications (i.e., Algorithm 2), leading to a total of 10 d-tuples and 4 d^2-tuples.
C4. The adversary additionally exploits the leakage of the field multiplications (i.e., Algorithm 3), leading to a total of 10 d-tuples and 8 d^2-tuples.

Furthermore, since a number of these d-tuples contain fresh randomness (e.g., the ones corresponding to multiplications algorithms) while other ones are deterministically related to each other, we will denote with $\delta = \lambda + \ell$ the number of d-tuples exploited, such that we have λ fresh ones and ℓ deterministic ones.

Note that our notations describe serial implementations where the adversary can observe the noisy leakage of each share in his d-tuples separately. This is a relevant choice since serial implementations are typically very expensive to analyze due to their large number of dimensions/leakage samples to consider. Yet, as recently discussed in [5], side-channel security for a serial implementation generally implies side-channel security for its parallel counterpart (as long as the ISL assumption remains fulfilled). So our conclusions apply in this case too.

3.2 Mutual Information Metric

In order to evaluate the worst-case security level of our different (unprotected and masked) simulated implementations, we will use the mutual information metric first put forward in [57]. The motivation of this choice is twofold. First, it was shown recently that this metric can be linked to the measurement complexity of the corresponding (worst-case) Bayesian adversary [20]. Second, it is significantly faster to estimate than the success rate, which is specially important/relevant in our context where we aim to minimize the evaluator's workload. We illustrate this fact with a simple example. Say an evaluator has 1000,000 measurements to estimate the security of an implementation with a worst-case Bayesian attack that is roughly successful after the collection of 1000 traces. In this case, it means that he can repeat 1000 independent experiments to estimate the success rate with 1000 traces (with good confidence). But say now that the implementation to evaluate can only be broken after (roughly) 1000,000 traces. Then it means that from his set of traces, the evaluator can only estimate the success rate based on a single experiment (which will not lead to any statistical confidence). By contrast, as discussed in [24], cross-validation allows him to exploit most of his 1000,000 evaluation traces to estimate the mutual information metric, which will then be correlated with the success rate (for any number of traces).[4]

Concretely, computing the mutual information for an unprotected implementation simply requires to estimate the following sum of log probabilities:

$$\mathrm{MI}(K;X,\boldsymbol{L}) = \mathrm{H}[K] + \sum_{k\in\mathcal{K}} \Pr[k]\cdot \sum_{x\in\mathcal{X}} \Pr[x]\cdot \underbrace{\sum_{\boldsymbol{l}\in\mathcal{L}^{\delta}} \Pr[\boldsymbol{l}|k,x]\cdot \log_2 \Pr[k|x,\boldsymbol{l}]}_{\delta-\text{dimension integral}}\,, \quad (2)$$

where the conditional probability $\Pr[k|x,\boldsymbol{l}]$ is computed from the Probability Density Function (PDF) $\mathsf{f}[\boldsymbol{l}|x,k]$ thanks to Bayes' theorem as: $\frac{\mathsf{f}[\boldsymbol{l}|x,k]}{\sum_{k^*}\mathsf{f}[\boldsymbol{l}|x,k^*]}$. This corresponds to performing δ-dimensional integrals over the leakage samples, for each combination of the key k and plaintext x, or each bitwise XOR between k and x if taking advantage of the Equivalence under Independent Subkeys (EIS) assumption formalized in [54]. There are numerous publications where this metric has been computed, via numerical integrals or sampling (e.g., [20] provides an open source code for it), so we do not detail its derivation further.

When moving to masked implementations, the computation of the metric remains essentially similar. The only difference is that we need to sum over the randomness vector $\bar{\boldsymbol{y}}$ (which may become computationally intensive as the number of shares increases, as discussed in the next sections):

$$\begin{aligned}\mathrm{MI}(K;X,\bar{\boldsymbol{L}}) = \mathrm{H}[K] + \sum_{k\in\mathcal{K}} \Pr[k]\cdot \sum_{x\in\mathcal{X}} \Pr[x]\,\cdot \\ \sum_{\bar{\boldsymbol{y}}\in\mathcal{Y}^{(d-1)\cdot\lambda}} \Pr[\bar{\boldsymbol{y}}]\cdot \underbrace{\sum_{\bar{\boldsymbol{l}}\in\mathcal{L}^{d\cdot\delta}} \Pr[\bar{\boldsymbol{l}}|k,x,\bar{\boldsymbol{y}}]\cdot \log_2 \Pr[k|x,\bar{\boldsymbol{l}}]}_{\delta-\text{dimension integral}}\,. \quad (3)\end{aligned}$$

The computation of the conditional probability $\Pr[k|x,\boldsymbol{l}]$ follows similar guidelines as in the unprotected case, where the PDF of masked implementations becomes a mixture that can be written as $\mathsf{f}[\boldsymbol{l}|x,k] = \sum_{\bar{\boldsymbol{y}}} \mathsf{f}[\boldsymbol{l}|x,k,\bar{\boldsymbol{y}}]$ [36,58].

Remark. In our experiments where the (simulated) noise is Gaussian, we use a Gaussian PDF in the unprotected case, and a Gaussian mixture PDF in the masked case. Since we know the PDF exactly in these cases, we can compute the MI metric exactly and perform worst-case security evaluations. However, we insist that our discussions relate to the *complexity* of side-channel security evaluations, not their *optimality*. More precisely, our goal is to show that we can significantly simplify the evaluation of a highly protected implementation. These efficiency gains and our methodological conclusions are independent of the

[4] Note that the mutual information metric is not the only one allowing to simplify the estimation of a security level for a leaking cryptographic implementation. However, it is the most generic one since it does not require assumptions on the leakage distribution, nor on the choice of concrete distinguisher chosen by the adversary. More specialized (and sometimes more efficient) solutions include [18,27,38,51].

leakage function and model used by a concrete adversary (which however impacts the numerical results obtained). The main difference, if a concrete adversarial model was used in place of the perfect one, is that the log probabilities in Eqs. 2 and 3 would be evaluated based on it. This implies that less information would be extracted in case of model estimation or assumption errors, which is again an orthogonal concern to ours. Leakage certification could then be used to test whether estimation and assumption errors are small enough [23, 24].

4 Unprotected Implementations

Evaluation complexity. Since our goal is to make side-channel security evaluations more efficient, a first question is to specify how we will evaluate complexity. Eventually we are interested in the measurement complexity of the attacks, which masking is expected to increase exponentially (in the number of shares). But of course, we also want to be able to evaluate the security of implementations of which the security is beyond what we can actually measure as evaluators. As just mentioned, computing the mutual information metric is an interesting tool for this purpose. Yet, it means that we still have to compute Eqs. 2 and 3, which are essentially made of a sum of δ-dimension integrals. Concretely, the (time) complexity for computing such a sum is highly dependent on the choice of PDF estimation tool chosen by the adversary/evaluator. In our case where we focus on attacks based on the exhaustive estimation of a mixture model, the number of integrals to perform is a natural candidate for the complexity of a (worst-case) side-channel evaluation, which we will next denote with $\mathsf{E}_{\$}$.[5]

In the case of an unprotected S-box implementation in $\mathrm{GF}(2^m)$, this leads to $\mathsf{E}_{\$} = 2^{2m}$ in general (since we sum over 2^m key bytes and 2^m plaintext bytes). This complexity is reduced to $\mathsf{E}_{\$} = 2^m$ if we take advantage of the EIS assumption. Since the latter assumption is generally correct in the "standard DPA" attack context we consider in this paper [39], we will always consider the complexity of evaluations taking advantage of EIS in the following (ignoring this simplification implies an additional 2^m factor in the evaluation complexities).

Practical evaluation results. As suggested by the previous formula, evaluating the security of an unprotected (8-bit) S-box is cheap. We now report on some exemplary results which we use to introduce an important assumption regarding our following simplifications. We consider different attack cases:

[5] Note that the only message this metric supports is that the evaluation complexity of an optimal side-channel attack can be reduced from unrealistic to easy by exploiting various assumptions. The integral count provides an intuitive solution for this purpose, but other metrics could be considered equivalently. Note also that heuristic attacks may approach worst-case ones more efficiently (we address this issue in Sect. 5.4). So we mostly use this metric to motivate the need of new tools for evaluating and attacking masked cryptographic implementations: for evaluations, it justifies why shortcut approaches are useful; for attacks, it justifies why heuristic appraoches such as outlined in Sect. 5.4 become necessary for large d's.

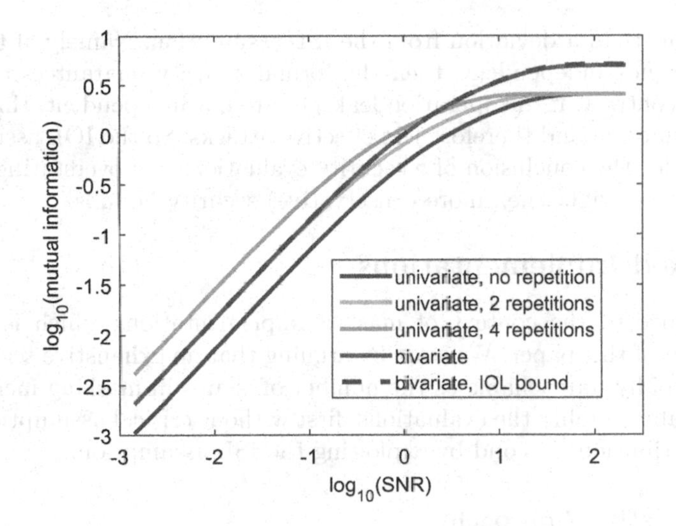

Fig. 2. Unprotected AES S-box evaluation results.

- Univariate, no repetition: the adversary observes the S-box output leakage.
- Univariate, with repetitions: the adversary observes the S-box output leakage several times with independent noise samples (e.g., 2 times, 4 times).
- Bivariate: the adversary observes the S-box input and output leakage.

Additionally, we consider a "bivariate attack bound" which is just the sum of two "univariate, no repetition" curves. In order to allow an easier interpretation of the results, we use the Signal-to-Noise Ratio (SNR) as X axis, defined as the variance of the noise-free traces (i.e., $m/4$ for Hamming weight leakages) divided by the variance of the noise. It better reflects the fact that the impact of the noise depends on the scaling of the signal. The results of these information evaluations are given in Fig. 2 from which two main conclusions can be extracted.

First, there is a difference between the impact of repeated observations, which just reduce the noise and therefore translate the information curves on the right, and bivariate attacks which (may) add information and shift these curves vertically. Interestingly, the latter observation is dependent on the S-boxes [49]: an identity S-box would lead to a repetition without information gain; a truly random one would lead to independent information for the two observations.

Second, the bivariate attack bound is tight in this case. This suggests that the AES S-box leads to quite independent information for the leakage samples L_a and L_b of our case study, which is consistent with the conclusions in [49]. Formally, we will say that this bound is tight if the Independent Operations' Leakages (IOL) assumption holds, which considers that the inputs/outputs of an operation (i.e., the AES S-box in our case study) are independent.

Note that as for the ISL assumption, the latter does not mean that the noise of the leakage samples has to be independent (which will be discussed in Sect. 5.2). Note also that the impact of a deviation from this IOL assumption is very different than a deviation from the ISL assumption. Namely, if the share's leakages are not independent, then the formal security guarantees of masking vanish. By contrast, if the operation leakages are not independent, this will lead to less information and therefore less effective attacks. So the IOL assumption is not critical for the conclusion of a security evaluation, and overstating IOL may only lead to less tight (i.e., more conservative) security bounds.

5 Masked Implementations

We now move to the context of masked implementations which is the main contribution of this paper. We start by arguing that an exhaustive security evaluation is rapidly unreachable as the number of shares in masking increases. We then gradually simplify the evaluations, first without critical assumptions on the leakage distributions, second by exploiting the ISL assumption.

5.1 Exhaustive Approach

By visual inspection of Eq. 3, we directly find that the evaluation complexity $\mathsf{E}_\$ = 2^{dm\lambda} + \ell \cdot 2^{dm}$, where we recall that λ is the number of fresh dimensions and ℓ the number of deterministic ones. For the case C1 in Sect. 3.1 with $d = 2$ shares, where the adversary targets only one 2-tuple of leakage samples corresponding to the masked S-box output y_8 in Fig. 1, this means a reachable 2^{2m} integrals. But as soon as we move to a (slightly) more powerful adversary, the complexity explodes. For example, the adversary of case C2 (who is still not optimal) with $m = 8$, $d = 2$, $\lambda = 6$ (due to the fresh intermediate values in Fig. 1) and $\ell = 4$ (due to the key addition and squarings), already leads to $\mathsf{E}_\$ = 2^{96}$ integrals which is by far too expensive for evaluation laboratories.

5.2 Reducing Dimensionality with the IOL Assumption

The first cause in the complexity explosion of the exhaustive approach is the number of fresh dimensions. In this respect, a natural simplification is to exploit the IOL assumption. Indeed, by considering the operations in the multiplication chain of Fig. 1 as independent, the evaluation complexity of the previous (C2) adversary can be reduced to $\mathsf{E}_\$ = \delta \cdot (2^{dm}) = 10 \cdot 2^{16}$ integrals. This is an interesting simplification since it corresponds to the strategy of an adversary willing to perform a multivariate attack against such a leaking masked implementation. Namely, he will identify the ten d-tuples of interest and combine their results via a maximum likelihood approach. We report the result of an information theoretic evaluation of this C2 adversary in Fig. 3, where we also plot the IOL bound provided by multiplying the information theoretic curve of the C1 adversary by ten. As for the case of unprotected implementations, the bound is tight.

Nevertheless, this simplification also implies two important technical questions. First, and since we assume the leakage of independent operations to be independent, what would be the impact of a dependent noise? Second, how to generalize this simplification to the adversaries C3 and C4 which imply the need of considering d^2-tuples jointly (rather than d-tuples jointly in the C2 case)?

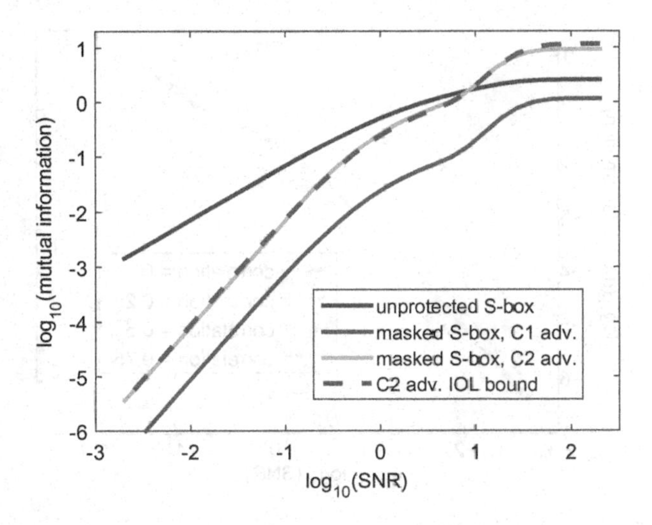

Fig. 3. Masked AES S-box evaluation results: cases C1 & C2 ($d = 2$).

5.3 The Dependent Noise Issue

To the best of our knowledge, this noise dependency issue has not been specifically discussed in the literature on masking, although the existence of correlated noise has been put forward in other contexts (e.g., see the discussion in [12], Chap. 6). We therefore launched an information theoretic evaluation of our masked S-box (case C1) with $d = 2$ and the covariance matrix such that the correlation between the noise samples of the two shares equals 0, 0.25, 0.5 and 0.75. The results of these evaluations are in Fig. 4. As expected, a correlated noise does not impact the security order of the countermeasure, defined as the lowest key-dependent moment in the leakage distribution $\Pr[k|x, \bar{l}]$ minus one, and reflected by the slope of the information theoretic curves in the high-noise region (i.e., where the curves are linear) minus one. By contrast, correlated noise implies a shift of the curves by a factor that can be significant (e.g., $\times 2$ for correlation 0.5 and $\times 8$ for correlation 0.75). Such large correlations typically vanish after a couple of clock cycles. Yet, our results highlight that estimating the non-diagonal elements of the noise covariance matrices in masked implementations is an important sanity check that could be part of a certification process.

5.4 Secure Multiplication Leakages

When also considering the leakages of the d^2 cross products involved in a secure multiplication (such as the ones of Eq. 1 in Sect. 3.1 for $d = 3$), an additional problem is that computing an integral of d^2 dimensions rapidly becomes computationally intensive. This is particularly true if one considers an optimal Gaussian mixture model for the PDF since in this case the computation of the integral requires summing over the randomness vector. In fact, already for small field

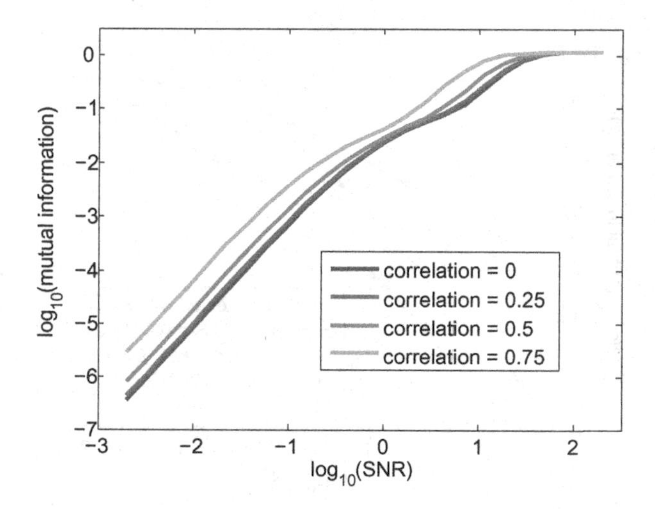

Fig. 4. Masked AES S-box evaluation results: case C1 with correlated noise ($d = 2$).

sizes and number of shares, the problem is hard. For example, for $d = 3$ and $m = 8$, the multiplication between two dependent values such as required in the multiplication chain of Fig. 1 requires performing 2^{24} integrals (corresponding to 8 bits of secret and twice 8 bits of randomness) of a 9-dimensional PDF.

In order to deal with this limitation, a solution is to look at masking proofs. In particular, Theorem 3 in [50] and Theorem 2 in [19] both provide bounds on the amount of information leaked by the multiplication of two secrets shared with Boolean masking, roughly corresponding to $(1.72d + 2.72)$ and $(28d + 16)$ times the information leakage of a single d-tuple. In this respect, there are again two important questions. First, are these bounds (and in particular the first one) tight? Second, given that the evaluation with an optimal attack becomes computationally intensive for large d values as just argued, does it mean that these bounds are unreachable by adversaries with realistic computing power?

We answer these questions in two steps. First, we investigate a simplified context with small d and m values such that the optimal attack is applicable. Second, we discuss heuristic attacks which approach the optimal attack efficiently.

Simplified case study. Figure 5 shows the information theoretic evaluation of a secure multiplication with $d = 3$ and $m = 2$.[6] We can clearly observe the larger leakage of optimal attack exploiting the $\delta = 9$ dimensions of the multiplication jointly, compared to the information provided by the encoding (i.e., the C1 adversary). As for the bounds, we first note that a simple (intuitive) bound is to assume that given two dependent values that are multiplied together, one

[6] Due to the large number of dimensions, the integrals were computed via sampling in this case, which also explains the lower noise variances that we could reach. However, we note that these lower noise levels were sufficient to reach the asymptotic (i.e., linear) regions of the information theoretic curves supporting our conclusions.

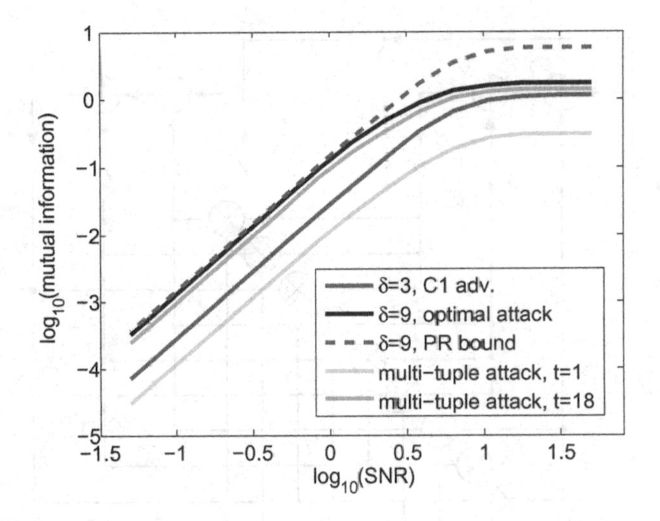

Fig. 5. Secure multiplication evaluation results ($d = 3$, $m = 2$).

leaks d horizontal d-tuples corresponding to one value (assuming the other to be known) and another d vertical d-tuples corresponding to the other value (under the same assumption). This leads to an estimation of the multiplication matrix leakage as $2d$ times the one of a single d-tuple, which is close to the $1.72d$ factor given by Prouff and Rivain in [50]. Hence, we added the latter bound on the figure (under the name PR bound). Concretely, it simply consists in multiplying the information of the encoding by $1.72d$ and turns out to be remarkably tight as soon as a sufficient amount of noise affects the measurements.[7]

Heuristic attacks. As the optimal attack in the previous paragraph becomes computationally intensive for large d and m values, we now consider alternatives that allow an adversary to exploit the information leakage of the multiplication matrix without summing over all the randomness and considering all the dimensions jointly. A first candidate is the recursive attack proposed by Battistello et al. at CHES 2016 [7]. In the following, we revisit and improve this attack by framing it as a Soft Analytical Side-Channel Attack (SASCA) [33,59].[8] In a SASCA, the adversary essentially describes all the leaking operations in his target implementation as a "factor graph" and then decodes the leakage information by exploiting the Belief Propagation (BP) algorithm. The main interest of this approach is that it allows combining the information of multiple leaking instructions (e.g., the cross products in a secure multiplication) locally, without the need to consider them jointly. Its time complexity depends on the diameter of the factor graph (which is constant when all target intermediate variables are directly connected as in the secure multiplication), the cost of the probabilities'

[7] Note that a parallel implementation would lead to a slightly better bound of $\approx d$ since reducing the amount of observable leakage samples by a factor d [5].

[8] Details about SASCA are provided in supplementary material for completeness.

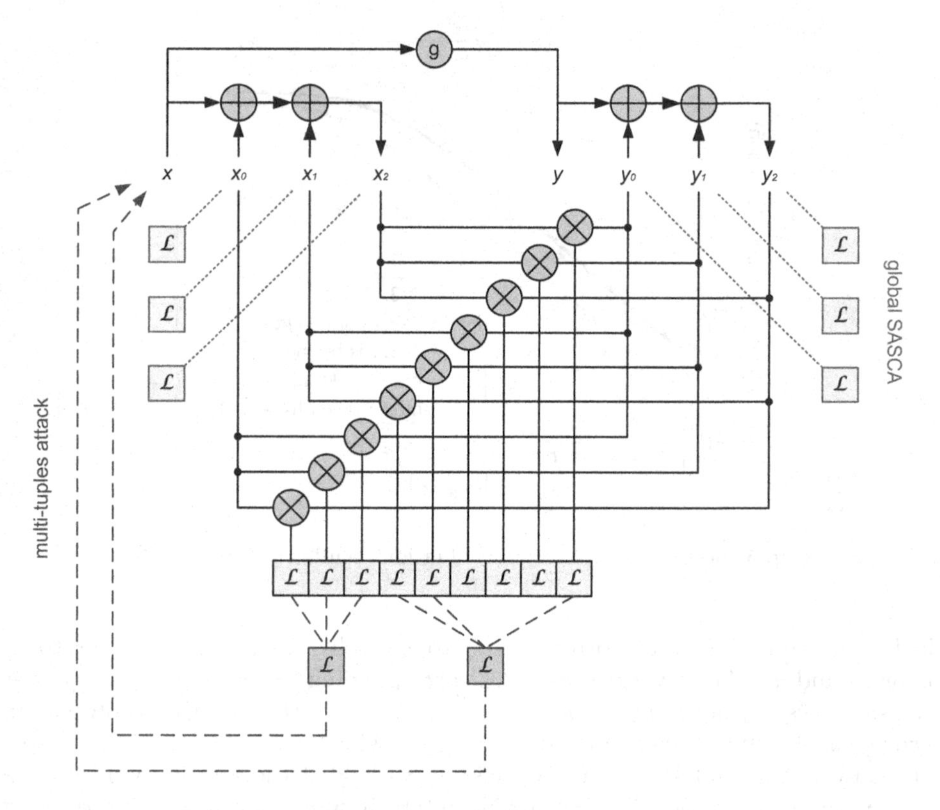

Fig. 6. Factor graph of a secure multiplication ($d = 3$).

updates (which is constant and depends on the bit size of the operations considered) and the number of these updates (which depends on the size of the factor graph and grows quadratically in d). The factor graph of a secure multiplication with $d = 3$ shares is pictured in Fig. 6. Its only specificity is that for the BP algorithm to succeed, we need to initialize the leakage on the shares x_0, x_1, x_2 and y_0, y_1, y_2, which means that a SASCA must consider the target operations more globally. In our experiments we added the leakage of these shares for this purpose, which can be obtained, e.g., when loading them into a register.

An alternative (and conceptually simple) approach allowing to get rid of the need of initialization is to always target d-tuples of informative leakage samples jointly. Such a "multi-tuple attack" can be viewed as an intermediate between the optimal attack targeting d^2 samples jointly and the previous SASCA targeting samples one by one, as illustrated in Fig. 6. More precisely, the optimal attack outlined in Sect. 3.2 exploits a leakage PDF $\Pr[\bar{\boldsymbol{l}}_{d^2}|k, x, \bar{\boldsymbol{y}}_{d^2}]$, where the d^2 subscripts of the vectors $\bar{\boldsymbol{l}}_{d^2}$ and $\bar{\boldsymbol{y}}_{d^2}$ now highlight their number of dimensions. In a multi-tuple attack, we simply select a number of d-tuples of which the combination depends on the target secret and approximate:

$$\Pr\left[\bar{\boldsymbol{l}}_{d^2}|k, x, \bar{\boldsymbol{y}}_{d^2}\right] \approx \Pr\left[\bar{\boldsymbol{l}}_d^1|k, x, \bar{\boldsymbol{y}}_d^1\right] \cdot \Pr\left[\bar{\boldsymbol{l}}_d^2|k, x, \bar{\boldsymbol{y}}_d^2\right] \cdot \ldots \cdot \Pr\left[\bar{\boldsymbol{l}}_d^t|k, x, \bar{\boldsymbol{y}}_d^t\right],$$

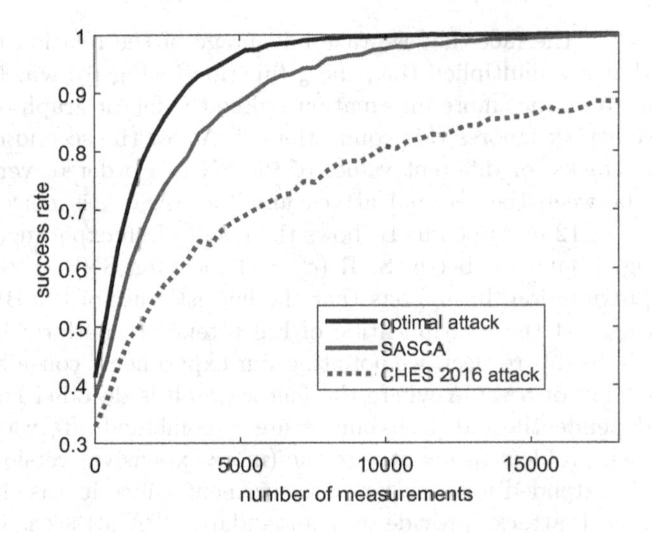

Fig. 7. Optimal attack vs. efficient heuristics ($d = 3$, $m = 2$, $\mathrm{SNR} = \frac{2}{10}$).

where t is the number of tuples exploited.[9] As illustrated in Fig. 5, an attack using a single d-tuple (e.g., here a matrix line) only leads to little exploitable information, which is consistent with the observations in [7]. By contrast, increasing t rapidly allows reaching a close-to-optimal attack.

Note that the multi-tuples attack still does not scale well since the total number of informative d-tuples in the matrix multiplications grows following a binomial rule. So the most appealing attacks to target the secure multiplication algorithm are the CHES 2016 iterative one and the SASCA. Unfortunately, in these cases we face the problem that the heuristic nature of the decoding algorithms (which both propagate information locally without formal guarantees of convergence) does not formally lead them to output probabilities. Typically, by iterating the CHES 2016 and BP algorithms more, it is possible to artificially crush the probabilities of the variable nodes in the factor graph. So formally, we cannot evaluate the mutual information metric in this case. As a result, and for this part of our experiments only, we directly evaluated the success rate of an optimal attack, a SASCA and the CHES 2016 iterative attack (using exactly the same leaking operations as the SASCA) for various noise levels.

For example, Fig. 7 contains the result of such an experiment (for $\sigma_n^2 = 10$ meaning $\mathrm{SNR} = \frac{2}{10}$) where we observe that (i) the SASCA outperforms the CHES 2016 iterative attack, and (ii) the SASCA leads to attack efficiencies that approach the optimal one. The first observation is easily explained since the CHES 2016 iterative attack can in fact be viewed as a modified version of SASCA. Namely, the main difference between the SASCA and the CHES 2016

[9] Note that whenever an imperfect model is used by the adversary/evaluator, the estimation of Eqs. 2 and 3 does not strictly converge towards the mutual information, but only to the so-called perceived information discussed in [24].

iterative attack is the fact that we take advantage of the relation between the two secrets that are multiplied (i.e., the g function in Fig. 6), which allows the BP algorithm to extract more information (while the factor graph of the CHES 2016 iterative attack ignores this connection).[10] As for the second observation, we launched attacks for different values of the SNR in order to verify whether the distance between the optimal attack and the SASCA is noise-dependent. For example, Fig. 12 in Appendix B shows the result of an experiment similar to the one of Fig. 7 but with better SNR ($\sigma_n^2 = 1$ meaning SNR $= 2$), leading to a tighter approximation. It suggests that the heuristic use of the BP algorithm in a SASCA against the multiplication of Fig. 6 tends to perform worse as the noise increases. In this respect, we note that our experiments consider the (more efficient) variation of SASCA where the factor graph is decoded for each measurement independently and probabilities are re-combined afterwards. It is an interesting open problem to investigate the (more expensive) version where the factor graph is extended for each new measurement (since it was shown in [59] that the gain such attacks provide over a standard DPA attack is independent of the noise in the context of an unprotected AES implementation).

Remark. As previously mentioned, extending these experiments to larger d and m values is not possible because the optimal attack becomes too expensive (computationally). By contrast, we could check that the success rate curves of the SASCA consistently outperform the ones of the CHES 2016 iterative attack by an approximate factor > 2 in measurement complexity, for larger m values. For example, we report the result of such a comparison for the relevant $m = 8$-bit case corresponding to the AES S-box in Appendix B, Fig. 13.

Overall, we conclude from this section that the IOL assumption and the PR bound for secure multiplications give rise to quite tight estimations of the information leakage of a masked implementation (at least for the leakage functions and noise levels considered experimentally). Furthermore, this leakage can also be exploited quite efficiently using heuristics such as the BP algorithm. We conjecture that these observations generally remain correct for most leakage functions, and when the number of shares in the masking schemes increases.

5.5 Reducing Cardinality with the ISL Assumption

Eventually, the previous experiments suggest that the evaluation of a masked implementation against multivariate attacks can boil down to the evaluation of the information leakage of a d-tuple. Yet, this still has evaluation cost proportional to 2^{dm}. Fortunately, at this stage we can use the ISL assumption and the bound discussed at Eurocrypt 2015 showing that this information can be (very efficiently) computed based on the information of a single share (essentially by raising this information to the security order), which has (now minimal) evaluation cost $\mathsf{E}_\$ = \delta \cdot 2^m$ (or even 2^m if one assumes that the leakage function

[10] Technically, the rules used for updating the probabilities in the CHES 2016 attack are also presented slightly differently than in SASCA, where the BP algorithm is explicitly invoked with variable to factors and factors to variable message passing.

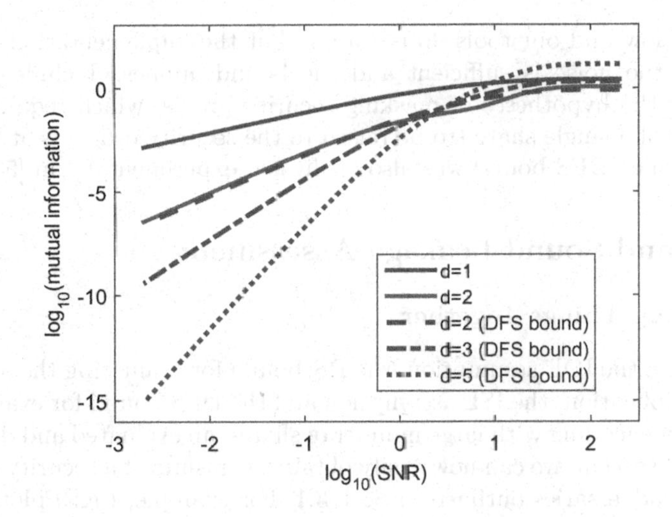

Fig. 8. Masked AES S-box evaluation results: case C1 with ISL assumption.

of the target implementation is similar for all operations, or if we bound the overall leakage based on the most informative d-tuple found) [20]. For completeness, we illustrate such a result in Fig. 8, where we compare the bound (denoted as DFS) and the true information leakage for $d = 2$, and only plot the bound for larger d's. As already mentioned, the big conceptual change at this step of our simplifications is that the ISL assumption is no longer a conservative one. If it turns out to be incorrect, then the security order of higher-order masking schemes may be less than predicted by the number of shares. Yet, as discussed in Sect. 2, this does not decrease the relevance of our method and bounds: it simply implies that applying them first requires to assess the security order of the target implementation (which we briefly discuss in Sect. 6.2).

Note also that as carefully discussed in [20] (Sect. 4.1, part c), the DFS bound is only conjectured and ignores a square root loss in the reduction from the mutual information to the statistical distance used in the proofs.[11] Yet, this square root loss vanishes when the noise increases, as per the upper bound in [50]. More precisely, this reference showed that the mutual information is (up to constants) lower than the statistical distance (not its square), and this inequality becomes an equality for low SNRs. In this respect, we recall that masking proofs are anyway only relevant for large enough noises (or low enough SNRs), which corresponds to the linear (left) parts of the information theoretic curves of Fig. 8 (i.e., where the DFS bound is tight). Intuitively, this is because masking can be viewed as a noise amplification mechanism. So without noise to amplify, the countermeasure does not provide any concrete benefits. In other words, either the

[11] Strictly speaking, it also ignores a small constant factor discussed in the optimal reduction given by [26], which is assumed to be a proof artifact and is at least not observed for the simple leakage functions considered in our experiments.

noise is too low and our tools do not apply but the implementation is insecure anyway, or the noise is sufficient and the bound applies. Technically, this is reflected by the hypotheses of masking security proofs, which require that the information of a single share (to be raised to the security order) is at least lower than one.[12] The DFS bound was also confirmed experimentally in [58].

6 Fast and Sound Leakage Assessment

6.1 Putting Things Together

By combining the IOL assumption, the PR bound for evaluating the leakage of a secure multiplication, the ISL assumption and the DFS bound for evaluating the leakage of an encoding with large number of shares, all evaluated and discussed in the previous section, we can now easily obtain the results of a security evaluation for the four adversaries outlined in Sect. 3.1. For example, Fig. 9 plots them for $d = 3, 5$ and 7 shares, for various noise levels. For readability, we only provide the results of the extreme attacks (C1 and C4). These curves are simply obtained by performing powers and sums of the information theoretic curve for the simplest possible case $d = 1$. In other words, we can evaluate the leakage of a masked implementation against optimal (highly multivariate) side-channel attacks at the cost of the evaluation of an unprotected implementation.

Note that the curves clearly highlight the need of a higher noise level when implementing higher-order masking schemes, in order to mitigate the noise reduction that is caused by the possibility to perform highly multivariate attacks (reflected by a shift of the curves towards the left of the figure). And quite naturally, they directly allow one to quantify the increasing impact of such attacks when the security order increases. For example, the factor between the measurement complexity of the adversary C1 (exploiting one tuple of leakage samples) and the optimal C4 ranges from 50 (for $d = 3$) to 100 (for $d = 7$).

In this respect, there is one final remark. In concrete implementations, it frequently happens that some of the target intermediate values appear several times (e.g., because they need to be reloaded for performing the cross products in a secure multiplication). In this case, the adversary can additionally average the noise for these target intermediate values, as proposed in [7]. As mentioned in Sect. 4, such an effect is also easy to integrate into our evaluations since it only corresponds to a shift of the information theoretic curves. However, it is worth emphasizing that this averaging process is applied to the shares (i.e., before their combination provides noise amplification), which implies that it is extremely damaging for the security of masking. Concretely, this means that averaging the leakage samples of a masked implementation with d shares by a factor d (because these shares are loaded d times to perform cross products) may lead to a reduction of the security level by a factor d^d. For illustration, Fig. 10

[12] Otherwise raising the information leakage of individual shares to some power may lead to larger values than the maximum m. For convenience, the following plots limit the mutual information to m when this happens (i.e., for too low noise levels).

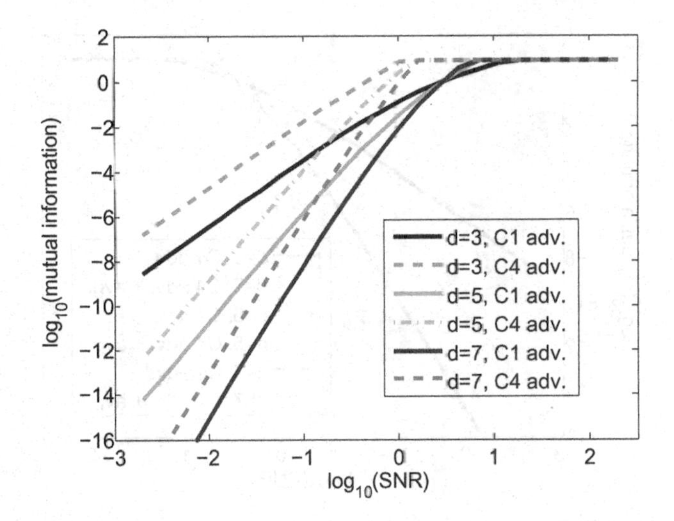

Fig. 9. Masked AES S-box evaluation results: cases C1 & C4 (with all assumptions).

shows the result of such a security evaluation in a context similar to Fig. 9, where the shares of each the masked multiplication are averaged d times, this times causing reductions of the security level by several orders of magnitude.

The difference between the multivariate attacks of Fig. 9 and the ones in Fig. 10 is easily explained by looking back at Fig. 8 where the leakage of a single d-tuple of shares is reported. First observe that in practice, any actual device has a fixed SNR: say 10^{-2} for illustration, leading to a mutual information of $\approx 10^{-13}$ for the $d = 5$ case. In case several (say ≈ 78) independent d tuples are combined in an attack (which is essentially what the C4 adversary of Fig. 9 achieves), the total amount of information available to the adversary is multiplied by a factor ≈ 78. This corresponds to a vertical shift in the information theoretic curves. Say now the adversary can average each of his leakage samples $d = 5$ times (which is essentially what the adversaries of Fig. 10 achieve). Then the SNR will be reduced 5 times, leading to an horizontal shift in the information theoretic curve, and reducing the averaged 5-tuples' information leakage by a factor $5^5 = 3125$, as per the curves' slope. This last observation suggests that each share in a masked implementation should be manipulated minimally.

Note that such an averaging is easy to perform in worst-case evaluations where the implementation is known to the adversary and the points in time where the shares are manipulated can be directly spot. But its exploitation with limited implementation knowledge may be more challenging. It is an interesting scope for further research to investigate whether some statistical tools can approach such worst-case attacks in this case (e.g., based on machine learning [37]).[13]

[13] Note also that traces averaging can be exploited constructively in the assessment of a security order. For example, in case the masks are known to the evaluator, he can average traces before evaluating the security order, leading to the efficiency gains [56].

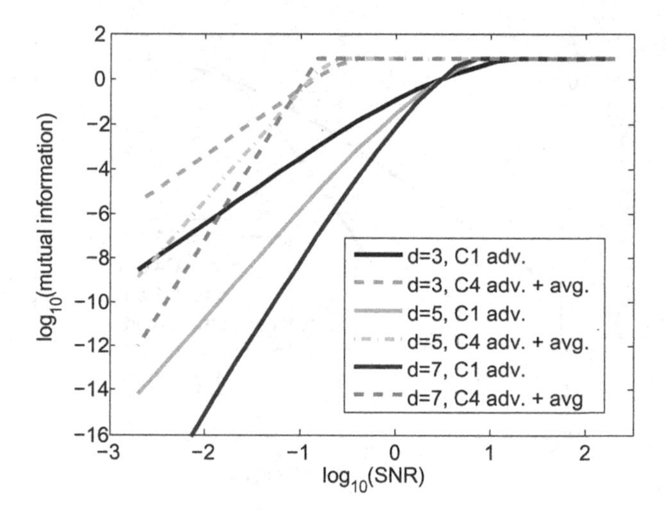

Fig. 10. Masked AES S-box evaluation results: cases C1 & C4 (with all assumptions & d-times averaging applied to the shares of the secure multiplications).

6.2 A Real-World Case Study

Our methodology has been recently applied (at CHES 2017) to a 32-bit shared implementation of the AES in an ARM Cortex M4 device [35]. As mentioned in introduction, such implementations are the typical targets for which our proof-based approach becomes necessary. This reference first assesses the security order by investigating reduced-shares versions of the algorithms (which can be viewed as an analogy to the reduced-rounds versions in block cipher cryptanalysis). Since overestimated orders are the main risk of overstated security (because they reflect the ISL assumption), the paper additionally considers a risk factor (i.e., evaluates the security for the best detected order and for its reduction by a conservative factor two). "Single tuple", "all tuples" and "all tuples + averaging" attacks are then analyzed, based on the number of leaking operations in the implementation code, and the number of shares' repetitions. It allows claiming so far unreported security levels, under well identified (and empirically falsifiable) assumptions.

6.3 Exploiting Computational Power

Eventually, and given some mutual information value extracted from the previous plots, we mention that one can easily insert this value in a metric-based bound in order to build a security graph, such as suggested in [21] and illustrated in Fig. 11. While such metric-based bounds only provide a conservative estimation of the impact of key enumeration in a side-channel attack [41,48], they are obtained withing seconds of computation on a desktop computer. We detail how to build such a graph and the heuristics we rely on in Appendix C.

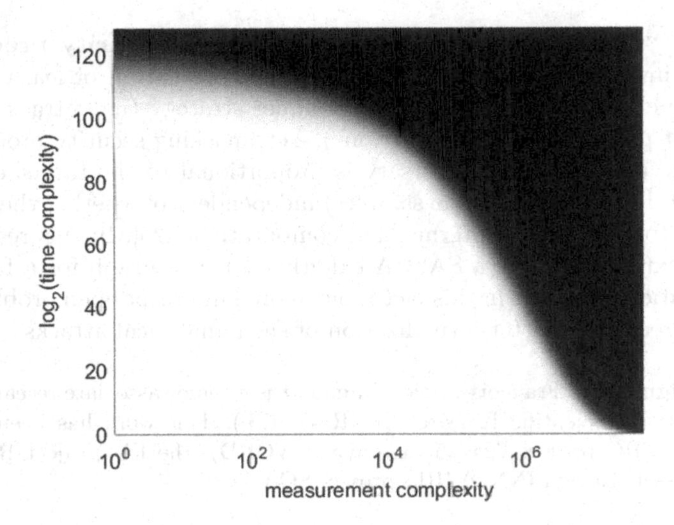

Fig. 11. Exemplary metric-based bound for a security graph (with $\mathrm{MI} = 10^{-7}$).

6.4 Conclusions

1. On too simple evaluation methodologies. Looking at the power of multivariate (aka horizontal) side-channel attacks taking advantage of all the leaking operations in the multiplicative chain of a masked AES S-box, an important conclusion is that simple (univariate) evaluation strategies become increasingly irrelevant as the number of shares in a masked implementation increases.

2. On the need of formal methods and security order detection. As made clear in Sect. 2, the tools we provide in this paper only solve the "noise" part of the security evaluation problem for masked implementations. Hence, their combination with formal methods and security order detection techniques is an interesting scope for further research. Typically, one could extend the tools put forward in [3] in order to detect all the leaking operations in an assembly code (possibly with repetitions), then use leakage detection methods such as [13,22,30,42,55] to assess the security order of actual measurements, and finally evaluate their informativeness as we suggest in this paper, in order to obtain a fast assessment of the worst-case security level of an implementation.

3. On how to reach high security levels. Our results show that ensuring high security levels against optimal adversaries taking advantage of all the information provided by a masked implementation is challenging. It requires many shares, high noise levels and independence. In this respect, the application of our progresses to alternative multiplication chains [16,32], to the optimized algorithms in [8], to new primitives allowing more efficient masking (e.g., the proposal in [25] of which the complexity scales linearly in the number of shares and is well suited to guarantee the ISL assumption), and the combination of these ideas with parallel and/or hardware implementations (which improve security against multivariate attacks), is another interesting research direction.

4. On the "circuit size parameter" of masking security proofs. Eventually, our investigations are still limited to the evaluation of leakage samples that can be exploited via a divide-and-conquer strategy (i.e., attacks targeting independent parts of the key one by one). Yet, masking security proofs suggest that the success rate of an adversary is proportional to the target circuit size (i.e., the total amount of leakage samples) independent of whether these samples correspond to enumerable intermediate computations [20]. In this respect, analyzing the extent to which a SASCA exloiting a factor graph for a full masked implementation can confirm this fact is one more important open problem, which would first require a better formalization of such analytical attacks.

Acknowledgments. François-Xavier Standaert is a senior associate researcher of the Belgian Fund for Scientific Research (FNRS-F.R.S.). This work has been funded in parts by the ERC project 724725 (acronym SWORD), the EU project REASSURE and the Brussels Region INNOVIRIS project SCAUT.

A Algorithms for the Masked S-box

Algorithm 1. Secure evaluation of a GF(2)-linear function g.

Require: Shares $x(i)$ such that $x = x(1) \oplus \cdots \oplus x(d)$.
Ensure: Shares $y(i)$ such that $\mathsf{g}(x) = y = y(1) \oplus \cdots \oplus y(d)$.
1: **for** i from 1 to d **do**
2: $\quad y(i) \leftarrow \mathsf{g}(x(i))$
3: **end for**

Algorithm 2. Multiplication of two masked secrets $\in \mathrm{GF}(2^m)$.

Require: Shares $x(i)$ and $y(i)$ such that $x = x(1) \oplus \cdots \oplus x(d)$ and $y = y(1) \oplus \cdots \oplus y(d)$.
Ensure: Shares $z(i)$ such that $x \otimes y = z = z(1) \oplus \cdots \oplus z(d)$.
1: **for** i from 1 to d **do**
2: $\quad$ **for** j from $i+1$ to d **do**
3: $\quad\quad r_{i,j} \overset{\mathrm{r}}{\leftarrow} \mathrm{GF}(2^m)$
4: $\quad\quad r_{j,i} \leftarrow (r_{i,j} \oplus x(i) \otimes y(j)) \oplus x(j) \otimes y(i)$
5: $\quad$ **end for**
6: **end for**
7: **for** i from 1 to d **do**
8: $\quad z(i) \leftarrow x(i) \otimes y(i)$
9: $\quad$ **for** j from 1 to $d, j \neq i$ **do**
10: $\quad\quad z(i) \leftarrow z(i) \oplus r_{i,j}$
11: $\quad$ **end for**
12: **end for**
13: **return** $(z(1), ..., z(d))$

Algorithm 3. Field multiplication of two elements $\in \mathrm{GF}(2^m)$.

Require: $x, y \in \mathrm{GF}(2^m)$.
Ensure: z such that $z = x \otimes y$.
1: $x' \leftarrow \mathsf{LogTab}[x]$
2: $y' \leftarrow \mathsf{LogTab}[y]$
3: $z' \leftarrow x' + y' \bmod 2^m - 1$
4: $z \leftarrow (x \neq 0 \wedge y \neq 0)\ \mathsf{aLogTab}[z']$
5: **return** z

B Additional Figures

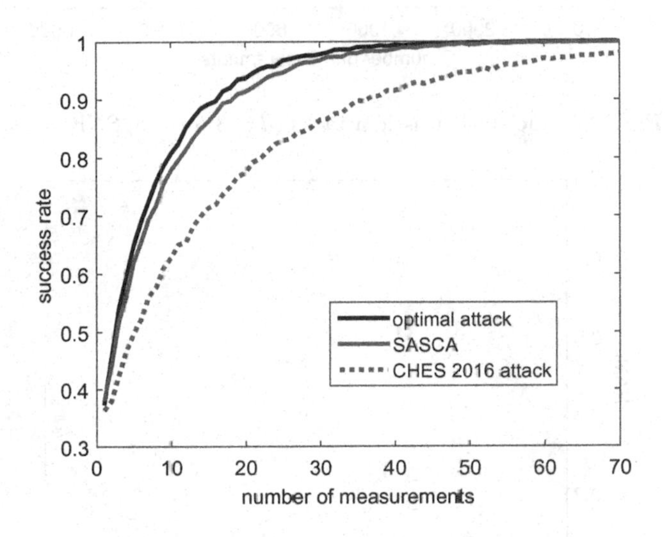

Fig. 12. Optimal attack vs. efficient heuristics ($d = 3$, $m = 2$, SNR $= 2$).

C Metric-Based Bound for the Key Rank

Very summarized, the two core ideas used in [21] to take the computational (enumeration) power of a divide-and-conquer adversary into account in a side-channel evaluation are: (i) to bound the success rate per S-box in function of the adversary's computational power thanks to the mutual information of an aggregated key variable K^c_{agg}, where c is an aggregation parameter (corresponding to the computational power), and (ii) to plug these success rate bounds into the metric-based rank-estimation algorithm of [48]. So technically, the only ingredient needed to exploit the same tools is the mutual information of the aggregated key variable (i.e., the so-called NAMI, for Normalized Aggregated Mutual Information). Unfortunately, the exact computation of the NAMI is impossible in our

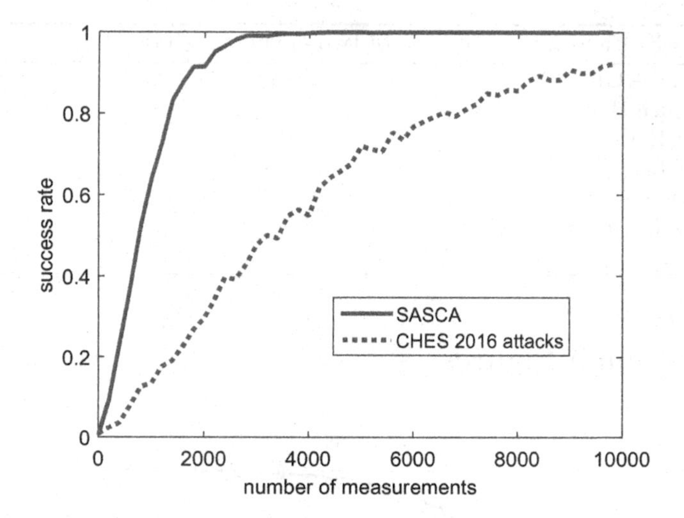

Fig. 13. Efficient heuristic attacks ($d = 3$, $m = 8$, $\text{SNR} = 2$).

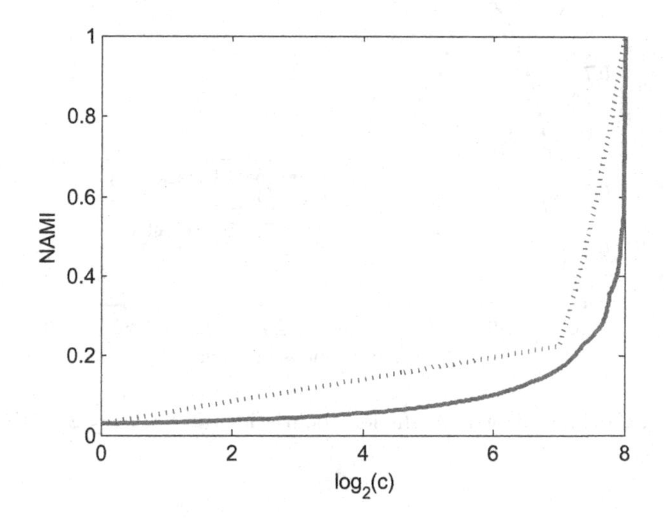

Fig. 14. Bound on the Normalized Aggregated Mutual Information.

case, since we do not have access to the probabilities of all the key candidates (that are combined during the aggregation process). So we need a way to bound the NAMI based on its first value $\text{NAMI}(c = 1) = \text{MI}(K; X, \bar{\boldsymbol{L}})$.

For this purpose, a simple observation is that for $c \leq 2^{m-1}$, aggregating $c = 2^q$ key candidates together can at most multiply the NAMI by $q+1$. The behavior of the NAMI for $c > 2^{m-1}$ is less intuitive (since in general, the definition of the NAMI is most intuitive when c is a power of two). Yet, as illustrated by the example in Fig. 14, a simple heuristic to bound it is then to connect the value of the NAMI at $c = 2^{m-1}$ and the maximum value of 1 that is reached at $c = 2^m$

by a straight line (which is obviously conservative as well, since the figure is in log-lin scale). Alternatively, when $\mathrm{MI}(K; X, \bar{\boldsymbol{L}}) < \frac{1}{2^m}$, an even simpler bound is to connect $\log(\mathrm{NAMI}(c = 1)) = \log(\mathrm{MI}(K; X, \bar{\boldsymbol{L}}))$ and $\log(\mathrm{NAMI}(c = 2^m)) = 0$ by a straight line. More accurate bounds are certainly reachable, yet not useful here since the general focus of the paper is on providing fast intuitions regarding the computational security of a key manipulated by a leaking device.

References

1. Archambeau, C., Peeters, E., Standaert, F.-X., Quisquater, J.-J.: Template attacks in principal subspaces. In: Goubin, L., Matsui, M. (eds.) CHES 2006. LNCS, vol. 4249, pp. 1–14. Springer, Heidelberg (2006). https://doi.org/10.1007/11894063_1
2. Balasch, J., Gierlichs, B., Grosso, V., Reparaz, O., Standaert, F.-X.: On the cost of lazy engineering for masked software implementations. In: Joye, M., Moradi, A. (eds.) CARDIS 2014. LNCS, vol. 8968, pp. 64–81. Springer, Cham (2015). https://doi.org/10.1007/978-3-319-16763-3_5
3. Barthe, G., Belaïd, S., Dupressoir, F., Fouque, P.-A., Grégoire, B., Strub, P.-Y.: Verified proofs of higher-order masking. In: Oswald and Fischlin [47], pp. 457–485
4. Barthe, G., Belaïd, S., Dupressoir, F., Fouque, P.-A., Grégoire, B., Strub, P.-Y., Zucchini, R.: Strong non-interference and type-directed higher-order masking. In: Weippl, E.R., Katzenbeisser, S., Kruegel, C., Myers, A.C., Halevi, S., (eds.) Proceedings of the 2016 ACM SIGSAC Conference on Computer and Communications Security, Vienna, Austria, 24–28 October, 2016, pp. 116–129. ACM (2016)
5. Barthe, G., Dupressoir, F., Faust, S., Grégoire, B., Standaert, F.-X., Strub, P.-Y.: Parallel implementations of masking schemes and the bounded moment leakage model. In: Coron and Nielsen [15], pp. 535–566
6. Batina, L., Robshaw, M. (eds.): CHES 2014. LNCS, vol. 8731. Springer, Heidelberg (2014). https://doi.org/10.1007/978-3-662-44709-3
7. Battistello, A., Coron, J.-S., Prouff, E., Zeitoun, R.: Horizontal side-channel attacks and countermeasures on the ISW masking scheme. In: Gierlichs and Poschmann [29], pp. 23–39
8. Belaïd, S., Benhamouda, F., Passelègue, A., Prouff, E., Thillard, A., Vergnaud, D.: Randomness complexity of private circuits for multiplication. In: Fischlin, M., Coron, J.-S. (eds.) EUROCRYPT 2016. LNCS, vol. 9666, pp. 616–648. Springer, Heidelberg (2016). https://doi.org/10.1007/978-3-662-49896-5_22
9. Brier, E., Clavier, C., Olivier, F.: Correlation power analysis with a leakage model. In: Joye, M., Quisquater, J.-J. (eds.) CHES 2004. LNCS, vol. 3156, pp. 16–29. Springer, Heidelberg (2004). https://doi.org/10.1007/978-3-540-28632-5_2
10. Chari, S., Jutla, C.S., Rao, J.R., Rohatgi, P.: Towards sound approaches to counteract power-analysis attacks. In: Wiener, M. (ed.) CRYPTO 1999. LNCS, vol. 1666, pp. 398–412. Springer, Heidelberg (1999). https://doi.org/10.1007/3-540-48405-1_26
11. Chari, S., Rao, J.R., Rohatgi, P.: Template attacks. In: Kaliski, B.S., Koç, K., Paar, C. (eds.) CHES 2002. LNCS, vol. 2523, pp. 13–28. Springer, Heidelberg (2003). https://doi.org/10.1007/3-540-36400-5_3
12. Choudary, M.O.: Efficient multivariate statistical techniques for extracting secrets from electronic devices. Ph.D. thesis, University of Cambridge (2014)

13. Cooper, J., De Mulder, E., Goodwill, G., Jaffe, J., Kenworthy, G., Rohatgi, P.: Test vector leakage assessment (TVLA) methodology in practice (extended abstract). In: ICMC 2013. http://icmc-2013.org/wp/wp-content/uploads/2013/09/goodwillkenworthtestvector.pdf
14. Coron, J.-S., Giraud, C., Prouff, E., Renner, S., Rivain, M., Vadnala, P.K.: Conversion of security proofs from one leakage model to another: a new issue. In: Schindler, W., Huss, S.A. (eds.) COSADE 2012. LNCS, vol. 7275, pp. 69–81. Springer, Heidelberg (2012). https://doi.org/10.1007/978-3-642-29912-4_6
15. Coron, J.-S., Nielsen, J.B. (eds.): EUROCRYPT 2017. LNCS, vol. 10210. Springer, Cham (2017). https://doi.org/10.1007/978-3-319-56620-7
16. Coron, J.-S., Prouff, E., Rivain, M., Roche, T.: Higher-order side channel security and mask refreshing. In: Moriai, S. (ed.) FSE 2013. LNCS, vol. 8424, pp. 410–424. Springer, Heidelberg (2014). https://doi.org/10.1007/978-3-662-43933-3_21
17. Daemen, J., Rijmen, V.: The wide trail design strategy. In: Honary, B. (ed.) Cryptography and Coding 2001. LNCS, vol. 2260, pp. 222–238. Springer, Heidelberg (2001). https://doi.org/10.1007/3-540-45325-3_20
18. Ding, A.A., Zhang, L., Fei, Y., Luo, P.: A statistical model for higher order DPA on masked devices. In: Batina and Robshaw [6], pp. 147–169
19. Duc, A., Dziembowski, S., Faust, S.: Unifying leakage models: from probing attacks to noisy leakage. In: Nguyen and Oswald [45], pp. 423–440
20. Duc, A., Faust, S., Standaert, F.-X.: Making masking security proofs concrete - or how to evaluate the security of any leaking device. In: Oswald and Fischlin [47], pp. 401–429
21. Duc, A., Faust, S., Standaert, F.-X.: Making masking security proofs concrete or how to evaluate the security of any leaking device (extended version). IACR Cryptology ePrint Archive 2015, 119 (2015)
22. Durvaux, F., Standaert, F.-X.: From improved leakage detection to the detection of points of interests in leakage traces. In: Fischlin, M., Coron, J.-S. (eds.) EUROCRYPT 2016. LNCS, vol. 9665, pp. 240–262. Springer, Heidelberg (2016). https://doi.org/10.1007/978-3-662-49890-3_10
23. Durvaux, F., Standaert, F.-X., Del Pozo, S.M.: Towards easy leakage certification. In: Gierlichs and Poschmann [29], pp. 40–60
24. Durvaux, F., Standaert, F.-X., Veyrat-Charvillon, N.: How to certify the leakage of a chip? In: Nguyen and Oswald [45], pp. 459–476
25. Dziembowski, S., Faust, S., Herold, G., Journault, A., Masny, D., Standaert, F.-X.: Towards sound fresh re-keying with hard (physical) learning problems. In: Robshaw, M., Katz, J. (eds.) CRYPTO 2016. LNCS, vol. 9815, pp. 272–301. Springer, Heidelberg (2016). https://doi.org/10.1007/978-3-662-53008-5_10
26. Dziembowski, S., Faust, S., Skorski, M.: Noisy leakage revisited. In: Oswald, E., Fischlin, M. (eds.) EUROCRYPT 2015. LNCS, vol. 9057, pp. 159–188. Springer, Heidelberg (2015). https://doi.org/10.1007/978-3-662-46803-6_6
27. Fei, Y., Luo, Q., Ding, A.A.: A statistical model for dpa with novel algorithmic confusion analysis. In: Prouff, E., Schaumont, P. (eds.) CHES 2012. LNCS, vol. 7428, pp. 233–250. Springer, Heidelberg (2012). https://doi.org/10.1007/978-3-642-33027-8_14
28. Gierlichs, B., Batina, L., Tuyls, P., Preneel, B.: Mutual information analysis. In: Oswald, E., Rohatgi, P. (eds.) CHES 2008. LNCS, vol. 5154, pp. 426–442. Springer, Heidelberg (2008). https://doi.org/10.1007/978-3-540-85053-3_27
29. Gierlichs, B., Poschmann, A.Y. (eds.): CHES 2016. LNCS, vol. 9813. Springer, Heidelberg (2016). https://doi.org/10.1007/978-3-662-53140-2

30. Goodwill, G., Jun, B., Jaffe, J., Rohatgi, P.: A testing methodology for side channel resistance validation. In: NIST Non-invasive Attack Testing Workshop (2011). http://csrc.nist.gov/news_events/non-invasive-attack-testing-workshop/papers/08_Goodwill.pdf
31. Goudarzi, D., Rivain, M.: How fast can higher-order masking be in software? In: Coron and Nielsen [15], pp. 567–597
32. Grosso, V., Prouff, E., Standaert, F.-X.: Efficient masked s-boxes processing – a step forward. In: Pointcheval, D., Vergnaud, D. (eds.) AFRICACRYPT 2014. LNCS, vol. 8469, pp. 251–266. Springer, Cham (2014). https://doi.org/10.1007/978-3-319-06734-6_16
33. Grosso, V., Standaert, F.-X.: ASCA, SASCA and DPA with enumeration: which one beats the other and when? In: Iwata, T., Cheon, J.H. (eds.) ASIACRYPT 2015. LNCS, vol. 9453, pp. 291–312. Springer, Heidelberg (2015). https://doi.org/10.1007/978-3-662-48800-3_12
34. Ishai, Y., Sahai, A., Wagner, D.: Private circuits: securing hardware against probing attacks. In: Boneh, D. (ed.) CRYPTO 2003. LNCS, vol. 2729, pp. 463–481. Springer, Heidelberg (2003). https://doi.org/10.1007/978-3-540-45146-4_27
35. Journault, A., Standaert, F.-X.: Very high order masking: efficient implementation and security evaluation. In: Fischer, W., Homma, N. (eds.) CHES 2017. LNCS, vol. 10529, pp. 623–643. Springer, Cham (2017). https://doi.org/10.1007/978-3-319-66787-4_30
36. Lemke-Rust, K., Paar, C.: Gaussian mixture models for higher-order side channel analysis. In: Paillier, P., Verbauwhede, I. (eds.) CHES 2007. LNCS, vol. 4727, pp. 14–27. Springer, Heidelberg (2007). https://doi.org/10.1007/978-3-540-74735-2_2
37. Lerman, L., Poussier, R., Bontempi, G., Markowitch, O., Standaert, F.-X.: Template attacks vs. machine learning revisited (and the curse of dimensionality in side-channel analysis). In: Mangard, S., Poschmann, A.Y. (eds.) COSADE 2014. LNCS, vol. 9064, pp. 20–33. Springer, Cham (2015). https://doi.org/10.1007/978-3-319-21476-4_2
38. Lomné, V., Prouff, E., Rivain, M., Roche, T., Thillard, A.: How to estimate the success rate of higher-order side-channel attacks. In: Batina and Robshaw [6], pp. 35–54
39. Mangard, S., Oswald, E., Standaert, F.-X.: One for all - all for one: unifying standard differential power analysis attacks. IET Inf. Secur. **5**(2), 100–110 (2011)
40. Mangard, S., Popp, T., Gammel, B.M.: Side-channel leakage of masked CMOS gates. In: Menezes, A. (ed.) CT-RSA 2005. LNCS, vol. 3376, pp. 351–365. Springer, Heidelberg (2005). https://doi.org/10.1007/978-3-540-30574-3_24
41. Martin, D.P., Mather, L., Oswald, E., Stam, M.: Characterisation and estimation of the key rank distribution in the context of side channel evaluations. In: Cheon, J.H., Takagi, T. (eds.) ASIACRYPT 2016. LNCS, vol. 10031, pp. 548–572. Springer, Heidelberg (2016). https://doi.org/10.1007/978-3-662-53887-6_20
42. Mather, L., Oswald, E., Bandenburg, J., Wójcik, M.: Does my device leak information? An *a priori* statistical power analysis of leakage detection tests. In: Sako, K., Sarkar, P. (eds.) ASIACRYPT 2013. LNCS, vol. 8269, pp. 486–505. Springer, Heidelberg (2013). https://doi.org/10.1007/978-3-642-42033-7_25
43. Mather, L., Oswald, E., Whitnall, C.: Multi-target DPA attacks: Pushing DPA beyond the limits of a desktop computer. In: Sarkar and Iwata [53], pp. 243–261
44. Matsui, M.: Linear cryptanalysis method for DES cipher. In: Helleseth, T. (ed.) EUROCRYPT 1993. LNCS, vol. 765, pp. 386–397. Springer, Heidelberg (1994). https://doi.org/10.1007/3-540-48285-7_33

45. Nguyen, P.Q., Oswald, E. (eds.): EUROCRYPT 2014. LNCS, vol. 8441. Springer, Heidelberg (2014). https://doi.org/10.1007/978-3-642-55220-5
46. Nikova, S., Rijmen, V., Schläffer, M.: Secure hardware implementation of nonlinear functions in the presence of glitches. J. Cryptol. **24**(2), 292–321 (2011)
47. Oswald, E., Fischlin, M. (eds.): EUROCRYPT 2015. LNCS, vol. 9056. Springer, Heidelberg (2015). https://doi.org/10.1007/978-3-662-46800-5
48. Poussier, R., Grosso, V., Standaert, F.-X.: Comparing approaches to rank estimation for side-channel security evaluations. In: Homma, N., Medwed, M. (eds.) CARDIS 2015. LNCS, vol. 9514, pp. 125–142. Springer, Cham (2016). https://doi.org/10.1007/978-3-319-31271-2_8
49. Prouff, E.: DPA attacks and s-boxes. In: Gilbert, H., Handschuh, H. (eds.) FSE 2005. LNCS, vol. 3557, pp. 424–441. Springer, Heidelberg (2005). https://doi.org/10.1007/11502760_29
50. Prouff, E., Rivain, M.: Masking against side-channel attacks: a formal security proof. In: Johansson, T., Nguyen, P.Q. (eds.) EUROCRYPT 2013. LNCS, vol. 7881, pp. 142–159. Springer, Heidelberg (2013). https://doi.org/10.1007/978-3-642-38348-9_9
51. Rivain, M.: On the exact success rate of side channel analysis in the Gaussian model. In: Avanzi, R.M., Keliher, L., Sica, F. (eds.) SAC 2008. LNCS, vol. 5381, pp. 165–183. Springer, Heidelberg (2009). https://doi.org/10.1007/978-3-642-04159-4_11
52. Rivain, M., Prouff, E.: Provably secure higher-order masking of AES. In: Mangard, S., Standaert, F.-X. (eds.) CHES 2010. LNCS, vol. 6225, pp. 413–427. Springer, Heidelberg (2010). https://doi.org/10.1007/978-3-642-15031-9_28
53. Sarkar, P., Iwata, T. (eds.): ASIACRYPT 2014. LNCS, vol. 8874. Springer, Heidelberg (2014). https://doi.org/10.1007/978-3-662-45608-8
54. Schindler, W., Lemke, K., Paar, C.: A stochastic model for differential side channel cryptanalysis. In: Rao, J.R., Sunar, B. (eds.) CHES 2005. LNCS, vol. 3659, pp. 30–46. Springer, Heidelberg (2005). https://doi.org/10.1007/11545262_3
55. Schneider, T., Moradi, A.: Leakage assessment methodology - extended version. J. Crypt. Eng. **6**(2), 85–99 (2016)
56. Standaert, F.-X.: How (not) to use Welch's t-test in side-channel security evaluations. IACR Cryptology ePrint Archive **2017**, 138 (2017)
57. Standaert, F.-X., Malkin, T.G., Yung, M.: A Unified Framework for the Analysis of Side-Channel Key Recovery Attacks. In: Joux, A. (ed.) EUROCRYPT 2009. LNCS, vol. 5479, pp. 443–461. Springer, Heidelberg (2009). https://doi.org/10.1007/978-3-642-01001-9_26
58. Standaert, F.-X., Veyrat-Charvillon, N., Oswald, E., Gierlichs, B., Medwed, M., Kasper, M., Mangard, S.: The world is not enough: another look on second-order DPA. In: Abe, M. (ed.) ASIACRYPT 2010. LNCS, vol. 6477, pp. 112–129. Springer, Heidelberg (2010). https://doi.org/10.1007/978-3-642-17373-8_7
59. Veyrat-Charvillon, N., Gérard, B., Standaert, F.-X.: Soft analytical side-channel attacks. In: Sarkar and Iwata [53], pp. 282–296

Best Young Researcher Paper Award

The Discrete-Logarithm Problem with Preprocessing

Henry Corrigan-Gibbs(✉) and Dmitry Kogan

Stanford University, Stanford, USA
henrycg@cs.stanford.edu

Abstract. This paper studies discrete-log algorithms that use *preprocessing*. In our model, an adversary may use a very large amount of precomputation to produce an "advice" string about a specific group (e.g., NIST P-256). In a subsequent online phase, the adversary's task is to use the preprocessed advice to quickly compute discrete logarithms in the group. Motivated by surprising recent preprocessing attacks on the discrete-log problem, we study the power and limits of such algorithms.

In particular, we focus on *generic* algorithms—these are algorithms that operate in every cyclic group. We show that any generic discrete-log algorithm with preprocessing that uses an S-bit advice string, runs in online time T, and succeeds with probability ϵ, in a group of prime order N, must satisfy $ST^2 = \widetilde{\Omega}(\epsilon N)$. Our lower bound, which is tight up to logarithmic factors, uses a synthesis of incompressibility techniques and classic methods for generic-group lower bounds. We apply our techniques to prove related lower bounds for the CDH, DDH, and multiple-discrete-log problems.

Finally, we demonstrate two new generic preprocessing attacks: one for the multiple-discrete-log problem and one for certain decisional-type problems in groups. This latter result demonstrates that, for generic algorithms with preprocessing, distinguishing tuples of the form $(g, g^x, g^{(x^2)})$ from random is much easier than the discrete-log problem.

1 Introduction

The problem of computing discrete logarithms in groups is fundamental to cryptography: it underpins the security of widespread cryptographic protocols for key exchange [31], public-key encryption [26,34], and digital signatures [46,53,68].

In the absence of an unconditional proof that computing discrete logarithms is hard, one fruitful research direction has focused on understanding the hardness of these problems against certain restricted classes of algorithms [6,61,71]. In particular, Shoup considered discrete-log algorithms that are *generic*, in the sense that they only use the group operation as a black box [71]. Generic algorithms are useful in practice since they apply to every group. In addition, lower bounds against generic algorithms are meaningful because, in popular elliptic-curve groups, generic attacks are the best known [38,51].

J. B. Nielsen and V. Rijmen (Eds.): EUROCRYPT 2018, LNCS 10821, pp. 415–447, 2018.
https://doi.org/10.1007/978-3-319-78375-8_14

The traditional notion of generic algorithms models *online-only attacks*, in which the adversary simultaneously receives the description of a cyclic group $\mathbb{G} = \langle g \rangle$ and a problem instance $g^x \in \mathbb{G}$. In this model, when the attack algorithm begins executing, the attacker has essentially no information about the group $\mathbb{G}$. Shoup [71] showed that, in this online-only setting, every generic discrete-log algorithm that succeeds with good probability in a group of prime order N must run in time at least $N^{1/2}$.

In practice, however, an adversary may have access to the description of the group $\mathbb{G}$ long before it has to solve a discrete-log problem instance. In particular, the vast majority of real-world cryptosystems use one of a handful of groups, such as NIST P-256, Curve25519 [12], or the DSA groups. In this setting, a real-world adversary could potentially perform a *preprocessing attack* [28,32,45] relative to a popular group: In an offline phase, the adversary would compute and store a data structure ("advice string") that depends on the group $\mathbb{G}$. In a subsequent online phase, the adversary could use its precomputed advice to solve the discrete-log problem in the group $\mathbb{G}$ much more quickly than would be possible in an online-only attack.

In recent work, Mihalcik [59] and Bernstein and Lange [13] demonstrated the surprising power of preprocessing attacks against the discrete-log problem. In particular, they construct a *generic* algorithm with preprocessing that computes discrete logarithms in every group of order N using $N^{1/3}$ bits of group-specific advice and roughly $N^{1/3}$ online time. Since their algorithm is generic, it applies to every group, including popular elliptic-curve groups. In contrast, Shoup's result shows that, without preprocessing, every generic discrete-log algorithm requires at least $N^{1/2}$ time. The careful use of a large amount of preprocessing—roughly $N^{2/3}$ operations—is what allows the attack of Mihalcik, Bernstein, and Lange to circumvent this lower bound.

As of now, there is no reason to believe that the attack of Mihalcik, Bernstein, and Lange is the best possible. For example, we know of no results ruling out a generic attack that uses $N^{1/2}$ precomputation to build an advice string of size $N^{1/8}$, which can be used to compute discrete logs in online time $N^{1/8}$.

The existence of such an attack would—at the very least—shake our confidence in 256-bit elliptic-curve groups. An attacker who wanted to break NIST P-256, for example, could perform a one-time 2^{128} precomputation to compute a 2^{32}-bit advice string. Given this advice string, an attacker could compute discrete logarithms on the P-256 curve in online time 2^{32}. The precomputed advice string would essentially be a "trapdoor" that would allow its holder to compute discrete-logs on the curve in seconds.

The possibility of such devastating discrete-log preprocessing attacks, and the lack of lower-bounds for such algorithms, leads us to ask:

How helpful can preprocessing be to generic discrete-log algorithms?

In this paper, we extend the classic model of generic algorithms to capture preprocessing attacks. To do so, we introduce the notion of *generic algorithms with preprocessing* for computational problems in cryptographic groups. These algorithms make only black-box use of the group operation, but may perform a large number of group operations during a preprocessing phase. Following prior work on preprocessing attacks [28,32,35,45], we measure the complexity of such algorithms by (a) the size of the advice string that the algorithm produces in the preprocessing phase, and (b) the running time of the algorithm's online phase.

These two standard cost metrics do not consider the preprocessing time required to compute the advice string. Ignoring the preprocessing cost only strengthens the resulting lower bounds, but it leaves open the question of how much preprocessing is really necessary to compute a useful advice string. Towards the end of this paper, we take up this question as well by extending our model to account for preprocessing time.

1.1 Our Results

We prove new lower bounds on generic algorithms with preprocessing that relate the time, advice, and preprocessing complexity of generic discrete-log algorithms, and algorithms for related problems. We also introduce new generic preprocessing attacks for the multiple-discrete-log problem and for certain distinguishing problems in groups.

Lower Bounds for Discrete Log and CDH. We prove in Theorem 2 that every generic algorithm that uses S bits of group-specific precomputed advice and that computes discrete logarithms in online time T with success probability ϵ must satisfy $ST^2 = \widetilde{\Omega}(\epsilon N)$, where the $\widetilde{\Omega}(\cdot)$ notation hides logarithmic factors in N. When $S = T$ the bound shows that, for constant ϵ, the best possible generic attack must use roughly $N^{1/3}$ bits of advice and runs in online time roughly $N^{1/3}$.

Our lower bound is tight, up to logarithmic factors, for the full range of parameters S, T, and ϵ, since the attack of Mihalcik [59] and Bernstein and Lange [13], which we summarize in Sect. 7.1, gives a matching upper bound. (These attacks sidestep Shoup's $N^{1/2}$-time lower bound for generic discrete-log algorithms [71] by using more than $N^{1/2}$ time in their preprocessing phase.) As a consequence, beating the preprocessing algorithm of Mihalcik, Bernstein, and Lange on the NIST P-256 curve, for example, would require developing a new non-generic attack.

Our lower bound extends naturally to the computational Diffie-Hellman problem, for which we also prove an $ST^2 = \widetilde{\Omega}(\epsilon N)$ lower bound (Theorem 6), and the M-instance multiple-discrete-log problem, for which we prove an $ST^2/M + T^2 = \widetilde{\Omega}(\epsilon^{1/M} MN)$ lower bound (Theorem 8). The attacks of Sect. 7 show that these lower bounds are tight.

Lower Bound for DDH with Preprocessing. We also look at the more subtle case of distinguishing attacks. We show in Theorem 9, that every generic distinguisher with preprocessing that achieves advantage ϵ against the decisional Diffie-Hellman problem (DDH) must satisfy $ST^2 = \widetilde{\Omega}(\epsilon^2 N)$. The quadratic dependence on the error probability makes this bound weaker than the previous ones. We know of no DDH distinguisher that matches this lower bound for all parameter ranges (e.g., for $\epsilon = N^{-1/4}$), and we leave the question of whether such a distinguisher exists as an open problem.

Lower Bound on Preprocessing Time. In addition, we prove lower bounds on the amount of computation required to produce the advice string in the preprocessing phase of a generic discrete-log algorithm. We show in Theorem 10 that any such algorithm that uses preprocessing time P, online time T, and achieves success probability ϵ must satisfy: $PT + T^2 = \Omega(\epsilon N)$. Our lower bound matches the preprocessing time used by the discrete-log preprocessing attack of Mihalcik, Bernstein, and Lange, and essentially rules out the existence of very fast generic algorithms that also use modest amounts of preprocessing. For example, any generic algorithm that runs in online time $T = N^{1/8}$ must use close to $N^{7/8}$ preprocessing time to succeed with good probability—no matter how large of an advice string it uses.

New Preprocessing Attacks. Finally, in Theorem 11, we introduce a new preprocessing algorithm for the multiple-discrete-log problem that shows that our lower bound is tight for constant ϵ. In addition, for the problem of distinguishing tuples of the form $(g, g^x, g^{(x^2)})$ from random, Theorem 13 gives a new algorithm that satisfies $ST^2 = \widetilde{O}(\epsilon^2 N)$. The existence of such an algorithm is especially surprising because solving the $(g, g^x, g^{(x^2)})$ distinguishing problem is *as hard as* computing discrete logarithms for *online-only* algorithms. In contrast, our algorithm shows that this problem is *substantially easier* than computing discrete logarithms for *preprocessing* algorithms: computing discrete logarithms requires $S = T = 1/\epsilon = N^{1/4}$ while our new distinguishing attack requires $S = T = 1/\epsilon = N^{1/5}$.

1.2 Our Techniques

The starting point of our lower bounds is an incompressibility argument, which is also at the heart of classic lower bounds against preprocessing algorithms (also known as "non-uniform algorithms") for inverting one-way permutations [42,76,77] and random functions [32]. At a high level, our approach is to show that if there exists a generic discrete-log algorithm $\mathcal{A}$ that (a) uses few bits of preprocessed advice and (b) uses few online group operations, then we can use such an algorithm $\mathcal{A}$ to compress a random permutation.

Incompressibility. The first technical challenge is that a straightforward application of incompressibility techniques does not suffice in the setting of generic groups. To explain the difficulty, let us sketch the argument that a random permutation oracle π is one-way, even against preprocessing adversaries [28,42,76,77]. The argument builds a compression scheme by invoking $\mathcal{A}(x)$ on some point x in the image of π and answering $\mathcal{A}$'s queries to π. The key observation is that when $\mathcal{A}$ produces its output $y = \pi^{-1}(x)$, we have learned some extra information about π beyond the information that the query responses contain. In this way, each invocation of $\mathcal{A}$ yields some "profit," in terms of our knowledge of π. We can use this profit to compress π.

To apply this argument to generic groups, we could replace the random permutation oracle π by an oracle that implements the group operation for a random group (We define the model precisely in Sect. 2.) The challenge is that a group-operation oracle has extra structure that a random permutation oracle does not. This extra structure fouls up the standard incompressibility argument, since the query responses that the compression routine must feed to $\mathcal{A}$ might themselves contain enough information to recover the discrete log that $\mathcal{A}$ will later output. If this happens, the compression scheme will not "profit" at all from invoking $\mathcal{A}$, and we will not be able to use $\mathcal{A}$ to compress the oracle.

To handle this case, we notice that this sort of compression failure only occurs when two distinct queries to the group oracle return the same string. By using a slightly more sophisticated compression routine, which notices and compensates for these "collision" events, we achieve compression even where the traditional incompressibility argument would have failed. (Dodis et al. [33] use a similar observation in their analysis of the RSA-FDH signature scheme.)

To keep track of when these collision events occur, we adopt an idea from Shoup's generic-group lower-bound proof [71], which does not use incompressibility at all. Shoup's idea is to keep a careful accounting of the information that the adversary's queries have revealed about the generic-group oracle at any point during the execution. Our compression scheme exploits a similar accounting strategy, which allows it to halt the adversary $\mathcal{A}$ as soon as the compressor notices that continuing to run $\mathcal{A}$ would be "unprofitable."

Handling Randomized Algorithms. The second technical challenge we face is in handling algorithms that succeed with arbitrarily small probability ϵ. The standard incompressibility methods invoke the algorithm $\mathcal{A}$ on many inputs, and the compression routine succeeds only if *all* of these executions succeed. If the algorithm $\mathcal{A}$ fails often, then we will fail to construct a useful compression scheme.

The naïve way around this problem would be to amplify $\mathcal{A}$'s success probability by having the compression scheme run the algorithm $\mathcal{A}$ many times on each input. The problem is that amplifying the success probability in this way decreases the "profit" that we gain from $\mathcal{A}$, since the compression scheme has to answer many more group-oracle queries in the amplified algorithm than in the unamplified algorithm. As a result, this naïve amplification strategy yields an $ST^2 = \widetilde{\Omega}(\epsilon^2 N)$ lower bound that is loose in its dependence on the success probability ϵ.

Our approach is to leverage the observation, applied fruitfully to the random-permutation model by De et al. [28], that it is without loss of generality to assume that the compression and decompression algorithms share a common string of independent random bits. Rather than amplifying the success probability of $\mathcal{A}$ by iteration, the compression scheme simply finds a set of random bits in the shared random string that cause $\mathcal{A}$ to produce the correct output. The compression scheme then writes this pointer out as part of the compressed representation of the group oracle. This optimization yields the tight $ST^2 = \widetilde{\Omega}(\epsilon N)$ lower bound.

Along the way, we exploit the random self-reducibility of the discrete-log problem to transform an average-case discrete-log algorithm, which succeeds on a *random* instance with probability ϵ, to a worst-case algorithm, which succeeds on *every* instance with probability ϵ. Using the random self-reduction substantially simplifies the incompressibility argument, since it allows the compression routine to invoke the algorithm $\mathcal{A}$ on arbitrary inputs.

Generalizing to Decisional Problems. The final technical challenge is to extend our core incompressibility argument to give lower bounds for the decisional Diffie-Hellman Problem (DDH). The difficulty with using a DDH algorithm to build a compression scheme is that each execution of the DDH distinguisher only produces a single bit of information. Furthermore, if the distinguishing advantage ϵ is small, the distinguisher produces only a fraction of a bit of information. The straightforward amplification would again work but would yield a very loose $ST^2 = \widetilde{\Omega}(\epsilon^4 N)$ bound.

To get around this issue, we execute the distinguisher on large batches of input instances. We judiciously choose the batch size to balance the profit from each batch with the probability that all runs in a batch succeed. Handling collision events in this case requires extra care. Putting these ingredients together, we achieve an $ST^2 = \widetilde{\Omega}(\epsilon^2 N)$ lower bound for the DDH problem.

1.3 Related Work

This paper builds upon two major lines of prior work: one on *preprocessing* lower bounds for *symmetric-key problems*, and the other on *online* lower bounds for *generic algorithms in groups.* We prove *preprocessing* lower bounds for *generic algorithms* and, indeed, our proofs use a combination of techniques from both prior settings.

Incompressibility Methods. One prominent related area of research puts lower bounds on the efficiency of *preprocessing algorithms* for inverting random functions and random permutations. An early motivation was Hellman's preprocessing algorithm ("Hellman tables") for inverting random functions [45]. Fiat and Naor [35] later extended the technique to allow inverting general functions and Oechslin [63] proposed practical improvements to Hellman's construction.

Yao [77] used an incompressibility argument to show the optimality of Hellman's method for inverting random permutations. Gennaro and Trevisan [42] and Wee [76] proved related lower bounds, also using incompressibility methods. Barkan et al. [9] showed that, in a restricted model of computation, Hellman's method is optimal for inverting random functions (not just permutations).

De et al. [28] demonstrated how to use *randomized encodings*, essentially an incompressibility argument augmented with random oracles, to give alternative proofs of preprocessing lower bounds on the complexity of inverting random permutations and breaking general pseudo-random generators. We adopt the powerful randomized encoding technique of De et al. in our proofs. Dodis et al. [32] applied this technique to show that salting [60] defeats preprocessing attacks against certain computational tasks (e.g., collision finding) in the random-oracle model [10]. Abusalah et al. [2] used the technique to construct proofs of space from random functions.

Unruh [74] gave an elegant framework for proving the hardness of computational problems in the random-oracle model against preprocessing adversaries (or against algorithms with "auxiliary input," in his terminology). He proves that if a computational problem is hard when a certain number of points of the random oracle are fixed ("presampled"), then the problem is hard in the random-oracle model against preprocessing adversaries using a certain amount of oracle-dependent advice. This presampling technique gives an often simpler alternative to incompressibility-based lower bounds. Coretti et al. [24] recently introduced new variants of Unruh's presampling technique that give tighter lower bounds against preprocessing adversaries for a broad set of problems.

Generic-Group Lower Bounds. All of the aforementioned work studies precomputation attacks on one-way permutations and one-way functions, which are essentially symmetric-key primitives. In the setting of public-key cryptography, a parallel—and quite distinct—line of work studies lower bounds on algorithms for the discrete-log problem and related problems in generic groups. All of these lower bounds study online-only algorithms (i.e., that do not use preprocessing).

In particular, Shoup [71] introduced the modern *generic-group model* to capture algorithms that make black-box use of a group operation. In Shoup's model, which draws on earlier treatments of black-box algorithms for groups [6,61], the discrete-logarithm problem in a group of prime order N requires time $\Omega(N^{1/2})$ to solve. Shoup's model captures many popular discrete-log algorithms, including Shanks' Baby-Step Giant-Step algorithm [70], Pollard's Rho and Kangaroo algorithms [67], and the Pohlig-Hellman algorithm [66]. For computing discrete logarithms on popular elliptic curves, variants of these algorithms are the best known [11,39,75,80].

Subsequent works used Shoup's model to prove lower bounds against generic algorithms for RSA-type problems [27], knowledge assumptions [30], the multiple-discrete-log problem [79], assumptions in groups with pairings [15], and for algorithms with access to additional oracles [57]. A number of works also prove the security of specific cryptosystems in the generic-group model [20,21, 29,36,49,69,72]. Other work studies computational problems in generic *rings*, to analyze generic algorithms for RSA-type problems [4,55].

Preprocessing Attacks in Generic Groups. The works most relevant to our new algorithms with preprocessing are Mihalcik's master's thesis [59], which surveys preprocessing attacks on the discrete-logarithm problem, and the paper of Bernstein and Lange [13], which demonstrated preprocessing attacks—both generic and non-generic—on a wide range of symmetric- and public-key primitives. We design new preprocessing attacks against the multiple-discrete-logarithm problem and against a large class of distinguishing problems in groups.

Non-generic discrete-log algorithms. In certain groups there are non-generic discrete-log attacks that dramatically outperform the generic ones. The landscape of non-generic discrete-log algorithms is vast, so we refer the reader to the

2000 survey of Odlyzko [62] and the 2014 survey of Joux et al. [47] for details. To give a taste of these results: when computing discrete logarithms in finite fields $\mathbb{F}_{p^n}$, the running time of the best discrete logarithms depend on the relative size of p and n. When $p \ll n$, a recent algorithm of Barbulescu et al. [8] computes discrete logarithms in quasi-polynomial time. When $p \gg n$, the best methods are based on "index calculus" techniques and run in sub-exponential time $e^{O((\log p)^{1/3}(\log\log p)^{2/3})}$ [44,56]. The analysis of these algorithms is heuristic, in that it relies on some unproved (but reasonable) number-theoretic assumptions.

In certain classes of elliptic-curve groups, there are non-generic algorithms for the discrete-log problem that outperform the generic algorithms [41]; some such algorithms run in sub-exponential time [58], or even in polynomial time [73]. In the standard elliptic-curve groups used for key exchange (e.g., NIST P-256) however, the generic preprocessing attacks discussed in this paper are still essentially the best known.

Non-generic discrete-log algorithms also benefit from preprocessing. Coppersmith demonstrated a sub-exponential-time preprocessing attack on the integer factorization problem [23] that also yields a non-generic sub-exponential-time preprocessing attack on the finite-field discrete-log problem [7,13]. Adrian et al. [3] show how to use such an attack compute discrete logs modulo a 512-bit prime in less than a minute of online time.

Organization of This Paper. In Sect. 2, we introduce notation, our model of computation, and a key lemma. In Sect. 3, we prove a lower bound on generic algorithms with preprocessing for the discrete-logarithm and CDH problems. In Sects. 4 and 5, we extend these bounds to the multiple-discrete-logarithm and DDH problems. In Sect. 6, we investigate the amount of precomputation such generic preprocessing algorithms require. In Sect. 7, we introduce new generic preprocessing attacks. In Sect. 8, we conclude with open questions.

2 Background

In this section, we recall the standard model of computation in generic groups, we introduce our model of generic algorithms with preprocessing, and we recall an incompressibility lemma that will be essential to our proofs.

Notation. We use $\mathbb{Z}_N$ to denote the ring of integers modulo N, $[N]$ indicates the set $\{1, \ldots, N\}$, and $\mathbb{Z}^+$ indicates the set of positive integers. Throughout this paper, we take N to be prime, so $\mathbb{Z}_N$ is also a field. We use the notation $x \leftarrow 5$ to indicate the assignment of a value to a variable and, when S is a finite set, the notation $x \xleftarrow{\text{R}} S$ indicates that x is a sample from the uniform distribution over S. For a probability distribution $\mathcal{D}$, $d \sim \mathcal{D}$ indicates that d is a random variable distributed according to $\mathcal{D}$. The statement $f(x) \stackrel{\text{def}}{=} x^2 - x$ indicates the definition of a function f. All logarithms are base two, unless otherwise noted.

We use the standard Landau notation $O(\cdot)$, $\Theta(\cdot)$, $\Omega(\cdot)$, and $o(\cdot)$ to indicate the asymptotics of a function. For example $f(N) = O(g(N))$ if there exists a constant $c > 0$ such that for all large enough N, $|f(N)| \le c \cdot g(N)$. When there are many variables inside the big-O, as in $f(N) = O(N/ST)$, all variables other than N are implicit functions of N. The tilde notation $\widetilde{O}(\cdot)$ and $\widetilde{\Omega}(\cdot)$ hides polylogarithmic factors in N. So, we can say for example that $S \log^2 N = \widetilde{O}(S)$.

Generic Algorithms. Following Shoup [71], we model a generic group using a random injective function σ that maps the integers in $\mathbb{Z}_N$ (representing the set of discrete logarithms) to a set of labels $\mathcal{L}$ (representing the set of group elements). We then write the elements of an order-N group as $\{\sigma(1), \sigma(2), \ldots, \sigma(N)\}$, instead of the usual $\{g, g^2, \cdots, g^N\}$. We often say that $i \in \mathbb{Z}_N$ is the "discrete log" of its label $\sigma(i) \in \mathcal{L}$.

The *generic group oracle* $\mathcal{O}_\sigma(\cdot, \cdot)$ for a labeling function σ takes as input two strings $s_i, s_j \in \mathcal{L}$ and responds as follows:

- If the arguments to the oracle are in the image of σ, then we can write $s_i = \sigma(i)$ and $s_j = \sigma(j)$. The oracle responds with $\sigma(i+j)$, where the addition is modulo the group order N.
- If either of the arguments to the oracle falls outside of the image of σ, the oracle returns $\perp$.

Given such an oracle and a label $\sigma(x)$, it is possible to compute $\sigma(\alpha x)$ for any constant $\alpha \in \mathbb{Z}_N$ using $O(\log N)$ oracle queries, by repeated squaring.

Some authors define the group oracle $\mathcal{O}_\sigma$ with a second functionality that maps labels $\sigma(x)$ to their inverses $\sigma(-x)$ in a single query. Our oracle can simulate this inversion oracle in at most $O(\log N)$ queries. To do so: given an element $\sigma(x)$, compute the element $\sigma((N-1)x) = \sigma(-x)$. Since providing an inversion oracle can decrease a generic algorithm's running time by at most a logarithmic factor, we omit it for simplicity.

A *generic algorithm* for $\mathbb{Z}_N$ on $\mathcal{L}$ is a probabilistic algorithm that takes as input a list of labels $(\sigma(x_1), \ldots, \sigma(x_L))$ and has oracle access to $\mathcal{O}_\sigma$. We measure the time complexity of a generic algorithm by counting the number of queries it makes to the generic group oracle.

Although the generic algorithms we consider may be probabilistic, we require that for every choice of σ, inputs, and random tapes, every algorithm halts after a finite number of steps. In this way, for every group order $N \in \mathbb{Z}^+$, we can compute an upper bound on the number of random bits the algorithm uses by iterating over all possible labelings, inputs, and random tapes. For this reason, we need only consider finite probability spaces in our discussion.

Generic Algorithms with Preprocessing. A *generic algorithm with preprocessing* is a pair of generic algorithms $(\mathcal{A}_0, \mathcal{A}_1)$ for $\mathbb{Z}_N$ on $\mathcal{L}$ such that:

- Algorithm $\mathcal{A}_0$ takes the label $\sigma(1)$ as input, makes some number of queries to the oracle $\mathcal{O}_\sigma$ ("preprocessing queries"), and outputs an advice string st_σ.
- Algorithm $\mathcal{A}_1$ takes as input the advice string st_σ and a list of labels $(\sigma(x_1), \ldots, \sigma(x_L))$, makes some number of queries to the oracle $\mathcal{O}_\sigma$ ("online queries"), and produces some output.

We typically measure the complexity of the algorithm $(\mathcal{A}_0, \mathcal{A}_1)$ by (a) the size of the advice string st_σ that $\mathcal{A}_0$ outputs, and (b) the number of oracle queries that algorithm $\mathcal{A}_1$ makes.

In Sect. 6, we consider generic algorithms with preprocessing for which the running time of $\mathcal{A}_0$ (i.e., the preprocessing time) is also bounded. In all other sections, we put no running time bound on $\mathcal{A}_0$, so without loss of generality, we may assume in these sections that $\mathcal{A}_0$ is deterministic.

Incompressibility Arguments. We use the following proposition of De et al. [28], which formalizes the notion that it is impossible to compress every element in a set $\mathcal{X}$ to a string less than $\log|\mathcal{X}|$ bits long, even relative to a random string.

Proposition 1 (De, Trevisan, and Tulsiani [28]). *Let $E : \mathcal{X} \times \{0,1\}^\rho \to \{0,1\}^m$ and $D : \{0,1\}^m \times \{0,1\}^\rho \to \mathcal{X}$ be randomized encoding and decoding procedures such that, for every $x \in \mathcal{X}$, $\Pr_{r \leftarrow \{0,1\}^\rho}\left[D(E(x,r),r) = x\right] \geq \delta$. Then $m \geq \log|\mathcal{X}| - \log 1/\delta$.*

Notice that the encoding and decoding algorithms of Proposition 1 take *the same* random string r as input. Additionally, that bound on the string length m is independent of the number of random bits that these routines take as input. As a consequence, Proposition 1 holds even when the algorithms E and D have access to a common random oracle.

3 Lower Bound for Discrete Logarithms

In this section we prove that every generic algorithm that uses S bits of group-specific precomputed advice and that computes discrete logs in online time T with probability ϵ must satisfy $ST^2 = \widetilde{\Omega}(\epsilon N)$.

Theorem 2. *Let N be a prime. Let $(\mathcal{A}_0, \mathcal{A}_1)$ be a pair of generic algorithms for $\mathbb{Z}_N$ on $\mathcal{L}$, such that $\mathcal{A}_0$ outputs an S-bit state, $\mathcal{A}_1$ makes at most T oracle queries, and*

$$\Pr_{\sigma, x, \mathcal{A}_1}\left[\mathcal{A}_1^{\mathcal{O}_\sigma}\left(\mathcal{A}_0^{\mathcal{O}_\sigma}(\sigma(1)), \sigma(x)\right) = x\right] \geq \epsilon,$$

where the probability is taken over the uniformly random choice of the labeling σ, the instance $x \in \mathbb{Z}_N$, and the coins of $\mathcal{A}_1$. Then $ST^2 = \widetilde{\Omega}(\epsilon N)$.

Remark. The statement of Theorem 2 models the case in which the group generator $\sigma(1)$ is fixed, and the online algorithm must compute the discrete-log of the instance $\sigma(x)$ with respect to the fixed generator. Using a fixed generator is essentially without loss of generality, since an algorithm that computes discrete logarithms with respect to one generator can also be used to compute discrete logarithms with respect to any generator by increasing its running time by a factor of two. Because of this, we treat the generator as fixed throughout this paper.

Remark. Theorem 2 treats only prime-order groups. In the more general case of composite-order groups a similar result holds, except that the bound is $ST^2 = \widetilde{\Omega}(\epsilon p)$, where p is the largest prime factor of the group order. Since the techniques needed to arrive at this more general result are essentially the same as in the proof of Theorem 2, we focus on the prime-order case for simplicity.

We first give the idea behind the proof of Theorem 2 and then present a detailed proof.

Proof Idea for Theorem 2. Our proof uses an incompressibility argument. The basic idea is to compress the random labeling function σ using a discrete-log algorithm with preprocessing $(\mathcal{A}_0, \mathcal{A}_1)$. To do so, we write $\mathcal{A}_0$'s S-bit advice about σ into the compressed string. We then run $\mathcal{A}_1$ on many discrete-log instances $\sigma(x)$ and we write the T responses to $\mathcal{A}_1$'s queries into the compressed string. For each execution of $\mathcal{A}_1$, we only need to write T values of σ into the compressed string, but we get $T+1$ values of σ back, since the output of $\mathcal{A}_1(\sigma(x))$ gives us the value of x "for free." If S and T are simultaneously small, then we can compress σ using this method, which yields a contradiction.

However, this naïve technique might never yield any compression at all. The problem is that the T responses to $\mathcal{A}_1$'s queries might contain "collision events," in which the response to one of $\mathcal{A}_1$'s queries is equal to a previously seen query response. For example, say that $\mathcal{A}_1$ makes a query of the form $\mathcal{O}_\sigma(\sigma(x), \sigma(3))$ and the oracle's response is a string $\sigma(7)$ that also appeared in response to a previous query. In this case, just seeing the queries of $\mathcal{A}_1$ and their responses is enough to conclude that $x+3 = 7 \bmod N$, which immediately yields the discrete log $x = 4$. This is problematic because even if $\mathcal{A}_1$ eventually halts and outputs $x = 4$, we have not received any "profit" from $\mathcal{A}_1$ since the T query responses themselves already contain all of the information we need to conclude that $x = 4$.

To profit in spite of these collisions, our compression scheme halts the execution of $\mathcal{A}_1$ as soon as it finds such a collision, since every collision event yields the discrete log being sought. The profit comes from the fact that, as long as the list of previous query responses is not too long, encoding a pointer to the collision-causing response requires many fewer bits than encoding an arbitrary element in the range of σ.

Our lower bound needs to handle randomized algorithms $\mathcal{A} = (\mathcal{A}_0, \mathcal{A}_1)$ that succeed with arbitrarily small probability ϵ. Yet to use $\mathcal{A}$ to compress σ, the algorithm $\mathcal{A}_1$ must succeed with very high probability. That is because the compression routine may invoke $\mathcal{A}_1$ as many as N times, and each execution must succeed for the compression scheme to succeed. The random self-reducibility of the discrete-log problem allows us to convert an average-case algorithm that succeeds on an ϵ fraction of instances (for a given labeling σ) to a worst-case algorithm that succeeds with probability ϵ on every instance (for a given labeling σ).

We still need to handle the fact that ϵ may be quite small. The straightforward way to amplify the success probability of $\mathcal{A}_1$ would be to construct an algorithm $\mathcal{A}'_1$ that runs R independent executions of $\mathcal{A}_1$ and that succeeds with probability at least $1-\epsilon^R$. We could then use the amplified algorithm $(\mathcal{A}_0, \mathcal{A}'_1)$ to compress σ.

The problem in our setting is that this simple amplification strategy yields a loose lower bound: if we run $\mathcal{A}_1$ for R iterations, and each iteration makes T queries, our compression scheme ends up "paying" for RT queries instead of T queries for each bit of "profit" it gets (i.e., for each output of $\mathcal{A}'_1$). Carrying this argument through yields an $ST^2 = \widetilde{\Omega}(\epsilon^2 N)$ bound, which is worse than our goal of $\widetilde{\Omega}(\epsilon N)$.

Our idea is to leverage the correlated randomness between the compressor and decompressor to our advantage. In our compression scheme, the compressor runs $\mathcal{A}_1$ using R sets of independent random coins, sampled from the random string shared with the decompressor. The compressor then writes into the compressed representation a $\log R$-bit pointer to a set of random coins (if one exists) that caused $\mathcal{A}_1$ to succeed. Using this strategy, instead of paying for RT queries per execution of $\mathcal{A}_1$, the compression scheme only pays for T queries, plus a small pointer. We can then choose R large enough to ensure that at least one of the R executions succeeds with extremely high probability. □

We now turn to the proof.

We say that a discrete-log algorithm succeeds in the *worst case* if it succeeds on every problem instance $\sigma(x)$ for $x \in \mathbb{Z}_N$. We say that a discrete-log algorithm succeeds in the *average case* if it succeeds on a random problem instance $\sigma(x)$ for $x \xleftarrow{R} \mathbb{Z}_N$.

We first use the random self-reducibility of the discrete-log problem to show that an average-case discrete-log algorithm implies a worst-case discrete-log algorithm. A lower bound on worst-case algorithms is therefore enough to prove Theorem 2. This is formalized in the next lemma.

Lemma 3 (Adapted from Abadi, Feigenbaum, and Kilian [1]). *Let N be a prime. Let $(\mathcal{A}_0, \mathcal{A}_1)$ be a pair of generic algorithms for $\mathbb{Z}_N$ on $\mathcal{L}$ such that $\mathcal{A}_0$ outputs an S-bit advice string and $\mathcal{A}_1$ makes at most T oracle queries. Then, there exists a generic algorithm $\mathcal{A}'_1$ that makes at most $T + O(\log N)$ oracle queries and, for every $\sigma : \mathbb{Z}_N \to \mathcal{L}$, if $\Pr_{x,\mathcal{A}_1}\left[\mathcal{A}_1^{\mathcal{O}_\sigma}(\mathcal{A}_0^{\mathcal{O}_\sigma}(\sigma(1)), \sigma(x)) = x\right] \geq \epsilon$, then for every $x \in \mathbb{Z}_N$, $\Pr_{\mathcal{A}'_1}\left[\mathcal{A}_1'^{\mathcal{O}_\sigma}(\mathcal{A}_0^{\mathcal{O}_\sigma}(\sigma(1)), \sigma(x)) = x\right] \geq \epsilon$.*

Proof. On input $(\mathsf{st}_\sigma, \sigma(x))$, algorithm $\mathcal{A}'_1$ executes the following steps: First, it samples a random $r \xleftarrow{R} \mathbb{Z}_N$ and computes $\sigma(x + r)$, using $O(\log N)$ group operations. Then, it runs $\mathcal{A}_1(\mathsf{st}_\sigma, \sigma(x + r))$. Finally, when $\mathcal{A}_1$ outputs a discrete log x', algorithm $\mathcal{A}'_1$ outputs $x = x' - r \bmod N$.

Notice that $\mathcal{A}'_1$ invokes $\mathcal{A}_1$ on $\sigma(x + r)$, which is the image of a uniformly random point in $\mathbb{Z}_N$. Since $\mathcal{A}_1$ succeeds with probability at least ϵ over the random choice of $x \xleftarrow{R} \mathbb{Z}_N$ and its coins, $\mathcal{A}'_1$ succeeds with probability ϵ, only over the choice of its coins. □

To prove Theorem 2, we will use the generic algorithms $(\mathcal{A}_0, \mathcal{A}_1)$ to construct a randomized encoding scheme that compresses a good fraction of the labeling functions σ. The following lemma gives us such a scheme.

Lemma 4. *Let N be a prime. Let $G = \{\sigma_1, \sigma_2, \dots\}$ be a subset of the labeling functions from $\mathbb{Z}_N$ to $\mathcal{L}$. Let $(\mathcal{A}_0, \mathcal{A}_1)$ be a pair of generic algorithms for $\mathbb{Z}_N$*

on $\mathcal{L}$ such that for every $\sigma \in G$ and every $x \in \mathbb{Z}_N$, $\mathcal{A}_0$ outputs an S-bit advice string, $\mathcal{A}_1$ makes at most T oracle queries, and $(\mathcal{A}_0, \mathcal{A}_1)$ satisfy

$$\Pr_{\mathcal{A}_1}\left[\mathcal{A}_1^{\mathcal{O}_\sigma}\left(\mathcal{A}_0^{\mathcal{O}_\sigma}(\sigma(1)), \sigma(x)\right) = x\right] \geq \epsilon.$$

Then, there exists a randomized encoding scheme that compresses elements of G to bitstrings of length at most

$$\log \frac{|\mathcal{L}|!}{(|\mathcal{L}| - N)!} + S + 1 - \frac{\epsilon N}{6T(T+1)(\log N + 1)},$$

and succeeds with probability at least 1/2.

We prove Lemma 4 in Sect. 3.1. Given the above two lemmas, we can prove Theorem 2.

Proof of Theorem 2. We say that a labeling σ is "good" if $(\mathcal{A}_0, \mathcal{A}_1)$ computes discrete logs with probability at least $\epsilon/2$ on σ. More precisely, a labeling σ is "good" if:

$$\Pr_{x, \mathcal{A}_1}\left[\mathcal{A}_1^{\mathcal{O}_\sigma}\left(\mathcal{A}_0^{\mathcal{O}_\sigma}(\sigma(1)), \sigma(x)\right) = x\right] \geq \epsilon/2,$$

where the probability is taken over the choice of $x \in \mathbb{Z}_N$ as well as over the random tape of $\mathcal{A}_1$. Let G be the set of good labelings. A standard averaging argument [5, Lemma A.12] guarantees that an $\epsilon/2$ fraction of injective mappings from $\mathbb{Z}_N$ to $\mathcal{L}$ are good. Then $|G| \geq \epsilon/2 \cdot |\mathcal{L}|!/(|\mathcal{L}| - N)!$, where we've used the fact that the number of injective functions from $\mathbb{Z}_N$ to $\mathcal{L}$ is $|\mathcal{L}|!/(|\mathcal{L}| - N)!$.

Lemma 3 then implies that there exists a pair of generic algorithms $(\mathcal{A}_0, \mathcal{A}_1')$ such that for every $\sigma \in G$ and every $x \in \mathbb{Z}_N$, $\mathcal{A}_1'^{\mathcal{O}_\sigma}(\mathcal{A}_0^{\mathcal{O}_\sigma}(\sigma(1)), \sigma(x))$ makes at most $T' = T + O(\log N)$ queries, and outputs x with probability at least $\epsilon/2$. Lemma 4 then implies that we can use $(\mathcal{A}_0, \mathcal{A}_1')$ to compress any labeling $\sigma \in G$ to a string of bitlength at most

$$\log \frac{|\mathcal{L}|!}{(|\mathcal{L}| - N)!} + S + 1 - \frac{(\epsilon/2)N}{6T'(T'+1)(\log N + 1)}, \tag{1}$$

where the encoding scheme works with probability at least 1/2. By Proposition 1, this length must be at least $\log |G| - \log 2$. Thus, it must hold that

$$\log \frac{|\mathcal{L}|}{(|\mathcal{L}| - N)!} + S + 1 - \frac{\epsilon N}{12T'(T'+1)(\log N + 1)} \geq \log \frac{|\mathcal{L}|!}{(|\mathcal{L}| - N)!} - \log \frac{4}{\epsilon}.$$

Rearranging, we obtain

$$S \geq \frac{\epsilon N}{O(T^2) \cdot \mathrm{polylog}(N)} - \log \frac{8}{\epsilon}.$$

We may assume without loss of generality that $\epsilon \geq 1/N$, since an algorithm that just guesses the discrete log achieves this advantage. Therefore, $\log \frac{8}{\epsilon} = O(\log N)$, and we get

$$(S + O(\log N))T^2 = \widetilde{\Omega}(\epsilon N),$$

which implies that $ST^2 = \widetilde{\Omega}(\epsilon N)$. □

3.1 Proof of Lemma 4

Recall that a randomized encoding scheme consists of an encoding and a decoding routine, such that both routines take the same string r of random bits as input. The encoding scheme we construct for the purposes of Lemma 4 operates on labelings σ. That is, the encoding routine takes a labeling $\sigma \in G$ and the random bits r, and constructs a compressed representation of σ. Correspondingly, the decoding routine takes this compressed representation and the *same* random bits r, and reconstructs σ.

While the encoding routine runs, it builds up a table of pairs $(f, \sigma(i)) \in (\mathbb{Z}_N[X] \times \mathcal{L})$. The decoder constructs a similar table during its execution. At any point during the encoding process, the table contains a representation of the information about σ that the encoder has communicated to the decoder up to the current point in the encoding process. The indeterminate X that appears in this table represents a discrete log value $x \in \mathbb{Z}_N$, which the decoder does not know. Once the decoder has enough information to determine x, each of the routines replaces every non-constant polynomial $f(X)$ in the table with its evaluation $f(x)$ at the point x. Subsequently, both routines can introduce a new variable X into the table, which represents a different unknown discrete logarithm in $\mathbb{Z}_N$. Therefore, at any point during the execution, there is at most a single indeterminate X in the table. Finally, when each of the routines completes, the table contains only constant polynomials, and the table fully determines σ.

We stress that the table is not part of the compressed representation of σ, but is part of the internal state of both routines.

Simulating $\mathcal{A}_1$'s Random Tape. Since the algorithm $\mathcal{A}_1$ is randomized, each time the encoder (or decoder) runs the algorithm $\mathcal{A}_1$, it must provide $\mathcal{A}_1$ with a fresh random tape. Both routines take as input a common random bitstring, and the encoder can reserve a substring of it to feed to each invocation of $\mathcal{A}_1$ as that algorithm's random tape. Since $\mathcal{A}_1$ always terminates, the encoder can determine an upper bound on the number of random bits that $\mathcal{A}_1$ will need for a given group size N and can partition the common random string accordingly.

The decoder follows the same process, and the fact that the encoder and decoder take the same random string r as input ensures that $\mathcal{A}_1$ behaves identically during the encoding and decoding processes.

Encoding Routine. The encoding routine, on input σ, uses two parameters $d, R \in \mathbb{Z}^+$, which we will set later, and proceeds as follows:

1. Compute $\mathsf{st}_\sigma \leftarrow \mathcal{A}_0(\sigma(1))$. The encoder can respond to all of the algorithm's oracle queries since the encoder knows all of σ. Write the S-bit output st_σ into the encoding.
2. Encode the image of σ as a subset of $\mathcal{L}$ using $\log \binom{|\mathcal{L}|}{N}$ bits, and append it to the encoding.
3. Initialize the table of pairs to an empty list.
4. Repeat d times:

(a) Choose the first string in the lexicographical order of the image of σ that does not yet appear in the table. Call this string $\sigma(x)$ and add the pair $(X, \sigma(x))$ to the table.
(b) Run $\mathcal{A}_1(\mathsf{st}_\sigma, \sigma(x))$ up to R times using independent randomness from the encoder's random string in each run. The encoder answers all of $\mathcal{A}_1$'s oracle queries using its knowledge of σ. If $\mathcal{A}_1$ fails on all R executions, abort the entire encoding routine. Otherwise, write into the encoding the index $r^* \in [R]$ of the successful execution (using $\log R$ bits).
(c) Write a placeholder of $\log T$ zeros into the encoding. (The routine overwrites these zeros with a meaningful value once this execution of $\mathcal{A}_1$ terminates.)
(d) Rerun $\mathcal{A}_1(\mathsf{st}_\sigma, \sigma(x))$ using the r^*-th random tape. While $\mathcal{A}_1$ is running, it makes a number of queries and then outputs its guess of the discrete log x. The encoding routine processes each of $\mathcal{A}_1$'s queries $(\sigma(i), \sigma(j))$ as follows:
   i. If either of the query arguments is outside of the range of σ, reply $\perp$ and continue to the next query.
   ii. If either (or both) of the arguments is missing from the table, then this is an "unexpected" query input. Add each such input, together with its discrete log, to the table, and append the discrete-log value i to the encoding, using $\log(N - |\mathsf{Table}|)$ bits.
   iii. Otherwise, look up the linear polynomials f_i, f_j representing $\sigma(i)$, $\sigma(j)$ in the table, and compute the linear polynomial $f_i + f_j$ representing the response $\sigma(i + j)$. We then distinguish between three cases:
      A. If $(f_i + f_j, \sigma(i + j))$ is already in the table, simply reply with $\sigma(i + j)$.
      B. If $\sigma(i + j)$ does not appear in the table, then add $\sigma(i + j)$ to the encoding, using $\log(N - |\mathsf{Table}|)$ bits, and reply with $\sigma(i + j)$.
      C. If $\sigma(i + j)$ appears in the table but its discrete log in the table is a polynomial f_k such that $f_k \neq f_i + f_j$, encode the reply to this query as a $(\log |\mathsf{Table}|)$-bit pointer to the table entry $(f_k, \sigma(i + j))$ and add this pointer the encoding. Stop this execution of $\mathcal{A}_1$, and indicate this "early stop" by writing the actual number of queries $t \leq T$ into its placeholder above.
(e) When the execution $\mathcal{A}_1(\mathsf{st}_\sigma, \sigma(x))$ outputs x, evaluate all of the polynomials in the table at the point x.

5. Append the remaining values that do not yet appear in the table to the encoding in lexicographic order.

Decoding Routine. The decoder proceeds analogously to the encoder. A key property of our randomized encoding scheme is that each position in the encoded string corresponds to the same state of the table in both the encoding and the decoding routines. In other words, when the decoding routine reads a certain position in the encoded string, its internal table is identical to the internal table the encoding routine had when it wrote to that position in the encoded string. The table allows the decoder to correctly classify each query to the correct category.

Note that in the case of a collision query (case C above), the decoder can use the collision to recover the value x of the indeterminate X. Specifically, for a query (u, v) where $u, v \in \mathcal{L}$, the decoder reads the reply $w \in \mathcal{L}$ from the encoding string, looks up the polynomials f_u, f_v, and f_w in the table, and solves for X the equation $f_w = f_u + f_v \bmod N$. This equation always has a unique solution, since N is a prime and f_u, f_v and f_w are linear polynomials in X such that $f_u + f_v$ is not identical to f_w.

The full description of the decoder appears in the full version of this paper [25].

Encoding Length. The encoding contains:

- the advice to the algorithm about the labeling σ (S bits),
- the encoding of the image of σ ($\log \binom{|\mathcal{L}|}{N}$ bits),
- for each of the d invocations of $\mathcal{A}_1$, the index r^* of the random tape on which it succeeded ($d \cdot \log R$ bits in total),
- for the i-th entry added to the table ($0 \le i < N$), if the entry was added
  - as the result of resolving a collision within the table, $\log i$ bits,
  - from the output of $\mathcal{A}_1$, 0 bits,
  - otherwise, $\log(N - i)$ bits,
- a counter indicating the number of queries for which to run each execution ($d \cdot \log T$ bits in total).

Observe that each of the d executions of $\mathcal{A}_1$ saves $\log(N - |\mathsf{Table}|)$ bits compared to the straightforward encoding (either due to $\mathcal{A}_1$ successfully computing the discrete log of its input, or finding a collision), but incurs an additional cost of at most $\log R + \log T + \log |\mathsf{Table}|$ bits. Since each execution of $\mathcal{A}_1$ adds at most $3T + 1$ rows to the table (T replies plus $2T$ unexpected inputs and either one collision or one output of $\mathcal{A}_1$) we have that $|\mathsf{Table}| \le d \cdot (3T + 1)$. Setting $d = \lfloor N/((2RT + 1)(3T + 1)) \rfloor$ guarantees that each of the d executions results in a net profit of

$$\log \frac{N - |\mathsf{Table}|}{RT|\mathsf{Table}|} \ge \log \frac{N - d(3T+1)}{RdT(3T+1)} \ge \log \frac{1 - \frac{1}{2RT+1}}{\frac{RT}{2RT+1}} = \log 2 = 1$$

bit. In this case, the total bitlength of the encoding is at most

$$\begin{aligned} S + \log \binom{|\mathcal{L}|}{N} + \sum_{i=0}^{N-1} \log(N - i) - d &= \log \frac{|\mathcal{L}|!}{(|\mathcal{L}| - N)!} + S - d \\ &\le \log \frac{|\mathcal{L}|!}{(|\mathcal{L}| - N)!} + S - \frac{N}{(2RT+1)(3T+1)} + 1 \\ &\le \log \frac{|\mathcal{L}|!}{(|\mathcal{L}| - N)!} + S - \frac{N}{6RT(T+1)} + 1\,. \end{aligned}$$

We need to choose R large enough to ensure that the encoding routine fails with probability at most 1/2. If we choose $R = (1+\log N)/\epsilon$, then the probability that R invocations of $\mathcal{A}_1$ all fail is, by a union bound, at most $(1-\epsilon)^R \le e^{-\epsilon R} \le 2^{-\epsilon R} \le 2^{-1-\log N} \le 1/(2N)$. The encoding scheme invokes $\mathcal{A}_1$ on at most N different inputs, so by a union bound, the probability that any invocation fails is at most 1/2. Overall, the encoding length is at most:

$$\log \frac{|\mathcal{L}|!}{(|\mathcal{L}| - N)!} + S + 1 - \frac{\epsilon N}{6T(T+1)(\log N + 1)} \text{ bits,}$$

which completes the proof of Lemma 4. □

3.2 Discrete Logarithms in Short Intervals

When working in groups of large order N, it is common to rely on the hardness of the *short-exponent discrete-log problem*, rather than the standard discrete-log problem [43,52,64,65]. In the usual discrete-log problem, a problem instance is a pair of the form $(g, g^x) \in \mathbb{G}^2$ for $x \xleftarrow{R} \mathbb{Z}_N$. The short-exponent problem is identical, except that x is sampled at random from $\{1, \ldots, W\} \subset \mathbb{Z}_N$, for some interval width parameter $W < N$. Using short exponents speeds up the Diffie-Hellman key-agreement protocol when it is feasible to set the interval width W to be much smaller than the group order N [64]. A variant of Pollard's "Lambda Method" [40,67] solves the short-exponent discrete-log problem in every group in time $O(W^{1/2})$, so W cannot be too small.

The following corollary of Theorem 2 shows that the short-exponent problem is no easier for generic algorithms with preprocessing than computing a discrete-logarithm in an order-W group.

Corollary 5 (Informal). *Let $\mathcal{A}$ be a generic algorithm with preprocessing that solves the short-exponent discrete-log problem in an interval of width W. If $\mathcal{A}$ uses S bits of group-specific advice, runs in online time T, and succeeds with probability ϵ, then $ST^2 = \widetilde{\Omega}(\epsilon W)$.*

Proof. We claim that the algorithm $\mathcal{A}$ of the corollary solves the standard discrete-log problem with probability $\epsilon' = \epsilon \cdot (W/N)$. The reason is that a standard discrete-log instance g^x for $x \xleftarrow{R} \mathbb{Z}_N$ has a short exponent (i.e., $x \in [W]$) with probability W/N. Algorithm $\mathcal{A}$ solves these short instances with probability ϵ. By Theorem 2, $ST^2 = \widetilde{\Omega}(\epsilon' N) = \widetilde{\Omega}(\epsilon W)$. □

As an application: decryption in the Boneh-Goh-Nissim cryptosystem [18] requires solving a short-exponent discrete-log problem in an interval of width W, for a polynomially large width W. The designers of that system suggest using a size-W table of precomputed discrete logs (i.e., $S = \widetilde{O}(W)$) to enable decryption in constant time. Corollary 5 shows that the best generic decryption algorithm that uses a size-S table requires roughly $\sqrt{W/S}$ time.

3.3 The Computational Diffie-Hellman Problem

A generic algorithm for the computational Diffie-Hellman problem takes as input a triple of labels $(\sigma(1), \sigma(x), \sigma(y))$ and must output the label $\sigma(xy)$. The following theorem demonstrates that in generic groups—even allowing for preprocessing—the computational Diffie-Hellman problem is as hard as computing discrete logarithms.

Theorem 6 (Informal). *Let $\mathcal{A} = (\mathcal{A}_0, \mathcal{A}_1)$ be a generic algorithm with preprocessing for the computational Diffie-Hellman problem in a group of prime order N. If $\mathcal{A}$ uses S bits of group-specific advice, runs in online time T, and succeeds with probability ϵ, then $ST^2 = \widetilde{\Omega}(\epsilon N)$.*

We present only the proof idea, since the structure of the proof is very similar to that of Theorem 2.

Proof Idea. The primary difference from the proof of Theorem 2, is that, we run $\mathcal{A}_1$ on pairs of labels $(\sigma(x), \sigma(y))$, and a successful run of $\mathcal{A}_1$ produces the CDH value $\sigma(xy)$. Since we run $\mathcal{A}_1$ on two labels at once, the encoder's table now has two formal variables: X and Y.

In this case, whenever the encoder encounters a collision, it gets a single linear relation on X and Y modulo the group order N. Since there are at most N solutions (x_0, y_0) to a linear relation in X and Y over $\mathbb{Z}_N$, the encoder can describe the solution to the decoder using $\log(N - |\mathsf{Table}|)$ bits. The encoder gets some profit, in terms of encoding length, since it will get two discrete logs for the cost of one discrete log and one pointer into the table (of length $\log |\mathsf{Table}|$ bits).

The rest of the proof is as in Theorem 2. □

3.4 Lower Bounds for Families of Groups

The lower bound of Theorem 2 suggests that one way to mitigate the risk of generic preprocessing attacks is to increase the group size. Doubling the size of group elements from $\log N$ to $2 \log N$ recovers the same level of security as if the attacker could not do any preprocessing. The downside of this mitigation strategy is that increasing the group size also increases the cost of each group operation and requires using larger cryptographic keys (e.g., when using the group for Diffie-Hellman key exchange [31]).

One might ask whether it would be possible to defend against preprocessing attacks without having to pay the price of using longer keys. One now-standard method to defend against preprocessing attacks when using a common cryptographic hash function H is to use "salts" [60]. When using salts, each user u of the hash function H chooses a random salt value s_u from a large space of possible salts. User u then uses the salted function $H_u(x) \stackrel{\text{def}}{=} H(s_u, x)$ as her hash function, and the salt value u can be made public. Chung et al. [22] showed that this approach can result in obtaining collision-resistant hashing against preprocessing attacks, and Dodis et al. [32] demonstrated the effectiveness of this approach for a variety of cryptographic primitives.

The analogue to salting in generic groups would be to have a large family of groups (e.g., of elliptic-curve groups) $\{\mathbb{G}_k\}_{k=1}^K$ indexed by a key k. Rather than having all users share a single group—as is the case today with NIST P-256—different users and systems could use different groups $\mathbb{G}_k$ sampled from this large family. In particular, pairs of users executing the Diffie-Hellman key-exchange protocol could first jointly sample a group $\mathbb{G}_k$ from this large family and then perform their key exchange in $\mathbb{G}_k$.

We show that using group families in this way effectively defends against generic preprocessing attacks, as long as the family contains a large enough number of groups.

To model group families, we replace the labeling function $\sigma : \mathbb{Z}_N \to \mathcal{L}$ with a keyed family of labeling functions $\sigma_{\mathsf{key}} : [K] \times \mathbb{Z}_N \to \mathcal{L}$. The keyed generic-group oracle $\mathcal{O}_{\sigma_{\mathsf{key}}}(\cdot,\cdot,\cdot)$ then takes a key k and two labels $\sigma_1, \sigma_2 \in \mathcal{L}$ and returns $\sigma_{\mathsf{key}}(k, x+y)$ if there exist $x, y \in \mathbb{Z}_N$ such that $\sigma_{\mathsf{key}}(k,x) = \sigma_1$ and $\sigma_{\mathsf{key}}(k,y) = \sigma_2$. The oracle returns $\perp$ otherwise. In addition, when fed the pair $(k, \star)$, for a key $k \in [K]$ and a special symbol $\star$, the oracle returns the identity element in the kth group: $\sigma(k, 1)$.

The following theorem demonstrates that using a large keyed family of groups effectively defends against generic preprocessing attacks:

Theorem 7. *Let N be a prime. Let $(\mathcal{A}_0, \mathcal{A}_1)$ be a pair of generic algorithms for $[K] \times \mathbb{Z}_N$ on $\mathcal{L}$, such that $\mathcal{A}_0$ outputs an S-bit state, $\mathcal{A}_1$ makes at most T oracle queries, and*

$$\Pr_{\sigma,k,x,\mathcal{A}_1}\left[\mathcal{A}_1^{\mathcal{O}_{\sigma_{\mathsf{key}}}}\left(\mathcal{A}_0^{\mathcal{O}_{\sigma_{\mathsf{key}}}}(), k, \sigma(k,x)\right) = x\right] \geq \epsilon,$$

where the probability is taken over the uniformly random choice of the labeling σ_{key}, the key $k \in [K]$, the instance $x \in \mathbb{Z}_N$, and the coins of $\mathcal{A}_1$. Then $ST^2 = \widetilde{\Omega}(\epsilon K N)$.

The proof of Theorem 7 appears in the full version of this paper [25]. The structure of the proof follows that of Theorem 2, except that we need some extra care to handle the fact that an adversary may query the oracle at many different values of k in a single execution.

4 Lower Bound for Computing Many Discrete Logarithms

A natural extension of the standard discrete-log problem is the *multiple*-discrete-log problem [37,48,54,78,79], in which the adversary's task is to solve M discrete-log problems at once. This problem arises in the setting of multiple-instance security of discrete-log-based cryptosystems. If an adversary has a list of M public keys $(g^{x_1}, \ldots, g^{x_M})$ in some group $\mathbb{G} = \langle g \rangle$ of prime order N, we would like to understand the cost to the adversary of recovering all M secret keys $x_1, \ldots, x_M \in \mathbb{Z}_N$.

Solving the multiple-discrete-log problem cannot be harder than solving M instances of the standard discrete-log problem independently using $\widetilde{O}(M\sqrt{N})$ time overall. One can however do better: generic algorithms due to Kuhn and Struik [54] and Fouque, Joux, and Mavromati [37] solve it in time $\widetilde{O}(\sqrt{MN})$. These algorithms achieve a speed-up over solving M discrete-log instances in sequence by reusing some of the work between instances. Yun [79] showed that in the generic-group model, these algorithms are optimal up to logarithmic factors by proving an $\Omega(\sqrt{NM})$-time lower bound for online-only algorithms, subject to the natural restriction that $M = o(N)$.

Our methods give the more general $ST^2 = \widetilde{\Omega}(\epsilon^{1/M}NM)$ generic lower bound for the M-instance multiple-discrete-log problem with preprocessing. For the special case of algorithms without preprocessing, our bound gives $T = \widetilde{\Omega}(\sqrt{NM})$, which matches the above upper and lower bounds. An additional benefit of our analysis it that it handles arbitrarily small success probabilities ϵ, whereas Yun's bound applies only to the $\epsilon = \Omega(1)$ case.

Let $\bar{x} = (x_1, \ldots, x_M) \in \mathbb{Z}_N^M$ and, for a labeling $\sigma : \mathbb{Z}_N \to \mathcal{L}$, define the vector $\sigma(\bar{x}) = (\sigma(x_1), \ldots, \sigma(x_M)) \in \mathcal{L}^M$. We restrict ourselves to the case of $M \leq T$, as otherwise the algorithm cannot even afford to perform a group operation on each of its inputs.

Theorem 8. *Let N be a prime. Let $(\mathcal{A}_0, \mathcal{A}_1)$ be a pair of generic algorithms for $\mathbb{Z}_N$ on $\mathcal{L}$ such that $\mathcal{A}_0$ outputs an S-bit advice string, $\mathcal{A}_1$ makes at most T oracle queries,*

$$\Pr_{\sigma, \bar{x}, \mathcal{A}_1}\left[\mathcal{A}_1^{\mathcal{O}_\sigma}\left(\mathcal{A}_0^{\mathcal{O}_\sigma}(\sigma(1)), \sigma(\bar{x})\right) = \bar{x}\right] \geq \epsilon,$$

where the probability is taken over the random choice of the labeling σ, a random input vector $\bar{x} \in \mathbb{Z}_N^M$ (for $M \leq T$), and the coins of $\mathcal{A}_1$. Then

$$ST^2/M + T^2 = \widetilde{\Omega}(\epsilon^{1/M}NM).$$

We prove this theorem in the full version of this paper [25].

The proof follows the proof of Theorem 2, except the encoder now runs $\mathcal{A}_1$ on M labels at a time. The encoder and decoder keep a table in M formal variables $(X_1, \ldots, X_M)$, representing the M discrete logs being sought. With every "collision event," we show that the number of formal variables in the table can decrease by one until either (a) $\mathcal{A}_1$ outputs the M discrete logs, or (b) the table has no more formal variables and the encoder halts $\mathcal{A}_1$.

5 The Decisional Diffie-Hellman Problem

The decisional Diffie-Hellman problem [14] (DDH) is to distinguish tuples of the form (g, g^x, g^y, g^{xy}) from tuples of the form (g, g^x, g^y, g^z), for random $x, y, z \in \mathbb{Z}_N$. In this section, we show that every generic distinguisher with preprocessing for the decisional Diffie-Hellman problem that achieves advantage ϵ must satisfy $ST^2 = \widetilde{\Omega}(\epsilon^2 N)$. More formally:

Theorem 9. *Let N be a prime. Let $(\mathcal{A}_0, \mathcal{A}_1)$ be a pair of generic algorithms for $\mathbb{Z}_N$ on $\mathcal{L}$, such that $\mathcal{A}_0$ outputs an S-bit state, $\mathcal{A}_1$ makes at most T oracle queries, and*

$$\Big| \Pr\left[\mathcal{A}_1^{\mathcal{O}_\sigma}\left(\mathcal{A}_0^{\mathcal{O}_\sigma}(\sigma(1)), \sigma(x), \sigma(y), \sigma(xy)\right) = 1\right] - \Pr\left[\mathcal{A}_1^{\mathcal{O}_\sigma}\left(\mathcal{A}_0^{\mathcal{O}_\sigma}(\sigma(1)), \sigma(x), \sigma(y), \sigma(z)\right) = 1\right]\Big| \geq \epsilon,$$

where the probabilities are over the choice of the label σ, the values $x, y, z \in \mathbb{Z}_N$, and the randomness of $\mathcal{A}_1$. Then $ST^2 = \widetilde{\Omega}(\epsilon^2 N)$.

The proof of Theorem 9 appears in the full version of this paper [25].

While the proof uses an incompressibility argument, extending the technique of Theorem 2 to give lower bounds for decisional-type problems requires overcoming additional technical challenges. Consider a DDH distinguisher with preprocessing $(\mathcal{A}_0, \mathcal{A}_1)$ that achieves advantage ϵ. The difficulty with using such an algorithm to build a scheme for compressing σ is that each execution of $\mathcal{A}_1$ only produces a single bit of output. When $\epsilon < 1$, each execution of $\mathcal{A}_1$ produces even less—a fraction of a bit of useful information.

To explain why getting only a single bit of output from $\mathcal{A}_1$ is challenging: the encoder of Theorem 2 derandomized $\mathcal{A}_1$ by writing a pointer $r^* \in [R]$ to a "good" set of random coins for $\mathcal{A}_1$ into the encoding, thus turning a faulty randomized algorithm into a correct deterministic algorithm at the cost of slightly increasing the encoding length. This derandomization technique does not apply immediately here, since the $\log R$-bit value required to point to the "good" set of random coins eliminates any profit in encoding length that we would have gained from the fraction of a bit that $\mathcal{A}_1$ produces as output.

A straightforward amplification strategy—building an algorithm $\mathcal{A}_1'$ that calls $\mathcal{A}_1$ many times and takes the majority output—would circumvent this problem, but would yield an $ST^2 = \widetilde{\Omega}(\epsilon^4 N)$ lower bound that is loose in ϵ.

To achieve a tighter $ST^2 = \widetilde{\Omega}(\epsilon^2 N)$ bound, our strategy is to use $\mathcal{A}_1$ to construct an algorithm $\mathcal{A}_1^{\times B}$ that executes $\mathcal{A}_1$ on a batch of B independent DDH problem instances (one at a time), for some batch size parameter $B \in \mathbb{Z}^+$. The algorithm $\mathcal{A}_1^{\times B}$ now produces B bits of output and succeeds with probability ϵ^B. If we now choose R such that $\log R < B$, we can now apply our prior derandomization technique, since each execution of $\mathcal{A}_1^{\times B}$ will yield some profit in our compression scheme.

Handling collisions in this case involves additional technicalities, since there might (or might not) be a collision in each of the B sub-executions of $\mathcal{A}_1^{\times B}$ and we need to be able to identify which execution encountered a collision without squandering the small profit that $\mathcal{A}_1^{\times B}$ yields.

Putting everything together, we achieve an $ST^2 = \widetilde{\Omega}(\epsilon^2 N)$ lower bound for the DDH problem.

6 Lower Bounds with Limited Preprocessing

Up to this point, we have measured the cost of a discrete-log algorithm with preprocessing by (a) number of bits of preprocessed advice it requires and (b) its online running time. In this section, we explore the preprocessing cost—the time required to compute the advice string—and we prove tight lower bounds on the preprocessing cost of generic discrete-log algorithms.

Let $(\mathcal{A}_0, \mathcal{A}_1)$ be a generic discrete-log algorithm with preprocessing, as defined in Sect. 2. For this section, we allow $\mathcal{A}_0$ to be randomized. We say that $(\mathcal{A}_0, \mathcal{A}_1)$ uses P preprocessing queries and T online queries if $\mathcal{A}_0$ makes P oracle queries and $\mathcal{A}_1$ makes T oracle queries. In this section, we do not put any restriction on the size of the state that $\mathcal{A}_0$ outputs—we are only interested in understanding the relationship between the preprocessing time P and the online time T.

Remark. When $P = \Theta(N)$, there is a trivial discrete-log algorithm with preprocessing $(\mathcal{A}_0, \mathcal{A}_1)$ that uses $T = 0$ online queries and succeeds with constant probability. In the preprocessing step, $\mathcal{A}_0$ computes a table of $\Theta(N)$ distinct pairs of the form $(i, \sigma(i)) \in \mathbb{Z}_N \times \mathcal{L}$. On receiving a discrete-log instance $\sigma(x)$, the online algorithm $\mathcal{A}_1$ looks to see if $\sigma(x)$ is already stored in its precomputed table and outputs the discrete log x if so. This algorithm succeeds with probability $\epsilon = P/N = \Omega(1)$.

Remark. When $P = o(\sqrt{N})$, we can rule out algorithms that run in online time $T = o(\sqrt{N})$ and succeed with constant probability. To do so, we observe that every generic discrete-log algorithm that uses P preprocessing queries and T online queries can be converted into an algorithm that uses *no* preprocessing queries and $T' = (P + T)$ online queries, such that both algorithms achieve the same success probability.

Shoup's lower bound [71] states that every generic discrete-log algorithm *without preprocessing* that runs in time T' succeeds with probability at most $\epsilon = O(T'^2/N)$. This implies that any algorithm *with preprocessing* P and online time T succeeds with probability at most $\epsilon = O((T + P)^2/N)$.

Put another way: Shoup's result implies a lower bound of $(T+P)^2 = \Omega(\epsilon N)$. So any algorithm that makes only $P = o(\sqrt{N})$ preprocessing queries must use $T = \Omega(\sqrt{N})$ online queries to succeed with constant probability. Thus, an algorithm that uses $o(\sqrt{N})$ preprocessing queries cannot asymptotically outperform an online algorithm.

Given these two remarks, the remaining parameter regime of interest is when $\sqrt{N} < P < N$. We prove:

Theorem 10. *Let $(\mathcal{A}_0, \mathcal{A}_1)$ be a generic discrete-log algorithm with preprocessing for $\mathbb{Z}_N$ on $\mathcal{L}$ that makes at most P preprocessing queries and T online queries. If $x \in \mathbb{Z}_N$ and a labeling function σ are chosen at random, then $\mathcal{A}$ succeeds with probability $\epsilon = O((PT + T^2)/N)$.*

As a corollary, we find that every algorithm that succeeds with probability ϵ must satisfy $PT + T^2 = \Omega(\epsilon N)$. For example, an algorithm that uses $P = O(N^{2/3})$ preprocessing queries must use online time at least $T = \Omega(N^{1/3})$ to succeed with constant probability.

The full proof appears in the full version of this paper [25], and we sketch the proof idea here.

Proof Idea for Theorem 10. We prove the theorem using a pair of probabilistic experiments, following the general strategy of Shoup's now-classic proof technique [71].

In both experiments, the adversary interacts with a challenger, who plays the role of the generic group oracle $\mathcal{O}_\sigma$. The challenger defines the labeling function $\sigma(\cdot)$ lazily in response to the adversary's queries. Both experiments follow similar steps:

1. The challenger sends a label $s_1 \in \mathcal{L}$, representing $\sigma(1)$, to the adversary.
2. The adversary makes P preprocessing group-oracle queries to the challenger.
3. The challenger sends the discrete-log instance $s_x \in \mathcal{L}$, representing $\sigma(x)$, to the adversary.
4. The adversary makes T online queries and outputs a guess x' of x.

The difference between the two experiments is in how the challenger defines the discrete log of the instance $s_x \in \mathcal{L}$.

In Experiment 0, the challenger chooses the discrete log $x \in \mathbb{Z}_N$ of s_x *before* the adversary makes any online queries. The challenger in Experiment 0 is thus a faithful (or honest) oracle.

In Experiment 1, the challenger chooses the discrete log x of σ_x *after* the adversary has made all of its online queries. In this latter case, the challenger is essentially "cheating" the adversary, since all of the challenger's query responses are independent of x and the adversary cannot recover x with probability better than random guessing. To complete the argument, we show that unless the adversary makes many queries, it can only rarely distinguish between the two experiments.

A detailed description of the experiments and their analysis appears in the full version of this paper [25]. □

The Lower Bound is Tight. From Theorem 2, we know that a discrete-log algorithm that succeeds with constant probability must use advice S and online time T such that $ST^2 = \widetilde{\Omega}(N)$. From Theorem 10, we know that any such algorithm must also use preprocessing P such that $PT + T^2 = \Omega(N)$. The best tradeoff we could hope for, ignoring the constants and logarithmic factors, is $PT + T^2 = ST^2$, or $P = ST$. Indeed, the known upper bound with preprocessing (see Sect. 7.1) matches this lower bound, disregarding low-order terms.

7 Preprocessing Attacks on Discrete-Log Problems

In this section, we recall the known generic discrete-log algorithm with preprocessing and we introduce two new generic attacks with preprocessing. Specifically, we show an attack on the multiple-discrete-log problem that matches the lower bound of Theorem 8, and we show an attack on certain decisional problems in groups that matches the lower bound of Theorem 9.

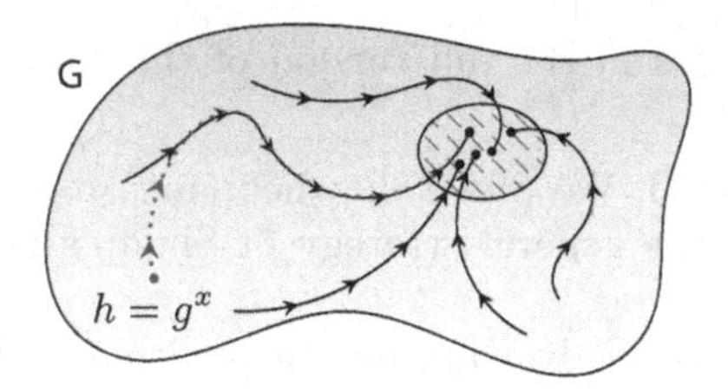

Fig. 1. The discrete-log algorithm with preprocessing of Sect. 7.1 uses a random function F to define a walk on the elements of $\mathbb{G}$. The preprocessed advice consists of the discrete logs of S points that lie at the end of length-$\Theta(T)$ disjoint paths on the walk. In the online phase, the algorithm walks from the input point until hitting a stored endpoint, which occurs with good probability.

These attacks are all generic, so they apply to every group, including popular elliptic-curve groups. Our preprocessing attacks are *not* polynomial-time attacks—indeed our lower bounds rule out such attacks—but they yield better-than-known exponential-time attacks on these problems.

The analysis of the algorithms in these sections rely on the attacker having access to a random function (i.e., a random oracle [10]), which the attacker could instantiate with a standard cryptographic hash function, such as SHA-256. Removing the attacks' reliance on a truly random function remains a useful task for future work.

7.1 The Existing Discrete-Log Algorithm with Preprocessing

For the reader's reference, we describe a variation of the discrete-log algorithm with preprocessing, introduced by Mihalcik [59] and Bernstein and Lange [13], with a slightly more detailed analysis. This discrete-log algorithm shows that the lower bound of Theorem 2 is tight. Our algorithms for the multiple-discrete-log problem (Sect. 7.2) and for distinguishing pseudo-random generators (Sect. 7.3) use ideas from this algorithm.

The algorithm computes discrete logs in a group $\mathbb{G}$ of prime order N with generator g. The algorithm takes as input parameters $S, T \in \mathbb{Z}^+$ such that $ST^2 \leq N$. The algorithm uses $\widetilde{O}(S)$ bits of precomputed advice about the group $\mathbb{G}$, uses $\widetilde{O}(T)$ group operations in the online phase, and succeeds with probability $\epsilon = \Omega(ST^2/N)$.

Let $F : \mathbb{G} \to \mathbb{Z}_N$ be a random function, which we can instantiate in practice using a standard hash function. We use the function F to define a walk on the elements of $\mathbb{G}$. Given a point $h \in \mathbb{G}$, the walk computes $\alpha \leftarrow F(h)$ and moves to the point $g^\alpha h \in \mathbb{G}$.

Given these preliminaries, the algorithm works as follows:

- *Preprocessing phase.* Repeat S times: pick $r \xleftarrow{R} \mathbb{Z}_N$ and, starting at $g^r \in \mathbb{G}$, take the walk defined by F for $T/2$ steps. Store the endpoint of the walk $g^{r'}$ and its discrete log r' in a table: $(r', g^{r'})$.

  At the end of the preprocessing phase, the algorithm stores this table of S group elements along with their discrete logs, using $O(S \log N)$ bits.
- *Online phase.* Given a discrete-log instance $h = g^x$, the algorithm takes T steps along the random walk defined by F, starting from the point h (see Fig. 1). If the walk hits one of the S points stored in the precomputed table, this collision yields a linear relation on x in the exponent: $g^{r'} = g^{x+\alpha_1+\cdots+\alpha_k} \in \mathbb{G}$. Solving this linear relation for $x \in \mathbb{Z}_N$ reveals the desired discrete log.

The algorithm uses $\widetilde{O}(S)$ bits of group-specific advice and runs in online time $\widetilde{O}(T)$. The remaining task is to analyze its success probability.

We first claim that, with good probability, the S walks in the preprocessing phase touch at least $ST/4$ distinct points. To this end, observe that for every walk in the preprocessing phase, the probability that it touches $T/2$ new points is at least $(1 - ST/(2N))^{T/2} \geq 1 - ST^2/(4N)$, by Bernoulli's inequality. Since $ST^2 \leq N$, we have that $1 - ST^2/(4N) \geq 1 - 1/4 = 3/4$. Therefore, in expectation, each walk touches at least $3T/8$ new points and by linearity of expectation, the overall expected number of touched points is at least $3ST/8$. The number of touched points is at most $ST/2$ and is at least $3ST/8$, in expectation. We can apply Markov's inequality to an auxiliary random variable to conclude that the number of touched points is greater than $ST/4$ with probability at least $1/2$.

Next, observe that if at any of its first $T/2$ steps, the online walk hits any of the points touched by one of the preprocessed walks, in the remaining $T/2$ steps it will hit the stored endpoint of that preprocessed walk. It will then successfully compute the discrete log. Moreover, as long as the online walk does not hit any of these points, its steps are independent random points in $\mathbb{G}$. If the number points touched during preprocessing is at least $ST/4$, then the online walk succeeds with probability at least $1-(1-(ST/(4N))^{T/2} \geq 1-\exp(-ST^2/(8N)) \geq ST^2/(16N)$. Overall, the probability of success ϵ is at least $1/2 \cdot ST^2/(16N) = \Omega(ST^2/N)$.

7.2 Multiple Discrete Logarithms with Preprocessing

We now demonstrate that a similar technique allows solving the multiple-discrete-log problem more quickly using preprocessing. The algorithm is a modification to the attack of Fouque et al. [37] to allow for precomputation, in the spirit of the algorithm of Sect. 7.1.

This upper bound matches the lower bound of Theorem 8 for a constant ϵ, up to logarithmic factors, which shows that the lower bound is tight for constant ϵ. To recall, an instance of the multiple-discrete-log problem is a vector $(g^{x_1}, \ldots, g^{x_M})$ for random $x_i \in \mathbb{Z}_N$. The solution is the vector $(x_1, \ldots, x_M)$. Then we have the following theorem:

Theorem 11. *There exists a generic algorithm with preprocessing for the M-instance multiple-discrete-log problem in a group of prime order N that makes use of a random function, uses $\widetilde{O}(S)$ bits of group-specific advice, runs in time $\widetilde{O}(T)$, succeeds with constant probability, and satisfies $ST^2/M + T^2 = O(MN)$.*

We prove the theorem in the full version of this paper [25].

7.3 Distinguishers with Preprocessing

In this section, we give a new distinguishing algorithm for certain decisional problems in groups.

For concreteness, we first demonstrate how to use preprocessing to attack the *square decisional Diffie-Hellman problem* (sqDDH) [50], which is the problem of distinguishing tuples of the form (g, g^x, g^y) from tuples of the form $(g, g^x, g^{(x^2)})$ for random $x, y \in \mathbb{Z}_N$. In groups for which DDH is hard, the best known attack against this assumption requires solving the discrete-log problem. Later on, we show how to generalize the attack to a larger family of natural decisional assumptions in groups.

Definition 12. We say that an oracle algorithm $\mathcal{A}^{\mathcal{O}}$ has *advantage* ϵ at distinguishing distributions $\mathcal{D}_1$ and $\mathcal{D}_2$ if $\left|\Pr[\mathcal{A}^{\mathcal{O}}(d_1) = 1] - \Pr[\mathcal{A}^{\mathcal{O}}(d_2) = 1]\right| = \epsilon$, where the probability is over the randomness of the oracle and samples $d_1 \sim \mathcal{D}_1$ and $d_2 \sim \mathcal{D}_2$.

Theorem 13. *There is a sqDDH distinguisher with preprocessing that makes use of a random function, uses $\widetilde{O}(S)$ bits of group-specific advice, runs in time $\widetilde{O}(T)$, and achieves distinguishing advantage ϵ whenever $ST^2 = \Omega(\epsilon^2 N)$.*

Remark. A simple sqDDH distinguisher takes as input a sample $(h_0, h_1) \in \mathbb{G}^2$, computes the discrete logarithm $x = \log_g(h_0)$ of the first group element and checks whether $h_1 = g^{(x^2)} \in \mathbb{G}$. Theorem 2 indicates that such a distinguisher using advice S and time T and achieving advantage ϵ must satisfy $ST^2 = \widetilde{\Omega}(\epsilon N)$. So, this attack allows the parameter setting $S = T = 1/\epsilon = N^{1/4}$. In contrast, the distinguisher of Theorem 13 allows the better running time and advice complexity roughly $S = T = 1/\epsilon = N^{1/5}$.

Remark. To see the cryptographic significance of Theorem 13, consider the pseudo-random generator $P(x) \stackrel{\text{def}}{=} (g^x, g^{(x^2)})$ that maps $\mathbb{Z}_N$ to $\mathbb{G}^2$. Theorem 13 shows that, for generic algorithms with preprocessing, it is significantly easier to distinguish this PRG from random than it is to compute discrete logs.

Proof Sketch of Theorem 13. The attack that proves the theorem combines two technical tools. The first tool is a general method for using preprocessing to distinguish PRG outputs from random, which we adopt from Bernstein and Lange [13] (De et al. [28] rigorously analyze a more nuanced PRG distinguisher with preprocessing.). The second tool, adopted from the attack of Sect. 7.1, is the idea of taking a walk on the elements of the group, and applying the PRG distinguisher only to the set of points that lie at the end of long walks.

The attack works because a walk that begins at a point of the form $(g^x, g^{(x^2)})$ is likely to hit one of the precomputed endpoints quickly and applying the PRG distinguisher yields an ϵ-biased output value. In contrast, an attack that begins at a point of the form (g^x, g^y) will never hit a precomputed point and applying the distinguisher yields a relatively unbiased output.

The algorithm (illustrated in Fig. 2) takes as input parameters $S, T \in \mathbb{Z}^+$.

As in the attack of Sect. 7.1, we use a random function to define a walk on a graph. In this case, the vertices of the graph are *pairs* of group elements—so every vertex is an element of $\mathbb{G}^2$. We also define the subset of vertices $\mathcal{Y} = \{(g^x, g^{(x^2)}) \mid x \in \mathbb{Z}_N\} \subset \mathbb{G}^2$ that correspond to "yes" instances of the sqDDH problem. The subset $\mathcal{Y}$ is very small relative to the set of all vertices $\mathbb{G}^2$, since $|\mathbb{G}^2| = N^2$, while $|\mathcal{Y}| = N$.

To define the walk on the vertices of this graph, we use a random function F that maps $\mathbb{G}^2 \to \mathbb{Z}_N$. Given a point $(h_0, h_1) \in \mathbb{G}^2$, the walk computes $\alpha \leftarrow F(h_0, h_1)$ and moves to the point $(h_0^\alpha, h_1^{(\alpha^2)}) \in \mathbb{G}^2$. Observe that if the walk starts in $\mathcal{Y}$ (i.e., at a "yes" point), the walk remains inside of $\mathcal{Y}$. If the walk starts at a point outside of $\mathcal{Y}$, the walk remains outside of $\mathcal{Y}$.

Out of the N^2 total vertices in the graph, we choose a set of distinguished or "marked" points $\mathcal{M}$, by marking each point independently at random with probability $1/T$. (In practice, we can choose the set of marked points using a hash function). To each point in $\mathcal{M}$, we assign one of S different "colors," again using a hash function. So there are roughly $N^2/(ST)$ points each with color $1, 2, \ldots, S$.

Given these preliminaries, the algorithm works as follows:

- *Preprocessing phase.* Choose $N/3T^2$ random points in $\mathcal{Y}$. From each of these points, take $2T$ steps of the walk on $\mathbb{G}^2$ that F defines. Halt the walk upon reaching a marked point $m \in \mathcal{M}$. If the walk hits a marked point, store the marked point along with its color c in a table.

  Group the endpoints of the walks by color. For each of the colors $c \in [S]$, find the prefix string $p_c \in \{0,1\}^{\log N}$ that maximizes the sum $\sum H(p_c, m)$, where $H : \{0,1\}^{\log N} \times \mathbb{G}^2 \to \{0,1\}$ is a random function and the sum is taken over the stored marked points m of color c.

  Store the prefix strings $(p_1, \ldots, p_S)$ as the distinguisher's advice.
- *Online phase.* Given a sqDDH challenge $(h_0, h_1) \in \mathbb{G}^2$ as input, perform at most $10T$ steps of the walk on $\mathbb{G}^2$ that the function F defines. As soon as the walk hits a marked point $m \in \mathcal{M}$ of color c, return the value $H(p_c, m)$ as output. If the walk never hits a marked point, output "0" or "1" with probability 1/2 each.

The distinguisher uses $\widetilde{O}(S)$ bits of group-specific advice and runs in time $\widetilde{O}(T)$ as desired. So all we must argue is that the algorithm achieves distinguishing advantage $\epsilon = \Omega(\sqrt{ST^2/N})$. We argue this last step in the full version of this paper [25].

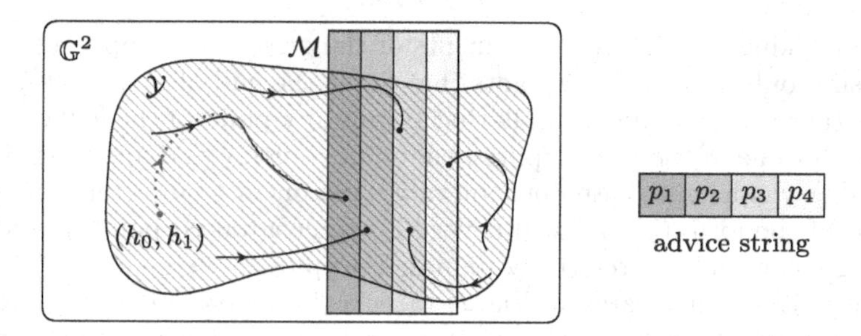

Fig. 2. The preprocessing phase of the sqDDH distinguisher takes walks on the elements of $\mathcal{Y} \subset \mathbb{G}^2$. Each walk terminates upon hitting the set of *marked points* $\mathcal{M}$, which we further partition into S "colors". The advice consists of a string p_c for each of the colors, such that the sum $\sum H(p_c, m)$ is maximized over all the endpoints of color c. In the online phase (in red), the algorithm walks from the input point until hitting a marked point. (Color figure online)

Attacking More-General Problems. The distinguishing attack of Theorem 13 applies to a general class of decisional problems in cyclic groups. Let $(f_1, \ldots, f_\ell)$ be k-variate polynomials and let $\bar{x} = (x_1, \ldots, x_k) \in \mathbb{Z}_N^k$. Then we can define the problem of distinguishing tuples of the form

$$(g^{x_1}, \ldots, g^{x_k}, g^{f_1(\bar{x})}, \ldots, g^{f_\ell(\bar{x})}) \quad \text{from} \quad (g^{x_1}, \ldots, g^{x_k}, g^{r_1}, \ldots, g^{r_\ell}),$$

for uniformly random $x_1, \ldots, x_k, r_1, \ldots, r_\ell \in \mathbb{Z}_N$.

The attack of Theorem 13 applies whenever there exists an index i, a linear function $L : \mathbb{G}^{k+\ell} \to \mathbb{G}$, and a constant $c > 1$ such that $L(\bar{x}, f_1(\bar{x}), \ldots, f_\ell(\bar{x})) = x_i^c$. To apply the attack, first apply $L(\cdot)$ "in the exponent" to the challenge to get a pair $(g^{x_i}, g^{x_i^c}) \in \mathbb{G}^2$ and then run the distinguisher on this pair of elements.

As an example, this attack can distinguish tuples of the form $(g^{x_1}, g^{x_2}, g^{(x_1^2)}, g^{x_1 x_2}, g^{(x_2^2)})$ from random. The attack uses $i = 1$, $L(z_1, z_2, z_3, z_4, z_5) = z_3$, and $c = 2$. Note that this assumption is very closely related to the standard DDH assumption, except that the challenge tuple includes the extra elements $g^{(x_1^2)}$ and $g^{(x_2^2)}$.

Remark. Somewhat surprising is that the distinguishing attack of Theorem 13 *does not* translate to an equivalently strong attack for the DDH problem. The immediate technical obstacle for this is the fact that the distinguishing advantage of the generic PRG distinguisher reduces as the size of the seed space of the PRG grows. That space is of size N in the sqDDH problem, but of size N^2 in the DDH case, which results in a weaker distinguisher.

8 Conclusion

We studied the limits of generic group algorithms with preprocessing for the discrete-logarithm problem and related computational tasks.

In almost all cases, our lower bounds match the best known attacks up to logarithmic factors in group order. The one exception is our lower bound for the decisional Diffie-Hellman problem, in which our lower bound is $ST^2 = \widetilde{\Omega}(\epsilon^2 N)$, but the attack requires computing a discrete logarithm with $ST^2 = \widetilde{O}(\epsilon N)$. When the success probability ϵ is constant, these bounds match. For intermediate values of ϵ, such as $\epsilon = N^{-1/4}$, it is not clear which bound is correct.

One useful task for future work would be to generalize our lower bounds to more complex assumptions, such as Diffie-Hellman assumptions on pairing-equipped groups [17], q-type assumptions [15], or the "uber" assumptions [16,19].

In addition, our upper bounds of Sect. 7 make use of a public random function. Making the attacks fully constructive by removing this heuristic, in the spirit of Fiat and Naor [35] and De et al. [28], would be valuable as well.

Acknowledgements. We would like to thank Dan Boneh for encouraging us to undertake this project and for his advice along the way. We thank Omer Reingold, David Wu, and Benedikt Bünz for fruitful discussions during the early stages of this work. Saba Eskandarian, Steven Galbraith, Sam Kim, and Florian Tramèr gave suggestions that improved the presentation. This work was supported by NSF, DARPA, the Stanford Cyber Initiative, the Simons foundation, a grant from ONR, and an NDSEG Fellowship.

References

1. Abadi, M., Feigenbaum, J., Kilian, J.: On hiding information from an oracle. In: STOC (1987). https://doi.org/10.1145/28395.28417
2. Abusalah, H., Alwen, J., Cohen, B., Khilko, D., Pietrzak, K., Reyzin, L.: Beyond Hellman's time-memory trade-offs with applications to proofs of space. In: Takagi, T., Peyrin, T. (eds.) ASIACRYPT 2017. LNCS, vol. 10625, pp. 357–379. Springer, Cham (2017). https://doi.org/10.1007/978-3-319-70697-9_13
3. Adrian, D., Bhargavan, K., Durumeric, Z., Gaudry, P., Green, M., Halderman, J.A., Heninger, N., Springall, D., Thomé, E., Valenta, L., et al.: Imperfect forward secrecy: how Diffie-Hellman fails in practice. In: CCS (2015). https://doi.org/10.1145/2810103.2813707
4. Aggarwal, D., Maurer, U.: Breaking RSA generically is equivalent to factoring. In: Joux, A. (ed.) EUROCRYPT 2009. LNCS, vol. 5479, pp. 36–53. Springer, Heidelberg (2009). https://doi.org/10.1007/978-3-642-01001-9_2
5. Arora, S., Barak, B.: Computational Complexity: A Modern Approach. Cambridge University Press, Cambridge (2009)
6. Babai, L., Szemeredi, E.: On the complexity of matrix group problems I. In: FOCS (1984). https://doi.org/10.1109/sfcs.1984.715919
7. Bărbulescu, R.: Improvements on the Discrete Logarithm Problem in GF(p). Master's thesis, Ècole Normale Supérieure de Lyon (2011). https://hal.inria.fr/inria-00588713
8. Barbulescu, R., Gaudry, P., Joux, A., Thomé, E.: A heuristic quasi-polynomial algorithm for discrete logarithm in finite fields of small characteristic. In: Nguyen, P.Q., Oswald, E. (eds.) EUROCRYPT 2014. LNCS, vol. 8441, pp. 1–16. Springer, Heidelberg (2014). https://doi.org/10.1007/978-3-642-55220-5_1

9. Barkan, E., Biham, E., Shamir, A.: Rigorous bounds on cryptanalytic time/memory tradeoffs. In: Dwork, C. (ed.) CRYPTO 2006. LNCS, vol. 4117, pp. 1–21. Springer, Heidelberg (2006). https://doi.org/10.1007/11818175_1
10. Bellare, M., Rogaway, P.: Random oracles are practical: a paradigm for designing efficient protocols. In: CCS (1993). https://doi.org/10.1145/168588.168596
11. Bernstein, D., Lange, T.: Two grumpy giants and a baby. In: The Open Book Series, vol. 1, no. 1, pp. 87–111 (2013). https://doi.org/10.2140/obs.2013.1.87
12. Bernstein, D.J.: Curve25519: new Diffie-Hellman speed records. In: Yung, M., Dodis, Y., Kiayias, A., Malkin, T. (eds.) PKC 2006. LNCS, vol. 3958, pp. 207–228. Springer, Heidelberg (2006). https://doi.org/10.1007/11745853_14
13. Bernstein, D.J., Lange, T.: Non-uniform cracks in the concrete: the power of free precomputation. In: Sako, K., Sarkar, P. (eds.) ASIACRYPT 2013. LNCS, vol. 8270, pp. 321–340. Springer, Heidelberg (2013). https://doi.org/10.1007/978-3-642-42045-0_17
14. Boneh, D.: The decision Diffie-Hellman problem. In: Buhler, J.P. (ed.) ANTS 1998. LNCS, vol. 1423, pp. 48–63. Springer, Heidelberg (1998). https://doi.org/10.1007/BFb0054851
15. Boneh, D., Boyen, X.: Short signatures without random oracles. In: Cachin, C., Camenisch, J.L. (eds.) EUROCRYPT 2004. LNCS, vol. 3027, pp. 56–73. Springer, Heidelberg (2004). https://doi.org/10.1007/978-3-540-24676-3_4
16. Boneh, D., Boyen, X., Goh, E.-J.: Hierarchical identity based encryption with constant size ciphertext. In: Cramer, R. (ed.) EUROCRYPT 2005. LNCS, vol. 3494, pp. 440–456. Springer, Heidelberg (2005). https://doi.org/10.1007/11426639_26
17. Boneh, D., Franklin, M.: Identity-based encryption from the Weil pairing. In: Kilian, J. (ed.) CRYPTO 2001. LNCS, vol. 2139, pp. 213–229. Springer, Heidelberg (2001). https://doi.org/10.1007/3-540-44647-8_13
18. Boneh, D., Goh, E.-J., Nissim, K.: Evaluating 2-DNF formulas on ciphertexts. In: Kilian, J. (ed.) TCC 2005. LNCS, vol. 3378, pp. 325–341. Springer, Heidelberg (2005). https://doi.org/10.1007/978-3-540-30576-7_18
19. Boyen, X.: The uber-assumption family. In: Galbraith, S.D., Paterson, K.G. (eds.) Pairing 2008. LNCS, vol. 5209, pp. 39–56. Springer, Heidelberg (2008). https://doi.org/10.1007/978-3-540-85538-5_3
20. Brown, D.: On the provable security of ECDSA. In: Advances in Elliptic Curve Cryptography, pp. 21–40. Cambridge University Press (2005). https://doi.org/10.1017/cbo9780511546570.004
21. Brown, D.R.L.: Generic groups, collision resistance, and ECDSA. Des. Codes Crypt. **35**(1), 119–152 (2005). https://doi.org/10.1007/s10623-003-6154-z
22. Chung, K.M., Lin, H., Mahmoody, M., Pass, R.: On the power of nonuniformity in proofs of security. In: ITCS (2013). http://doi.acm.org/10.1145/2422436.2422480
23. Coppersmith, D.: Modifications to the number field sieve. J. Cryptology **6**(3), 169–180 (1993)
24. Coretti, S., Dodis, Y., Guo, S., Steinberger, J.: Random oracles and non-uniformity. Cryptology ePrint Archive, Report 2017/937 (2017). https://eprint.iacr.org/2017/937
25. Corrigan-Gibbs, H., Kogan, D.: The discrete-logarithm problem with preprocessing. Cryptology ePrint Archive, Report 2017/1113 (2017). https://eprint.iacr.org/2017/1113
26. Cramer, R., Shoup, V.: A practical public key cryptosystem provably secure against adaptive chosen ciphertext attack. In: Krawczyk, H. (ed.) CRYPTO 1998. LNCS, vol. 1462, pp. 13–25. Springer, Heidelberg (1998). https://doi.org/10.1007/BFb0055717

27. Damgård, I., Koprowski, M.: Generic lower bounds for root extraction and signature schemes in general groups. In: Knudsen, L.R. (ed.) EUROCRYPT 2002. LNCS, vol. 2332, pp. 256–271. Springer, Heidelberg (2002). https://doi.org/10.1007/3-540-46035-7_17
28. De, A., Trevisan, L., Tulsiani, M.: Time space tradeoffs for attacks against one-way functions and PRGs. In: Rabin, T. (ed.) CRYPTO 2010. LNCS, vol. 6223, pp. 649–665. Springer, Heidelberg (2010). https://doi.org/10.1007/978-3-642-14623-7_35
29. Dent, A.W.: Adapting the weaknesses of the random oracle model to the generic group model. In: Zheng, Y. (ed.) ASIACRYPT 2002. LNCS, vol. 2501, pp. 100–109. Springer, Heidelberg (2002). https://doi.org/10.1007/3-540-36178-2_6
30. Dent, A.W.: The hardness of the DHK problem in the generic group model. Cryptology ePrint Archive, Report 2006/156 (2006). https://eprint.iacr.org/2006/156
31. Diffie, W., Hellman, M.: New directions in cryptography. IEEE Trans. Inf. Theory **22**(6), 644–654 (1976)
32. Dodis, Y., Guo, S., Katz, J.: Fixing cracks in the concrete: random oracles with auxiliary input, revisited. In: Coron, J.-S., Nielsen, J.B. (eds.) EUROCRYPT 2017. LNCS, vol. 10211, pp. 473–495. Springer, Cham (2017). https://doi.org/10.1007/978-3-319-56614-6_16
33. Dodis, Y., Haitner, I., Tentes, A.: On the instantiability of hash-and-sign RSA signatures. In: Cramer, R. (ed.) TCC 2012. LNCS, vol. 7194, pp. 112–132. Springer, Heidelberg (2012). https://doi.org/10.1007/978-3-642-28914-9_7
34. ElGamal, T.: A public key cryptosystem and a signature scheme based on discrete logarithms. In: Blakley, G.R., Chaum, D. (eds.) CRYPTO 1984. LNCS, vol. 196, pp. 10–18. Springer, Heidelberg (1985). https://doi.org/10.1007/3-540-39568-7_2
35. Fiat, A., Naor, M.: Rigorous time/space tradeoffs for inverting functions. In: STOC (1991). http://doi.acm.org/10.1145/103418.103473
36. Fischlin, M.: A note on security proofs in the generic model. In: Okamoto, T. (ed.) ASIACRYPT 2000. LNCS, vol. 1976, pp. 458–469. Springer, Heidelberg (2000). https://doi.org/10.1007/3-540-44448-3_35
37. Fouque, P.-A., Joux, A., Mavromati, C.: Multi-user collisions: applications to discrete logarithm, Even-Mansour and PRINCE. In: Sarkar, P., Iwata, T. (eds.) ASIACRYPT 2014. LNCS, vol. 8873, pp. 420–438. Springer, Heidelberg (2014). https://doi.org/10.1007/978-3-662-45611-8_22
38. Freeman, D., Scott, M., Teske, E.: A taxonomy of pairing-friendly elliptic curves. J. Cryptology **23**(2), 224–280 (2010). https://doi.org/10.1007/s00145-009-9048-z
39. Galbraith, S.D., Gaudry, P.: Recent progress on the elliptic curve discrete logarithm problem. Des. Codes Crypt. **78**(1), 51–72 (2016). https://doi.org/10.1007/s10623-015-0146-7
40. Galbraith, S.D., Ruprai, R.S.: Using equivalence classes to accelerate solving the discrete logarithm problem in a short interval. In: Nguyen, P.Q., Pointcheval, D. (eds.) PKC 2010. LNCS, vol. 6056, pp. 368–383. Springer, Heidelberg (2010). https://doi.org/10.1007/978-3-642-13013-7_22
41. Gaudry, P., Hess, F., Smart, N.P.: Constructive and destructive facets of Weil descent on elliptic curves. J. Cryptol. **15**(1), 19–46 (2002). https://doi.org/10.1007/s00145-001-0011-x
42. Gennaro, R., Trevisan, L.: Lower bounds on the efficiency of generic cryptographic constructions. In: FOCS (2000). https://doi.org/10.1109/SFCS.2000.892119
43. Gennaro, R.: An improved pseudo-random generator based on discrete log. In: Bellare, M. (ed.) CRYPTO 2000. LNCS, vol. 1880, pp. 469–481. Springer, Heidelberg (2000). https://doi.org/10.1007/3-540-44598-6_29

44. Gordon, D.M.: Discrete logarithms in GF(P) using the number field sieve. SIAM J. Discrete Math. **6**(1), 124–138 (1993). https://doi.org/10.1137/0406010
45. Hellman, M.: A cryptanalytic time-memory trade-off. IEEE Trans. Inf. Theory **26**(4), 401–406 (1980). https://doi.org/10.1109/TIT.1980.1056220
46. Johnson, D., Menezes, A., Vanstone, S.: The elliptic curve digital signature algorithm (ECDSA). Int. J. Inf. Secur. **1**(1), 36–63 (2001). https://doi.org/10.1007/s102070100002
47. Joux, A., Odlyzko, A., Pierrot, C.: The past, evolving present, and future of the discrete logarithm. In: Koç, Ç.K. (ed.) Open Problems in Mathematics and Computational Science, pp. 5–36. Springer, Cham (2014). https://doi.org/10.1007/978-3-319-10683-0_2
48. Kim, T.: Multiple discrete logarithm problems with auxiliary inputs. In: Iwata, T., Cheon, J.H. (eds.) ASIACRYPT 2015. LNCS, vol. 9452, pp. 174–188. Springer, Heidelberg (2015). https://doi.org/10.1007/978-3-662-48797-6_8
49. Koblitz, N., Menezes, A.: Another look at generic groups. Adv. Math. Commun. **1**(1), 13–28 (2007). https://doi.org/10.3934/amc.2007.1.13
50. Koblitz, N., Menezes, A.: Intractable problems in cryptography. In: Conference on Finite Fields and Their Applications (2010). https://doi.org/10.1090/conm/518/10212
51. Koblitz, N., Menezes, A., Vanstone, S.: The state of elliptic curve cryptography. Des. Codes Cryptogr. **19**(2–3), 173–193 (2000). https://doi.org/10.1023/A:1008354106356
52. Koshiba, T., Kurosawa, K.: Short exponent Diffie-Hellman problems. In: Bao, F., Deng, R., Zhou, J. (eds.) PKC 2004. LNCS, vol. 2947, pp. 173–186. Springer, Heidelberg (2004). https://doi.org/10.1007/978-3-540-24632-9_13
53. Kravitz, D.W.: Digital signature algorithm. US Patent 5,231,668 (1993)
54. Kuhn, F., Struik, R.: Random walks revisited: extensions of Pollard's Rho algorithm for computing multiple discrete logarithms. In: Vaudenay, S., Youssef, A.M. (eds.) SAC 2001. LNCS, vol. 2259, pp. 212–229. Springer, Heidelberg (2001). https://doi.org/10.1007/3-540-45537-X_17
55. Leander, G., Rupp, A.: On the equivalence of RSA and factoring regarding generic ring algorithms. In: Lai, X., Chen, K. (eds.) ASIACRYPT 2006. LNCS, vol. 4284, pp. 241–251. Springer, Heidelberg (2006). https://doi.org/10.1007/11935230_16
56. Matyukhin, D.V.: On asymptotic complexity of computing discrete logarithms over GF(p). Discrete Math. Appl. **13**(1), 27–50 (2003). https://doi.org/10.1515/156939203321669546
57. Maurer, U.: Abstract models of computation in cryptography. In: Smart, N.P. (ed.) Cryptography and Coding 2005. LNCS, vol. 3796, pp. 1–12. Springer, Heidelberg (2005). https://doi.org/10.1007/11586821_1
58. Menezes, A.J., Okamoto, T., Vanstone, S.A.: Reducing elliptic curve logarithms to logarithms in a finite field. IEEE Trans. Inf. Theory **39**(5), 1639–1646 (1993). https://doi.org/10.1109/18.259647
59. Mihalcik, J.P.: An analysis of algorithms for solving discrete logarithms in fixed groups. Master's thesis, Naval Postgraduate School (2010). https://calhoun.nps.edu/bitstream/handle/10945/5395/10Mar_Mihalcik.pdf
60. Morris, R., Thompson, K.: Password security: a case history. Commun. ACM **22**(11), 594–597 (1979). https://doi.org/10.1145/359168.359172
61. Nechaev, V.I.: Complexity of a determinate algorithm for the discrete logarithm. Math. Notes **55**(2), 165–172 (1994). https://doi.org/10.1007/bf02113297
62. Odlyzko, A.: Discrete logarithms: the past and the future. Des. Codes Cryptogr. **19**(2), 129–145 (2000). https://doi.org/10.1023/A:1008350005447

63. Oechslin, P.: Making a faster cryptanalytic time-memory trade-off. In: Boneh, D. (ed.) CRYPTO 2003. LNCS, vol. 2729, pp. 617–630. Springer, Heidelberg (2003). https://doi.org/10.1007/978-3-540-45146-4_36
64. van Oorschot, P.C., Wiener, M.J.: On Diffie-Hellman key agreement with short exponents. In: Maurer, U. (ed.) EUROCRYPT 1996. LNCS, vol. 1070, pp. 332–343. Springer, Heidelberg (1996). https://doi.org/10.1007/3-540-68339-9_29
65. Patel, S., Sundaram, G.S.: An efficient discrete log pseudo random generator. In: Krawczyk, H. (ed.) CRYPTO 1998. LNCS, vol. 1462, pp. 304–317. Springer, Heidelberg (1998). https://doi.org/10.1007/BFb0055737
66. Pohlig, S., Hellman, M.: An improved algorithm for computing logarithms over $GF(p)$ and its cryptographic significance (corresp.). IEEE Trans. Inf. Theory **24**(1), 106–110 (1978). https://doi.org/10.1109/tit.1978.1055817
67. Pollard, J.M.: Monte Carlo methods for index computation (mod p). Math. Comput. **32**(143), 918–924 (1978). https://doi.org/10.2307/2006496
68. Schnorr, C.P.: Efficient identification and signatures for smart cards. In: Brassard, G. (ed.) CRYPTO 1989. LNCS, vol. 435, pp. 239–252. Springer, New York (1990). https://doi.org/10.1007/0-387-34805-0_22
69. Schnorr, C.P., Jakobsson, M.: Security of signed ElGamal encryption. In: Okamoto, T. (ed.) ASIACRYPT 2000. LNCS, vol. 1976, pp. 73–89. Springer, Heidelberg (2000). https://doi.org/10.1007/3-540-44448-3_7
70. Shanks, D.: Class number, a theory of factorization, and genera (1971). https://doi.org/10.1090/pspum/020/0316385
71. Shoup, V.: Lower bounds for discrete logarithms and related problems. In: Fumy, W. (ed.) EUROCRYPT 1997. LNCS, vol. 1233, pp. 256–266. Springer, Heidelberg (1997). https://doi.org/10.1007/3-540-69053-0_18
72. Smart, N.P.: The exact security of ECIES in the generic group model. In: Honary, B. (ed.) Cryptography and Coding 2001. LNCS, vol. 2260, pp. 73–84. Springer, Heidelberg (2001). https://doi.org/10.1007/3-540-45325-3_8
73. Smart, N.P.: The discrete logarithm problem on elliptic curves of trace one. J. Cryptol. **12**(3), 193–196 (1999). https://doi.org/10.1007/s001459900052
74. Unruh, D.: Random oracles and auxiliary input. In: Menezes, A. (ed.) CRYPTO 2007. LNCS, vol. 4622, pp. 205–223. Springer, Heidelberg (2007). https://doi.org/10.1007/978-3-540-74143-5_12
75. Wang, P., Zhang, F.: Computing elliptic curve discrete logarithms with the negation map. Inf. Sci. **195**, 277–286 (2012). https://doi.org/10.1016/j.ins.2012.01.044
76. Wee, H.: On obfuscating point functions. In: STOC (2005). http://doi.acm.org/10.1145/1060590.1060669
77. Yao, A.C.C.: Coherent functions and program checkers. In: STOC (1990). http://doi.acm.org/10.1145/100216.100226
78. Ying, J.H.M., Kunihiro, N.: Bounds in various generalized settings of the discrete logarithm problem. In: Gollmann, D., Miyaji, A., Kikuchi, H. (eds.) ACNS 2017. LNCS, vol. 10355, pp. 498–517. Springer, Cham (2017). https://doi.org/10.1007/978-3-319-61204-1_25
79. Yun, A.: Generic hardness of the multiple discrete logarithm problem. In: Oswald, E., Fischlin, M. (eds.) EUROCRYPT 2015. LNCS, vol. 9057, pp. 817–836. Springer, Heidelberg (2015). https://doi.org/10.1007/978-3-662-46803-6_27
80. Zhang, F., Wang, P., Galbraith, S.: Computing elliptic curve discrete logarithms with improved baby-step giant-step algorithm. Adv. Math. Commun. **11**(3), 453–469 (2017). https://doi.org/10.3934/amc.2017038

Best Paper Awards

Simple Proofs of Sequential Work

Bram Cohen[1(✉)] and Krzysztof Pietrzak[2]

[1] Chia Network, San Francisco, USA
bram@chia.network
[2] IST Austria, Klosterneuburg, Austria
pietrzak@ist.ac.at

Abstract. At ITCS 2013, Mahmoody, Moran and Vadhan [MMV13] introduce and construct *publicly verifiable proofs of sequential work*, which is a protocol for proving that one spent sequential computational work related to some statement. The original motivation for such proofs included non-interactive time-stamping and universally verifiable CPU benchmarks. A more recent application, and our main motivation, are blockchain designs, where proofs of sequential work can be used – in combination with proofs of space – as a more ecological and economical substitute for proofs of work which are currently used to secure Bitcoin and other cryptocurrencies.

The construction proposed by [MMV13] is based on a hash function and can be proven secure in the random oracle model, or assuming *inherently sequential* hash-functions, which is a new standard model assumption introduced in their work.

In a proof of sequential work, a prover gets a "statement" χ, a time parameter N and access to a hash-function H, which for the security proof is modelled as a random oracle. Correctness requires that an honest prover can make a verifier accept making only N queries to H, while soundness requires that any prover who makes the verifier accept must have made (almost) N *sequential* queries to H. Thus a solution constitutes a proof that N time passed since χ was received. Solutions must be publicly verifiable in time at most polylogarithmic in N.

The construction of [MMV13] is based on "depth-robust" graphs, and as a consequence has rather poor concrete parameters. But the major drawback is that the prover needs not just N time, but also N space to compute a proof.

In this work we propose a proof of sequential work which is much simpler, more efficient and achieves much better concrete bounds. Most importantly, the space required can be as small as $\log(N)$ (but we get better soundness using slightly more memory than that).

An open problem stated by [MMV13] that our construction does not solve either is achieving a "unique" proof, where even a cheating prover can only generate a single accepting proof. This property would be extremely useful for applications to blockchains.

K. Pietrzak—Supported by the European Research Council (ERC), Horizon 2020, consolidator grant (682815 - TOCNeT).

J. B. Nielsen and V. Rijmen (Eds.): EUROCRYPT 2018, LNCS 10821, pp. 451–467, 2018.
https://doi.org/10.1007/978-3-319-78375-8_15

1 Introduction

1.1 Proofs of Sequential Work (PoSW)

Mahmoody, Moran and Vadhan [MMV13] introduce the notion of proofs of sequential work (PoSW), and give a construction in the random oracle model (ROM), their construction can be made non-interactive using the Fiat-Shamir methodology [FS87]. Informally, with such a non-interactive PoSW one can generate an efficiently verifiable proof showing that some computation was going on for N time steps since some statement χ was received. Soundness requires than one cannot generate such a proof in time much less than N even considering powerful adversaries that have a large number of processors they can use in parallel.

[MMV13] introduce a new standard model assumption called "inherently sequential" hash functions, and show that the random oracle in their construction can be securely instantiated with such hash functions.

Random Oracle Model (ROM). PoSW are easiest to define and prove secure in the ROM, as here we can identify a (potentially parallel) query to the RO as one time step. Throughout this paper we'll work in the ROM, but let us remark that everything can be lifted to the same standard model assumptions (collision resistant and sequential hash functions) used in [MMV13].

A proof of sequential work in the ROM is a protocol between a prover $\mathcal{P}$ and a verifier $\mathcal{V}$, both having access to a random oracle $\mathsf{H} : \{0,1\}^* \to \{0,1\}^w$. Figure 1 illustrates PoSW as constructed in [MMV13] and also here. We'll give a more formal definition in Sect. 1.2.

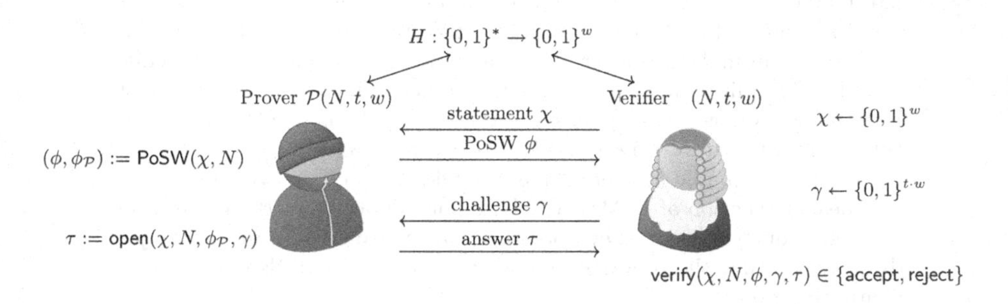

Fig. 1. Proofs of Sequential Work in the ROM as constructed in [MMV13] and this paper. N is the time parameter, i.e., $\mathsf{PoSW}(\chi, N)$ makes N queries to H computing ϕ, and any cheating prover $\tilde{\mathcal{P}}$ that makes $\mathcal{V}$ accept must make almost N sequential queries to H computing ϕ. t is a statistical security parameter, the larger t the better the soundness: any $\tilde{\mathcal{P}}$ making only $(1-\alpha)N$ sequential queries for some $\alpha > 0$, will succeed with probability at most $(1-\alpha)^t$ (e.g. with $t = 21$, a cheating prover making only $0.8 \cdot N$ sequential queries succeed with probability $< 1\%$). w is the output range of our hash function, which we need to be collision resistant and sequential, $w = 256$ is a typical value.

Non-interactive PoSW. The first message is sent from $\mathcal{V}$ to $\mathcal{P}$, and is just a uniformly random w bit string χ. In applications this first message is a "statement" for which we want to prove that N time has passed since it was received. The distribution and domain of this statement is not important, as long as it has sufficiently high min-entropy, because we can always first hash it down to a uniform w bit string using the RO.

As the prover is public-coin, we can make the protocol non-interactive using the Fiat-Shamir heuristic [FS87]: A non-interactive PoSW for statement χ and time parameter N is a tuple (χ, N, ϕ, τ) where the challenge $\gamma = (\mathsf{H}(\phi, 1), \ldots, \mathsf{H}(\phi, t))$ is derived from the proof ϕ by hashing with the RO.

1.2 PoSW Definition

The PoSW we consider are defined by a triple of oracle aided algorithms PoSW, open and verify as defined below.

Common Inputs $\mathcal{P}$ and $\mathcal{V}$ get as common input two statistical security parameters $w, t \in \mathbb{N}$ and a time parameter $N \in \mathbb{N}$. All parties have access to a random oracle $\mathsf{H} : \{0,1\}^* \rightarrow \{0,1\}^w$.

Statement $\mathcal{V}$ samples a random $\chi \leftarrow \{0,1\}^w$ and sends it to $\mathcal{P}$.

Compute PoSW $\mathcal{P}$ computes (ideally, making N queries to H sequentially) a proof $(\phi, \phi_{\mathcal{P}}) := \mathsf{PoSW}^{\mathsf{H}}(\chi, N)$. $\mathcal{P}$ sends ϕ to $\mathcal{V}$ and locally stores $\phi_{\mathcal{P}}$.

Opening Challenge $\mathcal{V}$ samples a random challenge $\gamma \leftarrow \{0,1\}^{t \cdot w}$ and sends it to $\mathcal{P}$.

Open $\mathcal{P}$ computes $\tau := \mathsf{open}^{\mathsf{H}}(\chi, N, \phi_{\mathcal{P}}, \gamma)$ and sends it to $\mathcal{V}$.

Verify $\mathcal{V}$ computes and outputs $\mathsf{verify}^{\mathsf{H}}(\chi, N, \phi, \gamma, \tau) \in \{\mathsf{accept}, \mathsf{reject}\}$.

We require perfect **correctness:** if $\mathcal{V}$ interacts with an honest $\mathcal{P}$, then it will output accept with probability 1. The **soundness** property requires that any potentially malicious prover $\widetilde{\mathcal{P}}$ who makes $\mathcal{V}$ accept with good probability must have queried H "almost" N times sequentially. This holds even if in every round $\widetilde{\mathcal{P}}$ can query H on many inputs in parallel, whereas the honest $\mathcal{P}$ just needs to make a small (in our construction 1, in [MMV13] 2) number of queries per round.

1.3 The [MMV13] and our Construction in a Nutshell

In the construction from [MMV13], the statement χ is used to sample a fresh random oracle H. Then $\mathcal{P}$ uses H to compute "labels" of a directed acyclic graph (DAG) G, where the label of a node is the hash of the labels of its parents. Next, $\mathcal{P}$ computes a Merkle tree commitment of those labels, sends it to $\mathcal{V}$, who then challenges $\mathcal{P}$ to open some of the labels together with its parents.

If $G = (V, E)$ is "depth-robust", which means it has a long path even after removing many vertices, a cheating prover can either (1) try to cheat and make up many of the labels, or (2) compute most of the labels correctly. The security proof now shows that in case (1) the prover will almost certainly not be able to correctly open the Merkle tree commitments, and in case (2) he must make a large number of *sequential* queries: if he cheats on labels of nodes $S \subseteq V$, then the number of sequential queries must be at least as large as the length of the longest path in the subgraph on $V - S$. As G is depth-robust and S is not too large, this path is long.

Our construction is conceptually similar, but our underlying graph is much simpler. We use the nodes in the tree underlying the Merkle commitment not just for the commitment, but also to enforce sequential computation. For this it suffices to add some edges as illustrated in Fig. 3.

Our graph has some convenient properties, for example the parents of a leaf node v are always a subset of the nodes whose labels one needs to provide for the opening of the Merkle tree commitment of the label of v, so checking that the labels are correctly computed and verifying the opening of a leaf label can be done simultaneously without increasing communication complexity and with only a little bit of extra computation.

But most importantly, the labels in our graph can be computed in topological order[1] while keeping only logarithmically many labels in memory at any point, whereas computing the labelling of a depth-robust graph is much more expensive. In fact, because of this property depth-robust graphs are used to build so called memory-hard functions. Concretely, [ABP17] show that if the labelling of a depth-robust graph on N nodes is done in time T using space S, then $T \cdot S \in \Omega(N^2)$. In particular, if one wants to compute the labels in time N, or even just some $O(N)$, then linear $\Omega(N)$ space is required.

There is a caveat. If using only logarithmic memory in our construction, the prover needs to recompute all the labels in the opening phase, whereas one wouldn't need any computation (just some lookups) in the opening phase if everything was stored. This is unfortunate, as it means we get a factor 2 difference in the sequential computation that is claimed, versus what has to actually be done, but some applications need this factor to be close to 1. Fortunately there is a simple trade-off, where using slightly more memory one can make the opening phase much more efficient. The basic idea, which we describe in detail in Sect. 5.4, is to store all the 2^m nodes at some level m of the tree. With this, one can compute any other node making just 2^{n-m} queries.

1.4 More Related Work

Time Release Cryptography. The idea of "time-release" cryptography goes back to [CLSY93, May93].

Most related to PoSW are time-lock puzzles, which were introduced by Rivest, Shamir and Wagner [RSW00]. They give a construction of such puzzles based on the assumption that exponentiation modulo an RSA integer is an "inherently sequential" computation. A recent treatment with new constructions is [BGJ+16].

[1] A topological ordering of the vertices of a DAG is an ordering $v_1, v_2, \ldots, v_{|V|}$ such that there's no path from v_j to v_i whenever $j > i$.

Time-lock puzzles allow a puzzle generator to generate a puzzle with a message of its choice encoded into it, such that this message can only be redeemed by a solver after t steps of sequential work. Such a scheme can be used as a PoSW as the decoded message constitutes a proof of sequential work, but as the puzzle generator has a trapdoor, this proof will not be convincing to anyone else and as it's not public-coin, it can't be made non-interactive by the Fiat-Shamir methodology. Although incomparable, time-lock puzzles seem to be more sophisticated objects than PoSW. Unlike for PoSW, we have no constructions based on random oracles, and [MMV11] give black-box separations showing this might be inherent (we refer to their paper for the exact statements). Existing PoSW (including ours) have another drawback, namely, that the proofs are not unique. We'll discuss this in more detail at the end of Sect. 6.

Proofs of Work. Proofs of work (PoW) – introduced by Dwork and Naor [DN93] – are defined similarly to proof of *sequential* work, but as the name suggests, here one does not require that the work has been done sequentially. Proofs of work are very easy to construct in the random oracle model. The simplest construction of a PoW goes as follows: given a statement χ and a work parameter t, find a nonce x s.t. $\mathsf{H}(\chi, x)$ starts with t zeros. If H is modelled as a random oracle, finding such an x requires an expected 2^t number of queries, while verifying that x is a valid solution just requires a single query. Proofs of work are used to secure several decentralised cryptocurrencies and other blockchain applications, most notably Bitcoin.

1.5 Basic Notation

We denote with $\{0,1\}^{\leq n} \stackrel{\mathsf{def}}{=} \bigcup_{i=0}^{n} \{0,1\}^i$ the set of all binary strings of length at most n, including the empty string ϵ. Concatenation of bitstrings is denoted with $\|$. For $x \in \{0,1\}^*$, $x[i]$ denotes its ith bit, $x[i \dots j] = x[i]\| \dots \|x[j]$ and $|x|$ denotes the bitlength of x.

2 Building Blocks

In Sect. 2.1 below, we define the basic properties of graphs used in this work. Then in Sect. 2.2 we summarize the properties of the random oracle model [BR93] used in our security proof.

2.1 Graphs Basics

To define the [MMV13] and our construction we'll need the following

Definition 1 (Graph Labelling). *Given a directed acyclic graph (DAG) $G = (V, E)$ on vertex set $V = \{0, \dots, N-1\}$ and a hash function $\mathsf{H} : \{0,1\}^* \rightarrow \{0,1\}^w$, the label $\ell_i \in \{0,1\}^w$ of $i \in V$ is recursively computed as (where u is a parent of v if there's a directed edge from u to v)*

$$\ell_i = \mathsf{H}(i, \ell_{p_1}, \dots, \ell_{p_d}) \text{ where } (p_1, \dots, p_d) = \mathsf{parents}(i). \quad (1)$$

Note that for any DAG the labels can be computed making N sequential queries to H by computing them in an arbitrary topological order. If the maximum indegree of G is δ, then the inputs will have length at most $\lceil\log(N)\rceil + \delta \cdot w$.

The PoSW by Mahmoody et al. [MMV13] is based on depth-robust graphs, a notion introduced by Erdős et al. in [EGS75].

Definition 2 (Depth-Robust DAG). *A DAG $G = (V, E)$ is (e, d) depth-robust if for any subset $S \subset V$ of at most $|S| \leq e$ vertices, the subgraph on $V - S$ has a path of length at least d.*

For example, the complete DAG $G = (V, E)$, $|V| = N$, $E = \{(i, j) : 0 \leq i < j \leq N - 1\}$ is $(e, N - e)$ depth-robust for any e, but for PoSW we need a DAG with small indegree. Already [EGS75] showed that $(\Theta(N), \Theta(N))$ depth-robust DAGs with indegree $O(\log(N))$ exist. Mahmoody et al. give an explicit construction with concrete constants, albeit with larger indegree $O(\log^2(N)\text{polyloglog}(N)) \in O(\log^3(N))$.

2.2 Random Oracles Basics

Salting the RO. In [MMV13] and also our construction, all three algorithms PoSW, open and verify described in Sect. 1.2 use the input χ only to sample a random oracle H_χ, for example by using χ as prefix to every input

$$\mathsf{H}_\chi(\cdot) \stackrel{\mathsf{def}}{=} \mathsf{H}(\chi, \cdot).$$

We will sometimes write e.g., $\mathsf{PoSW}^{\mathsf{H}_\chi}(N)$ instead $\mathsf{PoSW}^{\mathsf{H}}(\chi, N)$. Using the uniform χ like this implies that in the proof we can assume that to a cheating prover, the random oracle H_χ just looks like a "fresh" random oracle on which it has no auxiliary information [DGK17].

Random Oracles are Collision Resistant

Lemma 1 (RO is Collision Resistant). *Consider any adversary $\mathcal{A}^{\mathsf{H}}$ which is given access to a random function $\mathsf{H} : \{0,1\}^* \to \{0,1\}^w$. If $\mathcal{A}$ makes at most q queries, the probability it will make two colliding queries $x \neq x', \mathsf{H}(x) = \mathsf{H}(x')$ is at most $q^2/2^{w+1}$.*

Proof. The probability that the output of the i'th query collides with any of the $i - 1$ previous outputs is at most $\frac{i-1}{2^w}$. By the union bound, we get that the probability that any i hits a previous output is at most $\sum_{i=1}^{q} \frac{i-1}{2^w} < \frac{q^2}{2^{w+1}}$. □

Random Oracles are Sequential. Below we show that ROs are "sequential", this is already shown in [MMV13], except that we use concrete parameters instead of asymptotic notations.

Definition 3 (H-sequence). *An H sequence of length s is a sequence $x_0, \ldots, x_s \in \{0,1\}^*$ where for each $i, 1 \leq i < s$, $\mathsf{H}(x_i)$ is contained in x_{i+1} as continuous substring, i.e., $x_{i+1} = a \| \mathsf{H}(x_i) \| b$ for some $a, b \in \{0,1\}^*$.*

Lemma 2 (RO is Sequential). *Consider any adversary $\mathcal{A}^{\mathsf{H}}$ which is given access to a random function $\mathsf{H} : \{0,1\}^* \to \{0,1\}^w$ that it can query for at most $s-1$ rounds, where in each round it can make arbitrary many parallel queries. If $\mathcal{A}$ makes at most q queries of total length Q bits, then the probability that it outputs an H-sequence $x_0, \ldots, x_s$ (as defined above) is at most*

$$q \cdot \frac{Q + \sum_{i=0}^{s} |x_i|}{2^w}$$

Proof. There are two ways $\mathcal{A}$ can output an H sequence $x_0, \ldots, x_s$ making only $s-1$ sequential queries.

1. Lucky guess: It holds that for some i, $\mathsf{H}(x_i)$ is a substring of x_{i+1} and the adversary did not make the query $\mathsf{H}(x_i)$. As H is uniform, the probability of this event can be upper bounded by

$$q \cdot \frac{\sum_{i=0}^{s} |x_i|}{2^w}.$$

2. Collision: The x_i's were not computed sequentially. That is, it holds that for some $1 \leq i \leq j \leq s-1$, a query a_i is made in round i and query a_j in round j where $\mathsf{H}(a_j)$ is a substring of a_i. Again using that H is uniformly random, the probability of this event can be upper bounded by

$$q \cdot \frac{Q}{2^w}.$$

The claimed bound follows by a union bound over the two cases analysed above. □

Thus, whenever an adversary outputs an H-sequence of length s where $q \cdot (Q + \sum_{i=0}^{s} |x_i|)$ is much smaller than 2^w – which in practice will certainly be the case if we use a standard block length like $w = 256$ – we can assume that it made at least s *sequential* queries to H.

Merkle-Damgård. The inputs to our hash function H are of length up to $(\lceil \log N \rceil + 1)w$ bits (assuming $N \leq 2^w$, so the index of a node can be encoded into $\{0,1\}^w$). We can build a function for arbitrary input lengths from a compression function $\mathsf{h} : \{0,1\}^{2w} \to \{0,1\}^w$ using the classical Merkle-Damgård construction [Dam90,Mer90]. Concretely, let y_0 be the statement χ (used for salting as outlined above) and then recursively define

$$\mathsf{H}(x_1, \ldots, x_z) = y_z \text{ where } y_i = \mathsf{h}(x_i, y_{i-1}) \text{ for } i \geq 1.$$

One must be careful with this approach in our construction. As it's possible to compute y_i using only the prefix $x_1, \ldots, x_i$, an adversary might get an advantage by computing such intermediate y_i's before the entire input is known, and thus exploit parallelism to speed up the computation. This can be avoided by requiring that x_1 is always the label of the node that was computed right before the current node.

3 The [MMV13] Construction

In this section we informally describe the PoSW from [MMV13] using the high-level protocol layout from Sect. 1.2.

For any $N = 2^n$, the scheme is specified by a depth-robust DAG $G_n^{\mathsf{DR}} = (V, E)$ on $|V| = N$ vertices. Let $B_n = (V', E')$ denote the full binary tree with $N = 2^n$ leaves (and thus $2N - 1$ nodes) where the edges are directed towards the root. Let G_n^{MMV} be the DAG we get from B_n, by identifying the N leaves of this tree with the N nodes of G_n^{DR} as illustrated in Fig. 2.

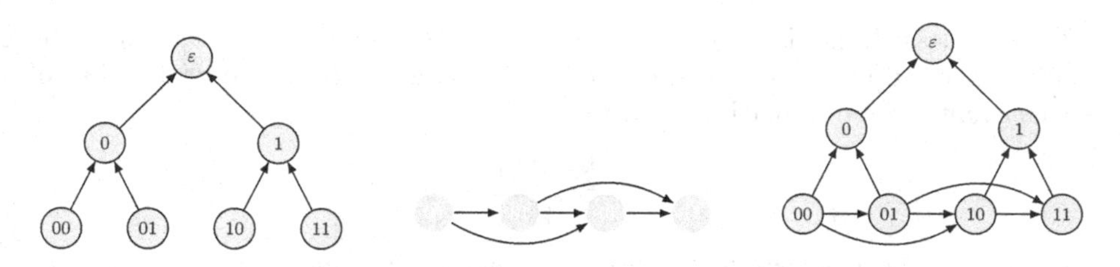

Fig. 2. Illustration of B_2 (left), a (toy example of a) depth-robust graph G_2^{DR} (middle) and the corresponding G_2^{MMV} graph.

Now $(\phi, \phi_{\mathcal{P}}) := \mathsf{PoSW}^{\mathsf{H}_\chi}(N)$ computes and stores the labels $\phi_{\mathcal{P}} = \{\ell_v\}_{v \in \{0,1\}^{\leq n}}$ (cf. Definition 1) of G_n^{MMV} using H_χ as hash function, and sends the label $\phi = \ell_\epsilon$ of the root to $\mathcal{V}$. We remark that in [MMV13] this is described as a two step process, where one first computes a labeling of G_n^{DR} (using a sequential hash function), and then a Merkle-tree commitment of the N labels (using a collision resistant hash function).

After receiving the challenge $\gamma = (\gamma_1, \ldots, \gamma_t)$ from $\mathcal{V}$, the prover $\mathcal{P}$ computes the answer $\tau := \mathsf{open}^{\mathsf{H}_\chi}(N, \phi_{\mathcal{P}}, \gamma)$ as follows: For any $i, 1 \leq i \leq t$, τ contains the opening of the Merkle commitments of the label ℓ_{γ_i}, and the labels of the parents of i, and moreover the labels labels required for the opening of the Merkle commitment of this label.[2]

Upon receiving the answer τ, $\mathcal{V}$ invokes $\mathsf{verify}^{\mathsf{H}_\chi}(N, \phi, \gamma, \tau)$ to check if the labels ℓ_{γ_i} were correctly computed as in Eq. (1), and if the Merkle openings of the labels ℓ_{γ_i} are all correct.

To argue soundness, one uses the fact that G_n^{DR} is (e, d) depth-robust with $e, d = \Theta(N)$. As H_χ is collision resistant, a cheating prover $\tilde{\mathcal{P}}$ must commit to unique labels $\{\ell'_v\}_{v \in \{0,1\}^n}$ of the leaves (that it can later open to). We say that a vertex i is inconsistent if it is not correctly computed from the other labels, i.e.,

$$\ell'_i \neq \mathsf{H}(i, \ell'_{p_1}, \ldots, \ell'_{p_d}) \text{ where } (p_1, \ldots, p_d) = \mathsf{parents}(i)$$

[2] That is, the labels of all siblings of the nodes on the path from this vertex to the root. E.g., for label ℓ_{01} (as in Fig. 2) that would be ℓ_{00} and ℓ_1. To verify, one checks if $\mathsf{H}_\chi(0, \ell_{00}, \ell_{01}) = \ell_0$ and $\mathsf{H}_\chi(\varepsilon, \ell_0, \ell_1) = \ell_\varepsilon = \phi$.

Let β be the number of inconsistent vertices. We make a case distinction:

- If $\beta \geq e$, then one uses the fact that the probability that a cheating prover $\tilde{\mathcal{P}}$ will be asked to open an inconsistent vertex is exponentially (in t) close to 1, namely $1 - \left(\frac{N-\beta}{N}\right)^t$, and thus $\tilde{P}$ will fail to make $\mathcal{V}$ accept except with exponentially small probability.
- if $\beta < e$, then there's a path of length $d = \Theta(N)$ of consistent verticies, which means the labels $\ell'_{i_0}, \ldots, \ell'_{i_{d-1}}$ on this path constitute an H_χ sequence (cf. Definition 3) of length $d-1$, and as H_χ is sequential, $\tilde{\mathcal{P}}$ must almost certainly have made $d-1 = \Theta(N)$ *sequential* queries to H_χ.

4 Definition and Properties of the DAG G_n^{PoSW}

In this section we describe the simple DAG underlying our construction, and prove state some simple combinatorial properties about it which we'll later need in the security proof and to argue efficiency.

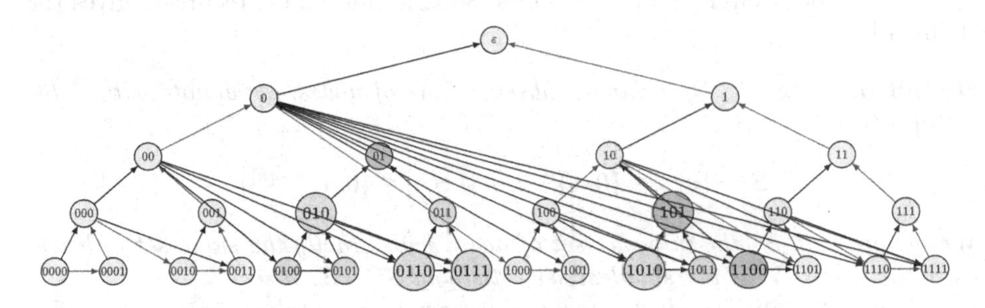

Fig. 3. Illustration of G_4^{PoSW}. The set $S^* = \{01, 101, 1100\}$ – which e.g. could be derived from $S = \{010, 0110, 0111, 101, 1010, 1100\}$ – is shown in red. D_{S^*} is the union of red and orange nodes. $\hat{S} = \hat{S}^*$ are the orange or red leaves. The path of length $2N - 1 - |B_S| = 32 - 1 - 11 = 19$ (as constructed in the proof of Lemma 4) is shown in blue. (Color figure online)

For $n \in \mathbb{N}$ let $N = 2^{n+1} - 1$ and $B_n = (V, E')$ be a complete binary tree of depth n. We identify the N nodes $V = \{0,1\}^{\leq n}$ with the binary strings of length at most n, the empty string ε being the root. We say vertex v is *above* u if $u = v\|a$ for some a (then u is *below* v). The directed edges go from the leaves towards the root

$$E' = \{(x\|b, x) \ : \ b \in \{0,1\}, x \in \{0,1\}^i, i < n\}.$$

We define the DAG $G_n^{\mathsf{PoSW}} = (V, E)$ by starting with B_n, and then adding some edges. Concretely $E = E' \cup E''$ where E'' contains, for all leaves $u \in \{0,1\}^n$, an edge (v, u) for any v that is a left sibling of a node on the path from u to the root ε. E.g., for $v = 1101$ we add the edges $(1100, 1101), (10, 1101), (0, 1101)$, formally

$$E'' = \{(v, u) \; : \; u \in \{0,1\}^n, u = a\|1\|a', v = a\|0\}.$$

Lemma 3. *The labels of* G_n^{PoSW} *can be computed in topological order using only* $w \cdot (n+1)$ *bits of memory.*

Proof. The proof of the lemma follows by induction: to compute the labels of G_n^{PoSW}, start by computing the labels of the left subtree L, which is isomorphic to G_{n-1}^{PoSW}. Once the last label ℓ_0 of L is computed, keep only this one label. Now compute the labels of the right subtree R, which is also isomorphic to G_{n-1}^{PoSW}, except that it has some incoming edges from the left subtree. As all these edges start at ℓ_0, one can compute the labeling of this graph with just w extra bits of space. Once the last label ℓ_1 of R is computed, delete all labels except ℓ_0, ℓ_1, and compute the label of the root $\ell_\varepsilon = \mathsf{H}(\varepsilon, \ell_0, \ell_1)$. The memory required to compute the labels of G_n^{PoSW} is thus the memory required for G_{n-1}^{PoSW} plus w bits (to store ℓ_0 while computing the right subtree). G_1^{PoSW} just has 3 nodes and so can trivially be computed with $3 \cdot w$ bits. Solving this simple recursion gives the claimed bound. □

Definition 4 $(\hat{S}, S^*, D_S)$**.** *For a subset* $S \subseteq V$ *of nodes, we denote with* $\hat{S}$ *the set of leaves below* S

$$\hat{S} = \{v\|u \in \{0,1\}^n \; : \; v \in S, u \in \{0,1\}^{n-|v|}\}$$

We denote with S^* *the minimal set of nodes with exactly the same set of leaves as* S*, i.e.,* $S^* \subseteq V$ *is the smallest set satisfying* $\hat{S}^* = \hat{S}$*.*

We denote with D_S *all the nodes which are in* S *or below some node in* S

$$D_S = \{v\|v' \; : \; v \in S, v' \in \{0,1\}^{\leq n-|v|}\}$$

Lemma 4. *The subgraph of* $G_n^{\mathsf{PoSW}} = (V, E)$ *on vertex set* $V - D_{S^*}$ *(for any* $S \subseteq V$*) has a directed path going through all the* $|V| - |D_{S^*}| = N - |D_{S^*}|$ *nodes.*

Proof. The proof is by induction on n, an example path is illustrated in Fig. 3. The lemma is trivially true for G_0^{PoSW}, which just contains a single node. Assume it's true for G_i^{PoSW}, now G_{i+1}^{PoSW} consists of a root ε, with a left and right subgraph L and R isomorphic to G_i^{PoSW}, with extra edges going from the root of L – which is 0 – to all the leaves in R. If $\varepsilon \in S^*$ the lemma is trivially true as $|V| - |D_{S^*}| = 0$. If $0 \in S^*$, then all of L is in D_{S^*}, in this case just apply the Lemma to $R \equiv G_i^{\mathsf{PoSW}}$, and add an extra last edge $1 \to \varepsilon$. If $0 \notin S^*$, apply the Lemma first to $L \equiv G_i^{\mathsf{PoSW}}$ to get a path that ends in its root 0, then – if $1 \notin S^*$ – apply the lemma to $R \equiv G_i^{\mathsf{PoSW}}$, to get a path that starts at a leaf v. Now add the edges $0 \to v$ and $1 \to \varepsilon$. If $1 \in S^*$ we just add the edge $0 \to \varepsilon$. □

Lemma 5 (trivial). *For any $S^*, S \subset V$, D_{S^*} contains*

$$|\{0,1\}^n \cap D_{S^*}| = \frac{|D_{S^*}| + |S^*|}{2}$$

many leaves.

Proof. Let $S^* = \{v_1, \ldots, v_k\}$, using that $D_{v_i} \cap D_{v_j} = \emptyset$ for all $i \neq j$ (as otherwise S^* would not be minimal), we can write

$$|\{0,1\}^n \cap D_{S^*}| = \sum_{i=1}^{k} |\{0,1\}^n \cap D_{v_i}|.$$

As each D_{v_i} is a full binary tree it has $(|D_{v_i}| + 1)/2$ many leaves, so

$$\sum_{i=1}^{k} |\{0,1\}^n \cap D_{v_i}| = \sum_{i=1}^{k} \frac{|D_{v_i}| + 1}{2} = \frac{|D_{S^*}| + |S^*|}{2}.$$

□

5 Our Construction

In this section we specify our PoSW based on the graphs G_n^{PoSW}.

5.1 Parameters

We have the following parameters:

N The time parameter which we assume is of the form $N = 2^{n+1} - 1$ for an integer $n \in \mathbb{N}$.

$\mathsf{H} : \{0,1\}^{\leq w(n+1)} \rightarrow \{0,1\}^w$ the hash function, which for the security proof is modelled as a random oracle, and which takes as inputs strings of length up to $w(n+1)$ bits.

t A statistical security parameter.

M Memory available to $\mathcal{P}$, we assume it's of the form

$$M = (t + n \cdot t + 1 + 2^{m+1})w$$

for some integer $m, 0 \leq m \leq n$.

5.2 The PoSW, open and verify Algorithms

Our PoSW follows the outline given in Sect. 1.2 using three algorithms PoSW, open and verify. Note that $n \approx \log N$ and $m \approx \log M$ are basically the logarithms of the time parameter N and the memory M (measured in w bit blocks) we allow $\mathcal{P}$ to use.

$(\phi, \phi_{\mathcal{P}}) := \mathsf{PoSW}^{\mathsf{H}_\chi}(N)$: computes the labels $\{\ell_i\}_{i \in \{0,1\}^{\leq n}}$ (cf. Definition 1) of the graph G_n^{PoSW} (as defined in Sect. 4) using H_χ. It stores the labels $\phi_{\mathcal{P}} = \{\ell_i\}_{i \in \{0,1\}^{\leq m}}$ of the m highest layers, and sends the root label $\phi = \ell_\varepsilon$ to $\mathcal{V}$.

$\tau := \mathsf{open}^{\mathsf{H}_\chi}(N, \phi_{\mathcal{P}}, \gamma)$: on challenge $\gamma = (\gamma_1 \ldots, \gamma_t)$, τ contains – for every $i, 1 \leq i \leq t$ – the label ℓ_{γ_i} of node $\gamma_i \in \{0,1\}^n$ and the labels of all siblings of the nodes on the path from γ_i to the root (as in an opening of a Merkle tree commitment), i.e.,

$$\{\ell_k\}_{k \in \mathcal{S}_{\gamma_i}} \text{ where } \mathcal{S}_{\gamma_i} \stackrel{\mathsf{def}}{=} \{\gamma_i[1 \ldots j-1] \| (1 - \gamma_i[j])\}_{j=1 \ldots n}$$

and

$$\tau \stackrel{\mathsf{def}}{=} \{\ell_{\gamma_i}, \{\ell_k\}_{k \in \mathcal{S}_{\gamma_i}}\}_{i=1 \ldots t}.$$

E.g., for $\gamma_i = 0101$ (cf. Fig. 3) τ contains the labels of 0101, 0100, 011, 00 and 1.

If $m = n$, $\mathcal{P}$ stores all labels in $\phi_{\mathcal{P}}$ and thus this needs no queries to H_χ. We'll discuss the case $0 < m < n$ in Sect. 5.4.

$\mathsf{verify}^{\mathsf{H}_\chi}(N, \phi, \gamma, \tau)$: Using that the graphs G_n^{PoSW} have the property that all the parents of a leaf γ_i are in $\mathcal{S}_{\gamma_i}$, for every $i, 1 \leq i \leq t$, one first checks that ℓ_{γ_i} was correctly computed from its parent labels (i.e., as in Eq. 1)

$$\ell_{\gamma_i} \stackrel{?}{=} \mathsf{H}_\chi(i, \ell_{p_1}, \ldots, \ell_{p_d}) \text{ where } (p_1, \ldots, p_d) = \mathsf{parents}(\gamma_i).$$

Then we verify the "Merkle tree like" commitment of ℓ_{γ_i}, by using the labels in τ to recursively compute, for $i = n-1, n-2, \ldots 0$

$$\ell_{\gamma_i[0 \ldots i]} := \mathsf{H}_\chi(\gamma_i[0 \ldots i], \ell_{\gamma_i[0 \ldots i] \| 0}, \ell_{\gamma_i[0 \ldots i] \| 1})$$

and then verifying that the computed root $\ell_{\gamma_i[0 \ldots 0]} = \ell_\varepsilon$ is equal to ϕ received before.

5.3 Security

Theorem 1. *Consider the PoSW from Sect. 5.2, with parameters t, w, N and a "soundness gap" $\alpha > 0$. If $\tilde{\mathcal{P}}$ makes at most $(1-\alpha)N$ sequential queries to H after receiving χ, and at most q queries in total, then $\mathcal{V}$ will output* reject *with probability*

$$1 - (1-\alpha)^t - \frac{2 \cdot n \cdot w \cdot q^2}{2^w}$$

So, for example setting the statistical security parameter to $t = 21$, means a $\tilde{\mathcal{P}}$ who makes only $0.8N$ *sequential* queries will be able to make $\mathcal{V}$ accept with $\leq$1% probability. This is sufficient for some applications, but if we want to use Fiat-Shamir to make the proof non-interactive, the error should be much smaller, say 2^{-50} which we get with $t = 150$.

Proof. The exponentially small $2 \cdot n \cdot w \cdot q^2/2^w$ loss accounts for the assumption we'll make, that $\tilde{\mathcal{P}}$, after receiving χ (1) won't find a collision in H_χ, and (2) whenever it outputs an H_χ-sequence of length s it must have made s sequential queries to H. The concrete bound follows from Lemmas 1 and 2 (recall that H only takes inputs of length $\leq (n+1)w$).

After sending ϕ, $\tilde{\mathcal{P}}$ is committed to the labels $\{\ell'_i\}_{i\in\{0,1\}^{\leq n}}$ it can open. We say a node i is inconsistent if its label ℓ'_i was not correctly computed, i.e.,

$$\ell'_i \neq \mathsf{H}(i, \ell'_{p_1}, \ldots, \ell'_{p_d}) \text{ where } (p_1, \ldots, p_d) = \mathsf{parents}(i).$$

Let us mention that i can be consistent even though $\ell'_i \neq \ell_i$ (ℓ_i denoting the label the honest $\mathcal{P}$ would compute), so being consistent is not the same as being correct. We can also determine these ℓ'_i from just looking at $\tilde{\mathcal{P}}$'s oracle queries, but for the proof we just need that they are unique.

Let $S \subseteq V = \{0,1\}^{\leq n}$ denote all inconsistent nodes. Then by Lemma 4 there's a path going through all the nodes in $V - D_{S^*}$. As all these nodes are consistent, the labels ℓ'_i on this path constitute an H_χ-sequence of length $N - |D_{S^*}|$. If $|D_{S^*}| \leq \alpha N$, $\tilde{\mathcal{P}}$ must have made at least $(1-\alpha)N$ sequential queries (recall we assume $\tilde{\mathcal{P}}$ did not break sequentiality of H_χ), so we now assume

$$|D_{S^*}| > \alpha N = \alpha(2^{n+1} - 1).$$

By Lemma 5 and the above equation

$$|\{0,1\}^n \cap D_{S^*}| = \frac{|D_{S^*}| + |S^*|}{2} > \alpha 2^n. \tag{2}$$

$\tilde{\mathcal{P}}$ will fail to produce a valid proof given t random challenges $\gamma = (\gamma_1, \ldots, \gamma_t)$ if there's at least one γ_i such that a node on the path from γ_i to the root is in S, i.e., $\gamma \cap \hat{S} \neq \emptyset$, or equivalently

$$\gamma \cap D_{S^*} = \gamma \cap \hat{S}^* = \gamma \cap \hat{S} \neq \emptyset.$$

By Eq. (2), and using that every γ_i is uniform

$$\Pr[\gamma_i \notin D_{S^*}] = 1 - |\{0,1\}^n \cap D_{S^*}|/2^n < 1 - \alpha$$

and as the γ_i are also independent

$$\Pr[\gamma \cap D_{S^*} = \emptyset] = \prod_{i=1}^{t} \Pr[\gamma_i \notin D_{S^*}] < (1-\alpha)^t$$

so $\tilde{\mathcal{P}}$ will fail to generate a valid proof with probability $> 1 - (1-\alpha)^t$ as claimed. □

5.4 Efficiency

We'll now discuss the efficiency of the scheme from Sect. 5.2 in terms of proof size, computation and memory requirements.

Proof Size. The exchanged messages χ, ϕ, γ, τ are of length (we need w bits to specify a label and n bits to specify a node)

$$|\chi| = w \qquad |\phi| = w \qquad |\gamma| = t \cdot n \qquad |\tau| \leq t \cdot w \cdot n$$

When we make the proof non-interactive using Fiat-Shamir (where γ is derived from ϕ) the length of a proof for a given statement χ becomes

$$|\phi| + |\tau| \leq w(t \cdot n + 1)$$

With $w = 256$ bit blocks, $t = 150$, which is sufficient to get 2^{-50} security for soundness gap $\alpha = 0.2$ (i.e., a cheating prover must make $0.8N$ sequential queries) and $n = 40$ (i.e., over a trillion steps) the size of the proof is less than 200 KB.

Prover Efficiency. $\mathcal{P}$'s efficiency is dominated by queries to H_χ for computing PoSW and open, so below we just count these.

$\mathsf{PoSW}^{\mathsf{H}_\chi}(N)$ can be computed making N (sequential) queries to H_χ, each input being of length at most $(n+1) \cdot w$ bits, and on average about 1/4 of that (for comparison, the construction from [MMV13] has inputs of length $n^2 \cdot \text{polylog}(n) \cdot w$).

$\mathsf{open}^{\mathsf{H}_\chi}(N, \phi, \gamma)$: Here the efficiency depends on m, which specifies the size of the memory $M = (n + 1 + n \cdot t + 2^{m+1})w$ we allow $\mathcal{P}$ to use. Here $w \cdot n \cdot t$ bits are used to store the values in τ to send back, $(n+1) \cdot w$ bits are used to compute the label (cf. Lemma 3), and $2^{m+1}w$ labels are used to store $\phi_\mathcal{P}$, which contains the labels of the m upmost levels $\{\ell_i\}_{i \in \{0,1\}^{\leq m}}$.

- If $m = n$, $\mathcal{P}$ stored all the labels computed by $\mathsf{PoSW}^{\mathsf{H}_\chi}(N)$, and thus needs no more queries.
- If $m = 0$, $\mathcal{P}$ needs to recompute all N labels. This is not very satisfying, as it means that we'll always have a soundness gap of at least 2: the honest prover needs a total of $2N$ sequential queries (N for each, PoSW and open), whereas (even an honest) prover with $m = n$ space will only require N sequential queries. Fortunately there is a nice trade-off, where already using a small memory means $\mathcal{P}$ just needs to make slightly more than N queries, as described next.
- In the general case $0 \leq m \leq n$, $\mathcal{P}$ needs to compute $2^{n-m+1} - 1$ labels for each of the t challenges, thus at most

  $$t \cdot (2^{n-m+1} - 1)$$

  in total (moreover this can be done making $2^{n-m+1} - 1$ queries sequentially, each with t inputs). E.g. if $m = n/2$, this means $\mathcal{P}$ uses around $\sqrt{N} \cdot w$ bits memory, and $\sqrt{N} \cdot t$ queries on top of the N for computing PoSW. For typical parameters $\sqrt{N} \cdot t$ will be marginal compared to N. More generally, for any $0 \leq \beta \leq 1$, given $N^{1-\beta} \cdot w$ memory means $\mathcal{P}$ needs $N^\beta \cdot t$ queries to compute open (or N^β sequential queries with parallelism t).

  For our example with $w = 256, n = 40, t = 150$, setting, say $m = 20$, means $\mathcal{P}$ uses 70 MB of memory, and the number of queries made by open is less than $N/1000$, which is marginal compared to the N queries made by PoSW.

5.5 Verifier Efficiency

The verifier is extremely efficient, it must only sample a random challenge γ (of length $t \cdot w$) and computing $\mathsf{verify}(\chi, N, \phi, \gamma, \tau)$ can be done making $t \cdot n$ queries to H_χ, each of length at most $n \cdot w$ bits. This is also basically the cost of verifying a non-interactive proof.

6 Conclusions and Open Problems

We constructed a proof of sequential work which is much simpler and enjoys much better parameters than the original construction from [MMV13]. They also state three open questions, two of which we answer in this work. Their first question is:

> **Space Complexity of the Solver.** In our construction of time stamping and time-lock puzzles for time N, the solver keeps the hash labels of a graph of N vertices. Is there any other solution that uses $o(N)$ storage? Or is there any inherent reason that $\Omega(N)$ storage is necessary?

We give a strong negative answer to this question, in our construction the storage of the prover is only $O(\log(N))$. Their second question is:

> **Necessity of Depth-Robust Graphs.** The efficiency and security of our construction is tightly tied to the parameters of depth-robust graph constructions: graphs with lower degree give more efficient solutions, while graphs with higher robustness (the lower bound on the length of the longest path remaining after some of the vertices are removed) give us puzzles with smaller adversarial advantage. An interesting open question is whether the converse also holds: do time-lock puzzles with better parameters also imply the existence of depth-robust graphs with better parameters?

Also here the answer is no. The graphs G_n^{PoSW} we use, as illustrated in Fig. 3, are basically as terrible in terms of depth-robustness as a simple path. For example just removing the vertex 0 cuts the depth in half. Or just removing the $2^{n/2} \approx \sqrt{N}$ vertices in the middle layer, will leave no paths of length more than $\sqrt{N}$. Maybe depth-robustness is the wrong notion to look at here, our graphs satisfy a notion of "weighted" depth-robustness: assign each leaf weight 1, the nodes one layer up weight 2, then 4 etc., doubling with every layer. The total weight of all nodes will be $n2^n$ (2^n for every layer), and one can show hat for any $0 \leq \alpha \leq 1$, removing nodes of weight $\alpha 2^n$, will leave a path of length $(1-\alpha)2^n$.

Apart from [MMV13], depth-robust graphs have been used for cryptographic applications in at least one other case, namely to construct memory-hard functions [ABP17]. Moreover the *proofs of space* protocol from [DFKP15] is quite similar to the PoSW from [MMV13], the main difference being that the underlying graph does not have to be depth-robust, but needs to have high space complexity. Due to this similarities, it seems conceivable that using ideas from this work one can get improved constructions for memory-hard functions and proofs of space.

Let us also mention the third open question asked by [MMV13]. It asks whether a PoSW based only on random oracles can be used to achieve fairness in protocols like coin tossing. We refer to their paper for the details, and just mention that to achieve this, it's sufficient to construct a PoSW with a "unique" proof (note that we already mention this problem in the related work Sect. 1.4). That is, we not only require that to generate a proof one needs to spend sequential time, but for every input (statement and time parameter), it should be hard to come up with two different valid proofs. Such a property would also be very useful in other contexts, like for constructing blockchains, which was the main motivation for this work.

Unfortunately, our construction also does not have unique proofs. It's an intriguing open problem to construct a PoSW with unique proofs and an exponential gap between proof generation and proof verification. Currently, the publicly verifiable function with the largest gap between computation and verification is the sloth function [LW17], which is based on the assumption that computing square roots in a field of size p takes $\log(p)$ times longer than the inverse operation, i.e., squaring. Under this assumption, the gap is $\log(p)$, in practice one would probably use something like $\log(p) \approx 1000$. Sloth is not a time-lock puzzle (as discussed in Sect. 1.4), as one can't sample an input together with its output. It's also not a good PoSW as there's no further speedup if we only want to verify that a lot of sequential time has been spend on the computation, not correctness. Let us also mention that sloth, as well as our PoSW (but not [MMV13]) allow for a speedup of q in verification time if parallelism q is allowed and the proof can be of size linear in q. Basically one adds q "checkpoints" to the proof. These are intermediate states that appear during the computation, and one can verify that each two consecutive checkpoints are consistent independently.

References

[ABP17] Alwen, J., Blocki, J., Pietrzak, K.: Depth-robust graphs and their cumulative memory complexity. In: Coron, J.-S., Nielsen, J.B. (eds.) EUROCRYPT 2017. LNCS, vol. 10212, pp. 3–32. Springer, Cham (2017). https://doi.org/10.1007/978-3-319-56617-7_1

[BGJ+16] Bitansky, N., Goldwasser, S., Jain, A., Paneth, O., Vaikuntanathan, V., Waters, B.: Time-lock puzzles from randomized encodings. In: Sudan, M. (ed.) ITCS 2016, pp. 345–356. ACM, January 2016

[BR93] Bellare, M., Rogaway, P.: Random oracles are practical: a paradigm for designing efficient protocols. In: Ashby, V. (ed.) ACM CCS 93, pp. 62–73. ACM Press, November 1993

[CLSY93] Cai, J.-Y., Lipton, R.J., Sedgewick, R., Yao, A.C.-C.: Towards uncheatable benchmarks. In: Proceedings of the Eighth Annual Structure in Complexity Theory Conference, San Diego, CA, USA, 18–21 May 1993, pp. 2–11 (1993)

[Dam90] Damgård, I.B.: A design principle for hash functions. In: Brassard, G. (ed.) CRYPTO 1989. LNCS, vol. 435, pp. 416–427. Springer, New York (1990). https://doi.org/10.1007/0-387-34805-0_39

[DFKP15] Dziembowski, S., Faust, S., Kolmogorov, V., Pietrzak, K.: Proofs of space. In: Gennaro, R., Robshaw, M. (eds.) CRYPTO 2015. LNCS, vol. 9216, pp. 585–605. Springer, Heidelberg (2015). https://doi.org/10.1007/978-3-662-48000-7_29

[DGK17] Dodis, Y., Guo, S., Katz, J.: Fixing cracks in the concrete: random oracles with auxiliary input, revisited. In: Coron, J.-S., Nielsen, J.B. (eds.) EUROCRYPT 2017. LNCS, vol. 10211, pp. 473–495. Springer, Cham (2017). https://doi.org/10.1007/978-3-319-56614-6_16

[DN93] Dwork, C., Naor, M.: Pricing via processing or combatting junk mail. In: Brickell, E.F. (ed.) CRYPTO 1992. LNCS, vol. 740, pp. 139–147. Springer, Heidelberg (1993). https://doi.org/10.1007/3-540-48071-4_10

[EGS75] Erdoes, P., Graham, R.L., Szemeredi, E.: On sparse graphs with dense long paths. Technical report, Stanford, CA, USA (1975)

[FS87] Fiat, A., Shamir, A.: How to prove yourself: practical solutions to identification and signature problems. In: Odlyzko, A.M. (ed.) CRYPTO 1986. LNCS, vol. 263, pp. 186–194. Springer, Heidelberg (1987). https://doi.org/10.1007/3-540-47721-7_12

[LW17] Lenstra, A.K., Wesolowski, B.: Trustworthy public randomness with sloth, unicorn, and trx. IJACT **3**(4), 330–343 (2017)

[May93] May, T.C.: Timed-release crypto (1993). http://www.hks.net/cpunks/cpunks-0/1460.html

[Mer90] Merkle, R.C.: A certified digital signature. In: Brassard, G. (ed.) CRYPTO 1989. LNCS, vol. 435, pp. 218–238. Springer, New York (1990). https://doi.org/10.1007/0-387-34805-0_21

[MMV11] Mahmoody, M., Moran, T., Vadhan, S.: Time-lock puzzles in the Random Oracle model. In: Rogaway, P. (ed.) CRYPTO 2011. LNCS, vol. 6841, pp. 39–50. Springer, Heidelberg (2011). https://doi.org/10.1007/978-3-642-22792-9_3

[MMV13] Mahmoody, M., Moran, T., Vadhan, S.P.: Publicly verifiable proofs of sequential work. In: Kleinberg, R.D. (ed.) ITCS 2013, pp. 373–388. ACM, January 2013

[RSW00] Rivest, R.L., Shamir, A., Wagner, D.A.: Time-lock puzzles and timed release crypto. Technical report (2000)

Two-Round Multiparty Secure Computation from Minimal Assumptions

Sanjam Garg[(✉)] and Akshayaram Srinivasan

University of California, Berkeley, USA
{sanjamg,akshayaram}@berkeley.edu

Abstract. We provide new two-round multiparty secure computation (MPC) protocols assuming the minimal assumption that two-round oblivious transfer (OT) exists. If the assumed two-round OT protocol is secure against semi-honest adversaries (in the plain model) then so is our two-round MPC protocol. Similarly, if the assumed two-round OT protocol is secure against malicious adversaries (in the common random/reference string model) then so is our two-round MPC protocol. Previously, two-round MPC protocols were only known under relatively stronger computational assumptions. Finally, we provide several extensions.

1 Introduction

Can a group of n mutually distrusting parties compute a joint function of their private inputs without revealing anything more than the output to each other? This is the classical problem of secure computation in cryptography. Yao [Yao86] and Goldreich et al. [GMW87] provided protocols for solving this problem in the two-party (2PC) and the multiparty (MPC) cases, respectively.

A remarkable aspect of the 2PC protocol based on Yao's garbled circuit construction is its simplicity and the fact that it requires only two-rounds of communication. Moreover, this protocol can be based just on the minimal assumption that two-round 1-out-of-2 oblivious transfer (OT) exists. Two-round OT can itself be based on a variety of computational assumptions such as the Decisional Diffie-Hellman Assumption [AIR01,NP01,PVW08], quadratic residuosity assumption [HK12,PVW08] or the learning-with-errors assumption [PVW08].

In contrast, much less is known about the assumptions that two-round MPC can be based on (constant-round MPC protocols based on any OT protocol are well-known [BMR90]). In particular, two-round MPC protocols are only known under assumptions such as indistinguishability obfuscation [GGHR14,GGH+13]

Research supported in part from AFOSR YIP Award, DARPA/ARL SAFEWARE Award W911NF15C0210, AFOSR Award FA9550-15-1-0274, and research grants by the Okawa Foundation, Visa Inc., and Center for Long-Term Cybersecurity (CLTC, UC Berkeley). The views expressed are those of the author and do not reflect the official policy or position of the funding agencies.

J. B. Nielsen and V. Rijmen (Eds.): EUROCRYPT 2018, LNCS 10821, pp. 468–499, 2018.
https://doi.org/10.1007/978-3-319-78375-8_16

(or, witness encryption [GLS15,GGSW13]), LWE [CM15,MW16,BP16,PS16], or bilinear maps [GS17,BF01,Jou04]. In summary, there is a significant gap between assumptions known to be sufficient for two-round MPC and the assumptions that known to be sufficient for two-round 2PC (or, two-round OT). This brings us to the following main question:

What are the minimal assumptions under which two-round MPC can be constructed?

1.1 Our Result

In this work, we give two-round MPC protocols assuming only the necessary assumption that two-round OT exists. In a bit more detail, our main theorem is:

Theorem 1 (Main Theorem). *Let $\mathcal{X} \in$ {semi-honest in plain model, malicious in common random/reference sting model}. Assuming the existence of a two-round $\mathcal{X}$-OT protocol, there exists a compiler that transforms any polynomial round, $\mathcal{X}$-MPC protocol into a two-round, $\mathcal{X}$-MPC protocol.*

Previously, such compilers [GGHR14,GLS15,GS17] were only known under comparatively stronger computational assumptions such as indistinguishability obfuscation [BGI+01,GGH+13], witness encryption [GGSW13], or using bilinear maps [GS17,BF01,Jou04]. Additionally, two-round MPC protocols assuming the learning-with-errors assumptions were known [MW16,PS16,BP16] in the CRS model satisfying semi-malicious security.[1] We now discuss instantiations of the above compiler with known protocols (with larger round complexity) that yield two-round MPC protocols in various settings under minimal assumptions.

Semi-honest Case. Plugging in the semi-honest secure MPC protocol by Goldreich, Micali, and Wigderson [GMW87], we get the following result:

Corollary 1. *Assuming the existence of semi-honest, two-round oblivious transfer in the plain model, there exists a semi-honest, two-round multiparty computation protocol in the plain model.*

Previously, two-round plain model semi-honest MPC protocols were only known assuming indistinguishability obfuscation [BGI+01,GGH+13], or witness encryption [GGSW13] or bilinear maps [GS17] or from DDH for a constant number of parties [BGI17]. Thus, using two-round plain model OT [NP01,AIR01, HK12] based on standard number theoretic assumptions such as DDH or QR, this work yields the first two-round semi-honest MPC protocol for polynomial number of parties in the plain model under the same assumptions.

[1] Semi-malicious security is a strengthening of the semi-honest security wherein the adversary is allowed to choose its random tape arbitrarily. Ashrov et al. [AJL+12] showed that any protocol satisfying semi-malicious security could be upgraded to one with malicious security additionally using Non-Interactive Zero-Knowledge proofs (NIZKs).

Malicious Case. Plugging in the maliciously secure MPC protocol by Kilian [Kil88] or by Ishai et al. [IPS08] based on any oblivious transfer, we get the following corollary:

Corollary 2. *Assuming the existence of UC secure, two-round oblivious transfer against static, malicious adversaries, there exists a UC secure, two-round multiparty computation protocol against static, malicious adversaries.*

Previously, all known two-round maliciously secure MPC protocols required additional use of non-interactive zero-knowledge proofs. As a special case, using a DDH based two-round OT protocol (e.g., [PVW08]), this work yields the first two-round malicious MPC protocol in the common random string model under the DDH assumption.

Extensions. In addition to the above main results we obtain several extensions and refer the reader to the main body for details.

Concurrent Work. In a concurrent and independent work, Benhamouda and Lin [BL18] also construct two-round multiparty computation from two-round oblivious transfer. Their construction against semi-honest adversaries is proven under the minimal assumption that two-round, semi-honest oblivious transfer exists. However, their construction against malicious adversaries additionally requires the existence of non-interactive zero-knowledge proofs. Additionally, in the plain model they give a construction of 5-round maliciously secure MPC from 5-round maliciously secure oblivious transfer.

2 Technical Overview

Towards demonstrating the intuition behind our result, in this section, we show how to squish the round complexity of a very simple "toy" protocol to two. Additionally, we sketch how these ideas extend to the general setting and also work in the malicious case. We postpone the details to later sections.

Background: "Garbled Circuits that talk." The starting point of this work is a recent work of Garg and Srinivasan [GS17] that obtains constructions of two-round MPC from bilinear maps. Building on [GGHR14, GLS15], the key idea behind [GS17] is a new method for enabling "garbled circuits to talk," which the authors call "garbled protocols." It is natural to imagine how "garbled circuits that can talk" might be useful for squishing the round complexity of any protocol. By employing this technique, a party can avoid multiple rounds of interaction just by sending a garbled circuit that interacts with the other parties on its behalf. At a technical level, a garbled circuit can "speak" by just outputting a value. However, the idea of enabling garbled circuits to "listen" without incurring any additional interaction poses new challenges. A bit more precisely, "listen" means that a garbled circuit can take as input a bit obtained via a joint computation on its secret state and the secret states of two or more other parties.

In [GS17], this idea was implemented by constructing a special purpose witness encryption [GGSW13,BH15,GOVW12,CDG+17,DG17] using specific algebraic properties of non-interactive zero-knowledge (NIZK) proofs by Gorth, Ostrovsky and Sahai [GOS06]. The key contribution of this work is a realization of the intuition of "garbled circuits that talk" using any two-round OT protocols rather than a specific NIZK proof system. In particular, we avoid using any specialized algebraic properties of the underlying primitives. At the heart of our construction is the following novel use of two-round OT protocols: in our MPC protocol multiple instances of the underlying two-round OT protocol are executed and the secret receiver's random coins used in some of these executed OT instances are revealed to the other parties. As we explain later, this is done carefully so that the security of the MPC protocol is not jeopardized.

A "toy" protocol for successive ANDs. Stripping away technical details, we highlight our core new idea in the context of a "toy" example, where a garbled circuit will need to listen to one bit. Later, we briefly sketch how this core idea can be used to squish the round complexity of any arbitrary round MPC protocol to two. Recall that, in one round, each party sends a message depending on its secret state and the messages received in prior rounds.

Consider three parties P_1, P_2 and P_3 with inputs α, β, and γ (which are single bits), respectively. Can we realize a protocol such that the parties learn $f(\alpha, \beta, \gamma) = (\alpha, \alpha \wedge \beta, \alpha \wedge \beta \wedge \gamma)$ and nothing more? Can we realize a two-round protocol for the same task? Here is a very simple three-round information theoretic protocol Φ (in the semi-honest setting) for this task: *In the first round,* P_1 sends its input α to P_2 and P_3. *In the second round,* P_2 computes $\delta = \alpha \wedge \beta$ and sends it to P_1 and P_3. Finally, *in the third round,* P_3 computes $\gamma \wedge \delta$ and sends it to P_1 and P_2.

Compiling Φ into a two-round protocol. The key challenge that we face is that the third party's message depends on the second party's message, and the second party's message depends on the first party's message. We will now describe our approach to overcome this three-way dependence using two-round oblivious transfer and thus squish this protocol Φ into a two-round protocol.

We assume the following notation for a two-round OT protocol. *In the first round,* the receiver with choice bit β generates $c = \mathsf{OT}_1(\beta; \omega)$ using ω as the randomness and passes c to the sender. Then *in the second round,* the sender responds with its OT response $d = \mathsf{OT}_2(c, s_0, s_1)$ where s_0 and s_1 are its input strings. *Finally,* using the OT response d and its randomness ω, the receiver recovers s_β. In our protocol below, we will use a circuit $C[\gamma]$ that has a bit γ hardwired in it and that on input a bit δ outputs $\gamma \wedge \delta$. At a high level in our protocol, we will have P_2 and P_3 send extra messages in the first and the second rounds, respectively, so that the third round can be avoided. Here is our protocol:

- **Round 1:** P_1 sends α to P_2 and P_3. P_2 prepares $c_0 = \mathsf{OT}_1(0 \wedge \beta; \omega_0)$ and $c_1 = \mathsf{OT}_1(1 \wedge \beta; \omega_1)$ and sends (c_0, c_1) to P_2 and P_3.

- **Round 2:** P_2 sends $(\alpha \wedge \beta, \omega_\alpha)$ to P_1 and P_3. P_3 garbles $\mathsf{C}[\gamma]$ obtaining $\tilde{\mathsf{C}}$ and input labels lab_0 and lab_1. It computes $d = \mathsf{OT}_2(c_\alpha, \mathsf{lab}_0, \mathsf{lab}_1)$ and sends $(\tilde{\mathsf{C}}, d)$ to P_1 and P_2.
- **Output Evaluation:** Every party recovers lab_δ where $\delta = \alpha \wedge \beta$ from d using ω_α. Next, it evaluates the garbled circuit $\tilde{\mathsf{C}}$ using lab_δ which outputs $\gamma \wedge \delta$ as desired.

Intuitively, in the protocol above P_2 sends two first OT messages c_0 and c_1 that are prepared assuming α is 0 and assuming α is 1, respectively. Note that P_3 does not know α at the beginning of the first round, but P_3 does know it at the end of the first round. Thus, P_3 just uses c_α while discarding $c_{1-\alpha}$ in preparing its messages for the second round. This achieves the three-way dependency while only using two-rounds. Furthermore, P_2's second round message reveals the randomness ω_α enabling all parties (and not just P_2 and P_3) to obtain the label lab_δ which can then be used for evaluation of $\tilde{\mathsf{C}}$. In summary, via this mechanism, the garbled circuit $\tilde{\mathsf{C}}$ was able to "listen" to the bit δ that P_3 did not know when generating the garbled circuit.

The above description highlights our ideas for squishing round complexity of an incredibly simple toy protocol where only one bit was being "listened to." Moreover, the garbled circuit "speaks" or outputs $\gamma \wedge \delta$, which is obtained by all parties. In the above "toy" example, P_3's garbled circuit computes a gate that takes only one bit as input. To compute a gate with two bit inputs, P_2 will need to send four first OT messages in the first round instead of two.

Squishing arbitrary protocols. Our approach to enable garbled circuits to "listen to" a larger number of bits with complex dependencies is as follows. We show that any MPC protocol Φ between parties $P_1, \cdots P_n$ can be transformed into one satisfying the following format. First, the parties execute a preprocessing step; namely, each party P_i computes some randomized function of its input x_i obtaining public value z_i which is shared with everyone else and private value v_i. z_i is roughly an encryption of x_i using randomness from v_i as a one-time pad. v_i also contains random bits that will be used as one-time pad to encrypt bits sent later by P_i. *Second*, each party sets its local state $\mathsf{st}_i = (z_1 \| \ldots \| z_n) \oplus v_i$. That places us at the beginning of the protocol execution phase. In our transformed protocol Φ can be written as a sequence of T *actions*. For each $t \in [T]$ the tth action $\phi_t = (i, f, g, h)$ involves party P_i computing *one* NAND gate; it sets $\mathsf{st}_{i,h} = \mathsf{NAND}(\mathsf{st}_{i,f}, \mathsf{st}_{i,g})$ and sends $v_{i,h} \oplus \mathsf{st}_{i,h}$ to all the other parties. Our transformed protocol is such that for any bit $\mathsf{st}_{i,h}$, the bit $v_{i,h}$ is unique and acts as the one-time pad to hide it from the other parties. (Some of the bits in v_i are set to 0. These bits do not need to be hidden from other parties.) *To complete this action*, each party P_j for $j \neq i$ sets $\mathsf{st}_{j,h}$ to be the received bit. After all the actions are completed, each party P_j outputs a function of its local state st_j. In this transformed MPC protocol, in any round only one bit is sent based on just one gate (i.e., the gate obtained as $v_{i,h} \oplus \mathsf{NAND}(\mathsf{st}_{i,f}, \mathsf{st}_{i,g})$ with inputs $\mathsf{st}_{i,f}$ and $\mathsf{st}_{i,g}$, where $v_{i,h}$ is hardwired inside it) computation on two bits. Thus, we can use the above "toy" protocol to achieve this effect.

To squish the round complexity of this transformed protocol, *in the first round*, we will have each party follow the pre-processing step from above along with a bunch of carefully crafted first OT messages as in our "toy" protocol. *In the second round*, parties will send a garbled circuit that is expected to "speak" and "listen" to the garbled circuits of the other parties. So when $\phi_1 = (i, f, g, h)$ is executed, we have that the garbled circuit sent by party P_i speaks and all the others listen. Each of these listening garbled circuits uses our "toy" protocol idea from above. After completion of the first action, all the garbled circuits will have read the transcript of communication (which is just the one bit communicated in the first action ϕ_1). Next, the parties need to execute action $\phi_2 = (i, f, g, h)$ and this is done like the first action, and the process continues. This completes the main idea of our construction. Building on this idea, we obtain a compiler that assuming semi-honest two-round OT transforms any semi-honest MPC protocol into a two-round semi-honest MPC protocol. Furthermore, if the assumed semi-honest two-round OT protocol is in the plain model then so will be the resulting MPC protocol.

Compilation in the Malicious Case. The protocol ideas described above only achieve semi-honest security and additional use of non-interactive zero-knowledge (NIZK) proofs [BFM88,FLS90] is required to upgrade security to malicious [AJL+12,MW16]. This has been the case for all known two-round MPC protocol constructions. In a bit more detail, by using NIZKs parties can (without increasing the round complexity) prove in zero-knowledge that they are following protocol specifications. The use of NIZKs might seem essential to such protocols. However, we show that this can be avoided. Our main idea is as follows: instead of proving that the garbled circuits are honestly generated, we require that the garbled circuits prove to each other that the messages they send are honestly generated. Since our garbled circuits can "speak" and "listen" over several rounds without increasing the round complexity of the squished protocol, therefore we can instead use interactive zero-knowledge proof system and avoid NIZKs. Building on this idea we obtain two-round MPC protocols secure against malicious adversaries. We elaborate on this new idea and other issues involved in subsequent sections.

3 Preliminaries

We recall some standard cryptographic definitions in this section. Let λ denote the security parameter. A function $\mu(\cdot) : \mathbb{N} \rightarrow \mathbb{R}^{+}$ is said to be negligible if for any polynomial $\mathsf{poly}(\cdot)$ there exists λ_0 such that for all $\lambda > \lambda_0$ we have $\mu(\lambda) < \frac{1}{\mathsf{poly}(\lambda)}$. We will use $\mathsf{negl}(\cdot)$ to denote an unspecified negligible function and $\mathsf{poly}(\cdot)$ to denote an unspecified polynomial function.

For a probabilistic algorithm A, we denote $A(x; r)$ to be the output of A on input x with the content of the random tape being r. When r is omitted, $A(x)$ denotes a distribution. For a finite set S, we denote $x \leftarrow S$ as the process of sampling x uniformly from the set S. We will use PPT to denote Probabilistic Polynomial Time algorithm.

3.1 Garbled Circuits

Below we recall the definition of garbling scheme for circuits [Yao86] (see Applebaum et al. [AIK04, AIK05], Lindell and Pinkas [LP09] and Bellare et al. [BHR12] for a detailed proof and further discussion). A garbling scheme for circuits is a tuple of PPT algorithms $(\mathsf{Garble}, \mathsf{Eval})$. Garble is the circuit garbling procedure and Eval is the corresponding evaluation procedure. More formally:

- $(\widetilde{\mathsf{C}}, \{\mathsf{lbl}_{w,b}\}_{w \in \mathsf{inp}(C), b \in \{0,1\}}) \leftarrow \mathsf{Garble}\left(1^\lambda, C\right)$: Garble takes as input a security parameter 1^λ, a circuit C, and outputs a *garbled circuit* $\widetilde{\mathsf{C}}$ along with labels $\mathsf{lbl}_{w,b}$ where $w \in \mathsf{inp}(C)$ ($\mathsf{inp}(C)$ is the set of input wires of C) and $b \in \{0,1\}$. Each label $\mathsf{lbl}_{w,b}$ is assumed to be in $\{0,1\}^\lambda$.
- $y \leftarrow \mathsf{Eval}\left(\widetilde{\mathsf{C}}, \{\mathsf{lbl}_{w,x_w}\}_{w \in \mathsf{inp}(C)}\right)$: Given a garbled circuit $\widetilde{\mathsf{C}}$ and a sequence of input labels $\{\mathsf{lbl}_{w,x_w}\}_{w \in \mathsf{inp}(C)}$ (referred to as the garbled input), Eval outputs a string y.

Correctness. For correctness, we require that for any circuit C and input $x \in \{0,1\}^{|\mathsf{inp}(C)|}$ we have that:

$$\Pr\left[C(x) = \mathsf{Eval}\left(\widetilde{\mathsf{C}}, \{\mathsf{lbl}_{w,x_w}\}_{w \in \mathsf{inp}(C)}\right)\right] = 1$$

where $(\widetilde{\mathsf{C}}, \{\mathsf{lbl}_{w,b}\}_{w \in \mathsf{inp}(C), b \in \{0,1\}}) \leftarrow \mathsf{Garble}\left(1^\lambda, C\right)$.

Security. For security, we require that there exists a PPT simulator Sim such that for any circuit C and input $x \in \{0,1\}^{|\mathsf{inp}(C)|}$, we have that

$$\left(\widetilde{\mathsf{C}}, \{\mathsf{lbl}_{w,x_w}\}_{w \in \mathsf{inp}(C)}\right) \stackrel{c}{\approx} \mathsf{Sim}\left(1^{|C|}, 1^{|x|}, C(x)\right)$$

where $(\widetilde{\mathsf{C}}, \{\mathsf{lbl}_{w,b}\}_{w \in \mathsf{inp}(C), b \in \{0,1\}}) \leftarrow \mathsf{Garble}\left(1^\lambda, C\right)$ and $\stackrel{c}{\approx}$ denotes that the two distributions are computationally indistinguishable.

3.2 Universal Composability Framework

We work in the Universal Composition (UC) framework [Can01] to formalize and analyze the security of our protocols. (Our protocols can also be analyzed in the stand-alone setting, using the composability framework of [Can00a], or in other UC-like frameworks, like that of [PW00].) We refer the reader to [Can00b] for details.

3.3 Oblivious Transfer

In this paper, we consider a 1-out-of-2 *oblivious transfer* protocol (OT), similar to [CCM98, NP01, AIR01, DHRS04, HK12] where one party, the *sender*, has input composed of two strings (s_0, s_1) and the input of the second party, the *receiver*, is a bit β. The receiver should learn s_β and nothing regarding $s_{1-\beta}$ while the sender should gain no information about β.

Security of the oblivious transfer (OT) functionality can be described easily by an ideal functionality $\mathcal{F}_{\mathsf{OT}}$ as is done in [CLOS02]. However, in our constructions the receiver needs to reveal the randomness (or a part of the randomness) it uses in an instance of two-round OT to other parties. Therefore, defining security as an ideal functionality raises issues require care and issues similar to one involved in defining ideal public-key encryption functionality [Can05, p. 96] arrise. Thus, in our context, it is much easier to directly work with a two-round OT protocol. We define the syntax and the security guarantees of a two-round OT protocol below.

Semi-honest Two-Round Oblivious Transfer. A two-round semi-honest OT protocol $\langle S, R\rangle$ is defined by three probabilistic algorithms $(\mathsf{OT}_1, \mathsf{OT}_2, \mathsf{OT}_3)$ as follows. The receiver runs the algorithm OT_1 which takes the security parameter 1^λ, and the receiver's input $\beta \in \{0,1\}$ as input and outputs ots_1 and ω.[2] The receiver then sends ots_1 to the sender, who obtains ots_2 by evaluating $\mathsf{OT}_2(\mathsf{ots}_1, (s_0, s_1))$, where $s_0, s_1 \in \{0,1\}^\lambda$ are the sender's input messages. The sender then sends ots_2 to the receiver who obtains s_β by evaluating $\mathsf{OT}_3(\mathsf{ots}_2, (\beta, \omega))$.

- **Correctness.** For every choice bit $\beta \in \{0,1\}$ of the receiver and input messages s_0 and s_1 of the sender we require that, if $(\mathsf{ots}_1, \omega) \leftarrow \mathsf{OT}_1(1^\lambda, \beta)$, $\mathsf{ots}_2 \leftarrow \mathsf{OT}_2(\mathsf{ots}_1, (s_0, s_1))$, then $\mathsf{OT}_3(\mathsf{ots}_2, (\beta, \omega)) = s_\beta$ with overwhelming probability.
- **Receiver's security.** We require that

$$\left\{\mathsf{ots}_1 : (\mathsf{ots}_1, \omega) \leftarrow \mathsf{OT}_1(1^\lambda, 0)\right\} \stackrel{c}{\approx} \left\{\mathsf{ots}_1 : (\mathsf{ots}_1, \omega) \leftarrow \mathsf{OT}_1(1^\lambda, 1)\right\}.$$

- **Sender's security.** We require that for any choice of $\beta \in \{0,1\}$, overwhelming choices of ω' and any strings $K_0, K_1, L_0, L_1 \in \{0,1\}^\lambda$ with $K_\beta = L_\beta$, we have that

$$\left\{\beta, \omega', \mathsf{OT}_2(1^\lambda, \mathsf{ots}_1, K_0, K_1)\right\} \stackrel{c}{\approx} \left\{\beta, \omega', \mathsf{OT}_2(1^\lambda, \mathsf{ots}_1, L_0, L_1)\right\}$$

where $(\mathsf{ots}_1, \omega) := \mathsf{OT}_1(1^\lambda, \beta; \omega')$.

Constructions of semi-honest two-round OT are known in the plain model under assumptions such as DDH [AIR01, NP01] and quadratic residuosity [HK12].

Maliciously Secure Two-Round Oblivious Transfer. We consider the stronger notion of oblivious transfer in the common random/reference string model. In terms of syntax, we supplement the syntax of semi-honest oblivious transfer with an algorithm K_{OT} that takes the security parameter 1^λ as input and outputs the common random/reference string σ. Also, the three algorithms $\mathsf{OT}_1, \mathsf{OT}_2$ and OT_3 additionally take σ as input. Correctness and receiver's security properties in the malicious case are the same as the semi-honest case. However, we strengthen the sender's security as described below.

[2] We note that ω in the output of OT_1 need not contain all the random coins used by OT_1. This fact will be useful in the stronger equivocal security notion of oblivious transfer.

- **Correctness.** For every choice bit $\beta \in \{0,1\}$ of the receiver and input messages s_0 and s_1 of the sender we require that, if $\sigma \leftarrow K_{\mathsf{OT}}(1^\lambda)$, $(\mathsf{ots}_1, \omega) \leftarrow \mathsf{OT}_1(\sigma, \beta)$, $\mathsf{ots}_2 \leftarrow \mathsf{OT}_2(\sigma, \mathsf{ots}_1, (s_0, s_1))$, then $\mathsf{OT}_3(\sigma, \mathsf{ots}_2, (\beta, \omega)) = s_\beta$ with overwhelming probability.
- **Receiver's security.** We require that

$$\{(\sigma, \mathsf{ots}_1) : \sigma \leftarrow K_{\mathsf{OT}}(1^\lambda), (\mathsf{ots}_1, \omega) \leftarrow OT_1(\sigma, 0)\}$$
$$\stackrel{c}{\approx} \{(\sigma, \mathsf{ots}_1) : \sigma \leftarrow K_{\mathsf{OT}}(1^\lambda), (\mathsf{ots}_1, \omega) \leftarrow OT_1(\sigma, 1)\}$$

- **Sender's security.** We require the existence of PPT algorithm $\mathsf{Ext} = (\mathsf{Ext}_1, \mathsf{Ext}_2)$ such that for any choice of $K_0, K_1 \in \{0,1\}^\lambda$ and PPT adversary $\mathcal{A}$ we have that

$$\left|\Pr[\mathrm{IND}_{\mathcal{A}}^{\mathrm{REAL}}(1^\lambda, K_0, K_1) = 1] - \Pr[\mathrm{IND}_{\mathcal{A}}^{\mathrm{IDEAL}}(1^\lambda, K_0, K_1) = 1]\right| \leq \frac{1}{2} + \mathsf{negl}(\lambda).$$

Experiment $\mathrm{IND}_{\mathcal{A}}^{\mathrm{REAL}}(1^\lambda, K_0, K_1)$:	**Experiment** $\mathrm{IND}_{\mathcal{A}}^{\mathrm{IDEAL}}(1^\lambda, K_0, K_1)$:
$\sigma \leftarrow K_{\mathsf{OT}}(1^\lambda)$	$(\sigma, \tau) \leftarrow \mathsf{Ext}_1(1^\lambda)$
$\mathsf{ots}_1 \leftarrow \mathcal{A}(\sigma)$	$\mathsf{ots}_1 \leftarrow \mathcal{A}(\sigma)$
	$\beta := \mathsf{Ext}_2(\tau, \mathsf{ots}_1)$
	$L_0 := K_\beta$ and $L_1 := K_\beta$
$\mathsf{ots}_2 \leftarrow \mathsf{OT}_1(\sigma, \mathsf{ots}_1, (K_0, K_1))$	$\mathsf{ots}_2 \leftarrow \mathsf{OT}_2(\sigma, \mathsf{ots}_1, (L_0, L_1))$
Output $\mathcal{A}(\mathsf{ots}_2)$	Output $\mathcal{A}(\mathsf{ots}_2)$

Constructions of maliciously secure two-round OT are known in the common random string model under assumptions such as DDH, quadratic residuosity, and LWE [PVW08].

Equivocal Receiver's Security. We also consider a strengthened notion of malicious receiver's security where we require the existence of a PPT simulator Sim_{Eq} such that the for any $\beta \in \{0,1\}$:

$$\left\{(\sigma, (\mathsf{ots}_1, \omega_\beta)) : (\sigma, \mathsf{ots}_1, \omega_0, \omega_1) \leftarrow \mathsf{Sim}_{Eq}(1^\lambda)\right\} \stackrel{c}{\approx} \left\{(\sigma, \mathsf{OT}_1(\sigma, \beta)) : \sigma \leftarrow K_{\mathsf{OT}}(1^\lambda)\right\}.$$

Using standard techniques in the literature (e.g., [CLOS02]) it is possible to add equivocal receiver's security to any OT protocol. We refer the reader to the full-version of our paper [GS18] for details.

4 Conforming Protocols

Our protocol compilers work for protocols satisfying certain syntactic structure. We refer to protocols satisfying this syntax as *conforming protocols*. In this subsection, we describe this notion and prove that any MPC protocol can be transformed into a conforming protocol while preserving its correctness and security properties.

4.1 Specifications for a Conforming Protocol

Consider an n party deterministic[3] MPC protocol Φ between parties $P_1, \ldots, P_n$ with inputs $x_1, \ldots, x_n$, respectively. For each $i \in [n]$, we let $x_i \in \{0,1\}^m$ denote the input of party P_i. A conforming protocol Φ is defined by functions pre, post, and computations steps or what we call *actions* $\phi_1, \cdots \phi_T$. The protocol Φ proceeds in three stages: the pre-processing stage, the computation stage and the output stage.

- **Pre-processing phase**: For each $i \in [n]$, party P_i computes

$$(z_i, v_i) \leftarrow \mathsf{pre}(1^\lambda, i, x_i)$$

  where pre is a randomized algorithm. The algorithm pre takes as input the index i of the party, its input x_i and outputs $z_i \in \{0,1\}^{\ell/n}$ and $v_i \in \{0,1\}^{\ell}$ (where ℓ is a parameter of the protocol). Finally, P_i retains v_i as the secret information and broadcasts z_i to every other party. We require that $v_{i,k} = 0$ for all $k \in [\ell] \backslash \{(i-1)\ell/n + 1, \ldots, i\ell/n\}$.
- **Computation phase**: For each $i \in [n]$, party P_i sets

$$\mathsf{st}_i := (z_1 \| \cdots \| z_n) \oplus v_i.$$

  Next, for each $t \in \{1 \cdots T\}$ parties proceed as follows:
  1. Parse action ϕ_t as (i, f, g, h) where $i \in [n]$ and $f, g, h \in [\ell]$.
  2. Party P_i computes *one* NAND gate as

$$\mathsf{st}_{i,h} = \mathsf{NAND}(\mathsf{st}_{i,f}, \mathsf{st}_{i,g})$$

     and broadcasts $\mathsf{st}_{i,h} \oplus v_{i,h}$ to every other party.
  3. Every party P_j for $j \neq i$ updates $\mathsf{st}_{j,h}$ to the bit value received from P_i.

  We require that for all $t, t' \in [T]$ such that $t \neq t'$, we have that if $\phi_t = (\cdot, \cdot, \cdot, h)$ and $\phi_{t'} = (\cdot, \cdot, \cdot, h')$ then $h \neq h'$. Also, we denote $A_i \subset [T]$ to be the set of rounds in with party P_i sends a bit. Namely, $A_i = \{t \in T \mid \phi_t = (i, \cdot, \cdot, \cdot)\}$.
- **Output phase**: For each $i \in [n]$, party P_i outputs $\mathsf{post}(\mathsf{st}_i)$.

4.2 Transformation for Making a Protocol Conforming

We show that any MPC protocol can made conforming by making only some syntactic changes. Our transformed protocols retains the correctness or security properties of the original protocol.

Lemma 1. *Any MPC protocol Π can be written as a conforming protocol Φ while inheriting the correctness and the security of the original protocol.*

[3] Randomized protocols can be handled by including the randomness used by a party as part of its input.

Proof. Let Π be any given MPC protocol. Without loss of generality we assume that in each round of Π, *one* party broadcasts *one* bit that is obtained by computing a circuit on its initial state and the messages it has received so far from other parties. Note that this restriction can be easily enforced by increasing the round complexity of the protocol to the communication complexity of the protocol. Let the round complexity (and also communication complexity) of Π be p. In every round $r \in [p]$ of Π, a single bit is sent by one of the parties by computing a circuit. Let the circuit computed in round r be C_r. Without loss of generality we assume that (i) these exists q such that for each $r \in [p]$, we have that $q = |C_r|$, (ii) each C_r is composed of just NAND gates with fan-in two, and (iii) each party sends an equal number of bits in the execution of Π. All three of these conditions can be met by adding dummy gates and dummy round of interaction.

We are now ready to describe our transformed conforming protocol Φ. The protocol Φ will have $T = pq$ rounds. We let $\ell = mn + pq$ and $\ell' = pq/n$ and depending on ℓ the compiled protocol Φ is as follows.

- $\mathsf{pre}(i, x_i)$: Sample $r_i \leftarrow \{0,1\}^m$ and $s_i \leftarrow (\{0,1\}^{q-1}\|0)^{p/n}$. (Observe that s_i is a pq/n bit random string such that its $q^{th}, 2q^{th} \cdots$ locations are set to 0.) Output $z_i := x_i \oplus r_i\|0^{\ell'}$ and $v_i := 0^{\ell/n}\|\ldots\|r_i\|s_i\|\ldots\|0^{\ell/n}$.
- We are now ready to describe the actions $\phi_1, \cdots \phi_T$. For each $r \in [p]$, round r in Π party is expanded into q actions in Φ — namely, actions $\{\phi_j\}_j$ where $j \in \{(r-1)q+1 \cdots rq\}$. Let P_i be the party that computes the circuit C_r and broadcast the output bit broadcast in round r of Π. We now describe the ϕ_j for $j \in \{(r-1)q+1 \cdots rq\}$. For each j, we set $\phi_j = (i, f, g, h)$ where f and g are the locations in st_i that the jth gate of C_r is computed on (recall that initially st_i is set to $z_i \oplus v_i$). Moreover, we set h to be the first location in st_i among the locations $(i-1)\ell/n + m + 1$ to $i\ell/n$ that has previously not been assigned to an action. (Note that this is ℓ' locations which is exactly equal to the number of bits computed and broadcast by P_i.)

  Recall from before than on the execution of ϕ_j, party P_i sets $\mathsf{st}_{i,h} := \mathsf{NAND}(\mathsf{st}_{i,f}, \mathsf{st}_{i,g})$ and broadcasts $\mathsf{st}_{i,h} \oplus v_{i,h}$ to all parties.
- $\mathsf{post}(i, \mathsf{st}_i)$: Gather the local state of P_i and the messages sent by the other parties in Π from st_i and output the output of Π.

Now we need to argue that Φ preserves the correctness and security properties of Π. Observe that Φ is essentially the same as the protocol Π except that in Φ some additional bits are sent. Specifically, in addition to the messages that were sent in Π, in Φ parties send z_i in the preprocessing step and $q-1$ additional bits per every bit sent in Π. Note that these additional bits sent are not used in the computation of Φ. Thus these bits do not affect the functionality of Π if dropped. This ensures that Φ inherits the correctness properties of Π. Next note that each of these bits is masked by a uniform independent bit. This ensures that Φ achieves the same security properties as the underlying properties of Π.

Finally, note that by construction for all $t, t' \in [T]$ such that $t \neq t'$, we have that if $\phi_t = (\cdot, \cdot, \cdot, h)$ and $\phi_{t'} = (\cdot, \cdot, \cdot, h')$ then $h \neq h'$ as required.

5 Two-Round MPC: Semi-honest Case

In this section, we give our construction of two-round multiparty computation protocol in the semi-honest case with security against static corruptions based on any two-round semi-honest oblivious transfer protocol in the plain model. This is achieved by designing a compiler that takes any conforming arbitrary (polynomial) round MPC protocol Φ and squishes it to two rounds.

5.1 Our Compiler

We give our construction of two-round MPC in Fig. 1 and the circuit that needs to be garbled (repeatedly) is shown in Fig. 2. We start by providing intuition behind this construction.

Overview. In the first round of our compiled protocol, each party runs the preprocessing phase of the protocol Φ and obtains z_i and v_i and broadcasts z_i to every other party. In the second round, each party sends a set of garbled circuits that "non-interactively" implement the entire computation phase of the protocol Φ. In other words, any party with the set of garbled circuits sent by every other party, can use them to compute the entire transcript of the computation phase of the protocol Φ. This allows each party to obtain the output of the protocol Φ. In the following paragraphs, we give more details on how this is achieved.

To understand the main idea, let us concentrate on a particular round (let us say the tth round) of the computation phase of the conforming protocol Φ and see how this step is implemented using garbled circuits. Recall that before starting the computation phase, each party locally computes $\mathsf{st}_i := (z_1 \| \ldots \| z_n) \oplus v_i$ using the first round messages sent by the other parties. This local state is updated (recall that only one bit location is updated) at the end of each round based on the bit that is sent in that round. We start with some notations.

Notations. Let us say that the party P_{i^*} is the designated party in round t. Let st_i^t be the updated local state of party P_i at the beginning of the tth round of the computation phase. In the tth round, the designated party P_{i^*} computes $\gamma := \mathsf{NAND}(\mathsf{st}_{i^*,f}^t, \mathsf{st}_{i^*,g}^t)$, writes this bit to position h of $\mathsf{st}_{i^*}^t$ and broadcasts $\gamma \oplus v_{i^*,h}$ to every other party. Every other party P_i (where $i \neq i^*$) updates its local state by writing the received bit at position h in its state st_i^t.

Implementing the Computation Phase. The tth round of the computation phase is implemented by the tth garbled circuit in each of these sequences. In a bit more details, the garbled circuit of party P_i takes as input st_i^t which is the state of the party P_i at the beginning of the t-th round and outputs or, aids the process of outputting the labels corresponding to the updated local state at the end of the tth round. These labels are then used to evaluate the garbled circuit corresponding to the $(t+1)^{th}$ round of the computation phase and this process continues. Finally, at the end each party can just compute output function on the final local state to obtain its output. Next, we describe how the tth garbled circuits in each of the n sequences can be used to complete the tth action of the computation phase.

The tth garbled circuit of party P_{i^*} is executed first and is the most natural one as in this round party P_{i^*} is the one that sends a bit to the other parties. Starting with the easy part, this garbled circuit takes as input $\mathsf{st}_{i^*}^t$, updates the local state by writing the bit γ in the position h of $\mathsf{st}_{i^*}^t$ and outputs the labels corresponding to its updated state. However, the main challenge is that this garbled circuit needs to communicate the bit $\gamma \oplus v_{i^*,h}$ to other garbled circuits of the other parties. Specifically, those garbled circuits also need to output the correct labels corresponding to the their updated local state. Note that only the hth bit of each of their local state needs to be updated. This was achieved in [GS17] by using specific properties of Groth, Ostrovsky and Sahai proofs and in this work, we only rely on oblivious transfer. This is our key new idea and we provide the details next.

Relying on Oblivious Transfer. In addition to broadcasting the encoded input z_i in the first round, the party P_i sends a set of 4 OT messages (acting as the receiver) for every round in the computation phase where P_i is the designated party. Thus, if the number of rounds in the computation phase where P_i is the designated party is a_i, then the party P_i sends $4a_i$ receiver OT messages. Specifically, in our running example from above P_{i^*} will generate 4 first OT messages to help in t^{th} round of Φ. In particular, for each value of $\alpha, \beta \in \{0,1\}$, P_{i^*} generates the first OT message with $v_{i^*,h} \oplus \mathsf{NAND}(v_{i^*,f} \oplus \alpha, v_{i^*,g} \oplus \beta)$ as its choice bit. Every other party P_i for $i \neq i^*$ acts as the sender and prepares four OT responses corresponding to each of the four OT messages using labels corresponding to the h-th input wire (say $(\mathsf{label}_{h,0}^{i,t+1}, \mathsf{label}_{h,1}^{i,t+1})$) of its next (i.e., $(t+1)^{th}$) garbled circuit. However, these values aren't sent to anyone yet! Because sending them all to P_{i^*} would lead to complete loss of security. Specifically, for every choice of $v_{i^*,f}, v_{i^*,g}, v_{i^*,h}$ there exists different choices of α, β such that $v_{i^*,h} \oplus \mathsf{NAND}(v_{i^*,f} \oplus \alpha, v_{i^*,g} \oplus \beta)$ is 0 and 1, respectively. Thus, if all these OT responses were reveled to P_{i^*} then P_{i^*} would learn both the input labels $\mathsf{label}_{h,0}^{i,t+1}, \mathsf{label}_{h,1}^{i,t+1}$ potentially breaking the security of garbled circuits. Our key idea here is that party P_i hardcodes these OT responses in its tth garbled circuit and only one of them is revealed to P_{i^*}. We now elaborate this.

The t-th garbled circuit of party P_i (where $i \neq i^*$) outputs the set of labels corresponding to the state bits $\{\mathsf{st}_{i,k}^t\}_{k \in [\ell] \setminus \{h\}}$ (as these bits do not change at the end of the t-th round) and additionally outputs the sender OT response for $\alpha = \mathsf{st}_{i,f}^t$ and $\beta = \mathsf{st}_{i,g}^t$ with the messages being set to the labels corresponding to h-th bit of st_i^t. It follows from the invariant of the protocol, that the choice bit in this OT_1 message is indeed $\gamma \oplus v_{i^*,h}$ which is exactly the bit P_{i^*} wants to communicate to the other parties. However, this leaves us with another problem. The OT responses only allow P_{i^*} to learn the labels of the next garbled circuits and it is unclear how a party $j \neq i^*$ obtains the labels of the garbled circuits generated by P_i.

Enabling all Parties to Compute. The party P_{i^*}'s t^{th} garbled circuit, in addition to outputting the labels corresponding to the updated state of P_{i^*}, outputs the randomness it used to prepare the first OT message for which all P_i for $i \neq i^*$ output OT responses; namely, $\alpha = \mathsf{st}_{i^*,f}^t \oplus v_{i^*,f}, \beta = \mathsf{st}_{i^*,g}^t \oplus v_{i^*,g}$. It again follows from the invariant of the protocol Φ that this allows every party P_j with $j \neq i^*$ to evaluate the recover $\mathsf{label}_{h,\gamma \oplus v_{i^*,h}}^{i,t+1}$ which is indeed the label corresponding to the correct updated state. Thus, using the randomness output by the garbled circuit of P_{i^*} all other parties can recover the label $\mathsf{label}_{h,\gamma \oplus v_{i^*,h}}^{i,t+1}$.

We stress that this process of revealing the randomness of the OT leads to complete loss of security for the particular instance OT. Nevertheless, since the randomness of only one of the four OT messages of P_{i^*} is reveled, overall security is ensured. In particular, our construction ensures that the learned choice bit is $\gamma \oplus v_{i^*,h}$ which is in fact the message that is broadcasted in the underlying protocol Φ. Thus, it follows from the security of the protocol Φ that learning this message does not cause any vulnerabilities.

Theorem 2. *Let Φ be a polynomial round, n-party semi-honest MPC protocol computing a function $f : (\{0,1\}^m)^n \to \{0,1\}^*$, $(\mathsf{Garble}, \mathsf{Eval})$ be a garbling scheme for circuits, and $(\mathsf{OT}_1, \mathsf{OT}_2, \mathsf{OT}_3)$ be a semi-honest two-round OT protocol. The protocol described in Fig. 1 is a two-round, n-party semi-honest MPC protocol computing f against static corruptions.*

This theorem is proved in the rest of this section.

5.2 Correctness

In order to prove correctness, it is sufficient to show that the label computed in Step 2(d)(ii) of the evaluation procedure corresponds to the bit $\mathsf{NAND}(\mathsf{st}_{i^*,f}, \mathsf{st}_{i^*,g}) \oplus v_{i^*,h}$. Notice that by the assumption on the structure of v_{i^*} (recall that v_{i^*} is such that $v_{i^*,k} = 0$ for all $k \in [\ell] \setminus \{(i^*-1)\ell/n + 1, \ldots, i^*\ell/n\}$) we deduce that for every $i \neq i^*$, $\mathsf{st}_{i,f} = \mathsf{st}_{i^*,f} \oplus v_{i^*,f}$ and $\mathsf{st}_{i,g} = \mathsf{st}_{i^*,g} \oplus v_{i^*,g}$. Thus, the label obtained by OT_2 corresponds to the bit $\mathsf{NAND}(v_{i^*,f} \oplus \underbrace{\mathsf{st}_{i^*,f} \oplus v_{i^*,f}}_{\alpha}, v_{i^*,g} \oplus \underbrace{\mathsf{st}_{i^*,g} \oplus v_{i^*,g}}_{\beta}) \oplus v_{i^*,h} = \mathsf{NAND}(\mathsf{st}_{i^*,f}, \mathsf{st}_{i^*,g}) \oplus v_{i^*,h}$ and correctness follows.

Let Φ be an n-party conforming semi-honest MPC protocol, $(\mathsf{Garble}, \mathsf{Eval})$ be a garbling scheme for circuits and $(\mathsf{OT}_1, \mathsf{OT}_2, \mathsf{OT}_3)$ be a semi-honest two-round oblivious transfer protocol.

Round-1: Each party P_i does the following:

1. Compute $(z_i, v_i) \leftarrow \mathsf{pre}(1^\lambda, i, x_i)$.
2. For each t such that $\phi_t = (i, f, g, h)$ (A_i is the set of such values of t), for each $\alpha, \beta \in \{0,1\}$

$$\mathsf{ots}_{1,t,\alpha,\beta} \leftarrow \mathsf{OT}_1(1^\lambda, v_{i,h} \oplus \mathsf{NAND}(v_{i,f} \oplus \alpha, v_{i,g} \oplus \beta); \omega_{t,\alpha,\beta}).$$

3. Send $\left(z_i, \{\mathsf{ots}_{1,t,\alpha,\beta}\}_{t \in A_i, \alpha,\beta \in \{0,1\}}\right)$ to every other party.

Round-2: In the second round, each party P_i does the following:

1. Set $\mathsf{st}_i := (z_1 \| \dots \| z_{i-1} \| z_i \| z_{i+1} \| \dots \| z_n) \oplus v_i$.
2. Set $\overline{\mathsf{lab}}^{i,T+1} := \{\mathsf{lab}^{i,T+1}_{k,0}, \mathsf{lab}^{i,T+1}_{k,1}\}_{k \in [\ell]}$ where for each $k \in [\ell]$ and $b \in \{0,1\}$ $\mathsf{lab}^{i,T+1}_{k,b} := 0^\lambda$.
3. **for** each t from T down to 1,
   (a) Parse ϕ_t as (i^*, f, g, h).
   (b) If $i = i^*$ then compute (where P is described in Figure 2)

$$\left(\widetilde{\mathsf{P}}^{i,t}, \overline{\mathsf{lab}}^{i,t}\right) \leftarrow \mathsf{Garble}(1^\lambda, \mathsf{P}[i, \phi_t, v_i, \{\omega_{t,\alpha,\beta}\}_{\alpha,\beta}, \bot, \overline{\mathsf{lab}}^{i,t+1}]).$$

   (c) If $i \neq i^*$ then for every $\alpha, \beta \in \{0,1\}$, set $\mathsf{ots}^i_{2,t,\alpha,\beta} \leftarrow \mathsf{OT}_2(\mathsf{ots}_{1,t,\alpha,\beta}, \mathsf{lab}^{i,t+1}_{h,0}, \mathsf{lab}^{i,t+1}_{h,1})$ and compute

$$\left(\widetilde{\mathsf{P}}^{i,t}, \overline{\mathsf{lab}}^{i,t}\right) \leftarrow \mathsf{Garble}(1^\lambda, \mathsf{P}[i, \phi_t, v_i, \bot, \{\mathsf{ots}^i_{2,t,\alpha,\beta}\}_{\alpha,\beta}, \overline{\mathsf{lab}}^{i,t+1}]).$$

4. Send $\left(\{\widetilde{\mathsf{P}}^{i,t}\}_{t \in [T]}, \{\mathsf{lab}^{i,1}_{k,\mathsf{st}_{i,k}}\}_{k \in [\ell]}\right)$ to every other party.

Evaluation: To compute the output of the protocol, each party P_i does the following:

1. For each $j \in [n]$, let $\widetilde{\mathsf{lab}}^{j,1} := \{\mathsf{lab}^{j,1}_k\}_{k \in [\ell]}$ be the labels received from party P_j at the end of round 2.
2. **for** each t from 1 to T do:
   (a) Parse ϕ_t as (i^*, f, g, h).
   (b) Compute $((\alpha, \beta, \gamma), \omega, \widetilde{\mathsf{lab}}^{i^*,t+1}) := \mathsf{Eval}(\widetilde{\mathsf{P}}^{i^*,t}, \widetilde{\mathsf{lab}}^{i^*,t})$.
   (c) Set $\mathsf{st}_{i,h} := \gamma \oplus v_{i,h}$.
   (d) **for** each $j \neq i^*$ do:
      i. Compute $(\mathsf{ots}_2, \{\mathsf{lab}^{j,t+1}_k\}_{k \in [\ell] \setminus \{h\}}) := \mathsf{Eval}(\widetilde{\mathsf{P}}^{j,t}, \widetilde{\mathsf{lab}}^{j,t})$.
      ii. Recover $\mathsf{lab}^{j,t+1}_h := \mathsf{OT}_3(\mathsf{ots}_2, \omega)$.
      iii. Set $\widetilde{\mathsf{lab}}^{j,t+1} := \{\mathsf{lab}^{j,t+1}_k\}_{k \in [\ell]}$.
3. Compute the output as $\mathsf{post}(i, \mathsf{st}_i)$.

Fig. 1. Two-round semi-honest MPC

P

Input. st_i.
Hardcoded. The index i of the party, the action $\phi_t = (i^*, f, g, h)$, the secret value v_i, the strings $\{\omega_{t,\alpha,\beta}\}_{\alpha,\beta}$, $\{\mathsf{ots}_{2,t,\alpha,\beta}\}_{\alpha,\beta}$ and a set of labels $\overline{\mathsf{lab}} = \{\mathsf{lab}_{k,0}, \mathsf{lab}_{k,1}\}_{k\in[\ell]}$.

1. **if** $i = i^*$ **then**:
   (a) Compute $\mathsf{st}_{i,h} := \mathsf{NAND}(\mathsf{st}_{i,f}, \mathsf{st}_{i,g})$, $\alpha := \mathsf{st}_{i,f} \oplus v_{i,f}$, $\beta := \mathsf{st}_{i,g} \oplus v_{i,g}$ and $\gamma := \mathsf{st}_{i,h} \oplus v_{i,h}$.
   (b) Output $((\alpha, \beta, \gamma), \omega_{t,\alpha,\beta}, \{\mathsf{lab}_{k,\mathsf{st}_{i,k}}\}_{k\in[\ell]})$.
2. **else**:
   (a) Output $(\mathsf{ots}_{2,t,\mathsf{st}_{i,f},\mathsf{st}_{i,g}}, \{\mathsf{lab}_{k,\mathsf{st}_{i,k}}\}_{k\in[\ell]\setminus\{h\}})$.

Fig. 2. The program P.

Via the same argument as above it is useful to keep in mind that for every $i, j \in [n]$ and $k \in [\ell]$, we have that $\mathsf{st}_{i,k} \oplus v_{i,k} = \mathsf{st}_{j,k} \oplus v_{j,k}$. Let us denote this shared value by st^*. Also, we denote the transcript of the interaction in the computation phase by $\mathsf{Z} \in \{0,1\}^t$.

5.3 Simulator

Let $\mathcal{A}$ be a semi-honest adversary corrupting a subset of parties and let $H \subseteq [n]$ be the set of honest/uncorrupted parties. Since we assume that the adversary is static, this set is fixed before the execution of the protocol. Below we provide the simulator.

Description of the Simulator. We give the description of the ideal world adversary $\mathcal{S}$ that simulates the view of the real world adversary $\mathcal{A}$. $\mathcal{S}$ will internally use the semi-honest simulator Sim_Φ for Φ and the simulator Sim_G for garbling scheme for circuits. Recall that $\mathcal{A}$ is static and hence the set of honest parties H is known before the execution of the protocol.

Simulating the interaction with $\mathcal{Z}$. For every input value for the set of corrupted parties that $\mathcal{S}$ receives from $\mathcal{Z}$, $\mathcal{S}$ writes that value to $\mathcal{A}$'s input tape. Similarly, the output of $\mathcal{A}$ is written as the output on $\mathcal{S}$'s output tape.

Simulating the interaction with $\mathcal{A}$: For every concurrent interaction with the session identifier sid that $\mathcal{A}$ may start, the simulator does the following:

- **Initialization**: $\mathcal{S}$ uses the inputs of the corrupted parties $\{x_i\}_{i\notin H}$ and output y of the functionality f to generate a simulated view of the adversary.[4]

[4] For simplicity of exposition, we only consider the case where every party gets the same output. The proof in the more general case where parties get different outputs follows analogously.

More formally, for each $i \in [n]\backslash H$ $\mathcal{S}$ sends $(\mathsf{input}, \mathsf{sid}, \{P_1 \cdots P_n\}, P_i, x_i)$ to the ideal functionality implementing f and obtains the output y. Next, it executes $\mathsf{Sim}_{\Phi}(1^\lambda, \{x_i\}_{i \notin H}, y)$ to obtain $\{z_i\}_{i \in H}$, the random tapes for the corrupted parties, the transcript of the computation phase denoted by $\mathsf{Z} \in \{0,1\}^t$ where Z_t is the bit sent in the tth round of the computation phase of Φ, and the value st^* (which for each $i \in [n]$ and $k \in [\ell]$ is equal to $\mathsf{st}_{i,k} \oplus v_{i,k}$). $\mathcal{S}$ starts the real-world adversary $\mathcal{A}$ with the inputs $\{z_i\}_{i \in H}$ and random tape generated by Sim_{Φ}.

- **Round-1 messages from $\mathcal{S}$ to $\mathcal{A}$:** Next $\mathcal{S}$ generates the OT messages on behalf of honest parties as follows. For each $i \in H, t \in A_i, \alpha, \beta \in \{0,1\}$, generate $\mathsf{ots}_{1,t,\alpha,\beta} \leftarrow \mathsf{OT}_1(1^\lambda, \mathsf{Z}_t; \omega_{t,\alpha,\beta})$. For each $i \in H$, $\mathcal{S}$ sends $(z_i, \{\mathsf{ots}_{1,t,\alpha,\beta}\}_{t \in A_i, \alpha, \beta \in \{0,1\}})$ to the adversary $\mathcal{A}$ on behalf of the honest party P_i.
- **Round-1 messages from $\mathcal{A}$ to $\mathcal{S}$:** Corresponding to every $i \in [n] \setminus H$, $\mathcal{S}$ receives from the adversary $\mathcal{A}$ the value $(z_i, \{\mathsf{ots}_{1,t,\alpha,\beta}\}_{t \in A_i, \alpha, \beta \in \{0,1\}})$ on behalf of the corrupted party P_i.
- **Round-2 messages from $\mathcal{S}$ to $\mathcal{A}$:** For each $i \in H$, the simulator $\mathcal{S}$ generates the second round message on behalf of party P_i as follows:
  1. For each $k \in [\ell]$ set $\mathsf{lab}_k^{i,T+1} := 0^\lambda$.
  2. **for** each t from T down to 1,
     (a) Parse ϕ_t as (i^*, f, g, h).
     (b) Set $\alpha^* := \mathsf{st}_f^*$, $\beta^* := \mathsf{st}_g^*$, and $\gamma^* := \mathsf{st}_h^*$.
     (c) If $i = i^*$ then compute
     $$\left(\widetilde{\mathsf{P}}^{i,t}, \{\mathsf{lab}_k^{i,t}\}_{k \in [\ell]}\right) \leftarrow \mathsf{Sim}_{\mathsf{G}}\left(1^\lambda, \left((\alpha^*, \beta^*, \gamma^*), \omega_{t,\alpha^*,\beta^*}, \{\mathsf{lab}_k^{i,t+1}\}_{k \in [\ell]}\right)\right).$$
     (d) If $i \neq i^*$ then set $\mathsf{ots}_{2,t,\alpha^*,\beta^*}^i \leftarrow \mathsf{OT}_2(\mathsf{ots}_{1,t,\alpha^*,\beta^*}, \mathsf{lab}_h^{i,t+1}, \mathsf{lab}_h^{i,t+1})$ and compute
     $$\left(\widetilde{\mathsf{P}}^{i,t}, \{\mathsf{lab}_k^{i,t}\}_{k \in [\ell]}\right) \leftarrow \mathsf{Sim}_{\mathsf{G}}\left(1^\lambda, \left(\mathsf{ots}_{2,t,\alpha^*,\beta^*}^i, \{\mathsf{lab}_k^{i,t+1}\}_{k \in [\ell] \setminus \{h\}}\right)\right).$$
  3. Send $\left(\{\widetilde{\mathsf{P}}^{i,t}\}_{t \in [T]}, \{\mathsf{lab}_k^{i,1}\}_{k \in [\ell]}\right)$ to every other party.
- **Round-2 messages from $\mathcal{A}$ to $\mathcal{S}$:** For every $i \in [n] \setminus H$, $\mathcal{S}$ obtains the second round message from $\mathcal{A}$ on behalf of the malicious parties. Subsequent to obtaining these messages, for each $i \in H$, $\mathcal{S}$ sends $(\mathsf{generateOutput}, \mathsf{sid}, \{P_1 \cdots P_n\}, P_i)$ to the ideal functionality.

5.4 Proof of Indistinguishability

We now show that no environment $\mathcal{Z}$ can distinguish whether it is interacting with a real world adversary $\mathcal{A}$ or an ideal world adversary $\mathcal{S}$. We prove this via an hybrid argument with $T+1$ hybrids.

- $\mathcal{H}_{Real}$: This hybrid is the same as the real world execution. Note that this hybrid is the same as hybrid $\mathcal{H}_t$ below with $t = 0$.

- $\mathcal{H}_t$ (where $t \in \{0, \ldots T\}$): Hybrid $\mathcal{H}_t$ (for $t \in \{1 \cdots T\}$) is the same as hybrid $\mathcal{H}_{t-1}$ except we change the distribution of the OT messages (both from the first and the second round of the protocol) and the garbled circuits (from the second round) that play a role in the execution of the tth round of the protocol Φ; namely, the action $\phi_t = (i^*, f, g, h)$. We describe the changes more formally below.

  We start by executing the protocol Φ on the inputs and the random coins of the honest and the corrupted parties. This yields a transcript $\mathsf{Z} \in \{0,1\}^T$ of the computation phase. Since the adversary is assumed to be semi-honest the execution of the protocol Φ with $\mathcal{A}$ will be consistent with Z. Let st^* be the local state of the end of execution of Faithful. Finally, let $\alpha^* := \mathsf{st}^*_f$, $\beta^* := \mathsf{st}^*_g$ and $\gamma^* := \mathsf{st}^*_h$. In hybrid $\mathcal{H}_t$ we make the following changes with respect to hybrid $\mathcal{H}_{t-1}$:
  - If $i^* \notin H$ then skip these changes. $\mathcal{S}$ makes two changes in how it generates messages on behalf of P_{i^*}. First, for all $\alpha, \beta \in \{0,1\}$, $\mathcal{S}$ generates $\mathsf{ots}_{1,t,\alpha,\beta}$ as $\mathsf{OT}_1(1^\lambda, \mathsf{Z}_t; \omega_{t,\alpha,\beta})$ (note that only one of these four values is subsequently used) rather than $\mathsf{OT}_1(1^\lambda, v_{i,h} \oplus \mathsf{NAND}(v_{i,f} \oplus \alpha, v_{i,g} \oplus \beta); \omega_{t,\alpha,\beta})$. Second, it generates the garbled circuit
  $$(\widetilde{\mathsf{P}}^{i^*,t}, \{\mathsf{lab}_k^{i^*,t}\}_{k\in[\ell]}) \leftarrow \mathsf{Sim}_{\mathsf{G}}\left(1^\lambda, \left((\alpha^*, \beta^*, \gamma^*), \omega_{t,\alpha^*,\beta^*}, \{\mathsf{lab}_{k,\mathsf{st}_{i,k}}^{i^*,t+1}\}_{k\in[\ell]}\right)\right),$$
  where $\{\mathsf{lab}_{k,\mathsf{st}_{i,k}}^{i^*,t+1}\}_{k\in[\ell]}$ are the honestly generates input labels for the garbled circuit $\widetilde{\mathsf{P}}^{i^*,t+1}$.
  - $\mathcal{S}$ makes the following two changes in how it generates messages for other honest parties P_i (i.e., $i \in H \setminus \{i^*\}$). $\mathcal{S}$ does not generate four $\mathsf{ots}^i_{2,t,\alpha,\beta}$ values but just one of them; namely, $\mathcal{S}$ generates $\mathsf{ots}^i_{2,t,\alpha^*,\beta^*}$ as $\mathsf{OT}_2(\mathsf{ots}_{1,t,\alpha^*,\beta^*}, \mathsf{lab}^{i,t+1}_{h,\mathsf{Z}_t}, \mathsf{lab}^{i,t+1}_{h,\mathsf{Z}_t})$ rather than $\mathsf{OT}_2(\mathsf{ots}_{1,t,\alpha^*,\beta^*}, \mathsf{lab}^{i,t+1}_{h,0}, \mathsf{lab}^{i,t+1}_{h,1})$. Second it generates the garbled circuit
  $$(\widetilde{\mathsf{P}}^{i,t}, \{\mathsf{lab}_k^{i,t}\}_{k\in[\ell]}) \leftarrow \mathsf{Sim}_{\mathsf{G}}\left(1^\lambda, \left(\mathsf{ots}^i_{2,t,\alpha^*,\beta^*}, \{\mathsf{lab}_{k,\mathsf{st}_{i,k}}^{i,t+1}\}_{k\in[\ell]\setminus\{h\}}\right)\right),$$
  where $\{\mathsf{lab}_{k,\mathsf{st}_{i,k}}^{i,t+1}\}_{k\in[\ell]}$ are the honestly generated input labels for the garbled circuit $\widetilde{\mathsf{P}}^{i,t+1}$.

  Indistinguishability between $\mathcal{H}_{t-1}$ and $\mathcal{H}_t$ is proved in Lemma 2.
- $\mathcal{H}_{T+1}$: In this hybrid we just change how the transcript Z, $\{z_i\}_{i\in H}$, random coins of malicious parties and value st^* are generated. Instead of generating these using honest party inputs we generate these values by executing the simulator Sim_Φ on input $\{x_i\}_{i\in[n]\setminus H}$ and the output y obtained from the ideal functionality.

  The indistinguishability between hybrids $\mathcal{H}_T$ and $\mathcal{H}_{T+1}$ follows directly from the semi-honest security of the protocol Φ. Finally note that $\mathcal{H}_{T+1}$ is same as the ideal execution (i.e., the simulator described in the previous subsection).

Lemma 2. *Assuming semi-honest security of the two-round OT protocol and the security of the garbling scheme, for all $t \in \{1 \ldots T\}$ hybrids $\mathcal{H}_{t-1}$ and $\mathcal{H}_t$ are computationally indistinguishable.*

Proof. Using the same notation as before, let $\phi_t = (i^*, f, g, h)$, st_{i^*} be the state of P_{i^*} at the end of round t, and $\alpha^* := \mathsf{st}_{i^*,f} \oplus v_{i^*,f}$, $\beta^* := \mathsf{st}_{i^*,g} \oplus v_{i^*,g}$ and $\gamma^* := \mathsf{st}_{i^*,h} \oplus v_{i^*,h}$. The indistinguishability between hybrids $\mathcal{H}_{t-1}$ and $\mathcal{H}_t$ follows by a sequence of three sub-hybrids $\mathcal{H}_{t,1}$, $\mathcal{H}_{t,2}$, and $\mathcal{H}_{t,3}$.

- $\mathcal{H}_{t,1}$: Hybrid $\mathcal{H}_{t,1}$ is same as hybrid $\mathcal{H}_{t-1}$ except that $\mathcal{S}$ now generates the garbled circuits $\widetilde{\mathsf{P}}^{i,t}$ for each $i \in H$ in a simulated manner (rather than generating them honestly). Specifically, instead of generating each garbled circuit and input labels $(\widetilde{\mathsf{P}}^{i,t}, \{\mathsf{lab}_k^{i,t}\}_{k\in[\ell]})$ honestly, they are generated via the simulator by hard coding the output of the circuit itself. In a bit more details, parse ϕ_t as (i^*, f, g, h).
  - If $i = i^*$ then
  $$(\widetilde{\mathsf{P}}^{i,t}, \{\mathsf{lab}_k^{i,t}\}_{k\in[\ell]}) \leftarrow \mathsf{Sim}_{\mathsf{G}}\left(1^\lambda, \left((\alpha^*, \beta^*, \gamma^*), \omega_{t,\alpha^*,\beta^*}, \{\mathsf{lab}_{k,\mathsf{st}_{i,k}}^{i,t+1}\}_{k\in[\ell]}\right)\right),$$
  where $\{\mathsf{lab}_{k,\mathsf{st}_{i,k}}^{i,t+1}\}_{k\in[\ell]}$ are the honestly generates input labels for the garbled circuit $\widetilde{\mathsf{P}}^{i,t+1}$.
  - If $i \neq i^*$ then
  $$(\widetilde{\mathsf{P}}^{i,t}, \{\mathsf{lab}_k^{i,t}\}_{k\in[\ell]}) \leftarrow \mathsf{Sim}_{\mathsf{G}}\left(1^\lambda, \left(\mathsf{ots}_{2,t,\alpha^*,\beta^*}^i, \{\mathsf{lab}_{k,\mathsf{st}_{i,k}}^{i,t+1}\}_{k\in[\ell]\setminus\{h\}}\right)\right),$$
  where $\{\mathsf{lab}_{k,\mathsf{st}_{i,k}}^{i,t+1}\}_{k\in[\ell]}$ are the honestly generated input labels for the garbled circuit $\widetilde{\mathsf{P}}^{i,t+1}$.

  The indistinguishability between hybrids $\mathcal{H}_{t,1}$ and $\mathcal{H}_{t-1}$ follows by $|H|$ invocations of security of the garbling scheme.
- $\mathcal{H}_{t,2}$: Skip this hybrid, if $i^* \notin H$. This hybrid is same as $\mathcal{H}_{t,1}$ except that we change how $\mathcal{S}$ generates the Round-1 message on behalf of P_{i^*}. Specifically, the simulator $\mathcal{S}$ generates $\mathsf{ots}_{1,t,\alpha,\beta}$ as is done in the $\mathcal{H}_t$. In a bit more detail, for all $\alpha, \beta \in \{0,1\}$, $\mathcal{S}$ generates $\mathsf{ots}_{1,t,\alpha,\beta}$ as $\mathsf{OT}_1(1^\lambda, \mathsf{Z}_t; \omega_{t,\alpha,\beta})$ rather than $\mathsf{OT}_1(1^\lambda, v_{i,h} \oplus \mathsf{NAND}(v_{i,f} \oplus \alpha, v_{i,g} \oplus \beta); \omega_{t,\alpha,\beta})$.

  Indistinguishability between hybrids $\mathcal{H}_{t,1}$ and $\mathcal{H}_{t,2}$ follows directly by a sequence of 3 sub-hybrids each one relying on the receiver's security of underlying semi-honest oblivious transfer protocol. Observe here that the security reduction crucially relies on the fact that $\widetilde{\mathsf{P}}^{i,t}$ only contains $\omega_{t,\alpha^*,\beta^*}$ (i.e., does not have $\omega_{t,\alpha,\beta}$ for $\alpha \neq \alpha^*$ or $\beta \neq \beta^*$).
- $\mathcal{H}_{t,3}$: Skip this hybrid if there does not exist $i \neq i^*$ such that $i \in H$. In this hybrid, we change how $\mathcal{S}$ generates the $\mathsf{ots}_{2,t,\alpha,\beta}^i$ on behalf of every honest party P_i such that $i \in H \setminus \{i^*\}$ for all choices of $\alpha, \beta \in \{0,1\}$. More specifically, $\mathcal{S}$ only generates one of these four values; namely, $\mathsf{ots}_{2,t,\alpha^*,\beta^*}^i$ which is now generated as $\mathsf{OT}_2(\mathsf{ots}_{1,t,\alpha^*,\beta^*}, \mathsf{lab}_{h,\mathsf{Z}_t}^{i,t+1}, \mathsf{lab}_{h,\mathsf{Z}_t}^{i,t+1})$ instead of $\mathsf{OT}_2(\mathsf{ots}_{1,t,\alpha^*,\beta^*}, \mathsf{lab}_{h,0}^{i,t+1}, \mathsf{lab}_{h,1}^{i,t+1})$.

  Indistinguishability between hybrids $\mathcal{H}_{t,2}$ and $\mathcal{H}_{t,3}$ follows directly from the sender's security of underlying semi-honest oblivious transfer protocol. Finally, observe that $\mathcal{H}_{t,3}$ is the same as hybrid $\mathcal{H}_t$.

5.5 Extensions

The protocol presented above is very general and can be extended in different ways to obtain several other additional properties. We list some of the simple extensions below.

Multi-round OT. We note that plugging in any multi-round (say, r-round) OT scheme with semi-honest security we obtain an r-round MPC for semi-honest adversaries. More specifically, this can be achieved as follows. We run the first $r-2$ rounds of the protocol as a pre-processing phase with the receiver's choice bits set as in the protocol and the sender's message being randomly chosen labels. We then run the first round of our MPC protocol with the $(r-1)^{th}$ round of OT from the receiver and run the second round using the last round message from the sender hardwired inside the garbled circuits. The proof of security follows identically to proof given above for a two-round OT. A direct corollary of this construction is a construction of three round MPC for semi-honest adversaries from enhanced trapdoor permutations.

Two-Round MPC for RAM programs. In the previous section, we described how protocol compilation can be done for the case of conforming MPC protocols for circuits. Specifically, the protocol communication depends on the lengths of the secret state of the parties. We note that we can extend this framework for securely evaluating RAM programs [OS97,GKK+12,GGMP16,HY16] in two-rounds. In this setting, each party has a huge database as its private input and the parties wishes to compute a RAM program on their private databases. We consider the persistent memory setting [LO13,GHL+14,GLOS15,GLO15] where several programs are evaluated on the same databases. We allow an (expensive) pre-processing phase where the parties communicate to get a shared garbled database and the programs must be evaluated with communication and computation costs that grow with the running time of the programs. In our construction of two-round MPC for RAM programs, the pre-processing phase involves the parties executing a two-round MPC to obtain garbled databases of all the parties using a garbled RAM scheme (say, [GLOS15]) along with the shared secret state. Next, when a program needs to be executed, then the parties execute our two-round MPC to obtain a garbled program. Finally, the obtained garbled program can be executed with the garbled database to obtain the output.

Reducing the Communication Complexity. Finally, we note that in our two-round protocol each party can reduce the communication complexity [Gen09,BGI16,CDG+17] of either one of its two messages (with size dependent just on the security parameter) using Laconic Oblivious Transfer (OT) [CDG+17]. Roughly, laconic OT allows one party to commit to a large message by a short *hash* string (depending just on the security parameter) such that the knowledge of the laconic hash suffices for generating a garbled circuit that can be executed on the large committed string as input. Next, we give simple transformations using which the first party in any two-round MPC protocol can make either its first message or its second message short, respectively. The general case can also be handled in a similar manner.

We start by providing a transformation by which the first party can make its first message short. The idea is that in the transformed protocol the first party now only sends a laconic hash of the first message of the underlying protocol, which is disclosed in the second round message of the transformed protocol. The first round of messages of all other parties in the transformed protocol remains unchanged. However, their second round messages are now obtained by sending garbled circuits that generate the second round message of the original protocol using the first round message of the first party as input. This can be done using laconic OT.

Using a similar transformation the first party can make its second message short. Specifically, in this case, the first party appends its first round message with a garbled circuit that generated its second round message given as input the laconic OT hash for the first round messages of all the other parties. Now in the second round, the first party only needs to disclose the labels for the garbled circuit corresponding to laconic OT hash of the first round messages of all the other parties. The messages of all the other parties remain unchanged.

6 Two-Round MPC: Malicious Case

In this section, we give our construction of two-round multiparty computation protocol in the malicious case with security against static corruptions based on any two-round malicious oblivious transfer protocol (with equivocal receiver security which as argued earlier can be added with need for any additional assumptions) This is achieved by designing a compiler that takes any conforming arbitrary (polynomial) round MPC protocol Φ and squishes it to two rounds.

6.1 Our Compiler

We give our construction of two-round MPC in Fig. 3 and the circuit that needs to be garbled (repeatedly) is shown in Fig. 2 (same as the semi-honest case). We start by providing intuition behind this construction. Our compiler is essentially the same as the semi-honest case. In addition to the minor syntactic changes, the main difference is that we compile malicious secure conforming protocols instead of semi-honest ones.

Another technical issue arises because the adversary may wait to receiver first round messages that $\mathcal{S}$ sends on the behalf of honest parties before the corrupted parties send out their first round messages. Recall that by sending the receiver OT messages in the first round, every party "commits" to all its future messages that it will send in the computation phase of the protocol. Thus, the ideal world simulator $\mathcal{S}$ must somehow commit to the messages generated on behalf of the honest party before extracting the adversary's effective input. To get around this issue, we use the equivocability property of the OT using which the simulator can equivocate its first round messages after learning the malicious adversary's effective input.

Let Φ be an n-party conforming malicious MPC protocol, $(\mathsf{Garble}, \mathsf{Eval})$ be a garbling scheme for circuits and $(K_{\mathsf{OT}}, \mathsf{OT}_1, \mathsf{OT}_2, \mathsf{OT}_3)$ be a malicious (with equivocal receiver security) two-round oblivious transfer protocol.

Common Random/Reference String: For each $t \in T, \alpha, \beta \in \{0,1\}$ sample $\sigma_{t,\alpha,\beta} \leftarrow K_{\mathsf{OT}}(1^\lambda)$ and output $\{\sigma_{t,\alpha,\beta}\}_{t\in[T],\alpha,\beta\in\{0,1\}}$ as the common random/reference string.

Round-1: Each party P_i does the following:

1. Compute $(z_i, v_i) \leftarrow \mathsf{pre}(1^\lambda, i, x_i)$.
2. For each t such that $\phi_t = (i, f, g, h)$ (A_i is the set of such values of t), for each $\alpha, \beta \in \{0,1\}$
$$\mathsf{ots}_{1,t,\alpha,\beta} \leftarrow \mathsf{OT}_1(\sigma_{t,\alpha,\beta}, v_{i,h} \oplus \mathsf{NAND}(v_{i,f} \oplus \alpha, v_{i,g} \oplus \beta); \omega_{t,\alpha,\beta}).$$
3. Send $\big(z_i, \{\mathsf{ots}_{1,t,\alpha,\beta}\}_{t\in A_i,\alpha,\beta\in\{0,1\}}\big)$ to every other party.

Round-2: In the second round, each party P_i does the following:

1. Set $\mathsf{st}_i := (z_1\|\ldots\|z_{i-1}\|z_i\|z_{i+1}\|\ldots\|z_n) \oplus v_i$.
2. Set $\overline{\mathsf{lab}}^{i,T+1} := \{\mathsf{lab}^{i,T+1}_{k,0}, \mathsf{lab}^{i,T+1}_{k,1}\}_{k\in[\ell]}$ where for each $k \in [\ell]$ and $b \in \{0,1\}$ $\mathsf{lab}^{i,T+1}_{k,b} := 0^\lambda$.
3. **for** each t from T down to 1,
   (a) Parse ϕ_t as (i^*, f, g, h).
   (b) If $i = i^*$ then compute (where P is described in Figure 2)
$$(\widetilde{\mathsf{P}}^{i,t}, \overline{\mathsf{lab}}^{i,t}) \leftarrow \mathsf{Garble}(1^\lambda, \mathsf{P}[i, \phi_t, v_i, \{\omega_{t,\alpha,\beta}\}_{\alpha,\beta}, \bot, \overline{\mathsf{lab}}^{i,t+1}]).$$
   (c) If $i \neq i^*$ then for every $\alpha, \beta \in \{0,1\}$, set $\mathsf{ots}^i_{2,t,\alpha,\beta} \leftarrow \mathsf{OT}_2(\sigma_{t,\alpha,\beta}, \mathsf{ots}_{1,t,\alpha,\beta}, \mathsf{lab}^{i,t+1}_{h,0}, \mathsf{lab}^{i,t+1}_{h,1})$ and compute
$$(\widetilde{\mathsf{P}}^{i,t}, \overline{\mathsf{lab}}^{i,t}) \leftarrow \mathsf{Garble}(1^\lambda, \mathsf{P}[i, \phi_t, v_i, \bot, \{\mathsf{ots}^i_{2,t,\alpha,\beta}\}_{\alpha,\beta}, \overline{\mathsf{lab}}^{i,t+1}]).$$
4. Send $\big(\{\widetilde{\mathsf{P}}^{i,t}\}_{t\in[T]}, \{\mathsf{lab}^{i,1}_{k,\mathsf{st}_{i,k}}\}_{k\in[\ell]}\big)$ to every other party.

Evaluation: To compute the output of the protocol, each party P_i does the following:

1. For each $j \in [n]$, let $\widetilde{\mathsf{lab}}^{j,1} := \{\mathsf{lab}^{j,1}_k\}_{k\in[\ell]}$ be the labels received from party P_j at the end of round 2.
2. **for** each t from 1 to T do:
   (a) Parse ϕ_t as (i^*, f, g, h).
   (b) Compute $((\alpha, \beta, \gamma), \omega, \widetilde{\mathsf{lab}}^{i^*,t+1}) := \mathsf{Eval}(\widetilde{\mathsf{P}}^{i^*,t}, \widetilde{\mathsf{lab}}^{i^*,t})$.
   (c) Set $\mathsf{st}_{i,h} := \gamma \oplus v_{i,h}$.
   (d) **for** each $j \neq i^*$ do:
      i. Compute $(\mathsf{ots}_2, \{\mathsf{lab}^{j,t+1}_k\}_{k\in[\ell]\setminus\{h\}}) := \mathsf{Eval}(\widetilde{\mathsf{P}}^{j,t}, \widetilde{\mathsf{lab}}^{j,t})$.
      ii. Recover $\mathsf{lab}^{j,t+1}_h := \mathsf{OT}_3(\sigma_{t,\alpha,\beta}, \mathsf{ots}_2, \omega)$.
      iii. Set $\widetilde{\mathsf{lab}}^{j,t+1} := \{\mathsf{lab}^{j,t+1}_k\}_{k\in[\ell]}$.
3. Compute the output as $\mathsf{post}(i, \mathsf{st}_i)$.

Fig. 3. Two-round malicious MPC.

Theorem 3. *Let Φ be a polynomial round, n-party malicious MPC protocol computing a function $f : (\{0,1\}^m)^n \rightarrow \{0,1\}^*$, $(\mathsf{Garble}, \mathsf{Eval})$ be a garbling scheme for circuits, and $(K_{\mathsf{OT}}, \mathsf{OT}_1, \mathsf{OT}_2, \mathsf{OT}_3)$ be a maliciously secure (with equivocal receiver security) two-round OT protocol. The protocol described in Fig. 3 is a two-round, n-party malicious MPC protocol computing f against static corruptions.*

We prove the security of our compiler in the rest of the section. The proof of correctness is the same as for the case of semi-honest security (see Sect. 5.2).

As in the semi-honest case Via the same argument as above it is useful to keep in mind that for every $i, j \in [n]$ and $k \in [\ell]$, we have that $\mathsf{st}_{i,k} \oplus v_{i,k} = \mathsf{st}_{j,k} \oplus v_{j,k}$. Let us denote this shared value by st^*. Also, we denote the transcript of the interaction in the computation phase by $\mathsf{Z} \in \{0,1\}^t$.

6.2 Simulator

Let $\mathcal{A}$ be a malicious adversary corrupting a subset of parties and let $H \subseteq [n]$ be the set of honest/uncorrupted parties. Since we assume that the adversary is static, this set is fixed before the execution of the protocol. Below we provide thenotion of faithful execution and then describe our simulator.

Faithful Execution. In the first round of our compiled protocol, $\mathcal{A}$ provides z_i for every $i \in [n] \setminus H$ and $\mathsf{ots}_{1,t,\alpha,\beta}$ for every $t \in \cup_{i \in [n] \setminus h}$ and $\alpha, \beta \in \{0,1\}$. These values act as "binding" commitments to all of the adversary's future choices. All these committed choices can be extracted using the extractor Ext_2. Let $b_{t,\alpha,\beta}$ be the value extracted from $\mathsf{ots}_{1,t,\alpha,\beta}$. Intuitively speaking, a faithful execution is an execution that is consistent with these extracted values.

More formally, we define an interactive procedure $\mathsf{Faithful}(i, \{z_i\}_{i \in [n]}, \{b_{t,\alpha,\beta}\}_{t \in A_i, \alpha, \beta})$ that on input $i \in [n]$, $\{z_i\}_{i \in [n]}$, $\{b_{t,\alpha,\beta}\}_{t \in A_i, \alpha, \beta \in \{0,1\}}$ produces protocol Φ message on behalf of party P_i (acting consistently/faithfully with the extracted values) as follows:

1. Set $\mathsf{st}^* := z_1 \| \dots \| z_n$.
2. For $t \in \{1 \cdots T\}$
   (a) Parse $\phi_t = (i^*, f, g, h)$.
   (b) If $i \neq i^*$ then it waits for a bit from P_{i^*} and sets st^*_h to be the received bit once it is received.
   (c) Set $\mathsf{st}^* := b_{t,\mathsf{st}^*_f,\mathsf{st}^*_g}$ and output it to all the other parties.

We will later argue that any deviation from the faithful execution by the adversary $\mathcal{A}$ on behalf of the corrupted parties (during the second round of our compiled protocol) will be be detected. Additionally, we prove that such deviations do not hurt the security of the honest parties.

Description of the Simulator. We give the description of the ideal world adversary $\mathcal{S}$ that simulates the view of the real world adversary $\mathcal{A}$. $\mathcal{S}$ will internally use the malicious simulator Sim_Φ for Φ, the extractor $\mathsf{Ext} = (\mathsf{Ext}_1, \mathsf{Ext}_2)$

implied by the sender security of two-round OT, the simulator Sim_{Eq} implied by the equivocal receiver's security and the simulator Sim_G for garbling scheme for circuits. Recall that $\mathcal{A}$ is static and hence the set of honest parties H is known before the execution of the protocol.

Simulating the interaction with $\mathcal{Z}$. For every input value for the set of corrupted parties that $\mathcal{S}$ receives from $\mathcal{Z}$, $\mathcal{S}$ writes that value to $\mathcal{A}$'s input tape. Similarly, the output of $\mathcal{A}$ is written as the output on $\mathcal{S}$'s output tape.

Simulating the interaction with $\mathcal{A}$: For every concurrent interaction with the session identifier sid that $\mathcal{A}$ may start, the simulator does the following:

- **Generation of the common random/reference string**: $\mathcal{S}$ generates the common random/reference string as follows:
  1. For each $i \in H, t \in A_i, \alpha, \beta \in \{0,1\}$ set $(\sigma_{t,\alpha,\beta}, (\mathsf{ots}_{1,t,\alpha,\beta}, \omega^0_{t,\alpha,\beta}, \omega^1_{t,\alpha,\beta})) \leftarrow \mathsf{Sim}_{Eq}(1^\lambda)$ (using equivocal simulator).
  2. For each $i \in [n] \setminus H, \alpha, \beta \in \{0,1\}$ and $t \in A_i$ generate $(\sigma_{t,\alpha,\beta}, \tau_{t,\alpha,\beta}) \leftarrow \mathsf{Ext}_1(1^\lambda)$ (using the extractor of the OT protocol).
  3. Output the common random/reference string as $\{\sigma_{t,\alpha,\beta}\}_{t,\alpha,\beta}$.
- **Initialization**: $\mathcal{S}$ executes the simulator (against malicious adversary's) $\mathsf{Sim}_\Phi(1^\lambda)$ to obtain $\{z_i\}_{i\in H}$. Moreover, $\mathcal{S}$ starts the real-world adversary $\mathcal{A}$. We next describe how $\mathcal{S}$ provides its messages to Sim_Φ and $\mathcal{A}$.
- **Round-1 messages from $\mathcal{S}$ to $\mathcal{A}$:** For each $i \in H$, $\mathcal{S}$ sends $(z_i, \{\mathsf{ots}_{1,t,\alpha,\beta}\}_{t\in A_i,\alpha,\beta\in\{0,1\}})$ to the adversary $\mathcal{A}$ on behalf of the honest party P_i.
- **Round-1 messages from $\mathcal{A}$ to $\mathcal{S}$:** Corresponding to every $i \in [n] \setminus H$, $\mathcal{S}$ receives from the adversary $\mathcal{A}$ the value $(z_i, \{\mathsf{ots}_{1,t,\alpha,\beta}\}_{t\in A_i,\alpha,\beta\in\{0,1\}})$ on behalf of the corrupted party P_i. Next, for each $i \in [n]\setminus H, t \in A_i, \alpha, \beta \in \{0,1\}$ extract $b_{t,\alpha,\beta} := \mathsf{Ext}_2(\tau_{t,\alpha,\beta}, \mathsf{ots}_{1,t,\alpha,\beta})$.
- **Completing the execution with the Sim_Φ:** For each $i \in [n] \setminus H$, $\mathcal{S}$ sends z_i to Sim_Φ on behalf of the corrupted party P_i. This starts the computation phase of Φ with the simulator Sim_Φ. $\mathcal{S}$ provides computation phase messages to Sim_Φ by following a faithful execution. More formally, for every corrupted party P_i where $i \in [n]\setminus H$, $\mathcal{S}$ generates messages on behalf of P_i for Sim_Φ using the procedure $\mathsf{Faithful}(i, \{z_i\}_{i\in[n]}, \{b_{t,\alpha,\beta}\}_{t\in A_i,\alpha,\beta})$. At some point during the execution, Sim_Φ will return the extracted inputs $\{x_i\}_{i\in[n]\setminus H}$ of the corrupted parties. For each $i \in [n]\setminus H$, $\mathcal{S}$ sends $(\mathsf{input}, \mathsf{sid}, \{P_1 \cdots P_n\}, P_i, x_i)$ to the ideal functionality implementing f and obtains the output y which is provided to Sim_Φ. Finally, at some point the faithful execution completes.

  Let $\mathsf{Z} \in \{0,1\}^t$ where Z_t is the bit sent in the tth round of the computation phase of Φ be output of this execution. And let st^* be the state value at the end of execution of one of the corrupted parties (this value is the same for all the parties). Also, set for each $t \in \cup_{i\in H} A_i$ and $\alpha, \beta \in \{0,1\}$ set $\omega_{t,\alpha,\beta} := \omega^{\mathsf{Z}_t}_{t,\alpha,\beta}$.
- **Round-2 messages from $\mathcal{S}$ to $\mathcal{A}$:** For each $i \in H$, the simulator $\mathcal{S}$ generates the second round message on behalf of party P_i as follows:
  1. For each $k \in [\ell]$ set $\mathsf{lab}^{i,T+1}_k := 0^\lambda$.
  2. **for** each t from T down to 1,

(a) Parse ϕ_t as (i^*, f, g, h).
(b) Set $\alpha^* := \mathsf{st}^*_f$, $\beta^* := \mathsf{st}^*_g$, and $\gamma^* := \mathsf{st}^*_h$.
(c) If $i = i^*$ then compute

$$(\widetilde{\mathsf{P}}^{i,t}, \{\mathsf{lab}_k^{i,t}\}_{k\in[\ell]}) \leftarrow \mathsf{Sim}_{\mathsf{G}}\left(1^\lambda, \left((\alpha^*, \beta^*, \gamma^*), \omega_{t,\alpha^*,\beta^*}, \{\mathsf{lab}_k^{i,t+1}\}_{k\in[\ell]}\right)\right).$$

(d) If $i \neq i^*$ then set $\mathsf{ots}^i_{2,t,\alpha^*,\beta^*} \leftarrow \mathsf{OT}_2(\sigma_{t,\alpha^*,\beta^*}, \mathsf{ots}_{1,t,\alpha^*,\beta^*}, \mathsf{lab}_h^{i,t+1}, \mathsf{lab}_h^{i,t+1})$ and compute

$$(\widetilde{\mathsf{P}}^{i,t}, \{\mathsf{lab}_k^{i,t}\}_{k\in[\ell]}) \leftarrow \mathsf{Sim}_{\mathsf{G}}\left(1^\lambda, \left(\mathsf{ots}^i_{2,t,\alpha^*,\beta^*}, \{\mathsf{lab}_k^{i,t+1}\}_{k\in[\ell]\setminus\{h\}}\right)\right).$$

3. Send $(\{\widetilde{\mathsf{P}}^{i,t}\}_{t\in[T]}, \{\mathsf{lab}_k^{i,1}\}_{k\in[\ell]})$ to every other party.

- **Round-2 messages from $\mathcal{A}$ to $\mathcal{S}$:** For every $i \in [n] \setminus H$, $\mathcal{S}$ obtains the second round message from $\mathcal{A}$ on behalf of the malicious parties. Subsequent to obtaining these messages, $\mathcal{S}$ executes the garbled circuits provided by $\mathcal{A}$ on behalf of the corrupted parties to see the execution of garbled circuits proceeds consistently with the expected faithful execution. If the computation succeeds then for each $i \in H$, $\mathcal{S}$ sends $(\mathsf{generateOutput}, \mathsf{sid}, \{P_1 \cdots P_n\}, P_i)$ to the ideal functionality.

6.3 Proof of Indistinguishability

We now show that no environment $\mathcal{Z}$ can distinguish whether it is interacting with a real world adversary $\mathcal{A}$ or an ideal world adversary $\mathcal{S}$. We prove this via an hybrid argument with $T + 2$ hybrids.

- $\mathcal{H}_{Real}$: This hybrid is the same as the real world execution.
- $\mathcal{H}_0$: In this hybrid we start by changing the distribution of the common random string. Specifically, the common random string is generated as is done in the simulation. More formally, $\mathcal{S}$ generates the common random/reference string as follows:
  1. For each $i \in H, t \in A_i, \alpha, \beta \in \{0,1\}$ set $(\sigma_{t,\alpha,\beta}, (\mathsf{ots}_{1,t,\alpha,\beta}, \omega^0_{t,\alpha,\beta}, \omega^1_{t,\alpha,\beta})) \leftarrow \mathsf{Sim}_{Eq}(1^\lambda)$ (using equivocal simulator).
     For all $t \in \cup_{i\in H} A_i$ and $\alpha, \beta \in \{0,1\}$ set $\omega_{t,\alpha,\beta} := \omega_{t,\alpha,\beta}^{v_{i,h}\oplus\mathsf{NAND}(v_{i,f}\oplus\alpha, v_{i,g}\oplus\beta)}$ where v_i is the secret value of party P_i generated in the pre-processing phase of Φ.
  2. For each $i \in [n] \setminus H, \alpha, \beta \in \{0,1\}$ and $t \in A_i$ generate $(\sigma_{t,\alpha,\beta}, \tau_{t,\alpha,\beta}) \leftarrow \mathsf{Ext}_1(1^\lambda)$ (using the extractor of the OT protocol).
     Corresponding to every $i \in [n] \setminus H$, $\mathcal{A}$ sends $(z_i, \{\mathsf{ots}_{1,t,\alpha,\beta}\}_{t\in A_i,\alpha,\beta\in\{0,1\}})$ on behalf of the corrupted party P_i as its first round message. For each $i \in [n] \setminus H, t \in A_i, \alpha, \beta \in \{0,1\}$ in this hybrid we extract $b_{t,\alpha,\beta} := \mathsf{Ext}(\tau_{t,\alpha,\beta}, \mathsf{ots}_{1,t,\alpha,\beta})$.

  Note that this hybrid is the same as hybrid $\mathcal{H}_t$ below with $t = 0$.

  The indistinguishability between hybrids $\mathcal{H}_{Real}$ and $\mathcal{H}_0$ follow from a reduction to the sender's security and the equivocal receiver's security of the two-round OT protocol.

- $\mathcal{H}_t$ (where $t \in \{0, \dots T\}$): Hybrid $\mathcal{H}_t$ (for $t \in \{1 \cdots T\}$) is the same as hybrid $\mathcal{H}_{t-1}$ except we change the distribution of the OT messages (both from the first and the second round of the protocol) and the garbled circuits (from the second round) that play a role in the execution of the tth round of the protocol Φ; namely, the action $\phi_t = (i^*, f, g, h)$. We describe the changes more formally below.

  For each $i \in [n] \setminus H$, in this hybrid $\mathcal{S}$ (in his head) completes an execution of Φ using honest party inputs and randomness. In this execution, the messages on behalf of corrupted parties are generated via faithful execution. Specifically, $\mathcal{S}$ sends $\{z_i\}_{i \in [n] \setminus H}$ to the honest parties on behalf of the corrupted party P_i in this mental execution of Φ. This starts the computation phase of Φ. In this computation phase, $\mathcal{S}$ generates honest party messages using the inputs and random coins of the honest parties and generates the messages of the each malicious party P_i by executing $\mathsf{Faithful}\left(i, \{z_i\}_{i \in [n] \setminus H}, \{b_{t,\alpha,\beta}\}_{t \in A_i, \alpha, \beta}\right)$. Let st^* be the local state of the end of execution of $\mathsf{Faithful}$. Finally, let $\alpha^* := \mathsf{st}^*_f$, $\beta^* := \mathsf{st}^*_g$ and $\gamma^* := \mathsf{st}^*_h$. In hybrid $\mathcal{H}_t$ we make the following changes with respect to hybrid $\mathcal{H}_{t-1}$:

  - If $i^* \notin H$ then skip these changes. $\mathcal{S}$ makes two changes in how it generates messages on behalf of P_{i^*}. First, for all $\alpha, \beta \in \{0,1\}$, $\mathcal{S}$ sets $\omega_{t,\alpha,\beta}$ as $\omega_{t,\alpha,\beta}^{\mathsf{Z}_t}$ rather than $\omega_{t,\alpha,\beta}^{v_{i,h} \oplus \mathsf{NAND}(v_{i,f} \oplus \alpha, v_{i,g} \oplus \beta)}$ (note that these two values are the same when using the honest party's input and randomness). Second, it generates the garbled circuit

    $$\left(\widetilde{\mathsf{P}}^{i^*,t}, \{\mathsf{lab}_k^{i^*,t}\}_{k \in [\ell]}\right) \leftarrow \mathsf{Sim}_{\mathsf{G}}\left(1^\lambda, \left((\alpha^*, \beta^*, \gamma^*), \omega_{t,\alpha^*,\beta^*}, \{\mathsf{lab}_{k,\mathsf{st}_{i,k}}^{i^*,t+1}\}_{k \in [\ell]}\right)\right),$$

    where $\{\mathsf{lab}_{k,\mathsf{st}_{i,k}}^{i^*,t+1}\}_{k \in [\ell]}$ are the honestly generates input labels for the garbled circuit $\widetilde{\mathsf{P}}^{i^*,t+1}$.
  - $\mathcal{S}$ makes the following two changes in how it generates messages for other honest parties P_i (i.e., $i \in H \setminus \{i^*\}$). $\mathcal{S}$ does not generate four $\mathsf{ots}_{2,t,\alpha,\beta}^i$ values but just one of them; namely, $\mathcal{S}$ generates $\mathsf{ots}_{2,t,\alpha^*,\beta^*}^i$ as $\mathsf{OT}_2(\sigma_{t,\alpha^*,\beta^*}, \mathsf{ots}_{1,t,\alpha^*,\beta^*}, \mathsf{lab}_{h,\mathsf{Z}_t}^{i,t+1}, \mathsf{lab}_{h,\mathsf{Z}_t}^{i,t+1})$ rather than $\mathsf{OT}_2(\sigma_{t,\alpha^*,\beta^*}, \mathsf{ots}_{1,t,\alpha^*,\beta^*}, \mathsf{lab}_{h,0}^{i,t+1}, \mathsf{lab}_{h,1}^{i,t+1})$. Second it generates the garbled circuit

    $$\left(\widetilde{\mathsf{P}}^{i,t}, \{\mathsf{lab}_k^{i,t}\}_{k \in [\ell]}\right) \leftarrow \mathsf{Sim}_{\mathsf{G}}\left(1^\lambda, \left(\mathsf{ots}_{2,t,\alpha^*,\beta^*}^i, \{\mathsf{lab}_{k,\mathsf{st}_{i,k}}^{i,t+1}\}_{k \in [\ell] \setminus \{h\}}\right)\right),$$

    where $\{\mathsf{lab}_{k,\mathsf{st}_{i,k}}^{i,t+1}\}_{k \in [\ell]}$ are the honestly generated input labels for the garbled circuit $\widetilde{\mathsf{P}}^{i,t+1}$.

  Indistinguishability between $\mathcal{H}_{t-1}$ and $\mathcal{H}_t$ is proved in Lemma 2.
- $\mathcal{H}_{T+1}$: In this hybrid we just change how the transcript Z, $\{z_i\}_{i \in H}$, random coins of malicious parties and value st^* are generated. Instead of generating these using honest party inputs in execution with a faithful execution of Φ, we generate it via the simulator Sim_Φ (of the maliciously secure protocol Φ). In other words, we execute the simulator Sim_Φ where messages on behalf of each

corrupted party P_i are generated using $\mathsf{Faithful}(i, \{z_i\}_{i \in [n] \setminus H}, \{b_{t,\alpha,\beta}\}_{t \in A_i, \alpha, \beta})$. (Note that Sim_Φ might rewind $\mathsf{Faithful}$. This can be achieved since $\mathsf{Faithful}$ is just a polynomial time interactive procedure that can also be rewound.)

The indistinguishability between hybrids $\mathcal{H}_T$ and $\mathcal{H}_{T+1}$ follows directly from the malicious security of the protocol Φ. Finally note that $\mathcal{H}_{T+1}$ is same as the ideal execution (i.e., the simulator described in the previous subsection).

Lemma 3. *Assuming malicious security of the two-round OT protocol and the security of the garbling scheme, for all $t \in \{1 \ldots T\}$ hybrids $\mathcal{H}_{t-1}$ and $\mathcal{H}_t$ are computationally indistinguishable.*

Proof. Using the same notation as before, let $\phi_t = (i^*, f, g, h)$, st_{i^*} be the state of P_{i^*} at the end of round t, and $\alpha^* := \mathsf{st}_{i^*,f} \oplus v_{i^*,f}$, $\beta^* := \mathsf{st}_{i^*,g} \oplus v_{i^*,g}$ and $\gamma^* := \mathsf{st}_{i^*,h} \oplus v_{i^*,h}$. The indistinguishability between hybrids $\mathcal{H}_{t-1}$ and $\mathcal{H}_t$ follows by a sequence of three sub-hybrids $\mathcal{H}_{t,1}$, $\mathcal{H}_{t,2}$, and $\mathcal{H}_{t,3}$.

- $\mathcal{H}_{t,1}$: Hybrid $\mathcal{H}_{t,1}$ is same as hybrid $\mathcal{H}_{t-1}$ except that $\mathcal{S}$ now generates the garbled circuits $\widetilde{\mathsf{P}}^{i,t}$ for each $i \in H$ in a simulated manner (rather than generating them honestly). Specifically, instead of generating each garbled circuit and input labels $(\widetilde{\mathsf{P}}^{i,t}, \{\mathsf{lab}_k^{i,t}\}_{k \in [\ell]})$ honestly, they are generated via the simulator by hard coding the output of the circuit itself. In a bit more details, parse ϕ_t as (i^*, f, g, h).
  - If $i = i^*$ then
  $$(\widetilde{\mathsf{P}}^{i,t}, \{\mathsf{lab}_k^{i,t}\}_{k \in [\ell]}) \leftarrow \mathsf{Sim}_{\mathsf{G}}\left(1^\lambda, \left((\alpha^*, \beta^*, \gamma^*), \omega_{t,\alpha^*,\beta^*}, \{\mathsf{lab}_{k,\mathsf{st}_{i,k}}^{i,t+1}\}_{k \in [\ell]}\right)\right),$$
  where $\{\mathsf{lab}_{k,\mathsf{st}_{i,k}}^{i,t+1}\}_{k \in [\ell]}$ are the honestly generates input labels for the garbled circuit $\widetilde{\mathsf{P}}^{i,t+1}$.
  - If $i \neq i^*$ then
  $$(\widetilde{\mathsf{P}}^{i,t}, \{\mathsf{lab}_k^{i,t}\}_{k \in [\ell]}) \leftarrow \mathsf{Sim}_{\mathsf{G}}\left(1^\lambda, \left(\mathsf{ots}_{2,t,\alpha^*,\beta^*}^i, \{\mathsf{lab}_{k,\mathsf{st}_{i,k}}^{i,t+1}\}_{k \in [\ell] \setminus \{h\}}\right)\right),$$
  where $\{\mathsf{lab}_{k,\mathsf{st}_{i,k}}^{i,t+1}\}_{k \in [\ell]}$ are the honestly generated input labels for the garbled circuit $\widetilde{\mathsf{P}}^{i,t+1}$.

  The indistinguishability between hybrids $\mathcal{H}_{t,1}$ and $\mathcal{H}_{t-1}$ follows by $|H|$ invocations of security of the garbling scheme.
- $\mathcal{H}_{t,2}$: Skip this hybrid, if $i^* \notin H$. This hybrid is same as $\mathcal{H}_{t,1}$ except that we change how $\mathcal{S}$ generates the Round-1 message on behalf of P_{i^*}. Specifically, the simulator $\mathcal{S}$ generates $\mathsf{ots}_{1,t,\alpha,\beta}$ as is done in the $\mathcal{H}_t$. In a bit more detail, for all $\alpha, \beta \in \{0,1\}$, $\mathcal{S}$ generates $\mathsf{ots}_{1,t,\alpha,\beta}$ as $\mathsf{OT}_1(\sigma_{t,\alpha,\beta}, \mathsf{Z}_t; \omega_{t,\alpha,\beta})$ rather than $\mathsf{OT}_1(\sigma t, \alpha, \beta, v_{i,h} \oplus \mathsf{NAND}(v_{i,f} \oplus \alpha, v_{i,g} \oplus \beta); \omega_{t,\alpha,\beta})$.

  Indistinguishability between hybrids $\mathcal{H}_{t,1}$ and $\mathcal{H}_{t,2}$ follows directly by a sequence of 3 sub-hybrids each one relying on the receiver's security of underlying semi-honest oblivious transfer protocol. Observe here that the security reduction crucially relies on the fact that $\widetilde{\mathsf{P}}^{i,t}$ only contains $\omega_{t,\alpha^*,\beta^*}$ (i.e., does not have $\omega_{t,\alpha,\beta}$ for $\alpha \neq \alpha^*$ or $\beta \neq \beta^*$).

- $\mathcal{H}_{t,3}$: Skip this hybrid if there does not exist $i \neq i^*$ such that $i \in H$. In this hybrid, we change how $\mathcal{S}$ generates the $\mathsf{ots}^i_{2,t,\alpha,\beta}$ on behalf of every honest party P_i such that $i \in H \setminus \{i^*\}$ for all choices of $\alpha, \beta \in \{0,1\}$. More specifically, $\mathcal{S}$ only generates one of these four values; namely, $\mathsf{ots}^i_{2,t,\alpha^*,\beta^*}$ which is now generated as $\mathsf{OT}_2(\sigma_{t,\alpha^*,\beta^*}, \mathsf{ots}_{1,t,\alpha^*,\beta^*}, \mathsf{lab}^{i,t+1}_{h,\mathsf{Z}_t}, \mathsf{lab}^{i,t+1}_{h,\mathsf{Z}_t})$ instead of $\mathsf{OT}_2(\sigma_{t,\alpha^*,\beta^*}, \mathsf{ots}_{1,t,\alpha^*,\beta^*}, \mathsf{lab}^{i,t+1}_{h,0}, \mathsf{lab}^{i,t+1}_{h,1})$.
  Indistinguishability between hybrids $\mathcal{H}_{t,2}$ and $\mathcal{H}_{t,3}$ follows directly from the sender's security of underlying malicious oblivious transfer protocol. Finally, observe that $\mathcal{H}_{t,3}$ is the same as hybrid $\mathcal{H}_t$.

6.4 Extensions

As in the semi-honest case, we discuss several extensions to the construction of two-round maliciously secure MPC.

Fairness. Assuming honest majority we obtain fairness in three rounds using techniques from [GLS15]. Specifically, we can change the function description to output a $n/2$-out-of-n secret sharing of the output. In the last round, the parties exchange their shares to reconstruct the output. Note that since the corrupted parties is in minority, it cannot learn the output of the function even if it obtains the second round messages from all the parties. Note that Gordon et al. [GLS15] showed that three rounds are necessary to achieve fairness. Thus this is optimal.

Semi-malicious security in Plain Model. We note that a simple modification of our construction in Fig. 3 can be made semi-maliciously secure in the plain model. The modification is to use a two-round OT secure against semi-malicious receiver and semi-honest sender (e.g., [NP01]) and achieve equivocability by sending two OT_1 messages in the first round having the same receiver's choice bit. Note that this is trivially equivocal since a simulator can use different choice bits in the OT_1 message. On the other hand, since a semi-malicious party is required to follow the protocol, it will always use the same choice bit in both the OT_1 messages.

References

[AIK04] Applebaum, B., Ishai, Y., Kushilevitz, E.: Cryptography in NC^0. In: 45th FOCS, Rome, Italy, 17–19 October 2004, pp. 166–175. IEEE Computer Society Press (2004)

[AIK05] Applebaum, B., Ishai, Y., Kushilevitz, E.: Computationally private randomizing polynomials and their applications. In: 20th Annual IEEE Conference on Computational Complexity (CCC 2005), San Jose, CA, USA, 11–15 June 2005, pp. 260–274 (2005)

[AIR01] Aiello, B., Ishai, Y., Reingold, O.: Priced oblivious transfer: how to sell digital goods. In: Pfitzmann, B. (ed.) EUROCRYPT 2001. LNCS, vol. 2045, pp. 119–135. Springer, Heidelberg (2001). https://doi.org/10.1007/3-540-44987-6_8

[AJL+12] Asharov, G., Jain, A., López-Alt, A., Tromer, E., Vaikuntanathan, V., Wichs, D.: Multiparty computation with low communication, computation and interaction via threshold FHE. In: Pointcheval, D., Johansson, T. (eds.) EUROCRYPT 2012. LNCS, vol. 7237, pp. 483–501. Springer, Heidelberg (2012). https://doi.org/10.1007/978-3-642-29011-4_29

[BF01] Boneh, D., Franklin, M.: Identity-based encryption from the weil pairing. In: Kilian, J. (ed.) CRYPTO 2001. LNCS, vol. 2139, pp. 213–229. Springer, Heidelberg (2001). https://doi.org/10.1007/3-540-44647-8_13

[BFM88] Blum, M., Feldman, P., Micali, S.: Non-interactive zero-knowledge and its applications (extended abstract). In: 20th ACM STOC, Chicago, IL, USA, 2–4 May 1988, pp. 103–112. ACM Press (1988)

[BGI+01] Barak, B., Goldreich, O., Impagliazzo, R., Rudich, S., Sahai, A., Vadhan, S., Yang, K.: On the (im)possibility of obfuscating programs. In: Kilian, J. (ed.) CRYPTO 2001. LNCS, vol. 2139, pp. 1–18. Springer, Heidelberg (2001). https://doi.org/10.1007/3-540-44647-8_1

[BGI16] Boyle, E., Gilboa, N., Ishai, Y.: Breaking the circuit size barrier for secure computation under DDH. In: Robshaw, M., Katz, J. (eds.) CRYPTO 2016. LNCS, vol. 9814, pp. 509–539. Springer, Heidelberg (2016). https://doi.org/10.1007/978-3-662-53018-4_19

[BGI17] Boyle, E., Gilboa, N., Ishai, Y.: Group-based secure computation: optimizing rounds, communication, and computation. In: Coron, J.-S., Nielsen, J.B. (eds.) EUROCRYPT 2017. LNCS, vol. 10211, pp. 163–193. Springer, Cham (2017). https://doi.org/10.1007/978-3-319-56614-6_6

[BH15] Bellare, M., Hoang, V.T.: Adaptive witness encryption and asymmetric password-based cryptography. In: Katz, J. (ed.) PKC 2015. LNCS, vol. 9020, pp. 308–331. Springer, Heidelberg (2015). https://doi.org/10.1007/978-3-662-46447-2_14

[BHR12] Bellare, M., Hoang, V.T., Rogaway, P.: Foundations of garbled circuits. In: Yu, T., Danezis, G., Gligor, V.D. (eds.) ACM CCS 12, Raleigh, NC, USA, 16–18 October 2012, pp. 784–796. ACM Press (2012)

[BL18] Benhamouda, F., Lin, H.: k-round multiparty computation from k-round oblivious transfer via garbled interactive circuits. In: Nielsen, J.B., Rijmen, V. (eds.) EUROCRYPT 2018. LNCS, vol. 10821, pp. 500–532. Springer, Cham (2018). https://eprint.iacr.org/2017/1125

[BMR90] Beaver, D., Micali, S., Rogaway, P.: The round complexity of secure protocols (extended abstract). In: 22nd ACM STOC, Baltimore, MD, USA, 14–16 May 1990, pp. 503–513. ACM Press (1990)

[BP16] Brakerski, Z., Perlman, R.: Lattice-based fully dynamic multi-key FHE with short ciphertexts. In: Robshaw, M., Katz, J. (eds.) CRYPTO 2016. LNCS, vol. 9814, pp. 190–213. Springer, Heidelberg (2016). https://doi.org/10.1007/978-3-662-53018-4_8

[Can00a] Canetti, R.: Security and composition of multiparty cryptographic protocols. J. Cryptol. **13**(1), 143–202 (2000)

[Can00b] Canetti, R.: Universally composable security: a new paradigm for cryptographic protocols. Cryptology ePrint Archive, Report 2000/067 (2000). http://eprint.iacr.org/2000/067

[Can01] Canetti, R.: Universally composable security: a new paradigm for cryptographic protocols. In: 42nd FOCS, Las Vegas, NV, USA, 14–17 October 2001, pp. 136–145. IEEE Computer Society Press (2001)

[Can05] Canetti, R.: Universally composable security: a new paradigm for cryptographic protocols, Version of December 2005 (2005). http://eccc.uni-trier.de/eccc-reports/2001/TR01-016

[CCM98] Cachin, C., Crépeau, C., Marcil, J.: Oblivious transfer with a memory-bounded receiver. In: 39th FOCS, Palo Alto, CA, USA, 8–11 November 1998, pp. 493–502. IEEE Computer Society Press (1998)

[CDG+17] Cho, C., Döttling, N., Garg, S., Gupta, D., Miao, P., Polychroniadou, A.: Laconic oblivious transfer and its applications. In: Katz, J., Shacham, H. (eds.) CRYPTO 2017. LNCS, vol. 10402, pp. 33–65. Springer, Cham (2017). https://doi.org/10.1007/978-3-319-63715-0_2

[CLOS02] Canetti, R., Lindell, Y., Ostrovsky, R., Sahai, A.: Universally composable two-party and multi-party secure computation. In: 34th ACM STOC, Montréal, Québec, Canada, 19–21 May 2002, pp. 494–503. ACM Press (2002)

[CM15] Clear, M., McGoldrick, C.: Multi-identity and multi-key leveled FHE from learning with errors. In: Gennaro, R., Robshaw, M. (eds.) CRYPTO 2015. LNCS, vol. 9216, pp. 630–656. Springer, Heidelberg (2015). https://doi.org/10.1007/978-3-662-48000-7_31

[DG17] Döttling, N., Garg, S.: Identity-based encryption from the Diffie-Hellman assumption. In: Katz, J., Shacham, H. (eds.) CRYPTO 2017. LNCS, vol. 10401, pp. 537–569. Springer, Cham (2017). https://doi.org/10.1007/978-3-319-63688-7_18

[DHRS04] Ding, Y.Z., Harnik, D., Rosen, A., Shaltiel, R.: Constant-round oblivious transfer in the bounded storage model. In: Naor, M. (ed.) TCC 2004. LNCS, vol. 2951, pp. 446–472. Springer, Heidelberg (2004). https://doi.org/10.1007/978-3-540-24638-1_25

[FLS90] Feige, U., Lapidot, D., Shamir, A.: Multiple non-interactive zero knowledge proofs based on a single random string (extended abstract). In: 31st FOCS, St. Louis, Missouri, 22–24 October 1990, pp. 308–317. IEEE Computer Society Press (1990)

[Gen09] Gentry, C.: Fully homomorphic encryption using ideal lattices. In: Mitzenmacher, M. (ed.) 41st ACM STOC, Bethesda, MD, USA, 31 May–2 June 2009, pp. 169–178. ACM Press (2009)

[GGH+13] Garg, S., Gentry, C., Halevi, S., Raykova, M., Sahai, A., Waters, B.: Candidate indistinguishability obfuscation and functional encryption for all circuits. In: 54th FOCS, Berkeley, CA, USA, 26–29 October 2013, pp. 40–49. IEEE Computer Society Press (2013)

[GGHR14] Garg, S., Gentry, C., Halevi, S., Raykova, M.: Two-round secure MPC from indistinguishability obfuscation. In: Lindell, Y. (ed.) TCC 2014. LNCS, vol. 8349, pp. 74–94. Springer, Heidelberg (2014). https://doi.org/10.1007/978-3-642-54242-8_4

[GGMP16] Garg, S., Gupta, D., Miao, P., Pandey, O.: Secure multiparty RAM computation in constant rounds. In: Hirt, M., Smith, A. (eds.) TCC 2016. LNCS, vol. 9985, pp. 491–520. Springer, Heidelberg (2016). https://doi.org/10.1007/978-3-662-53641-4_19

[GGSW13] Garg, S., Gentry, C., Sahai, A., Waters, B.: Witness encryption and its applications. In: Boneh, D., Roughgarden, T., Feigenbaum, J. (eds.) 45th ACM STOC, Palo Alto, CA, USA, 1–4 June 2013, pp. 467–476. ACM Press (2013)

[GHL+14] Gentry, C., Halevi, S., Lu, S., Ostrovsky, R., Raykova, M., Wichs, D.: Garbled RAM revisited. In: Nguyen, P.Q., Oswald, E. (eds.) EUROCRYPT 2014. LNCS, vol. 8441, pp. 405–422. Springer, Heidelberg (2014). https://doi.org/10.1007/978-3-642-55220-5_23

[GKK+12] Dov Gordon, S., Katz, J., Kolesnikov, V., Krell, F., Malkin, T., Raykova, M., Vahlis, Y.: Secure two-party computation in sublinear (amortized) time. In: Yu, T., Danezis, G., Gligor, V.D. (eds.) ACM CCS 2012, Raleigh, NC, USA, 16–18 October 2012, pp. 513–524. ACM Press (2012)

[GLO15] Garg, S., Lu, S., Ostrovsky, R.: Black-box garbled RAM. In: Guruswami, V. (ed.) 56th FOCS, Berkeley, CA, USA, 17–20 October 2015, pp. 210–229. IEEE Computer Society Press (2015)

[GLOS15] Garg, S., Lu, S., Ostrovsky, R., Scafuro, A.: Garbled RAM from one-way functions. In: Servedio, R.A., Rubinfeld, R. (eds.) 47th ACM STOC, Portland, OR, USA, 14–17 June 2015, pp. 449–458. ACM Press (2015)

[GLS15] Dov Gordon, S., Liu, F.-H., Shi, E.: Constant-round MPC with fairness and guarantee of output delivery. In: Gennaro, R., Robshaw, M. (eds.) CRYPTO 2015. LNCS, vol. 9216, pp. 63–82. Springer, Heidelberg (2015). https://doi.org/10.1007/978-3-662-48000-7_4

[GMW87] Goldreich, O., Micali, S., Wigderson, A.: How to play any mental game or a completeness theorem for protocols with honest majority. In: Aho, A. (ed.) 19th ACM STOC, New York City, NY, USA, 25–27 May 1987, pp. 218–229. ACM Press (1987)

[GOS06] Groth, J., Ostrovsky, R., Sahai, A.: Perfect non-interactive zero knowledge for NP. In: Vaudenay, S. (ed.) EUROCRYPT 2006. LNCS, vol. 4004, pp. 339–358. Springer, Heidelberg (2006). https://doi.org/10.1007/11761679_21

[GOVW12] Garg, S., Ostrovsky, R., Visconti, I., Wadia, A.: Resettable statistical zero knowledge. In: Cramer, R. (ed.) TCC 2012. LNCS, vol. 7194, pp. 494–511. Springer, Heidelberg (2012). https://doi.org/10.1007/978-3-642-28914-9_28

[GS17] Garg, S., Srinivasan, A.: Garbled protocols and two-round MPC from bilinear maps. In: 58th FOCS, pp. 588–599. IEEE Computer Society Press (2017)

[GS18] Garg, S., Srinivasan, A.: Two-round multiparty secure computation from minimal assumptions. In: Nielsen, J.B., Rijmen, V. (eds.) EUROCRYPT 2018. LNCS, vol. 10821, pp. 468–499. Springer, Cham (2018). https://eprint.iacr.org/2017/1156

[HK12] Halevi, S., Kalai, Y.T.: Smooth projective hashing and two message oblivious transfer. J. Cryptol. **25**(1), 158–193 (2012)

[HY16] Hazay, C., Yanai, A.: Constant-round maliciously secure two-party computation in the RAM model. In: Hirt, M., Smith, A. (eds.) TCC 2016. LNCS, vol. 9985, pp. 521–553. Springer, Heidelberg (2016). https://doi.org/10.1007/978-3-662-53641-4_20

[IPS08] Ishai, Y., Prabhakaran, M., Sahai, A.: Founding cryptography on oblivious transfer – efficiently. In: Wagner, D. (ed.) CRYPTO 2008. LNCS, vol. 5157, pp. 572–591. Springer, Heidelberg (2008). https://doi.org/10.1007/978-3-540-85174-5_32

[Jou04] Joux, A.: A one round protocol for tripartite Diffie-Hellman. J. Cryptol. **17**(4), 263–276 (2004)

[Kil88] Kilian, J.: Founding cryptography on oblivious transfer. In: 20th ACM STOC, Chicago, IL, USA, 2–4 May 1988, pp. 20–31. ACM Press (1988)

[LO13] Lu, S., Ostrovsky, R.: How to garble RAM programs? In: Johansson, T., Nguyen, P.Q. (eds.) EUROCRYPT 2013. LNCS, vol. 7881, pp. 719–734. Springer, Heidelberg (2013). https://doi.org/10.1007/978-3-642-38348-9_42

[LP09] Lindell, Y., Pinkas, B.: A proof of security of Yao's protocol for two-party computation. J. Cryptol. **22**(2), 161–188 (2009)

[MW16] Mukherjee, P., Wichs, D.: Two round multiparty computation via multi-key FHE. In: Fischlin, M., Coron, J.-S. (eds.) EUROCRYPT 2016. LNCS, vol. 9666, pp. 735–763. Springer, Heidelberg (2016). https://doi.org/10.1007/978-3-662-49896-5_26

[NP01] Naor, M., Pinkas, B.: Efficient oblivious transfer protocols. In: Rao Kosaraju, S. (ed.) 12th SODA, Washington, DC, USA, 7–9 January 2001, pp. 448–457. ACM-SIAM (2001)

[OS97] Ostrovsky, R., Shoup, V.: Private information storage (extended abstract). In: 29th ACM STOC, El Paso, TX, USA, 4–6 May 1997, pp. 294–303. ACM Press (1997)

[PS16] Peikert, C., Shiehian, S.: Multi-key FHE from LWE, revisited. In: Hirt, M., Smith, A. (eds.) TCC 2016. LNCS, vol. 9986, pp. 217–238. Springer, Heidelberg (2016). https://doi.org/10.1007/978-3-662-53644-5_9

[PVW08] Peikert, C., Vaikuntanathan, V., Waters, B.: A framework for efficient and composable oblivious transfer. In: Wagner, D. (ed.) CRYPTO 2008. LNCS, vol. 5157, pp. 554–571. Springer, Heidelberg (2008). https://doi.org/10.1007/978-3-540-85174-5_31

[PW00] Pfitzmann, B., Waidner, M.: Composition and integrity preservation of secure reactive systems. In: Jajodia, S., Samarati, P. (eds.) ACM CCS 2000, Athens, Greece, 1–4 November 2000, pp. 245–254. ACM Press (2000)

[Yao86] Yao, A.C.-C.: How to generate and exchange secrets (extended abstract). In: 27th FOCS, Toronto, Ontario, Canada, 27–29 October 1986, pp. 162–167. IEEE Computer Society Press (1986)

k-Round Multiparty Computation from k-Round Oblivious Transfer via Garbled Interactive Circuits

Fabrice Benhamouda[1(✉)] and Huijia Lin[2]

[1] IBM Research, Yorktown Heights, USA
fabrice.benhamouda@normalesup.org
[2] University of California, Santa Barbara, USA

Abstract. We present new constructions of *round-efficient*, or even *round-optimal*, Multi-Party Computation (MPC) protocols from Oblivious Transfer (OT) protocols. Our constructions establish a *tight* connection between MPC and OT: In the setting of semi-honest security, for any $k \geq 2$, k-round semi-honest OT is *necessary and complete* for k-round semi-honest MPC. In the round-optimal case of $k = 2$, we obtain 2-round semi-honest MPC from 2-round semi-honest OT, resolving the round complexity of semi-honest MPC assuming weak and necessary assumption. In comparison, previous 2-round constructions rely on either the heavy machinery of indistinguishability obfuscation or witness encryption, or the algebraic structure of bilinear pairing groups. More generally, for an arbitrary number of rounds k, all previous constructions of k-round semi-honest MPC require at least OT with k' rounds for $k' \leq \lfloor k/2 \rfloor$.

In the setting of malicious security, we show: For any $k \geq 5$, k-round malicious OT is *necessary and complete* for k-round malicious MPC. In fact, OT satisfying a weaker notion of *delayed-semi-malicious* security suffices. In the common reference string model, for any $k \geq 2$, we obtain k-round malicious Universal Composable (UC) protocols from any k-round semi-malicious OT and non-interactive zero-knowledge. Previous 5-round protocols in the plain model, and 2-round protocols in the common reference string model all require algebraic assumptions such as DDH or LWE.

At the core of our constructions is a new framework for *garbling interactive circuits*. Roughly speaking, it allows for garbling interactive machines that participates in interactions of a special form. The garbled machine can emulate the original interactions receiving messages sent in the *clear* (without being encoded using secrets), and reveals only the transcript of the interactions, provided that the transcript is *computationally uniquely defined*. We show that garbled interactive circuits for the purpose of constructing MPC can be implemented using OT. Along the way, we also propose a new primitive of *witness selector* that strengthens witness encryption, and a new notion of *zero-knowledge functional commitments*.

J. B. Nielsen and V. Rijmen (Eds.): EUROCRYPT 2018, LNCS 10821, pp. 500–532, 2018.
https://doi.org/10.1007/978-3-319-78375-8_17

1 Introduction

A *Multi-Party Computation (MPC) protocol* allows m mutually distrustful parties to securely compute a functionality $f(\bar{x})$ of their corresponding private inputs $\bar{x} = x_1, \ldots, x_m$, such that party P_i receives the i-th component of $f(\bar{x})$. The *semi-honest security* guarantees that *honest-but-curious* parties who follow the specification of the protocol learn nothing more than their prescribed outputs. The stronger *malicious security* guarantees that even malicious parties who may deviate from the protocol, cannot learn more information nor manipulate the outputs of the honest parties. MPC protocols for computing general functionalities are central primitives in cryptography and have been studied extensively. An important question is: "*how many rounds of interactions do general MPC protocols need, and under what assumptions?*"

The round complexity of *2-Party Computation* (2PC) was resolved more than three decades ago: Yao [44,45] gave a construction of general semi-honest 2PC protocols that have only *two rounds* of interaction (where parties have access to a simultaneous broadcast channel[1]), using garbled circuits and a 2-message semi-honest Oblivious Transfer (OT) protocol. The round complexity is optimal, as any one-round protocol is trivially broken. Moreover, the underlying assumption of 2-message semi-honest OT is weak and necessary.[2]

In contrast, constructing round-efficient MPC protocols turned out to be more challenging. The first general construction [32] requires a high number of rounds, $O(d)$, proportional to the depth d of the computation. Later, Beaver, Micali, and Rogaway (BMR) reduced the round complexity to a constant using garbled circuits [5]. However, the *exact* round complexity of MPC remained open until recently. By relying on specific algebraic assumptions, a recent line of works constructed *(i)* 2-round MPC protocols relying on trusted infrastructure (e.g., a common reference string) assuming LWE [2,14,21,39,41] or DDH [9–11], and *(ii)* 2-round protocols in the plain model from indistinguishability obfuscation or witness encryption with NIZK [16,22,24,28,35], or bilinear groups [29]. However, all these constructions heavily exploit the algebraic structures of the underlying assumptions, or rely on the heavy machinery of obfuscation or witness encryption.

The state-of-the-art for malicious security is similar. Garg et al. [27] showed that 4 round is optimal for malicious MPC. So far, there are constructions of *(i)* 5-round protocols from DDH [1], and *(ii)* 4-round protocols from subexponentially secure DDH [1], or subexponentially secure LWE and adaptive

[1] Using the simultaneous broadcast channel, every party can simultaneously broadcast a message to all other parties. A malicious adversary can rush in the sense that in every round it receives the messages broadcast by honest parties first before choosing its own messages. In the 2PC setting, if both parties receive outputs, Yao's protocols need simultaneous broadcast channel.

[2] A 2-round OT protocol consists of one message from the receiver, followed by another one from the sender. It is implied by 2-round 2PC protocols using the simultaneous broadcast channel.

commitments[3] [12]. In general, for any number of round k, all known constructions of semi-honest or malicious MPC require at least k' round OT for $k' \leq \lfloor k/2 \rfloor$. We ask the question,

Can we have round-optimal MPC protocols from weak and necessary assumptions?

We completely resolve this question in the semi-honest setting, constructing 2-round semi-honest MPC from 2-round semi-honest OT, and make significant progress in the malicious setting, constructing 5-round malicious MPC from 5-round delayed-semi-malicious OT, a weaker primitive than malicious OT. Our results are obtained via a new notion of *garbling interactive circuits.* Roughly speaking, classical garbling turns a computation, given by a circuit C and an input x, into another one $(\hat{C}, \hat{x})$ that reveals only the output $C(x)$. Our new notion considers garbling a machine participating in an interaction: Let C (with potentially hardcoded input x) be an interactive machine that interacts with an oracle $\mathcal{O}$, which is a *non-deterministic algorithm* that computes its replies to C's messages, *depending on some witnesses* $\bar{w}$. Garbling interactive machine turns C into $\hat{C}$, which can emulate the interaction between C and $\mathcal{O}$, given the witnesses $\bar{w}$ in the clear (without any secret encoding). It is guaranteed that $\hat{C}$ reveals only the transcript of messages in the interaction and nothing else, provided that the transcript is *computationally uniquely defined*, that is, it is computationally hard to find two different witnesses $\bar{w}, \bar{w}'$ that lead to different transcripts.

1.1 Our Contributions

<u>SEMI-HONEST SECURITY:</u> We construct 2-round semi-honest MPC protocols in the plain model from 2-round semi-honest OT. Our construction can be generalized to an arbitrary number of rounds, establishing a tight connection between MPC and OT: For any k, *k-round OT is necessary and complete for k-round MPC.*[4]

Theorem 1.1 (Semi-Honest Security). *For any $k \geq 2$, there is a k-*round *semi-honest MPC protocol for any functionality f, from any k-*round *semi-honest OT protocol.*

The above theorem resolves the exact round complexity of semi-honest MPC based on weak and necessary assumptions, closing the gap between the 2-party and multi-party case. In the optimal 2-round setting, by instantiating our construction with specific 2-round OT protocols, we obtain 2-round MPC protocols

[3] That is, CCA commitments introduced in [17].

[4] We recall that for MPC, we suppose that parties have access to a simultaneous broadcast channel. Furthermore a k-round OT with simultaneous broadcast channel can be transformed into a k-round OT where each round consists a single message or flow either from the receiver to the sender or the other way round. This is because in the last round there is no point for the receiver to send a message to the sender.

in the plain model from a wide range of number theoretic and algebraic assumptions, including CDH [6], factoring [6],[5] LWE [42],[6] and constant-noise LPN with a sub-exponential security [31,46]. This broadens the set of assumptions that round-optimal semi-honest MPC can be based on.

MALICIOUS SECURITY: Going beyond semi-honest security, we further strengthen our protocols to achieve the stronger notion of *semi-malicious security*, as a stepping stone towards *malicious security*. Semi-malicious security proposed by [2] considers semi-malicious attackers that follow the protocol specification, but may adaptively choose arbitrary inputs and random tapes for computing each of its messages. We enhance our semi-honest protocols to handle such attackers.

Theorem 1.2 (Semi-Malicious Security). *For any $k \geq 2$, there is a k-round* semi-malicious *MPC protocol for any functionality f, from any k-round* semi-malicious *OT protocol.*

Previous semi-malicious protocols have 3 rounds based on LWE [2,12], 2 rounds based on bilinear maps [29], or 2 rounds based on LWE but in the common reference string model [39]. We obtain the first 2-round construction from any 2-round semi-malicious OT, which is necessary and can be instantiated from a variety of assumptions, including DDH [40], QR, and N-th residuosity [36]. Furthermore, following the compilation paradigms in recent works [1,2,12], we immediately obtain maliciously secure Universal Composable (UC) protocols in the common reference string model [15,18], using non-interactive zero-knowledge (NIZK).

Corollary 1.3 (Malicious Security in the CRS Model). *For any $k \geq 2$, there is a k-round* malicious UC protocol *in the common reference string model for any functionality f, from any k-round* semi-malicious *OT protocol and NIZK.*

Moving forward to malicious MPC protocols in the plain model, we show that, for any $k \geq 5$, k-round malicious MPC protocols can be built from k-round delayed-semi-malicious OT, which is implied by k-round malicious OT.

Theorem 1.4 (Malicious Security in the Plain Model). *For any $k \geq 5$, there is a k-round* malicious *MPC protocol for every functionality f, from any k-round* delayed-semi-malicious *OT protocol.*

This theorem is obtained by first showing that our k-round semi-malicious MPC protocols satisfy a stronger notion of *delayed-semi-malicious* security, when instantiated with a k-round OT protocol satisfying the same notion. Here, delayed-semi-malicious security guards against a stronger variant of semi-malicious attackers, and is still significantly weaker than malicious security.

[5] This follows from the fact that CDH in the group of quadratic residues is as hard as factoring [8,38,43].

[6] The scheme in [42] uses a CRS, but in the semi-honest setting, the sender can generate the CRS and send it to the receiver.

For instance, delayed-semi-malicious OT provides only indistinguishability-based privacy guarantees, whereas malicious OT supports extraction of inputs and simulation. In the second step, we transform our k-round delayed-semi-malicious MPC protocols into k-round malicious MPC protocols, assuming only one-way functions. This transformation relies on specific structures of our protocols. In complement, we also present a generic transformation that starts with *any* $(k-1)$-round delayed semi-malicious MPC protocol.

Previous 5-round malicious protocols rely on LWE and adaptive commitments [12], or DDH [1]. Our construction weakens the assumptions, and in particular adds factoring-based assumptions into the picture. Our result is *one-step away* from constructing round-optimal malicious MPC from weak and necessary assumptions. So far, 4-round protocols can only be based on subexponential DDH [1] or subexponential LWE and adaptive commitments [12]. A clear open question is constructing 4-round malicious MPC from 4-round OT.

GARBLED INTERACTIVE CIRCUITS, AND MORE: Along the way of constructing our MPC protocols, we develop new techniques and primitives that are of independent interest: We propose a new notion of *garbling interactive circuits*, a new primitive of *witness selector* that strengthens witness encryption [26], and a new notion of *zero-knowledge functional commitment.* Roughly speaking,

- As mentioned above, garbling interactive machine transforms an interactive machine C talking to a *non-deterministic* oracle $\mathcal{O}(\bar{w})$ using some witnesses, into a garbled interactive machine $\hat{C}$ that upon receiving the witnesses $\bar{w}$ in the clear (*without* any secret encoding) reveals the transcript of the interaction between C and $\mathcal{O}(\bar{w})$ and nothing else, provided that the transcript is *computationally uniquely defined.*
- Witness selector strengthens witness encryption [26] in the dimension that hiding holds when it is *computationally* hard to find a witness that enables decryption, as opposed to when no such witnesses exist.
- Finally, we enhance standard (computationally binding and computationally hiding) commitment schemes with the capability of partially opening a commitment c to the output $f(v)$ of a function f evaluated on the committed value v, where the commitment and partial decommitment reveal nothing more than the output $f(v)$.

To construct 2-round MPC, we use garbled interactive circuits and functional commitments to collapse rounds of any multi-round MPC protocols down to 2, and implement garbled interactive circuits using witness selector and classical garbled circuits. Our technique generalizes the novel ideas in recent works on constructing laconic OT from DDH [19], identity based encryption from CDH or factoring [13,23], and 2-round MPC from bilinear pairing [29]. These works can be rephrased as implementing special-purpose garbled interactive circuits from standard assumptions, and applying them for their specific applications. In this work, we implement the garbled interactive circuits, witness selector, and functional commitments needed for our constructions of MPC, from OT. The generality of our notions gives a unified view of the techniques in this and prior works.

1.2 Organization

We start with an overview of our techniques in Sect. 2. Then, after some classical preliminaries in Sect. 3, we formally define garbled interactive circuit schemes in Sect. 4. In Sect. 5, we build 2-round semi-honest MPC protocols from any semi-honest MPC protocols and (zero-knowledge) functional commitment scheme with an associated garbled interactive circuit scheme. In Sect. 6, we define witness selector schemes and show that they imply garbled interactive circuit schemes. The construction of a functional commitment scheme with witness selector from any 2-round OT (which concludes the construction of 2-round semi-honest MPC protocols from 2-round OT), as well as the extensions to k-round OT and to the semi-malicious and malicious settings are in the full version [7].

1.3 Concurrent Work

In a concurrent and independent work [30], Garg and Srinivasan also built k-round semi-honest MPC from k-round semi-honest OT. In the malicious setting, they obtained a stronger result in the CRS model, constructing 2-round UC-secure MPC from 2-round UC-secure OT in the CRS model (without requiring NIZK contrary to us). On the other hand, they did not consider malicious MPC in the plain model, whereas we constructed k-round malicious MPC from k-round delayed-semi-malicous OT for any $k \geq 5$. While both works leverage the novel ideas in [13,19,23,29], the concrete techniques are different. In our language, if we see their protocols in the lens of garbled interactive circuits, each step of their garbled interactive circuit performs a NAND gate on the state of one of the parties, while each of our steps performs a full MPC round, thanks to the functional commitment. Our approach can also be seen as more modular by the introduction of garbled interactive circuits, witness selector, and functional commitments, which we believe are of independent interest.

2 Overview

Garg et al. [24] introduced a generic approach for collapsing any MPC protocol down to 2 rounds, using indistinguishability obfuscation [4,25]. Later et al. [35] showed how to perform round collapsing using garbled circuits, witness encryption, and NIZK. Very recently, Garg and Srinivasan [29] further showed how to do collapse rounds using *garbled protocols*, which can be implemented from bilinear pairing groups. In this work, we perform round collapsing using our new notion of *garbled interactive circuits*; this notion is general and enables us to weaken the assumption to 2-round OT. (See the full version [7] for a more detailed comparison with prior works.) Below, we give an overview of our construction in the 2-round setting; construction in the multi-round setting is similar.

2.1 Round-Collapsing via Obfuscation

The basic idea is natural and simple: To construct 2-round MPC protocols for a function f, take any multi-round MPC protocols for f, referred to as the *inner MPC protocols*, such as, the Goldreich-Micali-Wigderson protocol [32], and try to eliminate interaction. Garg, Gentry, Halevi, and Raykova (GGHR) [24] showed how to do this using indistinguishability obfuscation. The idea is to let each player P_i obfuscate their *next-step circuit* $\mathsf{Next}_i(x_i, r_i, \star)$ in an execution of the inner MPC protocol Π for computing f, where $\mathsf{Next}_i(x_i, r_i, \star)$ has P_i's private input x_i and random tape r_i hardcoded, and produces P_i's next message m_i^ℓ in round ℓ, on input the messages $\bar{m}^{<\ell} = \{m_j^{\ell'}\}_{j,\ell'<\ell}$ broadcast by all parties in the previous rounds,

$$\mathsf{Next}_i(x_i, r_i, \bar{m}^{<\ell}) = m_i^\ell \, . \tag{1}$$

Given all obfuscated circuits $\{i\mathcal{O}(\mathsf{Next}(x_i, r_i, \star)_j)\}$, each party P_i can emulate the execution of Π *in its head*, eliminating interaction completely.

The above idea achieves functionality, but not security. In fact, attackers, given the obfuscated next-step circuits of honest parties, can evaluate the residual function $f(\{x_i\}_{\text{honest } i}, \star)$ with the inputs of honest parties hardcoded, or even evaluate honest parties' next-step circuits on arbitrary "invalid" messages. To avoid this, the protocol requires each party to commit to its input and random tape in the first round, $c_i \xleftarrow{R} \mathsf{Com}(x_i, r_i)$. Then, in the second round, each party obfuscates an *augmented next-step circuit* $\mathsf{AugNext}_i$ that takes additionally a NIZK proof $\pi_j^{\ell'}$ for each message $m_j^{\ell'}$ it receives, and verifies the proof $\pi_j^{\ell'}$ that $m_j^{\ell'}$ is generated honestly from inputs and random tapes committed in c_j (it aborts otherwise). This way, only the *unique* sequence of honestly generated messages is accepted by honest parties' obfuscated circuits. In the security proof, by the security of indistinguishability obfuscation and NIZK, this unique sequence can even be hardcoded into honest parties' obfuscated circuits, enabling simulation using the simulator of the inner MPC protocol.

2.2 Garbled Interactive Circuits

The fact that it suffices and is necessary that the honest parties' obfuscated circuits only allow for a single meaningful "execution path" (determined by the unique sequence of honest messages), suggests that we should rather use garbling instead of obfuscation for hiding honest parties' next-step circuits. However, the challenge is that the next-step circuits Next_i are not plain circuits: They are *interactive* in the sense that they takes inputs (i.e., MPC messages) generated by other parties that cannot be fixed at time of garbling. To overcome the challenge, we formalize the MPC players as interactive circuits, and propose a new notion called *Garbled Interactive Circuits (GIC)*.

<u>INTERACTIVE CIRCUITS:</u> The interaction with an interactive circuit is captured via a *non-deterministic* (poly-size) oracle $\mathcal{O}$ that on inputs a *query* q and some *witness* w returns an *answer* $a = \mathcal{O}(q, w)$ (or $\perp$ if w is not accepting). (Note that $\mathcal{O}$ is non-deterministic in the sense that without a valid witness, one cannot

evaluate $\mathcal{O}$.) An interactive circuit $\mathsf{i}C$ consists of a list of L next-step circuits $\{\mathsf{i}C^\ell\}_{\ell\in[L]}$. Its execution with oracle $\mathcal{O}$ on input a list of witnesses $\bar{w} = \{\bar{w}^\ell\}$ proceeds in L iterations as depicted in Fig. 1: In round ℓ, $\mathsf{i}C^\ell$ on input the state $st^{\ell-1}$ output in the previous round, as well as the answers $\bar{a}^{\ell-1} = \{a_k^{\ell-1}\}$ from $\mathcal{O}$ to queries $\bar{q}^{\ell-1} = \{q_k^{\ell-1}\}$ produced in the previous round, outputs the new state st^ℓ and queries $\bar{q}^\ell = \{q_k^\ell\}$, and a (round) output o^ℓ.

$$\forall \ell, \qquad \mathsf{i}C^\ell(st^{\ell-1}, \bar{a}^{\ell-1}) = (st^\ell, \bar{q}^\ell, o^\ell) \text{ , where } \forall k,\ a_k^{\ell-1} = \mathcal{O}(q_k^{\ell-1}, w_k^{\ell-1}) \text{ .}$$

The *output* of the execution is the list of round outputs $\bar{o} = \{o^\ell\}_\ell$, and the *transcript* of the execution is the list of all queries, answers, and outputs $\mathsf{trans}(\mathsf{i}C, \bar{w}) = \{(\bar{q}^\ell, \bar{a}^\ell, o^\ell)\}_\ell$. In the case that any oracle answer is $a_k^\ell = \perp$, the execution is considered invalid. For simplicity of this high-level overview, we consider only valid executions and valid transcript; see Sect. 4 for more details.

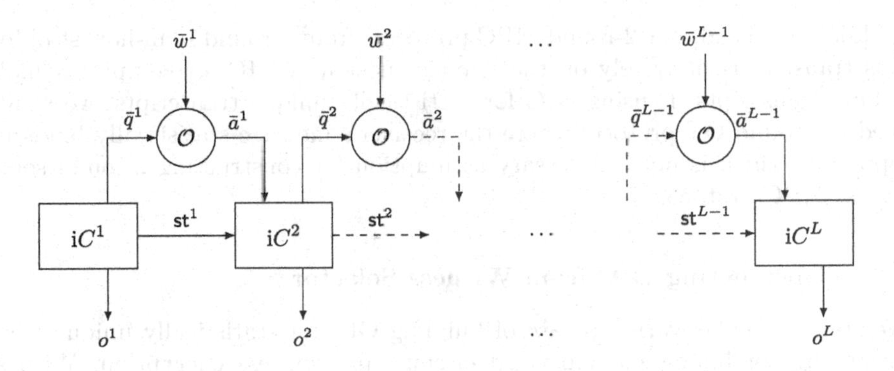

Fig. 1. Execution of an interactive circuit $\mathsf{i}C$ with witnesses $\bar{w}$

<u>GARBLED INTERACTIVE CIRCUIT SCHEME:</u> A Garbled Interactive Circuit (GIC) scheme GiC allows us to garble an interactive circuit $\widehat{\mathsf{i}C} \xleftarrow{R} \mathsf{GiC.Garble}(\mathsf{i}C)$, s.t.

Correctness: We can evaluate $\widehat{\mathsf{i}C}$ with the oracle $\mathcal{O}$ and a list $\bar{w}$ of witnesses (*in the clear*) to obtain each round output $o^\ell = \mathsf{GiC.Eval}(\widehat{\mathsf{i}C}, \bar{w}^{<\ell})$. This significantly differs from classical garbling techniques where inputs of the computation must be encoded using secrets (such as, mapping them to corresponding input keys or labels).

Simulation Security for Unique Transcripts Distribution: Security guarantees that $\widehat{\mathsf{i}C}$ reveals only the transcript of execution, including all outputs, queries, and answers, and nothing else, that is, it can be simulated by $\widehat{\mathsf{i}C} \xleftarrow{R} \mathsf{GiC.Sim}(\mathsf{trans})$, provided that there is a *unique* transcript of execution.

The requirement on unique transcript is necessary, otherwise, security is ill-defined as there may exist different transcripts produced by using different witnesses, and the simulator cannot hardcode them all. Furthermore, garbled interactive circuit schemes are meant to be different from obfuscation and hides only a single execution path. To formalize this, there are two options:

- STATISTICALLY UNIQUE TRANSCRIPT. The easier option is requiring simulation security only for interactive circuits $\mathrm{i}C$ that have unique transcript no matter what witnesses are used, that is, for all $\bar{w}, \bar{w}'$, $\mathsf{trans}(\mathrm{i}C, \mathcal{O}, \bar{w}) = \mathsf{trans}(\mathrm{i}C, \mathcal{O}, \bar{w}')$. This is, however, a strong requirement.
- (DEFAULT:) COMPUTATIONALLY UNIQUE TRANSCRIPT. The more general option is considering a distribution $\mathrm{i}\mathcal{D}$ over $(\mathrm{i}C, \bar{w})$ that has computationally unique transcripts, in the sense that given $(\mathrm{i}C, \bar{w})$, it is hard to find $\bar{w}'$ that leads to a different valid transcript, $\mathsf{trans}(\mathrm{i}C, \mathcal{O}, \bar{w}) \neq \mathsf{trans}(\mathrm{i}C, \mathcal{O}, \bar{w}')$.[7]

GIC for a computational or statistical unique-transcript distribution ensures:

$$\left\{\mathsf{GiC.Garble}(\mathrm{i}C) : (\mathrm{i}C, \bar{w}) \xleftarrow{R} \mathrm{i}\mathcal{D}\right\} \approx$$
$$\left\{\mathsf{GiC.Sim}(\mathsf{trans}(\mathrm{i}C, \mathcal{O}, \bar{w})) : (\mathrm{i}C, \bar{w}) \xleftarrow{R} \mathrm{i}\mathcal{D}\right\}$$

Looking ahead, our 2-round MPC protocols from 2-round semi-honest oblivious transfer crucially rely on the stronger notion of GIC for computationally unique transcripts. If using GIC for statistically unique transcripts, we would need a 2-round OT protocol where the receiver's message statistically binds its input bit, which is not a necessary assumption for constructing 2-round semi-honest MPC protocols.

2.3 Constructing GIC from Witness Selector

We start with the warm-up case of building GIC for statistically unique transcripts by combining plain garbled circuits and witness encryption. Witness Encryption (WE) proposed by Garg et al. [26], enables one to encrypt a message under an instance x of an NP language $\mathcal{L}$ to obtain a ciphertext $\mathsf{ct} \xleftarrow{R} \mathsf{WE.Enc}(\mathsf{x}, \mathsf{M})$; later this ciphertext can be decrypted using any witness w of x, $\mathsf{M} = \mathsf{WE.Dec}(\mathsf{ct}, \mathsf{w})$. The idea of combining garbled circuits and witness encryption has already appeared in three recent works by Gordon et al. [35], Cho et al. [19], and Döttling and Garg [23]. Our garbled interactive circuit scheme can be viewed as a generalization of their ideas for capturing the full power of this combination. As we explain shortly, to handle computationally unique transcripts, we need to rely on a new primitive called *Witness Selector*, which strengthens WE.[8]

<u>WARM-UP: GIC FOR STATISTICALLY UNIQUE TRANSCRIPT FROM WE:</u>
To garble an interactive circuit $\mathrm{i}C = \{\mathrm{i}C^\ell\}_\ell$, a natural first attempt is garbling each next-step circuit $\mathrm{i}C^\ell$ as a plain circuit, yielding L garbled circuits

[7] The distribution may output some additional auxiliary information, and it is hard to find witnesses that lead to a different valid transcript even given the auxiliary information. See Sect. 4 for more details.

[8] We mention that the work of Döttling and Garg [23] defined what is called *chameleon encryption scheme*, which can be viewed as a special case of our witness selector for a specific language.

$\{\widehat{iC}^\ell, \mathsf{key}^\ell\}_\ell$, where each input wire of $\widehat{iC}^\ell$ has two keys, $(\mathsf{key}^\ell[k,0], \mathsf{key}^\ell[k,1])$, one for this input bit being 0 and one for 1. The difficulty is that, to evaluate $\widehat{iC}^\ell$, the evaluator must obtain keys corresponding to the honestly generated state $st^{\ell-1}$ and answers $\bar{a}^{\ell-1}$ produced in the previous round; denote these keys as $\mathsf{key}^\ell[st^{\ell-1}]$ and $\mathsf{key}^\ell[\bar{a}^{\ell-1}]$.[9] We show how to enable this by modifying the garbled circuits $\{\widehat{iC}^\ell\}$ as follows.

- The first idea is embedding all keys key^ℓ for one garbled circuit $\widehat{iC}^\ell$ in the previous one $\widehat{iC}^{\ell-1}$, so that, $\widehat{iC}^{\ell-1}$ can output directly the keys $\mathsf{key}^\ell[st^{\ell-1}]$ for the state $st^{\ell-1}$ it produces. This idea, however, does not apply for selecting keys for answers $\bar{a}^{\ell-1}$, as $\widehat{iC}^{\ell-1}$ only computes queries $\bar{q}^{\ell-1}$ but not answers as it does not necessarily know the corresponding witnesses $\bar{w}^{\ell-1}$.
- The second idea is using WE as a "translator." To illustrate the idea, assume that there is a single query $q^{\ell-1}$ and it has a Boolean answer $a^{\ell-1}$. In this case, let $\widehat{iC}^{\ell-1}$ output a pair of WE ciphertexts $(\mathsf{ct}_0, \mathsf{ct}_1)$, where ct_b encrypts the key $\mathsf{key}^\ell[k,b]$ for the answer $a^{\ell-1}$ being b, under the statement x_b that the oracle outputs b, $\mathcal{O}(q^{\ell-1}, w_b') = b$, for some witness w_b'. Now, the evaluator after evaluating $\widehat{iC}^{\ell-1}$ obtains $\mathsf{ct}_0, \mathsf{ct}_1$. Using the witness w^ℓ it receives as input, it can decrypt the WE ciphertext $\mathsf{ct}^{\ell-1}_{a^{\ell-1}}$ for $a^{\ell-1} = \mathcal{O}(q^{\ell-1}, w^{\ell-1})$, obtaining the right key $\mathsf{key}^\ell[a^{\ell-1}]$ for evaluating the next garbled circuit.

To show security, it boils down to argue that for each garbled circuit $\widehat{iC}^\ell$, only one key for each input wire is revealed. The security of $\widehat{iC}^{\ell-1}$ ensures that only keys $\mathsf{key}^\ell[st^{\ell-1}]$ for the right state is revealed. On the other hand, to argue that only keys $\mathsf{key}^\ell[k, a^{\ell-1}]$ for the right answers are revealed, it crucially relies on the fact that the transcript including the answer is statistically unique. Thus, the ciphertext $\mathsf{ct}_{1-a^{\ell-1}}$ is encrypted under a false statement, and by security of WE, the label $\mathsf{key}^\ell[k, 1-a^{\ell-1}]$ is hidden. We emphasize that if the transcript were only computationally unique, both WE ciphertexts $\mathsf{ct}_0, \mathsf{ct}_1$ would potentially be encrypted under true statements, as there may exist two witnesses w_0, w_1 that make the oracle output 0 and 1, $\mathcal{O}(q^{\ell-1}, w_0) = 0$, $\mathcal{O}(q^{\ell-1}, w_1) = 1$, even though it is computationally hard to find them; and the security of WE would be vacuous.

GENERAL CASE: GIC FROM WITNESS SELECTOR: To handle computationally unique transcripts, WE is not the right tool. We propose a new primitive called *Witness Selective* (WS), which strengthens WE in two ways:

Correctness: WS is defined for a non-deterministic oracle $\mathcal{O}$. One can encrypt a set of keys $\mathsf{key} = \{\mathsf{key}[k,b]\}_{k\in[l], b\in\{0,1\}}$ under a query q, $\mathsf{ct} \leftarrow \mathsf{WS.Enc}(q, \mathsf{key})$, which can later be decrypted using a witness w revealing the keys selected according to the output $a = \mathcal{O}(q, w)$, that is, $\{\mathsf{key}[k, a_k]\}_k = \mathsf{WS.Dec}(\mathsf{ct}, w)$.

Semantic Security for Unique Answers: The security guarantee is that the WS ciphertext ct hides all the keys $\mathsf{key}[k, 1-a_k]$, provided that a is

[9] This is a slight abuse of notation, where $st^{\ell-1}$ and $\bar{a}^{\ell-1}$ denote both their actual values and the indices of the corresponding input wires.

the *computationally unique answer*. Clearly, if it were easy to find two witnesses w, w' such that, $(a = \mathcal{O}(q, w)) \neq (a' = \mathcal{O}(q, w'))$, the aforementioned semantic security cannot hold. Therefore, similarly to GIC, security is only required to hold for a distribution $\mathrm{w}\mathcal{D}$ over (q, w) that has computationally unique answers in the sense that given (q, w), it is hard to find w' that makes $\mathcal{O}$ output a different valid answer. Then,

$$\left\{\mathsf{WS.Enc}(q, \mathsf{key}) : (q, w) \xleftarrow{R} \mathrm{w}\mathcal{D}\right\} \approx$$
$$\left\{\mathsf{WS.Enc}(q, \mathsf{key}) : (q, w) \xleftarrow{R} \mathrm{w}\mathcal{D};\ a = \mathcal{O}(q, w);\ \forall k,\ \mathsf{key}[k, 1 - a_k] = 0\right\} .$$

We can construct general GIC scheme for computationally unique transcript by replacing WE in the warm-up construction with WS. Slightly more precisely, each garbled circuit $\widehat{\mathrm{i}C}^{\ell-1}$ outputs a WS ciphertext ct encrypting keys $\{\mathsf{key}[k, b]\}$ for all wires corresponding to the oracle answer $a^{\ell-1}$, under the query $q^{\ell-1}$ (if there are multiple queries, simply generate one WS ciphertext for each query); then, the evaluator can use the witness $w^{\ell-1}$ to decrypt and obtain keys $\{\mathsf{key}[k, a_k^{\ell-1}]\}$ selected according to the oracle answer $a^{\ell-1} = \mathcal{O}(q^{\ell-1}, w^{\ell-1})$. Since the oracle answer (as a part of the transcript) is computationally unique, semantic security of WS ensures that the other keys $\{\mathsf{key}[k, 1 - a_k^{\ell-1}]\}$ remain hidden, and hence we can invoke the security of the garbled circuits to argue the security of GIC.

<u>RELATION BETWEEN WS, WE, AND EXTRACTABLE WE:</u> As discussed above, WS is stronger than WE. For instance, one can use WS to encrypt a set of keys key under a query $q = (h, y = h(v))$ for a randomly sampled collision-resistant hash function h. With respect to the de-hashing oracle $\mathcal{O}(q, v')$ that outputs v' if $y = h(v')$, a WS ciphertext reveals only keys $\{\mathsf{key}[k, v_k]\}$ selected by v, and hides others. In contrast, WE provides no security in this case. On the other hand, WS is weaker than the notion of extractable WE [33]. Roughly speaking, extractable WE guarantees that for every attacker A, there is an extractor E, such that, if A can decrypt a ciphertext encrypted under statement x, then E can output a witness of x. Extractable WE implies WS, and is strictly stronger as it requires knowledge extraction.

We note that so far there is no construction of general-purpose WE, let alone WS or extractable WE, from standard assumptions. This is also not the goal of this work. Instead, we show below how to construct special-purpose WS that suffices to construct 2-round MPC protocols.

2.4 Round-Collapsing via Garbled Interactive Circuits

We now revisit the round-collapsing approach, by replacing obfuscation with garbled interactive circuits. First, we observe that each player P_i in the inner MPC protocol can be viewed as an interactive circuit $\{P_i^\ell\}$, interacting with an oracle $\mathcal{O}$ representing the other parties $\{P_j\}$, as described in Fig. 2.

The important details are: In each round ℓ, P_i^ℓ obtains through the oracle $\mathcal{O}$ all messages $\bar{m}^{\ell-1} = \{m_j^{\ell-1}\}_j$ output in the previous round, and additionally, it

P_i as an interactive circuit $\{P_i^\ell\}$

- The non-deterministic oracle $\mathcal{O}$ (representing all other parties) receives queries of form $q_j^\ell = (c_j, G_j^\ell)$, consisting of P_j's commitment and its next-step circuit with all messages in the first $\ell - 1$ rounds hardcoded, $G_j^\ell(\star, \star) = \mathsf{Next}_j(\star, \star, \bar{m}^{<\ell})$. On input such a query and a witness $w_j^\ell = (m_j^\ell, \pi_j^\ell)$, $\mathcal{O}$ computes:

$$a_j^\ell = \mathcal{O}(q_j^\ell, (m_j^\ell, \pi_j^\ell)) = \begin{cases} m_j^\ell & \text{if } \pi_j^\ell \text{ proves that the values } (x_j, r_j) \\ & \text{committed in } c_j \text{ satisfy } m_j^\ell = G_j^\ell(x_j, r_j) \\ \bot & \text{otherwise} \end{cases} .$$

- P_i^ℓ proceeds similarly as Next_i in Eq. (1) (page 7), except that, it additionally outputs one query $q_j^\ell = (c_j, G_j^\ell)$ for each player P_j's message m_j^ℓ, and a proof π_i^ℓ that its next message is indeed m_i^ℓ. (The proof system is described later.)

$$P_i^\ell(x_i, r_i, \overbrace{\bar{m}^{<\ell-1}}^{st^{\ell-1}}, \overbrace{\{m_j^{\ell-1}\}_j}^{\bar{a}^{\ell-1}}) = (\overbrace{\bar{m}^{<\ell}}^{st^\ell}, \overbrace{\{q_j^\ell\}_j}^{\bar{q}^\ell}, \overbrace{(m_i^\ell, \pi_i^\ell)}^{o^\ell}) ,$$

Fig. 2. Each player P_i can be formalized as an interactive circuit $P_i = \{P_i^\ell\}$.

outputs a proof π_i^ℓ that the message m_i^ℓ it outputs is generated honestly from its input x_i and random tape r_i committed in c_i. The message and proof are exactly the witness $w_i^\ell = (m_i^\ell, \pi_i^\ell)$ for the query q_i^ℓ that players P_j^ℓ make in round ℓ to the oracle $\mathcal{O}$ for obtaining P_i's message $a_i^\ell = m_i^\ell$ for the next round.

<u>OUR 2-ROUND MPC PROTOCOL:</u> Therefore, we can use a GIC scheme to garble the interactive circuit representing each player P_i to collapse round:

1. In the first round of MPC, each P_i broadcasts a commitment c_i to its input x_i and random tape r_i, and
2. in the second round, each P_i sends the garbled interactive circuit $\widehat{P}_i \xleftarrow{R} \mathsf{GiC.Garble}(\{P_i^\ell\})$, and
3. each P_i emulates the execution of inner MPC in its head, by evaluating all $\{\widehat{P}_j\}$ round by round: In round ℓ, it evaluates $o_j^\ell = (m_j^\ell, \pi_j^\ell) = \mathsf{GiC.Eval}(\widehat{P}_j, \bar{w}^{<\ell})$, using the outputs obtained in previous rounds as witnesses, $w^{<\ell} = o^{<\ell} = \{(m_k^{\ell'}, \pi_k^{\ell'})\}_{k, \ell' < \ell}$. P_i obtains its output when the inner MPC execution completes.

We observe that the transcript of execution of each $\{P_i^\ell\}$ is indeed *computationally unique*, as the commitments $\{c_j\}$ have unique committed values $\{x_j, r_j\}$ by the computational binding property, and lead to unique next messages $\{m_j^\ell\}$, by the soundness of proofs $\{\pi_j^\ell\}$. Therefore, the GIC scheme guarantees that the garbled interactive circuits reveals only their outputs, queries, and answers, summing up to all commitments $\{c_j\}$, inner MPC messages $\{m_j^\ell\}$, and proofs $\{\pi_j^\ell\}$, all of which can be made simulatable.

First Attempt of Instantiation: The MPC messages can be simulated by the simulator of the inner MPC protocol. To make commitments and proofs simulatable, the easiest way is using a standard non-interactive commitment scheme and a NIZK system, which however (1) requires a common reference string, and (2) makes the task of instantiating the associated WS scheme difficult. Recall that to instantiate the GIC scheme, we need a WS scheme for the oracle $\mathcal{O}$ described above, which internally verifies proofs. To solve this, we resort to a *zero-knowledge* Functional Commitment (FC) scheme that has a built-in special-purpose proof system. By minimizing the security requirements on this commitment, we manage to construct it, together with an associated WS scheme, from 2-message semi-honest OT (which is a necessary assumption). This gives 2-round MPC protocols in the plain model from 2-message semi-honest OT.

2.5 Functional Commitment with Witness Selector from OT

A zero-knowledge functional commitment scheme FC is computationally binding and computationally hiding, and additionally supports functional opening that is both *binding* and *zero-knowledge*. The notion of functional commitment was previously proposed by Libert et al. [37] for inner product functions, and later generalized to general functions in [3]. Here, we consider a stronger property, namely a *zero-knowledge* property. On the other hand, we do not require commitments nor functional decommitments to be of size constant in the length of the committed value, and our binding property only holds against semi-honest adversaries. Functional commitments were also implicitly and informally suggested by Gorbunov et al. in [34], as a way to interpret their new primitive: Homomorphic Trapdoor Functions (HTDFs). HTDFs could be used to construct our functional commitments (but the converse is not true). However, we do not know how to construct WS associated to an FC built from the HTDF proposed in [34].

Functional Opening: For a commitment $c = \mathsf{FC.Com}(v; \rho)$ and a circuit G, one can generate a *functional decommitment* d to the output of G evaluated on the committed value v, namely $m = G(v)$, using the randomness ρ of the commitment c,

$$d = \mathsf{FC.FOpen}(c, G, m, \rho), \quad \mathsf{FC.FVer}(c, G, m, d) = 1 \ .$$

We say that (m, d) is a decommitment to (c, G); here, d serves as a proof $\pi = d$ that the value committed in c evaluates to m through G in our 2-round MPC protocols.

(Semi-Honest) Functional Binding: For an honestly generated commitment $c = \mathsf{FC.Com}(v; \rho)$ with random tape ρ, it is hard to find a decommitment (m', d') to (c, G) for a different output $m' \neq m$, even given ρ. Note this is weaker than standard computational binding, as binding is only required for honestly generated commitments. This corresponds to *distributional soundness* of the proofs.

Simulation (i.e., Zero-Knowledge): An honestly generated commitment $c \xleftarrow{R} \mathsf{FC.Com}(v;\rho)$ (with random tape ρ) and decommitment d can be simulated *together*, using only the output m, $(\tilde{c},\tilde{d}) \xleftarrow{R} \mathsf{FC.Sim}(c,G,m)$. This property is weaker than standard zero-knowledge, as the statement is from a distribution and is also simulated; only a single decommitment d can be given for each commitment, or else simulation does not work.

A WS scheme associated with FC is for the oracle $\mathcal{O}^{\mathsf{FC}}$ that on input a query (c,G) and a witness $w=(m,d)$, outputs m if (m,d) is a valid decommitment to (c,G), and $\perp$ otherwise. The functional binding property ensures that for any v,G, the distribution $\mathrm{w}\mathcal{D}_{v,G}$ of query $q=(c,G)$ and decommitment $w=(m,d)$ for honestly generated $c=\mathsf{FC.Com}(v;\rho)$, produces computationally unique oracle answer m (even given the randomness ρ as auxiliary information). Despite the fact that functional commitments are only *semi-honestly* binding and *one-time* simulatable, we show that, together with an associated WS scheme, they suffice to instantiate our 2-round MPC protocols.

FC from Garbled Circuits and OT: We show how to construct a functional commitment, and its associated WS scheme, from garbled circuits and a 2-round string 2-to-1 semi-honest OT.

OT as semi-honest binding commitment: We start with observing that any string 2-to-1 semi-honest OT gives a commitment scheme that is *semi-honest binding*; that is, given an honestly generated commitment $c=\mathsf{Com}(v;\rho)$ using a uniformly random tape ρ, it is hard to find a decommitment (v',ρ') that opens c to a different value $v'\neq v$ even given ρ. To see this, consider the *parallelized* version of 2-to-1 string OT, where $\mathrm{ot}_1=\mathsf{pOT}_1(x;\rho)$ generates the first flows from OT receiver for every bit x_k, and $\mathrm{ot}_2=\mathsf{pOT}_2(\mathrm{ot}_1,\{\mathsf{key}[k,b]\})$ generates the second flows from OT sender for every pair of inputs $(\mathsf{key}[k,0],\mathsf{key}[k,1])$. Combining ot_2 with the randomness ρ used for generating the first flows, one can act as the OT receiver to recover exactly one input $\mathsf{key}[k,x_k]$ at each coordinate k. We argue that the first flow $\mathrm{ot}_1=\mathsf{pOT}_1(x;\rho)$ is a semi-honest commitment to x. Suppose that it is not the case and that it is easy to find a decommitment ρ' to a different value $x'\neq x$. Then a semi-honest attacker acting as OT receiver can violate the privacy of OT sender. (However, observe that $\mathsf{pOT}_1(x)$ is not necessarily computationally binding, as there is no security for maliciously generated first flows of OT.)

Functional Opening: We use garbled circuits and OT (as a semi-honest binding commitment scheme) to enable functional opening. To commit to a value v, garble a universal circuit $U_v(\star)=U(v,\star)$ with v hardcoded, and commit to all its input keys $\{\mathsf{key}[k,b]\}$ using pOT_1:

$$\mathsf{FC.Com}(v;\rho)=c=(\widehat{U}_v,\mathrm{ot}_1)\ ,\ \text{where}\ \ \mathrm{ot}_1[k,b]=\mathsf{pOT}_1(\mathsf{key}[k,b];\ \rho[k,b])\ .$$

To generate a decommitment (m,d) of (c,G), simply send the keys and randomness used for generating the OT first flows $\{\mathrm{ot}_1[k,G[k]]\}$ selected by G. More formally, if $G[k]$ is the k-th bit of the description of G which is used as input to U_v:

$$\mathsf{FC.FOpen}(c, G, m, \rho) = d = \{\mathsf{key}[k, G[k]],\ \rho[k, G[k]]\}.$$

Verifying a decommitment $d = \{\mathsf{key}', \rho'\}$ w.r.t. (c, G, m) involves checking that the keys and randomness contained in d' generate the OT first flows selected by G, and the garbled universal circuit $\widehat{U}_v$ evaluates to m on input these keys.

$$\mathsf{FC.FVer}(c, G, m, d) = 1 \quad \text{iff} \quad \begin{array}{l} (1)\ \forall k,\ \mathrm{ot}_1[k, G[k]] = \mathrm{p}\mathsf{OT}_1(\mathsf{key}'[k]; \rho'[k]) \text{ and} \\ (2)\ \widehat{U}_v(\mathsf{key}') = m. \end{array}$$

It is easy to see that the semi-honest binding property of $\mathrm{p}\mathsf{OT}_1$ implies the semi-honest functional binding of FC, and that a pair (c, d) can be simulated relying on the security of garbled circuits and the computational hiding property (i.e., receiver privacy) of $\mathrm{p}\mathsf{OT}_1$.

WS for FC: Next, to construct a WS scheme for the oracle $\mathcal{O}^{\mathsf{FC}}$ that verifies the functional decommitment of FC, we again use garbled circuits to "enforce and hide" this verification. To encrypt a set of messages $\mathsf{M}[i, b']$ under a query (c, G), our idea is to garble the following circuit V that acts as $\mathsf{FC.FVer}$ (without checking (1)), and selects messages according to the output m if verification passes,

$$V(\{\mathsf{key}'[k]\}) = \begin{cases} \{\mathsf{M}[i, m_i]\} & \text{if } \widehat{U}_v(\{\mathsf{key}'[k]\}) = m \\ \bot & \text{otherwise} \end{cases}. \tag{2}$$

Let $\widehat{V}$ be the garbled circuit, and $\{\mathsf{okey}_k[j, \beta]\}_j$ the set of keys for the input wires corresponding to $\mathsf{key}'[k]$. (For clarity, we denote keys for $\widehat{V}$ as okey.)

Given a decommitment $d = (\mathsf{key}', \rho')$, correct WS decryption should recover messages $\{\mathsf{M}[i, G(v)_i]\}$ selected according to the correct output $G(v)$ if the decommitment is valid, and $\bot$ if invalid. To enable this, what is missing is a "translation mechanism" that can achieve the following: For every k,

- Correctness: if $(\mathsf{key}'[k], \rho'[k])$ is a valid decommitment to $\mathrm{ot}_1[k, G[k]]$, it translates this pair into input keys of $\widehat{V}$ corresponding to $\mathsf{key}[k, G[k]]$, namely $\{\mathsf{okey}_k[j, \mathsf{key}[k, G[k]]_j]\}_j$.
- Security: the other keys $\{\mathsf{okey}_k[j, 1 - \mathsf{key}[k, G[k]]_j]\}_j$ are always hidden.

With such a translation mechanism, given a valid decommitment $d = \{\mathsf{key}[k, G[k]], \rho[k, G[k]]\}$, one can obtain all input keys corresponding to $\{\mathsf{key}[k, G[k]]\}$, and can evaluate $\widehat{V}$ with these keys to obtain the correct output,

$$\widehat{V}\left(\left\{\{\mathsf{okey}_k[j, \mathsf{key}[k, G[k]]_j]\}_j\right\}_k\right) = V(\{\mathsf{key}[k, G[k]]\}_k) = \{\mathsf{M}[i, G(v)_i]\}_i\ . \tag{3}$$

The security of the translation mechanism and garbled circuit $\widehat{V}$ guarantees that only the right messages $\{\mathsf{M}[i, G(v)_i]\}$ are revealed.

Our key observation is that the second flows of OT is exactly such a translation mechanism. For every OT first flows $\mathrm{ot}_1[k, G[k]]$ selected by G, generate the OT second flows using appropriate input keys of $\widehat{V}$ as sender's inputs,

$$\forall k, \qquad \mathrm{ot}_2[k] \xleftarrow{R} \mathsf{pOT}_2(\mathrm{ot}_1[k, G[k]], \{\mathsf{okey}_k[j,\beta]\}_{j,\beta}) \ . \tag{4}$$

Indeed, for every k, given a valid decommitment $(\mathsf{key}[k, G[k]], \rho')$ to $\mathrm{ot}_1[k, G[k]]$, one can act as an OT receiver to recover input keys $\{\mathsf{okey}_k[j, \mathsf{key}[k, G[k]]_j]\}_j$, achieving correct translation. On the other hand, the OT sender's security guarantees that the other keys $\{\mathsf{okey}_k[j, 1 - \mathsf{key}[k, G[k]]_j]\}_j$ remain hidden.

Summarizing the above ideas gives the following construction of WS for FC:

- $\mathsf{WS.Enc}((c, G), \mathsf{M})$: To encrypt M under (c, G), encryptor garbles the circuit V as in Eq. (2), and generates the second OT flows as in Eq. (4). The WS ciphertext is $\mathsf{ct} = (c, G, \widehat{V}, \{\mathrm{ot}_2[k]\})$.
- $\mathsf{WS.Dec}(\mathsf{ct}, d)$: To decrypt ct with a decommitment $d = \{\mathsf{key}', \rho'\}$, the decryptor first verifies that for every k $(\mathsf{key}'[k], \rho'[k])$ is a valid decommitment of $\mathrm{ot}_1[k, G[k]]$ in c; otherwise, abort. Then, for every k, it acts as an OT receiver with input $\mathsf{key}'[k]$, randomness $\rho'[k]$, and OT sender's message $\mathrm{ot}_2[k]$ to recover input keys of $\widehat{V}$ corresponding to $\mathsf{key}'[k]$. Finally, it evaluates $\widehat{V}$ with the obtained keys and output the messages output by $\widehat{V}$, as in Eq. (3).

The correctness and security of the WS scheme follows directly from the correctness and security of the translation mechanism, which are in turn implied by those of OT. See the full version [7] for more details.

Combining Sects. 2.1 to 2.5, we get a construction of a 2-round semi-honest MPC protocol from any 2-round semi-honest OT protocol using round collapsing for an inner MPC protocol.

2.6 Semi-Malicious and Malicious Security in the CRS Model

Toward achieving malicious security, we first achieve semi-malicious security. Roughly speaking, a semi-malicious party P_j generates its messages according to the protocol using arbitrarily and adaptively chosen inputs and random tapes. This is formalized by letting P_j "explain" each message m_j^ℓ it sends with a pair of input and random tape consistent with it, on a special witness tape. In the two-round setting, the challenge in simulating the view of P_j lies in simulating honest parties' first messages without knowing any secret information of P_j. This is because P_j may *rush* to see honest parties' first messages before outputting its own message, input, and random tape. (Observe that this is not an issue for semi-honest security, as the simulator learns the inputs and random tapes of corrupted parties first.)

Recall that in our 2-round protocols, each party P_i sends functional commitments c_i to its input and random tape (x_i, r_i) in the first round, which are later partially decommitted to reveal P_i's messages m in the inner MPC protocol. The simulation property of the functional commitment scheme FC ensures that the commitment and decommitment can be simulated together using just the message. However, this is insufficient for achieving semi-malicious security, as the simulator must simulate commitments in the first round with no information. To overcome this problem, we strengthen the simulatability of FC to *equivocability*, that is, simulation takes the following two steps: First, a commitment $\tilde{c}$ is simulated with no information, and later it is equivocated to open to any output m w.r.t. any circuit G. Instantiating our 2-round MPC protocols with such an *equivocal functional commitment scheme*, and other primitives that are semi-maliciously secure (e.g., using a semi-maliciously secure multi-round MPC protocol, and 2-round OT protocol), naturally "lift" semi-honest security to semi-malicious security.

With a simple idea, we can transform any *simulatable* functional commitment scheme FC into an equivocal one eFC: Let $(\mathsf{OT}_1, \mathsf{OT}_2)$ be the sender and receiver's algorithms of a 2-out-of-1 OT scheme.

- To commit to v, generate a FC commitment c to v, and then commit to each bit c_i twice using the algorithm OT_1, yielding the eFC commitment:

$$ec = \{\mathrm{ot}_1[i,0] = \mathsf{OT}_1(c_i;\ r[i,0]),\ \mathrm{ot}_1[i,1] = \mathsf{OT}_1(c_i;\ r[i,1])\}_i\,.$$

- An eFC decommitment $(ed, G(v))$ to (ec, G) contains the FC decommitment $(d, G(v))$ to (c, G), and the OT randomness $\{r[i, c_i]\}$ for generating the set of first flows $\{\mathrm{ot}_1[i, c_i]\}$ selected by c. Note that for any ec generated according to the above commitment algorithm, the revealed OT randomness determines the commitment c, and then the FC decommitment d determines $G(v)$.
- Now, a commitment can be simulated by committing to both 0 and 1 in ec,

$$\widetilde{ec} = \{\mathrm{ot}_1[i,0] = \mathsf{OT}_1(0;\ r[i,0]),\ \mathrm{ot}_1[i,1] = \mathsf{OT}_1(1;\ r[i,1])\}_i\,.$$

  To decommit $\widetilde{ec}$ to output $G(v)$, first simulate the FC commitment and decommitment $(\tilde{c}, \tilde{d})$ together using $G(v)$, and then reveal the set of randomness $\{r[i, \tilde{c}_i]\}$ selected according to the simulated commitment $\tilde{c}$.

The WS scheme associated with eFC can be constructed similarly as that for FC. The above idea is conceptually simple, but leads to nested calls of $\mathrm{p}\mathsf{OT}_1/\mathsf{OT}_1$, as a FC commitment c already contains OT first flows. This is not a problem when using 2-round OT, but does not extend to multi-round OT. In the full version [7], we present a more involved construction that avoids nested calls.

Malicious Security in the CRS Model. Given 2-round semi-maliciously secure protocols, in the CRS model, we can let each party prove using NIZK that each message is generated in a semi-malicious way (i.e., according to the protocol w.r.t. some input and random tape) as done in [2], which immediately gives Corollary 1.3 in the introduction. We refer the reader to [2] for more details.

Extension to k Rounds. Our 2-round semi-honest or semi-malicious constructions so far can be extended to k-round constructions, when replacing the underlying 2-round OT protocols with semi-honest or semi-malicious k-round OT protocols. See the full version [7] for more details.

2.7 Malicious Security in the Plain Model

<u>From General ($k-1$)-Round Delayed-Semi-Malicious MPC:</u> We first show a new compilation that turns *any* $(k-1)$-round MPC protocol for computing f satisfying a stronger variant of semi-malicious security, called *delayed-semi-malicious security*, into a k-round malicious MPC protocol for f, assuming only one-way functions, for any $k \geq 5$. Roughly speaking, a delayed-semi-malicious party P_j acts like a semi-malicious party, except that, it only "explains" *all* its messages *once*, *before the last round* (instead of explaining each of its messages after each round). This is formalized by letting P_j output a pair of input and random tape before the last round (on its special witness tape) which is required to be consistent with all P_j's messages. We say that a protocol is *delayed-semi-malicious secure* if it is secure against such adversaries. (For technical reasons, we require the protocols to have a *universal* simulator.) We observe that our k-round semi-malicious MPC protocols, when instantiated with a k-round delayed-semi-malicious OT become secure against delayed semi-malicious attackers (and admit a universal simulator).

To "lift" delayed-semi-malicious security to malicious security *generically*, our compilation builds on techniques of [1]. To illustrate the idea, consider compiling our 2-round delayed-semi-malicious MPC protocol Φ for f into a 5-round malicious MPC protocol Π for f. The basic idea is running Φ for computing f, and restricting a malicious adversary A to act as a delayed-semi-malicious one A' by requiring A to prove using zero-knowledge proof of knowledge (ZKPOK) that its messages in each round of Φ are generated correctly according to some input and random tape. Unlike the CRS model, ZKPOK in the plain model requires at least 4 rounds. Sequentializing the two ZKPOK leads to a *8-round* protocol. But if the ZKPOK allows for *delayed-input*, that is, only the last prover's message depends on the statement and witness, then the two ZKPOK can be partially parallelized, leading to a *5-round* protocol. In addition, in order to prevent mauling attacks, the ZKPOK must be *non-malleable*. Fortunately, Ciampi, Ostrovsky, Siniscalchi, and Visconti [20] (COSV) recently constructed a 4-round delayed-input non-malleable ZKPOK protocol from one-way functions, which suffice for our purpose. When starting from a 4-round (instead of 2-round) protocol Φ, to obtain a 5-round malicious protocol Π, we cannot afford to prove correctness of each round. But, if Φ is delayed-semi-malicious secure, then it suffices to prove correctness only at the last two rounds, keeping the round complexity at 5.

Though the high-level ideas are simple, there are subtleties in the construction and proof. We cannot use the non-malleable ZKPOK in a black-box. This is because simulation of non-malleable ZKPOK uses rewindings and may render the Φ instance running in parallel insecure. In addition, the COSV non-malleable ZKPOK is only many-many non-malleable in the *synchronous* setting, but in Π, the non-malleable ZKPOKs are not completely synchronized (ending either at the second last or the last round). Therefore, we use the COSV construction in a *non-black-box* way in Π (with some simplification) as done in [1]. The specific property of COSV non-malleable ZKPOK that we rely on is that simulation requires only rewinding the second and third rounds, while (witness) extraction requires only rewinding the third and forth rounds. This means Φ would be rewound at second/third and third/fourth rounds. The security of a generic delayed-semi-malicious protocol may not hold amid such rewinding. However, if we start with a *4-round* protocol, rewindings can be circumvented if Π contains no messages of Φ in its third round. This means, in the rewindings of second/third and third/fourth rounds, the simulator can simply *replay* messages of Φ in the main thread, keeping the instance of Φ secure. See the full version [7] for details.

FROM OUR SPECIFIC k-ROUND DELAYED-SEMI-MALICIOUS MPC: The above transformation is modular and general, but comes at a price—it only gives k-round malicious MPC from $(k-1)$-round delayed-semi-malicious OT, which is not necessary. To eliminate the gap, we leverage specific structures of our k-round delayed-semi-malicious protocols, to address the rewinding issue above. To illustrate the ideas, lets again examine the $k = 5$ case.

To handle rewindings at third/fourth rounds, we observe that at the end of fourth round, each party P_i proves using COSV non-malleable ZK that it has acted honestly in Φ according to some input and random tape (x_i, r_i). If in the malicious protocol Π, each party additionally commits to (x_i, r_i) in the first two rounds using a statistically binding commitment scheme (and prove that its messages are generated honestly using the committed value). Then, as long as the adversary cannot cheat in the non-malleable ZK proofs, its messages in the third/fourth rounds of Φ are determined by the commitments in the first two rounds. Therefore, the simulator can afford to continuously rewinding the adversary, until it *repeats* its messages in Φ in the main execution thread. In this case, the simulator can simply *replay* the honest parties' messages in Φ in the main thread.

To handle rewindings at second/third rounds, the specific property of our protocol that we rely on is that the first 2 rounds of Φ contains only instances of OT, whose messages do not depend on parties' inputs. The latter holds because of the random self-reducibility of OT (hence, the sender and receiver can only use their inputs for generating their last messages). To avoid rewinding these OT instances in Φ, our idea is modifying the malicious protocol Π as follows: In the first 2 rounds, for every OT instance OT_j in Φ, Π runs two independent OT instances OT_j^0 and OT_j^1. In the third round, an *random* instance $\mathsf{OT}_j^{b_j}$ for $b_j \leftarrow \{0, 1\}$ is chosen to be continued, and the other $\mathsf{OT}_j^{1-b_j}$ aborted—they are referred to as the *real* and *shadow* instances. Now in a rewinding of the second/third round, to avoid rewinding the real OT instances, the simulator *replays* the OT messages in the second round, and in the third round, continues the shadow instances $\mathsf{OT}_j^{1-b_j}$ and aborts the real instances $\mathsf{OT}_j^{b_j}$. Importantly, since for every pair $(\mathsf{OT}_j^0, \mathsf{OT}_j^1)$, the choice b_j of which is real and which is shadow is random and independent, the view of the adversary in a rewinding is identical to that in the main execution thread. This guarantees that rewindings would succeed.

We remark that this idea does not apply in general. This is because to continue a random instance of a general protocol Φ in the third round, parties may need to *agree* on that instance, which requires coin-tossing. In contrast, our protocol Φ consists of many OT instances OT_j, the decision of which of $(\mathsf{OT}_j^0, \mathsf{OT}_j^1)$ to continue can be made *locally* by the party who is supposed to send the third message of OT_j in Φ. In the full version [7], we put the above two ideas together, which gives k-round malicious OT from k-round delayed-semi-malicious OT.

A figure summarizing the results is provided in the full version [7].

3 Preliminaries

The security parameter is denoted λ. We recall the notion of polynomial-size circuit classes and families, together with the notion of statistical and computational indistinguishability in the full version [7].

For the sake of simplicity, we suppose that all circuits in a circuit class have the same input and output lengths. This can be achieved without loss of generality using appropriate paddings. We recall that for any S-size circuit class $\mathcal{C} = \{\mathcal{C}_\lambda\}_{\lambda \in \mathbb{N}}$, there exists a universal poly(S)-size circuit family $\{U_\lambda\}_{\lambda \in \mathbb{N}}$ such that for any $\lambda \in \mathbb{N}$, any circuit $C \in \mathcal{C}_\lambda$ with input and output lengths n, l, and any input $x \in \{0,1\}^n$, $U_\lambda(C, x) = C(x)$.

We make use of garbled circuit schemes. A *garbled circuit* scheme GC for a poly-size circuit class $\mathcal{C} = \{\mathcal{C}_\lambda\}_{\lambda \in \mathbb{N}}$ is defined by four polynomial-time algorithms $\mathsf{GC} = (\mathsf{GC.Gen}, \mathsf{GC.Garble}, \mathsf{GC.Eval}, \mathsf{GC.Sim})$: *(i)* $\mathsf{key} \xleftarrow{R} \mathsf{GC.Gen}(1^\lambda)$ generates input labels $\mathsf{key} = \{\mathsf{key}[i,b]\}_{i \in [n], b \in \{0,1\}}$; *(ii)* $\widehat{C} \xleftarrow{R} \mathsf{GC.Garble}(\mathsf{key}, C)$ garbles the circuit $C \in \mathcal{C}_\lambda$ into $\widehat{C}$; *(iii)* $y = \mathsf{GC.Eval}(\widehat{C}, \mathsf{key}')$ evaluates the garbled circuit $\mathsf{GC.Garble}$ using input labels $\mathsf{key}' = \{\mathsf{key}'[i]\}_{i \in [n]}$ and returns the output $y \in \{0,1\}^l$; *(iv)* $(\mathsf{key}', \widetilde{C}) \xleftarrow{R} \mathsf{GC.Sim}(1^\lambda, y)$ simulates input labels $\mathsf{key}' = \{\mathsf{key}'[i]\}_{i \in [n]}$ and a garbled circuit $\widetilde{C}$ corresponding to the output $y \in \{0,1\}^l$. The formal definition can be found in the full version [7]. We recall that garbled circuit schemes can be constructed from one-way functions.

4 Definition of Garbled Interactive Circuit Schemes

In this section, we define Garbled Interactive Circuit (GIC) schemes. An overview is provided in Sect. 2.2.

4.1 Interactive Circuits

We start by defining non-deterministic oracles and interactive circuits.

Definition 4.1 (Non-Deterministic Oracles). A non-deterministic oracle $\mathcal{O}$ is a circuit that takes as input a pair of bitstrings $(q, w) \in \{0,1\}^n \times \{0,1\}^m$, called *query* and *witness* respectively, and the output is a l-bit string or a special element $\bot$, called *answer*: $\mathcal{O}(q, w) \in \{0,1\}^l \cup \{\bot\}$. A *poly-size non-deterministic oracle family* is an ensemble of *poly-size* non-deterministic oracles $\mathcal{O} = \{\mathcal{O}_\lambda\}_{\lambda \in N}$.

Definition 4.2. Let $\mathcal{O}$ be a non-deterministic oracle. An L-round interactive circuit $\mathrm{i}C = \{\mathrm{i}C^\ell\}_{\ell \in [L]}$ with oracle $\mathcal{O}$ consists of a list of L next-step circuits.

Execution of $\mathrm{i}C$ with $\mathcal{O}$ on Witnesses $\bar{w}$: An execution of $\mathrm{i}C$ with $\mathcal{O}$ and a list of witnesses $\bar{w} = \{\bar{w}^\ell\}_{\ell \in [L]}$ proceeds in L iterations as follows: In round $\ell \in [L]$, the next-step circuit $\mathrm{i}C^\ell$ on input the state $st^{\ell-1}$ (output in the previous round) and answers $\bar{a}^{\ell-1} = \{a_k^{\ell-1}\}_k$ (to queries $\bar{q}^{\ell-1} = \{q_k^{\ell-1}\}_k$ produced in the previous round), outputs a new state st^ℓ, queries $\bar{q}^\ell = \{q_k^\ell\}_k$, and a (round) output o^ℓ,

$$(st^\ell, \bar{q}^\ell, o^\ell) = \begin{cases} \mathrm{i}C^\ell(st^{\ell-1}, \bar{a}^{\ell-1}) & \text{if } \forall k,\ a_k^{\ell-1} = \mathcal{O}(q_k^{\ell-1}, w_k^{\ell-1}) \neq \bot \\ (\bot, \bot, \bot) & \text{otherwise} \end{cases}.$$

The execution terminates after L rounds, or whenever $\bot$ is output. By convention, st^0 and $\bar{q}^0$ are empty strings.

We say that an execution is *valid* if it terminates after L rounds without outputting $\bot$. We call the list of witnesses $\bar{w}$ the *witnesses* of the execution. The *output* of the execution is the list of round outputs, denoted as $\mathsf{out}(\mathrm{i}C, \mathcal{O}, \bar{w}) = \bar{o} = \{o^\ell\}_{\ell \in [L]}$. The *transcript* of the execution is the list of queries, answers, and outputs, denoted as $\mathsf{trans}(\mathrm{i}C, \mathcal{O}, \bar{w}) = \{\bar{q}^\ell, \bar{a}^\ell, o^\ell\}_{\ell \in [L]}$. (If the execution outputs $\bot$ in round ℓ, $\bar{q}^{\ell'} = \bar{a}^{\ell'} = o^{\ell'} = \bot$ for all $\ell' \geq \ell$.) Finally, we say that $\mathrm{i}C$ has size S if the total size of all circuits are bounded by S. In the rest of the paper, when the oracle $\mathcal{O}$ is clear from the context, we often omit it in the notations and write $\mathsf{out}(\mathrm{i}C, \bar{w})$ and $\mathsf{trans}(\mathrm{i}C, \bar{w})$.

4.2 Garbling Interactive Circuits

As mentioned above, an important difference between GIC schemes and classical garbled circuit schemes is that to evaluate a garbled (plain) circuit, one must obtain encoded inputs, whereas a garble interactive circuit can be evaluated with its oracle $\mathcal{O}$ on input an arbitrary list of witnesses, without encoding. This provides a more powerful functionality, but poses an issue on security: There may exist different lists of witnesses $\bar{w}, \bar{w}'$ that lead to executions with completely different transcripts. In this case, it is unclear how simulation can be done. To circumvent this, we only require the security of the garbling scheme to hold for distributions $\mathrm{i}\mathcal{D}$ of interactive circuits $\mathrm{i}C$ and witnesses $\bar{w}$ (with potentially some auxiliary information aux) that have *computationally unique transcripts* $\mathsf{trans}(\mathrm{i}C, \mathcal{O}, \bar{w})$, in the sense that (given aux) it is hard to find another list of witnesses $\bar{w}'$ that leads to an *inconsistent* transcript $\mathsf{trans}(\mathrm{i}C, \mathcal{O}, \bar{w})$, where inconsistency means:

Definition 4.3 (Consistent Transcripts). We say that two transcripts $\{\bar{q}^\ell, \bar{a}^\ell, o^\ell\}_{\ell \in [L]}$ and $\{\bar{q}'^\ell, \bar{a}'^\ell, o'^\ell\}_{\ell \in [L]}$ are *consistent* if for every $\ell \in [L]$, $(\bar{q}^\ell, \bar{a}^\ell, o^\ell) = (\bar{q}'^\ell, \bar{a}'^\ell, o'^\ell)$ or $(\bar{q}^\ell, \bar{a}^\ell, o^\ell) = (\bot, \bot, \bot)$ or $(\bar{q}'^\ell, \bar{a}'^\ell, o'^\ell) = (\bot, \bot, \bot)$. Otherwise, we say that the two transcripts are *inconsistent*.

Note that one can always produce a list of invalid witnesses that lead to an invalid execution. Therefore, difference due to outputting $\perp$ does not count as inconsistency. Next, we formally define these distributions that produce unique transcripts.

Definition 4.4 (Unique-Transcript Distribution). Let $\mathcal{O} = \{\mathcal{O}_\lambda\}_{\lambda\in\mathbb{N}}$ be a non-deterministic oracle family. Let $\mathrm{i}\mathcal{D} = \{\mathrm{i}\mathcal{D}_{\lambda,\mathrm{id}}\}_{\lambda\in\mathbb{N},\mathrm{id}}$ be an ensemble of efficiently samplable distributions over tuples $(\mathrm{i}C, \bar{w}, \mathsf{aux})$. We say that $\mathrm{i}\mathcal{D}$ is a *(computationally) unique-transcript* distribution for $\mathcal{O}$, if

Valid Execution: For any $\lambda \in \mathbb{N}$, any index $\mathrm{id} \in \{0,1\}^{\mathrm{poly}(\lambda)}$, and any $(\mathrm{i}C, \bar{w}, \mathsf{aux})$ in the support of $\mathrm{i}\mathcal{D}_{\lambda,\mathrm{id}}$, the execution of $\mathrm{i}C$ with $\mathcal{O}_\lambda$ and $\bar{w}$ is valid.

Computationally Unique Transcript: For any poly-size circuit family $A = \{A_\lambda\}_\lambda$, any sequence of indices $\{\mathrm{id}_\lambda\}_\lambda$, there is a negligible function negl, such that for any λ:

$$\Pr\Big[\mathsf{trans}(\mathrm{i}C, \mathcal{O}_\lambda, \bar{w}') \text{ and } \mathsf{trans}(\mathrm{i}C, \mathcal{O}_\lambda, \bar{w}) \text{ are inconsistent} : \\ (\mathrm{i}C, \bar{w}, \mathsf{aux}) \xleftarrow{R} \mathrm{i}\mathcal{D}_{\lambda,\mathrm{id}_\lambda};\ \bar{w}' \xleftarrow{R} A_\lambda(\mathrm{i}C, \bar{w}, \mathsf{aux})\Big] \leq \mathrm{negl}(\lambda)\ .$$

It is a *statistically unique-transcript distribution* if the second property holds for any arbitrary-size circuit family $A = \{A_\lambda\}_\lambda$.

Now, we are ready to define GIC schemes.

Definition 4.5 (Garbled Interactive Circuit Schemes). Let $\mathcal{O} = \{\mathcal{O}_\lambda\}_{\lambda\in\mathbb{N}}$ be a non-deterministic oracle family, and $\mathrm{i}\mathcal{D} = \{\mathrm{i}\mathcal{D}_{\lambda,\mathrm{id}}\}_{\lambda\in\mathbb{N},\mathrm{id}}$ be a unique-transcript distribution for $\mathcal{O}$. A *garbled interactive circuit* scheme for $(\mathcal{O}, \mathrm{i}\mathcal{D})$ is a tuple of three polynomial-time algorithms $\mathsf{GiC} = (\mathsf{GiC.Garble}, \mathsf{GiC.Eval}, \mathsf{GiC.Sim})$:

Garbling: $\widehat{\mathrm{i}C} \xleftarrow{R} \mathsf{GiC.Garble}(1^\lambda, \mathrm{i}C)$ garbles an interactive circuit $\mathrm{i}C$ into a garbled interactive circuit $\widehat{\mathrm{i}C}$;

Evaluation: $o^\ell = \mathsf{GiC.Eval}(\widehat{\mathrm{i}C}, \bar{w}^{<\ell})$ evaluates a garbled interactive circuit $\widehat{\mathrm{i}C}$ with a partial list of witness $\bar{w}^{<\ell}$, and outputs the ℓ-th round output o^ℓ;

Simulation: $\widetilde{\mathrm{i}C} \xleftarrow{R} \mathsf{GiC.Sim}(1^\lambda, T)$ simulates a garbled circuit $\widetilde{\mathrm{i}C}$ from a transcript T of an execution;

satisfying the following properties:

Correctness: For any $\lambda \in \mathbb{N}$, any index $\mathrm{id} \in \{0,1\}^{\mathrm{poly}(\lambda)}$, any $(\mathrm{i}C, \bar{w}, \mathsf{aux})$ in the support of $\mathrm{i}\mathcal{D}_{\lambda,\mathrm{id}}$, it holds that

$$\Pr\Big[\{\mathsf{GiC.Eval}(\widehat{\mathrm{i}C}, \bar{w}^{<\ell})\}_{\ell\in[L]} = \mathsf{out}(\mathrm{i}C, \mathcal{O}_\lambda, \bar{w}) : \\ \widehat{\mathrm{i}C} \xleftarrow{R} \mathsf{GiC.Garble}(1^\lambda, \mathrm{i}C)\Big] = 1\ ;$$

Simulatability: The following two distributions are computationally indistinguishable:

$$\left\{(\mathrm{i}C, \bar{w}, \mathsf{aux}, \widehat{\mathrm{i}C}) \;:\; \begin{array}{l}(\mathrm{i}C, \bar{w}, \mathsf{aux}) \xleftarrow{R} \mathrm{i}\mathcal{D}_{\lambda,\mathrm{id}};\\ \widehat{\mathrm{i}C} \xleftarrow{R} \mathsf{GiC.Garble}(1^\lambda, \mathrm{i}C)\end{array}\right\}_{\lambda,\mathrm{id}},$$

$$\left\{(\mathrm{i}C, \bar{w}, \mathsf{aux}, \widetilde{\mathrm{i}C}) \;:\; \begin{array}{l}(\mathrm{i}C, \bar{w}, \mathsf{aux}) \xleftarrow{R} \mathrm{i}\mathcal{D}_{\lambda,\mathrm{id}};\\ \widetilde{\mathrm{i}C} \xleftarrow{R} \mathsf{GiC.Sim}(1^\lambda, \mathsf{trans}(\mathrm{i}C, \mathcal{O}_\lambda, \bar{w}))\end{array}\right\}_{\lambda,\mathrm{id}}.$$

Remark 4.6. In this paper, we always consider perfect correctness for all primitives for the sake of simplicity. We could relax this notion to correctness up to a negligible error probability if, in addition, we ask that no non-uniform poly-time adversary can generate inputs and randomness which would not satisfy the correctness property, with non-negligible probability. In other words, in the case of GIC schemes, semi-maliciously generated GIC should satisfy the correctness property (except with negligible probability). This additional property is not needed for our semi-honest constructions.

5 2-Round Semi-Honest MPC Protocols

In this section, we present our construction of 2-round semi-honest MPC protocols. For that purpose, we first introduce the notion of functional commitment. We then show the MPC construction.

5.1 New Tool: Functional Commitment

Definition 5.1 ((Zero-Knowledge) Functional Commitment). Let $\mathcal{G} = \{\mathcal{G}_\lambda\}_{\lambda\in\mathbb{N}}$ be a poly-size circuit class. A *(zero-knowledge) functional commitment* scheme FC for $\mathcal{G}$ is a tuple of four polynomial-time algorithms $\mathsf{FC} = (\mathsf{FC.Com}, \mathsf{FC.FOpen}, \mathsf{FC.FVer}, \mathsf{FC.Sim})$:

Commitment: $c = \mathsf{FC.Com}(1^\lambda, v; \rho)$ generates a commitment c of $v \in \{0,1\}^n$ using random tape $\rho \in \{0,1\}^\tau$, for the security parameter λ, where the random tape length τ is polynomial in λ;

Functional Opening: $d = \mathsf{FC.FOpen}(c, G, v, \rho)$ derives from the commitment c and the random tape ρ used to generate it, a functional decommitment d of c to $y = G(v)$ for $G \in \mathcal{G}_\lambda$;

Functional Verification: $b = \mathsf{FC.FVer}(c, G, y, d)$ outputs $b = 1$ if d is a valid functional decommitment of c to y for $G \in \mathcal{G}_\lambda$; and outputs $b = 0$ otherwise;

Simulation: $(c, d) \xleftarrow{R} \mathsf{FC.Sim}(1^\lambda, G, y)$ simulates a commitment c together with a functional decommitment d of c to y for $G \in \mathcal{G}_\lambda$;

satisfying the following properties:

Correctness: For any security parameter $\lambda \in \mathbb{N}$, for any $v \in \{0,1\}^n$, for any circuit $G \in \mathcal{G}_\lambda$, for any $\rho \in \{0,1\}^\tau$, it holds that if $c = \mathsf{FC.Com}(1^\lambda, v; \rho)$, then:

$$\mathsf{FC.FVer}(c, G, G(v), \mathsf{FC.FOpen}(c, G, v, \rho)) = 1 \;;$$

Semi-Honest Functional Binding: For any polynomial-time circuit family $A = \{A_\lambda\}_{\lambda \in \mathbb{N}}$, there exists a negligible function negl, such that for any $\lambda \in \mathbb{N}$, for any $v \in \{0,1\}^n$, for any circuit $G \in \mathcal{G}_\lambda$:

$$\Pr\Big[\mathsf{FC.FVer}(c, G, y, d) = 1 \text{ and } y \neq G(v) : \\ \rho \xleftarrow{R} \{0,1\}^\tau;\ c = \mathsf{FC.Com}(1^\lambda, v; \rho);\ (y, d) \xleftarrow{R} A_\lambda(1^\lambda, c, v, \rho)\Big] \leq \text{negl}(\lambda)\ ;$$

Simulatability: The following two distributions are computationally indistinguishable:

$$\left\{(c,d)\ :\ \begin{array}{l} \rho \xleftarrow{R} \{0,1\}^\tau;\ c \xleftarrow{R} \mathsf{FC.Com}(1^\lambda, v; \rho); \\ d = \mathsf{FC.FOpen}(c, G, v, \rho) \end{array}\right\}_{\lambda, G, v},$$
$$\left\{(c,d)\ :\ (c,d) \xleftarrow{R} \mathsf{FC.Sim}(1^\lambda, G, G(v))\right\}_{\lambda, G, v}.$$

Note that the simulatability property implies the standard hiding property of commitments, if each circuit class $\mathcal{G}_\lambda$ contains a constant circuit: Consider indeed any constant circuit $C(x) = \alpha$, the fact that (c, d) can be simulated from C and α implies that c hides the message committed inside.

Let us now define the non-deterministic oracle family associated to FC.

Definition 5.2. Let $\mathcal{G} = \{\mathcal{G}_\lambda\}_{\lambda \in \mathbb{N}}$ be a poly-size circuit class. Let $\mathsf{FC} = (\mathsf{FC.Com}, \mathsf{FC.FOpen}, \mathsf{FC.FVer}, \mathsf{FC.Sim})$ be a *functional commitment* scheme for $\mathcal{G}$. We define the following *associated non-deterministic oracle family* $\mathcal{O}^{\mathsf{FC}} = \{\mathcal{O}^{\mathsf{FC}}_\lambda\}_{\lambda \in \mathbb{N}}$:

$$\mathcal{O}^{\mathsf{FC}}_\lambda((c, G), (y, d)) = \begin{cases} y & \text{if } \mathsf{FC.FVer}(c, G, y, d) = 1; \\ \bot & \text{otherwise.} \end{cases}$$

5.2 Construction of 2-Round Semi-Honest MPC

Tools: Let f be an arbitrary N-party functionality.[10] To construct a 2-round semi-honest MPC protocol $\widetilde{\Pi}$ for f, we rely on the following tools:

- A semi-honestly secure L-round MPC protocol $\Pi = (\mathsf{Next}, \mathsf{Output})$ for f. We will refer to this protocol the "inner MPC protocol".
  Recall that Next is next message function that computes the message broadcasted by party P_i in round ℓ, $m_i^\ell = \mathsf{Next}_i(x_i, r_i, \bar{m}^{<\ell})$, on input x_i and random tape r_i, after receiving messages $\bar{m}^{<\ell} = \{m_j^{\ell'}\}_{j \in [N], \ell' < \ell}$ broadcasted by parties P_j on previous rounds. And Output is the output function that computes the output of party P_i, $y_i = \mathsf{Output}_i(x_i, r_i, \bar{m})$, after receiving all the messages $\bar{m} = \{m_j^\ell\}_{j \in [N], \ell \in [L]}$. The security parameter λ is an implicit parameter 1^λ of Next and Output.

[10] Formal definitions of MPC protocol and N-party functionality are provided in the full version [7].

- A functional commitment scheme $\mathsf{FC} = (\mathsf{FC.Com}, \mathsf{FC.FOpen}, \mathsf{FC.FVer}, \mathsf{FC.Sim})$ for the class of all S-size circuits with a sufficiently large polynomial bound S. We denote by $\mathcal{O}^{\mathsf{FC}}$ the associated non-deterministic oracle family defined in Definition 5.2.
- A GIC scheme $\mathsf{GiC} = (\mathsf{GiC.Garble}, \mathsf{GiC.Eval})$ for the oracle $\mathcal{O}^{\mathsf{FC}}$ and the unique-transcript distribution $\mathrm{i}\mathcal{D} = \{\mathrm{i}\mathcal{D}_{\lambda,\mathrm{id}}\}_{\lambda\in\mathbb{N},\mathrm{id}}$ that we define later.

We will show that using the constructions in Sect. 6 and in the full version [7], we can construct the two last tools from 2-round (semi-honest) OT. With the above tools, our 2-round MPC protocol $\widetilde{\Pi} = (\widetilde{\mathsf{Next}}, \widetilde{\mathsf{Output}})$ for f proceed as follows:

<u>The First Round:</u> Each party P_i computes its first message $\widetilde{m}_i^1 = \widetilde{\mathsf{Next}}_i(x_i, \tilde{r}_i, \emptyset)$, using security parameter λ, input x_i, random tape $\tilde{r}_i$, and no messages, as follows.

1. Take a sufficient long substring r_i of $\tilde{r}_i$ as the random tape for running the inner MPC protocol Π.
2. Commit L times to (x_i, r_i) using the functional commitment scheme FC: for $\ell \in [L]$, $c_i^\ell = \mathsf{FC.Com}(1^\lambda, (x_i, r_i); \rho_i^\ell)$, where all the ρ_i^ℓ's (and r_i) are non-overlapping substrings of $\tilde{r}_i$.
3. Broadcast the first message $\widetilde{m}_i^1 = \{c_i^\ell\}_{\ell\in[L]}$, and keep $\{\rho_i^\ell\}_{\ell\in[L]}$ secret.

<u>The Second Round:</u> Each party P_i computes its second message $\widetilde{m}_i^2 = \widetilde{\mathsf{Next}}_i(x_i, \tilde{r}_i, \{\widetilde{m}_j^1\}_{j\in N})$, using all first messages $\{\widetilde{m}_j^1\}_{j\in N}$ as follows:

1. Garble the interactive circuit $\mathrm{i}C_i = \{\mathrm{i}C_i^\ell\}_{\ell\in[L]}$ defined in Fig. 3: $\widehat{\mathrm{i}C}_i \xleftarrow{R} \mathsf{GiC.Garble}(1^\lambda, \mathrm{i}C_i)$.
2. Broadcast the second message $\widetilde{m}_i^2 = \widehat{\mathrm{i}C}_i$.

<u>The Output Function:</u> Each party P_i computes its output $y_i = \widetilde{\mathsf{Output}}_i(x_i, \tilde{r}_i, \{\widetilde{m}_j^1, \widetilde{m}_j^2\}_{j\in[N]})$, using all first and second messages $\{\widetilde{m}_j^1, \widetilde{m}_j^2\}_{j\in N}$ as follows. Proceed in L iterations to evaluate the N garbled circuits $\{\widehat{\mathrm{i}C}_j\}_{j\in[N]}$ in parallel. Before iteration $\ell \in [L]$ starts, the following invariant holds:

<u>*Invariant*</u>: After the first $(\ell - 1)$ iterations, P_i has obtained for every $j \in [N]$ and every $\ell' < \ell$:

- the inner MPC message $m_j^{\ell'}$ generated in the ℓ'-th round by party P_j, and
- the associated functional decommitment $d_j^{\ell'}$ of $c_j^{\ell'}$ for the circuit $G_j^{\ell'}(\star,\star) = \mathsf{Next}_j(\star, \star, \overline{m}^{<\ell'})$.

We define $\overline{w}^{<\ell} = \{w_j^{\ell'}\}_{j,\ell'<\ell} = \{(m_j^{\ell'}, d_j^{\ell'})\}_{\ell'<\ell}$.

In the first round $\ell = 1$, all these messages and functional decommitments are empty. Thus, the invariant holds initially. With the above, P_i does the following in iteration ℓ: for every $j \in [N]$: $(m_j^\ell, d_j^\ell) = o_j^\ell = \mathsf{GiC.Eval}(\widehat{\mathrm{i}C}_j, \overline{w}^{<\ell})$.

The Interactive Circuit iC_i

Constants: 1^λ, ℓ, x_i, r_i, the ℓ-th commitments c_j^ℓ for each party P_j (part of the first message $\widetilde{m}_j^1$), and the randomness ρ_i^ℓ used in commitment c_i^ℓ.

Inputs: $(st^{\ell-1}, \bar{a}^{\ell-1})$ where for $\ell > 1$:

- The state $st^{\ell-1} = \bar{m}^{<\ell-1}$ contains the inner MPC messages of the first $\ell - 1$ rounds.
- The answers $a_j^\ell = m_j^{\ell-1}$ are the answers of the non-deterministic oracle $\mathcal{O}^{\mathsf{FC}}$ to the queries $q_j^\ell = (c_j^{\ell-1}, G_j^{\ell-1})$, for $j \in [N]$, where the circuit $G_j^{\ell-1}$ is defined by $G_j^{\ell-1}(\star, \star) = \mathsf{Next}_j(\star, \star, \bar{m}^{<\ell-1})$.

These inputs define $\bar{m}^{<\ell}$.

Procedure:

1. Define the circuit G_j^ℓ as $G_j^\ell(\star, \star) = \mathsf{Next}_j(\star, \star, \bar{m}^{<\ell})$, for $j \in [N]$.
2. Compute the ℓ-th message of P_i in the inner MPC: $m_i^\ell = \mathsf{Next}_i\left(x_i, r_i, \bar{m}^{<\ell}\right)$.
3. Compute the associated functional decommitment of c_i^ℓ: $d_i^\ell = \mathsf{FC.FOpen}(c_i^\ell, G_i^\ell, (x_i, r_i), \rho_i^\ell)$.
4. Compute the next queries: for every $j \in [N]$, $q_j^\ell = (c_j^\ell, G_j^\ell)$.
5. Define the next state to be $st^\ell = \bar{m}^{<\ell}$ and the output to be $o_i^\ell = (m_i^\ell, d_i^\ell)$.

Output: $(st^\ell, \bar{q}^\ell, o_i^\ell)$.

Fig. 3. The interactive circuit iC_i

After all L iterations, P_i obtains the set of all messages $\bar{m}$, and computes the output by invoking the output function of the inner MPC protocol: $y_i = \mathsf{Output}_i\,(x_i, r_i, \bar{m})$.

<u>UNIQUE-TRANSCRIPT DISTRIBUTION:</u> We now define the unique-transcript distribution $\mathrm{i}\mathcal{D} = \{\mathrm{i}\mathcal{D}_{\lambda,\mathrm{id}}\}_{\lambda\in\mathbb{N},\mathrm{id}}$ (for the garbled interactive circuit iC_i) as follows: $\mathrm{id} = (i, \bar{x}, \bar{r}, \bar{m})$ and $\mathrm{i}\mathcal{D}_{\lambda,\mathrm{id}}$ is

$$\left\{ (\mathrm{i}C_i,\ \bar{w},\ \bar{\rho} = \{\rho_j^\ell\}_{j,\ell}) : \begin{array}{l} \forall j \in [N],\ \forall \ell \in [L], \\ \quad \rho_j^\ell \stackrel{R}{\leftarrow} \{0,1\}^{|\rho_j^\ell|};\ c_j^\ell = \mathsf{FC.Com}(1^\lambda, (x_j, r_j); \rho_j^\ell); \\ \quad G_j^\ell(\star, \star) = \mathsf{Next}_j(\star, \star, \bar{m}^{<\ell}); \\ \quad d_j^\ell = \mathsf{FC.FOpen}(c_j^\ell, G_j^\ell, (x_j, r_j), \rho_j^\ell); \\ \bar{w} = \{w_j^\ell = (m_j^\ell, d_j^\ell)\}_{j,\ell};\ \mathrm{i}C_i \text{ defined in Fig. 3} \end{array} \right\}.$$

The unique-transcript property follows from the semi-honest functional binding property of FC. See the full version [7] for details.

<u>SECURITY:</u> We have the following theorem proven in the full version [7].

Theorem 5.3. *If the inner MPC Π =* $(\mathsf{Next}, \mathsf{Output})$ *is correct and secure against semi-honest adversaries, if the functional scheme* FC *is correct, semi-honest functional binding, and simulatable, if the garbled interactive circuit*

scheme GiC *is correct and simulatable, then the MPC protocol defined above is correct and secure against semi-honest adversaries.*

6 Garbled Interactive Circuit from Witness Selector

In this section, we show how to construct GIC from another tool we call witness selector, which can be seen as generalization of witness encryption to languages defined by a non-deterministic oracle family $\mathcal{O}$. Contrary to witness encryption, each query to $\mathcal{O}$ may have multiple answers, as long as at most one can be found efficiently.

We first define the notion of computationally unique-answer distribution for $\mathcal{O}$ and the notion of witness selector for such a distribution. Then we show how to construct a garbled interactive circuit scheme for $(\mathcal{O}, \mathrm{i}\mathcal{D})$ from any witness selector for a unique-answer distribution for $\mathcal{O}$ which is *consistent* with the unique-transcript distribution $\mathrm{i}\mathcal{D}$.

6.1 Witness Selector

Definition 6.1 (Unique-Answer Distribution). Let $\mathcal{O}$ be a non-deterministic oracle family. Let $\mathrm{w}\mathcal{D} = \{\mathrm{w}\mathcal{D}_{\lambda,\mathsf{id}}\}_{\lambda\in\mathbb{N},\mathsf{id}}$ be an ensemble of efficiently samplable distributions over tuples (q, w, aux). We say that $\mathrm{w}\mathcal{D}$ is a *(computationally) unique-answer distribution* for $\mathcal{O}$ if

Non-$\perp$ Answer: For any $\lambda \in \mathbb{N}$, any index $\mathsf{id} \in \{0,1\}^{\mathrm{poly}(\lambda)}$, and any (q, w, aux) in the support of $\mathrm{w}\mathcal{D}_{\lambda,\mathsf{id}}$, $\mathcal{O}_\lambda(q, w) \neq \perp$.

Computationally Unique Answer: For any poly-size circuit family $A = \{A_\lambda\}_{\lambda\in\mathbb{N}}$, for any sequence of indices $\{\mathsf{id}_\lambda\}_\lambda$, there exists a negligible function negl, such that for any $\lambda \in \mathbb{N}$:

$$\Pr\Big[\mathcal{O}_\lambda(q, w') \neq \perp \text{ and } \mathcal{O}_\lambda(q, w') \neq \mathcal{O}_\lambda(q, w) : (q, w, \mathsf{aux}) \xleftarrow{R} \mathrm{w}\mathcal{D}_{\lambda,\mathsf{id}_\lambda};\ w' \xleftarrow{R} A_\lambda(q, w, \mathsf{aux})\Big] \leq \mathrm{negl}(\lambda) .$$

It is a *statistically unique-answer distribution* if the second property holds for any arbitrary-size circuit family $A = \{A_\lambda\}_\lambda$.

Definition 6.2 (Witness Selector). Let $\mathcal{O} = \{\mathcal{O}_\lambda\}_{\lambda\in N}$ be a non-deterministic oracle family, and $\mathrm{w}\mathcal{D} = \{\mathrm{w}\mathcal{D}_{\lambda,\mathsf{id}}\}_{\lambda\in\mathbb{N},\mathsf{id}}$ a unique-answer distribution for $\mathcal{O}$. A *witness selector* scheme for $(\mathcal{O}, \mathrm{w}\mathcal{D})$ is a tuple of two polynomial-time algorithms $\mathsf{WS} = (\mathsf{WS.Enc}, \mathsf{WS.Dec})$:

Encryption: $\mathsf{ct} \xleftarrow{R} \mathsf{WS.Enc}(1^\lambda, q, \mathsf{M})$ encrypts messages $\mathsf{M} = \{\mathsf{M}[i,b]\}_{i\in[l],b\in\{0,1\}}$ for a query q, into a ciphertext ct, where each message has the same length $|\mathsf{M}[i,b]| = \mathrm{poly}(\lambda)$;

Decryption: $\mathsf{M}' = \mathsf{WS.Dec}(\mathsf{ct}, w)$ decrypts a ciphertext ct into messages $\mathsf{M}' = \{\mathsf{M}'[i]\}_{i\in[l]}$ using a witness w;

satisfying the following properties:

Correctness: For any security parameter $\lambda \in \mathbb{N}$, for any index id, for any (q, w, aux) in the support of $\mathrm{w}\mathcal{D}_{\lambda,\mathsf{id}}$, for any messages $\mathsf{M} = \{\mathsf{M}[i,b]\}_{i,b}$, for $a = \mathcal{O}(q, w)$:

$$\Pr\Big[\mathsf{WS.Dec}(\mathsf{WS.Enc}(1^\lambda, q, \mathsf{M}),\ w) = \{\mathsf{M}[i, a_i]\}_{i\in[l]}\Big] = 1\,;$$

Semantic Security: The following two distributions are indistinguishable:

$$\big\{(q, w, \mathsf{aux}, \mathsf{WS.Enc}(1^\lambda, q, \mathsf{M}))\ :\ (q, w, \mathsf{aux}) \stackrel{R}{\leftarrow} \mathrm{w}\mathcal{D}_{\lambda,\mathsf{id}}\big\}_{\lambda,\mathsf{id},\mathsf{M}}\,,$$

$$\left\{ (q, w, \mathsf{aux}, \mathsf{WS.Enc}(1^\lambda, q, \mathsf{M}'))\ :\ \begin{array}{l} (q, w, \mathsf{aux}) \stackrel{R}{\leftarrow} \mathrm{w}\mathcal{D}_{\lambda,\mathsf{id}};\\ a = \mathcal{O}_\lambda(q, w);\\ \{\mathsf{M}'[i,b]\}_{i,b} = \{\mathsf{M}[i,a_i]\}_{i,b} \end{array} \right\}_{\lambda,\mathsf{id},\mathsf{M}}.$$

6.2 Garbled Interactive Circuit from Witness Selector

Let $\mathcal{O} = \{\mathcal{O}_\lambda\}_{\lambda\in\mathbb{N}}$ be a poly-size non-deterministic oracle family. Let $\mathrm{i}\mathcal{D} = \{\mathrm{i}\mathcal{D}_{\lambda,\mathsf{id}}\}_{\lambda\in\mathbb{N},\mathsf{id}}$ be an ensemble of efficiently samplable distributions over tuples $(\mathrm{i}C, \bar{w}, \mathsf{aux})$, where $\mathrm{i}C$ is an L-round interactive circuit. We suppose that $\mathrm{i}\mathcal{D}$ is a unique-transcript distribution for $\mathcal{O}$. To construct a garbled interactive circuit scheme $\mathsf{GiC} = (\mathsf{GiC.Garble}, \mathsf{GiC.Eval}, \mathsf{GiC.Sim})$ for $(\mathcal{O}, \mathrm{i}\mathcal{D})$, we rely on the following tools:

- A witness selector $\mathsf{WS} = (\mathsf{WS.Enc}, \mathsf{WS.Dec})$ for $(\mathcal{O}, \mathrm{w}\mathcal{D})$ where $\mathrm{w}\mathcal{D} = \{\mathrm{w}\mathcal{D}_{\lambda,\mathsf{id}}\}$ is a unique-answer distribution for $\mathcal{O}$, which is consistent with the unique-transcript distribution $\mathrm{i}\mathcal{D}$ (consistency is defined below).
- A garbled circuit scheme $\mathsf{GC} = (\mathsf{GC.Gen}, \mathsf{GC.Garble}, \mathsf{GC.Eval}, \mathsf{GC.Sim})$ for the class of all S-size circuits with a sufficiently large polynomial bound S.

The naive notion of consistence would be: $\mathrm{i}\mathcal{D}$ is consistent with $\mathrm{w}\mathcal{D}$ if each query q_k^ℓ and its associated witness w_k^ℓ follow the same distribution as $\mathrm{w}\mathcal{D}$. Unfortunately, this is not sufficient as the adversary may learn some auxiliary information. Instead, we require that for any ℓ and k, the distribution of $(\mathrm{i}C, \bar{w}, \mathsf{aux}) \stackrel{R}{\leftarrow} \mathrm{i}\mathcal{D}_{\lambda,\mathsf{id}}$ can be simulated from $(q, w, \mathsf{aux}) \stackrel{R}{\leftarrow} \mathrm{w}\mathcal{D}_{\lambda,\mathsf{id}'}$ (for some index id′ function of id) in such a way that q_k^ℓ and w_k^ℓ match q and w. A formal definition is provided in the full version [7].

The construction proceeds as follows:

Garbling: $\widehat{\mathrm{i}C} \stackrel{R}{\leftarrow} \mathsf{GiC.Garble}(1^\lambda, \mathrm{i}C)$ garbles the interactive circuit $\mathrm{i}C = \{\mathrm{i}C^\ell\}_{\ell\in[L]}$ into $\widehat{\mathrm{i}C}$ as follows: For ℓ from L to 1,

1. Generate input labels $\mathsf{key}^\ell \stackrel{R}{\leftarrow} \mathsf{GC.Gen}(1^\lambda)$.
2. Garble the circuit $\mathrm{i}C.\mathsf{AugNext}^\ell$ defined in Fig. 4:
   $\mathrm{i}C.\widehat{\mathsf{AugNext}}^\ell \stackrel{R}{\leftarrow} \mathsf{GC.Garble}(\mathsf{key}^\ell, \mathrm{i}C.\mathsf{AugNext}^\ell)$.

And output $\widehat{\mathrm{i}C} = \{\mathrm{i}C.\widehat{\mathsf{AugNext}}^\ell\}_{\ell\in[L]}$.

Evaluation: $o^{\ell'} = \mathsf{GiC.Eval}(\widehat{\mathrm{i}C}, \bar{w}^{<\ell'})$ evaluates the garbled interactive circuit $\widehat{\mathrm{i}C}$ with the partial list of witnesses $\bar{w}^{<\ell'}$ as follows. For $\ell \in [\ell']$, we denote by key'^{ℓ} the set of input labels that we actually use to evaluate $\widehat{\mathrm{i}C.\mathsf{AugNext}^{\ell}}$ (i.e., it contains one label per input wire; key'^{1} and key'^{L+1} are the empty strings). key'^{ℓ} is composed of two parts $\mathsf{key}'^{\ell}[[st^{\ell}]]$ and $\mathsf{key}'^{\ell}[[\bar{a}^{\ell}]] = \{\mathsf{key}'^{\ell}[[a_k^{\ell}]]\}_k$ corresponding to the input wires for st^{ℓ} and $\bar{a}^{\ell}$ respectively: $\mathsf{key}'^{\ell} = (\mathsf{key}'^{\ell}[[st^{\ell}]], \{\mathsf{key}'^{\ell}[[a_k^{\ell}]]\}_k)$. For ℓ from 1 to ℓ', the evaluator does the following:

1. Evaluate the garbled circuit $\widehat{\mathrm{i}C.\mathsf{AugNext}^{\ell}}$:
   $(\mathsf{key}'^{\ell+1}[[st^{\ell}]], \bar{q}^{\ell}, \bar{\mathsf{ct}}^{\ell}, o^{\ell}) = \mathsf{GC.Eval}(\widehat{\mathrm{i}C.\mathsf{AugNext}^{\ell}}, \mathsf{key}'^{\ell})$.
2. If $\ell < \ell'$, for each $k \in [|\bar{\mathsf{ct}}^{\ell}|]$, decrypt ct_k^{ℓ} using the witness w_k^{ℓ}:
   $\mathsf{key}'^{\ell+1}[[a_k^{\ell}]] = \mathsf{WS.Dec}(\mathsf{ct}_k^{\ell}, w_k^{\ell})$.

And output $o^{\ell'}$ (except if $o^{\ell} = \bot$ for some $\ell \leq \ell'$).

Simulation: $\widetilde{\mathrm{i}C} \xleftarrow{R} \mathsf{GiC.Sim}(1^{\lambda}, T)$ simulates a garbled interactive circuit $\widetilde{\mathrm{i}C}$ from a transcript $T = \{\bar{q}^{\ell}, \bar{a}^{\ell}, o^{\ell}\}_{\ell \in [L]}$ as follows. As for evaluation, for $\ell \in [L]$, we denote by $\mathsf{key}'^{\ell} = (\mathsf{key}'^{\ell}[[st^{\ell}]], \{\mathsf{key}'^{\ell}[[a_k^{\ell}]]\}_k)$ the set of input labels that we actually use as inputs to $\widehat{\mathrm{i}C.\mathsf{AugNext}^{\ell}}$ (i.e., it contains one label per input wire). For ℓ from L to 1, the simulator does the following:

1. Define $\mathsf{key}^{\ell+1}$ to be such that $\mathsf{key}^{\ell+1}[i, b] = \mathsf{key}'^{\ell+1}[i]$ for all input wire i and all bits $b \in \{0, 1\}$. key'^{L+1} and key^{L+1} are empty.

The Augmented Next Message Function $\mathrm{i}C.\mathsf{AugNext}^{\ell}$

Constants: 1^{λ}, ℓ, $\mathrm{i}C^{\ell}$, and the keys $\mathsf{key}_{i*}^{\ell+1}$ for the $(\ell+1)$-th garbled circuit.
Inputs: The previous state $st^{\ell-1}$ and the answers $\bar{a}^{\ell-1}$ (of the non-deterministic oracle $\mathcal{O}$ to the queries $\bar{q}^{\ell-1}$).
Procedure:

1. Compute $(st^{\ell}, \bar{q}^{\ell}, o^{\ell}) = \mathrm{i}C^{\ell}(st^{\ell-1}, \bar{a}^{\ell-1})$. If $o^{\ell} = \bot$, abort and output $(\bot, \bot, \bot, \bot)$. By convention, st^0 and $\bar{a}^0$ are empty strings.
2. For every k, generate using a hardcoded random tape:
$$\mathsf{ct}_k^{\ell} = \mathsf{WS.Enc}(1^{\lambda}, q_k^{\ell}, \mathsf{key}^{\ell+1}[[a_k^{\ell}]]) ,$$
   where $\mathsf{key}^{\ell+1}[[a_k^{\ell}]]$ is the tuple of input labels $\mathsf{key}^{\ell+1}[i, b]$ for all $b \in \{0, 1\}$ and for the input wires i corresponding to the input a_k^{ℓ} of $\mathrm{i}C.\mathsf{AugNext}^{\ell+1}$. Set $\bar{\mathsf{ct}}^{\ell} = \{\mathsf{ct}_k^{\ell}\}_k$. By convention, $\bar{q}^{\ell}$ is empty if $\ell = L$.
3. Select the input labels for the next step, corresponding to the new state st^{ℓ}: $\mathsf{key}^{\ell+1}[st^{\ell}] = \{\mathsf{key}^{\ell+1}[i, st_i^{\ell}]\}_i$. By convention, st^{ℓ} and $\mathsf{key}^{\ell+1}[st^{\ell}]$ are empty if $\ell = L$.

Output: $(\mathsf{key}^{\ell+1}[st^{\ell}], \bar{q}^{\ell}, \bar{\mathsf{ct}}^{\ell}, o^{\ell})$.

Fig. 4. The augmented next message function $\mathrm{i}C.\mathsf{AugNext}^{\ell}$

2. Encrypt the labels generated for the round $\ell + 1$ corresponding to the answer $\bar{a}^\ell$, using the witness selector scheme: for each k,
$\mathsf{ct}_k^\ell \xleftarrow{R} \mathsf{WS.Enc}(1^\lambda, \bar{q}^\ell, \mathsf{key}^{\ell+1}[[a_k^\ell]])$. (For $\ell = L$, $\bar{\mathsf{ct}}^\ell$ and $\mathsf{key}^{\ell+1}$ are empty.)
3. Simulate the garbling of $\mathrm{i}C.\widehat{\mathsf{AugNext}}^\ell$, using its outputs $\mathsf{key}'^{\ell+1}[[st^\ell]] = \mathsf{key}^{\ell+1}[st^\ell]$ (for $\ell = L$, this value is empty), $\bar{q}^{\ell+1}$, $\bar{\mathsf{ct}}^\ell$, and o^ℓ:
$\mathrm{i}C.\widehat{\mathsf{AugNext}}^\ell \xleftarrow{R} \mathsf{GC.Sim}(1^\lambda, (\mathsf{key}'^{\ell+1}[[st^\ell]], \bar{q}^\ell, \bar{\mathsf{ct}}^\ell, o^\ell))$.

<u>SECURITY:</u> We prove the following security theorem in the full version [7].

Theorem 6.3. *If* GC *is correct and simulatable, if* WS *is correct and semantically secure, if* w$\mathcal{D}$ *is unique-answer, and if* i$\mathcal{D}$ *and* w$\mathcal{D}$ *are consistent, then the garbled interactive circuit scheme* GiC *defined above is correct and simulatable.*

Acknowledgments. The authors thank Yuval Ishai, Antigoni Polychroniadou, and Stefano Tessaro for helpful discussions.

This work was supported by NSF grants CNS-1528178, CNS-1514526, CNS-1652849 (CAREER), a Hellman Fellowship, the Defense Advanced Research Projects Agency (DARPA) and Army Research Office (ARO) under Contract No. W911NF-15-C-0236, and a subcontract No. 2017-002 through Galois. The views expressed are those of the authors and do not reflect the official policy or position of the Department of Defense, the National Science Foundation, or the U.S. Government.

References

1. Ananth, P., Choudhuri, A.R., Jain, A.: A new approach to round-optimal secure multiparty computation. In: Katz, J., Shacham, H. (eds.) CRYPTO 2017. LNCS, vol. 10401, pp. 468–499. Springer, Cham (2017). https://doi.org/10.1007/978-3-319-63688-7_16
2. Asharov, G., Jain, A., López-Alt, A., Tromer, E., Vaikuntanathan, V., Wichs, D.: Multiparty computation with low communication, computation and interaction via threshold FHE. In: Pointcheval, D., Johansson, T. (eds.) EUROCRYPT 2012. LNCS, vol. 7237, pp. 483–501. Springer, Heidelberg (2012). https://doi.org/10.1007/978-3-642-29011-4_29
3. Badrinarayanan, S., Goyal, V., Jain, A., Sahai, A.: Verifiable functional encryption. In: Cheon, J.H., Takagi, T. (eds.) ASIACRYPT 2016. LNCS, vol. 10032, pp. 557–587. Springer, Heidelberg (2016). https://doi.org/10.1007/978-3-662-53890-6_19
4. Barak, B., Goldreich, O., Impagliazzo, R., Rudich, S., Sahai, A., Vadhan, S., Yang, K.: On the (im)possibility of obfuscating programs. In: Kilian, J. (ed.) CRYPTO 2001. LNCS, vol. 2139, pp. 1–18. Springer, Heidelberg (2001). https://doi.org/10.1007/3-540-44647-8_1
5. Beaver, D., Micali, S., Rogaway, P.: The round complexity of secure protocols (extended abstract). In: 22nd ACM STOC, pp. 503–513. ACM Press, May 1990
6. Bellare, M., Micali, S.: Non-interactive oblivious transfer and applications. In: Brassard, G. (ed.) CRYPTO 1989. LNCS, vol. 435, pp. 547–557. Springer, New York (1990). https://doi.org/10.1007/0-387-34805-0_48
7. Benhamouda, F., Lin, H.: k-round MPC from k-round OT via garbled interactive circuits. Cryptology ePrint Archive, Report 2017/1125 (2017). https://eprint.iacr.org/2017/1125

8. Biham, E., Boneh, D., Reingold, O.: Generalized Diffie-Hellman modulo a composite is not weaker than factoring. Cryptology ePrint Archive, Report 1997/014 (1997). http://eprint.iacr.org/1997/014
9. Boyle, E., Gilboa, N., Ishai, Y.: Function secret sharing: improvements and extensions. In: Weippl, E.R., Katzenbeisser, S., Kruegel, C., Myers, A.C., Halevi, S. (eds.) ACM CCS 16, pp. 1292–1303. ACM Press, October 2016
10. Boyle, E., Gilboa, N., Ishai, Y.: Group-based secure computation: optimizing rounds, communication, and computation. In: Coron, J.-S., Nielsen, J.B. (eds.) EUROCRYPT 2017. LNCS, vol. 10211, pp. 163–193. Springer, Cham (2017). https://doi.org/10.1007/978-3-319-56614-6_6
11. Boyle, E., Gilboa, N., Ishai, Y., Lin, H., Tessaro, S.: Foundations of homomorphic secret sharing. In: ITCS (2018, to appear)
12. Brakerski, Z., Halevi, S., Polychroniadou, A.: Four round secure computation without setup. In: Kalai, Y., Reyzin, L. (eds.) TCC 2017. LNCS, vol. 10677, pp. 645–677. Springer, Cham (2017). https://doi.org/10.1007/978-3-319-70500-2_22
13. Brakerski, Z., Lombardi, A., Segev, G., Vaikuntanathan, V.: Anonymous IBE, leakage resilience and circular security from new assumptions. Cryptology ePrint Archive, Report 2017/967 (2017). https://eprint.iacr.org/2017/967
14. Brakerski, Z., Perlman, R.: Lattice-based fully dynamic multi-key FHE with short ciphertexts. In: Robshaw, M., Katz, J. (eds.) CRYPTO 2016. LNCS, vol. 9814, pp. 190–213. Springer, Heidelberg (2016). https://doi.org/10.1007/978-3-662-53018-4_8
15. Canetti, R.: Universally composable security: a new paradigm for cryptographic protocols. In: 42nd FOCS, pp. 136–145. IEEE Computer Society Press, October 2001
16. Canetti, R., Goldwasser, S., Poburinnaya, O.: Adaptively secure two-party computation from indistinguishability obfuscation. In: Dodis, Y., Nielsen, J.B. (eds.) TCC 2015. LNCS, vol. 9015, pp. 557–585. Springer, Heidelberg (2015). https://doi.org/10.1007/978-3-662-46497-7_22
17. Canetti, R., Lin, H., Pass, R.: Adaptive hardness and composable security in the plain model from standard assumptions. In: 51st FOCS, pp. 541–550. IEEE Computer Society Press, October 2010
18. Canetti, R., Lindell, Y., Ostrovsky, R., Sahai, A.: Universally composable two-party and multi-party secure computation. In: 34th ACM STOC, pp. 494–503. ACM Press, May 2002
19. Cho, C., Döttling, N., Garg, S., Gupta, D., Miao, P., Polychroniadou, A.: Laconic oblivious transfer and its applications. In: Katz, J., Shacham, H. (eds.) CRYPTO 2017. LNCS, vol. 10402, pp. 33–65. Springer, Cham (2017). https://doi.org/10.1007/978-3-319-63715-0_2
20. Ciampi, M., Ostrovsky, R., Siniscalchi, L., Visconti, I.: Delayed-input non-malleable zero knowledge and multi-party coin tossing in four rounds. In: Kalai, Y., Reyzin, L. (eds.) TCC 2017. LNCS, vol. 10677, pp. 711–742. Springer, Cham (2017). https://doi.org/10.1007/978-3-319-70500-2_24
21. Clear, M., McGoldrick, C.: Multi-identity and multi-key leveled FHE from learning with errors. In: Gennaro, R., Robshaw, M. (eds.) CRYPTO 2015. LNCS, vol. 9216, pp. 630–656. Springer, Heidelberg (2015). https://doi.org/10.1007/978-3-662-48000-7_31
22. Dachman-Soled, D., Katz, J., Rao, V.: Adaptively secure, universally composable, multiparty computation in constant rounds. In: Dodis, Y., Nielsen, J.B. (eds.) TCC 2015. LNCS, vol. 9015, pp. 586–613. Springer, Heidelberg (2015). https://doi.org/10.1007/978-3-662-46497-7_23

23. Döttling, N., Garg, S.: Identity-based encryption from the Diffie-Hellman assumption. In: Katz, J., Shacham, H. (eds.) CRYPTO 2017. LNCS, vol. 10401, pp. 537–569. Springer, Cham (2017). https://doi.org/10.1007/978-3-319-63688-7_18
24. Garg, S., Gentry, C., Halevi, S., Raykova, M.: Two-round secure MPC from indistinguishability obfuscation. In: Lindell, Y. (ed.) TCC 2014. LNCS, vol. 8349, pp. 74–94. Springer, Heidelberg (2014). https://doi.org/10.1007/978-3-642-54242-8_4
25. Garg, S., Gentry, C., Halevi, S., Raykova, M., Sahai, A., Waters, B.: Candidate indistinguishability obfuscation and functional encryption for all circuits. In: 54th FOCS, pp. 40–49. IEEE Computer Society Press, October 2013
26. Garg, S., Gentry, C., Sahai, A., Waters, B.: Witness encryption and its applications. In: Boneh, D., Roughgarden, T., Feigenbaum, J. (eds.) 45th ACM STOC, pp. 467–476. ACM Press, June 2013
27. Garg, S., Mukherjee, P., Pandey, O., Polychroniadou, A.: The exact round complexity of secure computation. In: Fischlin, M., Coron, J.-S. (eds.) EUROCRYPT 2016. LNCS, vol. 9666, pp. 448–476. Springer, Heidelberg (2016). https://doi.org/10.1007/978-3-662-49896-5_16
28. Garg, S., Polychroniadou, A.: Two-round adaptively secure MPC from indistinguishability obfuscation. In: Dodis, Y., Nielsen, J.B. (eds.) TCC 2015. LNCS, vol. 9015, pp. 614–637. Springer, Heidelberg (2015). https://doi.org/10.1007/978-3-662-46497-7_24
29. Garg, S., Srinivasan, A.: Garbled protocols and two-round MPC from bilinear maps. In: 58th FOCS, pp. 588–599. IEEE Computer Society Press (2017)
30. Garg, S., Srinivasan, A.: Two-round multiparty secure computation from minimal assumptions. Cryptology ePrint Archive, Report 2017/1156 (2017). http://eprint.iacr.org/2017/1156
31. Gertner, Y., Kannan, S., Malkin, T., Reingold, O., Viswanathan, M.: The relationship between public key encryption and oblivious transfer. In: 41st FOCS, pp. 325–335. IEEE Computer Society Press, November 2000
32. Goldreich, O., Micali, S., Wigderson, A.: How to play any mental game or a completeness theorem for protocols with honest majority. In: Aho, A. (ed.) 19th ACM STOC, pp. 218–229. ACM Press, May 1987
33. Goldwasser, S., Kalai, Y.T., Popa, R.A., Vaikuntanathan, V., Zeldovich, N.: How to run turing machines on encrypted data. In: Canetti, R., Garay, J.A. (eds.) CRYPTO 2013. LNCS, vol. 8043, pp. 536–553. Springer, Heidelberg (2013). https://doi.org/10.1007/978-3-642-40084-1_30
34. Gorbunov, S., Vaikuntanathan, V., Wichs, D.: Leveled fully homomorphic signatures from standard lattices. In: Servedio, R.A., Rubinfeld, R. (eds.) 47th ACM STOC, pp. 469–477. ACM Press, June 2015
35. Dov Gordon, S., Liu, F.-H., Shi, E.: Constant-round MPC with fairness and guarantee of output delivery. In: Gennaro, R., Robshaw, M. (eds.) CRYPTO 2015. LNCS, vol. 9216, pp. 63–82. Springer, Heidelberg (2015). https://doi.org/10.1007/978-3-662-48000-7_4
36. Halevi, S., Kalai, Y.T.: Smooth projective hashing and two-message oblivious transfer. J. Cryptology **25**(1), 158–193 (2012)
37. Libert, B., Ramanna, S.C., Yung, M.: Functional commitment schemes: from polynomial commitments to pairing-based accumulators from simple assumptions. In: Chatzigiannakis, I., Mitzenmacher, M., Rabani, Y., Sangiorgi, D. (eds.) ICALP 2016. LIPIcs, vol. 55, pp. 30:1–30:14. Schloss Dagstuhl, July 2016
38. McCurley, K.S.: A key distribution system equivalent to factoring. J. Cryptol. **1**(2), 95–105 (1988)

39. Mukherjee, P., Wichs, D.: Two round multiparty computation via multi-key FHE. In: Fischlin, M., Coron, J.-S. (eds.) EUROCRYPT 2016. LNCS, vol. 9666, pp. 735–763. Springer, Heidelberg (2016). https://doi.org/10.1007/978-3-662-49896-5_26
40. Naor, M., Pinkas, B.: Efficient oblivious transfer protocols. In: Kosaraju, S.R. (ed.) 12th SODA, pp. 448–457. ACM-SIAM, January 2001
41. Peikert, C., Shiehian, S.: Multi-key FHE from LWE, revisited. In: Hirt, M., Smith, A. (eds.) TCC 2016. LNCS, vol. 9986, pp. 217–238. Springer, Heidelberg (2016). https://doi.org/10.1007/978-3-662-53644-5_9
42. Peikert, C., Vaikuntanathan, V., Waters, B.: A framework for efficient and composable oblivious transfer. In: Wagner, D. (ed.) CRYPTO 2008. LNCS, vol. 5157, pp. 554–571. Springer, Heidelberg (2008). https://doi.org/10.1007/978-3-540-85174-5_31
43. Shmuely, Z.: Composite Diffie-Hellman Public-Key Generating Systems are Hard to Break. Technical report, Technion (1985). http://www.cs.technion.ac.il/users/wwwb/cgi-bin/tr-info.cgi/1985/CS/CS0356
44. Yao, A.C.C.: Protocols for secure computations (extended abstract). In: 23rd FOCS, pp. 160–164. IEEE Computer Society Press, November 1982
45. Yao, A.C.C.: How to generate and exchange secrets (extended abstract). In: 27th FOCS, pp. 162–167. IEEE Computer Society Press, October 1986
46. Yu, Y., Zhang, J.: Cryptography with auxiliary input and trapdoor from constant-noise LPN. In: Robshaw, M., Katz, J. (eds.) CRYPTO 2016. LNCS, vol. 9814, pp. 214–243. Springer, Heidelberg (2016). https://doi.org/10.1007/978-3-662-53018-4_9

Theoretical Multiparty Computation

Adaptively Secure Garbling with Near Optimal Online Complexity

Sanjam Garg(✉) and Akshayaram Srinivasan

University of California, Berkeley, USA
{sanjamg,akshayaram}@berkeley.edu

Abstract. We construct an adaptively secure garbling scheme with an online communication complexity of $n + m + \mathsf{poly}(\log |C|, \lambda)$ where $C : \{0,1\}^n \rightarrow \{0,1\}^m$ is the circuit being garbled, and λ is the security parameter. The security of our scheme can be based on (polynomial hardness of) the Computational Diffie-Hellman (CDH) assumption, or the Factoring assumption or the Learning with Errors assumption. This is nearly the best achievable in the standard model (i.e., without random oracles) as the online communication complexity must be larger than both n and m. The online computational complexity of our scheme is $O(n+m) + \mathsf{poly}(\log |C|, \lambda)$. Previously known standard model adaptively secure garbling schemes had asymptotically worse online cost or relied on exponentially hard computational assumptions.

1 Introduction

Introduced in the seminal work of Yao [Yao86], garbling techniques are one of the main cornerstones of cryptography. Garbling schemes have found numerous applications in multiparty computation [Yao86, AF90, BMR90], parallel cryptography [AIK04, AIK05], one-time programs [GKR08], verifiable computation [GGP10, AIK10], functional encryption [SS10, GVW12, GKP+13], efficient zero-knowledge proofs [JKO13, FNO15] and program obfuscation [App14, LV16].

Garbling a circuit C and an input x yields a garbled circuit $\widetilde{C}$ and a garbled input $\widetilde{x}$ respectively. Next, using $\widetilde{C}$ and $\widetilde{x}$ anyone can efficiently compute $C(x)$ but security requires that $\widetilde{C}$ and $\widetilde{x}$ jointly reveal nothing about C or x beyond $C(x)$. Typical garbling schemes are only proved to satisfy the weaker notion of *selective security* where both the circuit C and the input x are chosen a priori. However, in certain applications, a stronger notion of *adaptive security* wherein the input x can be chosen adaptively based on the garbled circuit $\widetilde{C}$ is needed [BHR12a]. We refer to the size of $\widetilde{C}$ as the *offline* communication complexity and the size of $\widetilde{x}$ as the *online* communication complexity.

Research supported in part from 2017 AFOSR YIP Award, DARPA/ARL SAFEWARE Award W911NF15C0210, AFOSR Award FA9550-15-1-0274, and research grants by the Okawa Foundation, Visa Inc., and Center for Long-Term Cybersecurity (CLTC, UC Berkeley). The views expressed are those of the author and do not reflect the official policy or position of the funding agencies.

J. B. Nielsen and V. Rijmen (Eds.): EUROCRYPT 2018, LNCS 10821, pp. 535–565, 2018.
https://doi.org/10.1007/978-3-319-78375-8_18

Constructing such adaptively secure garbling schemes with better online communication cost has been an active area of investigation [BHR12a, BGG+14, HJO+16, JW16, AS16, JKK+17, JSW17]. Despite tremendous effort, all standard model constructions of adaptively secure garbling which are based on polynomially hard assumptions have online communication cost that grows with the width of the circuit.

1.1 Our Contributions

We obtain a new adaptive garbling scheme with online communication complexity of $n + m + \mathsf{poly}(\log |C|, \lambda)$ where n is the input length of the circuit C, m is its output length and λ is the security parameter. This almost matches the lower bounds of n and m due to Applebaum et al. [AIKW13].[1] Moreover, this complexity is very close to the best known constructions for the selective security setting [AIKW13]. More formally, our main result is:

Theorem 1. *Assuming either the Computational Diffie-Hellman assumption or the Factoring assumption or the Learning with Errors assumption, there exists a construction of adaptive garbling scheme with online communication complexity of* $n + m + \mathsf{poly}(\log |C|, \lambda)$ *with simulation security.*

All prior constructions of adaptively secure garbling schemes in the standard model had online communication complexity that grew with either the circuit depth/width. Moreover, several of these schemes suffered from an exponential loss in security reduction. We summarize the known constructions and our new results in Table 1.

Table 1. Constructions of known and new adaptive garbling schemes (with simulation security).

	Assumption	Online communication complexity	Security loss	Model
[BHR12a] Const. 1	OWF	$n\lambda$	$\mathsf{poly}(\|C\|, \lambda)$	RO
[BHR12a] Const. 2	OWF	$\|C\| + n\lambda$	$\mathsf{poly}(\|C\|, \lambda)$	Std.
[BGG+14] Const.1	LWE	$(n + m)\mathsf{poly}(\lambda, d)$	$2^{O(d)}$	Std.
[BGG+14] Const.2	LWE + MDDH	$O(n + m) + \mathsf{poly}(\lambda, d)$	$2^{O(d)}$	Std.
[HJO+16] Const. 1	OWF	$(n + m + w)\mathsf{poly}(\lambda)$	$\mathsf{poly}(\|C\|, \lambda)$	Std.
[HJO+16] Const. 2	OWF	$(n + m + d)\mathsf{poly}(\lambda)$	$2^{O(d)}$	Std.
[JW16]	OWF	$(n + m + d)\mathsf{poly}(\lambda)$	$2^{O(d)}$	Std.
[JKK+17]	OWF	$(n + m + d)\mathsf{poly}(\lambda)$	$2^{O(d)}$	Std.
This work	CDH/Factoring/LWE	$n + m + \mathsf{poly}(\lambda, \log \|C\|)$	$\mathsf{poly}(\|C\|, \lambda)$	Std.

[1] In this work, we consider the standard simulation based security notion. Indeed, if one considers the weaker notion of indistinguishablity based security this lower bound can be bypassed as shown in [AS16, JSW17].

Additionally, we note that as a special case, our result implies selectively secure garbling scheme with online cost $n + \mathsf{poly}(\lambda)$ from the same assumptions. Previously, this result was not known under CDH or Factoring. Specifically, constructions were known from DDH or RSA [AIKW13].

1.2 Applications

We now mention some of the applications of our result. These applications were already noted in the work of Hemenway et al. [HJO+16] and we improve their efficiency.

One-time Program and Verifiable Computation. Plugging our result in the one-time program construction of [GKR08], we get a construction of one-time program where the number of hardware tokens is $O(n + m + \mathsf{poly}(\lambda, \log |C|))$. Similarly, the running time of verification protocol in the work of [GGP10] can be improved to match our online complexity.

Compact Functional Encryption. Starting with a single-key, selective functional encryption scheme with weakly compact ciphertexts and using the transformations of [ABSV15, AS16, GS16, LM16] along with our construction of adaptively secure garbled circuits, we obtain a multi-key secure, adaptive functional encryption scheme whose ciphertext size grows only with the output size of the functions.

2 Our Techniques

In this section, we outline the main techniques and tools used in the construction of adaptively secure garbled circuits.

Adaptive Security Game. Before explaining our construction, let us first explain the adaptive security game in a bit more detail. In this game, the adversary provides the challenger with a circuit C and the challenger responds with a garbled circuit $\widetilde{C}$. The adversary later provides with an input x (that could potentially depend on $\widetilde{C}$) and the challenger responds with garbled input $\widetilde{x}$. In the real world, both the garbled circuit and the garbled input are generated honestly whereas in the ideal world, the garbled circuit $\widetilde{C}$ is generated by a simulator Sim_1 that is given the size of C as input and the garbled input $\widetilde{x}$ is generated by another simulator Sim_2 that is given $C(x)$ as input. The goal of the adversary is to distinguish between the real world and the ideal world distributions.

The reason why the proof of Yao's construction breaks down in the adaptive setting is because the distribution of the garbled circuit $\widetilde{C}$ in the intermediate hybrids depends on the value of the (adversarily chosen) input x. Naturally, Yao's approach is not feasible when the garbled circuit needs to be sent before the adversary gives its input x.

Prior Approaches. To solve the issue with Yao's construction, Bellare et al. [BHR12b] encrypted the garbled circuit by an (fully) equivocal encryption scheme and sent the ciphertext in the offline phase. Later, in the online phase, the key for decrypting this ciphertext was provided. Since an equivocal ciphertext can be opened to any value, the simulator in each intermediate hybrid opens the ciphertext sent in the offline phase to an appropriate simulated value (that depends on C and x). However, the key size for an equivocal encryption scheme in the standard model has to grow with the size of the message [Nie02] and in this case it grows with the size of the circuit. Thus, the online complexity of this approach has to grow with the size of the circuit.

The work of Hemenway, Jafargholi, Ostrovsky, Scafuro and Wichs [HJO+16] improved the online complexity by replacing the fully equivocal encryption scheme with a *somewhere equivocal encryption*. Roughly speaking, a somewhere equivocal encryption allows to generate a ciphertext encrypting a vector of messages with "holes" in some positions. Later, these "holes" could be filled with arbitrary message values by deriving a suitable decryption key. Intuitively, in each intermediate hybrid, "holes" are created in the garbled circuit in those positions that depend on the input and the simulator fills these "holes" in the online phase based on the input x. The crucial aspect of a somewhere equivocal encryption is that its key size is only proportional to number of holes which could be much smaller than the total length of the message vector. Thus to minimize the online complexity, it is sufficient to come up with a sequence of hybrids where the number of holes in each intermediate hybrid is minimized. Hemenway et al. provide two sequences of hybrid arguments: the first sequence where the number of "holes" in each hybrid is at most the width of the circuit and the second sequence of hybrids where the number of "holes" in each hybrid is at most the depth (with $2^{O(depth)}$ hybrids). However, even in this approach the online complexity could be as large as the circuit size as the circuit width or depth could be as large as the circuit itself.

Our approach. We follow the high level idea of Hemenway et al. [HJO+16] in encrypting the garbled circuit using a somewhere equivocal encryption but employ a *crucial trick* to minimize the number of "holes" in each intermediate hybrid. At a very high level, we use the recent construction of updatable laconic oblivious transfer [CDG+17, DG17, DGHM18, BLSV18] (which can be constructed based either on CDH/Factoring/LWE) to "linearize" the garbled circuit. Informally, a garbled circuit is "linearized" if the simulation of a garbled gate g depends *only* on simulating one additional gate. We note that all the prior approaches [HJO+16, JW16] resulted in "non-linearized" garbled circuits. In particular, in all the prior works, simulating the garbled gate g depended on simulating *all* gates that provide inputs to g (which are at least two in number). With this "linearization" in place, we design a sequence of hybrids (based on the pebbling strategy of [Ben89]) where the number of "holes" in each intermediate hybrid is $O(\log(|C|))$. This allows us to achieve nearly optimal online complexity. We elaborate on our approach in the next subsection.

2.1 Our Approach: "Linearizing" the Garbled Circuit

We now explain our construction of "linearized" garbled circuits.

Step Circuits. To understand our construction, it is best to view the circuit C as a *sequence of step circuits*. In more details, we will consider C as a sequence of step circuits along with a database/memory D. For simplicity, we consider a circuit with a single output bit. The i-th step circuit implements the i-th gate (with some topological ordering of the gates) in the circuit C. The database D is initially loaded with the input x and contents of the database represent the state of the computation. That is, the snapshot of the database before the evaluation of the i-th step circuit contains the output of every gate $g < i$ in the execution of C on input x. The i-th step circuit reads contents from two pre-determined locations in the database and writes a bit to location i. The bits that are read correspond to the values in the input wires for the i-th gate. The output of the circuit is easily derived from the contents of the database at the end of the computation. To garble the circuit C, we must garble each of the step circuits and the database D.

Garbling Step Circuits. Our approach of garbling the step circuits involves a primitive called as updatable laconic oblivious transfer [CDG+17]. To make the exposition easy, we first consider a simplistic setting where the database D is not protected i.e., it is revealed in the clear to the adversary. We will later explain how this restriction can be removed.

A laconic oblivious transfer is a protocol between two parties: sender and a receiver. The receiver holds a large database $D \in \{0,1\}^N$ and sends a short digest d (with length λ) of the database to the sender. The sender obtains as input a location $L \in [N]$ and two messages m_0, m_1. The sender computes a read-ciphertext c using his private inputs and the received digest d by running in time $\mathsf{poly}(\log N, |m_0|, |m_1|, \lambda)$ and sends c to the receiver. Note that the time required to compute the read-ciphertext c grows logarithmically with the size of the database. The receiver recovers the message $m_{D[L]}$ from the ciphertext c and the security requirement is that the message $m_{1-D[L]}$ is computationally hidden. A laconic oblivious transfer is said to be *updatable* if it additionally allows updates on the database. In particular, the sender on input a location $L \in [N]$, a bit b, digest d and a sequence of λ messages $\{m_{j,0}, m_{j,1}\}_{j \in [\lambda]}$ creates a write-ciphertext c_w (by running in time that grows logarithmically with the size of the database). The receiver on input c_w can recover $\{m_{j,\mathsf{d}^*_j}\}_{j \in [\lambda]}$ where d^* is the digest of the updated database with bit b written in location L. As in the previous case, the security requires that the messages $\{m_{j,1-\mathsf{d}^*_j}\}_{j \in [\lambda]}$ are computationally hidden. An updatable laconic oblivious transfer was first constructed in [CDG+17] from the Decisional Diffie-Hellman (DDH) problem and the assumptions were later improved to CDH/Factoring in [DG17] and to LWE in [DGHM18, BLSV18].

Let us now give details on how to use updatable laconic OT to garble the circuit C. At a very high level, the garbled circuit consists of a sequence of garbled augmented step circuits $\widetilde{\mathsf{SC}}'_1, \ldots, \widetilde{\mathsf{SC}}'_N$ and the garbled input consists of the labels for executing the first garbled step circuit $\widetilde{\mathsf{SC}}'_1$. These garbled step circuits are constructed in special way such that the output of the garbled step circuit $\widetilde{\mathsf{SC}}'_i$ can be used to derive the labels for executing the next garbled step circuit $\widetilde{\mathsf{SC}}'_{i+1}$. Thus, starting from $\widetilde{\mathsf{SC}}'_1$, we can evaluate every garbled step circuit in the sequence. Let us now give details on the internals of the augmented step circuits.

The i-th augmented step circuit SC'_i takes as input the digest d of the snapshot of database D before the evaluation of i-th gate and two bits α_i and β_i. The bits α_i and β_i correspond to the inputs to gate i in the evaluation of C. The augmented step circuit SC'_i additionally has the set of both labels for each input wire of $\widetilde{\mathsf{SC}}'_{i+1}$ hardwired in its description. We denote these labels by $\{\mathsf{lab}^{\mathsf{d}}_{j,0}, \mathsf{lab}^{\mathsf{d}}_{j,1}\}_{j\in[\lambda]}$ that correspond to the digest and $\{\mathsf{lab}^{\alpha}_0, \mathsf{lab}^{\alpha}_1\}$ and $\{\mathsf{lab}^{\beta}_0, \mathsf{lab}^{\beta}_1\}$ that correspond to the input bits of gate $i+1$. SC'_i first computes the output of the i-th gate (denoted by γ) using α_i and β_i. This bit must be written to the database and the updated hash value must be fed to the next circuit SC'_{i+1}. Towards this goal, SC'_i computes a write-ciphertext c_w using the digest d, location i, bit γ and $\{\mathsf{lab}^{\mathsf{d}}_{j,0}, \mathsf{lab}^{\mathsf{d}}_{j,1}\}_{j\in[\lambda]}$. This write-ciphertext will be used to derive the labels corresponding to the updated value of the digest which is fed to SC'_{i+1}. Recall that SC'_{i+1} must also take in the input values to the $(i+1)^{th}$ gate of the circuit C. For this purpose, SC'_i also computes two read ciphertexts c_α, c_β using the value of the (updated) digest d^* and labels $\{\mathsf{lab}^{\alpha}_0, \mathsf{lab}^{\alpha}_1\}$ and $\{\mathsf{lab}^{\beta}_0, \mathsf{lab}^{\beta}_1\}$ respectively. These read ciphertexts will be used to derive the labels corresponding to the values of the input wires to the gate $i+1$. It finally outputs c_w, c_α, c_β. An evaluator for this garbled circuit can recover the set of labels for evaluating $\widetilde{\mathsf{SC}}'_{i+1}$ from these ciphertexts using the decryption functionality of updatable laconic OT.

Notice that in order to simulate the garbled step-circuit $\widetilde{\mathsf{SC}}'_i$, it is sufficient to simulate the garbled step-circuit $\widetilde{\mathsf{SC}}'_{i-1}$. This is because the labels for evaluating $\widetilde{\mathsf{SC}}_i$ are only hardwired in the step-circuit SC'_{i-1} and are not available anywhere else. Once the garbled step circuit $\widetilde{\mathsf{SC}}'_{i-1}$ is simulated, we can use the security of updatable laconic oblivious transfer and (plain) garbled circuits to simulate $\widetilde{\mathsf{SC}}'_i$. This helps us to achieve the right "linearized" structure for simulating the garbled step circuits.

Protecting the Database. In the above exposition, the database D is revealed in the clear which is clearly insecure as database holds the values of all the intermediate wires in the evaluation of the circuit. To protect the database, we mask the contents of the database with a random string. To be more precise, each step circuit additionally has the masking bits for the two input wires and the masking bit for the output wire hardwired. When the step circuit is fed with the masked values of the input wires, it unmasks those values (using the hardwired masking bits) and computes the output of the gate. Finally, it uses the hardwired masking bit for the output wire to mask the output and uses this value to compute the updated digest. This trick of protecting the intermediate computation values using random masks is closely related to the "point and permute" construction of garbled circuits [BMR90, MNPS04].

Pebbling Game. As in the work of [HJO+16], we encrypt these garbled step circuits $\{\widetilde{\mathsf{SC}}'_i\}$ using a somewhere equivocal encryption scheme and send the ciphertext in the offline phase. Later in the online phase, we reveal the key for decrypting this ciphertext along with the labels for evaluating $\widetilde{\mathsf{SC}}'_1$. The task that remains is to come up with a sequence of hybrids such that the number of "holes" in each intermediate hybrid is minimized. Recall that a "hole" appears in a position that depends on the adaptively chosen input. To design a sequence of hybrids, we consider the following pebbling game.[2]

Consider a number line $1, 2, \ldots, N$. We are given some pebbles and we can place a pebble on the number line according the following rules:

- We can always place or remove a pebble from position 1.
- We can place or remove a pebble from position i if and only if there exists a pebble in position $i - 1$.

The goal is to be place at position N by minimizing the number of pebbles (denoted as the *pebbling complexity*) present in the graph at any point of time. A trivial strategy would be to consecutively place pebbles starting from position 1 upto N. The maximum number of pebbles used is N and the hope is to have a strategy that uses far less pebbles.

Intuitively, a pebble in the above game corresponds to a "hole" in the somewhere equivocal ciphertext. Alternatively, we can view the process of placing a pebble in position i as simulating the i-th garbled circuit.[3] The above two rules naturally correspond to rules for simulating a garbled step circuit i.e., the first garbled step circuit can always be simulated and we can simulate the i-th garbled step circuit if the $(i - 1)$-th garbled step circuit is simulated. Bennett [Ben89] showed an inductive pebbling strategy for the above game using $O(\log N)$ pebbles. This readily gives a sequence of hybrids to prove adaptive security where the number of "holes" in each intermediate hybrid is logarithmic in the size of the circuit. This helps us achieve nearly optimal online complexity.

Why is "linearization" important? The work of Hemenway et al. consider a pebbling game directly on the topology of the circuit rather than on a line graph. In more details, they interpreted the circuit C as a DAG with every gate in C being a node in the graph, the input gates represented as sources in the

[2] The pebble game we describe is a simplification of the actual pebbling game we design later. This simplification is sufficient to get the main intuition.

[3] These two views are equivalent since simulation of a garbled step circuit depends on the output of that step circuit which in turn depends on the adversarily chosen input x. Thus, if a garbled step circuit is simulated we must have a "hole" in the corresponding position of the somewhere equivocal encryption.

graph (nodes with in-degree 0) and output gate represented as sink (node with out-degree 0). The rules of the pebbling game are [4]:

1. A pebble can always be placed or removed from a source.
2. A pebble can be placed or removed from a node if all its predecessors have pebbles.

The goal is to place a pebble at the sink node by minimizing the number of pebbles placed in the graph at any point of time. Note that unlike our game, in order to place a pebble at a node it is required that pebbles are present on *all* the predecessors which are at least two in number. This makes the task of using logarithmic many pebbles extremely difficult and there are strong lower bounds [PTC76] concerning the pebbling complexity of the above game. In particular, the work of [PTC76] shows that existence of certain families of DAGs on n nodes (with in-degree 2 and out-degree more than 1) such that the pebbling complexity of those graphs is $\Omega(\frac{n}{\log n})$. This naturally corresponds to similar lower bounds on the pebbling complexity of circuits with fan-in 2 and fan-out greater than 1. Thus, to get around these lower bounds, the use of the "linearized" garbled circuit seems necessary.

Why Garbled RAM fails? To garble the step circuits and the database D, we could hope to use ideas from the garbled RAM literature [LO13, GHL+14, GLOS15, GLO15].[5] This would have given us a garbling scheme based on just one-way functions instead of requiring public-key assumptions. However, all known approaches of constructing garbled RAM introduce additional dependencies in garbling step circuits. This implies that in order to garble a particular step circuit, at least two other step circuits must be garbled. Thus, the graph to be pebbled is no longer a straight line and the known lower bounds apply.

3 Preliminaries

Let λ denote the security parameter. A function $\mu(\cdot) : \mathbb{N} \rightarrow \mathbb{R}^+$ is said to be negligible if for any polynomial $\mathsf{poly}(\cdot)$ there exists $\lambda_0 \in \mathbb{N}$ such that for all $\lambda > \lambda_0$ we have $\mu(\lambda) < \frac{1}{\mathsf{poly}(\lambda)}$. For a probabilistic algorithm A, we denote $A(x;r)$ to be the output of A on input x with the content of the random tape being r. When r is omitted, $A(x)$ denotes a distribution. For a finite set S, we denote $x \leftarrow S$ as the process of sampling x uniformly from the set S. We will use PPT to denote Probabilistic Polynomial Time. We denote $[a]$ to be the set $\{1, \ldots, a\}$ and $[a, b]$ to be the set $\{a, a+1, \ldots, b\}$ for $a \leq b$ and $a, b \in \mathbb{Z}$. For a binary string $x \in \{0,1\}^n$, we will denote the ith bit of x by x_i. We assume without loss of generality that the length of the random tape used by all cryptographic algorithms is λ. We will use $\mathsf{negl}(\cdot)$ to denote an unspecified negligible function and $\mathsf{poly}(\cdot)$ to denote an unspecified polynomial function.

[4] For the sake of exposition, we give a simplified version of the pebbling game considered in the work of [HJO+16]. We refer the reader to their work for the full description.

[5] We in fact do not require the full power of garbled RAM as the locations that are accessed by each step circuit are fixed a priori.

3.1 Garbled Circuits

Below we recall the definition of garbling scheme for circuits [Yao82] with selective security (see Lindell and Pinkas [LP09] and Bellare et al. [BHR12b] for a detailed proof and further discussion). A garbling scheme for circuits is a tuple of PPT algorithms $(\mathsf{GarbleCkt}, \mathsf{EvalCkt})$. Very roughly, $\mathsf{GarbleCkt}$ is the circuit garbling procedure and $\mathsf{EvalCkt}$ the corresponding evaluation procedure. We use a formulation where input labels for a garbled circuit are provided as input to the garbling procedure rather than generated as output. (This simplifies the presentation of our construction.) More formally:

- $\widetilde{\mathsf{C}} \leftarrow \mathsf{GarbleCkt}\left(1^\lambda, C, \{\mathsf{lab}_{w,b}\}_{w\in[n],b\in\{0,1\}}\right)$: $\mathsf{GarbleCkt}$ takes as input a security parameter λ, a circuit C, and input labels $\mathsf{lab}_{w,b}$ where $w \in [n]$ ($[n]$ is the set of input wires to the circuit C) and $b \in \{0,1\}$. This procedure outputs a *garbled circuit* $\widetilde{\mathsf{C}}$. We assume that for each w, b, $\mathsf{lab}_{w,b}$ is chosen uniformly from $\{0,1\}^\lambda$.
- $y \leftarrow \mathsf{EvalCkt}\left(\widetilde{\mathsf{C}}, \{\mathsf{lab}_{w,x_w}\}_{w\in[n]}\right)$: Given a garbled circuit $\widetilde{\mathsf{C}}$ and a sequence of input labels $\{\mathsf{lab}_{w,x_w}\}_{w\in[n]}$ (referred to as the garbled input), $\mathsf{EvalCkt}$ outputs a string y.

Correctness. For correctness, we require that for any circuit C, input $x \in \{0,1\}^{|[n]|}$ and input labels $\{\mathsf{lab}_{w,b}\}_{w\in[n],b\in\{0,1\}}$ we have that:

$$\Pr\left[C(x) = \mathsf{EvalCkt}\left(\widetilde{\mathsf{C}}, \{\mathsf{lab}_{w,x_w}\}_{w\in[n]}\right)\right] = 1$$

where $\widetilde{\mathsf{C}} \leftarrow \mathsf{GarbleCkt}\left(1^\lambda, C, \{\mathsf{lab}_{w,b}\}_{w\in[n],b\in\{0,1\}}\right)$.

Selective Security. For security, we require that there exists a PPT simulator $\mathsf{Sim}_{\mathsf{Ckt}}$ such that for any circuit C and input $x \in \{0,1\}^{|[n]|}$, we have that

$$\left\{\widetilde{\mathsf{C}}, \{\mathsf{lab}_{w,x_w}\}_{w\in[n]}\right\} \overset{c}{\approx} \left\{\mathsf{Sim}_{\mathsf{Ckt}}\left(1^\lambda, 1^{|C|}, C(x), \{\mathsf{lab}_{w,x_w}\}_{w\in[n]}\right), \{\mathsf{lab}_{w,x_w}\}_{w\in[n]}\right\}$$

where $\widetilde{\mathsf{C}} \leftarrow \mathsf{GarbleCkt}\left(1^\lambda, C, \{\mathsf{lab}_{w,b}\}_{w\in[n],b\in\{0,1\}}\right)$ and for each $w \in [n]$ and $b \in \{0,1\}$ we have $\mathsf{lab}_{w,b} \leftarrow \{0,1\}^\lambda$. Here $\overset{c}{\approx}$ denotes that the two distributions are computationally indistinguishable.

Theorem 2 ([Yao86,LP09]). *Assuming the existence of one-way functions, there exists a construction of garbling scheme for circuits.*

3.2 Updatable Laconic Oblivious Transfer

In this subsection, we recall the definition of updatable laconic oblivious transfer from [CDG+17].

Definition 1 **([CDG+17]).** *An updatable laconic oblivious transfer consists of the following algorithms:*

- $\mathsf{crs} \leftarrow \mathsf{crsGen}(1^\lambda)$: *It takes as input the security parameter* 1^λ *(encoded in unary) and outputs a common reference string* crs.
- $(\mathsf{d}, \widehat{D}) \leftarrow \mathsf{Hash}(\mathsf{crs}, D)$: *It takes as input the common reference string* crs *and database* $D \in \{0,1\}^*$ *as input and outputs a digest* d *and a state* $\widehat{D}$. *We assume that the state* $\widehat{D}$ *also includes the database* D.
- $e \leftarrow \mathsf{Send}(\mathsf{crs}, \mathsf{d}, L, m_0, m_1)$: *It takes as input the common reference string* crs, *a digest* d, *a location* $L \in \mathbb{N}$ *and two messages* $m_0, m_1 \in \{0,1\}^{p(\lambda)}$ *and outputs a ciphertext* e.
- $m \leftarrow \mathsf{Receive}^{\widehat{D}}(\mathsf{crs}, e, L)$: *This is a RAM algorithm with random read access to* $\widehat{D}$. *It takes as input a common reference string* crs, *a ciphertext* e, *and a database location* $L \in \mathbb{N}$ *and outputs a message* m.
- $e_w \leftarrow \mathsf{SendWrite}(\mathsf{crs}, \mathsf{d}, L, b, \{m_{j,0}, m_{j,1}\}_{j=1}^{|\mathsf{d}|})$: *It takes as input the common reference string* crs, *a digest* d, *a location* $L \in \mathbb{N}$, *a bit* $b \in \{0,1\}$ *to be written, and* $|\mathsf{d}|$ *pairs of messages* $\{m_{j,0}, m_{j,1}\}_{j=1}^{|\mathsf{d}|}$, *where each* $m_{j,c}$ *is of length* $p(\lambda)$ *and outputs a ciphertext* e_w.
- $\{m_j\}_{j=1}^{|\mathsf{d}|} \leftarrow \mathsf{ReceiveWrite}^{\widehat{D}}(\mathsf{crs}, L, b, e_w)$: *This is a RAM algorithm with random read/write access to* $\widehat{D}$. *It takes as input the common reference string* crs, *a location* L, *a bit* $b \in \{0,1\}$ *and a ciphertext* e_w. *It updates the state* $\widehat{D}$ *(such that* $D[L] = b$*) and outputs messages* $\{m_j\}_{j=1}^{|\mathsf{d}|}$.

We require an updatable laconic oblivious transfer to satisfy the following properties.

Correctness: *We require that for any database* D *of size at most* $M = \mathsf{poly}(\lambda)$, *any memory location* $L \in [M]$, *any pair of messages* $(m_0, m_1) \in \{0,1\}^{p(\lambda)}$ *where* $p(\cdot)$ *is a polynomial that*

$$\Pr\left[m = m_{D[L]} \;\middle|\; \begin{array}{rl} \mathsf{crs} & \leftarrow \mathsf{crsGen}(1^\lambda) \\ (\mathsf{d}, \widehat{D}) & \leftarrow \mathsf{Hash}(\mathsf{crs}, D) \\ e & \leftarrow \mathsf{Send}(\mathsf{crs}, \mathsf{d}, L, m_0, m_1) \\ m & \leftarrow \mathsf{Receive}^{\widehat{D}}(\mathsf{crs}, e, L) \end{array} \right] = 1,$$

Correctness of Writes: *Let database* D *be of size at most* $M = \mathsf{poly}(\lambda)$ *and let* $L \in [M]$ *be any memory location. Let* D^* *be a database that is identical to* D *except that* $D^*[L] = b$. *For any sequence of messages* $\{m_{j,0}, m_{j,1}\}_{j \in [\lambda]} \in \{0,1\}^{p(\lambda)}$ *we require that*

$$\Pr\left[\begin{array}{c} m'_j = m_{j,\mathsf{d}^*_j} \\ \forall j \in [|\mathsf{d}|] \end{array} \;\middle|\; \begin{array}{rl} \mathsf{crs} & \leftarrow \mathsf{crsGen}(1^\lambda) \\ (\mathsf{d}, \widehat{D}) & \leftarrow \mathsf{Hash}(\mathsf{crs}, D) \\ (\mathsf{d}^*, \widehat{D}^*) & \leftarrow \mathsf{Hash}(\mathsf{crs}, D^*) \\ e_w & \leftarrow \mathsf{SendWrite}\left(\mathsf{crs}, \mathsf{d}, L, b, \{m_{j,0}, m_{j,1}\}_{j=1}^{|\mathsf{d}|}\right) \\ \{m'_j\}_{j=1}^{|\mathsf{d}|} & \leftarrow \mathsf{ReceiveWrite}^{\widehat{D}}(\mathsf{crs}, L, b, e_w) \end{array} \right] = 1,$$

Sender Privacy: *There exists a PPT simulator* $\mathsf{Sim}_{\ell\mathsf{OT}}$ *such that the for any non-uniform PPT adversary* $\mathcal{A} = (\mathcal{A}_1, \mathcal{A}_2)$ *there exists a negligible function* $\mathsf{negl}(\cdot)$ *s.t.,*

$$\left|\Pr[\mathsf{SenPrivExpt}^{\mathsf{real}}(1^\lambda, \mathcal{A}) = 1] - \Pr[\mathsf{SenPrivExpt}^{\mathsf{ideal}}(1^\lambda, \mathcal{A}) = 1]\right| \leq \mathsf{negl}(\lambda)$$

where $\mathsf{SenPrivExpt}^{\mathsf{real}}$ *and* $\mathsf{SenPrivExpt}^{\mathsf{ideal}}$ *are described in Fig. 1.*

Sender Privacy for Writes: *There exists a PPT simulator* $\mathsf{Sim}_{\ell\mathsf{OTW}}$ *such that the for any non-uniform PPT adversary* $\mathcal{A} = (\mathcal{A}_1, \mathcal{A}_2)$ *there exists a negligible function* $\mathsf{negl}(\cdot)$ *s.t.,*

$$\left|\Pr[\mathsf{WriSenPrivExpt}^{\mathsf{real}}(1^\lambda, \mathcal{A}) = 1] - \Pr[\mathsf{WriSenPrivExpt}^{\mathsf{ideal}}(1^\lambda, \mathcal{A}) = 1]\right| \leq \mathsf{negl}(\lambda)$$

where $\mathsf{WriSenPrivExpt}^{\mathsf{real}}$ *and* $\mathsf{WriSenPrivExpt}^{\mathsf{ideal}}$ *are described in Fig. 2.*

Efficiency: *The algorithm Hash runs in time* $|D|\mathsf{poly}(\log|D|, \lambda)$. *The algorithms* Send, SendWrite, Receive, ReceiveWrite *run in time* $\mathsf{poly}(\log|D|, \lambda)$.

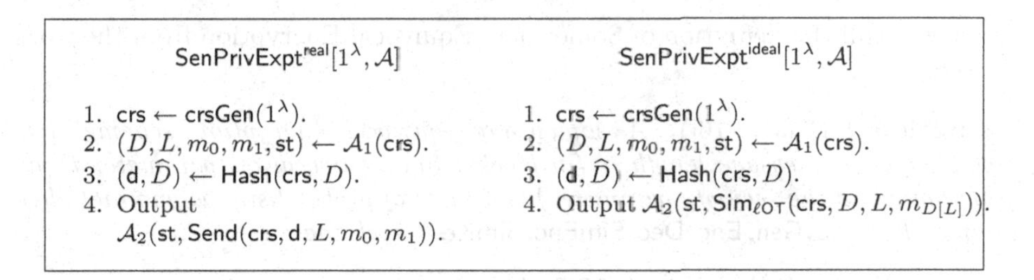

Fig. 1. Sender privacy security game

Theorem 3 ([CDG+17,DG17,BLSV18,DGHM18]). *Assuming either the Computational Diffie-Hellman assumption or the Factoring assumption or the Learning with Errors assumption, there exists a construction of updatable laconic oblivious transfer.*

Remark 1. We note that the security requirements given in Definition 1 is stronger than the one in [CDG+17] as we require the crs to be generated before the adversary provides the database D and the location L. However, the construction in [CDG+17] already satisfies this definition since in the proof, we can guess the location by incurring a $1/D$ loss in the security reduction.

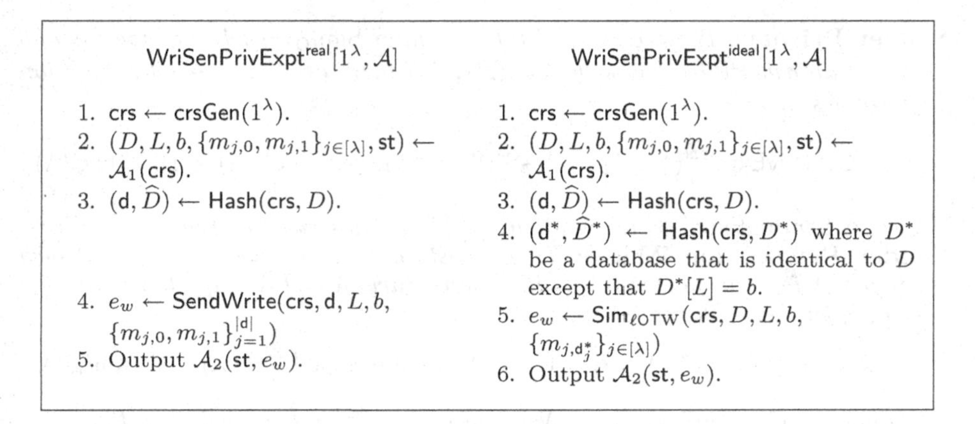

Fig. 2. Sender privacy for writes security game

3.3 Somewhere Equivocal Encryption

We now recall the definition of Somewhere Equivocal Encryption from the work of [HJO+16].

Definition 2 **([HJO+16]).** *A somewhere equivocal encryption scheme with block-length s, message length n (in blocks) and equivocation parameter t (all polynomials in the security parameter) is a tuple of probabilistic polynomial algorithms $\Pi = (\mathsf{KeyGen}, \mathsf{Enc}, \mathsf{Dec}, \mathsf{SimEnc}, \mathsf{SimKey})$ such that:*

- $\mathsf{key} \leftarrow \mathsf{KeyGen}(1^\lambda)$: *It is a PPT algorithm that takes as input the security parameter (encoded in unary) and outputs a key* key.
- $\overline{c} \leftarrow \mathsf{Enc}(\mathsf{key}, m_1 \dots m_n)$: *It is a PPT algorithm that takes as input a key* key *and a vector of messages $\overline{m} = m_1 \dots m_n$ with each $m_i \in \{0,1\}^s$ and outputs a ciphertext $\overline{c}$.*
- $\overline{m} \leftarrow \mathsf{Dec}(\mathsf{key}, \overline{c})$: *It is a deterministic algorithm that takes as input a key* key *and a ciphertext $\overline{c}$ and outputs a vector of messages $\overline{m} = m_1 \dots m_n$.*
- $(\mathsf{st}, \overline{c}) \leftarrow \mathsf{SimEnc}((m_i)_{i \notin I}, I)$: *It is a PPT algorithm that takes as input a set of indices $I \subseteq [n]$ and a vector of messages $(m_i)_{i \notin I}$ and outputs a ciphertext $\overline{c}$ and a state* st.
- $\mathsf{key}' \leftarrow \mathsf{SimKey}(\mathsf{st}, (m_i)_{i \in I})$: *It is a PPT algorithm that takes as input the state information* st *and a vector of messages $(m_i)_{i \in I}$ and outputs a key* key'.

and satisfies the following properties:

Correctness. *For every* $\mathsf{key} \leftarrow \mathsf{KeyGen}(1^\lambda)$, *for every* $\overline{m} \in \{0,1\}^{s \times n}$ *it holds that:*

$$\mathsf{Dec}(\mathsf{key}, \mathsf{Enc}(\mathsf{key}, \overline{m})) = \overline{m}$$

Simulation with No Holes. *We require that the distribution of* $(\overline{c}, \mathsf{key})$ *computed via* $(\mathsf{st}, \overline{c}) \leftarrow \mathsf{SimEnc}(\overline{m}, \emptyset)$ *and* $\mathsf{key} \leftarrow \mathsf{SimKey}(\mathsf{st}, \emptyset)$ *to be identical to*

$\mathsf{key} \leftarrow \mathsf{KeyGen}(1^\lambda)$ *and* $\overline{c} \leftarrow \mathsf{Enc}(\mathsf{key}, m_1 \ldots m_n)$. *In other words, simulation when there are no holes (i.e.,* $I = \emptyset$*) is identical to honest key generation and encryption.*

Security. *For any PPT adversary* $\mathcal{A}$*, there exists a negligible function* $\nu = \nu(\lambda)$ *such that:*

$$\left| \Pr[\mathsf{Exp}^{\mathsf{simenc}}_{\mathcal{A},\Pi}(1^\lambda, 0) = 1] - \Pr[\mathsf{Exp}^{\mathsf{simenc}}_{\mathcal{A},\Pi}(1^\lambda, 1) = 1] \right| \leq \nu(\lambda)$$

where the experiment $\mathsf{Exp}^{\mathsf{simenc}}_{\mathcal{A},\Pi}$ *is defined as follows:*

Experiment $\mathsf{Exp}^{\mathsf{simenc}}_{\mathcal{A},\Pi}$

1. *The adversary* $\mathcal{A}$ *on input* 1^λ *outputs a set* $I \subseteq [n]$ *s.t.* $|I| < t$*, a vector* $(m_i)_{i \notin I}$*, and a challenge* $j \in [n] \setminus I$*. Let* $I' = I \cup \{j\}$*.*
2. – *If* $b = 0$*, compute* $\overline{c}$ *as follows:* $(\mathsf{st}, \overline{c}) \leftarrow \mathsf{SimEnc}((m_i)_{i \notin I}, I)$*.*
   – *If* $b = 1$*, compute* $\overline{c}$ *as follows:* $(\mathsf{st}, \overline{c}) \leftarrow \mathsf{SimEnc}((m_i)_{i \notin I'}, I')$*.*
3. *Send* $\overline{c}$ *to the adversary* $\mathcal{A}$*.*
4. *The adversary* $\mathcal{A}$ *outputs the set of remaining messages* $(m_i)_{i \in I}$*.*
   – *If* $b = 0$*, compute* key *as follows:* $\mathsf{key} \leftarrow \mathsf{SimKey}(\mathsf{st}, (m_i)_{i \in I})$*.*
   – *If* $b = 1$*, compute* key *as follows:* $\mathsf{key} \leftarrow \mathsf{SimKey}(\mathsf{st}, (m_i)_{i \in I'})$
5. *Send* key *to the adversary.*
6. $\mathcal{A}$ *outputs* b' *which is the output of the experiment.*

Theorem 4 ([HJO+16]). *Assuming the existence of one-way functions, there exists a somewhere equivocal encryption scheme for any polynomial message-length* n*, black-length* s *and equivocation parameter* t*, having key size* $t \cdot s \cdot \mathsf{poly}(\lambda)$ *and ciphertext of size* $n \cdot s \cdot \mathsf{poly}(\lambda)$ *bits.*

3.4 Adaptive Garbled Circuits

We provide the definition of adaptive garbled circuits from [HJO+16].

Definition 3. *An adaptive garbling scheme for circuits is a tuple of PPT algorithms* $(\mathsf{AdaGarbleCkt}, \mathsf{AdaGarbleInp}, \mathsf{AdpEvalCkt})$ *such that:*

– $(\widetilde{C}, \mathsf{st}) \leftarrow \mathsf{AdaGarbleCkt}(1^\lambda, C)$: *It is a PPT algorithm that takes as input the security parameter* 1^λ *(encoded in unary) and a circuit* $C : \{0,1\}^n \to \{0,1\}^m$ *as input and outputs a garbled circuit* $\widetilde{C}$ *and state information* st*.*
– $\widetilde{x} \leftarrow \mathsf{AdaGarbleInp}(\mathsf{st}, x)$: *It is a PPT algorithm that takes as input the state information* st *and an input* $x \in \{0,1\}^n$ *and outputs the garbled input* $\widetilde{x}$*.*
– $y = \mathsf{AdpEvalCkt}(\widetilde{C}, \widetilde{x})$: *Given a garbled circuit* $\widetilde{C}$ *and a garbled input* $\widetilde{x}$*, it outputs a value* $y \in \{0,1\}^m$*.*

Correctness. *For every* $\lambda \in \mathbb{N}$*,* $C : \{0,1\}^n \to \{0,1\}^m$ *and* $x \in \{0,1\}^n$ *it holds that:*

$$\Pr\left[(\widetilde{C}, \mathsf{st}) \leftarrow \mathsf{AdaGarbleCkt}(1^\lambda, C); \widetilde{x} \leftarrow \mathsf{AdaGarbleInp}(\mathsf{st}, x) : C(x) = \mathsf{AdpEvalCkt}(\widetilde{C}, \widetilde{x})\right] = 1$$

Adaptive Security. *There exists a PPT simulator* $\mathsf{Sim} = (\mathsf{SimC}, \mathsf{SimIn})$ *such that, for any non-uniform PPT adversary* $\mathcal{A}$ *there exists a negligible function* ν *such that:*

$$\left|Pr[\mathsf{Exp}^{\mathsf{adaptive}}_{\mathcal{A},\mathsf{GC},\mathsf{Sim}}(1^\lambda, 0) = 1] - Pr[\mathsf{Exp}^{\mathsf{adaptive}}_{\mathcal{A},\mathsf{GC},\mathsf{Sim}}(1^\lambda, 1) = 1]\right| \leq \nu(\lambda)$$

where the experiment $\mathsf{Exp}^{\mathsf{adaptive}}_{\mathcal{A},\mathsf{GC},\mathsf{Sim}}$ *is defined as follows:*

1. *The adversary specifies the circuit* C *and obtains* $\widetilde{C}$ *where* $\widetilde{C}$ *is created as follows:*
   - *If* $b = 0$*:* $(\widetilde{C}, \mathsf{st}) \leftarrow \mathsf{AdaGarbleCkt}(1^\lambda, C)$.
   - *If* $b = 1$*:* $(\widetilde{C}, \mathsf{st}) \leftarrow \mathsf{SimC}(1^\lambda, 1^{|C|})$.
2. *The adversary* $\mathcal{A}$ *specifies the input* x *and gets* $\widetilde{x}$ *created as follows:*
   - *If* $b = 0$, $\widetilde{x} \leftarrow \mathsf{AdaGarbleInp}(\mathsf{st}, x)$.
   - *If* $b = 1$, $\widetilde{x} \leftarrow \mathsf{SimIn}(\mathsf{st}, C(x))$
3. *Finally, the adversary outputs a bit* b'*, which is the output of the experiment.*

Online Complexity. *The running time of* $\mathsf{AdaGarbleInp}$ *is called as the online computational complexity and* $|\widetilde{x}|$ *is called as the online communication complexity.*

4 Our Construction

In this section, we provide our construction of adaptive garbled circuits. The main theorem is:

Theorem 5. *Assuming the existence of updatable laconic oblivious transfer, somewhere equivocal encryption and garbling scheme for circuits with selective security, there exists a construction of adaptive garbling scheme for circuits. The online communication complexity of our scheme is* $n + m + \mathsf{poly}(\lambda, \log |C|)$ *and the online computational complexity is* $O(n + m + \mathsf{poly}(\lambda, \log |C|))$.

From Theorems 2, 3, 4 we obtain the following corollary:

Corollary 1. *Assuming either the Computational Diffie-Hellman assumption or the Factoring assumption or the Learning with Errors assumption, there exists a construction of adaptive garbling scheme for circuits with online communication complexity of* $n + m + \mathsf{poly}(\lambda, \log |C|)$ *and online computational complexity of* $O(n + m + \mathsf{poly}(\lambda, \log |C|))$.

We start with some notation on how we denote circuits. We choose this notation to simplify the description of our construction. In the rest of the paper, whenever we mention a circuit C, we implicitly mean the universal circuit $U[C]$ with the circuit C hardwired in it. This is done so that the topology of the circuit $U[C]$ does not reveal anything about C except its size.

$\mathsf{AdaGarbleCkt}(1^\lambda, C)$: On input a circuit $C : \{0,1\}^n \to \{0,1\}^m$ do:

1. Sample $\mathsf{crs} \leftarrow \mathsf{crsGen}(1^\lambda)$, $\mathsf{key} \leftarrow \mathsf{KeyGen}(1^\lambda)$ and $r \leftarrow \{0,1\}^N$.
2. For each $g \in [n+1, N+1]$, $k \in [\lambda]$ and $b \in \{0,1\}$ sample $\mathsf{lab}^g_{k,b} \leftarrow \{0,1\}^\lambda$. (We use $\{\mathsf{lab}^g_{k,b}\}$ to denote $\{\mathsf{lab}^g_{k,b}\}_{k\in[\lambda], b\in\{0,1\}}$.)
3. **for** each g from N down to $n+1$ **do**:
   (a) Let (i,j) be the description of the gate g.
   (b) Compute (where the step-circuit SC is described in Figure 4)
$$\widetilde{\mathsf{SC}}_g \leftarrow \mathsf{GarbleCkt}\left(1^\lambda, \mathsf{SC}[\mathsf{crs}, (r_i, r_j, r_g), (i,j), \{\mathsf{lab}^{g+1}_{k,b}\}, 0], \{\mathsf{lab}^g_{k,b}\}\right).$$
4. Compute $c \leftarrow \mathsf{Enc}(\mathsf{key}, \{\widetilde{\mathsf{SC}}_g\}_{g\in[n+1,N]})$.
5. Output $\widetilde{C} := (\mathsf{crs}, c)$ and $\mathsf{st} := (r, \mathsf{key}, \{\mathsf{lab}^{n+1}_{k,b}\})$.

$\mathsf{AdaGarbleInp}(\mathsf{st}, x)$: On input the state st and a string $x \in \{0,1\}^n$ do:

1. Parse st as $(r, \mathsf{key}, \{\mathsf{lab}^{n+1}_{k,b}\})$
2. Set $D := r_1 \oplus x_1 \| \dots \| r_n \oplus x_n \| 0^{N-n}$ and compute $(\mathsf{d}, \widehat{D}) := \mathsf{Hash}(\mathsf{crs}, D)$.
3. Output $\widetilde{x} := \left(\{\mathsf{lab}^{n+1}_{k,\mathsf{d}_k}\}_{k\in[\lambda]}, r_1 \oplus x_1 \| \dots \| r_n \oplus x_n, \mathsf{key}, r_{N-m+1}, \dots, r_N\right)$.

$\mathsf{AdpEvalCkt}(\widetilde{C}, \widetilde{x})$: On input garbled circuit $\widetilde{C}$, and garbled input $\widetilde{x}$ do:

1. Parse $\widetilde{C}$ as (crs, c) and $\widetilde{x}$ as $(\{\mathsf{lab}_k\}_{k\in[\lambda]}, s_1, \dots, s_n, \mathsf{key}, r_{N-m+1}, \dots r_N)$.
2. Set $D := s_1 \| \dots \| s_n \| 0^{N-n}$ and compute $(\mathsf{d}, \widehat{D}) := \mathsf{Hash}(\mathsf{crs}, D)$.
3. Compute $\{\widetilde{\mathsf{SC}}_g\}_{g\in[n+1,N]} := \mathsf{Dec}(\mathsf{key}, c)$.
4. Set $\overline{\mathsf{lab}} := \{\mathsf{lab}_k\}_{k\in[\lambda]}$.
5. **for** each g from $n+1$ to N **do**:
   (a) Let (i,j) be the description of gate g.
   (b) Compute $(\gamma, e) := \mathsf{Receive}^{\widehat{D}}(\mathsf{crs}, \mathsf{Receive}^{\widehat{D}}(\mathsf{crs}, \mathsf{EvalCkt}(\widetilde{\mathsf{SC}}_g, \overline{\mathsf{lab}}), i), j)$.
   (c) Set $\overline{\mathsf{lab}} := \mathsf{ReceiveWrite}^{\widehat{D}}(\mathsf{crs}, g, \gamma, e)$.
6. Recover the contents of the memory D from the final state $\widehat{D}$.
7. Output $D_{N-m+1} \oplus r_{N-m+1} \| \dots \| D_m \oplus r_N$.

Fig. 3. Adaptive garbling scheme for circuits

Notation. We model a circuit $C : \{0,1\}^n \to \{0,1\}^m$ as a set of $N - n$ NAND gates, each having fan-in 2. We number the gates of the circuit as follows. The input gates are given the numbers $\{1, \dots, n\}$. The intermediate gates are numbered $\{n+1, n+2, \dots, N-m\}$ such that a gate that receives its input from gates i and j is given a number greater than i and j. The output gates are numbered $\{N-m+1, \dots, N\}$. Each gate $g \in [n+1, N]$ is *described by a tuple* $(i,j) \in [g-1]^2$ where outputs of gates i and j serves as inputs to gate g.

Construction. Let (crsGen, Hash, Send, Receive, SendWrite, ReceiveWrite) be an updatable laconic oblivious transfer scheme, and (GarbleCkt, EvalCkt) be a garbling scheme for circuits. Moreover, let (KeyGen, Enc, Dec, SimEnc, SimKey) be a somewhere equivocal encryption with the block-length $s = |\widetilde{\mathsf{SC}}|$ (where $\widetilde{\mathsf{SC}}$ denotes a garbled version of the step-circuit SC defined in Fig. 4), the message-length equal to $N - n$ and the equivocation parameter $t = \log N$ (the choice of t comes from the security proof).

Step Circuit SC

Input: A digest d.
Hardcoded: The common reference string crs, a triplet of masking bits (r_i, r_j, r_g), a description (i, j) of gate g, a set of labels $\{\mathsf{lab}_{k,b}\}$ and a bit τ ($\tau = 1$ case is only relevant for the proof).

1. Compute $e_b \leftarrow \mathsf{SendWrite}(\mathsf{crs}, \mathsf{d}, g, b, \{\mathsf{lab}_{k,0}, \mathsf{lab}_{k,1}\}_{k\in[\lambda]})$ for $b \in \{0,1\}$.
2. Define for all $\alpha, \beta \in \{0,1\}$, $\gamma(\alpha,\beta) := \begin{cases} \mathsf{NAND}(\alpha \oplus r_i, \beta \oplus r_j) \oplus r_g & \text{if } \tau = 0 \\ r_g & \text{if } \tau = 1 \end{cases}$
3. Generate
$$f_0 \leftarrow \mathsf{Send}\left(\mathsf{crs}, \mathsf{d}, j, (\gamma(0,0), e_{\gamma(0,0)}), (\gamma(0,1), e_{\gamma(0,1)})\right),$$
$$f_1 \leftarrow \mathsf{Send}\left(\mathsf{crs}, \mathsf{d}, j, (\gamma(1,0), e_{\gamma(1,0)}), (\gamma(1,1), e_{\gamma(1,1)})\right).$$
4. Output
$$\mathsf{Send}\left(\mathsf{crs}, \mathsf{d}, i, f_0, f_1\right)$$

Fig. 4. Description of the step circuit

The formal description of our adaptive garbling scheme appears in Fig. 3. In this construction a selective secure garbling scheme is used to garble a step circuit SC repeatedly. This step circuit is described in Fig. 4. We now provide an informal overview of this construction. At a high level, the adaptive garbling of a circuit C simply consists of garbling of the $N - n$ intermediate step circuits using the standard selectively secure (Yao's) garbling scheme. The entire N bit state of the computation is stored in an external memory using laconic OT and each step circuit accesses two bits in this memory. These garbled step circuits are encrypted using a somewhere equivocal encryption scheme and the resulting ciphertext is sent in the offline phase. Later in the online phase, the key for decrypting this ciphertext is revealed. Note that in our description we use the string $r \in \{0,1\}^N$ as a one time pad to hide the state of the computation.

Communication Complexity of $\mathsf{AdaGarbleInp}$. It follows from the construction that the communication complexity of $\mathsf{AdaGarbleInp}$ is $\lambda^2 + n + m + |\mathsf{key}|$. From the parameters used in the somewhere equivocal encryption, we note that $|\mathsf{key}| = |\widetilde{\mathsf{SC}}|\mathsf{poly}(\log N, \lambda)$. It follows from the efficiency properties of updatable laconic oblivious transfer that $|\mathsf{SC}|$ is $\mathsf{poly}(\log N, \lambda)$. Thus, $|\mathsf{key}|$ is $\mathsf{poly}(\log N, \lambda)$.

Computational Complexity of $\mathsf{AdaGarbleInp}$. The running time of $\mathsf{AdaGarbleInp}$ described in Fig. 3 grows with the circuit size N. We note that the running time can be made independent of the circuit size N by analyzing the specific laconic OT construction of Cho et al. [CDG+17]. The construction of Cho et al. uses a Merkle tree to hash a large database into a short digest. Recall that Merkle hash is efficiently updatable. Specifically, let y and y' be two strings given as a sequence of blocks of λ bits and y, y' differ in only the first k blocks. Given the Merkle hash on y and a set of $\log|y|$ hash values, there is an efficient procedure running in time $O(\lambda(k + \log|y|))$ that computes the Merkle hash on y'. We use this property to reduce the online computational complexity of our construction.

Recall that in our construction, the contents of the database at the very beginning of our computation needs to be set to $((r_{[1,n]} \oplus x)||0^{N-n})$ where x is the n-bit input. However, note that the input x is specified in the online phase. So the goal is to compute the hash of $((r_{[1,n]} \oplus x)||0^{N-n})$ in the online phase efficiently. To do this, we compute the hash of 0^N in the offline phase and store the value of this hash along with the specific hash values for updating the first $\lceil n/\lambda \rceil$ blocks. Once x is specified in the online phase, we use the stored value to compute the hash on $((r_{[1,n]} \oplus x)||0^{N-n})$ by performing the Merkle hash update. The crucial point is that this update is efficient (i.e. grows only with $|x| + \log N$).

Also, the algorithm AdaGarbleInp does not need the entire input r as input. It suffices to provide the first n and the last m bits of r (i.e., $\{r_1 \dots r_n, r_{N-m+1} \dots r_N\}$ as input to AdaGarbleInp.

Correctness. Let D^{g^*} be the contents of the database at the end of Step 5.(c) of AdpEvalCkt in the g^*-th iteration of the **for** loop. We first argue via an inductive argument that for each gate $g^* \in [1, N]$, $D_g^{g^*}$ is the output of gate g masked with r_g for every $g \in [1, g^*]$. Given this, the correctness follows by setting $g^* := N$ and observing that the $\{D_k^N\}_{k \in [N-m+1,N]}$ is unmasked using $r_{[N-m+1,N]}$ in Step 7 of AdpEvalCkt.

The base case is $g^* = n$ which is clearly true since in the beginning D^n is set as $(r_{[1,n]} \oplus x||0^{N-n})$. In order to prove the inductive step for a gate g^* (with description (i,j)), we now argue that that the γ recovered in Step 4.(b) of AdpEvalCkt corresponds to $\mathsf{NAND}(D_i^{g^*-1} \oplus r_i, D_j^{g^*-1} \oplus r_j) \oplus r_{g^*}$ which by inductive hypothesis corresponds to output of the gate g^* masked with r_{g^*}. This is shown as follows.

$$
\begin{aligned}
(\gamma, e) &:= \mathsf{Receive}^{\widehat{D}}(\mathsf{crs}, \mathsf{Receive}^{\widehat{D}}(\mathsf{crs}, \mathsf{EvalCkt}(\widetilde{\mathsf{SC}}_g, \overline{\mathsf{lab}}), i), j) \\
&= \mathsf{Receive}^{\widehat{D}}(\mathsf{crs}, \mathsf{Receive}^{\widehat{D}}(\mathsf{crs}, \mathsf{Send}\,(\mathsf{crs}, \mathsf{d}, i, f_0, f_1)\,, i), j) \\
&= \mathsf{Receive}^{\widehat{D}}(\mathsf{crs}, f_{D_i^{g^*-1}}, j) \\
&= \mathsf{Receive}^{\widehat{D}}\left(\mathsf{crs}, \mathsf{Send}\left(\mathsf{crs}, \mathsf{d}, j, (\gamma(D_i^{g^*-1}, 0), e_{\gamma(D_i^{g^*-1}, 0)}), (\gamma(D_i^{g^*-1}, 1), e_{\gamma(D_i^{g^*-1}, 1)})\right), j\right) \\
&= \left(\gamma(D_i^{g^*-1}, D_j^{g^*-1}), e_{\gamma(D_i^{g^*-1}, D_j^{g^*-1})}\right) \\
&= \left(\mathsf{NAND}(D_i^{g^*-1} \oplus r_i, D_j^{g^*-1} \oplus r_j) \oplus r_{g^*}, e_{\mathsf{NAND}(D_i^{g^*-1} \oplus r_i, D_j^{g^*-1} \oplus r_j) \oplus r_{g^*}}\right)
\end{aligned}
$$

5 Proof of Security

In this section, we prove that the construction presented in the previous section satisfies adaptive security. In Subsect. 5.1, we start by defining circuit configurations. Next, in Subsect. 5.2 we show that both the real and ideal world executions are special cases of this circuit configuration (this will also provide a formal description of our simulator). Finally, in the rest of the subsection we show that the real and ideal world executions are indistinguishable. The indistinguishability argument proceeds by a sequence of hybrids over different configurations.

5.1 Circuit Configuration

Our proof of security proceeds via a hybrid argument over different *circuit configurations* which we describe in this section. A circuit configuration denoted by $\mathsf{conf} = (I, \{(g, \mathsf{mode}_g)\}_{g \in [n+1,N]})$ consists of a set $I \subseteq [n+1, N]$ and a set of tuples (g, mode_g) where for each gate $g \in [n+1, N]$ $\mathsf{mode}_g \in \{\mathsf{White}, \mathsf{Gray}, \mathsf{Black}\}$ describes the mode of operation of gate g.

The subset I denotes the set of indices in which there is a "hole" in the outer encryption layer. At an intuitive level, the White mode corresponds to the Real Garbling (as is done in the honest execution), the Gray mode corresponds to the Input Dependent Simulation (where the step circuit for this gate is in simulation but depends on the input), and the Black mode corresponds to the Input Independent Simulation (where simulation is done independent of the input). In other words, White mode matches the real execution, Black mode matches the ideal execution and Gray mode is an intermediate execution mode. Looking ahead, initially all the step circuit will be in White mode and the goal will be to convert all of them to Black in the simulation. Note that we refer to modes with color names as these modes will coincide with the pebbling game that we later describe.

Valid configurations. We say that a configuration $\mathsf{conf} = (I, \{(g, \mathsf{mode}_g)\}_{g \in [n+1,N]})$ is *valid* if and only if:

1. If $\mathsf{mode}_g = \mathsf{Black}$ then for every $k > g$, $\mathsf{mode}_k = \mathsf{Black}$.
2. If $\mathsf{mode}_g = \mathsf{Gray}$ then $g \in I$.

In other words, if gate g is in Black mode then we require that all the subsequent gates are also in Black mode. Moreover, if a gate g is in the Gray mode then there is a hole in positions g in the outer encryption layer.

Simulation in a valid configuration. In Fig. 5 we describe the simulated circuit garbling SimC and the simulated input garbling SimIn functions for any given valid configuration conf. Note that these simulated garbling functions take as input the circuit C and x respective as inputs which the ideal world simulation does not. We describe our simulator functions with these additional inputs so that it captures simulation in all of our intermediate hybrids. We note that final ideal world simulation does not uses these values.

5.2 Our Hybrids

For every valid circuit configuration $\mathsf{conf} = (I, \{(g, \mathsf{mode}_g)\}_{g\in[n+1,N]})$, we define $\mathsf{Hybrid}_{\mathsf{conf}}$ to be a distribution of $\widetilde{C}$ and $\widetilde{x}$ as given in Fig. 5. We start by observing that both real world and ideal distribution from Definition 3 can be seen as instance of $\mathsf{Hybrid}_{\mathsf{conf}}$ where $\mathsf{conf} = (\emptyset, \{(g, \mathsf{White})\}_{g\in[n+1,N]})$ and $\mathsf{conf} = (\emptyset, \{(g, \mathsf{Black})\}_{g\in[n+1,N]})$, respectively. In other words, the real world distribution corresponds to having all gates in White mode and the ideal world distribution corresponds to having all gates in Black mode. The goal is to move from the real world distribution to the ideal world distribution while minimizing the maximum number of gates in the Gray mode in any intermediate hybrid.

5.2.1 Rules of Indistinguishability

We will now describe the two rules (we call these rule A and rule B) to move from one valid circuit configuration conf to another valid configuration conf′ such that $\mathsf{Hybrid}_{\mathsf{conf}}$ is computationally indistinguishable from $\mathsf{Hybrid}_{\mathsf{conf}'}$.

Rule A: Very roughly, rule A says that for any valid configuration conf we can indistinguishably change gate g^* in White mode to Gray mode if it is the first gate or if its predecessor is also in Gray mode. More formally, let $\mathsf{conf} = (I, \{(g, \mathsf{mode}_g)\}_{g\in[n+1,N]})$ and $\mathsf{conf}' = (I', \{(g, \mathsf{mode}'_g)\}_{g\in[n+1,N]})$ be two valid circuit configurations and $g^* \in [n+1, N]$ be a gate such that:

- For all $g \in [n+1, N] \setminus g^*$ we have that $\mathsf{mode}_g = \mathsf{mode}'_g$.
- $g^* \notin I$, $I' = I \cup \{g^*\}$, and $|I'| \le t$ (where t is the equivocation parameter).
- Either $g^* = n+1$ or $(g^* - 1, \mathsf{Gray}) \in \mathsf{conf}$.
- $(g^*, \mathsf{White}) \in \mathsf{conf}$ and $(g^*, \mathsf{Gray}) \in \mathsf{conf}'$.

In Lemma 3 we show that for an valid configurations conf, conf′ satisfying the above constraints we have that $\mathsf{Hybrid}_{\mathsf{conf}} \overset{c}{\approx} \mathsf{Hybrid}_{\mathsf{conf}'}$. Note that we can also use this rule to move a gate g^* from Gray mode to White mode. We refer to those invocations of the rule as *inverse A rule*.

Rule B: Very roughly, rule B says that for any configuration for any valid configuration conf we can indistinguishably change gate g^* in Gray mode to Black mode if all gates subsequent to g^* is in Black mode and the predecessor is in Gray mode. More formally, let $\mathsf{conf} = (I, \{(g, \mathsf{mode}_g)\}_{g\in[n+1,N]})$ and $\mathsf{conf}' = (I', \{(g, \mathsf{mode}'_g)\}_{g\in[n+1,N]})$ be two valid circuit configurations and $g^* \in [n+1, N]$ be a gate such that:

- For all $g \in [n+1, N] \setminus g^*$ we have that $\mathsf{mode}_g = \mathsf{mode}'_g$.
- $g^* \in I$, $I' = I \setminus \{g^*\}$, and $|I| \le t$ (where t is the equivocation parameter).
- For all $g \in [g^*+1, N]$ we have that $(g, \mathsf{Black}) \in \mathsf{conf}$.
- Either $g^* = n+1$ or $(g^* - 1, \mathsf{Gray}) \in \mathsf{conf}$.
- $(g^*, \mathsf{Gray}) \in \mathsf{conf}$ and $(g^*, \mathsf{Black}) \in \mathsf{conf}'$.

In Lemma 4 we show that for an valid configurations conf, conf′ satisfying the above constraints we have that $\mathsf{Hybrid}_{\mathsf{conf}} \overset{c}{\approx} \mathsf{Hybrid}_{\mathsf{conf}'}$.

$\mathsf{SimC}(1^\lambda, C)$: On input a circuit $C : \{0,1\}^n \to \{0,1\}^m$ do:

1. Sample $\mathsf{crs} \leftarrow \mathsf{crsGen}(1^\lambda)$ and $r \xleftarrow{\$} \{0,1\}^N$.
2. For each $g \in [n+1, N+1]$, $k \in [\lambda]$ and $b \in \{0,1\}$ sample $\mathsf{lab}^g_{k,b} \leftarrow \{0,1\}^\lambda$. (We use $\{\mathsf{lab}^g_{k,b}\}$ to denote $\{\mathsf{lab}^g_{k,b}\}_{k\in[\lambda], b\in\{0,1\}}$.)
3. **for** each g from N down to $n+1$ such that $g \notin I$ **do**:
   (a) Let (i,j) be the description of the gate g.
   (b) If $\mathsf{mode}_g = \mathsf{White}$ then compute (where the step-circuit SC is described in Figure 4)
   $$\widetilde{\mathsf{SC}}_g \leftarrow \mathsf{GarbleCkt}\left(1^\lambda, \mathsf{SC}[\mathsf{crs}, (r_i, r_j, r_g), (i,j), \{\mathsf{lab}^{g+1}_{k,b}\}, 0], \{\mathsf{lab}^g_{k,b}\}\right).$$
   (c) If $\mathsf{mode}_g = \mathsf{Black}$ then compute
   $$\widetilde{\mathsf{SC}}_g \leftarrow \mathsf{GarbleCkt}\left(1^\lambda, \mathsf{SC}[\mathsf{crs}, (0, 0, r_g), (i,j), \{\mathsf{lab}^{g+1}_{k,b}\}, 1], \{\mathsf{lab}^g_{k,b}\}\right).$$
4. Compute $(\mathsf{st}_1, c) \leftarrow \mathsf{SimEnc}(I, \{\widetilde{\mathsf{SC}}_g\}_{g\notin I})$.
5. Output $\widetilde{C} := (\mathsf{crs}, c)$ and $\mathsf{st} := (r, \mathsf{st}_1, \{\mathsf{lab}^g_{k,b}\}_{k,b,g})$.

$\mathsf{SimIn}(\mathsf{st}, x, y)$: On input state $\mathsf{st} = (r, \mathsf{st}_1, \{\mathsf{lab}^g_{k,b}\}_{k,b,g})$, a string $x \in \{0,1\}^n$, and $y \in \{0,1\}^m$ do:

Notation: For $g \in [n+1, N+1]$ we let D_g be such that

$$D_{g,w} = \begin{cases} x_w \oplus r_w & w \le n, \\ E_w \oplus r_w & n+1 \le w < g, \\ 0 & \text{otherwise}, \end{cases}$$

where E_w is the bit assigned to wire w of the circuit C computed on input x. Finally, we let d_g be the digest of D_g (i.e., $(\mathsf{d}_g, \cdot) := \mathsf{Hash}(\mathsf{crs}, D_g)$) and $\mathsf{d}_{g,k}$ be the kth bit of d_g.

1. **for** each g from N down to $n+1$ such that $g \in I$:
   (a) Set $e \leftarrow \mathsf{Sim}_{\ell\mathsf{OTW}}(\mathsf{crs}, D_g, g, D_{g+1,g}, \{\mathsf{lab}^{g+1}_{k,\mathsf{d}_{g+1,k}}\}_{k\in[\lambda]})$.
   (b) Set $\mathsf{out} \leftarrow \mathsf{Sim}_{\ell\mathsf{OT}}(\mathsf{crs}, D_g, i, \mathsf{Sim}_{\ell\mathsf{OT}}(\mathsf{crs}, D_g, j, e))$
   (c) Compute $\widetilde{\mathsf{SC}}_g \leftarrow \mathsf{Sim}_{\mathsf{Ckt}}\left(1^\lambda, 1^{|\mathsf{SC}|}, \mathsf{out}, \{\mathsf{lab}^g_{k,\mathsf{d}_{g,k}}\}_{k\in[\lambda]}\right)$.
2. Compute $\mathsf{key} \leftarrow \mathsf{SimKey}(\mathsf{st}_1, \{\widetilde{\mathsf{SC}}_g\}_{g\in I})$.
3. **for** each $g \in [N-m+1, N]$ do:
   (a) If $\mathsf{mode}_g \in \mathsf{Black}$ then set $r'_g = r_g \oplus y_{w-N+m}$.
   (b) Else, $r'_g = r_g$.
4. Output $\widetilde{x} := \left(\{\mathsf{lab}^{n+1}_{k,\mathsf{d}_{n+1,k}}\}_{k\in[\lambda]}, r_1 \oplus x_1 \| \ldots \| r_n \oplus x_n, \mathsf{key}, r'_{N-m+1}, {}_N \cdots r'_N\right)$.

Fig. 5. Garbling in configuration $\mathsf{conf} = (I, \{(h, \mathsf{mode}_h)\}_{h\in[n+1,N]})$.

5.2.2 Interpreting the Rules of Indistinguishability as a Pebbling Game

Our sequence of hybrids from the real to the ideal world follow an optimal strategy for the following pebbling game. The two rules described above correspond to the rules of our pebbling game below.

Consider the positive integer line $n+1, n+2, \ldots N$. We are given pebbles of two colors: gray and black. A black pebble corresponds to a gate in the Black (i.e., input independent simulation) mode and a gray pebble corresponds to a gate in the Gray (i.e., input dependent simulation) mode. A position without any pebble corresponds to real garbling or in the White mode. We can place the pebbles on this positive integer line according to the following two rules:

Rule A: We can place or remove a gray pebble in position i if and only if there is a gray pebble in position $i-1$. This restriction does not apply to position $n+1$: we can always place or remove a gray pebble at position $n+1$.

Rule B: We can replace a gray pebble in position i with a black pebble as long as all the positions $> i$ have black pebbles and there is a gray pebble in position $i-1$ or if $i = n+1$.

Optimization goal of the pebbling game. The goal is to pebble the line $[n+1, N]$ such that every position has a black pebble while minimizing the number of gray pebbles that are present on the line at any point in time.

5.2.3 Optimal Pebbling Strategy

To provide some intuition, we start with the naïve pebbling strategy. The naïve pebbling strategy involves starting from position $n+1$ and placing a gray pebble at every position in $[n+1, N]$ and then replacing them with black pebbles from N to $n+1$. However, this strategy uses a total of $N-n$ gray pebbles. Using a more clever strategy it is actually possible to do the same using only $\log(N-n)$ gray pebbles. This is argued in the following lemmas.

Lemma 1. *For any integer $n+1 \leq p \leq n+2^k-1$, it is possible to make $O((p-n)^{\log_2 3}) \approx O((p-n)^{1.585})$ moves and get a gray pebble at position p using k gray pebbles.*

Proof. This proof is taken verbatim from [GPSZ17]. First we observe to get a gray pebble placed at p, for each $i \in [n+1, p-1]$ there must have been at some point a gray pebble placed at location i.

Next, we observe that it suffices to show we can get a gray pebble at position $p = n+2^k-1$ for every k using $O(3^k) = O((p-n)^{\log_2 3})$ steps. Indeed, for more general p, we run the protocol for $p' = n+2^k-1$ where $k = \lceil \log_2(p-n-1) \rceil$, but stop the first time we get a gray pebble at position p. Since $p'/p \leq 3$, the running time is at most $O((p-n)^{\log_2 3})$.

Now for the algorithm. The sequence of steps will create a fractal pattern, and we describe the steps recursively. We assume an algorithm A_{k-1} using $k-1$ gray pebbles that can get a gray pebble at position $n+2^{k-1}-1$. The steps are as follows:

- Run A_{k-1}. There is now a gray pebble at position $n + 2^{k-1} - 1$ on the line.
- Place the remaining gray pebble at position $n + 2^{k-1}$, which is allowed since there is a gray pebble at position $n + 2^{k-1} - 1$.
- Run A_{k-1} in reverse, recovering all of the $k - 1$ gray pebbles used by A. The result is that there is a single gray pebble on the line at position $n + 2^{k-1}$.
- Now associate the portion of the number line starting at $n + 2^{k-1} + 1$ with a new number line. That is, associate $n + 2^{k-1} + a$ on the original number line with $n' + a$ (where $n' = n + 2^{k-1}$) on the new number line. We now have $k - 1$ gray pebbles, and on this new number line, all of the same rules apply. In particular, we can always add or remove a gray pebble from the first position $n' + 1 = n + 2^{k-1} + 1$ since we have left a gray pebble at $n + 2^{k-1}$. Therefore, we can run A_{k+1} once more on the new number line starting at $n' + 1$. The end result is a pebble at position $n' + 2^{k-1} - 1 = n + 2^{k-1} + (2^{k-1} - 1) = n + 2^k - 1$.

It remains to analyze the running time. The algorithm makes 3 recursive calls to A_{k-1}, so by induction the overall running time is $O(3^k)$, as desired.

Using the above lemma, we now give an optimal strategy for our pebbling game.

Lemma 2. *For any $N \in \mathbb{N}$, there exists a strategy for pebbling the line graph $[n + 1, N]$ according to rules A and B by using at most* $\log N$ *gray pebbles and making* $\mathsf{poly}(N)$ *moves.*

Proof. The strategy is given below. For each g from N down to $n + 1$ **do**:

1. Use the strategy in Lemma 1 to place a gray pebble in position g. Note that there exists a gray pebble in position $g - 1$ as well.
2. Replace the gray pebble in position g with a black pebble. This replacement is allowed since all positions $> g$ have black pebbles and there is a gray pebble in position $g - 1$.
3. Recover all the gray pebbles by reversing the moves.

The correctness of this strategy follows by inspection and the number of moves is polynomial in N.

5.2.4 Completing the Hybrids

Using the strategy given in Lemma 2 yields a sequence of configurations $\mathsf{conf}_0 \ldots \mathsf{conf}_\ell$ for an appropriate polynomial ℓ with conf_0 and conf_ℓ being the real and the ideal world distributions and for each $i \in [\ell]$ we have that $\mathsf{Hybrid}_{\mathsf{conf}_{i-1}} \stackrel{c}{\approx} \mathsf{Hybrid}_{\mathsf{conf}_i}$ either using rule A (i.e., Lemma 3) or using rule B (i.e., Lemma 4). Finally note that the number of holes in the garbled circuit needed is the maximum size of I over the sequence of hybrids (i.e. the maximum number of gray pebbles used). Thus, it suffices to set the equivocation parameter t for somewhere equivocal encryption scheme to $\log N$. This completes the proof of security.

5.3 Proof of Indistinguishability for the Rules

In this subsection, we will use the security of underlying primitives to implement the two rules.

5.3.1 Implementing Rule A

Lemma 3 (Rule A). *Let* conf *and* conf' *be two valid circuit configurations satisfying the constraints of rule A, then assuming the security of somewhere equivocal encryption, garbling scheme for circuits and updatable laconic oblivious transfer, we have that* $\mathsf{Hybrid}_{\mathsf{conf}} \overset{c}{\approx} \mathsf{Hybrid}_{\mathsf{conf}'}$.

Proof. We prove this via a hybrid argument.

- $\mathsf{Hybrid}_{\mathsf{conf}}$: This is our starting hybrid and is distributed as $\mathsf{Hybrid}_{(I,\{(g,\mathsf{mode}_g)\}_{g\in[n+1,N]})}$.
- Hybrid_1 : In this hybrid, we change the configuration to $(I',\{(g,\mathsf{mode}_g)\}_{g\in[n+1,N]})$. This hybrid is distributed as $\mathsf{Hybrid}_{(I',\{(g,\mathsf{mode}_g)\}_{g\in[n+1,N]})}$.
  Computational indistinguishability between hybrid $\mathsf{Hybrid}_{\mathsf{conf}}$ and Hybrid_1 reduces directly to the security of somewhere equivocal encryption scheme. We give a formal reduction in Appendix A
- Hybrid_2 : By conditions of Rule A we have that $\mathsf{mode}_{g^*-1} = \mathsf{Gray}$. Thus, we have that $g^*-1 \in I'$. Therefore, we note that the input labels $\{\mathsf{lab}^{g^*}_{k,b}\}$ are not used in SimC but only in SimIn where it is used to generate $\widetilde{\mathsf{SC}}_{g^*-1}$ and $\widetilde{\mathsf{SC}}_{g^*}$. In this hybrid, we postpone the sampling of $\{\mathsf{lab}^{g^*}_{k,b}\}$ and the generation of $\widetilde{\mathsf{SC}}_{g^*}$ from SimC to SimIn.
  The change in hybrid Hybrid_2 from Hybrid_1 is just syntactic and the distributions are identical.
- Hybrid_3 : In this hybrid, we change the sampling of $\{\mathsf{lab}^{g^*}_{k,b}\}$ and the generation of $\widetilde{\mathsf{SC}}_{g^*}$. Specifically, we do not sample the entire set of labels $\{\mathsf{lab}^{g^*}_{k,b}\}$ but a subset namely $\{\mathsf{lab}^{g^*}_{k,\mathsf{d}_{g^*,k}}\}_k$ and we generate $\widetilde{\mathsf{SC}}_{g^*}$ from the simulated distribution. (Note that since g^*-1 is also in Gray mode. Thus we have that $\widetilde{\mathsf{SC}}_{g^*-1}$ is also simulated and only $\{\mathsf{lab}^{g^*}_{k,\mathsf{d}_{g^*,k}}\}_k$ are needed for its generation.) More formally, we generate

$$\widetilde{\mathsf{SC}}_{g^*} \leftarrow \mathsf{Sim}_{ckt}(1^\lambda, 1^{|\mathsf{SC}|}, \mathsf{out}, \{\mathsf{lab}^{g^*}_{k,\mathsf{d}_{g^*,k}}\}_{k\in[\lambda]})$$

  where $\mathsf{out} \leftarrow \mathsf{SC}[\mathsf{crs}, (r_i, r_j, r_g), (i,j), \{\mathsf{lab}^{g^*+1}_{k,b}\}, 0](\mathsf{d}_{g^*})$.
  The only change in hybrid Hybrid_3 from Hybrid_2 is in the generation of the garbled circuit $\widetilde{\mathsf{SC}}_{g^*}$ and the security follows directly from the selective security of the garbling scheme. We show this reduction in Appendix A.
- Hybrid_4 : In this hybrid, we change how the output value out hardwired in $\widetilde{\mathsf{SC}}_{g^*}$ is generated. Recall that in hybrid Hybrid_3 this value is generated by first computing f_0 and f_1 as in Fig. 4 and then generating out as $\mathsf{Send}(\mathsf{crs}, \mathsf{d}, i, f_0, f_1)$.

In this hybrid, we just generate $f_{D_{g^*,i}}$ and use the laconic OT simulator to generate out. More formally, out is generated as

$$\mathsf{out} \leftarrow \mathsf{Sim}_{\ell\mathsf{OT}}\left(\mathsf{crs}, D_{g^*}, i, f_{D_{g^*,i}}\right).$$

Computational indistinguishability between hybrids Hybrid_3 and Hybrid_4 follows directly from the sender privacy of the laconic OT scheme. The reduction is given in Appendix A.

- $\underline{\mathsf{Hybrid}_5}$: In this hybrid, we change how the value $f_{D_{g^*,i}}$ is generated. Recall from Fig. 4 that $f_{D_{g^*,i}}$ is set as $\mathsf{Send}(\mathsf{crs}, \mathsf{d}, j, (\gamma(D_{g^*,i}, 0), e_{\gamma(D_{g^*,i},0)}), (\gamma(D_{g^*,i}, 1), e_{\gamma(D_{g^*,i},1)}))$. We change the distribution of $f_{D_{g^*,i}}$ to $\mathsf{Sim}_{\ell\mathsf{OT}}(\mathsf{crs}, D_{g^*}, j, e_{D_{g^*+1,g^*}})$, where $e_{D_{g^*+1,g^*}}$ is sampled as in Fig. 4.

  Computational indistinguishability between hybrids Hybrid_4 and Hybrid_5 follows directly from the sender privacy of the laconic OT scheme. The argument is analogous to the argument of indistinguishability between Hybrid_3 and Hybrid_4.
- $\underline{\mathsf{Hybrid}_6}$: In this hybrid, we change how $e_{D_{g^*+1,g^*}}$ is generated. More specifically, we generate it using the simulator $\mathsf{Sim}_{\ell\mathsf{OTW}}$. In other words, $e_{D_{g^*+1,g}}$ is generated as

$$\mathsf{Sim}_{\ell\mathsf{OTW}}(\mathsf{crs}, D_{g^*}, g^*, D_{g^*+1,g^*}, \{\mathsf{lab}^{g^*+1}_{k,\mathsf{d}_{g^*+1,k}}\}_{k\in[\lambda]}).$$

  Computational indistinguishability between hybrids Hybrid_4 and Hybrid_5 follows directly from the sender privacy for writes of the laconic OT scheme.
- $\underline{\mathsf{Hybrid}_7}$: In this hybrid, we reverse the changes made earlier with respect to sampling of $\{\mathsf{lab}^{g^*}_{k,b}\}$. Specifically, we sample all values $\{\mathsf{lab}^{g^*}_{k,b}\}$ and not just $\{\mathsf{lab}^{g^*}_{k,\mathsf{d}_{g^*,k}}\}_k$. Additionally, this is done in SimC rather than SimIn.

  Note that this change is syntactic and the hybrids Hybrid_6 to Hybrid_7 are identical. Finally, observe that hybrid Hybrid_7 is the same as $\mathsf{Hybrid}_{\mathsf{conf}'}$.

This completes the proof of the lemma. We additionally note that the above sequence of hybrids is reversible. This implies the inverse rule A.

5.3.2 Implementing Rule B

Lemma 4 (Rule B). *Let* conf *and* conf' *be two valid circuit configurations satisfying the constraints of rule B, then assuming the security of somewhere equivocal encryption, garbling scheme for circuits and updatable laconic oblivious transfer, we have that* $\mathsf{Hybrid}_{\mathsf{conf}} \overset{c}{\approx} \mathsf{Hybrid}_{\mathsf{conf}'}$.

Proof. We prove this via a hybrid argument starting with $\mathsf{Hybrid}_{\mathsf{conf}'}$ and ending in hybrid $\mathsf{Hybrid}_{\mathsf{conf}}$. We follow this ordering of the hybrids as this keeps the proof very close to the proof of Lemma 3. In particular, we start with hybrid $\mathsf{Hybrid}_{\mathsf{conf}'}$ and make changes to get a hybrid Hybrid_7 which are almost the same as the hybrids in Lemma 3. One key difference is that we set the value $D_{g^*+1,g}$ differently than how it is set in the execution of SimIn in Fig. 5. Specifically, we set D_{g^*+1,g^*} as r_{g^*} rather than $E_{g^*} \oplus r_{g^*}$. Note that this also corresponding changes the value of d_{g^*+1} which is the hash of D_{g^*+1}. Finally we provide argument that hybrids Hybrid_7 and $\mathsf{Hybrid}_{\mathsf{conf}}$ are identical.

- $\underline{\mathsf{Hybrid}_{\mathsf{conf}'}}$: This is our starting hybrid and is distributed as $\mathsf{Hybrid}_{(I',\{(g,\mathsf{mode}'_g)\}_{g\in[n+1,N]})}$.
- $\underline{\mathsf{Hybrid}_1}$: In this hybrid, we change the configuration to $(I,\{(g,\mathsf{mode}'_g)\}_{g\in[n+1,N]})$. This hybrid is distributed as $\mathsf{Hybrid}_{(I,\{(g,\mathsf{mode}'_g)\}_{g\in[n+1,N]})}$. Computational indistinguishability between hybrid $\mathsf{Hybrid}_{\mathsf{conf}'}$ and Hybrid_1 reduces directly to the security of somewhere equivocal encryption.
- $\underline{\mathsf{Hybrid}_2}$: By conditions of Rule B we have that $\mathsf{mode}_{g^*-1} = \mathsf{Gray}$. Thus, we have that $g^*-1 \in I'$. Therefore, we note that the input labels $\{\mathsf{lab}^{g^*}_{k,b}\}$ are not used in SimC but only in SimIn where it is used to generate $\widetilde{\mathsf{SC}}_{g^*-1}$ and $\widetilde{\mathsf{SC}}_{g^*}$. In this hybrid, we postpone the sampling of $\{\mathsf{lab}^{g^*}_{k,b}\}$ and the generation of $\widetilde{\mathsf{SC}}_{g^*}$ from SimC to SimIn.

  The change in hybrid Hybrid_2 from Hybrid_1 is just syntactic and the distributions are identical.
- $\underline{\mathsf{Hybrid}_3}$: In this hybrid, we change the sampling of $\{\mathsf{lab}^{g^*}_{k,b}\}$ and the generation of $\widetilde{\mathsf{SC}}_{g^*}$. Specifically, we do not sample the entire set of labels $\{\mathsf{lab}^{g^*}_{k,b}\}$ but a subset namely $\{\mathsf{lab}^{g^*}_{k,\mathsf{d}_{g^*,k}}\}_k$ and we generate $\widetilde{\mathsf{SC}}_{g^*}$ from the simulated distribution. (Note that since g^*-1 is also in Gray mode. Thus we have that $\widetilde{\mathsf{SC}}_{g^*-1}$ is also simulated and only $\{\mathsf{lab}^{g^*}_{k,\mathsf{d}_{g^*,k}}\}_k$ are needed for its generation.) More formally, we generate

$$\widetilde{\mathsf{SC}}_{g^*} \leftarrow \mathsf{Sim}_{ckt}(1^\lambda, 1^{|\mathsf{SC}|}, \mathsf{out}, \{\mathsf{lab}^{g^*}_{k,\mathsf{d}_{g^*,k}}\}_{k\in[\lambda]})$$

  where $\mathsf{out} \leftarrow \mathsf{SC}[\mathsf{crs}, (0,0,r_g), (i,j), \{\mathsf{lab}^{g^*+1}_{k,b}\}, 1](\mathsf{d}_{g^*})$.

  The only change in hybrid Hybrid_3 from Hybrid_2 is in the generation of the garbled circuit $\widetilde{\mathsf{SC}}_{g^*}$ and the security follows directly from the selective security of the garbling scheme.
- $\underline{\mathsf{Hybrid}_4}$: In this hybrid, we set change how the output value out hardwired in $\widetilde{\mathsf{SC}}_{g^*}$ is generated. Recall that in hybrid Hybrid_3 this value is generated by first computing f_0 and f_1 as in Fig. 4 and then generating out as $\mathsf{Send}\left(\mathsf{crs}, \mathsf{d}, i, f_0, f_1\right)$. In this hybrid, we just generate $f_{D_{g^*,i}}$ and use the laconic OT simulator to generate out. More formally, out is generated as

$$\mathsf{out} \leftarrow \mathsf{Sim}_{\ell\mathsf{OT}}\left(\mathsf{crs}, D_{g^*}, i, f_{D_{g^*,i}}\right).$$

  Computational indistinguishability between hybrids Hybrid_3 and Hybrid_4 follows directly from the sender privacy of the laconic OT scheme.
- $\underline{\mathsf{Hybrid}_5}$: In this hybrid, we change how the how the value $f_{D_{g^*,i}}$ is generated in hybrid Hybrid_4. Recall from Fig. 4 that $f_{D_{g^*,i}}$ is set as $\mathsf{Send}\left(\mathsf{crs}, \mathsf{d}, j, e_{r_{g^*}}, e_{r_{g^*}}\right)$. We change the distribution of $f_{D_{g^*,i}}$ to $\mathsf{Sim}_{\ell\mathsf{OT}}\left(\mathsf{crs}, D_g, j, e_{r_{g^*}}\right)$, where $e_{r_{g^*}}$ is sampled as in Fig. 4.

Computational indistinguishability between hybrids Hybrid_3 and Hybrid_4 follows directly from the sender privacy of the laconic OT scheme. The argument is analogous to the argument of indistinguishability between Hybrid_2 and Hybrid_3.

- $\underline{\mathsf{Hybrid}_6}$: In this hybrid, we change how $e_{r_{g^*}}$ is generated. More specifically, we generate it using the simulator $\mathsf{Sim}_{\ell\mathsf{OTW}}$. In other words, $e_{r_{g^*}}$ is generated as

$$\mathsf{Sim}_{\ell\mathsf{OTW}}(\mathsf{crs}, D_{g^*}, g^*, r_{g^*}, \{\mathsf{lab}^{g^*+1}_{k,\mathsf{d}_{g^*+1,k}}\}_{k\in[\lambda]}).$$

Computational indistinguishability between hybrids Hybrid_4 and Hybrid_5 follows directly from the sender privacy for writes of the laconic OT scheme.
- $\underline{\mathsf{Hybrid}_7}$: In this hybrid, we reverse the changes made earlier with respect to sampling of $\{\mathsf{lab}^{g^*}_{k,b}\}$. Specifically, we sample all values $\{\mathsf{lab}^{g^*}_{k,b}\}$ and not just $\{\mathsf{lab}^{g^*}_{k,\mathsf{d}_{g^*,k}}\}_k$. Additionally, this is done in SimC rather than SimIn.

  Note that this change is syntactic and the hybrids Hybrid_6 to Hybrid_7 are identical.
- $\underline{\mathsf{Hybrid}_{\mathsf{conf}}}$: This corresponds to the hybrid $\mathsf{Hybrid}_{(I,\{(g,\mathsf{mode}_g)\}_{g\in[n+1,N]}}$. Observe that the only difference between Hybrid_7 and $\mathsf{Hybrid}_{\mathsf{conf}}$ is how D_{g^*+1,g^*} is set. Namely, in Hybrid_7 this value is set to be r_{g^*} while in $\mathsf{Hybrid}_{\mathsf{conf}}$ this value is set as $r_{g^*} \oplus \mathsf{NAND}(D_{g^*,i} \oplus r_i, D_{g^*,j} \oplus r_j,)$. We argue that the distributions Hybrid_7 and $\mathsf{Hybrid}_{\mathsf{conf}}$ are identical. Two cases arise:
  - $g^* \le N-m$: In this case, note that since r_{g^*} is not anywhere else we have that the distribution r_{g^*} and $r_{g^*} \oplus \mathsf{NAND}(D_{g^*,i} \oplus r_i, D_{g^*,j} \oplus r_j)$ are both uniform and identical.
  - $g^* > N-m$: In this case, we have that $r_{g^*} = y_{g^*-N+m} \oplus r'_{g^*}$ which is again identical to the distribution of r_{g^*} in $\mathsf{Hybrid}_{\mathsf{conf}}$.

This completes the proof of the lemma.

A Completing Proofs of Lemma 3

Claim. Assuming the security of somewhere equivocal encryption scheme, we have $\mathsf{Hybrid}_{\mathsf{conf}} \stackrel{c}{\approx} \mathsf{Hybrid}_1$.

Proof. We give a reduction to the security of somewhere equivocal encryption.

Generating the Garbled Circuit. To generate the garbled circuit $\widetilde{C}$:
1. Execute the steps 1, 2, 3 as described in Fig. 5.
2. Interact with the external challenger by giving $\{\widetilde{\mathsf{SC}}_g\}_{g\notin I}$ as the challenge messages, I as the challenge subset and g^* as the challenge index. Obtain c as the challenge ciphertext.
3. Output (crs, c) as the garbled circuit $\widetilde{C}$ and $(r, \mathsf{st}_1, \{\mathsf{lab}^g_{k,b}\}_{k,b,g})$ as the state st.

Generating the Garbled Input. To generate the garbled input $\widetilde{x}$:
1. Execute the steps 1, 2 as described in Fig. 5.

2. Interact with the external challenger by providing $\{\widetilde{\mathsf{SC}}_g\}_{g \in I}$ as the remaining messages and obtain key'.
3. Execute the rest of the steps as described in Fig. 5 and output $\widetilde{x}$ using the key key' obtained from the external challenger.

Notice that if the reduction is playing in the experiment $\mathsf{Exp}^{\mathsf{simenc}}_{\mathcal{B},\Pi}(1^\lambda, 0)$ then the distribution of inputs to the adversary is identical to $\mathsf{Hybrid}_{\mathsf{conf}}$. Else, it is distributed identically to Hybrid_1. Thus, the reduction breaks the security of somewhere equivocal encryption.

Claim. Assuming the selective security of garbling scheme for circuits, $\mathsf{Hybrid}_2 \overset{c}{\approx} \mathsf{Hybrid}_3$

Proof. We give a reduction to the security of garbling scheme for circuits.

Generating the Garbled Circuit. For each $g \in [n+1, N+1] \setminus \{g^*\}$, $k \in [\lambda]$ and $b \in \{0,1\}$ sample $\mathsf{lab}^g_{k,b} \leftarrow \{0,1\}^\lambda$. Generate the garbled circuit $\widetilde{C}$ as in Hybrid_2. Note that there is a "hole" in position g^* in the outer encryption layer and hence $\{\mathsf{lab}^{g^*}_{k,b}\}$ is not needed in the generation of $\widetilde{C}$. Also, recall that by assumption the gate $g^* - 1$ is Gray or $g^* = n+1$.

Generating the Garbled Input: To generate the garbled input $\widetilde{x}$ do:

1. Interact with the garbled circuits challenger and give $\mathsf{SC}[\mathsf{crs}, (r_i, r_j, r_g), (i,j), \{\mathsf{lab}^{g^*+1}_{k,b}\}, 0]$ (where the description of g^* is (i,j)) as the challenge circuit and d_{g^*} as the challenge input. Obtain $\widetilde{\mathsf{SC}}_{g^*}$ and $\{\mathsf{lab}^{g^*}_{k,\mathsf{d}_{g^*,k}}\}_{k \in [\lambda]}$.
2. For each $g \in I' \setminus \{g^*\}$, generate $\widetilde{\mathsf{SC}}_g$ as described in Fig. 5. Note that for generating $\widetilde{\mathsf{SC}}_{g^*-1}$ it is sufficient to only know $\{\mathsf{lab}^{g^*}_{k,\mathsf{d}_{g^*,k}}\}_{k \in [\lambda]}$.
3. Execute the rest of the steps as described in Fig. 5 and output $\widetilde{x}$.

Notice that if the garbling $\widetilde{\mathsf{SC}}_{g^*}$ is generated using the honest procedure then the inputs to the adversary are distributed identically to Hybrid_2. Else, they are distributed identically to Hybrid_3. Thus, the reduction breaks the selective security of garbling scheme for circuits which is a contradiction.

Claim. Assuming the sender privacy of updatable laconic oblivious transfer, $\mathsf{Hybrid}_3 \overset{c}{\approx} \mathsf{Hybrid}_4$

Proof. We show that $\mathsf{Hybrid}_3 \overset{c}{\approx} \mathsf{Hybrid}_4$ by giving a reduction to the sender privacy of updatable laconic oblivious transfer.

Generating the Garbled Circuit. Obtain crs from the external challenger and generate the garbled circuit $\widetilde{C}$ as in Hybrid_3.

Generating the Garbled Input:

1. To generate $\widetilde{\mathsf{SC}}_{g^*}$ with the description of g^* equal to (i,j):
   (a) Compute e_0, e_1, f_0, f_1 as described in Fig. 4.

   (b) Interact with the laconic OT challenger by giving D_{g^*} as the challenge database, i as the challenge locations and give f_0, f_1 as the challenge messages. Obtain the challenge ciphertext out.
   (c) Compute $\widetilde{\mathsf{SC}}_{g^*} \leftarrow \mathsf{Sim}_{ckt}(1^\lambda, 1^{|\mathsf{SC}|}, \mathsf{out}, \{\mathsf{lab}^{g^*}_{k,\mathsf{d}_{g^*,k}}\}_{k\in[\lambda]})$
2. Generate $\widetilde{\mathsf{SC}}_g$ for all $g \in I' \setminus \{g^*\}$ as in Hybrid_3.
3. Execute the rest of the steps as described in Fig. 5 to generate the garbled input $\widetilde{x}$.

Note that if out is generated using the honest procedure then the distribution of inputs to the adversary is identical to Hybrid_3. Else, it is identical to Hybrid_4. Thus, the above reduction breaks the sender privacy of updatable laconic oblivious transfer.

Claim. Assuming the sender privacy for writes of updatable laconic oblivious transfer, $\mathsf{Hybrid}_5 \overset{c}{\approx} \mathsf{Hybrid}_6$.

Proof. We give a reduction to the sender privacy for writes of updatable laconic oblivious transfer.

Generating the Garbled Circuit. Obtain crs from the external challenger and generate the garbled circuit $\widetilde{C}$ as in Hybrid_5.

Generating the Garbled Input:

1. To generate $\widetilde{\mathsf{SC}}_{g^*}$ with the description of g^* equal to (i, j):
   (a) Interact with the laconic OT challenger by giving D_{g^*} as the challenge database, g^* as the challenge location, D_{g^*+1,g^*} as the challenge bit and $\{\mathsf{lab}^{g^*+1}_{k,b}\}$ as the sequence of challenge messages. It obtains $e_{D_{g^*+1,g^*}}$ as the challenge ciphertext.
   (b) Generate $f_{D_{g^*+1,g^*}}$ and out as in Hybrid_5.
   (c) Compute $\widetilde{\mathsf{SC}}_{g^*} \leftarrow \mathsf{Sim}_{ckt}(1^\lambda, 1^{|\mathsf{SC}|}, \mathsf{out}, \{\mathsf{lab}^{g^*}_{k,\mathsf{d}_{g^*,k}}\}_{k\in[\lambda]})$
2. Generate $\widetilde{\mathsf{SC}}_g$ for all $g \in I' \setminus \{g^*\}$ as in Hybrid_5.
3. Execute the rest of the steps as described in Fig. 5 to generate the garbled input $\widetilde{x}$.

Note that if $e_{D_{g^*+1,g^*}}$ is generated using the honest procedure then the distribution of inputs to $\mathcal{A}$ is identical to Hybrid_5. Else, it is identical to Hybrid_6. Thus, the reduction breaks the sender privacy for writes of updatable laconic oblivious transfer.

References

[ABSV15] Ananth, P., Brakerski, Z., Segev, G., Vaikuntanathan, V.: From selective to adaptive security in functional encryption. In: Gennaro, R., Robshaw, M. (eds.) CRYPTO 2015. LNCS, vol. 9216, pp. 657–677. Springer, Heidelberg (2015). https://doi.org/10.1007/978-3-662-48000-7_32

[AF90] Abadi, M., Feigenbaum, J.: Secure circuit evaluation. J. Cryptol. **2**(1), 1–12 (1990)

[AIK04] Applebaum, B., Ishai, Y., Kushilevitz, E.: Cryptography in NC^0. In: 45th Annual Symposium on Foundations of Computer Science, Rome, Italy, 17–19 October 2004, pp. 166–175. IEEE Computer Society Press (2004)

[AIK05] Applebaum, B., Ishai, Y., Kushilevitz, E.: Computationally private randomizing polynomials and their applications. In: Proceedings of the 20th Annual IEEE Conference on Computational Complexity, CCC 2005, Washington, DC, USA, pp. 260–274. IEEE Computer Society (2005)

[AIK10] Applebaum, B., Ishai, Y., Kushilevitz, E.: From secrecy to soundness: efficient verification via secure computation. In: Abramsky, S., Gavoille, C., Kirchner, C., Meyer auf der Heide, F., Spirakis, P.G. (eds.) ICALP 2010. LNCS, vol. 6198, pp. 152–163. Springer, Heidelberg (2010). https://doi.org/10.1007/978-3-642-14165-2_14

[AIKW13] Applebaum, B., Ishai, Y., Kushilevitz, E., Waters, B.: Encoding functions with constant online rate or how to compress garbled circuits keys. In: Canetti, R., Garay, J.A. (eds.) CRYPTO 2013. LNCS, vol. 8043, pp. 166–184. Springer, Heidelberg (2013). https://doi.org/10.1007/978-3-642-40084-1_10

[App14] Applebaum, B.: Bootstrapping obfuscators via fast pseudorandom functions. In: Sarkar, P., Iwata, T. (eds.) ASIACRYPT 2014. LNCS, vol. 8874, pp. 162–172. Springer, Heidelberg (2014). https://doi.org/10.1007/978-3-662-45608-8_9

[AS16] Ananth, P., Sahai, A.: Functional encryption for turing machines. In: Kushilevitz, E., Malkin, T. (eds.) TCC 2016. LNCS, vol. 9562, pp. 125–153. Springer, Heidelberg (2016). https://doi.org/10.1007/978-3-662-49096-9_6

[Ben89] Bennett, C.H.: Time/space trade-offs for reversible computation. SIAM J. Comput. **18**(4), 766–776 (1989)

[BGG+14] Boneh, D., Gentry, C., Gorbunov, S., Halevi, S., Nikolaenko, V., Segev, G., Vaikuntanathan, V., Vinayagamurthy, D.: Fully key-homomorphic encryption, arithmetic circuit abe and compact garbled circuits. In: Nguyen, P.Q., Oswald, E. (eds.) EUROCRYPT 2014. LNCS, vol. 8441, pp. 533–556. Springer, Heidelberg (2014). https://doi.org/10.1007/978-3-642-55220-5_30

[BHR12a] Bellare, M., Hoang, V.T., Rogaway, P.: Adaptively secure garbling with applications to one-time programs and secure outsourcing. In: Wang, X., Sako, K. (eds.) ASIACRYPT 2012. LNCS, vol. 7658, pp. 134–153. Springer, Heidelberg (2012). https://doi.org/10.1007/978-3-642-34961-4_10

[BHR12b] Bellare, M., Hoang, V.T., Rogaway, P.: Foundations of garbled circuits. In: Yu, T., Danezis, G., Gligor, V.D. (eds.) 19th Conference on Computer and Communications Security, ACM CCS 12, Raleigh, NC, USA, 16–18 October 2012, pp. 784–796. ACM Press (2012)

[BLSV18] Brakerski, Z., Lombardi, A., Segev, G., Vaikuntanathan, V.: Anonymous IBE, leakage resilience and circular security from new assumptions. In: Nielsen, J.B., Rijmen, V. (eds.) Eurocrypt 2018. LNCS, vol. 10821, pp. 535–564. Springer, Cham (2018). https://eprint.iacr.org/2017/967

[BMR90] Beaver, D., Micali, S., Rogaway, P.: The round complexity of secure protocols (extended abstract). In: 22nd Annual ACM Symposium on Theory of Computing, Baltimore, MD, USA, 14–16 May 1990, pp. 503–513. ACM Press (1990)

[CDG+17] Cho, C., Döttling, N., Garg, S., Gupta, D., Miao, P., Polychroniadou, A.: Laconic oblivious transfer and its applications. In: Katz, J., Shacham, H. (eds.) CRYPTO 2017. LNCS, vol. 10402, pp. 33–65. Springer, Cham (2017). https://doi.org/10.1007/978-3-319-63715-0_2

[DG17] Döttling, N., Garg, S.: Identity-based encryption from the diffie-hellman assumption. In: Katz, J., Shacham, H. (eds.) CRYPTO 2017. LNCS, vol. 10401, pp. 537–569. Springer, Cham (2017). https://doi.org/10.1007/978-3-319-63688-7_18

[DGHM18] Döttling, N., Garg, S., Hajiabadi, M., Masny, D.: New constructions of identity-based and key-dependent message secure encryption schemes. In: PKC (2018, to appear). https://eprint.iacr.org/2017/978

[FNO15] Frederiksen, T.K., Nielsen, J.B., Orlandi, C.: Privacy-free garbled circuits with applications to efficient zero-knowledge. In: Oswald, E., Fischlin, M. (eds.) EUROCRYPT 2015. LNCS, vol. 9057, pp. 191–219. Springer, Heidelberg (2015). https://doi.org/10.1007/978-3-662-46803-6_7

[GGP10] Gennaro, R., Gentry, C., Parno, B.: Non-interactive verifiable computing: outsourcing computation to untrusted workers. In: Rabin, T. (ed.) CRYPTO 2010. LNCS, vol. 6223, pp. 465–482. Springer, Heidelberg (2010). https://doi.org/10.1007/978-3-642-14623-7_25

[GHL+14] Gentry, C., Halevi, S., Lu, S., Ostrovsky, R., Raykova, M., Wichs, D.: Garbled RAM revisited. In: Nguyen, P.Q., Oswald, E. (eds.) EUROCRYPT 2014. LNCS, vol. 8441, pp. 405–422. Springer, Heidelberg (2014). https://doi.org/10.1007/978-3-642-55220-5_23

[GKP+13] Goldwasser, S., Kalai, Y.T., Popa, R.A., Vaikuntanathan, V., Zeldovich, N.: Reusable garbled circuits and succinct functional encryption. In: Boneh, D., Roughgarden, T., Feigenbaum, J. (eds.) 45th Annual ACM Symposium on Theory of Computing, Palo Alto, CA, USA, 1–4 June 2013, pp. 555–564. ACM Press (2013)

[GKR08] Goldwasser, S., Kalai, Y.T., Rothblum, G.N.: One-time programs. In: Wagner, D. (ed.) CRYPTO 2008. LNCS, vol. 5157, pp. 39–56. Springer, Heidelberg (2008). https://doi.org/10.1007/978-3-540-85174-5_3

[GLO15] Garg, S., Lu, S., Ostrovsky, R.: Black-box garbled RAM. In: Guruswami, V. (ed.) 56th Annual Symposium on Foundations of Computer Science, Berkeley, CA, USA, 17–20 October 2015, pp. 210–229. IEEE Computer Society Press (2015)

[GLOS15] Garg, S., Lu, S., Ostrovsky, R., Scafuro, A.: Garbled RAM from one-way functions. In: Servedio, R.A., Rubinfeld, R. (eds.) 47th Annual ACM Symposium on Theory of Computing, Portland, OR, USA, 14–17 June 2015, pp. 449–458. ACM Press (2015)

[GPSZ17] Garg, S., Pandey, O., Srinivasan, A., Zhandry, M.: Breaking the sub-exponential barrier in obfustopia. In: Coron, J.-S., Nielsen, J.B. (eds.) EUROCRYPT 2017. LNCS, vol. 10212, pp. 156–181. Springer, Cham (2017). https://doi.org/10.1007/978-3-319-56617-7_6

[GS16] Garg, S., Srinivasan, A.: Single-key to multi-key functional encryption with polynomial loss. In: Hirt, M., Smith, A. (eds.) TCC 2016. LNCS, vol. 9986, pp. 419–442. Springer, Heidelberg (2016). https://doi.org/10.1007/978-3-662-53644-5_16

[GVW12] Gorbunov, S., Vaikuntanathan, V., Wee, H.: Functional encryption with bounded collusions via multi-party computation. In: Safavi-Naini, R., Canetti, R. (eds.) CRYPTO 2012. LNCS, vol. 7417, pp. 162–179. Springer, Heidelberg (2012). https://doi.org/10.1007/978-3-642-32009-5_11

[HJO+16] Hemenway, B., Jafargholi, Z., Ostrovsky, R., Scafuro, A., Wichs, D.: Adaptively secure garbled circuits from one-way functions. In: Robshaw, M., Katz, J. (eds.) CRYPTO 2016. LNCS, vol. 9816, pp. 149–178. Springer, Heidelberg (2016). https://doi.org/10.1007/978-3-662-53015-3_6

[JKK+17] Jafargholi, Z., Kamath, C., Klein, K., Komargodski, I., Pietrzak, K., Wichs, D.: Be adaptive, avoid overcommitting. In: Katz, J., Shacham, H. (eds.) CRYPTO 2017. LNCS, vol. 10401, pp. 133–163. Springer, Cham (2017). https://doi.org/10.1007/978-3-319-63688-7_5

[JKO13] Jawurek, M., Kerschbaum, F., Orlandi, C.: Zero-knowledge using garbled circuits: how to prove non-algebraic statements efficiently. In: Sadeghi, A.-R., Gligor, V.D., Yung, M. (eds.) 20th Conference on Computer and Communications Security, ACM CCS 13, Berlin, Germany, 4–8 November 2013, pp. 955–966. ACM Press (2013)

[JSW17] Jafargholi, Z., Scafuro, A., Wichs, D.: Adaptively indistinguishable garbled circuits. In: Kalai, Y., Reyzin, L. (eds.) TCC 2017. LNCS, vol. 10678, pp. 40–71. Springer, Cham (2017). https://doi.org/10.1007/978-3-319-70503-3_2

[JW16] Jafargholi, Z., Wichs, D.: Adaptive security of Yao's garbled circuits. In: Hirt, M., Smith, A. (eds.) TCC 2016. LNCS, vol. 9985, pp. 433–458. Springer, Heidelberg (2016). https://doi.org/10.1007/978-3-662-53641-4_17

[LM16] Li, B., Micciancio, D.: Compactness vs collusion resistance in functional encryption. In: Hirt, M., Smith, A. (eds.) TCC 2016. LNCS, vol. 9986, pp. 443–468. Springer, Heidelberg (2016). https://doi.org/10.1007/978-3-662-53644-5_17

[LO13] Lu, S., Ostrovsky, R.: How to garble RAM programs? In: Johansson, T., Nguyen, P.Q. (eds.) EUROCRYPT 2013. LNCS, vol. 7881, pp. 719–734. Springer, Heidelberg (2013). https://doi.org/10.1007/978-3-642-38348-9_42

[LP09] Lindell, Y., Pinkas, B.: A proof of security of Yao's protocol for two-party computation. J. Cryptol. **22**(2), 161–188 (2009)

[LV16] Lin, H., Vaikuntanathan, V.: Indistinguishability obfuscation from DDH-like assumptions on constant-degree graded encodings. In: Dinur, I. (ed.) 57th Annual Symposium on Foundations of Computer Science, New Brunswick, NJ, USA, 9–11 October 2016, pp. 11–20. IEEE Computer Society Press (2016)

[MNPS04] Malkhi, D., Nisan, N., Pinkas, B., Sella, Y.: Fairplay - secure two-party computation system. In: Proceedings of the 13th USENIX Security Symposium, San Diego, CA, USA, 9–13 August 2004, pp. 287–302 (2004)

[Nie02] Nielsen, J.B.: Separating random oracle proofs from complexity theoretic proofs: the non-committing encryption case. In: Yung, M. (ed.) CRYPTO 2002. LNCS, vol. 2442, pp. 111–126. Springer, Heidelberg (2002). https://doi.org/10.1007/3-540-45708-9_8

[PTC76] Paul, W.J., Tarjan, R.E., Celoni, J.R.: Space bounds for a game on graphs. Mathe. Syst. Theor. **10**(1), 239–251 (1976)

[SS10] Sahai, A., Seyalioglu, H.: Worry-free encryption: functional encryption with public keys. In: Al-Shaer, E., Keromytis, A.D., Shmatikov, V. (eds.) 17th Conference on Computer and Communications Security, ACM CCS 10, Chicago, Illinois, USA, 4–8 October 2010, pp. 463–472. ACM Press (2010)

[Yao82] Yao, A.C.-C.: Protocols for secure computations (extended abstract). In: 23rd Annual Symposium on Foundations of Computer Science, Chicago, Illinois, 3–5 November 1982, pp. 160–164. IEEE Computer Society Press (1982)

[Yao86] Yao, A.C.-C.: How to generate and exchange secrets (extended abstract). In: 27th Annual Symposium on Foundations of Computer Science, Toronto, Ontario, Canada, 27–29 October 1986, pp. 162–167. IEEE Computer Society Press (1986)

A New Approach to Black-Box Concurrent Secure Computation

Sanjam Garg[1(✉)], Susumu Kiyoshima[2], and Omkant Pandey[3]

[1] University of California, Berkeley, USA
sanjamg@berkeley.edu
[2] NTT Secure Platform Laboratories, Tokyo, Japan
kiyoshima.susumu@lab.ntt.co.jp
[3] Stony Brook University, Stony Brook, USA
omkant@cs.stonybrook.edu

Abstract. We consider the task of constructing concurrently composable protocols for general secure computation by making only *black-box* use of underlying cryptographic primitives. Existing approaches for this task first construct a black-box version of CCA-secure commitments which provide a strong form of concurrent security to the committed value(s). This strong form of security is then crucially used to construct higher level protocols such as concurrently secure OT/coin-tossing (and eventually all functionalities).

This work explores a fresh approach. We first aim to construct a concurrently-secure OT protocol whose concurrent security is proven directly using concurrent simulation techniques; in particular, it does not rely on the usual "non-polynomial oracles" of CCA-secure commitments. The notion of concurrent security we target is *super-polynomial simulation* (SPS). We show that such an OT protocol can be constructed from *polynomial* hardness assumptions in a *black-box* manner, and within a *constant* number of rounds. In fact, we only require the existence of (constant round) semi-honest OT and standard collision-resistant hash functions.

Next, we show that such an OT protocol is sufficient to obtain SPS-secure (concurrent) multiparty computation (MPC) for general functionalities. This transformation does not require any additional assumptions; it also maintains the black-box nature as well as the constant round feature of the original OT protocol. Prior to our work, the only known black-box construction of constant-round concurrently composable MPC required stronger assumptions; namely, verifiable perfectly binding homomorphic commitment schemes and PKE with oblivious public-key generation.

1 Introduction

Secure multiparty computation (MPC) protocols enable mutually distrustful parties to compute a joint functionality on their private inputs without compromising the correctness of the outputs and the privacy of their inputs. They

J. B. Nielsen and V. Rijmen (Eds.): EUROCRYPT 2018, LNCS 10821, pp. 566–599, 2018.
https://doi.org/10.1007/978-3-319-78375-8_19

have been studied in both two-party and multi-party cases. General constructions of such protocols for computing any functionality even when a majority of players are adversarial have been long known [16,49]. In this work, we are interested in MPC protocols that only make a black-box use of cryptographic primitives and maintain security in a concurrent environment with many simultaneous executions.

Black-box constructions. General purpose MPC protocols are often *non-black-box* in nature. They use the code of the underlying primitives at some stage of the computation, e.g., an NP reduction for general zero-knowledge proofs. Such non-black-use of the primitives is generally undesirable since not only it is computationally expensive, it also renders the protocol useless in situations where such code is not available (e.g., primitives based on hardware-tokens). One therefore seeks *black-box* constructions of such protocols which use the underlying primitives only in black-box way (i.e., only through their input/output interfaces).

Recently, a number of works have studied black-box constructions of general MPC protocols. Ishai et al. [26] presented the first black-box construction of general purpose MPC based on enhanced trapdoor permutations or homomorphic public-key encryption schemes. Combined with the subsequent work of Haitner [22] on black-box OT, this gives a black-box construction of general MPC based assuming only semi-honest OT [23]. Subsequently, Wee [48] reduced the round complexity of these constructions to $O(\log^* n)$, and Goyal [17] to only constant rounds. In the two-party setting, black-box construction were obtained by Pass and Wee [44] in constant-rounds and Ostrovsky et al. [39] in optimal 5-rounds.

Concurrent security. The standard notion of security for MPC, also called *stand-alone security* considers only a single execution of this protocol. While this is sufficient for many applications, other situations (such as protocol executions over the Internet) require stronger notions of security. Such a more demanding setting, where there may be many protocols executions at the same time, is called the *concurrent* setting. Unfortunately, it is known that stand-alone security does not necessarily imply security in the concurrent setting [12].

Secure computation in the concurrent setting is quite challenging to define. Canetti [4] proposed the notion of *universally composable* (UC) security where protocols maintain their strong simulation based security guarantees even in the presence of other arbitrary protocols. Achieving such strong notion of UC-security turned out to be impossible in the plain model [4,5]. Moreover, Lindell [34,35] proved that even in the special case where only instantiations of the *same* protocol are allowed, standard notion of polynomial-time simulation is impossible to achieve. (This is "self composition" and corresponds to the setting we are interested in.)

These strong negative results motivated the study of alternative notions of security; of these, most relevant to us are super-polynomial simulation (SPS) [41], angel-based security [6,46], and security with shielded oracles [3].

SPS Security. SPS security is similar to UC security except that the simulator is allowed to run in super-polynomial time. It guarantees that whatever an adversary can do in the real world can also be done in the ideal world *in super-polynomial time*. While SPS-security is a weaker guarantee, it is still meaningful security for many functionalities, and allows concurrent self-composition in the plain model. (In what follows, by SPS security we mean SPS-security under concurrent self-composition.) Prabhakaran and Sahai [46] provided the initial positive result for SPS security. Although, these early results [2,33,36,46] relied on non-standard/sub-exponential assumptions, Canetti, Lin and Pass achieved this notion under standard polynomial-time assumptions [6] in polynomially many rounds. Soon after, Garg et al. [14] presented a *constant round* construction. The works of [6,36,46] actually get angel-based security, discussed below.

Angel-Based Security. Angel-based UC security is the same as UC security except that the environment/adversary and the simulator have access to an additional entity—an *angel*—that allows some judicious use of super-polynomial resources. Angel-based UC security, though weaker than UC security, is meaningful for many settings and implies SPS security. Furthermore, like UC security, it also guarantees composability. As noted above, the works in [6,36,46] achieve angel-based security, though only [6] relies on standard polynomial hardness. Subsequently, Goyal et al. [20] presented a $\widetilde{O}(\log n)$ round construction under the same assumptions.

Black-box constructions of angel-based secure computation were first presented by Lin and Pass [31] assuming the existence of semi-honest OT, in $O(\max(n^{\epsilon}, R_{\mathrm{OT}}))$ rounds, where $\epsilon > 0$ is an arbitrary constant and R_{OT} is the round complexity of the underlying OT protocol. (Hence, if the underlying OT protocol has only constant round, the round complexity is $O(n^{\epsilon})$.) Subsequently, Kiyoshima [28] provided a $\widetilde{O}(\log^2 n)$-round construction under the same assumption.

Security with Shielded Oracles. Security with shielded oracles, proposed very recently by Broadnax et al. [3], is similar to angel-based security where the environment and the simulator have access to an additional entity—a *shielded oracle*—that can perform some super-polynomial computation. However, unlike angel-based security, the results of super-polynomial time computation are "shielded away" from the simulator, in the sense that the shielded oracle directly interacts with the ideal functionality; the simulator cannot observe their communication. This notion lies strictly between SPS and angel-based security, and guarantees composability. A constant-round MPC protocol satisfying this notion were also presented in [3]; one of their protocol is black-box and relies on standard polynomial hardness. More specifically, it requires (verifiable perfectly binding) homomorphic commitment schemes and PKE with oblivious public-key generation.

State-of-the-art. All of the constructions of concurrently-secure MPC protocols we have discussed so far, rely on first constructing non-malleable commitment schemes with strong concurrent or UC-security properties; in particular (robust)

"CCA-secure commitments" or "coin tossing" or UC-secure commitments. These schemes are then used to build higher level protocols such as OT and general secure computation. However, the concurrent security of these higher level protocols is proven indirectly, by relying on the strong concurrent security of the CCA-secure commitments. While this approach leads to (better) angel-based security, it is quite expensive in terms of rounds, requiring $\tilde{O}(\log^2 n)$ in [28] for *black-box* constructions. The work of Broadnax et al. [3] significantly improves this situation by relaxing the angel-based security requirement to SPS with shielded-oracles, and obtains a constant round construction. However, it still needs stronger assumptions (see above) and represents the only approach so far for obtaining constant round black-box constructions. In contrast, much of the results that make non-black-box use of the primitives, can rely on the minimal assumption of semi-honest OT. The approach of Broadnax et al. [3] is still based on first constructing a sufficiently strong commitment scheme with UC properties and using it to obtain OT and general functionalities. It is highly desirable to find new approaches to construct such protocols which have the potential to rely on minimal assumptions in constant rounds.

1.1 Our Contribution

In this work, we seek new approaches for constructing concurrently-secure black-box MPC protocols which can lead to qualitative improvements over existing constructions, such as minimal underlying assumptions, a constant number of rounds, and so on. Towards this goal, we deviate from the existing approaches which focus on incorporating both *concurrent security* and *non-malleability* into a single primitive such as (CCA-secure) commitment schemes or coin-tossing. Instead, we take a different approach and focus on incorporating concurrent security into the oblivious transfer functionality. We present a black-box OT protocol satisfying the SPS notion of concurrent-security. We achieve this by using concurrent simulation techniques and non-malleable commitments in a somewhat modular way where (roughly speaking) the former is primarily used for trapdoor extraction/simulation and the latter for independence of committed values. The protocol has constant rounds and relies only on the existence of (constant round) semi-honest OT and standard collision-resistant hash functions (CRHFs).

Having obtained concurrent security for OT, we proceed to construct SPS-secure MPC protocols for all functionalities. Our method does not require any additional assumptions, and maintains the black-box and constant round properties of the original OT protocol. Consequently, we obtain SPS-secure constant-round black-box MPC under much weaker assumptions than [3]. On the flip side, our work achieves a weaker security notion than [3].

Theorem 1 (Informal). *Assume the existence of constant-round semi-honest oblivious transfer protocols and collision-resistant hash functions. Then, there exists a constant-round black-box construction of concurrently secure MPC protocol that achieve SPS security.*

The formal statement is given as Theorem 3 in Sect. 5.

1.2 Other Related Work

Other than the works mentioned above, there are several works that study SPS security/angel-based security. For SPS-security, Pass et al. [43] present a constant-round non-black-box construction of MPC from constant-round semi-honest OT. Dachman-Soled et al. and Venkitasubramaniam [11,47] present a non-black-box construction that satisfies adaptive security. And very recently, Badrinarayanan et al. [1] present a non-black-box 3-round construction assuming sub-exponential hardness assumptions. For angel-based security, Kiyoshima et al. [29] present a constant-round *black-box* construction albeit under a sub-exponential hardness assumption, and Hazay and Venkitasubramaniam [25] present a non-black-box construction that achieves adaptive security.

We have not discussed several works that focus on other notions of concurrent security such as input-indistinguishable computation, bounded concurrent composition, and multiple ideal-query model [18,37,42].

Black-box constructions have been extensively explored for several other primitives such as non-malleable/CCA-secure encryption, non-malleable commitments, zero-knowledge proofs and so on (e.g., [9,10,19,21,40,45]). For concurrent OT, Garay and MacKenzie [13] presented a protocols for independent inputs under the DDH assumption, and Garg et al. [15] showed the impossibility of this task for general input distributions.

2 Overview of Our Techniques

We obtain our MPC protocol in two steps. First, we construct a constant-round black-box construction of a SPS-secure concurrent OT protocol. Second, we compose this OT protocol with an existing constant-round OT-hybrid UC-secure MPC protocol. We elaborate on each step below.

We remark that we consider concurrent security in the interchangeable-roles setting. So, in the case of OT, the adversary can participate a session as the sender while concurrently participating another session as the receiver.

2.1 Constant-Round Black-Box Concurrent OT

Our starting point is the (super-constant-round) black-box concurrent OT protocol of Lin and Pass [31], which is secure under angel-based security and makes only black-box use of semi-honest OT protocols. Our approach is to modify their protocol so that it has only constant number of rounds (while degrading security from angel-based security to SPS security).

Let us first recall the OT protocol of [31]. At a high level, it uses a semi-honest OT protocol in the black-box way in a similar manner to the stand-alone black-box OT of Haitner et al. [23] does. Specifically, the OT protocol of [31] proceeds roughly as follows.

1. First, the sender S and the receiver R execute many instances of a semi-honest OT protocol in parallel, where in each instance S and R use the inputs and the

randomness that are generated by a coin-tossing protocol. (S and R execute two instances of coin tossing for each instance of OT; the sender obtains random coin in the first coin tossing and the receiver obtains random coin in the second coin tossing.)

2. Next, S and R do a simple trick called *OT combiner*, which allows them to execute an OT with their real inputs securely when most of the OT instances in the previous step are correctly executed. To check that most of the OT instances in the previous step were indeed correctly executed, S and R do the well-known *cut-and-choose* trick, where S (resp., R) chooses a constant fraction of the OT instances randomly and R (resp., S) reveals the input and randomness that it used in those instances so that S (resp., R) can verity whether R executed those instances correctly.

(Actually, the underlying OT protocol is required to be secure against malicious senders, but we ignore this requirement in this overview.)

The OT protocol of [31] has more than constant number of rounds because it uses *CCA-secure commitment schemes* [6,7] in the coin-tossing part of the protocol and existing constructions of CCA-secure commitment schemes have more than constant number of rounds under standard assumptions.[1] Key observations by the authors of [31] are that CCA-secure commitment schemes can be used to obtain a "concurrently secure" coin tossing protocol,[2] and that their OT protocol is concurrently secure when its coin-tossing part is concurrently secure.

To obtain a constant-round protocol, we need to remove the CCA-secure commitments from the protocol of [31]. A naive approach is to simply replace the CCA-secure commitments with *(concurrent) non-malleable commitments*, which provide weaker security than CCA-secure ones but are known to have a constant-round black-box instantiation under the existence of one-way functions [19]. However, this approach does not work because, as mentioned by Lin and Pass [31], non-malleable commitment schemes only lead to "parallel secure" coin tossing protocols[3] and the parallel security of the coin tossing protocol is insufficient for proving the concurrent security of the OT protocol of [31].

At a high level, we remove the CCA-secure commitments from the protocol of [31] as follows. Our starting idea is to prove the concurrent security of the OT protocol of [31] without relying on the concurrent security of the coin tossing part (and therefore without using CCA-secure commitments there). To prove the concurrent security in this way, we modify the protocol of [31] so that it uses non-malleable commitments in a similar manner to the constant-round (non-black-box) SPS-secure concurrent MPC protocol of Garg et al. [14] does. Informally

---

[1] Roughly speaking, CCA-secure commitment schemes guarantee that the hiding property holds even when the adversary has access to the *committed-value oracle*, which computes the committed value of a given commitment by brute force.

[2] Concretely, the resultant coin-tossing protocol satisfies *simulation soundness*, which guarantees that any concurrent man-in-the-middle adversary cannot bias the outcome of a coin-tossing when it concurrently participates in simulated coin tossings.

[3] Very roughly speaking, this is because non-malleability allows the man-in-the adversary to obtain replies from the committed-value oracle only in parallel.

speaking, the protocol of Garg et al. [14] uses non-malleable commitments when each party commits to a witness for the fact that the "trapdoor statement" is false, where the trapdoor statement is a statement about the transcript and it is guaranteed that any adversary cannot "cheat" in the protocol when the trapdoor statement is false. With this use of non-malleable commitments, the concurrent security of the protocol of Garg et al. [14] is proven in two steps:

1. First, it is shown that in the real experiment (where an adversary interacts with honest parties in multiple sessions of the protocol concurrently), the non-malleable commitment from the adversary in each session is indeed a commitment of a valid witness for the fact that the trapdoor statement is false. (This is guaranteed by a zero-knowledge proof in the protocol).
2. Second, it is shown that if the non-malleable commitment in a session is indeed a commitment of a valid witness (which implies that the trapdoor statement is false in that session, which in turn implies that the adversary cannot "cheat" in that session), it is possible to switch the honest parties in that session with the simulator in an indistinguishable way, and furthermore this switch does not affect the non-malleable commitments in the other sessions (i.e., their committed values remain to be valid witnesses). (The latter is guaranteed by non-malleability of the non-malleable commitments[4].)
3. Now, the concurrent security follows from the above two since the honest parties can be switched to the simulator in all the sessions by repeatedly using what is shown in the second part.

Following this approach by Garg et al. [14], we first identify the trapdoor statement of the OT protocol of Lin and Pass [31] and then add non-malleable commitments to their protocol in such a way that the trapdoor statement is false whenever the committed values of the non-malleable commitments satisfy a specific condition. With this modification, we can prove the concurrent security of the OT protocol of [31] without relying on the concurrent security of coin tossing by following the approach of [14] outlined above.

Remark 1. It is not straightforward to use the approach of Garg et al. [14] in the OT protocol of Lin and Pass [31] since its trapdoor statement does not have a simple witness for the fact that the statement is false. Because of this difficulty, we do not use non-malleable commitments to commit to a witness; rather, we use them in such a way that there exists a condition on the committed values of the non-malleable commitments such that the trapdoor statement is false as long as this condition holds. For details, see Sect. 4 (in particular, Definitions 5 and 6 and Claims 3 and 4).

[4] Actually, *non-malleability w.r.t. other protocols* [30] is also required, where non-malleability w.r.t. a protocol Π guarantees non-malleability against man-in-the-middle adversaries that participates in the non-malleable commitment in the right interaction and Π in the left interaction.

2.2 Composition of OT with OT-Hybrid MPC

We next compose our OT protocol with a OT-hybrid UC-secure MPC protocol (i.e., replace each invocation of the ideal OT functionality in the latter with an execution of the former), thereby obtaining a MPC protocol in the plain model. A problem is that the security of the resultant MPC protocol cannot be derived trivially from those of the components since SPS security does not guarantee composability. Hence, we prove the security by analyzing the MPC protocol directly. In essence, what we do is to observe that the security proof for our OT protocol (which consists of a hybrid argument from the real world to the ideal world) still works even after the OT protocol is composed with a OT-hybrid MPC protocol, and in particular we observe that the condition on the committed values of the non-malleable commitments (which is mentioned in Sect. 2.1) remains to hold in each session even after switching the OT-hybrid MPC protocol in any session to simulation. Fortunately, this can be observed easily thanks to the non-malleability of the non-malleable commitments, so we can prove the concurrent security of our MPC protocol under SPS security easily.

3 Preliminaries

We denote the security parameter by n. We assume familiarity with basic cryptographic protocols (e.g., commitment schemes and oblivious transfer protocols).

3.1 Non-malleable Commitment Schemes

We recall the definition of non-malleable commitment schemes from [30]. Let $\langle C, R\rangle$ be a tag-based commitment scheme (i.e., a commitment scheme that takes a n-bit string—a *tag*—as an additional input). For any man-in-the-middle adversary $\mathcal{M}$, consider the following experiment. On input security parameter 1^n and auxiliary input $z \in \{0,1\}^*$, $\mathcal{M}$ participates in one left and one right interactions simultaneously. In the left interaction, $\mathcal{M}$ interacts with the committer of $\langle C, R\rangle$ and receives a commitment to value v using identity $\mathsf{id} \in \{0,1\}^n$ of its choice. In the right interaction, $\mathcal{M}$ interacts with the receiver of $\langle C, R\rangle$ and gives a commitment using identity $\widetilde{\mathsf{id}}$ of its choice. Let $\widetilde{v}$ be the value that $\mathcal{M}$ commits to on the right. If the right commitment is invalid or undefined, $\widetilde{v}$ is defined to be $\bot$. If $\mathsf{id} = \widetilde{\mathsf{id}}$, value $\widetilde{v}$ is also defined to be $\bot$. Let $\mathsf{mim}(\langle C, R\rangle, \mathcal{M}, v, z)$ be a random variable representing $\widetilde{v}$ and the view of $\mathcal{M}$ in the above experiment.

Definition 1. *A commitment scheme $\langle C, R\rangle$ is **non-malleable** if for any* PPT *adversary $\mathcal{M}$, the following are computationally indistinguishable.*

- $\{\mathsf{mim}(\langle C, R\rangle, \mathcal{M}, v, z)\}_{n\in\mathbb{N}, v\in\{0,1\}^n, v'\in\{0,1\}^n, z\in\{0,1\}^*}$
- $\{\mathsf{mim}(\langle C, R\rangle, \mathcal{M}, v', z)\}_{n\in\mathbb{N}, v\in\{0,1\}^n, v'\in\{0,1\}^n, z\in\{0,1\}^*}$

The above definition can be generalized naturally so that the adversary gives multiple commitments *in parallel* in the right interaction. The non-malleability in this generalized setting is called *parallel non-malleability*. (It is known that this "one-many" definition implies the "many-many" one, where the adversary receives multiple commitments in the left session [32].)

Robust Non-malleability. We next recall the definition of k-robust non-malleability (a.k.a. non-malleability w.r.t. k-round protocols) [30]. Consider a man-in-the-middle adversary $\mathcal{M}$ that participates in one left interaction—communicating with a machine B—and one right interaction—communicating with a receiver a commitment scheme $\langle C, R\rangle$. As in the standard definition of non-malleability, $\mathcal{M}$ can choose the identity in the right interaction. We denote by $\mathsf{mim}^{B,\mathcal{M}}_{\langle C,R\rangle}(y, z)$ the random variable consisting of the view of $\mathcal{M}(z)$ in a man-in-the-middle execution when communicating with $B(y)$ on the left and an honest receiver on the right, combined with the value $\mathcal{M}(z)$ commits to on the right. Intuitively, $\langle C, R\rangle$ is non-malleable w.r.t. B if $\mathsf{mim}^{B,\mathcal{M}}_{\langle C,R\rangle}(y_1, z)$ and $\mathsf{mim}^{B,\mathcal{M}}_{\langle C,R\rangle}(y_2, z)$ are indistinguishable whenever interactions with $B(y_1)$ and $B(y_2)$ are indistinguishable.

Definition 2. *Let $\langle C, R\rangle$ be a commitment scheme and B be a* PPT *ITM. We say that a commitment scheme $\langle C, R\rangle$ is* ***non-malleable w.r.t.*** *B if the following holds: For every two sequences $\{y^1_n\}_{n\in\mathbb{N}}$ and $\{y^2_n\}_{n\in\mathbb{N}}$ such that $y^1_n, y^2_n \in \{0,1\}^n$, if it holds that for any* PPT *ITM $\mathcal{A}$,*

$$\left\{\langle B(y^1_n), \mathcal{A}(z)\rangle(1^n)\right\}_{n\in\mathbb{N},z\in\{0,1\}^*} \approx \left\{\langle B(y^2_n), \mathcal{A}(z)\rangle(1^n)\right\}_{n\in\mathbb{N},z\in\{0,1\}^*},$$

it also holds that for any PPT *man-in-the-middle adversary $\mathcal{M}$,*

$$\left\{\mathsf{mim}^{B,\mathcal{M}}_{\langle C,R\rangle}(y_1, z)\right\}_{n\in\mathbb{N},z\in\{0,1\}^*} \approx \left\{\mathsf{mim}^{B,\mathcal{M}}_{\langle C,R\rangle}(y_2, z)\right\}_{n\in\mathbb{N},z\in\{0,1\}^*}.$$

$\langle C, R\rangle$ is *k-robust* if $\langle C, R\rangle$ is non-malleable w.r.t. any machine that interacts with the adversary in k rounds. We define parallel k-robustness naturally.

Black-Box Instantiation. There exists a constant-round black-box construction of a parallel (actually, concurrent) non-malleable commitment scheme based on one-way functions [19]. In the full version of this paper, we show that any parallel non-malleable commitment can be transformed into a parallel k-robust non-malleable one in the black-box way by using collision-resistant hash functions (more precisely, by using statistically hiding commitment schemes, which can be constructed from collision-resistant hash functions). If k is constant, the round complexity increases only by a constant factor in this transformation.

3.2 UC Security and Its SPS Variant

We next recall the definition of UC security [4] and its SPS variant [2,14,46]. A part of the text below is taken from [14]. (We assume that the readers are familiar with the UC framework. A brief overview can be found in, e.g., [7].)

UC Security. Recall that in the UC framework, the model for protocol execution consists of the environment $\mathcal{Z}$, the adversary $\mathcal{A}$, and the parties running protocol π. In this paper, we consider static adversaries and assume the existence of authenticated communication channels. Let $\mathrm{EXEC}_{\pi,\mathcal{A},\mathcal{Z}}(n, z)$ denote a random variable for the output of $\mathcal{Z}$ on security parameter $n \in \mathbb{N}$ and input $z \in \{0,1\}^*$ with a uniformly chosen random tape. Let $\mathrm{EXEC}_{\pi,\mathcal{A},\mathcal{Z}}$ denote the ensemble $\{\mathrm{EXEC}_{\pi,\mathcal{A},\mathcal{Z}}(n,z)\}_{n\in\mathbb{N},z\in\{0,1\}^*}$.

The security of a protocol π is defined using the *ideal protocol*. In the ideal protocol, all the parties simply hand their inputs to the *ideal functionality* $\mathcal{F}$, which carries out the desired task securely and gives outputs to the parties; the parties then forward these outputs to $\mathcal{Z}$. The adversary $\mathcal{S}im$ in the execution of the ideal protocol is often called the *simulator*. Let $\pi(\mathcal{F})$ denote the ideal protocol for functionality $\mathcal{F}$.

We say that a protocol π *emulates* protocol ϕ if for any adversary $\mathcal{A}$ there exists an adversary $\mathcal{S}im$ such that no environment $\mathcal{Z}$, on any input, can tell whether it is interacting with $\mathcal{A}$ and parties running π or it is interacting with $\mathcal{S}im$ and parties running ϕ. We say that π *securely realizes* an ideal functionality $\mathcal{F}$ if it emulates the ideal protocol $\Pi(\mathcal{F})$.

UC Security with Super-Polynomial Simulation. UC-SPS security is a relaxed notion of UC security where the simulator is given access to super-polynomial computational resources.

Definition 3. *Let π and ϕ be protocols. We say that π* UC-SPS-emulates *ϕ if for any adversary $\mathcal{A}$ there exists a super-polynomial-time adversary $\mathcal{S}im$ such that for any environment $\mathcal{Z}$ that obeys the rules of interaction for UC security, we have $\mathrm{EXEC}_{\phi,\mathcal{S}im,\mathcal{Z}} \approx \mathrm{EXEC}_{\pi,\mathcal{A},\mathcal{Z}}$.*

Definition 4. *Let $\mathcal{F}$ be an ideal functionality and let π be a protocol. We say that π* UC-SPS-realizes *$\mathcal{F}$ if π UC-SPS-emulates the ideal process $\Pi(\mathcal{F})$.*

The multi-session extension of an ideal functionality. When showing concurrent security of a protocol π under SPS security, we need to construct a simulator in a setting where parties execute π concurrently. To consider the simulator in this setting, we use a *multi-session extension* of an ideal functionality [8]. Roughly speaking, the multi-session extension $\hat{\mathcal{F}}$ of an ideal functionality $\mathcal{F}$ is a functionally that internally runs multiple copies of $\mathcal{F}$.

4 Our SPS Concurrent OT Protocol

In this section, we prove the following theorem.

Theorem 2. *Assume the existence of constant-round semi-honest oblivious transfer protocols and collision-resistant hash functions. Let $\mathcal{F}_{\mathrm{OT}}$ be the ideal oblivious transfer functionality (Fig. 1) and $\hat{\mathcal{F}}_{\mathrm{OT}}$ be its multi-session extension. Then, there exists a constant-round protocol that UC-SPS realizes $\hat{\mathcal{F}}_{\mathrm{OT}}$, and it uses the underlying primitives in the black-box way.*

The ideal OT functionality $\mathcal{F}_{OT}$ interacts with a sender S and a receiver R.

- Upon receiving a message $(\mathsf{sid}, \mathsf{sender}, v_0, v_1)$ from S, where each $v_i \in \{0,1\}^n$, store (v_0, v_1).
- Upon receiving a message $(\mathsf{sid}, \mathsf{receiver}, u)$ from R, where $u \in \{0,1\}$, check if a $(\mathsf{sid}, \mathsf{sender}, \ldots)$ message was previously sent. If yes, send (sid, v_u) to R and (sid) to the adversary $\mathcal{S}im$ and halt. If not, send nothing to R.

Fig. 1. The oblivious transfer functionality $\mathcal{F}_{OT}$.

4.1 Protocol Description

In our protocol, we use the following building blocks.

- A two-round statistically binding commitment Com and a four-round statistically binding extractable commitment ExtCom, both of which can be constructed from one-way functions in the black-box way [24,38,44].
- A $O(1)$-round OT protocol mS-OT that is secure against malicious senders and semi-honest receivers.[5] As shown in [23], such a OT protocol can be obtained from any semi-honest one in the black-box way.
- A $O(1)$-round parallel non-malleable commitment NMCom that is parallel k-robust for sufficiently large constant k. (Concretely, we require that k is larger than the round complexity of the above three building blocks.) As remarked in Sect. 3.1, we show in the full version of this paper that such a non-malleable commitment scheme can be constructed from collision-resistant hash functions in the black-box way.

Our OT protocol Π_{OT} is described below. As explained in Sect. 2.1, (1) our protocol is based on the OT protocol of Lin and Pass [31], which roughly consists of coin-tossing, semi-honest OT, OT combiner, and cut-and-choose, and (2) our protocol additionally uses non-malleable commitments, which will be used in the security proof to argue that the adversary cannot make the "trapdoor statement" true even in the concurrent setting. Below, we give intuitive explanations in italic.

Inputs: The input to the sender S is $v_0, v_1 \in \{0,1\}^n$. The input to the receiver R is $u \in \{0,1\}$.

Stage 1: (Preprocess for cut-and-choose)

1. S commits to a random subset $\Gamma_S \subset [11n]$ of size n by using Com.
2. R commits to a random subset $\Gamma_R \subset [11n]$ of size n by using Com.

COMMENT: *As in the OT protocol of Lin and Pass [31], the subsets that will be used in the cut-and-choose stages are committed in advance to prevent selective opening attacks.*

[5] We only requires mS-OT to be secure under a game-based definition (which is preserved under parallel composition). For details, see the full version of this paper.

Stage 2:

1. **(Coin tossing for S)** S commits to random strings $\boldsymbol{a}^S = (a_1^S, \ldots, a_{11n}^S)$ by using Com; let $d_1^S, \ldots, d_{11n}^S$ be the decommitments. R then sends random strings $\boldsymbol{b}^S = (b_1^S, \ldots, b_{11n}^S)$ to S. S then defines $\boldsymbol{r}^S = (r_1^S, \ldots, r_{11n}^S)$ by $r_i^S \stackrel{\text{def}}{=} a_i^S \oplus b_i^S$ for each $i \in [11n]$ and parses r_i^S as $s_{i,0} \,\|\, s_{i,1} \,\|\, \tau_i^S$ for each $i \in [11n]$.
2. **(Coin tossing for R)** R commits to random strings $\boldsymbol{a}^R = (a_1^R, \ldots, a_{11n}^R)$ by using Com; let $d_1^R, \ldots, d_{11n}^R$ be the decommitments. S then sends random strings $\boldsymbol{b}^R = (b_1^R, \ldots, b_{11n}^R)$ to R. R then defines $\boldsymbol{r}^R = (r_1^R, \ldots, r_{11n}^R)$ by $r_i^R \stackrel{\text{def}}{=} a_i^R \oplus b_i^R$ for each $i \in [11n]$ and parses r_i^R as $c_i \,\|\, \tau_i^R$ for each $i \in [11n]$.

Stage 3: (mS-OTs with random inputs)

S and R execute $11n$ instances of $\mathsf{mS\text{-}OT}$ in parallel. In the i-th instance, S uses $(s_{i,0}, s_{i,1})$ as the input and τ_i^S as the randomness, and R uses c_i as the input and τ_i^R as the randomness, where $\{s_{i,0}, s_{i,1}, \tau_i^S\}_i$ and $\{c_i, \tau_i^R\}_i$ are the random coins that were obtained in Stage 2. The output to R is denoted by $\widetilde{s}_1, \ldots, \widetilde{s}_{11n}$, which are supposed to be equal to $s_{1,c_1}, \ldots, s_{11n,c_{11n}}$.

Stage 4: (NMCom and ExtCom for checking honesty of R)

1. R commits to $(a_1^R, d_1^R), \ldots (a_{11n}^R, d_{11n}^R)$ using NMCom. Let $e_1^R, \ldots, e_{11n}^R$ be the decommitments.
2. R commits to $(a_1^R, d_1^R, e_1^R), \ldots (a_{11n}^R, d_{11n}^R, e_{11n}^R)$ using ExtCom.

COMMENT: *Roughly, the commitments in this stage, along with the cut-and-choose in the next stage, will be used in the security proof to argue that even cheating R must behave honestly in most instances of* $\mathsf{mS\text{-}OT}$ *in Stage 3. A key point is that given the values that are committed to in* NMCom *or* ExtCom *in this stage, one can obtain the random coins that R obtained in Stage 2 and thus can check whether R behaved honestly in Stage 3.*

Stage 5: (Cut-and-choose against R)

1. S reveals Γ_S by decommitting the Com commitment in Stage 1-1.
2. For every $i \in \Gamma_S$, R reveals (a_i^R, d_i^R, e_i^R) by decommitting the i-th ExtCom commitment in Stage 4.
3. For every $i \in \Gamma_S$, S checks the following.
   - $((a_i^R, d_i^R), e_i^R)$ is a valid decommitment of the i-th NMCom commitment in Stage 4.
   - (a_i^R, d_i^R) is a valid decommitment of the i-th Com commitment in Stage 2-2.
   - R executed the i-th mS-OT in Stage 3 honestly using $c_i \,\|\, \tau_i^R$, which is obtained from $r_i^R = a_i^R \oplus b_i^R$ as specified by the protocol.

COMMENT: *In other words, for each index that it randomly selected in Stage 1, S checks whether R behaved honestly in Stages 3 and 4 on that index.*

Stage 6: (OT combiner) Let $\Delta := [11n] \setminus \Gamma_S$.

1. R sends $\alpha_i := u \oplus c_i$ to S for every $i \in \Delta$.
2. S computes a $(6n+1)$-out-of-$10n$ secret sharing of v_0, denoted by $\boldsymbol{\rho}_0 = (\rho_{0,i})_{i \in \Delta}$, and computes a $(6n+1)$-out-of-$10n$ secret sharing of v_1, denoted by $\boldsymbol{\rho}_1 = (\rho_{1,i})_{i \in \Delta}$. Then, S sends $\beta_{b,i} := \rho_{b,i} \oplus s_{i,b \oplus \alpha_i}$ to R for every $i \in \Delta$, $b \in \{0,1\}$.
3. R computes $\widetilde{\rho}_i := \beta_{u,i} \oplus \widetilde{s}_i$ for every $i \in \Delta$. Let $\widetilde{\boldsymbol{\rho}} := (\widetilde{\rho}_i)_{i \in \Delta}$.

COMMENT: *In this stage, S and R execute OT with their true inputs by using the outputs of* mS-OT *in Stage 3. Roughly speaking, this stage is secure as long as most instances of* mS-OT *in Stage 3 are correctly executed.*

Stage 7: (NMCom and ExtCom for checking honesty of S)

1. S commits to $(a_1^S, d_1^S), \ldots (a_{11n}^S, d_{11n}^S)$ using NMCom. Let $e_1^S, \ldots, e_{11n}^S$ be the decommitments.
2. R commits to $(a_1^S, d_1^S, e_1^S), \ldots (a_{11n}^S, d_{11n}^S, e_{11n}^S)$ using ExtCom.

Stage 8: (Cut-and-choose against S)

1. R reveals Γ_R by decommitting the Com commitment in Stage 1-2.
2. For every $i \in \Gamma_R$, S reveals (a_i^S, d_i^S, e_i^S) by decommitting the i-th ExtCom commitment in Stage 7.
3. For every $i \in \Gamma_R$, R checks the following.
   - $((a_i^S, d_i^S), e_i^S)$ is a valid decommitment of the i-th NMCom commitment in Stage 7.
   - (a_i^S, d_i^S) is a valid decommitment of the i-th Com commitment in Stage 2-1.
   - S executed the i-th mS-OT in Stage 3 honestly using $s_{i,0} \parallel s_{i,1} \parallel \tau_i^S$, which is obtained from $r_i^S = a_i^S \oplus b_i^S$ as specified by the protocol.

Output: R outputs $\mathsf{Value}(\widetilde{\boldsymbol{\rho}}, \Gamma_R \cap \Delta)$, where $\mathsf{Value}(\cdot,\cdot)$ is the function that is defined in Fig. 2.

COMMENT: *As in the OT protocol of Lin and Pass [31], a carefully designed reconstruction procedure* $\mathsf{Value}(\cdot,\cdot)$ *is used here so that the simulator can extract correct implicit inputs from cheating S by obtaining sharing that is sufficiently "close" to* $\widetilde{\boldsymbol{\rho}}$.

> *Reconstruction procedure* $\mathsf{Value}(\cdot,\cdot)$: For a sharing $\boldsymbol{s} = (s_i)_{i\in\Delta}$ and a set $\Theta \subset \Delta$, the output of $\mathsf{Value}(\boldsymbol{s},\Theta)$ is computed as follows. If $\boldsymbol{s}$ is 0.9-close to a valid codeword $\boldsymbol{w} = (w_i)_{i\in\Delta}$ that satisfies $s_i = w_i$ for every $i \in \Theta$, then $\mathsf{Value}(\boldsymbol{s},\Theta)$ is the value decoded from $\boldsymbol{w}$; otherwise, $\mathsf{Value}(\boldsymbol{s},\Theta) = \bot$.

Fig. 2. The function $\mathsf{Value}(\cdot,\cdot)$.

4.2 Simulator $\mathcal{S}im_{\mathrm{OT}}$

To prove the security of Π_{OT}, we consider the following simulator $\mathcal{S}im_{\mathrm{OT}}$. Recall that our goal is to prove that Π_{OT} US-SPS realizes the multi-session extension of $\mathcal{F}_{OT}$. We therefore consider a simulator that works against adversaries that participate in multiple sessions of Π_{OT} both as senders and as receivers.

Let $\mathcal{Z}$ be any environment, $\mathcal{A}$ be any adversary that participates in multiple sessions of Π_{OT}. Our simulator $\mathcal{S}im_{\mathrm{OT}}$ internally invokes $\mathcal{A}$ and simulates each of the sessions for $\mathcal{A}$ as follows.

When R is corrupted: In a session where the receiver R is corrupted, $\mathcal{S}im_{\mathrm{OT}}$ simulates the sender S for $\mathcal{A}$ by extracting the implicit input $u^* \in \{0,1\}$ from $\mathcal{A}$. During the simulation, $\mathcal{S}im_{\mathrm{OT}}$ extracts the committed subset and random coins in Stages 1 and 2 by brute force; the former extraction is needed to execute most instances mS-OT in Stage 3 with true randomness (which is crucial to use their security in the analysis), and the latter extraction is needed to infer what information $\mathcal{A}$ obtained in the mS-OT instances in Stage 3 (which is crucial to extract the implicit input $u^* \in \{0,1\}$ from $\mathcal{A}$).

Concretely, $\mathcal{S}im_{\mathrm{OT}}$ simulates S for $\mathcal{A}$ as follows. From Stage 1 to Stage 5, $\mathcal{S}im_{\mathrm{OT}}$ interacts with $\mathcal{A}$ in the same way as an honest S except for the following.

- From the Com commitments from $\mathcal{A}$ in Stages 1 and 2, the committed subset Γ_R and the committed strings $\boldsymbol{a}^R = (a_1^R, \ldots, a_{11n}^R)$ are extracted by brute force.
  $\mathcal{S}im_{\mathrm{OT}}$ then defines $\boldsymbol{r}^R = (r_1^R, \ldots, r_{11n}^R)$ by $r_i^R \stackrel{\mathrm{def}}{=} a_i^R \oplus b_i^R$ for each $i \in [11n]$ and parses r_i^R as $c_i \,\|\, \tau_i^R$ for each $i \in [11n]$. (Notice that $\boldsymbol{r}^R$ is the outcome of the coin-tossing that $\mathcal{A}$ must have obtained.)
- In Stage 3, the i-th mS-OT is executed with a random input and true randomness rather than with $(s_{i,0}, s_{i,1})$ and τ_i^S for every $i \notin \Gamma_R$.

In Stage 6, $\mathcal{S}im_{\mathrm{OT}}$ interacts with $\mathcal{A}$ as follows.

1. Receive $\{\alpha_i\}_{i\in\Delta}$ from $\mathcal{A}$ in Stage 6-1.
2. Determine the implicit input u^* of $\mathcal{A}$ as follows. Let I_0, I_1 be the sets such that for $b \in \{0,1\}$ and $i \in \Delta$, we have $i \in I_b$ if and only if:
   - $i \in \Gamma_R$, or
   - $\mathcal{A}$ did not execute the i-th mS-OT in Stage 3 honestly using $c_i \,\|\, \tau_i^R$ as the input and randomness, or

- $c_i = b \oplus \alpha_i$, and $\mathcal{A}$ executed the i-th mS-OT in Stage 3 honestly using $c_i \,\|\, \tau_i^R$ as the input and randomness.

Then, define u^* by $u^* \stackrel{\text{def}}{=} 0$ if $|I_0| \geq 6n+1$ and $u^* \stackrel{\text{def}}{=} 1$ otherwise. (Roughly, $|I_b|$ is the number of strings that $\mathcal{A}$ can obtain out of $\{s_{i,b\oplus\alpha_i}\}_{i\in\Delta}$ by requiring S to reveal them in Stage 8, by cheating in mS-OT, or by executing mS-OT honestly with input $b \oplus \alpha_i$. We remind the readers that $\{s_{i,b\oplus\alpha_i}\}_{i\in\Delta}$ are the strings that are used to mask $\boldsymbol{\rho}_b = (\rho_{b,i})_{i\in\Delta}$ in Stage 6.)

3. Send u^* to the ideal functionality and obtains v^*.
4. Subsequently, interact with $\mathcal{A}$ in the same way as an honest S assuming that the inputs to S are $v_{u^*} = v^*$ and random v_{1-u^*}.

From Stage 7 to Stage 8, $\mathcal{S}im_{\mathrm{OT}}$ interacts with $\mathcal{A}$ in the same way as an honest S except that in Stage 7, an all-zero string is committed in the i-th NMCom rather than (a_i^S, d_i^S) for every $i \notin \Gamma_R$, and an all-zero string is committed in the i-th ExtCom rather than (a_i^S, d_i^S, e_i^S) for every $i \notin \Gamma_R$.

When S is corrupted: In a session where the sender S is corrupted, $\mathcal{S}im_{\mathrm{OT}}$ simulates the receiver R for $\mathcal{A}$ by extracting the implicit input v_0^*, v_1^* from $\mathcal{A}$. During the simulation, $\mathcal{S}im_{\mathrm{OT}}$ extracts the committed subset and random coins in Stages 1 and 2 by brute force; the former extraction is needed to execute most instances mS-OT in Stage 3 with true randomness (which is crucial to use their security in the analysis), and the latter extraction is needed to learn what input $\mathcal{A}$ used in the mS-OT instances in Stage 3 (which is crucial to extract the implicit input v_0^*, v_1^* from $\mathcal{A}$).

Concretely, $\mathcal{S}im_{\mathrm{OT}}$ simulates R for $\mathcal{A}$ as follows. $\mathcal{S}im_{\mathrm{OT}}$ interacts with $\mathcal{A}$ in the same way as an honest R in all the stages except for the following.

- From the Com commitment from $\mathcal{A}$ in Stage 1, the committed subset Γ_S is extracted by brute force.
- In Stage 3, the i-th mS-OT is executed with a random input and true randomness rather than with c_i and τ_i^R for every $i \notin \Gamma_S$.
- In Stage 4, an all-zero string is committed in the i-th NMCom rather than (a_i^S, d_i^S) for every $i \notin \Gamma_S$, and an all-zero string is committed in the i-th ExtCom rather than (a_i^S, d_i^S, e_i^S) for every $i \notin \Gamma_S$.
- In Stage 6, α_i is a random bit rather than $\alpha_i = u \oplus c_i$ for every $i \in \Delta$, and $\widetilde{\rho}_i$ is not computed for any $i \in \Delta$.

Then, $\mathcal{S}im_{\mathrm{OT}}$ determines the implicit inputs v_0^*, v_1^* of $\mathcal{A}$ as follows.

1. From the Com commitments from $\mathcal{A}$ in Stage 2, extract the committed strings $\boldsymbol{a}^S = (a_1^S, \ldots, a_{11n}^S)$ by brute force.
2. Define $\boldsymbol{r}^S = (r_1^S, \ldots, r_{11n}^S)$ by $r_i^S \stackrel{\text{def}}{=} a_i^S \oplus b_i^S$ for each $i \in [11n]$ and parse r_i^S as $s_{i,0} \,\|\, s_{i,1} \,\|\, \tau_i^S$ for each $i \in [11n]$. (Notice that $\boldsymbol{r}^S$ is the outcome of the coin-tossing that $\mathcal{A}$ must have obtained.)

3. Define $\rho_b^{\text{ext}} = (\rho_{b,i}^{\text{ext}})_{i\in\Delta}$ for each $b \in \{0,1\}$ as follows: $\rho_{b,i}^{\text{ext}} \stackrel{\text{def}}{=} \beta_{b,i} \oplus s_{i,b\oplus\alpha_i}$ if $\mathcal{A}$ executed the i-th mS-OT in stage 3 honestly using $s_{i,0} \parallel s_{i,1} \parallel \tau_i^S$, and $\rho_{b,i}^{\text{ext}} \stackrel{\text{def}}{=} \bot$ otherwise.
4. For each $b \in \{0,1\}$, define $v_b^* \stackrel{\text{def}}{=} \mathsf{Value}(\rho_b^{\text{ext}}, \Gamma_R \cap \Delta)$.

Then, $\mathcal{S}im_{\text{OT}}$ sends v_0^*, v_1^* to the ideal functionality.

4.3 Proof of Indistinguishability

We show the indistinguishability by using a hybrid argument. Before defining hybrid experiments, we define *special messages*, which we use in the definitions of the hybrid experiments. (Essentially, they are the messages on which the simulator applies brute-force extractions.)

- first special message is the Com commitment in Stage 1-1.
- second special message is the Com commitment in Stage 1-2.
- third special message is the Com commitments in Stage 2-1.
- fourth special message is the Com commitments in Stage 2-2.

Hybrid Experiments. Now, we define hybrid experiments. Let m denote an upper bound on the number of the sessions that $\mathcal{A}$ starts. Note that the number of special messages among m sessions can be bounded by $4m$. We order those $4m$ special messages by the order of their appearances; we use SM_k to denote the k-th special messages, and $s(k)$ to denote the session that SM_k belongs to.

We define hybrids H_0 and $H_{k:1}, \ldots, H_{k:7}$ ($k \in [4m]$) as follows. (For convenience, in what follows we occasionally denote H_0 as $H_{0:7}$.)

Remark 2 (Rough idea of the hybrids). In the sequence of the hybrid experiments, we gradually modify the read-world experiment to the ideal-world one. All the experiments (except for H_0) involve super-polynomial-time brute-force extraction, but we make sure that $H_{k:i}$ ($i \in [7]$) involves brute-force extraction only until SM_k, and it deviates from the previous hybrid only after SM_k. These properties help us prove the indistinguishability of each neighboring hybrids because we can think the results of brute-force extraction as non-uniform advice and use the non-uniform security of the underlying primitives to show the indistinguishability.[6] ◇

Hybrid H_0. H_0 is the same as the real experiment.

Hybrid $H_{k:1}$. $H_{k:1}$ is the same as $H_{k-1:7}$ except that in session $s(k)$, if S is corrupted and SM_k is first special message,

[6] We remark that, unlike Garg et al. [14] (who give a non-black-box constant-round SPS protocol), we cannot replace brute-force extraction with rewinding extraction for obtaining polynomial-time hybrids. This is because when considering black-box constructions, we cannot easily guarantee that brute-force extraction and rewinding one obtain the same value.

- the committed subset Γ_S is extracted by brute force in Stage 1-1,
- the value committed to in the i-th NMCom commitment in Stage 4 is switched to an all-zero string for every $i \notin \Gamma_S$, and
- the value committed to in the i-th ExtCom commitment in Stage 4 is switched to an all-zero string for every $i \notin \Gamma_S$.

Hybrid $H_{k:2}$. $H_{k:2}$ is the same as $H_{k:1}$ except that in session $s(k)$, if S is corrupted and SM_k is first special message, the i-th mS-OT in Stage 3 is executed with a random input and true randomness for every $i \notin \Gamma_S$.

Hybrid $H_{k:3}$. $H_{k:3}$ is the same as $H_{k:2}$ except that in session $s(k)$, if S is corrupted and SM_k is third special message, the following modifications are made.

1. The committed strings $\boldsymbol{a}^S = (a_1^S, \ldots, a_{11n}^S)$ are extracted by brute force in Stage 2-1. Define $\boldsymbol{r}^S = (r_1^S, \ldots, r_{11n}^S)$ by $r_i^S \stackrel{\text{def}}{=} a_i^S \oplus b_i^S$ for each $i \in [11n]$, and parse r_i^S as $s_{i,0} \| s_{i,1} \| \tau_i^S$ for each $i \in [11n]$. Define $\boldsymbol{\rho}_b^{\mathsf{ext}} = (\rho_{b,i}^{\mathsf{ext}})_{i \in \Delta}$ for each $b \in \{0,1\}$ as follows: $\rho_{b,i}^{\mathsf{ext}} \stackrel{\text{def}}{=} \beta_{b,i} \oplus s_{i,b \oplus \alpha_i}$ if $\mathcal{A}$ executed the i-th mS-OT in stage 3 honestly using $s_{i,0} \| s_{i,1} \| \tau_i^S$, and $\rho_{b,i}^{\mathsf{ext}} = \bot$ otherwise.
2. R outputs $\mathsf{Value}(\boldsymbol{\rho}_u^{\mathsf{ext}}, \Gamma_R \cap \Delta)$ rather than $\mathsf{Value}(\widetilde{\rho}, \Gamma_R \cap \Delta)$. (Recall that u is the real input to R.)

Hybrid $H_{k:4}$. $H_{k:4}$ is the same as $H_{k:3}$ except that in session $s(k)$, if S is corrupted and SM_k is third special message, α_i is a random bit rather than $\alpha_i = u \oplus c_i$ for every $i \in \Delta$ in Stage 6-1 and $\widetilde{\rho}_i$ is no longer computed for any $i \in \Delta$ in Stage 6-3.

Hybrid $H_{k:5}$. $H_{k:5}$ is the same as $H_{k:4}$ except that in session $s(k)$, if R is corrupted and SM_k is second special message,

- the committed subset Γ_R is extracted by brute force in Stage 1-2,
- the value committed in the i-th NMCom commitment in Stage 7 is switched to an all-zero string for every $i \notin \Gamma_R$, and
- the value committed in the i-th ExtCom commitment in Stage 7 is switched to an all-zero string for every $i \notin \Gamma_R$.

Hybrid $H_{k:6}$. $H_{k:6}$ is the same as $H_{k:5}$ except that in session $s(k)$, if R is corrupted and SM_k is second special message, the i-th mS-OT in Stage 3 is executed with a random input and true randomness for every $i \notin \Gamma_R$.

Hybrid $H_{k:7}$. $H_{k:7}$ is the same as $H_{k:6}$ except that in session $s(k)$, if R is corrupted and SM_k is fourth special message, the following modifications are made.

1. The committed strings $\boldsymbol{a}^R = (a_1^R, \ldots, a_{11n}^R)$ are extracted by brute force in Stage 2-2. Define $\boldsymbol{r}^R = (r_1^R, \ldots, r_{11n}^R)$ by $r_i^R \stackrel{\text{def}}{=} a_i^R \oplus b_i^R$ for each $i \in [11n]$, and parse r_i^R as $c_i \| \tau_i^R$ for each $i \in [11n]$. Define u^* as follows. Let I_0 and I_1 be the set such that for $b \in \{0,1\}$ and $i \in \Delta$, we have $i \in I_b$ if and only if:

- $i \in \Gamma_R$, or
- $\mathcal{A}$ did not execute the i-th mS-OT in Stage 3 honestly using $c_i \| \tau_i^R$ as the input and randomness, or
- $c_i = b \oplus \alpha_i$, and $\mathcal{A}$ executed the i-th mS-OT in Stage 3 honestly using $c_i \| \tau_i^R$ as the input and randomness.

Then, define u^* by $u^* \stackrel{\text{def}}{=} 0$ if $|I_0| \geq 6n + 1$ and $u^* \stackrel{\text{def}}{=} 1$ otherwise.

2. In Stage 6, $\boldsymbol{\rho}_{1-u^*}$ is a secret sharing of a random bit rather than that of v_{1-u^*}.

We remark that in $H_{4m:7}$, all the messages from the honest parties and their output are computed as in $\mathcal{S}im_{\mathrm{OT}}$.

Indistinguishability of Each Neighboring Hybrids. Below, we show that each neighboring hybrids are indistinguishable, and additionally show, for technical reasons, that an invariant condition holds in each session of every hybrid.

First, we define the invariant condition.

Definition 5 (Invariant Condition (when R is corrupted)). *For any session in which R is corrupted, we say that the invariant condition holds in that session if the following holds when the cut-and-choose in Stage 5 is accepted.*

1. Let $(\hat{a}_1^R, \hat{d}_1^R), \dots (\hat{a}_{11n}^R, \hat{d}_{11n}^R)$ be the values that are committed in NMCom *in Stage 4. Let $I_{\text{bad}} \subset [11n]$ be the set such that $i \in I_{\text{bad}}$ if and only if*
   (a) $(\hat{a}_i^R, \hat{d}_i^R)$ is not a valid decommitment of the i-th Com *commitment in Stage 2-2, or*
   (b) R does not execute the i-th mS-OT in Stage 3 honestly using $\hat{c}_i \| \hat{\tau}_i^R$ as the input and randomness, where $\hat{c}_i \| \hat{\tau}_i^R$ is obtained from $\hat{r}_i^R = \hat{a}_i^R \oplus b_i^R$.

   Then, it holds that $|I_{\text{bad}}| < n$.

Remark 3. Roughly speaking, this condition guarantees that most of the mS-OTs in Stage 3 are honestly executed using the outcome of the coin tossing, which in turn guarantees that the cheating receiver's input can be extracted by extracting the outcome of the coin tossing. ◊

Remark 4. When Stage 5 is accepted, we also have $I_{\text{bad}} \cap \Gamma_S = \emptyset$ from the definition of I_{bad}. ◊

Definition 6 (Invariant Condition (when S is corrupted)). *For any session in which S is corrupted, we say that the invariant condition holds in that session if the following hold when the cut-and-choose in Stage 8 is accepted.*

1. Let $(\hat{a}_1^S, \hat{d}_1^S), \dots (\hat{a}_{11n}^S, \hat{d}_{11n}^S)$ be the values that are committed in NMCom *in Stage 7. Let $I_{\text{bad}} \subset [11n]$ be the set such that $i \in I_{\text{bad}}$ if and only if*

   (a) $(\hat{a}_i^S, \hat{d}_i^S)$ is not a valid decommitment of the i-th Com *commitment in Stage 2-1, or*

(b) *S does not execute the i-th mS-OT in Stage 3 honestly using* $\hat{s}_{i,0} \| \hat{s}_{i,1} \| \hat{\tau}_i^S$ *as the input and randomness, where* $\hat{s}_{i,0} \| \hat{s}_{i,1} \| \hat{\tau}_i^S$ *is obtained from* $\hat{r}_i^S = \hat{a}_i^S \oplus b_i^S$.

Then, it holds that $|I_{\mathrm{bad}}| < 0.1n$.

2. *For each* $b \in \{0,1\}$*, define* $\boldsymbol{\rho}_b^{\mathsf{nm}} = (\rho_{b,i}^{\mathsf{nm}})_{i \in \Delta}$ *as follows:* $\rho_{b,i}^{\mathsf{nm}} \stackrel{\mathrm{def}}{=} \beta_{b,i} \oplus \hat{s}_{i,b\oplus\alpha_i}$ *if* $i \notin I_{\mathrm{bad}}$ *and* $\rho_{b,i}^{\mathsf{nm}} \stackrel{\mathrm{def}}{=} \perp$ *otherwise. Then, for each* $b \in \{0,1\}$, $\boldsymbol{\rho}_b^{\mathsf{nm}}$ *is either 0.9-close to a valid codeword* $\boldsymbol{w} = (w_i)_{i \in \Delta}$ *that satisfies* $w_i = \rho_{b,i}^{\mathsf{nm}}$ *for every* $i \in \Gamma_R$ *or 0.15-far from any such valid codeword.*

Remark 5. Roughly speaking, this condition guarantees that the cheating sender's input can be extracted from the outcome of the coin tossing. In particular, it guarantees that the sharing that is computed from the outcome of mS-OTs (i.e., the sharing that is computed by the honest receiver) and the sharing that is computed from the outcome of the coin tossing (i.e., the sharing that is computed by the simulator) are very "close" (see Claim 3 below). ◊

Remark 6. When Stage 8 is accepted, we also have $I_{\mathrm{bad}} \cap \Gamma_R = \emptyset$ from the definition of I_{bad}. ◊

Next, we show that the invariant condition holds in every session in H_0 (i.e., the real experiment).

Definition 7. *We say that $\mathcal{A}$* ***cheats*** *in a session if the invariant condition does not hold in that session.*

Lemma 1. *In H_0, $\mathcal{A}$ does not cheat in every session except with negligible probability.*

Proof. Assume for contradiction that in H_0, $\mathcal{A}$ cheats in a session with non-negligible probability. Since the number of the sessions is bounded by a polynomial, there exists a function $i^*(\cdot)$ and a polynomial $p(\cdot)$ such that for infinitely many n, $\mathcal{A}$ cheats in the $i^*(n)$-th session with probability at least $1/p(n)$; furthermore, since $\mathcal{A}$ cheats only when either R or S is corrupted, in the $i^*(n)$-th session either R is corrupted for infinitely many such n or S is corrupted for infinitely many such n. In both cases, we derive contradiction by using $\mathcal{A}$ to break the hiding property of Com.

Case 1. R is corrupted in the $i^*(n)$-th session. We show that when $\mathcal{A}$ cheats, we can break the hiding property of the Com commitment in Stage 1-1 (i.e., the commitment by which Γ_S is committed to). From the definition of the invariant condition (Definition 5), when $\mathcal{A}$ cheats, we have $|I_{\mathrm{bad}}| \geq n$ even though the cut-and-choose in Stage 5 is accepting (and hence $I_{\mathrm{bad}} \cap \Gamma_S = \emptyset$ as remarked in Remark 4), where $I_{\mathrm{bad}} \subseteq [11n]$ is the set defined from the committed values of the NMCom commitments in Stage 4. If we can compute I_{bad} efficiently, we can use it to distinguish Γ_S from a random subset of size n (this is because a random subset Γ of size n satisfies $I_{\mathsf{bad}} \cap \Gamma = \emptyset$ only with negligible probability when $|I_{\mathsf{bad}}| \geq n$), so we can use it to break the hiding property of the commitment to Γ_S. However, I_{bad} is not efficiently computable since the committed values of the NMCom commitments are not efficiently computable. We thus first show that we can "approximate" I_{bad} by extracting the committed values of the ExtCom commitments in Stage 4. Details are given below.

First, we observe that if we extract the committed values of the ExtCom commitments in Stage 4 of the $i^*(n)$-th session, the extracted values, $(\hat{a}_1^R, \hat{d}_1^R, \hat{e}_1^R), \ldots, (\hat{a}_{11n}^R, \hat{d}_{11n}^R, \hat{e}_{11n}^R)$, satisfy the following condition.

- Let $\hat{I}_{\mathsf{bad}} \subset [11n]$ be a set such that $i \in \hat{I}_{\mathsf{bad}}$ if and only if
  1. $((\hat{a}_i^R, \hat{d}_i^R), \hat{e}_i^R)$ is not a valid decommitment of the i-th NMCom commitment in Stage 4, or
  2. $(\hat{a}_i^R, \hat{d}_i^R)$ is not a valid decommitment of the i-th Com commitment in Stage 2-2, or
  3. R does not execute the i-th mS-OT in Stage 3 honestly using $\hat{c}_i \,\|\, \hat{\tau}_i^R$ as the input and randomness, where $\hat{c}_i \,\|\, \hat{\tau}_i^R$ is obtained from $\hat{r}_i^R = \hat{a}_i^R \oplus b_i^R$.

  Then, $|\hat{I}_{\mathsf{bad}}| \geq n$ and $\hat{I}_{\mathsf{bad}} \cap \Gamma_S = \emptyset$ with probability at least $1/2p(n)$.

The extracted values satisfy this condition because when $\mathcal{A}$ cheats, we have $|\hat{I}_{\mathsf{bad}}| \geq n$ and $\hat{I}_{\mathsf{bad}} \cap \Gamma_S = \emptyset$ except with negligible probability. (We have $|\hat{I}_{\mathsf{bad}}| \geq n$ since we have $I_{\mathsf{bad}} \subset \hat{I}_{\mathsf{bad}}$ from the definitions of $I_{\mathsf{bad}}, \hat{I}_{\mathsf{bad}}$ and the binding property of NMCom. We have $\hat{I}_{\mathsf{bad}} \cap \Gamma_S = \emptyset$ since when the cut-and-choose in Stage 5 is accepting, for every $i \in \Gamma_S$ the i-th ExtCom commitment is a valid decommitment of the i-th NMCom commitment, and $I_{\mathsf{bad}} \cap \Gamma_S = \emptyset$.)

Based on this observation, we derive contradiction by considering the following adversary $\mathcal{A}_{\mathsf{Com}}$ against the hiding property of Com.

> $\mathcal{A}_{\mathsf{Com}}$ receives a Com commitment c^* in which either Γ_S^0 or Γ_S^1 is committed, where $\Gamma_S^0, \Gamma_S^1 \subset [11n]$ are random subsets of size n.
> Then, $\mathcal{A}_{\mathsf{Com}}$ internally executes the experiment H_0 honestly except that in the $i^*(n)$-th session, $\mathcal{A}_{\mathsf{Com}}$ uses c^* as the commitment in Stage 1-1 (i.e., as the Com commitment in which S commits to a subset Γ_S). When the experiment H_0 reaches Stage 4 of the $i^*(n)$-th session, $\mathcal{A}_{\mathsf{Com}}$ extracts the committed values of the ExtCom commitments in this stage by using its extractability.[7] Let $\hat{I}_{\mathsf{bad}} \subset [11n]$ be the set that is defined as above from the extracted values. Then, $\mathcal{A}_{\mathsf{Com}}$ outputs 1 if and only if $|\hat{I}_{\mathsf{bad}}| \geq n$ and $\hat{I}_{\mathsf{bad}} \cap \Gamma_S^1 = \emptyset$.

If $\mathcal{A}_{\mathsf{Com}}$ receives a commitment to Γ_S^1, $\mathcal{A}_{\mathsf{Com}}$ outputs 1 with probability at least $1/2p(n)$ (this follows from the above observation). In contrast, if $\mathcal{A}_{\mathsf{Com}}$ receives a commitment to Γ_S^0, $\mathcal{A}_{\mathsf{Com}}$ outputs 1 with exponentially small probability (this is because when no information about Γ_S^1 is fed into H_0, the probability that $|\hat{I}_{\mathsf{bad}}| \geq n$ but $\hat{I}_{\mathsf{bad}} \cap \Gamma_S^1 = \emptyset$ is exponentially small). Hence, $\mathcal{A}_{\mathsf{Com}}$ breaks the hiding property of Com.

[7] This extraction involves rewinding the execution of the whole experiment, i.e., the executions of the environment, the adversary, and all the other parties.

Case 2. S is corrupted in the $i^*(n)$-th session. The proof for this case is similar to (but a little more complex than) the one for Case 1. Specifically, we show that if the invariant condition does not hold, we can break the hiding property of Com in Stage 1-2 by approximating I_{bad} using the extractability of ExtCom. We give a formal proof in the full version of this paper. (A somewhat similar proof is given as the proof of Claim 4 later.) □

Finally, we show the indistinguishability between each neighboring hybrids.

Lemma 2. *Assume that in $H_{k-1:7}$ ($k \in [4m]$), $\mathcal{A}$ does not cheat in sessions $s(k), \ldots, s(4m)$ except with negligible probability. Then,*

- *$H_{k-1:7}$ and $H_{k:1}$ are indistinguishable, and*
- *in $H_{k:1}$, $\mathcal{A}$ does not cheat in sessions $s(k), \ldots, s(4m)$ except with negligible probability.*

Proof. We prove the lemma by using a hybrid argument. Specifically, we consider the following intermediate hybrid $H'_{k-1:7}$.

Hybrid $H'_{k-1:7}$. $H'_{k-1:7}$ is the same as $H_{k-1:7}$ except that in session $s(k)$, if S is corrupted and SM_k is first special message,

- the committed subset Γ_S is extracted by brute force in Stage 1-1, and
- the value committed to in the i-th ExtCom commitment in Stage 4 is switched to an all-zero string for every $i \notin \Gamma_S$.

Claim 1. *Assume that in $H_{k-1:7}$, $\mathcal{A}$ does not cheat in sessions $s(k), \ldots, s(4m)$ except with negligible probability. Then,*

- *$H_{k-1:7}$ and $H'_{k-1:7}$ are indistinguishable, and*
- *in $H'_{k-1:7}$, $\mathcal{A}$ does not cheat in sessions $s(k), \ldots, s(4m)$ except with negligible probability.*

Proof. We first show the indistinguishability between $H_{k-1:7}$ and $H'_{k-1:7}$. Assume for contradiction that $H_{k-1:7}$ and $H'_{k-1:7}$ are distinguishable. From an average argument, we can fix the execution of the experiment up until SM_k (inclusive) in such a way that even after being fixed, $H_{k-1:7}$ and $H'_{k-1:7}$ are still distinguishable. As remarked in Remark 2, no brute-force extraction is performed after SM_k in $H_{k-1:7}$ and $H'_{k-1:7}$; hence, by considering the transcript (including the inputs and randomness of all the parties) and the extracted values up until SM_k as non-uniform advice, we can break the hiding property of ExtCom as follows.

> The adversary $\mathcal{A}_{\mathsf{ExtCom}}$ internally executes $H_{k-1:7}$ from SM_k using the non-uniform advice. In Stage 4 of session $s(k)$, $\mathcal{A}_{\mathsf{ExtCom}}$ sends $(a_i^R, d_i^R, e_i^R)_{i \notin \Gamma_S}$ and $(0,0,0)_{i \notin \Gamma_S}$ to the external committer, receives back ExtCom commitments (in which either $(a_i^R, d_i^R, e_i^R)_{i \notin \Gamma_S}$ or $(0,0,0)_{i \notin \Gamma_S}$ are committed to), and feeds them into $H_{k-1:7}$. After the execution of $H_{k-1:7}$ finishes, $\mathcal{A}_{\mathsf{ExtCom}}$ outputs whatever $\mathcal{Z}$ outputs in the experiment.

When $\mathcal{A}_{\mathsf{ExtCom}}$ receives commitments to $(a_i^R, d_i^R, e_i^R)_{i \notin \Gamma_S}$, the internally executed experiment is identical with $H_{k-1:7}$, whereas when $\mathcal{A}_{\mathsf{ExtCom}}$ receives a commitments to $(0,0,0)_{i \notin \Gamma_S}$, the internally executed experiment is identical with $H'_{k-1:7}$. Hence, from the assumption that $H_{k-1:7}$ and $H'_{k-1:7}$ are distinguishable (even after being fixed up until SM_k), $\mathcal{A}_{\mathsf{ExtCom}}$ distinguishes ExtCom commitments.

We next show that in $H'_{k-1:7}$, $\mathcal{A}$ does not cheat in sessions $s(k), \ldots, s(4m)$. Assume for contradiction that in $H'_{k-1:7}$, $\mathcal{A}$ cheats in one of those sessions, say, session $s(j)$, with non-negligible probability. Then, from an average argument, we can fix the execution of the experiment up until SM_k (inclusive) in such a way that even after being fixed, $\mathcal{A}$ cheats in session $s(j)$ only with negligible probability in $H_{k-1:7}$ but with non-negligible probability in $H'_{k-1:7}$. Then, by considering the transcript and the extracted values up until SM_k as non-uniform advice, we can break the robust non-malleability of NMCom as follows. (Note that the ExtCom commitments in sessions $s(k), \ldots, s(4m)$ starts only after SM_k.)

The man-in-the-middle adversary $\mathcal{A}_{\mathsf{NMCom}}$ internally executes $H_{k-1:7}$ from SM_k using the non-uniform advice. In Stage 4 of session $s(k)$, $\mathcal{A}_{\mathsf{NMCom}}$ sends $(a_i^R, d_i^R, e_i^R)_{i \notin \Gamma_S}$ and $(0,0,0)_{i \notin \Gamma_S}$ to the external committer, receives back ExtCom commitments (in which either $(a_i^R, d_i^R, e_i^R)_{i \notin \Gamma_S}$ or $(0,0,0)_{i \notin \Gamma_S}$ are committed to), and feeds them into $H_{k-1:7}$. Also, in session $s(j)$, $\mathcal{A}_{\mathsf{NMCom}}$ forwards the NMCom commitments from $\mathcal{A}$ to the external receiver (specifically, the NMCom commitments in Stage 4 if R is corrupted and in Stage 7 if S is corrupted). After the execution of $H_{k-1:7}$ finishes, $\mathcal{A}_{\mathsf{NMCom}}$ outputs its view.
The distinguisher $\mathcal{D}_{\mathsf{NMCom}}$ takes as input the view of $\mathcal{A}_{\mathsf{NMCom}}$ and the values committed by $\mathcal{A}_{\mathsf{NMCom}}$ (which are equal to the values committed to by $\mathcal{A}$ in session $s(j)$ in the internally executed experiment). $\mathcal{D}_{\mathsf{NMCom}}$ then outputs 1 if and only if $\mathcal{A}$ cheated in session $s(j)$. (Notice that given the committed values of the NMCom commitments, $\mathcal{D}_{\mathsf{NMCom}}$ can check whether $\mathcal{A}$ cheated or not in polynomial time.)
When $\mathcal{A}_{\mathsf{NMCom}}$ receives commitments to $(a_i^R, d_i^R, e_i^R)_{i \notin \Gamma_S}$, the internally executed experiment is identical with $H_{k-1:7}$, whereas when $\mathcal{A}_{\mathsf{NMCom}}$ receives a commitments to $(0,0,0)_{i \notin \Gamma_S}$, the internally executed experiment is identical with $H'_{k-1:7}$. Hence, from the assumption that $\mathcal{A}$ cheats in session $s(j)$ with negligible probability in $H_{k-1:7}$ but with non-negligible probability in $H'_{k-1:7}$, $\mathcal{A}_{\mathsf{NMCom}}$ breaks the robust non-malleability of NMCom.

This completes the proof of Claim 1. □

Claim 2. *Assume that in $H'_{k-1:7}$, $\mathcal{A}$ does not cheat in sessions $s(k), \ldots, s(4m)$ except with negligible probability. Then,*

- *$H'_{k-1:7}$ and $H_{k:1}$ are indistinguishable, and*
- *in $H_{k:1}$, $\mathcal{A}$ does not cheat in sessions $s(k), \ldots, s(4m)$ except with negligible probability.*

This claim can be proven very similarly to Claim 1 (the only difference is that we use the hiding property of NMCom rather than that of ExtCom in the first part, and use the non-malleability of NMCom rather than its robust non-malleability in the second part). We give a proof in the full version of this paper.

This completes the proof of Lemma 2. □

Lemma 3. *Assume that in $H_{k:1}$ ($k \in [4m]$), $\mathcal{A}$ does not cheat in sessions $s(k), \ldots, s(4m)$ except with negligible probability. Then,*

- *$H_{k:1}$ and $H_{k:2}$ are indistinguishable, and*
- *in $H_{k:2}$, $\mathcal{A}$ does not cheat in sessions $s(k), \ldots, s(4m)$ except with negligible probability.*

Recall that hybrids $H_{k:1}, H_{k:2}$ differ only in the input and the randomness that are used in some of the mS-OTs in Stage 3, where those that are derived from the outcomes of the coin tossing is used in $H_{k:1}$ and random inputs and true randomness are used in $H_{k:2}$. Intuitively, we prove this lemma by using the security of the coin tossing (which is guaranteed by the hiding property of Com) because it guarantees that the outcome of the coin tossing is pseudorandom. The proof is quite similar to the proof of Claim 1 (we use the hiding of Com rather than that of ExtCom), and given in the full version of this paper.

Lemma 4. *Assume that in $H_{k:2}$ ($k \in [4m]$), $\mathcal{A}$ does not cheat in sessions $s(k), \ldots, s(4m)$ except with negligible probability. Then,*

- *$H_{k:2}$ and $H_{k:3}$ are indistinguishable, and*
- *in $H_{k:3}$, $\mathcal{A}$ does not cheat in sessions $s(k), \ldots, s(4m)$ except with negligible probability.*

Proof. Recall that $H_{k:2}$ and $H_{k:3}$ differ only in that in session $s(k)$ of $H_{k:3}$, if S is corrupted and SM_k is third special message, R outputs $\mathsf{Value}(\boldsymbol{\rho}_u^{\mathsf{ext}}, \Gamma_R \cap \Delta)$ rather than $\mathsf{Value}(\widetilde{\boldsymbol{\rho}}, \Gamma_R \cap \Delta)$.

For proving the lemma, it suffices to show that in $H_{k:3}$, we have $\mathsf{Value}(\boldsymbol{\rho}_u^{\mathsf{ext}}, \Gamma_R \cap \Delta) = \mathsf{Value}(\widetilde{\boldsymbol{\rho}}, \Gamma_R \cap \Delta)$ except with negligible probability. This is because if $\mathsf{Value}(\boldsymbol{\rho}_u^{\mathsf{ext}}, \Gamma_R \cap \Delta) = \mathsf{Value}(\widetilde{\boldsymbol{\rho}}, \Gamma_R \cap \Delta)$ holds in $H_{k:3}$ except with negligible probability, $H_{k:2}$ and $H_{k:3}$ are statistically close, which implies that in $H_{k:3}$, $\mathcal{A}$ does not cheat in sessions $s(k), \ldots, s(4m)$ except with negligible probability.

Hence, we show that in $H_{k:3}$, we have $\mathsf{Value}(\boldsymbol{\rho}_u^{\mathsf{ext}}, \Gamma_R \cap \Delta) = \mathsf{Value}(\widetilde{\boldsymbol{\rho}}, \Gamma_R \cap \Delta)$ except with negligible probability. Since $H_{k:2}$ and $H_{k:3}$ proceed identically until the end of session $s(k)$, we have that in $H_{k:3}$, $\mathcal{A}$ does not cheat in sessions $s(k)$ except with negligible probability. It thus suffices to show the following two claims.

Claim 3. *For any $\boldsymbol{x} = (x_i)_{i \in \Delta}, \boldsymbol{y} = (y_i)_{i \in \Delta}$ and a set Θ, we have $\mathsf{Value}(\boldsymbol{x}, \Theta) = \mathsf{Value}(\boldsymbol{y}, \Theta)$ if the following conditions hold.*

1. *$\boldsymbol{x}$ and $\boldsymbol{y}$ are 0.99-close, and $x_i = y_i$ holds for every $i \in \Theta$.*
2. *If $x_i \neq \bot$, then $x_i = y_i$.*

3. *x is either 0.9-close to a valid codeword $\boldsymbol{w} = (w_i)_{i\in\Delta}$ that satisfies $w_i = x_i$ for every $i \in \Theta$ or 0.14-far from any such valid codeword.*

Claim 4. *In $H_{k:3}$, if in session $s(k)$ the sender S is corrupted, $\mathcal{A}$ does not cheat, and the session is accepting, the following hold.*

1. *$\boldsymbol{\rho}_u^{\mathsf{ext}}$ and $\widetilde{\boldsymbol{\rho}}$ are 0.99-close, and $\rho_{u,i}^{\mathsf{ext}} = \tilde{\rho}_i$ holds for every $i \in \Gamma_R \cap \Delta$.*
2. *If $\rho_{u,i}^{\mathsf{ext}} \neq \bot$, then $\rho_{u,i}^{\mathsf{ext}} = \tilde{\rho}_i$.*
3. *$\boldsymbol{\rho}_u^{\mathsf{ext}}$ is either 0.9-close to a valid codeword $\boldsymbol{w} = (w_i)_{i\in\Delta}$ that satisfies $w_i = \rho_{u,i}^{\mathsf{ext}}$ for every $i \in \Gamma_R \cap \Delta$ or 0.14-far from any such valid codeword.*

We prove each of the claims below.

Proof (of Claim 3). We consider the following two cases.

Case 1. $\boldsymbol{x}$ is 0.9-close to a valid codeword $\boldsymbol{w} = (w_i)_{i\in\Delta}$ that satisfies $w_i = x_i$ for every $i \in \Theta$: First, we observe that $\boldsymbol{y}$ is 0.9-close to $\boldsymbol{w}$. Since $\boldsymbol{w}$ is a valid codeword, we have $w_i \neq \bot$ for every $i \in \Delta$; thus, for every i such that $x_i = w_i$, we have $x_i \neq \bot$. Recall that from the assumed conditions, for every i such that $x_i \neq \bot$, we have $x_i = y_i$. Therefore, for every i such that $x_i = w_i$, we have $y_i = w_i$, which implies that $\boldsymbol{y}$ is 0.9-close to $\boldsymbol{w}$.
Next, we observe that $\boldsymbol{w}$ satisfies $w_i = y_i$ for every $i \in \Theta$. From the assumed conditions, we have $x_i = y_i$ for every $i \in \Theta$. Also, from the condition of this case, $\boldsymbol{w}$ satisfies $w_i = x_i$ for every $i \in \Theta$. From these two, we have that $\boldsymbol{w}$ satisfies $w_i = y_i$ for every $i \in \Theta$.
Now, from the definition of $\mathsf{Value}(\cdot,\cdot)$, we have $\mathsf{Value}(\boldsymbol{x},\Theta) = \mathsf{Value}(\boldsymbol{y},\Theta) = \mathsf{Decode}(\boldsymbol{w})$.

Case 2. $\boldsymbol{x}$ is 0.14-far from any valid codeword $\boldsymbol{w} = (w_i)_{i\in\Delta}$ that satisfies $w_i = x_i$ for every $i \in \Theta$: For any valid codeword $\boldsymbol{w}' = (w'_i)_{i\in\Delta}$ that satisfies $w'_i = y_i$ for every $i \in \Theta$, we observe that $\boldsymbol{y}$ is 0.1-far from $\boldsymbol{w}'$. Since we assume that $x_i = y_i$ holds for every $i \in \Theta$, we have $w'_i = x_i$ for every $i \in \Theta$. Therefore, from the assumption of this case, $\boldsymbol{x}$ is 0.14-far from $\boldsymbol{w}'$. Now, since we assume that $\boldsymbol{x}$ and $\boldsymbol{y}$ are 0.99-close, $\boldsymbol{y}$ is 0.1-far from $\boldsymbol{w}'$.
Now, from the definition of $\mathsf{Value}(\cdot,\cdot)$, we conclude that $\mathsf{Value}(\boldsymbol{x},\Theta) = \mathsf{Value}(\boldsymbol{y},\Theta) = \bot$.

Notice that from the assumed conditions, either Case 1 or Case 2 is true. This concludes the proof of Claim 3. □

Proof (of Claim 4). Recall that if $\mathcal{A}$ does not cheat in an accepting session in which S is corrupted, we have the following.

1. Let $(\hat{a}_1^S, \hat{d}_1^S), \ldots (\hat{a}_{11n}^S, \hat{d}_{11n}^S)$ be the values committed in NMCom in Stage 7. Let $I_{\mathsf{bad}} \subset [11n]$ be the set that is defined as follows: $i \in I_{\mathsf{bad}}$ if and only if

   (a) $(\hat{a}_i^S, \hat{d}_i^S)$ is not a valid decommitment of the i-th Com commitment in Stage 2-1, or

(b) S does not execute the i-th mS-OT in Stage 3 honestly using $\hat{s}_{i,0} \parallel \hat{s}_{i,1} \parallel \hat{\tau}_i^S$ as the input and randomness, where $\hat{s}_{i,0} \parallel \hat{s}_{i,1} \parallel \hat{\tau}_i^S$ is obtained from $\hat{r}_i^S = \hat{a}_i^S \oplus b_i^S$.

Then, it holds that $|I_{\text{bad}}| < 0.1n$.

2. For each $b \in \{0,1\}$, define $\boldsymbol{\rho}_b^{\mathsf{nm}} = (\rho_{b,i}^{\mathsf{nm}})_{i\in\Delta}$ as follows: $\rho_{b,i}^{\mathsf{nm}} \stackrel{\text{def}}{=} \beta_{b,i} \oplus \hat{s}_{i,b\oplus\alpha_i}$ if $i \notin I_{\text{bad}}$ and $\rho_{b,i}^{\mathsf{nm}} \stackrel{\text{def}}{=} \bot$ otherwise. Then, for each $b \in \{0,1\}$, $\boldsymbol{\rho}_b^{\mathsf{nm}}$ is either 0.9-close to a valid codeword $\boldsymbol{w} = (w_i)_{i\in\Delta}$ that satisfies $w_i = \rho_{b,i}^{\mathsf{nm}}$ for every $i \in \Gamma_R$ or 0.15-far from any such valid codeword.

We show that the above two imply all the three conditions in the claim statement.

First, we show that $\boldsymbol{\rho}_u^{\mathsf{ext}}$ and $\widetilde{\boldsymbol{\rho}}$ are 0.99-close and that $\rho_{u,i}^{\mathsf{ext}} = \tilde{\rho}_i$ holds for every $i \in \Gamma_R \cap \Delta$. From the definition of I_{bad}, we have $\rho_{u,i}^{\mathsf{ext}} = \tilde{\rho}_i$ for every $i \notin I_{\text{bad}}$ (this is because for every $i \notin I_{\text{bad}}$, $\mathcal{A}$ executed the i-th mS-OT in Stage 3 honestly using the coin obtained in Stage 2-1, which implies that the value $\widetilde{s}_i$ that was obtained from the i-th mS-OT is equal to the value s_{i,c_i} that was obtained by extracting the coin in Stage 2-1 by brute-force). Then, since $|I_{\text{bad}}| < 0.1n$ and $I_{\text{bad}} \cap \Gamma_R = \emptyset$ (the latter holds since the session would be rejected otherwise), we have that $\boldsymbol{\rho}_u^{\mathsf{ext}}$ and $\widetilde{\boldsymbol{\rho}}$ are 0.99-close and that $\rho_{u,i}^{\mathsf{ext}} = \tilde{\rho}_i$ holds for every $i \in \Gamma_R \cap \Delta$.

Next, we show that if $\rho_{u,i}^{\mathsf{ext}} \neq \bot$ then $\rho_{u,i}^{\mathsf{ext}} = \tilde{\rho}_i$. From the definition of $\boldsymbol{\rho}_u^{\mathsf{ext}}$, if $\rho_{u,i}^{\mathsf{ext}} \neq \bot$, $\mathcal{A}$ executed the i-th mS-OT in Stage 3 honestly using the coin obtained in Stage 2-1, so we have $\rho_{u,i}^{\mathsf{ext}} = \tilde{\rho}_i$ from the argument same as above.

Finally, we show that $\boldsymbol{\rho}_u^{\mathsf{ext}}$ is either 0.9-close to a valid codeword $\boldsymbol{w} = (w_i)_{i\in\Delta}$ that satisfies $w_i = \rho_{u,i}^{\mathsf{ext}}$ for every $i \in \Gamma_R \cap \Delta$ or 0.14-far from any such valid codeword. From the assumption that $\mathcal{A}$ does not cheat, it suffices to consider the following two cases.

Case 1. $\boldsymbol{\rho}_u^{\mathsf{nm}}$ is 0.9-close to a valid codeword $\boldsymbol{w} = (w_i)_{i\in\Delta}$ that satisfies $w_i = \rho_{b,i}^{\mathsf{nm}}$ for every $i \in \Gamma_R \cap \Delta$: In this case, $\boldsymbol{\rho}_u^{\mathsf{ext}}$ is 0.9-close to $\boldsymbol{w}$, and $w_i = \rho_{b,i}^{\mathsf{ext}}$ holds for every $i \in \Gamma_R$. This is because for every i such that $\rho_{u,i}^{\mathsf{nm}} = w_i$, we have $\rho_{u,i}^{\mathsf{nm}} \neq \bot$ and thus we have $\rho_{u,i}^{\mathsf{nm}} = \rho_{u,i}^{\mathsf{ext}}$ from the definition of $\boldsymbol{\rho}_u^{\mathsf{nm}}$.

Case 2. $\boldsymbol{\rho}_u^{\mathsf{nm}}$ is 0.15-far from any valid codeword $\boldsymbol{w} = (w_i)_{i\in\Delta}$ that satisfies $w_i = \rho_{b,i}^{\mathsf{nm}}$ for every $i \in \Gamma_R \cap \Delta$: In this case, $\boldsymbol{\rho}_u^{\mathsf{ext}}$ is 0.14-far from any valid codeword $\boldsymbol{w}'$ that satisfies $w'_i = \rho_{b,i}^{\mathsf{ext}}$ for every $i \in \Gamma_R \cap \Delta$. This can be seen by observing the following: (1) for every $i \in \Gamma_R \cap \Delta$, we have $i \notin I_{\text{bad}}$ (this is because the session is accepting) and hence $\rho_{u,i}^{\mathsf{ext}} = \rho_{u,i}^{\mathsf{nm}}$; (2) therefore, for any valid codeword $\boldsymbol{w}'$ that satisfies $w'_i = \rho_{b,i}^{\mathsf{ext}}$ for every $i \in \Gamma_R \cap \Delta$, we have that $\boldsymbol{w}'$ also satisfies $w'_i = \rho_{b,i}^{\mathsf{nm}}$ for every $i \in \Gamma_R \cap \Delta$; (3) then, from the assumption of this case, $\boldsymbol{\rho}_u^{\mathsf{nm}}$ is 0.15-far from $\boldsymbol{w}'$; (4) now, since $\boldsymbol{\rho}_u^{\mathsf{nm}}$ and $\boldsymbol{\rho}_u^{\mathsf{ext}}$ are 0.99-close, $\boldsymbol{\rho}_u^{\mathsf{ext}}$ is 0.14-far from $\boldsymbol{w}'$.

This concludes the proof of Claim 4. □

This concludes the proof of Lemma 4. □

Lemma 5. *Assume that in $H_{k:3}$ ($k \in [4m]$), $\mathcal{A}$ does not cheat in sessions $s(k), \ldots, s(4m)$ except with negligible probability. Then,*

- *$H_{k:3}$ and $H_{k:4}$ are indistinguishable, and*
- *in $H_{k:4}$, $\mathcal{A}$ does not cheat in sessions $s(k), \ldots, s(4m)$ except with negligible probability.*

Recall that $H_{k:3}$ and $H_{k:4}$ differ only in that in session $s(k)$ of $H_{k:4}$, if S is corrupted and SM_k is third special message, α_i is a random bit rather than $\alpha_i = u \oplus c_i$ for every $i \in \Delta$ in Stage 6-1. Intuitively, we can prove this lemma by using the security of mS-OT: For every $i \notin \Gamma_S$, the choice bit c_i of the i-th mS-OT in Stage 3 is hidden from $\mathcal{A}$ and hence $\alpha_i = u \oplus c_i$ in $H_{k:3}$ is indistinguishable from a random bit. Formally, we prove this Lemma in the same way as we do for Claim 1 (we use the security of mS-OT rather than the hiding of ExtCom); the proof is given in the full version of this paper.

Lemma 6. *Assume that in $H_{k:4}$ ($k \in [4m]$), $\mathcal{A}$ does not cheat in sessions $s(k), \ldots, s(4m)$ except with negligible probability. Then,*

- *$H_{k:4}$ and $H_{k:5}$ are indistinguishable, and*
- *in $H_{k:5}$, $\mathcal{A}$ does not cheat in sessions $s(k), \ldots, s(4m)$ except with negligible probability.*

Since hybrids $H_{k:4}, H_{k:5}$ differ only in the values committed to in NMCom and ExtCom for the indices outside of Γ_R, this lemma can be proven identically with Lemma 2. We give a prof in the full version of this paper.

Lemma 7. *Assume that in $H_{k:5}$ ($k \in [4m]$), $\mathcal{A}$ does not cheat in sessions $s(k), \ldots, s(4m)$ except with negligible probability. Then,*

- *$H_{k:5}$ and $H_{k:6}$ are indistinguishable, and*
- *in $H_{k:6}$, $\mathcal{A}$ does not cheat in sessions $s(k), \ldots, s(4m)$ except with negligible probability.*

Since hybrids $H_{k:5}, H_{k:6}$ differ only in the inputs and the randomness that are used in some of the mS-OTs in Stage 3, this lemma can be proven identically with Lemma 3 (which in turn can be proven quite similarly to Lemma 2). We give a prof in the full version of this paper.

Lemma 8. *Assume that in $H_{k:6}$ ($k \in [4m]$), $\mathcal{A}$ does not cheat in sessions $s(k), \ldots, s(4m)$ except with negligible probability. Then,*

- *$H_{k:6}$ and $H_{k:7}$ are indistinguishable, and*
- *in $H_{k:7}$, $\mathcal{A}$ does not cheat in sessions $s(k), \ldots, s(4m)$ except with negligible probability.*

Proof. We prove the lemma by considering the following intermediate hybrids $H'_{k:6}$, $H''_{k:6}$, and $H'''_{k:6}$.

Hybrid $H'_{k:6}$**.** $H'_{k:6}$ is the same as $H_{k:6}$ except that in session $s(k)$, if R is corrupted and SM_k is fourth special message, the following modifications are made.

1. As in $H_{k:7}$, the committed strings $\boldsymbol{a}^R = (a_1^R, \ldots, a_{11n}^R)$ are extracted by brute force in Stage 2-2, $\boldsymbol{r}^R = (r_1^R, \ldots, r_{11n}^R)$ is defined by $r_i^R \stackrel{\text{def}}{=} a_i^R \oplus b_i^R$ for each $i \in [11n]$, and r_i^R is parsed as $c_i \,\|\, \tau_i^R$ for each $i \in [11n]$. Also, I_0, I_1, and u^* are defined as in $H_{k:7}$.
2. In Stage 6, $\beta_{b,i}$ is a random bit rather than $\beta_{b,i} = \rho_{b,i} \oplus s_{i,b\oplus\alpha_i}$ for every $b \in \{0,1\}$ and $i \in \Delta \backslash I_b$. (Recall that, roughly, $I_b \subset \Delta$ is the set of indices on which $\mathcal{A}$ could have obtained $s_{i,b\oplus\alpha_i}$.)

Hybrid $H''_{k:6}$**.** $H''_{k:6}$ is the same as $H'_{k:6}$ except that in session $s(k)$, if R is corrupted and SM_k is fourth special message, the following modification is made.

1. In Stage 6, $\boldsymbol{\rho}_{1-u^*} = \{\rho_{1-u^*,i}\}_{i\in\Delta}$ is a secret sharing of a random bit rather than that of v_{1-u^*}.

Hybrid $H'''_{k:6}$**.** $H'''_{k:6}$ is the same as $H''_{k:6}$ except that in session $s(k)$, if R is corrupted and SM_k is fourth special message, the following modification is made.

1. In Stage 6, $\beta_{b,i}$ is $\beta_{b,i} = \rho_{b,i} \oplus s_{i,b\oplus\alpha_i}$ rather than a random bit for every $b \in \{0,1\}$ and $i \in \Delta \backslash I_b$.

Notice that $H'''_{k:6}$ is identical with $H_{k:7}$.

Claim 5. *Assume that in* $H_{k:6}$*,* $\mathcal{A}$ *does not cheat in sessions* $s(k), \ldots, s(4m)$ *except with negligible probability. Then,*

- $H_{k:6}$ *and* $H'_{k:6}$ *are indistinguishable, and*
- *in* $H'_{k:6}$*,* $\mathcal{A}$ *does not cheat in sessions* $s(k), \ldots, s(4m)$ *except with negligible probability.*

Recall that $H_{k:6}$ and $H'_{k:6}$ differ only in that in session $s(k)$ of $H'_{k:6}$, if R is corrupted and SM_k is fourth special message, $\beta_{b,i}$ is a random bit rather than $\beta_{b,i} = \rho_{b,i} \oplus s_{i,b\oplus\alpha_i}$ for every $b \in \{0,1\}$ and $i \in \Delta \backslash I_b$. Intuitively, we can prove this claim by using the security of mS-OT: For every $i \in \Delta \backslash I_b$, $\mathcal{A}$ executed the i-th mS-OT honestly with choice bit $(1-b) \oplus \alpha_i$, and the sender's input and randomness of this mS-OT are not revealed in Stage 8; therefore, the value of $s_{i,b\oplus\alpha_i}$ is hidden from $\mathcal{A}$ and thus $\beta_{b,i} = \rho_{b,i} \oplus s_{i,b\oplus\alpha_i}$ is indistinguishable from a random bit. Formally, we prove this claim in the same way as we do for Claim 1 (we use the security of mS-OT rather than the hiding of ExtCom); a formal proof is given in the full version of this paper.

Claim 6. *Assume that in* $H'_{k:6}$*,* $\mathcal{A}$ *does not cheat in sessions* $s(k), \ldots, s(4m)$ *except with negligible probability. Then,*

- $H'_{k:6}$ *and* $H''_{k:6}$ *are indistinguishable, and*
- *in* $H''_{k:6}$*,* $\mathcal{A}$ *does not cheat in sessions* $s(k), \ldots, s(4m)$ *except with negligible probability.*

Proof. Recall that hybrids $H'_{k:6}, H''_{k:6}$ differ only in that in Stage 6, $\rho_{1-u^*} = \{\rho_{1-u^*,i}\}_{i\in\Delta}$ is a secret sharing of a random bit rather than that of v_{1-u^*}.

For proving the lemma, it suffices to show that we have $|I_{1-u^*}| \le 6n$ in $H'_{k:6}$ except with negligible probability. This is because if $|I_{1-u^*}| \le 6n$ in $H'_{k:6}$, then $\beta_{1-u^*,i}$ is a random bit on at least $4n$ indices and thus $\rho_{1-u^*,i}$ is hidden on at least $4n$ indices, which implies that $H'_{k:6}$ and $H''_{k:6}$ are statistically indistinguishable. (Statistical indistinguishability between $H'_{k:6}$ and $H''_{k:6}$ implies that in $H''_{k:6}$, $\mathcal{A}$ does not cheat in sessions $s(k), \ldots, s(4m)$ except with negligible probability.)

Hence, we show that we have $|I_{1-u^*}| \le 6n$ in $H'_{k:6}$ except with negligible probability. Since we assume that $\mathcal{A}$ does not cheat in session $s(k)$ except with negligible probability, it suffices to show that we have either $|I_0| \le 6n$ or $|I_1| \le 6n$ whenever $\mathcal{A}$ does not cheat in session $s(k)$. Assume that $\mathcal{A}$ does not cheat in session $s(k)$. Then, since $|\Gamma_R| = n$ and the number of indices on which $\mathcal{A}$ does not execute mS-OT using the outcome of coin-tossing is at most n, we have $|I_0 \cap I_1| \le 2n$. Now, since $I_0, I_1 \subset \Delta$ and thus $|I_0 \cup I_1| \le |\Delta| = 10n$, we have $|I_0| + |I_1| \le 12n$, and hence, we have either $|I_0| \le 6n$ or $|I_1| \le 6n$. □

Claim 7. *Assume that in $H''_{k:6}$, $\mathcal{A}$ does not cheat in sessions $s(k), \ldots, s(4m)$ except with negligible probability. Then,*

- *$H''_{k:6}$ and $H'''_{k:6}$ are indistinguishable, and*
- *in $H'''_{k:6}$, $\mathcal{A}$ does not cheat in sessions $s(k), \ldots, s(4m)$ except with negligible probability.*

Proof. This claim can be proven identically with Claim 5. □

This completes the proof of Lemma 8. □

From Lemmas 2–8, we conclude that the output of H_0 and that of $H_{4m:7}$ are indistinguishable, i.e., the output of the real world and that of the ideal world are indistinguishable. This concludes the proof of Theorem 2.

5 Our SPS Concurrent MPC Protocol

In this section, we prove the following theorem.

Theorem 3. *Assume the existence of constant-round semi-honest oblivious transfer protocols and collision-resistant hash functions. Let $\mathcal{F}$ be any well-formed functionality and $\hat{\mathcal{F}}$ be its multi-session extension. Then, there exists a constant-round protocol that UC-SPS realizes $\hat{\mathcal{F}}$, and it uses the underlying primitives in the black-box way.*

We focus on the two-party case below (the MPC case is analogues).

Protocol Description. Roughly speaking, we obtain our SPS 2PC protocol by composing our SPS OT protocol in Sect. 4 with a UC-secure OT-hybrid 2PC protocol. Concretely, let Π_{OT} be our SPS OT protocol in Sect. 4, and $\Pi_{\mathrm{2PC}}^{\mathcal{F}_{\mathrm{OT}}}$ be a UC-secure OT-hybrid 2PC protocol with the following property: The two parties use the OT functionality $\mathcal{F}_{OT}$ only at the beginning of the protocol, and they send only randomly chosen inputs to $\mathcal{F}_{OT}$. Then, we obtain our SPS 2PC protocol Π_{2PC} by replacing each invocation of $\mathcal{F}_{OT}$ in $\Pi_{\mathrm{2PC}}^{\mathcal{F}_{\mathrm{OT}}}$ with an execution of Π_{OT} (i.e., the two parties execute Π_{OT} instead of calling to $\mathcal{F}_{OT}$), where all the executions of Π_{OT} are carried out in a synchronous manner, i.e., in a manner that the first message of all the executions are sent before the second message of any execution is sent etc.

As the UC-secure OT-hybrid 2PC protocol, we use the constant-round 2PC (actually, MPC) protocol of Ishai et al. [27], which makes only black-box use of pseudorandom generators (which in turn can be obtained in the black-box way from any semi-honest OT protocol). (The protocol of Ishai et al. [27] itself does not satisfy the above property, but it can be modified easily to satisfy them; see the full version of this paper.) Since the OT-hybrid protocol of Ishai et al. [27] is a black-box construction and has only constant number of rounds, our protocol Π_{2PC} is also a black-box construction and has only constant number of rounds.

Simulator $\mathcal{S}im$. As in Sect. 4.2, we consider a simulator that works against adversaries that participate in multiple sessions of Π_{2PC}. Let $\mathcal{Z}$ be any environment, $\mathcal{A}$ be any adversary that participates in multiple sessions of Π_{2PC}. Our simulator $\mathcal{S}im_{\mathrm{OT}}$ internally invokes the adversary $\mathcal{A}$, and simulates each of the sessions by using the simulator of Π_{OT} (Sect. 4.2) and that of $\Pi_{\mathrm{2PC}}^{\mathcal{F}_{\mathrm{OT}}}$ as follows.

1. In each execution of Π_{OT} at the beginning of Π_{2PC}, $\mathcal{S}im$ simulates the honest party's messages for $\mathcal{A}$ in the same way as $\mathcal{S}im_{\mathrm{OT}}$.
   Recall that $\mathcal{S}im_{\mathrm{OT}}$ makes a query to $\mathcal{F}_{OT}$ during the simulation. When $\mathcal{S}im_{\mathrm{OT}}$ makes a query to $\mathcal{F}_{OT}$, $\mathcal{S}im$ sends those queries to the simulator of $\Pi_{\mathrm{2PC}}^{\mathcal{F}_{\mathrm{OT}}}$ in order to simulate the answer from $\mathcal{F}_{OT}$. (Recall that the simulator of $\Pi_{\mathrm{2PC}}^{\mathcal{F}_{\mathrm{OT}}}$ simulates $\mathcal{F}_{OT}$ for the adversary.)
2. In the execution of $\Pi_{\mathrm{2PC}}^{\mathcal{F}_{\mathrm{OT}}}$ during Π_{2PC}, $\mathcal{S}im$ simulates the honest party's messages for $\mathcal{A}$ by using the simulator of $\Pi_{\mathrm{2PC}}^{\mathcal{F}_{\mathrm{OT}}}$, who obtained the queries to $\mathcal{F}_{OT}$ as above.

We remark that here we use the simulator of $\Pi_{\mathrm{2PC}}^{\mathcal{F}_{\mathrm{OT}}}$ in the setting where multiple sessions of $\Pi_{\mathrm{2PC}}^{\mathcal{F}_{\mathrm{OT}}}$ are concurrently executed and some super-polynomial-time computation is performed. However, the use of it in this setting does not cause any problem because it runs in the black-box straight-line manner.

Proof of Indistinguishability. We show that the output of the environment in the real world and that in the ideal world are indistinguishable. The proof proceeds very similarly to the proof for our SPS OT protocol (Sect. 4). To simplify the exposition, below we assume that $\Pi_{\mathrm{2PC}}^{\mathcal{F}_{\mathrm{OT}}}$ makes only a single call to

$\mathcal{F}_{OT}$. (The proof can be modified straightforwardly when $\Pi_{2PC}^{\mathcal{F}_{OT}}$ makes multiple calls to $\mathcal{F}_{OT}$.)

Recall that Π_{2PC} is obtained by composing our OT protocol Π_{OT} with a OT-hybrid 2PC protocol $\Pi_{2PC}^{\mathcal{F}_{OT}}$. Roughly, we consider a sequence of hybrid experiments in which:

- Each execution of Π_{OT} is gradually changed to simulation as in the sequence of hybrid experiments that we considered in the proof of Π_{OT} (Sect. 4.3).
- Once the execution of Π_{OT} in a session of Π_{2PC} is changed to simulation completely, the execution of $\Pi_{2PC}^{\mathcal{F}_{OT}}$ in that session is changed to simulation.

More concretely, we consider hybrids H_0 and $H_{k:1}, \ldots, H_{k:9}$ ($k \in [4m]$), where H_0 and $H_{k:1}, \ldots, H_{k:7}$ are defined as in Sect. 4.3, and $H_{k:8}$ and $H_{k:9}$ are defined as follows.

Hybrid $H_{k:8}$. $H_{k:8}$ is the same as $H_{k:7}$ except that in session $s(k)$, if S is corrupted and SM_k is third special message, all the messages of $\Pi_{2PC}^{\mathcal{F}_{OT}}$ from R are generated by the simulator of $\Pi_{2PC}^{\mathcal{F}_{OT}}$. More concretely, the messages of $\Pi_{2PC}^{\mathcal{F}_{OT}}$ from R are generated as follows. Recall that from the definition of Hybrid $H_{k:3}$, the implicit input $v_b^* \stackrel{\text{def}}{=} \mathsf{Value}(\boldsymbol{\rho}_b^{\mathsf{ext}}, \Gamma_R \cap \Delta)$ ($b \in \{0,1\}$) to Π_{OT} is extracted from the adversary in session $s(k)$ (as $\boldsymbol{\rho}_b^{\mathsf{ext}}$ are computed for both $b \in \{0,1\}$). Now, the messages of $\Pi_{2PC}^{\mathcal{F}_{OT}}$ from R are simulated by feeding those extracted implicit input and the subsequent messages to the simulator of $\Pi_{2PC}^{\mathcal{F}_{OT}}$.

Hybrid $H_{k:9}$. $H_{k:9}$ is the same as $H_{k:8}$ except that in session $s(k)$, if R is corrupted and SM_k is fourth special message, all the messages of $\Pi_{2PC}^{\mathcal{F}_{OT}}$ from S are generated by the simulator of $\Pi_{2PC}^{\mathcal{F}_{OT}}$.

Lemma 9. *Assume that in $H_{k:7}$ ($k \in [4m]$), $\mathcal{A}$ does not cheat in sessions $s(k), \ldots, s(4m)$ except with negligible probability. Then,*

- *$H_{k:7}$ and $H_{k:8}$ are indistinguishable, and*
- *in $H_{k:8}$, $\mathcal{A}$ does not cheat in sessions $s(k), \ldots, s(4m)$ except with negligible probability.*

Lemma 10. *Assume that in $H_{k:8}$ ($k \in [4m]$), $\mathcal{A}$ does not cheat in sessions $s(k), \ldots, s(4m)$ except with negligible probability. Then,*

- *$H_{k:8}$ and $H_{k:9}$ are indistinguishable, and*
- *in $H_{k:9}$, $\mathcal{A}$ does not cheat in sessions $s(k), \ldots, s(4m)$ except with negligible probability.*

Lemma 10 can be proven identically with Lemma 9, and Lemma 9 can be proven quite similarly to Claim 1 (Sect. 4.3); the only difference is that we use the security of $\Pi_{2PC}^{\mathcal{F}_{OT}}$ rather than the hiding of ExtCom. We give a proof of Lemma 9 in the full version of this paper.

By combining Lemmas 9 and 10 with Lemmas 2–8 in Sect. 4.3, we conclude that the output of H_0 and that of $H_{4m:9}$ are indistinguishable, i.e., the output of the real world and that of the ideal world are indistinguishable. This concludes the proof of Theorem 3.

References

1. Badrinarayanan, S., Goyal, V., Jain, A., Khurana, D., Sahai, A.: Round optimal concurrent MPC via strong simulation. Cryptology ePrint Archive, Report 2017/597 (2017), http://eprint.iacr.org/2017/597
2. Barak, B., Sahai, A.: How to play almost any mental game over the net - concurrent composition via super-polynomial simulation. In: 46th FOCS, pp. 543–552. IEEE Computer Society Press, October 2005
3. Broadnax, B., Döttling, N., Hartung, G., Müller-Quade, J., Nagel, M.: Concurrently composable security with shielded super-polynomial simulators. In: Coron, J.-S., Nielsen, J.B. (eds.) EUROCRYPT 2017. LNCS, vol. 10210, pp. 351–381. Springer, Cham (2017). https://doi.org/10.1007/978-3-319-56620-7_13
4. Canetti, R.: Universally composable security: a new paradigm for cryptographic protocols. In: 42nd FOCS, pp. 136–145. IEEE Computer Society Press, October 2001
5. Canetti, R., Kushilevitz, E., Lindell, Y.: On the limitations of universally composable two-party computation without set-up assumptions. In: Biham, E. (ed.) EUROCRYPT 2003. LNCS, vol. 2656, pp. 68–86. Springer, Heidelberg (2003). https://doi.org/10.1007/3-540-39200-9_5
6. Canetti, R., Lin, H., Pass, R.: Adaptive hardness and composable security in the plain model from standard assumptions. In: 51st FOCS, pp. 541–550. IEEE Computer Society Press, October 2010
7. Canetti, R., Lin, H., Pass, R.: Adaptive hardness and composable security in the plain model from standard assumptions. SIAM J. Comput. **45**(5), 1793–1834 (2016)
8. Canetti, R., Lindell, Y., Ostrovsky, R., Sahai, A.: Universally composable two-party and multi-party secure computation. In: 34th ACM STOC, pp. 494–503. ACM Press, May 2002
9. Choi, S.G., Dachman-Soled, D., Malkin, T., Wee, H.: A black-box construction of non-malleable encryption from semantically secure encryption. J. Cryptol. (2017)
10. Cramer, R., Hanaoka, G., Hofheinz, D., Imai, H., Kiltz, E., Pass, R., Shelat, A., Vaikuntanathan, V.: Bounded CCA2-secure encryption. In: Kurosawa, K. (ed.) ASIACRYPT 2007. LNCS, vol. 4833, pp. 502–518. Springer, Heidelberg (2007). https://doi.org/10.1007/978-3-540-76900-2_31
11. Dachman-Soled, D., Malkin, T., Raykova, M., Venkitasubramaniam, M.: Adaptive and concurrent secure computation from new adaptive, non-malleable commitments. In: Sako, K., Sarkar, P. (eds.) ASIACRYPT 2013. LNCS, vol. 8269, pp. 316–336. Springer, Heidelberg (2013). https://doi.org/10.1007/978-3-642-42033-7_17
12. Feige, U., Shamir, A.: Witness indistinguishable and witness hiding protocols. In: 22nd ACM STOC, pp. 416–426. ACM Press, May 1990
13. Garay, J.A., MacKenzie, P.D.: Concurrent oblivious transfer. In: 41st FOCS, pp. 314–324. IEEE Computer Society Press, November 2000
14. Garg, S., Goyal, V., Jain, A., Sahai, A.: Concurrently secure computation in constant rounds. In: Pointcheval, D., Johansson, T. (eds.) EUROCRYPT 2012. LNCS, vol. 7237, pp. 99–116. Springer, Heidelberg (2012). https://doi.org/10.1007/978-3-642-29011-4_8

15. Garg, S., Kumarasubramanian, A., Ostrovsky, R., Visconti, I.: Impossibility results for static input secure computation. In: Safavi-Naini, R., Canetti, R. (eds.) CRYPTO 2012. LNCS, vol. 7417, pp. 424–442. Springer, Heidelberg (2012). https://doi.org/10.1007/978-3-642-32009-5_25
16. Goldreich, O., Micali, S., Wigderson, A.: How to play any mental game or a completeness theorem for protocols with honest majority. In: Aho, A. (ed.) 19th ACM STOC, pp. 218–229. ACM Press, May 1987
17. Goyal, V.: Constant round non-malleable protocols using one way functions. In: Fortnow, L., Vadhan, S.P. (eds.) 43rd ACM STOC, pp. 695–704. ACM Press, June 2011
18. Goyal, V., Jain, A.: On concurrently secure computation in the multiple ideal query model. In: Johansson, T., Nguyen, P.Q. (eds.) EUROCRYPT 2013. LNCS, vol. 7881, pp. 684–701. Springer, Heidelberg (2013). https://doi.org/10.1007/978-3-642-38348-9_40
19. Goyal, V., Lee, C.K., Ostrovsky, R., Visconti, I.: Constructing non-malleable commitments: a black-box approach. In: 53rd FOCS, pp. 51–60. IEEE Computer Society Press, October 2012
20. Goyal, V., Lin, H., Pandey, O., Pass, R., Sahai, A.: Round-efficient concurrently composable secure computation via a robust extraction lemma. In: Dodis, Y., Nielsen, J.B. (eds.) TCC 2015. LNCS, vol. 9014, pp. 260–289. Springer, Heidelberg (2015). https://doi.org/10.1007/978-3-662-46494-6_12
21. Goyal, V., Ostrovsky, R., Scafuro, A., Visconti, I.: Black-box non-black-box zero knowledge. In: Shmoys, D.B. (ed.) 46th ACM STOC, pp. 515–524. ACM Press, May/June 2014
22. Haitner, I.: Semi-honest to malicious oblivious transfer—the black-box way. In: Canetti, R. (ed.) TCC 2008. LNCS, vol. 4948, pp. 412–426. Springer, Heidelberg (2008). https://doi.org/10.1007/978-3-540-78524-8_23
23. Haitner, I., Ishai, Y., Kushilevitz, E., Lindell, Y., Petrank, E.: Black-box constructions of protocols for secure computation. SIAM J. Comput. **40**(2), 225–266 (2011)
24. Håstad, J., Impagliazzo, R., Levin, L.A., Luby, M.: A pseudorandom generator from any one-way function. SIAM J. Comput. **28**(4), 1364–1396 (1999)
25. Hazay, C., Venkitasubramaniam, M.: Composable adaptive secure protocols without setup under polytime assumptions. In: Hirt, M., Smith, A. (eds.) TCC 2016. LNCS, vol. 9985, pp. 400–432. Springer, Heidelberg (2016). https://doi.org/10.1007/978-3-662-53641-4_16
26. Ishai, Y., Kushilevitz, E., Lindell, Y., Petrank, E.: Black-box constructions for secure computation. In: Kleinberg, J.M. (ed.) 38th ACM STOC, pp. 99–108. ACM Press, May 2006
27. Ishai, Y., Prabhakaran, M., Sahai, A.: Founding cryptography on oblivious transfer – efficiently. In: Wagner, D. (ed.) CRYPTO 2008. LNCS, vol. 5157, pp. 572–591. Springer, Heidelberg (2008). https://doi.org/10.1007/978-3-540-85174-5_32
28. Kiyoshima, S.: Round-efficient black-box construction of composable multi-party computation. In: Garay, J.A., Gennaro, R. (eds.) CRYPTO 2014, Part II. LNCS, vol. 8617, pp. 351–368. Springer, Heidelberg (2014). https://doi.org/10.1007/s00145-018-9276-1
29. Kiyoshima, S., Manabe, Y., Okamoto, T.: Constant-round black-box construction of composable multi-party computation protocol. In: Lindell, Y. (ed.) TCC 2014. LNCS, vol. 8349, pp. 343–367. Springer, Heidelberg (2014). https://doi.org/10.1007/978-3-642-54242-8_15
30. Lin, H., Pass, R.: Non-malleability amplification. In: Mitzenmacher, M. (ed.) 41st ACM STOC, pp. 189–198. ACM Press, May/June 2009

31. Lin, H., Pass, R.: Black-box constructions of composable protocols without set-up. In: Safavi-Naini, R., Canetti, R. (eds.) CRYPTO 2012. LNCS, vol. 7417, pp. 461–478. Springer, Heidelberg (2012). https://doi.org/10.1007/978-3-642-32009-5_27
32. Lin, H., Pass, R., Venkitasubramaniam, M.: Concurrent non-malleable commitments from any one-way function. In: Canetti, R. (ed.) TCC 2008. LNCS, vol. 4948, pp. 571–588. Springer, Heidelberg (2008). https://doi.org/10.1007/978-3-540-78524-8_31
33. Lin, H., Pass, R., Venkitasubramaniam, M.: A unified framework for concurrent security: universal composability from stand-alone non-malleability. In: Mitzenmacher, M. (ed.) 41st ACM STOC, pp. 179–188. ACM Press, May/June 2009
34. Lindell, Y.: Bounded-concurrent secure two-party computation without setup assumptions. In: 35th ACM STOC, pp. 683–692. ACM Press, June 2003
35. Lindell, Y.: Lower bounds for concurrent self composition. In: Naor, M. (ed.) TCC 2004. LNCS, vol. 2951, pp. 203–222. Springer, Heidelberg (2004). https://doi.org/10.1007/978-3-540-24638-1_12
36. Malkin, T., Moriarty, R., Yakovenko, N.: Generalized environmental security from number theoretic assumptions. In: Halevi, S., Rabin, T. (eds.) TCC 2006. LNCS, vol. 3876, pp. 343–359. Springer, Heidelberg (2006). https://doi.org/10.1007/11681878_18
37. Micali, S., Pass, R., Rosen, A.: Input-indistinguishable computation. In: 47th FOCS, pp. 367–378. IEEE Computer Society Press, October 2006
38. Naor, M.: Bit commitment using pseudorandomness. J. Cryptol. **4**(2), 151–158 (1991)
39. Ostrovsky, R., Richelson, S., Scafuro, A.: Round-optimal black-box two-party computation. In: Gennaro, R., Robshaw, M. (eds.) CRYPTO 2015. LNCS, vol. 9216, pp. 339–358. Springer, Heidelberg (2015). https://doi.org/10.1007/978-3-662-48000-7_17
40. Ostrovsky, R., Scafuro, A., Venkitasubramanian, M.: Resettably sound zero-knowledge arguments from OWFs - the (semi) black-box way. In: Dodis, Y., Nielsen, J.B. (eds.) TCC 2015. LNCS, vol. 9014, pp. 345–374. Springer, Heidelberg (2015). https://doi.org/10.1007/978-3-662-46494-6_15
41. Pass, R.: Simulation in quasi-polynomial time, and its application to protocol composition. In: Biham, E. (ed.) EUROCRYPT 2003. LNCS, vol. 2656, pp. 160–176. Springer, Heidelberg (2003). https://doi.org/10.1007/3-540-39200-9_10
42. Pass, R.: Bounded-concurrent secure multi-party computation with a dishonest majority. In: Babai, L. (ed.) 36th ACM STOC, pp. 232–241. ACM Press, June 2004
43. Pass, R., Lin, H., Venkitasubramaniam, M.: A unified framework for UC from only OT. In: Wang, X., Sako, K. (eds.) ASIACRYPT 2012. LNCS, vol. 7658, pp. 699–717. Springer, Heidelberg (2012). https://doi.org/10.1007/978-3-642-34961-4_42
44. Pass, R., Wee, H.: Black-box constructions of two-party protocols from one-way functions. In: Reingold, O. (ed.) TCC 2009. LNCS, vol. 5444, pp. 403–418. Springer, Heidelberg (2009). https://doi.org/10.1007/978-3-642-00457-5_24
45. Peikert, C., Waters, B.: Lossy trapdoor functions and their applications. SIAM J. Comput. **40**(6), 1803–1844 (2011)
46. Prabhakaran, M., Sahai, A.: New notions of security: achieving universal composability without trusted setup. In: Babai, L. (ed.) 36th ACM STOC, pp. 242–251. ACM Press, June 2004

47. Venkitasubramaniam, M.: On adaptively secure protocols. In: Abdalla, M., De Prisco, R. (eds.) SCN 2014. LNCS, vol. 8642, pp. 455–475. Springer, Cham (2014). https://doi.org/10.1007/978-3-319-10879-7_26
48. Wee, H.: Black-box, round-efficient secure computation via non-malleability amplification. In: 51st FOCS, pp. 531–540. IEEE Computer Society Press, October 2010
49. Yao, A.C.C.: How to generate and exchange secrets (extended abstract). In: 27th FOCS, pp. 162–167. IEEE Computer Society Press, October 1986

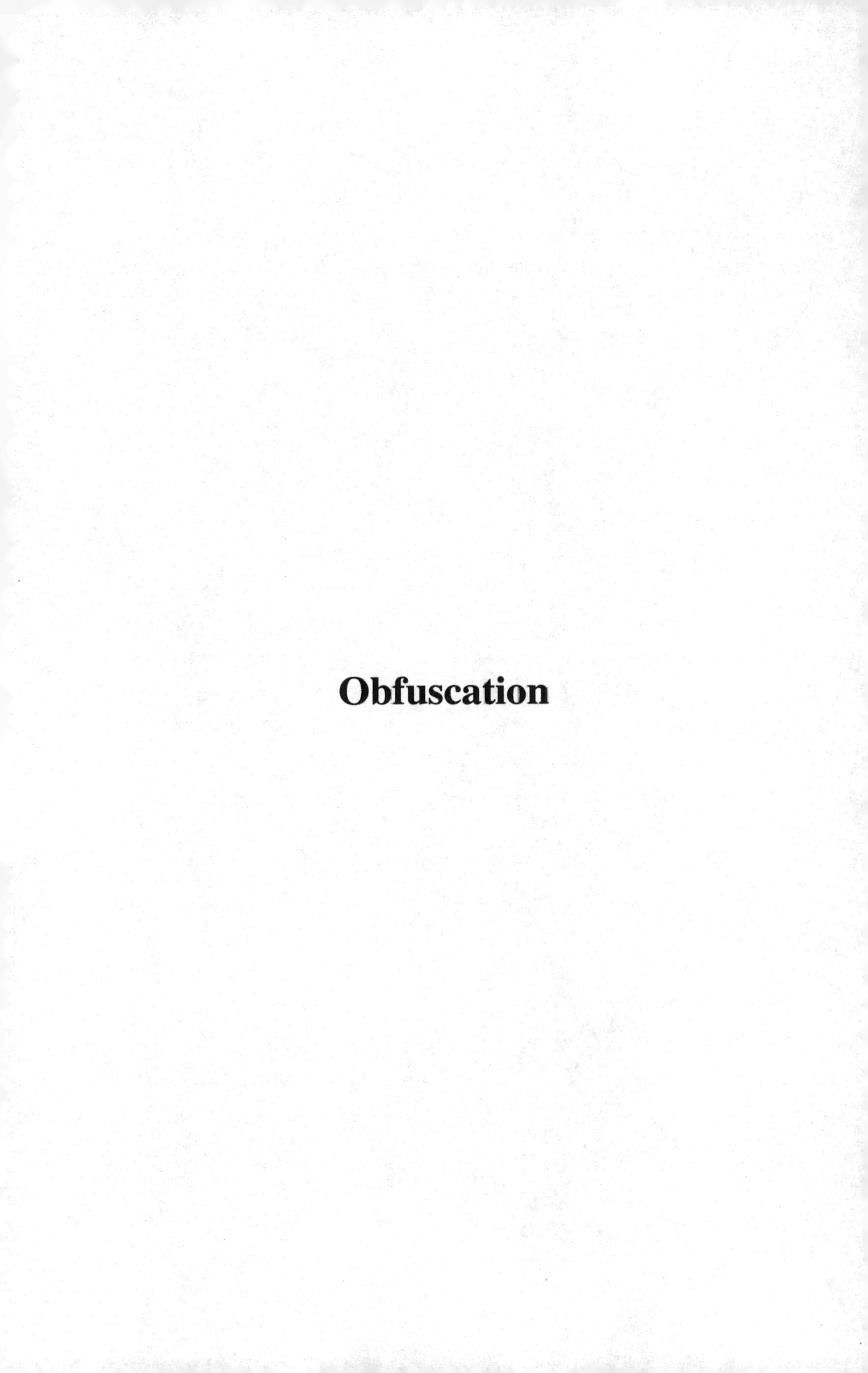

Obfuscation

Obfustopia Built on Secret-Key Functional Encryption

Fuyuki Kitagawa[1(✉)], Ryo Nishimaki[2], and Keisuke Tanaka[1]

[1] Tokyo Institute of Technology, Tokyo, Japan
{kitagaw1,keisuke}@is.titech.ac.jp

[2] Secure Platform Laboratories, NTT Corporation, Tokyo, Japan
nishimaki.ryo@lab.ntt.co.jp

Abstract. We show that indistinguishability obfuscation (IO) for all circuits can be constructed solely from secret-key functional encryption (SKFE). In the construction, SKFE need to be able to issue a-priori unbounded number of functional keys, that is, collusion-resistant. Our strategy is to replace public-key functional encryption (PKFE) in the construction of IO proposed by Bitansky and Vaikuntanathan (FOCS 2015) with *puncturable SKFE*. Bitansky and Vaikuntanathan introduced the notion of puncturable SKFE and observed that the strategy works. However, it has not been clear whether we can construct puncturable SKFE without assuming PKFE. In particular, it has not been known whether puncturable SKFE is constructed from ordinary SKFE. In this work, we show that a relaxed variant of puncturable SKFE can be constructed from collusion-resistant SKFE. Moreover, we show that the relaxed variant of puncturable SKFE is sufficient for constructing IO.

In addition, we also study the relation of collusion-resistance and succinctness for SKFE. Functional encryption is said to be weakly-succinct if the size of its encryption circuit is sub-linear in the size of functions. We show that collusion-resistant SKFE can be constructed from weakly-succinct SKFE supporting only one functional key.

By combining the above two results, we show that IO for all circuits can be constructed from weakly-succinct SKFE supporting only one functional key.

1 Introduction

1.1 Backgrounds

Program obfuscation is now one of the central topics in cryptography. Program obfuscation aims to turn programs "unintelligible" while preserving its functionality. The theoretical study of program obfuscation was initiated by Barak et al. [12]. They introduced *virtual-black-box* obfuscation as a formal definition of obfuscation. The definition of virtual black-box obfuscation is intuitive and naturally captures the requirement that obfuscators hide information about programs. However, Barak et al. showed that it is impossible to achieve virtual

J. B. Nielsen and V. Rijmen (Eds.): EUROCRYPT 2018, LNCS 10821, pp. 603–648, 2018.
https://doi.org/10.1007/978-3-319-78375-8_20

black-box obfuscation for all circuits. In order to avoid the impossibility result, they also defined an weaker variant of obfuscation called *indistinguishability obfuscation (IO)*. Impossibility of IO for all circuits is not known.

Garg et al. [33] proposed the first candidate construction of IO for all circuits. Subsequently, many works have shown that IO is powerful enough in the sense that we can achieve a wide variety of cryptographic primitives based on IO though it is weaker than virtual-black-box obfuscation [14,16,18,26,29,33,39,40,49,62].

While we know the usefulness of IO well, we know very little about how to achieve IO. Although the first candidate construction was demonstrated, we are still at the embryonic stage for constructing IO. All known constructions of IO are based on a little-studied cryptographic tool called multi-linear maps [4,5,8,10,11,24,32–34,51,52,56,60,65]. Moreover, security flaws were discovered in some IO constructions [7,28,30,31,58].

Thus, constructing IO based on a standard assumption is still standing as a major open question in the study of cryptography. As a stepping-stone for solving the question, it is important to find a seemingly weaker primitive that implies IO. As such a cryptographic primitive, we already have *functional encryption*.

Functional encryption is one of the most advanced cryptographic primitives which enable a system having flexibility in controlling encrypted data [20,59,63]. In functional encryption, an owner of a master secret key MSK can generate a functional decryption key sk_f for a function f belonging to a function family $\mathcal{F}$. By decrypting a ciphertext of a message m using sk_f, a holder of sk_f can learn only a value $f(m)$. No information about x except $f(m)$ is revealed from the ciphertext of m. This feature enables us to construct a cryptographic system with fine-grained access control. In addition, it is known that functional encryption is a versatile building block to construct other cryptographic primitives. In particular, we can construct IO for all circuits by using functional encryption that satisfies certain security notions and efficiency requirements [2,3,15,17].

Bitansky and Vaikuntanathan [17] and Ananth and Jain [2] independently showed that we can construct IO based on public-key functional encryption (PKFE) which supports a single functional key and whose encryption circuit size is sub-linear in the size of functions. A functional encryption scheme that supports a single key is called a *single-key* scheme. A functional encryption scheme that satisfies the efficiency property above is said to be *weakly-succinct.*

Bitansky et al. [15] subsequently showed that *collusion-resistant* secret-key functional encryption (SKFE) is powerful enough to yield IO if we additionally assume plain public key encryption. Collusion-resistant functional encryption is functional encryption that can securely issue a-priori unbounded number of functional keys.

From these results, we see that the combination of functional encryption with some property and a public-key cryptographic primitive is sufficient for achieving IO. This fact is a great progress as a stepping-stone for achieving IO based on a standard assumption.

However, one natural question arises for this situation. The question is whether we really need public-key primitives to construct IO or not. In other words, we have the following fundamental question:

Is it possible to achieve IO for all circuits based solely on secret-key primitives?

SKFE is the best possible candidate for a secret-key cryptographic primitive that gives an affirmative answer to this question. However, Asharov and Segev [9] gave a somewhat negative answer to the question. Their result can be seen as a substantial evidence that SKFE is somewhat unlikely to imply IO as long as we use *black-box* techniques.[1] Although Komargodski and Segev [48] already showed that we can construct IO for somewhat restricted class of circuits based on SKFE via non-black-box construction, it is still open whether we can construct IO for all circuits from SKFE bypassing the barrier with a non-black-box technique.

The real power of IO appears in the fact that it can transform secret-key primitives into public-key ones. Therefore, solving the above problem is a key advancement to discover the exact requirements for achieving IO.

1.2 Our Results

We give an affirmative answer to the question above. More precisely, we prove the following theorem.

Theorem 1 (Informal). *Assuming there exists sub-exponentially secure collusion-resistant SKFE for all circuits. Then, there exists IO for all circuits.*

Since our construction of IO is *non-black-box*, we can circumvent the impossibility result shown by Asharov and Segev [9].

The security loss of our construction of IO is exponential in the input length of circuits, but is independent of the size of circuits. Thus, if the input length of circuits is poly-logarithmic in the security parameter, our construction of IO incurs only quasi-polynomial security loss regardless of the size of circuits. Therefore, we can obtain IO for circuits of *polynomial size* with input of poly-logarithmic length from *quasi-polynomially secure* collusion-resistant SKFE for all circuits. This is an improvement over the IO construction by Komargodski and Segev [48]. They showed that IO for circuits of *sub-polynomial size* with input of poly-logarithmic length is constructed from quasi-polynomially secure collusion-resistant SKFE for all circuits.

We show Theorem 1 by using *puncturable SKFE*. The notion of puncturable SKFE was introduced by Bitansky and Vaikuntanathan [17]. They showed that in their construction of IO, the building block PKFE can be replaced with puncturable SKFE. However, it has been an open issue whether we can achieve puncturable SKFE without assuming the existence of PKFE.

[1] More precisely, Asharov and Segev [9] introduced an extended model for black-box reductions to include a limited class of non-black-box reductions into their impossibility results.

We show how to construct puncturable SKFE that is sufficient for constructing IO, based solely on SKFE. More precisely, we show the following theorem.

Theorem 2 (Informal). *Assuming there exists collusion-resistant SKFE for all circuits. Then, there exists single-key weakly-succinct puncturable SKFE for all circuits.*

Note that our definition of puncturable SKFE is slightly different from that proposed by Bitansky and Vaikuntanathan. Our requirement for puncturable SKFE looks weaker than that of Bitansky and Vaikuntanathan. However, they are actually incomparable. In fact, we show that puncturable SKFE defined in this paper is also sufficient for a building block of IO. See Sect. 2 for the details of the notion of puncturable SKFE and the difference between our definition and that of Bitansky and Vaikuntanathan.

This result makes a progress on the study of IO and functional encryption as we note in the next paragraph.

Impacts on the hierarchy of cryptographic primitives. It is known that we can classify cryptographic primitives into two hierarchies MINICRYPT and CRYPTOMANIA since the beautiful work of Impagliazzo and Rudich [42] showed that public-key encryption is not implied by one-way functions via black-box reductions. The terminologies, MINICRYPT and CRYPTOMANIA, were introduced by Impagliazzo [41]. In MINICRYPT, one-way functions exist, but public-key encryption does not. In CRYPTOMANIA, public-key encryption also exists.

We have recently started to consider a new hierarchy called OBFUSTOPIA. Garg et al. [35] introduced the term OBFUSTOPIA, which seems to indicate the "world" where there exists IO. Garg et al. did not give a formal definition of OBFUSTOPIA. In this paper, we explicitly define OBFUSTOPIA as the "world" where there exists efficient IO for all circuits and one-way functions.[2] It is known that we can construct almost all existing cryptographic primitives which are stronger than public-key encryption by using IO. This is the reason why we consider the new hierarchy beyond CRYPTOMANIA.[3]

The landscape of OBFUSTOPIA is not clear while those of MINICRYPT and CRYPTOMANIA are. In particular, we do not know how to construct IO based on standard assumptions. There has been significant effort to find out cryptographic primitives that are in OBFUSTOPIA. That is, we have been asking what kind of cryptographic primitive implies the existence of IO. We know that sub-exponentially-secure succinct PKFE exists in OBFUSTOPIA [2,17].

[2] Komargodski et al. [47] proved that IO implies one-way functions under a mild complexity theoretic assumption. More specifically, the complexity assumption is $\mathsf{NP} \not\subseteq \mathsf{io\text{-}BPP}$, where $\mathsf{io\text{-}BPP}$ is the class of languages that is decided by probabilistic polynomial-time algorithms for infinitely many input sizes. Therefore, under the assumption, we say that OBFUSTOPIA is the complexity spectrum where efficient IO for all circuits exists.

[3] Strictly speaking, it was known that there are stronger primitives than public-key encryption before the candidate of obfuscation appeared. For example, public-key encryption does not imply identity-based encryption [19].

It is natural to ask whether SKFE is also in OBFUSTOPIA or not since SKFE seems to be a strong primitive as PKFE. Asharov and Segev [9] gave a somewhat negative answer to this question. They showed that SKFE is unlikely to imply IO as long as we use black-box techniques. They also showed that SKFE does not imply any primitive in CRYPTOMANIA via black-box reductions. Moreover, it was not known whether SKFE implies any primitive outside MINICRYPT even if we use it in a non-black-box manner before the work of Bitansky et al. [15].

Bitansky et al. showed that the combination of sub-exponentially secure collusion-resistant SKFE and exponentially secure one-way functions implies quasi-polynomially secure public-key encryption. This also implies that the above combination yields quasi-polynomially secure succinct PKFE from their main result showing that the combination of collusion-resistant SKFE and public-key encryption implies succinct PKFE.

Komargodski and Segev [48] showed that quasi-polynomially secure IO for circuits of sub-polynomial size with input of poly-logarithmic length can be constructed from quasi-polynomially secure collusion-resistant SKFE for all circuits. In addition, they showed that by combining quasi-polynomially secure collusion-resistant SKFE and sub-exponentially secure one-way functions, we can construct quasi-polynomially secure succinct PKFE. However, in this construction, the resulting PKFE supports only circuits of sub-polynomial size with input of poly-logarithmic length though the building block SKFE supports all polynomial size circuits.

These two results surely demonstrated that SKFE is stronger than we thought. Nevertheless, we see that both two results involve degradation of security level or functionality. Thus, it is still open whether SKFE implies a cryptographic primitive other than those in MINICRYPT without such degradation, and especially SKFE is in OBFUSTOPIA or not.

We gives an affirmative answer to this question. More concretely, we can construct sub-exponentially secure IO for all circuits from sub-exponentially secure collusion-resistant SKFE for all circuits through our transformation by setting security parameter appropriately. This result means that sub-exponentially secure collusion-resistant SKFE exists in OBFUSTOPIA. In addition, by combining this result and the result by Garg et al. [33], we see that the existence of sub-exponentially secure collusion-resistant PKFE for all circuits is equivalent to that of sub-exponentially secure collusion-resistant SKFE for all circuits.

Collusion-resistance versus succinctness for SKFE. We also study the relation of collusion-resistance and succinctness for SKFE.

Collusion-resistance and succinctness for functional encryption are seemingly incomparable notions and implications between them are non-trivial. Therefore, it is also a major concern whether we can transform a scheme satisfying one of the two properties into a collusion-resistant and succinct one.

Such a transformation is already known for PKFE. Ananth et al. [3] showed how to construct collusion-resistant and succinct PKFE from collusion-resistant one. In addition, Garg and Srinivasan [36] and Li and Micciancio [50] showed a transformation from single-key weakly-succinct PKFE to collusion-resistant one with polynomial security loss. Their transformations preserve succinctness of the building block scheme.[4] From these results, collusion-resistance and succinctness are equivalent for PKFE.

On the other hand, the situation is different for SKFE. While we know how to construct collusion-resistant and succinct schemes from collusion-resistant ones [3] similarly to PKFE, we do not know how to construct such schemes from succinct ones *even if sub-exponential security loss is permitted.*

As stated above, some recent results including Theorem 1 show that SKFE is a strong cryptographic primitive beyond MINICRYPT if we consider non-black-box reductions. However, one natural question arises for this situation. All of those results assume collusion-resistant SKFE as a building block. Thus, while we see that collusion-resistant SKFE is outside MINICRYPT, it is still open whether succinct SKFE is also a strong cryptographic primitive beyond MINICRYPT since we do not know how to construct collusion-resistant SKFE from succinct one.

Succinctness seems to be as powerful as collusion-resistance from the equivalence of them in the PKFE setting. Therefore, it is natural to ask whether succinct SKFE is also outside MINICRYPT. If we have a transformation from succinct SKFE to collusion-resistant one without assuming public-key primitives, we can solve the question affirmatively. Solving the question is an advancement to understand the complexity of SKFE.

We solve the question by showing the following result.

Theorem 3 (Informal). *Assume that there exists quasi-polynomially (resp. sub-exponentially) secure single-key weakly-succinct SKFE for all circuits. Then, there also exists quasi-polynomially (resp. sub-exponentially) secure collusion-resistant SKFE for all circuits.*

We note that our transformation incurs quasi-polynomial security loss. However, we can transform any quasi-polynomially secure single-key weakly-succinct SKFE into quasi-polynomially secure collusion-resistant one, if we know the security bound of the underlying single-key SKFE. In addition, if the underlying single-key scheme is sub-exponentially secure, then so does the resulting collusion-resistant one.[5]

Our transformation preserves the succinctness of the underlying scheme. In other words, if the building block single-key scheme is succinct (resp. weakly succinct), the resulting collusion-resistant scheme is also succinct (resp. weakly succinct).

Analogous to PKFE, we can transform collusion-resistant SKFE into collusion-resistant and succinct one [3]. From this fact and Theorem 3, we discover that the existence of collusion-resistant SKFE and that of succinct one

[4] The resulting scheme of the transformation proposed by Garg and Srinivasan is succinct even if the building block scheme is only weakly-succinct. The transformation proposed by Li and Micciancio preserves succinctness of the building block scheme.

[5] When transforming a sub-exponentially secure scheme, our transformation incurs sub-exponentially security loss. However, we can transform any sub-exponentially secure single-key scheme into a sub-exponentially secure collusion-resistant one.

are actually equivalent if we allow quasi-polynomial security loss. Due to this equivalence, we see that succinct SKFE is also a strong cryptographic primitive beyond MINICRYPT similarly to collusion-resistant SKFE. Especially, we obtain the following corollary from Theorems 1 and 3.

Corollary 1 (Informal). *Assume that there exists sub-exponentially secure single-key weakly-succinct SKFE for all circuits. Then, there exists IO for all circuits.*

From this result, we can remove the learning with errors (LWE) assumption from recent state-of-the-art constructions of IO based on multi-linear maps and (block-wise) local pseudorandom generators [52,55].

These works first construct single-key weakly-succinct *SKFE* based on multi-linear maps and (block-wise) local pseudorandom generators. Then, assuming the LWE assumption, they transform it into IO using the result by Bitansky et al. [15]. By relying on Corollary 1 instead of the result by Bitansky et al. [15] in their construction, we can obtain IO based only on multi-linear maps and (block-wise) local pseudorandom generators.

1.3 Organization

We provide the overview of this work using the majority of the remaining part of this paper. For Theorem 1, we show only constructions and omit its security proofs. See [45] for those omitted proofs. For Theorem 3, we provide only the overview. See [44] for details of this result. The detailed organization is as follows.

In Sect. 2, we provide the overview of our construction of IO based on collusion-resistant SKFE via puncturable SKFE. In Sect. 3, we also provide the overview of how collusion-resistant SKFE is constructed based on weakly-succinct SKFE. In Sect. 4, we provide notations and definitions of cryptographic primitives. In Sect. 5, we formally define puncturable SKFE, and introduce security and efficiency notions for it. We also discuss the difference between our definition of puncturable SKFE and that of Bitansky and Vaikuntanathan [17] in Sect. 5. In Sect. 6, we show the construction of single-key non-succinct puncturable SKFE. In Sect. 7, we show how to transform single-key non-succinct puncturable SKFE into single-key weakly-succinct one. In Sect. 8, we then show how to construct IO based on SKFE.

2 Overview: IO from Collusion-Resistant SKFE

We give an overview of our construction of IO based on SKFE in this section.

Our basic strategy is to replace PKFE in the construction of Bitansky and Vaikuntanathan [17] with puncturable SKFE. Bitansky and Vaikuntanathan observed that this strategy works. However, it is not known whether puncturable SKFE is constructed from cryptographic primitives other than PKFE or IO.

In this work, we show that we can construct a relaxed variant of puncturable SKFE that is a single-key scheme and weakly-succinct from collusion-resistant

SKFE. Moreover, we show that such a relaxed variant of puncturable SKFE is sufficient for constructing IO.

We give an overview of the construction of Bitansky and Vaikuntanathan [17] in Sect. 2.1 and explain why SKFE must be "puncturable" when we replace PKFE with SKFE in their construction in Sect. 2.2. Next, we give an overview of how to construct our puncturable SKFE scheme and IO in Sects. 2.3 and 2.4, respectively.

2.1 Construction of IO Based on PKFE

The main idea of Bitansky and Vaikuntanathan is to design an obfuscator $i\mathcal{O}_i$ for circuits with i-bit input from an obfuscator $i\mathcal{O}_{i-1}$ for circuits with $(i-1)$-bit input. If we can design such a bit extension construction, for any polynomial n, we can construct an obfuscator $i\mathcal{O}_n$ for circuits with n-bit input since we can easily achieve $i\mathcal{O}_1$ for circuits with 1-bit input by outputting an entire truth table of a circuit with 1-bit input. If you are familiar with the construction of Bitansky and Vaikuntanathan [17], then you can skip this section.

When we construct IO based on the bit extension construction above, it is important to avoid a circuit-size blow-up of circuits to be obfuscated at each recursive step. In fact, if we allow a circuit-size blow-up, we can obtain the bit extension construction by defining

$$i\mathcal{O}_i(C(x_1 \cdots x_i)) := i\mathcal{O}_{i-1}(C(x_1 \cdots x_{i-1} \| 0)) \| i\mathcal{O}_{i-1}(C(x_1 \cdots x_{i-1} \| 1)) \ .$$

However, this construction obviously incurs an exponential blow-up and thus we cannot rely on this solution. Bitansky and Vaikuntanathan showed how to achieve the bit extension construction without an exponential blow-up using *weakly-succinct* PKFE.

In their construction, a functional key of PKFE should hide information about the corresponding circuit. Such security notion is called function privacy. However, it is not known how to achieve function private PKFE. Then, Bitansky and Vaikuntanathan explicitly accommodated the technique for function private SKFE used by Brakerski and Segev [25] to their IO construction based on PKFE.

We review their construction based on PKFE. For simplicity, we ignore the issue of the randomness for encryption algorithms. It is generated by puncturable pseudorandom function (PRF) in the actual construction.

$i\mathcal{O}_i$ based on $i\mathcal{O}_{i-1}$ and PKFE works as follows. The construction additionally uses plain secret key encryption (SKE) to implement the technique used by Brakerski and Segev [25]. To obfuscate a circuit C with i-bit input, it first generates a key pair $(\mathsf{PK}_i, \mathsf{MSK}_i)$ of PKFE. Then, using MSK_i, it generates a functional key sk_{C^*} tied to the following circuit C^*. C^* has hardwired two SKE ciphertexts $\mathsf{CT}_0^{\mathsf{ske}}$ and $\mathsf{CT}_1^{\mathsf{ske}}$ of plaintext C under independent keys K_0 and K_1, respectively. C^* expects as an input not only an i-bit string $\boldsymbol{x}_i$ but also an SKE key K_b. On those inputs, C^* first obtains C by decrypting $\mathsf{CT}_b^{\mathsf{ske}}$ by K_b and outputs $U(C, \boldsymbol{x}_i) = C(\boldsymbol{x}_i)$, where $U(\cdot, \cdot)$ is an universal circuit. Finally, the construction obfuscates the following encryption circuit E_{i-1} by $i\mathcal{O}_{i-1}$. E_{i-1} has hardwired PK_i and K_b. On input $(i-1)$-bit string $\boldsymbol{x}_{i-1}$, it outputs ciphertexts $\mathsf{Enc}(\mathsf{PK}_i, (\boldsymbol{x}_{i-1}\|0, K_b))$ and $\mathsf{Enc}(\mathsf{PK}_i, (\boldsymbol{x}_{i-1}\|1, K_b))$, where Enc is the encryption algorithm of PKFE. The resulting obfuscation of C is a tuple $(\mathsf{sk}_{C^*}, i\mathcal{O}_{i-1}(\mathsf{E}_{i-1}))$. Note that we always set the value of b as 0 in the actual construction. We set b as 1 only in the security proof.

// Description of (simplified) C^*	// Description of (simplified) E_{i-1}
Hard-Coded Constants: $\mathsf{CT}_0^{\mathsf{ske}}$, $\mathsf{CT}_1^{\mathsf{ske}}$.	**Hard-Coded Constants**: PK_i, K_b.
Input: $\boldsymbol{x}_i$, K_b	**Input:** $\boldsymbol{x}_{i-1} \in \{0,1\}^{i-1}$
1. Compute $C = \mathsf{D}(K_b, \mathsf{CT}_b^{\mathsf{ske}})$.	1. Compute $\mathsf{CT}_{i,x_i} \xleftarrow{\mathsf{r}} \mathsf{Enc}(\mathsf{PK}_i, (\boldsymbol{x}_{i-1}\|x_i, K_b))$.
2. Return $U(C, \boldsymbol{x}_i)$.	2. Output $\mathsf{CT}_{i,0}$ and $\mathsf{CT}_{i,1}$.

When evaluating the obfuscated C on input $\boldsymbol{x}_i = x_1 \cdots x_{i-1}x_i \in \{0,1\}^i$, we first invoke $i\mathcal{O}(\mathsf{E}_{i-1})$ on input $\boldsymbol{x}_{i-1} = x_1 \cdots x_{i-1}$ and obtain $\mathsf{Enc}(\mathsf{PK}_i, (\boldsymbol{x}_{i-1}\|0, K_b))$ and $\mathsf{Enc}(\mathsf{PK}_i, (\boldsymbol{x}_{i-1}\|1, K_b))$. Then, by decrypting $\mathsf{Enc}(\mathsf{PK}_i, (\boldsymbol{x}_{i-1}\|x_i, K_b))$ using sk_{C^*}, we obtain $C(\boldsymbol{x}_i)$.

Consequently, by using this bit extension construction, the obfuscation of a circuit C with n-bit input consists of n functional keys $\mathsf{sk}_1, \cdots, \mathsf{sk}_n$ each of which is generated under a different master secret key MSK_i, and pair of ciphertexts of 0 and 1 under PK_1 corresponding to MSK_1. For any $\boldsymbol{x}_n = x_1 \cdots x_n \in \{0,1\}^n$, we can first compute a ciphertext of $\boldsymbol{x}_n$ by repeatedly decrypting a ciphertext of $\boldsymbol{x}_{i-1} = x_1 \cdots x_{i-1}$ by sk_{i-1} and obtaining a ciphertext of $\boldsymbol{x}_i = x_1 \cdots x_i$ for every $i \in \{2, \cdots, n\}$. We can finally obtain $C(\boldsymbol{x}_n)$ by decrypting the ciphertext of $\boldsymbol{x}_n$ by sk_n.

In this construction, each instance of PKFE needs to issue only one functional key. This is a minimum requirement for functional encryption. However, for efficiency, PKFE in the construction above should satisfy a somewhat strong requirement, that is, weak-succinctness to avoid a circuit-size blow-up of circuits to be obfuscated at each recursive step. Therefore, we need to use a single-key weakly-succinct PKFE scheme in the IO construction above.

We can prove the security of the construction recursively. More precisely, we can prove the security of $i\mathcal{O}_i$ based on those of $i\mathcal{O}_{i-1}$, PKFE, and SKE. Note that it is sufficient that PKFE satisfies a mild selective-security to complete the proof. Their security proof relies on the argument of probabilistic IO formalized by Canneti et al. [27], and thus the security loss of each recursive step is exponential in i, that is 2^i. This is the reason their building block PKFE must be sub-exponentially secure.

2.2 Replacing PKFE with SKFE: Need of Puncturable SKFE

The security proof of Bitansky and Vaikuntanathan relies on the fact that we can use the security of PKFE even when its encryption circuit is publicly available. Concretely, PK_i is hardwired into obfuscated encryption circuit $i\mathcal{O}_{i-1}(\mathsf{E}_{i-1})$ and this encryption circuit is public when we use the security of PKFE under the key pair $(\mathsf{PK}_i, \mathsf{MSK}_i)$.

The above security argument might not work if ordinary SKFE is used instead of PKFE. This intuition comes from the impossibility result shown by Barak et al. [12]. In fact, Bitansky and Vaikuntanathan showed that it is impossible to instantiate their IO by using SKFE. More precisely, they showed that there exists a secure SKFE scheme such that their transformation results in insecure IO if the SKFE scheme is used as the building block. This is why they adopted PKFE as their building block. Therefore, in order to replace PKFE with SKFE in the construction above, we need SKFE whose security holds even when its encryption circuit is publicly available. As one of such primitives, Bitansky and Vaikuntanathan proposed *puncturable SKFE*.

In puncturable SKFE defined by Bitansky and Vaikuntanathan, there are a puncturing algorithm Punc and a punctured encryption algorithm PEnc in addition to algorithms of ordinary SKFE. We can generate a punctured master secret key $\mathsf{MSK}^*\{m_0, m_1\}$ at two messages m_0 and m_1 from a master secret key MSK by using Punc. Puncturable SKFE satisfies the following two properties: *functionality preserving under puncturing* and *semantic security at punctured point*. Functionality preserving under puncturing requires that

$$\mathsf{Enc}(\mathsf{MSK}, m; r) = \mathsf{PEnc}(\mathsf{MSK}^*\{m_0, m_1\}, m; r)$$

holds for any message m other than m_0 and m_1 and for any randomness r. Semantic security at punctured point requires that

$$(\mathsf{MSK}^*\{m_0, m_1\}, \mathsf{Enc}(\mathsf{MSK}, m_0) \stackrel{c}{\approx} (\mathsf{MSK}^*\{m_0, m_1\}, \mathsf{Enc}(\mathsf{MSK}, m_1))$$

holds for all adversaries, where $\stackrel{c}{\approx}$ denotes computational indistinguishability.

Bitansky and Vaikuntanathan showed that single-key weakly-succinct puncturable SKFE is also a sufficient building block for their IO construction while ordinary SKFE is not. Note that weak-succinctness of puncturable SKFE requires that not only the encryption circuit but also the punctured encryption circuit should be weakly-succinct. However, as stated earlier, there was no instantiation of puncturable SKFE other than regarding PKFE as puncturable SKFE at that point. In particular, it was not clear whether we can construct puncturable SKFE based on ordinary SKFE.

2.3 Puncturable SKFE from SKFE

In this work, we show we can construct single-key weakly-succinct puncturable SKFE from collusion-resistant SKFE. More specifically, we show the following two results. First, we show how to construct single-key non-succinct puncturable SKFE based only on one-way functions. In addition, we show that we can transform it into single-key weakly-succinct one using collusion-resistant SKFE. Our formalization of puncturable SKFE is different from that of Bitansky and Vaikuntanathan [17] in several aspects. Nevertheless, we show that our puncturable SKFE is also sufficient for constructing IO.

Below, we give the overview of these two constructions.

Single-Key Non-Succinct Puncturable SKFE Based on One-Way Functions. Our starting point is the SKFE variant of the single-key non-succinct PKFE scheme proposed by Sahai and Seyalioglu [61]. It is constructed from garbled circuit and SKE, which are implied by one-way functions. Their construction is as follows.

Setup: A master secret key consists of $2s$ secret keys $\{K_{j,\alpha}\}_{j\in[s],\alpha\in\{0,1\}}$ of SKE, where s is the length of a binary representation of functions supported by the resulting SKFE scheme.

Enc: When we encrypt a message m, we first generates a garbled circuit $\widetilde{U}_m$ with labels $\{L_{j,\alpha}\}_{j\in[s],\alpha\in\{0,1\}}$ by garbling an universal circuit $U(\cdot,m)$ into which m is hardwired. Then, we encrypt $L_{j,\alpha}$ under $K_{j,\alpha}$ and obtain an SKE ciphertext $c_{j,\alpha}$ for every $j\in[s]$ and $\alpha\in\{0,1\}$. The resulting ciphertext of the scheme is $(\widetilde{U}_m,\{c_{j,\alpha}\}_{j\in[s],\alpha\in\{0,1\}})$.

KeyGen: A functional key sk_f for a function f consists of $\{K_{j,f[j]}\}_{j\in[s]}$, where $f[1]\cdots f[s]$ is the binary representation of f and each $f[j]$ is a single bit.

Dec: A decryptor who has a ciphertext $(\widetilde{U}_m,\{c_{j,\alpha}\}_{j\in[s],\alpha\in\{0,1\}})$ and a functional key $\{K_{j,f[j]}\}_{j\in[s]}$ can compute $\{L_{j,f[j]}\}_{j\in[s]}$ by decrypting each $c_{j,f[j]}$ by $K_{j,f[j]}$ and obtain $\widetilde{U}_m(\{L_{j,f[j]}\}_{j\in[s]})=U(f,m)=f(m)$.

In the construction above, we observe that if we use puncturable PRF instead of SKE, the resulting scheme is puncturable in some sense. More specifically, a master secret key now consists of $2s$ puncturable PRF keys $\{S_{j,\alpha}\}_{j\in[s],\alpha\in\{0,1\}}$. When we encrypt a message m, we first generate $(\widetilde{U}_m,\{L_{j,\alpha}\}_{j\in[s],\alpha\in\{0,1\}})$ and encrypt each label by using a puncturable PRF value. That is, $c_{j,\alpha}\leftarrow L_{j,\alpha}\oplus \mathsf{F}_{S_{j,\alpha}}(\mathsf{tag})$, where F is puncturable PRF and tag is a public tag chosen in some way.

In this case, we can generate a punctured master secret key $\mathsf{MSK}^*\{\mathsf{tag}\}$ at a tag tag. Thus, we define an encryption algorithm in a tag-based manner. The encryption algorithm Enc, given MSK, tag, and m, outputs a ciphertext of m under the tag tag. That is, $\mathsf{Enc}(\mathsf{MSK},\mathsf{tag},m)=(\widetilde{U}_m,\{L_{j,\alpha}\oplus \mathsf{F}_{S_{j,\alpha}}(\mathsf{tag})\}_{j\in[s],\alpha\in\{0,1\}})$. A punctured master secret key $\mathsf{MSK}^*\{\mathsf{tag}\}$ consists of $2s$ puncturable PRF keys $\{S^*_{j,\alpha}\{\mathsf{tag}\}\}_{j\in[s],\alpha\in\{0,1\}}$ all of which are punctured at tag.

By using $\mathsf{MSK}^*\{\mathsf{tag}\}$, we can generate a ciphertext of any message m under a tag tag' different from tag, that is, $\mathsf{PEnc}(\mathsf{MSK}^*\{\mathsf{tag}\},\mathsf{tag}',m)=(\widetilde{U}_m,\{L_{j,\alpha}\oplus \mathsf{F}_{S^*_{j,\alpha}\{\mathsf{tag}\}}(\mathsf{tag}')\}_{j\in[s],\alpha\in\{0,1\}})$. Then, we have

$$\mathsf{Enc}(\mathsf{MSK},\mathsf{tag}',m;r)=\mathsf{PEnc}(\mathsf{MSK}^*\{\mathsf{tag}\},\mathsf{tag}',m;r)$$

for any tag tag and tag' such that $\mathsf{tag}\neq\mathsf{tag}'$, message m, and randomness r due to the functionality preserving property of puncturable PRF. Namely, this scheme satisfies functionality preserving under puncturing.

In addition, we can prove that $\mathsf{Enc}(\mathsf{MSK},\mathsf{tag},m_0)$ and $\mathsf{Enc}(\mathsf{MSK},\mathsf{tag},m_1)$ are indistinguishable for adversaries that have $\mathsf{MSK}^*\{\mathsf{tag}\}$ based on the security of puncturable PRF. In other words, it satisfies semantic security at punctured tag.

This formalization is different from that proposed by Bitansky and Vaikuntanathan. Nevertheless, our formalization of puncturable SKFE is sufficient for constructing IO. In fact, when we construct IO, we set the tag same as the message to be encrypted itself. Then, our formalization is conceptually the same as that of Bitansky and Vaikuntanathan. Our tag-based definition is well-suited for our constructions.

Achieving Weak-Succinctness via Collusion-Succinctness. We cannot directly use the puncturable SKFE scheme above as a building block of IO since it is non-succinct. We need to transform it into an weakly-succinct scheme while preserving security and functionality.

We extend the work by Kitagawa et al. [46] that showed how to transform non-succinct PKFE into weakly-succinct one using collusion-resistant SKFE. They accomplished the transformation via a *collusion-succinct* scheme. We try to accommodate their transformation techniques into the context of puncturable SKFE.

Collusion-succinctness requires that each size of the encryption circuit and punctured encryption circuit is sub-linear in the number of functional keys that the scheme can issue. Note that when we consider collusion-succinctness, the size of these circuits can be polynomial of the size of functions.

We first show that we can construct collusion-succinct puncturable SKFE based on single-key non-succinct puncturable SKFE and collusion-resistant SKFE. Then, we transform the collusion-succinct scheme into an weakly-succinct scheme via a transformation based on decomposable randomized encoding. The latter transformation based on decomposable randomized encoding is similar to that proposed by Bitansky and Vaikuntanathan [17] and that proposed by Ananth et al. [3]. We give an illustration of our construction path in Fig. 1.

The general picture is similar to that of Kitagawa et al. [46] and we can accomplish the latter transformation based on a known technique, but there is a technical hurdle in the former transformation. The most biggest issue is how to define punctured master secret keys and the punctured encryption algorithm. We show the overview of the former transformation and explain the technical hurdle below.

Construction of collusion-succinct scheme. Our goal of this step is to construct a collusion-succinct scheme, that is, a scheme which supports q functional keys and the size of whose encryption and punctured encryption circuits are sub-linear in q, where q is an a-priori fixed polynomial. The key tool for achieving this goal is strong exponentially-efficient IO (SXIO) proposed by Lin et al. [53].

SXIO is a relaxed variant of IO. SXIO is required that, given a circuit C with n-bit input, it runs in $2^{\gamma n} \cdot \mathrm{poly}(\lambda, |C|)$-time, where γ is a constant smaller than 1, poly is some polynomial, and λ is the security parameter. We call γ the compression factor since it represents how SXIO can compress the truth table of the circuit to be obfuscated. SXIO with arbitrarily small constant compression factor can be constructed from collusion-resistant SKFE [15].

We show how to construct collusion-succinct puncturable SKFE from single-key non-succinct one and SXIO. To achieve a collusion-succinct scheme, we need to increase the number of functional keys to some polynomial q while compressing the size of its encryption circuits into sub-linear in q.

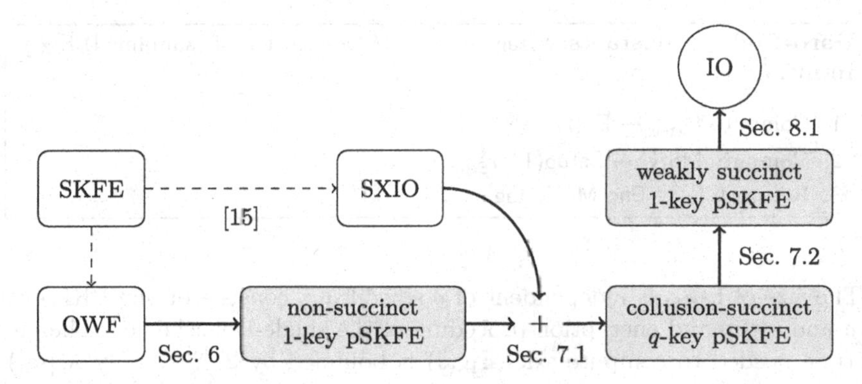

Fig. 1. Illustration of our construction path. pSKFE denotes puncturable SKFE. Dashed lines denote known or trivial implications. White boxes denote our ingredients or goal. Purple boxes denote our core schemes. A transformation from an object in a rectangle to one in a rectangle incurs only polynomial security loss. A transformation from an object in a rectangle to one in a circle incurs super-polynomial security loss. (Color figure online)

The most naive way to increase the number of functional keys is to run multiple instances of the single-key scheme. If we have q master secret keys $\mathsf{MSK}_1, \cdots, \mathsf{MSK}_q$, we can generate q functional keys since we can generate one functional key under each master secret key. In this case, to ensure that we can decrypt a ciphertext using every functional key under different master secret keys MSK_i for every $i \in [q]$, a ciphertext should be composed of q ciphertexts each of which is generated under MSK_i for every $i \in [q]$. In addition, when we generate a punctured master secret key punctured at tag, we generate q punctured master secret keys $\mathsf{MSK}_i^*\{\mathsf{tag}\}$ for every $i \in [q]$ all of which are punctured at tag.

In the naive construction above, we see that if the single-key scheme satisfies functionality preserving under puncturing and semantic security at punctured tag, then so does the resulting scheme since a ciphertext of the resulting scheme consists of only those of the single-key scheme. However, if a ciphertext of the resulting scheme consists of q ciphertexts of the single-key scheme, the encryption time is obviously at least linear in q. Therefore, we cannot construct a collusion-succinct scheme based on this naive idea.

We then consider to compress the encryption time by using SXIO. We extend the technique used in some previous results [15,46,53]. Let $\mathsf{sxi}\mathcal{O}$ be SXIO. We set a ciphertext as a circuit computing q ciphertexts obfuscated by $\mathsf{sxi}\mathcal{O}$ instead of setting it as q ciphertexts themselves. Concretely, we obfuscate the following circuit $\mathsf{E}_{1\mathsf{Key}}$ using $\mathsf{sxi}\mathcal{O}$. $\mathsf{E}_{1\mathsf{Key}}$ has hardwired message m, tag tag, and puncturable PRF key S, and on input $i \in [q]$, it first generates MSK_i pseudorandomly from S and i, and then outputs a ciphertext of m under MSK_i and tag. Note that the master secret key of this scheme is now one puncturable PRF key S. In other words, the scheme generates q master secret keys of the single-key scheme from one puncturable PRF key. For the formal description of $\mathsf{E}_{1\mathsf{Key}}$, see Fig. 4 in Sect. 7.1.

Hard-Coded Constants: S, tag, m. // Description of (simplified) $\mathsf{E}_{1\mathsf{Key}}$
Input: $i \in [q]$

1. Compute $r^i_{\mathsf{Setup}} \leftarrow \mathsf{F}_S(i)$.
2. Compute $\mathsf{MSK}_i \leftarrow \mathsf{Setup}(1^\lambda; r^i_{\mathsf{Setup}})$.
3. Return $\mathsf{CT}_i \leftarrow \mathsf{Enc}(\mathsf{MSK}_i, \mathsf{tag}, m)$.

The size of $\mathsf{E}_{1\mathsf{Key}}$ is independent of q since $\mathsf{E}_{1\mathsf{Key}}$ consists of one PRF evaluation and setup and encryption procedure of the single-key scheme.[6] Therefore, the time needed to compute $\mathsf{sxi}\mathcal{O}(\mathsf{E}_{1\mathsf{Key}})$ is bounded by $2^{\gamma \log q} \cdot \mathrm{poly}(\lambda, |m|) = q^\gamma \cdot \mathrm{poly}(\lambda, |m|)$ for some constant $\gamma < 1$ and polynomial poly, that is, sub-linear in q. Namely, we succeeds in reducing the encryption time from linear to sub-linear in q.

However, we need more complicated structure to compress the running-time of a punctured encryption algorithm into sub-linear in q. The main reason is that we cannot give master secret key S in the clear in the punctured encryption circuit to reduce the security to that of the building block single-key scheme.

We first argue how to set a punctured master secret key. We cannot rely on the trivial way that sets q punctured master secret keys of the single-key scheme as a punctured master secret key since the size of the punctured encryption circuit becomes linear in q in this trivial way.

Our solution is to set a punctured master secret key as also an obfuscated circuit under SXIO. More precisely, we obfuscate the following circuit $\mathsf{P}_{1\mathsf{Key}}$. $\mathsf{P}_{1\mathsf{Key}}$ has hardwired tag tag and puncturable PRF key S. Note that S is the master secret key thus is the same puncturable PRF key as that hardwired into $\mathsf{E}_{1\mathsf{Key}}$. On input $i \in [q]$, $\mathsf{P}_{1\mathsf{Key}}$ first generates MSK_i pseudorandomly from S and i, and then outputs a punctured master secret key $\mathsf{MSK}^*_i\{\mathsf{tag}\}$ of the single-key scheme. For the formal description of $\mathsf{P}_{1\mathsf{Key}}$, see Fig. 5 in Sect. 7.1.

// Description of (simplified) $\mathsf{P}_{1\mathsf{Key}}$
Hard-Coded Constants: S, tag.
Input: $i \in [q]$

1. Compute $r^i_{\mathsf{Setup}} \leftarrow \mathsf{F}_S(i)$.
2. Compute $\mathsf{MSK}_i \leftarrow \mathsf{Setup}(1^\lambda; r^i_{\mathsf{Setup}})$.
3. Return $\mathsf{MSK}^*_i\{\mathsf{tag}\} \leftarrow \mathsf{Punc}(\mathsf{MSK}_i, \mathsf{tag})$.

// Description of (simplified) $\mathsf{PE}_{1\mathsf{Key}}$
Hard-Coded Constants: $\mathsf{MSK}^*\{\mathsf{tag}\}, \mathsf{tag}', m$.
Input: $i \in [q]$

1. Parse $\mathsf{sxi}\mathcal{O}(\mathsf{P}_{1\mathsf{Key}}) \leftarrow \mathsf{MSK}^*\{\mathsf{tag}\}$.
2. Compute $\mathsf{MSK}^*_i\{\mathsf{tag}\} \leftarrow \mathsf{sxi}\mathcal{O}(\mathsf{P}_{1\mathsf{Key}})(i)$.
3. Return $\mathsf{CT}_i \leftarrow \mathsf{PEnc}(\mathsf{MSK}^*_i\{\mathsf{tag}\}, \mathsf{tag}', m)$.

In addition, we define the punctured encryption algorithm as follows. On input $\mathsf{MSK}^*\{\mathsf{tag}\}$ that is $\mathsf{sxi}\mathcal{O}(\mathsf{P}_{1\mathsf{Key}})$, tag tag', and message m, the punctured encryption algorithm obfuscates the following circuit $\mathsf{PE}_{1\mathsf{Key}}$ using $\mathsf{sxi}\mathcal{O}$ and outputs the obfuscated circuit. $\mathsf{PE}_{1\mathsf{Key}}$ has hardwired $\mathsf{MSK}^*\{\mathsf{tag}\}$, tag', and m, and on input

[6] Strictly speaking, the domain of PRF is $[q]$, and thus the size of $\mathsf{E}_{1\mathsf{Key}}$ depends on q in logarithmic. However, it does not matter since logarithmic factor is absorbed by sub-linear factor. We ignore this issue here for simplicity.

$i \in [q]$, it first generates the i-th punctured key $\mathsf{MSK}_i^*\{\mathsf{tag}\}$ by feeding i into $\mathsf{MSK}^*\{\mathsf{tag}\} = \mathsf{sxi}\mathcal{O}(\mathsf{PE}_{\mathsf{1Key}})$, and then outputs a ciphertext of m under $\mathsf{MSK}_i^*\{\mathsf{tag}\}$ and tag' using the punctured encryption algorithm of the single-key scheme. If the compression factor of $\mathsf{sxi}\mathcal{O}$ is *sufficiently small*, we ensure that the running time of this punctured encryption algorithm is sub-linear in q. For the formal description of $\mathsf{PE}_{\mathsf{1Key}}$, see Fig. 6 in Sect. 7.1.

We can prove the semantic security at punctured tag by the punctured programming technique proposed by Sahai and Waters [62]. However, the construction above does not satisfy functionality preserving under puncturing. This is because ciphertexts output by the encryption and punctured encryption algorithms are different. The ciphertexts are obfuscation of different circuits $\mathsf{E}_{\mathsf{1Key}}$ and $\mathsf{PE}_{\mathsf{1Key}}$, respectively.

In fact, it seems difficult to avoid this problem as long as we use SXIO to gain succinctness. To the best of our knowledge, how to achieve succinctness in a generic way without using SXIO is not known.

Indistinguishability of functionality under puncturing. To overcome the problem above, we introduce a relaxed variant functionality preserving property that is compatible with the construction based on SXIO. We call it *indistinguishability of functionality under puncturing.* Informally speaking, the property requires that

$$(\mathsf{MSK}, \mathsf{MSK}^*\{\mathsf{tag}\}, \mathsf{Enc}(\mathsf{MSK}, \mathsf{tag}', m)) \stackrel{\mathsf{c}}{\approx} (\mathsf{MSK}, \mathsf{MSK}^*\{\mathsf{tag}\}, \mathsf{PEnc}(\mathsf{MSK}^*\{\mathsf{tag}\}, \mathsf{tag}', m))$$

holds for any tag tag and tag' such that $\mathsf{tag} \neq \mathsf{tag}'$, and message m, where $\stackrel{\mathsf{c}}{\approx}$ denotes computational indistinguishability. In other words, it requires that no distinguisher can distinguish ciphertexts output by Enc and PEnc even given both the master secret key and punctured master secret key.

We see that the collusion-succinct construction based on SXIO above satisfies indistinguishability of functionality under puncturing. This comes from the security guarantee of SXIO and the fact that $\mathsf{E}_{\mathsf{1Key}}$ and $\mathsf{PE}_{\mathsf{1Key}}$ are functionally equivalent as long as the above tag and tag' are different.

Overall, we can construct collusion-succinct puncturable SKFE with indistinguishability of functionality under puncturing from a single-key non-succinct scheme and SXIO.

Transforming into an weakly-succinct scheme. As stated earlier, we can in turn transform a collusion-succinct scheme into an weakly-succinct one using decomposable randomized encoding. This transformation is based on those proposed by Bitansky and Vaikuntanathan [17] and Ananth et al. [3].

In this transformation, a ciphertext of the weakly-succinct scheme is a ciphertext of the collusion-succinct scheme itself. Thus, if the collusion-succinct scheme satisfies semantic security at punctured tag and indistinguishability of functionality under puncturing, then so does the weakly-succinct scheme. Therefore, we can construct a single-key weakly-succinct puncturable SKFE with indistinguishability of functionality under puncturing.

Indistinguishability of functionality under puncturing looks to be insufficient for constructing IO. Nevertheless, we show that we can replace PKFE in the construction of IO proposed by Bitansky and Vaikuntanathan with our puncturable SKFE that satisfies only indistinguishability of functionality under puncturing if we allow more but asymptotically the same security loss.

2.4 IO from Puncturable SKFE

Finally, we give an overview of our IO construction below.

The construction of IO based on puncturable SKFE is almost the same as that based on PKFE proposed by Bitansky and Vaikuntanathan [17]. It does not depend on which functionality preserving property puncturable SKFE satisfies. Recall that, in their construction, a key pair $(\mathsf{PK}_i, \mathsf{MSK}_i)$ of PKFE is generated and the circuit E_{i-1} that has hardwired PK_i is obfuscated at every recursive step. In our construction based on puncturable SKFE, a master secret key MSK_i of puncturable SKFE is generated and E_{i-1} that has hardwired MSK_i is obfuscated at each recursive step. Concretely, we construct E_{i-1} as a circuit that has hardwired MSK_i and an SKE key K, and on $(i-1)$-bit input $\boldsymbol{x}_{i-1}$, it outputs a ciphertext of $(\boldsymbol{x}_{i-1} \| x_i, K)$ for $x_i \in \{0,1\}$ under MSK_i and a tag $\boldsymbol{x}_{i-1}$, that is, $\mathsf{Enc}(\mathsf{MSK}_i, \boldsymbol{x}_{i-1}, (\boldsymbol{x}_{i-1} \| x_i, K))$ for $x_i \in \{0,1\}$. In the proof, we replace MSK_i hardwired into E_{i-1} with the tuple of a punctured master secret key $\mathsf{MSK}_i^*\{\boldsymbol{j}\}$ punctured at $\boldsymbol{j} \in \{0,1\}^{i-1}$ and a ciphertext of $(\boldsymbol{j} \| x_i, K)$ for $x_i \in \{0,1\}$, where $\boldsymbol{j}$ is a string in $\{0,1\}^{i-1}$ that we focus on at that time.

Outline of Security Proof. We give an overview of the security proof of IO based on puncturable SKFE. If the building block puncturable SKFE satisfies functionality preserving under puncturing, the security proof is almost the same as that of Bitansky and Vaikuntanathan. However, our puncturable SKFE satisfies only indistinguishability of functionality under puncturing, and thus we need more complicated arguments. The first half of the following overview is similar to that of Bitansky and Vaikuntanathan. The rest is an overview of proofs that we additionally need due to indistinguishability of functionality under puncturing.

Analogous to IO based on PKFE, we can accomplish this proof recursively. More precisely, we can prove the security of $i\mathcal{O}_i$ based on those of $i\mathcal{O}_{i-1}$, puncturable SKFE, and plain SKE. We proceed the proof as follows. Note again that, we ignore the issue of the randomness for the encryption algorithm and punctured encryption algorithm for simplicity. It is generated by puncturable PRF in the actual construction.

Suppose that we have two functionally equivalent circuits C_0 and C_1 both of which expect an i-bit input. We show that no efficient distinguisher $\mathcal{D}$ can distinguish $i\mathcal{O}_i(C_0)$ and $i\mathcal{O}_i(C_1)$. We consider the following sequence of hybrid experiments. Below, for two hybrids $\mathcal{H}$ and $\mathcal{H}'$, we write $\mathcal{H} \sim \mathcal{H}'$ to denote that the behavior of $\mathcal{D}$ does not change between $\mathcal{H}$ and $\mathcal{H}'$.

In the first hybrid $\mathcal{H}_0$, $\mathcal{D}$ is given $i\mathcal{O}_i(C_0)$. Recall that $i\mathcal{O}_i(C_0)$ consists of sk_{C^*} and $i\mathcal{O}_{i-1}(\mathsf{E}_{i-1})$. C^* has hardwired two SKE ciphertexts $\mathsf{CT}_0^{\mathsf{ske}}$ and $\mathsf{CT}_1^{\mathsf{ske}}$ of C_0

under independent keys K_0 and K_1. On i-bit input $\boldsymbol{x}_i$ and SKE key K_b, C^* first obtains C by decrypting $\mathsf{CT}_b^{\mathsf{ske}}$ by K_b and outputs $C(\boldsymbol{x}_i)$.

In the next hybrid $\mathcal{H}_1$, we change how $\mathsf{CT}_1^{\mathsf{ske}}$ hardwired in C^* is generated. Concretely, we generate $\mathsf{CT}_1^{\mathsf{ske}}$ as a ciphertext of C_1 under the key K_1. It holds that $\mathcal{H}_0 \sim \mathcal{H}_1$ due to the security of SKE. Then, in the next hybrid $\mathcal{H}_2$, we change the circuit E_{i-1} so that, on $(i-1)$-bit input $\boldsymbol{x}_{i-1}$, it outputs a ciphertext of $(\boldsymbol{x}_{i-1} \| x_i, K_1)$ instead of $(\boldsymbol{x}_{i-1} \| x_i, K_0)$ for $x_i \in \{0,1\}$ under MSK_i and a tag $\boldsymbol{x}_{i-1}$.

If we prove $\mathcal{H}_1 \sim \mathcal{H}_2$, we also prove $\mathcal{H}_0 \sim \mathcal{H}_2$ and almost complete the security proof. This is because we can argue that the behavior of $\mathcal{D}$ does not change between $\mathcal{H}_2$ and the hybrid where $\mathcal{D}$ is given $i\mathcal{O}_i(C_1)$ by a similar argument for $\mathcal{H}_0 \sim \mathcal{H}_2$.

Therefore, the main part of the proof is how we change the circuit E_{i-1} from encrypting K_0 in $\mathcal{H}_1$ to encrypting K_1 in $\mathcal{H}_2$. As mentioned earlier, we accomplish this task by relying on the argument of probabilistic IO formalized by Canneti et al. [27].

Concretely, we consider $2^{i-1}+1$ intermediate hybrid experiments $\mathcal{H}_{1,j}$ for $j \in \{0, \cdots, 2^{i-1}\}$ between $\mathcal{H}_1$ and $\mathcal{H}_2$. Between $\mathcal{H}_{1,j}$ and $\mathcal{H}_{1,j+1}$, we change E_{i-1} so that on input $\boldsymbol{j} \in \{0,1\}^{i-1}$, it outputs ciphertexts of $(\boldsymbol{j} \| x_i, K_1)$ instead of $(\boldsymbol{j} \| x_i, K_0)$ for $x_i \in \{0,1\}$, where $\boldsymbol{j}$ is the binary representation of j. More precisely, we construct E_{i-1} in $\mathcal{H}_{1,j}$ as follows. E_{i-1} has hardwired MSK_i, K_0, and K_1. On $(i-1)$-bit input $\boldsymbol{x}_{i-1}$,

- if $\boldsymbol{x}_{i-1} < \boldsymbol{j}$, it outputs a ciphertext of $(\boldsymbol{x}_{i-1} \| x_i, K_1)$ for $x_i \in \{0,1\}$ under MSK_i and a tag $\boldsymbol{x}_{i-1}$.
- Otherwise, it outputs a ciphertext of $(\boldsymbol{x}_{i-1} \| x_i, K_0)$ for $x_i \in \{0,1\}$ under MSK_i and a tag $\boldsymbol{x}_{i-1}$.

We see that E_{i-1} in $\mathcal{H}_1$ has the same functionality as E_{i-1} in $\mathcal{H}_{1,0}$. In addition, E_{i-1} in $\mathcal{H}_2$ has the same functionality as E_{i-1} in $\mathcal{H}_{1,2^{i-1}}$. Therefore, we have $\mathcal{H}_1 \sim \mathcal{H}_{1,0}$ and $\mathcal{H}_2 \sim \mathcal{H}_{1,2^{i-1}}$ from the security guarantee of $i\mathcal{O}_{i-1}$.

We show how to prove $\mathcal{H}_{1,j} \sim \mathcal{H}_{1,j+1}$. For simplicity, we first assume that puncturable SKFE satisfies functionality preserving under puncturing. In this case, we show $\mathcal{H}_{1,j} \sim \mathcal{H}_{1,j+1}$ by the following three steps.

(1) In the first step, we hardwire ciphertexts of $(\boldsymbol{j} \| x_i, K_0)$ under MSK_i and a tag $\boldsymbol{j}$ for $x_i \in \{0,1\}$ in E_{i-1}. In addition, we replace hardwired MSK_i in E_{i-1} with $\mathsf{MSK}_i^*\{\boldsymbol{j}\}$ that is a master secret key punctured at a tag $\boldsymbol{j}$. On $(i-1)$-bit input $\boldsymbol{x}_{i-1}$,
   - if $\boldsymbol{x}_{i-1} = \boldsymbol{j}$, E_{i-1} outputs hardwired ciphertexts of $(\boldsymbol{j} \| x_i, K_0)$ for $x_i \in \{0,1\}$.
   - if $\boldsymbol{x}_{i-1} \neq \boldsymbol{j}$, it generates ciphertexts of $(\boldsymbol{x}_{i-1} \| x_i, K_\beta)$ under $\mathsf{MSK}_i^*\{\boldsymbol{j}\}$ and a tag $\boldsymbol{x}_{i-1}$ and outputs them, where $\beta = 1$ if $\boldsymbol{x}_{i-1} < \boldsymbol{j}$ and $\beta = 0$ otherwise.

   We see that this change does not affect the functionality of E_{i-1} if puncturable SKFE satisfies functionality preserving under puncturing. Thus, this step is done by the security of $i\mathcal{O}_{i-1}$.

(2) In the second step, we change the hardwired ciphertexts to ciphertexts of $(\boldsymbol{j}\|x_i, K_1)$ for $x_i \in \{0,1\}$. This is done by the semantic security at punctured tag of puncturable SKFE.

(3) In the final step, we change E_{i-1} so that it does not have hardwired ciphertexts of $(\boldsymbol{j}\|x_i, K_1)$ for $x_i \in \{0,1\}$. Moreover, we change E_{i-1} so that E_{i-1} has hardwired MSK_i and use it to generate the output ciphertexts. This change also does not affect the functionality of E_{i-1}, and thus we can accomplish this step by relying on the security of $i\mathcal{O}_{i-1}$ again.

From the above, if puncturable SKFE satisfies functionality preserving under puncturing, we have $\mathcal{H}_{1,j} \sim \mathcal{H}_{1,j+1}$ for every $j \in \{0, \cdots, 2^{i-1}-1\}$. By combining $\mathcal{H}_1 \sim \mathcal{H}_{1,0}$ and $\mathcal{H}_{1,2^{i-1}} \sim \mathcal{H}_2$, we obtain $\mathcal{H}_1 \sim \mathcal{H}_2$.

Therefore, we complete the entire proof. In fact, in this case, the proof is essentially the same as that for the case where PKFE is used as a building block shown by Bitansky and Vaikuntanathan.

Additional hybrids for the case of indistinguishability of functionality under puncturing. Recall that our puncturable SKFE satisfies only indistinguishability of functionality under puncturing. Thus, the above argument for steps 1 and 3 do not work straightforwardly. This is because if puncturable SKFE satisfies only indistinguishability of functionality under puncturing, the functionality of E_{i-1} might change at each step of 1 and 3. Therefore, we cannot directly use the security of $i\mathcal{O}_{i-1}$.

Nevertheless, even if puncturable SKFE satisfies only indistinguishability of functionality under puncturing, we can proceed steps 1 and 3 by introducing more additional hybrids. Since steps 1 and 3 are symmetric, we focus on proceeding the step 1. We can apply the following argument for the step 3. Below, we let $\mathcal{H}^0_{1,j}$ denote the hybrid experiment after applying the step 1 to $\mathcal{H}_{1,j}$.

To accomplish the step 1, we introduce the additional intermediate hybrids $\mathcal{H}_{1,j,k}$ for every $k \in \{0, \cdots, 2^{i-1}\} \setminus \{j\}$ between $\mathcal{H}_{1,j}$ and $\mathcal{H}^0_{1,j}$. Between $\mathcal{H}_{1,j,k}$ and $\mathcal{H}_{1,j,k+1}$, we change E_{i-1} so that, on input $\boldsymbol{k} \in \{0,1\}^{i-1}$, it outputs ciphertexts under $\mathsf{MSK}^*_i\{\boldsymbol{j}\}$ instead of ciphertexts under MSK_i, where $\boldsymbol{k}$ is the binary representation of k. More precisely, we construct E_{i-1} in $\mathcal{H}_{1,j,k}$ as follows. E_{i-1} has hardwired $\mathsf{MSK}^*_i\{\boldsymbol{j}\}$ in addition to MSK_i, K_0, and K_1. On $(i-1)$-bit input $\boldsymbol{x}_{i-1}$, it runs as follows.

- If $\boldsymbol{x}_{i-1} < \boldsymbol{j}$, it sets $\beta = 1$ and $\beta = 0$ otherwise.
- If $\boldsymbol{x}_{i-1} < k$ and $\boldsymbol{x}_{i-1} \neq j$, it outputs a ciphertext of $(\boldsymbol{x}_{i-1}\|x_i, K_\beta)$ under $\mathsf{MSK}^*_i\{\boldsymbol{j}\}$ and a tag $\boldsymbol{x}_{i-1}$, that is, $\mathsf{PEnc}\,(\mathsf{MSK}^*_i\{\boldsymbol{j}\}, \boldsymbol{x}_{i-1}, (\boldsymbol{x}_{i-1}\|x_i, K_\beta))$ for $x_i \in \{0,1\}$.
- Otherwise ($\boldsymbol{x}_{i-1} \geq k$ or $\boldsymbol{x}_{i-1} = j$), it outputs a ciphertext of $(\boldsymbol{x}_{i-1}\|x_i, K_\beta)$ under MSK_i and a tag $\boldsymbol{x}_{i-1}$, that is, $\mathsf{Enc}\,(\mathsf{MSK}_i, \boldsymbol{x}_{i-1}, (\boldsymbol{x}_{i-1}\|x_i, K_\beta))$ for $x_i \in \{0,1\}$.

We see that E_{i-1} in $\mathcal{H}_{1,j}$ and $\mathcal{H}^0_{1,j}$ have the same functionality as that in $\mathcal{H}_{1,j,0}$ and $\mathcal{H}_{1,j,2^{i-1}}$, respectively. In addition, E_{i-1} in $\mathcal{H}_{1,j,j}$ has the same functionality as that in $\mathcal{H}_{1,j,j+1}$. Therefore, we have $\mathcal{H}_{1,j} \sim \mathcal{H}_{1,j,0}$, $\mathcal{H}^0_{1,j} \sim \mathcal{H}_{1,j,2^{i-1}}$, and $\mathcal{H}_{1,j,j} \sim \mathcal{H}_{1,j,j+1}$ from the security guarantee of $i\mathcal{O}_{i-1}$.

We can prove $\mathcal{H}_{1,j,k} \sim \mathcal{H}_{1,j,k+1}$ for every $k \in \{0, \cdots, 2^{i-1}\} \setminus \{j\}$ by three steps again based on indistinguishability of functionality under puncturing.

(1) We hardwire ciphertexts of $(\boldsymbol{k}\|x_i, K_\beta)$ under MSK_i and a tag $\boldsymbol{k}$, that is, $\mathsf{Enc}(\mathsf{MSK}_i, \boldsymbol{k}, (\boldsymbol{k}\|x_i, K_\beta))$ for $x_i \in \{0,1\}$ in E_{i-1} in the first step. In addition, we change E_{i-1} so that it outputs the hardwired ciphertext of $(\boldsymbol{k}\|x_i, K_0)$ for $x_i \in \{0,1\}$ if the input is $\boldsymbol{k}$. We see that this change does not affect the functionality of E_{i-1}. Thus, this step is done by the security of $i\mathcal{O}_{i-1}$.
(2) In the second step, we change the hardwired ciphertexts to a ciphertext of $(\boldsymbol{k}\|x_i, K_\beta)$ under $\mathsf{MSK}^*_i\{j\}$, that is $\mathsf{PEnc}(\mathsf{MSK}^*_i\{j\}, \boldsymbol{k}, (\boldsymbol{k}\|x_i, K_\beta))$ for $x_i \in \{0,1\}$. This is done by the indistinguishability of functionality under puncturing of puncturable SKFE.
(3) In the final step, we change E_{i-1} so that it does not have hardwired ciphertexts of $(\boldsymbol{k}\|x_i, K_1)$ for $x_i \in \{0,1\}$. Namely, we change E_{i-1} so that on input $\boldsymbol{k}$, E_{i-1} generates ciphertexts of $\boldsymbol{k}$ under $\mathsf{MSK}^*_i\{j\}$ and outputs them. This change does not affect the functionality of E_{i-1}, and thus we can accomplish this step by relying on the security of $i\mathcal{O}_{i-1}$ again.

From these, $\mathcal{H}_{1,j,k} \sim \mathcal{H}_{1,j,k+1}$ holds for every $k \in \{0, \cdots, 2^{i-1}\} \setminus \{j\}$. By combining $\mathcal{H}_{1,j} \sim \mathcal{H}_{1,j,0}$, $\mathcal{H}^0_{1,j} \sim \mathcal{H}_{1,j,2^{i-1}}$, and $\mathcal{H}_{1,j,j} \sim \mathcal{H}_{1,j,j+1}$, we obtain $\mathcal{H}_{1,j} \sim \mathcal{H}^0_{1,j}$.

Therefore, we obtain $\mathcal{H}_{1,j} \sim \mathcal{H}^0_{1,j}$ even if puncturable SKFE satisfies only indistinguishability of functionality under puncturing. Overall, we can complete the entire security proof.

We note that our security proof incurs more security loss than those of Bitansky and Vaikuntanathan [17] and the case where puncturable SKFE satisfies functionality preserving under puncturing. Our security proof incurs roughly $2^{2 \cdot i}$ security loss while the latter proofs incurs 2^i security loss when we prove the security of $i\mathcal{O}_i$ based on that of $i\mathcal{O}_{i-1}$. Nevertheless, this difference is not an issue in the sense that if the building block primitives are roughly $2^{\Omega(n^2)}$-secure, we can prove the security of our indistinguishability obfuscator, where n is the input length of circuits to be obfuscated. This requirement is the same as that of Bitansky and Vaikuntanathan.

3 Overview: Collusion-Resistant SKFE from Weakly-Succinct One

In this section, we give a high-level overview of our technique for increasing the number of functional decryption keys that an SKFE scheme supports. The basic idea behind our proposed construction is that we combine multiple instances of a functional encryption scheme and use functional decryption keys tied to a function that outputs a re-encrypted ciphertext under a different encryption key. Several re-encryption techniques have been studied in the context of functional encryption [2, 17, 23, 36, 50], but we cannot directly use such techniques as we see below.

3.1 First Attempt: Applying Re-encryption Techniques in the Public-Key Setting

It is natural to try using the techniques in the public-key setting. In particular, it was shown that single-key weakly succinct PKFE implies collusion-resistant PKFE by Garg and Srinivasan [36] and Li and Micciancio [50]. Their techniques are different, but the core idea seems to be the same. Both techniques use functional decryption keys for a re-encryption function that outputs a ciphertext under a different encryption key.

We give more details of the technique by Li and Micciancio since it is our starting point. It is unclear whether the technique by Garg and Srinivasan is applicable in the secret-key setting since it seems that they use plain public-key encryption in an essential way.

The main technical tool of Li and Micciancio is the PRODUCT construction. Given two PKFE schemes, the PRODUCT construction combines them into a new PKFE scheme. The most notable feature of the PRODUCT construction is that the number of functional decryption keys of the resulting scheme is the product of those of the building block schemes. For example, if we have a λ-key PKFE scheme, by combining two instances of it via the PRODUCT construction, we can construct a λ^2-key PKFE scheme, where λ is the security parameter.

By applying the PRODUCT construction k times iteratively, we can construct a λ^k-key PKFE scheme from a λ-key PKFE scheme. Note that we can in turn construct a λ-key PKFE scheme by simply running λ instances of a single-key PKFE scheme in parallel. Moreover, if the underlying single-key scheme is weakly succinct, the running time of the λ^k-key scheme depends only on k instead of λ^k. Thus, by setting $k = \omega(1)$, we can construct a $\lambda^{\omega(1)}$-key PKFE scheme and achieve collusion-resistance from a single-key weakly succinct one.

Li and Micciancio proceeded with the above series of transformations via a stateful variant of PKFE, and thus the resulting collusion-resistant scheme is also a stateful scheme. Therefore, after achieving collusion-resistance, they converted the stateful PKFE scheme into a standard PKFE scheme. For simplicity, we ignore the issue here.

One might think that we can construct a collusion-resistant SKFE scheme from a single-key SKFE scheme by using the PRODUCT construction. However, we encounter several difficulties in the SKFE setting.

The PRODUCT construction involves *the chaining of re-encryption by functional decryption keys* used in many previous works [2,17,23,36]. This technique causes several difficulties when we adopt the PRODUCT construction in the SKFE setting. This is also the reason why the building block single-key PKFE scheme must satisfy (weak) succinctness property.

We now look closer at the technique of Li and Micciancio to see difficulties in the SKFE setting. Let PKFE be a 2-key PKFE scheme. As stated above, for functional key generation in this construction, we need state information called index, which indicates how many functional keys generated so far and which master secret and public key should be used to generate the next functional key. A simplified version of the PRODUCT construction proposed by Li and Micciancio is as follows.

<u>(2×2)**-key scheme from 2-key scheme.**</u>

Setup: Generates PKFE key pairs $(\mathsf{MPK}, \mathsf{MSK}) \leftarrow \mathsf{Setup}(1^\lambda)$ and $(\mathsf{MPK}_i, \mathsf{MSK}_i) \leftarrow \mathsf{Setup}(1^\lambda)$ for $i \in [2]$. MPK is the master public key and $(\mathsf{MSK}, \mathsf{MSK}_1, \mathsf{MSK}_2, \mathsf{MPK}_1, \mathsf{MPK}_2)$ is the master secret key of this scheme, respectively. In the actual construction, we maintain $(\mathsf{MPK}_i, \mathsf{MSK}_i)$ for $i \in [2]$ as one PRF key to avoid blow-ups.[7]

Functional Key: For n-th functional key generation, a positive integer $n \in [4]$ is interpreted as a pair of index $(i, j) \in [2] \times [2]$. Generates two keys $\mathsf{sk}^i_{\mathcal{E}[\mathsf{MPK}_i]} \leftarrow \mathsf{KG}(\mathsf{MSK}, \mathcal{E}[\mathsf{MPK}_i], i)$ and $\mathsf{sk}^{(i,j)}_f \leftarrow \mathsf{KG}(\mathsf{MSK}_i, f, j)$ where $\mathcal{E}$ is a re-encryption circuit described below. A functional key is $(\mathsf{sk}^i_{\mathcal{E}[\mathsf{MPK}_i]}, \mathsf{sk}^{(i,j)}_f)$.

Encryption: A ciphertext is $\mathsf{ct}_{\mathsf{pre}} \leftarrow \mathsf{Enc}(\mathsf{MPK}, m)$.

Decryption: First, applies the decryption algorithm with MPK, that is, $\mathsf{ct}_{\mathsf{post}} \leftarrow \mathsf{Dec}(\mathsf{sk}^i_{\mathcal{E}[\mathsf{MPK}_i]}, \mathsf{ct}_{\mathsf{pre}})$. Next, applies it with MPK_i, $f(m) \leftarrow \mathsf{Dec}(\mathsf{sk}^{(i,j)}_f, \mathsf{ct}_{\mathsf{post}})$.

The description of $\mathcal{E}$ defined at the functional key generation phase is as in the figure below. Re-encryption circuit $\mathcal{E}[\mathsf{MPK}_i]$ takes as an input a message m and outputs $\mathsf{ct}_{\mathsf{post}} \leftarrow \mathsf{Enc}(\mathsf{MPK}_i, m)$ by using a hard-wired master public-key MPK_i.

Hard-Coded Constants: MPK_i. // Description of (simplified) $\mathcal{E}$
Input: m

1. Return $\mathsf{ct}_{\mathsf{post}} \leftarrow \mathsf{Enc}(\mathsf{MPK}_i, m)$.

Using the master secret-key MSK_1, we can generate two functional keys $\mathsf{sk}^{1,1}_{f_1}, \mathsf{sk}^{1,2}_{f_2}$ since PKFE is a 2-key scheme. Similarly, we can generate two functional keys using MSK_2. Moreover, since MSK is also a master secret-key of the 2-key scheme, we can generate two functional keys $\mathsf{sk}_{\mathcal{E}[\mathsf{MPK}_1]}$ and $\mathsf{sk}_{\mathcal{E}[\mathsf{MPK}_2]}$ using MSK at the functional key generation step. By these combinations, we can generate 2×2 keys

$$(\mathsf{sk}_{\mathcal{E}[\mathsf{MPK}_1]}, \mathsf{sk}^{1,1}_{f_1}), (\mathsf{sk}_{\mathcal{E}[\mathsf{MPK}_1]}, \mathsf{sk}^{1,2}_{f_2}), (\mathsf{sk}_{\mathcal{E}[\mathsf{MPK}_2]}, \mathsf{sk}^{2,1}_{f_3}), (\mathsf{sk}_{\mathcal{E}[\mathsf{MPK}_2]}, \mathsf{sk}^{2,2}_{f_4}).$$

This is generalized to the case where the underlying schemes are a p-key scheme and q-key scheme for any p and q. That is, for n-th functional key generation where $n \leq p \cdot q$, n is interpreted as $(i, j) \in [p] \times [q]$. Thus, by applying the PRODUCT construction to a λ-key scheme k times iteratively, we can obtain a λ^k-key scheme. Note again that we can construct a λ-key weakly succinct SKFE scheme from a single-key weakly succinct one by simple parallelization.

[7] In fact, $(\mathsf{MPK}_i, \mathsf{MSK}_i)$ for $i \in [2]$ are generated at the functional key generation phase by computing $r_i \leftarrow \mathsf{PRF}(K, i)$ and $(\mathsf{MPK}_i, \mathsf{MSK}_i) \leftarrow \mathsf{Setup}(1^\lambda; r_i)$, where K is a PRF key and is stored as a part of the master secret key.

While such a re-encryption technique is widely used in the context of PKFE, it is difficult to use it directly in the SKFE setting. The main cause of the difficulty is the fact that we have to release a functional key implementing the encryption circuit in which a *master secret key* of an SKFE scheme is hardwired to achieve the re-encryption by functional decryption keys. The fact seems to be a crucial problem for the security proof since sk_f might leak information about f. In the PKFE setting, this issue does not arise since an encryption key is publicly available.

3.2 Second Attempt: Applying Techniques in a Different Context of SKFE

To solve the above issue, we try using a technique in the secret-key setting but in a different context from the collusion-resistance.

Brakerski et al. [23] introduced a new re-encryption technique by functional decryption keys in the context of multi-input SKFE [38]. They showed that we can overcome the difficulty above by using the security notion of function privacy [25].

By function privacy, we can hide the information about a master-secret key embedded in a re-encryption circuit $\mathcal{E}[\mathsf{MSK}^*]$. With their technique, we embed a post-re-encrypted ciphertext $\mathsf{ct}_{\mathsf{post}}$ as a trapdoor into a pre-re-encrypted ciphertext $\mathsf{ct}_{\mathsf{pre}}$ in advance in the simulation for the security proof. By embedding this trapdoor, we can remove MSK^* from the re-encryption circuit $\mathcal{E}$ when we reduce the security of the resulting scheme to that of the underlying scheme corresponding to MSK^*.

Their technique is useful, but it incurs a polynomial blow-up of the running time of the encryption circuit for each application of a construction with the re-encryption procedure by a functional decryption key. This is because it embeds a ciphertext into another ciphertext (we call this nested-ciphertext-embedding).

Such a nest does not occur with the technique of Li and Micciancio in the PKFE setting since a post-re-encrypted ciphertext as a trapdoor is embedded in a functional decryption key. One might think we can avoid the issue of nested-ciphertext embedding by embedding ciphertexts in a functional key. However, this is not the case because the number of ciphertext queries is not a-priori bounded in the secret-key setting.

In fact, we obtain a new PRODUCT construction by accommodating the function privacy and nested-ciphertext-embedding technique to the PRODUCT construction of Li and Micciancio. However, if we use our new PRODUCT construction in a naive way, each application of the new PRODUCT construction incurs a polynomial blow-up of the encryption time. In general, k applications of our new PRODUCT construction with nested-ciphertext-embedding incur a double exponential blow-up $\lambda^{2^{O(k)}}$.

Thus, in a naive way, we can apply our new PRODUCT construction iteratively *only constant times*. This is not sufficient for our goal since we must apply our new PRODUCT construction $\omega(1)$ times to achieve collusion-resistant SKFE.

3.3 Our Solution: Sandwiched Size-Shifting

To solve the difficulty of size blow-up, we propose a new construction technique called "sandwiched size-shifting". In this new technique, we use a *hybrid encryption methodology* to reduce the exponential blow-up of the encryption time caused by our new PRODUCT construction with nested-ciphertext-embedding.

A hybrid encryption methodology is used in many encryption schemes. In particular, Ananth et al. [1] showed that a hybrid encryption construction is useful in designing adaptively secure functional encryption from selectively secure one without any additional assumption. In fact, Brakerski et al. [23] also used a hybrid encryption construction to achieve an input aggregation mechanism for multi-input SKFE.

In this study, we propose a new hybrid encryption construction for functional encryption to *reduce the encryption time* of a functional encryption scheme without any additional assumption. Our key tool is a single-ciphertext collusion-resistant SKFE scheme called $\mathsf{1CT}$, which is constructed only from one-way functions. The notable features of $\mathsf{1CT}$ are as follows.

1. The size of a master secret key of $\mathsf{1CT}$ is independent of the length of a message to be encrypted.
2. The encryption is fully succinct.
3. The size of a functional decryption key is only linear in the size of a function.

The drawback of $\mathsf{1CT}$ is that we can release *only one ciphertext*. However, this is not an issue for our purpose since a master secret key of $\mathsf{1CT}$ is freshly chosen at each ciphertext generation in our hybrid construction.

$\mathsf{1CT}$ is based on a garbled circuit [64]. A functional decryption key is a garbled circuit of f with encrypted labels by a standard secret-key encryption scheme.[8] A ciphertext consists of a randomly masked message and keys of the secret-key encryption scheme that corresponds to the randomly masked message. Thus, we can generate only one ciphertext since if two ciphertexts are generated, then labels for both bits are revealed and the security of the garbled circuit is completely broken. Note that $\mathsf{1CT}$ is selectively secure. In fact, this construction is a flipped variant of the single-key SKFE by Sahai and Seyalioglu [61].

We then modify the SKFE variant of the hybrid construction proposed by Ananth et al. [1].[9] We use $\mathsf{1CT}$ as data encapsulation mechanism and a q-key weakly succinct SKFE scheme SKFE as key encapsulation mechanism. In our hybrid construction, the encryption algorithm of SKFE encrypts only short values (concretely, a one-time master secret-key of $\mathsf{1CT}$), which are independent of the length of a message to be encrypted. A one-time encryption key (short and fixed length) of $\mathsf{1CT}$ is encrypted by SKFE.

---

[8] Each pair of labels is shuffled by a random masking.

[9] Their goal is to construct an adaptively secure scheme. They used *adaptively secure* single-ciphertext functional encryption that is *non-succinct* as data encapsulation mechanism.

That is, by this hybrid construction, a real message part is shifted onto $\mathsf{1CT}$, whose ciphertext has the full succinctness property. In other words, we can separate the blow-up due to recursion from nested-ciphertext-embedding part. Therefore, we call our new hybrid construction technique "size-shifting".

The third property of $\mathsf{1CT}$ is also important. The size of a functional key of $\mathsf{1CT}$ affects the encryption time of the hybrid construction. This is because a functional key for f of the hybrid construction consists of a functional key of SKFE for a function G, which generates a functional key of $\mathsf{1CT}$ for f. A simplified description of G is below. Due to the third property of $\mathsf{1CT}$, the hybrid construction preserves weak succinctness.

Hard-Coded Constants: f. // Description of (simplified) G
Input: $\mathsf{1CT.MSK}$

1. Return $\mathsf{1CT.sk}_f \leftarrow \mathsf{1CT.KG}(\mathsf{1CT.MSK}, f)$.

Moreover, from the above construction of the key generation algorithm, the number of issuable functional keys of the resulting scheme is minimum of those of building block SKFE and $\mathsf{1CT}$. Therefore, since $\mathsf{1CT}$ is collusion-resistant, if SKFE supports q functional keys, then so does the resulting scheme, where q is any fixed polynomial of λ.

Thus, we can apply the hybrid construction after each application of our new PRODUCT construction, preserving the weak succinctness and the number of functional keys that can be released.

The size-shifting procedure is "sandwiched" by each our new PRODUCT construction. As a result, we can reduce the blow-up of the encryption time after k iterations to $k \cdot \lambda^{O(1)}$ if the underlying single-key scheme is weakly succinct while the naive k iterated applications of our new PRODUCT construction incurs $\lambda^{2^{O(k)}}$ size blow-up. Therefore, we can iterate our new PRODUCT construction $\omega(1)$ times via the size-shifting and construct a collusion-resistant SKFE scheme based only on a single-key (weakly) succinct SKFE scheme.[10]

Our analysis is highly non-trivial though our transformation consists of relatively simple transformations. We believe that it is better to achieve non-trivial results by using simple techniques than complex ones.

Figure 2 illustrates how to construct our building blocks. An illustration of our sandwiched size-shifting procedure is described in Fig. 3.

[10] While we can reduce the blow-up of the encryption time, we cannot reduce the security loss caused by each iteration step. As a result, $\lambda^{\omega(1)}$ security loss occurs after $\omega(1)$ times iterations. This is the reason our transformation incurs quasi-polynomial security loss.

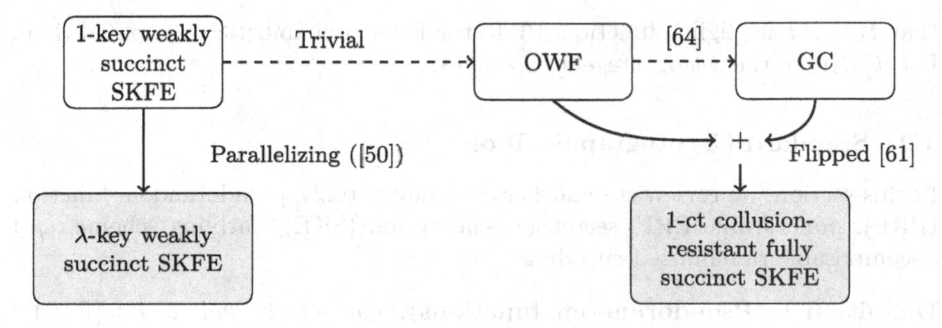

Fig. 2. Our building blocks. Green boxes denote our core schemes used in our iterated construction in Fig. 3. (Color figure online)

4 Preliminaries

We define some notations and cryptographic primitives.

4.1 Notations

We write $x \xleftarrow{r} X$ to denote that an element x is chosen from a finite set X uniformly at random and $y \leftarrow \mathsf{A}(x;r)$ to denote that the output of an algorithm A on an input x and a randomness r is assigned to y. When there is no need to write the randomness explicitly, we omit it and simply write $y \leftarrow \mathsf{A}(x)$.

Throughout this paper, λ denotes a security parameter. poly denotes an unspecified polynomial. A function $f(\lambda)$ is a negligible function if $f(\lambda)$ tends to 0 faster than $\frac{1}{\lambda^c}$ for every constant $c > 0$. We write $f(\lambda) = \mathsf{negl}(\lambda)$ to denote

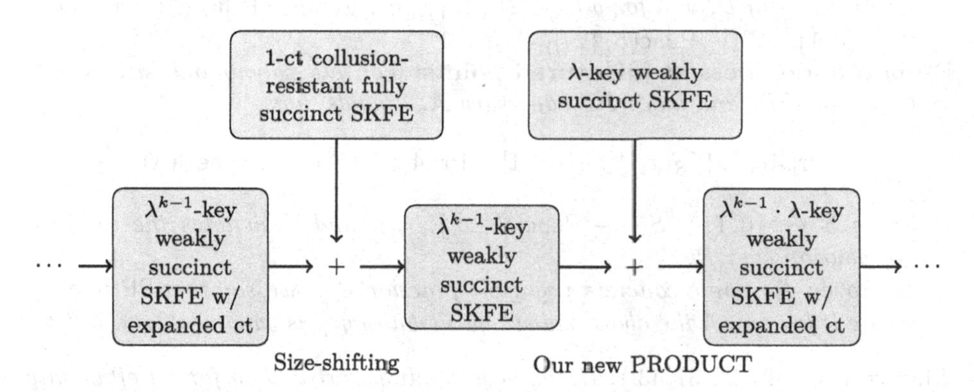

Fig. 3. An illustration of our iteration technique, in which our size-shifting procedure is sandwiched. For k-th iteration, first, we apply the size-shifting procedure to a λ^{k-1}-key weakly succinct SKFE scheme with expanded ciphertexts incurred by nested-ciphertext-embedding (the result of $(k-1)$-th iteration). Second, we apply our new PRODUCT construction to increase the number of issuable keys.

that $f(\lambda)$ is a negligible function. PPT stands for probabilistic polynomial time. Let $[\ell]$ denote the set of integers $\{1,\cdots,\ell\}$.

4.2 Standard Cryptographic Tools

In this section, we review standard cryptographic tools, pseudorandom function (PRF), puncturable PRF, secret-key encryption (SKE), garbling scheme, and decomposable randomized encoding.

Definition 1 (Pseudorandom functions). *For sets $\mathcal{D}$ and $\mathcal{R}$, let $\{\mathsf{F}_S(\cdot) : \mathcal{D} \to \mathcal{R} | S \in \{0,1\}^\lambda\}$ be a family of polynomially computable functions. We say that F is pseudorandom if for any PPT adversary $\mathcal{A}$, it holds that*

$$\begin{aligned}\mathsf{Adv}^{\mathsf{prf}}_{\mathsf{F},\mathcal{A}}(\lambda) = |\Pr[\mathcal{A}^{\mathsf{F}_S(\cdot)}(1^\lambda) = 1 : S \xleftarrow{\mathsf{r}} \{0,1\}^\lambda] \\ - \Pr[\mathcal{A}^{\mathsf{R}(\cdot)}(1^\lambda) = 1 : \mathsf{R} \xleftarrow{\mathsf{r}} \mathcal{U}]| = \mathsf{negl}(\lambda) \enspace ,\end{aligned}$$

where $\mathcal{U}$ is the set of all functions from $\mathcal{D}$ to $\mathcal{R}$. Moreover, for some concrete negligible function $\epsilon(\cdot)$, we say that F is ϵ-secure if for any PPT $\mathcal{A}$ the above indistinguishability gap is smaller than $\epsilon(\lambda)^{\Omega(1)}$.

Theorem 4 ([37]). *If one-way functions exist, then for all efficiently computable functions $n(\lambda)$ and $m(\lambda)$, there exists a pseudorandom function that maps $n(\lambda)$ bits to $m(\lambda)$ bits (i.e., $\mathcal{D} := \{0,1\}^{n(\lambda)}$ and $\mathcal{R} := \{0,1\}^{m(\lambda)}$).*

Definition 2 (Puncturable pseudorandom function). *For sets $\mathcal{D}$ and $\mathcal{R}$, a puncturable pseudorandom function PPRF consists of a tuple of algorithms $(\mathsf{F}, \mathsf{Punc})$ that satisfies the following two conditions.*

Functionality preserving under puncturing: *For all polynomial size subset $\{x_i\}_{i\in[k]}$ of $\mathcal{D}$, and for all $x \in \mathcal{D} \setminus \{x_i\}_{i\in[k]}$, we have $\Pr[\mathsf{F}_S(x) = \mathsf{F}_{S^*}(x) : S \leftarrow \{0,1\}^\lambda, S^* \leftarrow \mathsf{Punc}(S, \{x_i\}_{i\in[k]})] = 1$.*

Pseudorandomness at punctured points: *For all polynomial size subset $\{x_i\}_{i\in[k]}$ of $\mathcal{D}$, and any PPT adversary $\mathcal{A}$, it holds that*

$$\Pr[\mathcal{A}(S^*, \{\mathsf{F}_S(x_i)\}_{i\in[k]}) = 1] - \Pr[\mathcal{A}(S^*, U^k) = 1] = \mathsf{negl}(\lambda) \enspace ,$$

where $S \xleftarrow{\mathsf{r}} \{0,1\}^\lambda$, $S^* \leftarrow \mathsf{Punc}(S, \{x_i\}_{i\in[k]})$, and U denotes the uniform distribution over $\mathcal{R}$.

Moreover, for some concrete negligible function $\epsilon(\cdot)$, we say that PPRF is ϵ-secure if for any $\mathcal{A}$ the above indistinguishability gap is smaller than $\epsilon(\lambda)^{\Omega(1)}$.

Theorem 5 ([21,22,37,43]). *If one-way functions exist, then for all efficiently computable functions $n(\lambda)$ and $m(\lambda)$, there exists a puncturable pseudorandom function that maps $n(\lambda)$ bits to $m(\lambda)$ bits (i.e., $\mathcal{D} := \{0,1\}^{n(\lambda)}$ and $\mathcal{R} := \{0,1\}^{m(\lambda)}$).*

Definition 3 (Secret key encryption). *An SKE scheme SKE is a two tuple (E, D) of PPT algorithms.*

- *The encryption algorithm* E*, given a key* $K \in \{0,1\}^\lambda$ *and a message* $m \in \mathcal{M}$*, outputs a ciphertext* c*, where* $\mathcal{M}$ *is the plaintext space of* SKE*.*
- *The decryption algorithm* D*, given a key* K *and a ciphertext* c*, outputs a message* $\tilde{m} \in \{\bot\} \cup \mathcal{M}$*. This algorithm is deterministic.*

Correctness: *We require* $\mathsf{D}(K, \mathsf{E}(K, m)) = m$ *for every* $m \in \mathcal{M}$ *and key* $K \in \{0,1\}^\lambda$*.*

CPA security: *We define the security game between a challenger and an adversary* $\mathcal{A}$ *as follows.*

1. *The challenger generates* $K \xleftarrow{r} \{0,1\}^\lambda$ *and chooses the challenge bit* $b \xleftarrow{r} \{0,1\}$*. Then, the challenger sends* 1^λ *to* $\mathcal{A}$*.*
2. $\mathcal{A}$ *may make polynomially many encryption queries adaptively.* $\mathcal{A}$ *sends* $(m_0, m_1) \in \mathcal{M} \times \mathcal{M}$ *to the challenger. Then, the challenger returns* $c \leftarrow \mathsf{E}(K, m_b)$*.*
3. $\mathcal{A}$ *outputs* $b' \in \{0,1\}$*.*

In this game, we define the advantage of the adversary $\mathcal{A}$ *as*

$$\mathsf{Adv}^{\mathsf{cpa}}_{\mathsf{SKE},\mathcal{A}}(\lambda) = 2|\Pr[b = b'] - \frac{1}{2}| = |\Pr[b' = 1|b = 0] - \Pr[b' = 1|b = 1]| \ .$$

For a negligible function $\epsilon(\cdot)$*, We say that* SKE *is* ϵ*-secure if for any PPT* $\mathcal{A}$*, we have* $\mathsf{Adv}^{\mathsf{cpa}}_{\mathsf{SKE},\mathcal{A}}(\lambda) < \epsilon(\lambda)^{\Omega(1)}$*.*

Theorem 6 ([54])**.** *If there exist one-way functions, there exists CPA-secure SKE.*

Definition 4 (Garbling scheme). *Let* $\{\mathcal{C}_n\}_{n\in\mathbb{N}}$ *be a family of circuits where each circuit in* $\mathcal{C}_n$ *takes an* n*-bit input. A circuit garbling scheme* GC *is a two tuple* $(\mathsf{Grbl}, \mathsf{Eval})$ *of PPT algorithms.*

- *The garbling algorithm* Grbl*, given a security parameter* 1^λ *and a circuit* $C \in \mathcal{C}_n$*, outputs a garbled circuit* $\widetilde{C}$*, together with* $2n$ *labels* $\{L_{j,\alpha}\}_{j\in[n],\alpha\in\{0,1\}}$*.*
- *The evaluation algorithm, given a garbled circuit* $\widetilde{C}$ *and* n *labels* $\{L_j\}_{j\in[n]}$*, outputs* y*.*

Correctness: *We require* $\mathsf{Eval}(\widetilde{C}, \{L_{j,x_j}\}_{j\in[n]}) = C(x)$ *for every* $n \in \mathbb{N}$*,* $C \in \mathcal{C}_n$*, and* $x \in \{0,1\}^n$*, where* $(\widetilde{C}, \{L_{j,\alpha}\}_{j\in[n],\alpha\in\{0,1\}}) \leftarrow \mathsf{Grbl}(1^\lambda, C)$ *and* x_j *is the* j*-th bit of* x *for every* $j \in [n]$*.*

Security: *Let* Sim *be a PPT simulator. We define the following game between a challenger and an adversary* $\mathcal{A}$ *as follows.*

1. *The challenger chooses the challenge bit* $b \xleftarrow{r} \{0,1\}$ *and sends security parameter* 1^λ *to* $\mathcal{A}$*.*
2. $\mathcal{A}$ *sends a circuit* $C \in \mathcal{C}_n$ *and an input* $x \in \{0,1\}^n$ *for the challenger.*
3. *If* $b = 0$*, the challenger computes* $(\widetilde{C}, \{L_{j,\alpha}\}_{j\in[n],\alpha\in\{0,1\}}) \leftarrow \mathsf{Grbl}(1^\lambda, C)$ *and returns* $(\widetilde{C}, \{L_{j,x_j}\}_{j\in[n]})$ *to* $\mathcal{A}$*. Otherwise, the challenger returns* $(\widetilde{C}, \{L_j\}_{j\in[n]}) \leftarrow \mathsf{Sim}(1^\lambda, |C|, C(x))$*.*
4. $\mathcal{A}$ *outputs* $b' \in \{0,1\}$*.*

In this game, we define the advantage of $\mathcal{A}$ *as*

$$\mathsf{Adv}^{\mathsf{gc}}_{\mathsf{GC},\mathcal{A},\mathsf{Sim}}(\lambda) = 2|\Pr[b = b'] - \frac{1}{2}| = |\Pr[b' = 1|b = 0] - \Pr[b' = 1|b = 1]| \ .$$

For a concrete negligible function $\epsilon(\cdot)$, *We say that* GC *is* ϵ*-secure if there exists a PPT* Sim *such that for any PPT* $\mathcal{A}$, *we have* $\mathsf{Adv}^{\mathsf{gc}}_{\mathsf{GC},\mathcal{A},\mathsf{Sim}}(\lambda) < \epsilon(\lambda)^{\Omega(1)}$.

Theorem 7 ([13,57,64]). *If there exist one-way functions, there exists secure garbling scheme for any polynomial size circuits.*

Definition 5 (Decomposable randomized encoding). *Let* $c \geq 1$ *be an integer constant. A c-local decomposable randomized encoding* RE, *given security parameter* 1^λ *and a function* f *of size* s *and* n*-bit input, outputs a function* $\widehat{f} : \{0,1\}^n \times \{0,1\}^\rho \rightarrow \{0,1\}^\mu$ *with the following properties.* ρ *and* μ *are polynomials bounded by* $s \cdot \mathrm{poly}_{\mathsf{RE}}(\lambda, n)$, *where* $\mathrm{poly}_{\mathsf{RE}}$ *is a fixed polynomial.*

Correctness: *There is a polynomial time decoder that, given* $\widehat{f}(x; r)$, *outputs* $f(x)$ *for any* $x \in \{0,1\}^n$ *and* $r \in \{0,1\}^\rho$.

Decomposability: *Computation of* $\widehat{f}$ *can be decomposed into computation of* μ *functions. That is, there exist* μ *functions* $\widehat{f}_1, \cdots, \widehat{f}_\mu$ *such that* $\widehat{f}(x; r) = (\widehat{f}_1(x; r), \cdots, \widehat{f}_\mu(x; r))$. *Each* $\widehat{f}_i$ *depends on a single bit of* x *at most and* c *bits of* r. *We write* $\widehat{f}(x; r) = (\widehat{f}_1(x; r_{S_1}), \cdots, \widehat{f}_\mu(x; r_{S_\mu}))$, *where* S_i *denotes the subset of bits of* r *that* $\widehat{f}_i$ *depends on.*

Security: *Let* Sim *be a PPT simulator. We define the following game between a challenger and an adversary* $\mathcal{A}$ *as follows.*

1. *The challenger chooses a bit* $b \xleftarrow{r} \{0,1\}$ *and sends security parameter* 1^λ *to* $\mathcal{A}$.
2. $\mathcal{A}$ *sends a function* f *of size* s *and* n*-bit input and an input* $x \in \{0,1\}^n$ *to the challenger.*
3. *If* $b = 0$, *the challenger computes* $\widehat{f} \leftarrow \mathsf{RE}(1^\lambda, f)$, *generates* $r \leftarrow \{0,1\}^\rho$, *and returns* $\widehat{f}(x; r)$ *to* $\mathcal{A}$. *Otherwise, the challenger returns* $\mathsf{Sim}(1^\lambda, s, f(x))$.
4. $\mathcal{A}$ *outputs* $b' \in \{0,1\}$.

In this game, we define the advantage of $\mathcal{A}$ *as*

$$\mathsf{Adv}^{\mathsf{re}}_{\mathsf{RE},\mathsf{Sim},\mathcal{A}}(\lambda) = |\Pr[b' = 1|b = 0] - \Pr[b' = 1|b = 1]| \ .$$

For a negligible function $\epsilon(\cdot)$, *we say that* RE *is* ϵ*-secure if there exists a PPT* Sim *such that for any PPT* $\mathcal{A}$, *we have* $\mathsf{Adv}^{\mathsf{re}}_{\mathsf{RE},\mathsf{Sim},\mathcal{A}}(\lambda) < \epsilon(\lambda)^{\Omega(1)}$.

It is known that a decomposable randomized encoding can be based on one-way functions.

Theorem 8 ([6,64]). *If there exist one-way functions, there exists secure decomposable randomized encoding for all polynomial size functions.*

4.3 Secret-Key Functional Encryption

We review the definition of ordinary secret-key functional encryption (SKFE).

Definition 6 (Secret-key functional encryption). *An SKFE scheme* SKFE *is a four tuple of PPT algorithms* (Setup, KG, Enc, Dec). *Below, let* $\mathcal{M}$ *and* $\mathcal{F}$ *be the message space and function space of* SKFE, *respectively.*

- *The setup algorithm* Setup, *given a security parameter* 1^λ, *outputs a master secret key* MSK.
- *The key generation algorithm* KG, *given a master secret key* MSK *and a function* $f \in \mathcal{F}$, *outputs a functional decryption key* sk_f.
- *The encryption algorithm* Enc, *given a master secret key* MSK *and a message* $m \in \mathcal{M}$, *outputs a ciphertext* CT.
- *The decryption algorithm* Dec, *given a functional decryption key* sk_f *and a ciphertext* CT, *outputs a message* $\tilde{m} \in \{\bot\} \cup \mathcal{M}$.

Correctness: *We require* $\mathsf{Dec}(\mathsf{KG}(\mathsf{MSK}, f), \mathsf{Enc}(\mathsf{MSK}, m)) = f(m)$ *for every* $m \in \mathcal{M}$, $f \in \mathcal{F}$, *and* $\mathsf{MSK} \leftarrow \mathsf{Setup}(1^\lambda)$.

Next, we introduce selective-message message privacy for SKFE schemes.

Definition 7 (Selective-message message privacy). *Let* SKFE *be an SKFE scheme whose message space and function space are* $\mathcal{M}$ *and* $\mathcal{F}$, *respectively. Let* q *be a polynomial of* λ. *We define the selective-message message privacy game between a challenger and an adversary* $\mathcal{A}$ *as follows.*

1. *The challenger generates a master secret key* $\mathsf{MSK} \leftarrow \mathsf{Setup}(1^\lambda)$ *and chooses the challenge bit* $b \xleftarrow{r} \{0,1\}$. *Then, the challenger sends security parameter* 1^λ *to* $\mathcal{A}$.
2. $\mathcal{A}$ *sends* $\{(m_0^\ell, m_1^\ell)\}_{\ell \in [p]}$ *to the challenger, where* p *is an a-priori unbounded polynomial of* λ.
3. *The challenger generates ciphertexts* $\mathsf{CT}^{(\ell)} \leftarrow \mathsf{Enc}(\mathsf{MSK}, m_b^\ell) (\ell \in [p])$ *and sends them to* $\mathcal{A}$.
4. $\mathcal{A}$ *may adaptively make key queries* q *times at most. For a key query* $f \in \mathcal{F}$ *from* $\mathcal{A}$, *the challenger generates* $sk_f \leftarrow \mathsf{KG}(\mathsf{MSK}, f)$, *and returns* sk_f *to* $\mathcal{A}$. *Here,* f *needs to satisfy* $f(m_0^\ell) = f(m_1^\ell)$ *for all* $\ell \in [p]$.
5. $\mathcal{A}$ *outputs* $b' \in \{0,1\}$.

In this game, we define the advantage of $\mathcal{A}$ *as*

$$\mathsf{Adv}_{\mathsf{SKFE},\mathcal{A}}^{\mathsf{sm\text{-}mp}}(\lambda) = 2|\Pr[b = b'] - \frac{1}{2}| = |\Pr[b' = 1|b = 0] - \Pr[b' = 1|b = 1]| \ .$$

$\mathcal{A}$ *is said to be valid if each function query* f *made by* $\mathcal{A}$ *satisfies that* $f(m_0^\ell) = f(m_1^\ell)$ *for all* $\ell \in [p]$ *in the above game. For a negligible function* $\epsilon(\cdot)$, *We say that* SKFE *is* (q, ϵ)*-selective-message message private if for any valid PPT* $\mathcal{A}$, *we have* $\mathsf{Adv}_{\mathsf{SKFE},\mathcal{A}}^{\mathsf{sm\text{-}mp}}(\lambda) < \epsilon(\lambda)^{\Omega(1)}$.

We further say that an SKFE scheme is ϵ-secure collusion-resistant SKFE if it is (q, ϵ)-selective-message message private for any polynomial q.

Next, we define the succinctness for SKFE.

Definition 8 (Succinctness). *Let $\mathcal{F}$ be a function family. Let s and n be the maximum size and input length of functions contained in $\mathcal{F}$, respectively. We say that SKFE for $\mathcal{F}$ is weakly succinct if the size of the encryption circuit is bounded by $s^{\gamma} \cdot \mathrm{poly}(\lambda, n)$, where $\gamma < 1$ is a fixed constant.*

4.4 Indistinguishability Obfuscation

We review the definition of indistinguishability obfuscation (IO).

Definition 9 (Indistinguishability obfuscation). *A PPT algorithm $i\mathcal{O}$ is an indistinguishability obfuscator (IO) for a circuit class $\{\mathcal{C}_\lambda\}_{\lambda \in \mathbb{N}}$ if it satisfies the following two conditions.*

Functionality: *for all security parameters $\lambda \in \mathbb{N}$, for all $C \in \mathcal{C}_\lambda$, for all inputs x, we have that $\Pr[C'(x) = C(x) : C' \leftarrow i\mathcal{O}(1^\lambda, C)] = 1$.*

Indistinguishability: *for any PPT distinguisher $\mathcal{D}$, there exists a negligible function $\mathsf{negl}(\cdot)$ such that the following holds: for all security parameters $\lambda \in \mathbb{N}$, for all pairs of circuits $C_0, C_1 \in \mathcal{C}_\lambda$ of the same size and such that $C_0(x) = C_1(x)$ for all inputs x, then*

$$|\Pr\left[\mathcal{D}(i\mathcal{O}(1^\lambda, C_0)) = 1\right] - \Pr\left[\mathcal{D}(i\mathcal{O}(1^\lambda, C_1)) = 1\right]| = \mathsf{negl}(\lambda) \ .$$

We further say that $i\mathcal{O}$ is ϵ-secure, for some concrete negligible function $\epsilon(\cdot)$, if for any PPT distinguisher the above advantage is smaller than $\epsilon(\lambda)^{\Omega(1)}$.

4.5 Strong Exponentially-Efficient Indistinguishability Obfuscation

We next define strong exponentially-efficient IO (SXIO).

Definition 10 (Strong exponentially-efficient indistinguishability obfuscation). *Let $\gamma < 1$ be a constant. A PPT algorithm $\mathsf{sxi}\mathcal{O}$ is a γ-compressing strong exponentially-efficient indistinguishability obfuscator (SXIO) for a circuit class $\{\mathcal{C}\}_{\lambda \in \mathbb{N}}$ if it satisfies the functionality and indistinguishability in Definition 9 and the following efficiency requirement:*

Non-trivial time efficiency *We require that the running time of $\mathsf{sxi}\mathcal{O}$ on input $(1^\lambda, C)$ is at most $2^{n\gamma} \cdot \mathrm{poly}(\lambda, |C|)$ for every $\lambda \in \mathbb{N}$ and circuit $C \in \{\mathcal{C}_\lambda\}_{\lambda \in \mathbb{N}}$ with input length n.*

We have the following theorem.

Theorem 9 ([15]). *Assuming there exists ϵ-secure collusion-resistant SKFE for all circuits, where $\epsilon(\cdot)$ is a negligible function. Then, for any constant $\gamma < 1$, there exists ϵ-secure γ-compressing SXIO for polynomial-size circuits with logarithmic size input.*

5 Punctarable Secret-Key Functional Encryption

We introduce puncturable secret-key functional encryption (puncturable SKFE).

The notion of puncturable SKFE was introduced by Bitansky and Vaikuntanathan [17]. They showed that in their construction of IO, the building block PKFE can be replaced with puncturable SKFE. However, it has been open whether we can achieve puncturable SKFE without assuming PKFE.

In this work, we answer the question affirmatively. We show how to construct a relaxed variant of puncturable SKFE scheme that is single-key weakly-succinct. Our relaxed variant is sufficient for constructing IO. Our construction consists of two steps.

1. We prove that a single-key non-succinct puncturable SKFE scheme is constructed only from one-way functions.
2. We prove that we can transform the non-succinct scheme into an weakly-succinct one by using SXIO.

We can construct SXIO based on standard (i.e., not puncturable) SKFE by Theorem 9. Thus, we can construct our puncturable SKFE from standard SKFE.

5.1 Syntax

Our definition of puncturable SKFE introduced below is slightly different from that proposed by Bitansky and Vaikuntanathan [17]. However, we show that puncturable SKFE defined in this paper is also a sufficient building block of IO. We state differences between our definition and theirs after describing the syntax and security of our puncturable SKFE.

Definition 11 (Puncturable secret-key functional encryption). *A puncturable SKFE scheme* pSKFE *is a tuple* $(\mathsf{Setup}, \mathsf{KG}, \mathsf{Enc}, \mathsf{Dec}, \mathsf{Punc}, \mathsf{PEnc})$ *of six PPT algorithms. Below, let* $\mathcal{M}$, $\mathcal{F}$, *and* $\mathcal{T}$ *be the message space, function space, and tag space of* pSKFE, *respectively. In addition, let* q *be a polynomial denoting the upper bound of the number of issuable functional keys.*

- *The setup algorithm* Setup, *given a security parameter* 1^λ, *outputs a master secret key* MSK.
- *The key generation algorithm* KG, *given a master secret key* MSK, *function* $f \in \mathcal{F}$, *and an index* $i \in [q]$, *outputs a functional key* sk_f.
- *The encryption algorithm* Enc, *given a master secret key* MSK, *a tag* tag, *and a message* $m \in \mathcal{M}$, *outputs a ciphertext* CT.
- *The decryption algorithm* Dec, *given a functional key* sk_f, *a tag* tag, *and a ciphertext* CT, *outputs a message* $\tilde{m} \in \{\bot\} \cup \mathcal{M}$.
- *The puncturing algorithm* Punc, *given a master secret key* MSK *and a tag* tag, *outputs a punctured master secret key* $\mathsf{MSK}^*\{\mathsf{tag}\}$
- *The punctured encryption algorithm* PEnc, *given a punctured master secret key* MSK^*, *a tag* tag', *and a message* m, *outputs a ciphertext* CT.

Correctness: *For every* $m \in \mathcal{M}, f \in \mathcal{F}, i \in [q], \mathsf{tag} \in \mathcal{T}$, *and* $\mathsf{MSK} \leftarrow \mathsf{Setup}(1^\lambda)$, *we require that* $\mathsf{Dec}(\mathsf{KG}(\mathsf{MSK}, f, i), \mathsf{tag}, \mathsf{Enc}(\mathsf{MSK}, \mathsf{tag}, m)) = f(m)$.

5.2 Security

In this section, we introduce two variants of security. Their difference is the functionality of punctured encryption algorithms.

Definition 12 (Secure puncturable SKFE). *Let* $\mathsf{pSKFE} = (\mathsf{Setup}, \mathsf{KG}, \mathsf{Enc}, \mathsf{Dec}, \mathsf{Punc}, \mathsf{PEnc})$ *be puncturable SKFE. Below, let* $\mathcal{M}$, $\mathcal{F}$, *and* $\mathcal{T}$ *be the message space, function space, and tag space of* pSKFE, *respectively. In addition, let* q *be a polynomial denoting the upper bound of the number of issuable functional keys. We say that* pSKFE *is secure puncturable SKFE if it satisfies the following properties.*

Functionality preserving under puncturing:
For every $m \in \mathcal{M}$, $(\mathsf{tag}, \mathsf{tag}') \in \mathcal{T} \times \mathcal{T}$ *such that* $\mathsf{tag} \neq \mathsf{tag}'$, *randomness* r, $\mathsf{MSK} \leftarrow \mathsf{Setup}(1^\lambda)$, *and* $\mathsf{MSK}^*\{\mathsf{tag}\} \leftarrow \mathsf{Punc}(\mathsf{MSK}, \mathsf{tag})$, *it holds that*

$$\mathsf{PEnc}(\mathsf{MSK}^*\{\mathsf{tag}\}, \mathsf{tag}', m; r) = \mathsf{Enc}(\mathsf{MSK}, \mathsf{tag}', m; r) \ .$$

Semantic security at punctured tag: *We define punctured semantic security game between a challenger and an adversary* $\mathcal{A}$ *as follows.*

1. *The challenger generates a master secret key* $\mathsf{MSK} \leftarrow \mathsf{Setup}(1^\lambda)$ *and chooses a challenge bit* $b \xleftarrow{\mathsf{r}} \{0,1\}$. *The challenger sends security parameter* 1^λ *to* $\mathcal{A}$.
2. $\mathcal{A}$ *sends* $(m_0, m_1) \in \mathcal{M} \times \mathcal{M}$, $\mathsf{tag} \in \mathcal{T}$, *and* $\{f_i\}_{i \in [q]} \in \mathcal{F}^q$ *to the challenger. We require that for every* $i \in [q]$ *it holds that* $f_i(m_0) = f_i(m_1)$.
3. *The challenger computes* $\mathsf{CT} \leftarrow \mathsf{Enc}(\mathsf{MSK}, \mathsf{tag}, m_b)$, $\mathsf{sk}_{f_i} \leftarrow \mathsf{KG}(\mathsf{MSK}, f_i, i)$ *for every* $i \in [q]$, *and* $\mathsf{MSK}^*\{\mathsf{tag}\} \leftarrow \mathsf{Punc}(\mathsf{MSK}, \mathsf{tag})$.
Then, the challenger returns $(\mathsf{MSK}^*\{\mathsf{tag}\}, \mathsf{CT}, \{\mathsf{sk}_{f_i}\}_{i \in [q]})$ *to* $\mathcal{A}$.
4. $\mathcal{A}$ *outputs* $b' \in \{0,1\}$.

In this game, we define the advantage of the adversary $\mathcal{A}$ *as*

$$\mathsf{Adv}^{\mathsf{ss}}_{\mathsf{pSKFE},\mathcal{A}}(\lambda) = 2|\Pr[b = b'] - \frac{1}{2}| = |\Pr[b' = 1 | b = 0] - \Pr[b' = 1 | b = 1]| \ .$$

$\mathcal{A}$ *is said to be valid if* $f_i(m_0) = f_i(m_1)$ *holds for every* $i \in [q]$ *in the above game. We say that* pSKFE *satisfies semantic security at punctured tag if for any valid PPT* $\mathcal{A}$, *we have* $\mathsf{Adv}^{\mathsf{ss}}_{\mathsf{pSKFE},\mathcal{A}}(\lambda) = \mathsf{negl}(\lambda)$.
We further say that pSKFE *satisfies* ϵ*-semantic security at punctured tag, for some concrete negligible function* $\epsilon(\cdot)$, *if for any valid PPT* $\mathcal{A}$ *the above advantage* $\mathsf{Adv}^{\mathsf{ss}}_{\mathsf{pSKFE},\mathcal{A}}(\lambda)$ *is smaller than* $\epsilon(\lambda)^{\Omega(1)}$.

In addition, we say that pSKFE *is* ϵ*-secure puncturable SKFE if it satisfies functionality preserving under puncturing and* ϵ*-semantic security at punctured tag.*

Instead of functionality preserving under puncturing, we can consider a relaxed variant which we call *indistinguishability of functionality under puncturing*. This property requires that any PPT distinguisher cannot distinguish ciphertexts output by Enc and PEnc even given both master secret key and punctured master secret key. The formal definition is as follows.

Definition 13 (Indistinguishability of functionality under puncturing). *Let* $\mathsf{pSKFE} = (\mathsf{Setup}, \mathsf{KG}, \mathsf{Enc}, \mathsf{Dec}, \mathsf{Punc}, \mathsf{PEnc})$ *be puncturable SKFE whose message space and tag space are* $\mathcal{M}$ *and* $\mathcal{T}$*, respectively. We define indistinguishability of functionality game between a challenger and an adversary* $\mathcal{A}$ *as follows.*

1. *The challenger generates a master secret key* $\mathsf{MSK} \leftarrow \mathsf{Setup}(1^\lambda)$ *and chooses a challenge bit* $b \xleftarrow{\mathsf{r}} \{0,1\}$*. The challenger sends security parameter* 1^λ *to* $\mathcal{A}$*.*
2. $\mathcal{A}$ *sends* $m \in \mathcal{M}$ *and* $(\mathsf{tag}, \mathsf{tag}') \in \mathcal{T} \times \mathcal{T}$ *such that* $\mathsf{tag} \neq \mathsf{tag}'$ *to the challenger.*
3. *The challenger first computes* $\mathsf{MSK}^*\{\mathsf{tag}\} \leftarrow \mathsf{Punc}(\mathsf{MSK}, \mathsf{tag})$*. Then, the challenger computes* $\mathsf{CT} \leftarrow \mathsf{Enc}(\mathsf{MSK}, \mathsf{tag}', m)$ *if* $b = 0$*, and otherwise* $\mathsf{CT} \leftarrow \mathsf{PEnc}(\mathsf{MSK}^*\{\mathsf{tag}\}, \mathsf{tag}', m)$*.*
   Then, the challenger returns $(\mathsf{MSK}, \mathsf{MSK}^*\{\mathsf{tag}\}, \mathsf{CT})$ *to* $\mathcal{A}$*.*
4. $\mathcal{A}$ *outputs* $b' \in \{0,1\}$*.*

In this game, we define the advantage of the adversary $\mathcal{A}$ *as*

$$\mathsf{Adv}^{\mathsf{if}}_{\mathsf{pSKFE},\mathcal{A}}(\lambda) = 2|\Pr[b = b'] - \frac{1}{2}| = |\Pr[b' = 1|b = 0] - \Pr[b' = 1|b = 1]| \ .$$

We say that pSKFE *satisfies indistinguishability of functionality under puncturing if for any PPT* $\mathcal{A}$*, we have* $\mathsf{Adv}^{\mathsf{if}}_{\mathsf{pSKFE},\mathcal{A}}(\lambda) = \mathsf{negl}(\lambda)$*.*

We further say that pSKFE *satisfies* ϵ*-indistinguishability of functionality under puncturing, for some concrete negligible function* $\epsilon(\cdot)$*, if for any PPT* $\mathcal{A}$ *the above advantage* $\mathsf{Adv}^{\mathsf{if}}_{\mathsf{pSKFE},\mathcal{A}}(\lambda)$ *is smaller than* $\epsilon(\lambda)^{\Omega(1)}$*.*

Definition 14 (Secure puncturable SKFE with indistinguishability of functionality). *Let* pSKFE *be puncturable SKFE. Let* $\epsilon_1(\cdot)$ *and* $\epsilon_2(\cdot)$ *be some negligible functions. If* pSKFE *satisfies* ϵ_1*-semantic security at punctured tag and* ϵ_2*-indistinguishability of functionality under puncturing, then we say that* pSKFE *is* (ϵ_1, ϵ_2)*-secure puncturable SKFE with indistinguishability of functionality.*

5.3 Efficiency

We introduce the notion of succinctness for puncturable SKFE.

Definition 15 (Succinctness). *Let* $\mathcal{F}$ *be a function family. Let* s *and* n *be the maximum size and input length of functions contained in* $\mathcal{F}$*, respectively.*

Weakly-succinct: *Puncturable SKFE for* $\mathcal{F}$ *is said to be weakly-succinct if the size of both the encryption circuit and punctured encryption circuit are bounded by* $s^\gamma \cdot \mathrm{poly}(\lambda, n)$*, where* $\gamma < 1$ *is a fixed constant. We call* γ *the compression factor.*

Collusion-succinct: *Puncturable SKFE for* $\mathcal{F}$ *is said to be collusion-succinct if the size of both the encryption circuit and punctured encryption circuit are bounded by* $q^\gamma \cdot \mathrm{poly}(n, \lambda, s)$*, where* q *is the upper bound of issuable functional decryption keys and* $\gamma < 1$ *is a fixed constant. We call* γ *the compression factor.*

5.4 Difference from the Definition of Bitansky and Vaikuntanathan

There are three main differences between our definition of puncturable SKFE and that of Bitansky and Vaikuntanathan [17]. Two are about syntax. The other is about security.

Syntactical differences are as follows.

Tag-based encryption and decryption: In the definition of Bitansky and Vaikuntanathan, a master secret key is punctured at *two messages*. Their semantic security requires that no PPT adversary can distinguish ciphertexts of these two messages given the punctured master secret key.
We adopt the tag based syntax for the encryption and decryption algorithms while Bitansky and Vaikuntanathan do not. A tag-based definition is well-suited for our non-succinct puncturable SKFE scheme. When our non-succinct scheme encrypts a message, it generates a garbled circuit of an universal circuit into which the message is hardwired, and then masks labels of the garbled circuit by a string generated by puncturable PRF. A tag fed to the encryption algorithm is used as an input to puncturable PRF. See Sect. 6 for details.
In our construction of IO in Sect. 8, we use an input to an obfuscated circuit as a tag for ciphertexts of puncturable SKFE. Therefore, our IO construction is not significantly different from the IO construction based on puncturable SKFE by Bitansky and Vaikuntanathan from the syntactical point of view though ours is based on tag-based puncturable SKFE.

Index based key generation: We define the key generation algorithm as a stateful algorithm. In other words, for the i-th invocation, we need to feed an index i to the key generation algorithm in addition to a master secret key and a function. This is because we transform a non-succinct scheme into an weakly-succinct one *via a collusion-succinct scheme whose key generation algorithm is stateful* in Sect. 7.
We note that our stateful collusion-succinct scheme is just an intermediate scheme to achieve IO. We also emphasize the fact that the index-based key generation is not an issue to construct IO because our main building block is a *single-key* weakly-succinct puncturable SKFE scheme. For a single-key scheme, we do not need any state for key generation because it can issue only a single functional key.
Below, we omit the index of single-key schemes in the syntax for simplicity.

Functionality under puncturing. In addition to the syntactic differences above, there is a difference about security. We defined indistinguishability of functionality under puncturing in Definition 13. The reason why we introduce the relaxed notion of functionality preserving property is that our weakly-succinct scheme does not satisfy functionality preserving under puncturing in Definition 12 but the relaxed one. Our *non-succinct* scheme satisfies functionality preserving under puncturing.

One might think that puncturable SKFE satisfying indistinguishability of functionality under puncturing is not sufficient to construct IO. This is not the case. We show that indistinguishability of functionality under puncturing suffices for constructing IO and our weakly-succinct scheme satisfies the property.

6 Single-Key Non-Succinct Puncturable SKFE

We show we can construct a single-key (non-succinct) puncturable SKFE scheme assuming only one-way functions. This construction is similar to that of single-key non-succinct PKFE proposed by Sahai and Seyalioglu [61]. Their construction is based on garbling scheme and public-key encryption. In our construction, we use puncturable PRF instead of public-key encryption, and, as a result, achieve the puncturable property. We recall that we can realize both garbling scheme and puncturable PRF assuming only one-way functions. We give the construction below.

Let $\mathsf{GC} = (\mathsf{Grbl}, \mathsf{Eval})$ be garbling scheme, and $\mathsf{PPRF} = (\mathsf{F}, \mathsf{Punc_F})$ be puncturable PRF. Using GC and PPRF, we construct puncturable SKFE $\mathsf{OneKey} = (\mathsf{1Key.Setup}, \mathsf{1Key.KG}, \mathsf{1Key.Enc}, \mathsf{1Key.Dec}, \mathsf{1Key.Punc}, \mathsf{1Key.PEnc})$ supporting only one functional key as follows. Note that the tag space of OneKey is the same as the domain of PPRF. In addition, the index space of OneKey is $[1]$, and thus we omit the index from the description by assuming the index is always fixed to 1. Below, we assume that we can represent every function f by an s-bit string $(f[1], \cdots, f[s])$.

Construction. The scheme consists of the following algorithms.

$\mathsf{1Key.Setup}(1^\lambda)$:
- Generate $S_{j,\alpha} \xleftarrow{r} \{0,1\}^\lambda$ for every $j \in [s]$ and $\alpha \in \{0,1\}$.
- Return $\mathsf{MSK} \leftarrow \{S_{j,\alpha}\}_{j\in[s],\alpha\in\{0,1\}}$.

$\mathsf{1Key.KG}(\mathsf{MSK}, f)$:
- Parse $\{S_{j,\alpha}\}_{j\in[s],\alpha\in\{0,1\}} \leftarrow \mathsf{MSK}$ and $(f[1], \cdots, f[s]) \leftarrow f$.
- Return $\mathsf{sk}_f \leftarrow (f, \{S_{j,f[j]}\}_{j\in[s]})$.

$\mathsf{1Key.Enc}(\mathsf{MSK}, \mathsf{tag}, m)$:
- Parse $\{S_{j,\alpha}\}_{j\in[s],\alpha\in\{0,1\}} \leftarrow \mathsf{MSK}$.
- Compute $(\widetilde{U}, \{L_{j,\alpha}\}_{j\in[s],\alpha\in\{0,1\}}) \leftarrow \mathsf{Grbl}(1^\lambda, U(\cdot, m))$.
- For every $j \in [s]$ and $\alpha \in \{0,1\}$, compute $R_{j,\alpha} \leftarrow \mathsf{F}(S_{j,\alpha}, \mathsf{tag})$ and $c_{j,\alpha} \leftarrow L_{j,\alpha} \oplus R_{j,\alpha}$.
- Return $\mathsf{CT} \leftarrow (\widetilde{U}, \{c_{j,\alpha}\}_{j\in[s],\alpha\in\{0,1\}})$.

$\mathsf{1Key.Dec}(\mathsf{sk}_f, \mathsf{tag}, \mathsf{CT})$:
- Parse $(f, \{S_j\}_{j\in[s]}) \leftarrow \mathsf{sk}_f$ and $(\widetilde{U}, \{c_{j,\alpha}\}_{j\in[s],\alpha\in\{0,1\}}) \leftarrow \mathsf{CT}$.
- For every $j \in [s]$, compute $R_j \leftarrow \mathsf{F}(S_j, \mathsf{tag})$ and $L_j \leftarrow c_{j,f[j]} \oplus R_j$.
- Return $y \leftarrow \mathsf{Eval}(\widetilde{U}, \{L_j\}_{j\in[s]})$.

$\mathsf{1Key.Punc}(\mathsf{MSK}, \mathsf{tag})$:
- Parse $\{S_{j,\alpha}\}_{j\in[s],\alpha\in\{0,1\}} \leftarrow \mathsf{MSK}$.
- For every $j \in [s]$ and $\alpha \in \{0,1\}$, compute $S^*_{j,\alpha}\{\mathsf{tag}\} \leftarrow \mathsf{Punc_F}(S_{j,\alpha}, \mathsf{tag})$.
- Return $\mathsf{MSK}^*\{\mathsf{tag}\} \leftarrow \{S^*_{j,\alpha}\{\mathsf{tag}\}\}_{j\in[s],\alpha\in\{0,1\}}$.

$\mathsf{1Key.PEnc}(\mathsf{MSK}^*, \mathsf{tag}', m)$

- Parse $\{S^*_{j,\alpha}\}_{j\in[s],\alpha\in\{0,1\}} \leftarrow \mathsf{MSK}^*$.
- Compute $(\widetilde{U}, \{L_{j,\alpha}\}_{j\in[s],\alpha\in\{0,1\}}) \leftarrow \mathsf{Grbl}(1^\lambda, U(\cdot, m))$.
- For every $j \in [s]$ and $\alpha \in \{0,1\}$, compute $R_{j,\alpha} \leftarrow \mathsf{F}_{S^*_{j,\alpha}}(\mathsf{tag}')$ and $c_{j,\alpha} \leftarrow L_{j,\alpha} \oplus R_{j,\alpha}$.
- Return $\mathsf{CT} \leftarrow (\widetilde{U}, \{c_{j,\alpha}\}_{j\in[s],\alpha\in\{0,1\}})$.

Then, we have the following theorem.

Theorem 10. *Let* GC *be δ-secure garbling scheme, and* PPRF *δ-secure puncturable PRF, where $\delta(\cdot)$ is some negligible function. Then,* OneKey *is δ-secure single-key puncturable SKFE.*

See [45] for the formal proof of this theorem.

7 From Non-Succinct Puncturable SKFE to Weakly-Succinct One

In this section, we show how to transform single-key non-succinct puncturable SKFE into single-key weakly-succinct one using SXIO. Note that the resulting scheme satisfies only indistinguishability of functionality under puncturing property even if we start the transformation with a non-succinct scheme satisfying functionality preserving under puncturing property.

The transformation consists of 2 steps. First, we show how to construct collusion-succinct puncturable SKFE from single-key non-succinct puncturable SKFE and SXIO. Then, we give the transformation from collusion-succinct puncturable SKFE to weakly-succinct one.

In fact, the intermediate collusion-succinct scheme satisfies only indistinguishability of functionality under puncturing property. This is because we adopt a construction technique similar to that proposed by Lin et al. [53] (and extended by Bitansky et al. [15] and Kitagawa et al. [46]), and thus we use an obfuscated encryption circuit of the building block scheme by SXIO as a ciphertext of the resulting scheme. This fact is the reason the resulting weakly-succinct scheme satisfies only indistinguishability of functionality under puncturing property.

7.1 From Non-Succinct to Collusion-Succinct by Using SXIO

For any q which is a fixed polynomial of λ, we show how to construct a puncturable SKFE scheme whose index space is $[q]$ based on a single-key puncturable SKFE scheme. The resulting scheme is collusion-succinct, that is, the running time of both the encryption algorithm and the punctured encryption algorithm are sub-linear in q. We show the construction below.

Let $\mathsf{OneKey} = (\mathsf{1Key.Setup}, \mathsf{1Key.KG}, \mathsf{1Key.Enc}, \mathsf{1Key.Dec}, \mathsf{1Key.Punc}, \mathsf{1Key.PEnc})$ be puncturable SKFE that we constructed in Sect. 6. Let $\mathsf{sxi}\mathcal{O}$ be SXIO and $\mathsf{PPRF} = (\mathsf{F}, \mathsf{Punc}_\mathsf{F})$ puncturable PRF. Using OneKey, $\mathsf{sxi}\mathcal{O}$, and PPRF, we construct puncturable SKFE $\mathsf{CollSuc} = (\mathsf{CS.Setup}, \mathsf{CS.KG}, \mathsf{CS.Enc}, \mathsf{CS.Dec}, \mathsf{CS.Punc}, \mathsf{CS.PEnc})$ as follows. We again note that q is a fixed polynomial of λ. Let the tag of CollSuc be $\mathcal{T}$. Then, the tag space of OneKey is also $\mathcal{T}$.

Construction. The scheme consists of the following algorithms.

$\mathsf{CS.Setup}(1^\lambda)$:
- Generate $S \xleftarrow{r} \{0,1\}^\lambda$ and return $\mathsf{MSK} \leftarrow S$.

$\mathsf{CS.KG}(\mathsf{MSK}, f, i)$:
- Parse $S \leftarrow \mathsf{MSK}$.
- Compute $r^i_{\mathsf{Setup}} \leftarrow \mathsf{F}_S(i)$ and $\mathsf{MSK}_i \leftarrow \mathsf{1Key.Setup}(1^\lambda; r^i_{\mathsf{Setup}})$.
- Compute $\mathsf{1Key.sk}_f \leftarrow \mathsf{1Key.KG}(\mathsf{MSK}_i, f)$ and return $\mathsf{sk}_f \leftarrow (i, \mathsf{1Key.sk}_f)$.

$\mathsf{CS.Enc}(\mathsf{MSK}, \mathsf{tag}, m)$:
- Parse $S \leftarrow \mathsf{MSK}$.
- Generate $S_{\mathsf{Enc}} \xleftarrow{r} \{0,1\}^\lambda$ and return $\mathsf{CT} \leftarrow \mathsf{sxi}\mathcal{O}(\mathsf{E}_{\mathsf{1Key}}[S, S_{\mathsf{Enc}}, \mathsf{tag}, m])$. The circuit $\mathsf{E}_{\mathsf{1Key}}$ is defined in Fig. 4.

$\mathsf{CS.Dec}(\mathsf{sk}_f, \mathsf{tag}, \mathsf{CT})$:
- Parse $(i, \mathsf{1Key.sk}_f) \leftarrow \mathsf{sk}_f$.
- Compute $\mathsf{CT}_i \leftarrow \mathsf{CT}(i)$ and return $y \leftarrow \mathsf{1Key.Dec}(\mathsf{1Key.sk}_f, \mathsf{tag}, \mathsf{CT}_i)$.

$\mathsf{CS.Punc}(\mathsf{MSK}, \mathsf{tag})$:
- Parse $S \leftarrow \mathsf{MSK}$.
- Generate $S_{\mathsf{Punc}} \xleftarrow{r} \{0,1\}^\lambda$ and compute $\widetilde{\mathsf{P}} \leftarrow \mathsf{sxi}\mathcal{O}(\mathsf{P}_{\mathsf{1Key}}[S, S_{\mathsf{Punc}}, \mathsf{tag}])$. The circuit $\mathsf{P}_{\mathsf{1Key}}$ is defined in Fig. 5.
- Return $\mathsf{MSK}^*\{\mathsf{tag}\} \leftarrow \widetilde{\mathsf{P}}$.

$\mathsf{CS.PEnc}(\mathsf{MSK}^*, \mathsf{tag}', m)$:
- Parse $\widetilde{\mathsf{P}} \leftarrow \mathsf{MSK}^*$.
- Generate $S_{\mathsf{Enc}} \xleftarrow{r} \{0,1\}^\lambda$ and return $\mathsf{CT} \leftarrow \mathsf{sxi}\mathcal{O}(\mathsf{PE}_{\mathsf{1Key}}[\widetilde{\mathsf{P}}, S_{\mathsf{Enc}}, \mathsf{tag}', m])$. The circuit $\mathsf{PE}_{\mathsf{1Key}}$ is defined in Fig. 6.

Encryption circuit $\mathsf{E}_{\mathsf{1Key}}[S, S_{\mathsf{Enc}}, \mathsf{tag}, m](i)$:

Hardwired: Two PRF keys S and S_{Enc}, a tag tag, and a message m.
Input: An index $i \in [q]$.
Padding: This circuit is padded to size $\mathsf{pad}_{\mathsf{E}} := \mathsf{pad}_{\mathsf{E}}(\lambda, n, s)$, which is determined in analysis.

1. Compute $r^i_{\mathsf{Setup}} \leftarrow \mathsf{F}_S(i)$ and $r_{\mathsf{Enc}} \leftarrow \mathsf{F}_{S_{\mathsf{Enc}}}(i)$.
2. Compute $\mathsf{MSK}_i \leftarrow \mathsf{1Key.Setup}(1^\lambda; r^i_{\mathsf{Setup}})$.
3. Return $\mathsf{CT}_i \leftarrow \mathsf{1Key.Enc}(\mathsf{MSK}_i, \mathsf{tag}, m; r_{\mathsf{Enc}})$.

Fig. 4. The description of $\mathsf{E}_{\mathsf{1Key}}$.

Then, we have the following theorem.

Theorem 11. *Let $\delta(\cdot)$ be some negligible function. Let* OneKey *be δ-secure single-key puncturable SKFE constructed in Sect. 6. Let* $\mathsf{sxi}\mathcal{O}$ *be δ-secure γ-compressing SXIO, where γ is a sufficiently small constant such that $\gamma < 1$. Let* PPRF *be δ-secure puncturable PRF. Then,* CollSuc *is (δ, δ)-secure puncturable SKFE with indistinguishability of functionality that is collusion-succinct with compression factor $\hat{\gamma}$, which is a constant smaller than* 1.

Punctured key generation circuit $\mathsf{P}_{\mathsf{1Key}}[S, S_{\mathsf{Punc}}, \mathsf{tag}](i)$:

Hardwired: Two PRF keys S and S_{Punc}, and a tag tag.
Input: An index $i \in [q]$.
Padding: This circuit is padded to size $\mathsf{pad}_{\mathsf{P}} := \mathsf{pad}_{\mathsf{P}}(\lambda, n, s)$, which is determined in analysis.

1. Compute $r^i_{\mathsf{Setup}} \leftarrow \mathsf{F}_S(i)$ and $r_{\mathsf{Punc}} \leftarrow \mathsf{F}_{S_{\mathsf{Punc}}}(i)$.
2. Compute $\mathsf{MSK}_i \leftarrow \mathsf{1Key.Setup}(1^\lambda; r^i_{\mathsf{Setup}})$.
3. Return $\mathsf{MSK}^*_i\{\mathsf{tag}\} \leftarrow \mathsf{1Key.Punc}(\mathsf{MSK}_i, \mathsf{tag}; r_{\mathsf{Punc}})$.

Fig. 5. The description of $\mathsf{P}_{\mathsf{1Key}}$.

Punctured encryption circuit $\mathsf{PE}_{\mathsf{1Key}}[\widetilde{\mathsf{P}}, S_{\mathsf{Enc}}, \mathsf{tag}, m](i)$:

Hardwired: A circuit $\widetilde{\mathsf{P}}$, a PRF key S_{Enc}, a tag tag, and a message m.
Input: An index $i \in [q]$.
Padding: This circuit is padded to size $\mathsf{pad}_{\mathsf{E}} := \mathsf{pad}_{\mathsf{E}}(\lambda, n, s)$, which is determined in analysis.

1. Compute $\mathsf{MSK}^*_i \leftarrow \widetilde{\mathsf{P}}(i)$ and $r_{\mathsf{Enc}} \leftarrow \mathsf{F}_{S_{\mathsf{Enc}}}(i)$.
2. Return $\mathsf{CT}_i \leftarrow \mathsf{1Key.PEnc}(\mathsf{MSK}^*_i, \mathsf{tag}, m; r_{\mathsf{Enc}})$.

Fig. 6. The description of $\mathsf{PE}_{\mathsf{1Key}}$.

See [45] for the formal proof of this theorem.

The requirement for γ and the concrete value of $\hat{\gamma}$ is determined in the efficiency analysis in the proof of Theorem 11. We can make $\hat{\gamma}$ smaller than 1 by using SXIO with sufficiently small compression factor γ as the building block. Such SXIO is constructed from collusion-resistant SKFE [15].

7.2 From Collusion-Succinct to Weakly-Succinct

In this section, we show how to construct a single-key weakly-succinct puncturable SKFE scheme from a collusion-succinct one.

This transformation is based on those proposed by Bitansky and Vaikuntanathan [17] and Ananth et al. [3], and thus utilizes a decomposable randomized encoding. The difference is that we must consider puncturing and punctured encryption algorithms since we construct a puncturable SKFE scheme. In fact, we show their construction works for puncturable SKFE schemes. In addition, we consider semantic security defined in the weakly selective security manner while they considered selective security. Below, we give the construction.

We construct single-key puncturable SKFE $\mathsf{WeakSuc} = (\mathsf{WS.Setup}, \mathsf{WS.KG}, \mathsf{WS.Enc}, \mathsf{WS.Dec}, \mathsf{WS.Punc}, \mathsf{WS.PEnc})$. Let s and n be the maximum size and input length of functions supported by $\mathsf{WeakSuc}$. Let RE be c-local decomposable randomized encoding, where c is a constant. We suppose that the number of decomposed encodings of RE for a function of size s is μ. Then, μ is a polynomial bounded by $s \cdot \mathrm{poly}_{\mathsf{RE}}(\lambda, n)$, where $\mathrm{poly}_{\mathsf{RE}}(\lambda, n)$ is a fixed polynomial. We also suppose that the randomness space of RE is $\{0,1\}^{\rho}$, where ρ is a polynomial bounded by $s \cdot \mathrm{poly}_{\mathsf{RE}}(\lambda, n)$. Let $\mathsf{CollSuc} = (\mathsf{CS.Setup}, \mathsf{CS.KG}, \mathsf{CS.Enc}, \mathsf{CS.Dec}, \mathsf{CS.Punc}, \mathsf{CS.PEnc})$ be puncturable SKFE whose index space and tag space are $[\mu]$ and $\mathcal{T}$, respectively. Let $\mathsf{SKE} = (\mathsf{E}, \mathsf{D})$ be SKE and PRF PRF. In the scheme, we use $\mathsf{PRF} : \{0,1\}^{\lambda} \times (\{0,1\}^{\lambda} \times [\rho]) \rightarrow \{0,1\}$. Using $\mathsf{CollSuc}$, RE, SKE, and PRF, we construct $\mathsf{WeakSuc}$ as follows. The tag space of $\mathsf{WeakSuc}$ is $\mathcal{T}$.

$\mathsf{WS.Setup}(1^{\lambda})$:
- Return $\mathsf{MSK} \leftarrow \mathsf{CS.Setup}(1^{\lambda})$.

$\mathsf{WS.KG}(\mathsf{MSK}, f)$:
- Generate $K \xleftarrow{r} \{0,1\}^{\lambda}$ and $t \leftarrow \{0,1\}^{\lambda}$.
- Compute $\widehat{f} \leftarrow \mathsf{RE}(1^{\lambda}, f)$ and decomposed encodings $\widehat{f}_1, \cdots \widehat{f}_{\mu}$ together with sets of integers $(R_1, \cdots, R_{\mu})$. R_i indicates which bit of a randomness $\widehat{f}_i$ depends on for every $i \in [\mu]$. Note that $R_i \subseteq [\rho]$ and $|R_i| = c$ for every $i \in [\mu]$.
- Generate $\mathsf{CT}_i^{\mathsf{ske}} \leftarrow \mathsf{E}(K, 0^{|\widehat{f}_i(\cdot,\cdot)|})$, and compute $\mathsf{sk}_{\mathsf{En}_i} \leftarrow \mathsf{CS.KG}(\mathsf{MSK}, \mathsf{En}_{\mathsf{dre}}[\widehat{f}_i, R_i, t, \mathsf{CT}_i^{\mathsf{ske}}], i)$ for every $i \in [\mu]$. $\mathsf{En}_{\mathsf{dre}}$ defined in Fig. 7.
- Return $\mathsf{sk}_f \leftarrow (\mathsf{sk}_{\mathsf{En}_1}, \cdots, \mathsf{sk}_{\mathsf{En}_{\mu}})$.

$\mathsf{WS.Enc}(\mathsf{MSK}, \mathsf{tag}, m)$:
- Generate $S_{\mathsf{encd}} \leftarrow \{0,1\}^{\lambda}$.
- Return $\mathsf{CT} \leftarrow \mathsf{CS.Enc}(\mathsf{MSK}, \mathsf{tag}, (m, S_{\mathsf{encd}}, \bot))$.

$\mathsf{WS.Dec}(\mathsf{sk}_f, \mathsf{tag}, \mathsf{CT})$:
- Parse $(\mathsf{sk}_{\mathsf{En}_1}, \cdots, \mathsf{sk}_{\mathsf{En}_{\mu}}) \leftarrow \mathsf{sk}_f$.
- For every $i \in [\mu]$, compute $e_i \leftarrow \mathsf{CS.Dec}(\mathsf{sk}_{\mathsf{En}_i}, \mathsf{tag}, \mathsf{CT})$.
- Decode y from $(e_1, \cdots, e_{\mu})$.
- Return y.

$\mathsf{WS.Punc}(\mathsf{MSK}, \mathsf{tag})$:
- Return $\mathsf{MSK}^*\{\mathsf{tag}\} \leftarrow \mathsf{CS.Punc}(\mathsf{MSK}, \mathsf{tag})$.

$\mathsf{WS.PEnc}(\mathsf{MSK}^*, \mathsf{tag}', m)$:
- Generate $S_{\mathsf{encd}} \leftarrow \{0,1\}^{\lambda}$.
- Return $\mathsf{CT} \leftarrow \mathsf{CS.PEnc}(\mathsf{MSK}^*, \mathsf{tag}', (m, S_{\mathsf{encd}}, \bot))$.

Then, we have the following theorem.

Theorem 12. *Let $\delta(\cdot)$ be negligible function. Let* $\mathsf{CollSuc}$ *be (δ, δ)-secure puncturable SKFE with indistinguishability of functionality that can issue μ functional keys and is collusion-succinct with compression factor γ, where $\gamma < 1$ is a constant. Let* RE, SKE, *and* PRF *be δ-secure decomposable randomized encoding, SKE, and PRF, respectively. Then,* $\mathsf{WeakSuc}$ *be (δ, δ)-secure single-key puncturable SKFE with indistinguishability of functionality that is weakly-succinct with compression factor γ', where γ' is a constant such that $\gamma < \gamma' < 1$.*

See [45] for the formal proof of this theorem.

Decomposable Randomized Encoding Circuit
$\mathsf{En}_{\mathsf{dre}}[\widehat{f_i}, R_i, t, \mathsf{CT}_i^{\mathsf{ske}}](m, S_{\mathsf{encd}}, K)$

Hardwired: A randomized encoding $\widehat{f_i}$, a set R_i, a string t, and a ciphertext $\mathsf{CT}_i^{\mathsf{ske}}$.

Input: A message m, a PRF key S_{encd}, and an SKE secret key K.

1. If $m = \bot$, return $e_i \leftarrow \mathsf{D}(K, \mathsf{CT}_i^{\mathsf{ske}})$.
2. Else, compute as follows:
   - For $j \in R_i$, compute $r_j \leftarrow \mathsf{PRF}(S_{\mathsf{encd}}, t \| j)$, set $r_{R_i} \leftarrow \{r_j\}_{j \in R_i}$.
   - Return $e_i \leftarrow \widehat{f_i}(m; r_{R_i})$.

Fig. 7. The description of $\mathsf{En}_{\mathsf{dre}}$.

8 Indistinguishability Obfuscation from SKFE

We show how to obtain IO based on SKFE via puncturable SKFE.

8.1 IO from Collusion-Resistant SKFE

We construct IO from puncturable SKFE satisfying only indistinguishability of functionality under puncturing. Formally, we have the following theorem.

Theorem 13. *Let $\delta(\lambda) = 2^{-\lambda^{\epsilon}}$, where $\epsilon < 1$ is a constant. Assuming there exists (δ, δ)-secure single-key weakly-succinct puncturable SKFE with indistinguishability of functionality for all circuits. Then, there exists secure IO for all circuits.*

We omit the formal proof of it. See Sect. 2.4 for the overview of it. In [45], we formally prove it by first providing the construction of IO based on puncturable SKFE, and then analyzing its security and efficiency.

In addition, by combining Theorems 9, 10, 11, and 12, we also obtain the following theorem.

Theorem 14. *Assuming there exists δ-secure collusion-resistant SKFE for all circuits, where $\delta(\cdot)$ is a negligible function. Then, there exists (δ, δ)-secure single-key weakly-succinct puncturable SKFE with indistinguishability of functionality for all circuits.*

In order to obtain Theorem 14, we also use δ-secure PRF, puncturable PRF, plain SKE, garbling scheme, and decomposable randomized encoding as building blocks. From Theorems 4, 5, 6, 7, and 8, all of these primitives are implied by δ-secure one-way functions thus implied by δ-secure collusion-resistant SKFE for all circuits.

From Theorems 13 and 14, we obtain the following main theorem.

Theorem 15. *Let $\delta(\lambda) = 2^{-\lambda^{\epsilon}}$, where $\epsilon < 1$ is a constant. Assuming there exists δ-secure collusion-resistant SKFE for all circuits. Then, there exists secure IO for all circuits.*

Remark 1 (IO for circuits with input of poly-logarithmic length). The security loss of our IO construction is exponential in the input length of circuits, but is independent of the size of circuits. Thus, if the input length of circuits is poly-logarithmic in the security parameter, our IO construction incurs only quasi-polynomial security loss regardless of the size of circuits. Therefore, we can obtain IO for circuits of *polynomial size* with input of poly-logarithmic length from *quasi-polynomially secure* collusion-resistant SKFE for all circuits. This is an improvement over the IO construction by Komargodski and Segev [48]. They showed that IO for circuits of *sub-polynomial size* with input of poly-logarithmic length is constructed from quasi-polynomially secure collusion-resistant SKFE for all circuits.

Komargodski and Segev also showed that the combination of their IO and sub-exponentially secure one-way functions yields succinct and collusion-resistant PKFE for circuits of *sub-polynomial size* with input of poly-logarithmic length. We observe that our IO for circuits of polynomial size with input of poly-logarithmic length leads to succinct and collusion-resistant PKFE for circuits of *polynomial size* with input of poly-logarithmic length by combining sub-exponentially secure one-way functions from the result of Komargodski and Segev.

8.2 Collusion-Resistant SKFE from Weakly-Succinct One

We also show that collusion-resistant SKFE is constructed from single-key weakly-succinct SKFE. Formally, we have the following theorem.

Theorem 16. *Let $\delta(\lambda) = \lambda^{-\zeta}$, where $\zeta = \omega(1)$. Assuming there exists $(1, \delta)$-selective-message message private SKFE for all circuits that is weakly succinct. Then, there exists δ'-secure collusion-resistant SKFE for all circuits, where $\delta'(\lambda) = \lambda^{-\zeta^{1/2}}$.*[11]

In [44], we formally show Theorem 16. See Sect. 3 for the overview for this result.

Theorem 16 states that if the underlying single-key scheme is sub-exponentially secure, then so is the resulting scheme. Therefore, from Theorems 15 and 16, we have the following corollary.

Corollary 2. *Assuming there exists sub-exponentially secure single-key weakly-succinct SKFE for all circuits. Then, there exists IO for all circuits.*

Acknowledgement. The first and third authors are supported by NTT Secure Platform Laboratories, JST CREST JPMJCR14D6, JST OPERA, JSPS KAKENHI JP16H01705, JP16J10322, JP17H01695.

[11] We can slightly generalize the result. By setting $\eta = \zeta^{1/c}$ in the construction for any constant $c > 1$, we can achieve $\delta'(\lambda) = \lambda^{-\zeta^{1/c}}$.

References

1. Ananth, P., Brakerski, Z., Segev, G., Vaikuntanathan, V.: From selective to adaptive security in functional encryption. In: Gennaro, R., Robshaw, M. (eds.) CRYPTO 2015. LNCS, vol. 9216, pp. 657–677. Springer, Heidelberg (2015). https://doi.org/10.1007/978-3-662-48000-7_32
2. Ananth, P., Jain, A.: Indistinguishability obfuscation from compact functional encryption. In: Gennaro, R., Robshaw, M. (eds.) CRYPTO 2015. LNCS, vol. 9215, pp. 308–326. Springer, Heidelberg (2015). https://doi.org/10.1007/978-3-662-47989-6_15
3. Ananth, P., Jain, A., Sahai, A.: Indistinguishability obfuscation from functional encryption for simple functions. Cryptology ePrint Archive, Report 2015/730
4. Ananth, P., Sahai, A.: Projective arithmetic functional encryption and indistinguishability obfuscation from degree-5 multilinear maps. In: Coron, J.-S., Nielsen, J.B. (eds.) EUROCRYPT 2017. LNCS, vol. 10210, pp. 152–181. Springer, Cham (2017). https://doi.org/10.1007/978-3-319-56620-7_6
5. Ananth, P.V., Gupta, D., Ishai, Y., Sahai, A.: Optimizing obfuscation: avoiding Barrington's theorem. In: ACM CCS 2014, pp. 646–658 (2014)
6. Applebaum, B., Ishai, Y., Kushilevitz, E.: Computationally private randomizing polynomials and their applications. Comput. Complex. **15**(2), 115–162 (2006)
7. Apon, D., Döttling, N., Garg, S., Mukherjee, P.: Cryptanalysis of indistinguishability obfuscations of circuits over GGH13. In: 44rd International Colloquium on Automata, Languages, and Programming, ICALP 2017, Warsaw, Poland, 10–14 July 2017 (2017, to appear)
8. Applebaum, B., Brakerski, Z.: Obfuscating circuits via composite-order graded encoding. In: Dodis, Y., Nielsen, J.B. (eds.) TCC 2015. LNCS, vol. 9015, pp. 528–556. Springer, Heidelberg (2015). https://doi.org/10.1007/978-3-662-46497-7_21
9. Asharov, G., Segev, G.: Limits on the power of indistinguishability obfuscation and functional encryption. In: 56th FOCS, pp. 191–209 (2015)
10. Badrinarayanan, S., Miles, E., Sahai, A., Zhandry, M.: Post-zeroizing obfuscation: new mathematical tools, and the case of evasive circuits. In: Fischlin, M., Coron, J.-S. (eds.) EUROCRYPT 2016. LNCS, vol. 9666, pp. 764–791. Springer, Heidelberg (2016). https://doi.org/10.1007/978-3-662-49896-5_27
11. Barak, B., Garg, S., Kalai, Y.T., Paneth, O., Sahai, A.: Protecting obfuscation against algebraic attacks. In: Nguyen, P.Q., Oswald, E. (eds.) EUROCRYPT 2014. LNCS, vol. 8441, pp. 221–238. Springer, Heidelberg (2014). https://doi.org/10.1007/978-3-642-55220-5_13
12. Barak, B., Goldreich, O., Impagliazzo, R., Rudich, S., Sahai, A., Vadhan, S., Yang, K.: On the (im)possibility of obfuscating programs. In: Kilian, J. (ed.) CRYPTO 2001. LNCS, vol. 2139, pp. 1–18. Springer, Heidelberg (2001). https://doi.org/10.1007/3-540-44647-8_1
13. Bellare, M., Hoang, V.T., Rogaway, P.: Foundations of garbled circuits. In: ACM CCS 2012, pp. 784–796 (2012)
14. Bitansky, N., Garg, S., Lin, H., Pass, R., Telang, S.: Succinct randomized encodings and their applications. In: 47th ACM STOC, pp. 439–448 (2015)

15. Bitansky, N., Nishimaki, R., Passelègue, A., Wichs, D.: From cryptomania to obfustopia through secret-key functional encryption. In: Hirt, M., Smith, A. (eds.) TCC 2016. LNCS, vol. 9986, pp. 391–418. Springer, Heidelberg (2016). https://doi.org/10.1007/978-3-662-53644-5_15
16. Bitansky, N., Paneth, O., Wichs, D.: Perfect structure on the edge of chaos. In: Kushilevitz, E., Malkin, T. (eds.) TCC 2016. LNCS, vol. 9562, pp. 474–502. Springer, Heidelberg (2016). https://doi.org/10.1007/978-3-662-49096-9_20
17. Bitansky, N., Vaikuntanathan, V.: Indistinguishability obfuscation from functional encryption. In: 56th FOCS, pp. 171–190 (2015)
18. Boneh, D., Gupta, D., Mironov, I., Sahai, A.: Hosting services on an untrusted cloud. In: Oswald, E., Fischlin, M. (eds.) EUROCRYPT 2015. LNCS, vol. 9057, pp. 404–436. Springer, Heidelberg (2015). https://doi.org/10.1007/978-3-662-46803-6_14
19. Boneh, D., Papakonstantinou, P.A., Rackoff, C., Vahlis, Y., Waters, B.: On the impossibility of basing identity based encryption on trapdoor permutations. In: 49th FOCS, pp. 283–292 (2008)
20. Boneh, D., Sahai, A., Waters, B.: Functional encryption: definitions and challenges. In: Ishai, Y. (ed.) TCC 2011. LNCS, vol. 6597, pp. 253–273. Springer, Heidelberg (2011). https://doi.org/10.1007/978-3-642-19571-6_16
21. Boneh, D., Waters, B.: Constrained pseudorandom functions and their applications. In: Sako, K., Sarkar, P. (eds.) ASIACRYPT 2013. LNCS, vol. 8270, pp. 280–300. Springer, Heidelberg (2013). https://doi.org/10.1007/978-3-642-42045-0_15
22. Boyle, E., Goldwasser, S., Ivan, I.: Functional signatures and pseudorandom functions. In: Krawczyk, H. (ed.) PKC 2014. LNCS, vol. 8383, pp. 501–519. Springer, Heidelberg (2014). https://doi.org/10.1007/978-3-642-54631-0_29
23. Brakerski, Z., Komargodski, I., Segev, G.: Multi-input functional encryption in the private-key setting: stronger security from weaker assumptions. In: Fischlin, M., Coron, J.-S. (eds.) EUROCRYPT 2016. LNCS, vol. 9666, pp. 852–880. Springer, Heidelberg (2016). https://doi.org/10.1007/978-3-662-49896-5_30
24. Brakerski, Z., Rothblum, G.N.: Virtual black-box obfuscation for all circuits via generic graded encoding. In: Lindell, Y. (ed.) TCC 2014. LNCS, vol. 8349, pp. 1–25. Springer, Heidelberg (2014). https://doi.org/10.1007/978-3-642-54242-8_1
25. Brakerski, Z., Segev, G.: Function-private functional encryption in the private-key setting. In: Dodis, Y., Nielsen, J.B. (eds.) TCC 2015. LNCS, vol. 9015, pp. 306–324. Springer, Heidelberg (2015). https://doi.org/10.1007/978-3-662-46497-7_12
26. Canetti, R., Holmgren, J., Jain, A., Vaikuntanathan, V.: Succinct garbling and indistinguishability obfuscation for RAM programs. In: 47th ACM STOC, pp. 429–437 (2015)
27. Canetti, R., Lin, H., Tessaro, S., Vaikuntanathan, V.: Obfuscation of probabilistic circuits and applications. In: Dodis, Y., Nielsen, J.B. (eds.) TCC 2015. LNCS, vol. 9015, pp. 468–497. Springer, Heidelberg (2015). https://doi.org/10.1007/978-3-662-46497-7_19
28. Chen, Y., Gentry, C., Halevi, S.: Cryptanalyses of candidate branching program obfuscators. In: Coron, J.-S., Nielsen, J.B. (eds.) EUROCRYPT 2017. LNCS, vol. 10212, pp. 278–307. Springer, Cham (2017). https://doi.org/10.1007/978-3-319-56617-7_10

29. Cohen, A., Holmgren, J., Nishimaki, R., Vaikuntanathan, V., Wichs, D.: Watermarking cryptographic capabilities. In: 48th ACM STOC, pp. 1115–1127 (2016)
30. Coron, J.-S., Gentry, C., Halevi, S., Lepoint, T., Maji, H.K., Miles, E., Raykova, M., Sahai, A., Tibouchi, M.: Zeroizing without low-level zeroes: new MMAP attacks and their limitations. In: Gennaro, R., Robshaw, M. (eds.) CRYPTO 2015. LNCS, vol. 9215, pp. 247–266. Springer, Heidelberg (2015). https://doi.org/10.1007/978-3-662-47989-6_12
31. Coron, J.-S., Lee, M.S., Lepoint, T., Tibouchi, M.: Zeroizing attacks on indistinguishability obfuscation over CLT13. In: Fehr, S. (ed.) PKC 2017. LNCS, vol. 10174, pp. 41–58. Springer, Heidelberg (2017). https://doi.org/10.1007/978-3-662-54365-8_3
32. Fernando, R., Rasmussen, P.M.R., Sahai, A.: Preventing CLT attacks on obfuscation with linear overhead. Cryptology ePrint Archive, Report 2016/1070
33. Garg, S., Gentry, C., Halevi, S., Raykova, M., Sahai, A., Waters, B.: Candidate indistinguishability obfuscation and functional encryption for all circuits. In: 54th FOCS, pp. 40–49 (2013)
34. Garg, S., Miles, E., Mukherjee, P., Sahai, A., Srinivasan, A., Zhandry, M.: Secure obfuscation in a weak multilinear map model. In: Hirt, M., Smith, A. (eds.) TCC 2016. LNCS, vol. 9986, pp. 241–268. Springer, Heidelberg (2016). https://doi.org/10.1007/978-3-662-53644-5_10
35. Garg, S., Pandey, O., Srinivasan, A., Zhandry, M.: Breaking the sub-exponential barrier in obfustopia. In: Coron, J.-S., Nielsen, J.B. (eds.) EUROCRYPT 2017. LNCS, vol. 10212, pp. 156–181. Springer, Cham (2017). https://doi.org/10.1007/978-3-319-56617-7_6
36. Garg, S., Srinivasan, A.: Single-key to multi-key functional encryption with polynomial loss. In: Hirt, M., Smith, A. (eds.) TCC 2016. LNCS, vol. 9986, pp. 419–442. Springer, Heidelberg (2016). https://doi.org/10.1007/978-3-662-53644-5_16
37. Goldreich, O., Goldwasser, S., Micali, S.: How to construct random functions. J. ACM **33**(4), 792–807 (1986)
38. Goldwasser, S., Gordon, S.D., Goyal, V., Jain, A., Katz, J., Liu, F.-H., Sahai, A., Shi, E., Zhou, H.-S.: Multi-input functional encryption. In: Nguyen, P.Q., Oswald, E. (eds.) EUROCRYPT 2014. LNCS, vol. 8441, pp. 578–602. Springer, Heidelberg (2014). https://doi.org/10.1007/978-3-642-55220-5_32
39. Hofheinz, D., Jager, T., Khurana, D., Sahai, A., Waters, B., Zhandry, M.: How to generate and use universal samplers. In: Cheon, J.H., Takagi, T. (eds.) ASIACRYPT 2016. LNCS, vol. 10032, pp. 715–744. Springer, Heidelberg (2016). https://doi.org/10.1007/978-3-662-53890-6_24
40. Hohenberger, S., Sahai, A., Waters, B.: Replacing a random oracle: full domain hash from indistinguishability obfuscation. In: Nguyen, P.Q., Oswald, E. (eds.) EUROCRYPT 2014. LNCS, vol. 8441, pp. 201–220. Springer, Heidelberg (2014). https://doi.org/10.1007/978-3-642-55220-5_12
41. Impagliazzo, R.: A personal view of average-case complexity. In: Proceedings of the Tenth Annual Structure in Complexity Theory Conference, Minneapolis, Minnesota, USA, 19–22 June 1995, pp. 134–147 (1995)
42. Impagliazzo, R., Rudich, S.: Limits on the provable consequences of one-way permutations. In: 21st ACM STOC, pp. 44–61 (1989)
43. Kiayias, A., Papadopoulos, S., Triandopoulos, N., Zacharias, T.: Delegatable pseudorandom functions and applications. In: ACM CCS 2013, pp. 669–684 (2013)
44. Kitagawa, F., Nishimaki, R., Tanaka, K.: From single-key to collusion-resistant secret-key functional encryption by leveraging succinctness. Cryptology ePrint Archive, Report 2017/638

45. Kitagawa, F., Nishimaki, R., Tanaka, K.: Indistinguishability obfuscation for all circuits from secret-key functional encryption. Cryptology ePrint Archive, Report 2017/361
46. Kitagawa, F., Nishimaki, R., Tanaka, K.: Simple and generic constructions of succinct functional encryption. Cryptology ePrint Archive, Report 2017/275 (to appear in PKC 2018)
47. Komargodski, I., Moran, T., Naor, M., Pass, R., Rosen, A., Yogev, E.: One-way functions and (im)perfect obfuscation. In: 55th FOCS, pp. 374–383 (2014)
48. Komargodski, I., Segev, G.: From minicrypt to obfustopia via private-key functional encryption. In: Coron, J.-S., Nielsen, J.B. (eds.) EUROCRYPT 2017. LNCS, vol. 10210, pp. 122–151. Springer, Cham (2017). https://doi.org/10.1007/978-3-319-56620-7_5
49. Koppula, V., Lewko, A.B., Waters, B.: Indistinguishability obfuscation for turing machines with unbounded memory. In: 47th ACM STOC, pp. 419–428 (2015)
50. Li, B., Micciancio, D.: Compactness vs collusion resistance in functional encryption. In: Hirt, M., Smith, A. (eds.) TCC 2016. LNCS, vol. 9986, pp. 443–468. Springer, Heidelberg (2016). https://doi.org/10.1007/978-3-662-53644-5_17
51. Lin, H.: Indistinguishability obfuscation from constant-degree graded encoding schemes. In: Fischlin, M., Coron, J.-S. (eds.) EUROCRYPT 2016. LNCS, vol. 9665, pp. 28–57. Springer, Heidelberg (2016). https://doi.org/10.1007/978-3-662-49890-3_2
52. Lin, H.: Indistinguishability obfuscation from SXDH on 5-linear maps and locality-5 PRGs. In: Katz, J., Shacham, H. (eds.) CRYPTO 2017. LNCS, vol. 10401, pp. 599–629. Springer, Cham (2017). https://doi.org/10.1007/978-3-319-63688-7_20
53. Lin, H., Pass, R., Seth, K., Telang, S.: Indistinguishability obfuscation with non-trivial efficiency. In: Cheng, C.-M., Chung, K.-M., Persiano, G., Yang, B.-Y. (eds.) PKC 2016. LNCS, vol. 9615, pp. 447–462. Springer, Heidelberg (2016). https://doi.org/10.1007/978-3-662-49387-8_17
54. Luby, M., Rackoff, C.: How to construct pseudorandom permutations from pseudorandom functions. SIAM J. Comput. **17**(2), 373–386 (1988)
55. Lin, H., Tessaro, S.: Indistinguishability obfuscation from trilinear maps and block-wise local PRGs. In: Katz, J., Shacham, H. (eds.) CRYPTO 2017. LNCS, vol. 10401, pp. 630–660. Springer, Cham (2017). https://doi.org/10.1007/978-3-319-63688-7_21
56. Lin, H., Vaikuntanathan, V.: Indistinguishability obfuscation from DDH-like assumptions on constant-degree graded encodings. In: 57th FOCS, pp. 11–20 (2016)
57. Lindell, Y., Pinkas, B.: A proof of security of yao's protocol for two-party computation. J. Cryptol. **22**(2), 161–188 (2009)
58. Miles, E., Sahai, A., Zhandry, M.: Annihilation attacks for multilinear maps: cryptanalysis of indistinguishability obfuscation over GGH13. In: Robshaw, M., Katz, J. (eds.) CRYPTO 2016. LNCS, vol. 9815, pp. 629–658. Springer, Heidelberg (2016). https://doi.org/10.1007/978-3-662-53008-5_22
59. O'Neill, A.: Definitional issues in functional encryption. Cryptology ePrint Archive, Report 2010/556 (2010)
60. Pass, R., Seth, K., Telang, S.: Indistinguishability obfuscation from semantically-secure multilinear encodings. In: Garay, J.A., Gennaro, R. (eds.) CRYPTO 2014. LNCS, vol. 8616, pp. 500–517. Springer, Heidelberg (2014). https://doi.org/10.1007/978-3-662-44371-2_28
61. Sahai, A., Seyalioglu, H.: Worry-free encryption: functional encryption with public keys. In: ACM CCS 2010, pp. 463–472 (2010)

62. Sahai, A., Waters, B.: How to use indistinguishability obfuscation: deniable encryption, and more. In: 46th ACM STOC, pp. 475–484 (2014)
63. Sahai, A., Waters, B.: Fuzzy identity-based encryption. In: Cramer, R. (ed.) EUROCRYPT 2005. LNCS, vol. 3494, pp. 457–473. Springer, Heidelberg (2005). https://doi.org/10.1007/11426639_27
64. Yao, A.C.-C.: How to generate and exchange secrets (extended abstract). In: 27th FOCS, pp. 162–167 (1986)
65. Zimmerman, J.: How to obfuscate programs directly. In: Oswald, E., Fischlin, M. (eds.) EUROCRYPT 2015. LNCS, vol. 9057, pp. 439–467. Springer, Heidelberg (2015). https://doi.org/10.1007/978-3-662-46803-6_15

Limits on Low-Degree Pseudorandom Generators (Or: Sum-of-Squares Meets Program Obfuscation)

Boaz Barak[1(✉)], Zvika Brakerski[2], Ilan Komargodski[3], and Pravesh K. Kothari[4,5]

[1] Harvard University, Cambridge, MA, USA
b@boazbarak.org
[2] Weizmann Institute of Science, Rehovot, Israel
zvika.brakerski@weizmann.ac.il
[3] Cornell Tech, New York, NY, USA
komargodski@cornell.edu
[4] Princeton University, Princeton, NJ, USA
kothari@cs.princeton.edu
[5] IAS, Princeton, NJ, USA

Abstract. An m output pseudorandom generator $\mathcal{G}\colon (\{\pm 1\}^b)^n \to \{\pm 1\}^m$ that takes input n blocks of b bits each is said to be ℓ-block local if every output is a function of at most ℓ blocks. We show that such ℓ-block local pseudorandom generators can have output length at most $\tilde{O}(2^{\ell b} n^{\lceil \ell/2 \rceil})$, by presenting a polynomial time algorithm that distinguishes inputs of the form $\mathcal{G}(x)$ from inputs where each coordinate is sampled from the uniform distribution on m bits.

As a corollary, we refute some conjectures recently made in the context of constructing provably secure indistinguishability obfuscation (iO). This includes refuting the assumptions underlying Lin and Tessaro's [47] recently proposed candidate iO from bilinear maps. Specifically, they assumed the existence of a secure pseudorandom generator $\mathcal{G}\colon \{\pm 1\}^{nb} \to \{\pm 1\}^{2^{cb} n}$ as above for large enough $c > 3$ and $\ell = 2$. (Following this

B. Barak—Supported by NSF awards CCF 1565264 and CNS 1618026, and the Simons Foundation. Work done while the author visited Weizmann Institute of Science during Spring 2017.

Z. Brakerski—Supported by the Israel Science Foundation (Grant No. 468/14) and Binational Science Foundation (Grants No. 2016726, 2014276), ERC Project 756482 REACT and European Union PROMETHEUS Project (Horizon 2020 Research and Innovation Program, Grant 780701).

I. Komargodski—Supported in part by a Packard Foundation Fellowship and AFOSR grant FA9550-15-1-0262. Most work done while the author was a Ph.D. student at the Weizmann Institute of Science, supported in part by a grant from the Israel Science Foundation (no. 950/16) and by a Levzion Fellowship.

P. K. Kothari—Work done while the author visited Weizmann Institute of Science in March 2017.

J. B. Nielsen and V. Rijmen (Eds.): EUROCRYPT 2018, LNCS 10821, pp. 649–679, 2018.
https://doi.org/10.1007/978-3-319-78375-8_21

work, and an independent work of Lombardi and Vaikuntanthan [49], Lin and Tessaro retracted the bilinear maps based candidate from their manuscript.)

Our results actually hold for the much wider class of low-degree, non-binary valued pseudorandom generators: if every output of $\mathcal{G}\colon \{\pm 1\}^n \rightarrow \mathbb{R}^m$ ($\mathbb{R}$ = reals) is a polynomial (over $\mathbb{R}$) of degree at most d with at most s monomials and $m \geq \tilde{\Omega}(sn^{\lceil d/2 \rceil})$, then there is a polynomial time algorithm for distinguishing the output $\mathcal{G}(x)$ from z where each coordinate z_i is sampled independently from the marginal distribution on $\mathcal{G}_i$. Furthermore, our results continue to hold under arbitrary *pre-processing* of the seed. This implies that any such map $\mathcal{G}$, with arbitrary seed pre-processing, cannot be a pseudorandom generator in the mild sense of fooling a product distribution on the output space. This allows us to rule out various natural modifications to the notion of generators suggested in other works that still allow obtaining indistinguishability obfuscation from bilinear maps.

Our algorithms are based on the Sum of Squares (SoS) paradigm, and in most cases can even be defined more simply using a canonical semidefinite program. We complement our algorithm by presenting a class of candidate generators with block-wise locality 3 and constant block size, that resists both Gaussian elimination and sum of squares (SOS) algorithms whenever $m = n^{1.5-\varepsilon}$. This class is extremely easy to describe: Let $\mathbb{G}$ be any simple non-abelian group with the group operation "$*$", and interpret the blocks of x as elements in $\mathbb{G}$. The description of the pseudorandom generator is a sequence of m triples of indices (i, j, k) chosen at random and each output of the generator is of the form $x_i * x_j * x_k$.

1 Introduction

Understanding how "simple" a pseudorandom generator can be has been of great interest in cryptography and computational complexity. In particular, researchers have studied the question of whether there exist pseudorandom generators with *constant input locality*, in the sense that every output bit only depends on a constant number of the input bits. Applebaum et al. [9] showed that, assuming the existence of one-way functions computable by log-depth circuits, there is such a generator mapping n bits to $n + n^{\varepsilon}$ bits for a small constant $\varepsilon > 0$. Goldreich [36] gave a candidate pseudorandom generator of constant locality that could potentially have even *polynomially large* stretch (e.g. map n bits to n^s bits for some $s > 1$).[1] The possibility of such "ultra simple" high-stretch pseudorandom generators has attracted significant attention recently with applications including:

- Public key cryptography from "combinatorial" assumptions [8].
- Highly efficient multiparty computation [40].

[1] While Goldreich originally only conjectured that his function is a one-way function, followup work has considered the conjecture that it is a pseudorandom generator, and also linked the two questions (see e.g., [6,11]; see also Applebaum's survey [7]).

- Reducing the assumptions needed for constructing *indistinguishability obfuscators* (iO) [4,5,45–48].

The last application is perhaps the most exciting, as it represents the most promising pathway for basing this important cryptographic primitive on more standard assumptions. Furthermore, this application provides motivation for considering qualitatively different notions of "simplicity" of a generator. For example, it is possible to relax the condition of having small input locality to that of just having small algebraic *degree* (over the rationals), as well as allow other features such as preprocessing of the input and admitting non-Boolean outputs.

At the same time, the application to obfuscation emphasizes a fine-grained understanding of the quantitative relationship between the "simplicity" of a generator (such as its locality, or algebraic degree) and its *stretch* (i.e., ratio of output and input lengths). For example, works of Lin and Ananth and Sahai [5,46] show that a generator mapping n bits to $n^{1+\varepsilon}$ bits with locality 2 implies an obfuscation candidate based on standard cryptographic assumptions – a highly desired goal, but it is known that it is impossible to achieve super-linear stretch with locality four (let alone two) generator [52].

Very recently, Lin and Tessaro [47] proposed bypassing this limitation by considering a relaxation of locality to a notion they referred to as *block locality*. They also proposed a candidate generator with the required properties. If such secure PRGs exist, this would imply obfuscators whose security is based on standard cryptographic assumptions, a highly desirable goal. Ananth et al. [3] observed that the conditions can be relaxed further to allow generators without a block structure, and even allow non-Boolean outputs, but their method requires (among other restrictions) that each output is computed by a sparse polynomial of small degree.

In this paper we give strong limitations on this approach, in particular giving negative answers to some of the questions raised in prior works. While a priori, questions of algebraic flavor, such as the difference between the power of bilinear vs trilinear maps, and those of combinatorial essence such as the difficulty of refuting random constraint satisfaction instances might seem unrelated, it turns out that techniques useful in the study of CSP refutation yield a barrier that, somewhat surprisingly, seems to exactly correspond to what is needed to bypass the "trilinear map barrier" for obfuscation constructions.

We complement our negative results with a simple construction of a candidate degree *three* pseudorandom generator which resists known attacks (Gaussian elimination and sum-of-squares algorithms) even for output length $n^{1+\Omega(1)}$.

1.1 Our Results

To state our results, let us define the notion of the *image refutation problem* for a map $\mathcal{G}$ that takes n inputs into m outputs (e.g., a purported pseudorandom generator). Looking ahead, we will allow maps to have non-Boolean outputs.[2] Informally, the image refutation problem asks for a efficiently computable certificate for a random string *not* being in the image of a purported generator $\mathcal{G}$.

Definition 1.1 (Refutation problem). *Let $\mathcal{G}\colon \{\pm 1\}^n \to \mathbb{R}^m$ and Z be a distribution over $\mathbb{R}^m$. An algorithm A is said to solve the $\mathcal{G}$-*image refutation prob-lem w.r.t *Z if on input $z \in \mathbb{R}^m$, A outputs either* "`refuted`" *or* "?" *and satisfies:*

- *If $z = \mathcal{G}(x)$ for some $x \in \{\pm 1\}^n$ then $A(z) =$"?".*
- $\mathbb{P}_{z\sim Z}[A(z) =$"`refuted`"$] \geq 0.5$

Note that in particular if Z is the uniform distribution over $\{0,1\}^m$, then the existence of an efficient algorithm that solves the $\mathcal{G}$ image refutation problem with respect to Z means that $\mathcal{G}$ is not a pseudorandom generator - in fact, an image refutation algorithm, with probability at least $1/2$, shows that a random string from $\{\pm 1\}^m$ is not in the image of $\mathcal{G}$.

Remark 1.2 (Refutation vs Distinguishing). It is instructive to contrast the algorithmic tasks of image refutation with the easier task of distinguishing the output of a pseudorandom generator from a uniformly random string. In the latter case, we are typically concerned with distinguishing the output distribution of a generator $\mathcal{G}\colon \{\pm 1\}^n \to \{\pm 1\}^m$ when the input is chosen according to the uniform distribution on $\{\pm 1\}^m$. It's easy to see that a refutation algorithm immediately yields a distinguisher. In general, refutation, however can be more powerful. For example, a refutation algorithm can distinguish between the uniform distribution on $\{\pm 1\}^m$ from the output distribution of the generator even under *arbitrary* distributions on the seed. Thus, an image refutation algorithm rules out not only the natural PRG construction but also natural modifications that involve using some non-trivial pre-processing on the seed before inputting it into the generator, thus modifying the input distribution. Such modifications were in fact suggested for candidate constructions of iO from bilinear maps in the concurrent work of [49]. While a distinguisher for the original PRG may fail after this modification, a refutation algorithm continues to work. As we discuss later, this is one of the key differences in our approach from that of [49].

Our first result is a limitation on generators with "block locality" two:

Theorem 1.3 (Limitations of two block local generators). *For every n, b, let $\mathcal{G}\colon \{\pm 1\}^{nb} \to \{\pm 1\}^m$ be such that, if we partition the input into n blocks of size b, then every output of G depends only on variables inside two blocks. Then, there is an absolute constant K such that if $m > K \cdot 2^{2b} n \log^2 n$, then there is an efficient algorithm for the $\mathcal{G}$-image-refutation problem w.r.t. the uniform distribution over $\{\pm 1\}^m$.*

[2] Allowing non-Boolean output can make a significant difference. For example, [50, Theorem 6.1] show that every degree two Boolean-valued function on $\{\pm 1\}^n$ depends on at most four variables, which in particular means that it cannot be used as the basis for a pseudorandom generator with super-linear output length. It also allows us to consider polynomials that only take the values in $\{\pm 1\}$ on a subset of their inputs.

Theorem 1.3 yields an attack on the aforementioned candidate pseudorandom generator proposed by Lin and Tessaro [47] towards basing indistinguishability obfuscator on bilinear maps, as well as any other candidate of block-locality 2 compatible with their construction.

A special case that has been of considerable interest in literature is one where all outputs of the PRG are computed by the same two-block-local predicate $P\colon \{\pm1\}^b \to \{\pm1\}^b \to \{\pm1\}$. For this case, we give an image refutation algorithm that works whenever the stretch $m = \tilde{\Omega}(n2^b)$.[3]

Theorem 1.4 (Limitations of two block local generators with a single predicate, Theorem 5.3). *For every n, b, let $\mathcal{G}\colon \{\pm1\}^{nb} \to \{\pm1\}^m$ be such that, if we partition the input into n blocks of size b, then every output of $\mathcal{G}$ is the same predicate P applied to two b-bit blocks. Then, there is an absolute constant K such that if $m > K \cdot 2^b n \log^2 n$, then there is an efficient algorithm for the $\mathcal{G}$-image-refutation problem w.r.t. the uniform distribution over $\{\pm1\}^m$.*

Yet another special case of interest is where the candidate generator obtained is chosen at random: that is, the m pairs of blocks used to compute the output are chosen at random and, further, each predicate computing an output is chosen randomly and independently conditioned on being balanced. For this case, we show (in Theorem 5.4, Sect. 5.3) that we can again improve our bound on the output length from $\tilde{O}(2^{2b}n)$ to $\tilde{O}(2^b n)$:

Our next result applies to any degree d map, and even allows maps with non-Boolean output. For the refutation problem to make sense, the probability distribution Z must be non-degenerate or have large entropy, as otherwise it may well be the case that $z \sim Z$ is in the image of $\mathcal{G}$ with high probability. For real-valued distributions, a reasonable notion of non-degeneracy is that the distribution does not fall inside any small interval with high probability. Specifically, if we consider *normalized* product distributions (where $\mathbb{E}Z_i = 0$ and $\mathbb{E}Z_i^2 = 1$ for every i and the Z_i are independent), then we say that Z is *c-spread* (see Definition 4.1) if it is a product distribution and $\mathbb{P}[Z_i \notin I] \geq 0.1$ for every interval $I \subseteq \mathbb{R}$ of length at most $1/c$ (where we can think of c as a large constant or even a poly-logarithmic or small polynomial factor).

If Z is supposed to be indistinguishable from $\mathcal{G}(U)$, where U is the uniform distribution over $\{\pm1\}^n$, then these two distributions should agree on the marginals and in particular at least on their first and second moments. Hence, we can assume that the map $\mathcal{G}$ has the same normalization as Z, meaning that $\mathbb{E}\mathcal{G}(U)_i = 0$ and $\mathbb{E}\mathcal{G}(U)_i^2 = 1$.[4] Our result for general low degree generators is the following:

Theorem 1.5 (Limitations on degree d generators). *Suppose that $\mathcal{G}\colon \{\pm1\}^n \to \mathbb{R}^m$ is such that for every $i \in [m]$ the map $x \mapsto \mathcal{G}(x)_i$ is a normalized polynomial of degree at most d with at most s monomials. Let Z be a*

[3] Unlike the other results in this paper, Theorem 1.4 builds upon the concurrent work [49]. See Sect. 1.3 for a detailed comparison between this work and [49].

[4] We say that $\mathcal{G}$ is *normalized* if it satisfies these conditions. Clearly, any map can be normalized by appropriate shifting and scaling.

c-spread product distribution over $\mathbb{R}^m$*. Then, there is some absolute constant* K *such that if* $m \geq Kc^2 sn^{\lceil d/2 \rceil} \log^2 n$*, then there is an efficient algorithm for the* $\mathcal{G}$*-image-refutation problem w.r.t.* Z*.*

We believe the dependence on the degree d can be improved in the odd case from $\lceil d/2 \rceil$ to $d/2$. Resolving this is related to some problems raised in the CSP refutation literature (e.g., see [60, Questions 5.2.3, 5.2.7, 5.2.8]).

While for arbitrary polynomials we do not know how to remove the restriction on sparsity (i.e., number of non-zero monomials s), we show in Sect. 4 that we can significantly relax it in several settings. Moreover, the applications to obfuscation require generators that are both low degree and sparse; see Sect. 2. Nevertheless, we view eliminating the dependence on the sparsity as the main open question left by this work. We conjecture that this can be done, at least in the pseudorandom generator setting, as paradoxically, it seems that the only case where our current algorithm fails is when the pseudorandom generator exhibits some "non-random" behavior. Improving this is related to obtaining better upper bound on the stretch of block-local generators.

Up to the dependence on sparsity, Theorem 1.5 answers negatively a question of Lombardi and Vaikuntanathan [50, Question 7.2], who asked whether it is possible to have a degree d pseudorandom generator with stretch $n^{\lceil \frac{3}{4} d \rceil + \varepsilon}$. It was already known by the work of Mossel et al. [52] that such output length cannot be achieved by *d-local* generators; our work shows that, at least for $n^{o(1)}$-sparse polynomials, relaxing locality to the notion of algebraic degree does not help achieve a better dependency.

All of our results are based on the same algorithm: the *sum of squares* (SOS) semidefinite program ([44,55,59]; see the lecture notes [16]). This is not surprising as for refuting CSPs, semidefinite programs in general and the sum-of-squares semi-definite programming hierarchy in particular are the strongest known general tools [43,56]. This suggests that for future candidate generators, it will be useful to prove resilience at least with respect to this algorithm. Fortunately, there is now a growing body of techniques to prove such lower bounds.

Here, we establish that the sum-of-squares algorithm cannot be used to give an attack on PRGs with stretch $O(n2^b)$. Note that the sum of squares algorithm captures all the techniques in literature for efficiently refuting (non-linear) random CSPs including the algorithms in this paper and the work of [49]. Our lower bound on the sum of squares algorithm below shows that using such techniques, one cannot hope to attack two-block-local PRGs with stretch at most $O(n2^b)$ - for the case of identical predicates computing all outputs of the generator, this, in particular, establishes the optimality of our analyis of any technique captured by the sum of squares framework.

Concretely, in Sect. 6, we show that there is a natural sum-of-squares resistant construction with a stretch of $\tilde{\Theta}(n2^b)$. We stress that this PRG is only secure against a sum-of-square algorithm, and is actually *insecure* outside the sum-of-squares framework.

Theorem 1.6 (See Theorem 6.1 for a formal version). *For any $b \geq 10 \log\log(n)$, there is a construction of a two-block-local PRG $\mathcal{G}\colon (\{\pm 1\}^b)^n \to \{\pm 1\}^m$ for $m = \Omega(n2^b)$ such that degree-$\Theta(n/2^{4b})$ sum of squares algorithm cannot solve the refutation problem for $\mathcal{G}$.*

For example, for $b < \varepsilon/4 \log(n)$, the above results rules out an attack on $\Omega(n2^b)$-stretch PRGs using SoS algorithm that runs in time $\sim 2^{n^{1-\varepsilon}}$.

While our results give strong barriers for degree *two* pseudorandom generators, they do not rule out a degree *three* pseudorandom generator with output length $n^{1+\Omega(1)}$. Indeed, we show a very simple candidate generator that might satisfy this property. This is the generator $\mathcal{G}$ mapping $\mathbb{G}^n$ to $\mathbb{G}^m$ where $\mathbb{G}$ is some finite *non-abelian* simple group (e.g., the size 60 group A_5), where for every $\ell \in [m]$, the ℓth output of $\mathcal{G}(x)$ is obtained as

$$\mathcal{G}(x)_\ell = x_i * x_j * x_k$$

for randomly chosen indices i, j, k and $*$ is the group operation. This generator has block locality three with constant size blocks and also (using the standard representation of group elements as matrices) has algebraic degree three as well. Yet, it is a hard instance for the SOS algorithm which encapsulates all the techniques used in this paper. While more study of this candidate's security is surely needed, there are results suggesting that it resists algebraic attacks such as Gaussian elimination [35]. See Sect. 7 for details.

1.2 Prior Work

Most prior work on limitations of "simple" pseudorandom generators focused on providing upper bounds on the output length in terms of the *locality*. Cryan and Miltersen [27] observe that there is no PRG with locality 2 and proved that there is no PRG with locality 3 achieving super linear stretch (i.e., having input length n and output length $n + \omega(n)$ bits). Mossel, Shpilka, and Trevisan [52] extended this result to locality 4 PRGs and constructed (non-cryptographic) small-biased locality 5 generators with linear stretch and exponentially-small bias. They also showed that a k local generator cannot have output length better than $O(2^k n^{\lceil k/2 \rceil})$. Applebaum, Ishai, and Kushilevitz [9] showed that, under standard cryptographic assumptions, there are locality 4 PRGs with sublinear-stretch. Applebaum and Raykov [6,11] related the pseudorandomness and one-wayness of Goldreich's proposed one-way function [36] in some regime of parameters.

We focus on (algebraic) *degree* instead of locality of the predicate that is used. There were few works in the past with this property (for example [10, 31]). Apart from this, another feature that distinguishes our work from much of the prior works on pseudorandom generators is the focus on the *refutation* problem (certifying that a random string is *not* in the image of the generator) as opposed to the *decision* problem (given the output of a uniformly random seed, distinguish from a random string) or the *search* problem (given the output of a uniformly random seed, recover the seed). This is important for us since we do not want to make the typical assumption that the input (i.e., seed) to the pseudorandom generator is uniformly distributed, as to allow the possibility of preprocessing for it.

The refutation problem was extensively studied in the context of random *constraint satisfaction problems* (CSPs). The refutation problem for a k-local generator with n inputs and m outputs corresponds to refuting a CSP with n variables and m constraints. Thus, the study of limitations for local generators is tightly connected to the study of refutation algorithm for CSPs. Most well studied in this setting is the problem of refuting random CSPs - given a random CSP instance with a predicate P, certify that it is far from satisfiable with high probability. There is a large body of works on the study of refuting *random* and *semirandom* CSPs, starting with the work of Feige [28].[5]

In particular, we now know tight relations between the *arity* (or locality) of the predicates and the number of constraints required to refute random instances [1,43,56] using the sum-of-squares semidefinite programming hierarchy - the algorithm of choice for the problem.

Most relevant to the current paper are works from this literature that deal with predicates that have large arity but have small degree d (or the related notion of not supporting $(d+1)$-wise independent distribution). Allen, O'Donnell, and Witmer [1] showed that random instances of such predicates can be refuted when the number of constraints m is larger than $\tilde{O}(k^d n^{d/2})$. In his thesis proposal, Witmer [60] sketched how to generalize this to the *semirandom* setting, though only for the case of *even* degree d. This is related to the questions considered in this work for higher degree, though our model is somewhat more general, considering not just CSPs but arbitrary low-degree maps.

The notion of ℓ *block locality* is equivalent to the notion of CSPs of arity ℓ over a *large alphabet* (specifically, exponential in the block size). Though much of the CSP refutation and approximation literature deals with CSPs over a binary alphabet, there have been works dealing with larger alphabet (see e.g., [1]). The work of [15] gives an SOS based algorithm for 2-local CSPs over large alphabet (or equivalently, 2 block-local CSPs) as long as the underlying constraint graph is a sufficiently good expander. However, their algorithm (at least their analysis) has an *exponential* dependence in the running time on the alphabet size which is unsuitable for our applications.

The main technical difference between our work and prior results in the CSP literature, is that since for CSPs we often think as the arity as constant, these works often had poor dependence on this parameter, whereas we want to handle the case that it can be as large as n^{ε} or in some cases even unrestricted. Another difference is that in the cryptographic setting, we wish to allow the designer of a pseudorandom generator significant freedom, and this motivates studying more

[5] In a *random* CSP the graph of dependence between variables and constraints is random, and we also typically consider adding a random pattern of negations or shifts to either the inputs or the outputs of the predicates. In *semirandom* instances [29,30], the graph is arbitrary and only this pattern of negations or shifts is random.

challenging semirandom models than those typically used in prior works. We discuss these technical issues in more depth in Sect. 3.

The algorithms in almost all the refutation works in the CSP literature can be encapsulated by the *sum of squares* semidefinite programming hierarchy. Some lower bounds for this hierarchy, showing tightness of these analysis, were given in [13,43,54]. For the alphabet-size sensitive setting of block-local PRGs, we give a lower bound in Sect. 6.

1.3 Comparison with [49]

In a concurrent and independent work, Lombardi and Vaikuntanathan [49] also analyzed the possibility of a secure block-wise local PRG motivated by the work of Lin and Tessaro [47]. They show that there exists an efficient polynomial-time distinguisher with the following property: for any $m \geq \tilde{\Omega}(n2^b)$ and any predicate $P\colon \{\pm 1\}^b \times \{\pm 1\}^b \to \{\pm 1\}$ in two blocks of size b, there's an efficient distinguishing algorithm for the following two distributions over $\{\pm 1\}^m$: (1) the uniform distribution on $\{\pm 1\}^m$ and (2) the output distribution of Goldreich's PRG $G_H\colon (\{\pm 1\}^b)^n \to \{0,1\}^m$ instantiated with a random graph H and the single predicate P computing all m outputs when given a uniformly random nb bit string as input.[6]

We point out the major differences between our results on block-local PRGs and that of [49] here.

1. *Distinguishing vs Refutation:* As discussed in Remark 1.2, our approach yields the stronger refutation guarantees while that of [49] yields a distinguisher. This allows us to show that reinforcing the block-local (or low-degree, more generally) PRGs by allowing arbitrary input preprocessing cannot lead to a larger stretch. This is important, as preprocessing is OK to do in the context of the applications for obfuscation, and in fact this was one of the avenues suggested for bypassing these general type of negative results.
2. *Single Predicates vs Multiple Predicates:* The work of [49] only applies to the PRGs where each output is computed using the *same* predicate. Our approach shows that block-local (or low-degree) PRGs cannot achieve large enough stretch even if each output is computed using a different predicate - a priori, one could hope that using different predicates for different outputs could add significantly to the stretch of the PRG. This bottleneck is in fact inherent in the technical approach of [49]. In particular, our approach allows us to analyze the natural candidate for 2-block-local generator obtained by applying independently chosen multiple random predicates to randomly chosen pairs of input blocks and yields an $\tilde{O}(2^b n)$ upper bound on their stretch, see Sect. 5.3.
3. *Random Graph vs Arbitrary Graphs:* The work of [49] only handles block-local PRGs when the underlying graph G defining the generator is chosen

[6] We learned that in an updated version of [49], they use a refutation algorithm from our work to extend their distinguisher to the case when the graph H is arbitrarily chosen.

at random. This was because [49] relied on CSP refutation results that work under the assumption of the instance being random.
4. *Special Case of Single Predicate Block-Local PRGs:* For the PRGs with all outputs computed by a single predicate, [49] show a distinguisher that works whenever the stretch of the PRG is $\Omega(n2^b)$. For this case, we show that our algorithm in fact guarantees image *refutation* at the same stretch requirement. (A previous version of our work didn't include this result on PRGs with single predicate.) Our refutation algorithm (Theorem 1.4) is in fact inspired by the application of the Chor-Goldreich Lemma in the work of [49].

We note that the three first differences: image refutation as opposed to distinguishing, allowing different predicates as opposed to a single predicate, and using arbitrary graphs as opposed to random graphs, exactly correspond to the open questions raised by [49].[7] Thus, our results block all the approaches that [49] identified as potential strategies for repairing the iO candidate. This suggests that, rather than a "patchable problem", there is perhaps a fundamental barrier to this approach of obtaining iO from bilinear maps.

1.4 Paper Organization

Section 2 explains the connection between simple generators and the construction of indistinguishability obfuscator. This explanation allows us to draw the conclusion that our algorithm renders recently proposed methods ineffective for constructing obfuscation from standard cryptographic assumptions. For those interested in additional details, the full version [12, Appendix B] contains more information about constructing obfuscators and in particular on the result of [47]. In Sect. 3, we provide a high level overview of our algorithmic techniques. Section 4 contains our main algorithm and analysis, and in particular proves Theorem 1.5. We use standard tools from the SDP/SOS literature that can be found in Appendix A. In Sect. 5 we focus our attention on pseudorandom generators with small block-locality and show tighter results than those achieved by our general analysis, in particular we prove Theorem 1.3 as well as an even tighter result for generators with single predicates (Theorem 5.3) and random two-block-local PRGs (Theorem 5.4). In Sect. 6, we show that sum-of-squares algorithm cannot be used to prove sharper upper bounds on the stretch than $\sim$n2^b. Finally, in Sect. 7 we present our class of candidate block-local generators.

2 Relating Simple Generators and Program Obfuscators

A program obfuscator [14,38] is a compiler that given a program (say represented as a Boolean circuit) transforms it into another "scrambled" program which is functionally equivalent but its implementation details are "hidden", making it hard to reverse-engineer. The study of *indistinguishability obfuscation* (iO) stands at the forefront of cryptographic research in recent years due to two

[7] See Sect. 5 on page 12 of https://eprint.iacr.org/2017/301/20170409:183008.

main developments. Firstly, Garg et al. [33] suggested that this notion might be achievable given sufficiently strong *cryptographic multilinear maps*, for which a candidate construction was given by [32]. Secondly, it was shown by Sahai and Waters [58] and numerous follow-up works that iO is extremely useful for constructing a wide variety of cryptographic objects, many of which are unknown to exist under any other assumption.

A fundamental question in the construction of iO from multilinear maps is the *level of multilinearity*. Without going into details, this essentially corresponds to the highest degree of polynomials that can be evaluated by this object. Whereas multilinear maps of level 2, a.k.a *bilinear maps*, can be constructed based on pairing on elliptic curves [17,41] and have been used in cryptographic literature for over 15 years, the first obfuscation candidates required *polynomial* level (in the "security parameter" of the scheme). Proposed constructions of multilinear maps for level >2 have only started to emerge recently [25,26,32,34] and their security is highly questionable. Indeed, many concrete security assumptions were shown to be broken w.r.t all known candidates with level >2 [18,20–22,24,39,51].

A beautiful work of Lin [45], followed by [5,46,48], showed that the required level of multilinearity can be reduced to a constant (ultimately 5 in [5,46]). These works show a relation between the required multilinearity level and the existence of "simple" pseudorandom generators (PRGs). At a rudimentary level, the PRGs are used to "bootstrap" simple obfuscation-like objects into full-fledged obfuscators. This approach requires PRGs mapping $\{0,1\}^n$ to $\{0,1\}^m$ with $m = n^{1+\Omega(1)}$, which can be represented as low-degree polynomials over $\mathbb{R}$.

More accurately, for a security parameter λ and large enough n, the required output length is $m = n^{1+\varepsilon} \cdot \text{poly}(\lambda)$, for some fixed polynomial $\text{poly}(\cdot)$ which is related to the computational complexity of evaluating the underlying cryptographic primitives. One can ensure this condition as long as the output length is at least $n^{1+\Omega(1)}$ by setting n to be a sufficiently large polynomial in λ. The situation complicates further when trying to optimize the concrete constant corresponding to the level of multilinearity by means of preprocessing as in [5,46,47]. The stretch bound needs to hold even with respect to the preprocessed seed length (see the full version [12, Appendix B] for more details).

Lin [46] and Ananth and Sahai [5] instantiated this approach with locality-5 PRGs, which can trivially be represented as degree 5 polynomials. Their main insight was that for constant locality PRGs, preprocessing only blows up the seed by a constant factor. However, even so, the required stretch is impossible to achieve with locality smaller than 5 [52].

Implications of Our Work to Candidate Bilinear-Maps-Based Constructions. Very recently, Lin and Tessaro [47] proposed an approach to overcome the locality barrier and possibly get all the way to an instantiation of iO based on bilinear maps. This could be a major breakthrough in cryptographic research, allowing to base "fantasy" cryptography on well studied hardness assumptions. Lin and Tessaro showed that it is sufficient if the PRG has low *block-wise locality* for blocks of logarithmic size. Namely, if we consider the seed of the PRG as an $b \times n$ matrix for $b = O(\log n)$, then each output bit can be allowed to depend on ℓ columns of this matrix. The required output length is $m = 2^{c \cdot b} n^{1+\Omega(1)}$ for some constant c. An explicit value for c is not given, but the construction requires $c > 3$ which seems to be essential for this approach (see the full version [12, Appendix B]). Block-wise locality allows a possible way to bypass the impossibility results for standard (i.e., bitwise) locality, and indeed Lin and Tessaro conjectured that there is a pseudorandom generator with output length $n^{1+\Omega(1)}$ and block-wise locality $\ell = 2$, and proposed a candidate construction.

Theorem 1.3 shows that generators with block-wise locality 2 cannot have the stretch required by the [47] construction, thus suggesting that their current techniques are insufficient for achieving obfuscation from bilinear maps. While our worst-case result leaves a narrow margin for possible improvement of the obfuscation reduction to work with $1 < c < 2$, our improved analysis for random graphs and predicates (see Theorem 5.4 in Sect. 5.3) suggests that our methods may be effective, at least heuristically, for generators with *any* $c > 1$.

Ananth et al. [3] observed that there is a way to generalize the [47] approach, so that it is sufficient that the range of the PRG is not $\{0, 1\}$, but rather some small specified set, so long as the degree (as a polynomial over the rationals) is bounded by the level of multilinearity. Furthermore, pseudorandomness was no longer a requirement, but rather it is only required that the output of the generator is indistinguishable from some product distribution (in particular, the one where each output entry is distributed according to its marginal). This suggests that perhaps a broader class of generators than ones that have been considered in the literature so far are useful for reducing the degree of multilinearity. However, their approach imposes a number of restrictions on such generators in order to be effective. In particular, it requires preprocessing which increases the seed length by a factor of s^c, for some $c > 1$, where s is the number of monomials in each output coordinate of the generator. Therefore, Theorem 1.5 rules out the applicability of this technique for degree 2 generators, as well.

Supporting Evidence for Block-Wise Locality 3. We show that while the Lin-Tessaro approach might not yet bring us all the way to level 2, it is quite plausible that it implies a construction from tri-linear maps. Namely, that any improvement on the state of the art would imply full-fledged program obfuscators. Specifically, as explained in Sect. 1.1, we present a candidate generator of block-wise locality 3, with *constant* size blocks. We show that this candidate is robust against algorithms such as ours, as well as other algorithmic methods. See Sect. 7 for more details.

3 Our Techniques

In this section we give an informal overview of the proof of our main result, Theorem 1.5 (i.e., limitations of low degree generators), focusing mostly on the degree two case, and making some simplifying assumptions. For the full proof see Sect. 4. We also describe at a high level, the ideas involved in the improved algorithm for the special cases of single-predicate generators (Theorem 1.4), random block-local generators (Theorem 5.4) and sum-of-squares lower bound

(Theorem 1.6) that shows a generator with stretch $m = \Omega(n2^b)$ that is resistant to sum-of-squares based attacks (an algorithm that encapsulates all our techniques).

As we observe in Sect. 3.1 below, Theorem 1.5 can be used in a black-box way to obtain a slightly weaker variant of Theorem 1.3, showing limitations of two block-local (and more generally ℓ block-local) generators. The full proof of Theorem 1.3, with the stated parameters, appears in Sect. 5.

Our work builds on some of the prior tools used for analyzing local pseudorandom generators and refuting constraint satisfaction problems, and in particular relies on *semidefinite programming*. The key technical difference is that while prior work mostly focused on generators/predicates with *constant* input locality or arity, we consider functions that could have much larger input locality, but have small degree. The fact that (due to our motivations in the context of obfuscation) we consider mappings with *non-Boolean* output also induces an extra layer of complexity.

We now describe our results in more detail. For simplicity, we focus on the degree two case, which is the case that is of greatest interest in the application for obfuscation. Recall that a *degree-two map of* $\mathbb{R}^n$ to $\mathbb{R}^m$ is a tuple of m degree two polynomials $\bar{p} = (p_1, \ldots, p_m)$. We will assume that the polynomials are *normalized* in the sense that $\mathbb{E}p_i(U) = 0$ and $\mathbb{E}p_i(U)^2 = 1$ for every i. Let Z be some "nice" (e.g., $O(1)$-spread) distribution over $\mathbb{R}^m$. (For starters, one can think of the case that Z is the uniform distribution over $\{\pm 1\}^n$, though we will want to consider more general cases as well.) The *image refutation problem* for the map $\bar{p}$ and the distribution Z is the task of certifying, given a random element z from Z, that $z \notin \bar{p}(\{\pm 1\}^n)$.

A natural approach is to use an approximation or refutation algorithm for the constraint satisfaction problem obtained from the constraints $\{p_i(x) = z_i\}$ for every i. The problem in our case is that while each of these predicates is "simple" in the sense of having quadratic degree, it can have very large locality or arity. In particular, the locality can be as large as s— the number of monomials of p_i— which we typically think of as equal to n^ε for some small $\varepsilon > 0$.

Much of the CSP refutation literature (e.g., see [1]) followed the so called "XOR principle" which reduces the task of refuting a CSP with arbitrary predicates, to the task of refuting a CSP where all constraints involve XORs (or products, when the input is thought of as ± 1 valued) of the input variables. Generally, applying this principle to arity s predicates leads to a 2^s multiplicative loss in the number of constraints, and also yields XORs that can involve up to s variables, which is unacceptable in our setting. However, as shown by [1], the situation is much better when the original predicate has small degree d (which, in particular, means it does not support a $(d+1)$-wise-independent distribution). In this case, utilizing the XOR principle results in a d-XOR instance, and only yields roughly an s^d loss in the number of constraints.

However, there are two issues with this approach. First, this reduction is not directly applicable in the non-Boolean setting, which is relevant to potential applications in obfuscation. Second, reducing to an XOR inherently leads to a loss in the output length that is related to the sparsity s, while, as we'll see, it may be sometimes possible to avoid losing such factors altogether.

Thus, our algorithm takes a somewhat different approach. Given the variables $z_1, \ldots, z_m$, we consider the quadratic program

$$\max_{x \in \{\pm 1\}^n} \sum_{i=1}^{m} z_i p_i(x). \tag{3.1}$$

The value of this program can be approximated to within a $O(\log n)$ factor using semidefinite relaxation via the *symmetric Grothendieck inequality* of Charikar and Wirth [19]. Thus, it is sufficient to show a gap in the value of this program between the "planted" case, where there is some x such that $p_i(x) = z_i$ for every i, and the case where the values z_i are sampled from Z.

If there is some x such that $p_i(x) = z_i$ for every i, then the value of the program (3.1) is at least $\sum_{i=1}^{m} z_i^2$ which (using the fact that $\mathbb{E} z_i^2 = 1$ and standard concentration bounds) we can assume to be very close to m.[8]

On the other hand, consider the case where $(z_1, \ldots, z_m)$ is chosen from Z. For every fixed $x \in \{\pm 1\}^n$, we can define m random variables $Y_1^x, \ldots, Y_m^x$ such that $Y_i^x = z_i p_i(x)$ and let $Y^x = \sum_{i=1}^{m} Y_i^x$. Since Z is a product distribution, the random variables Y_i^x are independent, and hence we can use the Chernoff bound to show that with all but $0.01 \cdot 2^{-n}$ probability, the value of Y^x will be at most $O(\sqrt{nBm})$, where B is a bound on the magnitude of $z_i p_i(x)$. We can then apply the union bound over all possible x's to show that the value of the quadratic program (3.1) is at most $O(\sqrt{nBm})$ with probability 0.99.

For example, if each z_i is a uniform element in $\{\pm 1\}$, and $|p_i(x)| \leq O(1)$ for every x (as is the case when p_i is a *predicate*), then $B = O(1)$ and so in this case the value of (3.1) will be at most m/c as long as $m \gg c^2 n$. Setting c to the aforementioned approximation factor $O(\log n)$, we get a successful refutation.

The resulting algorithm does the following. On input $z_1, \ldots, z_m$, run the SDP relaxation for (3.1) and if the value is smaller than $m/2$, then output "`refuted`" and declare that z is not in the image of G. In the case where $z = \mathcal{G}(x)$ the value of the quadratic program, and so also its SDP relaxation, will be at least $0.9m$.[9] On the other hand, if $m = \omega(n \log n)$, then with high probability the value of the quadratic program will be $o(m/\log n)$ and hence the relaxation will have value $o(m)$.

In the discussion above we made two key assumptions:

- $|p_i(x)| \leq O(1)$ for every $x \in \{\pm 1\}^n$
- $|z_i| \leq O(1)$ for $x \in \{\pm 1\}^n$

In general both of these might be false. If p_i has at most s non-zero monomials, and satisfies $\mathbb{E} p_i(U)^2 = 1$, then we can show that $|p_i(x)| \leq \sqrt{s}$ for every x,

[8] Formally, in the case that $p_i(x) = z_i$ we do not assume anything about the distribution of z. However, if $\sum_{i=1}^{m} z_i^2 < 0.9m$, we can simply choose to output "?".

[9] We ignore here the case where $\sum z_i^2 < 0.9m$, in which case our algorithm will halt with the output "?".

using the known relations between the ℓ_1 and ℓ_2 norms of p_i's Fourier transform. The second condition can be a little more tricky. If the z_i's are *subgaussian*, then we can use Hoeffding's inequality in place of the Chernoff bound, but in general we cannot assume that this is the case. Luckily, it turns out that in our application we can use a simple trick of rejecting outputs in which z_i has unusually large magnitude to reduce to the bounded case. The bottom line is that we get an efficient algorithm for the image-refutation problem of an s-sparse quadratic map whenever $m \gg sn \log n$.

The higher degree case reduces to the degree 2 by "quadratisizing" polynomials. That is, we can consider a degree d polynomial on n variables as a degree 2 polynomial on the $n^{\lceil d/2 \rceil}$ variables obtained by considering all degree $\lceil d/2 \rceil$ monomials. Using this approach, we can generalize our results (at a corresponding loss in the bound on the output) to higher degree maps.

3.1 Distinguishing Generators with Block-Locality 2

A priori the notions of *block locality* and *algebraic degree* seem unrelated to one another. After all, a two block local generator on size b blocks could have degree that is as large as $2b$. However, we can *pre-process* a length bn input $x \in \{\pm 1\}^{bn}$, by mapping it to an input $x' \in \{\pm 1\}^{n'}$ for $n' = 2^b n$ where for every $i \in [n]$, the ith block of x' will consist of the values of all the 2^b monomials on the ith block of x. Note that a map of block locality ℓ in x becomes a map of *degree* ℓ in x'. Moreover, since every output bit depends on at most ℓ blocks, each containing 2^b variables, the number of monomials in this degree ℓ polynomial is at most $2^{\ell b}$.

In this way, we can transform a candidate two block-local pseudorandom generator $\mathcal{G}\colon \{\pm 1\}^{bn} \to \{\pm 1\}^m$ into a degree-2 sparsity-2^{2b} map $\mathcal{G}'\colon \{\pm 1\}^{n'} \to \mathbb{R}^m$. Note that even if $\mathcal{G}$ is a secure pseudorandom generator, it is *not* necessarily the case that $\mathcal{G}'$ is also a pseudorandom generator, as the uniform distribution on $x \in \{\pm 1\}^{bn}$ does not translate to the uniform distribution over $x' \in \{\pm 1\}^{2^b n}$. However, the image of $\mathcal{G}'$ contains the image of $\mathcal{G}$, and hence if we can solve the image refutation problem for $\mathcal{G}'$, then we can do so for $\mathcal{G}$ as well. Applying the above result as a black-box gives an efficient algorithm to break a two block-local generator of block size b as long as the output length m satisfies

$$m \gg 2^{2b} n' \log^2 n = 2^{3b} n \log^2 n \, .$$

This is already enough to break the concrete candidate of Lin and Tessaro [47], but a more refined analysis shows that we can improve the 2^{3b} factor to 2^{2b}. Furthermore, if we initialize the construction with a random predicate on an expanding constraint graph we can bring this factor down to 2^b. Both improvements still use the same algorithm, only providing a tighter analysis of it in these cases. We do not know if our analysis can be improved even further. Mapping out the various trade-offs for block-local generators (or, equivalently, refuting very large alphabet CSPs), is a very interesting open question.

The first improvement, described in Sect. 5.1, yields a better bound on the output of any two-block-wise generator. As mentioned above, it uses the same algorithm. That is, we take a candidate two-block-local generator $\mathcal{G}\colon \{\pm1\}^{bn} \to \{\pm1\}^m$ and transform it into a degree two mapping $\mathcal{G}'\colon \{\pm1\}^{2^b n} \to \mathbb{R}^m$ by "expanding out" the monomials in each block. We then run the same algorithm as before on the generator $\mathcal{G}'$, but the key idea is that because $\mathcal{G}'$ arose out of the expansion of a two-block-local generator, we can show a better upper bound on the objective value of the quadratic program (3.1). Specifically, we can express each of these polynomials as a function of the Fourier transform of the predicate that the original block local generator applied to each pair of blocks. We can then change the order of summations, which enables us to reduce bounding (3.1) to bounding 2^{2b} "simpler" sums, for which we able to obtain, in the random case, tighter bounds with sufficiently high probability that allows to take a union bound over these 2^{2b} options. See Sect. 5.1 for the full detail.

3.2 Improving the Stretch to $n2^b$ for the Single Predicate Case

The second improvement (Theorem 5.3), considers the special case where each output of the generator is computed using the same predicate (as discussed before, this case is the principle focus of [49]). In this case, we show that our image refutation algorithm works whenever m (the number of outputs) of the generator satisfies $m = \tilde{\Omega}(n2^b)$. This matches the stretch required for the *distinguisher* of [49] to work.

We now describe at a high level, how our refutation algorithm works. The refutation algorithm is given a string $z \in \{\pm1\}^m$ and description of the generator $\mathcal{G}$ that includes the underlying graph G on n vertices and the predicate $P\colon \{\pm1\}^b \times \{\pm1\}^b \to \{\pm1\}$. As a first step, we will reduce the problem of image refuting $\mathcal{G}$ to image refuting a somewhat simpler $\mathcal{G}'$ where the predicate P will be replaced by a "product-predicate" P'. A predicate $P'\colon [q] \times [q] \to \{\pm1\}$ is a *product* predicate if it can be written as a product of two functions $f\colon [q] \to \{\pm1\}$ and $g\colon [q] \to \{\pm1\}$ applied to each of the inputs to P. In the second step, we will give an efficient algorithm for image-refuting two-block-local, single product predicate PRG.

We now describe the first step. Here, the algorithm wishes to certify that there's no $x \in (\{\pm1\}^b)^n$ such that $\mathcal{G}(x) = z$. Fix any $x \in (\{\pm1\}^b)^n$. For this fixed x, consider the distribution $\mathcal{D}$ on inputs to P, generated by taking a random edge $\{i, j\}$ in G and outputting (x_i, x_j). We will show, using a result of Linial and Schraibman shown in the context of relating marginal complexity to various measures of communication complexity, that on $\mathcal{D}$ (more generally, any distribution on inputs to P), there's a product predicate $F(\alpha, \beta) = f(\alpha) \cdot g(\beta)$ such that $\mathbb{E}_{(\alpha,\beta)\sim\mathcal{D}}[P(\alpha, \beta) \cdot F(\alpha, \beta)] \geq \Theta(2^{-b/2})$. Thus, if there is an $x \in (\{\pm1\}^b)^n$ such that $\mathcal{G}(x) = z$, then for the same x, $\mathbb{E}_{i\sim[m]}[\mathcal{G}'(x)_i \cdot z_i] \geq \Theta(2^{-b/2})$. If we can now certify an upper bound of $\ll 2^{-b/2}$ on $\mathbb{E}_{i\sim[m]}[\mathcal{G}'(x)_i \cdot z_i]$ for every x and with high probability over the draw of z, we'd obtain an image refutation algorithm. This latter question turns out to be simpler because of the product nature of the predicate defining $\mathcal{G}'$.

This step in our algorithm is inspired by the use of a result of Chor-Goldreich in the work of [49]. This lemma says[10] that for the uniform distribution on the inputs to P, there's a product predicate that has a correlation of $\Theta(2^{-b/2})$ with P. In the work of [49] this observation is used to replace P by a *constant-alphabet* predicate (obtained by massaging the constituents of the product predicate given by Chor-Goldreich lemma above) to obtain a simplified PRG on constant-alphabet size such that when the seed is chosen according to the uniform distribution on $(\{\pm 1\}^b)^n$, the modified PRG's output distribution correlates well with that of the original one. Thus, a strong enough refutation algorithm (they use one due to [1]) applied to the modified PRG is enough to give a distinguisher. Observe that this approach doesn't give a refutation algorithm because the key step of replacing P with $f \cdot g$ relies on x being drawn uniformly from $[q]^n$.

Instead of using off-the-shelf refutation algorithms (such as that of [2]), we solve the image refutation problem for single product predicate block-local PRGs by giving a direct, simple algorithm – this algorithm crucially works without the knowledge of the product predicate itself or even the block size parameter b. This is important, as our argument that obtains $\mathcal{G}'$ is not constructive, in particular, the distribution that the product predicate approximates P on is a complicated function of the (purported) arbitrary assignment x and the graph G. Thus, our product-predicate refutation algorithm must work without the explicit knowledge of the underlying product predicate.

Indeed, we show (in the full version [12]) that given a graph G on n vertices with $m \gg n$ edges and any string z, we can (in one shot) show that z (w.h.p) is not in the image of *any* of the (infinitely many!) generators obtained by using any two-block-local product predicate of arbitrarily large block size with the same underlying graph G. In particular, our refutation algorithm does not need to know the predicate itself or even the number of bits in each block of the seed for the generator!

3.3 Random Block Local Generators

We analyze the natural candidate of multiple-predicate, block-local generators, where both the underlying graph and each of the predicates are chosen uniformly at random (conditioned on the predicates being balanced), and show (see Sect. 5.3) that our refutation algorithm works whenever $m = \Omega(n2^b)$. As before, our idea to consider the problem of maximizing the polynomial $\sum_i z_i p_i(x)$. We work with the *matrix* M such that our target polynomial $\sum_i z_i p_i(x)$ is a bilinear form of M. To obtain a certificate for the upper bound on the polynomial, it then suffices to show a strong enough upper bound on the *spectral* norm of the matrix M – which we show is small enough (w.h.p) because of the randomness involved in defining the generator. M has some dependencies between its various entries that preclude the use of standard bounds to upper bound the spectral norm. So we compute an upper bound on the spectral norm using the standard

[10] We use a somewhat different way to describe the use Chor-Goldreich lemma by [49] in order to show how it inspires our approach.

trace method that reduces the problem to some combinatorial properties that are simple to reason about.

4 Image Refutation for Low Degree Maps

In this section we will prove our main technical theorem, which is an algorithm for the image refutation problem for every low degree map and "nice" or "non-degenerate" product distributions. We start by defining the notion of non-degenerate distributions, which amounts to distributions that do not put almost all their probability mass on a small (compared to their standard deviation) interval.

Definition 4.1 (c-spread distributions). *Let Z be a product distribution over $\mathbb{R}^m$ with $\mathbb{E}Z_i = 0$ and $\mathbb{E}Z_i^2 = 1$ for every i. We say that Z is c-spread if for every interval $I \subseteq \mathbb{R}$ of length $1/c$, the probability that $Z_i \in I$ is at most 0.9.*

Normalized low-degree maps are polynomials over $\{\pm 1\}^n$ - we use the standard Fourier basis (e.g., see [53]) to represent them:

Definition 4.2 (Fourier notation). *For any $S \subseteq [n]$, let $\chi_S(x) = \Pi_{i \in S} x_i$ for any $x \in \{\pm 1\}^n$. A function p: $\{\pm 1\}^n \to \mathbb{R}$ can be uniquely expanded as $\sum_{S \subseteq [n]} \hat{p}(S)\chi_S$ where the "Fourier coefficients" $\hat{p}(S) = \mathbb{E}_{x \sim \{\pm 1\}^n}[\chi_S(x)p(x)]$ and the expectation is over the uniform distribution over the hypercube $\{\pm 1\}^n$. Fourier coefficients satisfy the Parseval's theorem: $\mathbb{E}_{x \sim \{\pm 1\}^n} p(x)^2 = \sum_{S \subseteq [n]} \hat{p}(S)^2$.*

We define a normalized degree d map to be a collection of degree d polynomials $\bar{p} = (p_1, \ldots, p_m)$ mapping $\{\pm 1\}^n$ to $\mathbb{R}^m$ such that $\mathbb{E}p_i(U) = 0$ and $\mathbb{E}p_i(U)^2 = 1$ for every i where U is the uniform distribution.[11]

Our main technical theorem is the following:

Theorem 4.3 (Main theorem). *There is an efficient algorithm that solves the refutation problem for every normalized degree d map $\bar{p}$ and c-spread probability distribution Z as long as*

$$m > K \cdot c^2 s(\bar{p}) n^{\lceil d/2 \rceil} \log^2(n) \tag{4.1}$$

for some global constant K.

To state the result in a stronger form, we use a somewhat technical definition for the parameter $s(\bar{p})$, which is deferred till later (see Eq. (4.5) and Definition 4.9 below). However, one important property of it is that for every

[11] Note that we are using the same normalization for the Z_i's and $p_i(U)$, which makes sense in the context of a pseudorandom generator applied to the uniform distribution over the seed. If we wanted to consider other distributions D over the seed, we would need to require that $\mathbb{E}p_i(D)^2$ is not much smaller than $\mathbb{E}p_i(U)^2$. This condition is satisfied by many natural distributions.

normalized polynomial map $\bar{p} = (p_1, \ldots, p_m)$, $s(\bar{p})$ is smaller than the maximum *sparsity* (i.e., number of monomials) of the polynomials. Hence, Theorem 4.3 implies Theorem 1.5 from Sect. 1.1. The fact that we only require a factor of $s(\bar{p})$ as opposed to the sparsity makes our result stronger, and in some cases this difference can be very significant.

The algorithm for proving Theorem 4.3 is fairly simple:

Refutation algorithm
Input: $z \in \mathbb{R}^m$, $p_1, \ldots, p_m$ normalized polynomials of degree d in $\{\pm 1\}^n$.
Output: "`refuted`" or "?".
Operation:

1. Let $I = \{i \in [m] : z_i^2 \leq 100\}$. Let μ_i be the conditional expectation of z_i conditioned on $z_i^2 \leq 100$.
2. If $\sum_{i \in I}(z_i - \mu_i)^2 < m/(10c)$ return "?".
3. Let θ be the value of the degree $\lceil d/2 \rceil$ SOS relaxation for the degree d polynomial optimization problem

$$\max_{x \in \{\pm 1\}^n} \sum_{i \in I}(z_i - \mu_i)p_i(x) \tag{4.2}$$

4. Return "`refuted`" if $\theta - \sum_{i \in I} \mu_i(z_i - \mu_i) < m/(10c)$ otherwise return "?".

The *degree d sum of squares program* is a semidefinite programming relaxation to a polynomial optimization problem, which means that the value θ is always an upper bound on (4.2). The most important fact we will use about this program is the *symmetric Grothendieck Inequality* of Charikar and Wirth [19], which states that in the important case where $d = 2$, the *integrality gap* of this program (i.e., ratio between its value and the true maximum) is $O(\log n)$.

For this case, where $d = 2$, this program is equivalent to the semidefinite program known as the *basic SDP* relaxation for the corresponding quadratic program. This means that θ can also be computed as

$$\max_{\substack{X \in \mathbb{R}^{(n+1)\times(n+1)} \\ X \succeq 0,\, X_{ii}=1\ \forall i}} \operatorname{tr}(A \cdot X)\,, \tag{4.3}$$

where A is an $(n+1) \times (n+1)$ matrix that *represents* the quadratic polynomial $\sum_{i \in I}(z_i - \mu_i)p_i$, in the sense that for every $i, j \in [n]$, $A_{i,j}$ corresponds to the coefficient of $x_i x_j$ in this polynomial, and for every $i \in [n]$, $A_{i,n+1} = A_{n+1,i}$ is the coefficient of x_i.

We now turn to proving Theorem 4.3. We start by showing the case that $d = 2$. The proof for general degree will follow by a reduction to that case.

4.1 Degree 2 Image Refutation

In this section, we prove Theorem 4.3 for the case $d = 2$, which is restated below as the following lemma:

Lemma 4.4 (Image refutation for degree 2). *There is an efficient algorithm that solves the refutation problem for every normalized degree 2 map $\bar{p}$ and c-spread probability distribution Z as long as*

$$m > K \cdot c^2 s(\bar{p}) n \log^2 n \tag{4.4}$$

for some absolute constant $K > 0$.

In this case, the parameter $s(\bar{p})$ is defined as follows:

$$s(p_1, \ldots, p_m) = \tfrac{1}{m} \max_{x \in \{\pm 1\}^n} \sum_{i=1}^{m} p_i(x)^2 \tag{4.5}$$

By expanding each p_i in the Fourier basis as $p_i = \sum \hat{p}_i(S)\chi_S$, we can see that $\max_{x\in\{\pm1\}^n} |p_i(x)| \leq \sum |\hat{p}_i|$. Hence, in particular, $s(\bar{p})$ is smaller than the average of the ℓ_1 norm squared of the p_i's Fourier coefficients. Using the fact that $\mathbb{E}p_i(U)^2 = 1$, and the standard relations between the ℓ_1 and ℓ_2 norms, we can see that if every one of the p_i polynomials has at most s monomials (i.e., non-zero Fourier coefficients), then $s(\bar{p}) \leq s$.

We now prove Lemma 4.4. To do so, we need to show two statements:

- If $z = \bar{p}(x)$, then the algorithm will never output "`refuted`".
- If z is chosen at random from Z, then the algorithm will output "`refuted`" with high probability.

We start with the first and easiest fact, which in fact holds for *every* degree d.

Lemma 4.5. *Let $z \in \mathbb{R}^m$ be such that there exists an x^* such that $p_i(x^*) = z_i$. Then, the algorithm does not output* "`refuted`".

Proof. Suppose otherwise. We can assume that $\sum_{i\in I}(z_i - \mu_i)^2 \geq m/(10c)$ as otherwise we will output "?". Since the SDP is a relaxation, in particular, the value θ is larger than $\sum_{i\in I}(z_i - \mu_i)p_i(x^*) = \sum_{i\in I}(z_i - \mu_i)z_i$ under our assumption. Hence, $\theta - \sum_{i\in I}(z_i - \mu_i)\mu_i \geq \sum_{i\in I}(z_i - \mu_i)^2 \geq m/(10c)$.

We now turn to the more challenging part, which is to show that the algorithm outputs "`refuted`" with high probability when z is sampled from Z. We start by observing that by Markov's inequality, for every i, the probability that $z_i^2 > 100\mathbb{E}z_i^2 = 100$ is at most 0.99. Hence, the expected size of the set I defined by the algorithm is at least $0.99m$ and using Chernoff's bound it follows with very high probability that $|I| > 0.9m$. Let Z_i' be the random variable Z_i conditioned on the (probability ≥ 0.99) event that $Z_i^2 \leq 100$, and $\mu_i = \mathbb{E}Z_i'$. Note that by definition $(Z_i')^2 \leq 100$ with probability 1, i.e. $|Z_i'| \leq 10$ with probability 1, which in turn implies that $|\mu_i| \leq 10$. By the "spread-out-ness" condition on Z_i and the union bound, $\mathbb{P}[Z_i' \notin [\mu_i - \frac{1}{2c}, \mu_i + \frac{1}{2c}]] \geq 0.1 - 0.01$ and hence, in particular, $\mathbb{E}[(Z_i' - \mu_i)^2] \geq \frac{1}{500c^2}$.

We can consider the process of sampling the z_i values from the algorithm as being obtained by first choosing the set I, and then sampling z_i independently from the random variable Z'_i for every coordinate $i \in I$. The following lemma says that there will not be an *integral* (i.e., $\{\pm 1\}$-valued) solution to the SDP with large value.

Lemma 4.6. *With probability at least* 0.99 *it holds that for every* $x \in \{\pm 1\}^n$,

$$\sum_{i \in I} (z'_i - \mu_i) p_i(x) \leq O(\sqrt{nms(\bar{p})}) \tag{4.6}$$

Proof. We use the union bound. For every fixed $x \in \{\pm 1\}^n$, we let $\alpha_i = p_i(x)$. We know that $\sum_{i \in I} \alpha_i^2 \leq \sum_{i=1}^m \alpha_i^2 \leq \max_{x \in \{\pm 1\}^n} \sum p_i(x)^2 = ms(\bar{p})$. Since $|z'_i - \mu_i| \leq 20$, it follows that $(z'_i - \mu_i)$ is sub-gaussian with constant standard deviation. Therefore, $\sum_{i \in I} (z'_i - \mu_i)\alpha_i$ is sub-gaussian with zero expectation standard deviation $O(\sqrt{ms(\bar{p})})$. Therefore, there exists a value $O(\sqrt{nms(\bar{p})})$ s.t. the probability that $\sum_{i \in I} (z'_i - \mu_i)\alpha_i$ exceeds it is smaller than $0.001 \cdot 2^{-n}$. Applying the union bound implies the lemma.

Lemma 4.4 will follow from Lemma 4.6 using the fact that the SDP gives $O(\log n)$ approximation factor for true maximum. In particular the symmetric version of Grothendieck inequality shown by [19] implies that the value θ computed by the algorithm is at most a factor of $O(\log n)$ larger than the true maximum of the integer program (4.2), see Theorem A.3 in Appendix A.

To finish the proof, we need to ensure that (after multiplying by $O(\log n)$) the bound on the RHS of (4.6) will be smaller than $m/(100c) + \sum_{i \in I} (z_i - \mu_i)\mu_i$. Indeed, since $|\mu_i| \leq 10$, with high probability over the choice of the z_i's (which are chosen from Z'_i), the quantity $\sum_i (z_i - \mu_i)\mu_i$ is at most, say, 10 times the standard deviation, which is $O(\sqrt{m}) \ll m/c$. (Here no union bound is needed.) So, by plugging in (4.6) what we really need is to ensure that

$$m/(20c \log n) \geq O(\sqrt{nms(\bar{p})})$$

or that

$$m \geq O(ns(\bar{p})c^2 \log^2 n)$$

which exactly matches the conditions of Lemma 4.4 hence concluding its proof (and hence the proof Theorem 4.3 for the $d = 2$ case).

4.2 Refutation for $d > 2$

In this section, we show how to reduce the general degree d case to the case $d = 2$, hence completing the proof of Theorem 4.3. The main tool we use is the notion of "quadratizing" a polynomial. That is, we can convert a degree d polynomial p on n variables into a degree two polynomial $\tilde{p}$ on $(n+1)^{\lceil d/2 \rceil}$ variables by simply encoding every monomial of degree up to $\lceil d/2 \rceil$ of the input as a separate variable.

Definition 4.7 (Quadratization). *Let p be a degree d polynomial on $\mathbb{R}^n$ which we write in Fourier notation (see Definition 4.2) as $p = \sum_{|S| \le d} \hat{p}(S)\chi_S$. Let $d' = \lceil d/2 \rceil$ Then the quadratization of p is the degree two polynomial q on $\binom{n}{\le d'}$ variables defined as:*

$$q(y) = \sum_{S,T} \hat{p}(S \cup T) y_S y_T,$$

where the elements of the $\binom{n}{\le d'}$ dimensional vector y are indexed by sets of size at most d', and this sum is taken over all sets $S, T \subseteq [n]$ of size at most d' such that every element in S is smaller than every element of T, $|S| = \max\{|S \cup T|, d'\}$.

The following simple properties ensured by quadratization are easy to verify:

Lemma 4.8. *Let q be the quadratization of a degree d polynomial p on $\binom{n}{\le d'}$ variables for $d' = \lceil d/2 \rceil$. Then,*

1. *For any $x \in \{\pm 1\}^n$ there exists $y \in \{\pm 1\}^{\binom{n}{\le d'}}$ such that $q(y) = p(x)$.*
2. *$\sum_{S,S'} \hat{q}(\{S, S'\})^2 = \sum_T \hat{p}(T)^2$.*
3. *$\max_{y \in \{\pm 1\}^{\binom{n}{\le d'}}} q(y) \le \sum_{|T| \le d} |\hat{p}(T)|$.*

Proof (sketch). For 1, we let $y_S = \chi_S(x)$ for every $|S| \le d'$. For 2 and 3, we note that the set of nonzero Fourier coefficients of p and q is identical because for every set $|U| \le d$ there is a unique way to split it into disjoint sets S, T of size at most d' where S is the first $\min\{|U|, d'\}$ coordinates of U, and $\hat{q}(\{S, T\}) = \hat{U}$. For all other pairs S, T that do not arise in this manner, it will hold that $\hat{q}(\{S, T\}) = 0$. This means that both the ℓ_1 and ℓ_2 norms of the vector $\hat{q}$ are the same as that of the vector $\hat{p}$, implying both 2 and 3.

We define the complexity of the degree d normalized map $\bar{p}$ as the complexity of the degree 2 normalized map of the quadratizations of p_is:

Definition 4.9 (Complexity of degree d normalized maps). *Let $\bar{p}$ be a normalized degree d map and let $\bar{q}$ be its quadratization. Then, we define $s(\bar{p})$ as $s(\bar{q})$ from* (4.5).

Remark 4.10. Part 2 of Lemma 4.8 shows that if $\bar{p}$ is normalized the so is its quadratization $\bar{q}$. Part 3 of Lemma 4.8 shows that $s(\bar{p}) \le \text{sparsity}(p)$ for any normalized degree d map p.

We can now complete the proof of Theorem 4.3.

Proof (of Theorem 4.3). Let $\bar{p} = (p_1, \ldots, p_m)$ be a normalized degree d polynomial map and let $z_1, \ldots, z_m$ be the inputs given to the algorithm. If there is an x such that $p_i(x) = z_i$ for every i, then by Lemma 4.5 (which did not assume that $d = 2$), the algorithm will return "?".

Suppose otherwise, that $z_1, \ldots, z_m$ are chosen from the distribution Z. Recall that our algorithm computes θ to be the value of the degree $2d'$ SOS relaxation for the quadratic program (4.2). This value satisfies

$$\theta = \max_{\mu(x)} \tilde{\mathbb{E}}_{\mu} \left[\sum_{i \in I} (z_i - \mu_i) p_i(x) \right] ,$$

where the maximum is over all degree $2d'$ pseudo-distributions satisfying $\{x_i^2 = 1\}$ for every $i \leq n$.

If μ is a degree $2d'$ pseudodistribution over $\{\pm1\}^n$ then we can define a degree 2 pseudodistribution μ' over $\{\pm1\}^{\binom{n}{d'}}$ by having $y \sim \mu$ be defined as $y_S = \chi_S(x)$ for $x \sim \mu$.[12] Let $\bar{q} = (q_1, \ldots, q_m)$ be the quadratization of $\bar{p} = (p_1, \ldots, p_m)$. Then the distribution μ' above demonstrates that $\theta \leq \theta'$ where

$$\theta' = \max_{\mu'(y)} \tilde{\mathbb{E}}_{\mu'} \left[\sum_{i \in I} (z_i - \mu_i) q_i(x) \right] .$$

But since this is the value of a degree two SDP relaxation for a quadratic program, we know by Theorem A.3 that it provides an $O(\log n)$ approximation factor, or in other words that

$$\theta' \leq O(\log n) \max_{y \in \{\pm1\}^{\binom{n}{d'}}} \sum_{i \in I} (z_i - \mu_i) q_i(y) . \tag{4.7}$$

Since the q_i's are degree two polynomials over $O(n^{d'})$ variables, Lemma 4.6 implies that when $z_1, \ldots, z_m$ are randomly chosen from Z, w.h.p. the RHS of (4.7) is at most $O((\log n)\sqrt{n^{d'} ms(\bar{q})}) = O((\log n)\sqrt{n^{d'} ms(\bar{p})})$. Setting this to be smaller than $(m/10c^2)$ recovers Theorem 4.3.

5 Block Local Generators

Recall that a map $\mathcal{G} \colon \{\pm1\}^{bn} \to \{\pm1\}^m$ is ℓ *block-local* if the input can be separated into n blocks of b bits each[13], such that every output of $\mathcal{G}$ depends on at most ℓ blocks.

In this section we will show tighter bounds for block-local generators than those derived from the theorem in Sect. 4. Of particular interest is the case of block-locality 2 due to its applications for obfuscation from bilinear maps. In Sect. 5.1 we show a tighter analysis of our algorithm from Sect. 4 for any block-local generator. This yields a distinguisher for any block-locality 2 generator with $m \gg 2^{2b} n \log n$. In Sect. 5.3, we analyze a particularly natural instantiation

[12] While it is clear that this operation makes sense for actual distributions, it turns out to be not hard to verify that it also holds for pseudodistributions, see the lecture notes [16].

[13] Our algorithm works even if the blocks intersect arbitrarily. The construction in [47] uses only non-intersecting blocks.

for 2-block-local PRGs - a random predicate and random constraint graph and show that our distinguisher works for an even smaller $m \gg 2^b n$. In fact, we show that one can even use a simpler distinguisher that computes the largest singular value of a certain matrix arising out of the input instead of running a semidefinite program.

5.1 Bounds on General Block-Local Generators

In this subsection we prove the following result:

Theorem 5.1 (Limitations of block local generators). *For every ℓ-block-local $\mathcal{G}\colon \{\pm 1\}^{bn} \to \{\pm 1\}^m$ there is an efficient algorithm for the $\mathcal{G}$ image refutation problem w.r.t. the uniform distribution over $\{\pm 1\}^m$ as long as*

$$m > (K \log n) 2^{\ell b} (n + 2\ell b)^{\lceil \ell/2 \rceil},$$

where K is a constant depending only on ℓ.

If ℓ is constant and $b = o(n)$ (as is mostly the case), the above translates to refutation for $m > (K \log n) 2^{\ell b} n^{\lceil \ell/2 \rceil}$.

The proof of this theorem can be found in the full version [12].

Theorem 1.3 from the introduction is the special case of Theorem 5.1 for the case $\ell = 2$, and so in particular Theorem 5.1 breaks any 2 block local pseudorandom generator with stretch $\tilde{\Omega}(n 2^{2b})$ to instantiate the bilinear-map based construction of iO of [47].

Remark 5.2. A slightly weaker bound can be obtained by a direct application of Theorem 4.3. We sketch the argument in the full version [12].

5.2 Sharper Bounds on the Stretch of Block-Local PRGs with a Single Predicate

Next, we prove a tighter upper bound of $\tilde{\Theta}(n 2^b)$ on the stretch of a block local PRGs with a *single* predicate P (instead of a different predicate for each output) with block-locality 2. The following is the main result of this section:

Theorem 5.3. *For $b \in \mathbb{N}$, let $\mathcal{G}\colon \{\pm 1\}^{bn} \to \{\pm 1\}^m$ be a two block-local PRG defined by an instance graph $G([n], E)$ with $m = |E|$ edges and an arbitrary predicate $P\colon \{\pm 1\}^b \to \{\pm 1\}^b \to \{\pm 1\}$ such that for any seed $x \in (\{\pm 1\}^b)^n$, for every $e \in E$, $\mathcal{G}_e = P(x_{e_1}, x_{e_2})$. Let $z \in \{\pm 1\}^m$.*

Then, for any $m > O(\log^2(n)) n 2^b$, there exists a poly(m, n) *time algorithm that takes input G, z and P and outputs "`refuted`" or "?" with the following guarantees:*

1. *If the output is "`refuted`", then,*

$$\max_{x \in (\{\pm 1\}^b)^n} \sum_{(i,j) \in E} P(x_{e_1}, x_{e_2}) z_e < 0.99m.$$

2. *When $z \in \{\pm 1\}^m$ is chosen uniformly at random, then $\mathbb{P}[$ Algorithm outputs "`refuted`"$] > 1 - 1/n$.*

The proof of this theorem can be found in the full version [12].

5.3 Image Refutation for Random Block-Local PRGs

A particularly appealing construction of block local PRGs is obtained by instantiating them with a random graph with ~m edges and a random and independent predicate for every edge. A priori, the randomness in this construction could appear to *aid* the security of the PRG. Indeed, such instantiations are in fact suggested by [47]. We show that in this case, as in the previous section where all predicates are identical, we can show a *stronger* upper bound on the stretch of the local PRG in terms of the block size b. Whereas in Sect. 5.1, for general block-local PRGs with non-identical predicates, we lost a factor of $2^{2b}\log(n)$ in the output length, for the special case of a random graphs and random, independent predicates, this can be improved to $\Theta(2^b)$ as we show in this section. We note that the only property of random graphs that we use is expansion.

More concretely, in this section, we analyze the stretch of the following candidate construction of a block-local PRG.

- We choose a graph $G([n], q)$ where every edge is present in G with probability $q = \frac{m}{\binom{n}{2}}$. Thus, with high probability, the number of edges in the graph is $m \pm \sqrt{m}$.
- For every edge $\{i, j\}$ in G, we choose a uniformly random predicate $P_{i,j}(x, y) = \pm 1$ conditioned on $P_{i,j}$s being balanced, i.e. $\mathbb{E}_{x,y\sim\{\pm 1\}^b} P_{i,j}(x, y) = 0$.
- On input (seed) $x \in \{\pm 1\}^{bn}$, which we think of as partitioned into blocks $x_1, \ldots, x_n \in \{\pm 1\}^b$, the generator outputs $h_{i,j}(x_i, x_j)$ for every edge (i, j) of G.

Theorem 5.4 (Limitations of random block-local generators). *There is some constant K such that if $\mathcal{G}\colon \{\pm 1\}^{bn} \to \{\pm 1\}^m$ is a generator sampled according to the above model and $m \geq K 2^b n \log^3(n)$, then w.h.p. there is a polynomial-time algorithm for the $\mathcal{G}$ image refutation problem w.r.t. the uniform distribution over $\{\pm 1\}^m$.*

The proof of this theorem can be found in the full version [12].

6 Lower Bound for Refuting Two-Block-Local PRGs

In this section, we establish that if $b > 10 \log\log(n)$, then there's no $2^{O(n/2^{4b})}$-time algorithm for image refutation of block-local PRG of stretch $\Omega(n2^b)$ based on the sum-of-squares method.

The main goal of this section is summarized in the following theorem.

Theorem 6.1. *For any $b > 10 \log\log(n)$, there's a construction $\mathcal{G}\colon \{\pm 1\}^n \to \{\pm 1\}^m$ for $m = \Omega(n2^b)$ such that for any $z \in \{\pm 1\}^m$, there's a feasible solution for the degree $\Theta(n/2^{4b})$ sum-of-squares relaxation of the constraints $\{\mathcal{G}_i = z_i\}$. In particular, sum of squares algorithm of degree $\Theta(n/2^{4b})$ cannot accomplish image refutation for $\mathcal{G}$.*

The proof of this theorem can be found in the full version [12].

7 A Class of Block-Local Candidate Pseudorandom Generators

In this section we outline a simple candidate pseudorandom generator of degree d that has potentially output length as large as $n^{d/2-\varepsilon}$. We have not conducted an extensive study of this candidate's security, but do believe it's worthwhile example as a potential counterpoint to our results on limitations for pseudorandom generator, demonstrating that they might be tight.

The idea is simple: for a finite group $\mathbb{G}$ that does not have any abelian quotient group (for example, a non-abelian simple group will do), we choose dm random indices $\{i_{j,k}\}_{j\in[m],k\in[d]}$ and let $\mathcal{G}$ be the generator mapping $\mathbb{G}^n$ to $\mathbb{G}^m$ where

$$\mathcal{G}(x)_j = x_{i_{j,1}} * x_{i_{j,2}} * \cdots * x_{i_{j,d}} \tag{7.1}$$

If want to output m bits rather than m elements of $\mathbb{G}$, then we use a group $\mathbb{G}$ of even order and apply to each coordinate some balanced map $f\colon \mathbb{G} \to \{0,1\}$. For every group element $g \in \mathbb{G}$, the predicate

$$x_1 * \cdots * x_d = g \tag{7.2}$$

supports a $d-1$ wise independent distribution. Hence, using the results of [43] we can show that as long $m < n^{d/2-\varepsilon}$, for a random $z \in \mathbb{G}^m$, the SOS algorithm cannot be used to efficiently refute the statement that $z = \mathcal{G}(x)$ for some x.

Ruling out Gaussian-elimination type attacks is trickier. For starters, solving a linear system over a non-abelian group is NP-hard [35,42]. Also, Applebaum and Lovett [10, Theorem 5.5] showed that at least for the large d case, because the predicate (7.2) has rational degree d, the image-refutation problem for this generator is hard with respect to algebraic attacks (that include Gaussian elimination) for $m = n^{\Omega(d)}$. Nevertheless, there are non trivial algorithms in the group theoretic settings (such as the low index subgroup algorithm, see [23] and [57, Sect. 6]). A more extensive study of algebraic attacks against this predicate is needed to get better justifications of its security, and we leave such study for future work.

We remark that the condition that the group $\mathbb{G}$ does not have abelian normal subgroups is crucial. Otherwise, we can write $\mathbb{G}$ as the direct product $\mathbb{H} \times \mathbb{H}'$ where $\mathbb{H}$ is abelian, and project all equations to their component in $\mathbb{H}$. We will get m random equations in n variables over the abelian group $\mathbb{H}$, and hence we can use Gaussian elimination to refute those.

Acknowledgements. We thank Prabhanjan Ananth, Dakshita Khurana and Amit Sahai for discussions regarding the class of generators needed for obfuscation. Thanks to Rachel Lin and Stefano Tessaro for discussing the parameters of their construction with us. We thank Avi Wigderson and Andrei Bulatov for references regarding Gaussian elimination in non-abelian groups.

A Analysis of the Basic SDP Program

The degree d SOS program [16] for a polynomial optimization problem of the form

$$\max_{x \in \{\pm 1\}^n} p(x)$$

corresponds to

$$\max_{\mu} \tilde{\mathbb{E}} p$$

where $\tilde{\mathbb{E}}$ ranges over the set of degree d expectation operators that satisfy the constraints $\{x_i^2 = 1\}_{i=1}^n$. These are defined as follows:

Definition A.1 (Pseudo-expectation). *Let $\mathcal{P}_{n,d}$ denote the space of all degree $\leq d$ polynomials on n variables. A linear operator $\tilde{\mathbb{E}} : \mathcal{P}_{n,d}$ is a degree d* pseudo-expectation *if it satisfies the following conditions:*

1. $\tilde{\mathbb{E}}[1] = 1$.
2. $\tilde{\mathbb{E}}[p^2] \geq 0$ *for every polynomial p of degree at most $d/2$.*

A pseudo-expectation is said to satisfy a constraint $\{q = 0\}$ if for every polynomial p of degree at most $d - deg(q)$, $\tilde{\mathbb{E}}[pq] = 0$. We say that $\tilde{\mathbb{E}}$ satisfies the constraint $\{q \geq 0\}$ if for every polynomial p of degree at most $d/2 - deg(q)/2$, $\tilde{\mathbb{E}}[p^2 q] \geq 0$.

If μ is any distribution on $\mathbb{R}^n$, then the associated expectation is a pseudo-expectation operator of all degrees. The above definition can be thought of as a relaxation of the notion of an actual expectation.

Key to the utility of the definition above is the following theorem that shows one can efficiently search over the space of all degree d pseudo-expectations.

Theorem A.2 ([44,55,59]). *For any n, and integer d, the following set has an $n^{O(d)}$ time weak separation oracle (in the sense of [37]):*

$$\{\tilde{\mathbb{E}}[(1, x_1, x_2, \ldots, x_n,)^{\otimes d}] \mid \tilde{\mathbb{E}} \text{ is a degree } d \text{ pseudo-expectation}\}$$

In this appendix we expand on how Charikar and Wirth's work [19] implies the the following theorem:

Theorem A.3. *For every degree two polynomial p: $\mathbb{R}^n \to \mathbb{R}$ with no constant term, the value of the degree two SOS program for*

$$\max_{x \in \{\pm 1\}^n} p(x) \tag{A.1}$$

is larger than the true value of (A.1) by a factor of at most $O(\log n)$.

Theorem A.3 is a direct implication of the following result of [19]:

Theorem A.4 (Symmetric Grothendieck Inequality, [19]**, Theorem 1).** *Let A be any $m \times m$ matrix such that $A_{i,i} = 0$ for every i. Then,*

$$\max_{X \succeq 0, X_{i,i}=1 \forall i} Tr(AX) \leq O(\log n) \max_{x \in \{\pm 1\}^n} x^\top Ax$$

Proof (of Theorem A.3 *from Theorem* A.4). Suppose that there is a degree 2 pseudo-distribution $\{x\}$ such that $\tilde{\mathbb{E}}p(x) \geq \theta$, and let X be the $n+1 \times n+1$ matrix corresponding to $\tilde{\mathbb{E}}(x,1)(x,1)^\top$. That is, $X_{i,j} = \tilde{\mathbb{E}}x_i x_j$ and $X_{n+1,i} = X_{i,n+1} = \tilde{\mathbb{E}}x_i$. Note that X is a psd matrix with 1's on the diagonal.

Then $\mathrm{Tr}(AX) \geq \theta$ if A be the $(n+1) \times (n+1)$ matrix that represents the polynomial p. In this case Theorem A.4 implies that there is an $n+1$ dimensional vector $(x, \sigma) \in \{\pm 1\}^{n+1}$ such that $(x,\sigma)^\top A(x,\sigma) \geq \Omega(\theta/\log n)$. If we write $p(x) = q(x) + l(x)$, where q is the homogeneous degree two and l is linear, then we can see by direct inspection that

$$(x,\sigma)^\top A(x,\sigma) = q(x) + \sigma l(x) = p(\sigma x)$$

with the last equality following from the fact that $q(-x) = q(x)$ and $l(-x) = -l(x)$. Hence the vector $\sigma x \in \{\pm 1\}^n$ demonstrates that the value of (A.1) is at least $\Omega(\theta/\log n)$.

References

1. Allen, S.R., O'Donnell, R., Witmer, D.: How to refute a random CSP. In: FOCS, pp. 689–708. IEEE Computer Society (2015)
2. Allen, S.R., O'Donnell, R., Witmer, D.: How to refute a random CSP. In: 2015 IEEE 56th Annual Symposium on Foundations of Computer Science—FOCS 2015, pp. 689–708. IEEE Computer Society, Los Alamitos, CA (2015)
3. Ananth, P., Brakerski, Z., Khurana, D., Sahai, A.: Private communication (2017)
4. Ananth, P., Jain, A., Sahai, A.: Indistinguishability obfuscation from functional encryption for simple functions. IACR Cryptology ePrint Archive 2015, 730 (2015)
5. Ananth, P., Sahai, A.: Projective arithmetic functional encryption and indistinguishability obfuscation from degree-5 multilinear maps. In: Coron, J.-S., Nielsen, J.B. (eds.) EUROCRYPT 2017. LNCS, vol. 10210, pp. 152–181. Springer, Cham (2017). https://doi.org/10.1007/978-3-319-56620-7_6
6. Applebaum, B.: Pseudorandom generators with long stretch and low locality from random local one-way functions. SIAM J. Comput. **42**(5), 2008–2037 (2013)
7. Applebaum, B.: Cryptographic hardness of random local functions - survey. Comput. Complex. **25**(3), 667–722 (2016)
8. Applebaum, B., Barak, B., Wigderson, A.: Public-key cryptography from different assumptions. In: Proceedings of the 42nd ACM Symposium on Theory of Computing, STOC, pp. 171–180. ACM (2010)
9. Applebaum, B., Ishai, Y., Kushilevitz, E.: Cryptography in NC^0. SIAM J. Comput. **36**(4), 845–888 (2006)
10. Applebaum, B., Lovett, S.: Algebraic attacks against random local functions and their countermeasures. In: STOC, pp. 1087–1100. ACM (2016)
11. Applebaum, B., Raykov, P.: Fast pseudorandom functions based on expander graphs. In: Hirt, M., Smith, A. (eds.) TCC 2016. LNCS, vol. 9985, pp. 27–56. Springer, Heidelberg (2016). https://doi.org/10.1007/978-3-662-53641-4_2

12. Barak, B., Brakerski, Z., Komargodski, I., Kothari, P.K.: Limits on low-degree pseudorandom generators (or: Sum-of-squares meets program obfuscation). IACR Cryptology ePrint Archive 2017, 312 (2017)
13. Barak, B., Chan, S.O., Kothari, P.K.: Sum of squares lower bounds from pairwise independence [extended abstract]. In: Proceedings of the 2015 ACM Symposium on Theory of Computing, STOC 2015, pp. 97–106. ACM, New York (2015)
14. Barak, B., Goldreich, O., Impagliazzo, R., Rudich, S., Sahai, A., Vadhan, S., Yang, K.: On the (im)possibility of obfuscating programs. In: Kilian, J. (ed.) CRYPTO 2001. LNCS, vol. 2139, pp. 1–18. Springer, Heidelberg (2001). https://doi.org/10.1007/3-540-44647-8_1
15. Barak, B., Raghavendra, P., Steurer, D.: Rounding semidefinite programming hierarchies via global correlation. In: 2011 IEEE 52nd Annual Symposium on Foundations of Computer Science, FOCS 2011, pp. 472–481. IEEE Computer Society, Los Alamitos, CA (2011). http://dx.doi.org/10.1109/FOCS.2011.95
16. Barak, B., Steurer, D.: Proofs, beliefs, and algorithms through the lens of sum-of-squares (2017). http://sumofsquares.org
17. Boneh, D., Franklin, M.K.: Identity-based encryption from the weil pairing. SIAM J. Comput. **32**(3), 586–615 (2003)
18. Brakerski, Z., Gentry, C., Halevi, S., Lepoint, T., Sahai, A., Tibouchi, M.: Cryptanalysis of the quadratic zero-testing of GGH. Cryptology ePrint Archive, Report 2015/845 (2015)
19. Charikar, M., Wirth, A.: Maximizing quadratic programs: extending grothendieck's inequality. In: FOCS, pp. 54–60. IEEE Computer Society (2004)
20. Cheon, J.H., Fouque, P.-A., Lee, C., Minaud, B., Ryu, H.: Cryptanalysis of the new CLT multilinear map over the integers. In: Fischlin, M., Coron, J.-S. (eds.) EUROCRYPT 2016. LNCS, vol. 9665, pp. 509–536. Springer, Heidelberg (2016). https://doi.org/10.1007/978-3-662-49890-3_20
21. Cheon, J.H., Han, K., Lee, C., Ryu, H., Stehlé, D.: Cryptanalysis of the multilinear map over the integers. In: Oswald, E., Fischlin, M. (eds.) EUROCRYPT 2015. LNCS, vol. 9056, pp. 3–12. Springer, Heidelberg (2015). https://doi.org/10.1007/978-3-662-46800-5_1
22. Cheon, J.H., Jeong, J., Lee, C.: An algorithm for NTRU problems and cryptanalysis of the GGH multilinear map without an encoding of zero. Cryptology ePrint Archive, Report 2016/139 (2016)
23. Conder, M., Dobcsányi, P.: Applications and adaptations of the low index subgroups procedure. Math. Comput. **74**(249), 485–497 (2005)
24. Coron, J.-S., et al.: Zeroizing without low-level zeroes: new MMAP attacks and their limitations. In: Gennaro, R., Robshaw, M. (eds.) CRYPTO 2015. LNCS, vol. 9215, pp. 247–266. Springer, Heidelberg (2015). https://doi.org/10.1007/978-3-662-47989-6_12
25. Coron, J.-S., Lepoint, T., Tibouchi, M.: Practical multilinear maps over the integers. In: Canetti, R., Garay, J.A. (eds.) CRYPTO 2013. LNCS, vol. 8042, pp. 476–493. Springer, Heidelberg (2013). https://doi.org/10.1007/978-3-642-40041-4_26
26. Coron, J.-S., Lepoint, T., Tibouchi, M.: New multilinear maps over the integers. In: Gennaro, R., Robshaw, M. (eds.) CRYPTO 2015. LNCS, vol. 9215, pp. 267–286. Springer, Heidelberg (2015). https://doi.org/10.1007/978-3-662-47989-6_13
27. Cryan, M., Miltersen, P.B.: On pseudorandom generators in NC^0. In: Sgall, J., Pultr, A., Kolman, P. (eds.) MFCS 2001. LNCS, vol. 2136, pp. 272–284. Springer, Heidelberg (2001). https://doi.org/10.1007/3-540-44683-4_24

28. Feige, U.: Relations between average case complexity and approximation complexity. In: Proceedings of the Thirty-Fourth Annual ACM Symposium on Theory of Computing, pp. 534–543. ACM, New York (2002). http://dx.doi.org/10.1145/509907.509985. (electronic)
29. Feige, U.: Refuting smoothed 3CNF formulas. In: FOCS, pp. 407–417. IEEE Computer Society (2007)
30. Feige, U., Ofek, E.: Easily refutable subformulas of large random 3CNF formulas. Theory Comput. **3**, 25–43 (2007). https://doi.org/10.4086/toc.2007.v003a002
31. Feldman, V., Perkins, W., Vempala, S.: On the complexity of random satisfiability problems with planted solutions. In: STOC, pp. 77–86. ACM (2015)
32. Garg, S., Gentry, C., Halevi, S.: Candidate multilinear maps from ideal lattices. In: Johansson, T., Nguyen, P.Q. (eds.) EUROCRYPT 2013. LNCS, vol. 7881, pp. 1–17. Springer, Heidelberg (2013). https://doi.org/10.1007/978-3-642-38348-9_1
33. Garg, S., Gentry, C., Halevi, S., Raykova, M., Sahai, A., Waters, B.: Candidate indistinguishability obfuscation and functional encryption for all circuits. In: 54th Annual IEEE Symposium on Foundations of Computer Science, FOCS 2013, pp. 40–49 (2013). http://dx.doi.org/10.1109/FOCS.2013.13
34. Gentry, C., Gorbunov, S., Halevi, S.: Graph-induced multilinear maps from lattices. In: Dodis, Y., Nielsen, J.B. (eds.) TCC 2015. LNCS, vol. 9015, pp. 498–527. Springer, Heidelberg (2015). https://doi.org/10.1007/978-3-662-46497-7_20
35. Goldmann, M., Russell, A.: The complexity of solving equations over finite groups. Inf. Comput. **178**(1), 253–262 (2002). https://doi.org/10.1006/inco.2002.3173
36. Goldreich, O.: Candidate one-way functions based on expander graphs. In: Electronic Colloquium on Computational Complexity (ECCC), vol. 7, no. 90 (2000)
37. Grötschel, M., Lovász, L., Schrijver, A.: The ellipsoid method and its consequences in combinatorial optimization. Combinatorica **1**(2), 169–197 (1981). https://doi.org/10.1007/BF02579273
38. Hada, S.: Zero-knowledge and code obfuscation. In: Okamoto, T. (ed.) ASIACRYPT 2000. LNCS, vol. 1976, pp. 443–457. Springer, Heidelberg (2000). https://doi.org/10.1007/3-540-44448-3_34
39. Hu, Y., Jia, H.: Cryptanalysis of GGH map. In: Fischlin, M., Coron, J.-S. (eds.) EUROCRYPT 2016. LNCS, vol. 9665, pp. 537–565. Springer, Heidelberg (2016). https://doi.org/10.1007/978-3-662-49890-3_21
40. Ishai, Y., Kushilevitz, E., Ostrovsky, R., Prabhakaran, M., Sahai, A.: Efficient non-interactive secure computation. In: Paterson, K.G. (ed.) EUROCRYPT 2011. LNCS, vol. 6632, pp. 406–425. Springer, Heidelberg (2011). https://doi.org/10.1007/978-3-642-20465-4_23
41. Joux, A.: A one round protocol for tripartite Diffie–Hellman. In: Bosma, W. (ed.) ANTS 2000. LNCS, vol. 1838, pp. 385–393. Springer, Heidelberg (2000). https://doi.org/10.1007/10722028_23
42. Klíma, O., Tesson, P., Thérien, D.: Dichotomies in the complexity of solving systems of equations over finite semigroups. Theory Comput. Syst. **40**(3), 263–297 (2007)
43. Kothari, P.K., Mori, R., O'Donnell, R., Witmer, D.: Sum of squares lower bounds for refuting any CSP. In: Proceedings of the 49th Annual ACM SIGACT Symposium on Theory of Computing, STOC, pp. 132–145. ACM (2017)
44. Lasserre, J.B.: New positive semidefinite relaxations for nonconvex quadratic programs. In: Hadjisavvas, N., Pardalos, P.M. (eds.) Advances in Convex Analysis and Global Optimization. Nonconvex Optimization and Its Applications, vol. 54, pp. 319–331. Kluwer Academic Publishers, Dordrecht (2001). https://doi.org/10.1007/978-1-4613-0279-7_18. (Pythagorion, 2000)

45. Lin, H.: Indistinguishability obfuscation from constant-degree graded encoding schemes. In: Fischlin, M., Coron, J.-S. (eds.) EUROCRYPT 2016. LNCS, vol. 9665, pp. 28–57. Springer, Heidelberg (2016). https://doi.org/10.1007/978-3-662-49890-3_2
46. Lin, H.: Indistinguishability obfuscation from SXDH on 5-Linear maps and locality-5 PRGs. In: Katz, J., Shacham, H. (eds.) CRYPTO 2017. LNCS, vol. 10401, pp. 599–629. Springer, Cham (2017). https://doi.org/10.1007/978-3-319-63688-7_20
47. Lin, H., Tessaro, S.: Indistinguishability obfuscation from bilinear maps and block-wise local PRGs. IACR Cryptology ePrint Archive, p. 250 (2017)
48. Lin, H., Vaikuntanathan, V.: Indistinguishability obfuscation from DDH-like assumptions on constant-degree graded encodings. In: IEEE 57th Annual Symposium on Foundations of Computer Science, FOCS, pp. 11–20. IEEE Computer Society (2016)
49. Lombardi, A., Vaikuntanathan, V.: Limits on the locality of pseudorandom generators and applications to indistinguishability obfuscation. In: Kalai, Y., Reyzin, L. (eds.) TCC 2017. LNCS, vol. 10677, pp. 119–137. Springer, Cham (2017). https://doi.org/10.1007/978-3-319-70500-2_5
50. Lombardi, A., Vaikuntanathan, V.: Minimizing the complexity of Goldreich's pseudorandom generator. IACR Cryptology ePrint Archive, p. 277 (2017)
51. Miles, E., Sahai, A., Zhandry, M.: Annihilation attacks for multilinear maps: cryptanalysis of indistinguishability obfuscation over GGH13. In: Robshaw, M., Katz, J. (eds.) CRYPTO 2016. LNCS, vol. 9815, pp. 629–658. Springer, Heidelberg (2016). https://doi.org/10.1007/978-3-662-53008-5_22
52. Mossel, E., Shpilka, A., Trevisan, L.: On epsilon-biased generators in NC^0. Random Struct. Algorithms **29**(1), 56–81 (2006)
53. O'Donnell, R.: Analysis of Boolean Functions. Cambridge University Press, Cambridge (2014)
54. O'Donnell, R., Witmer, D.: Goldreich's PRG: evidence for near-optimal polynomial stretch. In: IEEE 29th Conference on Computational Complexity–CCC 2014, pp. 1–12. IEEE Computer Society, Los Alamitos, CA (2014). http://dx.doi.org/10.1109/CCC.2014.9
55. Parrilo, P.A.: Structured semidefinite programs and semialgebraic geometry methods in robustness and optimization. Ph.D. thesis, Citeseer (2000)
56. Raghavendra, P., Rao, S., Schramm, T.: Strongly refuting random CSPs below the spectral threshold. In: Proceedings of the 49th Annual ACM SIGACT Symposium on Theory of Computing, STOC 2017, pp. 121–131. ACM (2017)
57. Rozenman, E., Shalev, A., Wigderson, A.: Iterative construction of cayley expander graphs. Theory Comput. **2**(5), 91–120 (2006)
58. Sahai, A., Waters, B.: How to use indistinguishability obfuscation: deniable encryption, and more. In: Symposium on Theory of Computing, STOC, pp. 475–484. ACM (2014)
59. Shor, N.Z.: Quadratic optimization problems. Izv. Akad. Nauk SSSR Tekhn. Kibernet. **222**(1), 128–139 (1987)
60. Witmer, D.: On refutation of random constraint satisfaction problems (thesis proposal) (2017). http://www.cs.cmu.edu/~dwitmer/papers/proposal.pdf

Symmetric Cryptanalysis

Boomerang Connectivity Table: A New Cryptanalysis Tool

Carlos Cid[1(✉)], Tao Huang[2(✉)], Thomas Peyrin[2,3,4(✉)], Yu Sasaki[5(✉)], and Ling Song[2,3,6(✉)]

[1] Information Security Group, Royal Holloway, University of London, Egham, UK
carlos.cid@rhul.ac.uk
[2] School of Physical and Mathematical Sciences, Nanyang Technological University, Singapore, Singapore
{huangtao,thomas.peyrin}@ntu.edu.sg
[3] Temasek Laboratories, Nanyang Technological University, Singapore, Singapore
[4] School of Computer Science and Engineering, Nanyang Technological University, Singapore, Singapore
[5] NTT Secure Platform Laboratories, Tokyo, Japan
sasaki.yu@lab.ntt.co.jp
[6] State Key Laboratory of Information Security, Institute of Information Engineering, Chinese Academy of Sciences, Beijing, China
songling@iie.ac.cn

Abstract. A boomerang attack is a cryptanalysis framework that regards a block cipher E as the composition of two sub-ciphers $E_1 \circ E_0$ and builds a particular characteristic for E with probability p^2q^2 by combining differential characteristics for E_0 and E_1 with probability p and q, respectively. Crucially the validity of this figure is under the assumption that the characteristics for E_0 and E_1 can be chosen independently. Indeed, Murphy has shown that independently chosen characteristics may turn out to be incompatible. On the other hand, several researchers observed that the probability can be improved to p or q around the boundary between E_0 and E_1 by considering a positive dependency of the two characteristics, e.g. the ladder switch and S-box switch by Biryukov and Khovratovich. This phenomenon was later formalised by Dunkelman et al. as a sandwich attack that regards E as $E_1 \circ E_m \circ E_0$, where E_m satisfies some differential propagation among four texts with probability r, and the entire probability is p^2q^2r. In this paper, we revisit the issue of dependency of two characteristics in E_m, and propose a new tool called *Boomerang Connectivity Table (BCT)*, which evaluates r in a systematic and easy-to-understand way when E_m is composed of a single S-box layer. With the BCT, previous observations on the S-box including the incompatibility, the ladder switch and the S-box switch are represented in a unified manner. Moreover, the BCT can detect a new switching effect, which shows that the probability around the boundary may be even higher than p or q. To illustrate the power of the

J. B. Nielsen and V. Rijmen (Eds.): EUROCRYPT 2018, LNCS 10821, pp. 683–713, 2018.
https://doi.org/10.1007/978-3-319-78375-8_22

BCT-based analysis, we improve boomerang attacks against Deoxys-BC, and disclose the mechanism behind an unsolved probability amplification for generating a quartet in SKINNY. Lastly, we discuss the issue of searching for S-boxes having good BCT and extending the analysis to modular addition.

Keywords: Boomerang attack · Differential distribution table S-box · Incompatibility · Ladder switch · S-box switch · Deoxys SKINNY

1 Introduction

Differential cryptanalysis, proposed by Biham and Shamir in the early 1990s [BS93], remains one of the most fundamental cryptanalytic approaches for assessing the security of block ciphers. For iterated ciphers based on predefined substitution tables (S-box), resistance against differential cryptanalysis is highly dependent on the non-linearity features of the S-box.

For an n-bit S-box $S : \{0,1\}^n \mapsto \{0,1\}^n$, the properties for differential propagations of S are typically represented in the $2^n \times 2^n$ table $\mathcal{T}$, called the *Difference Distribution Table (DDT)*. For any pair (Δ_i, Δ_o), the value

$$\#\{x \in \{0,1\}^n | S(x) \oplus S(x \oplus \Delta_i) = \Delta_o\}$$

is stored in the corresponding entry $\mathcal{T}(\Delta_i, \Delta_o)$ of the DDT, representing that the input difference Δ_i propagates to the output difference Δ_o with probability

$$\frac{\mathcal{T}(\Delta_i, \Delta_o)}{2^n}. \tag{1}$$

The maximum entry in the table $\mathcal{T}$ (outside the first row and column) is called the *differential uniformity* of S.

As an example, the DDT for the 4-bit S-box used in PRESENT [BKL+07] and LED [GPPR11] is shown in Table 1. We can observe that the differential uniformity of the S-box is 4.

While Eq. (1) represents the differential propagation property for a single S-box, in order to derive the differential properties of the entire cipher, a trail of high-probability differentials is searched through the cipher iteration, by assuming that the S-boxes and other operations applied in different rounds behave independently.

In many cases, it may not be possible to find a high-probability trail for the entire cipher. In such cases, the *Boomerang attack* framework, proposed by Wagner [Wag99], may be applied to exploit the differential properties of different segments of the cipher. In a boomerang attack, the target cipher E is regarded as a composition of two sub-ciphers E_0 and E_1, i.e. $E = E_1 \circ E_0$. Then suppose that the input difference α is propagated to the difference β by E_0 with probability p, while the difference γ is propagated to δ by E_1 with

Table 1. Difference Distribution Table (DDT) of the PRESENT S-box

Δ_i \ Δ_o	0	1	2	3	4	5	6	7	8	9	a	b	c	d	e	f
0	16	0	0	0	0	0	0	0	0	0	0	0	0	0	0	0
1	0	0	0	4	0	0	0	4	0	4	0	0	0	4	0	0
2	0	0	0	2	0	4	2	0	0	0	2	0	2	2	2	0
3	0	2	0	2	2	0	4	2	0	0	2	2	0	0	0	0
4	0	0	0	0	0	4	2	2	0	2	2	0	2	0	2	0
5	0	2	0	0	2	0	0	0	0	2	2	2	4	2	0	0
6	0	0	2	0	0	0	2	0	2	0	0	4	2	0	0	4
7	0	4	2	0	0	0	2	0	2	0	0	0	2	0	0	4
8	0	0	0	2	0	0	0	2	0	2	0	4	0	2	0	4
9	0	0	2	0	4	0	2	0	2	0	0	0	2	0	4	0
a	0	0	2	2	0	4	0	0	2	0	2	0	0	2	2	0
b	0	2	0	0	2	0	0	0	4	2	2	2	0	2	0	0
c	0	0	2	0	0	4	0	2	2	2	2	0	0	0	2	0
d	0	2	4	2	2	0	0	2	0	0	2	2	0	0	0	0
e	0	0	2	2	0	0	2	2	2	2	0	0	2	2	0	0
f	0	4	0	0	4	0	0	0	0	0	0	0	0	0	4	4

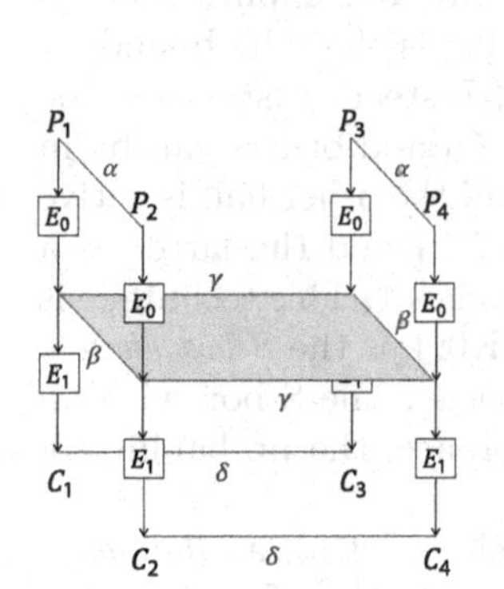

Fig. 1. Boomerang attack

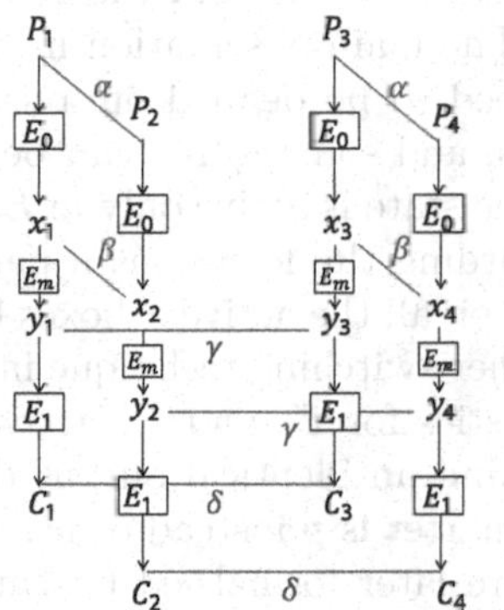

Fig. 2. Sandwich attack

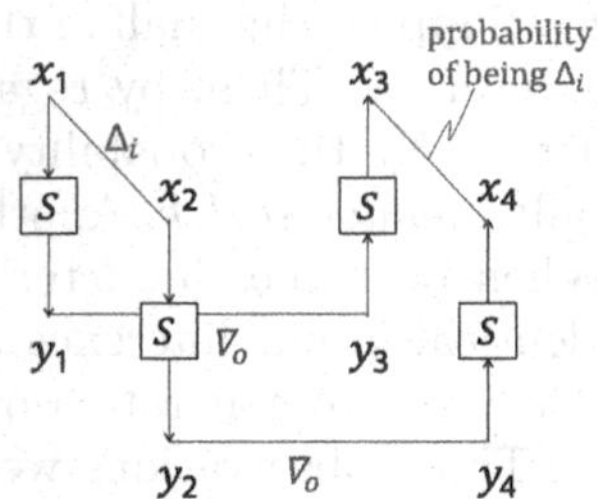

Fig. 3. Computation of r when E_m is an S-box layer

probability q. The boomerang attack exploits the expected probability of the following differential (depicted in Fig. 1):

$$\Pr\big[E^{-1}\big(E(x) \oplus \delta\big) \oplus E^{-1}\big(E(x \oplus \alpha) \oplus \delta\big) = \alpha\big] = p^2q^2. \tag{2}$$

Then, on making around $(pq)^{-2}$ adaptive chosen plaintext/ciphertext queries, E can be distinguished from an ideal cipher.

Variants of the boomerang attack were later proposed: the amplified boomerang attack (also called 'the rectangle attack') works in a chosen-plaintext scenario and a right quartet is expected to be obtained with probability $p^2q^22^{-n}$ [KKS00]. Further, it was pointed out in [BDK01,BDK02] that any values of β and γ are allowed as long as $\beta \neq \gamma$. As a result, the probability of the right quartet increases to $2^{-n}\hat{p}^2\hat{q}^2$, where $\hat{p} = \sqrt{\Sigma_i \Pr^2(\alpha \rightarrow \beta_i)}$ and $\hat{q} = \sqrt{\Sigma_j \Pr^2(\gamma_j \rightarrow \delta)}$.

In boomerang-style attacks, the most important part of the attack is selecting suitable differential characteristics for E_0 and E_1. Initially, the standard assumption used in boomerang-style attacks was that two characteristics independently chosen for E_0 and E_1 could be used; as a result the typical attacker's strategy was to optimise the best characteristics independently for the sub-ciphers E_0 and E_1. However, Murphy [Mur11] pointed out that, for S-box based ciphers, two independently chosen characteristics can be *incompatible*, thus the probability of generating a right quartet can be zero. He also showed that the dependency between two characteristics could give advantages for the attacker, giving an example that the probability of generating a quartet was pq instead of p^2q^2 when E_0 and E_1 are composed of a single S-box. The same phenomenon was observed by Biryukov et al. as the *middle round S-box trick* [BCD03].

Another improvement, proposed by Biryukov and Khovratovich [BK09], was named the *boomerang switch*. Suppose that the cipher state is composed of several words (typically 8 bits or 4 bits) and the round function applies S-boxes to each word in parallel. The main observation in [BK09] is that the boundary of E_0 and E_1 does not need to be defined on a state. Instead, a state can be further divided into words, and some words can be in E_0 and others can be in E_1. Suppose that half of the state is active only in E_0 and the other half is active only in E_1. Then, by regarding the former as a part of E_1 and the latter as a part of E_0, the probability on all the active S-boxes becomes 1. This technique is called *ladder switch*. Another switching technique in [BK09], is the *S-box switch*. When both the characteristics for E_0 and E_1 activate the same S-box with an identical input difference and an identical output difference, the probability of this S-box to generate a quartet is p instead of p^2.

Those observations were later formalised by Dunkelman et al. as *the sandwich attack* [DKS10, DKS14] depicted in Fig. 2, that regards E as $E_1 \circ E_m \circ E_0$, where E_m is a relatively short operation satisfying some differential propagation among four texts with probability r, and the entire probability is p^2q^2r. Let (x_1, x_2, x_3, x_4) and (y_1, y_2, y_3, y_4) be input and output quartet values for E_m, where $y_i = E_m(x_i)$. The differential characteristics for E_0 specify the input differences α to E_m, namely $x_1 \oplus x_2 = x_3 \oplus x_4 = \alpha$, and E_1 specifies the output differences β to each S-box, namely $y_1 \oplus y_3 = y_2 \oplus y_4 = \beta$. Dunkelman et al. define r as follows [DKS10, Eq. (4)].

$$r = \Pr\Big[(x_3 \oplus x_4) = \beta | (x_1 \oplus x_2 = \beta) \wedge (y_1 \oplus y_3 = \gamma) \wedge (y_2 \oplus y_4 = \gamma)\Big] \tag{3}$$

Boomerang-style attacks have become an ever more popular cryptanalytic method for assessing the security of block ciphers. Yet, considering the research results above, we note the following questions that arise in their context:

- the probability r of the middle part E_m is for a quartet. Then, how can we evaluate r in an efficient and systematic way? The only known approach is to run experiments as in [DKS10,DKS14,BDK01,KHP+12].
- are there other switching techniques that can be used to improve boomerang-style attacks? In particular, can we find switching techniques that connect two characteristics with even higher probability than the S-box switch and Murphy's examples?

Answer to these questions would be of course of great interest to researchers working on block cipher cryptanalysis, but also significant to provide a deeper understanding of the subtleties of boomerang-style attacks. Besides, it also contributes to block ciphers designers by taking into account this property as a criterion to choose a good S-box.

Our Contributions. This paper positively answers the above questions by proposing a new tool for evaluating the probability that boomerang-style quartets are generated. While we focus mainly on explaining the effects against ciphers employing S-boxes, we also present the extension to analyse ciphers based on modular addition.

Suppose that the middle layer E_m of the sandwich attack is composed of a single S-box layer. Then, for a given pair of (Δ_i, ∇_o), the probability that a right quartet is generated in each S-box in the middle S-box layer is given by:

$$\frac{\#\{x \in \{0,1\}^n | S^{-1}(S(x) \oplus \nabla_o) \oplus S^{-1}(S(x \oplus \Delta_i) \oplus \nabla_o) = \Delta_i\}}{2^n}, \quad (4)$$

where $S : \{0,1\}^n \mapsto \{0,1\}^n$ is an n-bit to n-bit S-box and S^{-1} is its inverse. What Eq. (4) evaluates is illustrated in Fig. 3, which is exactly r in Eq. (3) when E_m is a single S-box layer. Note that the differences for E_0 and E_1 are defined between different paired values, thus we use Δ and ∇ to denote the differences of E_0 and E_1, respectively. We also note that in the figures we mainly use red and blue colours to describe Δ and ∇, respectively. The denominator is 2^n instead of 2^{2n}, which shows the implication of the sandwich attack that the probability r of generating a right quartet in E_m is at least 2^{-n} (if not 0).

Similar to the DDT, we can of course evaluate Eq. (4) for all pairs of (Δ_i, ∇_o), storing the results (in fact the numerator) in a table. We call this table the *Boomerang Connectivity Table (BCT)*. The BCT for the PRESENT S-box is shown in Table 2.

The BCT represents the observations by [Mur11,BK09] in a unified manner.

Incompatibility. (Δ_i, ∇_o) is incompatible when the corresponding entry in the BCT is 0.

Ladder switch. It corresponds to the first row and the first column of the BCT, in which either one of the input or output difference is zero, while the other is non-zero. As suggested by Table 2, Eq. (4) gives probability 1.

S-box switch. It corresponds to the claim that a DDT entry with non-zero value v would imply that the corresponding BCT entry is v. While this is

Table 2. Boomerang Connectivity Table (BCT) of the PRESENT S-box

Δ_i \ ∇_o	0	1	2	3	4	5	6	7	8	9	a	b	c	d	e	f
0	16	16	16	16	16	16	16	16	16	16	16	16	16	16	16	16
1	16	0	4	4	0	16	4	4	4	4	0	0	4	4	0	0
2	16	0	0	6	0	4	6	0	0	0	2	0	2	2	2	0
3	16	2	0	6	2	4	4	2	0	0	2	2	0	0	0	0
4	16	0	0	0	0	4	2	2	0	6	2	0	6	0	2	0
5	16	2	0	0	2	4	0	0	0	6	2	2	4	2	0	0
6	16	4	2	0	4	0	2	0	2	0	0	4	2	0	4	8
7	16	4	2	0	4	0	2	0	2	0	0	4	2	0	4	8
8	16	4	0	2	4	0	0	2	0	2	0	4	0	2	4	8
9	16	4	2	0	4	0	2	0	2	0	0	4	2	0	4	8
a	16	0	2	2	0	4	0	0	6	0	2	0	0	6	2	0
b	16	2	0	0	2	4	0	0	4	2	2	2	0	6	0	0
c	16	0	6	0	0	4	0	6	2	2	2	0	0	0	2	0
d	16	2	4	2	2	4	0	6	0	0	2	2	0	0	0	0
e	16	0	2	2	0	0	2	2	2	2	0	0	2	2	0	0
f	16	8	0	0	8	0	0	0	0	0	0	8	0	0	8	16

correct in some cases, as it can be observed from the two tables, we show that the value of the BCT can in fact be larger than v owing to the new switching effect. However, at least the effect of S-box switch is always guaranteed.

Study of the BCT can also present more advantages to the attacker compared to the previously known switching techniques. With respect to versatility, the BCT shows that the switching effect can be applied even when Δ_i cannot be propagated to Δ_o in the DDT. With respect to strength, the maximum probability in the BCT is usually higher than that of the DDT. For example, the DDT in Table 1 has entry 0 for $(\Delta_i, \Delta_o) = (\mathtt{1}, \mathtt{5})$, while the BCT in Table 2 for $(\Delta_i, \nabla_o) = (\mathtt{1}, \mathtt{5})$ gives us probability 1. As far as the authors are aware, such an event has never been pointed out in previous works, and we expect that many existing boomerang attacks can be improved by considering superior switching effects represented in the BCT. To illustrate this point, we show in this paper how to improve the boomerang attack against 10-round `Deoxys-BC-384` which was recently presented [CHP+17]. We also use the BCT with related-tweakey boomerang characteristics of `SKINNY-64` and `SKINNY-128` presented by [LGL17]. The BCT allows us to accurately evaluate the amplification of the probability of forming distinguishers. As a result, we detect flaws on the experimentally evaluated probability in [LGL17], and probabilities for `SKINNY-64` are improved.

To better understand the relationship between the DDT and the BCT, we consider the problem of finding an S-box such that the maximum probability in the BCT is the same as one in the DDT. We show that while 2-uniform DDT always derives 2-uniform BCT, finding such an S-box with 4-uniform DDT is hard especially when the size of the S-box increases. Finally, we discuss the application of our idea to the modular addition operation. We show that the ladder switch observed for the S-box based designs can be applied to the modular addition, while the S-box switch cannot be applied. We also find a new switching mechanism called *MSB-switch* for modular addition which generates a right quartet with probability 1.

Finally, we would like to emphasise that the BCT should not be considered only from the attackers' point-of-view. One major feature of our approach is that the BCT can (and should) also be considered by designers. A block-cipher designer need to evaluate many S-box choices according to various criteria. The simple form of the BCT, which allows one to measure the strength of the S-box against boomerang-style attacks independently from the other components (not too dissimilar to the relation between differential cryptanalysis and an S-box DDT) will be of great benefit to designers as well.

Outline. In Sect. 2, we give a brief overview of related work. Section 3 introduces the boomerang connectivity table as a new method to evaluate the probability of two differential characteristics, and explains the mechanisms based on which our improved switching technique can work. The BCT is applied to `Deoxys` and `SKINNY` in Sects. 4 and 5. We then discuss difficulties in finding 4-uniform BCT and extends the analysis to modular addition in Sect. 6. We present our conclusions in Sect. 7.

2 Previous Work

The boomerang attack, originally proposed by Wagner [Wag99], was extended to the related-key setting and was formalised in [BDK05] by using four related-key oracles K_1, $K_2 = K_1 \oplus \Delta K$, $K_3 = K_1 \oplus \nabla K$ and $K_4 = K_1 \oplus \Delta K \oplus \nabla K$.

Let $E_K(P)$ and $D_K(C)$ denote the encryption of P and the decryption of C under a key K, respectively. In the framework, a pair (P_1, P_2) with plaintext difference Δ_i is first queried to E_{K_1} and E_{K_2} to produce (C_1, C_2). Then (C_3, C_4) is computed from (C_1, C_2) by xoring ∇_o, and queried to decryption oracles D_{K_3} and D_{K_4} to produce (P_3, P_4). With probability p^2q^2, where $p^2q^2 > 2^{-n}$, the pair (P_3, P_4) will have difference Δ_i, and the cipher may be distinguished. The pseudo-code of the related-key boomerang attack is given below.

1. $\kappa_1 \leftarrow random(), \kappa_2 \leftarrow \kappa_1 \oplus \Delta K, \kappa_3 \leftarrow \kappa_1 \oplus \nabla K, \kappa_4 \leftarrow \kappa_1 \oplus \Delta K \oplus \nabla K$.
2. Repeat the following steps $\mathcal{N}$ times, where $\mathcal{N} \geq (pq)^{-2}$.
3. $P_1 \leftarrow random()$ and $P_2 \leftarrow P_1 \oplus \Delta_P$.
4. $C_1 \leftarrow E_{\kappa_1}(P_1)$ and $C_2 \leftarrow E_{\kappa_2}(P_2)$.
5. $C_3 \leftarrow C_1 \oplus \nabla_C$ and $C_4 \leftarrow C_2 \oplus \nabla_C$.
6. $P_3 \leftarrow D_{\kappa_3}(C_3)$ and $P_4 \leftarrow D_{\kappa_4}(C_4)$.
7. Check if $P_3 \oplus P_4 = \Delta_P$.

Boomerang-style attacks have been widely considered in symmetric-key cryptanalysis, and thus we refrain from providing a complete list of previous works that apply the technique. A few noticeable examples of boomerang-style attacks against block ciphers include [BDK05,BK09,LGL17,CHP+17].

3 BCT – Boomerang Connectivity Table

In this section we introduce our novel idea, the Boomerang Connectivity Table (BCT), which can be used to more accurately evaluate the probability of generating a right quartet in boomerang-style attacks. As briefly explained in Sect. 1, the BCT is constructed by directly computing the probabilities for generating boomerang quartets at the local level (Eq. 4), and thus provides more useful information for boomerang attacks when compared to the DDT, which was typically used in previous works.

3.1 Definition of the BCT

As illustrated in Fig. 3, we consider the case where the input difference to the S-box, Δ_i, is defined by the sub-cipher E_0 and the output difference from the S-box, ∇_o, is defined by E_1. The important observation is that when one of the input values to the S-box is fixed, all the values in the quartet are fixed. Hence, the generation of the right quartet is a probabilistic event, which we can compute as:

$$\frac{\#\{x \in \{0,1\}^n | S^{-1}(S(x) \oplus \nabla_o) \oplus S^{-1}(S(x \oplus \Delta_i) \oplus \nabla_o) = \Delta_i\}}{2^n}.$$

The table that stores the results of this equation for all (Δ_i, ∇_o) is useful in the analysis of the target cipher. We call it "Boomerang Connectivity Table (BCT)".

Definition 3.1 (Boomerang Connectivity Table). *Let $S : \{0,1\}^n \to \{0,1\}^n$ be an invertible function, and $\Delta_i, \nabla_o \in \{0,1\}^n$. The Boomerang Connectivity Table (BCT) of S is given by a $2^n \times 2^n$ table $\mathcal{T}$, in which the entry for the (Δ_i, ∇_o) position is given by*

$$\mathcal{T}(\Delta_i, \nabla_o) = \#\{x \in \{0,1\}^n | S^{-1}(S(x) \oplus \nabla_o) \oplus S^{-1}(S(x \oplus \Delta_i) \oplus \nabla_o) = \Delta_i\}.$$

The BCT for the PRESENT S-box is shown in Table 2. We note that the complexity for generating the BCT for an n-bit to n-bit S-box is $O(2^{3n})$, which is higher than $O(2^{2n})$ for the DDT.

The BCT provides a unified representation of existing observations on quartet generation/probabilities for boomerang-style attacks, which can be easily detected on analysis of the cipher's S-box BCT.

Incompatibility. In previous works, the compatibility or incompatibility of (Δ_i, ∇_o) noted in [Mur11] would be typically checked experimentally. This can however be observed directly in the BCT: the difference pair (Δ_i, ∇_o) is incompatible if the corresponding entry of the BCT is 0.

Ladder switch. The value in any entry in the first row and the first column of the BCT is 2^n. This corresponds to the ladder switch proposed in [BK09]. This probability 1 transition can also be explained in the way of Fig. 3.

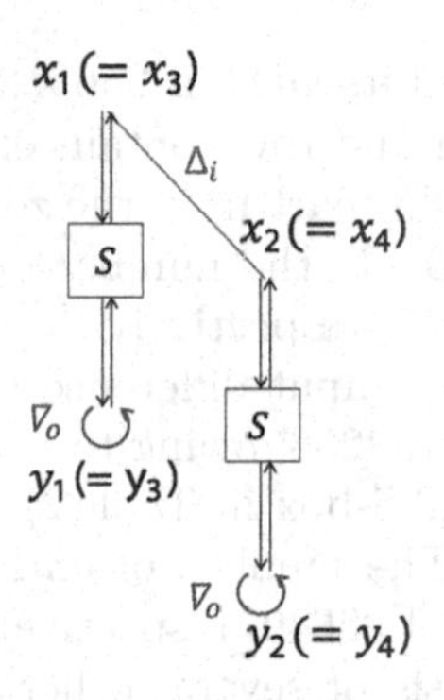

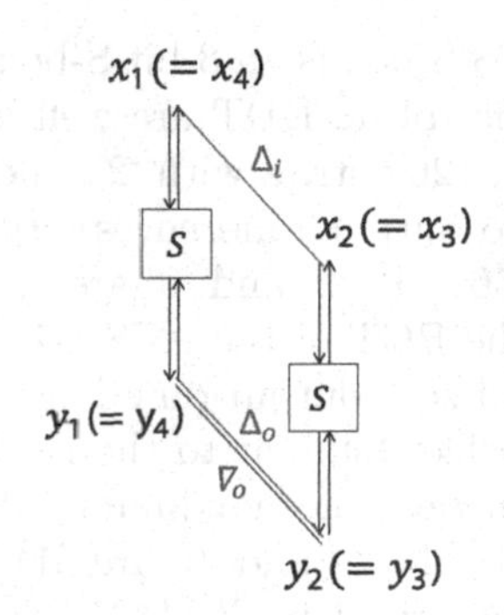

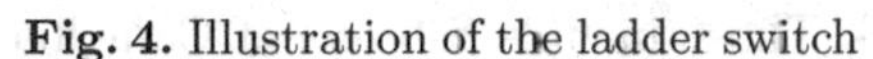

Fig. 4. Illustration of the ladder switch

Fig. 5. Illustration of the S-box switch

The case with $\Delta_i \neq 0$ and $\nabla_o = 0$ is illustrated in Fig. 4. As we can observe, for any choice of x_1 and $x_2(= x_1 \oplus \Delta_i)$, we have their images y_1 and y_2 after the S-box application. Now since $\nabla_o = 0$, no modification is made to create y_3 and y_4, and thus after S^{-1} is applied, the paired values get back to (x_1, x_2) with probability 1, and the second pair will always satisfy Δ_i. The same holds when $\Delta_i = 0$ and $\nabla_o \neq 0$.

S-box switch. The S-box switch can be explained in the context of the BCT as follows: if the DDT entry for (Δ_i, Δ_o) is non-zero, then by setting $\nabla_o = \Delta_o$, the BCT entry for (Δ_i, ∇_o) will take the same value. The mechanism of the S-box switch is the same as explained in [BK09], but here we explain it in the way of Fig. 3, which will be useful to understand our new switching effects presented later. As illustrated in Fig. 5, suppose that two input values x_1 and $x_2(= x_1 \oplus \Delta_i)$ are mapped to y_1 and y_2 satisfying $y_1 \oplus y_2 = \Delta_o$ with probability p. By setting $\nabla_o = \Delta_o$, y_3 and y_4 are computed by $y_1 \oplus \Delta_o$ and $y_2 \oplus \Delta_o$. This merely switches y_1 and y_2, and after S^{-1} is applied, the paired values become (x_2, x_1) with probability 1, and thus the second pair always satisfies Δ_i.

The above analysis, especially for the S-box switch, can be summarised as the following lemma about the relationship between the DDT and the BCT.

Lemma 1. *For any choice of (Δ_i, Δ_o), the value in the BCT is greater than or equal to the one in the DDT.*

Proof. The lemma is trivially valid when the value in the DDT is 0, or when $(\Delta_i, \Delta_o) = (0, 0)$. For the other non-zero DDT entries, the lemma follows from the discussion for the S-box switch above. □

BCT of the AES S-box. Because the PRESENT S-box does not offer the strongest resistance against maximum differential and linear probabilities, it may be interesting to study the properties of the BCT of the AES S-box, for example.

The AES S-box is an 8-bit S-box, and thus the size of its DDT is 256×256. The properties of its DDT are well known: each column and row contain one entry with '4', 126 entries with '2', and the remaining is '0' (apart from the zero input and zero output differences). Hence in the entire DDT, the number of entries with '256', '4', '2' and '0' are 1, 255, 32130 and 33150, respectively.

In the BCT of the AES S-box, all entries for zero input difference (the first row) and zero output difference (the first column) are '256' owing to the ladder switch effect (similar to the BCT for the PRESENT S-box in Table 2). For the other entries, the maximum value of BCT is '6'. The number of entries with '256', '6', '4', '2' and '0' are 511, 510, 255, 31620 and 32640, respectively; these are summarised in Table 3. We also list the analysis of several other S-boxes having the same DDT structure in Table 3. Those include S-boxes of Camellia [AIK+00], TWINE [SMMK12], and Lilliput [BFMT16].

Table 3. Number of entries for each value for the DDT and BCT for the S-boxes in AES, Camellia, TWINE and Lilliput

Cipher	Table	256	6	4	2	0
AES	DDT	1	-	255	32130	33150
	BCT	511	510	255	31620	32640
Camellia	DDT	1	-	255	32130	33150
	BCT	511	510	255	31620	32640

Cipher	Table	16	6	4	2	0
TWINE	DDT	1	-	15	90	150
	BCT	31	30	15	60	120
Lilliput	DDT	1	-	15	90	150
	BCT	31	30	15	60	120

In Table 3, the following two facts deserve careful attention.

- The maximum non-trivial value in the BCT is '6', which is higher than the one in DDT. It means that for some Δ_i and $\Delta_o = \nabla_o$, generating a right quartet against an S-box can be easier than satisfying a differential transition for a pair.
- The number of zero entries in the BCT is smaller than in DDT. This means that even if DDT for (Δ_i, Δ_o) is 0, by setting $\nabla_o = \Delta_o$, a right quartet can be generated with (Δ_i, ∇_o).

The mechanisms behind these properties will be explained in the next subsection.

3.2 Increased Probability with Generalized Switching Effect

As shown in Lemma 1, each BCT entry may have a higher value than the corresponding entry in the DDT. This is caused by a new switching effect, but can be easily detected by considering the BCT.

Let us focus on the DDT entry for (Δ_i, Δ_o) whose value is '4,' namely Δ_i is propagated to Δ_o with probability 2^{-n+2}. In this case, there are two paired values such that the input difference is Δ_i and the difference after the S-box is Δ_o. This situation is illustrated in Fig. 6.

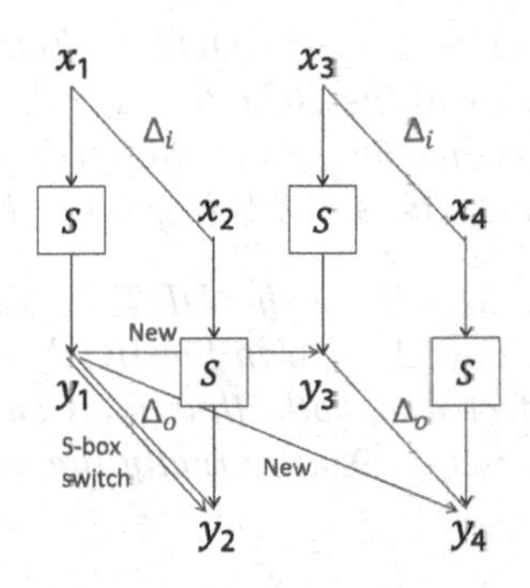

Fig. 6. Generalized switching effect: S-box switch and new switch

Let $\mathcal{X}_{DDT}(\Delta_i, \Delta_o)$ and $\mathcal{Y}_{DDT}(\Delta_i, \Delta_o)$ be a set of paired values satisfying the differential transition from Δ_i to Δ_o.

$$\mathcal{X}_{\mathrm{DDT}}(\Delta_i, \Delta_o) \triangleq \{(a, b) \in \{0,1\}^n \times \{0,1\}^n : S(a) \oplus S(b) = \Delta_o, a \oplus b = \Delta_i\},$$
$$\mathcal{Y}_{\mathrm{DDT}}(\Delta_i, \Delta_o) \triangleq \{(S(a), S(b)) \in \{0,1\}^n \times \{0,1\}^n : S(a) \oplus S(b) = \Delta_o, a \oplus b = \Delta_i\}.$$

In the example in Fig. 6, we have $\mathcal{X}_{\mathrm{DDT}}(\Delta_i, \Delta_o) = \{(x_1, x_2), (x_3, x_4)\}$ and $\mathcal{Y}_{\mathrm{DDT}}(\Delta_i, \Delta_o) = \{(y_1, y_2), (y_3, y_4)\}$.

Recall the strategy of the S-box switch, which sets $\nabla_o = \Delta_o$. Then for any $\mathcal{Y}_{\mathrm{DDT}}(\Delta_i, \Delta_o)$, $\mathcal{Y}_{\mathrm{DDT}} \oplus \nabla_o = \mathcal{Y}_{\mathrm{DDT}}$. Thus after the application of the inverse S-box, they will map back to $\mathcal{X}_{\mathrm{DDT}}(\Delta_i, \Delta_o)$. The essence of the S-box switch is finding a ∇_o for which $\mathcal{Y}_{\mathrm{DDT}} \oplus \nabla_o = \mathcal{Y}_{\mathrm{DDT}}$. Our observation of the generalized switching effect is that from two pairs in $\mathcal{Y}_{\mathrm{DDT}}(\Delta_i, \Delta_o)$, there are three ways to define such ∇_o:

$$\nabla_o \in \{y_1 \oplus y_2, y_1 \oplus y_3, y_1 \oplus y_4\}. \tag{5}$$

While one corresponds to the known S-box switch, the other two are new. Those choices of ∇_o are illustrated in Fig. 6.

Thus, one entry of value '4' for Δ_i in the DDT will increase the value of two entries in the BCT, namely $(\Delta_i, y_1 \oplus y_3)$ and $(\Delta_i, y_1 \oplus y_4)$ by 4. Note that the BCT entry for $(\Delta_i, y_1 \oplus y_2)$ becomes '4', but the DDT of this entry is already '4' and we do not get an increase by 4. Let $y_{new} \in \{y_3, y_4\}$ and ℓ be a non-negative integer. The generalized switching effect can thus be summarised as follows:

$$\text{DDT for } (\Delta_i, y_1 \oplus y_{new}) \text{ is } 2\ell \Rightarrow \text{BCT for } (\Delta_i, y_1 \oplus y_{new}) \text{ is } 2\ell + 4.$$

From the above analysis, we obtain the following lemma about the relationship between the DDT and the BCT of an S-box.

Lemma 2. *For any fixed Δ_i, for each entry with '4' in the DDT, the value of two positions in the BCT will increase by 4.*

We omit the proof (it follows from the discussion above). We use instead the examples below to illustrate the lemma.

Example 1. *The row for $\Delta_i = 2$ in the DDT in Table 1 contains an entry with '4.' This increases two entries of the BCT for $\Delta_i = 2$. In fact, values for $\Delta_o = 3$ and $\Delta_o = 6$ in the BCT increase by 4 from the DDT, while the other non-trivial entries for $\Delta_i = 2$ are exactly the same between the DDT and the BCT.*

Example 2. *The row for $\Delta_i = 9$ in the DDT in Table 1 contains two entries with '4.' Values for $\Delta_o = 1$ and $\Delta_o = \mathtt{b}$ in the BCT increase by 4 from the DDT. The value for $\Delta_o = \mathtt{f}$ is affected by both, thus increases by 8 from the DDT. The other non-trivial entries for $\Delta_i = 9$ are exactly the same between the DDT and BCT.*

Note that Lemma 2 is about fixed Δ_i, but considering the symmetry, the same applies to any fixed ∇_o. In this paper, we omit lemmas for fixed ∇_o.

For 4-bit S-boxes, we propose a sufficient condition such that the S-box is free (has probability 1) with non-zero input and output differences using BCT.

Lemma 3. *For any 4-bit S-box, if the DDT has a row for some input difference Δ_i such that there are 4 entries of '4', then there exists an output difference ∇_o, such that (Δ_i, ∇_o) has probability 1 in the boomerang switch of this 4-bit S-box.*

Proof. Since the DDT has a row with 4 entries of '4' for some input difference Δ_i, we divide the input values into 4 sets $V_j = \{a_j, b_j, c_j, d_j\}$, s.t. $a_j \oplus b_j = c_j \oplus d_j = \Delta_i$ and $S(a_j) \oplus S(b_j) = S(c_j) \oplus S(d_j)$ for $j = 1, 2, 3, 4$. Each V_j corresponds to a boomerang quartet. Let $T_j = \{x_j, y_j, z_j, w_j\}$ for $j = 1, 2, 3, 4$ be the sets of output after S-box corresponding to V_j. Then $x_j \oplus y_j = z_j \oplus w_j = \Delta_{o,j}$ holds. Note that the row for Δ_i of DDT will have 4 non-zero entries in the columns for $\Delta_{o,j}$ for $j = 1, 2, 3, 4$. We define the 4 sets $D_j = \{x_j \oplus y_j, x_j \oplus z_j, x_j \oplus w_j\}$ for $j = 1, 2, 3, 4$ to store the XOR difference of T_j. Set $\nabla_o = D_1 \cap D_2 \cap D_3 \cap D_4$. Then if $\nabla_o \neq \emptyset$, (Δ_i, ∇_o) generates a quartet with probability 1. In fact, for any input value $x \in V_j$ with difference (Δ_i, ∇_o), we can verify that the second pair in the quartet will be exactly V_j. For example, suppose that the input pair $(x, x \oplus \Delta_i)$ is (a_1, b_1), the output will be (x_1, y_1). When ∇_o is applied, (x_1, y_1) must be changed to one of the $\{(y_1, x_1), (z_1, w_1), (w_1, z_1)\}$, which will have difference Δ_i after inverse S-box.

Then we only need to prove that $\nabla_o \neq \emptyset$ under the assumption that the DDT has 4 entries of '4' for Δ_i. Here we prove it experimentally as the mathematical proof is not trivial. While the number of all possible output of 4-bit bijective S-box is 16!, only those satisfying the condition imposed on T_j need to be checked. This greatly reduces the search space. We first choose 4 numbers from 0 to 15 as $\{x_1, y_1, z_1, w_1\}$. There are $Perm(16, 4) = 43680$ possible choices. But only 3360 choices satisfy $x_1 \oplus y_1 = z_1 \oplus w_1$, which are the valid choices of T_1. Similarly we can generate T_j for $j = 2, 3, 4$. The total number of valid $(T_1, ..., T_4)$ is around 2^{30}. Then we can compute D_j for $j = 1, 2, 3, 4$ and verify if ∇_o is empty. It takes less than 1 hour on a desktop to check all possible valid $(T_1, ..., T_4)$. The result confirms that ∇_o is always non-empty. Therefore, we can conclude that the (Δ_i, ∇_o) has probability 1 with the generalized switching effect. □

Lemma 3 implies that having a row with four entries of '4' in the DDT may increase the power of the boomerang attack on those designs. This is an important observation since 4-bit S-boxes are widely used in lightweight designs.

Another observation is that the mechanism of the generalized switching effect requires the existence of a differential transition through the S-box with probability 2^{-n+2} or higher. In other words, the generalized switching effect does not exist in any 2-uniform DDT, which results in the following lemma.

Lemma 4. *For any S-box with 2-uniform DDT, the BCT is the same as the DDT but for the first row and the first column.*

We again omit the proof, and provide several examples.

Example 3. *The row for Δ_i = e in the DDT in Table 1 does not contain any entries with '4.' All the non-trivial entries for Δ_i = e are exactly the same between the DDT and BCT.*

Example 4. *When n is an odd number, n-bit S-boxes achieving 2-uniformity can be found easily. An example of such a 3-bit S-box is $S^{(3)}$ = [1, 7, 6, 3, 0, 2, 5, 4]. The DDT and the BCT of $S^{(3)}$ are shown in Tables 4 and 5, respectively, which clearly shows that besides the ladder switch, no generalized switching effect is available.*

Table 4. 2-uniform DDT of $S^{(3)}$

Δ_i \ Δ_o	0	1	2	3	4	5	6	7
0	8	0	0	0	0	0	0	0
1	0	2	2	0	0	2	2	0
2	0	0	0	0	2	2	2	2
3	0	2	2	0	2	0	0	2
4	0	2	0	2	0	2	0	2
5	0	0	2	2	0	0	2	2
6	0	2	0	2	2	0	2	0
7	0	0	2	2	2	2	0	0

Table 5. 2-uniform BCT of $S^{(3)}$

Δ_i \ ∇_o	0	1	2	3	4	5	6	7
0	8	8	8	8	8	8	8	8
1	8	2	2	0	0	2	2	0
2	8	0	0	0	2	2	2	2
3	8	2	2	0	2	0	0	2
4	8	2	0	2	0	2	0	2
5	8	0	2	2	0	0	2	2
6	8	2	0	2	2	0	2	0
7	8	0	2	2	2	2	0	0

3.3 Extension of Generalized Switching Effect to General DDT

The analysis in Sect. 3.2 applies only for the DDT whose maximum value is '4.' Although most of the existing S-boxes used in block ciphers were designed to satisfy this criterion, there has been a recent trend to weaken this criterion in order to achieve higher efficiency. For example, the 4-bit S-box of GIFT [BPP+17] and the 8-bit S-box of SKINNY [BJK+16] have DDT whose maximum entry is higher than '4.' Motivated by these designs, we further extend the analysis in Sect. 3.2 to any 2ℓ-uniform DDT for a non-negative integer ℓ.

Recall that in the previous section, we explained that from one quartet there are two new ways to define ∇_o such that the BCT entry for (Δ_i, ∇_o) is higher than the corresponding one in the DDT by 4. When the DDT contains an entry of 2ℓ, where $\ell \geq 2$, there are ℓ paired values that satisfy the differential propagation. Then, $\binom{\ell}{2}$ distinct quartets can be constructed from ℓ paired values, which is illustrated in Fig. 7 for $\ell = 3$.

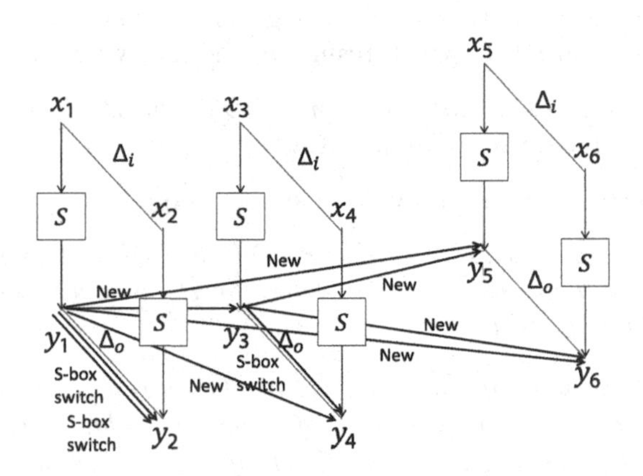

Fig. 7. Generalization of new switching effect. In total $\binom{3}{2} = 3$ distinct quartets are defined: $y_1y_2y_3y_4$ in blue, $y_1y_2y_5y_6$ in yellow, and $y_3y_4y_5y_6$ in green. Each quartet produces two new ways to define ∇_o. (Color figure online)

Each of the $\binom{\ell}{2}$ quartets gives two new ways to define ∇_o such that the BCT entry for (Δ_i, ∇_o) is higher than the DDT by 4. Thus Lemma 2 is generalised as follows.

Lemma 5. *For any fixed Δ_i, for each entry with '2ℓ' in the DDT, the value of $2 \cdot \binom{\ell}{2}$ non-trivial positions in the BCT increase by 4.*

Example 5. *A single row $\Delta_i = 4$ of the DDT and BCT for the GIFT S-box is shown in Table 6. The DDT contains a single entry of '6' and '4'. Lemma 5 can be used to predict the sum of all the entries of the same row in the BCT. Namely, '6' in the DDT increases the sum in the BCT by $4 \cdot 2 \cdot \binom{3}{2} = 24$ and '4' in the DDT increases the sum in the BCT by $4 \cdot 2 \cdot \binom{2}{2} = 8$. Along with the ladder switch for the first column, the sum of entries for the BCT should be higher than the one in the DDT by $48 (= 24 + 8 + 16)$, which matches the actual BCT.*

Example 6. *Application of Lemma 5 to the 8-bit S-box of SKINNY-128 is discussed in Appendix A.*

Table 6. DDT and BCT of the `GIFT` S-box for $\Delta_i = 4$

	Δ_o																
	0	1	2	3	4	5	6	7	8	9	a	b	c	d	e	f	sum
DDT	0	0	0	2	0	4	0	6	0	2	0	0	0	2	0	0	16
BCT	16	4	4	10	4	8	8	6	0	2	0	0	0	2	0	0	64

Lemma 5 shows that the impact from the DDT entry with 'x' to the BCT is large, on the order of x^2. Thus block-cipher designers adopting S-boxes with weak differential resistance need to be careful about how their choice will impact the corresponding BCT.

4 Applications to `Deoxys-BC`

In this section, we apply our BCT-based analysis to improve the recently proposed related-tweakey boomerang attacks against `Deoxys-BC` [CHP+17]. The specification of `Deoxys-BC` is briefly given in Sect. 4.1. The improved boomerang distinguishers are presented in Sect. 4.2, and the results of our experimental verification are reported in Sect. 4.3.

4.1 Specification

`Deoxys-BC` is an `AES`-based tweakable block cipher [JNPS16], which is based on the `TWEAKEY` framework [JNP14]. It is the underlying tweakable block cipher of the `Deoxys` authenticated encryption scheme submitted to the CAESAR competition (and one of the 15 candidates still being considered in the competition's third round). The `Deoxys` authenticated encryption scheme makes use of two versions of the cipher as its internal primitive: `Deoxys-BC-256` and `Deoxys-BC-384`. Hereafter, we mainly focus on the specification of `Deoxys-BC-384`, which is a target in this paper. `Deoxys-BC` is a dedicated 128-bit tweakable block cipher which besides the two standard inputs, a plaintext P (or a ciphertext C) and a key K, also takes an additional input called a *tweak* T. The concatenation of the key and tweak states is called the *tweakey* state. For `Deoxys-BC-384` the tweakey size is 384 bits. We assume that the reader is familiar with the `AES` block cipher [Nat01].

The round function of `Deoxys-BC` is exactly the same as that of the `AES`, except that the operation AddRoundKey is renamed as AddRoundTweakey. The internal state is viewed as a 4×4 matrix of bytes, and is updated by applying the following round function 14 times and 16 times for `Deoxys-BC-256` and `Deoxys-BC-384`, respectively.

- AddRoundTweakey – XOR the 128-bit round subtweakey to the state.
- SubBytes – Apply the `AES` S-box S to each byte of the state.
- ShiftRows – Rotate the 4-byte in the i-th row left by i positions.
- MixColumns – Multiply the state by the 4×4 MDS matrix of `AES`.

After the last round, a final AddRoundTweakey operation is performed to produce the ciphertext.

Subtweakeys. The size of tweakey for `Deoxys-BC-384` is 384 bits. Those are separated into three 128-bit words, and loaded into the initial tweakey states TK_0^1, TK_0^2, and TK_0^3. The 128-bit *subtweakey* used in the AddRoundTweakey operation is extracted from three tweakey states as $STK_i = TK_i^1 \oplus TK_i^2 \oplus TK_i^3 \oplus RC_i$, where RC_i is a round constant. Here, we omit the details of RC_i. Please refer to the original design document [JNPS16] for the exact specification.

In each round, the 128-bit words TK_i^1, TK_i^2, TK_i^3 are updated with the *tweakey schedule* algorithm, which is defined as

$$TK_{i+1}^1 = h(TK_i^1),$$
$$TK_{i+1}^2 = h(LFSR_2(TK_i^2)),$$
$$TK_{i+1}^3 = h(LFSR_3(TK_i^3)),$$

where the byte permutation h is

$$\begin{pmatrix} 0 & 1 & 2 & 3 & 4 & 5 & 6 & 7 & 8 & 9 & 10 & 11 & 12 & 13 & 14 & 15 \\ 1 & 6 & 11 & 12 & 5 & 10 & 15 & 0 & 9 & 14 & 3 & 4 & 13 & 2 & 7 & 8 \end{pmatrix},$$

with the 16 bytes numbered by the usual `AES` byte ordering.

The $LFSR_2$ and $LFSR_3$ functions are simply the application of an LFSR to each on the 16 bytes of a 128-bit tweakey word. The two LFSRs used are given in Table 7 (x_0 stands for the LSB of the cell).

Table 7. The two LFSRs used in `Deoxys-BC` tweakey schedule

$LFSR_2$	$(x_7\|\|x_6\|\|x_5\|\|x_4\|\|x_3\|\|x_2\|\|x_1\|\|x_0) \rightarrow (x_6\|\|x_5\|\|x_4\|\|x_3\|\|x_2\|\|x_1\|\|x_0\|\|x_7 \oplus x_5)$
$LFSR_3$	$(x_7\|\|x_6\|\|x_5\|\|x_4\|\|x_3\|\|x_2\|\|x_1\|\|x_0) \rightarrow (x_0 \oplus x_6\|\|x_7\|\|x_6\|\|x_5\|\|x_4\|\|x_3\|\|x_2\|\|x_1)$

A schematic diagram of the instantiation of the `TWEAKEY` framework for `Deoxys-BC-384` is shown in Fig. 8.

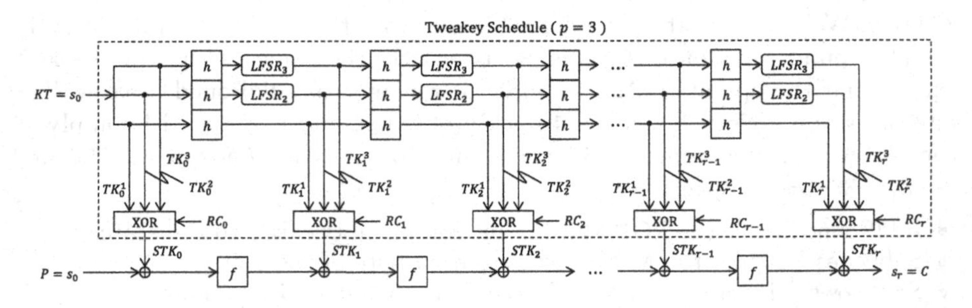

Fig. 8. A schematic diagram of `Deoxys-BC-384` with `TWEAKEY` framework

4.2 Improved 10-Round Boomerang Attack

Cid et al. have recently presented in [CHP+17] several boomerang attacks in the related-tweakey setting, including 8-round, 9-round and 10-round boomerang distinguishers against Deoxys-BC-384 having probability 2^{-6}, 2^{-18}, and 2^{-42}, respectively. They proposed an MILP-based automated search method of differential characteristics that takes into account linear incompatibility in truncated differentials and the ladder switch effect in the boomerang attack. Among all the possible differential characteristics, the authors chose the ones that exploit the S-box switch effect. Owing to the very detailed and careful optimisation, it seemed very unlikely that one could improve their proposed boomerang attacks; in other words, Cid et al. [CHP+17] picked the optimal choice under their assumptions on the search range.

However, our novel idea to use the BCT in boomerang-style attacks motivated us to improve their attacks by enlarging the search space when taking into account the generalised switching effect observed in the BCT of the AES S-box. In particular, their 8-round distinguisher includes only one active S-box that exploits the S-box switch effect, and hence an improvement by using BCT should be observed very clearly.

Our Goal. Recall that the maximum differential probability of the AES S-box is 2^{-6}, which is a reason why the probability of the 8-round distinguisher in [CHP+17] is 2^{-6}. As shown in Table 3, we observed in the BCT that the maximum probability of generating a quartet is $6/256 \approx 2^{-5.4}$ for the AES S-box. Hence, our goal here is to search for differential characteristics that achieve the probability of $2^{-5.4}$ and experimentally verify the correctness of the theory explained in Sect. 3.

In our analysis, we noticed that the authors of [CHP+17] interpreted the byte permutation h in the reverse order, thus their original analysis and results are in fact for a Deoxys-BC variant. Because our purpose here is demonstrate the possibility of improving existing attacks by use of the BCT, we analyse the same Deoxys-BC variant as in [CHP+17].

Searching for Differential Characteristics. We borrow the idea of the differential characteristic search proposed in [CHP+17]. Because the main focus of this paper is the generalised switching effect, we only briefly explain the search method.

The search in [CHP+17] is a two-stage approach. The first stage is searching for truncated differentials with the minimum number of active S-boxes using MILP. At this stage, there is no guarantee that each discovered truncated differentials can be instantiated with actual differences. Here, the authors in [CHP+17] introduced two levels of tradeoff between the accuracy of truncated differentials and the assumption of the search range:

1. It only assumed independence between subtweakeys in different rounds, while the real different tweakeys are linearly related in the real cipher's algorithm,

thus the truncated differentials detected in this approach may contain contradiction (often called "linear incompatibility").
2. Degrees of freedom (the number of differences that can be chosen independently of the other part of the trail) and the number of constraints for a valid trail (e.g. linear relations between subtweakeys mentioned above) were counted, and it was assumed that truncated differentials could be instantiated only if the degrees of freedom were higher than the degrees of consumption. Instead, truncated differentials that are detected in this way do not include contradiction about the linear incompatibility.

We refer to [CHP+17] for the exact MILP modelling for searching truncated differentials.

The second stage is searching for differences satisfying the given active-byte positions. This is done by listing all linear constraints in the truncated differential to build a system of linear equations, and by solving the system. We again refer to [CHP+17] for the exact method for generating the system.

The 10-round boomerang attack against `Deoxys-BC-384` uses 5-round differential characteristics for both E_0 and E_1. Active byte positions are chosen so that the ladder switch effect can be optimally exploited in the middle two rounds. Then the differential value is fixed to one of E_0 and E_1 and finally the differential value for the other half is fixed to exploit the S-box switch. The 10-round distinguisher [CHP+17] is given in Table 8. Cid *et al.* showed the differential propagation of E_0 in round 6 and of E_1 in round 5 to explicitly show that the ladder switch is applied. Both characteristics activate the S-box at position (1,1) in round 6 and both characteristics specify the same input and output difference (from `9e` to `68`), namely $\Delta_o = \nabla_o$, which is the condition to apply the S-box switch. The S-box is highlighted in red in Table 8. Note that in the DDT of the AES S-box, `9e` propagates to `68` with the highest probability of 2^{-6}.

We now replace the differential characteristic for the attack. Because of the optimisations done in [CHP+17], we use exactly the same differential characteristic for E_1, and only replace the difference of E_0. The characteristic for E_1 fixes the ∇_o of the target S-box to `68`. We confirmed that there exist two choices of Δ_i such that the BCT entry for $(\Delta_i, \texttt{68})$ is '6.' Those Δ_i are `2a` and `b4`. Hence, we added the linear equation $\Delta_i = \texttt{2a}$ or $\Delta_i = \texttt{b4}$ to the system of linear equations and solved the system to obtain the corresponding characteristics. The obtained differential characteristic for E_0 with $\Delta_i = \texttt{2a}$ is shown in Table 9.

4.3 Experimental Verification and Summary

As done in [CHP+17], we drop the first round and the last round of the 10-round boomerang characteristic, which leads to the 8-round boomerang characteristic only with a single active S-box now with the generalised switching effect. Our experiments clearly verify this effect.

Let κ_i, where $i \in \{1, 2, 3, 4\}$, be a 384-bit master tweakey for the first, second, third, and fourth oracles, respectively. Our experiments follow the pseudo-code in Sect. 2. The exact value of the master tweakey difference for E_1 denoted by ∇K is given in [CHP+17, Table 6]. We set $\mathcal{N}$ to 2^{15} and the number of

Table 8. 10-round distinguisher of Deoxys-BC-384 [CHP+17]. † denotes the probability of the rounds that are evaluated for the boomerang switch. The probability is counted in the other half of the characteristic, thus the probability with † can be ignored.

rounds	initial Δ	tweakey Δ	before SB	after SR	p_r
1	00 00 8e 00 a3 00 00 10 9e 00 00 00 00 8e 00 00	00 00 8e 00 00 00 00 10 9e 00 00 00 00 8e 00 00	00 00 00 00 a3 00 00 00 00 00 00 00 00 00 00 00	00 00 00 00 00 00 00 69 00 00 00 00 00 00 00 00	$(2^{-6})^2$
2	00 00 00 bb 00 00 00 d2 00 00 00 69 00 00 00 69	00 00 00 bb 00 00 00 d2 00 00 00 69 00 00 00 69	00 00 00 00 00 00 00 00 00 00 00 00 00 00 00 00	00 00 00 00 00 00 00 00 00 00 00 00 00 00 00 00	1
3	00 00 00 00 00 00 00 00 00 00 00 00 00 00 00 00	00 00 00 00 00 00 00 00 00 00 00 00 00 00 00 00	00 00 00 00 00 00 00 00 00 00 00 00 00 00 00 00	00 00 00 00 00 00 00 00 00 00 00 00 00 00 00 00	1
4	00 00 00 00 00 00 00 00 00 00 00 00 00 00 00 00	00 00 00 00 00 00 00 00 00 00 00 00 00 00 00 00	00 00 00 00 00 00 00 00 00 00 00 00 00 00 00 00	00 00 00 00 00 00 00 00 00 00 00 00 00 00 00 00	1
5	00 00 00 00 00 00 00 00 00 00 00 00 00 00 00 00	69 00 00 00 00 bb 00 00 00 00 d2 00 00 00 00 69	69 00 00 00 00 bb 00 00 00 00 d2 00 00 00 00 69	** 00 00 00 ** 00 00 00 ** 00 00 00 ** 00 00 00	1
6	** 00 00 00 ** 00 00 00 ** 00 00 00 ** 00 00 00	00 10 00 00 00 9e 00 00 00 8e 00 00 00 8e 00 00	** 10 00 00 ** 9e 00 00 ** 8e 00 00 ** 8e 00 00	** ** 00 00 68 00 00 ** 00 00 ** ** 00 ** ** 00	2^{-6}
5	00 ** ** ** ** 00 ** ** ** ** 00 ** ** ** ** **	00 ee 00 00 00 00 00 00 00 00 00 00 00 00 00 11	00 ** ** ** ** 00 ** ** ** ** 00 ** ** ** ** 00	00 ** ** ** 00 ** ** ** 00 ** ** ** 00 ** ** **	1 †
6	00 00 00 00 00 9e 00 00 00 0a ab 00 00 00 93 7a	00 00 00 00 00 00 00 00 00 0a 00 00 00 00 93 00	00 00 00 00 00 9e 00 00 00 00 ab 00 00 00 00 7a	00 00 00 00 68 00 00 00 01 00 00 00 b9 00 00 00	2^{-6} †
7	00 00 00 00 6a 00 00 00 ba 00 00 00 00 00 00 00	00 00 00 00 6a 00 00 00 ba 00 00 00 00 00 00 00	00 00 00 00 00 00 00 00 00 00 00 00 00 00 00 00	00 00 00 00 00 00 00 00 00 00 00 00 00 00 00 00	1
8	00 00 00 00 00 00 00 00 00 00 00 00 00 00 00 00	00 00 00 00 00 00 00 00 00 00 00 00 00 00 00 00	00 00 00 00 00 00 00 00 00 00 00 00 00 00 00 00	00 00 00 00 00 00 00 00 00 00 00 00 00 00 00 00	1
9	00 00 00 00 00 00 00 00 00 00 00 00 00 00 00 00	00 00 00 00 00 00 00 00 00 00 00 00 00 00 00 00	00 00 00 00 00 00 00 00 00 00 00 00 00 00 00 00	00 00 00 00 00 00 00 00 00 00 00 00 00 00 00 00	1
10	00 00 00 00 00 00 00 00 00 00 00 00 00 00 00 00	00 00 00 00 00 00 00 00 00 00 00 6a ba 00 00 00	00 00 00 00 00 00 00 00 00 00 00 6a ba 00 00 00	00 00 00 00 00 00 00 00 00 61 00 00 00 97 00 00	$(2^{-12})^2$

attempts satisfying the last equation is counted. The test was iterated for 1,000 randomly chosen tweakeys; the average number of successes was 763. Hence, the probability of generating a right quartet is $763/2^{15} \approx 2^{-5.42}$, which closely matches and confirms the generalised switching effect.

Table 9. Improved differential characteristic for E_0 of `Deoxys-BC-384`.

rounds	initial Δ	tweakey Δ	before SB	after SR	p_r
1	00 00 15 00	00 00 15 00	00 00 00 00	00 00 00 00	$(2^{-6})^2$
	b3 00 00 3f	00 00 00 3f	b3 00 00 00	00 00 00 0e	
	2a 00 00 00	2a 00 00 00	00 00 00 00	00 00 00 00	
	00 15 00 00	00 15 00 00	00 00 00 00	00 00 00 00	
2	00 00 00 12	00 00 00 12	00 00 00 00	00 00 00 00	1
	00 00 00 1c	00 00 00 1c	00 00 00 00	00 00 00 00	
	00 00 00 0e	00 00 00 0e	00 00 00 00	00 00 00 00	
	00 00 00 0e	00 00 00 0e	00 00 00 00	00 00 00 00	
3	00 00 00 00	00 00 00 00	00 00 00 00	00 00 00 00	1
	00 00 00 00	00 00 00 00	00 00 00 00	00 00 00 00	
	00 00 00 00	00 00 00 00	00 00 00 00	00 00 00 00	
	00 00 00 00	00 00 00 00	00 00 00 00	00 00 00 00	
4	00 00 00 00	00 00 00 00	00 00 00 00	00 00 00 00	1
	00 00 00 00	00 00 00 00	00 00 00 00	00 00 00 00	
	00 00 00 00	00 00 00 00	00 00 00 00	00 00 00 00	
	00 00 00 00	00 00 00 00	00 00 00 00	00 00 00 00	
5	00 00 00 00	0e 00 00 00	0e 00 00 00	** 00 00 00	1
	00 00 00 00	00 12 00 00	00 12 00 00	** 00 00 00	
	00 00 00 00	00 00 1c 00	00 00 1c 00	** 00 00 00	
	00 00 00 00	00 00 00 0e	00 00 00 0e	** 00 00 00	
6	** 00 00 00	00 3f 00 00	** 3f 00 00	** ** 00 00	$2^{-5.4}$
	** 00 00 00	00 2a 00 00	** 2a 00 00	68 00 00 **	
	** 00 00 00	00 15 00 00	** 15 00 00	00 00 ** **	
	** 00 00 00	00 15 00 00	** 15 00 00	00 ** ** 00	
	Master tweakey differences (ΔK)				
	00 00 ac 00	00 00 00 f4	58 00 00 00	00 ac 00 00	
	00 00 66 00	00 00 00 ab	cd 00 00 00	00 66 00 00	
	00 00 df 00	00 00 00 60	bf 00 00 00	00 df 00 00	

We also derived the differential characteristic for $\Delta_i =$ `b4` and implemented the 8-round distinguisher for verification. In the experiments, the average number of successes over 1,000 different choices of keys was $775/2^{15} \approx 2^{-5.40}$, which again demonstrates the validity of the generalised switching effect.

Thus using the BCT for the AES S-box and the generalised switching effect, we were able to improve the probability of the boomerang distinguishers against `Deoxys-BC-384` by a factor of $2^{-0.6}$; namely to $2^{-5.4}$, $2^{-17.4}$, and $2^{-41.4}$ for 8 rounds, 9 rounds and 10 rounds, respectively. Although the improved factor in this particular case is small, the relevant point is that the effect of the generalised switch represented by the BCT could be experimentally verified against the AES S-box. This indicates that the probability of boomerang distinguishers presented in previous works, which did not make use of the BCT, is unlikely to be optimal.

5 Applications to SKINNY

In [LGL17] Liu *et al.* proposed related-tweakey rectangle attacks against the `SKINNY` tweakable block cipher. The attacks evaluated the probability of generating a right quartet by taking into account the amplified probability, but did not consider the boomerang switch effect. In this section, we accurately evaluate the probability of generating the right quartet by applying the BCT. By doing so, we detect flaws in the experimentally evaluated probability in [LGL17] and show that the actual probabilities are higher than reported in [LGL17]. We first briefly review the specification of SKINNY in Sect. 5.1. The previous distinguishers and improved probabilities are then presented in Sects. 5.2 and 5.3, respectively.

5.1 Specification of SKINNY-128

SKINNY [BJK+16] is another family of lightweight tweakable block ciphers, based on the TWEAKEY framework [JNP14], which was introduced by Beierle *et al.* at CRYPTO 2016. The block size can be $n \in \{64, 128\}$ and the tweakey size can be $t \in \{n, 2n, 3n\}$. The 64-bit block version adopts a nibble-oriented SPN structure and is called SKINNY-64, while the 128-bit block version adopts a byte-oriented SPN structure and is called SKINNY-128.

An n-bit plaintext is loaded into the state represented by a 4×4-cell array, and the round function is then applied N_r times, where N_r is 40, 48 and 56 for n-bit, $2n$-bit and $3n$-bit tweakeys, respectively.

The round function consists of five operations: SubCells, AddRoundConstant, AddRoundTweakey, ShiftRowsand MixColumns.

SubCells. A 4-bit (resp. 8-bit) S-box whose maximum differential probability is 2^{-2} is applied to all cells in SKINNY-64 (resp. SKINNY-128).

AddRoundConstant. A 7-bit constant updated by an LFSR in every round is added to three cells of the state. Details of the LFSR can be found in [BJK+16].

AddRoundTweakey. A $n/2$-bit value is extracted from the n, $2n$ or $3n$-bit tweakey state, and is XORed to the upper half of the state. We omit the details of the tweakey schedule.

ShiftRows. Each cell in row j is rotated to the right (i.e. opposite to AES) by j positions.

MixColumns. Four cells in each column are multiplied by a binary matrix $\mathcal{M}$. When (i_0, i_1, i_2, i_3) is the 4-cell value input to $\mathcal{M}$, the output (o_0, o_1, o_2, o_3) is computed by $o_0 = i_0 \oplus i_2 \oplus i_3$, $o_1 = i_0$, $o_2 = i_1 \oplus i_2$, and $o_3 = i_0 \oplus i_2$. This is illustrated in Fig. 9.

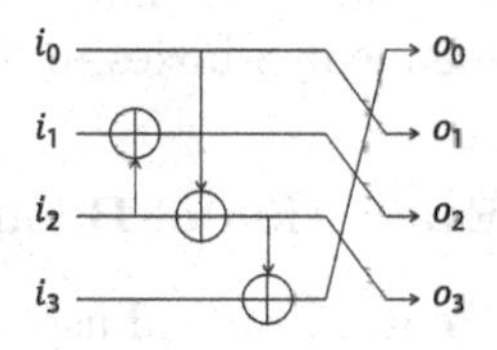

Fig. 9. A schematic representation of MixColumns of SKINNY

5.2 Previous Related-Tweakey Rectangle Attacks

Liu *et al.* [LGL17], among several cryptanalytic results, proposed 17-round, 18-round, 22-round and 23-round boomerang distinguishers against SKINNY-64-128, SKINNY-64-192, SKINNY-128-256, and SKINNY-128-384, respectively. The probabilities of those distinguishers are however too small to practically implement the verification experiments. Instead, the authors of [LGL17] implemented only the middle two rounds, the last round of E_0 and the first round of E_1, to experimentally verify that the proposed characteristics did not contain an incompatibility as pointed out by Murphy [Mur11]. If only the middle two rounds are evaluated, the probability including the amplified effect is calculated by $2^{-8.42}$, $2^{-16.30}$, $2^{-15.98}$ and $2^{-19.04}$ for the above four targets respectively, while their experimental verification implied that the probability should be $2^{-4.01}$, $2^{-7.53}$, $2^{-1.86}$, and $2^{-4.89}$, respectively. Those probabilities are summarised in Table 10.

Table 10. Previous boomerang distinguishers on SKINNY and our correction

Versions	$(\hat{p}\hat{q})^2$	Probability by Experiment	Our Corrected Probability
SKINNY-64-128	$2^{-8.42}$	$2^{-4.01}$	2^{-2}
SKINNY-64-192	$2^{-16.30}$	$2^{-7.53}$	$2^{-5.31}$
SKINNY-128-256	$2^{-15.98}$	$2^{-1.86}$	$2^{-1.86}$
SKINNY-128-384	$2^{-19.04}$	$2^{-4.89}$	0

Liu *et al.* mentioned in [LGL17] that one reason why the probabilities observed were higher than expected may be that some active Sboxes can be "saved", as the authors of [BK09] explained (the ladder switch and the S-box switch). They concluded that it is unlikely for the authors to overestimate the probability of the distinguishers.

This motivates us to apply the generalized switching effect of the BCT to explain the reasons behind their experimental results, and to improve their $\hat{p}^2\hat{q}^2$ probabilities to match the experimentally observed ones. We show that, while their experimental results cannot be explained only with the ladder switch and S-box switch from [BK09], they can be explained rigorously by using the BCT along with the analysis for dependent S-boxes in [CLN+17].[1]

5.3 Precise Probability Evaluation of Boomerang Distinguishers

To explain the observed probabilities, we will use the attack against SKINNY-64-128. The last round (round 8) of E_0 and the first round (round 9) of E_1 are shown in Fig. 10.

[1] Our experiments and theoretical explanation discovered different probabilities from the experiments by Liu *et al.* [LGL17]. We contacted the authors and confirmed that our evaluation is correct.

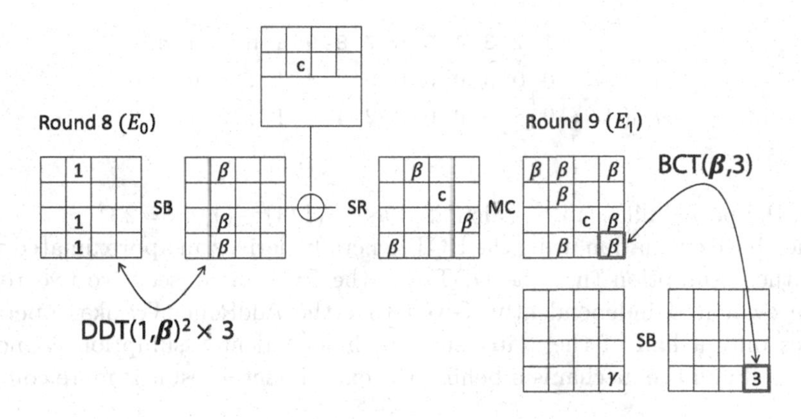

Fig. 10. Two rounds of 18-round distinguishers against SKINNY-64-128. Round 8 is covered by the characteristic in E_0 and round 9 is covered by the characteristic in E_1.

E_0 starts with three active nibbles with difference 1. Those will change into some difference in $\{0,1\}^4$ denoted by β. Then the difference c is introduced from the subtweakey difference.

In E_1, the differential propagation through the linear computations after the S-box is established with probability 1, thus omitted from Fig. 10. In the end, E_1 consists only of a single S-box layer. It specifies that there is only one active S-box in round 9, with the output difference of the S-box 3 and the input difference that can be some value in $\{0,1\}^4$ denoted by γ.

In the straightforward evaluation with amplified probability, $\hat{p}$ is computed as $\left(4 \cdot (2^{-2})^2\right)^3 = 2^{-6}$, while $\hat{q}$ is calculated as $2 \cdot (2^{-2})^2 + 4 \cdot (2^{-3})^2 \approx 2^{-2.42}$. Thus $\hat{p}\hat{q} \approx 2^{-8.42}$ which matches the evaluation by the authors of [LGL17].

A careful analysis shows that the active S-boxes from E_0 in round 9 and the active S-boxes from E_1 in round 9 overlap each other in only one byte. E_0 specifies that the input difference Δ_i to the active S-box is β, while E_1 specifies that the output difference ∇_o from the S-box is 3. This is exactly the situation in which the BCT can be applied to evaluate the probability of the active S-box and the other active S-boxes can be satisfied with probability 1 thanks to the ladder switch. Hence, we compute the probability of those two rounds as

$$\sum_{\beta \in \{0,1\}^4, \beta \neq 0} \left(\frac{\mathcal{T}_{\mathrm{DDT}}(1,\beta)}{16}\right)^2 \cdot \frac{\mathcal{T}_{\mathrm{BCT}}(\beta,3)}{16}, \tag{6}$$

where $\mathcal{T}_{\mathrm{DDT}}(\Delta_i, \Delta_o)$ and $\mathcal{T}_{\mathrm{BCT}}(\Delta_i, \nabla_o)$ are the values of the DDT and the BCT for the input difference Δ_i and the output difference Δ_o or ∇_o, respectively. Those values for SKINNY's 4-bit S-box are summarised below.

β	1	2	3	4	5	6	7	8	9	a	b	c	d	e	f
$\mathcal{T}_{\mathrm{DDT}}(1,\beta)$	0	0	0	0	0	0	0	4	4	4	4	0	0	0	0
$\mathcal{T}_{\mathrm{BCT}}(\beta,3)$	0	8	0	0	0	2	2	4	4	4	4	0	0	2	2

Hence, the probability can be calculated as $4 \cdot (1/4)^2 \cdot (1/4) = 2^{-4}$.

The above evaluation using the BCT generally derives an approximated value under the assumption that the DDT and the BCT in consecutive two rounds can be evaluated independently. Given that the AddRoundTweakey operation updates only a half of the state, such an independent assumption cannot be established and the mechanism behind the experimental result is more complex.

Analysis Including Dependency of Consecutive S-box Applications. Here, an analysis involving several dependent S-boxes in [CLN+17] can be applied. By following [CLN+17] we introduce the notation:

$$\mathcal{X}_{\mathrm{DDT}}(\Delta_i, \Delta_o) \triangleq \{x : S(x) \oplus S(x \oplus \Delta_i) = \Delta_o\},$$
$$\mathcal{Y}_{\mathrm{DDT}}(\Delta_i, \Delta_o) \triangleq \{S(x) : S(x) \oplus S(x \oplus \Delta_i) = \Delta_o\}.$$

And similarly for the BCT:

$$\mathcal{X}_{\mathrm{BCT}}(\Delta_i, \nabla_o) \triangleq \{x : S^{-1}(S(x) \oplus \nabla_o) \oplus S^{-1}(S(x \oplus \nabla_i) \oplus \nabla_o) = \Delta_i\},$$
$$\mathcal{DX}_{\mathrm{BCT}}(\Delta_i, \nabla_o) \triangleq \{x \oplus S^{-1}(S(x) \oplus \nabla_o) : S^{-1}(S(x) \oplus \nabla_o) \oplus S^{-1}(S(x \oplus \Delta_i) \oplus \nabla_o) = \Delta_i\}.$$

In the first S-box in round 8, the input difference 1 can change into one of {8, 9, a, b} with equal probability. We first consider the case for 8.

Case 1: $1 \to 8$.

$$\mathcal{Y}_{\mathrm{DDT}}(1,8) = \{5,7,\mathtt{d},\mathtt{f}\}, \quad \mathcal{X}_{\mathrm{BCT}}(8,3) = \{4,6,\mathtt{c},\mathtt{e}\}, \quad \mathcal{DX}_{\mathrm{BCT}}(8,3) = \{2\}.$$

After the first S-box application in round 8, the paired values can take $\{5,7,\mathtt{d},\mathtt{f}\}$. They change to $\{4,6,\mathtt{c},\mathtt{e}\}$ with probability 2^{-2} after AddRoundTweakey and MixColumns. Here the source of randomness are subtweakey values xored to 8 nibbles of the state and other nibble values during the MixColumns operation. Then, after going through the propagation of the BCT with probability 1, the paired values $S^{-1}(C_3)$ and $S^{-1}(C_4)$ become $2 \oplus \{4,6,\mathtt{c},\mathtt{e}\} = \{4,6,\mathtt{c},\mathtt{e}\}$. This is a heavily S-box-dependent feature that the set of paired values does not change after the application of BCT.

During the backward computation for the second pair, ∇_o only impacts to one active nibble value, but as explained above, the set of possible values does not change. Thus during the inverse of MixColumns, the source of randomness does not change from the first pair. Hence all the values can return to the paired values with the same difference as the first pair with probability 1. In summary, in Case 1 a right quartet is generated with probability 2^{-2}.

Other cases. The analysis for the other three cases is similar.

Case 2 : $\mathcal{X}_{\mathrm{BCT}}(\mathtt{9},\mathtt{3}) = \{\mathtt{1},\mathtt{3},\mathtt{8},\mathtt{a}\}, \mathcal{DX}_{\mathrm{BCT}}(\mathtt{9},\mathtt{3}) = \{\mathtt{b}\}$.
Case 3 : $\mathcal{X}_{\mathrm{BCT}}(\mathtt{a},\mathtt{3}) = \{\mathtt{4},\mathtt{6},\mathtt{c},\mathtt{e}\}, \mathcal{DX}_{\mathrm{BCT}}(\mathtt{a},\mathtt{3}) = \{\mathtt{2}\}$.
Case 4 : $\mathcal{X}_{\mathrm{BCT}}(\mathtt{b},\mathtt{3}) = \{\mathtt{1},\mathtt{3},\mathtt{8},\mathtt{a}\}, \mathcal{DX}_{\mathrm{BCT}}(\mathtt{b},\mathtt{3}) = \{\mathtt{b}\}$.

Thus, for any $\beta : \mathcal{T}(\mathtt{1},\beta) \neq 0$, $u \in \mathcal{X}_{\mathrm{BCT}}(\beta,\mathtt{3})$ and $v \in \mathcal{DX}_{\mathrm{BCT}}(\beta,\mathtt{3})$, the core property that $u \oplus v \in \mathcal{X}_{\mathrm{BCT}}(\beta,\mathtt{3})$ is established. Hence after falling into each case, a right quartet is generated with probability 2^{-2}.

Finally, considering that each case occurs with probability $\frac{1}{4}$, the entire probability of generating a right quartet is $4 \cdot \frac{1}{4} \cdot 2^{-2} = 2^{-2}$. We implemented those two rounds and verified that the results match the above theory.

A similar analysis can be applied to other members of the SKINNY family to verify the experimental results in Table 10. We omit the details of those evaluation in this paper.

6 Discussion

The paper has so far mainly focused on the use of the BCT to improve previously proposed boomerang attacks. In this section, we considered further properties and aspects of the BCT. The hardness of finding S-boxes achieving 4-uniform BCT is explained in Sect. 6.1. Moreover, the boomerang switch for modular addition is discussed in Sect. 6.2, where we show that the switching effect is quite different for that operation.

6.1 Difficulties of Achieving 4-Uniform BCT

If the BCT provides the opportunity for attackers to improve their attack, as shown earlier in this paper, a natural question is therefore whether it is possible to find an S-box with minimum boomerang switching effect. As discussed in Example 4, finding such S-boxes for n-bit to n-bit S-box is easy when n is odd, while in practice $n = 4$ and $n = 8$ are the most popular choices. In particular, most differentially strong S-boxes are designed to have 4-uniform DDT. Hence it is interesting to investigate whether 4-uniform DDT and BCT can be achieved simultaneously. Unfortunately, as we argue below, achieving 4-uniform BCT appears to be hard, especially as the size of the S-box increases, e.g. 8 bits.

Here, it is assumed that the differential spectrum of an AES-like S-box is used, i.e. the analysed S-box is an n-bit to n-bit S-box, and for each input and output difference of its DDT, there exist exactly one entry of '4' and $(2^{n/2}) - 2$ entries of '2.'

As in Lemma 2, each entry of '4' in the DDT increases two positions in the BCT for the same input or output difference by 4. To generate a 4-uniform BCT, the increased entries would have to have '0' in the DDT. Assume that the increased positions are chosen uniformly at random from all but zero. Then, the probability that the maximum value of the BCT in that row or column is '4' is

$$\frac{2^n/2}{2^n-1} \cdot \frac{2^n/2-1}{2^n-2} = \frac{2^{n-2}}{2^n-1}, \tag{7}$$

where the first term is the probability that the first increased position is chosen from '0' entries in the DDT, and the second term is for the second increased position. This must hold for $2^n - 1$ non-zero input or output differences, thus the probability is

$$\left(\frac{2^{n-2}}{2^n - 1}\right)^{2^n - 1}. \quad (8)$$

By setting $n = 4$ and 8, the probabilities that a randomly chosen S-box with a 4-uniform DDT simultaneously achieves 4-uniform BCT for 4-bit S-box and 8-bit S-box are $(4/15)^{15} \approx 2^{-28.6}$ and $(64/255)^{255} \approx 2^{-508.6}$, respectively.

For $n = 4$, if we consider the number of all 4-bit S-boxes with the optimal differential spectrum like the AES S-box, then it is unlikely that we find one that also achieves a 4-uniform BCT. Regarding $n = 8$, such an S-box may exist, but it is computationally hard to search for it.

6.2 Boomerang Switch for Modular Addition

The early analysis in this paper considers the BCT for S-boxes. A natural extension is to study how to apply a BCT-type analysis to other non-linear operations. In this section, we consider the boomerang switch for modular addition.

While an S-box is an n-bit to n-bit mapping, modular addition maps $2n$-bit inputs to n-bit outputs. Thus the previous definition of BCT cannot be directly applied to modular addition, and we need a different way to define the BCT for modular addition.

Suppose that the target cipher is divided into E_0, a middle modular addition step, and E_1. Let $((x_1, x'_1), (x_2, x'_2), (x_3, x'_3), (x_4, x'_4))$ be a quartet of modular addition inputs, and (y_1, y_2, y_3, y_4) be the corresponding output quartet. In order to make the modular addition invertible, one of the addends needs to be fixed. Here we let x'_i for $i = 1, ..., 4$ be the fixed addends of the quartet. Thus $x'_1 = x'_3$ and $x'_2 = x'_4$. The input difference of modular addition specified by E_0 is (Δ_i, Δ'_i), namely $x_1 \oplus x_2 = x_3 \oplus x_4 = \Delta_i$ and $x'_1 \oplus x'_2 = x'_3 \oplus x'_4 = \Delta'_i$. The output difference specified by E_1 is ∇_o, namely $y_1 \oplus y_3 = y_2 \oplus y_4 = \nabla_o$. Figure 11 shows a valid boomerang quartet for modular addition.

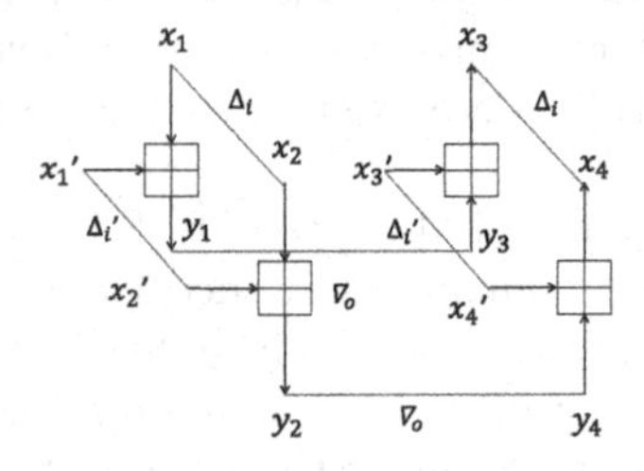

Fig. 11. A valid boomerang quartet for modular addition. Note that $x'_3 = x'_1$, $x'_4 = x'_2$.

The BCT for modular addition counts the number of inputs (x_i, x'_i) such that the corresponding quartet with input difference (Δ_i, Δ'_i) and output difference ∇_o is valid. Let '$\boxplus$' denote the modular addition and '$\boxminus$' the modular subtraction. The BCT for modular addition can then be defined in Eq. (9). Tables 11 and 12 give an example of the DDT and BCT for 3-bit modular addition when Δ_i is set to 0.

$$\begin{aligned}\mathcal{T}(\Delta_i, \Delta'_i, \nabla_o) =\#\Big\{&(x, x') \in (\{0,1\}^n, \{0,1\}^n) | ((x \boxplus x') \oplus \nabla_o \boxminus x') \\ &\oplus \Big(((x \oplus \Delta_i) \boxplus (x' \oplus \Delta'_i) \oplus \nabla_o) \boxminus (x' \oplus \Delta'_i)\Big) = \Delta_i\Big\}\end{aligned} \quad (9)$$

Like the BCT for an S-box, it is easy to verify that the BCT for modular addition has a similar property in representing the *ladder switch* (see the first row and the first column in Table 12). Moreover, another interesting property, which we call *most significant bit (MSB) switch*, can also be observed for modular addition.

MSB switch. Suppose the output difference ∇_o specified by E_1 is on the most significant bit. Then the modular addition, with probability 1, generates a right boomerang quartet. This property can be derived by replacing the 'xor' of ∇_o with 'modular addition' in Eq. (9). It can be observed in the column $\nabla_o = 4$ in Table 12.

Table 11. DDT of 3-bit modular addition with $\Delta_i = 0$

Δ'_i \ Δ_o	0	1	2	3	4	5	6	7
0	64	0	0	0	0	0	0	0
1	0	32	0	16	0	0	0	16
2	0	0	32	0	0	0	32	0
3	0	16	0	16	0	16	0	16
4	0	0	0	0	64	0	0	0
5	0	0	0	16	0	32	0	16
6	0	0	32	0	0	0	32	0
7	0	16	0	16	0	16	0	16

Table 12. BCT of 3-bit modular addition with $\Delta_i = 0$

Δ'_i \ ∇_o	0	1	2	3	4	5	6	7
0	64	64	64	64	64	64	64	64
1	64	0	32	0	64	0	32	0
2	64	64	0	0	64	64	0	0
3	64	0	32	0	64	0	32	0
4	64	64	64	64	64	64	64	64
5	64	0	32	0	64	0	32	0
6	64	64	0	0	64	64	0	0
7	64	0	32	0	64	0	32	0

On the other hand, the *S-box switch* does not work for modular addition. We can observe that in Table 11 the entry (1, 1) is 32 while the corresponding entry in Table 12 is 0, which contradicts the result of S-box switch (Lemma 1).

The reason is that in the S-box switch, when the first pair of values are (x_1, x_2), the condition $\nabla_o = \Delta_o$ implies the S-box output (y_1, y_2) are swapped to (y_2, y_1). The paired output (y_2, y_1) are exactly the input of the inverse S-box to compute the second pair. However, for the modular addition with first pair of input $((x_1, x'_1), (x_2, x'_2))$, although the output of modular addition (y_1, y_2) are swapped to (y_2, y_1) under the condition $\nabla_o = \Delta_o$, the values x'_1 and x'_2 are not swapped. Thus, $((y_2, x'_1), (y_1, x'_2))$ will be the input of the inverse modular addition. Since y_2 and x'_1 are not related, the original input difference is not guaranteed by the S-box switch.

Applications in Actual Ciphers. The analysis in Fig. 11 can be directly applied to particular differential trails in ARX ciphers. As an example, we show the application in the SPECK32/64 cipher [BSS+13], in which the internal state in round i is composed of two 16-bit words l_{i-1} and r_{i-1} and the round function updates those values as $l_i \leftarrow (l_{i-1} \ggg 7) \boxplus r_{i-1} \oplus k_i$ and $r_i \leftarrow (r_{i-1} \lll 2) \oplus l_i$. Then, the above BCT corresponds to the probability of the modular addition in a single round of SPECK with $\Delta l_{i-1} = 0$, $\Delta r_{i-1} = \Delta'_i$, and $\Delta l_i = \nabla_o$ (as shown in Fig. 12). Note that the ladder switch can be applied to the right word as long as active bit positions in $\Delta r_{i-1} \lll 2$ and $\Delta r_i \oplus \Delta l_i$ do not overlap.

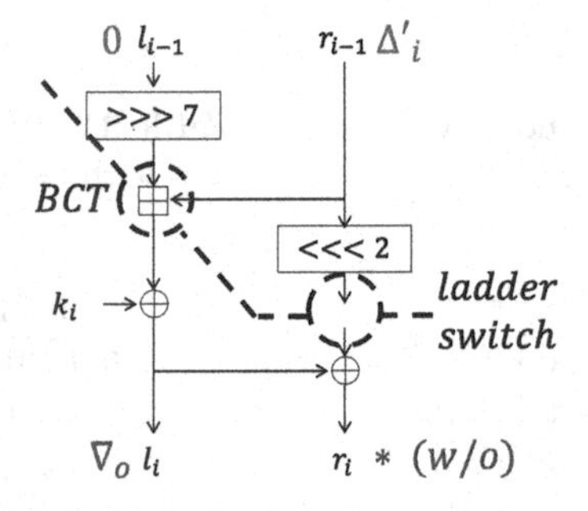

Fig. 12. Application of BCT for SPECK

For example, with $\Delta r_{i-1} = 8000$ and with any choice of Δl_i and Δr_i such that $(\Delta l_i \oplus \Delta r_i) \wedge 0002 = 0$, the MSB switch is applied to the modular addition and the ladder switch is applied to the right word. Hence, the probability r for one middle round is 1. Similarly, incompatible choices of $(\Delta r_{i-1}, \Delta l_i)$ with respect to the modular addition can be easily checked by using the BCT.

7 Concluding Remarks

In this paper, we introduced the BCT as a generalised method to analyse the dependency of two differential characteristics in boomerang distinguishers. The BCT includes the existing observations of incompatibility between two characteristics, as well as the ladder switch and the S-box switch. Moreover, the BCT offers stronger switching effect than previous ones, and we analysed the mechanism why such an effect is generated. The larger the bias in the DDT becomes, the more advantages the BCT provides. Future primitive designers who wish to adopt differentially weak S-boxes should take into account the impact of their choices on the BCT.

The effect of the BCT-based analysis was demonstrated by improving the boomerang attacks against `Deoxys-BC` and by precisely evaluating the probability of previous boomerang distinguishers against `SKINNY`.

We also discussed the issue of searching for S-boxes having good BCT, and showed that the S-boxes having 2-uniform DDT always have 2-uniform BCT, while S-boxes having 4-uniform DDT usually cannot ensure 4-uniform BCT. Lastly, we extended the analysis to modular addition along with an application to SPECK, and explained the different behaviours between the BCT for a S-box and the BCT for the modular addition.

Acknowledgements. We thank the anonymous reviewers for their valuable comments. We also thank attendees of the 2018 Dagstuhl seminar for Symmetric Cryptography, who provided us with various comments. The last author is supported by the Fundamental Theory and Cutting Edge Technology Research Program of Institute of Information Engineering, CAS (Grant No. Y7Z0341103), Youth Innovation Promotion Association CAS and the National Natural Science Foundation of China (Grants No. 61472415, 61732021 and 61772519). We also thank the ASK2016 organisers for providing us an opportunity for the initial discussion.

References

[AIK+00] Aoki, K., Ichikawa, T., Kanda, M., Matsui, M., Moriai, S., Nakajima, J., Tokita, T.: *Camellia*: a 128-bit block cipher suitable for multiple platforms — design and analysis. In: Stinson, D.R., Tavares, S. (eds.) SAC 2000. LNCS, vol. 2012, pp. 39–56. Springer, Heidelberg (2001). https://doi.org/10.1007/3-540-44983-3_4

[BCD03] Biryukov, A., De Cannière, C., Dellkrantz, G.: Cryptanalysis of Safer++. In: Boneh, D. (ed.) CRYPTO 2003. LNCS, vol. 2729, pp. 195–211. Springer, Heidelberg (2003). https://doi.org/10.1007/978-3-540-45146-4_12

[BDK01] Biham, E., Dunkelman, O., Keller, N.: The rectangle attack—rectangling the serpent. In: Pfitzmann, B. (ed.) EUROCRYPT 2001. LNCS, vol. 2045, pp. 340–357. Springer, Heidelberg (2001). https://doi.org/10.1007/3-540-44987-6_21

[BDK02] Biham, E., Dunkelman, O., Keller, N.: New results on boomerang and rectangle attacks. In: Daemen, J., Rijmen, V. (eds.) FSE 2002. LNCS, vol. 2365, pp. 1–16. Springer, Heidelberg (2002). https://doi.org/10.1007/3-540-45661-9_1

[BDK05] Biham, E., Dunkelman, O., Keller, N.: Related-key boomerang and rectangle attacks. In: Cramer, R. (ed.) EUROCRYPT 2005. LNCS, vol. 3494, pp. 507–525. Springer, Heidelberg (2005). https://doi.org/10.1007/11426639_30

[BFMT16] Berger, T.P., Francq, J., Minier, M., Thomas, G.: Extended generalized feistel networks using matrix representation to propose a new lightweight block cipher: Lilliput. IEEE Trans. Comput. **65**(7), 2074–2089 (2016)

[BJK+16] Beierle, C., Jean, J., Kölbl, S., Leander, G., Moradi, A., Peyrin, T., Sasaki, Y., Sasdrich, P., Sim, S.M.: The SKINNY family of block ciphers and its low-latency variant MANTIS. In: Robshaw, M., Katz, J. (eds.) CRYPTO 2016. LNCS, vol. 9815, pp. 123–153. Springer, Heidelberg (2016). https://doi.org/10.1007/978-3-662-53008-5_5

[BK09] Biryukov, A., Khovratovich, D.: Related-key cryptanalysis of the full AES-192 and AES-256. In: Matsui, M. (ed.) ASIACRYPT 2009. LNCS, vol. 5912, pp. 1–18. Springer, Heidelberg (2009). https://doi.org/10.1007/978-3-642-10366-7_1

[BKL+07] Bogdanov, A., Knudsen, L.R., Leander, G., Paar, C., Poschmann, A., Robshaw, M.J.B., Seurin, Y., Vikkelsoe, C.: PRESENT: an ultra-lightweight block cipher. In: Paillier, P., Verbauwhede, I. (eds.) CHES 2007. LNCS, vol. 4727, pp. 450–466. Springer, Heidelberg (2007). https://doi.org/10.1007/978-3-540-74735-2_31

[BPP+17] Banik, S., Pandey, S.K., Peyrin, T., Sasaki, Y., Sim, S.M., Todo, Y.: GIFT: A small present - towards reaching the limit of lightweight encryption. Cryptology ePrint Archive, Report 2017/622 (2017)

[BS93] Biham, E., Shamir, A.: Differential Cryptanalysis of the Data Encryption Standard. Springer, New York (1993). https://doi.org/10.1007/978-1-4613-9314-6

[BSS+13] Beaulieu, R., Shors, D., Smith, J., Treatman-Clark, S., Weeks, B., Wingers, L.: The simon and speck families of lightweight block ciphers. Cryptology ePrint Archive, Report 2013/404 (2013)

[CHP+17] Cid, C., Huang, T., Peyrin, T., Sasaki, Y., Song, L.: A security analysis of deoxys and its internal tweakable block ciphers. IACR Trans. Symmetric Cryptol. **2017**(3), 73–107 (2017)

[CLN+17] Canteaut, A., Lambooij, E., Neves, S., Rasoolzadeh, S., Sasaki, Y., Stevens, M.: Refined probability of differential characteristics including dependency between multiple rounds. IACR Trans. Symmetric Cryptol. **2017**(2), 203–227 (2017)

[DKS10] Dunkelman, O., Keller, N., Shamir, A.: A practical-time related-key attack on the KASUMI cryptosystem used in GSM and 3G telephony. In: Rabin, T. (ed.) CRYPTO 2010. LNCS, vol. 6223, pp. 393–410. Springer, Heidelberg (2010). https://doi.org/10.1007/978-3-642-14623-7_21

[DKS14] Dunkelman, O., Keller, N., Shamir, A.: A practical-time related-key attack on the KASUMI cryptosystem used in GSM and 3G telephony. J. Cryptology **27**(4), 824–849 (2014)

[GPPR11] Guo, J., Peyrin, T., Poschmann, A., Robshaw, M.: The LED block cipher. In: Preneel, B., Takagi, T. (eds.) CHES 2011. LNCS, vol. 6917, pp. 326–341. Springer, Heidelberg (2011). https://doi.org/10.1007/978-3-642-23951-9_22

[JNP14] Jean, J., Nikolić, I., Peyrin, T.: Tweaks and keys for block ciphers: the TWEAKEY framework. In: Sarkar, P., Iwata, T. (eds.) ASIACRYPT 2014. LNCS, vol. 8874, pp. 274–288. Springer, Heidelberg (2014). https://doi.org/10.1007/978-3-662-45608-8_15

[JNPS16] Jean, J., Nikolić, I., Peyrin, T., Seurin, Y.: Deoxys v1.41. Submitted to CAESAR, October 2016

[KHP+12] Kim, J., Hong, S., Preneel, B., Biham, E., Dunkelman, O., Keller, N.: Related-key boomerang and rectangle attacks: theory and experimental analysis. IEEE Trans. Inf. Theor. **58**(7), 4948–4966 (2012)

[KKS00] Kelsey, J., Kohno, T., Schneier, B.: Amplified boomerang attacks against reduced-round MARS and serpent. In: Goos, G., Hartmanis, J., van Leeuwen, J., Schneier, B. (eds.) FSE 2000. LNCS, vol. 1978, pp. 75–93. Springer, Heidelberg (2001). https://doi.org/10.1007/3-540-44706-7_6

[LGL17] Liu, G., Ghosh, M., Ling, S.: Security analysis of SKINNY under related-tweakey settings (long paper). IACR Trans. Symmetric Cryptol. **2017**(3), 37–72 (2017)

[Mur11] Murphy, S.: The return of the cryptographic boomerang. IEEE Trans. Inf. Theor. **57**(4), 2517–2521 (2011)

[Nat01] National Institute of Standards and Technology. Federal Information Processing Standards Publication 197: Advanced Encryption Standard (AES). NIST, November 2001

[SMMK12] Suzaki, T., Minematsu, K., Morioka, S., Kobayashi, E.: *TWINE*: a lightweight block cipher for multiple platforms. In: Knudsen, L.R., Wu, H. (eds.) SAC 2012. LNCS, vol. 7707, pp. 339–354. Springer, Heidelberg (2013). https://doi.org/10.1007/978-3-642-35999-6_22

[Wag99] Wagner, D.: The boomerang attack. In: Knudsen, L. (ed.) FSE 1999. LNCS, vol. 1636, pp. 156–170. Springer, Heidelberg (1999). https://doi.org/10.1007/3-540-48519-8_12

A Demonstration of Lemma 5 for SKINNY-128

The size of the DDT for the 8-bit S-box of SKINNY-128 is 256×256. Each entry can take one of 13 different values but for 0 and 256: 2, 4, 6, 8, 12, 16, 20, 24, 28, 32, 40, 48 and 64. Hence, the impact to the BCT is much bigger than for many other S-boxes, making it a good target for verifying the correctness of Lemma 5.

Each row of Table 13 shows the number of entries with the designated value in the DDT. For example, when $\Delta_i = \texttt{01}$, there are 6, 3 and 1 entries that take 16, 32 and 64, respectively. The column of "sum" shows the sum of the values of the BCT entries that were computed experimentally. The column of "Lemma 5" shows that value calculated by applying Lemma 5. Due to the limited space we only list the data for $\Delta_i = 1$ to 100.

As in Table 13, for any Δ_i, the relationships between the DDT and the BCT are correctly simulated by Lemma 5.

Table 13. Relationships between the DDT and the BCT simulated by Lemma 5

Δ_i	2	4	6	8	12	16	20	24	28	32	40	48	64	sum	Lem. 5	Δ_i	2	4	6	8	12	16	20	24	28	32	40	48	64	sum	Lem. 5
01	0	0	0	0	0	6	0	0	0	3	0	0	1	8704	8704	33	0	24	0	14	0	3	0	0	0	0	0	0	0	2048	2048
02	0	0	0	0	0	0	0	0	0	4	0	0	2	12288	12288	34	0	16	0	16	0	4	0	0	0	0	0	0	0	2304	2304
03	0	0	0	18	0	5	0	0	0	1	0	0	0	3456	3456	35	0	0	0	16	0	8	0	0	0	0	0	0	0	3072	3072
04	0	0	0	12	0	2	0	0	0	2	0	0	1	7424	7424	36	0	24	0	14	0	3	0	0	0	0	0	0	0	2048	2048
05	0	0	0	0	0	8	0	0	0	2	0	0	1	8192	8192	37	0	32	0	16	0	0	0	0	0	0	0	0	0	1536	1536
06	0	0	0	6	0	7	0	0	0	3	0	0	0	5248	5248	38	0	12	0	10	0	8	0	0	0	0	0	0	0	2880	2880
07	0	0	0	12	0	6	0	0	0	2	0	0	0	4352	4352	39	0	12	0	10	0	8	0	0	0	0	0	0	0	2880	2880
08	0	0	0	6	0	3	0	0	0	3	0	0	1	8320	8320	3a	0	12	0	16	0	1	0	0	0	2	0	0	0	3520	3520
09	0	0	0	6	0	3	0	0	0	3	0	0	1	8320	8320	3b	0	12	0	14	0	4	0	0	0	1	0	0	0	3136	3136
0a	0	0	0	12	0	2	0	0	0	2	0	0	1	7424	7424	3c	16	16	0	20	0	0	0	0	0	0	0	0	0	1600	1600
0b	0	0	0	6	0	9	0	0	0	2	0	0	0	4736	4736	3d	16	16	0	20	0	0	0	0	0	0	0	0	0	1600	1600
0c	0	12	0	12	0	5	0	0	0	1	0	0	0	3264	3264	3e	16	32	0	4	0	4	0	0	0	0	0	0	0	1856	1856
0d	0	12	0	12	0	5	0	0	0	1	0	0	0	3264	3264	3f	16	32	0	4	0	4	0	0	0	0	0	0	0	1856	1856
0e	0	12	0	12	0	5	0	0	0	1	0	0	0	3264	3264	40	0	0	0	4	0	4	0	0	0	3	0	0	1	8448	8448
0f	0	12	0	12	0	5	0	0	0	1	0	0	0	3264	3264	41	0	8	0	16	0	6	0	0	0	0	0	0	0	2688	2688
10	0	0	0	0	0	4	0	0	0	2	0	0	2	11264	11264	42	0	8	0	8	0	5	0	0	0	1	0	1	0	5248	5248
11	0	0	0	12	0	6	0	0	0	2	0	0	0	4352	4352	43	16	27	0	11	1	1	0	0	0	0	0	0	0	1600	1600
12	0	0	0	16	0	8	0	0	0	0	0	0	0	3072	3072	44	0	27	0	9	1	2	0	0	0	1	0	0	0	2688	2688
13	0	24	0	14	0	3	0	0	0	0	0	0	0	2048	2048	45	0	16	0	16	0	2	0	0	0	1	0	0	0	2816	2816
14	0	16	0	16	0	4	0	0	0	0	0	0	0	2304	2304	46	16	20	0	11	0	2	0	1	0	0	0	0	0	2176	2176
15	0	0	0	16	0	8	0	0	0	0	0	0	0	3072	3072	47	16	23	0	10	1	1	0	1	0	0	0	0	0	2048	2048
16	0	24	0	14	0	3	0	0	0	0	0	0	0	2048	2048	48	8	20	0	14	0	3	0	0	0	0	0	0	0	2016	2016
17	0	32	0	16	0	0	0	0	0	0	0	0	0	1536	1536	49	8	20	0	14	0	3	0	0	0	0	0	0	0	2016	2016
18	0	12	0	14	0	2	0	0	0	2	0	0	0	3648	3648	4a	8	15	0	13	1	4	0	0	0	0	0	0	0	2272	2272
19	0	12	0	14	0	2	0	0	0	2	0	0	0	3648	3648	4b	8	20	0	11	0	1	0	1	0	1	0	0	0	2912	2912
1a	0	12	0	12	0	7	0	0	0	0	0	0	0	2752	2752	4c	35	22	1	8	1	1	0	0	0	0	0	0	0	1440	1440
1b	0	12	0	10	0	8	0	0	0	0	0	0	0	2880	2880	4d	35	22	1	8	1	1	0	0	0	0	0	0	0	1440	1440
1c	16	32	0	4	0	4	0	0	0	0	0	0	0	1856	1856	4e	27	30	1	6	1	1	0	0	0	0	0	0	0	1408	1408
1d	16	32	0	4	0	4	0	0	0	0	0	0	0	1856	1856	4f	27	30	1	6	1	1	0	0	0	0	0	0	0	1408	1408
1e	16	16	0	20	0	0	0	0	0	0	0	0	0	1600	1600	50	0	0	0	4	0	4	0	0	0	3	0	0	1	8448	8448
1f	16	16	0	20	0	0	0	0	0	0	0	0	0	1600	1600	51	0	8	0	16	0	6	0	0	0	0	0	0	0	2688	2688
20	0	0	0	0	0	0	0	0	0	4	0	0	2	12288	12288	52	0	8	0	8	0	5	0	0	0	1	0	1	0	5248	5248
21	0	0	0	0	0	6	0	0	0	3	0	0	1	8704	8704	53	16	27	0	11	1	1	0	0	0	0	0	0	0	1600	1600
22	0	0	0	0	0	16	0	0	0	0	0	0	0	4096	4096	54	0	27	0	9	1	2	0	0	0	1	0	0	0	2688	2688
23	0	0	0	18	0	5	0	0	0	1	0	0	0	3456	3456	55	0	16	0	16	0	2	0	0	0	1	0	0	0	2816	2816
24	0	0	0	12	0	10	0	0	0	0	0	0	0	3328	3328	56	16	20	0	11	0	2	0	1	0	0	0	0	0	2176	2176
25	0	0	0	8	0	10	0	0	0	1	0	0	0	4096	4096	57	16	23	0	10	1	1	0	1	0	0	0	0	0	2048	2048
26	0	0	0	22	0	3	0	0	0	1	0	0	0	3200	3200	58	8	20	0	14	0	3	0	0	0	0	0	0	0	2016	2016
27	0	0	0	28	0	2	0	0	0	0	0	0	0	2304	2304	59	8	20	0	14	0	3	0	0	0	0	0	0	0	2016	2016
28	0	0	0	18	0	5	0	0	0	1	0	0	0	3456	3456	5a	8	15	0	13	1	4	0	0	0	0	0	0	0	2272	2272
29	0	0	0	18	0	5	0	0	0	1	0	0	0	3456	3456	5b	8	20	0	11	0	1	0	1	0	1	0	0	0	2912	2912
2a	0	0	0	24	0	4	0	0	0	0	0	0	0	2560	2560	5c	35	22	1	8	1	1	0	0	0	0	0	0	0	1440	1440
2b	0	0	0	18	0	5	0	0	0	1	0	0	0	3456	3456	5d	35	22	1	8	1	1	0	0	0	0	0	0	0	1440	1440
2c	0	28	0	16	0	1	0	0	0	0	0	0	0	1728	1728	5e	27	30	1	6	1	1	0	0	0	0	0	0	0	1408	1408
2d	0	28	0	16	0	1	0	0	0	0	0	0	0	1728	1728	5f	27	30	1	6	1	1	0	0	0	0	0	0	0	1408	1408
2e	0	28	0	16	0	1	0	0	0	0	0	0	0	1728	1728	60	0	0	0	4	0	8	0	0	0	3	0	0	0	5376	5376
2f	0	28	0	16	0	1	0	0	0	0	0	0	0	1728	1728	61	0	8	0	16	0	6	0	0	0	0	0	0	0	2688	2688
30	0	0	0	0	0	4	0	0	0	2	0	0	2	11264	11264	62	0	8	0	16	0	6	0	0	0	0	0	0	0	2688	2688
31	0	0	0	12	0	6	0	0	0	2	0	0	0	4352	4352	63	16	27	0	11	1	1	0	0	0	0	0	0	0	1600	1600
32	0	0	0	16	0	8	0	0	0	0	0	0	0	3072	3072	64	0	27	0	17	1	0	0	0	0	0	0	0	0	1664	1664

Correlation Cube Attacks: From Weak-Key Distinguisher to Key Recovery

Meicheng Liu(✉), Jingchun Yang, Wenhao Wang, and Dongdai Lin

State Key Laboratory of Information Security, Institute of Information Engineering, Chinese Academy of Sciences, Beijing 100093, People's Republic of China
meicheng.liu@gmail.com

Abstract. In this paper, we describe a new variant of cube attacks called *correlation cube attack*. The new attack recovers the secret key of a cryptosystem by exploiting conditional correlation properties between the superpoly of a cube and a specific set of low-degree polynomials that we call a *basis*, which satisfies that the superpoly is a zero constant when all the polynomials in the basis are zeros. We present a detailed procedure of correlation cube attack for the general case, including how to find a basis of the superpoly of a given cube. One of the most significant advantages of this new analysis technique over other variants of cube attacks is that it converts from a weak-key distinguisher to a key recovery attack.

As an illustration, we apply the attack to round-reduced variants of the stream cipher Trivium. Based on the tool of numeric mapping introduced by Liu at CRYPTO 2017, we develop a specific technique to efficiently find a basis of the superpoly of a given cube as well as a large set of potentially good cubes used in the attack on Trivium variants, and further set up deterministic or probabilistic equations on the key bits according to the conditional correlation properties between the superpolys of the cubes and their bases. For a variant when the number of initialization rounds is reduced from 1152 to 805, we can recover about 7-bit key information on average with time complexity 2^{44}, using 2^{45} keystream bits and preprocessing time 2^{51}. For a variant of Trivium reduced to 835 rounds, we can recover about 5-bit key information on average with the same complexity. All the attacks are practical and fully verified by experiments. To the best of our knowledge, they are thus far the best known key recovery attacks for these variants of Trivium, and this is the first time that a weak-key distinguisher on Trivium stream cipher can be converted to a key recovery attack.

Keywords: Cryptanalysis · Cube attack · Numeric mapping
Stream cipher · Trivium

This work was supported by the National Natural Science Foundation of China (Grant No. 61672516), the Strategic Priority Research Program of the Chinese Academy of Sciences (Grant No. XDA06010701), and the Fundamental Theory and Cutting Edge Technology Research Program of Institute of Information Engineering, CAS (Grant No. Y7Z0331102).

J. B. Nielsen and V. Rijmen (Eds.): EUROCRYPT 2018, LNCS 10821, pp. 715–744, 2018.
https://doi.org/10.1007/978-3-319-78375-8_23

1 Introduction

In recent years, cube attacks [11] and their variants [2,12,18] have been proven powerful in the security analysis of symmetric cryptosystems, such as Trivium [2,8,11,15], Grain-128 [9,12,16] and Keccak sponge function [3,10,18], producing the best cryptanalytic results for these primitives up to the present. Cube attacks were introduced by Dinur and Shamir at EUROCRYPT 2009 [11]. They are a generalization of chosen IV statistical attacks on stream ciphers [13,14,27], as well as an extension of higher order differential cryptanalysis [21] and AIDA [32]. The attacks treat a cryptosystem as a black-box polynomial. An attacker evaluates the sum of the output of polynomials system with a fixed private key over a subset of public variables, called a *cube*, in the hope of finding a linear coefficient of the term with maximum degree over the cube, referred to as a *superpoly*. The basic idea of cube attacks is that the symbolic sum of all the derived polynomials obtained from the black-box polynomial by assigning all the possible values to the cube variables is exactly the superpoly of the cube. The target of cube attacks is to find a number of linear superpolys on the secret variables and recover the secret information by solving a system of linear equations. In [11], the techniques was applied to a practical full key recovery on a variant of Trivium reduced to 767 rounds.

Since the seminal work of Dinur and Shamir, several variants of cube attacks have been proposed, including cube testers [2], dynamic cube attacks [12] and conditional cube attacks [18]. A cube tester [2] can detect the nonrandomness in cryptographic primitives by extracting the testable properties of the superpoly, such as unbalance, constantness and low degree, with the help of property testers. However a cube tester does not directly lead to key recovery attacks. Dynamic cube attacks [12] improve upon cube testers by introducing *dynamic variables*. When a set of conditions involving both the key bits and the dynamic variables are satisfied, the intermediate polynomials can be simplified, and cube testers (with assigned values to satisfy the conditions) are used to extract the nonrandomness of the cipher output. In this respect, a system of equations in the key bits and the dynamic variables are established. The discovery of the conditions mostly attributes to the manual work of analyzing the targeted cipher structure. Conditional cube attacks [18] work by introducing *conditional cube variables* and imposing conditions to restrain the propagation of conditional cube variables. Similar to dynamic cube attacks, the conditions used in conditional cube attacks are required to be dependent on both public bits and secret bits.

A key step to a successful cube-like attack is the search of good cubes and the corresponding superpolys during a precomputation phase. When such cubes are found, the attacker simply establishes and solves a polynomial system regarding the private key during the online phase. When cube attacks were first introduced in [11], the cryptosystems were regarded as black-box, and the authors used random walk to search for cubes experimentally. As the sum over a cube of size d involves 2^d evaluations under the fixed key, the search of cubes is time-consuming and the size of the cube is typically around 30, which restricts the capability of the attacker for better cubes. In [1] Aumasson *et al.* proposed an evolutionary algorithm to search for good cubes. Greedy bit set algorithm was applied by Stankovski in [28] to finding cubes in distinguishers of stream ciphers. The authors of [15] and [23] both used the union of two subcubes to generate larger cube candidates. With the improved cubes of size between 34 and 37, a key recovery attack on TRIVIUM reduced to 799 rounds [15] and a distinguisher on TRIVIUM reduced to 839 rounds [23] are proposed.

Recently two works on cube attacks using large cubes of size greater than 50 were presented at CRYPTO 2017. Both of them treat the cryptosystems as non-blackbox polynomials. The one by Todo *et al.* [30] uses the propagation of the bit-based division property (see also [29,31]) of stream ciphers, and presents possible key recovery attacks on 832-round TRIVIUM, 183-round `Grain`-128a and 704-round ACORN with the cubes of sizes 72, 92 and 64 respectively. The other one by Liu [22] uses numeric mapping to iteratively obtain the upper bound on the algebraic degree of an NFSR-based cryptosystem. Based on the tool of numeric mapping, cube testers are found for 842-round TRIVIUM, 872-round KREYVIUM, 1035-round TRIVIA-SC (v1) and 1047-round TRIVIA-SC (v2) with the cubes of sizes 37, 61, 63 and 61 resepectively [22].

Our Contributions. In this paper, we propose a new variant of cube attacks, named *correlation cube attack*. The general idea of this new attack is to exploit conditional correlation properties between the superpoly of a cube and a specific set of low-degree polynomials that we call a *basis*. The basis satisfies that the superpoly is a zero constant when all the polynomials in the basis are zeros. If the basis involves secret bits and has non-zero correlation with the superpoly, we can recover the secret information by solving probabilistic equations.

The attack consists of two phases: the preprocessing phase and online phase. The preprocessing phase tries to find a basis of a superpoly and its conditional correlation properties with the superpoly. The online phase targets at recovering the key by setting up and solving systems of probabilistic equations. We give a detailed procedure of both phases for the general case, including how to find a basis of the superpoly of a given cube.

As an illustration, we apply the attack to two reduced variants of the well-known stream cipher TRIVIUM [8], and obtain the best known key recovery results for these variants. TRIVIUM uses an 80-bit key and an 80-bit initial value (IV). We present two attacks for a variant of TRIVIUM when the number of initialization rounds is reduced from 1152 to 805. The first attack recovers about 7 equations on the key bits by 2^4 trials on average, *i.e.*, 3-bit key information, using 2^{37}-bit operations and 2^{37}-bit data, at the expense of preprocessing time 2^{47}. In the second attack, we can recover about 14 equations on the key bits by 2^7 trials on average, *i.e.*, 7-bit key information, with time complexity 2^{44} and 2^{45} keystream bits, at the expense of preprocessing time 2^{51}. For a variant of TRIVIUM reduced to 835 rounds, we can recover about 11 equations by 2^6 trials on average with the same complexity, that is, we can recover about 5-bit key information on average. The equations we recovered are linear or quadratic, and the quadratic ones can be easily linearized after guessing a few bits of the key.

All the attacks are directly valid for more than 30% of the keys in our experiments, and it also works for most of the other keys at the cost of recovering less key information.

Our results are summarized in Table 1 with the comparisons of the previous key recovery attacks on TRIVIUM. In this table, by "Time" we mean the time complexity of a full key recovery. The attack time $2^{99.5}$ [24] of the full cipher is measured by bit operations, while the others are measured by cipher operations. The previous best known practical partial key recovery is applicable to a variant of Trivium reduced to 799 rounds, proposed by Fouque and Vannet [15]. The previous best known impractical (and possible) partial key recovery that is faster than an exhaustive search is applicable to a variant of Trivium reduced to 832 rounds, presented by Todo *et al.* [30]. This was shown by recovering the superpoly of a cube of size 72 with preprocessing time 2^{77}. It is possible to extract at most one key bit expression (if the superpoly depends on the key). At the same time, it is also possible that it is a distinguisher rather than a key recovery (when the superpoly does not depend on the key). In this paper, we convert from a practical weak-key distinguisher to a practical partial key recovery attack, which is applicable to a variant of Trivium reduced to 835 rounds.

Table 1. Key recovery attacks on round-reduced TRIVIUM

#Rounds	Preproc	Data	Time	Ref
576	-	2^{12}	2^{33}	[32]
672	-	2^{15}	2^{55}	[14]
735	-	2^{29}	2^{30}	[11]
767	-	2^{34}	2^{36}	[11]
784	-	2^{39}	2^{38}	[15]
799	-	2^{40}	2^{62}	[15]
805	2^{47}	2^{37}	2^{77}	Section 4.3
805	2^{51}	2^{44}	2^{73}	Section 4.5
832	2^{77}	2^{72}	N.A	[30]
835	2^{51}	2^{44}	2^{75}	Section 4.4
Full	-	$2^{61.5}$	$2^{99.5}$	[24]
Full	-	-	2^{80}	Brute Force

The first and most critical steps in our attack are how to find good cubes and their bases. Benefited from the tool of numeric mapping [22], one can evaluate an upper bound on the algebraic degree of internal state of TRIVIUM in linear running time. Based on this tool, we specialize the techniques to efficiently find a basis of the superpoly of a given cube as well as a large set of potentially good cubes. After this, we evaluate the conditional probability $\Pr(g = 0|f_c(key, \cdot) \equiv 0)$ and $\Pr(g = 1|f_c(key, \cdot) \not\equiv 0)$ for a random fixed key, where g is a function depending on key bits in the basis of the superpoly f_c of a cube c and $f_c(key, \cdot)$ denotes the function f_c restricted at a fixed key. Finally, we record all the equations with high probability, and use them to recover the key. In the attacks, we use up to 54 cubes of sizes 28, 36 or 37. While we have found a thousand potentially favorite cubes with sizes 36 and 37 for TRIVIUM reduced to from 833 to 841 rounds, we can only make use of a small number of them in our attacks due to a limited computation resource.

Besides, we also partially apply our techniques to the stream ciphers TRIVIA-SC [5,6] and KREYVIUM [4]. We have found some cubes whose superpolys after 1047 and 852 rounds have a low-degree basis with a few elements for TRIVIA-SC and KREYVIUM respectively. The cubes for TRIVIA-SC have size larger than 60, and for KREYVIUM the size is at least 54. Though we are unable to fully verify the validity of the attack on TRIVIA-SC and KREYVIUM, we believe that there is a high chance of validness due to their similar structures with TRIVIUM.

Related Work. Similar to dynamic cube attacks and conditional cube attacks, correlation cube attacks recover the key by exploiting cube testers with constraints. Dynamic cube attacks [9,12] was applied to the full Grain-128 [16], while conditional cube attacks [18] was applied to round-reduced variants of KECCAK sponge function [3]. Unlike these attacks, however, the new attacks do not require the conditions to be dependent on public bits. The conditions imposed on conditional cube variables in conditional cube attacks also form a basis of the superpoly of a cube. Therefore, correlation cube attacks can be considered as a generalization of conditional cube attacks.

Actually, the idea of assigning (dynamic) constraints to public variables and using them to recover key bits was earlier appeared in conditional differential attacks, which was introduced by Knellwolf, Meier and Naya-Plasencia at ASIACRYPT 2010 [19]. The authors classified the conditions into three types:

- Type 0 conditions only involve public bits;
- Type 1 conditions involve both public bits and secret bits;
- Type 2 conditions only involve secret bits.

They exploited type 2 conditions to derive key recovery attacks based on hypothesis tests, as well as type 1 conditions to recover the key in another different way. This technique was applied to reduced variants of a few ciphers, including Grain-v1 [17], Grain-128 [16] and the block cipher family KATAN/KTANTAN [7]. Correlation cube attacks also exploit type 1 and type 2 conditions to derive key recovery attacks, while the underlying idea is very different from the work of [19]. Our techniques for finding a basis of the superpoly of a cube are more related to the automatic strategies for analyzing the conditions of higher order derivatives [20], which were exploited to derive weak-key distinguishing attacks on reduced variants of TRIVIUM. Nevertheless, the ideas are still different, and our strategies are more customized and suitable for key recovery attacks.

Organization. The rest of this paper is structured as follows. In Sect. 2, the basic definitions, notations, and background are provided. Section 3 shows the general framework of correlation cube attack, while its applications to Trivium are given in Sect. 4. Section 5 concludes the paper.

2 Preliminaries

Boolean Functions and Algebraic Degree. Let $\mathbb{F}_2$ denote the binary field and $\mathbb{F}_2^n$ the n-dimensional vector space over $\mathbb{F}_2$. An n-variable Boolean function is a mapping from $\mathbb{F}_2^n$ into $\mathbb{F}_2$. Denote by $\mathbb{B}_n$ the set of all n-variable Boolean functions. An n-variable Boolean function f can be uniquely represented as a multivariate polynomial over $\mathbb{F}_2$,

$$f(x_1, x_2, \cdots, x_n) = \bigoplus_{c=(c_1,\cdots,c_n)\in\mathbb{F}_2^n} a_c \prod_{i=1}^{n} x_i^{c_i},\ a_c \in \mathbb{F}_2,$$

called the algebraic normal form (ANF). The algebraic degree of f, denoted by $\deg(f)$, is defined as $\max\{wt(c) \mid a_c \neq 0\}$, where $wt(c)$ is the Hamming weight of c.

Decomposition and Basis of Boolean Functions. Given a Boolean function f, we call $f = \bigoplus_{i=1}^{u} g_i \cdot f_i$ a decomposition of f, and $G = \{g_1, g_2, \cdots, g_u\}$ a basis of f. It is clear that $g = \prod_{i=1}^{u}(g_i + 1)$ is an annihilator of f, that is, $g \cdot f = 0$.

Cube Attacks and Cube Testers. Given a Boolean function f and a term t_I containing variables from an index subset I that are multiplied together, the function can be written as the sum of terms which are supersets of I and terms that miss at least one variable from I,

$$f(x_1, x_2, \cdots, x_n) = f_S(I) \cdot t_I \oplus q(x_1, x_2, \cdots, x_n),$$

where $f_S(I)$ is called the superpoly of I in f. The basic idea of cube attacks [11] and cube testers [2] is that the symbolic sum of all the derived polynomials obtained from the function f by assigning all the possible values to the subset of variables in the term t_I is exactly $f_S(I)$. The target of cube attacks is finding a set of linear (or low-degree) functions f_S's on the secret key and recovering the key by solving this linear (or low-degree) system. Cube testers work by evaluating superpolys of carefully selected terms t_I's which are products of public variables (*e.g.*, IV bits), and trying to distinguish them from a random function. Especially, the superpoly $f_S(I)$ is equal to a zero constant, if the algebraic degree of f in the variables from I is smaller than the size of I.

Numeric Mapping. Let $f(x) = \bigoplus_{c=(c_1,\cdots,c_n)\in\mathbb{F}_2^n} a_c \prod_{i=1}^{n} x_i^{c_i}$ be a Boolean function on n variables. The *numeric mapping* [22], denoted by DEG, is defined as

$$\texttt{DEG}:\ \mathbb{B}_n \times \mathbb{Z}^n \to \mathbb{Z},$$

$$(f, D) \mapsto \max_{a_c \neq 0}\{\sum_{i=1}^{n} c_i d_i\},$$

where $D = (d_1, d_2, \cdots, d_n)$ and a_c's are coefficients of the ANF of f. Let $g_i (1 \le i \le m)$ be Boolean functions on m variables, and denote $\deg(G) = (\deg(g_1), \deg(g_2), \cdots, \deg(g_n))$ for $G = (g_1, g_2, \cdots, g_n)$. The numeric degree of the composite function $h = f \circ G$ is defined as $\mathtt{DEG}(f, \deg(G))$, denoted by $\mathtt{DEG}(h)$ for short. The algebraic degree of h is always less than or equal to the numeric degree of h. The algebraic degrees of the output bits and the internal states can be estimated iteratively for NFSR-based cryptosystems by using numeric mapping [22].

3 Correlation Cube Attacks

In this section, we propose a new model for cube attacks, called *correlation cube attack*. It is a hybrid of correlation attacks [25] and cube attacks [11]. The attacked cryptosystem is supposed to be a modern symmetric-key cryptosystem. The general idea is to find a low-degree decomposition of the superpoly over a given cube, evaluate the correlation relations between the low-degree basis and the superpoly, and recover the key by solving systems of probabilistic equations. The low-degree decomposition is based on an upper bound on the algebraic degree and determines whether the superpoly is a zero constant when imposing some conditions.

The attack consists of two phases, the preprocessing phase and online phase. The preprocessing phase tries to find a basis of a superpoly and its correlation properties with the superpoly. The online phase targets at recovering the key by setting up and solving systems of probabilistic equations. In the following, we will give the details of the attack.

3.1 Preprocessing Phase

The procedure of preprocessing phase of the attack is depicted as Algorithm 1. In this phase, we can choose the input to the cipher, including the secret and public bits. First we generate a set of cubes which are potentially good in the attacks. Then for each cube c, we use a procedure `Decomposition` to find a low-degree basis of the superpoly f_c of c in the output bits of the cipher. The details of this procedure will be discussed later. If a basis of f_c is found, we calculate the conditional probability $\Pr(g = b | f_c)$ for each function g in the basis. More exactly, we compute the values of the superpoly f_c by choosing random keys and random values of free non-cube public bits, and evaluate the conditional probability $\Pr(g = 0 | f_c(key, \cdot) \equiv 0)$ and $\Pr(g = 1 | f_c(key, \cdot) \not\equiv 0)$ for a random key, where $f_c(key, \cdot)$ denotes the function f_c restricted at a fixed *key*. Finally, we record the set of (c, g, b) that satisfies $\Pr(g = b | f_c) > p$, *i.e.*,

$$\Omega = \{(c, g, b) | \Pr(g = b | f_c) > p\}.$$

Note that if g depends on both key bits and public bits, the attack will become more efficient, at least not worse, than the case that g depends only on key bits (which can naturally be used to mount weak-key distinguishing attacks). A distinguishing attack is a much weaker attack compared to a key recovery attack, while a weak-key distinguishing attack is even weaker than a normal distinguishing attack. To illustrate how to convert from a weak-key distinguishing attack to a key recovery attack, we assume the weak case: g only depends on key bits (if not, we can set the public bits in g to constants).

Algorithm 1. Correlation Cube Attacks (Preprocessing Phase)

1: Generate a cube set C;
2: For each cube c in C do:
3: $Q_c \leftarrow$ `Decomposition`(c), and goto next c if Q_c is empty; `/* try to find a basis of the superpoly` f_c `of` c `in the output bits of the cipher */`
4: Estimate the conditional probability $\Pr(g = b|f_c)$ for each function g in the basis Q_c of the superpoly f_c, and select (c, g, b) that satisfies $\Pr(g = b|f_c) > p$.

Example 1. Given a Boolean polynomial f on five public variables $v = (v_1, v_2, v_3, v_4, v_5)$ and five secret variables $x = (x_1, x_2, x_3, x_4, x_5)$,

$$\begin{aligned} f(v, x) = & f_7(v_5, x)v_1v_2v_3v_4 + f_6(v_5, x)v_1v_2v_4 \\ & + f_5(v_5, x)v_2v_3v_4 + f_4(v_5, x)v_1v_4 \\ & + f_3(v_5, x)v_2v_4 + f_2(v_5, x)v_3 \\ & + f_1(v_5, x)v_4 + f_0(v_5, x) \end{aligned}$$

and

$$f_7(v_5, x) = h_1(v_5, x_2, x_3, x_4, x_5)x_1 + h_2(v_5, x_1, x_2, x_3, x_4)x_5,$$

where h_1, h_2 and $f_i (0 \le i \le 6)$ are arbitrary Boolean functions. We can build a weak-key cube tester for the polynomial f, by using the cube $\{v_1, v_2, v_3, v_4\}$ under the conditions $x_1 = x_5 = 0$, while it seems to be immune to cube or dynamic cube attacks. To convert from a weak-key cube tester to a key recovery, we test the correlation properties between the superpoly f_7 and its basis $\{x_1, x_5\}$. We observe the values of $f_7(v_5, x)$ for $v_5 = 0, 1$, and estimate the conditional probability

$$\Pr(x_i = 0|f_7(0, x) = f_7(1, x) = 0)$$

and

$$\Pr(x_i = 1|f_7(0, x) \neq 0 \text{ or } f_7(1, x) \neq 0)$$

for $i = 1, 5$. Noting that $(x_1 + 1)(x_5 + 1)f_7 = 0$, we also have

$$(x_1 + 1)(x_5 + 1) = 0 \text{ if } f_7(0, x) \neq 0 \text{ or } f_7(1, x) \neq 0.$$

This allows us to derive information regarding the secret key.

Now we explain how to find a basis of the superpoly f_c for a given cube c. The procedure `Decomposition` is described in Algorithm 2. The main idea is to make use of the coefficients Q_t of the terms with maximum degree on cube variables in the bits s_t of the internal state at the first rounds of the attacked cipher ($t \leq N_0$). Note that it is highly possible that f_c depends on these coefficients. To find a set Q such that $f_c = \bigoplus_{g \in Q} g \cdot f_g$, we first annihilate all the coefficients in Q_t with $1 \notin Q_t$ for all $t \leq N_0$, and then determine whether f_c is a zero constant. Once we detect that the algebraic degree of an output bit on cube variables is less than the size of c, which implies $f_c = 0$, we obtain a basis $Q = \bigcup_{t \leq N_0 | 1 \notin Q_t} Q_t$ for f_c. Then we minimize the number of elements in Q by removing redundant equations one by one. Finally, the procedure returns the minimum basis Q.

Algorithm 2. `Decomposition`

Require: a cube c of size n

1: Set Q to the empty set and X to the variable set $\{x_i | i \in c\}$;

/* find a basis Q */

2: For t from 0 to N_0 do:
3: Compute the ANF of s_t and set $d_t = \deg(s_t, X)$;
4: Set Q_t to the set of the coefficients of all the terms with degree d_t in the ANF of s_t;
5: If $d_t \geq 1$ and $1 \notin Q_t$, then set $Q = Q \cup Q_t$ and $d_t = \deg(s'_t, X)$, where s'_t is the function formed by removing all the terms with degree d_t from s_t;
6: Given $\{d_t\}$ and under the conditions that $g = 0$ for each $g \in Q$, find an upper bound $d(Q)$ on the degree of the N-round output bit;
7: If $d(Q) \geq n$, then **return** $\emptyset$;

/* minimize the basis Q */

8: Minimize N_0 such that $d(Q) < n$, and generate a new Q;
9: For each g in Q do:
10: Set $Q' = Q \setminus \{g\}$;
11: For $t \leq N_0$, if zero(Q') $\subseteq$ zero(Q_t) then set $d_t = \deg(s'_t, X)$, otherwise set $d_t = \deg(s_t, X)$, where zero(Q) is the solution set of $\{g = 0 | g \in Q\}$;
12: If $d(Q') < n$, then set $Q = Q'$;
13: **return** Q.

For explanation of Algorithm 2, we give an example on a nonlinear feedback shift register (NFSR) in the following.

Example 2. Let $s_t = s_{t-6}s_{t-7} + s_{t-8}$ be the update function of an NFSR with size 8. Let $(s_0, s_1, \cdots, s_7) = (x_1, x_2, x_3, x_4, v_1, v_2, v_3, 0)$, and $X = \{v_1, v_2, v_3\}$ be the cube variables. Taking $t = 10$ for example, we compute $s_{10} = s_4 s_3 + s_2 = v_1 x_4 + x_3$, then have $d_{10} = 1$, $Q_{10} = \{x_4\}$ and $s'_{10} = x_3$. Since $1 \notin Q_{10}$, we set $Q = Q \cup Q_{10}$ and $d_{10} = 0$. After computations for $t \leq N_0 = 17$, we obtain

$$Q = Q_{10} \cup Q_{16} \cup Q_{17} = \{x_4, x_2 x_4 + x_3 x_4, x_3 + x_4\},$$
$$(d_0, d_1, \cdots, d_{17}) = (0, 0, 0, 0, 1, 1, 1, -\infty, 0, 0, 0, 2, 2, 1, 1, 0, 0, 1).$$

For $N = 29$, we find an upper bound $d(Q) = 2$ on the algebraic degree of s_N by applying the numeric mapping. We can check that 17 is the minimum N_0 such that $d(Q) < n = 3$. After minimizing the basis Q, we obtain $Q = \{x_4, x_3 + x_4\}$.

Actually, the ANF of s_{29} is $v_1v_2v_3(x_2x_3x_4 + x_1x_3 + x_1x_4 + x_2x_4) + v_1v_3(x_2x_3x_4 + x_1x_4) + v_3(x_2x_3x_4 + x_1x_2) + v_2$, and the coefficient of the maximum term $v_1v_2v_3$ is $f_c = x_2x_3x_4 + x_1x_3 + x_1x_4 + x_2x_4$, which will be annihilated when $x_4 = x_3 + x_4 = 0$. We can see that Q is a basis of the superpoly f_c.

Complexity. It is hard to evaluate the complexity of the step for generating good cubes. How to find favorite cubes is still an intractable problem in cube attacks. The time complexity of `Decomposition` is $T_{N_0} + n_Q \cdot T_N$, where n_Q is the size of the primary basis $\bigcup_{t \le N_0 | 1 \notin Q_t} Q_t$, T_{N_0} is the time for computing this basis (Line 2–5 in Algorithm 2), and T_N is the time complexity of finding an upper bound on the algebraic degree of the N-round output bits. The estimation of the conditional probability $\Pr(g = b|f_c)$ for a cube c of size n needs about $\alpha \cdot 2^n$ cipher encryption operations, when using α values of f_c in the estimation. The total time complexity of preprocessing phase is thus about

$$n_C(T_{N_0} + n_Q \cdot T_N + \alpha \cdot 2^n),$$

where n_C is the number of cubes in C, not taking into account the time for generating the cube set C.

3.2 Online Phase

The procedure of online phase of the attack is depicted in Algorithm 3. In the online phase, the key is unknown, and we can only control the public bits. We first derive two sets of probabilistic equations according to the ciphertexts (or keystream bits), and then repeatedly solve a system consisting of a part of these equations until the correct key is found. In preprocessing phase, we have obtained a set Ω of (c, g, b) that satisfies $\Pr(g = b|f_c) > p$. In online phase, for each cube c, we test whether its superpoly f_c is a zero constant by computing α values of f_c over the cube with different non-cube public bits. If f_c is not a zero constant, we derive new equations $g = 1$ with $(c, g, 1) \in \Omega$; otherwise, we record the equations $g = 0$ with $(c, g, 0) \in \Omega$. After all the cubes are handled, we derive two sets of equations, $G_0 = \{g = 0|(c, g, 0) \in \Omega, f_c = 0\}$ and $G_1 = \{g = 1|(c, g, 1) \in \Omega, f_c \neq 0\}$. For the case that $\{g|g = 0 \in G_0 \text{ and } g = 1 \in G_1\}$ is not empty, we can use the one with higher probability between $g = 0$ and $g = 1$ or neither of them. We then randomly choose r_0 equations from G_0 and r_1 equations from G_1, solve these $r_0 + r_1$ equations and check whether the solutions are correct. Repeat this step until the correct key is found.

Complexity. The loop for deriving the equation sets G_0 and G_1 requires at most $n_C\alpha 2^n$ bit operations, where n_C is the number of cubes in C. Step 7 runs in time $2^{\ell_{key} - (r_0 + r_1)}$, where ℓ_{key} is the size of the key, when the equations are

Algorithm 3. Correlation Cube Attacks (Online Phase)

Require: a cube set C and $\Omega = \{(c, g, b) | \Pr(g = b | f_c) > p\}$
1: Set G_0 and G_1 to empty sets;
2: For each cube c in C do:
3: Randomly generate α values from free non-cube public bits, and request $\alpha 2^n$ keystream bits (or ciphertexts) corresponding to the cube c of size n and these non-cube public values;
4: Compute the α values of the superpoly f_c over the cube c;
5: If all the values of f_c equal 0, then set $G_0 = G_0 \cup \{g = 0 | (c, g, 0) \in \Omega\}$, otherwise set $G_1 = G_1 \cup \{g = 1 | (c, g, 1) \in \Omega\}$;
6: Deal with the case that $\{g | g = 0 \in G_0 \text{ and } g = 1 \in G_1\}$ is not empty;
7: Randomly choose r_0 equations from G_0 and r_1 equations from G_1, solve these $r_0 + r_1$ equations and check whether the solutions are correct;
8: Repeat Step 7 if none of the solutions is correct.

balanced and easy to be solved. We can estimate the probability $q > p^{(r_0+r_1)}$ that a trial successes, so the expected number of trials is $q^{-1} < p^{-(r_0+r_1)}$. Here we require that the total number of equations in G_0 and G_1 is greater than $p^{-(r_0+r_1)}$. The expected time of online phase is thus less than

$$n_C \alpha 2^n + p^{-(r_0+r_1)} 2^{\ell_{key}-(r_0+r_1)}.$$

3.3 Discussion

The crux point of the attack is finding a low-degree basis of the superpoly over a given cube, that is, finding Q with low degree such that

$$f_c = \bigoplus_{g \in Q} g \cdot f_g.$$

Theoretical Bound on the Probability. We first discuss the conditional probability $\Pr(g = 1 | f_c \neq 0)$, *i.e.*, for the case that f_c is not a zero constant for a fixed key. For any x such that $f_c(x) \neq 0$, there is at least one g such that $g(x) = 1$, that is, $\prod_{g \in Q}(g(x) + 1) = 0$. Specially, if Q contains only one function g, then $g(x) = 1$ holds with probability 1. If Q contains two functions g_1 and g_2, then we have $g_1(x) = 1$ or $g_2(x) = 1$, and thus at least one of them holds with probability $\geq \frac{2}{3}$. Generally, if Q contains n_Q functions, then there is at least one $g(x) = 1$ that holds with probability $\geq \frac{2^{n_Q-1}}{2^{n_Q}-1}$, under the condition $f_c \neq 0$.

The conditional probability $p_0 = \Pr(g = 0 | f_c = 0)$ can be computed according to $p_1 = \Pr(g = 1 | f_c \neq 0)$ and the probability γ that $f_c \neq 0$, *i.e.*, $p_0 = \frac{\frac{1}{2} - (1-p_1)\gamma}{1-\gamma}$.

When the upper bound $d(Q)$ on algebraic degree of f_c restricted to $\{g = 0 | g \in Q\}$ is tight in Algorithm 2, we expect that f_g is not a zero constant for $g \in Q$. If all the functions f_g's depend on the free non-cube bits, then f_c is a zero constant for a fixed key if and only if $g = 0$ holds with probability 1 (or close to 1) for all $g \in Q$.

Assuming that the event of (g, f_1) and the event of (g, f_2) are statistically independent, we have

$$\begin{aligned}\Pr(g = b|f_1, f_2) &= \frac{\Pr(g = b, f_1, f_2)}{\Pr(g = b, f_1, f_2) + \Pr(g = b + 1, f_1, f_2)} \\ &= \frac{\Pr(g = b, f_1)\Pr(g = b, f_2)}{\Pr(g = b, f_1)\Pr(g = b, f_2) + \Pr(g = b + 1, f_1)\Pr(g = b + 1, f_2)} \\ &= \frac{\Pr(g = b|f_1)\Pr(g = b|f_2)}{\Pr(g = b|f_1)\Pr(g = b|f_2) + \Pr(g = b + 1|f_1)\Pr(g = b + 1|f_2)}.\end{aligned}$$

Denote by ε_1 and ε_2 the correlation coefficients of $g = b$ given f_1 and f_2 respectively, *i.e.*, $\Pr(g = b|f_i) = \frac{1}{2}(1 + \varepsilon_i)$ for $i = 1, 2$. Then

$$\varepsilon = \frac{\varepsilon_1 + \varepsilon_2}{1 + \varepsilon_1\varepsilon_2}$$

is the correlation coefficient of the event that $g = b$ given both f_1 and f_2. Specially, if ε_1 and ε_2 have the same sign, then

$$|\varepsilon| = |\frac{\varepsilon_1 + \varepsilon_2}{1 + \varepsilon_1\varepsilon_2}| \geq \max\{|\varepsilon_1|, |\varepsilon_2|\}.$$

Our experiments on Trivium show that the assumption is reasonable. In fact, we do not expect that the assumption is perfectly true. The independence assumption is used to guarantee a bound on correlation coefficient. We believe the bound is sound, at least for the case that the correlations have the same sign, even if the assumption is not true in general.

Modifications of the Attack. We may slightly modify the online phase of the attack if necessary. As mentioned above, for the case that f_c is not a zero constant for a fixed key, we have $\prod_{g \in Q}(g(x) + 1) = 0$. We may make use of this kind of equations at Step 7 in Algorithm 3. Another modification is to use two different threshold probabilities separately for the equation sets G_0 and G_1 rather than the same one p.

4 Applications to Trivium Stream Cipher

In this section, we first give a brief description of the stream cipher Trivium [8], as well as recall the technique for estimating the degree of Trivium based on numeric mapping, and then apply the correlation cube attack to two variants of Trivium when the number of initialization rounds is reduced from 1152 to 805 and 835. At the end of this section, we will discuss the possible improvements, and partially apply our analysis techniques to the stream ciphers TriviA-SC [5,6] and Kreyvium [4].

4.1 Description of Trivium

A Brief Description of Trivium-Like Ciphers. Let A, B and C be three registers with sizes of n_A, n_B and n_C, denoted by A_t, B_t and C_t their corresponding states at clock t,

$$A_t = (x_t, x_{t-1}, \cdots, x_{t-n_A+1}), \quad (1)$$
$$B_t = (y_t, y_{t-1}, \cdots, y_{t-n_B+1}), \quad (2)$$
$$C_t = (z_t, z_{t-1}, \cdots, z_{t-n_C+1}), \quad (3)$$

and respectively updated by the following three quadratic functions,

$$x_t = z_{t-r_C} \cdot z_{t-r_C+1} + \ell_A(s^{(t-1)}), \quad (4)$$
$$y_t = x_{t-r_A} \cdot x_{t-r_A+1} + \ell_B(s^{(t-1)}), \quad (5)$$
$$z_t = y_{t-r_B} \cdot y_{t-r_B+1} + \ell_C(s^{(t-1)}), \quad (6)$$

where $1 \leq r_\lambda < n_\lambda$ for $\lambda \in \{A, B, C\}$ and ℓ_A, ℓ_B and ℓ_C are linear functions. We denote $A_t[i] = x_i$, $B_t[i] = y_i$ and $C_t[i] = z_i$, and define $g_A^{(t)} = z_{t-r_C} \cdot z_{t-r_C+1}$, $g_B^{(t)} = x_{t-r_A} \cdot x_{t-r_A+1}$ and $g_C^{(t)} = y_{t-r_B} \cdot y_{t-r_B+1}$. The internal state, denoted by $s^{(t)}$ at clock t, consists of the three registers A, B, C, that is, $s^{(t)} = (A_t,\ B_t,\ C_t)$. Let f be the output function. After an initialization of N rounds, in which the internal state is updated for N times, the cipher generates a keystream bit by $f(s^{(t)})$ for each $t \geq N$.

The stream ciphers Trivium (designed by De Cannière and Preneel [8]) and TriviA-SC (designed by Chakraborti *et al.* [5,6]) exactly fall into this kind of ciphers. Kreyvium [4] is a variant of Trivium with 128-bit security, designed by Canteaut *et al.* at FSE 2016 for efficient homomorphic-ciphertext compression. Compared with Trivium, Kreyvium uses two extra registers (K^*, V^*) without updating but shifting, *i.e.*, $s^{(t)} = (A_t,\ B_t,\ C_t, K^*, V^*)$, and add a single bit of (K^*, V^*) to each of ℓ_A and ℓ_B, where K^* and V^* only involve the key bits and IV bits respectively. Trivium uses an 80-bit key and an 80-bit initial value (IV), while Kreyvium and TriviA-SC both use a 128-bit key and a 128-bit IV. All these ciphers have 1152 rounds.

A Brief Description of Trivium. Trivium contains a 288-bit internal state with three NFSRs of different lengths. The key stream generation consists of an iterative process which extracts the values of 15 specific state bits and uses them both to update 3 bits of the state and to compute 1 bit of key stream. The algorithm is initialized by loading an 80-bit key and an 80-bit IV into the 288-bit initial state, and setting all remaining bits to 0, except for three bits. Then, the state is updated for $4 \times 288 = 1152$ rounds, in the same way as explained above, but without generating key stream bits. This is summarized in the pseudo-code below.

$(x_0, x_{-1}, \cdots, x_{-92}) \leftarrow (k_0, k_1, \cdots, k_{79}, 0, \cdots, 0)$
$(y_0, y_{-1}, \cdots, y_{-83}) \leftarrow (iv_0, iv_1, \cdots, iv_{79}, 0, \cdots, 0)$
$(z_0, z_{-1}, \cdots, z_{-110}) \leftarrow (0, \cdots, 0, 1, 1, 1)$
for i **from** 1 **to** N **do**
$x_i = z_{i-66} + z_{i-111} + z_{i-110} \cdot z_{i-109} + x_{i-69}$
$y_i = x_{i-66} + x_{i-93} + x_{i-92} \cdot x_{i-91} + y_{i-78}$
$z_i = y_{i-69} + y_{i-84} + y_{i-83} \cdot y_{i-82} + z_{i-87}$
if $N > 1152$ **then**
$ks_{i-1152} = z_{i-66} + z_{i-111} + x_{i-66} + x_{i-93} + y_{i-69} + y_{i-84}$
end if
end for

4.2 Degree Estimation of Trivium

In this section, we recall the algorithm proposed by Liu [22] for estimating algebraic degree of the output of f after N rounds for a Trivium-like cipher, as described in Algorithm 4.

This algorithm first computes the exact algebraic degrees of the internal states for the first N_0 rounds, where the degrees of the functions $g_A^{(t)}$, $g_B^{(t)}$ and $g_C^{(t)}$ are also recorded, then iteratively compute $D^{(t)}$ for $t = N_0+1, N_0+2, \cdots, N$, and finally apply the numeric mapping to calculate an estimated degree for the first bit of the keystream. In Algorithm 4, three sequences, denoted by d_A, d_B and d_C, are used to record the estimated degrees of the three registers A, B, C. In each step of a Trivium-like cipher, three bits are updated. Accordingly, the estimated degrees for these three bits in each step t are calculated, denoted by $d_A^{(t)}, d_B^{(t)}$ and $d_C^{(t)}$. Then update $D^{(t)}$ from $D^{(t-1)}$. For estimating the algebraic degrees of x_t, y_t, z_t, the two procedures DegMul* and DEG deal with their "quadratic" and "linear" parts separately. The procedure DegMul* is used to compute an upper bound on the algebraic degree of $g_A^{(t)} = z_{t-r_C} \cdot z_{t-r_C+1}$, $g_B^{(t)} = x_{t-r_A} \cdot x_{t-r_A+1}$ and $g_C^{(t)} = y_{t-r_B} \cdot y_{t-r_B+1}$. It has been demonstrated in [22] that for all t with $1 \leq t \leq N$ the estimated degrees $d_A^{(t)}, d_B^{(t)}, d_C^{(t)}$ for x_t, y_t, z_t are greater than or equal to their corresponding algebraic degrees, and therefore the output $\texttt{DEG}(f, D^{(N)})$ of Algorithm 4 gives an upper bound on algebraic degree of the N-round output bit of a Trivium-like cipher.

The algorithm has a linear time and space complexity on N, if we do not take into account the time and memory used for computing the exact algebraic degrees of the internal states for the first N_0 rounds.

4.3 The Attack on 805-Round TRIVIUM

Generating a Candidate Set of Favorite Cubes. A favorite cube of size 37 was found in [22] for distinguishing attacks on TRIVIUM. We exhaustively search the subcubes with size 28 of this cube, and pick up the subcubes such that the corresponding superpolys after 790 rounds are zero constants (*i.e.*, the output bits after 790 rounds do not achieve maximum algebraic degree over the subcube variables), by using Algorithm 4 with $N_0 = 0$. Then we find 5444 such subcubes.

Algorithm 4. Estimation of Degree of TRIVIUM-Like Ciphers [22]

Require: Given the ANFs of all internal states (A_t, B_t, C_t) with $t \leq N_0$, and the set of variables X.
1: For λ in $\{A, B, C\}$ do:
2: For t from $1 - n_\lambda$ to 0 do:
3: $d_\lambda^{(t)} \leftarrow \deg(\lambda_0[t], X)$; // $A_i[t] = x_t$, $B_i[t] = y_t$ and $C_i[t] = z_t$
4: $D^{(0)} \leftarrow (d_A^{(1-n_A)}, \cdots, d_A^{(0)}, d_B^{(1-n_B)}, \cdots, d_B^{(0)}, d_C^{(1-n_C)}, \cdots, d_C^{(0)})$;
/* Compute the exact algebraic degrees of the internal states for the first N_0 rounds */
5: For t from 1 to N_0 do:
6: For λ in $\{A, B, C\}$ do:
7: $dm_\lambda^{(t)} \leftarrow \deg(g_\lambda^{(t)}, X)$;
8: $d_\lambda^{(t)} \leftarrow \deg(\lambda_t[t], X)$;
9: $D^{(t)} \leftarrow (d_A^{(t-n_A+1)}, \cdots, d_A^{(t)}, d_B^{(t-n_B+1)}, \cdots, d_B^{(t)}, d_C^{(t-n_C+1)}, \cdots, d_C^{(t)})$;
/* Iteratively compute the upper bounds on algebraic degrees of the internal states for the remaining rounds */
10: For t from $N_0 + 1$ to N do:
11: For λ in $\{A, B, C\}$ do:
12: $dm_\lambda^{(t)} \leftarrow \mathtt{DegMul}^*(g_\lambda^{(t)})$;
13: $d_\lambda^{(t)} \leftarrow \max\{dm_\lambda^{(t)}, \mathtt{DEG}(\ell_\lambda, D^{(t-1)})\}$;
14: $D^{(t)} \leftarrow (d_A^{(t-n_A+1)}, \cdots, d_A^{(t)}, d_B^{(t-n_B+1)}, \cdots, d_B^{(t)}, d_C^{(t-n_C+1)}, \cdots, d_C^{(t)})$;
15: **return** $\mathtt{DEG}(f, D^{(N)})$.
/* Description of the procedure $\mathtt{DegMul}^*(g_\lambda^{(t)})$ for $\lambda \in \{A, B, C\}$ */
procedure $\mathtt{DegMul}^*(g_\lambda^{(t)})$
16: $t_1 \leftarrow t - r_{\rho(\lambda)}$; // $\rho(A) = C, \rho(C) = B, \rho(B) = A$
17: If $t_1 \leq 0$ then:
return $d_{\rho(\lambda)}^{(t_1)} + d_{\rho(\lambda)}^{(t_1+1)}$.
18: $t_2 \leftarrow t_1 - r_{\rho^2(\lambda)}$;
19: $d_1 \leftarrow \min\{d_{\rho^2(\lambda)}^{(t_2)} + dm_{\rho(\lambda)}^{(t_1+1)}, d_{\rho^2(\lambda)}^{(t_2+2)} + dm_{\rho(\lambda)}^{(t_1)}, d_{\rho^2(\lambda)}^{(t_2)} + d_{\rho^2(\lambda)}^{(t_2+1)} + d_{\rho^2(\lambda)}^{(t_2+2)}\}$;
20: $d_2 \leftarrow \mathtt{DEG}(\ell_{\rho(\lambda)}, D^{(t_1)}) + dm_{\rho(\lambda)}^{(t_1)}$;
21: $d_3 \leftarrow \mathtt{DEG}(\ell_{\rho(\lambda)}, D^{(t_1-1)}) + d_{\rho(\lambda)}^{(t_1+1)}$;
22: $d \leftarrow \max\{d_1, d_2, d_3\}$;
23: **return** d.
end procedure

Finding the Basis and Free Non-cube IV Bits. We apply the procedure `Decomposition` to each cube c from the 5444 candidates, setting all the non-cube IV bits to zeros. Note here that we use Algorithm 4 in the procedure `Decomposition` to find an upper bound $d(Q)$ on the algebraic degree. Once a non-trivial basis of f_c is found, we set one of the non-cube IV bits to a parameter variable, and apply `Decomposition` again. This bit is considered as a free IV bit if it does not affect the basis, and otherwise we set it to 0. Then we add another non-cube IV bit to be a parameter variable, and do this again. By this way, we obtain a set of free non-cube IV bits.

Using this method, we get 47 cubes of size 28 satisfying that a basis of the superpoly after 805 rounds can be found. To make the attack more efficient, we further search the cubes whose superpolys after 805 rounds have a basis containing at most two elements. Once such cubes are found, we modify them by randomly shifting and changing some indexes, and test them by the same method. After computations within a dozen hours on a desktop computer, we are able to find more than 100 cubes whose superpolys after 805 rounds have a basis with one or two elements.

Computing the Probability. We test 32 out of these cubes, each of which has a different basis after 805 rounds, according to Step 4 of Algorithm 1. In each test, we compute the values of the superpoly f_c for 1000 random keys and at most $\alpha = 16$ non-cube IVs for each key, and evaluate the conditional probability $\Pr(g = 0|f_c(key, \cdot) \equiv 0)$ and $\Pr(g = 1|f_c(key, \cdot) \not\equiv 0)$ for a random fixed key, where g is a function depending on key bits in the basis of f_c and $f_c(key, \cdot)$ denotes the function f_c restricted at a fixed key. Our experiment shows that all the computations need about $13.5 \cdot 1000 \cdot 32 \cdot 2^{28} \approx 2^{47}$ cipher operations. We remind the readers that, in our experiment, we take all the possible values of the first $\log_2(\alpha) = 4$ free non-cube IV bits and set random values for the other free non-cube IV bits. Once we observe a non-zero value of the superpoly f_c, we skip the remaining IVs and continue to compute for the next key. On average, we need to compute 13.5 IVs for each key.

The results are listed in Table 3 in Appendix, together with the cubes, free non-cube IV bits and the equations. Note that 4 out of the 32 cubes are excluded from the table due to their little impact in our attack. In Table 3, by $p(0|0)$ (resp., $p(1|1)$) we mean the conditional probability of $g = 0$ (resp., $g = 1$) when the superpoly f_c is a zero constant (resp., not a zero constant) for a fixed key, by $p_{f_c \neq 0}$ we denote the probability that the superpoly f_c is not a zero constant for a fixed key, and #Rds is the number of rounds. We set the estimate threshold value of the probability to $\sigma = \frac{1+\sqrt{10/N_s}}{2}$, where N_s is the number of the samples, and set the attack threshold value of the probability, *i.e.*, the minimum probability used in the attack, to $p_0 = 0.6$ and $p_1 = 0.7$ for $\Pr(g = 0|f_c = 0)$ and $\Pr(g = 1|f_c \neq 0)$ respectively. The probability below the estimate threshold value σ is marked with slash throughs, *e.g.*, 0.314 (with slash throughs), and will never be used in the attack. The probability with a strikethrough, *e.g.*, ~~0.568~~, is below the attack threshold value p_0 or p_1. From the experimental results, we derive two sets

$$\Omega_0 = \{(c, g, 0) | \Pr(g = 0 | f_c = 0) > p_0\}$$

and

$$\Omega_1 = \{(c, g, 1) | \Pr(g = 1 | f_c \neq 0) > p_1\}.$$

All the functions g's are either linear or quadratic. We also record all the equations with probability 1,

$$\Lambda = \{(c, g, b) | \Pr(g = b | f_c = 0) = 1 \text{ or } \Pr(g = b | f_c \neq 0) = 1\}.$$

As shown in Table 3, there are 11 cubes with a basis that contains only one linear or quadratic function. As discussed in Sect. 3.3, if their superpolys are not zero constants for a fixed key, then the sole function in the basis is always equal to one. In addition, for the 19th cube in the table, we observe that one of the functions in its basis is always equal to one given $f_c \neq 0$, *i.e.*, the conditional probability $\Pr(g_4 = 1 | f_c \neq 0) = 1$. The remaining 16 cubes have a basis that contains two linear or quadratic functions. The number of rounds ranges from 805 to 808.

We have also verified for 100 random keys, each with 16 IVs, that the superpolys of the cubes listed in the table are zero constants when imposing all the functions in their bases to zeros.

Recovering the Key in Online Phase. In this phase, we set C to the set of the 28 cubes listed in Table 3, $\Omega = \Omega_0 \cup \Omega_1$, and $\alpha = 16$. Then execute Algorithm 3 as described in Sect. 3.2. For avoidance of repetition, here we only show some necessary details that are not included in Algorithm 3. Remind that, in Step 3 of the algorithm, we take all the possible values of the first $\log_2(\alpha) = 4$ free non-cube IV bits, and set the other free non-cube IV bits to random values. The non-free non-cube IV bits are set to zeros. In Step 5, we update the equation sets G_0 and G_1 according to the values of f_c, and use an extra set E to collect the equations with probability 1 according to Λ. In Step 6, for the case that $\{g | g = 0 \in G_0 \text{ and } g = 1 \in G_1\}$ is not empty, we retain the one with higher probability between $g = 0$ and $g = 1$, and remove the other one. Meanwhile, we remove the equations in E from G_0 and G_1. In Step 7, we set r_i to the maximum r_i such that $p_i^{-r_i} < \binom{|G_i|}{r_i}$, where $|G_i|$ is the cardinality of G_i, $i = 0, 1$. Then randomly choose r_0 equations from G_0 and r_1 equations from G_1, solve these $r_0 + r_1$ equations together with E and check whether the solutions are correct.

Note that all the equations are linearly independent and can be linearized after guessing the values of some key bits. The expected time complexity of the online phase is less than

$$28 \times 13.5 \times 2^{28} + p_0^{-r_0} p_1^{-r_1} 2^{80-(r_0+r_1+|E|)} \approx 2^{37} + 2^{80-(\frac{1}{4}r_0+\frac{1}{2}r_1+|E|)}.$$

As shown in Table 3, the probability $p_{f_c \neq 0}$ of non-zero superpoly ranges from 0.036 to 0.113 for the 12 cubes that can generate deterministic equations.

Our experiments show that, for about 45% keys, there is at least one cube c out of these 12 cubes such that $f_c \neq 0$. For such 45% keys, the average number of such non-zero superpolys is around 2, and the maximum number is 7. In our experiments for 1000 keys, the average values of r_0 and r_1 are respectively 3.8 and 2.4, and the average value of $\frac{1}{4}r_0 + \frac{1}{2}r_1 + |E|$ is about 3. In other words, we can recover 7 equations on key bits by 2^4 trials on average. The average attack time is thus around 2^{77}, using 2^{37} keystream bits and at the expense of preprocessing time 2^{47}. The attack time on 805-round Trivium can be cut down by using more cubes and at the expense of more preprocessing time and higher data complexity. Our attack is valid for more than half of 1000 random keys in our experiments. The attack fails when none of the systems of $r_0+r_1+|E|$ equations derived from G_0 and G_1 are correct. We stress here that the success probability of the attack can be increased by using smaller systems of equations (smaller r_0 and r_1) or larger probability thresholds (larger p_0 and p_1), at the cost of more attack time.

Next we give an example of the attack procedure. Note here that in the example the time complexity is better than the average case.

Example 3. Given that the 80-bit secret key is 71 DB 8B B3 21 CD AE F9 97 84 in hexadecimal, where the most significant bit is k_0 and the least significant bit is k_{79}. For each of the 28 cubes in Table 3, we generate 16 different non-cube IVs according to its free IV bits, and request 16×2^{28} keystream bits corresponding to this cube and the non-cube IVs, then compute the values of its superpoly. Taking the 8-th cube as an instance, we set the four free IV bits $0, 8, 53, 54$ to all possible values, the other free IV bits to random values, and the remaining non-cube IV bits to zeros; we then request 2^{32} keystream bits of 806 rounds accordingly, and sum these bits over the cube (module 2); finally we find a non-zero sum and get a deterministic equation $g_6 = k_{63} = 1$. We request $28 \times 16 \times 2^{28} \approx 2^{37}$ keystream bits in total, and find that there are 9 cubes having zero superpolys,

$$1, 2, 4, 10, 11, 16, 19, 24, 25,$$

and 19 cubes whose superpolys are not zero constants,

$$3, 5, 6, 7, 8, 9, 12, 13, 14, 15, 17, 18, 20, 21, 22, 23, 26, 27, 28.$$

From Table 3, we obtain 6 deterministic equations by the cubes $3, 5, 6, 7, 8, 9$.

$$E = \{g_i = 1 | i \in \{2, 6, 7, 11, 12, 13\}\},$$

where

$$\begin{aligned}
g_2 &= k_{59}, \\
g_6 &= k_{63}, \\
g_7 &= k_{64}, \\
g_{11} &= k_{66} \cdot k_{67} + k_{41} + k_{68}, \\
g_{12} &= k_{67} \cdot k_{68} + k_{42} + k_{69}, \\
g_{13} &= k_{68} \cdot k_{69} + k_{43} + k_{70}.
\end{aligned}$$

Further, we derive two equation sets G_0 and G_1

$$G_0 = \{g_5 = k_{63} = 0\},\ G_1 = \{g_i = 1 | i \in \{3, 4, 8, 14\}\},$$

where

$$\begin{aligned} g_3 &= k_{60}, \\ g_4 &= k_{61}, \\ g_8 &= k_{34} \cdot k_{35} + k_9 + k_{36}, \\ g_{14} &= k_{69} \cdot k_{70} + k_{44} + k_{71}. \end{aligned}$$

We can see that all the equations are linear after guessing the values of three bits k_{35}, k_{67} and k_{69}. The equation $g_5 = 0$ in G_0 holds with probability 0.643, and the equations $g_i = 1$ in G_1 hold with probability $0.888, 0.735, 0.907, 0.799$ respectively for $i = 3, 4, 8, 14$. Accordingly, we have $r_0 = 0$ and $r_1 = 3$. Then we randomly choose 3 equations from G_1, solve a system of $3 + 6$ equations (together with the 6 equations in E), and repeat this step until the correct solution is found. In theory, the expected number of trials for finding the correct solution is less than 3. As a matter of fact, all the equations in G_1 but $g_4 = 1$ are true for the secret key, which means that we could find the correct key by at most 4 trials of solving a system of 9 equations. Therefore we can recover the key with time complexity of $2^{37} + 4 \times 2^{71} \approx 2^{73}$. The time complexity can be cut down to 2^{72} if we set $r_0 + r_1 = 4$ and exploit the equations in G_0 and G_1 together.

4.4 The Attack on 835-Round Trivium

Generating a Candidate Set of Favorite Cubes. In [22], an exhaustive search was done on the cubes of size $37 \le n \le 40$ that contain no adjacent indexes, by using a simplified version of Algorithm 4. Similarly, we exhaustively search the cubes of size $36 \le n \le 40$ that contain no adjacent indexes, and pick up the cubes such that the corresponding superpolys after 815 rounds are zero constants. Then we find 37595 and 3902 cubes of sizes 36 and 37 respectively that satisfy the requirement. There are also a number of such cubes of size higher than 37. This step is done in a few hours on a desktop computer.

Finding the Basis and Free Non-cube IV Bits. As done before, we apply the procedure `Decomposition` to each cube c from the candidate set, and also obtain a set of free non-cube IV bits. We then get 1085 and 99 cubes of sizes 36 and 37 such that a basis of the superpoly after 833 rounds can be found. The maximum number of rounds after which we can still find a basis is 841. No basis is found for the superpoly after 833 rounds of the cubes with size higher than 37 in the candidate set. The results are found in several hours on a desktop computer.

Computing the Probability. Computing the value of the superpoly f_c over a big cube is time consuming. We test 13 cubes of size 37 and 28 cubes of size 36, each of which has a different basis with less than 8 elements after 835 rounds. In each test, we compute the values of the superpoly f_c for 128 random keys with at most $\alpha = 8$ non-cube IVs, and evaluate the conditional probability $\Pr(g = 0|f_c(key, \cdot) \equiv 0)$ and $\Pr(g = 1|f_c(key, \cdot) \not\equiv 0)$ for a random fixed key. The values of non-cube IVs are taken in the same manner as done in Sect. 4.3. Our experiment shows that all the computations need about $6 \cdot 128 \cdot (13 \cdot 2^{37} + 28 \cdot 2^{36}) \approx 2^{51}$ cipher operations. On average, we need to compute 6 IVs for each key.

The results are listed in Tables 4 and 5 in Appendix, together with the cubes, free non-cube IV bits and the equations. We set the attack threshold value of the probability to $p = \frac{2}{3}$ for both $\Pr(g = 0|f_c = 0)$ and $\Pr(g = 1|f_c \neq 0)$. The probability below the estimate threshold value σ is marked with slash throughs, *e.g.*, 0.514, and will never be used in the attack. The probability with a strikethrough, *e.g.*, ~~0.654~~, is below the attack threshold value p. From the experimental results, we derive one set

$$\Omega = \{(c, g, b)|\Pr(g = b|f_c = 0) > p \text{ or } \Pr(g = b|f_c \neq 0) > p\}.$$

All the functions g's are either linear or quadratic. We also record all the equations with probability 1,

$$\Lambda = \{(c, g, b)|\Pr(g = b|f_c = 0) = 1 \text{ or } \Pr(g = b|f_c \neq 0) = 1\}.$$

As shown in Table 4 for the cubes of size 37, there are 2 cubes having a basis that contains only one function, while there are 5 cubes from which it is possible to set up an equation with probability 1. The third and 11-th cubes have no qualified equations, and will be discarded in online phase. The 13-th and 14-th cubes are the same, while the keystream bits of two different numbers of rounds, 835 and 840, are used.

The results for the cubes of size 36 are listed in Table 5, and there are 7 cubes that have no qualified equations and will be discarded in online phase.

We have also verified for 32 random keys, each with 4 IVs, that the superpolys of the cubes listed in the table sum to zeros when imposing all the functions in their bases to zeros.

Recovering the Key in Online Phase. In this phase, we set $\alpha = 8$, and then execute Algorithm 3. Remind that, in Step 3 of the algorithm, we take all the possible values of the first $\log_2(\alpha) = 3$ free non-cube IV bits, and set the other free non-cube IV bits to random values. The non-free non-cube IV bits are set to zeros. In Step 5, we update the equation sets $G = G_0 \cup G_1$ according to the values of f_c, and use an extra set E to collect the equations with probability 1 according to Λ. In Step 6, if G has two incompatible equations $g = 0$ and $g = 1$, we remove them both from G. Meanwhile, we remove the equations in E from G. In Step 7, we set r to the maximum r such that $p^{-r} < \binom{|G|}{r}$. Then randomly choose r equations from G, solve these r equations together with E and check whether the solutions are correct.

Note that all the equations are linearly independent and can be linearized after guessing the values of some key bits. The expected time complexity of this phase is less than

$$6 \times (12 \times 2^{37} + 21 \times 2^{36}) + p^{-r} 2^{80-(r+|E|)} \approx 2^{44} + 2^{80-(\frac{2}{5}r+|E|)}.$$

As shown in Table 4, the probability for the 5 cubes that can generate equations with probability 1 ranges from 0.008 to 0.297. Our experiments show that, for about half keys, we can generate from one to three equations with probability 1. In our experiments for 128 random keys, the average values of r is larger than 10, and the average value of $\frac{2}{5}r + |E|$ is about 5. In other words, we can recover 11 equations on key bits by 2^6 trials on average. The average attack time is thus around 2^{75}, using $8 \times (12 \times 2^{37} + 21 \times 2^{36}) \approx 2^{45}$ keystream bits and at the expense of preprocessing time 2^{51}. The attack is valid for more than 44% out of 128 random keys in our experiments. The attack time can be cut down by using more cubes and at the expense of more preprocessing time and more data. On the other hand, the success probability of the attack can be increased at the cost of more attack time.

4.5 Discussion

Improvements of the Attack. A natural method to cut down the attack time is to use more cubes with keystream bits of different numbers of rounds. While we have found a thousand potentially favorite cubes for Trivium reduced to from 833 to 841 rounds, we can make use of a small number of them due to a limited computation resource. Increasing the number α in online phase gives a higher chance to find deterministic equations. Testing more random keys with larger α in preprocessing phase gives a more accurate estimate of the conditional probability $\Pr(g|f_c)$, as well as generates more valid probabilistic equations. One may also exploit one of the two equations $g = 0$ and $g = 1$ by carefully computing the probability $\Pr(g = b|f_1, f_2)$ as discussed in Sect. 3.3, when both of them appear in the equation set G.

For the attack on Trivium reduced to less than 835 rounds, it is possible to cut down the attack time by using cubes of size less than 36 and combining the equations retrieved in Sect. 4.4. For instance, using 54 out of the 69 cubes in Tables 3, 4 and 5 gives an improved key recovery attack on 805-round Trivium. In this improved attack, we adopt the same strategy that was used for analysis of 835-round Trivium, and find that we can recover 14 equations by 2^7 trials on average. Thus the attack on 805-round Trivium is faster than an exhaustive search by a factor of around 2^7, using 2^{45} keystream bits and at the expense of preprocessing time 2^{51}. The attack is directly valid for 31% out of 128 random keys in our experiments. The attack also works for most of the remaining keys after increasing the probability threshold p and repeating the attack again.

Table 2. Success probability of the attack

805 rounds:	#key bits	7.2	6.9	6.5	6.1	5.7
	Success rate	31%	60%	77%	86%	93%
835 rounds:	#key bits	5.0	4.6	4.2	3.8	3.4
	Success rate	44%	72%	83%	95%	98%

Success Probability of the Attack. In the above attack, we maximize the system of probabilistic equations. This is achieved by setting r to the maximum r such that $p^{-r} < \binom{|G|}{r}$ in Algorithm 3. The attack works for more keys when a smaller system with fewer equations is used, *i.e.*, a smaller r is adopted. We have verified this by supplementary experiments on round-reduced TRIVIUM.

As shown in Table 2, for 805-round variant of TRIVIUM, we can deduce 7.2, 6.9, 6.5, 6.1 and 5.7 key bits on average for 31%, 60%, 77%, 86% and 93% of the keys, respectively; and for 835-round variant, we can deduce 5.0, 4.6, 4.2, 3.8 and 3.4 key bits on average for 44%, 72%, 83%, 95% and 98% of the keys, respectively. Actually, our experiments for 128 random keys show that we can always set up equations. As shown in Tables 3 and 4, there are many cubes (e.g., Cube 5 in Table 3) such that we can set up probabilistic equations from both sides, which implies that the attack works for a random key.

Applications to TRIVIA-SC and KREYVIUM. We apply our techniques to TRIVIA-SC and KREYVIUM, and can find some cubes whose superpolys after 1047 and 852 rounds have a low-degree basis with a few elements for TRIVIA-SC and KREYVIUM respectively. The cubes for TRIVIA-SC have size larger than 60, and for KREYVIUM the size is at least 54. Computing the conditional probability $\Pr(g|f_c)$ for such large cubes is infeasible for us. Though we are unable to fully verify the validity of the attack on TRIVIA-SC and KREYVIUM, we believe that there is a high chance of validness due to their similar structures with TRIVIUM.

5 Conclusions

In this paper, we have shown a general framework of a new model of cube attacks, called *correlation cube attack*. It is a generalization of conditional cube attack, as well as a variant of conditional differential attacks. As an illustration, we applied it to TRIVIUM stream cipher, and gained the best key recovery attacks for TRIVIUM. To the best of our knowledge, this is the first time that a weak-key distinguisher on TRIVIUM stream cipher can be converted to a key recovery attack. We believe that this new cryptanalytic tool is useful in both cryptanalysis and design of symmetric cryptosystems. In the future, it is worthy of working on its applications to more cryptographic primitives, such as the `Grain` family of stream ciphers, block cipher SIMON and hash function KECCAK.

Acknowledgments. We are grateful to the anonymous reviewers for their valuable comments.

A The Cubes, Equations and Probabilities

Table 3. The cubes, equations and probabilities in the attack on 805-round Trivium (16 IVs and 1000 keys for cube size 28)

No.	Cube Indexes	Free IV bits	Eqs.	$p(0\|0)$	$p(1\|1)$	$p_{f_c \neq 0}$	#Rds
1	2, 4, 6, 8, 10, 12, 14, 17, 19, 21, 23, 25, 27, 29, 32, 34, 36, 38, 40, 42, 44, 47, 49, 51, 53, 57, 62, 79	0, 1, 9, 54, 55, 58, 59, 60, 64, 65, 66, 67, 68, 69, 72, 73, 74, 77	g_{14}	~~0.533~~	1	0.078	805
2	2, 4, 6, 8, 10, 12, 14, 17, 19, 21, 23, 25, 27, 29, 32, 34, 36, 38, 40, 42, 44, 47, 49, 51, 53, 62, 73, 79	0, 9, 54, 55, 56, 64, 65, 66, 67, 68, 69, 70, 71, 75, 77	g_3	~~0.534~~	1	0.036	805
3	2, 4, 6, 8, 10, 12, 14, 17, 19, 21, 23, 25, 27, 29, 32, 34, 36, 38, 40, 42, 44, 47, 49, 51, 53, 62, 77, 79	0, 1, 9, 54, 55, 56, 57, 58, 59, 60, 63, 64, 65, 66, 67, 68, 69, 70, 71, 72, 73, 74	g_7	~~0.508~~	1	0.071	805
4	1, 3, 5, 7, 9, 11, 13, 16, 18, 20, 22, 24, 26, 28, 31, 33, 35, 37, 39, 41, 43, 46, 48, 50, 52, 61, 74, 78	53, 54, 55, 56, 57, 63, 64, 65, 66, 67, 68, 69, 70, 71, 76	g_4	~~0.550~~	1	0.061	805
5	1, 3, 5, 7, 9, 11, 13, 16, 18, 20, 22, 24, 26, 28, 31, 33, 35, 37, 39, 41, 43, 46, 48, 50, 52, 56, 61, 78	0, 8, 53, 54, 57, 58, 59, 63, 64, 65, 66, 67, 68, 71, 72, 73, 76	g_{13}	~~0.534~~	1	0.093	806
6	1, 3, 5, 7, 9, 11, 13, 16, 18, 20, 22, 24, 26, 28, 31, 33, 35, 37, 39, 41, 43, 46, 48, 50, 52, 54, 61, 78	0, 8, 55, 56, 57, 58, 59, 62, 63, 64, 65, 66, 69, 70, 71, 72, 73, 76	g_{11}	~~0.514~~	1	0.056	806
7	1, 3, 5, 7, 9, 11, 13, 16, 18, 20, 22, 24, 26, 28, 31, 33, 35, 37, 39, 41, 43, 46, 48, 50, 52, 61, 72, 78	8, 53, 54, 55, 63, 64, 65, 66, 67, 68, 69, 70, 74, 76	g_2	~~0.509~~	1	0.041	806
8	1, 3, 5, 7, 9, 11, 13, 16, 18, 20, 22, 24, 26, 28, 31, 33, 35, 37, 39, 41, 43, 46, 48, 50, 52, 61, 76, 78	0, 8, 53, 54, 55, 56, 57, 58, 59, 62, 63, 64, 65, 66, 67, 68, 69, 70, 71, 72, 73	g_6	~~0.547~~	1	0.080	806
9	0, 2, 4, 6, 8, 10, 12, 15, 17, 19, 21, 23, 25, 27, 30, 32, 34, 36, 38, 40, 42, 45, 47, 49, 51, 55, 60, 79	7, 52, 53, 56, 57, 58, 61, 62, 64, 65, 66, 67, 70, 71, 72, 75, 76	g_{12}	~~0.568~~	1	0.113	807
10	0, 2, 4, 6, 8, 10, 12, 15, 17, 19, 21, 23, 25, 27, 30, 32, 34, 36, 38, 40, 42, 45, 47, 49, 51, 53, 60, 79	7, 54, 55, 56, 57, 58, 61, 62, 64, 65, 68, 69, 70, 71, 72, 75, 76	g_{10}	~~0.520~~	1	0.061	807
11	0, 2, 4, 6, 8, 10, 12, 15, 17, 19, 21, 23, 25, 27, 30, 32, 34, 36, 38, 40, 42, 45, 47, 49, 51, 60, 75, 79	7, 52, 53, 54, 55, 56, 57, 58, 61, 62, 64, 65, 66, 67, 68, 69, 70, 71, 72, 76	g_5	~~0.556~~	1	0.097	807
12	2, 6, 8, 10, 12, 14, 17, 19, 21, 23, 25, 27, 29, 32, 34, 36, 38, 40, 42, 44, 47, 49, 51, 53, 57, 62, 77, 79	0, 1, 3, 4, 9, 54, 55, 58, 59, 60, 64, 65, 66, 67, 68, 69, 72, 73, 74	g_7 g_{14}	~~0.594~~ ~~0.587~~	0.832 0.748	0.286	805
13	2, 4, 8, 10, 12, 14, 17, 19, 21, 23, 25, 27, 29, 32, 34, 36, 38, 40, 42, 44, 47, 49, 51, 53, 55, 57, 62, 79	0, 1, 5, 6, 9, 58, 59, 60, 64, 65, 66, 67, 70, 72, 73, 74, 77	g_{12} g_{14}	~~0.575~~ 0.621	~~0.654~~ 0.799	0.309	805
14	2, 6, 8, 10, 12, 14, 17, 19, 21, 23, 25, 27, 29, 32, 34, 36, 38, 40, 42, 44, 47, 49, 51, 53, 55, 62, 77, 79	0, 1, 3, 4, 9, 56, 57, 58, 59, 60, 63, 64, 65, 66, 67, 70, 71, 72, 73, 74	g_7 g_{12}	~~0.573~~ ~~0.587~~	0.803 0.721	0.269	805
15	1, 5, 7, 9, 11, 13, 16, 18, 20, 22, 24, 26, 28, 31, 33, 35, 37, 39, 41, 43, 46, 48, 50, 52, 61, 72, 74, 78	2, 3, 53, 54, 55, 56, 57, 63, 64, 65, 66, 67, 68, 69, 70, 76	g_2 g_4	~~0.533~~ ~~0.573~~	0.708 0.735	0.185	805
16	2, 4, 6, 8, 10, 12, 15, 17, 19, 21, 23, 25, 27, 30, 32, 34, 36, 38, 40, 42, 45, 47, 49, 51, 60, 62, 75, 79	0, 1, 9, 52, 53, 54, 55, 56, 57, 58, 64, 65, 66, 67, 68, 69, 70, 71, 72, 77	g_5 g_{16}	~~0.558~~ ~~0.522~~	0.902 ~~0.623~~	0.122	805
17	1, 5, 7, 9, 11, 13, 16, 18, 20, 22, 24, 26, 28, 31, 33, 35, 37, 39, 41, 43, 46, 48, 50, 52, 56, 61, 76, 78	0, 2, 3, 8, 53, 54, 57, 58, 59, 63, 64, 65, 66, 67, 68, 71, 72, 73	g_6 g_{13}	0.622 ~~0.580~~	0.778 0.744	0.297	806
18	1, 3, 7, 9, 11, 13, 16, 18, 20, 22, 24, 26, 28, 31, 33, 35, 37, 39, 41, 43, 46, 48, 50, 52, 54, 56, 61, 78	0, 4, 5, 8, 57, 58, 59, 63, 64, 65, 66, 69, 71, 72, 73, 76	g_{11} g_{13}	~~0.580~~ 0.635	~~0.697~~ 0.805	0.343	806

Table 3. (*continued*)

No.	Cube Indexes	Free IV bits	Eqs.	$p(0\|0)$	$p(1\|1)$	$p_{f_c \neq 0}$	#Rds
19	1, 3, 5, 7, 9, 11, 14, 16, 18, 20, 22, 24, 26, 29, 31, 33, 35, 37, 39, 41, 44, 46, 48, 50, 61, 70, 74, 78	0, 8, 51, 52, 53, 59, 62, 63, 64, 65, 66, 67, 68, 72, 76	g_1 g_4	~~0.501~~ ~~0.568~~	~~0.511~~ 1	0.092	806
20	1, 5, 7, 9, 11, 13, 16, 18, 20, 22, 24, 26, 28, 31, 33, 35, 37, 39, 41, 43, 46, 48, 50, 52, 54, 61, 76, 78	0, 2, 3, 8, 55, 56, 57, 58, 59, 62, 63, 64, 65, 66, 69, 70, 71, 72, 73	g_6 g_{11}	0.607 ~~0.547~~	0.792 ~~0.692~~	0.260	806
21	0, 3, 5, 7, 9, 11, 13, 16, 18, 20, 22, 24, 26, 28, 31, 33, 35, 37, 39, 41, 43, 46, 48, 50, 52, 54, 61, 78	1, 8, 14, 55, 56, 57, 58, 59, 62, 63, 64, 65, 66, 69, 70, 71, 72, 73, 76	g_8 g_{11}	0.694 ~~0.542~~	0.907 ~~0.629~~	0.334	806
22	2, 4, 6, 8, 10, 12, 15, 17, 19, 21, 23, 25, 27, 30, 32, 34, 36, 38, 40, 42, 45, 47, 49, 51, 55, 60, 75, 79	0, 7, 52, 53, 56, 57, 58, 61, 62, 64, 65, 66, 67, 70, 71, 72, 73, 76	g_5 g_{12}	0.660 0.705	~~0.691~~ 0.742	0.450	807
23	2, 4, 6, 8, 10, 12, 15, 17, 19, 21, 23, 25, 27, 30, 32, 34, 36, 38, 40, 42, 45, 47, 49, 51, 53, 55, 60, 79	0, 7, 56, 57, 58, 61, 62, 64, 65, 68, 70, 71, 72, 73, 74, 75, 76	g_{10} g_{12}	0.636 0.762	~~0.665~~ 0.762	0.492	807
24	2, 4, 6, 8, 10, 12, 15, 17, 19, 21, 23, 25, 27, 30, 32, 34, 36, 38, 40, 42, 45, 47, 49, 51, 53, 60, 75, 79	0, 7, 54, 55, 56, 57, 58, 61, 62, 64, 65, 66, 67, 68, 69, 70, 71, 72, 76	g_5 g_{10}	0.643 ~~0.558~~	0.815 ~~0.669~~	0.308	807
25	0, 2, 4, 6, 8, 10, 12, 15, 17, 19, 21, 23, 25, 27, 30, 32, 34, 36, 38, 40, 42, 45, 47, 49, 51, 60, 64, 79	7, 52, 53, 54, 55, 56, 57, 58, 61, 62, 65, 66, 67, 68, 69, 70, 71, 72, 75, 76	g_{17} g_{18}	~~0.553~~ ~~0.496~~	0.801 ~~0.585~~	0.246	807
26	0, 2, 4, 6, 8, 10, 13, 15, 17, 19, 21, 23, 25, 28, 30, 32, 34, 36, 38, 40, 43, 45, 47, 49, 58, 60, 73, 79	7, 50, 51, 52, 53, 54, 55, 56, 61, 62, 64, 65, 66, 67, 68, 69, 70, 75, 76	g_3 g_{15}	~~0.592~~ ~~0.499~~	0.888 ~~0.650~~	0.160	807
27	1, 3, 5, 7, 9, 11, 14, 16, 18, 20, 22, 24, 26, 29, 31, 33, 35, 37, 39, 41, 44, 46, 48, 50, 54, 59, 74, 78	6, 51, 52, 55, 56, 57, 60, 61, 63, 64, 65, 66, 69, 70, 71, 72, 75	g_4 g_{11}	0.652 0.696	~~0.641~~ 0.760	0.463	808
28	1, 3, 5, 7, 9, 11, 14, 16, 18, 20, 22, 24, 26, 29, 31, 33, 35, 37, 39, 41, 44, 46, 48, 50, 52, 54, 59, 78	6, 55, 56, 57, 60, 61, 63, 64, 67, 69, 70, 71, 72, 73, 74, 75	g_9 g_{11}	0.644 0.766	~~0.636~~ 0.787	0.508	808

$g_1 = k_{57}$
$g_2 = k_{59}$
$g_3 = k_{60}$
$g_4 = k_{61}$
$g_5 = k_{62}$
$g_6 = k_{63}$
$g_7 = k_{64}$
$g_8 = k_{34} \cdot k_{35} + k_9 + k_{36}$
$g_9 = k_{64} \cdot k_{65} + k_{39} + k_{66}$
$g_{10} = k_{65} \cdot k_{66} + k_{40} + k_{67}$
$g_{11} = k_{66} \cdot k_{67} + k_{41} + k_{68}$
$g_{12} = k_{67} \cdot k_{68} + k_{42} + k_{69}$
$g_{13} = k_{68} \cdot k_{69} + k_{43} + k_{70}$
$g_{14} = k_{69} \cdot k_{70} + k_{44} + k_{71}$
$g_{15} = k_{70} \cdot k_{71} + k_{45} + k_{72}$
$g_{16} = k_{72} \cdot k_{73} + k_{47} + k_{74}$
$g_{17} = k_{76} \cdot k_{77} + k_{51} + k_{78}$
$g_{18} = k_{67} \cdot k_{68} + k_{76} \cdot k_{77} + k_0 + k_{42} + k_{51} + k_{68} + k_{78}$

Table 4. The cubes, equations and probabilities in the attack on 835-round Trivium (8 IVs and 128 keys for cube size 37)

No.	Cube Indexes	Free IV bits	Eqs.	$p(0\|0)$	$p(1\|1)$	$p_{f_c \neq 0}$	#Rds
1	2, 4, 6, 8, 10, 12, 14, 17, 19, 21, 23, 25, 27, 29, 32, 34, 36, 38, 40, 42, 44, 47, 49, 51, 53, 55, 57, 59, 62, 64, 66, 68, 70, 72, 74, 77, 79	0, 1, 3, 5, 7, 9	g_{16}	~~0.472~~	1	0.008	836
2	1, 3, 5, 7, 9, 11, 13, 16, 18, 20, 22, 24, 26, 28, 31, 33, 35, 37, 39, 41, 43, 46, 48, 50, 52, 54, 56, 58, 61, 63, 65, 67, 69, 71, 73, 76, 78	0, 2, 4, 6, 8, 79	g_{15}	~~0.524~~	1	0.016	837
3	0, 2, 4, 7, 9, 11, 13, 15, 17, 19, 22, 24, 26, 28, 30, 32, 34, 37, 39, 41, 43, 45, 47, 49, 52, 54, 56, 58, 60, 62, 64, 67, 69, 71, 73, 75, 79	1, 3, 5, 6, 8, 10, 77	g_5 g_{47}	~~0.534~~ ~~0.653~~	~~0.714~~ ~~0.857~~	0.055	835
4	0, 2, 4, 6, 8, 11, 13, 15, 17, 19, 21, 23, 26, 28, 30, 32, 34, 36, 38, 41, 43, 45, 47, 49, 51, 53, 56, 58, 60, 62, 64, 66, 68, 71, 73, 75, 79	1, 3, 5, 7, 9, 10, 77	g_{24} g_{26}	~~0.500~~ ~~0.459~~	~~0.667~~ 1	0.047	835
5	1, 3, 5, 7, 10, 12, 14, 16, 18, 20, 22, 25, 27, 29, 31, 33, 35, 37, 40, 42, 44, 46, 48, 50, 52, 55, 57, 59, 61, 63, 65, 67, 70, 72, 74, 76, 78	0, 2, 4, 6, 8, 9, 11	g_{17} g_{25} g_{27}	0.696 ~~0.478~~ ~~0.543~~	0.972 ~~0.444~~ ~~0.694~~	0.281	835
6	0, 2, 4, 6, 9, 11, 13, 15, 17, 19, 21, 24, 26, 28, 30, 32, 34, 36, 39, 41, 43, 45, 47, 49, 51, 54, 56, 58, 60, 62, 64, 66, 69, 71, 73, 75, 79	1, 3, 5, 7, 8, 10, 77	g_{16} g_{24} g_{26}	0.667 ~~0.478~~ ~~0.389~~	1 ~~0.474~~ ~~0.447~~	0.297	836
7	0, 2, 5, 7, 9, 11, 13, 15, 17, 20, 22, 24, 26, 28, 30, 32, 34, 37, 39, 41, 43, 45, 47, 49, 52, 54, 56, 58, 60, 62, 64, 67, 69, 71, 73, 75, 79	1, 3, 4, 6, 8, 10, 33, 76, 77	g_{31} g_{33} g_{46} g_{47} g_{49}	0.962 0.962 ~~0.615~~ ~~0.808~~ ~~0.346~~	~~0.627~~ ~~0.627~~ ~~0.529~~ ~~0.422~~ ~~0.480~~	0.797	835
8	0, 2, 4, 7, 9, 11, 13, 15, 17, 19, 21, 24, 26, 28, 30, 32, 34, 36, 39, 41, 43, 45, 47, 49, 51, 54, 56, 58, 60, 62, 64, 66, 69, 71, 73, 75, 79	1, 3, 5, 6, 8, 10, 20, 77	g_{20} g_{21} g_{33} g_{35} g_{36}	0.968 0.968 ~~0.419~~ ~~0.452~~ ~~0.419~~	~~0.619~~ 0.691 ~~0.485~~ ~~0.485~~ ~~0.515~~	0.758	835
9	0, 2, 4, 6, 9, 11, 13, 15, 17, 19, 22, 24, 26, 28, 30, 32, 34, 37, 39, 41, 43, 45, 47, 49, 52, 54, 56, 58, 60, 62, 64, 67, 69, 71, 73, 75, 79	3, 5, 7, 8, 10, 20, 76, 77	g_{31} g_{33} g_{35} g_{36} g_{46} g_{47}	~~0.610~~ ~~0.644~~ ~~0.627~~ ~~0.458~~ ~~0.492~~ 0.729	~~0.594~~ ~~0.638~~ ~~0.609~~ ~~0.522~~ ~~0.493~~ ~~0.464~~	0.539	835
10	0, 2, 5, 7, 9, 11, 13, 15, 17, 19, 22, 24, 26, 28, 30, 32, 34, 37, 39, 41, 43, 45, 47, 49, 52, 54, 56, 58, 60, 62, 64, 67, 69, 71, 73, 75, 79	3, 4, 6, 8, 10, 18, 76, 77	g_{20} g_{31} g_{33} g_{35} g_{36} g_{46} g_{47}	0.716 ~~0.597~~ ~~0.582~~ ~~0.612~~ ~~0.522~~ ~~0.537~~ 0.776	~~0.689~~ ~~0.607~~ ~~0.607~~ ~~0.623~~ ~~0.590~~ ~~0.541~~ ~~0.541~~	0.477	835
11	0, 2, 4, 7, 9, 11, 13, 15, 17, 20, 22, 24, 26, 28, 30, 32, 34, 37, 39, 41, 43, 45, 47, 49, 52, 54, 56, 58, 60, 62, 64, 67, 69, 71, 73, 75, 79	5, 6, 8, 10, 18, 33, 76, 77	g_{19} g_{20} g_{31} g_{33} g_{46} g_{47} g_{49}	~~0.568~~ ~~0.703~~ ~~0.703~~ ~~0.703~~ ~~0.486~~ ~~0.730~~ ~~0.486~~	~~0.484~~ ~~0.549~~ ~~0.582~~ ~~0.593~~ ~~0.495~~ ~~0.418~~ ~~0.516~~	0.711	835
12	2, 4, 6, 8, 10, 12, 15, 17, 19, 21, 23, 25, 27, 30, 32, 34, 36, 38, 40, 42, 45, 47, 49, 51, 53, 55, 57, 60, 62, 64, 66, 68, 70, 72, 75, 77, 79	0, 1, 3, 5, 7, 13	g_8 g_{10} g_{12} g_{14} g_{16} g_{18} g_{45}	~~0.569~~ ~~0.538~~ ~~0.585~~ ~~0.600~~ ~~0.508~~ 0.708 ~~0.615~~	~~0.571~~ ~~0.603~~ ~~0.651~~ ~~0.619~~ ~~0.571~~ 0.778 ~~0.524~~	0.492	839

Table 4. (*continued*)

No.	Cube Indexes	Free IV bits	Eqs.	$p(0\|0)$	$p(1\|1)$	$p_{f_c \neq 0}$	#Rds
13	1, 3, 5, 7, 9, 11, 14, 16, 18, 20, 22, 24, 26, 29, 31, 33, 35, 37, 39, 41, 44, 46, 48, 50, 52, 54, 56, 59, 61, 63, 65, 67, 69, 71, 74, 76, 78	0, 2, 4, 6, 8, 12	g_{15}	~~0.500~~	~~0.438~~	0.125	835
			g_{17}	~~0.580~~	1		
14	1, 3, 5, 7, 9, 11, 14, 16, 18, 20, 22, 24, 26, 29, 31, 33, 35, 37, 39, 41, 44, 46, 48, 50, 52, 54, 56, 59, 61, 63, 65, 67, 69, 71, 74, 76, 78	0, 2, 4, 6, 12	g_7	~~0.561~~	~~0.606~~	0.555	840
			g_9	~~0.632~~	~~0.493~~		
			g_{13}	~~0.614~~	~~0.535~~		
			g_{15}	0.754	~~0.676~~		
			g_{17}	0.754	0.690		
			g_{44}	~~0.561~~	~~0.507~~		
			g_{50}	~~0.649~~	~~0.521~~		

$g_1 = k_{54}$
$g_2 = k_{55}$
$g_3 = k_{56}$
$g_4 = k_{57}$
$g_5 = k_{58}$
$g_6 = k_{59}$
$g_7 = k_{61}$
$g_8 = k_{62}$
$g_9 = k_{63}$
$g_{10} = k_{64}$
$g_{11} = k_{65}$
$g_{12} = k_{78} \cdot k_{79} + k_{53}$
$g_{13} = k_{30} \cdot k_{31} + k_5 + k_{32}$
$g_{14} = k_{31} \cdot k_{32} + k_6 + k_{33}$
$g_{15} = k_{32} \cdot k_{33} + k_7 + k_{34}$
$g_{16} = k_{33} \cdot k_{34} + k_8 + k_{35}$
$g_{17} = k_{34} \cdot k_{35} + k_9 + k_{36}$
$g_{18} = k_{35} \cdot k_{36} + k_{10} + k_{37}$
$g_{19} = k_{38} \cdot k_{39} + k_{13} + k_{40}$
$g_{20} = k_{40} \cdot k_{41} + k_{15} + k_{42}$
$g_{21} = k_{42} \cdot k_{43} + k_{17} + k_{44}$
$g_{22} = k_{44} \cdot k_{45} + k_{19} + k_{46}$
$g_{23} = k_{45} \cdot k_{46} + k_{20} + k_{47}$
$g_{24} = k_{46} \cdot k_{47} + k_{21} + k_{48}$
$g_{25} = k_{47} \cdot k_{48} + k_{22} + k_{49}$
$g_{26} = k_{48} \cdot k_{49} + k_{23} + k_{50}$
$g_{27} = k_{49} \cdot k_{50} + k_{24} + k_{51}$
$g_{28} = k_{50} \cdot k_{51} + k_{25} + k_{52}$
$g_{29} = k_{51} \cdot k_{52} + k_{26} + k_{53}$
$g_{30} = k_{52} \cdot k_{53} + k_{27} + k_{54}$
$g_{31} = k_{53} \cdot k_{54} + k_{28} + k_{55}$
$g_{32} = k_{54} \cdot k_{55} + k_{29} + k_{56}$
$g_{33} = k_{55} \cdot k_{56} + k_{30} + k_{57}$
$g_{34} = k_{56} \cdot k_{57} + k_{31} + k_{58}$
$g_{35} = k_{57} \cdot k_{58} + k_{32} + k_{59}$
$g_{36} = k_{59} \cdot k_{60} + k_{34} + k_{61}$
$g_{37} = k_{61} \cdot k_{62} + k_{36} + k_{63}$
$g_{38} = k_{62} \cdot k_{63} + k_{37} + k_{64}$
$g_{39} = k_{63} \cdot k_{64} + k_{38} + k_{65}$
$g_{40} = k_{64} \cdot k_{65} + k_{39} + k_{66}$
$g_{41} = k_{65} \cdot k_{66} + k_{40} + k_{67}$
$g_{42} = k_{66} \cdot k_{67} + k_{41} + k_{68}$
$g_{43} = k_{67} \cdot k_{68} + k_{42} + k_{69}$
$g_{44} = k_{68} \cdot k_{69} + k_{43} + k_{70}$
$g_{45} = k_{69} \cdot k_{70} + k_{44} + k_{71}$
$g_{46} = k_{72} \cdot k_{73} + k_{47} + k_{74}$
$g_{47} = k_{74} \cdot k_{75} + k_{49} + k_{76}$
$g_{48} = k_{75} \cdot k_{76} + k_{50} + k_{77}$
$g_{49} = k_{76} \cdot k_{77} + k_{51} + k_{78}$
$g_{50} = k_{77} \cdot k_{78} + k_{52} + k_{79}$

Table 5. The cubes, equations and probabilities in the attack on 835-round Trivium (8 IVs and 128 keys for cube size 36)

No.	Cube Indexes	Free IV bits	Eqs.	$p(0\|0)$	$p(1\|1)$	$p_{f_c \neq 0}$	#Rds
1	1, 3, 5, 7, 9, 11, 13, 16, 18, 20, 22, 24, 26, 28, 31, 33, 35, 37, 39, 41, 43, 46, 48, 50, 52, 54, 56, 58, 61, 63, 65, 67, 69, 71, 76, 78	0, 2, 4, 6, 73	g_3	~~0.530~~	~~0.536~~	0.219	835
			g_5	~~0.630~~	~~0.786~~		
2	1, 3, 5, 7, 9, 11, 13, 16, 18, 20, 22, 24, 26, 28, 31, 33, 35, 37, 41, 43, 46, 48, 50, 52, 54, 56, 58, 61, 63, 65, 67, 69, 71, 73, 76, 78	0, 2, 4, 6, 8, 38, 39	g_1	~~0.431~~	~~0.375~~	0.438	836
			g_3	~~0.653~~	~~0.661~~		
			g_5	~~0.639~~	~~0.589~~		
			g_{50}	~~0.556~~	~~0.446~~		
3	1, 3, 5, 7, 9, 11, 13, 16, 18, 20, 22, 24, 26, 28, 31, 33, 35, 37, 39, 41, 43, 46, 48, 50, 52, 54, 56, 58, 61, 63, 65, 67, 69, 73, 76, 78	0, 2, 4, 6, 8, 71	g_1	~~0.523~~	~~0.500~~	0.328	836
			g_3	0.698	0.857		
			g_{40}	~~0.535~~	~~0.619~~		
			g_{50}	~~0.593~~	~~0.524~~		
4	1, 3, 5, 7, 9, 11, 13, 16, 18, 20, 22, 24, 26, 28, 31, 33, 35, 37, 39, 41, 43, 46, 48, 50, 52, 54, 56, 58, 61, 63, 65, 69, 71, 73, 76, 78	0, 2, 4, 6, 8, 67	g_3	~~0.570~~	~~0.595~~	0.328	836
			g_5	~~0.616~~	~~0.619~~		
			g_{40}	~~0.523~~	~~0.595~~		
			g_{50}	~~0.605~~	~~0.548~~		
5	0, 2, 4, 6, 8, 10, 12, 15, 17, 19, 21, 23, 25, 27, 30, 32, 34, 36, 40, 42, 45, 47, 49, 51, 53, 55, 57, 60, 62, 64, 66, 68, 70, 72, 75, 79	1, 3, 5, 7, 37, 38, 77	g_2	~~0.554~~	0.762	0.492	837
			g_4	~~0.585~~	~~0.683~~		
			g_{12}	~~0.431~~	~~0.492~~		
			g_{49}	~~0.492~~	~~0.524~~		

Table 5. (*continued*)

No.	Cube Indexes	Free IV bits	Eqs.	$p(0\|0)$	$p(1\|1)$	$p_{f_c \neq 0}$	#Rds
6	0, 2, 4, 6, 8, 10, 12, 15, 17, 19, 21, 23, 25, 27, 30, 32, 34, 36, 38, 40, 42, 45, 47, 49, 51, 53, 55, 57, 60, 62, 66, 68, 70, 72, 75, 79	1, 3, 5, 7, 64, 77	g_2 g_4 g_{12} g_{39}	~~0.500~~ ~~0.554~~ ~~0.473~~ ~~0.486~~	0.741 ~~0.685~~ ~~0.537~~ ~~0.648~~	0.422	837
7	0, 2, 4, 6, 8, 10, 12, 15, 17, 19, 21, 23, 25, 27, 30, 32, 34, 36, 38, 40, 42, 45, 47, 49, 51, 53, 55, 57, 60, 62, 64, 68, 70, 72, 75, 79	1, 3, 5, 7, 66, 77	g_2 g_4 g_{39} g_{49}	~~0.507~~ ~~0.573~~ ~~0.493~~ ~~0.480~~	0.755 ~~0.717~~ ~~0.660~~ ~~0.509~~	0.414	837
8	0, 2, 4, 6, 8, 10, 12, 15, 17, 19, 21, 23, 25, 27, 30, 32, 34, 36, 38, 40, 42, 45, 47, 49, 51, 53, 55, 57, 60, 62, 64, 66, 68, 72, 75, 79	1, 3, 5, 7, 70, 77	g_2 g_{12} g_{39} g_{49}	~~0.613~~ ~~0.480~~ ~~0.480~~ ~~0.533~~	0.906 ~~0.547~~ ~~0.642~~ ~~0.585~~	0.414	837
9	0, 2, 4, 6, 8, 10, 12, 15, 17, 19, 21, 23, 25, 27, 30, 32, 34, 36, 38, 40, 42, 45, 47, 49, 51, 53, 55, 57, 60, 62, 64, 66, 70, 72, 75, 79	1, 3, 5, 7, 68, 77	g_4 g_{12} g_{39} g_{49}	~~0.602~~ ~~0.477~~ ~~0.489~~ ~~0.466~~	0.875 ~~0.550~~ ~~0.700~~ ~~0.475~~	0.312	837
10	1, 3, 5, 7, 9, 11, 13, 16, 18, 20, 22, 24, 26, 28, 31, 33, 35, 37, 39, 41, 43, 46, 48, 50, 52, 56, 58, 61, 63, 65, 67, 69, 71, 73, 76, 78	0, 2, 4, 6, 8, 54	g_1 g_3 g_5 g_{40} g_{44}	~~0.423~~ ~~0.641~~ ~~0.641~~ ~~0.538~~ ~~0.564~~	~~0.340~~ ~~0.680~~ ~~0.620~~ ~~0.600~~ ~~0.540~~	0.391	836
11	0, 4, 6, 8, 10, 12, 15, 17, 19, 21, 23, 25, 27, 30, 32, 34, 36, 38, 40, 42, 45, 47, 49, 51, 53, 55, 57, 60, 62, 64, 66, 68, 70, 72, 75, 79	1, 2, 3, 5, 7, 77	g_2 g_4 g_{12} g_{39} g_{49}	~~0.493~~ ~~0.565~~ ~~0.406~~ ~~0.507~~ ~~0.536~~	0.712 ~~0.678~~ ~~0.458~~ ~~0.661~~ ~~0.576~~	0.461	837
12	0, 2, 4, 6, 8, 10, 15, 17, 19, 21, 23, 25, 28, 30, 32, 34, 36, 38, 40, 43, 45, 47, 49, 51, 53, 55, 58, 60, 62, 64, 66, 68, 70, 73, 75, 79	1, 3, 5, 11, 12, 13, 77	g_2 g_{12} g_{37} g_{47} g_{49}	~~0.529~~ ~~0.588~~ ~~0.441~~ ~~0.588~~ ~~0.544~~	0.750 ~~0.667~~ ~~0.583~~ ~~0.333~~ ~~0.583~~	0.469	839
13	2, 6, 8, 10, 12, 14, 17, 19, 21, 23, 25, 27, 29, 32, 34, 36, 38, 40, 42, 44, 47, 49, 51, 53, 55, 57, 59, 62, 64, 66, 68, 70, 72, 74, 77, 79	0, 1, 3, 4, 5, 7, 9	g_2 g_4 g_6 g_{41} g_{43} g_{45}	~~0.453~~ ~~0.562~~ ~~0.562~~ ~~0.578~~ ~~0.656~~ ~~0.641~~	~~0.656~~ ~~0.656~~ 0.703 0.703 ~~0.562~~ ~~0.547~~	0.500	835
14	1, 4, 6, 8, 10, 12, 14, 16, 19, 21, 23, 25, 27, 29, 31, 34, 36, 38, 40, 42, 44, 49, 51, 53, 55, 57, 59, 62, 64, 66, 68, 70, 72, 74, 77, 79	0, 2, 3, 5, 7, 9, 45, 46, 47	g_2 g_4 g_6 g_{12} g_{34} g_{41}	~~0.463~~ ~~0.683~~ 0.780 ~~0.488~~ ~~0.732~~ ~~0.561~~	~~0.632~~ ~~0.655~~ 0.736 ~~0.540~~ 0.678 ~~0.621~~	0.680	835
15	0, 2, 4, 7, 9, 11, 13, 15, 17, 19, 22, 24, 26, 28, 30, 32, 34, 37, 39, 41, 43, 45, 47, 49, 52, 54, 56, 58, 60, 62, 67, 69, 71, 73, 75, 79	5, 6, 8, 10, 63, 64, 76	g_{11} g_{33} g_{35} g_{36} g_{46} g_{47}	0.821 ~~0.692~~ ~~0.692~~ ~~0.410~~ ~~0.564~~ 0.795	0.753 ~~0.596~~ ~~0.584~~ ~~0.506~~ ~~0.528~~ ~~0.449~~	0.695	835
16	0, 2, 4, 6, 8, 10, 13, 15, 17, 19, 21, 23, 25, 28, 30, 32, 36, 38, 40, 43, 45, 47, 49, 51, 53, 55, 58, 60, 62, 64, 66, 68, 70, 73, 75, 79	1, 3, 5, 7, 9, 11, 33, 34, 77	g_2 g_4 g_{12} g_{26} g_{28} g_{30}	~~0.590~~ ~~0.590~~ ~~0.718~~ ~~0.615~~ ~~0.487~~ ~~0.487~~	0.685 ~~0.607~~ ~~0.640~~ ~~0.640~~ ~~0.584~~ ~~0.506~~	0.695	835
17	0, 2, 4, 6, 8, 10, 13, 15, 17, 19, 21, 23, 25, 28, 30, 32, 34, 36, 38, 43, 45, 47, 49, 51, 53, 55, 58, 60, 62, 64, 66, 68, 70, 73, 75, 79	1, 3, 5, 7, 9, 11, 39, 40, 41, 77	g_2 g_4 g_{12} g_{24} g_{26} g_{28}	~~0.533~~ ~~0.533~~ ~~0.533~~ ~~0.700~~ ~~0.733~~ ~~0.667~~	~~0.643~~ ~~0.571~~ ~~0.551~~ ~~0.571~~ ~~0.653~~ ~~0.633~~	0.766	835

Table 5. (*continued*)

No.	Cube Indexes	Free IV bits	Eqs.	$p(0\|0)$	$p(1\|1)$	$p_{f_c \neq 0}$	#Rds
18	0, 2, 4, 6, 8, 10, 13, 15, 17, 19, 21, 23, 25, 28, 30, 32, 34, 36, 40, 43, 45, 47, 49, 51, 53, 55, 58, 60, 62, 64, 66, 68, 70, 73, 75, 79	1, 3, 5, 7, 9, 11, 37, 38, 77	g_2	~~0.429~~	~~0.616~~	0.672	835
			g_4	~~0.595~~	~~0.616~~		
			g_{12}	~~0.643~~	~~0.616~~		
			g_{24}	~~0.619~~	~~0.570~~		
			g_{26}	~~0.714~~	0.698		
			g_{30}	~~0.643~~	~~0.581~~		
19	0, 2, 4, 6, 8, 10, 13, 15, 17, 19, 21, 23, 25, 28, 30, 32, 34, 36, 38, 40, 43, 45, 47, 49, 51, 53, 55, 58, 60, 62, 64, 68, 70, 73, 75, 79	1, 3, 5, 7, 9, 11, 66, 77	g_2	~~0.551~~	0.696	0.617	835
			g_4	~~0.469~~	~~0.537~~		
			g_{24}	~~0.592~~	~~0.570~~		
			g_{26}	~~0.592~~	~~0.658~~		
			g_{28}	~~0.551~~	~~0.620~~		
			g_{30}	~~0.653~~	~~0.608~~		
20	0, 2, 4, 6, 8, 10, 13, 15, 17, 19, 21, 23, 25, 28, 30, 32, 34, 36, 38, 40, 43, 45, 47, 49, 51, 53, 55, 58, 60, 62, 64, 66, 68, 73, 75, 79	1, 3, 5, 7, 9, 11, 70, 77	g_2	~~0.564~~	0.674	0.695	835
			g_{12}	~~0.538~~	~~0.562~~		
			g_{24}	~~0.615~~	~~0.562~~		
			g_{26}	~~0.692~~	0.674		
			g_{28}	~~0.564~~	~~0.618~~		
			g_{30}	~~0.641~~	~~0.573~~		
21	0, 3, 5, 7, 9, 11, 13, 15, 18, 20, 22, 24, 26, 28, 30, 33, 35, 37, 39, 41, 43, 48, 50, 52, 54, 56, 58, 61, 63, 65, 67, 69, 71, 73, 76, 78	1, 2, 4, 6, 8, 44, 45, 46, 79	g_1	~~0.486~~	~~0.473~~	0.727	836
			g_3	~~0.629~~	~~0.527~~		
			g_5	~~0.714~~	~~0.527~~		
			g_{33}	0.829	~~0.634~~		
			g_{40}	~~0.514~~	~~0.527~~		
			g_{50}	~~0.657~~	~~0.484~~		
22	2, 4, 6, 8, 10, 12, 14, 17, 19, 21, 23, 25, 27, 29, 32, 34, 36, 38, 40, 42, 47, 49, 51, 53, 55, 57, 60, 62, 64, 66, 68, 70, 72, 75, 77, 79	0, 1, 3, 5, 7, 43, 44, 45	g_2	~~0.620~~	0.744	0.609	837
			g_4	~~0.620~~	~~0.654~~		
			g_{12}	~~0.380~~	~~0.474~~		
			g_{32}	~~0.720~~	~~0.641~~		
			g_{39}	~~0.440~~	~~0.577~~		
			g_{49}	~~0.500~~	~~0.526~~		
23	1, 3, 5, 7, 9, 11, 13, 16, 18, 20, 22, 24, 26, 28, 31, 33, 35, 37, 39, 41, 46, 48, 50, 52, 54, 56, 59, 61, 63, 65, 67, 69, 71, 74, 76, 78	0, 2, 4, 6, 42, 43, 44	g_1	~~0.673~~	~~0.603~~	0.570	838
			g_3	~~0.709~~	~~0.630~~		
			g_{31}	0.818	0.740		
			g_{38}	~~0.582~~	~~0.413~~		
			g_{48}	~~0.436~~	~~0.384~~		
			g_{50}	~~0.636~~	~~0.507~~		
24	0, 2, 4, 6, 8, 10, 12, 15, 17, 19, 21, 23, 25, 27, 30, 32, 34, 36, 38, 40, 45, 47, 49, 51, 53, 55, 58, 60, 62, 64, 66, 68, 70, 73, 75, 79	1, 3, 5, 41, 42, 43, 76	g_2	~~0.590~~	0.685	0.695	839
			g_{12}	~~0.615~~	~~0.596~~		
			g_{30}	~~0.744~~	~~0.618~~		
			g_{37}	~~0.487~~	~~0.596~~		
			g_{47}	~~0.692~~	~~0.404~~		
			g_{49}	~~0.513~~	~~0.528~~		
25	0, 4, 6, 8, 10, 13, 15, 17, 19, 21, 23, 25, 28, 30, 32, 34, 36, 38, 40, 43, 45, 47, 49, 51, 53, 55, 58, 60, 62, 64, 66, 68, 70, 73, 75, 79	1, 2, 3, 5, 7, 9, 11, 77	g_2	~~0.522~~	~~0.671~~	0.641	835
			g_4	~~0.522~~	~~0.585~~		
			g_{12}	~~0.543~~	~~0.573~~		
			g_{24}	~~0.609~~	~~0.573~~		
			g_{26}	~~0.565~~	~~0.634~~		
			g_{28}	~~0.543~~	~~0.622~~		
			g_{30}	~~0.652~~	~~0.598~~		
26	1, 5, 7, 9, 11, 14, 16, 18, 20, 22, 24, 26, 29, 31, 33, 35, 37, 39, 41, 44, 46, 48, 50, 52, 54, 56, 59, 61, 63, 65, 67, 69, 71, 74, 76, 78	0, 3, 4, 6, 8, 12	g_{23}	~~0.577~~	~~0.596~~	0.445	835
			g_{25}	~~0.634~~	~~0.667~~		
			g_{27}	~~0.535~~	~~0.596~~		
			g_{29}	~~0.521~~	~~0.491~~		
			g_{31}	~~0.493~~	~~0.491~~		
			g_{42}	~~0.535~~	~~0.456~~		
			g_{44}	~~0.521~~	~~0.474~~		
27	0, 2, 4, 7, 9, 11, 13, 15, 17, 19, 22, 24, 26, 28, 30, 32, 34, 37, 39, 41, 43, 45, 47, 49, 52, 54, 56, 58, 60, 62, 64, 67, 69, 71, 73, 79	3, 5, 6, 8, 75, 77	g_{20}	~~0.592~~	~~0.519~~	0.617	835
			g_{21}	~~0.551~~	~~0.582~~		
			g_{31}	~~0.592~~	~~0.557~~		
			g_{33}	0.837	0.722		
			g_{35}	0.755	~~0.658~~		
			g_{36}	~~0.408~~	~~0.494~~		
			g_{44}	~~0.592~~	~~0.519~~		
28	0, 2, 4, 6, 8, 11, 13, 15, 17, 19, 21, 23, 26, 28, 30, 32, 34, 36, 38, 41, 43, 45, 47, 49, 51, 53, 56, 58, 60, 62, 64, 66, 68, 71, 73, 79	1, 3, 5, 7, 9, 75, 77	g_2	~~0.581~~	~~0.609~~	0.758	837
			g_{12}	~~0.548~~	~~0.537~~		
			g_{22}	~~0.710~~	~~0.507~~		
			g_{24}	0.839	~~0.619~~		
			g_{26}	~~0.613~~	~~0.619~~		
			g_{28}	~~0.581~~	~~0.608~~		
			g_{49}	~~0.581~~	~~0.546~~		

References

1. Aumasson, J., Dinur, I., Henzen, L., Meier, W., Shamir, A.: Efficient FPGA implementations of high-dimensional cube testers on the stream cipher Grain-128. IACR Cryptology ePrint Archive 2009:218 (2009)
2. Aumasson, J.-P., Dinur, I., Meier, W., Shamir, A.: Cube testers and key recovery attacks on reduced-round MD6 and Trivium. In: Dunkelman, O. (ed.) FSE 2009. LNCS, vol. 5665, pp. 1–22. Springer, Heidelberg (2009)
3. Bertoni, G., Daemen, J., Peeters, M., Van Assche, G.: The Keccak reference, January 2011. http://keccak.noekeon.org, Version 3.0
4. Canteaut, A., Carpov, S., Fontaine, C., Lepoint, T., Naya-Plasencia, M., Paillier, P., Sirdey, R.: Stream ciphers: a practical solution for efficient homomorphic-ciphertext compression. In: Peyrin, T. (ed.) FSE 2016. LNCS, vol. 9783, pp. 313–333. Springer, Heidelberg (2016)
5. Chakraborti, A., Chattopadhyay, A., Hassan, M., Nandi, M.: TriviA: a fast and secure authenticated encryption scheme. In: Güneysu, T., Handschuh, H. (eds.) CHES 2015. LNCS, vol. 9293, pp. 330–353. Springer, Heidelberg (2015)
6. Chakraborti, A., Nandi, M.: TriviA-ck-v2. CAESAR Submission (2015). http://competitions.cr.yp.to/round2/triviackv2.pdf
7. De Cannière, C., Dunkelman, O., Knežević, M.: KATAN and KTANTAN — a family of small and efficient hardware-oriented block ciphers. In: Clavier, C., Gaj, K. (eds.) CHES 2009. LNCS, vol. 5747, pp. 272–288. Springer, Heidelberg (2009). https://doi.org/10.1007/978-3-642-04138-9_20
8. De Cannière, C., Preneel, B.: Trivium. In: Robshaw, M., Billet, O. (eds.) New Stream Cipher Designs. LNCS, vol. 4986, pp. 244–266. Springer, Heidelberg (2008). https://doi.org/10.1007/978-3-540-68351-3_18
9. Dinur, I., Güneysu, T., Paar, C., Shamir, A., Zimmermann, R.: An experimentally verified attack on Full Grain-128 using dedicated reconfigurable hardware. In: Lee, D.H., Wang, X. (eds.) ASIACRYPT 2011. LNCS, vol. 7073, pp. 327–343. Springer, Heidelberg (2011)
10. Dinur, I., Morawiecki, P., Pieprzyk, J., Srebrny, M., Straus, M.: Cube attacks and cube-attack-like cryptanalysis on the Round-Reduced Keccak Sponge Function. In: Oswald and Fischlin [26], pp. 733–761
11. Dinur, I., Shamir, A.: Cube attacks on tweakable black box polynomials. In: Joux, A. (ed.) EUROCRYPT 2009. LNCS, vol. 5479, pp. 278–299. Springer, Heidelberg (2009)
12. Dinur, I., Shamir, A.: Breaking Grain-128 with dynamic cube attacks. In: Joux, A. (ed.) FSE 2011. LNCS, vol. 6733, pp. 167–187. Springer, Heidelberg (2011)
13. Englund, H., Johansson, T., Sönmez Turan, M.: A framework for chosen IV statistical analysis of stream ciphers. In: Srinathan, K., Rangan, C.P., Yung, M. (eds.) INDOCRYPT 2007. LNCS, vol. 4859, pp. 268–281. Springer, Heidelberg (2007)
14. Fischer, S., Khazaei, S., Meier, W.: Chosen IV statistical analysis for key recovery attacks on stream ciphers. In: Vaudenay, S. (ed.) AFRICACRYPT 2008. LNCS, vol. 5023, pp. 236–245. Springer, Heidelberg (2008)
15. Fouque, P.-A., Vannet, T.: Improving key recovery to 784 and 799 rounds of Trivium using optimized cube attacks. In: Moriai, S. (ed.) FSE 2013. LNCS, vol. 8424, pp. 502–517. Springer, Heidelberg (2014)
16. Hell, M., Johansson, T., Maximov, A., Meier, W.: A stream cipher proposal: grain-128. In: 2006 IEEE International Symposium on Information Theory, pp. 1614–1618. IEEE (2006)

17. Hell, M., Johansson, T., Maximov, A., Meier, W.: The grain family of stream ciphers. In: Robshaw, M., Billet, O. (eds.) New Stream Cipher Designs. LNCS, vol. 4986, pp. 179–190. Springer, Heidelberg (2008)
18. Huang, S., Wang, X., Xu, G., Wang, M., Zhao, J.: Conditional cube attack on reduced-round Keccak sponge function. In: Coron, J.-S., Nielsen, J.B. (eds.) EUROCRYPT 2017. LNCS, vol. 10211, pp. 259–288. Springer, Cham (2017)
19. Knellwolf, S., Meier, W., Naya-Plasencia, M.: Conditional differential cryptanalysis of NLFSR-based cryptosystems. In: Abe, M. (ed.) ASIACRYPT 2010. LNCS, vol. 6477, pp. 130–145. Springer, Heidelberg (2010)
20. Knellwolf, S., Meier, W., Naya-Plasencia, M.: Conditional differential cryptanalysis of Trivium and KATAN. In: Miri, A., Vaudenay, S. (eds.) SAC 2011. LNCS, vol. 7118, pp. 200–212. Springer, Heidelberg (2012)
21. Lai, X.: Higher order derivatives and differential cryptanalysis. In: Proceedings Symposium in Communications, Coding Cryptography, pp. 227–233. Kluwer Academic Publishers (1994)
22. Liu, M.: Degree evaluation of NFSR-based cryptosystems. In: Katz, J., Shacham, H. (eds.) CRYPTO 2017. LNCS, vol. 10403, pp. 227–249. Springer, Cham (2017)
23. Liu, M., Lin, D., Wang, W.: Searching cubes for testing Boolean functions and its application to Trivium. In: IEEE International Symposium on Information Theory, ISIT 2015, Hong Kong, China, 14–19 June 2015, pp. 496–500. IEEE (2015)
24. Maximov, A., Biryukov, A.: Two trivial attacks on Trivium. In: Adams, C., Miri, A., Wiener, M. (eds.) SAC 2007. LNCS, vol. 4876, pp. 36–55. Springer, Heidelberg (2007)
25. Meier, W., Staffelbach, O.: Fast correlation attacks on certain stream ciphers. J. Cryptol. **1**(3), 159–176 (1989)
26. Oswald, E., Fischlin, M. (eds.): EUROCRYPT 2015. LNCS, vol. 9056. Springer, Heidelberg (2015). https://doi.org/10.1007/978-3-662-46800-5
27. Saarinen, M.O.: Chosen-IV statistical attacks on estream ciphers. In: Malek, M., Fernández-Medina, E., Hernando, J. (eds.) SECRYPT 2006, Proceedings of the International Conference on Security and Cryptography, Setúbal, Portugal, 7–10 August 2006, SECRYPT is part of ICETE - The International Joint Conference on e-Business and Telecommunications, pp. 260–266. INSTICC Press (2006)
28. Stankovski, P.: Greedy distinguishers and nonrandomness detectors. In: Gong, G., Gupta, K.C. (eds.) INDOCRYPT 2010. LNCS, vol. 6498, pp. 210–226. Springer, Heidelberg (2010)
29. Todo, Y.: Structural evaluation by generalized integral property. In: Oswald and Fischlin [26], pp. 287–314
30. Todo, Y., Isobe, T., Hao, Y., Meier, W.: Cube attacks on non-blackbox polynomials based on division property. In: Katz, J., Shacham, H. (eds.) CRYPTO 2017. LNCS, vol. 10403, pp. 250–279. Springer, Cham (2017)
31. Todo, Y., Morii, M.: Bit-based division property and application to Simon family. In: Peyrin, T. (ed.) FSE 2016. LNCS, vol. 9783, pp. 357–377. Springer, Heidelberg (2016)
32. Vielhaber, M.: Breaking ONE.FIVIUM by AIDA an algebraic IV differential attack. IACR Cryptology ePrint Archive, 2007:413 (2007)

The Missing Difference Problem, and Its Applications to Counter Mode Encryption

Gaëtan Leurent(✉) and Ferdinand Sibleyras(✉)

Inria, Paris, France
{gaetan.leurent,ferdinand.sibleyras}@inria.fr

Abstract. The counter mode (CTR) is a simple, efficient and widely used encryption mode using a block cipher. It comes with a security proof that guarantees no attacks up to the birthday bound (*i.e.* as long as the number of encrypted blocks σ satisfies $\sigma \ll 2^{n/2}$), and a matching attack that can distinguish plaintext/ciphertext pairs from random using about $2^{n/2}$ blocks of data.

The main goal of this paper is to study attacks against the counter mode beyond this simple distinguisher. We focus on message recovery attacks, with realistic assumptions about the capabilities of an adversary, and evaluate the full time complexity of the attacks rather than just the query complexity. Our main result is an attack to recover a block of message with complexity $\tilde{\mathcal{O}}(2^{n/2})$. This shows that the actual security of CTR is similar to that of CBC, where collision attacks are well known to reveal information about the message.

To achieve this result, we study a simple algorithmic problem related to the security of the CTR mode: the missing difference problem. We give efficient algorithms for this problem in two practically relevant cases: where the missing difference is known to be in some linear subspace, and when the amount of data is higher than strictly required.

As a further application, we show that the second algorithm can also be used to break some polynomial MACs such as GMAC and Poly1305, with a universal forgery attack with complexity $\tilde{\mathcal{O}}(2^{2n/3})$.

Keywords: Modes of operation · CTR · GCM · Poly1305
Cryptanalysis

1 Introduction

Block ciphers (such as DES or the AES) are probably the most widely used cryptographic primitives. Formally, a block cipher is just a keyed family of permutations over n-bit blocks, but when combined with a mode of operation, it can provide confidentiality (*e.g.* using CBC, or CTR), authenticity (*e.g.* using CBC-MAC, CMAC, or GMAC), or authenticated encryption (*e.g.* using GCM, CCM, or OCB). A mode of operation defines how to divide a message into blocks, and how to process the blocks one by one with some chaining rule.

J. B. Nielsen and V. Rijmen (Eds.): EUROCRYPT 2018, LNCS 10821, pp. 745–770, 2018.
https://doi.org/10.1007/978-3-319-78375-8_24

The security of block ciphers is studied with cryptanalysis, with classical techniques such as differential [8] and linear [27] cryptanalysis, dedicated techniques like the SQUARE attack [9], and ad-hoc improvements for specific targets. This allows to evaluate the security margin of block ciphers, and today we have a high confidence that AES or Blowfish are as secure as a family of pseudo-random permutations with the same parameters (key size and block size).

On the other hand, modes of operation are mostly studied with security proofs, in order to determine conditions where using a particular mode of operation is safe. However, exceeding those conditions doesn't imply that there is an attack, and even when there is one, it can range from a weak distinguisher to a devastating key recovery. In order to get a better understanding of the security of modes of operations, we must combine lower bound on the security from security proofs, and upper bounds from attacks.

In particular, most of the modes used today are sensible to birthday attacks because of collisions; those attacks can even be practical with 64-bit block ciphers, as shown in [7], but security proofs don't tell us how dangerous the attacks are. For instance, the CBC and CTR modes have been proven secure against chosen plaintext attacks up to $\sigma \ll 2^{n/2}$ blocks of encrypted data [5,35]. Formally, the security statements bound the maximum advantage of an attacker against the modes as follows:

$$\begin{aligned}\mathbf{Adv}^{\mathrm{CPA}}_{\mathrm{CBC}-E} &\leq \mathbf{Adv}^{\mathrm{prp}}_{E} + \sigma^2/2^n,\\ \mathbf{Adv}^{\mathrm{CPA}}_{\mathrm{CTR}-E} &\leq \mathbf{Adv}^{\mathrm{prp}}_{E} + \sigma^2/2^{n+1}.\end{aligned}$$

Both statement are essentially the same, and become moot when σ reaches $2^{n/2}$, but attacks can actually be quite different.

More precisely, the CBC mode is defined as $c_i = E(m_i \oplus c_{i-1})$, with E a block cipher. A collision between two ciphertext blocks $c_i = c_j$ is expected after $2^{n/2}$ blocks, and reveals the xor of two plaintext blocks: $m_i \oplus m_j = c_{i-1} \oplus c_{j-1}$. On the other hand, the counter mode is defined as $c_i = E(i) \oplus m_i$. There are no collisions in the inputs/outputs of E, but this can actually be used by a distinguisher. Indeed, if an adversary has access to $2^{n/2}$ known plaintext/ciphertext pairs, he can recover $E(i) = c_i \oplus m_i$ and detect that the values are unique (because E is a permutation), while collisions would be expected with a random ciphertext. Both attacks have the same complexity, and show that the corresponding proofs are tight. However, the loss of security is quite different: the attack against CBC lets an attacker recover message blocks from collisions (as shown in practice in [7]), but the attack against the counter mode hardly reveals any useful information.

In general, there is a folklore belief that the leakage of the CTR mode is not as bad as the leakage of the CBC mode. For instance, Ferguson et al. wrote [15, Sect. 4.8.2] (in the context of a 128-bit block cipher):

> CTR leaks very little data. [...] It would be reasonable to limit the cipher mode to 2^{60} blocks, which allows you to encrypt 2^{64} bytes but restricts the leakage to a small fraction of a bit.
> When using CBC mode you should be a bit more restrictive. [...] We suggest limiting CBC encryption to 2^{32} blocks or so.

Our Contribution. The main goal of this paper is to study attacks against the counter mode beyond the simple distinguisher given above. This is an important security issue, because uses of the CTR mode with 64-bit block ciphers could be attacked in practice. We consider generic attacks that work for any instance of the block cipher E, and assume that E behaves as a pseudo-random permutation. The complexity of the attacks will be determined by the block size n, rather than the key size, and we focus on the asymptotic complexity, using the Big-O notation $\mathcal{O}()$, and the Soft-O notation $\tilde{\mathcal{O}}()$ (ignoring logarithmic factors).

We consider message recovery attacks, where an attacker tries to recover secret information contained in the message, rather than recovering the encryption key k. Following recent attacks against HTTPS [2,7,12], we assume that a fixed message containing both known blocks and secret blocks is encrypted multiple times (this is common with web cookies, for instance). As shown by McGrew [28], this kind of attack against the CTR mode can be written as a simple algorithmic problem: the *missing difference problem*, defined as follows: given two functions $f, g : X \rightarrow \{0,1\}^n$, with the promise that there exists a unique $S \in \{0,1\}^n$ such that $\forall(x,y),\ f(x) \oplus g(y) \neq S$, recover S. We further assume that f and g behave like random functions, and that we are given a set $\mathcal{S} \subseteq \{0,1\}^n$, such that $S \in \mathcal{S}$ ($\mathcal{S}$ represents prior knowledge about the secret). In an attack against the counter mode, f outputs correspond to known keystream blocks, while g outputs correspond to encryptions of S.

In the information theoretic setting, this problem can be solved with $\tilde{\mathcal{O}}(2^{n/2})$ queries for any set $\mathcal{S}$, and requires at least $\Omega(2^{n/2})$ queries when $|\mathcal{S}| \geq 2$. However, the analysis is more complex when taking into account the cost of the computations required to recover S. McGrew introduces two algorithms for this problem: a sieving algorithm with $\tilde{\mathcal{O}}(2^{n/2})$ queries and time $\tilde{\mathcal{O}}(2^n)$, and a searching algorithm that can be optimized to time and query complexity $\tilde{\mathcal{O}}(2^{n/2}\sqrt{|\mathcal{S}|})$. Our main contribution is to give better algorithms for this problem:

1. An algorithm with $\tilde{\mathcal{O}}(2^{n/2})$ queries and time $\tilde{\mathcal{O}}(2^{n/2} + 2^{\dim\langle\mathcal{S}\rangle})$, in the case where $\mathcal{S}$ is (a subset of) a linear subspace of $\{0,1\}^n$. In particular, when $\mathcal{S}$ is a linear subspace of dimension $n/2$, we reach a time and query complexity of $\tilde{\mathcal{O}}(2^{n/2})$, while the searching algorithm of McGrew has a time and query complexity of $\tilde{\mathcal{O}}(2^{3n/4})$.
2. An algorithm with time and query complexity $\tilde{\mathcal{O}}(2^{2n/3})$ for any $\mathcal{S}$. In particular, with $\mathcal{S} = \{0,1\}^n$, the best previous algorithm had a time complexity of $\tilde{\mathcal{O}}(2^n)$.

We also show new applications of these algorithms. The first algorithm leads to an efficient message recovery attack with complexity $\tilde{\mathcal{O}}(2^{n/2})$ against the CTR mode, assuming that the adversary can control the position of the secret, by splitting it across block boundaries (following ideas of [12,32]). The second algorithm can be used to recover the polynomial key in some polynomial based MACs such as GMAC and Poly1305, leading to a universal forgery attack with complexity $\tilde{\mathcal{O}}(2^{2n/3})$. As far as we know, this is the first universal forgery attack against those MACs with complexity below 2^n.

Related Works. There are several known results about the security of mode of operation beyond the birthday bound, when the proof is not applicable. For encryption modes, the security of the CBC mode beyond the birthday bound is well understood: collision attacks reveal the XOR of two message blocks, and can exploited in practice [7]. Other modes that allow collisions (eg. CFB) have the same properties. The goal of this paper is to study the security of modes that don't have collisions, to get a similar understanding of their security.

Many interesting attacks have also been found against authentication modes. In 1995, Preneel and van Oorschot [31] gave a generic collision attack against all deterministic iterated message authentication codes (MACs), leading to existential forgeries with complexity $\mathcal{O}(2^{n/2})$. Later, a number of more advanced generic attacks have been described, with stronger outcomes than existential forgeries, starting with a key-recovery attack against the envelop MAC by the same authors [32]. In particular, a series of attack against hash-based MAC [11,18,25,30] led to universal forgery attacks against long challenges, and key-recovery attacks when the hash function has an internal checksum (like the GOST family). Against PMAC, Lee *et al.* showed a universal forgery attack in 2006 [24]. Later, Fuhr et al. gave a key-recovery attack against the PMAC variant used in AEZv3 [17]. Issues with GCM authentication with truncated tags were also pointed out by Ferguson [14].

None of these attacks contradict the proof of security of the scheme they target, but they are important results to understand the security degradation after the birthday bound.

Organization of the Paper. We introduce the CTR mode and the missing difference problem in Sect. 2, and present our algorithmic contributions in Sect. 3. Then we describe concrete attacks against the CTR mode in Sect. 4, and attacks against Carter-Wegman MACs in Sect. 5. At last we show detailed proofs and simulation results in Sect. 6.

2 Message Recovery Attacks on CTR Mode

The CTR mode was first proposed by Diffie and Hellman in 1979 [10]. It was not included in the first series of standardized modes by NIST [16], but was added later [13]. The CTR mode essentially turns a block cipher into a stream cipher, by encrypting some non-repeating counter. It is now a popular mode of operation, thanks to its parallelizability, speed, and simple design. This led Phillip Rogaway to write in an evaluation of different privacy modes of operation talking about CTR [35]: "Overall, usually the best and most modern way to achieve privacy-only encryption". In particular, CTR is used as the basic of the authenticated encryption mode GCM, the most widely used mode in TLS today.

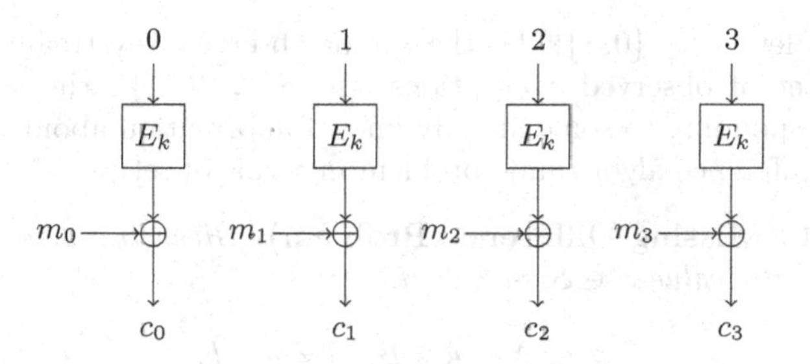

Fig. 1. CTR mode

2.1 Setting and Notations

In the following we assume that the counter mode is implemented such that the input to the block cipher never repeats. For simplicity we consider a stateful variant of the counter mode with a global counter that is maintained across messages and initialized as 0 (as shown in Fig. 1):

$$c_i = E_k(i) \oplus m_i,$$

where E_k is an n-bit block cipher, m_i an n-bit block of plaintext and c_i an n-bit block of ciphertext.

Our attacks do not depend on the details of how the input to the block cipher is constructed, and can also be applied to nonce-based variants[1]; we only require that all inputs are different. Note that some variants of the counter mode can have repetitions in the block cipher input[2], but this gives easy attacks because repetitions leak the xor of two plaintext blocks (as in the CBC mode).

We consider a message recovery attack, where the attacker tries to recover some secret message block S. Throughout the attack, the key k will be invariant so we will write $E_k(i)$ as a_i to represent the ith block of CTR keystream. We can immediately notice that if we have partial knowledge of the plaintext, for every known block m_i we can recover the associated a_i as $c_i \oplus m_i = a_i$. Assume further that we have access to the repeated encryption b_j of the secret S so that $b_j = a_j \oplus S$. The first property of the CTR mode is that $E_k(\cdot)$ being a permutation, the keystream a_i never repeats, thus we have the following inequalities:

$$i \neq j \Rightarrow a_i \neq a_j \Rightarrow a_i \oplus a_j \oplus S \neq S \Rightarrow a_i \oplus b_j \neq S.$$

From now on we will always assume that we can observe and collect lists of many a_i and b_j and use them with the previous inequality to recover S. This setting is similar to the practical attack Sweet32 on the CBC mode mounted by Bhargavan and Leurent, using repeated encryptions of an authentication token to obtain many different ciphertext blocks for the same secret information [7].

[1] For instance, GCM concatenates a per-message nonce and a counter within a message.
[2] For instance, the treatment of non-default-length nonces in GCM can lead to collisions [23].

Formally, let $\mathcal{A} \subseteq \{0,1\}^n$ be the set of observed keystream blocks, $\mathcal{B} \subseteq \{0,1\}^n$ the set of observed encryptions and $\mathcal{S} \subseteq \{0,1\}^n$ the set of possible secrets (corresponding to some already known information about S). We define the missing difference algorithmic problem in terms of set:

Definition 1 (Missing Difference Problem). *Given two sets $\mathcal{A}$ and $\mathcal{B}$, and a hint $\mathcal{S}$, find the value $S \in \mathcal{S}$ such that:*

$$\forall (a,b) \in \mathcal{A} \times \mathcal{B},\ S \neq a \oplus b.$$

Alternatively, we can consider that the attacker is given oracle access to $\mathcal{A}$ and $\mathcal{B}$ though some functions f and g, so that its running time includes calls to f and g, and computations to recover S. This presentation corresponds to a more active attack, where the adversary can optimize the size of the sets.

Definition 2 (Missing Difference Problem with Functions). *Given two functions $f, g : X \to \{0,1\}^n$, and a hint $\mathcal{S}$, find the value $S \in \mathcal{S}$ such that:*

$$\forall (x,y),\ S \neq f(x) \oplus g(y).$$

2.2 Previous Work

An attack can only be carried to the end if the secret S is the only value in $\mathcal{S}$ such that $\forall (a,b) \in \mathcal{A} \times \mathcal{B},\ S \neq a \oplus b$, or else it will be indistinguishable from the other values that satisfy the same condition (those values could have produced the same sets with same probability). The coupon collector's problem predicts that N out of N different coupons are found after $N \cdot H_N \simeq N \ln N$ draws (with H_N the N-th harmonic number), assuming uniform distribution of the draws. In our case we will assume that all the differences $a \oplus b$ are independent and uniformly distributed over $\{0,1\}^n \setminus S$, which is a reasonable approximation validated by our experiments. To carry the attack to the end we require to collect $N = |\mathcal{S}| - 1$ differences thus we will need $\mathcal{O}(|\mathcal{S}| \ln |\mathcal{S}|)$ "draws". A draw is a couple (a,b) s.t. $a \oplus b \in |\mathcal{S}|$, otherwise we discard it; it happens with probability $(|\mathcal{S}| - 1)/(2^n - 1)$. Therefore we need to observe enough data to have $|\mathcal{A}| \cdot |\mathcal{B}|$ in the order of $\mathcal{O}(2^n \ln |\mathcal{S}|)$; this may be achieved by having both sets in the order of $\mathcal{O}(2^{n/2}\sqrt{\ln |\mathcal{S}|})$. This size of the observed sets can be understood as the query complexity, that is the number of encrypted messages the attacker will have to intercept in order to carry out the attack. Notice that even for $|\mathcal{S}| = \mathcal{O}(2^n)$, $|\mathcal{A}| = |\mathcal{B}| = \mathcal{O}(\sqrt{n} \cdot 2^{n/2})$ is quite close to the theoretical lower bound of $\mathcal{O}(2^{n/2})$ given by the distinguishing attack and the security proof for the CTR mode. Therefore, message recovery attacks are possible with an (almost) optimal data complexity. The next question is to study the time complexity, *i.e.* how to efficiently recover S.

A first approach consists in computing all the impossible values of S from the large set of $\mathcal{A} \times \mathcal{B}$ and discard any new value we encounter as impossible until there's only one possible plaintext left. This is Algorithm 1. This approach works but requires to actually compute $\mathcal{O}(2^n \ln |\mathcal{S}|)$ values and maintain in memory a

Algorithm 1. Simple sieving algorithm

```
Input: A, B, S
Output: {s ∈ S | ∀(a,b) ∈ A × B, a ⊕ b ≠ s}
  for a in A do
    for b in B do
      Remove (a ⊕ b) from S;
    end for
  end for
  return S
```

Algorithm 2. Searching algorithm

```
Input: A, B, S
Output: {s ∈ S | ∀(a,b) ∈ A × B, a ⊕ b ≠ s}
  Store B so that operation ∈ is efficient.
  for s in S do
    for a in A do
      if (s ⊕ a) ∈ B then
        Remove s from S;
      end if
    end for
  end for
  return S
```

sieve of size $|\mathcal{S}|$. In the case where the key size is equal to the block size n, like AES-128, this attack is actually worse than a simple exhaustive search of the key. In a 2012 work, McGrew [28] described this sieving algorithm and noticed that when the set $\mathcal{S}$ is small, the sieving wastes a lot of time computing useless values. Therefore he proposed a second algorithm, Algorithm 2, to test and eliminate values of $\mathcal{S}$ one by one. This algorithm loops over $\mathcal{S}$ and $\mathcal{A}$ to efficiently test whether $s \oplus a \in \mathcal{B}$; if yes then we sieve the value s out of $\mathcal{S}$.

Both algorithms act on a sieving set $\mathcal{S}$ to reduce it, so McGrew proposed a hybrid algorithm switching from one algorithm to the other in order to reduce the searching space as quickly as possible. This improves the attack when $\mathcal{A}$ and $\mathcal{B}$ are fixed, but if the adversary can choose the sizes of $\mathcal{A}$ and $\mathcal{B}$ (in particular, if he actually has oracle access to functions f and g), then the searching algorithm allows better trade-offs. Indeed, the searching algorithm has a complexity of $\mathcal{O}(|\mathcal{B}| + |\mathcal{A}| \cdot |\mathcal{S}|)$, and is successful as soon as $|\mathcal{A}| \cdot |\mathcal{B}| = \Omega(2^n \ln |\mathcal{S}|)$. To optimize the complexity, we use $|\mathcal{B}| = |\mathcal{A}| \cdot |\mathcal{S}|$ to obtain an overall complexity of $\mathcal{O}(2^{n/2}\sqrt{|\mathcal{S}| \ln |\mathcal{S}|})$ in both time and queries. In particular for small $\mathcal{S}$ (of size polynomial in n) this algorithm is (almost) optimal, reaching the birthday bound $\tilde{\mathcal{O}}(2^{n/2})$.

Starting from these observations we will show improved algorithms to recover a block of secret information without big exhaustive searches in the next section.

3 Efficient Algorithms for the Missing Difference Problem

We now propose two new algorithms to solve the missing difference algorithmic problem more efficiently in two practically relevant different settings. Our first algorithm requires that the set $\mathcal{S}$ — or its linear span $\langle\mathcal{S}\rangle$ — is a vector space of relatively small dimension, and has complexity $\tilde{\mathcal{O}}(2^{n/2} + |\langle\mathcal{S}\rangle|)$. The second algorithm uses a larger query complexity of $\tilde{\mathcal{O}}(2^{2n/3})$, to reduce the computation and memory usage to $\tilde{\mathcal{O}}(2^{2n/3})$.

3.1 Known Prefix Sieving

In many concrete attack scenarios, an attacker knows some bits of the secret message in advance. For instance, an HTTP cookie typically uses ASCII printable characters, whose high order bit is always set to zero. More generally, we assume that $\mathcal{S}$ is (included in) an affine subspace of $\{0,1\}^n$ of dimension $n-z$ for some natural $z < n$. In order to simplify the attack, we use a bijective affine function ϕ that maps $\mathcal{S}$ unto $\{0\}^z \times \{0,1\}^{n-z}$, and rewrite the problem as follows:

$$\begin{aligned} S \neq a \oplus b &\Leftrightarrow \phi(S) \neq \phi(a \oplus b), && \text{as } \phi \text{ is a bijection.} \\ &\Leftrightarrow \phi(S) \neq \phi(a) \oplus \phi(b) \oplus \phi(0), && \text{as } \phi \text{ is affine} \end{aligned}$$

Therefore, we can reduce the missing difference problem on $\mathcal{A}$, $\mathcal{B}$, $\mathcal{S}$ with $\dim(\langle\mathcal{S}\rangle) = n - z$ to the missing difference problem on $\mathcal{A}'$, $\mathcal{B}'$, $\mathcal{S}'$, where the secret is known to start with z zeroes:

$$\begin{aligned} \mathcal{S}' &:= \{0\}^z \times \{0,1\}^{n-z} \\ \mathcal{A}' &:= \{\phi(a) \mid a \in \mathcal{A}\} \\ \mathcal{B}' &:= \{\phi(b) \oplus \phi(0) \mid b \in \mathcal{B}\} \end{aligned}$$

We now introduce a known prefix sieving algorithm (Algorithm 3) to solve this problem efficiently. The algorithm is quite straightforward; it looks for a prefix collision before sieving in the same way as before to recover S. The complexity depend on the dimension $n-z$; the sieving requires $\mathcal{O}(2^{n-z})$ memory and $\mathcal{O}((n-z) \cdot 2^{n-z})$ XOR computations in expectation, while looking for collisions only requires to store the prefix keys and to go through one of the set. Looking for collisions allows us to skip the computations of many pairs (a, b) that would be irrelevant as $a \oplus b \notin \mathcal{S}$.

The expected number of collisions required to isolate the secret is given by the coupon collector problem as $\ln(2^{n-z})2^{n-z} = \ln 2 \cdot (n-z) \cdot 2^{n-z}$. Therefore the total optimized complexity (with balanced sets $\mathcal{A}$ and $\mathcal{B}$) to recover an $n-z$ bits secret with this algorithm is:

$$\begin{aligned} &\mathcal{O}\left(\sqrt{n-z} \cdot 2^{n/2}\right) && \text{queries} \\ &\mathcal{O}\left(2^{n-z} + n\sqrt{n-z} \cdot 2^{n/2}\right) && \text{bits of memory (sieving \& queries)} \\ &\mathcal{O}\left((n-z) \cdot 2^{n-z} + \sqrt{n-z} \cdot 2^{n/2}\right) && \text{operations (sieving \& collisions searching)} \end{aligned}$$

As we can see from the complexity, when $z = 0$ this is the naive algorithm with its original complexity. When z nears n, this performs similarly to McGrew's searching algorithm *i.e.* the cost of looking for collisions (or storing $\mathcal{B}$ so that the search is efficient) will dominate the overall cost of the algorithm therefore the time and query complexity will match. Actually, this algorithm improves over previous works for intermediate values of z. With $z = n/2$, we have an algorithm with complexity $\tilde{\mathcal{O}}(2^{n/2})$, while McGrew's searching algorithm would require $\tilde{\mathcal{O}}(2^{3n/4})$ computations in the same setting. The complexity therefore becomes tractable and we could implement and run this algorithm for $n = 64$ bits with success, as shown in Sect. 6.2.

Algorithm 3. Known prefix sieving algorithm

Input: n, $z < n$, $\mathcal{A}, \mathcal{B}$, $\mathcal{S} \subseteq \{0\}^z \times \{0,1\}^{n-z}$
Output: $\{s \in \mathcal{S} \mid \forall (a,b) \in \mathcal{A} \times \mathcal{B},\ a \oplus b \neq s\}$
  $h_B \leftarrow$ Empty hash table.
  for b **in** $\mathcal{B}$ **do**
    $h_B[b_{[0...(z-1)]}] \xleftarrow{\cup} \{b_{[z...(n-1)]}\}$
  end for

  for a **in** $\mathcal{A}$ **do**
    $v_a \leftarrow a_{[z...(n-1)]}$
    for v_b **in** $h_B[a_{[0...(z-1)]}]$ **do**
      Remove $\overline{0} \parallel (v_a \oplus v_b)$ from $\mathcal{S}$;
    end for
  end for
  return $\mathcal{S}$

3.2 Fast Convolution Sieving

Alternatively, we can reduce the complexity of the sieving algorithm by using sets $\mathcal{A}$ and $\mathcal{B}$ of size $2^m \gg 2^{n/2}$, rather than $\tilde{\mathcal{O}}(2^{n/2})$ as required to uniquely identify S. If we consider all the values $a \oplus b$ for (a,b) in $\mathcal{A} \times \mathcal{B}$, we expect that they are close to uniformly distributed over $\{0,1\}^n \setminus S$, so that every value except S is reached about 2^{2m-n} times, while S is never hit. Increasing m makes the gap more visible than with sets of size only $\tilde{\mathcal{O}}(2^{n/2})$. Therefore, we can consider buckets of several candidates s, and accumulate the number of $a \oplus b$ in each bucket. If we consider buckets of 2^t values, each bucket receives 2^{2m+t-n} values on average, but the bucket containing S receives only $2^{2m+t-n} - 2^{2m-n}$ values. If we model this number with random variables following a binomial distribution, the variance σ^2 is about $2^{m+t/2-n/2}$. Therefore, the bias will be detectable when: $\sigma \ll 2^{2m-n}$, *i.e.* when $t \ll 2m - n$.

Concretely, we use a truncation function T that keeps only $n - t$ bits of an n-bit word. We consider the values $T(a \oplus b)$ for all $(a,b) \in \mathcal{A} \times \mathcal{B}$, and count how many times each value is reached. If m is large enough, the value with the lowest counter corresponds to $T(S)$. This attack does not require any prior information on the secret; it can be used with $\mathcal{S} = \{0,1\}^n$, and once $T(S)$ is known, we can use known-prefix sieving to recover the remaining bits (looking for S in an affine space of dimension t).

We now show an algorithm to quickly count the number of occurrences for each combination. For a given multi-set $\mathcal{X}$, we consider an array of counters $C_{\mathcal{X}}$, to represent how many times each value $T(x)$ is reached:

$$C_{\mathcal{X}}[i] = \left|\{x \in \mathcal{X} \mid T(x) = i\}\right|.$$

Our goal is to compute $C_{\mathcal{A} \oplus \mathcal{B}}$ efficiently from $\mathcal{A}$ and $\mathcal{B}$, where $\mathcal{A} \oplus \mathcal{B}$ is the multi-set $\{a \oplus b \mid (a,b) \in \mathcal{A} \times \mathcal{B}\}$. We observe that:

$$\begin{aligned}
C_{\mathcal{A} \oplus \mathcal{B}}[i] &= |\{(a,b) \in \mathcal{A} \times \mathcal{B} \mid T(a \oplus b) = i\}| \\
&= \sum_{a \in \mathcal{A}} |\{b \in \mathcal{B} \mid T(a \oplus b) = i\}| \\
&= \sum_{a \in \mathcal{A}} |\{b \in \mathcal{B} \mid T(b) = i \oplus T(a)\}| \\
&= \sum_{a \in \mathcal{A}} C_{\mathcal{B}}[i \oplus T(a)] \\
&= \sum_{j \in \{0,1\}^{n-t}} C_{\mathcal{A}}[j] C_{\mathcal{B}}[i \oplus j]
\end{aligned}$$

This is a form of convolution that can be computed efficiently only using the Fast Walsh-Hadamard Transform (Algorithm 4), in the same way we use the Fast Fourier Transform to compute circular convolutions (see Algorithm 5). Therefore the full attack (shown in Algorithm 6) takes time $\tilde{\mathcal{O}}(2^{n-t})$ using lists of size 2^m with $m \gg (n+t)/2$ and a sieve of 2^{n-t} elements.

In order to optimize the attack, we select $t = n/3$ such that the time complexity, data complexity, and memory usage are all roughly $2^{2n/3}$. A detailed analysis in Sect. 6.1 shows that we reach a constant success rate with $t = n/3$ using lists of size $\mathcal{O}(\sqrt{n} \cdot 2^{2n/3})$. This gives the following complexity for the full attack:

$\mathcal{O}(\sqrt{n} \cdot 2^{2n/3})$	queries
$\mathcal{O}(n \cdot 2^{2n/3}) + \mathcal{O}(n\sqrt{n} \cdot 2^{n/2})$	bits of memory (counters + sieving)
$\mathcal{O}(n \cdot 2^{2n/3}) + \mathcal{O}(n\sqrt{n} \cdot 2^{n/2})$	computations (fast Walsh-Hadamard + sieving)

As seen in Sect. 6.1, we performed experiments with $n = 12, 24, 48$, and the correct S was found with the lowest counter in at least 70% of our experiments, using list of size $\sqrt{n}2^{2n/3}$. This validates our approach and shows that the constant in the $\mathcal{O}$ notation is small. We could run this algorithm over $n = 48$ bits in a matter of minutes.

Optimizations. In order to increase the success rate of the algorithm, one can test several candidates for $T(S)$ (using the lowest remaining counters), and

Algorithm 4. Fast Walsh-Hadamard Transform

Input: $C_{\mathcal{A}}$, $|C_{\mathcal{A}}| = 2^m$
Output: The Walsh-Hadamard transform of $C_{\mathcal{A}}$
  for $d = m$ **downto** 0 **do**
    for $i = 0$ **to** 2^{m-d} **do**
      for $j = 0$ **to** 2^{d-1} **do**
        $C_{\mathcal{A}}[i \cdot 2^d + j] \leftarrow C_{\mathcal{A}}[i \cdot 2^d + j] + C_{\mathcal{A}}[i \cdot 2^d + j + 2^{d-1}]$
        $C_{\mathcal{A}}[i \cdot 2^d + j + 2^{d-1}] \leftarrow C_{\mathcal{A}}[i \cdot 2^d + j] - 2 \cdot C_{\mathcal{A}}[i \cdot 2^d + j + 2^{d-1}]$
      end for
    end for
  end for
  return $C_{\mathcal{A}}$

Algorithm 5. Fast convolution

Input: $C_{\mathcal{A}}, C_{\mathcal{B}}$
Output: $C_{\mathcal{A}\oplus\mathcal{B}}$
  {Perform fast Walsh-Hadamard transform in-place}
  FWHT($C_{\mathcal{A}}$); FWHT($C_{\mathcal{B}}$);
  for $c = 0$ **to** 2^{n-t} **do**
    $C_{\mathcal{A}\oplus\mathcal{B}}[c] \leftarrow C_{\mathcal{A}}[c] \cdot C_{\mathcal{B}}[c]$
  end for
  {Perform fast Walsh-Hadamard transform in-place}
  FWHT($C_{\mathcal{A}\oplus\mathcal{B}}$);
  return $C_{\mathcal{A}\oplus\mathcal{B}}$

Algorithm 6. Sieving with fast convolution

Input: $\mathcal{A}, \mathcal{B}, t \leq n$
Output: S s.t. $\forall (a,b) \in \mathcal{A} \times \mathcal{B}$, $a \oplus b \neq S$
  $C_{\mathcal{A}}, C_{\mathcal{B}}, C_{\mathcal{A}\oplus\mathcal{B}} \leftarrow$ arrays of 2^{n-t} integers initialized to 0;
  for a **in** $\mathcal{A}$ **do**
    Increment $C_{\mathcal{A}}[a_{0..(n-t-1)}]$
  end for
  for b **in** $\mathcal{B}$ **do**
    Increment $C_{\mathcal{B}}[b_{0..(n-t-1)}]$
  end for
  $C_{\mathcal{A}\oplus\mathcal{B}} \leftarrow$ FastConvolution($C_{\mathcal{A}}, C_{\mathcal{B}}$)
  $u \leftarrow \text{argmin}_i C_{\mathcal{A}\oplus\mathcal{B}}[i]$
  Run known prefix sieving (Algorithm 3), knowing that $T(S) = u$

use the known-prefix sieving to detect whether the candidate is correct. Another option is to run multiple independent runs of the algorithm with different choices of the $n/3$ truncated bits. This would avoid some bad cases we have observed in simulations, where the right counter grows abnormally high and gets hidden in all of the other counters.

For the memory complexity, notice that we don't need to store all the data but simply to increment a counter. We only need to keep enough blocks for the second part of the algorithm so that the sieving yields a unique result. Initially the counters for $C_{\mathcal{A}}$ and $C_{\mathcal{B}}$ are quite small, $\sqrt{n}$ in expectation. However, $C_{\mathcal{A}\oplus\mathcal{B}}$ will have much bigger entries, $n \cdot 2^{2n/3}$ in expectation, so that we need $\mathcal{O}(n)$ bits to store each entry.

4 Application to the CTR Mode

We know show how to mount attacks against the counter mode using the new algorithms for the missing difference problem.

4.1 Attack Using Fast Convolution

Use of the fast convolution algorithm to recover one block of CTR mode plaintext is straightforward. The attacker is completely passive and observes encryptions of S (gathered in set $\mathcal{B}$), and keystream blocks recovered from the encryption of known message blocks (gathered in set $\mathcal{A}$). When the lists are large enough, he runs the fast convolution algorithm on $\mathcal{A}$ and $\mathcal{B}$ to recover S.

4.2 Attacks Using Known Prefix Sieving

Direct Attack. There are many settings where unknown plaintext will naturally lie in some known affine subspace, and the known prefix sieving algorithm can be used directly. For instance a credit card number (or any number) could be encoded in 16 bytes of ASCII then encrypted. Because in ASCII the encoding of any digit starts by `0x3` (`0x30` to `0x39`), we know half of the bits of the plaintext, and we can use the known-prefix sieving with $z = n/2$. Other examples are information encoded by `uuencode` that uses ASCII values `0x20` to `0x5F` (corresponding to two known bits) or HTML authentication cookies that are typically encoded to some subset of ASCII numbers and letters[3].

Block Splitting. We often assume that the secret is encrypted in its own block, but when the secret is part of the message, it can also be split across block boundaries, depending on how the plaintext is constructed and encrypted by the protocol. In particular, if a message block contains both known bytes and secret bytes, we can apply the known prefix sieving algorithm to this block and recover the secret bytes.

In many protocols, messages start with some low entropy header that can be guessed by an attacker. Moreover, the attacker often has some degree of control over those headers. For instance, in the BEAST attack [12] against HTTPS, an attacker uses Javascript code to generate HTTPS requests, and he can choose the URL corresponding to the requests. Using this control of the length of the header, block splitting attacks have been shown in the BEAST model [12,20].

[3] For example, wikipedia.org encodes cookies with lower case letters and digits, this corresponds to two known bits.

Table 1. Example of an attack on two blocks secret $S = S_1\|S_2\|S_3\|S_4$. Each step performs the known prefix sieving algorithm. Known information in blue, unknown information in red, attacked information in yellow.

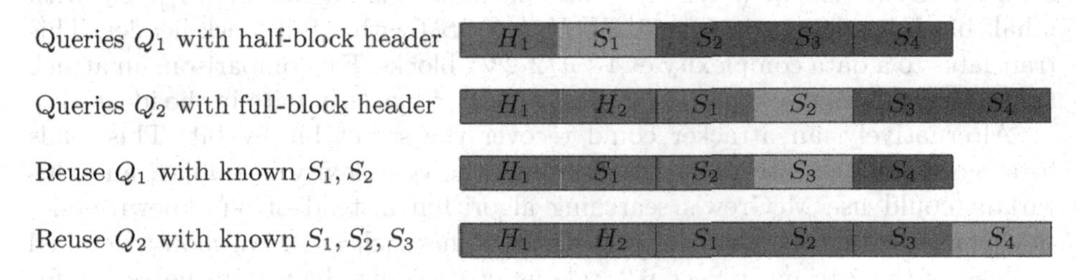

The attacker starts with a header length so that a small chunk of the secret message is encrypted together with known information, and recovers this secret chunk. Then he changes the length of the header to recover a second chunk of the message, using the fact that the first chunk is now known. Eventually, the full secret can be recovered iteratively.

In our case, the easiest choice is to recover chunks of $n/2$ bits of secret one by one, using the known-prefix sieving algorithm with $z = n/2$. We illustrate this attack in Table 1, assuming a two-block secret $S = S_1\|S_2\|S_3\|S_4$, and a protocol that lets the adversary query an encryption of the secret with an arbitrary chosen prefix:

1. The attacker makes two kind of queries
   - Q_1 with a known half-block header H_1 ($\mathcal{E}([H_1\|S_1]\|[S_2\|S_3]\|[S_4])$);
   - Q_2 with a known full-block header $H_1\|H_2$ ($\mathcal{E}([H_1\|H_2]\|[S_1\|S_2]\|[S_3\|S_4])$).
2. He first recovers S_1 using the known-prefix sieving with the first block of each type of query. More precisely, he uses $\mathcal{A} = \{\mathcal{E}(H_1\|H_2)\}$ and $\mathcal{B} = \{\mathcal{E}(H_1\|S_1)\}$, so that the missing difference is $0 \,\|\, (S_1 \oplus H_2)$.
3. When S_1 is known, he can again use known prefix sieving to recover S_2, with the first and second blocks of Q_2 queries: $\mathcal{A} = \{\mathcal{E}(H_1\|H_2)\}$ and $\mathcal{B} = \{\mathcal{E}(S_1\|S_2)\}$, so that the missing difference is $(S_1 \oplus H_1)\|(S_2 \oplus H_2)$. To improve the success rate of this step, he can also consider the first block of Q_1 queries as known keystream.
4. When S_2 is known, another round of known prefix sieving reveals S_3, *e.g.* with $\mathcal{A} = \{\mathcal{E}(H_1\|H_2)\}$ and $\mathcal{B} = \{\mathcal{E}(S_2\|S_3)\}$, the missing difference is $(S_2 \oplus H_1)\|(S_3 \oplus H_2)$.
5. Finally, S_4 is recovered with a last round of known prefix sieving using $\mathcal{A} = \{\mathcal{E}(H_1\|H_2)\}$ and $\mathcal{B} = \{\mathcal{E}(S_3\|S_4)\}$, with missing difference is $(S_3 \oplus H_1)\|(S_4 \oplus H_2)$.

This gives an algorithm with query complexity of $\mathcal{O}(\sqrt{n}2^{n/2})$ to recover repeated encryption of a secret over multiple blocks in the BEAST attacker model. In Sect. 6.2, we analyze the constants in the $\mathcal{O}()$ and run experiments with $n = 64$ using locally encrypted data. In particular, we have a success probability higher than 80% using two lists of 5×2^{32} queries with $n = 64$.

More generally, we show that for $n \geq 32$ the success probability of this attack is at least 99% with lists of size $\sqrt{n/2} \cdot 2^{n/2}$. With a one block secret, an optimal attack uses two lists of $\sqrt{n/2} \cdot 2^{n/2}$ two-block queries: queries $[H_1 \| S_1] \| [S_2]$ with a half-block header, and queries $[H_1 \| H_2] \| [S_1 \| S_2]$ with a full-block header. This translates to a data complexity of $4\sqrt{n/2} \cdot 2^{n/2}$ blocks. For comparison, an attack against the CBC mode requires on average $2 \cdot 2^{n/2}$ blocks of data in the ideal case.

Alternatively, an attacker could recover the secret bit by bit. This leads to a more complex attack in practice, but the complexity is similar, and this variant could use McGrew's searching algorithm instead of our known-prefix sieving algorithm (because in this scenario, we have $|\mathcal{S}| = 2$). We show a detailed analysis of this variant in Sect. 6.2, taking into account the n steps necessary for this attack.

4.3 Use of CTR Mode in Communication Protocols

The CTR mode is widely used in internet protocols, in particular as part of the GCM authenticated encryption mode [29], with the AES block cipher. For instance, Mozilla telemetry data show that more than 90% of HTTPS connections from Firefox 58 use AES-GCM[4]. While attacks against modes with a 128-bit block cipher are not practical yet, it is important to limit the amount of data processed with a given key, in order to keep the probability of a successfull attack negligible, following the guidelines of Luykx and Paterson [26].

Surprisingly, there are also real protocols that use 64-bit block ciphers with the CTR mode (or variants of the CTR mode), as shown below. Attacks against those protocols would be (close to) practical, assuming a scenario where an attacker can generate the encryption of a large number of messages with some fixed secret.

SSH. Ciphersuites based on the CTR mode were added to SSHv2 in 2006 [4]. In particular, 3DES-CTR is one of the recommended ciphers, but actual usage of 3DES-CTR seems to be rather low [1]. In practice, 3DES-CTR is optionally supported by the dropbear server, but it is not implemented in OpenSSH. According to a scan of the full IPv4 space by Censys.io[5], around 9% of SSH servers support 3DES-CTR, but actual usage is hard to estimate because it depends on client configuration.

The SSH specification requires to rekey after 1 GB of data, but an attack is still possible, although the complexity increases.

3G Telephony. The main encryption algorithm in UMTS telephony is based on the 64-bit blockcipher Kasumi. The mode of operation, denoted as f8, is represented in Fig. 2. While this mode in not the CTR mode and was designed to avoid its weaknesses, our attack can be applied to the first block of ciphertext.

---

[4] https://mzl.la/2GY53Mc, accessed February 8, 2018.

[5] https://censys.io/data/22-ssh-banner-full_ipv4, scan performed July 5, 2017.

Indeed the first block of message i is encrypted as $c_{i,0} = m_{i,0} \oplus E_k(E_{k'}(i))$, where the value $E_k(E_{k'}(i))$ is unique for all the messages encrypted with a given key.

There is a maximum of 2^{32} messages encrypted with a given key in 3G, but this only has a small effect on the complexity of attacks.

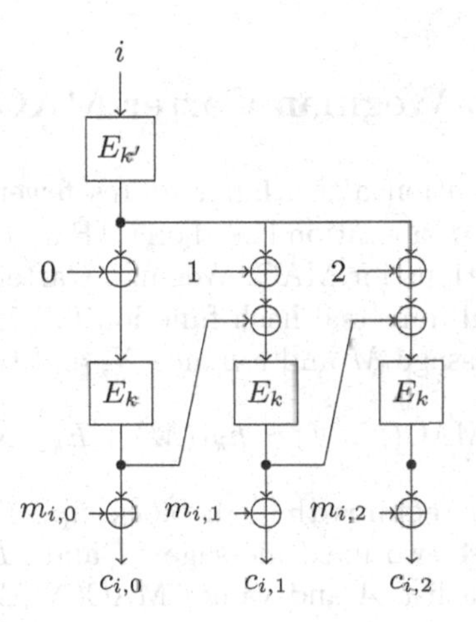

Fig. 2. f8 mode (i is a message counter)

Because of the low usage of 3DES-CTR in SSH, and the difficulty of mounting an attack against 3G telephony in practice, we did not attempt to demonstrate the attack in practice, but the setting and complexity of our attacks are comparable to recent results on the CBC mode with 64-bit ciphers [7].

4.4 Counter-Measures

As for many modes of operation, the common wisdom to counter this kind of attacks asks for rekeying before the birthday bound, *i.e.* before $2^{n/2}$ blocks. However rekeying too close to the birthday bound may not be enough. For example let's consider an implementation of a CTR based mode of operation that rekeys every $2^{n/2}$ blocks, Using the same model as previously, and a one-block secret, an optimal attack uses queries $[H_1 \| S_1] \| [S_2]$ with a half-block header, and queries $[H_1 \| H_2] \| [S_1 \| S_2]$ with a full-block header, where rekeying occurs after $2^{n/2-2}$ queries of each type. To recover S_1, we use the known prefix sieving algorithm as previously, but we can only use relations between ciphertext blocks encrypted with the same key. In each session of $2^{n/2}$ blocks, we consider 2^{n-4} pairs of ciphertext blocks; on average there are $2^{n/2-4}$ pairs with the correct prefix used for sieving. Since we need $n/2 \cdot 2^{n/2}$ draws to reduce the sieve to a single element with high probability, we use $8n$ sessions, *i.e.* $8n \cdot 2^{n/2}$ blocks of data in total. The same data can be reused to recover S_2 when S_1 is known. This should be compared with the previous data complexity of $4\sqrt{n/2} \cdot 2^{n/2}$ in the absence of rekeying.

However, rekeying every $2^{n/2-16}$ blocks makes the data complexity goes up to $2^{35}n$ sessions or $n \cdot 2^{19+n/2}$ blocks to recover the secret block. Notice that the security gain of rekeying is comparable with what is gained in CBC, where rekeying every $2^{n/2-16}$ blocks forces increases the data complexity from $2 \cdot 2^{n/2}$ to $2^{18} \cdot 2^{n/2}$.

5 Application to Wegman-Carter MACs

Because the fast convolution algorithm requires fewer assumptions, it can be adapted to other modes of operation based on CTR and particularly to Wegman-Carter type of constructions for MAC. Wegman-Carter MACs use a keyed permutation E and a keyed universal hash function h, with $k1$ and $k2$ two private keys. The input is a message M and a nonce N, and the MAC is defined as:

$$\mathrm{MAC}(N, M) = h_{k1}(M) + E_{k2}(N)$$

Again, the construction requires that all block cipher inputs are different. To apply our attack, we use two fixed message M and M', and we capture many values $\mathrm{MAC}(N, M)$ in a list $\mathcal{A}$ and values $\mathrm{MAC}(N', M')$ in a list $\mathcal{B}$, all using unique nonces. Then we solve the missing difference problem to recover $h_{k1}(M) - h_{k1}(M')$ as we know that $\forall N \neq N' : E_{k2}(N) - E_{k2}(N') \neq 0$. It is often sufficient to know this difference and the two messages M and M' to recover the key k_1. We give two examples with concrete MAC algorithms.

Galois/Counter Mode. GCM is an authenticated encryption mode with associated data, combining the CTR mode for encryption and a Wegman-Carter MAC based on polynomial evaluation in a Galois field for authentication. It takes as input a message M that is encrypted and authenticated, and some associated data A that is authenticated but not encrypted. When used with an empty message, the resulting MAC is known as GMAC. In our attack, we use an empty message with one block of authenticated data A, so that the tag is computed as:

$$\mathrm{MAC}(N, A) = A \cdot H^2 \oplus H \oplus E_k(N),$$

with H the hash key and $(\cdot)$ the multiplication in a Galois Field defined by a public polynomial. So, for two different blocks of authenticated data A and A' we collect $\mathcal{O}(\sqrt{n} \cdot 2^{2n/3})$ MACs and perform the fast convolution algorithm to recover $A \cdot H^2 \oplus H \oplus A' \cdot H^2 \oplus H = (A \oplus A') \cdot H^2$. We known $A \oplus A'$ and the field is known so we invert that value and recover H^2 then compute the square root and recover the hash key H.

Comparison with previous attacks against GMAC. There are several known attacks against GCM and GMAC, but none of them seems to allow universal forgery with just $2^{2n/3}$ blocks of data and $2^{2n/3}$ computations. In particular, Handschuh and Preneel [19] gave a weak-key attack, that can also be used to recover the hash key without weak key assumptions, using roughly $2^{n/2}$ messages of $2^{n/2}$ blocks. Later work extended these weak key properties [33,36] but an attack still requires about 2^n blocks in total when no assumptions are made about the key. We also note that these attacks require access to a verification oracle, while our attack only uses a MAC oracle.

Some earlier attacks use specific options of the GCM specifications to reach a lower complexity, but cannot be applied with standard-length IV, and tag: Ferguson [14] showed an attack when the tag is truncated, and Joux [23] gave an attack based on non-default IV lengths.

Poly1305. Poly1305 [6] is a MAC scheme following the Wegman-Carter construction, using polynomial evaluation modulo the prime number $2^{130} - 5$. It uses a keyed 128-bit permutation (usually AES), and the hash function key, r, has 106 free bits (22 bits of the key are set to 0, including in particular the 4 most significant ones). The message blocks are first padded to 129-bit values c_i. Then the MAC of a q-block message M with nonce N is defined as:

$$T(M,N) = (((c_1 r^q + c_2 r^{q-1} + ... + c_q r) \mod 2^{130} - 5) + E_k(N)) \mod 2^{128}.$$

With the same strategy as above, using two different messages M and M' we recover the missing difference

$$(((c_1 - c'_1)r^q + (c_2 - c'_2)r^{q-1} + ... + (c_q - c'_q)r) \mod 2^{130} - 5) \mod 2^{128}.$$

Moreover, we chose M and M' such that $c_i - c'_i = 0$ and $c_q - c'_q = 1$; since by design, $r < 2^{124}$ the value recovered is simply the hash key r.

Notice that Poly1305 doesn't use the XOR operation but a modular addition, and we have to adapt our algorithms to this case. Luckily, the fast convolution algorithm can easily be tweaked. First, we keep the $2n/3$ least significant bits to avoid issues the carry, something the XOR operation doesn't have. Then, when the lists of counters are up, we need to compute their cyclic convolution, which is done with a fast convolution algorithm based on the fast Fourier transform (instead of fast Walsh-Hadamard). Then we verify the value suggested by the lowest counter by running the known prefix algorithm looking for collisions on the least significant bits and sieving the modular subtraction of the most significant bits. This adaptation has similar complexities and proofs than the one described earlier. Moreover, in the case of Poly1305, one can further adapt the algorithms to take into account the fact that 22 bits of the key r are fixed at 0 effectively reducing the dimension of $\mathcal{S}$.

6 Proofs and Simulations

In this section we give some theoretical and simulation results that further support the claims we made thus far.

6.1 About the Fast Convolution Algorithm

Proof of query complexity for the claim made in Sect. 3.2. Consider, without loss of generality and for blocks of size n, that we possess $a \cdot 2^{2n/3}$ blocks of keystream and the same number of blocks of encrypted secret S with a a function of n. So in this setting we have $a^2 \cdot 2^{4n/3}$ different XORed-values possible between the two lists, that we will consider as independent and uniformly distributed over $2^n - 1$ values. We will then focus on the $2n/3$ bits truncation, $T(\cdot)$, and ignore the rest. We count the number of occurrences for every truncated values and store them in two lists of size $2^{2n/3}$. Using the fast Walsh-Hadamard transform 3 times, Algorithm 5, we can therefore compute the same counters but for all the XORed-values. We hope that the counter for $T(S)$, the good counter, will be lower than all of the other counters, the bad counters, with probability $\Omega(1)$. In which case we say the algorithm succeeds.

Let X_i^c represents the fact that the ith value truncates to c, so that X_i^c follows a Bernoulli distribution and any counter can be written as $X^c = \sum_{i=1}^{a^2 2^{4n/3}} X_i^c$. Now we have to discriminate between the distributions of the good and bad counters:

$$\text{Good case } c = T(S): \quad \Pr(X_i^{T(S)} = 1) = (2^{n/3} - 1)/2^n = 2^{-2n/3} - 2^{-n}$$
$$\implies \mathbf{E}[X^{T(S)}] = 2^{2n/3}a^2 - 2^{n/3}a^2$$

$$\text{Bad case } c \neq T(S): \quad \Pr(X_i^c = 1) = (2^{n/3})/2^n = 2^{-2n/3}$$
$$\implies \mathbf{E}[X^c] = 2^{2n/3}a^2$$

Now we are interested by the probability that a bad counter gets a value below $\mathbf{E}[X^{T(S)}]$ as a measure of how distinct the distributions are. Using Chernov Bound we get for all $c \neq T(S)$:

$$\begin{aligned}\Pr(X^c < \mathbf{E}[X^{T(S)}]) &= \Pr(X^c < (1 - 2^{-n/3})2^{2n/3}a^2) \\ &= \Pr(X^c < (1 - 2^{-n/3})\mathbf{E}[X^c]) \\ &\leq e^{-((2^{-n/3})^2 \cdot 2^{2n/3}a^2)/2)} = e^{-a^2/2}\end{aligned}$$

And to compute the probability that no bad counter gets below $\mathbf{E}[X^{T(S)}]$ we will have to assume their independence, which is wrong, but we will come back later to discuss this assumption.

$$\begin{aligned}\Pr(\forall c \neq T(S) : X^c \geq \mathbf{E}[X^{T(S)}]) &= \prod_{c \neq T(S)} \left(1 - \Pr(X^c < \mathbf{E}[X^{T(S)}])\right) \\ &\geq \left(1 - e^{-a^2/2}\right)^{2^{2n/3}}\end{aligned}$$

To conclude, we need to find an $a = a(n)$ such that this probability remains greater than some positive value as n grows. This is clearly achieved with $a = \mathcal{O}(\sqrt{n})$ as for example taking $a = \frac{2\sqrt{n}}{\sqrt{3 \cdot \log_2(e)}} \simeq 0.96\sqrt{n}$ we get:

$$\begin{aligned}\Pr(\forall c \neq T(S) : X^c \geq \mathbf{E}[X^{T(S)}]) &\geq (1 - e^{-a^2/2})^{2^{2n/3}} \\ &\geq (1 - 2^{-2n/3})^{2^{2n/3}} \\ &\geq 0.25, \qquad\qquad \forall n \geq 3/2\end{aligned}$$

Therefore we can bound the probability of success by the events '$X^{T(S)} < \mathbf{E}[X^{T(S)}]$', probability $\simeq 1/2$, and '$\forall c \neq T(S) : X^c \geq \mathbf{E}[X^{T(S)}]$', probability at least 1/4. Then we indeed have a probability of at least 1/8 of having a successful algorithm. We can conclude that with $\mathcal{O}(n \cdot 2^{4n/3})$ XORed-values the algorithm has probability $\Omega(1)$ of succeeding.

Notice that this requires lists of size $\mathcal{O}(\sqrt{n} \cdot 2^{2n/3})$ but for the proof we only need the total number of pairs between the two lists. So we can break the requirement that the two lists are of comparable sizes as long as the product of their sizes sum up to the order of required values.

On the independence of the counters, this is obviously wrong as they are bound by the relation $\sum_c X^c = a^2 2^{4n/3}$. However this relation becomes looser and looser as n grows so the approximation obtained should still be correct asymptotically. Moreover, the covariances implied are negative *i.e.* knowing one draw is big makes the other draws smaller in expectation to compensate. Small negative covariances will make the distribution look more evenly distributed in the sense that we can't observe too many extreme events in a particular direction which is good for the success rate of the algorithm. So the assumption of independence may be a conservative one for this complexity analysis.

Simulation Results. We ran simulations for block sizes $n = 12$, 24, 32 and 48 bits, so that we could do some statistical estimations of the success probability for this attack. We first create two lists of same size, one of raw keystream output and one XORed with an n-bit secret S. Then we pass the two lists in Algorithm 5 counting over $n' = 2n/3$ bits (unless specified otherwise) to get a list of counters for each possible XOR outputs on those n' bits. Then the expected behaviour of the attack would be to look for a solution whose n' first bits correspond to the position of the lowest counter and test this hypothesis with Algorithm 3. If it returns a unique value then this is S and we are done, if it returns an empty set then test with the position of the second lowest counter, etc. We can therefore know the number of key candidates that would be required to recover S and, over many trials, have an estimation of the probability of success after a given number of candidates in these parameters.

For block sizes of 12 and 24 we simulated a permutation simply by shuffling a range into a list. For bigger sizes of 32 and 48 we used the Simon lightweight cipher from the NSA [3] as that is one of the rare block cipher who can act on 48-bit blocks. We could quickly gather 10 000 runs for each setting except for the 48-bit blocks simulation where we gathered 756 runs.

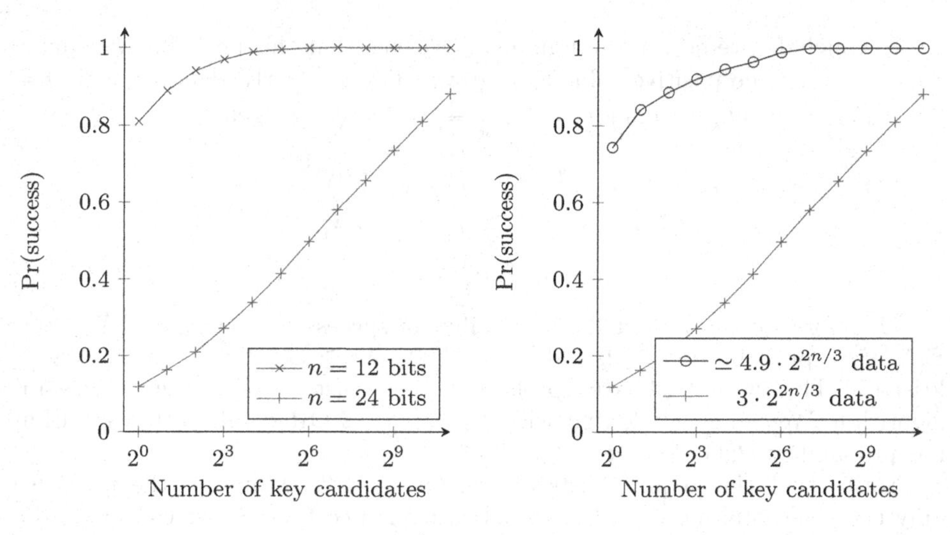

Fig. 3. Results for lists size of $3 \cdot 2^{2n/3}$

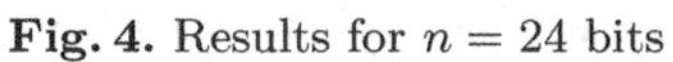

Fig. 4. Results for $n = 24$ bits

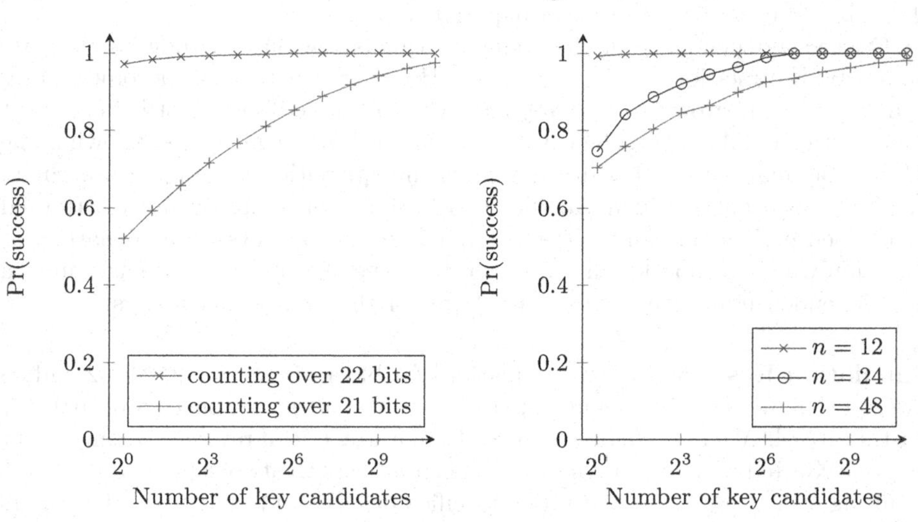

Fig. 5. Results for $n = 32$ bits; $\sqrt{n}2^{2n/3} \simeq 5.66 \cdot 2^{2n/3}$ data

Fig. 6. Results for $\sqrt{n}2^{2n/3}$ data; counting over $2n/3$ bits

In general we observe in Fig. 6 that the algorithm has a good chance of success with the first few candidates when using the suggested parameters. Moreover the sensibility with respect to the data complexity (Fig. 4) and to the number of bits counted over (Fig. 5) is fairly high. These results back up our complexity analysis and are a good indication that no big constant is ignored by the $\mathcal{O}()$ notation.

On the speed at which the probability increases we realized that, despite the log scale on the x axis, the curves take a straight (Fig. 3) or concave shape (Figs. 5 and 6). That means that the probability of success with the next key candidate decreases very quickly with the number of key candidates already tested and proved wrong. For example for $n = 48$ bits (Fig. 6) over 756 trials the right key candidate was in the 2048 lowest counters in 98.1% of the time but the worst case found was 1 313 576 and these "very bad" cases push the mean rank of the right key candidate to 2287 and its sample variance to 2 336 937 008.

For $n = 48$ bits, one simulation took us 40 min over 10 cores (each step is highly parallelizable), and 64 gibibytes of RAM for the counters lists.

6.2 About the Known Prefix Sieving Algorithm

We consider two particular settings for the known prefix sieving algorithm and the corresping block splitting attack, with $z = n/2$ and $z = 1$.

Theoretical Bound. We first give a theoretical lower bound to the probability of success of the sieving when $\dim(\mathcal{S}) = n/2$ (*i.e.* $z = n/2$), depending on the query complexity. Every partial collision found helps us to sieve. After collecting many blocks of keystream and encryption of S let $|\mathcal{A}| \cdot |\mathcal{B}| =: \alpha 2^n$ for some α. Thus we get $\alpha 2^n / 2^{n/2} = \alpha 2^{n/2}$ partial collisions in expectation. More precisely, the Chernoff bound gives us a lower bound for the probability of finding at least $(1-\delta)\alpha 2^{n/2}$ collisions:

$$p \geq 1 - \left(\frac{e^{-\delta}}{(1-\delta)^{(1-\delta)}} \right)^{\alpha 2^{n/2}}$$

for any $\delta > 0$.

We see one partial collision as a draw in the coupon collector problem. One can use the formula in [34] for the tail of coupon collector problem probability distribution to estimate the chance of success after obtaining $\beta \cdot 2^{n/2}$ partial collisions:

$$p \geq 1 - 2^{-\beta/\ln(2)+n/2}$$

which is positive whenever $\beta \geq n/2 \cdot \ln(2)$.

Therefore we bound the probability of success when collecting $|\mathcal{A}| \cdot |\mathcal{B}| = \alpha 2^n$ pairs as the probability of obtaining at least $(1-\delta)\alpha 2^{n/2}$ partial collisions multiplied by the probability of success after sieving $(1-\delta)\alpha 2^{n/2}$ values:

$$p \geq \left(1 - \left(\frac{e^{-\delta}}{(1-\delta)^{(1-\delta)}} \right)^{\alpha 2^{n/2}} \right) \cdot \left(1 - 2^{-(1-\delta)\cdot\alpha/\ln(2)+n/2} \right)$$

In particular, with two lists of size $\sqrt{n/2} \cdot 2^{n/2}$ (*i.e.* $\alpha = n/2$), we get $p \geq 0.99$ as long as $n \geq 32$ (using $\delta = 2^{-8}$).

Simulation Results. We ran simulations with a block size $n = 64$ bits, and a secret S of size $n/2 = 32$ bits, using the Tiny Encryption Algorithm (TEA [37])

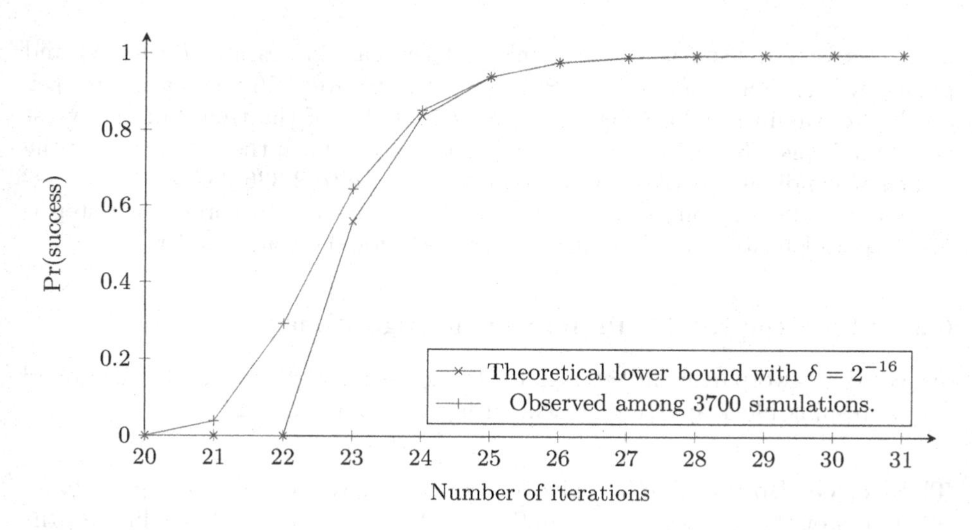

Fig. 7. Probability of success of the known prefix sieving knowing 2^{32} encryptions of a 32-bit secret against the number of chunks of 2^{32} keystream blocks of size $n = 64$ bits used.

in CTR mode to encrypt the data. We create two lists, the keystream output list $a_i \in \mathcal{A}$, and the encryptions $b_j = a_j \oplus (\bar{0}\|S) \in \mathcal{B}$. We first produce and sort a list $\mathcal{B}$ with 2^{32} elements then produce, sort and sieve iteratively several lists $\mathcal{A}$ with 2^{32} elements, until the secret S is the only one remaining in the sieve.

One simulation runs in around 20 min over 36 cores, as every steps are trivially parallelizable: encryption, sorting and sieving. We ran 3700 simulations and tracked how many chunks of $2^{n/2} = 2^{32}$ keystream outputs were needed for sieving. The coupon collector problem predicts that one will need on average $n/2 \cdot \ln(2) \cdot 2^{n/2}$ partial collisions which will be obtained after $n/2 \cdot \ln(2) \simeq 22.18 < 23$ rounds in expectation. And indeed the simulations showed a 64.5% probability of success after 23 iterations. Figure 7 shows the convergence between the theoretical lower bound and the simulated probabilities. We also noticed that the discrepancy in the number of rounds required is largely due to the last few candidates remaining in the sieve. If we decided the attack is successful when we are left with less than 1000 potential candidates for the secret then the algorithm successfully finishes after 16 rounds every time. In fact after 16 rounds the number of candidates left varies from 419 to 560 in all the simulations we have run.

Bit by Bit Secret Recovery. We also want to study the complexity of recovering the secret S bit by bit as an extreme case of the block splitting scenario described in Sect. 4.2. For simplicity, we consider a setting where one query returns a block of keystream and the encryption of $0\,\|\,s_i$ with an unknown bit s_i. We are interested in the query complexity for recovering n bits of secret one bit at a time; that is we need to know the first bit to ask for the second one, etc.

Clearly this can be done in $\mathcal{O}(n \cdot 2^{n/2})$ queries by repeating n times the attack on one bit. But the intuition is that we may need less and less queries to uncover the next bit as we go forward and accumulate blocks of keystream.

Let:

$U_i \leftarrow$The expected number of encryption of $0 \,\|\, s_i$ to recover s_i.

$K_i \leftarrow$The expected number of raw keystream outputs to recover s_i.

From the definition of a query, the above description and because each time we find a bit of secret we can deduce a range of keystream blocks for the next step we have the relations:

$$K_1 = U_1 \tag{1}$$

$$K_{i+1} = K_i + U_i + U_{i+1} \quad \text{for } i \geq 1 \tag{2}$$

$$K_i \cdot U_i = 2^n \quad \text{(in expectation)} \tag{3}$$

We consider the following proposition:

$$P_i : U_i = 2^{n/2}(\sqrt{i} - \sqrt{i-1}),$$

and, using (2), when P_k true for all $k \leq i$ we have:

$$K_i = 2\sum_{k=1}^{i-1} U_k + U_i = 2^{n/2}(\sqrt{i} + \sqrt{i-1}).$$

Moreover (1) and (3) imply $K_1 = U_1 = 2^{n/2}$ so P_1 is true. Now suppose P_k true for all $k \leq i$, let's prove it holds for P_{i+1}:

$$\begin{aligned}
& K_{i+1} \cdot U_{i+1} = 2^n && \text{by (3)} \\
\Longrightarrow\; & U_{i+1}^2 + (K_i + U_i) \cdot U_{i+1} - 2^n = 0 && \text{by (2)} \\
\Longrightarrow\; & U_{i+1}^2 + 2^{n/2} \cdot 2\sqrt{i} \cdot U_{i+1} - 2^n = 0 && \text{by } P_i \\
\Longrightarrow\; & U_{i+1} = 2^{n/2}(\sqrt{i+1} - \sqrt{i}) && \text{as } U_{i+1} \geq 0 \\
\Longrightarrow\; & P_{i+1} \text{ is true.}
\end{aligned}$$

Now that we have a closed form for U_i we can deduce the expected number of queries needed to recover n bits of secret by summing over as $\sum_{i=1}^{n} U_i = 2^{n/2}\sqrt{n}$.

Therefore the query complexity is really $\mathcal{O}(\sqrt{n} \cdot 2^{n/2})$ ignoring a constant depending on the length of a query. Notice that this complexity is the same as when sieving S as a whole showing that we don't grow the query complexity by more than a constant with this strategy.

7 Conclusion

In this work, we have studied the missing difference problem and its relation to the security of the CTR mode. We have given efficient algorithms for the missing difference problem in two practically relevant cases: with an arbitrary missing difference, and when the missing difference is known to be in some low-dimension vector space. These algorithms lead to a message-recovery attack against the CTR mode with complexity $\tilde{\mathcal{O}}(2^{n/2})$, and a universal forgery attack against some Carter-Wegman MACs with complexity $\tilde{\mathcal{O}}(2^{2n/3})$.

In particular, we show that message-recovery attacks against the CTR mode can be mounted with roughly the same requirements and the same complexity as attacks against the CBC mode. While both modes have similar security proofs, there was a folklore assumption that the security loss of the CTR mode with large amounts of data is slower than in the CBC mode, because the absence of collision in the CTR keystream is harder to exploit than CBC collisions [15, Sect. 4.8.2]. Our results show that this is baseless, and use of the CTR mode with 64-bit block ciphers should be considered unsafe (unless strict data limits are in place). As a counter-measure, we recommend to use larger block sizes, and to rekey well before $2^{n/2}$ blocks of data. Concrete guidelines for 128-bit block ciphers have been given by Luykx and Paterson [26]. Alternatively, if the use of small block is required, we suggest using a mode with provable security beyond the birthday bound, such as CENC [21,22].

Our missing difference attacks against CTR and the collision attacks against CBC are two different possible failure of block cipher modes beyond the birthday bound. They exploit different properties of the modes but result in similar attacks. These techniques can be used against other modes of operations (OFB, CFB, ...), and most of them will be vulnerable to at least one the attacks, unless they have been specially designed to provide security beyond the birthday bound.

Acknowledgement. Part of this work was supported by the French DGA, and the authors are partially supported by the French Agence Nationale de la Recherche through the BRUTUS project under Contract ANR-14-CE28-0015.

References

1. Albrecht, M.R., Degabriele, J.P., Hansen, T.B., Paterson, K.G.: A surfeit of SSH cipher suites. In: Weippl, E.R., Katzenbeisser, S., Kruegel, C., Myers, A.C., Halevi, S. (eds.) ACM CCS 2016, pp. 1480–1491. ACM Press, October 2016
2. AlFardan, N.J., Bernstein, D.J., Paterson, K.G., Poettering, B., Schuldt, J.C.N.: On the security of RC4 in TLS. In: King, S.T. (ed.) USENIX Security 2013, pp. 305–320. USENIX Association (2013)
3. Beaulieu, R., Shors, D., Smith, J., Treatman-Clark, S., Weeks, B., Wingers, L.: SIMON and SPECK: Block ciphers for the internet of things. Cryptology ePrint Archive, Report 2015/585 (2015). http://eprint.iacr.org/2015/585
4. Bellare, M., Kohno, T., Namprempre, C.: The Secure Shell (SSH) Transport Layer Encryption Modes. IETF RFC 4344 (2006)
5. Bellare, M., Desai, A., Jokipii, E., Rogaway, P.: A concrete security treatment of symmetric encryption. In: 38th FOCS, pp. 394–403. IEEE Computer Society Press, October 1997
6. Bernstein, D.J.: The Poly1305-AES message-authentication code. In: Gilbert, H., Handschuh, H. (eds.) FSE 2005. LNCS, vol. 3557, pp. 32–49. Springer, Heidelberg (2005). https://doi.org/10.1007/11502760_3

7. Bhargavan, K., Leurent, G.: On the practical (in-)security of 64-bit block ciphers: collision attacks on HTTP over TLS and OpenVPN. In: Weippl, E.R., Katzenbeisser, S., Kruegel, C., Myers, A.C., Halevi, S. (eds.) ACM CCS 2016, pp. 456–467. ACM Press, October 2016
8. Biham, E., Shamir, A.: Differential cryptanalysis of DES-like cryptosystems. J. Cryptol. **4**(1), 3–72 (1991). https://doi.org/10.1007/BF00630563
9. Daemen, J., Knudsen, L., Rijmen, V.: The block cipher Square. In: Biham, E. (ed.) FSE 1997. LNCS, vol. 1267, pp. 149–165. Springer, Heidelberg (1997). https://doi.org/10.1007/BFb0052343
10. Diffie, W., Hellman, M.E.: Privacy and authentication: an introduction to cryptography. Proc. IEEE **67**(3), 397–427 (1979)
11. Dinur, I., Leurent, G.: Improved generic attacks against hash-based MACs and HAIFA. In: Garay, J.A., Gennaro, R. (eds.) CRYPTO 2014, Part I. LNCS, vol. 8616, pp. 149–168. Springer, Heidelberg (2014). https://doi.org/10.1007/978-3-662-44371-2_9
12. Duong, T., Rizzo, J.: Here come the $\oplus$ ninjas (2011)
13. Dworkin, M.: Recommendation for Block Cipher Modes of Operation: Methods and Techniques. NIST Special Publication 800–38A, National Institute for Standards and Technology, December 2001
14. Ferguson, N.: Authentication weaknesses in GCM. Comment to NIST (2005). http://csrc.nist.gov/groups/ST/toolkit/BCM/documents/comments/CWC-GCM/Ferguson2.pdf
15. Ferguson, N., Schneier, B., Kohno, T.: Cryptography Engineering: Design Principles and Practical Applications. Wiley, New York (2011)
16. DES Modes of Operation. NIST Special Publication 81, National Institute for Standards and Technology, December 1980
17. Fuhr, T., Leurent, G., Suder, V.: Collision attacks against CAESAR candidates. In: Iwata, T., Cheon, J.H. (eds.) ASIACRYPT 2015, Part II. LNCS, vol. 9453, pp. 510–532. Springer, Heidelberg (2015). https://doi.org/10.1007/978-3-662-48800-3_21
18. Guo, J., Peyrin, T., Sasaki, Y., Wang, L.: Updates on generic attacks against HMAC and NMAC. In: Garay, J.A., Gennaro, R. (eds.) CRYPTO 2014, Part I. LNCS, vol. 8616, pp. 131–148. Springer, Heidelberg (2014). https://doi.org/10.1007/978-3-662-44371-2_8
19. Handschuh, H., Preneel, B.: Key-recovery attacks on universal hash function based MAC algorithms. In: Wagner, D. (ed.) CRYPTO 2008. LNCS, vol. 5157, pp. 144–161. Springer, Heidelberg (2008). https://doi.org/10.1007/978-3-540-85174-5_9
20. Hoang, V.T., Reyhanitabar, R., Rogaway, P., Vizár, D.: Online authenticated-encryption and its nonce-reuse misuse-resistance. In: Gennaro, R., Robshaw, M. (eds.) CRYPTO 2015, Part I. LNCS, vol. 9215, pp. 493–517. Springer, Heidelberg (2015). https://doi.org/10.1007/978-3-662-47989-6_24
21. Iwata, T.: New blockcipher modes of operation with beyond the birthday bound security. In: Robshaw, M. (ed.) FSE 2006. LNCS, vol. 4047, pp. 310–327. Springer, Heidelberg (2006). https://doi.org/10.1007/11799313_20
22. Iwata, T., Mennink, B., Vizár, D.: CENC is optimally secure. Cryptology ePrint Archive, Report 2016/1087 (2016). http://eprint.iacr.org/2016/1087
23. Joux, A.: Authentication failures in NIST version of GCM. Comment to NIST (2006). http://csrc.nist.gov/groups/ST/toolkit/BCM/documents/comments/800-38_Series-Drafts/GCM/Joux_comments.pdf

24. Lee, C., Kim, J., Sung, J., Hong, S., Lee, S.: Forgery and key recovery attacks on PMAC and Mitchell's TMAC variant. In: Batten, L.M., Safavi-Naini, R. (eds.) ACISP 2006. LNCS, vol. 4058, pp. 421–431. Springer, Heidelberg (2006). https://doi.org/10.1007/11780656_35
25. Leurent, G., Peyrin, T., Wang, L.: New generic attacks against hash-based MACs. In: Sako, K., Sarkar, P. (eds.) ASIACRYPT 2013, Part II. LNCS, vol. 8270, pp. 1–20. Springer, Heidelberg (2013). https://doi.org/10.1007/978-3-642-42045-0_1
26. Luykx, A., Paterson, K.G.: Limits on authenticated encryption use in TLS, March 2016. http://www.isg.rhul.ac.uk/~kp/TLS-AEbounds.pdf
27. Matsui, M.: Linear cryptanalysis method for DES cipher. In: Helleseth, T. (ed.) EUROCRYPT 1993. LNCS, vol. 765, pp. 386–397. Springer, Heidelberg (1994). https://doi.org/10.1007/3-540-48285-7_33
28. McGrew, D.: Impossible plaintext cryptanalysis and probable-plaintext collision attacks of 64-bit block cipher modes. Cryptology ePrint Archive, Report 2012/623. Accepted to FSE 2013 (2012). http://eprint.iacr.org/2012/623
29. McGrew, D.A., Viega, J.: The security and performance of the Galois/counter mode (GCM) of operation. In: Canteaut, A., Viswanathan, K. (eds.) INDOCRYPT 2004. LNCS, vol. 3348, pp. 343–355. Springer, Heidelberg (2004). https://doi.org/10.1007/978-3-540-30556-9_27
30. Peyrin, T., Wang, L.: Generic universal forgery attack on iterative hash-based MACs. In: Nguyen, P.Q., Oswald, E. (eds.) EUROCRYPT 2014. LNCS, vol. 8441, pp. 147–164. Springer, Heidelberg (2014). https://doi.org/10.1007/978-3-642-55220-5_9
31. Preneel, B., van Oorschot, P.C.: MDx-MAC and building fast MACs from hash functions. In: Coppersmith, D. (ed.) CRYPTO 1995. LNCS, vol. 963, pp. 1–14. Springer, Heidelberg (1995). https://doi.org/10.1007/3-540-44750-4_1
32. Preneel, B., van Oorschot, P.C.: On the security of two MAC algorithms. In: Maurer, U. (ed.) EUROCRYPT 1996. LNCS, vol. 1070, pp. 19–32. Springer, Heidelberg (1996). https://doi.org/10.1007/3-540-68339-9_3
33. Procter, G., Cid, C.: On weak keys and forgery attacks against polynomial-based MAC schemes. J. Cryptol. **28**(4), 769–795 (2015)
34. Rajeev, M., Prabhakar, R.: Randomized Algorithms. Cambridge University Press, New York (1995)
35. Rogaway, P.: Evaluation of some blockcipher modes of operation (2011)
36. Saarinen, M.-J.O.: Cycling attacks on GCM, GHASH and other polynomial MACs and hashes. In: Canteaut, A. (ed.) FSE 2012. LNCS, vol. 7549, pp. 216–225. Springer, Heidelberg (2012). https://doi.org/10.1007/978-3-642-34047-5_13
37. Wheeler, D.J., Needham, R.M.: TEA, a tiny encryption algorithm. In: Preneel, B. (ed.) FSE 1994. LNCS, vol. 1008, pp. 363–366. Springer, Heidelberg (1995). https://doi.org/10.1007/3-540-60590-8_29

Fast Near Collision Attack on the Grain v1 Stream Cipher

Bin Zhang[1,2,3,4(✉)], Chao Xu[1,2], and Willi Meier[5]

[1] TCA Laboratory, SKLCS, Institute of Software, Chinese Academy of Sciences, Beijing, China
{zhangbin,xuchao}@tca.iscas.ac.cn
[2] State Key Laboratory of Cryptology, P.O. Box 5159, Beijing 100878, China
[3] University of Chinese Academy of Sciences, Beijing 100049, China
[4] State Key Laboratory of Information Security, Institute of Information Engineering, Chinese Academy of Sciences, Beijing, China
[5] FHNW, Windisch, Switzerland
willi.meier@fhnw.ch

Abstract. Modern stream ciphers often adopt a large internal state to resist various attacks, where the cryptanalysts have to deal with a large number of variables when mounting state recovery attacks. In this paper, we propose a general new cryptanalytic method on stream ciphers, called fast near collision attack, to address this situation. It combines a near collision property with the divide-and-conquer strategy so that only subsets of the internal state, associated with different keystream vectors, are recovered first and merged carefully later to retrieve the full large internal state. A self-contained method is introduced and improved to derive the target subset of the internal state from the partial state difference efficiently. As an application, we propose a new key recovery attack on Grain v1, one of the 7 finalists selected by the eSTREAM project, in the single-key setting. Both the pre-computation and the online phases are tailored according to its internal structure, to provide an attack for any fixed IV in $2^{75.7}$ cipher ticks after the pre-computation of $2^{8.1}$ cipher ticks, given 2^{28}-bit memory and about 2^{19} keystream bits. Practical experiments on Grain v1 itself whenever possible and on a 80-bit reduced version confirmed our results.

Keywords: Cryptanalysis · Stream ciphers · Grain · Near collision

1 Introduction

As a rule of thumb, the internal state size of modern stream ciphers is at least twice as large as the key size, e.g., all the eSTREAM finalists follow this principle, which considerably complicates cryptanalysis. As a typical case, Grain v1, designed by Hell et al. [8,10], has an internal state size of 160 bits with a 80-bit key. Grain v1 has successfully withstood huge cryptanalytic efforts thus far in the single key model [2,5,15,18].

J. B. Nielsen and V. Rijmen (Eds.): EUROCRYPT 2018, LNCS 10821, pp. 771–802, 2018.
https://doi.org/10.1007/978-3-319-78375-8_25

In this paper, we propose a general new cryptanalytic framework on stream ciphers, called fast near collision attack (FNCA). Given a keystream prefix, finding a corresponding internal state of the stream cipher that generates it is equivalent to determining a preimage of the output of a specific function F. For our purpose, it is assumed that each component function of F can be rewritten in a way that depends on only few variables or combinations of the original variables, which is indeed the case for many stream ciphers. The new strategy is based on a combination of the birthday paradox with respect to near collisions and local differential properties of the component functions. To deal with the situation that F has a large number of variables, the near collision property is combined with a divide-and-conquer strategy so that only subsets of the internal state, associated with different keystream vectors, are restored first and merged carefully later to retrieve the full large internal state. The subset of the internal state associated with a specified keystream vector is called the restricted internal state of the corresponding keystream vector. It is observed that the keystream segment difference (KSD) of a given keystream vector *only* depends on the internal state difference (ISD) and the value of the restricted internal state, i.e., only the differences and the values in the restricted internal state can affect the KSD of a specified keystream vector, whatever the difference distribution and state values in the other parts of the full internal state. Thus, we could apply the near collision idea to this restricted internal state, rather than to the whole internal state. Then a self-contained method [16] is introduced and improved to derive the target subset of the internal state from the partial state difference efficiently. The observation here is that instead of collecting two keystream vector sets to find a near collision state pair, we only collect one set and virtualize the other by directly computing it. An efficient distilling technique is suggested to properly maintain the size of the candidates subset so that the correct candidate is contained in this subset with a higher probability than in a purely random situation. The attack consists of two phases. In the pre-computation phase, we prepare a list of differential look-up tables which are often quite small because of the local differential properties of F. Thus the preprocessing complexity is significantly reduced due to the relatively small number of involved variables. These small tables are carefully exploited in the online phase to determine a series of local candidate sets, which are merged carefully to cover a larger partial state. From this partial state, we aim to recover the full internal state according to the concrete structure of the primitive.

As our main application, we mount a fast near collision attack against Grain v1, one of the 3 finalists selected by the eSTREAM project for restricted hardware environments. In addition to the above general strategies, we further reduce the number of variables associated with a specified keystream vector by rewriting variables according to the internal structure of Grain v1 and making some state variables linearly dependent on the others, similar to the linear enumeration procedure in the BSW sampling in [4]. We first focus on the state recovery of the LFSR together with some partial information on the NFSR. Then a new property in the keystream generation of Grain v1 is exploited in the proposed

Z-technique: given the keystream and the LFSR, it is possible to construct a number of keystream chains to efficiently find out some linear equations on the original NFSR variables, which further reduce the number of unknown variables in the NFSR initial state. Given the LFSR part, Grain v1 degrades into a dynamically linearly filtered NFSR in forward direction and a pure linearly filtered NFSR in backward direction. In both cases, all the NFSR internal state variables can be formally expressed as a linear combination of the initial state variables and of some keystream bits [3]. Taking into account that the best linear approximation of the NFSR feedback function in Grain v1 has a bias of $\frac{41}{512}$, we could construct a system of parity-checks of weight 2 on an even smaller number of the initial NFSR variables with a low complexity. These parity-checks need not to be solved, but can be used as a distinguisher via the Fast Walsh Transform (FWT), called the Walsh distinguisher. The correct LFSR candidate could be identified directly from a glance at the distribution of the Walsh spectrum. Thus, we determine the LFSR part in Grain v1 *independent of* the NFSR state, which releases the complexity issue if the whole internal state is treated together. Finally, the left NFSR state could be restored easily by an algebraic attack with a complexity much lower than the above dominated step and the list of remaining candidates could be tested with the consistency of the available keystream to yield the correct one. As a result, both the pre-computation and the online attack phases are tailored to provide a state/key recovery attack[1] on Grain v1 in the single-key setting with an arbitrary known IV in $2^{75.7}$ cipher ticks after a pre-computation of $2^{8.1}$ cipher ticks, given 2^{28}-bit memory and around 2^{19} keystream bits, which is the best key recovery attack against Grain v1 so far and manages to remove the two unresolved assumptions in complexity manipulation in the previous near collision attack at FSE 2013. This attack is about $2^{11.7}$ times faster than the exhaustive search[2]. Our results have been verified both on Grain v1 itself whenever possible and on a reduced version of Grain v1 with a 40-bit LFSR and a 40-bit NFSR in experiments. A comparison of our attack with the exhaustive search is depicted in Table 1. In summary, though the whole structure of Grain v1 is sound, here we list some properties that facilitate our attack.

- The state size is exactly 160 bits with respect to the 80-bit security.
- The whole system will degrade into a linearly filtered NFSR after knowing the LFSR.
- There is a good linear approximation of the updating function of the NFSR.
- The 2-bit keystream vector depends on a relatively small number of variables after rewriting variables.

[1] Due to the invertible state updating, a state recovery attack on Grain v1 could be converted into a key recovery attack directly.

[2] The brute force attack with an expected complexity of $2^{87.4}$ cipher ticks is shown in [18]. Besides, NCA-2.0 [18] requires a huge pre-computation and memory complexities; while NCA-3.0 [18] is based on two assumptions which remains to be verified on Grain v1 itself.

Table 1. Comparison with the best previous attack on the full Grain v1

Attack	Complexities			
	Pre-comp	Data	Memory	Time
Brute force	-	$2^{7.4}$	$2^{7.4}$	$2^{87.4}$
NCA-2.0 [18]	$2^{83.4}$	2^{62}	$2^{65.9}$	$2^{76.1}$
This paper	$2^{8.1}$	2^{19}	2^{28}	$2^{75.7}$

The time complexity unit here is 1 cipher tick as in [18] and the data/memory complexity unit is 1 bit.

Outline. A brief description of the Grain v1 stream cipher is presented in Sect. 2. Then, some preliminaries relevant to our work are presented in Sect. 3 together with a brief review of the previous near collision attack in [18]. The framework of FNCA is established with the theoretical analysis in Sect. 4 and then applied to Grain v1 in Sect. 5, respectively. In Sect. 6, practical simulations both on Grain v1 itself and on the reduced version are provided. Finally, some conclusions are drawn and future work is pointed out in Sect. 7.

2 Description of Grain v1

Grain v1 is a bit-oriented stream cipher, which consists of a pair of linked 80-bit shift registers, one is a linear feedback shift register (LFSR) and another is a non-linear feedback shift register (NFSR), whose states are denoted by $(l_i, l_{i+1}, ..., l_{i+79})$ and $(n_i, n_{i+1}, ..., n_{i+79})$ respectively. The updating function of the LFSR is $l_{i+80} = l_{i+62} \oplus l_{i+51} \oplus l_{i+38} \oplus l_{i+23} \oplus l_{i+13} \oplus l_i$ and the updating function of the NFSR is

$$\begin{aligned} n_{i+80} = \; & l_i \oplus n_{i+62} \oplus n_{i+60} \oplus n_{i+52} \oplus n_{i+45} \oplus n_{i+37} \oplus n_{i+33} \oplus n_{i+28} \oplus n_{i+21} \\ & \oplus n_{i+14} \oplus n_{i+9} \oplus n_i \oplus n_{i+63}n_{i+60} \oplus n_{i+37}n_{i+33} \oplus n_{i+15}n_{i+9} \\ & \oplus n_{i+60}n_{i+52}n_{i+45} \oplus n_{i+33}n_{i+28}n_{i+21} \oplus n_{i+63}n_{i+45}n_{i+28}n_{i+9} \\ & \oplus n_{i+60}n_{i+52}n_{i+37}n_{i+33} \oplus n_{i+63}n_{i+60}n_{i+21}n_{i+15} \\ & \oplus n_{i+63}n_{i+60}n_{i+52}n_{i+45}n_{i+37} \oplus n_{i+33}n_{i+28}n_{i+21}n_{i+15}n_{i+9} \\ & \oplus n_{i+52}n_{i+45}n_{i+37}n_{i+33}n_{i+28}n_{i+21}. \end{aligned}$$

The keystream generation phase, shown in Fig. 1, works as follows. The combined NFSR-LFSR internal state is filtered by a non-linear boolean function $h(x) = x_1 \oplus x_4 \oplus x_0x_3 \oplus x_2x_3 \oplus x_3x_4 \oplus x_0x_1x_2 \oplus x_0x_2x_3 \oplus x_0x_2x_4 \oplus x_1x_2x_4 \oplus x_2x_3x_4$, which is chosen to be balanced and correlation immune of the first order with the variables x_0, x_1, x_2, x_3 and x_4 corresponding to the tap positions l_{i+3}, l_{i+25}, l_{i+46}, l_{i+64} and n_{i+63} respectively. The output z_i is taken as $z_i = \bigoplus_{k \in \mathcal{A}} n_{i+k} \oplus h(l_{i+3}, l_{i+25}, l_{i+46}, l_{i+64}, n_{i+63})$, where $\mathcal{A} = \{1, 2, 4, 10, 31, 43, 56\}$. The details of the initialization phase are omitted here, the only property relevant to our work is that the initialization phase is invertible.

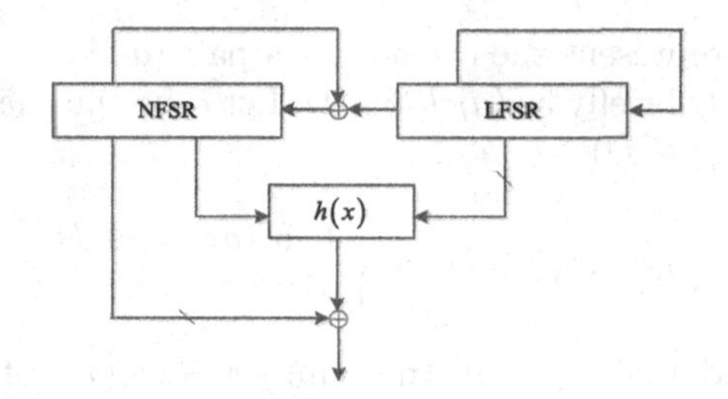

Fig. 1. Keystream generation of Grain v1

3 Preliminaries

In this section, some basic definitions and lemmas are presented with a brief review of the previous near collision attack on Grain v1 in [18]. The following notations are used hereafter.

- $w_H(\cdot)$: the Hamming weight of the input argument.
- d: the Hamming weight of the internal state difference (ISD).
- l: the bit length of the keystream vector.
- n: the bit length of the internal state, whether restricted or not.
- Δx: the value of the ISD, whether restricted or not.
- $V(n, d)$: the total number of the ISDs with $w_H(\Delta x) \leq d$.
- Ω: the number of CPU-cycles to generate 1 bit keystream in Grain v1 in software.

3.1 Basic Conceptions and Lemmas

Let $B_d = \{\Delta x \in \mathbb{F}_2^n | w_H(\Delta x) \leq d\} = \{\Delta x_1, \Delta x_2, ..., \Delta x_{V(n,d)}\}$ and $|B_d| = V(n, d) = \sum_{i=0}^{d} \binom{n}{i}$, where $|\cdot|$ denotes the cardinality of a set. Two n-bit strings s, s' are said to be d-near-collision, if $w_H(s \oplus s') \leq d$ holds. Similar to the birthday paradox, which states that two random subsets of a space with 2^n elements are expected to intersect when the product of their sizes exceeds 2^n, we present the following generalized lemma, which includes the d-near-collision Lemma in [18] as a special case.

Lemma 1. *Given two random sets A and B consisting of n-bit elements and a condition set D, then there exists a pair $(a, b) \in A \times B$ satisfying one of the conditions in D if*

$$|A| \cdot |B| \geq \frac{c \cdot 2^n}{|D|} \tag{1}$$

holds, where c is a constant that determines the existence probability of one good pair (a, b).

Proof. We regard each $a_i \in A$ and $b_j \in B$ as an uniformly distributed random variable with the realization values in $\mathbb{F}_2^n$. Let $A = \{a_1, a_2, ..., a_{|A|}\}$ and $B =$

$\{b_1, b_2, ..., b_{|B|}\}$, we represent the event that a pair $(a_i, b_j) \in A \times B$ satisfies one of the conditions in D briefly as $(a_i, b_j) \in D$. Let ϕ be the characteristic function of the event $\phi((a_i, b_j) \in D)$, i.e.,

$$\phi((a_i, b_j) \in D) = \begin{cases} 1 & \text{if } (a_i, b_j) \in D \\ 0 & \text{otherwise.} \end{cases}$$

For $1 \leq i \leq |A|$ and $1 \leq j \leq |B|$, the number $N_{A,B}(D)$ of good pairs (a_i, b_j) satisfying $(a_i, b_j) \in D$ is $N_{A,B}(D) = \sum_{i=1}^{|A|} \sum_{j=1}^{|B|} \phi((a_i, b_j) \in D)$. Thus, the expected value of $N_{A,B}(D)$ of the pairwise independent random variables can be computed as $E(N_{A,B}(D)) = |A| \cdot |B| \cdot \frac{|D|}{2^n}$. Therefore, if we choose the sizes of A and B satisfying Eq. (1), the expected number of good pairs is at least c. □

While when $D = B_d$, Lemma 1 reduces to the d-near-collision Lemma in [18], Lemma 1 itself is much more general in the sense that D could be an arbitrary condition set chosen by the adversary, which provides a lot of freedom for cryptanalysis. Another issue is the choice of c. In [16], the relation between the choice of the constant c and the existence probability of a d-near-collision pair is illustrated as follows for random samples:

$$\Pr(\text{d-near-collision}) = \begin{cases} 0.606 & \text{if } c = 1 \\ 0.946 & \text{if } c = 3 \\ 0.992 & \text{if } c = 5. \end{cases}$$

As stated in [16], these relations are obtained from the random experiments with a modest size, i.e., for each c value, 100 strings of length 40 to 49 for d-values from 10 to 15 are generated, *not* in a real cipher setting.

Remarks. In a concrete primitive scenario, it is found that the constant c sometimes needs to be even larger to assure a high existence probability of near collision good pairs. In our experiments, we find that the above relation does not hold for Grain v1 and its reduced versions. In these cases, we have to set $c = 8$ or even $c = 10$ to have an existence probability as high as desirable for the subsequent attack procedures. We believe that for each cipher, the choice of the constant c and its correspondence to the existence probability of a near collision pair is a fundamental measure related to the security of the primitive. The following fact is used in our new attack.

Corollary 1. *For a specified cipher and a chosen constant c, let A and B be the internal state subsets associated with the observable keystream vectors, where each element of A and B is of n-bit length. If we choose* $|A| = 1$ *and* $|B| \geq c \cdot \frac{2^n}{|D|}$*, then there exists an element* $b_i \in B$ *such that the pair* (a, b_i) *with the only element* $a \in A$ *forms a d-near collision pair with a probability dependent on c.*

Note that in the near collision setting, the bits in the restricted internal state may be nonadjacent. These bit positions are determined by the tap positions of the composition mapping of the output function if only a short prefix is chosen by the adversary and also dependent on the state updating function, whether linear or non-linear, if some inconsecutive, or even far away, keystream bits are chosen to be considered together.

In [18], the two sets A and B are chosen to be of equal size, i.e., $|A| = |B|$ to minimize the data complexity, in which case the adversary has to deal with all the candidate state positions one-by-one. Instead, Corollary 1 is used in our new attack on Grain v1 via the self-contained method introduced later in Sect. 4.3, to restore the restricted internal state defined below, at a *specified* chosen position along the keystream segment under consideration.

Definition 1. *For a specified cipher, the subset* $\mathbf{x} = (x_{i_0}, x_{i_1}, \ldots, x_{i_{n-1}})$ *of the full internal state associated with a given keystream vector* $\mathbf{z} = (z_{j_0}, z_{j_1}, \ldots, z_{j_{l-1}})$ *is called the restricted internal state associated with* $\mathbf{z}$.

We choose the following definition of the restricted BSW sampling resistance in stream ciphers.

Definition 2. *Let* $\mathbf{z} = (z_{j_0}, z_{j_1}, \ldots, z_{j_{l-1}})$ *be the known keystream vector selected by the adversary, if* l *internal state bits in the restricted internal state* $\mathbf{x}$ *associated with* $\mathbf{z}$ *could be represented explicitly by* $\mathbf{z}$ *and the other bits in* $\mathbf{x}$, l *is called the restricted BSW sampling resistance corresponding to* $(\mathbf{x}, \mathbf{z})$.

It is well known that Grain v1 has a sampling resistance of at least 18 [18], thus from Definition 2, we have $l \leq 18$. Actually, we prefer to consider small values of l in our analysis to reduce the memory complexity and to facilitate the verification of theoretical predictions. Note that here the indices $j_0, j_1, \ldots, j_{l-1}$, either consecutive or inconsecutive, could be chosen arbitrarily by the adversary. The restricted BSW sampling inherits the linear enumeration nature of the classical BSW sampling in [4], but unlike the classical BSW sampling, the new sampling does not try to push this enumeration procedure as far as possible, it just enumerates a suitable number of steps and then terminates.

3.2 The Previous Near Collision Attack

At FSE 2013, a near collision attack on Grain v1 was proposed in [18], trying to identify a near collision in the whole internal state at different time instants and to restore the two involved states accordingly. For such an inner state pair, the keystream prefixes they generate will be similar to each other and the distribution of the KSDs are non-uniform.

The adversary first intends to store the mapping from the specific KSDs to the possible ISDs of the *full* inner state with the sorted occurring probabilities in the pre-computed tables. Then in the online phase, he/she tries to recover the correct ISD by utilizing the pre-computed tables, then the two internal states from the determined ISD. The crucial problem here is to examine a large number of possible pairs whether they are truly near collision or not. In this process, strong wrong-candidate filter with a low complexity is needed, while in its form in [18], the reducing effect is not so satisfactory. In order to overcome this problem, the BSW sampling property and the special table techniques are briefly outlined based on two assumptions, which are essential for the complexity manipulation from the 64-bit reduced version experiments to the full version theoretical attack. In [18], the examination of the candidate state pairs is executed by first recovering the two internal states from the specified ISD. For the LFSR part, this is of no problem since the LFSR updates independently; but for the NFSR part, it is really a problem in [18]. Though the adversary knows the two specified keystream vectors and their corresponding ISD, it is still difficult to restore the full 80-bit NFSR state in such an efficient way that this routine could be invoked a large number of times. Besides, the special table technique assumes that about 50% of all the possible ISDs could be covered on average, which is very hard to verify for the full Grain v1, thus the successful probability of this attack cannot be guaranteed.

In the following, we will show that the adversary need not to recover the full internal state at once when making the examination, actually specified subsets of the internal state could be restored more efficiently than previously thought to be possible, thus the time/memory complexities of the new attack can be considerably reduced with an assured success probability and *without* any assumption in the complexity manipulation.

4 Fast Near Collision Attacks

In this section, we will describe the new framework for fast near collision attacks, including both the pre-computation phase and the online attack phase, with the theoretical justifications.

4.1 General Description of Fast Near Collision Attacks

The new framework is based on the notion of the restricted internal state corresponding to a fixed keystream vector, which is presented in Definition 1 above. Given $\mathbf{z} = (z_{j_0}, z_{j_1}, \ldots, z_{j_{l-1}})$ with z_{j_i} $(0 \leq i \leq l-1)$ not necessarily being consecutive in the real keystream, the corresponding restricted internal state $\mathbf{x}$ for $\mathbf{z}$ is determined by the output function f together with its tap positions, and the state updating function g of the cipher, i.e., induced by the intrinsic structure of the cipher. Besides, from the keystream vector $\mathbf{z}$, it is natural to look at the augmented function for $\mathbf{z}$.

Definition 3. *For a specified cipher with the output function f and the state updating function g, which outputs one keystream bit in one tick, the lth-order augmented function $Af : \mathbb{F}_2^{|\mathbf{x}|} \rightarrow \mathbb{F}_2^l$ for a given $(\mathbf{x}, \mathbf{z})$ pair is defined as $Af(\mathbf{x}) = (f(\mathbf{x}), f(g^{i_1}(\mathbf{x})), \ldots, f(g^{i_{l-1}}(\mathbf{x})))$.*

Note that the definition of the augmented function here is different from the previous ones in [1,11]. In Definition 3, the augmented function is defined on a *subset* of the whole internal state; while in [1,11], similar functions are usually defined on the full internal state. As can be seen later, this difference will make sense in launching a near collision attack, i.e., we need not target the full internal state, which is usually quite large, at once any more, now we just look for near collision in the sub-states chosen by the attacker. A high-level description of fast near collision attack is depicted in Algorithm 1.

Algorithm 1. FNCA

Parameters: `index`: the concrete value of a KSD
`prefix`: the concrete value of a keystream vector

Offline: **for** each combination of (`index`, `prefix`) **do**
Construct the table T[`index`, `prefix`], projecting from the KSD `index` to all the possible ISDs sorted by the occurring rates
end for

Input: A keystream segment $\mathbf{z}_{\text{total}} = (z_{j_0}, z_{j_1}, \ldots, z_{j_{l-1}}, z_{j_l}, \ldots, z_{j_{l+\gamma}})$

Online: Recover the full internal state $\mathbf{x}_{\text{full}}$ matching with $\mathbf{z}_{\text{total}}$

1: Divide $\mathbf{z}_{\text{total}}$ into α overlapping parts $\mathbf{z}_i$ $(1 \leq i \leq \alpha)$ and a suffix $\mathbf{z}_\mu$
2: **for** $i = 1$ **to** α **do**
3: get the candidates list L_i of the restricted internal state $\mathbf{x}_i$ for $\mathbf{z}_i$
4: **end for**
5: Merge L_is to get a candidate list for the possible partial state $\mathbf{x}_{\text{merge}}$
6: **for** each candidate of $\mathbf{x}_{\text{merge}}$ **do**
7: restore $\mathbf{x}_{\text{full}}$ and test the consistency with the suffix $\mathbf{z}_\mu$

The following proposition provides new insights on what influence the whole internal state size has on the feasibility of a near collision attack.

Proposition 1. *For a specified cipher and two keystream vectors* $\mathbf{z}$ *and* $\mathbf{z}'$, *the KSD* $\Delta\mathbf{z} = \mathbf{z} \oplus \mathbf{z}'$ *only depends on the ISD* $\Delta\mathbf{x} = \mathbf{x} \oplus \mathbf{x}'$ *and the values of* $\mathbf{x}$ *and* $\mathbf{x}'$, *whatever the difference and the values in* $\bar{\mathbf{x}}$, *the other parts of the whole internal state.*

Proof. It suffices to see the algebraic expressions of the keystream bits under consideration. By taking a look at the input variables, we have the claim. □

Offline. Proposition 1 makes the pre-computation phase in FNCA quite different from and much more efficient than that in the previous NCA in [18]. Now we need not to exhaustively search through all the possible ISDs over the *full* internal state, which is usually quite large, instead we just search through all the possible ISDs over a specified restricted internal state corresponding to a given keystream vector, which is usually much shorter compared to the full internal state. In Algorithm 1, we use two parameters `index` and `prefix` to characterise this difference with `index` being the KSD and `prefix` being the *value* of one of the two specified keystream vectors. For each possible combination of (`index`, `prefix`), we construct an individual table for the pair. Thus, many relatively small tables are built instead of one large pre-computed table, which greatly reduces the time/memory complexities of the offline phase, improves the accuracy of the pre-computed information, and finally assures a high success rate of the new attack.

Online. With the differential tables prepared in the offline phase, the adversary first tries to get some candidates of the target restricted internal state $\mathbf{x}_1$ and to filter out as much as possible wrong candidates of $\mathbf{x_1}$ in a reasonable time complexity. Then he/she moves to another restricted internal state $\mathbf{x_2}$, possibly overlapped with $\mathbf{x_1}$, but not coincident with $\mathbf{x_1}$, and get some candidates of $\mathbf{x_2}$. This process is repeated until enough internal state bits are recovered in an acceptable time/memory complexities and merge the candidate lists L_i together to get a candidate list of possible partial state $\mathbf{x}_{\text{merge}}$. Finally, from $\mathbf{x}_{\text{merge}}$, he/she tries to retrieve the full internal state and check the candidates by using the consistency with the keystream segment.

There are three essential problems that have to be solved in this process. The first one is how to efficiently get the candidates for each restricted internal state and further to filter out those wrong values as much as possible in each case? The second is how to efficiently merge these partial states together without the overflowing of the number of possible internal state candidates, i.e., we need to carefully control the increasing speed of the possible candidates during the merging phase. At last, we need to find some very efficient method to restore the other parts of the full internal state given $\mathbf{x}_{\text{merge}}$, which lies at the core of the routine. We will provide our solutions to these problems in the following sections.

4.2 Offline Phase: Parameterizing the Differential Tables

Now we explain how to pre-compute the differential tables $\text{T}[\texttt{index}, \texttt{prefix}]$, conditioned on the event that the value of one of the two keystream vectors is `prefix` when the KSD is `index`. Let $\mathbf{x} = (x_{i_0}, x_{i_1}, \ldots, x_{i_{n-1}})$ be the restricted internal state associated with $\mathbf{z} = (z_{j_0}, z_{j_1}, \ldots, z_{j_{l-1}})$, for such a chosen $(\mathbf{x}, \mathbf{z})$ pair, Algorithm 2 fulfills this task. Algorithm 2 is the inner routine of the pre-computation phase of FNCA in the general case, where N_1 and N_2 are the two random sampling sizes when determining whether a given ISD Δx of the restricted internal state $\mathbf{x}$ could generate the KSD `index` and what the occurring probability is. Algorithm 2 is interesting in its own right, though not adopted in the state recovery attack on Grain v1 in Sect. 5. It can be applied in the most general case with the theoretical justification when dedicated pre-computation is impossible.

Algorithm 2. Constructing the differential table $\text{T}[\texttt{index}, \texttt{prefix}]$

1: **for** each ISD Δx s.t. $w_H(\Delta x) \leq d$ **do**
2: **for** $i = 1$ **to** N_1 **do**
3: determine whether Δx could generate the specified KSD `index`
4: **if** yes **then**
5: **for** $j = 1$ **to** N_2 **do**
6: generate random $\mathbf{x}$ s.t. $Af(\mathbf{x}) = \texttt{prefix}$ and form the pair $(\mathbf{x}, \mathbf{x} \oplus \Delta x)$
7: compute $\mathbf{z} = Af(\mathbf{x})$ and $\mathbf{z}' = Af(\mathbf{x} \oplus \Delta x)$
8: count the number of times `counter` that $\Delta z = \mathbf{z} \oplus \mathbf{z}' = \texttt{index}$
9: store the ratio $\texttt{counter}/N_2$ with Δx in $\text{T}[\texttt{index}, \texttt{prefix}]$
10: Sort the ISDs according to the occurring rates

From $V(n,d) = \sum_{i=0}^{d} \binom{n}{i}$, the time complexity of Algorithm 2 is $P = 2 \cdot V(n,d) \cdot (N_1 + N_2 \cdot (j_{l-1} + 1))$ cipher ticks and the memory requirement is at most $V(n,d) \cdot (\lceil \log_2 n \rceil \cdot d + (7+7))$ bits, where a $14 = 7 + 7$-bit string is used to store the percentage number, e.g., for 76.025%, we use 7 bits to store the integer part $76 < 128$ and another 7 bits to save the fractional part 0.025 as 0.0000011. In Table 2, the global information of the 16 T tables for Grain v1 with the 23 original variables when $l = 2$ is shown, where Pr_{divs} is defined as follows.

Table 2. The summary of the pre-computation phase of Grain v1 for 2-bit keystream vector with the 23 original variables

(index, prefix)	|T|	Pr_{divs}	(index, prefix)	|T|	Pr_{divs}
(0x0, 0x0)	16126	0.314426	(0x2, 0x0)	16106	0.319008
(0x0, 0x1)	16126	0.314434	(0x2, 0x1)	16106	0.318892
(0x0, 0x2)	16126	0.314504	(0x2, 0x2)	16106	0.318934
(0x0, 0x3)	16126	0.314504	(0x2, 0x3)	16106	0.318955
(0x1, 0x0)	16106	0.318958	(0x3, 0x0)	16044	0.311827
(0x1, 0x1)	16106	0.319050	(0x3, 0x1)	16044	0.311839
(0x1, 0x2)	16106	0.318896	(0x3, 0x2)	16044	0.311979
(0x1, 0x3)	16106	0.318928	(0x3, 0x3)	16044	0.311833

Definition 4. *For each T[*`index, prefix`*], let $|T|$ be the number of ISDs in the table, the diversified probability of this table is defined as* $Pr_{divs} = \frac{\sum_{\Delta x \in T} Pr_{\Delta x}}{|T|}$*, where Δx ranges over all the possible ISDs in the table.*

The diversified probability of a T[`index, prefix`] table measures the average reducing effect of this table that for a random restricted internal state $\mathbf{x}$ such that $Af(\mathbf{x}) = \texttt{prefix}$, flip the bits in $\mathbf{x}$ according to a $\Delta x \in \mathrm{T}$ and get $\mathbf{x}'$, then with probability Pr_{divs}, $Af(\mathbf{x} \oplus \mathbf{\Delta x}) = \texttt{prefix} \oplus \texttt{index}$. From Definition 4, the success rate of the new FNCA is quite high, for we have taken each possible ISD into consideration in the attack.

Corollary 2. *From Table 2, if the* `index` *is fixed, then the 4 Pr_{divs}s corresponding to different* `prefix`*es are approximately the same, i.e., the 4 T tables have almost the same reducing effect for filtering out wrong candidates.*

Corollary 2 is the basis of the merging operation in the online attack phase, which assures that with the restricted BSW sampling resistance of Grain v1 and the self-contained method introduced later, the partial state recovery procedure will not be affected when the keytream vector under consideration has changed its value along the actual keystream. If the keystream vector under consideration has the value `prefix`, the adversary uses the value `prefix` $\oplus$ `index` in the computing stage of the self-contained method to have the KSD remaining the same.

4.3 Online Phase: Restoring and Distilling the Candidates

Now we come to the online phase of FNCA. The aim is to restore the overlapping restricted internal states one-by-one, merge them together, and finally from the already covered state $\mathbf{x}_{\text{merge}}$ to retrieve the correct full internal state. For a chosen $(\mathbf{x}, \mathbf{z})$ pair, the refined self-contained method is depicted in Algorithm 3.

Algorithm 3. The refined self-contained method

```
1: Initialize i = 0
2: while i ≤ c · 2^n/|D| do
3:     load x with a new random value so that it generates z ⊕ index
4:     for each possible ISD Δx in T[index, z ⊕ index] do
5:         compute x' = x ⊕ Δx
6:         if x' generates z then
7:             put x' into the candidates list L
8:         end if
9:     end for
10:    i = i + 1
11: end while
```

The original self-contained method was proposed in [16], whose idea is to make a tradeoff between the data and the online time complexity in such a way that the second set B of keystream vectors in Lemma 1 is generated by the adversary himself, thus he also knows the actual value of the corresponding internal state that generates the keystream vector. Therefore, given the ISD from the pre-computed tables, the adversary could just xor the ISD with the internal state matching with the keystream vector in B to get the candidate internal state for the keystream vectors in A. In Algorithm 3, it is quite possible that although obtained from a different new starting value for the restricted internal state $\mathbf{x}$, some candidates $\mathbf{x}'$ will collide with the already existing element in the list, thus the final number of hitting values in the list is not so much as the number of invoking times $c \cdot \frac{2^n}{|D|}$. The following theorem gives the expected value of the actual hitting numbers.

Theorem 1. *Let b be the number of all the values that can be hit and $a = c \cdot \frac{2^n}{|D|} \cdot |T| \cdot P_{divs}$, then after one invoking of Algorithm 3, the mathematical expectation of the final number r of hitting values in the list is*

$$E[r] = \sum_{r=1}^{a} \frac{\binom{b}{r} \cdot r! \cdot \left\{ {a \atop r} \right\} \cdot r}{b^a}, \tag{2}$$

where $\left\{ {a \atop r} \right\}$ is the Stirling number of the second kind, $\binom{b}{r}$ is the binomial coefficient and $r!$ is the factorial.

Proof. Note that the Stirling number of the second kind [17] $\left\{ {a \atop r} \right\}$ counts the number of ways to partition a set of a objects into r non-empty subsets, i.e., each way is a partition of the a subjects, which coincides with our circumstance. Thus we can model the process of Algorithm 3 as follows.

We throw a balls into b different boxes, and we want to know the probability that there are exactly r boxes having some number of balls in. From this converted model, we can see that the size of the total sample space is b^a, while the number of samples in our expected event can be calculated in the following steps.

1. Choose r boxes to hold the thrown balls, there are $\binom{b}{r}$ ways to fulfill this step.
2. Permute these r boxes, there are $r!$ ways to fulfill this step.
3. Partition the a balls into r non-empty sets, this is just the Stirling number of the second kind $\left\{ {a \atop r} \right\}$.

Following the multiplication principle in combinatorics, the size of our expected event is just the product of the above three. This completes the proof. □

We have made extensive experiments to verify Theorem 1 and the simulation results match the theoretical predictions quite well. Back to the self-contained method setting, a is just the number of valid candidates satisfying the conditions that the KSD is `index` and one of the keystream prefix is `prefix`, of which some may be identical due to the flipping according to the ISDs in T.

In general, the opponent can build a table for the function f, mapping the partial sub-states to the keystream vector z, to get a full list of inputs that map to a given z. In order to reduce the candidate list size in a search, he may somehow choose a smaller list of inputs that map to z, and hope that the correct partial state is still in the list with some probability p, depending on the size of the list. The aim of the distilling phase is to exploit the birthday paradox regarding d-near collisions, local differential properties of f, and the self-contained method to derive smaller lists of input sub-states so that the probability that the correct state is in the list is at least p. From Lemma 1, with a properly chosen constant c, the correct internal state $\mathbf{x}$ will be in the list L with a high probability, e.g., 0.8 or 0.95. For a chosen $(\mathbf{x}, \mathbf{z})$ pair, the candidates reduction process is depicted in Algorithm 4.

Algorithm 4. Distilling the candidates

Parameter: a well chosen constant β
1: **for** $i = 1$ to β **do**
2: run Algorithm 3 to get the candidates list L_i
3: **end for**
4: Initialize a list U and let $U = L_1$
5: **for** $i = 2$ to β **do**
6: intersect U with L_i, i.e., $U \leftarrow U \cap L_i$

The next theorem characterizes the number of candidates passing through the distilling process in Algorithm 4.

Theorem 2. *The expected number of candidates in the list U in Algorithm 4 after $\beta - 1$ steps of intersection is $|U_1| \cdot (\frac{E[r]}{b})^{\beta-1}$, where $|U_1| = |L_1|$ is the number of candidates present in the first list L_1 and $E[r]$ is the expected number of hitting values in one single invoking of Algorithm 3.*

Proof. For simplicity, let $|U_i| = f_i$ denote the cardinality of the candidates list after $i-1$ steps of intersection for $1 \leq i \leq \beta - 1$. Note that in the intersection process, if there are f_i candidates in the current list U, then at the next intersection operation, an element in U has the probability $\frac{E[r]}{b}$ to remain, and the probability $1 - \frac{E[r]}{b}$ to be filtered out.

Let $f_i \rightarrow f_{i+1}$ denote the event that there are f_{i+1} elements left after one intersection operation on the f_i elements in the current U. The expected value of f_{i+1} is $E[f_{i+1}] = \sum_{j=0}^{f_i} \binom{f_i}{j} \cdot (\frac{E[r]}{b})^j \cdot (1 - \frac{E[r]}{b})^{f_i - j} \cdot j = f_i \cdot \frac{E[r]}{b}$. Thus we have the following recursion

$$f_{\beta-1} = f_{\beta-t-1} \cdot \prod_{i=1}^{t} (\frac{E[r]}{b}) = |U_1| \cdot \underbrace{(\frac{E[r]}{b}) \cdot (\frac{E[r]}{b}) \cdot \cdots \cdot (\frac{E[r]}{b})}_{\beta-1} = |U_1| \cdot (\frac{E[r]}{b})^{\beta-1},$$

which completes the proof. □

Algorithm 5. Improving the existence probability of the correct $\mathbf{x}$

Parameter: a well chosen constant γ
1: **for** $i = 1$ to γ **do**
2: run Algorithm 4 to get the candidates list U_i
3: **end for**
4: Initialize a list V and let $V = U_1$
5: **for** $i = 2$ to γ **do**
6: union V with U_i, i.e., $V \leftarrow V \cup U_i$

Theorem 2 partially characterizes the distilling process in theory. Now the crucial problem is what the *reduction effect* of this process is, which is determined by the choice of the constant c intrinsic to each primitive and the number of variables involved in the current augmented function. From the cryptanalyst's point of view, the larger β is, the lower the probability that the correct $\mathbf{x}$ is involved in each generated list, thus it is better for the adversary to make some tradeoff between β and this existence probability. Algorithm 5 provides a way to exploit this tradeoff to get some higher existence probability of the correct restricted internal state $\mathbf{x}$.

In Algorithm 5, several candidate lists are first generated by Algorithm 4 with a number of intersection operations for each list, then these lists are unified together to form a larger list so that the existence probability of the correct $\mathbf{x}$ becomes higher compared to that of each component list.

Theorem 3. *Let the expected number of candidates in list V in Algorithm 5 after i $(1 \leq i \leq \gamma)$ steps of union be F_i, then the following relation holds*

$$F_{i+1} = F_i + |U_{i+1}| - \sum_{j=0}^{|U_{i+1}|} \frac{\binom{F_i}{j} \cdot \binom{F_{i+1} - F_i}{|U_{i+1}| - j}}{\binom{F_{i+1}}{|U_{i+1}|}} \cdot j \, , \; 1 \leq i \leq \gamma - 1 \quad (3)$$

where $|F_1| = |U_1|$.

Proof. Note that when a new U_{i+1} is unified into V, we have $F_{i+1} = F_i + |U_{i+1}| - |F_i \cap U_{i+1}|$. It suffices to note that Eq.(3) can be derived from the hypergeometric distribution for the $|F_i \cap U_{i+1}|$ part, which completes the proof. □

Theorem 3 provides a theoretical estimate of $|V|$, which is quite close to the experimental results. After getting V for $\mathbf{x}_1$, we move to the next restricted internal state $\mathbf{x}_2$, as depicted in Algorithm 1 until we recovered all the α restricted internal states. Then we merge the restored partial states $\mathbf{x}_i$ for $1 \leq i \leq \alpha$ to cover a larger part of the full internal state. We have to recover the full internal state conditioned on $\mathbf{x}_{\text{merge}}$. The next theorem describes the reduction effect when merging the candidate lists of two restricted internal states.

Theorem 4. *Let the candidates list for $\mathbf{x}_i$ be V_i, then when merging the candidates list V_i for $\mathbf{x}_i$ and V_{i+1} for $\mathbf{x}_{i+1}$ to cover an union state $\mathbf{x}_i \cup \mathbf{x}_{i+1}$, the expected number of candidates for the union state $\mathbf{x}_i \cup \mathbf{x}_{i+1}$ is*

$$E[|V_{\mathbf{x}_i \cup \mathbf{x}_{i+1}}|] = \frac{|V_i| \cdot |V_{i+1}|}{|V_i \cap V_{i+1}|},$$

where $V_{\mathbf{x}_i \cup \mathbf{x}_{i+1}}$ is the candidates list for the union state $\mathbf{x}_i \cup \mathbf{x}_{i+1}$.

Proof. Denote the bits in $\mathbf{x}_i \cap \mathbf{x}_{i+1}$ by $\mathbf{I} = I_0, I_1, \cdots, I_{|\mathbf{x}_i \cap \mathbf{x}_{i+1}|-1}$ when merging the two adjacent restricted internal states $\mathbf{x}_i$ and $\mathbf{x}_{i+1}$, then we can group V_i and V_{i+1} according to the $|\mathbf{x}_i \cap \mathbf{x}_{i+1}|$ concrete values of $\mathbf{I}$. For the same value pattern of the common bits in $\mathbf{I}$, we can just merge the two states $\mathbf{x}_i$ and $\mathbf{x}_{i+1}$ by concatenating the corresponding candidate states together directly. Thus, the expected number of candidates for the union state is

$$E[|V_{\mathbf{x}_i \cup \mathbf{x}_{i+1}}|] = \frac{|V_i|}{|V_i \cap V_{i+1}|} \cdot \frac{|V_{i+1}|}{|V_i \cap V_{i+1}|} \cdot |V_i \cap V_{i+1}| = \frac{|V_i| \cdot |V_{i+1}|}{|V_i \cap V_{i+1}|},$$

which completes the proof. □

Corollary 3. *In the merging process of Algorithm 1, let M_A and M_B be two partial internal states, each merged from possibly several restricted internal states respectively, then when merging M_A and M_B together, the expected number of candidates for the union state $M_A \cup M_B$ is $E[|M_A \cup M_B|] = \frac{|M_A| \cdot |M_B|}{|M_A \cap M_B|}$.*

Proof. It suffices to note the statistical independence of each invoking of Algorithm 5 and Theorem 4. □

Finally, we present the theorem on the success probability of Algorithm 5.

Theorem 5. *Let the probability that the correct value of the restricted internal state $\mathbf{x}$ will exist in V be $Pr_{\mathbf{x}}$, then we have $Pr_{\mathbf{x}} = 1 - (1 - (P_c)^{\beta})^{\gamma}$, where P_c is the probability that the correct value of the restricted internal state $\mathbf{x}$ exist in U for one single invoking of Algorithm 3.*

Proof. From Algorithms 4 and 5, the probability that for all the γ U_is, the correct value of $\mathbf{x}$ does not exist in the list is $(1 - (P_c)^{\beta})^{\gamma}$, thus the opposite event has the probability given above. This completes the proof. □

Based on the above theoretical framework of FNCA, we will develop a state recovery attack against Grain v1 in the next section, taking into account the dedicated internal structure of the primitive.

5 State Recovery Attack on Grain v1

Now we demonstrate a state recovery attack on the full Grain v1. The new attack is based on the FNCA framework described in Sect. 4 with some techniques to control the attack complexities.

5.1 Rewriting Variables and Parameter Configuration

From the keystream generation of Grain v1, we have $z_i = \bigoplus_{k \in \mathcal{A}} n_{i+k} \oplus h(l_{i+3}, l_{i+25}, l_{i+46}, l_{i+64}, n_{i+63})$, where $\mathcal{A} = \{1, 2, 4, 10, 31, 43, 56\}$, i.e., one keystream bit z_i is dependent on 12 binary variables, of which 7 bits from the NFSR form the linear masking bit $\bigoplus_{k \in \mathcal{A}} n_{i+k}$, 4 bits from the LFSR and n_{i+63} from the NFSR are involved in the filter function h.

For a straightforward FNCA on Grain v1, even considering two consecutive keystream bits, we have to deal with 23 binary variables simultaneously at the beginning of the attack. Thus the number of involved variables will grow rapidly with the running of the attack, and probably overflow at some intermediate point. To overcome this difficulty, we introduce the following two techniques to reduce the number of free variables involved in the keystream vectors.

Let $x_i = n_{i+1} \oplus n_{i+2} \oplus n_{i+4} \oplus n_{i+10} \oplus n_{i+31} \oplus n_{i+43} \oplus n_{i+56}$, then we have

$$z_i = x_i \oplus h(l_{i+3}, l_{i+25}, l_{i+46}, l_{i+64}, n_{i+63}). \tag{4}$$

There are only 6 binary variables $x_i, l_{i+3}, l_{i+25}, l_{i+46}, l_{i+64}, n_{i+63}$ involved in Eq.(4) and if we consider a keystream vector $\mathbf{z} = (z_i, z_{i+1})$, there are only 12 variables now, almost reduced by half compared to the previous number 23. Note that the rewriting technique is known to be useful in [9] before in algebraic attacks on stream ciphers.

Besides, we can still use the linear enumeration procedure as in the BSW sampling case to reduce the variables further. Precisely, from Eq.(4), we have $x_i = z_i \oplus h(l_{i+3}, l_{i+25}, l_{i+46}, l_{i+64}, n_{i+63})$, thus for the above keystream vector $\mathbf{z} = (z_i, z_{i+1})$, we could actually deal with 10 binary variables only, making x_i and x_{i+1} dependent on the other 10 variables and (z_i, z_{i+1}).

Algorithm 6. The pre-computation after rewriting variables

Parameter: matrix P_1 of size $2^l \times V(n, d)$ with $P_1[i][j] \neq 0$ if the ISD j could generate the KSD i and 0 otherwise

1: Initialize the table T[`index`, `prefix`]
2: **for** each possible value of $\mathbf{x}$ **do**
3:     **for** each ISD Δx s.t. $w_H(\Delta x) \leq d$ **do**
4:         determine whether $f_{sr}(\mathbf{x}) =$ `prefix` and $f_{sr}(\mathbf{x} \oplus \Delta x) =$ `prefix` $\oplus$ `index`
5:         **if** yes **then** P_1[`index`][Δx] $= P_1$[`index`][Δx] $+ 1$
6: **for** each ISD Δx s.t. $w_H(\Delta x) \leq d$ **do**
7:     set P_1[`index`][Δx]$/|\mathbf{x}|$ as the occurring rate of Δx
8: Sort the ISDs according to the occurring rates

There is an extra advantage of the above strategy. That is we could now exhaustively search the full input variable space when preparing the differential tables T[index, prefix] for the chosen attack parameters shown in Algorithm 6, which results in the accurately computed occurring probabilities compared to Algorithm 2 in Sect. 4.2, where $f_{sr}(\cdot)$ is the evaluation of the underlying stream cipher. The complete pre-computation table of Grain v1 is listed in Table 3 for $d = 3$, where $*$ indicates that the `prefix` could take any value from `0x0` to `0x3` due to the same distribution for different prefix values and `number` denotes the number of ISDs having the corresponding occurring probability.

Table 3. The full pre-computation information of Grain v1 after rewriting variables when $d = 3$

(index, prefix)	(0x0, $*$)					(0x1, $*$) (0x2, $*$)						(0x3, $*$)		
prob.	1	$\frac{1}{2}$	$\frac{1}{4}$	$\frac{1}{8}$	$\frac{1}{16}$	$\frac{3}{4}$	$\frac{1}{2}$	$\frac{3}{8}$	$\frac{1}{4}$	$\frac{3}{16}$	$\frac{1}{8}$	$\frac{9}{16}$	$\frac{3}{8}$	$\frac{1}{4}$
number	1	44	69	54	8	3	22	27	63	8	27	8	54	63

From Table 3 and Definition 4, we have the following corollary on the diversified probabilities of different pre-computed tables.

Corollary 4. *For the pre-computation table of Grain v1 after rewriting variables, we have*

$$Pr_{divs} = \begin{cases} 0.269886, & \textit{if } \texttt{index}{=}0x0 \\ 0.293333, & \textit{if } \texttt{index}{=}0x1 \\ 0.293333, & \textit{if } \texttt{index}{=}0x2 \\ 0.324000, & \textit{if } \texttt{index}{=}0x3. \end{cases}$$

Proof. From Definition 4, we have $Pr_{divs} = \frac{\sum_{\Delta x \in T} \Pr_{\Delta x}}{|T|}$, it suffices to substitute the variables with the values from Table 3 to get the results. □

From Corollary 4, we choose the KSD to be `0x0` in our attack, for in this case the reduction effect is maximized with the minimum $\Pr_{divs}$. Under this condition, we have run extensive experiments to determine the constant c for Grain v1, which is shown in the following table, where P_c is the probability that the correct value of the restricted internal state exists in the resultant list after one single invoking of

Table 4. The correspondence between the constant c and the existence probability for $index = 0x0$

c	5	6	7	8	9	10
P_c	0.757137	0.816551	0.860638	0.89502	0.92114	0.94644
c	11	12	13	14	15	16
P_c	0.95423	0.96573	0.97567	0.98021	0.985524	0.989411

Algorithm 3. Based on Table 4, we have run a number of numerical experiments to determine the appropriate configuration of attack parameters and found that $c = 10$ provides a balanced tradeoff between various complexities.

Precisely, under the condition that $c = 10$ and the $l = 2$-bit keystream vector with 12 variables (either consecutive or non-consecutive to construct the augmented function Af), we find that if $\beta = 21$ and $\gamma = 6$, then we get $\Pr_{\mathbf{x}_i} = 1 - (1 - 0.94644^{21})^6 = 0.896456$ from Theorem 5. We have tested this fact in experiments for 10^6 times, and found that the *average* value of the success rate well matches to the theoretical prediction. Besides, we have also found that under this parameter configuration, the number of candidates in the list V for the current restricted internal state $\mathbf{x}$ is $848 \approx 2^{9.73}$, which is also quite close to the theoretical value $2^{9.732}$ got from Theorem 3.

Corollary 5. *For Grain v1 when $c = 10$ and $l = 2$, the configuration that the resultant candidate list V is of size 848 with the average probability of 0.896456 for the correct restricted internal state being in V is non-random.*

Proof. Note that in the pure random case, the list V should have a size of $2^{10} \cdot 0.896456 = 917.971$ with the probability 0.896456; now in Grain v1, we get a list V of size 848 with the same probability. In the pure random case, we have

$$E[|V|] = \mu = 2^{10} \cdot 0.896456 = 917.971, \ \sigma = \sqrt{2^{10} \cdot \frac{1}{4} \cdot \frac{3}{4}} = 13.8564.$$

Further, $\frac{\mu - 848}{\sigma} = \frac{917.971 - 848}{13.8564} = 5.0497$; from Chebyshev's inequality, the configuration $(848, 0.896456)$ is far from random with the probability around 0.99. □

Now we are ready to describe the attack in details based on the above attack parameter configuration.

5.2 Concrete Attack: Strategy and Profile

First note that if we just run Algorithm 1 along a randomly known keystream segment to retrieve the overlapping restricted internal states one-by-one without considering the concrete internal structure of Grain v1, then we will probably meet the complexity overflow problem in the process when the restored internal state $\mathbf{x}_{\text{merge}}$ does not cover a large enough internal state, and at the same time, the number of candidates and the complexity needed to check these candidates will exceed the security bound already. Instead, we proceed as follows to have a more efficient attack. First observe that if we target the keystream vector $\mathbf{z} = (z_i, z_{i+1})$ through rewriting variables in Table 5 and restore the variables therein by our method, then for such a 2-bit keystream vector, we can obtain 8 LFSR variables involved in the h function and 2 NFSR bits n_{i+63}, n_{i+64}, together with 2 linear equations $x_i = \bigoplus_{k \in \mathcal{A}} n_{i+k}$ and $x_{i+1} = \bigoplus_{k \in \mathcal{A}} n_{i+k+1}$ on the NFSR variables. If we repeat this procedure for the time instants from 0 to 19, then from $z_i = x_i \oplus h(l_{i+3}, l_{i+25}, l_{i+46}, l_{i+64}, n_{i+63})$, we will have l_{i+3+j}, l_{i+25+j}, l_{i+46+j}, l_{i+64+j} and n_{i+63+j} for $0 \leq j \leq 19$ involved in Table 5.

Table 5. The target keystream equations first exploited in our attack

	output	output
eqns.	$1 : x_0 \oplus h(l_3, l_{25}, l_{46}, l_{64}, n_{63}) = z_0$	$2 : x_1 \oplus h(l_4, l_{26}, l_{47}, l_{65}, n_{64}) = z_1$
	$3 : x_2 \oplus h(l_5, l_{27}, l_{48}, l_{66}, n_{65}) = z_2$	$4 : x_3 \oplus h(l_6, l_{28}, l_{49}, l_{67}, n_{66}) = z_3$
	$5 : x_4 \oplus h(l_7, l_{29}, l_{50}, l_{68}, n_{67}) = z_4$	$6 : x_5 \oplus h(l_8, l_{30}, l_{51}, l_{69}, n_{68}) = z_5$
	$7 : x_6 \oplus h(l_9, l_{31}, l_{52}, l_{70}, n_{69}) = z_6$	$8 : x_7 \oplus h(l_{10}, l_{32}, l_{53}, l_{71}, n_{70}) = z_7$
	$9 : x_8 \oplus h(l_{11}, l_{33}, l_{54}, l_{72}, n_{71}) = z_8$	$10 : x_9 \oplus h(l_{12}, l_{34}, l_{55}, l_{73}, n_{72}) = z_9$
	$11 : x_{10} \oplus h(l_{13}, l_{35}, l_{56}, l_{74}, n_{73}) = z_{10}$	$12 : x_{11} \oplus h(l_{14}, l_{36}, l_{57}, l_{75}, n_{74}) = z_{11}$
	$13 : x_{12} \oplus h(l_{15}, l_{37}, l_{58}, l_{76}, n_{75}) = z_{12}$	$14 : x_{13} \oplus h(l_{16}, l_{38}, l_{59}, l_{77}, n_{76}) = z_{13}$
	$15 : x_{14} \oplus h(l_{17}, l_{39}, l_{60}, l_{78}, n_{77}) = z_{14}$	$16 : x_{15} \oplus h(l_{18}, l_{40}, l_{61}, l_{79}, n_{78}) = z_{15}$
	$17 : x_{16} \oplus h(l_{19}, l_{41}, l_{62}, l_{80}, n_{79}) = z_{16}$	$18 : x_{17} \oplus h(l_{20}, l_{42}, l_{63}, l_{81}, n_{80}) = z_{17}$
	$19 : x_{18} \oplus h(l_{21}, l_{43}, l_{64}, l_{82}, n_{81}) = z_{18}$	$20 : x_{19} \oplus h(l_{22}, l_{44}, l_{65}, l_{83}, n_{82}) = z_{19}$

Let $\mathbf{x}^*$ be the restricted internal state consisting of the input variables involved in Table 5, the details of how to restore the restricted internal state $\mathbf{x}^*$ is presented in the following Tables 6 and 7. We first use FNCA to restore $\mathbf{x}^*$, then we know nearly 80 bits of the LFSR internal state with the corresponding positions, from which we can easily recover the initial internal state of the LFSR with a quite low complexity. Algorithm 7 presents the sketch of our online attack against Grain v1.

Algorithm 7. The online attack on the full Grain v1

1: Apply FNCA to $\mathbf{x}^*$ to restore the input variables
2: **for** each candidate of $\mathbf{x}^*$ **do**
3: use the statistical test in Section 5.4 to check the candidate
4: **for** the passed ones **do**
5: recover the remaining NFSR state, shown in Section 5.4
6: **for** each candidate of $\mathbf{x}_{\text{full}}$ **do**
7: check the consistency with the available keystream

After knowing the LFSR part and more than half of the NFSR, we could first identify the correct LFSR state by the Walsh distinguisher, then the remaining NFSR state could easily be retrieved with an algebraic attack, both shown in the following Sect. 5.4. Note that in Tables 6 and 7, the list size for each merging operation is listed in the middle column, based on Theorem 4 and Corollary 3. For example, let us look at the 1st step. The reason that 2^5 is used instead of 2^6 in denominator is that the x_i variables are not freely generated random variables, for we have made them linearly dependent on the 5 variables in h function and the corresponding keystream bits to fulfill our criterion on the pre-computed tables. $2^{14.4558}$ is the list size when merging the two restricted internal states corresponding to (z_0, z_1) and (z_1, z_2), respectively. After merging (z_0, z_1) and (z_1, z_2), we get a list for the restricted internal state of the 3-bit keystream vector (z_0, z_1, z_2). Now we further invoke the self-contained method for the keystream vector (z_0, z_2), which consists of only z_0 and the non-consecutive z_2. Since there are now 10 free common variables between the restricted internal state of (z_0, z_1, z_2) and that of (z_0, z_2), thus the denominator becomes 2^{10}. During this merging procedure, we use 3 keystream vectors (z_0, z_1), (z_1, z_2) and (z_0, z_2), and 3 times of the union result to form 3 independent candidate lists of size 848 with the probability 0.896456 that the corresponding correct partial state is indeed therein. Thus we have the probability 0.896456^3. The subsequent procedures in Tables 6 and 7 are deduced in a similar way as the above. The key point here is to the count the number of freely chosen variables between the corresponding internal state subsets, not including the linearly dependent variables. This process is repeated until merging the 20th equation in Table 5.

Table 6. The attack process for recovering $\mathbf{x}^*$ (1)

	$\mathbf{z}$	*List merging*	*Probability*
1.	(z_0, z_1) (z_1, z_2) (z_0, z_2)	$\frac{848 \cdot 848}{2^5} = 2^{14.4558}$ $\frac{848 \cdot 2^{14.4558}}{2^{10}} = 2^{14.1837}$	$0.896456^3 = 2^{-0.473086}$
2.	(z_1, z_2) (z_2, z_3) (z_1, z_3)	$\frac{2^{14.1837} \cdot 2^{14.1837}}{2^{10}} = 2^{18.3674}$ $\frac{2^{18.3674} \cdot 848}{2^{10}} = 2^{18.0953}$	$0.896456^{2 \cdot 3+1} = 2^{-1.10387}$
3.	$(z_0, \cdots, z_3)$ $(z_1, \cdots, z_4)$ (z_0, z_4)	$\frac{2^{18.0953} \cdot 2^{18.0953}}{2^{15}} = 2^{21.1906}$ $\frac{2^{21.1906} \cdot 848}{2^{10}} = 2^{20.9185}$	$0.896456^{2 \cdot 7+1} = 2^{-2.36543}$
4.	$(z_0, \cdots, z_4)$ $(z_1, \cdots, z_5)$ (z_0, z_5)	$\frac{2^{20.9185} \cdot 2^{20.9185}}{2^{20}} = 2^{21.837}$ $\frac{2^{21.837} \cdot 848}{2^{10}} = 2^{21.5649}$	$0.896456^{2 \cdot 15+1} = 2^{-4.88855}$
5.	$(z_0, \cdots, z_5)$ $(z_1, \cdots, z_6)$ (z_0, z_6)	$\frac{2^{21.5649} \cdot 2^{21.5649}}{2^{25}} = 2^{18.1298}$ $\frac{2^{18.1298} \cdot 848}{2^{10}} = 2^{17.8577}$	$0.896456^{2 \cdot 31+1} = 2^{-9.93481}$
6.	$(z_0, \cdots, z_6)$ $(z_1, \cdots, z_7)$ (z_0, z_7)	$\frac{2^{17.8577} \cdot 2^{17.8577}}{2^{30}} = 2^{5.7154}$ $\frac{2^{5.7154} \cdot 848}{2^{10}} = 2^{5.44332}$	$0.896456^{2 \cdot 63+1} = 2^{-20.0273}$
7.	$(z_0, \cdots, z_7)$ $(z_3, \cdots, z_8)$ (z_0, z_8)	$\frac{2^{5.44332} \cdot 2^{21.5649}}{2^{25}} = 2^{2.00822}$ $\frac{2^{2.00822} \cdot 848}{2^{10}} = 2^{1.73614}$	$0.896456^{127+31+1} = 2^{-25.0736}$
8.	$(z_0, \cdots, z_8)$ $(z_4, \cdots, z_9)$ (z_0, z_9)	$\frac{2^{1.73614} \cdot 2^{21.5649}}{2^{25}} = 2^{-1.69896}$ $\frac{2^{-1.69896} \cdot 848}{2^{10}} = 2^{-1.97104}$	$0.896456^{159+31+1} = 2^{-30.1198}$
9.	$(z_0, \cdots, z_9)$ $(z_5, \cdots, z_{10})$ (z_0, z_{10})	$\frac{2^{-1.97104} \cdot 2^{21.5649}}{2^{25}} = 2^{-5.40614}$ $\frac{2^{-5.40614} \cdot 848}{2^{10}} = 2^{-5.67822}$	$0.896456^{191+31+1} = 2^{-35.1661}$

Table 7. The attack process for recovering $\mathbf{x}^*$ (2)

	$\mathbf{z}$	*List merging*	*Probability*
10.	$(z_0, \cdots, z_{10})$	$\frac{2^{-5.67262} \cdot 2^{21.5649}}{2^{25}} = 2^{-9.11332}$	
	$(z_6, \cdots, z_{11})$		$0.896456^{223+31+1} = 2^{-40.2123}$
	(z_0, z_{11})	$\frac{2^{-9.11332} \cdot 848}{2^{10}} = 2^{-9.3854}$	
11.	$(z_0, \cdots, z_{11})$	$\frac{2^{-9.3854} \cdot 2^{20.9185}}{2^{20}} = 2^{-8.4669}$	
	$(z_8, \cdots, z_{12})$		$0.896456^{255+15+1} = 2^{-42.7354}$
	(z_0, z_{12})	$\frac{2^{-8.4669} \cdot 848}{2^{10}} = 2^{-8.73898}$	
12.	$(z_0, \cdots, z_{12})$	$\frac{2^{-8.73898} \cdot 2^{20.9185}}{2^{20}} = 2^{-7.82048}$	
	$(z_9, \cdots, z_{13})$		$0.896456^{271+15+1} = 2^{-45.2586}$
	(z_0, z_2)	$\frac{2^{-7.82048} \cdot 848}{2^{10}} = 2^{-8.09256}$	
13.	$(z_0, \cdots, z_{13})$	$\frac{2^{-8.09256} \cdot 2^{20.9185}}{2^{20}} = 2^{-7.17406}$	
	$(z_{10}, \cdots, z_{14})$		$0.896456^{287+15+1} = 2^{-47.7817}$
	(z_0, z_{14})	$\frac{2^{-7.17406} \cdot 848}{2^{10}} = 2^{-7.44614}$	
14.	$(z_0, \cdots, z_{14})$	$\frac{2^{-7.44614} \cdot 2^{18.0953}}{2^{15}} = 2^{-4.35084}$	
	$(z_{12}, \cdots, z_{15})$		$0.896456^{303+7+1} = 2^{-49.0432}$
	(z_0, z_{15})	$\frac{2^{-4.35084} \cdot 848}{2^{10}} = 2^{-4.62292}$	
15.	$(z_0, \cdots, z_{15})$	$\frac{2^{-4.62292} \cdot 2^{18.0953}}{2^{15}} = 2^{-1.52762}$	
	$(z_{13}, \cdots, z_{16})$		$0.896456^{311+7+1} = 2^{-50.3048}$
	(z_0, z_{16})	$\frac{2^{-1.52762} \cdot 848}{2^{10}} = 2^{-1.7997}$	
16.	$(z_0, \cdots, z_{16})$	$\frac{2^{-1.7997} \cdot 2^{18.0953}}{2^{15}} = 2^{1.2956}$	
	$(z_{14}, \cdots, z_{17})$		$0.896456^{319+7+1} = 2^{-51.5664}$
	(z_0, z_{17})	$\frac{2^{1.2956} \cdot 848}{2^{10}} = 2^{1.02352}$	
17.	$(z_0, \cdots, z_{17})$	$\frac{2^{1.02352} \cdot 2^{18.0953}}{2^{15}} = 2^{4.11882}$	
	$(z_{15}, \cdots, z_{18})$		$0.896456^{327+7+1} = 2^{-52.8279}$
	(z_0, z_{18})	$\frac{2^{4.11882} \cdot 848}{2^{10}} = 2^{3.84674}$	
18.	$(z_0, \cdots, z_{18})$	$\frac{2^{3.84674} \cdot 2^{18.0953}}{2^{15}} = 2^{6.94204}$	
	$(z_{16}, \cdots, z_{19})$		$0.896456^{335+7+1} = 2^{-54.0895}$
	(z_0, z_{19})	$\frac{2^{6.94204} \cdot 848}{2^{10}} = 2^{6.66996}$	

5.3 Restoring the Internal State of the LFSR

From Table 5, $\mathbf{x}^*$ involves 78 LFSR bits in total, it seems that we need to guess 2 more LFSR bits to have a linear system covering the 80 initial LFSR variables. To have an efficient attack, first note that both l_{64} and l_{65} are used 2 times in these equations, thus the candidate values should be consistent on l_{64} and l_{65}, which will provide a reduction factor of $\frac{1}{2^2} = \frac{1}{4}$ on the total number of candidates. Further, from $l_{83} = l_{65} \oplus l_{54} \oplus l_{41} \oplus l_{26} \oplus l_{16} \oplus l_3$, we have a third linear consistency check on the candidates. Hence, the number of candidates after going through Tables 6 and 7 is $\frac{1}{2^{-54.0895}} \cdot 2^{6.66996} \cdot 2^{-3} = 2^{57.7595}$. By guessing 2 more bits l_0, l_1, we can get l_{23}, l_{24} from the recursion $l_{80+j} = l_{62+j} \oplus l_{51+j} \oplus l_{38+j} \oplus l_{23+j} \oplus l_{13+j} \oplus l_j$ for $j = 0, 1$. In addition, we can derive l_2 from $l_{82} = l_{64} \oplus l_{53} \oplus l_{40} \oplus l_{25} \oplus l_{15} \oplus l_2$.

Note that the LFSR updates independently in the keystream generation phase and we also know the positions of the restored LFSR bits either from FNCA or from guessing, thus we could make a pre-computation to store the inverse of the corresponding linear systems with an off-line complexity of $\frac{80^{2.8}}{\Omega}$ cipher ticks and a online complexity of $\frac{80^2}{\Omega}$ to find the corresponding unique solution, where 2.8 is the exponent for Gauss reduction. This complexity is negligible compared to those of the other procedures. The total number of candidates for the LFSR part and the accompanying partial NFSR state, $2^2 \cdot 2^{57.7595} = 2^{59.7595}$, will dominate the complexity.

Remarks. Note that the gain in our attack mainly comes from the following two aspects. First, we exploit the first 20-bit keystream information in this procedure in a *probabilistic* way, not in a deterministic way, which is depicted later in Theorem 7. Now we target $78 + 20 + 20 = 118$ variables, not 160 variables, in a tradeoff-like manner. Here only 98 variables can be freely chosen. This cannot be interpreted in a straightforward information-theoretical way, which is usually evaluated in a deterministic way. Second, we use the pre-computed tables which also contain quite some information on the internal structure of Grain v1 in an *implicit* way in the attack.

5.4 Restoring the Internal State of the NFSR

After obtaining the candidate list for the LFSR part, the adversary could run the LFSR individually forwards and backwards to get all the necessary values and thus peel off the non-linearity of the h function. Now there are 2 choices in front of us, one is to efficiently restore the 80 NFSR variables with a low complexity that allows to be invoked many times, for there are probably many candidates of the restricted internal state $\mathbf{x}^*$ to be checked; the other is to check the correctness of the LFSR candidate first, then the NFSR could be restored afterwards independently. Fortunately, we find the latter way feasible in this scenario, which is shown below.

From the above step, the FNCA method has provided the adversary with the NFSR bits n_{63+i} and $x_i = n_{i+1} \oplus n_{i+2} \oplus n_{i+4} \oplus n_{i+10} \oplus n_{i+31} \oplus n_{i+43} \oplus n_{i+56}$ for $0 \leq i \leq 19$, i.e., now there are $20 + 20 = 40$-bit information available on the NFSR initial state. We proceed as follows to get more information.

Collecting More Linear Equations on NFSR. First note that if we go back 1 step, we get $x_{-1} = n_0 \oplus n_1 \oplus n_3 \oplus n_9 \oplus n_{30} \oplus n_{42} \oplus n_{55}$, i.e., we get 1 more linear equation for free. If we go back further, we could get a series of variables that can be expressed as the linear combination of the known values and the target initial NFSR state variables. On the other side, if we go forwards and take a look at the coefficient polynomial of x_4 in the h function, i.e., $1 \oplus x_3 \oplus x_0x_2 \oplus x_1x_2 \oplus x_2x_3$, we find it is a balanced Boolean function. Thus, the n_{82+i} variables have a probability of 0.5 to vanish in the resultant keystream bit and the adversary could directly collect a linear equation through the corresponding x_{20+i} variable at the beginning time instants from 20.

To get more linear equations on the NFSR initial state, we can use the following Z-technique, which is based on the index difference of the involving variables in the keystream bit. Precisely, if n_{82+i} appears at the z_{19+i} position, let us look at the end of the keystream equation z_{26+i} to see whether n_{82+i} exists there or not. If it is not there, then this will probably give us one more linear equation on the NFSR initial variables due to the index difference $56 - 43 = 13 > 7$; if it is there, we could just xor the two keystream equations to cancel out the n_{82+i} variable to get a linear equation on the NFSR initial variables. Then increase i by 1 and repeat the above process for the new i. Since the trace of the equations looks like the capital letter 'Z', we call this technique Z-technique. An illustrative example is provided in Appendix A.

It can be proved by induction that the Z-technique can also be used to express the newly generated NFSR variables as linear combinations of the keystream bits and of the initial state variables in forward direction. In backward direction, it is trivial to do the same task. We have run extensive experiments to see the average number of linear equations that the adversary could collect using the Z-technique, it turns out that the average number is 8, i.e., we could reduce the number of unknown variables in the initial NFSR state to around $80 - 40 - (8 - 3) = 35$, which facilitates the following linear distinguisher.

The Walsh Distinguisher. First note that the NFSR updating function in Grain v1 has a linear approximation with bias $\frac{41}{512}$, shown below.

$$n_{80+i} = n_{62+i} \oplus n_{60+i} \oplus n_{52+i} \oplus n_{45+i} \oplus n_{37+i} \oplus n_{28+i} \oplus n_{21+i} \oplus n_{14+i} \oplus n_i \oplus e,$$

where e is the binary noise variable satisfying $\Pr(e = 0) = \frac{1}{2} + \frac{41}{512}$. Since now there are only around 35 unknown variables left, we could collect a system of probabilistic linear equations on the left 35 NFSR variables by iteratively expressing the NFSR variables with indices larger than 80 by the corresponding linear combinations of keystream bits and the known information from the LFSR part and the partial NFSR state. If there are δ NFSR variables represented in this process, the complexity is just $35 \cdot \delta$. As a result, we have a system of the following form

$$\begin{cases} c_0^0 n_{i_0} \oplus c_1^0 n_{i_1} \oplus \cdots \oplus c_{34}^0 n_{i_{34}} = kz_0 \oplus e_0 \\ c_0^1 n_{i_0} \oplus c_1^1 n_{i_1} \oplus \cdots \oplus c_{34}^1 n_{i_{34}} = kz_1 \oplus e_1 \\ \quad \vdots \qquad\qquad \vdots \qquad\qquad\qquad \vdots \\ c_0^{\delta-1} n_{i_0} \oplus c_1^{\delta-1} n_{i_1} \oplus \cdots \oplus c_{34}^{\delta-1} n_{i_{34}} = kz_{\delta-1} \oplus e_{\delta-1}, \end{cases} \tag{5}$$

where $c_j^i \in \mathbb{F}_2$ for $0 \leq i \leq \delta - 1$ and $0 \leq j \leq 34$ is the coefficient of the remaining NFSR variable n_{i_j} $(0 \leq j \leq 34)$, kz_i $(0 \leq i \leq \delta - 1)$ is the accumulated linear combination of the keystream bits and the known information from the LFSR part and the partial NFSR state derived before and e_i $(0 \leq i \leq \delta - 1)$ is the binary noise variable with the distribution $\Pr(e_i = 0) = \frac{1}{2} + \frac{41}{512}$.

To further reduce the number of unknown NFSR variables, we construct the parity checks of weight 2 from the above system as follows. First note that the bias of the parity checks is $2 \cdot (\frac{41}{512})^2 = 2^{-6.2849}$ from the Piling-up lemma in [14].

Second, this problem is equivalent to the LF2 reduction in LPN solving problems [12], which can be solved in a sort-and-merge manner with a complexity of at most δ using pre-computed small tables. We have tuned the attack parameters in this procedure and found that if $\delta = 2^{19}$ and $y = 15$, we could collect $\binom{2^{19-15}}{2} \cdot 2^{15} = 2^{21.9069}$ parity-checks on $35 - y = 20$ NFSR variables of the bias $2^{-6.2849}$. Note that we could further cancel out 4 more NFSR variables in these parity-checks by only selecting those equations that the corresponding coefficient of the assigned variable is 0, in this way we could easily get $\frac{2^{21.9069}}{2^4} = 2^{17.9069}$ parity-checks on $20 - 4 = 16$ NFSR variables. On the other side, from the unique solution distance in correlation attacks [6,13], we have

$$\frac{8 \cdot 16 \cdot \ln 2}{1 - h(p)} = 2^{17.5121} < 2^{17.9069},$$

where $p = \frac{1}{2} + 2^{-6.2849}$ and $h(p) = -p \cdot \log p - (1-p) \cdot \log(1-p)$ is the binary entropy function. Thus, we can have the success probability very close to 1 given $2^{17.9069}$ parity-checks to identify the correct value of the 16 NFSR variables under consideration. That is, we reach the following theorem.

Theorem 6. *If both the LFSR candidate and the partial NFSR state are correct, we can distinguish the correct value of the remaining 16 NFSR variables from the wrong ones with a success probability very close to* 1.

Proof. It suffices to note that if either the LFSR or the partial NFSR state is wrong, there exists no bias in the system (5), thus following the classical reasoning from correlation attacks in [6,13], we have the claim. □

Precisely, for each parity-check of weight 2 for the system (3), we have

$$(c_0^{j_1} \oplus c_0^{j_2})n_{i_0} \oplus (c_1^{j_1} \oplus c_1^{j_2})n_{i_1} \oplus \cdots \oplus (c_{34-y}^{j_1} \oplus c_{34-y}^{j_2})n_{i_0} = \bigoplus_{t=1}^{2} kz_{j_t} \oplus \bigoplus_{t=1}^{2} e_{j_t}. \quad (6)$$

Let $(n'_{i_0}, n'_{i_1}, \cdots, n'_{i_{34-y}})$ be the guessed value of $(n_{i_0}, n_{i_1}, \cdots, n_{i_{34-y}})$, we rewrite Eq.(6) as follows.

$$\bigoplus_{t=1}^{2} kz_{j_t} \oplus \bigoplus_{t=1}^{34-y} (c_t^{j_1} \oplus c_t^{j_2})n'_{i_t} = \bigoplus_{t=1}^{34-y} (c_t^{j_1} \oplus c_t^{j_2})(n'_{i_t} \oplus n_{i_t}) \oplus \bigoplus_{t=1}^{2} e_{j_t}. \quad (7)$$

From (7), let $\Delta(j_1, j_2) = \bigoplus_{t=1}^{34-y} (c_t^{j_1} \oplus c_t^{j_2})(n'_{i_t} \oplus n_{i_t}) \oplus \bigoplus_{t=1}^{2} e_{j_t}$, it is obvious if $(n'_{i_0}, n'_{i_1}, \cdots, n'_{i_{34-y}})$ coincides with the correct value, we get $\Delta(j_1, j_2) = \bigoplus_{t=1}^{2} e_{j_t}$; otherwise, we have $\Delta(j_1, j_2) = \bigoplus_{t: n'_{i_t} \oplus n_{i_t} = 1} (c_t^{j_1} \oplus c_t^{j_2}) \oplus \bigoplus_{t=1}^{2} e_{j_t}$. Since $c_t^{j_1} \oplus c_t^{j_2}$ is the xor of 2 independent uniformly distributed variables, we have $\Pr(c_t^{j_1} \oplus c_t^{j_2} = 0) = \frac{1}{2}$. Hence, when $(n'_{i_0}, n'_{i_1}, \cdots, n'_{i_{34-y}})$ is wrongly guessed, $\Delta(j_1, j_2)$ has the distribution $\Pr(\Delta(j_1, j_2) = 0) = \frac{1}{2}$, which is quite different from the correct case, i.e., $\Pr(\Delta(j_1, j_2) = 0) = \frac{1}{2} + 2^{-6.2849}$. For $2^{17.9069}$ such parity-checks of the system (3), $\sum_{t=1}^{2^{17.9069}}(\Delta(j_1, j_2) \oplus 1)$ should follow the binomial distribution $(2^{17.9069}, \frac{1}{2} + 2^{-6.2849})$ if $(n'_{i_0}, n'_{i_1}, \cdots, n'_{i_{34-y}})$ is correctly guessed; otherwise this sum should have the binomial distribution $(2^{17.9069}, \frac{1}{2})$. Now the situation is the same as that in binary correlation attacks. Thus, we can use the FWT technique to speed up the whole process as follows. Denote the set of the constructed parity-checks by P_t. First regroup the $2^{17.9069}$ parity-checks according to the pattern of $x = (c_0^{j_1} \oplus c_0^{j_2}, c_1^{j_1} \oplus c_1^{j_2}, \cdots, c_{34-y}^{j_1} \oplus c_{34-y}^{j_2})$ and define $f_{\text{NFSR}}(x) = \sum_{(c_0^{j_1} \oplus c_0^{j_2}, c_1^{j_1} \oplus c_1^{j_2}, \cdots, c_{34-y}^{j_1} \oplus c_{34-y}^{j_2})}(-1)^{kz_{j_1} \oplus kz_{j_2}}$ for all the values of the coefficient vector appearing in the $2^{17.9069}$ parity-checks; if some value of $(c_0^{j_1} \oplus c_0^{j_2}, c_1^{j_1} \oplus c_1^{j_2}, \cdots, c_{34-y}^{j_1} \oplus c_{34-y}^{j_2})$ is not hit in these equations, just let $f_{NFSR} = 0$ at that point. For this well-defined function f_{NFSR}, consider the Walsh transform $F(\omega) = \sum_{x \in \mathbb{F}_2^{35-y}} f_{\text{NFSR}}(x) \cdot (-1)^{\omega \cdot x} = \sum_{P_t}(-1)^{kz_{j_1} \oplus kz_{j_2} \oplus \bigoplus_{t=0}^{34-y} \omega_t (c_t^{j_1} \oplus c_t^{j_2})} = F_0 - F_1$, where $\omega = (\omega_0, \omega_1, \cdots, \omega_{34-y}) \in \mathbb{F}_2^{35-y}$, F_0 and F_1 are the number of 0s and 1s, respectively. It is easy to see that if $\omega = (n_{i_0}, n_{i_1}, \cdots, n_{i_{34-y}})$, we have $\sum_{t=1}^{2^{17.9069}}(\Delta(j_1, j_2) \oplus 1) = \frac{F(\omega)+F_1}{2}$. If we set a threshold value T and accept only those guesses of x satisfying $\frac{F(\omega)+F_1}{2} \geq T$, then the probability that the correct value of $(c_0^{j_1} \oplus c_0^{j_2}, c_1^{j_1} \oplus c_1^{j_2}, \cdots, c_{34-y}^{j_1} \oplus c_{34-y}^{j_2})$ will pass the test is $P_1 = \sum_{t=T}^{2^{17.9069}} \binom{2^{17.9069}}{t} (\frac{1}{2} + 2^{-6.2849})^t (\frac{1}{2} - 2^{-6.2849})^{2^{17.9069}-t}$ and the probability that a wrong guess will be accepted is $P_2 = \sum_{t=T}^{2^{17.9069}} \binom{2^{17.9069}}{t} (\frac{1}{2})^{2^{17.9069}}$. Set $T = 2^{16.9306}$, we find that $P_1 = 0.999996$ and $P_2 \approx 2^{-53}$, i.e., the correct LFSR candidate and the correct partial NFSR state will pass almost certainly; while about $2^{59.7595} \cdot 2^{-53} = 2^{6.7595}$ wrong cases will survive in the above statistical test.

Hence, the time complexity of this Walsh distinguisher for one invoking is $\frac{2^{19} \cdot 35 + 2^{19} + 2^{17.9069} + 2^{16} \cdot 16}{\Omega} = \frac{2^{24.2656}}{\Omega}$ cipher ticks. Hence, by observing the Walsh spectrum of the function, the adversary could identify the correct LFSR and the correct partial NFSR states if they survived through the first step.

Restoring the Remaining NFSR State. After the Walsh distinguisher step, we could use an algebraic attack as that in [3] to restore the remaining NFSR state, which has a complexity much lower than the previous step. Precisely, the adversary could exploit the non-linear feedback function, say g, of the NFSR in Grain v1 to establish algebraic equations. Note that the algebraic degree of g in Grain v1 is 6 and the multiple $(n_{28} \oplus 1)(n_{60} \oplus 1) \cdot g$ has the algebraic degree 4, thus if the linearization method is adopted for solving the algebraic system, there are now $\sum_{i=0}^{4} \binom{35}{i} \approx 2^{15.8615}$ monomials in the system. If we take the same complexity metric as that in [7] in complexity estimate and taking into account that we have to repeat the solving routine for restoring the remaining NFSR state for each candidate survived through the above statistical test, the time complexity of this step is $T_{solving} = \frac{2^{6.7595} \cdot \frac{7 \cdot (2^{15.8615})^{\log_2 7}}{64}}{\Omega} \doteq \frac{2^{48.0957}}{\Omega}$ cipher ticks. Finally, the overall time complexity of all the procedures in Sect. 5.4 is $2^{59.7595} \cdot \frac{2^{24.2656}}{\Omega} + \frac{2^{48.0957}}{\Omega}$ cipher ticks.

5.5 Final Complexity Analysis

Now we analyze the final complexity of the above attack against Grain v1. First note that quite some complexity analysis have already been involved in the above sections, here we just focus on the total complexity, which is stated in the following theorem.

Theorem 7. *Let T_{Alg5} and λ_{Alg5} be the time complexity and the number of invoking times of Algorithm 5, then the time complexity of our attack is*

$$T_{Alg6}+\xi\cdot(\frac{1}{Pr_{\mathbf{x}}^{\lambda_{Alg5}}}\cdot(T_{Alg5}\cdot\lambda_{Alg5}+\sum_{i=1}^{18}T_{merg}^{i})+\frac{|L_{18}|}{Pr_{\mathbf{x}}^{\lambda_{Alg5}}}\cdot T_{walsh}+T_{solving}+T_{cst})$$

cipher ticks, where T_{Alg6} is the pre-computation complexity of Algorithm 6, T_{merg}^{i} $(1 \leq i \leq 18)$ is the list merging complexity at step i in Tables 6 and 7, T_{walsh} is the complexity for the Z-technique and Walsh distinguisher in Sect. 5.4, $T_{solving}$ is the complexity for restoring the remaining NFSR state in Sect. 5.4, T_{cst} is the complexity of the final consistency examination and we repeat the online attack ξ times to ensure a high success probability. The memory complexity of our attack is at most $2^{2l}\cdot V(n,d)\cdot(\lceil log_2 n\rceil\cdot d+14)+max_{1\leq i\leq 18}|L_i|$ bits, where L_i $(1 \leq i \leq 18)$ are the lists generated during the process in Tables 6 and 7, and the data complexity is $2^{19}+20+160=2^{19.0005}$ keystream bits.

Proof. Note that in our attack, to assure the existence of each correct restricted internal state in the corresponding candidate list, we have to assure its existence in the generated list when invoking Algorithm 5. This contributes to the factor $\frac{1}{\mathrm{Pr}_{\mathbf{x}}^{\lambda_{Alg5}}}$ since our attack is a dynamically growing process with the assumption that the probability $\mathrm{Pr}_{\mathbf{x}}$ is stable. $\mathrm{T}_{Alg5}\cdot\lambda_{Alg5}$ stands for the complexity of invoking Algorithm 5 during the evolution process in Tables 6 and 7 and $\sum_{i=1}^{18}T_{merg}^{i}$ comes from the sorting and merging complexities when merging the candidate lists for the steps from 1 to 18 in Tables 6 and 7. When the adversary goes out of all the steps in Tables 6 and 7, there exists only a resultant list L_{18} for each time, all the other lists generated in the intermediate process have been erased and overwritten. Thus, the total number of candidates for the LFSR part and partial NFSR state is $\frac{|L_{18}|}{\mathrm{Pr}_{\mathbf{x}}^{\lambda_{Alg5}}}$, and we have to use the Z-technique and Walsh distinguisher in Sect. 5.4 to check the correctness of these candidates. Since we could identify the correct LFSR candidate and the partial NFSR state with high probability independent of the remaining unknown NFSR state, the factor $T_{solving}$ is added in, not multiplied by. Finally, we have to find out the real correct internal state by the consistency test with the available keystream.

For the memory complexity, we invoke Algorithm 6 in the pre-processing phase with the l-bit KSD/keystream prefix, thus the factor 2^{2l} comes. That is, the adversary constructed 2^{2l} relatively small pre-computed tables, each of which consists of at most $V(n,d)$ items. Among all these tables, the adversary chooses one to be used in the online phase. It is worth noting that we could pre-compute the most inner routine of the self-contained method by storing all the possible xors between all the ISDs and each value of the restricted internal state. Taking into account on the concrete storage data structure in Sect. 4.2 of each item, we have the first item in the expression. During the process illustrated in Tables 6 and 7, we only have to allocate a memory space that fits the largest memory consumption among all the intermediate lists L_i for $1 \leq i \leq 18$. By checking the list sizes in Tables 6 and 7 and the corresponding number of variables in Table 5, we have $2 \cdot 2^{21.5649} \cdot 42 < 2^{28}$. For different iterations, the same memory is reused.

For the data complexity, only the first 20 keystream bits are exploited when recovering the LFSR and the partial NFSR state. Note that $2^{15.8615} < 2^{19}$ and 2^{19} keystream bits are used by the Walsh distinguisher in Sect. 5.4. The last 160 bits are needed in consistency test for the surviving candidates. □

Since our attack is dynamically executed, the above formula can also depict the time consuming in the intermediate process. Here the dominated factors are the complexity cost by the invoking of Algorithm 5 and that of checking the surviving candidates. When $l = 2$, $n = 10$, $c = 10$ and $d = 3$ with the chosen pre-computed Table of size $|\mathrm{T}[0x0, \texttt{prefix}]| = 176$, the time complexity can be computed as

$$(343 \cdot \frac{2^{20.3436}}{\Omega} \cdot 2^{54.0895} + 2^{59.7595} \cdot \frac{2^{24.2656}}{\Omega} + \frac{2^{48.0957}}{\Omega} + 2^{54.0895} \cdot \frac{\sum_{i=1}^{18} T_{merg}^{i}}{\Omega} + T_{cst}) \cdot 2^2 \approx 2^{75.7}$$

cipher ticks, where $\frac{21 \cdot 6 \cdot 60 \cdot 176}{\Omega} = \frac{2^{20.3436}}{\Omega}$ is the complexity for one invoking of Algorithm 5 and $\frac{\sum_{i=1}^{18} T_{merg}^{i}}{\Omega}$ is the sorting and merging complexity in Tables 6 and 7. If we adopt the same $\Omega = 2^{10.4}$ as that in [18] and let $\xi = 2^2$ to stabilize a success rate higher than 0.9, we have the above result. The memory complexity is around $2 \cdot 2^{21.5649} \cdot 42 \approx 2^{28}$ bits. The pre-computation complexity is $\frac{2^{10} \cdot 176 \cdot 2 + 2^4 \cdot 176 \cdot 2}{\Omega} = \frac{2^{18.4818}}{\Omega} = 2^{8.1}$ cipher ticks.

Remarks. First note that the time complexity actually depends on the parameter Ω, which may be different for different implementations. Second, in the existence of a faster hardware implementation of Grain v1 that could generate 16 keystream bits in one cipher tick, our attack still holds, for such a hardware will also speed up the attack 16 times.

6 Experimental Results

6.1 The Experiments on Grain v1

Note that a large proportion of practical experiments are already presented in the previous sections, here we just provide the remaining simulations. We have run extensive experiments on Grain v1 to check the correctness of our attack. Since the total complexity is too large to be implemented on a single PC, we have verified the beginning steps of Tables 6 and 7 practically. Note that the remaining steps in the evolution process are just the repetition process of the first ones, we have enough confidence that the whole process for recovering the inner state of the LFSR part is correct.

The profile of our experiments is as follows. We first generate the inner states of the LFSR and NFSR in Grain v1 randomly. Here the RC4 stream cipher is adopted by discarding the first 8192 output bytes as the main random source. Then we run the cipher forwards from this random state and generate the corresponding keystream. After that, we apply the FNCA to generate the possible states following the steps in Tables 6 and 7. At each step, we use the concrete data from simulations to verify the complexity and probability predictions in the Tables. We have done a large number of experiments to recover the restricted internal state corresponding to the keystream segment $(z_0, z_1, \cdots, z_6)$, and almost all the experimental results conform to our theoretical predictions in Table 6. For example, if the states of the LFSR and NFSR in Grain v1 are `0xB038f07C133370269B6C` and `0xC7F5B36FF85C13249603`, respectively, then the first 20-bit keystream are $(z_0, z_1, \cdots, z_{19})$ = `11100000111000000100`. Set $l = 2$, $d \leq 3$ and $c = 10$, let $\beta = 21$ and $\gamma = 6$, run Algorithm 5 to generate the union list of size 848. In 10^6 repetitions, the average probability of the correct restricted internal state being in the final list is quite close to the theoretical value 0.896456, which confirmed the correctness of the theoretical prediction. We also verified the list merging procedure of FNCA in experiments in the beginning steps 1 to 5 of Table 6, though the complexity of this procedure does not dominate. In general, the list sizes got in simulations match well with the theoretical estimates when represented in terms of the power of 2. Further, we have implemented the Walsh distinguisher to check its validity and got the confirmed results as well.

6.2 Simulations on the Reduced Version

For the reduced version of Grain v1 in Appendix B, we rewrite the keystream bit as $z'_i = x_i \oplus h(l'_{i+1}, l'_{i+21}, n'_{i+23})$, where $x_i = n'_{i+1} \oplus n'_{i+7} \oplus n'_{i+15}$. We have established a similar evolution process for restoring the restricted internal state of $(z_0, z_1, \cdots, z_{19})$, which consists of 40 LFSR state bits, 20 NFSR variables n_{i+23} and 20 x_i variables for $(0 \leq i \leq 19)$. In simulations, we first randomly loaded the internal states of the LFSR and NFSR as `0x9b97284782` and `0xb20027ea7d`, respectively. Then we got the first 20-bit keystream, `00010010010000 001100`. Let $l = 2$, $d \leq 2$ and $c = 8$, the probability that the correct inner state is in the candidate list is 0.923 after one call of Algorithm 3. Now we adopt a group of attack parameters similar to the case of Grain v1 in the distilling phase. We invoked the Algorithm 3 10 times to generate the corresponding 10 lists, and then intersect these lists to have a smaller list. Repeat this process 4 times to acquire 4 similar intersection lists. At last, combine these 4 lists to form the union list and the existence probability of the correct state in the list is around 0.91. For (z_0, z_1), we can recover an 8-bit restricted internal state, i.e., $(n_{22}, l_1, l_{21}, n_{23}, l_2, l_{22}, x_0, x_1)$. Note that there are 6 free variables in this case and 64 possibilities in total. After distilling, the candidate number of this partial inner state is reduced to 50, while the value of expectation in theory is 53. Next, for (z_1, z_2), we got 51 candidates. Since the inner states of (z_0, z_1) and (z_1, z_2) have 3 common free variables, the expected number of inner states of (z_0, z_1, z_2) is $50 \cdot 51/2^3 \approx 319$. In the experiments, the practical number is 308. The same method can also be applied to (z_1, z_2, z_3) to recover the candidates list of size 312. There are 6 common free variables between the inner state of (z_0, z_1, z_2) and that of (z_1, z_2, z_3). We can reduce the number of inner states associated with (z_0, z_1, z_2, z_3) to $308 \cdot 319/2^6 \approx 2^{10.9}$, and the number got in experiments is $1921 \approx 2^{10.9}$. We continue this process to z_{19} until we have recovered the target inner state. In the experiments, we repeat the above whole process for $2^{24.26}$ times until the correct inner state is indeed in the candidate list of size $2^{12.01}$. In theory, we need about $2^{23.94}$ repetitions of the whole process and get a list with $2^{11.64}$ candidates. Therefore, on average we could restore the internal state with a complexity of about $2^{37.58}$ reduced version cipher ticks, and currently, it took several hours for our non-optimized C implementation to have the candidate list with the correct candidate in, which verified the theoretical analysis of FNCA. For the reduced version, there is no need to use the Z-technique and Walsh distinguisher to deal with the LFSR independently. The codes for the reduced version experiments are available via https://github.com/martinzhangbin/nca_reducedversion.

7 Conclusions

In this paper, we have tried to develop a new cryptanalytic method, called fast near collision attack, on modern stream ciphers with a large internal state. The new attack utilizes the basic, yet often ignored fact in the primitives that each keystream vector actually depends on only a subset of the internal state bits, not on the full internal state. Thus it is natural to combine the near collision property with the divide-and-conquer strategy to mount the new kind of state recovery attacks. In the process, a self-contained method is introduced and improved to derive the partial internal state from the partial state difference efficiently. After the recovery of certain subsets of the whole internal state, a careful merging and further retrieval step is conducted to restore the full large internal state. As an application of the new methodology, we demonstrated a key recovery attack against Grain v1, one of the 7 finalists in the European eSTREAM project. Combined with the rewriting variables technique, it is shown that the internal state of Grain v1, thus the secret key, can be reliably restored in $2^{75.7}$ cipher ticks after the pre-computation of $2^{8.1}$ cipher ticks, given 2^{28}-bit memory and around 2^{19} keystream bits in the single key model, which is the best key recovery attack against Grain v1 so far. It is suggested to strengthen Grain v1 with a new NFSR that eliminates the existence of good linear approximations for the feedback function. It is our future work to study fast near collision attacks against other NFSR-based primitives.

Acknowledgements. We would like to thank the anonymous reviewers for very helpful comments. This work is supported by the National Key R&D Research programm

(Grant No. 2017YFB0802504), the program of the National Natural Science Foundation of China (Grant No. 61572482), National Cryptography Development Fund (Grant No. MMJJ20170107) and National Grand Fundamental Research 973 Programs of China (Grant No. 2013CB338002).

A An Example to Illustrate the Z-technique

Example 1. Assume the adversary collects the following linear equations of z_{20+i} for $i \geq 0$. Now he could use the Z-technique as follows to derive more linear equations on the initial NFSR state.

$$
\begin{aligned}
1: &\ z_{20} = n_{21} \oplus n_{22} \oplus n_{24} \oplus n_{30} \oplus n_{51} \oplus n_{63} \oplus n_{76} \oplus n_{83} \\
2: &\ z_{21} = n_{22} \oplus n_{23} \oplus n_{25} \oplus n_{31} \oplus n_{52} \oplus n_{64} \oplus n_{77} \oplus n_{84} \\
3: &\ z_{22} = n_{23} \oplus n_{24} \oplus n_{26} \oplus n_{32} \oplus n_{53} \oplus n_{65} \oplus n_{78} \\
4: &\ z_{23} = n_{24} \oplus n_{25} \oplus n_{27} \oplus n_{33} \oplus n_{54} \oplus n_{66} \oplus n_{79} \\
5: &\ z_{24} = n_{25} \oplus n_{26} \oplus n_{28} \oplus n_{34} \oplus n_{55} \oplus n_{67} \oplus n_{80} \oplus n_{87} \\
6: &\ z_{25} = n_{26} \oplus n_{27} \oplus n_{29} \oplus n_{35} \oplus n_{56} \oplus n_{68} \oplus n_{81} \\
7: &\ z_{26} = n_{27} \oplus n_{28} \oplus n_{30} \oplus n_{36} \oplus n_{57} \oplus n_{69} \oplus n_{82} \\
8: &\ z_{27} = n_{28} \oplus n_{29} \oplus n_{31} \oplus n_{37} \oplus n_{58} \oplus n_{70} \oplus n_{83} \\
9: &\ z_{28} = n_{29} \oplus n_{30} \oplus n_{32} \oplus n_{38} \oplus n_{59} \oplus n_{71} \oplus n_{84} \oplus n_{91} \\
10: &\ z_{29} = n_{30} \oplus n_{31} \oplus n_{33} \oplus n_{39} \oplus n_{60} \oplus n_{72} \oplus n_{85} \oplus n_{92} \\
11: &\ z_{30} = n_{31} \oplus n_{32} \oplus n_{34} \oplus n_{40} \oplus n_{61} \oplus n_{73} \oplus n_{86} \oplus n_{93} \\
12: &\ z_{31} = n_{32} \oplus n_{33} \oplus n_{35} \oplus n_{41} \oplus n_{62} \oplus n_{74} \oplus n_{87} \oplus n_{94} \\
13: &\ z_{32} = n_{33} \oplus n_{34} \oplus n_{36} \oplus n_{42} \oplus n_{63} \oplus n_{75} \oplus n_{88} \oplus n_{95} \\
14: &\ z_{33} = n_{34} \oplus n_{35} \oplus n_{37} \oplus n_{43} \oplus n_{64} \oplus n_{76} \oplus n_{89} \\
15: &\ z_{34} = n_{35} \oplus n_{36} \oplus n_{38} \oplus n_{44} \oplus n_{65} \oplus n_{77} \oplus n_{90} \\
16: &\ z_{35} = n_{36} \oplus n_{37} \oplus n_{39} \oplus n_{45} \oplus n_{66} \oplus n_{78} \oplus n_{91} \\
17: &\ z_{36} = n_{37} \oplus n_{38} \oplus n_{40} \oplus n_{46} \oplus n_{67} \oplus n_{79} \oplus n_{92} \\
18: &\ z_{37} = n_{38} \oplus n_{39} \oplus n_{41} \oplus n_{47} \oplus n_{68} \oplus n_{80} \oplus n_{93} \oplus n_{100} \\
19: &\ z_{38} = n_{39} \oplus n_{40} \oplus n_{42} \oplus n_{48} \oplus n_{69} \oplus n_{81} \oplus n_{94} \oplus n_{101} \\
20: &\ z_{39} = n_{40} \oplus n_{41} \oplus n_{43} \oplus n_{49} \oplus n_{70} \oplus n_{82} \oplus n_{95} \oplus n_{102} \\
21: &\ z_{40} = n_{41} \oplus n_{42} \oplus n_{44} \oplus n_{50} \oplus n_{71} \oplus n_{83} \oplus n_{96} \\
22: &\ z_{41} = n_{42} \oplus n_{43} \oplus n_{45} \oplus n_{51} \oplus n_{72} \oplus n_{84} \oplus n_{97} \oplus n_{104} \\
23: &\ z_{42} = n_{43} \oplus n_{44} \oplus n_{46} \oplus n_{52} \oplus n_{73} \oplus n_{85} \oplus n_{98} \\
24: &\ z_{43} = n_{44} \oplus n_{45} \oplus n_{47} \oplus n_{53} \oplus n_{74} \oplus n_{86} \oplus n_{99} \\
25: &\ z_{44} = n_{45} \oplus n_{46} \oplus n_{48} \oplus n_{54} \oplus n_{75} \oplus n_{87} \oplus n_{100} \\
26: &\ z_{45} = n_{46} \oplus n_{47} \oplus n_{49} \oplus n_{55} \oplus n_{76} \oplus n_{88} \oplus n_{101} \\
27: &\ z_{46} = n_{47} \oplus n_{48} \oplus n_{50} \oplus n_{56} \oplus n_{77} \oplus n_{89} \oplus n_{102}
\end{aligned}
$$

From Eqs.1 ⇀ 8, $z_{20} \oplus z_{27}$ is a linear equation on the initial NFSR state. From Eqs.2 ⇀ 9 ⇀ 16, $z_{21} \oplus z_{28} \oplus z_{35}$ is also a linear equation on the initial NFSR state. Eqs.3 and 4 provide 2 more such equations directly. From Eqs.5 ⇀ 12 ⇀ 19 ⇀ 26 ⇀ 13 ⇀ 20 ⇀ 27 ⇀ 14, another linear equation on the initial NFSR state is established. Finally, as n_{80}, n_{81} and n_{82} are known from the previous step, Eqs.6 and 7 are also linear equations on the initial NFSR state. Thus, the adversary could collect 7 linear equations in forwards direction and 1 linear equation in backwards direction. In total, he could collect 8 linear equations on the initial NFSR state with a negligible complexity. □

B The Reduced Version of Grain v1

This reduced cipher generates the keystream from a 40-bit key and a 32-bit IV. Precisely, let $f'(x) = 1 + x^7 + x^{22} + x^{31} + x^{40}$ be the primitive polynomial of degree 40, then the updating function of the LFSR is defined as $l'_{i+40} = l'_{i+33} + l'_{i+18} + l'_{i+9} + l'_i$ for $i \geq 0$. Similar to the original Grain v1, the updating function of the NFSR is

$$\begin{aligned} n'_{i+40} = \; & l'_i \oplus n'_{i+33} \oplus n'_{i+29} \oplus n'_{i+23} \oplus n'_{i+17} \oplus n'_{i+11} \oplus n'_{i+9} \oplus n'_{i+33}n'_{i+29} \\ & \oplus n'_{i+23}n'_{i+17} \oplus n'_{i+33}n'_{i+9} \oplus n'_{i+33}n'_{i+29}n'_{i+23} \oplus n'_{i+29}n'_{i+23}n'_{i+17} \\ & \oplus n'_{i+33}n'_{i+29}n'_{i+23}n'_{i+17} \oplus n'_{i+29}n'_{i+23}n'_{i+17}n'_{i+11}n'_{i+9}. \end{aligned}$$

The filter function $h'(x)$ is defined as $h'(x) = x_1 \oplus x_0x_2 \oplus x_1x_2 \oplus x_0x_1x_2$ with the different tap positions, which are provided below. The output function is $z'_i = \sum_{k \in \mathcal{A}'} n'_{i+k} \oplus h'(l'_{i+1}, l'_{i+21}, n'_{i+22})$, where $\mathcal{A}' = \{1, 7, 15\}$. Its key/IV initialization is similar to that of Grain v1 with 80 initialization rounds. The actual complexity of the brute force attack for a fixed IV on the reduced version is $(2^{40} - 1) \cdot (80 + \sum_{i=1}^{40} i \cdot \frac{1}{2^{i-1}}) \approx 2^{46}$ cipher ticks.

References

1. Anderson, R.: Searching for the optimum correlation attack. In: Preneel, B. (ed.) FSE 1994. LNCS, vol. 1008, pp. 137–143. Springer, Heidelberg (1995). https://doi.org/10.1007/3-540-60590-8_11
2. Berbain, C., Gilbert, H., Maximov, A.: Cryptanalysis of grain. In: Robshaw, M. (ed.) FSE 2006. LNCS, vol. 4047, pp. 15–29. Springer, Heidelberg (2006). https://doi.org/10.1007/11799313_2
3. Berbain, C., Gilbert, H., Joux, A.: Algebraic and correlation attacks against linearly filtered non linear feedback shift registers. In: Avanzi, R.M., Keliher, L., Sica, F. (eds.) SAC 2008. LNCS, vol. 5381, pp. 184–198. Springer, Heidelberg (2009). https://doi.org/10.1007/978-3-642-04159-4_12
4. Biryukov, A., Shamir, A.: Cryptanalytic time/memory/data tradeoffs for stream ciphers. In: Okamoto, T. (ed.) ASIACRYPT 2000. LNCS, vol. 1976, pp. 1–13. Springer, Heidelberg (2000). https://doi.org/10.1007/3-540-44448-3_1
5. De Cannière, C., Küçük, Ö., Preneel, B.: Analysis of grain's initialization algorithm. In: Vaudenay, S. (ed.) AFRICACRYPT 2008. LNCS, vol. 5023, pp. 276–289. Springer, Heidelberg (2008). https://doi.org/10.1007/978-3-540-68164-9_19

6. Chepyzhov, V.V., Johansson, T., Smeets, B.: A simple algorithm for fast correlation attacks on stream ciphers. In: Goos, G., Hartmanis, J., van Leeuwen, J., Schneier, B. (eds.) FSE 2000. LNCS, vol. 1978, pp. 181–195. Springer, Heidelberg (2001). https://doi.org/10.1007/3-540-44706-7_13
7. Courtois, N.T., Meier, W.: Algebraic attacks on stream ciphers with linear feedback. In: Biham, E. (ed.) EUROCRYPT 2003. LNCS, vol. 2656, pp. 345–359. Springer, Heidelberg (2003). https://doi.org/10.1007/3-540-39200-9_21
8. http://www.ecrypt.eu.org/stream/e2-grain.html
9. Hawkes, P., Rose, G.G.: Rewriting variables: the complexity of fast algebraic attacks on stream ciphers. In: Franklin, M. (ed.) CRYPTO 2004. LNCS, vol. 3152, pp. 390–406. Springer, Heidelberg (2004). https://doi.org/10.1007/978-3-540-28628-8_24
10. Hell, M., Johansson, T., Meier, W.: Grain: a stream cipher for constrained environments. Int. J. Wirel. Mob. Comput. (IJWMC) **2**(1), 86–93 (2007)
11. Fischer, S., Meier, W.: Algebraic immunity of S-boxes and augmented functions. In: Biryukov, A. (ed.) FSE 2007. LNCS, vol. 4593, pp. 366–381. Springer, Heidelberg (2007). https://doi.org/10.1007/978-3-540-74619-5_23
12. Guo, Q., Johansson, T., Löndahl, C.: Solving LPN using covering codes. In: Sarkar, P., Iwata, T. (eds.) ASIACRYPT 2014. LNCS, vol. 8873, pp. 1–20. Springer, Heidelberg (2014). https://doi.org/10.1007/978-3-662-45611-8_1
13. Lu, Y., Vaudenay, S.: Faster correlation attack on bluetooth keystream generator E0. In: Franklin, M. (ed.) CRYPTO 2004. LNCS, vol. 3152, pp. 407–425. Springer, Heidelberg (2004). https://doi.org/10.1007/978-3-540-28628-8_25
14. Matsui, M.: Linear cryptanalysis method for DES cipher. In: Helleseth, T. (ed.) EUROCRYPT 1993. LNCS, vol. 765, pp. 386–397. Springer, Heidelberg (1994). https://doi.org/10.1007/3-540-48285-7_33
15. Knellwolf, S., Meier, W., Naya-Plasencia, M.: Conditional differential cryptanalysis of NLFSR-based cryptosystems. In: Abe, M. (ed.) ASIACRYPT 2010. LNCS, vol. 6477, pp. 130–145. Springer, Heidelberg (2010). https://doi.org/10.1007/978-3-642-17373-8_8
16. Koch, P.C.: Cryptanalysis of stream ciphers-analysis and application of the near collision attack for stream ciphers, Technical University of Denmark, Master Thesis-Supervisor: Christian Rechberger, November 2013
17. http://en.wikipedia.org/wiki/Stirling_numbers_of_the_second_kind
18. Zhang, B., Li, Z., Feng, D., Lin, D.: Near collision attack on the grain v1 stream cipher. In: Moriai, S. (ed.) FSE 2013. LNCS, vol. 8424, pp. 518–538. Springer, Heidelberg (2014). https://doi.org/10.1007/978-3-662-43933-3_27

Author Index

Printed in the United States
by Bookmasters

Printed in the United States
By Bookmasters